Transactions on Computational Science and Computational Intelligence

Series Editor

Hamid R. Arabnia
Department of Computer Science
The University of Georgia
Athens, GA, USA

Computational Science (CS) and Computational Intelligence (CI) both share the same objective: finding solutions to difficult problems. However, the methods to the solutions are different. The main objective of this book series, "Transactions on Computational Science and Computational Intelligence", is to facilitate increased opportunities for cross-fertilization across CS and CI. This book series will publish monographs, professional books, contributed volumes, and textbooks in Computational Science and Computational Intelligence. Book proposals are solicited for consideration in all topics in CS and CI including, but not limited to, Pattern recognition applications; Machine vision; Brain-machine interface; Embodied robotics; Biometrics; Computational biology; Bioinformatics; Image and signal processing; Information mining and forecasting; Sensor networks; Information processing; Internet and multimedia; DNA computing; Machine learning applications; Multi-agent systems applications; Telecommunications; Transportation systems; Intrusion detection and fault diagnosis; Game technologies; Material sciences; Space, weather, climate systems, and global changes; Computational ocean and earth sciences; Combustion system simulation; Computational chemistry and biochemistry; Computational physics; Medical applications; Transportation systems and simulations; Structural engineering; Computational electro-magnetic; Computer graphics and multimedia; Face recognition; Semiconductor technology, electronic circuits, and system design; Dynamic systems; Computational finance; Information mining and applications; Astrophysics; Biometric modeling; Geology and geophysics; Nuclear physics; Computational journalism; Geographical Information Systems (GIS) and remote sensing; Military and defense related applications; Ubiquitous computing; Virtual reality; Agent-based modeling; Computational psychometrics; Affective computing; Computational economics; Computational statistics; and Emerging applications. For further information, please contact Mary James, Senior Editor, Springer, mary.james@springer.com.

More information about this series at http://www.springer.com/series/11769

Hamid R. Arabnia • Ken Ferens • David de la Fuente
Elena B. Kozerenko • José Angel Olivas Varela
Fernando G. Tinetti
Editors

Advances in Artificial Intelligence and Applied Cognitive Computing

Proceedings from ICAI'20 and ACC'20

Editors
Hamid R. Arabnia
Department of Computer Science
University of Georgia
Athens, GA, USA

David de la Fuente
Business Administration
University of Oviedo
Oviedo, Asturias, Spain

José Angel Olivas Varela
Technology and Information systems
Universidad de Castilla La Mancha
Ciudad Real, Ciudad Real, Spain

Ken Ferens
Department of Electrical and Computer
Engineering
University of Manitoba
Winnipeg, MB, Canada

Elena B. Kozerenko
Institute of Informatics Problems
The Russian Academy of Sciences
Moscow, Russia

Fernando G. Tinetti
Facultad de Informática - CIC PBA
Universidad Nacional de La Plata
La Plata, Argentina

ISSN 2569-7072 ISSN 2569-7080 (electronic)
Transactions on Computational Science and Computational Intelligence
ISBN 978-3-030-70298-4 ISBN 978-3-030-70296-0 (eBook)
https://doi.org/10.1007/978-3-030-70296-0

© Springer Nature Switzerland AG 2021
This work is subject to copyright. All rights are reserved by the Publisher, whether the whole or part of
the material is concerned, specifically the rights of translation, reprinting, reuse of illustrations, recitation,
broadcasting, reproduction on microfilms or in any other physical way, and transmission or information
storage and retrieval, electronic adaptation, computer software, or by similar or dissimilar methodology
now known or hereafter developed.
The use of general descriptive names, registered names, trademarks, service marks, etc. in this publication
does not imply, even in the absence of a specific statement, that such names are exempt from the relevant
protective laws and regulations and therefore free for general use.
The publisher, the authors, and the editors are safe to assume that the advice and information in this book
are believed to be true and accurate at the date of publication. Neither the publisher nor the authors or
the editors give a warranty, expressed or implied, with respect to the material contained herein or for any
errors or omissions that may have been made. The publisher remains neutral with regard to jurisdictional
claims in published maps and institutional affiliations.

This Springer imprint is published by the registered company Springer Nature Switzerland AG
The registered company address is: Gewerbestrasse 11, 6330 Cham, Switzerland

Preface

It gives us great pleasure to introduce this collection of papers that were presented at the following international conferences: Artificial Intelligence (ICAI 2020) and Applied Cognitive Computing (ACC 2020). These two conferences were held simultaneously (same location and dates) at Luxor Hotel (MGM Resorts International), Las Vegas, USA, July 27–30, 2020. This international event was held using a hybrid approach, that is, "in-person" and "virtual/online" presentations and discussions.

This book is composed of nine parts. Parts 1 through 8 (composed of 78 chapters) include articles that address various challenges in the area of artificial intelligence (ICAI). Part 9 (composed of 12 chapters) includes a collection of research papers in the area of applied cognitive computing (ACC).

An important mission of the World Congress in Computer Science, Computer Engineering, and Applied Computing, CSCE (a federated congress to which this event is affiliated with), includes *"Providing a unique platform for a diverse community of constituents composed of scholars, researchers, developers, educators, and practitioners. The Congress makes concerted effort to reach out to participants affiliated with diverse entities (such as: universities, institutions, corporations, government agencies, and research centers/labs) from all over the world. The congress also attempts to connect participants from institutions that have **teaching** as their main mission with those who are affiliated with institutions that have **research** as their main mission. The congress uses a quota system to achieve its institution and geography diversity objectives."* By any definition of diversity, this congress is among the most diverse scientific meeting in the USA. We are proud to report that this federated congress had authors and participants from 54 different nations representing variety of personal and scientific experiences that arise from differences in culture and values.

The program committees (refer to subsequent pages for the list of the members of committees) would like to thank all those who submitted papers for consideration. About 50% of the submissions were from outside the USA. Each submitted paper was peer-reviewed by two experts in the field for originality, significance, clarity, impact, and soundness. In cases of contradictory recommendations, a member of the conference program committee was charged to make the final decision; often, this involved seeking help from additional referees. In addition, papers whose authors included a member of the conference program committee were evaluated using the double-blind review process. One exception to the above evaluation process was for papers that were submitted directly to chairs/organizers of pre-approved sessions/workshops; in these cases, the chairs/organizers were responsible for the evaluation of such submissions. The overall paper acceptance rate for regular papers was 20%; 18% of the remaining papers were accepted as short and/or poster papers.

We are grateful to the many colleagues who offered their services in preparing this book. In particular, we would like to thank the members of the Program Committees of individual research tracks as well as the members of the Steering Committees of ICAI 2020 and ACC 2020; their names appear in the subsequent pages. We would also like to extend our appreciation to over 500 referees.

As sponsors-at-large, partners, and/or organizers, each of the followings (separated by semicolons) provided help for at least one research track: Computer Science Research, Education, and Applications (CSREA); US Chapter of World Academy of Science; American Council on Science and Education & Federated Research council; and Colorado Engineering Inc. In addition, a number of university faculty members and their staff, several publishers of computer science and computer engineering books and journals, chapters and/or task forces of computer science associations/organizations from 3 regions, and developers of high-performance machines and systems provided significant help in organizing the event as well as providing some resources. We are grateful to them all.

We express our gratitude to all authors of the articles published in this book and the speakers who delivered their research results at the congress. We would also like to thank the following: UCMSS (Universal Conference Management Systems & Support, California, USA) for managing all aspects of the conference; Dr. Tim Field at APC for coordinating and managing the printing of the programs; the staff at Luxor Hotel (MGM Convention) for the professional service they provided; and Ashu M. G. Solo for his help in publicizing the congress. Last but not least, we would like to thank Ms. Mary James (Springer Senior Editor in New York) and Arun Pandian KJ (Springer Production Editor) for the excellent professional service they provided for this book project.

Book Co-editors and Chapter Co-editors: ICAI 2020 and ACC 2020

Athens, GA, USA	Hamid R. Arabnia
Winnipeg, MB, Canada	Ken Ferens
Oviedo, Asturias, Spain	David de la Fuente
Moscow, Russia	Elena B. Kozerenko
Ciudad Real, Ciudad Real, Spain	José Angel Olivas Varela
La Plata, Argentina	Fernando G. Tinetti
Seoul, South Korea	Charlie (Seungmin) Rho
Houston, TX, USA	Xiaokun Yang

Artificial Intelligence: ICAI 2020 – Program Committee

- Prof. Abbas M. Al-Bakry (Steering Committee); University of IT and Communications, Baghdad, Iraq
- Prof. Emeritus Nizar Al-Holou (Steering Committee); Electrical and Computer Engineering Department; Vice Chair, IEEE/SEM-Computer Chapter; University of Detroit Mercy, Detroit, Michigan, USA
- Prof. Emeritus Hamid R. Arabnia (Steering Committee); University of Georgia, USA; Editor-in-Chief, Journal of Supercomputing (Springer); Fellow, Center of Excellence in Terrorism, Resilience, Intelligence & Organized Crime Research (CENTRIC).
- Prof. Mehran Asadi; Department of Business and Entrepreneurial Studies, The Lincoln University, Pennsylvania, USA
- Prof. Juan Jose Martinez Castillo; Director, The Acantelys Alan Turing Nikola Tesla Research Group and GIPEB, Universidad Nacional Abierta, Venezuela
- Dr. Arianna D'Ulizia; Institute of Research on Population and Social Policies, National Research Council of Italy (IRPPS), Rome, Italy
- Prof. Emeritus Kevin Daimi (Steering Committee); Department of Mathematics, Computer Science and Software Engineering, University of Detroit Mercy, Detroit, Michigan, USA
- Prof. Zhangisina Gulnur Davletzhanovna; Vice-Rector of the Science, Central-Asian University, Kazakhstan, Almaty, Republic of Kazakhstan; Vice President of International Academy of Informatization, Kazskhstan, Almaty, Republic of Kazakhstan
- Prof. Leonidas Deligiannidis (Steering Committee); Department of Computer Information Systems, Wentworth Institute of Technology, Boston, Massachusetts, USA
- Dr. Roger Dziegiel; US Air Force Research Lab, AFRL/RIEA, USA
- Prof. Mary Mehrnoosh Eshaghian-Wilner (Steering Committee); Professor of Engineering Practice, University of Southern California, California, USA; Adjunct Professor, Electrical Engineering, University of California Los Angeles, Los Angeles (UCLA), California, USA

- Prof. Ken Ferens (Steering Committee); Department of Electrical and Computer Engineering, University of Manitoba, Winnipeg, Canada
- Dr. David de la Fuente (Chapter Editor); University of Oviedo, Spain
- Hindenburgo Elvas Goncalves de Sa; Robertshaw Controls (Multi-National Company), System Analyst, Brazil; Information Technology Coordinator and Manager, Brazil
- Prof. George A. Gravvanis (Steering Committee); Director, Physics Laboratory & Head of Advanced Scientific Computing, Applied Math & Applications Research Group; Professor of Applied Mathematics and Numerical Computing and Department of ECE, School of Engineering, Democritus University of Thrace, Xanthi, Greece.
- Prof. George Jandieri (Steering Committee); Georgian Technical University, Tbilisi, Georgia; Chief Scientist, The Institute of Cybernetics, Georgian Academy of Science, Georgia; Ed. Member, International Journal of Microwaves and Optical Technology, The Open Atmospheric Science Journal, American Journal of Remote Sensing, Georgia
- Prof. Byung-Gyu Kim (Steering Committee); Multimedia Processing Communications Lab.(MPCL), Department of CSE, College of Engineering, SunMoon University, South Korea
- Prof. Tai-hoon Kim; School of Information and Computing Science, University of Tasmania, Australia
- Dr. Elena B. Kozerenko (Chapter Editor); Institute of Informatics Problems of the Russian Academy of Sciences, Moscow, Russia
- Prof. Louie Lolong Lacatan; Chairperson, Computer Engineerig Department, College of Engineering, Adamson University, Manila, Philippines; Senior Member, International Association of Computer Science and Information Technology (IACSIT), Singapore; Member, International Association of Online Engineering (IAOE), Austria
- Prof. Dr. Guoming Lai; Computer Science and Technology, Sun Yat-Sen University, Guangzhou, P. R. China
- Dr. Peter M. LaMonica; US Air Force Research Lab, AFRL/RIEBB, USA
- Prof. Hyo Jong Lee; Director, Center for Advanced Image and Information Technology, Division of Computer Science and Engineering, Chonbuk National University, South Korea
- Dr. Changyu Liu; College of Mathematics and Informatics, South China Agricultural University, Guangzhou, P. R. China and Visiting scientist, School of Computer Science, Carnegie Mellon University, USA
- Dr. Muhammad Naufal Bin Mansor; Faculty of Engineering Technology, Department of Electrical, Universiti Malaysia Perlis (UniMAP), Perlis, Malaysia
- Dr. Andrew Marsh (Steering Committee); CEO, HoIP Telecom Ltd (Healthcare over Internet Protocol), UK; Secretary General of World Academy of BioMedical Sciences and Technologies (WABT) a UNESCO NGO, The United Nations
- Dr. Mohamed Arezki Mellal; Faculty of Engineering Sciences (FSI), M'Hamed Bougara University, Boumerdes, Algeria

Artificial Intelligence: ICAI 2020 – Program Committee

- Dr. Ali Mostafaeipour; Industrial Engineering Department, Yazd University, Yazd, Iran
- Dr. Houssem Eddine Nouri; Informatics Applied in Management, Institut Superieur de Gestion de Tunis, University of Tunis, Tunisia
- Prof. Dr., Eng. Robert Ehimen Okonigene (Steering Committee); Department of EEE, Faculty of Engineering and Technology, Ambrose Alli University, Edo State, Nigeria
- Dr. Jose A. Olivas (Chapter Editor); University of Castilla - La Mancha, Spain
- Prof. James J. (Jong Hyuk) Park (Steering Committee); DCSE, SeoulTech, Korea; President, FTRA, EiC, HCIS Springer, JoC, IJITCC; Head of DCSE, SeoulTech, Korea
- Dr. Xuewei Qi; Research Faculty & PI, Center for Environmental Research and Technology, University of California, Riverside, California, USA
- Dr. Charlie (Seungmin) Rho (Chapter Editor); Department Software, Sejong University, Gwangjin-gu, Seoul, Republic of Korea
- Prof. Abdel-Badeeh M. Salem; Head of Artificial Intelligence and Knowledge Engineering Research Labs and Professor of Computer Science, Faculty of Computer and Information Sciences, Ain Shams University, Cairo, Egypt; Editor-In-Chief, Egyptian Computer Science Journal; Editor-In-Chief, International Journal of Bio-Medical Informatics and e-Health (IJBMIeH); Associate-Editor-In-Chief, International Journal of Applications of Fuzzy Sets and Artificial Intelligence (IJAFSAI)
- Dr. Akash Singh (Steering Committee); IBM Corporation, Sacramento, California, USA; Chartered Scientist, Science Council, UK; Fellow, British Computer Society; Member, Senior IEEE, AACR, AAAS, and AAAI; IBM Corporation, USA
- Ashu M. G. Solo (Publicity), Fellow of British Computer Society, Principal/R&D Engineer, Maverick Technologies America Inc.
- Dr. Tse Guan Tan; Faculty of Creative Technology and Heritage, Universiti Malaysia Kelantan, Malaysia
- Prof. Fernando G. Tinetti (Steering Committee); School of Computer Science, Universidad Nacional de La Plata, La Plata, Argentina; also at Comision Investigaciones Cientificas de la Prov. de Bs. As., Argentina
- Prof. Hahanov Vladimir (Steering Committee); Vice Rector, and Dean of the Computer Engineering Faculty, Kharkov National University of Radio Electronics, Ukraine and Professor of Design Automation Department, Computer Engineering Faculty, Kharkov; IEEE Computer Society Golden Core Member; National University of Radio Electronics, Ukraine
- Prof. Shiuh-Jeng Wang (Steering Committee); Director of Information Cryptology and Construction Laboratory (ICCL) and Director of Chinese Cryptology and Information Security Association (CCISA); Department of Information Management, Central Police University, Taoyuan, Taiwan; Guest Ed., IEEE Journal on Selected Areas in Communications.
- Dr. Todd Waskiewicz; US Air Force Research Lab, AFRL/RIED, USA

- Prof. Layne T. Watson (Steering Committee); Fellow of IEEE; Fellow of The National Institute of Aerospace; Professor of Computer Science, Mathematics, and Aerospace and Ocean Engineering, Virginia Polytechnic Institute & State University, Blacksburg, Virginia, USA
- Dr. Xiaokun Yang (Chapter Editor); College of Science and Engineering, University of Houston Clear Lake, Houston, Texas, USA
- Prof. Jane You (Steering Committee); Associate Head, Department of Computing, The Hong Kong Polytechnic University, Kowloon, Hong Kong

Applied Cognitive Computing: ACC 2020 – Program Committee

- Prof. Emeritus Hamid R. Arabnia (Steering Committee); University of Georgia, USA; Editor-in-Chief, Journal of Supercomputing (Springer); Editor-in-Chief, Transactions of Computational Science & Computational Intelligence (Springer); Fellow, Center of Excellence in Terrorism, Resilience, Intelligence & Organized Crime Research (CENTRIC).
- Prof. Juan Jose Martinez Castillo; Director, The Acantelys Alan Turing Nikola Tesla Research Group and GIPEB, Universidad Nacional Abierta, Venezuela
- Prof. Leonidas Deligiannidis (Steering Committee); Department of Computer Information Systems, Wentworth Institute of Technology, Boston, Massachusetts, USA
- Dr. Ken Ferens (Co-Chair); Department of Electrical and Computer Engineering, University of Manitoba, Winnipeg, MB, Canada
- Prof. George A. Gravvanis (Steering Committee); Director, Physics Laboratory & Head of Advanced Scientific Computing, Applied Math & Applications Research Group; Professor of Applied Mathematics and Numerical Computing and Department of ECE, School of Engineering, Democritus University of Thrace, Xanthi, Greece.
- Prof. Byung-Gyu Kim (Steering Committee); Multimedia Processing Communications Lab.(MPCL), Department of CSE, College of Engineering, SunMoon University, South Korea
- Prof. Tai-hoon Kim; School of Information and Computing Science, University of Tasmania, Australia
- Prof. Louie Lolong Lacatan; Chairperson, Computer Engineerig Department, College of Engineering, Adamson University, Manila, Philippines; Senior Member, International Association of Computer Science and Information Technology (IACSIT), Singapore; Member, International Association of Online Engineering (IAOE), Austria
- Dr. Changyu Liu; College of Mathematics and Informatics, South China Agricultural University, Guangzhou, P. R. China and Visiting scientist, School of Computer Science, Carnegie Mellon University, USA

- Dr. Muhammad Naufal Bin Mansor; Faculty of Engineering Technology, Department of Electrical, Universiti Malaysia Perlis (UniMAP), Perlis, Malaysia
- Prof. Dr., Eng. Robert Ehimen Okonigene (Steering Committee); Department of Electrical & Electronics Engineering, Faculty of Engineering and Technology, Ambrose Alli University, Edo State, Nigeria
- Prof. James J. (Jong Hyuk) Park (Steering Committee); DCSE, SeoulTech, Korea; President, FTRA, EiC, HCIS Springer, JoC, IJITCC; Head of DCSE, SeoulTech, Korea
- Prof. Fernando G. Tinetti (Steering Committee); School of Computer Science, Universidad Nacional de La Plata, La Plata, Argentina; also at Comision Investigaciones Cientificas de la Prov. de Bs. As., Argentina
- Prof. Hahanov Vladimir (Steering Committee); Vice Rector, and Dean of the Computer Engineering Faculty, Kharkov National University of Radio Electronics, Ukraine and Professor of Design Automation Department, Computer Engineering Faculty, Kharkov; IEEE Computer Society Golden Core Member; National University of Radio Electronics, Ukraine
- Prof. Shiuh-Jeng Wang (Steering Committee); Director of Information Cryptology and Construction Laboratory (ICCL) and Director of Chinese Cryptology and Information Security Association (CCISA); Department of Information Management, Central Police University, Taoyuan, Taiwan; Guest Ed., IEEE Journal on Selected Areas in Communications.
- Prof. Layne T. Watson (Steering Committee); Fellow of IEEE; Fellow of The National Institute of Aerospace; Professor of Computer Science, Mathematics, and Aerospace and Ocean Engineering, Virginia Polytechnic Institute & State University, Blacksburg, Virginia, USA
- Prof. Jane You; Associate Head, Department of Computing, The Hong Kong Polytechnic University, Kowloon, Hong Kong
- Dr. Farhana H. Zulkernine; Coordinator of the Cognitive Science Program, School of Computing, Queen's University, Kingston, ON, Canada

Contents

Part I Deep Learning, Generative Adversarial Network, CNN, and Applications

Fine Tuning a Generative Adversarial Network's Discriminator for Student Attrition Prediction 3
Eric Stenton and Pablo Rivas

Automatic Generation of Descriptive Titles for Video Clips Using Deep Learning 17
Soheyla Amirian, Khaled Rasheed, Thiab R. Taha, and Hamid R. Arabnia

White Blood Cell Classification Using Genetic Algorithm–Enhanced Deep Convolutional Neural Networks 29
Omer Sevinc, Mehrube Mehrubeoglu, Mehmet S. Guzel, and Iman Askerzade

Deep Learning–Based Constituency Parsing for Arabic Language 45
Amr Morad, Magdy Nagi, and Sameh Alansary

Deep Embedded Knowledge Graph Representations for Tactic Discovery 59
Joshua Haley, Ross Hoehn, John L. Singleton, Chris Ballinger, and Alejandro Carbonara

Pathways to Artificial General Intelligence: A Brief Overview of Developments and Ethical Issues via Artificial Intelligence, Machine Learning, Deep Learning, and Data Science 73
Mohammadreza Iman, Hamid R. Arabnia, and Robert Maribe Branchinst

Brain Tumor Segmentation Using Deep Neural Networks and Survival Prediction 89
Xiaoxu Na, Li Ma, Mariofanna Milanova, and Mary Qu Yang

xiii

Combination of Variational Autoencoders and Generative Adversarial Network into an Unsupervised Generative Model 101
Ali Jaber Almalki and Pawel Wocjan

Long Short-Term Memory in Chemistry Dynamics Simulation 111
Heng Wu, Shaofei Lu, Colmenares-Diaz Eduardo, Junbin Liang,
Jingke She, and Xiaolin Tan

When Entity Resolution Meets Deep Learning, Is Similarity Measure Necessary? 127
Xinming Li, John R. Talburt, Ting Li, and Xiangwen Liu

Generic Object Recognition Using Both Illustration Images and Real-Object Images by CNN 141
Hirokazu Watabe, Misako Imono, and Seiji Tsuchiya

A Deep Learning Approach to Diagnose Skin Cancer Using Image Processing 147
Roli Srivastava, Musarath Jahan Rahamathullah, Siamak Aram,
Nathaniel Ashby, and Roozbeh Sadeghian

Part II Learning Strategies, Data Science, and Applications

Effects of Domain Randomization on Simulation-to-Reality Transfer of Reinforcement Learning Policies for Industrial Robots 157
C. Scheiderer, N. Dorndorf, and T. Meisen

Human Motion Recognition Using Zero-Shot Learning 171
Farid Ghareh Mohammadi, Ahmed Imteaj, M. Hadi Amini,
and Hamid R. Arabnia

The Effectiveness of Data Mining Techniques at Estimating Future Population Levels for Isolated Moose Populations 183
Charles E. Knadler

Unsupervised Classification of Cell-Imaging Data Using the Quantization Error in a Self-Organizing Map 201
Birgitta Dresp-Langley and John M. Wandeto

Event-Based Keyframing: Transforming Observation Data into Compact and Meaningful Form 211
Robert Wray, Robert Bridgman, Joshua Haley, Laura Hamel,
and Angela Woods

An Incremental Learning Scheme with Adaptive Earlystopping for AMI Datastream Processing 223
Yungi Ha, Changha Lee, Seong-Hwan Kim, and Chan-Hyun Youn

Contents

Traceability Analysis of Patterns Using Clustering Techniques 235
Jose Aguilar, Camilo Salazar, Julian Monsalve-Pulido, Edwin Montoya,
and Henry Velasco

**An Approach to Interactive Analysis of StarCraft: BroodWar
Replay Data** 251
Dylan Schwesinger, Tyler Stoney, and Braden Luancing

**Merging Deep Learning and Data Analytics for Inferring
Coronavirus Human Adaptive Transmutability and Transmissibility** ... 263
Jack Y. Yang, Xuesen Wu, Gang Chen, William Yang, John R. Talburt,
Hong Xie, Qiang Fang, Shiren Wang, and Mary Qu Yang

Activity Recognition for Elderly Using Machine Learning Algorithms .. 277
Heba Elgazzar

**Machine Learning for Understanding the Relationship Between
Political Participation and Political Culture** 297
A. Hannibal Leach and Sajid Hussain

**Targeted Aspect-Based Sentiment Analysis for Ugandan Telecom
Reviews from Twitter** 311
David Kabiito and Joyce Nakatumba-Nabende

A Path-Based Personalized Recommendation Using Q Learning 323
Hyeseong Park and Kyung-Whan Oh

Reducing the Data Cost of Machine Learning with AI: A Case Study ... 335
Joshua Haley, Robert Wray, Robert Bridgman, and Austin Brehob

Judging Emotion from EEGs Using Pseudo Data 345
Seiji Tsuchiya, Misako Imono, and Hirokazu Watabe

Part III   Neural Networks, Genetic Algorithms, Prediction Methods, and Swarm Algorithms

**Using Neural Networks and Genetic Algorithms for Predicting
Human Movement in Crowds** 353
Abdullah Alajlan, Alaa Edris, Robert B. Heckendorn, and Terence Soule

**Hybrid Car Trajectory by Genetic Algorithms with Non-Uniform
Key Framing** 369
Dana Vrajitoru

Which Scaling Rule Applies to Artificial Neural Networks 381
János Végh

Growing Artificial Neural Networks 409
John Mixter and Ali Akoglu

Neural-Based Adversarial Encryption of Images in ECB Mode with 16-Bit Blocks 425
Pablo Rivas and Prabuddha Banerjee

Application of Modified Social Spider Algorithm on Unit Commitment Solution Considering the Uncertainty of Wind Power in Restructured Electricity Market 437
Heidar Ali Shayanfar, Hossein Shayeghi, and L. Bagherzadeh

Predicting Number of Personnel to Deploy for Wildfire Containment ... 449
John Carr, Matthew Lewis, and Qingguo Wang

An Evaluation of Bayesian Network Models for Predicting Credit Risk on Ugandan Credit Contracts 461
Peter Nabende, Samuel Senfuma, and Joyce Nakatumba-Nabende

Part IV   Artificial Intelligence – Fundamentals, Applications, and Novel Algorithms

Synthetic AI Nervous/Limbic-Derived Instances (SANDI) 477
Shelli Friess, James A. Crowder, and Michael Hirsch

Emergent Heterogeneous Strategies from Homogeneous Capabilities in Multi-Agent Systems 491
Rolando Fernandez, Erin Zaroukian, James D. Humann,
Brandon Perelman, Michael R. Dorothy, Sebastian S. Rodriguez,
and Derrik E. Asher

Artificially Intelligent Cyber Security: Reducing Risk and Complexity 499
John N. Carbone and James A. Crowder

Procedural Image Generation Using Markov Wave Function Collapse 525
Pronay Peddiraju and Corey Clark

Parallel Algorithms to Detect and Classify Defects in Surface Steel Strips 543
Khaled R. Ahmed, Majed Al-Saeed, and Maryam I. Al-Jumah

Lightweight Approximation of Softmax Layer for On-Device Inference 561
Ihor Vasyltsov and Wooseok Chang

A Similarity-Based Decision Process for Decisions' Implementation 571
Maryna Averkyna

Dynamic Heuristics for Surveillance Mission Scheduling with Unmanned Aerial Vehicles in Heterogeneous Environments 583
Dylan Machovec, James A. Crowder, Howard Jay Siegel, Sudeep Pasricha,
and Anthony A. Maciejewski

Contents

Would You Turn on Bluetooth for Location-Based Advertising? 607
Heng-Li Yang, Shiang-Lin Lin, and Jui-Yen Chang

Adaptive Chromosome Diagnosis Based on Scaling Hierarchical Clusters 619
Muhammed Akif Ağca, Cihan Taştan, Kadir Üstün, and Ibrahim Halil Giden

Application of Associations to Assess Similarity in Situations Prior to Armed Conflict 641
Ahto Kuuseok

A Multigraph-Based Method for Improving Music Recommendation .. 651
James Waggoner, Randi Dunkleman, Yang Gao, Todd Gary, and Qingguo Wang

A Low-Cost Video Analytics System with Velocity Based Configuration Adaptation in Edge Computing 667
Woo-Joong Kim and Chan-Hyun Youn

Hybrid Resource Scheduling Scheme for Video Surveillance in GPU-FPGA Accelerated Edge Computing System 679
Gyusang Cho, Seong-Hwan Kim and Chan-Hyun Youn

Artificial Psychosocial Framework for Affective Non-player Characters 695
Lawrence J. Klinkert and Corey Clark

A Prototype Implementation of the NNEF Interpreter 715
Nakhoon Baek

A Classifier of Popular Music Online Reviews: Joy Emotion Analysis .. 721
Qing-Feng Lin and Heng-Li Yang

Part V Hardware Acceleration in Artificial Intelligence (Chair: Dr. Xiaokun Yang)

A Design on Multilayer Perceptron (MLP) Neural Network for Digit Recognition 729
Isaac Westby, Hakduran Koc, Jiang Lu, and Xiaokun Yang

An LSTM and GAN Based ECG Abnormal Signal Generator 743
Han Sun, Fan Zhang, and Yunxiang Zhang

An IoT-Edge-Server System with BLE Mesh Network, LBPH, and Deep Metric Learning 757
Archit Gajjar, Shivang Dave, T. Andrew Yang, Lei Wu, and Xiaokun Yang

An Edge Detection IP of Low-Cost System on Chip for Autonomous Vehicles 775
Xiaokun Yang, T. Andrew Yang, and Lei Wu

Advancing AI-aided Computational Thinking in STEM (Science, Technology, Engineering & Math) Education (*Act*-STEM) 787
Lei Wu, Alan Yang, Anton Dubrovskiy, Han He, Hua Yan, Xiaokun Yang, Xiao Qin, Bo Liu, Zhimin Gao, Shan Du, and T. Andrew Yang

Realistic Drawing & Painting with AI-Supported Geometrical and Computational Method (*Fun-Joy*) 797
Lei Wu, Alan Yang, Han He, Xiaokun Yang, Hua Yan, Zhimin Gao, Xiao Qin, Bo Liu, Shan Du, Anton Dubrovskiy, and T. Andrew Yang

Part VI   Artificial Intelligence for Smart Cities (Chair: Dr. Charlie (Seungmin) Rho)

Training-Data Generation and Incremental Testing for Daily Peak Load Forecasting 807
Jihoon Moon, Sungwoo Park, Seungmin Jung, Eenjun Hwang, and Seungmin Rho

Attention Mechanism for Improving Facial Landmark Semantic Segmentation 817
Hyungjoon Kim, Hyeonwoo Kim, Seongkuk Cho, and Eenjun Hwang

Person Re-identification Scheme Using Cross-Input Neighborhood Differences 825
Hyeonwoo Kim, Hyungjoon Kim, Bumyeon Ko, and Eenjun Hwang

Variational AutoEncoder-Based Anomaly Detection Scheme for Load Forecasting 833
Sungwoo Park, Seungmin Jung, Eenjun Hwang, and Seungmin Rho

Prediction of Clinical Disease with AI-Based Multiclass Classification Using Naïve Bayes and Random Forest Classifier 841
V. Jackins, S. Vimal, M. Kaliappan, and Mi Young Lee

A Hybrid Deep Learning Approach for Detecting and Classifying Breast Cancer Using Mammogram Images 851
K. Lakshminarayanan, Y. Harold Robinson, S. Vimal, and Dongwann Kang

Food-Type Recognition and Estimation of Calories Using Neural Network 857
R. Dinesh Kumar, E. Golden Julie, Y. Harold Robinson, and Sanghyun Seo

Progression Detection of Glaucoma Using K-means and GLCM Algorithm 863
S. Vimal, Y. Harold Robinson, M. Kaliappan, K. Vijayalakshmi, and Sanghyun Seo

Contents

Trend Analysis Using Agglomerative Hierarchical Clustering Approach for Time Series Big Data 869
P. Subbulakshmi, S. Vimal, M. Kaliappan, Y. Harold Robinson, and Mucheol Kim

Demand Response: Multiagent System Based DR Implementation 877
Faisal Saeed, Anand Paul, Seungmin Rho, and Muhammad Jamal Ahmed

t-SNE-Based K-NN: A New Approach for MNIST 883
Muhammad Jamal Ahmed, Faisal Saeed, Anand Paul, and Seungmin Rho

Short- to Mid-Term Prediction for Electricity Consumption Using Statistical Model and Neural Networks 889
Malik Junaid Jami Gul, Malik Urfa Gul, Yangsun Lee, Seungmin Rho, and Anand Paul

BI-LSTM-LSTM Based Time Series Electricity Consumption Forecast for South Korea 897
Malik Junaid Jami Gul, M. Hafid Firmansyah, Seungmin Rho, and Anand Paul

Part VII   XX Technical Session on Applications of Advanced AI Techniques to Information Management for Solving Company-Related Problems (Co-Chairs: Dr. David de la Fuente and Dr. Jose A. Olivas)

Inside Blockchain and Bitcoin 905
Simon Fernandez-Vazquez, Rafael Rosillo, Paolo Priore, Isabel Fernandez, Alberto Gomez, and Jose Parreño

Smart Marketing on Audiovisual Content Platforms: Intellectual Property Implications 913
Elisa Gutierrez, Cristina Puente, Cristina Velasco, and José Angel Olivas Varela

Priority Management in a Cybernetic Organization: A Simulation-Based Support Tool 923
J. C. Puche-Regaliza, J. Costas, B. Ponte, R. Pino, and D. de la Fuente

A Model for the Strategic Management of Innovation and R&D Based on Real Options Valuation: Assessing the Options to Abandon and Expand Clinical Trials in Pharmaceutical Firms 927
J. Puente, S. Alonso, F. Gascon, B. Ponte, and D. de la Fuente

Part VIII International Workshop – Intelligent Linguistic Technologies; ILINTEC'20 (Chair: Dr. Elena B. Kozerenko)

The Contrastive Study of Spatial Constructions *na* NP_{loc} in the Russian Language and 在NP上 in the Chinese Language in the Cognitive Aspect 935
Irina M. Kobozeva and Li Dan

Methods and Algorithms for Generating Sustainable Cognitive Systems Based on Thematic Category Hierarchies for the Development of Heterogeneous Information Resources in Technological and Social Spheres 951
Michael M. Charnine and Elena B. Kozerenko

Mental Model of Educational Environments............... 963
Natalia R. Sabanina and Valery S. Meskov

Part IX Applied Cognitive Computing

An Adaptive Tribal Topology for Particle Swarm Optimization 981
Kenneth Brezinski and Ken Ferens

The Systems AI Thinking Process (SATP) for Artificial Intelligent Systems............... 999
James A. Crowder and Shelli Friess

Improving the Efficiency of Genetic-Based Incremental Local Outlier Factor Algorithm for Network Intrusion Detection............... 1011
Omar Alghushairy, Raed Alsini, Xiaogang Ma, and Terence Soule

Variance Fractal Dimension Feature Selection for Detection of Cyber Security Attacks............... 1029
Samilat Kaiser and Ken Ferens

A Grid Partition-Based Local Outlier Factor for Data Stream Processing............... 1047
Raed Alsini, Omar Alghushairy, Xiaogang Ma, and Terrance Soule

A Cognitive Unsupervised Clustering for Detecting Cyber Attacks...... 1061
Kaiser Nahiyan, Samilat Kaiser, and Ken Ferens

A Hybrid Cognitive System for Radar Monitoring and Control Using the Rasmussen Cognition Model............... 1071
James A. Crowder and John N. Carbone

Assessing Cognitive Load via Pupillometry 1087
Pavel Weber, Franca Rupprecht, Stefan Wiesen, Bernd Hamann, and Achim Ebert

Contents

A Hybrid Chaotic Activation Function for Artificial Neural Networks .. 1097
Siobhan Reid and Ken Ferens

**Defending Aviation Cyber-Physical Systems from DDOS Attack
Using NARX Model** 1107
Abdulaziz A. Alsulami and Saleh Zein-Sabatto

**Simulated Annealing Embedded Within Personal Velocity
Update of Particle Swarm Optimization** 1123
Ainslee Heim and Ken Ferens

Cognitive Discovery Pipeline Applied to Informal Knowledge 1145
Nicola Severini, Pietro Leo, and Paolo Bellavista

Index 1153

Part I
Deep Learning, Generative Adversarial Network, CNN, and Applications

Fine Tuning a Generative Adversarial Network's Discriminator for Student Attrition Prediction

Eric Stenton and Pablo Rivas ⓘ

1 Introduction

Most colleges want to retain the number of freshman students enrolled and do what they can to prevent them from leaving within the first year. We will use the word "attrition" to describe students who have either dropped out or transferred to another college. A strong tool in lowering the amount of student attrition is the ability to predict who will leave as well as determine a trend or commonality between those who do leave. An inevitable problem with developing a good manner of prediction is the small amount of data that is available as a result of a typically small incoming class and the even smaller amount of those who leave. In other words, predicting student attrition in the first year can be proposed as an anomaly detection problem with a very limited amount of data to use in creating prediction models. In this paper, the freshman population of Marist College of years 2016 and 2017 will be examined using a GAN architecture in order to predict attrition in 2018. First, the neural network model learns the characteristics of a first-year student through adversarial learning. Second, the model is fine-tuned to classify students as either those who will stay or those who will leave. Third, the latent space of the layer directly before the final one that gives the final prediction is inspected for comparing three versions of the model. The versions are the following: The model traditionally trained without a GAN (the control), one adversarially trained without tuning, and one adversarially trained with tuning. The hypothesis is that the model that is adversarially trained with tuning will have a latent space more representative of the freshman population producing a higher accuracy when predicting student attrition.

E. Stenton (✉) · P. Rivas
Computer Science, Baylor University, Waco, TX, USA
e-mail: eric.stenton1@marist.edu; pablo.rivas@marist.edu

© Springer Nature Switzerland AG 2021
H. R. Arabnia et al. (eds.), *Advances in Artificial Intelligence and Applied Cognitive Computing*, Transactions on Computational Science and Computational Intelligence, https://doi.org/10.1007/978-3-030-70296-0_1

The following section will provide a brief background of the concepts in this paper. Following this section will be a description of the methodology used to test the models and how the models were built. The next section will be an overview of the three experiments performed, their accompanying diagrams, and a short explanation of the results. Finally, the last section will be a concluding paragraph on the findings of the experiments.

2 Background and Other Work

It is important to note this paper serves as an extension of research carried out by Dr. Eitel Lauria and colleagues in which the same population of students was used to predict attrition using multiple machine learning algorithms, the primary one being XGBoost [3]. Dr. Lauria's research produced models with accurate predictions of student attrition despite minimal amounts of data. This research extends the knowledge of neural models for student attrition introduced by E. Lauria et al. [8].

Current insights in GAN architectures originated in a paper by Dr. Ian Goodfellow et al. where the concept of a discriminator model and generator model playing a minimax game first arose [5]. Their paper shows the following value function for how the GAN operates:

$$\min_{G} \max_{D} V(D, G) = \mathbb{E}_{x \sim p_{\text{data}}(x)}[\log D(x)] + \mathbb{E}_{z \sim p_z(z)}[\log(1 - D(G(z)))]. \quad (1)$$

In the value function $V(D, G)$, G is a differential function representing the generator model that takes noise input $p_z(z)$ and maps it to a data space. This data space is meant to represent possible values that can mimic variables $p_{\text{data}}(x)$, real data, when inputted into another function represented by the discriminator model and denoted as D that outputs a prediction of whether the input was generated or not. D is trained to maximize the probability of correctly labeling generated and real samples while G is trained to minimize $\log(1 - D(G(z)))$, or lower the probability of D predicting correctly.

Shortly after Dr. Goodfellow's paper, the structure of the GAN training python code and the calculation of both the Wasserstein loss and gradient penalty for the training of the discriminator originated in an experiment from a paper by Martin Arjovsky, Soumith Chintala, and Léon Bottou [1]. The formula for the Wasserstein distance which is described in further detail in the referenced paper is the following:

$$W\left(\mathbb{P}_r, \mathbb{P}_g\right) = \inf_{\gamma \in \Pi(\mathbb{P}_r, \mathbb{P}_g)} \mathbb{E}_{(x,y) \sim \gamma}[\|x - y\|]. \quad (2)$$

In the Wasserstein distance equation, $\Pi(\mathbb{P}_r, \mathbb{P}_g)$ represents the set of all joint distributions $\gamma(x, y)$ with marginals $\mathbb{P}_r$ and $\mathbb{P}_g$, respectively. In order to transform distributions $\mathbb{P}_r$ into distribution $\mathbb{P}_g$, $\gamma(x, y)$ denotes the amount of "mass" to be transported from x to y while the Wasserstein distance describes the "cost" of the optimal method of transport.

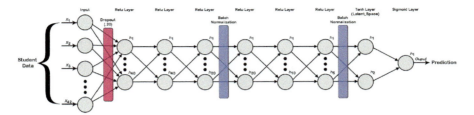

Fig. 1 Discriminator architecture diagram

The next section will describe the methodology for building the discriminator and generator models as well as how the Wasserstein distance equation will be utilized.

3 Methodology

The main pieces of GAN architectures are the discriminator and generator models as shown in Eq. 2. These models will be explained in this section in detail.

3.1 Discriminator

The discriminator is a neural model composed of 12 layers as shown in Fig. 1. These layers are: dropout, ReLU, batch normalization, tanh, and sigmoid. First, in order to prevent any one feature of the input data becoming heavily weighted, the dropout layer disconnects about 20% of the features randomly on each training step [9]. Second, batch normalization layers are placed intermittently to prevent the outputs of the ReLU layers from becoming too large and slowing or preventing convergence [4]. Third, the Python implementation of our model is based on Keras' functional model due to its ability to work with the tanh layer separately as this will serve as a view into the latent space of the model directly before an output is computed. Fourth, the discriminator's loss is based on weighing two Wasserstein loss calculations with a weight of one and a gradient penalty with a weight of ten.

3.2 Generator

The generator is a sequential model made up of 9 layers with a similar layout to the discriminator in which it has ReLU layers with intermittent batch normalization layers and an output consisting of a sigmoid layer as shown in Fig. 2. The most notable difference the generator has from the discriminator is the nature of its input

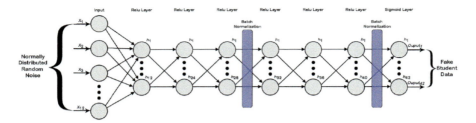

Fig. 2 Generator architecture diagram

which is 12 normally distributed random values between 0 and 1. These values are "noise" or values within a latent dimension defining different vectors that will eventually become generated data mimicking the input to the discriminator. This latent dimension should not be confused with the latent space referenced in this paper describing the output of the tanh layer in the discriminator. Furthermore, the generator's loss function is simpler than the discriminator's as it only consists of a single Wasserstein calculation.

The following section will present the three experiments conducted using the aforementioned models in detail as well as expound on the results of each.

4 Experiments and Results

Before getting into the details of the experiment, let us take a look at the input data. Table 1 describes the features and corresponding data types.

Some of the most noteworthy predictors in Table 1 are the following: "HSGPA," "DistanceInMiles," "MeritScholAmt," and "APCourses." "HSGPA" is a student's GPA from high school measured with a 4.0 scale. "DistanceInMiles" is the distance from a student's hometown to the college measured in miles. "MeritScholAmt" is the amount of money awarded to the student through a merit scholarship. Finally, the "APCourses" feature is a binary value where 1 means the student has taken AP courses and 0 means they have not. It is important to note that the majority of the aforementioned predictors relate to how well the student has done academically in high school. Furthermore, the "DistanceInMiles" predictor may indirectly relate to the student's emotional well-being as a larger distance away from their hometown may limit visits home. However, due to the difficulty in measuring the importance of predictors in a neural network, the speculation on the impact each feature has on predicting student attrition is rooted in the work by E. Lauria et al. where many of the same predictors are used and measured based on their importance in multiple machine learning models [8].

Besides the most important predictors, it is also imperative to point out the most "noisy" predictors, or those that have a large number of null values, which are the following: "DistanceInMiles," "OccupantsBuilding," "OccupantsRoom," and

Fine Tuning a GAN's Discriminator

Table 1 Description of predictors

Feature	Description	Data type
EarlyAction	Applied for early action	Binary (1/0)
EarlyDecision	Applied for early decision	Binary (1/0)
MeritScholAmt	Merit scholarship amount awarded	Binary (1/0)
FinAidRating	Financial aid rating	Categorical encoded as binary (1,0)
HSTier	High school tier	Categorical encoded as binary (1,0)
Foreign	Foreign student	Binary (1/0)
FAFSA	Applied for federal student aid	Binary (1/0)
APCourses	Took AP courses	Binary (1/0)
Sex	The sex of the student	Binary (1/0)
Athlete	Is a student athlete	Binary (1/0)
EarlyDeferral	Applied for early deferral	Binary (1/0)
WaitlistYN	Was waitlisted	Binary (1/0)
Commute	Is a commuter student	Binary (1/0)
HSGPA	High School GPA	Integer
DistanceInMiles	Distance from home (miles)	Integer
School	Member of a certain school, e.g., CC (ComSci & Math)	Categorical encoded as binary (1,0)
IsPellRecipient	Is recipient of Pell Grant	Binary (1/0)
IsDeansList	Joined Dean's List	Binary (1/0)
IsProbation	Is on probation	Binary (1/0)
OccupantsBuilding	Number of occupants in dorm	Integer
OccupantsRoom	Number of occupants in dorm room	Integer
IsSingleRoom	Uses a single room	Binary (1/0)
IsUnlimitedMealPlan	Has unlimited meal plan	Binary (1/0)
PercentHigherEd	Percent of those with higher education in home area	Float
GiniIndex	Gini Index value of home area	Float
MedianIncome	Median income of home area	Float
PercentWithInternet	Percent with internet in home area	Float
Attrited (Target)	Left the college	Binary (1/0)

"GiniIndex." As mentioned previously, "DistanceInMiles" is the amount of miles between the college and the student's hometown. "OccupantsBuilding" is the number of students that live in a student's dorm building. Similarly, "OccupantsRoom" is the number of students that live within the student's dorm room including themselves. Last, "GiniIndex" is the Gini coefficient of the student's hometown which is a measurement of income distribution in the area where a high value indicates greater inequality. In order to handle these features, the data is cleaned.

Our method of preprocessing the data includes removing any feature that is comprised of more than 30% of nulls and imputing the remaining features with missing values using K nearest neighbors (KNN). Additionally, the preprocessing step also included normalizing values between 0 and 1 for all integer and float type features. All categorical features mentioned in Table 1 are dummified.

After preprocessing the data, it is used to perform three experiments as described in the next few sections.

4.1 Experiment 1

In the first experiment, the GAN model was trained for 10,000 epochs. The weights were then transferred to two models, one that is tuned for 500 epochs to classify student attrition and the other that is left alone. This transference of weights is an example of transfer learning where the knowledge gained through adversarial training is applied to predicting student attrition (further details can be found in the referenced work) [6]. A control model was made from the same architecture as the GAN one, but trained separately on only the data previously used to tune for classification for 500 epochs. From the Receiver Operating Characteristic (ROC) diagrams, the control model performed marginally better with an accuracy of 0.68 than the tuned GAN model with only 0.64 accuracy. A ROC curve is a plot of the true positive rate against the false positive rate across various thresholds that determine the dividing line between classifications for a given model (more info in the provided reference) [2]. The accuracy of the GAN model, before tuning, is extremely low at 0.42. It is important to also note the discriminator and generator loss converging at about 10,000 epochs, or around the amount of epochs this experiment ran.

Directing our attention to the Cohen's kappa statistic, we observed that the relationship between the control and tuned model shows a kappa value of 0.5301 when a threshold resulting in about a 5% error rate is used. This value could be in the range of -1 to 1 and shows how close the model's outputs are where 1 is identical and anything 0 or below is akin to equivalent by chance [7]. The formula for the Cohen's kappa coefficient is the following:

$$\kappa = (p_o - p_e) / (1 - p_e).$$ (3)

The p_o variable in the equation is the observed agreement of the labels applied to a sample by the models while p_e is the probability of chance agreement. The aforementioned value 0.5301 demonstrates that the control and tuned models for this experiment are outputting predictions that are similar but also having a good number of discrepancies. The fact that they are different suggests that the models are fundamentally different in their output distributions which is desired. In the second experiment, we will see how the Cohen's kappa coefficients change.

4.2 Experiment 2

In the second experiment, the GAN model trained for 15,000 epochs. We observed that the discriminator and generator losses converged and begun separating again though on inverse sides. The GAN model, before tuning, still demonstrates a low accuracy and a latent space with a similar linear relationship as in experiment 1. The control model's accuracy remains at about 0.68 with 500 epochs of training. It is here that we see an improvement in the accuracy of the tuned model boasting a 0.69 which is 0.05 higher than its previous. Last, the kappa value for the control and tuned model is 0.4426 which is lower than in the first experiment when ran with a threshold resulting in about a 5% false positive rate despite the overall accuracy of the two models being different by a 0.01 margin. This means that despite their close accuracies, the two models are providing differing outputs which suggests the two models are correctly classifying students the other is misidentifying. The third and final experiment will demonstrate what happens to the Cohen's kappa coefficient when the accuracy of the tuned model is higher than the control model.

4.3 Experiment 3

In the third experiment, the GAN model trained for 20,000 epochs. We chose 20,000 as the largest amount of epochs for an experiment due to the losses converging at about 10,000 epochs and to see how well the model performed with a large number of epochs at about double the point of loss convergence. The loss and kappa statistic results are shown in Fig. 3. As shown in (a), the discriminator and generator losses converged, separated, and continued to grow apart though on inverse sides to where they began. When we take a look at the kappa score in (b), where the control and tuned models are predicting at a threshold resulting in about a 5% false positive rate, it is higher than the previous two experiments. Here, we see that their output similarity is measured to be a 0.6358 kappa score. This increase in the kappa score is expected since both models have an increased accuracy from the previous two experiments which naturally leads to their outputs being similar as they both are making more correct predictions. While this value is higher than in experiment 1 and 2, it still demonstrates the predictions of the two models show a noteworthy degree of discrepancy and produce different output distributions.

Figure 4 shows the GAN model before tuning. In (a) observe a low accuracy though now with a noticeably different latent space, (b), that seems to still have some semblance of a linear relationship with a high amount of data clumping at the bottom left corner and some at the top right corner. This can be explained by the nature of hyperbolic tangent activation function which aims to pull separate classes into opposite sides of the quadrants.

Figure 5 shows the control model, which was able to reach an accuracy of 0.69 with 500 epochs of training (a). However, it is still 0.01 below the tuned model in

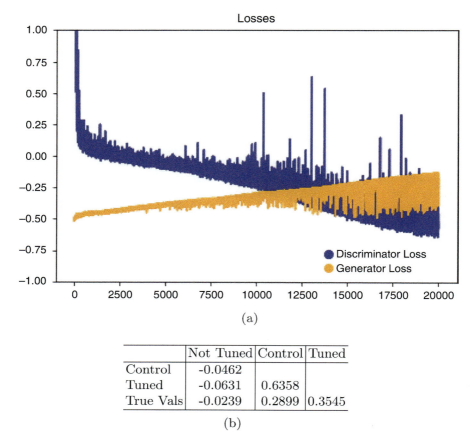

Fig. 3 Experiment 3 results: loss and Kappa statistic. (**a**) Discriminator and generator loss. (**b**) Cohen's Kappa statistics

this experiment as it reached a 0.70 accuracy. In (b) we observe that the data points are more spread-out in the latent space while still pushing to have separate classes in opposite sides of the quadrants.

Finally, Fig. 6 depicts tuned model ROC (a) and its corresponding latent space (b). In comparison to the control model in Fig. 5 (b) we see groupings of attrited students in the upper right corner suggesting there may be a correlation in the predictors for these cases. The AUC and ROC are similar in both the control and tuned models; however, it is evident that the tuned model gives an advantage over the traditional approach.

In the next paragraphs we will discuss the results of the three experiments in more detail.

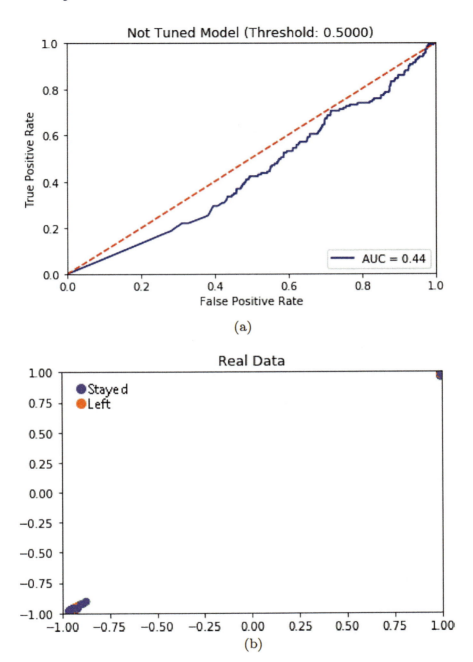

Fig. 4 Experiment 3 results: not tuned GAN model ROC and latent space. (**a**) Not tuned GAN model ROC. (**b**) Not tuned GAN model latent space

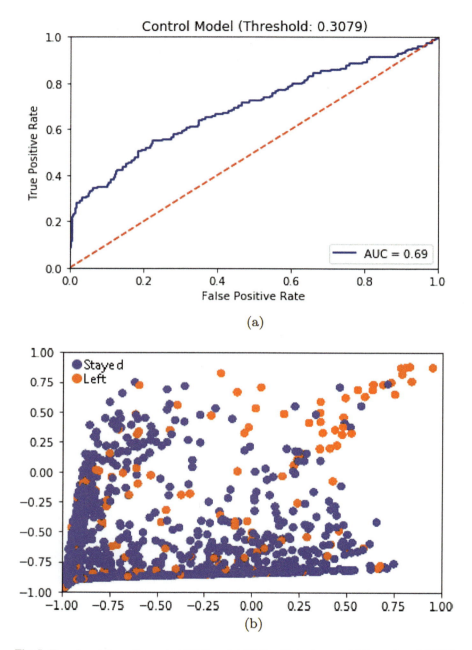

Fig. 5 Experiment 3 results: control GAN model ROC and latent space. (**a**) Control model ROC. (**b**) Control model latent space

Fine Tuning a GAN's Discriminator

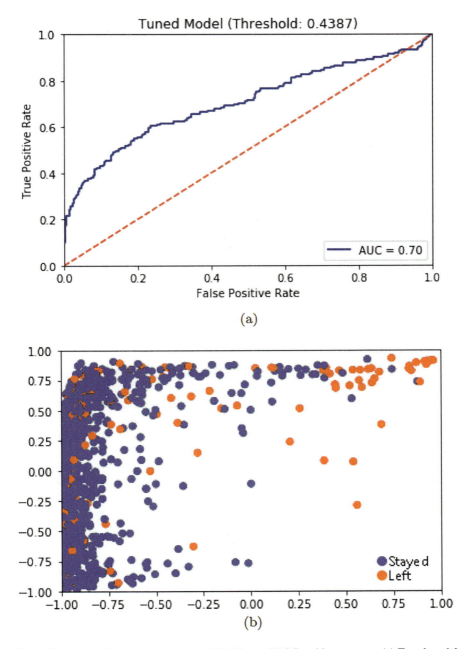

Fig. 6 Experiment 3 training results: tuned GAN model ROC and latent space. (**a**) Tuned model ROC. (**b**) Tuned model latent space

```
Accuracy.........: 87.8347
Precision........: 45.1923
Recall...........: 31.5436
FP Rate..........:4.9223
Balanced Accuracy: 45.1923
F1-score.........: 31.5436
ROC AUC (probs)..: 0.6936
Confusion matrix.:
 [[1101    57]
  [ 102    47]]
```

```
Accuracy.........: 88.2938
Precision........: 48.0769
Recall...........: 33.5570
FP Rate..........:4.6632
Balanced Accuracy: 48.0769
F1-score.........: 33.5570
ROC AUC (probs)..: 0.7048
Confusion matrix.:
 [[1104    54]
  [ 99    50]]
```

(a)

(b)

Fig. 7 Confusion matrices for experiment 3. (**a**) Control model C.M. (**b**) Tuned model C.M.

4.4 Results

The above three experiments demonstrate the effectiveness of adversarial training despite limited data and detecting a target between two very unbalanced classes. As the number of epochs of adversarial training increased, the accuracy of the tuned model is able to predict student attrition with a higher accuracy and a lower false positive rate. This can prominently be seen in experiment 3; using a threshold of about 0.31 and 0.44 for the control model and the tuned model, respectively.

With this model we are able to make more accurate predictions while remaining at about a 5% false positive rate as can be seen in the confusion matrices in Fig. 7. Furthermore, the Cohen's kappa values that relate the control and tuned models show that their outputs differ in each experiment to some degree suggesting the two models are predicting differently for a number of students.

Increasing the amount of epochs for the GAN training may produce higher accuracy for the tuned model, but examining the loss diagram shows the discriminator and generator losses converging and later diverging at about 10,000 epochs which may reveal a problem in the GAN training on which we comment next.

5 Conclusions

As can be seen in the experiments, the classifier model of the GAN with tuning increases its accuracy the more epochs it trains and eventually is more accurate than the traditionally trained model. While the overall accuracy of either model is low, any increase in the ability to detect anomalous students who leave during freshmen year in such a small sample decreases the amount of false positives, which is important if the model is to be used in any official capacity. Thus, utilizing a GAN for better accuracy and a smaller false positive rate is a step in the right direction.

If we are to compare the latent space of the GAN trained model that is tuned and the traditionally trained model, the layout of the data in the latent space in one model seems to form a semblance of a reflection along the diagonal of the layout in the other. The difference between the aforementioned latent spaces suggests the tuned model uses what it has learned about the input data in its prediction which may be the reason behind its higher accuracy in experiment 2 and experiment 3. This is further exemplified by the kappa values in each experiment comparing the control model to the tuned model displaying the two are making predictions that differ to a notable degree. It should also be noted that the loss diagrams show the discriminator and generator losses converging at about 10,000 epochs where the discriminator loss then continues to decrease and the generator's loss increases until both reach a point of stability with no major changes in their loss. This is most likely due to the generator suffering from mode collapse. The generator outputs data with most of its values hovering around 0.5 for the columns containing binary values which is likely the reason behind the immediate divergence in the loss diagrams after 10,000 epochs. Looking to the future, the adversarial training may produce better results using some degree of reinforcement training in order to introduce a penalty in the generator for values that are not 1 or 0 in binary columns rather than solely relying on unsupervised training to avoid mode collapse. Furthermore, a paper by Akash Srivastava and others shows promise in reducing mode collapse using implicit variational learning which is explained in detail in the referenced paper [10]. Nonetheless, using GANs in situations of limited data and anomaly detection shows promising results that should be explored further.

Acknowledgments We want to thank the subject matter experts at Marist College, in particular, professor Eitel Lauria whose seminal work in the area inspired and motivated this experimental work. This work was supported in part by the New York State Cloud Computing and Analytics Center, and by the VPAA's office at Marist College.

References

1. M. Arjovsky, S. Chintala, L. Bottou, Wasserstein gan (2017, preprint). arXiv:1701.07875
2. V. Bewick, L. Cheek, J. Ball, Statistics review 13: receiver operating characteristic curves. Crit. Care **8**(6), 508 (2004)
3. T. Chen, C. Guestrin, Xgboost: a scalable tree boosting system, in *Proceedings of the 22nd ACM SIGKDD International Conference on Knowledge Discovery and Data Mining* (2016), pp. 785–794
4. G.E. Dahl, T.N. Sainath, G.E. Hinton, Improving deep neural networks for LVCSR using rectified linear units and dropout, in *2013 IEEE International Conference on Acoustics, Speech and Signal Processing* (IEEE, Piscataway, 2013), pp. 8609–8613
5. I. Goodfellow, J. Pouget-Abadie, M. Mirza, B. Xu, D. Warde-Farley, S. Ozair, A. Courville, Y. Bengio, Generative adversarial nets, in *Advances in Neural Information Processing Systems* (2014)
6. Z. Huang, Z. Pan, B. Lei, Transfer learning with deep convolutional neural network for SAR target classification with limited labeled data. Remote Sens. **9**(9), 907 (2017)
7. T.O. Kvålseth, Note on Cohen's kappa. Psychol. Rep. **65**(1), 223–226 (1989)

8. E.J.M. Lauria, E. Stenton, E. Presutti, Boosting early detection of spring semester freshmen attrition: a preliminary exploration, in *International Conference on Computer Supported Education* (2020)
9. N. Srivastava, G. Hinton, A. Krizhevsky, I. Sutskever, R. Salakhutdinov, Dropout: a simple way to prevent neural networks from overfitting. J. Mach. Learn. Res. **15**(1), 1929–1958 (2014)
10. A. Srivastava, L. Valkov, C. Russell, M.U. Gutmann, C. Sutton, Veegan: reducing mode collapse in GANs using implicit variational learning, in *Advances in Neural Information Processing Systems* (2017), pp. 3308–3318

Automatic Generation of Descriptive Titles for Video Clips Using Deep Learning

Soheyla Amirian, Khaled Rasheed, Thiab R. Taha, and Hamid R. Arabnia

1 Introduction

The use of very large neural networks as Deep Learning methods that are inspired by the human brain system has recently dominated most of the researchers' work in several domains to help in improving the results and make it more desirable for people. Machine Translation, Self-driving cars, Robotics [23], Digital Marketing, Customer Services, and Better Recommendations are some applications for deep learning. In more recent years, deep learning [5] has positively and significantly impacted the field of image recognition specifically, allowing much more flexibility. In this research, we attempt to utilize image/video captioning [18, 24] methods and Natural Language Processing systems to generate a sentence as a title for a long video that could be useful in many ways. Using an automated system instead of watching many videos to get titles could be time-saving. It can also be used in the cinema industry, search engines, and supervision cameras to name a few. We present an example of the overall process in Fig. 1.

Image and video captioning with deep learning are used for the difficult task of recognizing the objects and actions in an image or video and creating a succinct meaningful sentence based on the contents found. Text summarization [2] is the task of generating a concise and fluent summary for a document(s) while preserving key information content. This paper proposes an architecture by utilizing the image/video captioning system and text summarization methods to make a title and an abstract for a long video. For constructing a story about a video, we extract the key-frames of the video which give more information, and then we feed those key-frames to the captioning system to make a caption or a document for them. For the

---

S. Amirian (✉) · K. Rasheed · T. R. Taha · H. R. Arabnia
Department of Computer Science, University of Georgia, Athens, GA, USA
e-mail: amirian@uga.edu

© Springer Nature Switzerland AG 2021
H. R. Arabnia et al. (eds.), *Advances in Artificial Intelligence and Applied Cognitive Computing*, Transactions on Computational Science and Computational Intelligence, https://doi.org/10.1007/978-3-030-70296-0_2

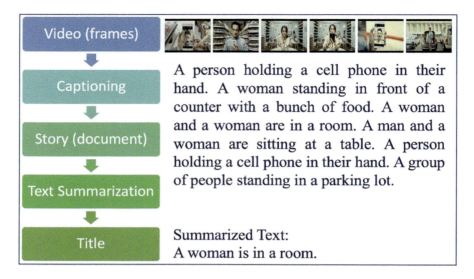

Fig. 1 This is an overall example of the proposed system. The key-frames of the video are selected and captioned. The resulting document is processed with the text summarization method and the output is a possible title for the corresponding video

captioning system, different methods as encoder-decoder or generative adversarial networks exist that propose different object detection methods. Also for the text summarization, we use both Extractive and abstractive methods [19] to generate the title and the abstract, respectively. We provide more details in the next sections.

The main contribution of this research is to explore the possibility of making a title and a concise abstract for a long video by utilizing deep learning technologies to save time through automation in many application domains. In the rest of the article, we describe the different parts of our proposed architecture which are image/video captioning and text summarization methods and provide a literature review for each component. Then, we explain the methodology of the proposed architecture and how it works. We present a proof of concept through experiments using publicly available datasets. The article is concluded with a discussion of the results and our future work.

2 Definition and Related Work

The proposed architecture in this paper consists of two main components, namely Image/Video captioning and Text summarization methods, each having different parts. In this section, we dissect each component. Also, we review some previous work about image/video captioning and text summarization that support parts of our proposed architecture.

2.1 Image/Video Captioning

Describing a short video in natural language is a trivial task for most people, but a challenging one for machines. From the methodological perspective, categorizing the models or algorithms is challenging because it is difficult to assert the contributions of the visual features and the adopted language model to the final description. Automatically generating natural language sentences describing a video clip generally has two components: extracting the visual information, as Encoder; and describing it in a grammatically correct natural language sentence as Decoder. With a convolutional neural network, the objects and features are extracted from the video frames, then a neural network is used to generate a natural sentence based on the available information, on which an image captioning method would be utilized for captioning the frames [4].

In the field of image captioning, Aneja et al. [6] developed a convolutional image captioning technique with existing LSTM techniques and also analyzed the differences between RNN based learning and their method. This technique contains three main components. The first and the last components are input/output word embeddings, respectively. However, while the middle component contains LSTM or GRU units in the RNN case, masked convolutions are employed in their CNN-based approach. This component is feed-forward without any recurrent function. Their CNN with attention (Attn) achieved comparable performance. They also experimented with an attention mechanism and attention parameters using the conv-layer activations. The results of the CNN+Attn method were increased relative to the LSTM baseline. For better performance on the MSCOCO they used ResNet features and the results show that ResNet boosts their performance on the MSCOCO. The results on MSCOCO with ResNet101 and ResNet152 were impressive.

In video captioning, Krishna et al. [15], however, presented Dense-captioning, which focuses on detecting multiple events that occur in a video by jointly localizing temporal proposals of interest and then describing each with natural language. This model introduced a new captioning module that uses contextual information from past and future events to jointly describe all events. They implemented the model on the ActivityNet Captions dataset. The captions that came out of ActivityNet shift sentence descriptions from being object-centric in images to action-centric in videos. Ding et al. [10] proposed novel techniques for the application of long video segmentation, which can effectively shorten the retrieval time. Redundant video frame detection based on the Spatio-temporal interest points (STIPs) and a novel super-frame segmentation are combined to improve the effectiveness of video segmentation. After that, the super-frame segmentation of the filteblue long video is performed to find an interesting clip. Key-frames from the most impactful segments are converted to video captioning by using the saliency detection and LSTM variant network. Finally, the attention mechanism is used to select more crucial information to the traditional LSTM. Generative Adversarial Networks help to have more flexibility in these methods [3]. Therefore, we can see that Sung Park et al. [24] applied Adversarial Networks in their framework. They propose to apply adversarial techniques during inference, designing a discriminator which encourages multi-sentence video description. They decouple a discriminator to evaluate visual relevance to the video, language diversity and fluency, and coherence across sentences on the ActivityNet Captions dataset.

Sequence models like recurrent neural network (RNN) [9] have been widely utilized in speech recognition, natural language processing, and other areas. Sequence models can address supervised learning problems like machine translation [8], name entity recognition, DNA sequence analysis, video activity recognition, and sentiment classification. *LSTM*, as a special RNN structure, has proven to be stable and powerful for modeling long-range dependencies in various studies. LSTM can be adopted as a building block for complex structures. The complex unit in Long Short-Term Memory is called a memory cell. Each memory cell is built around a central linear unit with a fixed self-connection [13]. LSTM is historically proven to be more powerful and more effective than a regular RNN since it has three gates (forget, update, and output). Long Short-Term Memory recurrent neural networks can be used to generate complex sequences with long-range structure [14, 26].

2.2 Text Summarization

Automatic text summarization is the task of producing a concise and fluent summary while preserving key information content and overall meaning [2]. Extractive and Abstractive are the two main categories of summarization algorithms.

Extractive summarization systems form summaries by copying parts of the input. Extractive summarization is implemented by identifying the important sections of the text, processing, and combining them to form a meaningful summary. Abstractive summarization systems generate new phrases, possibly rephrasing or using words that were not in the original text. Abstractive summaries are generated by interpreting the raw text and generating the same information in a different and concise form by using complex neural network-based architectures such as RNNs and LSTMs. Paulus et al. [19] proposed a neural network model with a novel intra-attention that attends over the input and continuously generated output separately and a new training method that combines standard supervised word prediction and reinforcement learning (RL). Also, Roul et al. [20] introduced the landscape of transfer learning techniques for NLP with a unified framework that converts every language problem into a text-to-text format.

Text summarization can be further divided into two categories: single and multi-text summarization. In single text summarization [21], the text is summarized from one document, whereas Multi-document text summarization systems are able to generate reports that are rich in important information, and present varying views that span multiple documents.

3 Methodology

The proposed architecture consists of two different, complementary processes: Video Captioning and Text Summarization. In the first process for video captioning, the system gets a video as an input, then generates a story for the video. The generated story will feed to the second process as a document and it summarizes the document to a sentence and an abstract. Figure 2 shows the complete process of the suggested architecture. Further, we explain the details of each part.

3.1 Video to Document Process

Image/video description is the automatic generation of meaningful sentences that describes the events in an image/video (frames). A video consists of many frames each representing an image. Some of the images/frames give much information and some are just basically repeating a scene. Therefore, we select some key-frames that include more information. The in-between frames are just repeating with subtle changes. A sequence of key-frames defines which movement the viewer will see. Therefore, the order of the key-frames on the video or animation defines the timing of the movement.

One of our contributions in this research is doing some experiments by selecting different key-frames to have a story for long videos to see if we can have the same extracted information. So, one task is to get the key-frames and process them to be captioned, instead of using all the frames of the video to save time and resources for getting the same result. See Fig. 3 for an illustration of the frames, key-frames, and in-between frames.

The captioning part consists of two phases: Encoder and Decoder. The Encoder part extracts the image information using convolutional neural networks like object detection methods to extract the objects and actions and then put them in a vector. ResNet, DenseNet, RCNN series, Yolo, and ... [5] can be used as object detection methods. Then the vector enters the decoder phase. The Decoder gets the vector and then with RNN methods generate a meaningful caption for the image. These

Fig. 2 This is the overall architecture of the proposed method that parts to two separate process for the video captioning and text summarization

Fig. 3 Video frames: in-between frames and the key-frames. We observe that many frames are repeating

two phases could work simultaneously. Figure 4 illustrates the captioning process. Captions are evaluated using the BLEU, METEOR, CIDEr, and other metrics [7, 25, 27]. These metrics are common for comparing the different image and video captioning models and have varying degrees of similarity with human judgment [22].

3.2 Document to Title Process

For generating and assigning a title to the video clip, we use an extractive text summarization technique. To keep it simple, we are using an unsupervised learning approach to find the sentence similarity and rank them [11]. The process is that we give the produced document as an input, it splits the whole document into sentences, then it removes stop words, builds a similarity matrix, generates rank based on the matrix, and at the end, it picks the top N sentences for a descriptive title. Figure 5 shows an example of the extractive text summarization system. Also for having an abstract, we implemented and used the abstractive text summarization method for the video [12]. Abstractive summarization methods interpret and examine the text by using advanced natural language techniques in order to generate a new shorter text that conveys the most important information from the original text [2].

Descriptive Titles for Video Clips

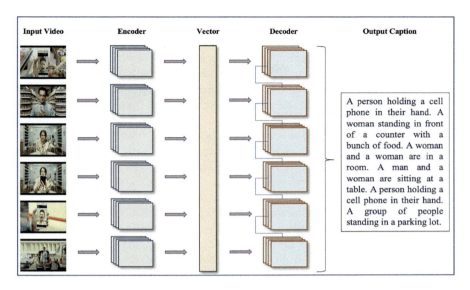

Fig. 4 The task of video captioning can be divided logically into two modules: one module is based on an image-based model which extracts important features and nuances from video frames; another module is based on a language-based model, which translates the features and objects produced with the image-based model to meaningful sentences

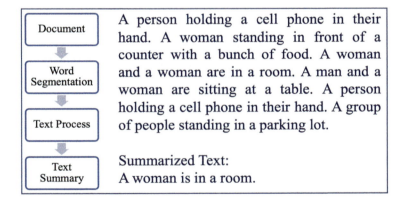

Fig. 5 The output of the captioning system is a document(s) that is an input to the extractive text summarization method. Then the text would be processed to weight the words, and it shows the most likely descriptive title that the text could have

4 Experiments

The main goal of our experiments is to evaluate the utility of the proposed architecture as a proof of concept. For implementing our idea, first, we need to get a story as a document from the image/video captioning model. So, for a given video we explore the frames, then feed the selected key-frames to the system to get the description. Implementing this part, captions have been generated by the by Luo et al. [18] method. The encoder has been trained with the COCO dataset. In fact, we utilized an image captioning system for this part. Some of the videos have been selected from the YouTube-8M dataset composed of almost 8 million videos totaling 500K hours of video [1] and some from the COCO dataset [17]. The captioning method has been trained and evaluated on the COCO dataset, which includes 113,287 images for training, 5,000 images for validation, and another 5,000 held out for testing. Each image is associated with five human captions.

For the image encoder in the retrieval and FC captioning model, Resnet-101 is used. For each image, the global average pooling of the final convolutional layer output is used, results in a vector of dimension 2048. The spatial features are extracted from the output of a Faster R-CNN with ResNet-101 [5], trained with the object and attribute annotations from Visual Genome [16]. Both the FC features and Spatial features are pre-extracted, and no fine-tuning is applied to image encoders. For captioning models, the dimension of LSTM hidden state, image feature embedding, and word embedding are all set to 512. The retrieval model uses GRU-RNN to encode text. The word embedding has 300 dimensions and the GRU hidden state size and joint embedding size are 1024 [18]. The captions generated with this model describe valuable information about the frames. However, richer and more diverse sources of training signal may further improve the training of caption generators.

The Text Summarization method that has been used in the first experiment is extractive and single summarization. First, we read the generated document from the previous process. Then, we generate a Similarly Matrix across the sentences. We then rank the sentences in the similarity matrix. And at the end, we sort the rank and pick the top sentence. Figure 6 shows some experiments that have been done. The videos are selected from the YouTube-8M dataset [1] and some from the COCO dataset [17]. The reader can find all the guidance and code here[1] for replicating the experiments for each part of the process.

In another experiment, we implement the abstractive text summarization method [20] to generate an abstract for the video clips, instead of assigning a title. Figure 7 shows the results by using Simple abstractive text summarization with pre-trained T5 (Text-To-Text Transfer Transformer) code [12].

---

[1] https://github.com/sohamirian/VideoTitle.

Descriptive Titles for Video Clips 25

A group of people on a beach near the water. A group of people sitting around a table. A man with a tennis racket in his hand. A group of people sitting around a table. A group of people standing around a table with food. A group of people standing around a table with food ...

Possible title:
A group of people sitting around a table.

A church with a clock tower in the middle. A group of dogs standing next to each other. A man and a woman standing in a kitchen. A black and white photo of a person sitting on a bed. A group of people standing around a fire hydrant ...

Possible title:
A man and a woman standing in a kitchen.

A close up of a tree in a forest. A couple of people that are sitting on a bench. A man is riding on the back of a truck. A group of people riding bikes down a street ...

Possible title:
A man is riding on the back of a truck.

A man is holding a skateboard in his hand. A group of men playing a game of soccer. A woman holding a tennis racquet on a tennis court. A man and a woman sitting on a couch. A man walking down a street with an umbrella. A man is standing in front of a building.
A street scene with focus on a street. A man riding a skateboard down the side of a ramp. A group of people walking down a street. A city street with a stop sign on the corner. A living room with a couch and a window ...

Possible title:
A man is holding a skateboard in his hand.

Fig. 6 These are different videos of YouTube-8M and ActivityNet dataset that we captioned with [18] and made a document. Then, the possible title is generated with an extractive text summarization algorithm for each document

A person holding a cell phone in their hand. A woman standing in front of a counter with a bunch of food. A woman and a woman are in a room. A man and a woman are sitting at a table. A person holding a cell phone in their hand. A group of people standing in a parking lot ...   Abstract: A person is holding cell phone in their hand. he stands in front of counter with food and eats food from the counter - and is sitting at table in the parking lot.
A group of people on a beach near the water. A group of people sitting around a table. A man with a tennis racket in his hand. A group of people sitting around a table. A group of people standing around a table with food. A group of people standing around a table with food ...   Abstract: A group of people sitting around table sits around the table with food. man with tennis racket in his hand. he is able to play tennis and swam in the water with his tv show on Thursday night.
A church with a clock tower in the middle. A group of dogs standing next to each other. A man and a woman standing in a kitchen. A black and white photo of a person sitting on a bed. A group of people standing around a fire hydrant ...   Abstract: Church with a clock tower in the middle. dogs standing next to each other. man and woman standing in kitchen. black and white photo of person sitting on bed.
A close up of a tree in a forest. A couple of people that are sitting on a bench. A man is riding on the back of a truck. A group of people riding bikes down a street ...   Abstract: A man is riding on the back of the truck. the group of people riding bikes down the street... and biking down street.
A man is holding a skateboard in his hand. A group of men playing a game of soccer. A woman holding a tennis racquet on a tennis court. A man and a woman sitting on a couch. A man walking down a street with an umbrella. A man is standing in front of a building.   A street scene with focus on a street. A man riding a skateboard down the side of a ramp. A group of people walking down a street. A city street with a stop sign on the corner. A living room with a couch and a window ...   Abstract: A man is holding skateboard in his hand. he is standing in front of an umbrella and walking down the side of the ramp.

Fig. 7 Here are the results with the abstractive text summarization method, which generates a summary for each document

5 Conclusion

The purpose of this research is to propose an architecture that could generate an appropriate title and a concise abstract for a video by utilizing an image/video caption systems and text summarization methods to help in several domains such as search engines, supervision cameras, and the cinema industry. We utilized deep learning systems as captioning methods to generate a document describing a video. We then use extractive text summarization methods to assign a title and abstractive text summarization methods to create a concise abstract to the video. We explained the components of the proposed framework and conducted experiments using videos from different datasets. The results prove that the concept is valid. However, the results could become better by applying more improved frameworks of image/video captioning and text summarization methods.

In our future work, we plan to explore more recent techniques of image/video captioning systems to generate a more natural story to describe the video clips.

Therefore, the text summarization system could generate a better title using the extractive text summarization algorithms and a better abstract using the abstractive text summarization algorithms.

Acknowledgments We gratefully acknowledge the support of NVIDIA Corporation with the donation of the Titan V GPU used for this research.

References

1. S. Abu-El-Haija, N. Kothari, J. Lee, P. Natsev, G. Toderici, B. Varadarajan, S. Vijaya-narasimhan, Youtube-8m: a large-scale video classification benchmark (2016, preprint). arXiv:1609.08675
2. M. Allahyari, S. Pouriyeh, M. Assefi, S. Safaei, E.D. Trippe, J.B. Gutierrez, K. Kochut, Text summarization techniques: a brief survey (2017, preprint). arXiv:1707.02268
3. S. Amirian, K. Rasheed, T.R. Taha, H.R. Arabnia, Image captioning with generative adversarial network, in *2019 International Conference on Computational Science and Computational Intelligence (CSCI)* (2019), pp. 272–275
4. S. Amirian, K. Rasheed, T.R. Taha, H.R. Arabnia, A short review on image caption generation with deep learning, in *The 23rd International Conference on Image Processing, Computer Vision and Pattern Recognition (IPCV'19), World Congress in Computer Science, Computer Engineering and Applied Computing (CSCE'19)* (IEEE, Piscataway, 2019), pp. 10–18
5. S. Amirian, Z. Wang, T.R. Taha, H.R. Arabnia, Dissection of deep learning with applications in image recognition, in *2018 International Conference on Computational Science and Computational Intelligence; "Artificial Intelligence" (CSCI-ISAI)* (IEEE, 2018), pp. 1132–1138
6. J. Aneja, A. Deshpande, A.G. Schwing, Convolutional image captioning, in *Proceedings of the IEEE Conference on Computer Vision and Pattern Recognition* (2018), pp. 5561–5570
7. X. Chen, H. Fang, T.Y. Lin, R. Vedantam, S. Gupta, P. Dollár, C.L. Zitnick, Microsoft coco captions: data collection and evaluation server (2015, preprint). arXiv:1504.00325
8. K. Cho, B. Van Merriënboer, C. Gulcehre, D. Bahdanau, F. Bougares, H. Schwenk, Y. Bengio, Learning phrase representations using RNN encoder-decoder for statistical machine translation (2014, preprint). arXiv:1406.1078
9. J. Chung, C. Gulcehre, K. Cho, Y. Bengio, Empirical evaluation of gated recurrent neural networks on sequence modeling (2014, preprint). arXiv:1412.3555
10. S. Ding, S. Qu, Y. Xi, S. Wan, A long video caption generation algorithm for big video data retrieval. Futur. Gener. Comput. Syst. **93**, 583–595 (2019)
11. P. Dubey, Text summarization. https://github.com/edubey/text-summarizer. Accessed 04 April 2020
12. R. Goutham, Simple abstractive text summarization with pretrained T5- text-to-text transfer transformer. https://towardsdatascience.com/simple-abstractive-text-summarization-with-pretrained-t5-text-to-text-transfer-transformer-10f6d602c426. Accessed 06 June 2020
13. S. Hochreiter, J. Schmidhuber, Long short-term memory. Neural Comput. **9**(8), 1735–1780 (1997)
14. R. Kiros, R. Salakhutdinov, R.S. Zemel, Unifying visual-semantic embeddings with multi-modal neural language models (2014, preprint). arXiv:1411.2539
15. R. Krishna, K. Hata, F. Ren, L. Fei-Fei, J. Carlos Niebles, Dense-captioning events in videos, in *Proceedings of the IEEE International Conference on Computer Vision* (2017), pp. 706–715
16. R. Krishna, Y. Zhu, O. Groth, J. Johnson, K. Hata, J. Kravitz, S. Chen, Y. Kalantidis, L.J. Li, D.A. Shamma, et al., Visual genome: connecting language and vision using crowdsourced dense image annotations. Int. J. Comput. Vis. **123**(1), 32–73 (2017)

17. T.Y. Lin, M. Maire, S. Belongie, J. Hays, P. Perona, D. Ramanan, P. Dollár, C.L. Zitnick, Microsoft coco: common objects in context, in *European Conference on Computer Vision* (Springer, Berlin, 2014), pp. 740–755

18. R. Luo, B. Price, S. Cohen, G. Shakhnarovich, Discriminability objective for training descriptive captions, in *Proceedings of the IEEE Conference on Computer Vision and Pattern Recognition* (2018), pp. 6964–6974

19. R. Paulus, C. Xiong, R. Socher, A deep reinforced model for abstractive summarization (2017, preprint). arXiv:1705.04304

20. C. Raffel, N. Shazeer, A. Roberts, K. Lee, S. Narang, M. Matena, Y. Zhou, W. Li, P.J. Liu, Exploring the limits of transfer learning with a unified text-to-text transformer (2019, preprint). arXiv:1910.10683

21. R.K. Roul, S. Mehrotra, Y. Pungaliya, J.K. Sahoo, A new automatic multi-document text summarization using topic modeling, in *International Conference on Distributed Computing and Internet Technology* (Springer, Berlin, 2019), pp. 212–221

22. G.A. Sigurdsson, G. Varol, X. Wang, A. Farhadi, I. Laptev, A. Gupta, Hollywood in homes: crowdsourcing data collection for activity understanding, in *European Conference on Computer Vision* (Springer, Berlin, 2016), pp. 510–526

23. N. Soans, E. Asali, Y. Hong, P. Doshi, Sa-net: robust state-action recognition for learning from observations, in *IEEE International Conference on Robotics and Automation (ICRA)* (2020), pp. 2153–2159

24. J. Sung Park, M. Rohrbach, T. Darrell, A. Rohrbach, Adversarial inference for multi-sentence video description, in *Proceedings of the IEEE Conference on Computer Vision and Pattern Recognition Workshops* (2019)

25. R. Vedantam, C. Lawrence Zitnick, D. Parikh, Cider: consensus-based image description evaluation, in *Proceedings of the IEEE Conference on Computer Vision and Pattern Recognition* (2015), pp. 4566–4575

26. Y. Wu, M. Schuster, Z. Chen, Q.V. Le, M. Norouzi, W. Macherey, M. Krikun, Y. Cao, Q. Gao, K. Macherey, et al., Google's neural machine translation system: bridging the gap between human and machine translation (2016, preprint). arXiv:1609.08144

27. K. Xu, J. Ba, R. Kiros, K. Cho, A. Courville, R. Salakhudinov, R. Zemel, Y. Bengio, Show, attend and tell: neural image caption generation with visual attention, in *International Conference on Machine Learning* (2015), pp. 2048–2057

White Blood Cell Classification Using Genetic Algorithm–Enhanced Deep Convolutional Neural Networks

Omer Sevinc, Mehrube Mehrubeoglu, Mehmet S. Guzel, and Iman Askerzade

1 Introduction

White blood cells (WBC), also referred to as leukocytes, play an important role in the diagnosis of many diseases such as allergies, leukemia, AIDS, and cancer [21]. The number or density of the white blood cells in the blood flow provides clues about the state of a person's immune system as well as potential health risks. White blood cells are divided into five main types: eosinophils, lymphocytes, monocytes, neutrophils [49], and basophils [8]. A large change in the white blood cell count is often a sign that the body is affected by an antigen. Furthermore, a change in a WBC type is generally associated with an antigen. Devices exist to count WBC in each blood sample [36]. However, these devices are expensive and require a few hours to complete the WBC count in a blood sample [22].

The proposed deep convolutional neural network (D-CNN) algorithmic model offers an alternative solution to the WBC counting and classification problem. Along with the developing technologies, the emergence of artificial intelligence has paved the way for important developments in the field of healthcare and in almost every other field. Nowadays, with the advances made in hardware, such as processors and graphics cards, and the increase in the amount of recorded data, artificial intelligence models have come to the forefront of solutions with very successful classification

---

O. Sevinc (✉)
Ondokuz Mayis University, Samsun, Turkey
e-mail: osevinc@omu.edu.tr

M. Mehrubeoglu
Texas A&M University-Corpus Christi, Corpus Christi, TX, USA
e-mail: ruby.mehrubeoglu@tamucc.edu

M. S. Guzel · I. Askerzade
Ankara University, Computer Engineering, Ankara, Turkey

© Springer Nature Switzerland AG 2021
H. R. Arabnia et al. (eds.), *Advances in Artificial Intelligence and Applied Cognitive Computing*, Transactions on Computational Science and Computational Intelligence, https://doi.org/10.1007/978-3-030-70296-0_3

and analysis capabilities. Among effective classification techniques [23], the layered artificial intelligence model has become the star method of deep learning [13].

In previous studies, researchers were working on creating the best sets of attributes to express the problem and determine and select attributes with the highest representation of the data to solve the problem [35]. However, the design structure must address issues such as how one should design the multilayered artificial neural network with the deep learning approach; how many layers to use in the neural network; how many neurons to include in each layer; and which optimization algorithm or activation function to choose to solve the problem. Thus, solving problems with deep learning came to be equivalent to determining the optimal combination of the layers, neurons, algorithms, and activation functions when designing the multilayered network [3]. These parameters are called hyperparameters in deep learning and constitute the determinants of the model; therefore, hyperparameters of the model must be designed in an optimal way to achieve the highest accuracy and performance in the results.

Deep learning models, for example, for painting classification, have developed rapidly since 2010. Thus, data set hosts such as ImageNet and Kaggle are set to run machine learning and high-powered virtual machine services [28]. They create a competitive environment to create the most efficient model. Models such as LeNet [29], GoogleNet [15], VGG-19 [43], and ResNet [44] have come to the forefront in the image classification competitions. ImageNet [15, 27] hosts a popular competition called ILSVRC. The deep convolutional neural network models learn the parts obtained from the paintings in the layers and make inferences from the details to the whole, so that they can finally make a classification decision on the painting [17, 24]. Depth in D-CNN is an important parameter. Increasing the number of layers in the model increases the capacity to express features of image theoretically, and, therefore, increased layers are expected to create more successful network architectures. In practice, however, this is not the case, since the backpropagation structure is used and the weight values are multiplied by taking partial derivatives, thus losing the effect of direct multiplication in the first layers and resetting the weights. This causes some neurons, which are ineffective in the calculations, to "die." This is called the vanishing gradient problem. The opposite is also possible due to exploding gradient multipliers [39]. As the number of layers increases, the number of parameters increases, so both the calculation speed and learning rate may slow down. Of course, improvements are made in the solution by appropriate arrangement of hyperparameters. For example, each input is taken as batch normalization in groups of numbers, rather than individual number processing, thus avoiding excessive learning [18]. In addition, some of the neurons which do not contribute much to the training of the network are excluded, thus increasing both the performance and accuracy. [9].

Pretrained models are also used to reduce computation time and increase accuracy of the network. Data sets such as ImageNet and COCO provide millions of images and contain the best results created by weights that are stored. Thus, while training a new data set, the pre-identified weights can be run by loading them into the new model [7]. This provides the advantage of shortening the training period. For example, VGG-19 and ResNet have pretrained models. These models are focused on the last three layers by making changes only to these layers [10].

In this study, the deep convolutional neural network model is developed for the Kaggle white blood cell data set which is open to the public. The parameters affecting the operation of the model were evaluated to analyze the different results obtained. Optimum performance was targeted by improving the hyperparameters of the proposed GA-enhanced D-CNN model. Performance measurements were then compared with other deep CNN models.

The proposed model's performance was compared with those obtained in other studies [12, 49], using the Kaggle white blood cell data set [30, 31]. Here, it is shown that the proposed model has performed better than the existing models for this data set. Thus, it would be possible to achieve the white blood cell classification at incredibly low cost using the GA-enhanced D-CNN model via software, and to provide almost instant results. As deep learning models as well as hardware infrastructures and architectures evolve, the data sets will be improved, and the model will be renewed.

The rest of this chapter includes more details on white blood cells used in this study (Sect. 2), deep convolutional neural networks and Kaggle data set (Sect. 3), and finally the proposed GA-enhanced D-CNN with results (Sect. 4). Conclusions are summarized in Sect. 5.

2 White Blood Cells

White blood cells (WBC) are one of the most important indicators of the body's immune system. The number of eosinophils, lymphocytes, monocytes, and neutrophils seen in the total number and subgroups of the body's WBC, as shown in Fig. 1, are important indicators of the immune system and with what kind of diseases a person is affected [5].

For example, lymphocyte counts suggest chronic infections and leukemia. Monocytes represent acute, chronic infections, leukemia, as well as collagen tissue and granulomatous diseases. Neutrophil counts indicate acute bacterial infections. Finally, eosinophils refer to allergic, parasitic, hematological diseases, in addition to intestinal diseases [4, 16, 42].

To emphasize, leukemia patients often have a significantly higher level of lymphocytes in their blood flow due to the weakness of their immune system. Likewise, people with allergies often see an increase in the number of eosinophils because these WBC are the key to combating allergens; therefore, determining and analyzing the number of white blood cells in the blood stream can offer a strong quantitative picture of a person's health.

The devices used for white blood cell counting are measured by impedance, radio wave, and optical scattering methods [25]. For this reason, the studies to be carried out in this field will make very important contributions to both the healthcare sector

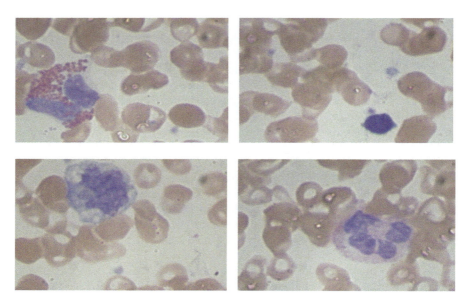

Fig. 1 (Top left) eosinophil; (top right) lymphocyte; (bottom left) monocyte; (bottom right) neutrophil

and the national budget by offering fast, efficient, and low-cost alternative solutions based on imaging and D-CNN models [14].

3 Deep Convolutional Network Model and Kaggle Data Set

With deep learning, a transition to design engineering is made directly by combining feature extraction and selection with classification, rather than feature extraction and selection first as in traditional machine learning. In machine learning, it is important to choose attributes with the highest ability to represent the data, and to remove the attributes that are redundant. However, with the development of deep learning, high performance is based on the successful design of the multilayer artificial neural network architecture to solve the problem at hand. The choice of parameters, such as the number of layers and neurons, the optimization algorithms to be used, and the activation function, all directly affect the operation of the model. The speed of training, the neural network that stands out as the main challenge with deep learning, as well as the adaptation of the network to solve the problems that require deepening the number of layers of the network have brought on new approaches and methods to feature extraction and classification. The development of graphic card processors in recent years has allowed researchers and other developers to model layered webs to achieve successful results. One of the methods implemented is the "death" of some neurons during the training of artificial neural networks. The death of neurons is allowed for those neurons whose weight values approach zero and, therefore, dissolve between the layer strands of the neurons. The associated neurons, thus, do not contribute to the training of the network since the network is multiplied by a partial derivative when fed back into the network. Another challenge related to deep neural network design is that even though the number of parameters does not increase as much as the expansion of the network, the number of parameters does increase. Consequently, more parameters mean increased processing and computation time, lengthening the learning time of the model. Pretrained models have emerged, and relatively accurate results have been obtained in general picture classification problems when analyzing unknown images. Pretrained models have previously been run on data sets containing millions of images, with the best results, weight values, activation and improvement functions already known. By adapting the final layers according to an individual problem, one can design a customized model and test the results on a new data set.

One of the most prominent approaches to solving the local optima and increasing the number of parameters is the ResNet model, which was first implemented in the 2015 ImageNet ILSVCR competition. ImageNet has data sets containing millions of images. ImageNet organizes a competition every year to achieve the best classification and reveal the best deep learning models. In this competition, deep convolutional neural network models were designed and developed as LeNet in 2008, as AlexNet in 2012, as GoogLeNet and VGGNet in 2014, and as ResNet in 2015, as shown in Fig. 2. Each network model implemented new approaches in the competition. In AlexNet model, for example, there are innovative ideas such as the links, in parallel to two deep convolutional neural networks that are trained in two separate graphics processors (GPUs), which are then designed to be modular in GoogLeNet, and can produce results in different parts of the web. Although one of the most important contributions of ResNet is increasing the depth of the network, values coming from the previous layers are added to the advanced layers with a short path (Fig. 3), reducing the process complexity and computation time. Even when the depth of the network is increased, the model is still successfully trained.

In the healthcare field, there are many studies in which very different types of images, such as chest, brain tumors, and lung cancer, are resolved with deep convolutional neural networks, with highly accurate results ([1, 2, 14]. In the work presented in this paper, a deep convolutional neural network model has been developed for white blood cell data set on the Kaggle website. The model has worked very successfully, resulting in higher performance than most of the existing models [12]. The proposed approach achieved the hyperparameters' optimization by using the genetic algorithm [32], and embedded these parameters to the D-CNN model to output the best results. The results of some of the studies showed higher accuracy while others displayed reduced number of parameters and computational time.

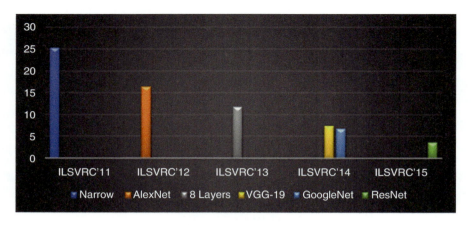

Fig. 2 The models and error rates in ILSVRC competition [40]

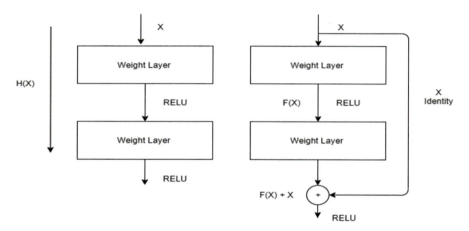

Fig. 3 ResNet shortcut approach [20]. X: input features; H(X): output; F(X): residual knowledge; F(X) = H(X) − X)

3.1 Genetic Algorithm

The Genetic Algorithm (GA) is one of the most effective and widely used optimization tools [45]. GA is modeled by mimicking nature, specifically evolution. GA uses evolution concepts, like reproduction and survival of the fittest, to solve a problem [32]. The main processes are crossover and mutation that include data being exchanged between two parent chromosomes to create two children. GA then proceeds to changing the children's genes to create diversity in the population. Crossover provides new solutions, as most of the information still stays. Crossover can be single-point, two-point, or multi-point crossover. The mutation consists of bit-flip, inversion, scramble, and immigration. Crossover and mutation are applied

genotype				
integer	integer	Integer	integer	float

Feature Maps	FM Length	FM Width	Batch Size	Dropout
		phenotype		

Fig. 4 Solution representation for the CNN using GA [19] (FM: Feature Map)

based on probabilities [41]. The probability of mutation default is set to 0.01 where the probability of crossover is mostly set to 0.80 based on existing literature [37, 38]. In the work presented here, GA is used as an optimization algorithm for the deep convolutional neural network model created to classify and count white blood cell types.

3.2 Chromosome Representation of CNN Optimization Using GA

To handle the hyperparameter values via GA, the chromosome which represents the hyperparameters is illustrated in Fig. 4. Five CNN features are presented, namely, the number of feature maps in each convolutional layer, the width for the number of neurons in layers, the depth for the number of the layers, the batch size for specifying the number of images to be fed as clusters in each epoch, and the dropout percentage to prevent overfitting. The feature map, batch size, feature map length, and width were encoded as integers whereas the dropout percentage was encoded as floating points.

3.3 Data Preprocessing

Image preprocessing was used in this study by resizing the images to a 90×120 canvas to speed up the training process of the CNN. After resizing, the images were then transformed to gray-level images in which they were flattened into a list of raw pixel intensities with 8-bit pixel depth.

3.4 Genetic Algorithm

GA was used to optimize the hyperparameters of the CNN, as described above. The hyperparameters selected for GA were crossover, which was set to a probability of 0.80 due to its importance, and mutation which was set to a probability of 0.01. Tournament selection and elitism was used. The model was run for 200 generations with a population size of 100.

3.5 Convolutional Neural Network Model

Through the optimized CNN hyperparameters achieved by GA, the deep CNN model was trained on 9957 images and tested on another dataset of 2478 images of WBC. The optimized number of convolutional layers were six. Two pooling layers were used in the developed model. The number of feature maps used was selected as 32 for a good starting point. The feature map length and width for features were 3 and 3, respectively. The dropout percentage, which is a regularization technique, was set to 0.10 to get optimum results, and the model was run for 25 epochs as there were no significant improvements beyond 25 epochs.

3.6 Kaggle Data Set

Kaggle is a highly respected website that has been widely used worldwide as a data set source for artificial intelligence, machine learning, and deep learning. Kaggle also allows developers to encode their model and run it on the on-site Kaggle servers [11]. Kaggle stands out with its data sets, especially in the field of healthcare, and is the only platform to open prize-winning competitions to develop the best artificial intelligence models on many different data sets. For this reason, the model presented in this chapter has been developed through Kaggle, and the developed model has been operated on the white blood cell data set available at Kaggle. Since other groups have worked with Kaggle data, it is possible to review and compare any studies with others on Kaggle.

4 GA-Enhanced D-CNN Model and Results

This study is designed to develop a deep convolutional neural network (D-CNN) model whose hyperparameters self-adjust with a genetic algorithm in order to optimize the results. The overall goal is to increase the classification accuracy of the white blood cells using the described genetic algorithm-optimized D-CNN model

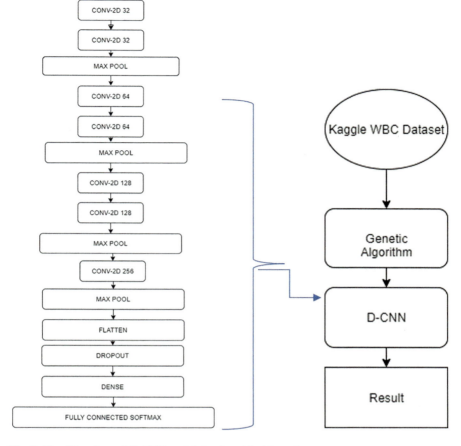

Fig. 5 The GA-enhanced D-CNN model developed in this study

that will result in a significant contribution to the healthcare field through correct diagnosis of many diseases, and, in turn, contribute to the economy by reducing the cost of such diagnoses. The model is designed as shown in Fig. 5, and its parameters can be adapted and created in an add-on architecture.

To implement the developed model, the white blood cell images obtained from Kaggle were run and tried separately in 60 × 80, 120 × 160, and 224 × 224 sizes, and the most successful size of 60x80 was chosen. While the model design was carried out, the depth of the CNN was increased starting from a shallow model structure. The model was finalized when the optimal results were achieved, as seen in Table 1. Since there exists a blurring problem [6] when using images from the microscope, or microscopic images, a 3 × 3 core filter was used and the number of filters was increased from 32 to 256, as the network progressively deepened [46].

Table 1 Determination of the number of layers of the GA-enhanced D-CNN Model

# of Layers	Parameter Count	Accuracy (%)
3	88,548	79.85
5	262,308	84.47
7	**584,228**	**93.3**
9	2,385,188	60.55

In convolutional neural networks, the convolution layer identifies the areas to be removed from the image, and after this process, the dimensions of the image are reduced. While the size of the image is reduced, the lines in the picture are highlighted by maximizing the 2×2 frames each time by jumping the pool two times. This results in decreased image size. As the layers become deeper, different attributes emerge, and the features are derived from the details. Images can be treated as separate inputs or as batch entries. In addition, groups are often prevented from overlearning and a shortened processing time is achieved. The proposed model is composed of two layers, which are formed by using three expansion pool layers and then completed with flattening, dropout, densification, and fully connected layers. Since the cell nuclei expected to be detected from the images of the white blood cells [31] taken as input are relatively small in the images, the convolution was achieved using 3×3 filters. In the convolution process, starting with 32 channels, the number of channels is increased to 64, then 128, and then to 256 to extract the detail properties. The L2 regularization technique was tested to improve the learning rate, but it was not used in the end due to the lack of successful group normalization (Batch Normalization) [48]. In addition, dilution (dropout) was applied to neurons with little effect on learning.

The activation function has been proven to be successful [47] and has been selected as the ReLU, which works very well with deep CNN. Adam learning rate has been chosen as the most effective function for adapting the learning rate to its own [26]. For classification decision, Softmax, which is one of the functions that result in successful convergence, is used. While the images were classified into four categories of white blood cells, the categorical cross-linking function was used to compute the error.

After the model was designed as described, it was trained and then tested. To improve the results, changes were made to the hyperparameters according to their purpose. The image size was reduced to 60×80 to simplify the process. The batch size was selected to be 32 in the presented model where the inputs were given in batches. Again, all the data were passed through the model and feedback error values were calculated by weighting the numbers determined during 25 epochs.

The created model was run on the Kaggle server where the graphics processor support was utilized. The proposed model was compared to existing models previously studied here. The proposed model resulted in 93.3% accuracy when compared to Inception V3 model, which produced 84.08% accuracy, and ResNet-50 model, which produced 87.62% accuracy. Figure 6 demonstrates the training and test error graphs over 25 epochs. Figure 7 provides the obtained accuracy. The accuracy values were calculated using Eqs. (1) and (2) below, where TP is true

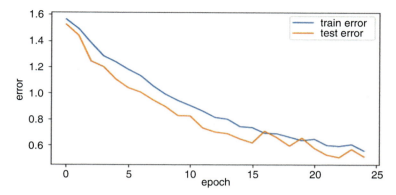

Fig. 6 Training and test data error graphs for 25 epochs. Vertical axis shows error values for each epoch

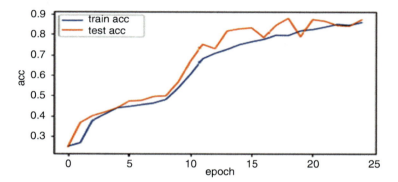

Fig. 7 Training and test data accuracy for 25 epochs (acc: accuracy)

positives, TN is true negatives, FP is false positives, and FN is false negatives.

$$\text{Accuracy} = \frac{TP + TN}{TP + FP + TN + FN} \qquad (1)$$

$$\text{Precision} = \frac{TP}{TP + FP} \qquad (2)$$

Figure 8 represents the training set and test set confusion matrices. In the confusion matrices, true (dark diagonal) and false (off diagonal) estimates are given by comparing the estimated or predicted values to the actual or real values.

Table 2 shows the comparative results from the proposed model and other convolutional deep learning models. The results demonstrate that the proposed

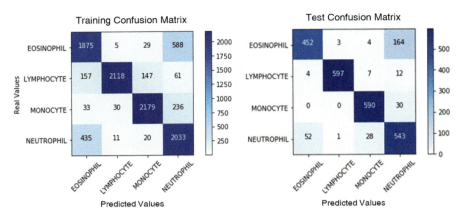

Fig. 8 Training (left) and test (right) dataset confusion matrices

Table 2 Comparison of performance of proposed and other models

Model	Accuracy (%)
Inception V3	84.08
ResNet-50	87.62
Proposed GA-DCNN	**93.30**

Table 3 Comparison of proposed model with other studies

Benchmarking with Other Studies		
Study	Model	Accuracy (%)
[33]	DCNN	78.33
[34]	Tiny YOLO - CNN	87.10
[22]	DCNN	83.84
Proposed model	**GA-DCNN**	**93.30**

method of GA-enhanced D-CNN model produced the optimal results with seven layers (Table 1) among the three models compared. All are ready models that can be trained on ImageNet images, and adapted to the current problem of white blood cell identification. Again, the performance of the proposed model surpasses the other two models studied on Kaggle as well as the literature, as seen in Table 3.

5 Conclusions

In this study, a GA-enhanced D-CNN model, based on artificial intelligence and deep learning methods, has been tested in the field of healthcare, which is perhaps today's most prominent technological solution that can rewrite the rules of medical research. The counts of white blood cells provide very important clues about individuals' health and are used to diagnose important diseases. Traditional devices used for counting blood cells are both costly and time consuming since the analysis of samples takes time. This study offers an alternative deep learning model that can be used to classify four types of white blood cells. This engineering study resulted in the design of a hyperparametric neural network architecture in deep learning, determining how to set hyperparameters such as number of layers, optimization, activation functions, and leading to a suggested model that will make a significant contribution to the diagnosis of potential diseases. On the other hand, the developed model is designed for the optimal performance by implementing a shallow structure towards a deeper structure and tested with different parameter values. The developed model was compared with other known architectures as well as other studies which have investigated and analyzed white blood cells. The model presented here is self-parameter adapting with the genetic algorithm (GA) resulting in successful evaluation results of 93.3% accuracy in classification of white blood cells. This GA-enhanced deep learning model can provide instant results with microscopic images as the input. Thus, it will be possible to produce more accurate and faster analysis of white blood cells using the developed model. The GA-enhanced deep convolutional neural network model successfully classifies the four categories of white blood cells using deep convolutional learning. To further improve the results, a general model can be produced and applied to many problems in healthcare by means of learning transfer method by working on other cell sample information like white blood cell structure. The developed model presented here uses GA-enhanced deep convolutional neural networks. In the future, this design can be combined with models such as long-term memory learning and repetitive neural networks to test improvements in its performance.

Acknowledgments Images of human white blood cells were selected from publicly available databases without any identifying information on any individuals. This work has been supported in part by Ondokuz Mayis University Research Center.

References

1. A. Akselrod-Ballin, L. Karlinsky, Alpert, et al., A region based convolutional network for tumor detection and classification in breast mammography, in *Deep Learning and Data Labeling for Medical Applications*, (Springer, Cham, 2016), pp. 197–205
2. Y. Bar, I. Diamant, L. Wolf, et al., Chest pathology detection using deep learning with non-medical training. In *2015 IEEE 12th Int. Symp. Biomedical Imaging*, pp. 294–97, 2015, Apr.
3. J. Bergstra, Y. Bengio, Random search for hyper-parameter optimization. J. Mach. Learn. Res. **13**, 281–305 (2012)
4. T. Bhat, S. Teli, J. Rijal, et al., Neutrophil to lymphocyte ratio and cardiovascular diseases: a review. Expert. Rev. Cardiovasc. Ther. **11**(1), 55–59 (2013)
5. C. Briggs, Quality counts: new parameters in blood cell counting. Int. J. Lab. Hematol. **31**(3), 277–297 (2009)
6. J.A. Conchello, J.W. Lichtman, Optical sectioning microscopy. Nat. Methods **2**(12), 920–931 (2005) https://doi.org/10.1038/nmeth815/
7. G.E. Dahl, D. Yu, L. Deng, A. Acero, Context-dependent pre-trained deep neural networks for large-vocabulary speech recognition. IEEE Trans. Audio Speech Lang. Process. **20**(1), 30–42 (2011)

8. L. Dean, *Blood Groups and Red Cell Antigens*, Chapter 1. Bethesda, MD. https://www.ncbi.nlm.nih.gov/books/NBK2261/ (2005)
9. Y. Gal, Z. Ghahramani, Dropout as a Bayesian approximation: Representing model uncertainty in deep learning. In *Int.l Conf. on Machine Learning*, pp. 1050–1059 (2016, June)
10. X. Glorot, Y. Bengio, Understanding the difficulty of training deep feedforward neural networks. *Proc. 13th Int'l Conf. Artificial Intelligence and Statistics*, pp. 249–256 (2010, March)
11. B. Graham, *Kaggle Diabetic Retinopathy Detection Competition Report* (University of Warwick press, Coventry, UK, 2015)
12. M. Habibzadeh, A. Krzyżak, T. Fevens, White blood cell differential counts using convolutional neural networks for low resolution images. In *Int. Conf. Artificial Intelligence and Soft Computing*, pp. 263–274. Springer, Berlin, Heidelberg (2013, June)
13. M.A. Hall, Correlation-based feature selection for machine learning. PhD Thesis, University Waikato, Hamilton (1999)
14. S. Hamidian, B. Sahiner, N. Petrick, A. Pezeshk, 3D convolutional neural network for automatic detection of lung nodules in chest CT. In *Medical Imaging 2017: Computer-Aided Diagnosis*, Vol. 10134, p. 1013409. Int.l Society for Optics and Photonics (2017)
15. K. He, X. Zhang, S. Ren, J. Sun, Delving deep into rectifiers: Surpassing human-level performance on imageNet classification. *Proc. IEEE Int.l Conf. Comp. Vision*, pp. 1026–1034 (2015)
16. B.D. Horne, J.L. Anderson, J.M. John, et al., Which white blood cell subtypes predict increased cardiovascular risk? J. Am. Coll. Cardiol. **45**(10), 1638–1643 (2005)
17. V. Iglovikov, S. Mushinskiy, V. Osin, Satellite imagery feature detection using deep convolutional neural network: a Kaggle competition. arXiv preprint arXiv:1706.06169 (2017)
18. S. Ioffe, C. Szegedy, Batch normalization: Accelerating deep network training by reducing internal covariate shift. arXiv preprint arXiv:1502.03167 (2015)
19. N.S. Jaddi, S. Abdullah, A.R. Hamdan, A solution representation of genetic algorithm for neural network weights and structure. Inf. Process. Lett. **116**(1), 22–25 (2016)
20. V. Jain, S. Patnaik, F. P. Vlădicescu, I. K. Sethi (eds.), *Recent Trends in Intelligent Computing, Communication and Devices* (Springer Nature (Singapore/multi-national) Springer, 2018)
21. M.D. Joshi, A.H. Karode, S.R. Suralkar, White blood cells segmentation and classification to detect acute leukemia. Int. J. Emerging Trends Tech. Comp. Sci. **2**(3), 147–151 (2013)
22. D. Kansara, S. Sompura, S. Momin, M. D'Silva, Classification of WBC for blood cancer diagnosis using deep convolutional neural networks. Int. J. Res. Advent Technol. **6**(12), 3576–3581 (2018)
23. A.M. Karim, M.S. Güzel, M.R. Tolun, H. Kaya, F.V. Çelebi, A new generalized deep learning framework combining sparse autoencoder and Taguchi method for novel data classification and processing. Math. Probl. Eng. **2018**, 1–13 (2018)
24. A.M. Karim, M.S. Güzel, M.R. Tolun, H. Kaya, F.V. Çelebi, A new framework using deep autoencoder and energy spectral density for medical waveform data classification and processing. Biocybern. Biomed. Eng. **39**(1), 148–159 (2019)
25. T.S. Kickler, Clinical analyzers. Advances in automated cell counting. Anal. Chem. **71**(12), 363–365 (1999)
26. D.P. Kingma, J. Ba, Adam: a method for stochastic optimization. arXiv preprint arXiv:1412.6980 (2014)
27. A. Krizhevsky, I. Sutskever, G.E. Hinton, ImageNet classification with deep convolutional neural networks. In *Advances in Neural Information Processing Systems (1097–05)* (2012)
28. K. Kuan, M. Ravaut, G. Manek, et al., Deep learning for lung cancer detection: tackling the Kaggle data science bowl 2017 challenge. arXiv preprint arXiv:1705.09435 (2017)
29. Y. LeCun, Y. Bengio, G. Hinton, Deep learning. Nature **521**(7553), 436 (2015)
30. G. Liang, H. Hong, W. Xie, L. Zheng, Combining convolutional neural network with recursive neural network for blood cell image classification. IEEE Access **6**, 36188–36197 (2018)
31. J.S. Loudon, *Detecting and Localizing Cell Nuclei in Medical Images*. MSc thesis, NTNU (2018)

32. N. Metawa, M.K. Hassan, M. Elhoseny, Genetic algorithm-based model for optimizing bank lending decisions. Expert Syst. Appl. **80**, 75–82 (2017)
33. Mooney Paul, Identify Blood Cell Subtypes from Images, Kaggle (2018). https://www.kaggle.com/paultimothymooney/identify-blood-cell-subtypes-from-images
34. S. Newman, T. Persson, *White Blood Cell Differential Counting in Blood Smears Via Tiny YOLO* (Stanford University Press, Stanford, California, USA, 2018)
35. S. Ohlsson, *Deep Learning: How the Mind Overrides Experience* (Cambridge University Press, UK, 2011)
36. A. Osei-Bimpong, C. Jury, R. McLean, S.M. Lewis, Point-of-care method for total white cell count: an evaluation of the HemoCue WBC device. Int J. Laboratory Hematol. **31**(6), 657–664 (2009)
37. S.K. Pal, P.P. Wang, *Genetic Algorithms for Pattern Recognition* (CRC Press (USA/multi-national), 2017)
38. N.K. Pareek, V. Patidar, Medical image protection using genetic algorithm operations. Soft. Comput. **20**(2), 763–772 (2016)
39. R. Pascanu, T. Mikolov, Y. Bengio, Understanding the exploding gradient problem. CoRR, abs/1211.5063, 2, 1–11 (2012)
40. O. Russakovsky, J. Deng, Su, et al., ImageNet Large Scale Visual Recognition Challenge. IJCV 2015 http://www.image-net.org/challenges/LSVRC/ (2015)
41. O. Roeva, S. Fidanova, M. Paprzycki, Population size influence on the genetic and ant algorithms performance in case of cultivation process modeling, in *Recent Advances in Computational Optimization*, (Springer, Cham, 2015), pp. 107–120
42. J.D. Seebach, R. Morant, R. Rüegg, et al., The diagnostic value of the neutrophil left shift in predicting inflammatory and infectious disease. Am. J. Clin. Pathol. **107**(5), 582–591 (1997)
43. M. Simon, E. Rodner, J. Denzler, ImageNet pre-trained models with batch normalization. arXiv preprint arXiv:1612.01452 (2016)
44. C. Szegedy, S. Ioffe, V. Vanhoucke, A.A. Alemi, Inception-v4, inception-resNet and the impact of residual connections on learning. In *Thirty-First AAAI Conf. on AI* (2017, Feb.)
45. D. Tigkas, V. Christelis, G. Tsakiris, Comparative study of evolutionary algorithms for the automatic calibration of the Medbasin-D conceptual hydrological model. Environ. Proc. **3**(3), 629–644 (2016)
46. Q. Wu, F. Merchant, K. Castleman, *Microscope image processing* (Elsevier, 2010)
47. C. Zhang, P.C. Woodland Parameterised sigmoid and ReLU hidden activation functions for DNN acoustic modelling. *16th Annual Conf. Int.l Speech Communication Assoc.* (2015)
48. K. Zhang, W. Zuo, Y. Chen, et al., Beyond a gaussian denoiser: Residual learning of deep CNN for image denoising. IEEE Trans. Image Process. **26**(7), 3142–3155 (2017)
49. J. Zhao, M. Zhang, Z. Zhou, et al., Automatic detection and classification of leukocytes using convolutional neural networks. Med. Biol. Eng. Comput. **55**(8), 1287–1301 (2017)

Deep Learning–Based Constituency Parsing for Arabic Language

Amr Morad, Magdy Nagi, and Sameh Alansary

1 Introduction

Developing an effective algorithm to generate constituency parse tree is a major challenge in Natural Language Processing (NLP). Constituency parse tree is the first and most important step, prior to several tasks in NLP such as word processors grammar checking [1], language translation [2], Arabic diacritization, and many more.

The past few years have witnessed a significant increase in adopting neural network–based approaches for generating constituency parse tree [3, 4]. These approaches, which have proved to be performant and adaptive, are called data-driven ones as they extract the rules from the data itself, hence they are adaptive to any language, provided that the dataset is changed. Moreover, most of these techniques rely on dense input representations [5] for words in the given sentences. In such representation, semantic meaning of the words is disclosed by making the representation of similar words closer according to the task at hand. These

---

A. Morad (✉)
Bibliotheca Alexandrina, Information and Communication Technologies (ICT), Alexandria, Egypt
e-mail: amr.morad@bibalex.org

M. Nagi
Bibliotheca Alexandrina, Faculty of Eng., Alexandria University, Computer and Systems Department, Alexandria, Egypt
e-mail: magdy.nagi@bibalex.org

S. Alansary
Bibliotheca Alexandrina, Arabic Computational Linguistic Center, Faculty of Arts, Alexandria University, Phonetics and Linguistics Department, Alexandria, Egypt
e-mail: sameh.alansary@bibalex.org

© Springer Nature Switzerland AG 2021
H. R. Arabnia et al. (eds.), *Advances in Artificial Intelligence and Applied Cognitive Computing*, Transactions on Computational Science and Computational Intelligence, https://doi.org/10.1007/978-3-030-70296-0_4

word representations are typically achieved by deep learning analysis based on the input sentences. This representation of words is called "dense input representation," which in turn needs deep learning–based technique, and the dataset used in training of these models could be in different languages; however, it needs different resources for the required language in order to find this similarity.

In this chapter, a neural network syntactic distance-based model [6] is applied on Arabic sentences. The Arabic language is challenging because of its history, culture, and literary heritage, as well as being very complex language because of its linguistic structure [7]. Some words spell the same, but have different meanings, depending on their context and diacritization. Moreover, there are multiple Arabic words spellings corresponding to one word in other languages, especially in case of known entities, for example, the city of Washington could be spelled 'واشنطن', 'وشنطن', 'واشنجطن'. As a result, novel challenges are considered when applying deep learning techniques to Arabic language. Arabic is not only a complex language, but also has very limited linguistic resources, specially parsed corpora (Fig. 1).

The model, adopted in this chapter, is based on the concept of syntactic distance [6], which is defined for each split point of the sentence. For implementation, dense input representation is achieved by using Glove [8] (see Sect. 3 below), for Arabic corpus. Several experiments are carried out in this work, with various split points, based on linguistic factors and length in Arabic sentences (see Sect. 6).

In order to predict the correct label (nonterminal tag) accurately, a list of these labels is determined, and its dimensionality should be affordable, meaning that this list should not be too long to avoid the problem of dimensionality curse. The curse of dimensionality refers to a well-known problem encountered during data analysis, in high-dimension space, however, as previously mentioned, our Arabic dataset is limited. Therefore, the full list of Arabic labels, nonterminal tags, is reviewed by linguists and then grouped. This happens by grouping similar labels together in one label. For example, label "NOUN.VN + NSUFF_FEM_SG + CASE_ DEF_GEN" is turned into "NOUN." The rest of the label linguistic features, denoting the gender of the noun, whether it is definite or not and so on, is stored in the node displayed in the resultant constituency parse tree. More explanation will be further illustrated in the experiments section (VI).

Fig. 1 A constituent parse tree

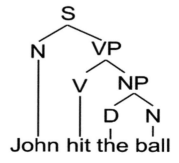

The model is fully parallel [6], which means that it is performant and capable of handling long sentences efficiently. Mapping from syntactic distance to the parse tree and vice versa can be done in O (n log n), which makes the decoding computationally efficient.

The obtained model is used in a complete workflow for linguists and editors, predicting the constituency parse tree for the Arabic sentence. A constituency parse tree viewer is implemented as a web-based application for linguists, to display the predicted tree. Tree viewer is user-friendly, enabling the linguists to edit the resultant displayed parse tree easily, and save the modified one. These valid and curated sentences will be used later to retrain the model, assisting in the existing challenge of lack of Arabic language resources.

2 Survey of Related Work

In Natural Language Processing (NLP), designing fast and accurate parsing algorithms for constituencies is a major challenge. Constituency parsing is necessary, as a first step toward the interpretation of the semantic meaning, which in turn is used as a foundation for other tasks, like translation [2] and grammar checking in word processors [1].

Traditionally, syntactic parsers represent the sentence as one-hot vector of a collection of unique features. This is accomplished by having only "1" in a location unique to the given word; "0" otherwise. However, enormous drawbacks were discovered for this representation technique. Apart from the length of this representation, it mainly handles all words as independent entities, with no relation to each other [8].

Transition-based dependency parsing [10] represents a step toward word similarity, which assigns similar words to the same buffer, in order to provide contextual meaning for these words. The problem in the case of getting word vectors is the necessity of a certain knowledge to the language of the task at hand, in order to extract the appropriate rules. For each language, various rules should be applied which makes rule-based techniques harder in application, as a general solution to achieve the words vectors representation. Hence, deep learning approaches are the most appropriate in tackling this issue.

GloVe (Global Vectors for Word Representation) [8] is a well-known unsupervised deep learning algorithm to generate vectors representation for words, called word vectors. It defines some sort of similarity between words by using deep learning approach and creates word vectors by using matrix factorization, with local window methods. In brief, the output word vectors represent the semantics of the word, which is relevant to the current task. These vectors are dense, meaning that their entries are typically nonzero, according to the features extracted from given sentences. GloVe could be considered a general technique for word embeddings for any language where a general deep learning model could be trained with different resources in various languages [9].

In this chapter, we trained GloVe model with multiple Arabic resources such as Arabic corpus, Elwatan news, and many others. As a result, GloVe is a global technique for any language, which is unsupervised learning of word representations.

The challenge represented by the Arabic language to researchers is mainly the inherent linguistic structure. Some Arabic words have different POS tags according to their place in the sentence or according to the vocalization. Another problem is when Arabic texts include many translated and transliterated known places, whose spelling might be inconsistent in Arabic texts. Moreover, Arabic language is one of the non-Latin languages, written from right to left, so the Arabic characters need different encoding techniques. In order to overcome some of these challenges, Buckwalter for Arabic characters transliteration [11] was introduced. Although Buckwalter addresses some of Arabic language challenges, it adds more complexities because it requires more encoding and decoding steps.

Another challenge in Arabic language labels is represented in the dimensionality of labels selected. Each Arabic word could have up to three prefixes and two suffixes. Accordingly, a label could be a combination depending on the context, whether it is singular or plural and so on. For example, label "NOUN.VN + NSUFF_FEM_ SG + CASE_DEF_GEN" refers to a NOUN with a suffix for a singular female, it also denotes that this word has definite and in genitive case. Hence, given these different components in each single label, it could be concluded that the number of possible labels will be huge.

Designing a model which has to predict a label among the large number of choices is really hard, whether this model is rule based or data driven (deep learning based). A short list, in cooperation with linguists, should be determined while preserving the original one, and without sacrificing the context of the word.

Recently, deep learning–based techniques relying on dense word vectors representation for constituency parsing are widely adopted [3, 4]. Some of them decompose the problem of generating the parse tree into local decisions sequentially. However, this makes the accuracy of such models low, because they were never exposed to their mistakes during the training phase. In order to solve this problem, a technique such as spa-based constituency parsing with a structure-label system [12] is proposed. The drawback of this technique is that the training phase is complicated.

One of these techniques is accomplished by adopting the concept of syntactic distance [6]. Syntactic distance is a vector of real valued scalars for each split point in the given sentence (e.g., white space). A fully parallel model for constituency parser is introduced for English and Chinese languages using syntactic distance notion [6]. Mainly, it uses a top-down approach to construct the parse tree by splitting larger constituents into smaller ones. A combined neural network is built to estimate the syntactic distances vector for a sentence. Algorithms for converting syntactic distances into binary parse trees and vice versa are explained with average running time of O (n log n) [6].

In this chapter, a combined neural network model is built to generate the parse tree for Arabic sentences without the need to use Buckwalter, because there is no known word embeddings for transliterated words. This step requires preprocessing and postprocessing as intermediate stages. Therefore, those two steps are eliminated in our solution as it deals with Arabic language directly. Arabic language sentences impose some complications such as the length of sentences and the ambiguity of words. As previously mentioned, words with the same spelling may have multiple different POS tags according to the context of this word or its diacritization. These challenges are addressed in this chapter by conducting several experiments to show the effect of these factors on the accuracy of the given model.

The model is fully parallel, which can make use of efficient and powerful Graphical Processing Units (GPU) capabilities. Several experiments have been established to discover the effect, in terms of the length of Arabic sentences, as well as the effect of split points, on the model's accuracy, running times, and accuracy. In addition, the model predicts the constituency parse-tree efficiently. This is important when the model is considered a step in a larger workflow, as presented in this chapter (see Sect. 5).

3 Dense Input Representation

Generating the constituency parse tree depends on the given words, Part Of Speech (POS) tags, and the semantic meaning of given phrases. Traditionally, words representation depends on technique called "one-hot vector" in which "1" is added in a unique location for the word and "0" in the remaining vector bits. Enormous drawbacks are faced in this representation, because it requires huge vectors for the handled vocabulary, as well as treating all words as independent entities. "One-hot vector" is not the ideal representation technique [8], because parse tree depends also on the semantic meaning and the context of words.

Other techniques are developed to overcome the mentioned complexities based on the use of pretrained embedding vectors, as additional features to the model. One of the most common embedding word vectors known is "word embeddings." In "word embeddings," each word is linked to a vector representation in a way that captures the semantic meaning of relationships of words. This is used as in transition-based dependency parsing, which defines concept of buffer to denote the contextual meaning of the words [10].

GloVe (Global Vectors for Word Representation) [6] is one of the well-known deep learning–based techniques, used for generating word vector representations. It is an unsupervised method for generating word representations, based on the word occurrences statistics in a given context. GloVe is proven that it outperforms other models on word similarity tasks.

Using precalculated word vectors in a deep learning solution to generate constituency parse tree is proven to be efficient and accurate [9]. The precalculated word vectors, called pretrained, are used as initial values and additional features, for a network within the model neural network. Such approaches depend on the input dataset, making it extensible and applicable to any language, to calculate both the unique vocabulary in the given dataset, as well as the word vectors representation. These pertained values are injected then to the main model which is used to generate the constituency parse tree. As the dataset for generating the parse tree could be different from the dataset used to calculate the word vectors representations, these vectors could be fine-tuned during the training of the constituency parse tree model, using the dataset for that model. In other words, the whole model network could allow these word vectors to be changed, according to the dataset of the given task.

In this chapter, GloVe is chosen to obtain the dense word representations. GloVe model is trained using Arabic resources, such as Arabic Corpus, Arabic Wikipedia articles, Arabic tweets, and Elwatan newspaper. These resources result into a huge number of words (1.9 billion words), in addition to comprising large number of unique Arabic vocabulary (1.5 million unique vocabulary). As this dataset contains numerous collection of unique words, more parameters are required to be learned. Therefore, the model is more prone to overfitting, which is one of the known problems in Arabic language called lexical sparsity. To overcome sparsity problem, a large amount of tokens or words is required.

GloVe model is modified in order to handle Arabic characters and overcome some encoding problems. After that, the modified model is used to calculate word vectors representations from the dataset mentioned previously.

GloVe model is trained using the mentioned dataset for 20 epochs, which results into getting the similarity between Arabic words. The model is slightly changed in order to handle the non-Latin nature of Arabic language. Changes mainly take place in encoding and memory management for the Arabic corpus.

The obtained Arabic vocabulary and Arabic words vectors, after the training process, are used within the constituency parse tree generator model network (as illustrated in the next section). The model itself fine-tunes these word representations according to the new dataset used to train the constituency parse tree generator model. This is done using RNN network, which inputs the word representation vector (256d vector) as initial values resulting in obtaining the best-predicted parse tree from the model faster and more accurate (Fig. 2).

4 Parse Tree Generator Model

The model used to generate the constituency parse tree depends mainly on syntactic distance [6]. To explain the meaning of a syntactic distance, the top-down scheme for construction the parse tree from a sentence needs more elaboration. One can process the sentence by splitting larger constituent into smaller constituents, in which the order of the split points defines the hierarchical structure. Hence, the syntactic distance is defined for the split points. The model uses neural networks to estimate the vector of syntactic distances for a given sentence. Figure 2 shows an overview of the model.

As mentioned previously, Arabic GloVe with dimension of 256 is used for Arabic word embeddings. The model fine-tunes word vectors. Not only the words are represented as dense vectors, but also tags, which helps in increasing the accuracy of the predicted parse trees. The result of these embeddings (fine-tuned) is fed to

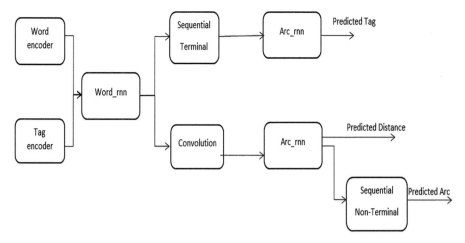

Fig. 2 Constituency parse tree generator overview

a BiLSTM (Bidirectional Long Short-Term Memory) word RNN. This makes the model captures a long-term syntactical relations between the words. Predicted tags could be calculated using feed forward network. Another bidirectional long short-term memory network called "arc_rnn" is used to calculate the syntactic distances. Constituent labels are then generated using the mentioned "arc_rnn," in addition to feed forward layer.

The model is fully parallel allowing it to be performed using a GPU. The model computes the distances in batch, and afterwards, it is decoded using an algorithm with complexity time of O (n log n). Moreover, the encoding and decoding algorithms are naturally adopted for execution in parallel environment, which can further reduce the time to O (log n). The estimated distance for each split point in the sentence is calculated independently from the others, allowing efficient parallelism.

For Arabic dataset, the model is slightly changed to handle the new dimensionality of dense Arabic words representations obtained from GloVe. Moreover, the model is now modified so as to handle non-Latin nature of Arabic language. The model is performant and is able to predict constituency parse tree even for very long Arabic sentences fast.

5 Workflow

In this chapter, a complete workflow is built upon the constituency parse tree model. The workflow typically enables linguists to predict the parse tree of a sentence from International Corpus of Arabic (ICA) and then they can edit the resultant tree. Moreover, these modified sentences could be used later to retrain the model. This section could be divided into two parts for explaining the workflow and how the dataset for retraining the model is extracted.

5.1 Workflow

The workflow consists of a series of stages. First, a list of sentences is shown to linguists using a web portal application. Linguists, users for this system, have accounts in this web portal. There are three main types of users to this web portal: a linguist, editorial reviewer (who is a more experienced linguist), and scientific reviewer.

Firstly, linguists need to be authenticated to be able to log in through the portal and once logged-in, a list of sentences is shown to the linguists by an administrator. The linguist selects the desired sentence in order to generate its constituency parse tree and then the sentence is sent to the model, pretrained with Elnahar data, so as to predict the constituent of this sentence.

Constituency parse tree is shown to the linguist in a viewer that enables him to modify the resultant tree, if and only if the final order of words remains the same as the given one. In brief, the linguist could edit a constituent label (nonterminal label, predicted by the model), modify the order of the parse tree nodes, or even delete a node. After amending the needed changes, the linguist could save the parse tree. As previously mentioned, the modified parse tree will be kept only on condition that the order of the sentence words is still the same as the input sentence.

Progressively, a new group of modified parse trees, which are curated by group of specialists (linguist, editorial, or even managerial layer), will be obtained. Since these sentences are verified, they will be used for model's incremental training. Incremental training means retraining the model using the new verified sentences. Initially, the saved model weights, obtained from previous training times, are loaded, followed by feeding the verified sentences to the model as a new dataset to be trained with. Thereafter, the model is trained, not only with Elnahar dataset, but also by the curated sentences resulting in the creation of a trusted model, which can generate and Arabic constituency tree efficiently. Figure 3 exhibits the workflow steps.

As a conclusion, the workflow can be easily operated and deployed in any hardware configuration. The main challenge tackled by this workflow, is how linguists could alter a deep learning model in order to benefit from it. The linguist uses the model to generate an initial constituency parse tree, which he could either accept or modify. In case of modifications, as long as it is curated, considerations are made to obtain a more accurate model. The resultant model is not only trained on the original Arabic dataset given (Elnahar) with limited number of sentences, but also a list of curated sentences obtained from a trusted dataset of International Corpus of Arabic (ICA).

Deep Learning–Based Constituency Parsing for Arabic Language 53

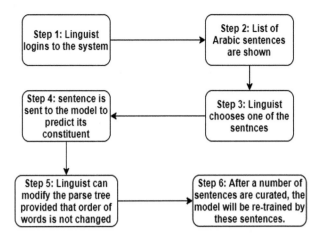

Fig. 3 Workflow

5.2 Dataset

As mentioned previously, linguists choose sentence from a list of Arabic sentences in order to predict its constituency parse tree and do the needed modifications. These sentences are extracted from International Corpus of Arabic (ICA) [13], using a script to build the parenthesized pair of word and Part Of Speech (POS) tag. This corpus is initiated by Bibliotheca Alexandrina (BA), in order to fulfill research demands on Arabic language. ICA currently contains 100 million Arabic words from four main sources; press, net articles, books, and academics. Press source, previously mentioned, comprises newspapers, magazines, and electronic press (Fig. 3).

ICA not only covers these sources but also it has numerous genres, such as strategic sciences, sports, religions, and many others. These genres are divided into 24 subgenres such as economy, law, politics, and others. Moreover, these subgenres are divided into four main sub-subgenres, namely short stories, novels, children's stories, and plays. The effect of these sources or genres to the Arabic corpus is not equal. It depends on how common this source or genre is. Hence, balance is not measured by the amount of texts.

Most of these sources and genres are got from all over the Arab world. Egypt covered about 13 million words out of the 100 million words of the corpus. Saudi Arabia comes in the second place after Egypt covering about eight million words. Other countries like Kuwait and UAE covered about five million words. Not only Arab world contributes to these words, but also some countries outside the Arab region.

The International Corpus of Arabic analysis is achieved automatically, using rule-based and statistical approaches, in which BASMA [14] Arabic morphological analyzer is used. This analysis lists information like list of prefixes, suffixes, word class, stem, lemma, root along with the number, gender, and definiteness of the word, depending on the context of this word within the sentence in the corpus.

6 Experiments

The constituency parse tree generator model, along with pretrained Arabic GloVe, is trained on Arabic dataset of Elnahar, which contains more than 12.5 K sentences, varying from short to very long ones, with varying data contexts between politics, sport, health, and others. Three different (unique) sets are created from the previously mentioned dataset: training, validation, and testing. Training data contains 10 K statements, in different context, with varying lengths, while validation data comprises 1 K sentences, varying between long and short sentences.

The validation dataset is utilized to address overfitting problem, which occurs when the model tightly understands the training data, but fails in understanding and generating parse trees for data with either different vocabulary, lengths, or context, thereby allowing the parse tree generator model to self-correct itself. Finally, the test dataset contains more than 1.6 K sentences, with similar characteristics to both training and validation datasets.

The validation and testing datasets are chosen in a way, such that they are different and unbiased. In order to validate this claim, the model is used to generate the constituency parse trees for both of validation and test datasets. The model outputs almost the same result on both the validation and test datasets (F-measure = 86%).

In order to evaluate the accuracy of the model, F-measure [15] is used. F-measure could be considered a ratio of precision and recall; therefore, it captures efficiently both of them during the calculations. Standard Evalb tool, in its Python version, is used for this evaluation. This version, after some modifications in this work, assists in creating a cross-environment model, as well as helping in solving the encoding problems encountered, due to complexities in Arabic language (non-Latin nature).

F-measure is calculated by:

$$F - \text{measure} = 2 * \frac{\text{Precision} * \text{Recall}}{\text{Precision} + \text{Recall}}$$

where

- Precision is considered the positive predictive value. It is calculated as follows:

$$\text{Precision} = \frac{\text{tp}}{\text{tp} + \text{fp}}$$

  - tp is truly predicted result.
  - fp is a false alarm (predicted to be true, though it is false – Type I error).

Deep Learning–Based Constituency Parsing for Arabic Language

- Recall is also called sensitivity.

$$\text{Recall} = \frac{tp}{tp + fn}$$

- tp is truly predicted result.
- fn (predicted to be false, though it is true – Type II error).

Experiments are carried out on Bibliotheca Alexandrina (BA) High Performance Computing (HPC) cluster, which provides both CPU and GPU capabilities. We choose to use the GPU unit to achieve the results from the model faster since the problem can be handled in parallel nature. After the training process of the constituency parse tree generator model is complete, the model can run in any environment, with in any hardware configurations.

Labels, nonterminal tags, are eliminated by grouping similar labels into one label. For example, "NOUN.VN + NSUFF_FEM_SG + CASE_DEF_GEN," is reduced to the POS "NOUN" tag. In this work, number of labels is reduced from around 600 labels to only 20 labels. Moreover, POS tags are reduced from around 400 tags to only 38. This reduction minimizes the number of labels to be predicted, resulting in a lower dimensionality, hence more accurate model.

In this chapter, four main experiments are evaluated, considering the length of the sentence, in addition to the different split points in it (Al-taareef position), as shown below. The result of them proves that the length of the sentence along with the split points does not have strong effect on the final result. These experiments show that the usage of word embeddings technique (GloVe) has a very important effect to the results and helps into getting the final result fast and more accurate.

The following shows the results obtained from the experiments.

6.1 Short Sentences with Minimum Split Points

Elnahar dataset contains very long lines, forming multiple sentences. In this experiment, very long sentences are split, if there is no direct association with the original sentence, which can occur in case of opening parenthesis while the closing one is not located at the end of it. Therefore, these types of sentences are split into multiple ones, according to the opening and closing parenthesis. Moreover, prefix such as Al-taareef (the) is not split from the word, to minimize the split points.

From the table below, it is noticed that at the first loop, F-measure was 70% for the testing dataset, which is better than the corresponding one without GloVe. As a result, this assures that using word embeddings techniques, along with the neural network approaches, for constituency parse tree generators could affect the accuracy (F-measure) of the model, especially at the beginning of training iterations (epochs). In addition, word embeddings techniques help in minimizing the number of epochs needed, in order to reach the final accuracy value, hence minimizing processing cost. Also, it is noticed that the F-measure rate increased rapidly for the first five epochs as shown in Table 1, to slow down until reaching the best accuracy (F-measure = 86%).

The following table shows the number of epochs needed, in order to achieve different percentages from the final result.

6.2 Short Sentences with Maximum Split Points

This experiment is similar to the previous one, however the only difference is that Al-taareef (the) is separated from the word, thus creating more split points to be used in the model measurements for distance and constituent.

It is noticed that the results are very similar to what we obtained from the previous experiment, concluding that adding or removing the Al-taareef (the) in short sentences has almost no difference noticed. The following Table 2 shows the number of epochs needed in order to achieve different percentages from the final result. It can be concluded from the table that the rate in which the model achieves 95% from the final result is almost the same between the current experiment and the previous one.

6.3 Long Sentences with Minimum Split Points

This experiment maintains long sentences as they are given in their original dataset without splitting, leading into very long sentences, but this time, Al-taareef (the) is not split from the original word. Consequently, one line, which might contain multiple sentences, is fed as a single entity to the model for training, validation, and testing.

Table 3 indicates the results obtained from this experiment. As observed, we can achieve 95% from the best accuracy obtained (F-measure = 86%) rapidly, when compared to the previous two experiments. This means that the length of the Arabic sentence does not cause an effect to the final results nor when these results are achieved.

Table 1 F-measure percentage from the best result

F-measure *(percentage from the final result which is 86%)*	81%	85%	90%	95%
Epochs	1	2	4	12

Table 2 F-measure percentage from the best result

F-measure *(percentage from the final result which is 86%)*	81%	85%	90%	95%
Epochs	1	3	4	13

Deep Learning–Based Constituency Parsing for Arabic Language 57

Table 3 F-measure percentage from the best result

F-measure *(percentage from the final result which is 86%)*	82%	85%	90%	95%
Epochs	1	2	4	10

Table 4 F-measure percentage from the best result

F-measure *(percentage from the final result which is 86%)*	77%	85%	90%	95%
Epochs	1	2	5	11

6.4 Long Sentences with Maximum Split Points

This experiment is similar to the last one, yet adding the Al-taareef (the) to the words. In Table 4, which illustrates the obtained results, we can note that the result achieves 95% from the best accuracy (F-measure = 86%) in almost the same manner as the previous experiment. However, currently the first epoch shows that the starting accuracy is less than its corresponding one, from experiment number 3.

7 Conclusion

Constituency parse tree is the backbone of many Natural Language Processing (NLP) tasks, such as language translation and Arabic diacritization. This is why valuable research efforts have been exerted in order to find an algorithm which generates parse tree efficiently. Deep learning approaches are adopted, because they are efficient and can generate the parse tree using dataset without any predefined rules. In this chapter, deep leaning approaches are used to generate constituency parse tree for Arabic sentences. First, dense words representations are accomplished by GloVe, which is trained using various Arabic resources with billions of words for multiple epochs. Afterwards, this dense representation is fed into deep learning technique, which is responsible for generating parse trees for Arabic sentences. This model is trained initially by Elnahar dataset. This dataset contains more than 12 K Arabic sentences with varying data contexts between politics, sport, health, and others. This is considered a step in a complete workflow, which enables linguists to choose sentences and get the corresponding parse trees. The Arabic sentences are extracted from the International Corpus of Arabic (ICA) initiated by Bibliotheca Alexandrina, which contains 100 million words from multiple sources and genres. Linguists could accept the parse trees predicted or modify them, by editing the constituent, reorder the resultant constituent, or deleting nodes, provided that modified words order in the sentence remains the same as the input one. Finally, these sentences are used to retrain the model to obtain better results eventually and enrich the dataset. The used model is fully parallel and capable of predicting constituency parse tree for long sentences. High-performance computers could be used to facilitate the training process and cut down the training time.

References

1. C. Callison-Burch, Syntactic constraints on paraphrases extracted from parallel corpora, in *Proceedings of the Conference on Empirical Methods in Natural Language Processing (NLP)*, (2008), pp. 196–205
2. J.C. Maxwell, *A Treatise on Electricity and Magnetism*, vol Vol. 2, 3rd edn. (Clarendon, Oxford, 1892), pp. 68–73
3. A. Eriguchi, Y. Tsuruoka, K. Cho, Learning to parse and translate improves neural machine translation, in *Proceedings of the 55th Annual Meeting of the Association for Computational Linguistics*, (2017), pp. 72–78
4. O. Vinyals, Ł. Kaiser, T. Koo, S. Petrov, I. Sutskever, G. Hinton, Grammar as a foreign language, in *In Advances in Neural Information Processing Systems*, (2015), pp. 2773–2781
5. D. Chen, C.D. Manning, *A Fast and Accurate Dependency Parser Using Neural Networks* (Association for Computational Linguistics, USA, 2015), pp. 740–750
6. Y. Shen, Z. Lin, A. Paul Jacob, A. Sordoni, A. Courville, Y. Bengio, Straight to the Tree: Constituency Parsing with Neural Syntactic Distance. arXiv:1806.04168, 2018
7. A. Farghaly, K. Shaalan, Arabic natural language processing, in *ACM Transactions on Asian Language Information Processing*, (2009)
8. J. Pennington, R. Socher, C.D. Manning, GloVe: Global vectors for word representation, in *EMNLP*, (2014)
9. D. Vilares, M. Strzyz, A. Søgaard, C. Gomez-Rodrıguez, Parsing as Pretraining, arXiv: 2002.01685v1, 2020
10. C. Dyer, M. Ballesteros, W. Ling, A. Matthews, N.A. Smith, Transition-based dependency parsing with stack long short-term memory, in *Proceedings of the 53rd Annual Meeting of the Association for Computational Linguistics and the 7th International Joint Conference on Natural Language Processing (Volume 1: Long Papers)*, (2015), pp. 334–343
11. A. Bakar, O. Khairuddin, M. Faidzul, M. Zamrim, Implementation of Buckwalter transliteration to Malay corpora, in *International Conference on Intelligent Systems Design and Applications, ISDA ER*, (2013)
12. J. Cross, L. Huang, Span-based constituency parsing with a structure-label system and provably optimal dynamic oracles, in *Proceedings of the 2016 Conference on Empirical Methods in Natural Language Processing. Association for Computational Linguistics, Pp 1–âA ̦S11*, (2016)
13. S. Alansary, M. Nagi, The International Corpus of Arabic: Compilation, Analysis and Evaluation. In *Proceedings of the EMNLP 2014 Workshop on Arabic Natural Language Processing (ANLP)*, pp 8–17, October 25, 2014, Doha, Qatar
14. S. Alansary, BASMA: BibAlex Standard Arabic Morphological Analyzer. In *The 14th Egyptian Society of Language Engineering Conference*, December 2015, Cairo, Egypt
15. L. Derczynski, Complementarity, F-score, and NLP Evaluation. In *Proceedings of the Tenth International Conference on Language Resources and Evaluation (LREC'16)*, pp. 261–266, 2016

Deep Embedded Knowledge Graph Representations for Tactic Discovery

Joshua Haley, Ross Hoehn, John L. Singleton, Chris Ballinger, and Alejandro Carbonara

1 Introduction

Tactics within a broadly defined game are actions aimed at the achievement of some goal—or subgoal—that is either global or local to a portion of the game; the selection of individual tactics is typically driven by strategies seeking to optimize the outcome of an overall game [1]. The use of tactics to achieve a strategic goal applies to a large number of fields, including finance [2], business [3], and military operations, as well as games [4, 5] and sports [6, 7]. The ability to train Machine Learning (ML) systems to recognize, classify, and predict successful tactics within a data structure comprised of context, execution, and result is a critical step toward developing systems that not only learn the scoring of board/play states, but can recommend and execute both novel and familiar tactics within a game [8, 9].

Tactics identification problems can be viewed as being either recognition or discovery. The recognition task takes place when the tactics are known and must be recognized within unlabeled data, lending itself to supervised ML techniques. The tactics discovery task evidences itself when available tactics are not known a priori and must be discovered from principles inferred directly from the data. Existing AI and ML approaches for game tactics discovery and execution exist within state evaluation [1], self-play [10], tree searches, or hybrid methods [8, 11]; yet, they do not yield the labels of the tactics themselves. Tactics discovery is an unsupervised data analysis task in which the game play tactics are discovered in order to inform decision-making and downstream consideration. Previous works have shown

---

J. Haley (✉) · R. Hoehn · J. L. Singleton · C. Ballinger · A. Carbonara
Intelligent Systems Division, Soar Technology Inc., Ann Arbor, MI, USA
e-mail: Joshua.Haley@soartech.com; Ross.Hoehn@soartech.com;
John.Singleton@soartech.com; Chris.Ballinger@soartech.com;
Alejandro.Carbonara@soartech.com

© Springer Nature Switzerland AG 2021
H. R. Arabnia et al. (eds.), *Advances in Artificial Intelligence and Applied Cognitive Computing*, Transactions on Computational Science and Computational Intelligence, https://doi.org/10.1007/978-3-030-70296-0_5

the utility of unsupervised ML methods for the tactics discovery problem when deployed in conjunction with semi-supervised deep embeddings [12]. However, these joint approaches require a significant amount of feature engineering to ensure that the relations of interest are present and retained in the data. This upfront feature engineering task becomes even more difficult when relations of interest need to be discovered. Instead, we encode game play logs within a semantic graph and then use graph embedding techniques to construct a vectorization with minimal supervision. Semantic graphs (*e.g.*, Wordnet [13]) are representational structures that encode facts and relations over which reasoning can be conducted [14]. We exploit this scheme by placing relatively simple domain facts into the semantic graph structures; the emergent graph structure and connectivity of each datum facilitate additional input in a downstream tactics discovery ML implementation. Overall, the transformation of a game play log into a semantic graph form is a representational transform requiring fewer design and data transformations and preliminary calculations than does the rote feature engineering approach exemplified by Haley et al. [12].

Herein we examine several methods for constructing semantic graph representations of game play and embedding the resultant graphs, where each graph construction method incorporates a different degree of domain knowledge. These methods include the route inclusion of all features in an unstructured manner, the *ab initio* generation of graphical representations, domain-driven feature selection, and domain-driven generation of graph representations. Within Sect. 1.1 we describe the target domain: professional-level American Football. In Sect. 2 we detail the production of each knowledge embedding method and the procedure used to evaluate these embeddings. Sections 3 and 4 are reserved for the interpretation of results and provision of concluding remarks.

1.1 Exploration Domain: NFL Football

American Football is a timed, two-team athletic sport where each team alternates between playing in an offensive or defensive manner. The game is initiated by an event where the defensive team commits a long-distance kick of the game ball so that the offensive team may recover the ball, setting the initial location for play (line of scrimmage). During a team's offensive phase, they focus on conveying the game ball to a terminal location within the opposing side of the field; this location is referred to as the end zone. The conveyance of this ball is facilitated by tactical plays primarily composed of runs and passes. The offensive team may attempt four run/pass plays during their turn, each being called a down; the down count is reset for each conveyance of the ball by 10 yards. Additionally, the offensive team may opt to manage either time or distance by conducting a punt, spike, or kneel. Once the offensive team reaches the end zone they are awarded six points and may choose to conduct an action that may award additional points (a field goal or conversion). While the offensive team is conducting these tactical events, the defensive team focuses on preventing the conveyance of the ball. Should the offensive team fail to reach 10 yards within four downs, the teams change roles. This is an extremely brief introduction to football game play, yet is sufficient to introduce most of the events, tactics, and quantities to be discussed herein.

As previously discussed, we employed the Detailed NFL Play-by-Play Data 2009–2018 dataset while evaluating methods [15]. This dataset offers a large number of tactical instances from within a game environment, and these plays have tactically relevant labels provided for each play/down. This dataset contains a record of each play conducted during 2009–2018 in all professional National Football League (NFL) games. The cardinality of the dataset is roughly 450,000, where each entry comprises 255 data fields. Each cardinal entry represents a particular play that is contextualized, described, and labelled. As we are concerned with the (re-)discovery of tactics, the labels provided for each play describe the tactical action taken by the ball-holding team, if a play is completed; these labels include `kickoff`, `run`, `pass`, `punt`, `qb_spike`, `qb_kneel`, `field_goal`, `extra_point`, `no_play`, `NA`. The `no_play` and `NA` labels describe plays that were not completed due to referee intervention and plays that were insufficiently described within the dataset; as these two classes may obfuscate the tactics discovery task and confuse the prediction task, data instances of the `no_play` and `NA` classes were removed along with any fully duplicated entries, reducing the cardinality of the dataset to roughly 391,000.

2 Methodology

All instances (plays) in our dataset were encoded into a numeric representation and then run through a supervised task of tactic recognition and an unsupervised task of tactic discovery, as described in Testing Protocol. Herein we describe the methods used to encode game play logs with a naive vectorization method, typical in ML approaches, serving as a baseline comparison.

2.1 Naive Vectorization

The feature engineering tasks undertaken during the naive vectorization are intended to replicate the typical *in natura* data analyst problem. These steps require no domain-specific information, except that which is necessary to disregard information developed through post-observation computation.

Machine learning tasks fundamentally require all incoming information to be organized into a vector containing a numerical representation of information. This requirement necessitates the transformation of categorical and string-based data into corresponding numerical formats. Game play features corresponding to categorical data were transformed into a one-hot encoding representation. Unique identifiers regarding the specific play, game, date, or player were dropped from the dataset as they would lead to learned correlations that were meaningless. Additionally, the dataset contained a large array of post-processed statistics concerning each play and its probabilistic outcome on the season; these data fields were also removed. After all of the aforementioned actions, the resulting number of feature fields relating to each play are represented by a feature vector of length 112. Due to the large magnitude of differences between specific continuous valued features, a feature-wise normalization was conducted to ensure each value is between [0–1].

2.2 Knowledge Graph Construction

The naive vectorization development method described above provides a baseline comparison against which more advanced methods will be benchmarked. The more sophisticated approaches being undertaken for evaluation still utilize all features of the learning methods in some capacity, yet each feature may no longer appear in every embedded data instance. The intention is to provide additional information—driven either by domain knowledge or relationships highlighted within the data—in the form of connectivity between features. Herein we describe several methodologies by which knowledge graphs may be created from game-play logs, each leveraging a unique level of domain knowledge.

2.2.1 Domain-Specific Semantic Graph

Voluminous domain-specific overhead is typically expended during feature engineering and problem formulation; we sought to reduce the overall overhead, to generalize the domain knowledge across tasks, and to determine a domain knowledge imbued structure that can operate across specific ML tasks. This structure would utilize the same features as the naive vectorization, yet provide domain knowledge through the order and connectivity between single features. This emerging graph structure was stored within a non-edge attributed NetworkX Graph structure within Python. The graph trees conditionally included prefabricated branches containing features if and when those features were observed within the dataset. This permitted a generalized procedure for generating graphs that was blind to specific play classification when being formed.

The procedure for generating these conditional graphs involves several steps. We began by leveraging pre-existing domain knowledge to identify information within the dataset that is a contextual predicate of the play being described. This information is graphically ordered in the traditional subject-predicate-object order common to RDF triples. Then, actions undertaken by the Quarterback (a central player and on-field decision maker) are identified and placed as nodes in a fully connected structure. These nodes are conditional in nature and appear if the binary values present in the dataset indicate that this action/event took place; therefore not all nodes will appear in the final graph. (Edges to conditional nodes are also

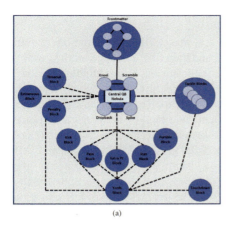 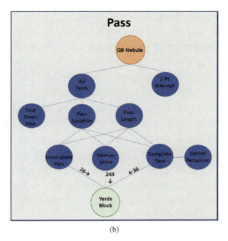

Fig. 1 Example structures conveying the organization of the domain-specific semantic graphs. (**a**) The overall connectivity of game-related blocks, each being a conditional graph related to a specific event feature of game play. (**b**) A more detailed example of the Pass Block contained with the Overall structure (see subfigure a) showing the connectivity of nodes/features within a block

conditional.) All remaining features in the dataset were organized by relevancy into the following categorical blocks: Tackles, Timeout, Extraneous, Penalty, Kick, Pass, Run, Fumble, Extra Point, Yards, and Touchdown. These blocks are each a graph whose nodes and edges are both domain knowledge driven and conditional in nature. The presence of each categorical block and the features contained within are conditional on those events having occurred in any particular play. As a result of this procedure, graphs of the same play class may have different structures as they may or may not have included (for example) a time out, penalty/flag on play, or an interception, or resulted in a first down.

A graph was generated per play via the above conditional procedure, where each graph possessed an identically formatted predicate section, referred to as the Front Matter. This Front Matter section can be thought of as the Predicate block in a semantic triple statement and contains information that contextualizes the play prior to the execution of the play (e.g., down, yard line, scores). This predicate block is attached to a small, fully connected block of conditionally populated actions undertaken by the offensive team's Quarterback. From the Quarterback Action block, there are conditional and interconnected blocks for each event and outcome observable in the dataset. This general, yet dynamic, procedure is intended to result in a series of diverse graphs with emergent play class similarities informed by domain knowledge. Figure 1a shows the general connectivity of the conditional blocks, whereas Fig. 1b shows the internal structure of the Pass block in greater detail.

2.2.2 Specification Graph

Exemplified by the methods used in Sect. 2.2.1, many ML representations require large amounts of domain-specific knowledge, and therefore do not generalize well to other classes or problems [16–20]. This undesirable attribute is highlighted ancillary to the NFL play-by-play domain as game play structure varies widely between types of game, preventing the method from being applied elsewhere. To address both the domain-specific overhead requirements and lack of domain transference, we have developed a new knowledge graph formulation inspired from behavioral interface specification languages [21] such as Eiffel [22], SPARK [23, 24], and JML [25–27]. This new representation—called a *specification graph* (spec graph)—uses the description of each play in the form of a Hoare triple as a conceptual tenant through which to generalize gameplay [28, 29]. The general form of a Hoare triple is provided as Eq. 1.

$$\{\phi\}\,p\,\{\psi\}\tag{1}$$

In Eq. 1, ϕ is a pre-state predicate said to hold prior to the execution of play p. Similarly, ψ is a post-state predicate, said to hold after the successful execution of play p.

An example of a valid Hoare triple for a given play p:

$$\{\texttt{yards\_to\_go} = 10\}\; p\; \{\texttt{yards\_to\_go} = 5\}\tag{2}$$

Within Eq. 2, the pre-state assumes a value of `yards_to_go` equal to 10 and in the post-state evolves to a value of 5; this evolution indicates that the offensive team—starting with 10 `yards_to_go`—has gained 5 yards during the course of play p.

Our Spec Graph method sequences each play into pre- and post-play pairs rather than representing each play as an individual network. This sequencing allows for rich information to be added to our representation that would otherwise be unavailable. An example of this additional embedded information that specifies properties as in Eq. 2, our representation supports using *relational* operators that relate the pre- and post-play game states. A logically equivalent way to represent the post-state of Eq. 2 is as follows:

$$\{\texttt{yards\_to\_go} < old(\texttt{yards\_to\_go})\}\tag{3}$$

In Eq. 3, we introduce the function *old* that operates on the pre-state value of its argument. In Eq. 3, the meaning of the expression `yardstogo` $<$ *old*(`yardstogo`) is that the value of `yards_to_go` in the post-state is less than its value in the pre-state.

In the context of machine learning, the Spec Graph representation presents a notable advantage for use in ML systems since a common problem for ML algorithms arises in continuous value domains. For example, $\mathbb{R}$ values may range over many subtle and difficult-to-separate ranges, where the meaning of these ranges may be difficult to distinguish (where exactly in the 100-yard range of the field should a player run, pass, or punt?). The common technique of "binning" data is typically necessary in order to work with these values. The Spec Graph representation lessens this requirement as it describes values relationally, i.e., without respect to their concrete values.

In Fig. 3, we give the abstract syntax of the specification graph representation. A key component of our approach is a translation from this abstract syntax onto graph structures. We elide these semantics in our description of specification graphs in this discussion. The building blocks of our representation are as follows:

Actions In the spec graph representation, an action represents an event that happens in the course of play. In the domain of the NFL dataset, examples of actions include timeouts and shotguns. Note that these values are strictly encoded as binary values within the initial dataset.

Properties Similar to actions, properties pertain to information about a specific game play. Unlike actions, properties are not binary and may be either categorical or continuous in nature. In our evaluation we infer properties by analyzing the set of all valued features and first extracting those that range over $\mathbb{B}$. The resulting set of features are considered to be properties. Examples of properties include "air yards" and "score."

Pre-Play Predicates Similar to the behavioral interface specification languages on which the Spec Graph representation was modeled, the pre-play predicates are Boolean formulas that are assumed to be True prior to the play. In terms of a Hoare triple, pre-play predicates make up the first component of the triple, i.e., the pre-state predicate. Note that pre-play predicates are always written in terms of properties.

Post-Play Predicates In the context of a Hoare triple, post-play predicates make up the third component of the triple. After the conclusion of a play, certain property values of the state may have changed. In a Spec Graph, one specifies this change via the post-play predicates of the graph. As previously mentioned, it is possible to relationally describe properties in a post-play predicate while also referencing the concrete values of the property. Herein we make use of the *old* function, which operates on the pre-state value of its argument; see Eq. 3.

Play Actions In the context of a Hoare triple, actions are modeled via the second component of the Hoare triple. Unlike pre- and post-play predicates, play actions are assertions about actions that have occurred during a play. Therefore, play actions are always expressed in terms of actions, not properties.

As described in Fig. 2a, the first step in constructing a Spec Graph is to sequence plays into pre- and post-play pairs. Once these pairs have been constructed, in order to construct a Spec Graph, the abstract syntax shown in Fig. 3 is used to represent the semantic building blocks consisting of actions, properties, pre-play predicates, post-play predicates, and play actions. Once this representation is constructed, it is translated into an embedding for subsequent analysis. In this discussion, we translate our Spec Graph into a `networkx` [30] graph for further analysis. An example of

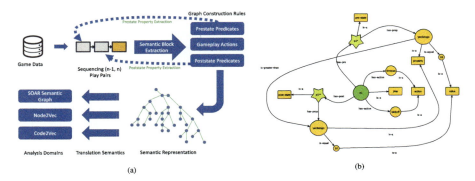

Fig. 2 Overview of the specification graph processing technique and a resulting graph. In Fig. 2a, specification graphs start with raw game play data that is then converted into pre- and post-play pairs. These pairs are then analyzed and the semantic building blocks (Sect. 2.2.2) are extracted from them. These building blocks are then converted into a semantic representation. The abstract syntax for this representation is given in Fig. 3. Finally, our translation semantics are responsible for converting our semantic representation into a target form for analysis. An example of the resulting network is shown in Fig. 2b. In this discussion, we focus on translation semantics that produce a `networkx` graph suitable for analysis by `word2vec`

such a graph is shown in Fig. 2b. The analysis we perform on our resulting graphs is described later in Sects. 2.3 and 2.4.

2.3 Embedding Techniques

Once graphical structures have been designed and all data instances have been represented in that fashion, we must settle the problem of embedding the graphs within a vectorized structure comprised of numerical values so that standard ML techniques can be employed. A typical method for translating the graph embedding problem is the Node2Vec method [31], which is an extension of the readily used Word2Vec linguistic embedding procedure by Gensim [32]. This method learns to embed the graphical representations by repeatedly conducting random walks over the graphs to develop a dataset for a semi-supervised task. Hyper-parameters related to the probability of revisiting nodes balance the tension between global verses local connectivity. The semi-supervised prediction task essentially predicts a given node's value based upon the connected nodes as enumerated by the random walks. The intermediate output of this predictor yields an ML-appropriate node-level embedding that incorporates much of the critical graph-embedded relationships through a generalized procedure. In order to embed the whole graph, every node is embedded and averaged.

$$
\begin{array}{lll}
p & \in \mathbb{P} = \{\textit{the set of properties}\} & \text{(domains)} \\
a & \in \mathbb{A} = \{\textit{the set of actions}\} \\
b & \in \mathbb{B} = \{\textit{true}, \textit{false}\} \\
op & \in \{\leq, =, \ldots\} \subseteq \mathbb{Z} \times \mathbb{Z} \to \mathbb{B} \\
v & \in \mathbb{Z} \\
id & \in \mathbb{Z} \\
l & \in \{\textit{the set of play classes}\} \\
G & ::= \textbf{game } id \ \{ \ \textit{Play} \ \} & \text{(games)} \\
\textit{Play} & ::= \epsilon \mid p_1; \ p_2 \mid \textbf{play } l \ \{ \ \textit{PrePlay Action} \\
& \quad \textit{PostPlay} \ \} \\
\textit{Preplay} & ::= \epsilon \mid \textit{Preplay}_1; \ \textit{Preplay}_2 \\
& \quad \mid \textbf{preplay } \psi \\
\textit{Action} & ::= \epsilon \mid \textit{Action}_1; \ \textit{Action}_2 \\
& \quad \mid \textbf{action } a \\
\textit{Postplay} & ::= \epsilon \mid \textit{PostPlay}_1; \ \textit{Postplay}_2 \\
& \quad \mid \textbf{postplay } \psi \\
\psi & ::= v \mid p & \text{(property expressions)} \\
& \quad \mid \psi_1 \ (op) \ \psi_2
\end{array}
$$

Fig. 3 The abstract syntax for our specification graph representation

2.4 Testing Protocol

In order to test the efficacy and utility of different representations, two ML tasks were devised. First, a supervised classification task was used to ensure that a decision boundary exists to differentiate between play types. Second, an unsupervised task was used to discover any differences in play type without a priori knowledge of class labels. All representation methods were tested using the same supervised and unsupervised tasks developed in Python using Scikit-learn [33], TensorFlow [34], and Keras [35] libraries.

2.4.1 Supervised Classification Task

A prerequisite for the tactics discovery task is the existence of separations between classes. In order to test that such separations exist, we used a deep neural network to test if a decision boundary could be learned. In order to handle the class imbalance present within the dataset (as `run` and `pass` instances far out-weighed other classes), we used weighted samples to prevent over-prediction of larger classes. A k-fold Cross Validation technique with k=5 was used to prevent skewing of results while reporting the average Recall, Precision, and F-Score.

After trying several different neural net configurations, a model architecture that worked experimentally well on our naive representation was chosen. The neural net contained eight fully connected dense layers, containing 1024, 1024, 512, 264, 128, 64, 32, and 16 units with a rectilinear activation function, and an output layer comprising a softmax activation function.

2.4.2 Unsupervised Discovery Task

To test the separability of the data to facilitate tactic Discovery, a k-means clustering approach was devised. In order to not bias the selection in the numbers of clusters, a ranged sweep from k=[2,32] was performed using the silhouette score and the knee method to select the optimal number of clusters (k) [36].

Computational metrics, such as silhouette scores, provide evidence that separations in the embedded space were exploited, but they do not necessarily mean that a useful clustering was uncovered. For this reason, we defined a set of metrics to indicate how much manual analysis would be required to make use of the clusters to support discovery. Completeness is the average proportion of how many of the same class (play type) exist within the same cluster. Homogeneity is the average portion of each cluster dominated by a single class, and Average Spread is the number of clusters on average each class is spread among. Taken together, these metrics define how well plays were grouped together (completeness), how much of each cluster would need to be viewed for an analyst to understand cluster meaning (homogeneity), and how many clusters would need analysis before a tactic is derived (spread).

3 Results

Herein we present the results of the supervised and unsupervised discovery tasks.

3.1 *Supervised Classification Task*

As can be seen from Table 1, the neural network method only yielded sufficient performance for the Naive Featurization case. Follow-up testing using simple logistic regression was able to yield metric results greater than .9 for all embedded representations. This is indicative of a useful structure to the data that enables prediction, yet the particular NN architecture was optimized only for the Naive representation.

Deep Embedded Knowledge Graph Representations for Tactic Discovery

Table 1 Supervised classification task results

Graph method	Embedding	Recall	Precision	F1
Domain informed	node2vec	0.103	0.103	0.103
Spec graph	node2vec	0.266	0.266	0.266
Naive featurization		0.99	0.99	0.99

(a)

Play Type\Cluster	0	1	2	3	4	5	6	7
play_type_extra_point	1149	1054	1099	1055	1081	1097	1066	1074
play_type_field_goal	69	63	71	76	68	70	58	66
play_type_kickoff	966	984	946	974	948	985	1000	963
play_type_pass	2282	2265	2350	2346	2223	2200	2330	2279
play_type_punt	18549	18678	18626	18451	18436	18661	18563	18628
play_type_qb_kneel	2384	2356	2349	2377	2442	2366	2407	2349
play_type_qb_spike	417	363	346	396	361	391	396	391
play_type_run	13257	13102	13101	13301	13298	13232	13216	13089

(b)

Play Type\Cluster	0	1	2	3	4	5	6
play_type_extra_point	0	0	0	0	0	0	10863
play_type_field_goal	0	9727	0	0	0	0	0
play_type_kickoff	0	17964	7441	0	0	0	0
play_type_pass	55995	0	0	0	55993	28571	45128
play_type_punt	0	23815	0	0	0	0	0
play_type_qb_kneel	0	3811	0	0	0	0	0
play_type_qb_spike	0	682	0	0	0	0	0
play_type_run	0	0	48557	55997	0	27424	0

(c)

Graph Method	Embedding	K	Completeness	Homogeneity	Average Spread	Silhouette Score
Domain Informed	node2vec	7	0.56	0.77	1.62	0.55
Spec Graph	node2vec	8	0	0	8	0.54
Naive Featurization		8	.41	.64	5	0.07

Fig. 4 Clustering contingency matrices for the embedded specification graph (**a**), Embedded domain-informed graph (**b**) and (**c**) the clustering metrics

3.2 Unsupervised Discovery Task

The results in Fig. 4c indicate that the Naive Featurization did not yield computationally meaningful results if judged by the silhouette scores in comparison to either embedded graph approach. However, the other clustering metrics indicate that the Naive Featurization marginally out-performed the embedded Spec Graph. This is why computational features do not necessarily indicate useful performance. Among all three representations, the domain-informed graph is the most computationally meaningful, and is the only representation that reduced absolute spread among clusters.

Exploring the contingency matrix in Fig. 4, we can see that in Fig. 4a the classes were evenly distributed among clusters, yielding no utility for the tactic discovery use case. In Fig. 4b, many of the clusters only contain specific plays, or groups of plays, indicating that the clustering was meaningful and useful for the tactic discovery use case. In particular, Passes and Runs can be discovered from minimal inspection of clusters 0 & 3. The classes with the highest amount of overlap are those whose features in the domain are more difficult to disambiguate and would require additional analysis (i.e., kickoff, punts, and field goals).

4 Conclusions

Based upon the results of the Unsupervised discovery task, we can conclude that embedded representations take advantage of similarities of data to produce computationally meaningful clusters as noted through higher silhouette scores.

However, the embedded approach does not negate the need to encode domain knowledge into the graph representation to have domain meaningful clusters. The graph representation, even one regarding domain knowledge in its construction, still saves more time than that of traditional feature engineering. While traditional feature engineering may involve computation of higher order features, the graph representation was more of a rote transcoding of the game play data.

While this work viewed the tactics discovery problem from a static data analysis prospective, an online approach can be imagined to use the a priori discovery to identify novel tactics as they are performed. Such a system would take advantage of the data analysis discussed here, but allow for real time uses such as employment of tactics algorithmically.

Acknowledgments The authors thank their teammates for conceiving and implementing many of the concepts reported here. In particular, Dr. Mike van Lent who led an earlier effort that facilitated this research and Mr. Cameron Copland, an AI Engineer who was instrumental to earlier research efforts. The opinions expressed here are not necessarily those of the Department of Defense nor the sponsor of this effort: The Combat Capabilities Development Command Ground Vehicle Systems Center (CCDC GVSC). This work was funded under contract W56HZV-17-C-0012.[1]

References

1. M. Ponsen, H. Munoz-Avila, P. Spronck, D.W. Aha, Automatically generating game tactics through evolutionary learning. AI Mag. **27**(3), 75 (2006). https://www.aaai.org/ojs/index.php/aimagazine/article/view/1894
2. C.K.-S. Leung, R.K. MacKinnon, Y. Wang, A machine learning approach for stock price prediction, in *Proceedings of the 18th International Database Engineering & Applications Symposium (IDEAS)* (2014)
3. G.P. Lawrence, B.W. Jones, L.J. Hardy, From wall street to expertise development: predicting the rise and demise of talent investment by using machine learning to identify 'game changers', in *Journal of Sport and Exercise Psychology: SUPPLEMENT: North American Society for the Psychology of Sport and Physical Activity* (2018)
4. J.A. Brown, A. Cuzzocrea, M. Kresta, K.D.L. Kristjanson, C.K. Leung, T.W. Tebinka, A machine learning tool for supporting advanced knowledge discovery from chess game data, in *2017 16th IEEE International Conference on Machine Learning and Applications (ICMLA)* (2017), pp. 649–654
5. L. Wu, A. Markham, Evolutionary machine learning for RTS game StarCraft (2017). https://www.aaai.org/ocs/index.php/AAAI/AAAI17/paper/view/14790/14245
6. F. Thabtah, L. Zhang, N. Abdelhamid, NBA game result prediction using feature analysis and machine learning. Ann. Data Sci. **6**(1), 103–116 (2019). https://doi.org/10.1007/s40745-018-00189-x
7. K. Kapadia, H. Abdel-Jaber, F. Thabtah, W. Hadi, Sport analytics for cricket game results using machine learning: an experimental study. Appl.Comput. Inf. (2019). http://www.sciencedirect.com/science/article/pii/S2210832719302868

---

[1]DISTRIBUTION STATEMENT A. Approved for public release; distribution unlimited. (OPSEC #4189).

Deep Embedded Knowledge Graph Representations for Tactic Discovery

8. D. Silver, A. Huang, C.J. Maddison, A. Guez, L. Sifre, G. van den Driessche, J. Schrittwieser, I. Antonoglou, V. Panneershelvam, M. Lanctot, S. Dieleman, D. Grewe, J. Nham, N. Kalchbrenner, I. Sutskever, T. Lillicrap, M. Leach, K. Kavukcuoglu, T. Graepel, D. Hassabis, Mastering the game of go with deep neural networks and tree search. Nature **529**(7587), 484–489 (2016). https://doi.org/10.1038/nature16961

9. D. Draskovic, M. Brzakovic, B. Nikolic, A comparison of machine learning methods using a two player board game, in *IEEE EUROCON 2019-18th International Conference on Smart Technologies* (2019), pp. 1–5

10. J. Heinrich, D. Silver, *Deep Reinforcement Learning from Self-Play in Imperfect-Information Games* (Open Access Archive of Cornell University, New York, USA, 2016) https://arxiv.org/abs/1603.01121. Accessed June 15, 2021

11. J. Haley, R.D. Hoehn, J. Folsom-Kovarik, R. Wray, R. Pazda, B. Stensrud, Approaches for deep learning in data sparse environments (2019), Proceedings of Interservice/industry training, simulation & education conference (I/ITSEC), NTSA (USA) , (2016), p. 9

12. J. Haley, A. Wearne, C. Copland, E. Ortiz, A. Bond, M. van Lent, R. Smith, Cluster analysis of deep embeddings in real-time strategy games, in *Artificial Intelligence and Machine Learning for Multi-Domain Operations Applications II*, ed. by T. Pham, L. Solomon, K. Rainey. International Society for Optics and Photonics, vol. 11413 (SPIE, Bellingham, 2020), pp. 271–280. https://doi.org/10.1117/12.2558105

13. G.A. Miller, Wordnet: a lexical database for English. Commun. ACM **38**(11), 39–41 (1995)

14. J.F. Sowa, Semantic Networks., CiteSeer archival system, (1987). http://citeseerx.ist.psu.edu/viewdoc/summary?doi=10.1.1.694.700. Accessed: June 15, 2021

15. M. Horowitz, R. Yurko, S. Ventura, Detailed NFL play-by-play data 2009–2018 (2020). https://www.kaggle.com/maxhorowitz/nflplaybyplay2009to2016

16. K. Muandet, D. Balduzzi, B. Schölkopf, Domain generalization via invariant feature representation, in *30th International Conference on Machine Learning, ICML 2013*, vol. 28, no. PART 1 (2013), pp. 10–18

17. M. Ghifary, W.B. Kleijn, M. Zhang, D. Balduzzi, Domain generalization for object recognition with multi-task autoencoders, in *Proceedings of the IEEE International Conference on Computer Vision, ICCV 2015*, vol. 2015 (2015), pp. 2551–2559

18. D. Li, Y. Yang, Y.Z. Song, T.M. Hospedales, Deeper, broader and artier domain generalization, in *Proceedings of the IEEE International Conference on Computer Vision*, vol. 2017, pp. 5543–5551 (2017)

19. C. Gan, T. Yang, B. Gong, Learning attributes equals multi-source domain generalization, in *Proceedings of the IEEE Computer Society Conference on Computer Vision and Pattern Recognition*, vol. 2016 (2016), pp. 87–97

20. A.M. Dunn, O.S. Hofmann, B. Waters, E. Witchel, Cloaking malware with the trusted platform module. USENIX, USA, (2011), pp. 395-410. Archival. available at: https://www.usenix.org/legacy/event/sec11/tech/full_papers/Dunn.pdf. Accessed June 15, 2021
A.M. Dunn, O.S. Hofmann, B.Waters, E.Witchel, Cloaking malware with the trusted platform module. USENIX, USA, pp. 395.410 (2011); Archival available at: https://www.usenix.org/legacy/event/sec11/tech/full_papers/Dunn.pdf (Accessed: June 15, 2021)

21. J. Hatcliff, G.T. Leavens, K.R.M. Leino, P. Müller, M. Parkinson, Behavioral interface specification languages. ACM Comput. Surv. **44**(3), 1–58 (2012). http://doi.acm.org/10.1145/2187671.2187678. http://dl.acm.org/ft_gateway.cfm?id=2187678&type=pdf

22. E. ECMA, 367: eiffel analysis, design and programming language, in *ECMA (European Association for Standardizing Information and Communication Systems), pub-ECMA: adr* (2005)

23. J.G.P. John, *High Integrity Software: The Spark Approach to Safety and Security* (Addison-Wesley, Boston, 2003)

24. R. Chapman, Industrial experience with spark. Ada Lett. **20**(4), 64–68 (2000). https://doi.org/10.1145/369264.369270

25. G.T. Leavens, Y. Cheon, C. Clifton, C. Ruby, D.R. Cok, How the design of JML accommodates both runtime assertion checking and formal verification. Sci. Comput. Program. **55**(1–3), 185–208 (2005). https://doi.org/10.1016/j.scico.2004.05.015
26. G.T. Leavens, A.L. Baker, C. Ruby, Preliminary design of JML. ACM SIGSOFT Softw. Eng. Notes **31**(3), 1 (2006)
27. L. Burdy, Y. Cheon, D.R. Cok, M.D. Ernst, J.R. Kiniry, G.T. Leavens, K.R.M. Leino, E. Poll, An overview of JML tools and applications. Int. J. Softw. Tools Technol. Transf. **7**(3), 212–232 (2005)
28. C.A.R. Hoare, An axiomatic basis for computer programming. Commun. ACM **12**(10), 576–580 (1969). https://doi.org/10.1145/363235.363259
29. R.W. Floyd, Assigning meanings to programs, in *Proceedings Symposium on Applied Mathematics*, vol. 19 (1967), pp. 19–32
30. A.A. Hagberg, D.A. Schult, P.J. Swart, Exploring network structure, dynamics, and function using networkx, in *Proceedings of the 7th Python in Science Conference, Pasadena*, ed. by G. Varoquaux, T. Vaught, J. Millman (2008), pp. 11–15
31. A. Grover, J. Leskovec, Node2vec: scalable feature learning for networks, in *Proceedings of the 22nd ACM SIGKDD International Conference on Knowledge Discovery and Data Mining, KDD'16* (Association for Computing Machinery, New York, 2016), pp. 855–864. https://doi.org/10.1145/2939672.2939754
32. R. Řehůřek, P. Sojka, Software Framework for Topic Modelling with Large Corpora, in *Proceedings of the LREC 2010 Workshop on New Challenges for NLP Frameworks* (ELRA, Valletta, 2010), pp. 45–50. http://is.muni.cz/publication/884893/en.
33. F. Pedregosa, G. Varoquaux, A. Gramfort, V. Michel, B. Thirion, O. Grisel, M. Blondel, P. Prettenhofer, R. Weiss, V. Dubourg, J. Vanderplas, A. Passos, D. Cournapeau, M. Brucher, M. Perrot, E. Duchesnay, Scikit-learn: machine learning in Python. J. Mach. Learn. Res. **12**, 2825–2830 (2011)
34. M. Abadi, A. Agarwal, P. Barham, E. Brevdo, Z. Chen, C. Citro, G.S. Corrado, A. Davis, J. Dean, M. Devin, S. Ghemawat, I. Goodfellow, A. Harp, G. Irving, M. Isard, Y. Jia, R. Jozefowicz, L. Kaiser, M. Kudlur, J. Levenberg, D. Mané, R. Monga, S. Moore, D. Murray, C. Olah, M. Schuster, J. Shlens, B. Steiner, I. Sutskever, K. Talwar, P. Tucker, V. Vanhoucke, V. Vasudevan, F. Viégas, O. Vinyals, P. Warden, M. Wattenberg, M. Wicke, Y. Yu, X. Zheng, TensorFlow: large-scale machine learning on heterogeneous systems (2015). Software available from tensorflow.org. http://tensorflow.org/
35. F. Chollet, et al., Keras (2015). https://github.com/fchollet/keras
36. P.J. Rousseeuw, Silhouettes: a graphical aid to the interpretation and validation of cluster analysis. J. Comput. Appl. Math. **20**, 53–65 (1987)

Pathways to Artificial General Intelligence: A Brief Overview of Developments and Ethical Issues via Artificial Intelligence, Machine Learning, Deep Learning, and Data Science

Mohammadreza Iman, Hamid R. Arabnia, and Robert Maribe Branchinst

1 Introduction

It is nearly impossible to move around modern society without encountering a device or application powered by artificial intelligence (AI). Weather forecasts, traffic signals, airplanes, factory lines, home appliances, and mobile applications are just a few examples of areas likely to encounter elements controlled by AI. Yet, there is even more happening under the surface with AI managing countless applications including internet traffic, gene-related research, and medical image and history analyzation. For most people today, deep learning, machine learning, and AI are all terms for which they are at least familiar.

Another body of work that most people will have heard of is data science and data analytics. Technological advances over the past few decades have transferred the possibility of generating, storing, sharing, and analyzing data to nearly everyone. With data now being a true commodity, some have said that data is the new oil or gold. For example, retailers are now able to gather information about their sales as well as their customers habits and preferences to greatly benefit both parties. Retailers can then use this information to intelligently predict customer shopping habits during other times of the year as well as control their supplies based on projected demands, thus, not wasting time and money on unnecessary storage or

---

M. Iman (✉) · H. R. Arabnia
Department of Computer Science, Franklin College of Arts and Sciences, University of Georgia, Athens, GA, USA
e-mail: hra@uga.edu

R. M. Branchinst
Learning, Design, and Technology Mary Frances Early College of Education, University of Georgia, Athens, GA, USA
e-mail: rbranch@uga.edu

© Springer Nature Switzerland AG 2021
H. R. Arabnia et al. (eds.), *Advances in Artificial Intelligence and Applied Cognitive Computing*, Transactions on Computational Science and Computational Intelligence, https://doi.org/10.1007/978-3-030-70296-0_6

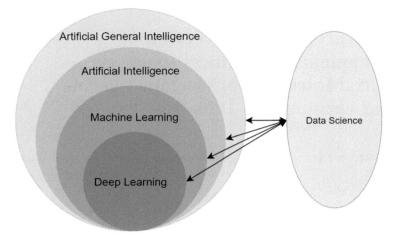

Fig. 1 Domain hierarchy

creating shortages. This is just one example of the great advances made possible by data science and its varying applications. With advances such as autonomous driving now available, there is no telling where data science and AI might take us.

In this article, we briefly review the history of these developments in artificial general intelligence, artificial intelligence, machine learning, deep learning, and data science (see Fig. 1), tracing the history from the first mechanical computer in 1850 to the current state of deep learning in 2020. We overview the many evolutions in AI and discuss possible future directions as well as some of the ethical dilemmas posed by such advances. Ultimately, our goal is to overview these processes for a lay audience who may not have intimate knowledge of AI and data science at large.

2 Artificial Intelligence

The first mechanical computer was invented in the 1850s by Charles Babbage [1]. In 1950, Alan Turing, renown for advancing the general-purpose programmable computer, asked the big question for the first time: "Can machines think?" [2]. Alan Turing proposed an operational test for machine intelligence. A machine "passes the test if a human interrogator, after posing some written questions, cannot tell whether the written responses come from a human or a machine." [2].

In 1956, the term "Artificial Intelligence" (AI) was used for the first time in a proposal for a summer research workshop at Dartmouth College in New Hampshire. The goal of AI was to "[make] a machine behave in ways that would be called intelligent if a human were so behaving" [3]. The aim of the workshop was to develop AI such that it might pass the Turing test. To pass the Turing test the AI needs to [1]:

- Understand speech; *natural language processing (NLP)*
- Store the information and data; *knowledge representation*
- Use the stored data to draw conclusions; *automated reasoning/decision-making*
- Detect new patterns and adapt to new circumstances; *machine learning (ML)*

To fully pass the Turing test, two additional capabilities are needed [4, 5]:

- Extract knowledge and/or comprehension from images or videos (e.g., face recognition); *computer vision*
- Mimic human physical behaviors corresponding with the senses to interact with the environment (e.g., touch, motor functioning); *physical interaction*

Overall, the AI could be divided into two categories of **Artificial General Intelligence** (AGI) or strong AI (actual thinking), and narrow or weak AI (simulated thinking) [1, 6]. Some scholars argue that achieving AGI may be decades into the future, and that the emergence of AGI will bring with it an "intelligence explosion" leading to "profound changes in human civilization" [6]. Yet, the field of computer science has already begun to develop the narrower form of AI. In fact, the ability to have devices such as sensors and robots, and intelligent Decision Support Systems (DSS), such as the autocorrect software analyzing the words on this page are the result of already existing forms of AI [6]. These technologies have already evolved beyond what many people could have imagined, and yet the future of AGI has the potential to transform the experience of generations to come in ways we cannot yet predict.

Early (late twentieth century) AI approaches were rule-based and focused on attending to all possible solutions for a specific and identifiable problem [7]. Some board games and various types of robots used in factory lines are just two examples of this type of rule-based AI. Going forward, decision making systems began to advance these types of approaches [8]. Specifically, decision making attended to the fact that real-life problems are rarely contained within such specific and rule-based features. Even board games must contend with players who make unpredictable decisions. Although decision making may follow some of the same patterns of rule-based prediction systems, it began to extend the bounds of these rules by accounting for uncertainty [9]. The Boltzmann machine research line during the 1990s through the early 2000s delivered a well-known example of this type of AI, which utilizes probability and statistics predicting behavior patterns in various settings [10–12]. Sum-Products Networks (SPNs) are another advancement in AI that began to incorporate networks able to compete with deep learning models in many applications by taking into account the probability distributions of features [13].

The boom of digital storage development in the 1990s and 2000s—delivering cloud storage and advanced data collection methods—brought with it a new era of "big data." Big data refers to the vast amount of easily accessible consumer data including images, texts, audio, transactions, and human and environmental sensing data from electronic devices. This surge in available data required new methods for analyzing it, translating to Data Science (DS) and a new chapter in machine learning (ML).

3 Data Science

Data Science is divided into three main areas including collecting, storing, and analyzing data (structured or unstructured) [14]. Data collection methods were advanced through the spread of high-speed internet world wide—including less costly wireless connections—as well as increased variety of cheaper electronic connections and sensors, such as smartwatches, exercise trackers, and cameras [15]. Data storage was advanced through cloud storage, which further influenced big data collection by offering these services at a reduced cost and to an increasing proportion of the population.

Data analysis consists of two major components: preprocessing and processing. Preprocessing refers to various aspects of raw data management including unbalanced data, imputation techniques for missing data, detecting and addressing outliers, and data labeling procedures. Processing refers to extracting information and knowledge from preprocessed data to identify patterns, make predictions, and/or classify data [14]. One of the promising methodological categories for processing big data is Machine Learning (ML), a subdivision of AI.

4 Machine Learning (ML)

Machine learning (ML) is a subdivision of AI that consists of statistics, mathematics, and logical techniques to extract patterns (i.e., information) from a set of training data and apply the inferences to unseen data. Again, these recent advances in ML were made possible by the new era of big data and the vast advancement in computational capacity. Importantly, ML differs from other forms of AI in that it does not require extensive and complicated programming, but rather, has the ability to learn patterns and later apply them. Thus, ML does not need to consider every possible solution (i.e., be deterministic) and can manage noise and uncertainty [16].

Innovation in ML brought with its exponential advancement in earlier techniques—some of them developed before the 1970s—such as Linear Regressions, decision trees, Random forest, K-nearest neighbor (KNN), Support Vector Machine (SVM), Artificial Neural Networks (ANNs). For example, early ML ANN models for autonomous driving [17] and facial recognition [18] were developed in the 1980s but lacked access to the data and computation capacity needed to apply them [16].

Like any method, ML brings with it its own unique techniques and challenges. Common types of ML include supervised learning, unsupervised learning, semi-supervised learning, and reinforcement learning. Each of these is discussed in brief below, followed by some of the challenges associated with ML such as overfitting and dealing with extraneous features.

Supervised learning refers to the use of an ML training data set that has been labeled, typically by humans, and the goal of which is to categorize or label the unseen data [19]. The process of categorization or labeling often occurs through **classification** or **regression** techniques, which has value for making predictions—using regression—such as predicting stock market values or classifying objects in an image, such as identifying tumors in a medical x-ray.

Unsupervised learning refers to categorizing data by analyzing patterns and shared features without utilizing a pre-labeled training dataset [19]. In unsupervised learning, **clustering** is often used to detect patterns and anomalies, such as in grouping customers for marketing strategies or marking emails from unknown sources as "spam."

In addition, a smaller (i.e., limited) labeled ML dataset may be used to improve the categorization of a larger, unlabeled dataset. This is known as *semi-supervised learning*. Semi-supervised learning may be a more cost-effective option of labeling large datasets, in addition to allowing for greater accuracy by limiting human error [19]. For example, speech recognition errors may be reduced by 22% when human-labeled data are combined with machine-labeled data using semi-supervised learning [20].

Reinforcement learning (RL) operates using a reward-based system. Reinforcement learning attempts to select the best possible action that would maximize the final reward (or conversely, minimize the punishment), all while keeping track of these actions to improve the choice-selection of the following round. Thus, it is a trial-and-error process that works through the system's ability to learn improvement strategies and decisions through the success or failure of previous attempts. There are many different types of RL algorithms, each designed to address a specific problem [16]. Examples of RL applications include some types of board games (e.g., chess and Go), robots, and various elements of autonomous driving systems.

Although Machine Learning algorithms demonstrate immense accuracy in identifying training dataset patterns, a common problem in these models is **overfitting** the data [21, 22]. Overfitting occurs when the ML network has been trained using all labeled (i.e., training) data and cannot deal with the noise (i.e., uncertainty) in the unseen data. It also occurs when patterns observed in the limited training data are not accurate of the existing patterns in the larger data. Overfitting may occur when using unbalanced or biased datasets, indicating that the training set does not include all possible samples within the domain[16, 21, 22].

There are several ways of correcting for the risk of overfitting. One of these is to divide the training dataset into two parts: training and **validation** [6, 16]. The size of the validation set will depend on the size of the overall training set, but typically ranges from about 10 to 30% of the full set. The validation set is not used for training purposes, but is instead verified against the final dataset to ensure accuracy. In this procedure, **cross-validation** is used to correct for the risk of selecting a biased validation set [23]. For example, a 10-fold cross-validation procedure would involve

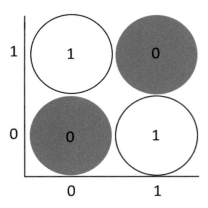

Fig. 2 Not linearly separable pattern, known as XOR problem

dividing a training set into 10 separate sets and then training the ML model 10 times using only nine of those sets each time. The final model would then be validated against the remaining set (1/10th of the original), with the accuracy being equal to the average of the 10 validation runs.

Another challenge that may arise in ML models is the issue of extraneous features, such as the vast number of potentially uncorrelated features present in some big data sets. In many cases, not all of the features present in a dataset will be related to the objective of the ML model and, thus, are not useful. For example, to predict the seasonal sales of an online store, customers' employment status and income may be related to the outcome, but their specific job title may not be. There are several known processes for responding to unrelated features in a dataset including **feature selection, combination, and extraction** [24]. These are performed through techniques like correlation analyzation, principal component analyzation (PCA), and dimensionality reduction techniques. These techniques work mainly by validating the correlation of each feature to the target [24].

Machine Learning techniques and data sets can be categorized into two groups: linear and non-linear. A linear data pattern is the simplest data pattern and can be categorized using a linear function to perform regression or classification. Many algorithms had been developed to fit linear models such as linear regression, logistic regression, classification and regression trees, K-nearest neighbors, and support vector machine [16]. Non-linear functions are those that cannot be classified using linear methods. Like other models of data analysis and management, non-linear data associations may pose additional challenges to ML [16]. The non-linear problem in ML is known as the **XOR (i.e., "exclusive or") problem**, which refers to a mixed pattern of data that cannot be categorized using linear functions, Fig. 2 [25].

Although many algorithms have been developed to manage linear data (as mentioned above), the non-linear nature of many data sets remained a challenge for ML. For example, the decision trees, k-nearest neighbors, and support vector machine mentioned above are functions that can manage some non-linear data problems; yet, they do this imperfectly, and issues remain. Artificial Neural Networks (described in the next section) began to address these issues [16].

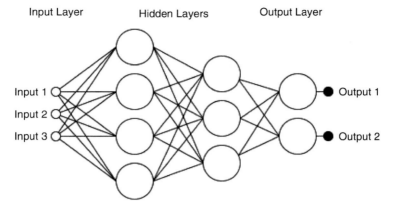

Fig. 3 An example of MLP network architecture

5 Artificial Neural Network

In 1958, the first artificial neuron was introduced—attempting to mimic the neural pathways of the human brain. Named Perceptron, it used a sigmoid function and performed linear functions with great success [26]. To advance this then new technology, several Perceptrons were later clustered into a layer, allowing for linear patterns to be detected through the use of input data connecting into the Perceptron layer. Training happens by feedforwarding the data while backpropagating the labels to tune the weights of each node. Thus, the first artificial neural network (ANN) was born [27]. Perceptron remained at the height of ANN mechanisms until 1969, when rigorous reviews demonstrated its shortcomings—namely, that Perceptron could not address the issue of non-linearity; it had hit a dead-end [28].

By the 1980s, scientists again attempted to address the issue of non-linearity (i.e., the XOR problem) by using hidden layer(s) of Perceptron, known as Multilayer Perceptron (MLP). MLP is a type of ANN consisting of one or more layers of varying nodes—the network architecture (see Fig. 3) [27]. Using an activation function, such as sigmoid, on the front end of the nodes, again combined with backpropagation techniques, allowed for increasingly advanced classification and regression models—including those for non-linear patterns [27]. These advances greatly improved the accuracy of some of the advanced technologies we enjoy today, such as autonomous driving and facial recognition.

The early ANN designs were fully connected, with each node tied to the next, and each connection having a weight. Each node contained an activation function and uses the value of prior nodes multiplied by the weight of the connection to calculate the next node in a recursive loop. The simultaneous backpropagation by means of the training dataset serves to update and fine-tune the node weights and thresholds. Similar to the reinforcement learning process described above, cost/loss functions are used as additional metrics by which to measure the compatibility between the training data (i.e., ground truth) and the network predictions [27].

Despite vast advances in theorizing, the ANNs (MLP) of the 1980s faced several challenges. Specifically, the limited number of available nodes in each MLP layer, combined with the limited number of layers, produced a heavy burden for the computers of the day. In short, advanced theorizing was limited to the computational capacity of the 1980s machines. However, by the year 2000, significant advances were made in computational capacity. These advances, paired with the ability to replace the nodes' sigmoid activation function with more efficient functions such as sign, linear, tanh [29, 30], and more recently ReLU and leaky-ReLU [31], allowed for the creation of a larger network of nodes, including more hidden layers. This led to the creation and advancement of deep neural networks (DNN), also known as deep learning (DL).

6 Deep Learning (DL)

As mentioned, the vast improvements in the computational capacity of the 2000s helped shape the development of deep neural networks (DNN) or deep learning (DL). Another shaping factor in the development of DL was the arrival of big data sets, which offered the opportunity to improve the training process and thus the performance of DNN.

Similar to ANNs, the learning in DNNs occurs through the optimization of the weights throughout the entire network. One of the well-known algorithms for handling this type of optimization problem is Stochastic Gradient Descent (SGD) [32]. There are several other methods based on the SGD algorithm such as Momentum, Nesterov Momentum, and Adam [32]. Each of these methods works by tracing the error surface of the error calculation function (known as loss function) with the goal of finding the global minima, as shown in Fig. 4 [32, 33]. The loss function is based on the adjustments of the weights of each of the connections in the network [32].

Other parameters that need to be taken into consideration in order to maximize DNNs' accuracy include data preprocessing, hyperparameter adjusting such as learning rate adjustments, weight initialization, initializing biases, and batch normalization [34, 35].

Several modifications of DNNs have vastly improved the implementation of these models. The modifications aim to reduce the models' generalization error by regularizing the weights. There are several methods to do such regularizations including considering the noise robustness, stop learning point (i.e., early stopping), parameter sharing, and dropout [34, 35]. The following section overviews some of the main events and advancements in determining the current state of DLs.

In 2007, Fei Fei Li and colleagues introduced ImageNet, the largest database of labeled images with over 14 million images categorized into nearly 22,000 indexed synsets (categories) as of 2020 [image-net.org]. These images can be used

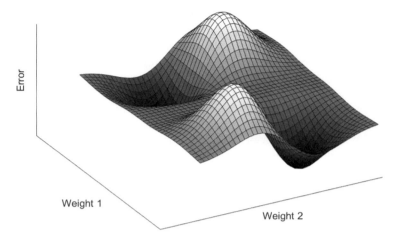

Fig. 4 An example of error surface in an ANN/DNN

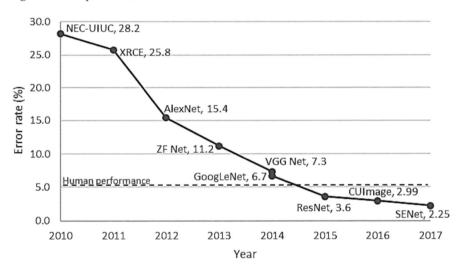

Fig. 5 ILSVRC winners

for technologies such as object location, detection, and classification in videos and other image-related media [image-net.org]. Since 2010, ImageNet has led an annual challenge—the ImageNet Large Scale Visual Recognition Challenge (ILSVRC). The challenge brings bright minds from around the globe together to explore new ideas in DL by allowing them to use a large collection of image-data they would not otherwise have access to. This challenge has brought great success in minimizing error as demonstrated in Fig. 5. Remarkably, the classification error of 28% in 2010 was reduced to less than 3% by 2017 [36].

By 2011, the Convolutional Neural Network (CNN) was beginning to grow in popularity. CNN was able to outperform humans in recognition of traffic signs with an accuracy of 99.46% (compared to humans at 99.22%) [37]. First introduced in 1997, CNN was inspired by the visual cortex of animals and attempted to regularize input data to find hierarchical patterns within image data—called self-organized map [38]. By 2020, nearly all DLs utilized CNN layer(s) for visual-related tasks. Going back to the ILSVRC, CNN was utilized by the majority of champions, but it has been used for many other applications as well [36]. Interesting research conducted in 2015 demonstrates how a DNN network functions using CNN layers (i.e., https://youtu.be/AgkfIQ4IGaM) [39].

Another network architecture from the year 1997, Long Short-Term Memory (LSTM) [40], also saw vast improvements in the 2010 decade [41]. LSTM, also known as Recurrent Neural Network (RNN), is a modified neural network that utilized feedback connections. LSTM allowed for the subsequent advancements in DL such as speech and handwriting recognition applications, as well as anomaly detection in data series (e.g., network traffic) [41].

In 2014, Ian Goodfellow and colleagues invented the Generative Adversarial Network (GAN) [42]. GAN is comprised of two neural networks competing against each other; the first is a generative network which generates new data while the second is a discriminative network that evaluates the generated data. This advanced network can generate new data based on the characteristics of the inputted training data. For example, if the training data were to be of a human face, the network would generate a new face, looking entirely human but never having previously existed [43]. A vast amount of applications benefit from GANs, such as imaginary fashion models and scientific simulations [43]. Despite their many advances, GANs raise some concerns, specifically regarding the production of falsified voice or video records [44].

By 2016, Google announced the Tensor Processing Unit (TPU), followed by the Google TensorFlow framework of open source libraries [45]. TensorFlow touts well-tailored hardware and software to be used for neural network computations and applications [45]. Using this technology, Google's DeepMind AlphaGo defeated the Go champion in 2016 by combining DL and RL in a new mechanism named Deep Reinforcement Learning [46].

Another contemporary topic in ML that was also initiated in the 1990s is transfer learning or domain adaptation, first published by Lorien Pratt [47]. Transfer learning works by using knowledge garnered through the ML model during the training phase and literally transferring the learning to another task in a similar domain [48]. For example, a DL model trained to classify flowers can be used to also classify leaves by modifying the trained model using a transfer learning technique (e.g., fine-tuning some of the layers). Since 2014, transfer learning has been used to adapt deep learning models, such as in domains like medical imaging. This is known as deep transfer learning and has been used to reduce the often-long training time as well as to handle the small training samples of some deep learning [48]. Progressive learning introduced by Google's DeepMind Project in 2016 is another specific type of deep transfer learning that is attempting to build on previous, related knowledge, similar to human learning capabilities [49].

In summary, this overview summarized a few notable types of DL that are on the rise. It is important to note that the aforementioned advancements in DL are vast topics in and of themselves, each carrying with them a research line with hundreds or even thousands of relevant articles that could not be overviewed here. In addition, there are many more DL advances not discussed here, including the autoencoders used for image segmentation models such as U-Net and new types of data compressors [50], among many others.

7 Discussion

As mentioned, the era of big data, spurred by drastic advancements in computational capacity, has brought a new chapter to machine learning (ML) and artificial intelligence (AI) since the turn of the century. In the past decade alone, the movement toward artificial general intelligence (AGI) has grown exponentially, and there is no telling where it might take us.

A common example of the great innovations of AGI is IBM's Watson, first introduced in 2011. Watson is a natural language processing (NLP) platform, whose architecture benefits from a variety of developments in AI and ML [51]. In 2011, Watson defeated the champions of the popular quiz show Jeopardy—a feat spurred by its ability to "process 500 gigabytes, the equivalent of a million books, per second" [52].

The use of even narrow AIs to mimic human cognition is opening the pathway to AGI, and AGI is a force that future humans will have to contend with. The competition is likely to be intense given that computer programs do not suffer from fatigue, boredom, or other common human ailments—and impediments to work and/or output. For example, Google's well-known program, AlphaGo, managed to train itself in one night to rise from an amateur to a champion player by the next morning [46].

There are countless other examples of the ways in which AGI is looming closer. Present-day Artificial Neural Networks (ANN) are already a simple mimic of human brain cells, and Convolutional Neural Networks (CNN) mime the human visual cortex. Generative Adversarial Networks (GAN) work like the human imagination—generating new data from observed data—which can be used for better understanding facts by imaging-related data. All this is done without needing to access all or even a vast amount of data. Long-Short Term Memory (LSTM) works similarly to the human memory and is able to solve problems related to sequential data analyzing. Transfer learning, followed by progressive learning, is attempting to mimic human skill-learning abilities, a task that is endless for human beings. However, human beings must rely on previous knowledge and skills—oftentimes garnered over a lifetime—that AI programs can learn in a matter of hours. All of this evidence suggests that AGI is close to becoming reality, and the implications of this have yet to be explored.

8 Ethics

With any scientific advance—particularly one that travels so quickly—ethical issues and considerations are unavoidable. Recent years have seen an increase in considerations of the potential ethical pitfalls of AI and the use of personal data raised by data scientists and other scholars (including social scientists, historians, and others) [53–55]. Unfortunately, the ethical consequences of many advances are difficult to assess in real-time. For example, although it is relatively obvious to see the issues with falsifying evidence through GANs [44], the impact of wrongly classifying a disease through X-ray images is more difficult to project, not to mention the social implications of these advanced changes. This section discusses some primary areas of concern in the ethics debate surrounding AI and offers some additional points for consideration.

One primary area of ethics currently involves **data privacy** surrounding sensitive data and personal information such as credit card transactions and medical data. Issues related to data privacy are complicated by the need for access to personal information in order to move many fields forward. For example, not using medical information means that patients miss out on the opportunity to have their diagnoses made by more accurate AI programs. On the other hand, there are also consequences to this data being made available. Doctors risk being sued if later AI advances point to something they missed, and patients are at risk of having their private information shared elsewhere. Like most ethical issues, it is imperative to consider both sides of this debate and to seek solutions that maximize benefits while limiting risks. Data anonymization mechanisms begin to address these issues by making data unidentifiable, which allow for the positive usage of private information without risking patient or physician privacy [56].

A second ethical implication includes the social impact of **human job loss** if AI automates such jobs. For example, the rise of autonomous driving semis and other transport vehicles has the potential to contribute to the unemployment of a large proportion of middle-class workers in the USA. This process is similar to the transition of farming to factory jobs across Europe and other parts of the world during the industrial era as well as the subsequent automation of factory lines. During these times, many workers lost their jobs, thereby moving from middle-class into poverty. Although these changes are unavoidable, the social impact on families and communities must be considered. However, existing data and insight from the industrial and automation booms can help data scientists and social science researchers better predict and prepare for the implications of AI-automation for employment. Planning for these potential consequences may help to ease the transition for society and future generations.

A final ethical implication worth noting here is the broader **impact on society**. Such impacts are often difficult to observe in real-time. One poignant example of the effects of AI is political polarization. With the rise of social media and worldwide connectivity via cell phones, tablets, smartwatches, and other devices, the implications of AI automating what information people have access to is more evident than ever. Some have suggested that the targeted marketing and information brought by AI has contributed to political and social polarization, with many people having access only to the information they already agree with. Opinions are constantly being validated, and the ability to be objective in any topic is becoming limited. This process has also occurred with consumer branding, as AI approaches target consumers with ads for products they are more likely to buy based on their previous purchasing behaviors. In fact, it has been said that these mechanisms know people better than they know themselves. The implications of the human mind becoming inundated with certain types of data remain to be seen, and the ethical considerations have yet to be examined. However, such implications must be on our radar as they have the potential to change society for generations.

Ethical dilemmas are just that: dilemmas. The vast majority are not easily solvable or even identifiable. However, their elusiveness cannot be the reason that data scientists fail to consider these issues—potential or currently a reality. Rather, it is the responsibility of the data scientist community to partner with other disciplines (e.g., social and behavioral sciences) to consider the effects of their creations on society, no matter how far into the future they may reach.

References

1. S.J. Russell, P. Norvig, *Artificial Intelligence: A Modern Approach* (Pearson, Kuala Lumpur, 2016)
2. I.B.Y.A.M. Turing, Computing machinery and intelligence-AM Turing. Mind **59**(236), 433 (1950)
3. J. McCarthy, M.L. Minsky, N. Rochester, C.E. Shannon, A proposal for the Dartmouth summer research project on artificial intelligence, August 31, 1955. AI Mag. **27**(4), 12 (2006)
4. S. Harnad, Other bodies, other minds: a machine incarnation of an old philosophical problem. Minds Mach. **1**(1), 43–54 (1991)
5. P. Schweizer, The truly total Turing test. Minds Mach. **8**(2), 263 (1998)
6. J. Howard, Artificial intelligence: implications for the future of work. Am. J. Ind. Med. **62**, 917–926 (2019)
7. F. Hayes-Roth, Rule-based systems. Commun. ACM **28**(9), 921 (1985)
8. E.J. Horvitz, J.S. Breese, M. Henrion, Decision theory in expert systems and artificial intelligence. Int. J. Approx. Reason. **2**(3), 247–302 (1988)
9. P.L. Jones, I. Graham, *Expert Systems: Knowledge, Uncertainty and Decision* (Chapman and Hall, London, 1988)
10. M.C. Moed, G.N. Saridis, A Boltzmann machine for the organization of intelligent machines. IEEE Trans. Syst. Man. Cybern. **20**(5), 1094 (1990)
11. S. Ruslan, and G. Hinton, Deep Boltzmann machines., In *Artificial intelligence and statistics*, PMLR (2009), pp. 448–455

12. K.H. Cho, T. Raiko, A. Ilin, Gaussian-Bernoulli deep Boltzmann machine, in *The 2013 International Joint Conference on Neural Networks (IJCNN)* (2013), pp. 1–7
13. H. Poon, P. Domingos, Sum-product networks: a new deep architecture, in *2011 IEEE International Conference on Computer Vision Workshops (ICCV Workshops)* (2011), pp. 689–690
14. W. Van Der Aalst, Data science in action, in *Process Mining* (Springer, Berlin, 2016), pp. 3–23
15. M. Iman, F.C. Delicato, C.M. De Farias, L. Pirmez, I.L. Dos Santos, P.F. Pires, THESEUS: a routing system for shared sensor networks, in *Proceedings - 15th IEEE International Conference on Computer and Information Technology, CIT 2015* (2015), pp. 108–115
16. E. Alpaydin, *Introduction to Machine Learning* (MIT Press, Cambridge, 2014)
17. D.A. Pomerleau, Alvinn: an autonomous land vehicle in a neural network, in *Advances in Neural Information Processing Systems* (1989), pp. 305–313
18. E.L. Hines, R.A. Hutchinson, Application of multi-layer perceptrons to facial feature location, in *Third International Conference on Image Processing and its Applications, 1989* (1989), pp. 39–43
19. O. Chapelle, B. Scholkopf, A. Zien, Semi-supervised learning (chapelle, o. et al., eds.; 2006)[book reviews]. IEEE Trans. Neural Netw. **20**(3), 542 (2009)
20. A. Sapru, and Sri Garimella, Leveraging Unlabeled Speech for Sequence Discriminative Training of Acoustic Models., In INTERSPEECH (2020), pp. 3585–3589
21. T. Dietterich, Overfitting and undercomputing in machine learning. ACM Comput. Surv. **27**(3), 326–327 (1995)
22. T. Poggio, K. Kawaguchi, Q. Liao, B. Miranda, L. Rosasco, X. Boix, J. Hidary, H. Mhaskar, Theory of deep learning III: the non-overfitting puzzle. CBMM Memo. 073 (2018)
23. J.D. Rodriguez, A. Perez, J.A. Lozano, Sensitivity analysis of k-fold cross validation in prediction error estimation. IEEE Trans. Pattern Anal. Mach. Intell. **32**(3), 569–575 (2009)
24. S. Khalid, T. Khalil, S. Nasreen, A survey of feature selection and feature extraction techniques in machine learning, in *2014 Science and Information Conference* (2014), pp. 372–378
25. T.L. Clarke, T.M. Ronayne, Categorial approach to machine learning, in *Conference Proceedings 1991 IEEE International Conference on Systems, Man, and Cybernetics* (1991), pp. 1563–1568
26. F. Rosenblatt, The perceptron: a probabilistic model for information storage and organization in the brain. Psychol. Rev. **65**(6), 386 (1958)
27. D. Anderson, G. McNeill, Artificial neural networks technology. Kaman Sci. Corp. **258**(6), 1–83 (1992)
28. M. Minsky, S. Papert, An introduction to computational geometry, in Cambridge tiass., HIT (1969)
29. B. Karlik, A.V. Olgac, Performance analysis of various activation functions in generalized MLP architectures of neural networks. Int. J. Artif. Intell. Exp. Syst. **1**(4), 111–122 (2011)
30. P. Sibi, S.A. Jones, P. Siddarth, Analysis of different activation functions using back propagation neural networks. J. Theor. Appl. Inf. Technol. **47**(3), 1264–1268 (2013)
31. B. Xu, N. Wang, T. Chen, M. Li, Empirical evaluation of rectified activations in convolutional network (2015, preprint). arXiv1505.00853
32. S. Ruder, An overview of gradient descent optimization algorithms (2016, preprint). arXiv1609.04747
33. M.W. Gardner, S.R. Dorling, Artificial neural networks (the multilayer perceptron)—a review of applications in the atmospheric sciences. Atmos. Environ. **32**(14–15), 2627–2636 (1998)
34. J. Bergstra, Y. Bengio, Random search for hyper-parameter optimization. J. Mach. Learn. Res. **13**, 281–305 (2012)
35. D. Maclaurin, D. Duvenaud, R. Adams, Gradient-based hyperparameter optimization through reversible learning, in *International Conference on Machine Learning*, (2015), pp. 2113–2122
36. J.W. Tweedale, An application of transfer learning for maritime vision processing using machine learning, in *International Conference on Intelligent Decision Technologies* (2018), pp. 87–97

37. D. Cireşan, U. Meier, J. Masci, J. Schmidhuber, A committee of neural networks for traffic sign classification, in *The 2011 International Joint Conference on Neural Networks* (2011), pp. 1918–1921
38. S. Lawrence, C.L. Giles, A.C. Tsoi, A.D. Back, Face recognition: a convolutional neural-network approach. IEEE Trans. Neural Netw. **8**(1), 98–113 (1997)
39. J. Yosinski, J. Clune, A. Nguyen, T. Fuchs, H. Lipson, Understanding neural networks through deep visualization (2015, preprint). arXiv1506.06579
40. S. Hochreiter, J. Schmidhuber, Long short-term memory. Neural Comput. **9**(8), 1735–1780 (1997)
41. Y. Yu, X. Si, C. Hu, J. Zhang, A review of recurrent neural networks: LSTM cells and network architectures. Neural Comput. **31**(7), 1235–1270 (2019)
42. I. Goodfellow, J. Pouget-Abadie, M. Mirza, B. Xu, D. Warde-Farley, S. Ozair, A. Courville, Y. Bengio, Generative adversarial nets, in *Advances in Neural Information Processing Systems* (2014), pp. 2672–2680
43. T. Karras, S. Laine, T. Aila, A style-based generator architecture for generative adversarial networks, in *Proceedings of the IEEE Conference on Computer Vision and Pattern Recognition* (2019), pp. 4401–4410
44. T. Zubair, A. Raquib, J. Qadir, Combating fake news, misinformation, and machine learning generated fakes: insight's from the Islamic ethical tradition. Islam Civilisational Renew. **10**(2), 189–212 (2019)
45. K. Sato, C. Young, D. Patterson, An in-depth look at google's first tensor processing unit (TPU). *Google Cloud Big Data Machine Learning Blog*, vol. 12 (2017)
46. H.S. Chang, M.C. Fu, J. Hu, S.I. Marcus, Google deep mind's alphago. OR/MS Today **43**(5), 24–29 (2016)
47. L.Y. Pratt, *Transferring Previously Learned Back-Propagation Neural Networks to New Learning Tasks* (Rutgers University, New Brunswick, 1993)
48. F. Zhuang, Z. Qi, K. Duan, D. Xi, Y. Zhu, H. Zhu, H. Xiong, Q. He, A comprehensive survey on transfer learning (2019, preprint). arXiv:1911.02685
49. A.A. Rusu, N.C. Rabinowitz, G. Desjardins, H. Soyer, J. Kirkpatrick, K. Kavukcuoglu, R. Pascanu, R. Hadsell, Progressive neural networks (2016, preprint). arXiv:1606.04671
50. S. Zebang, K. Sei-ichiro, Densely connected AutoEncoders for image compression, in *Proceedings of the 2nd International Conference on Image and Graphics Processing* (2019), pp. 78–83
51. D.A. Ferrucci, Introduction to 'this is Watson'. IBM J. Res. Dev. **56**(3–4), 1 (2012)
52. W.-D.J. Zhu, B. Foyle, D. Gagné, V. Gupta, J. Magdalen, A.S. Mundi, T. Nasukawa, M. Paulis, J. Singer, M. Triska, *IBM Watson Content Analytics: Discovering Actionable Insight from Your Content.* (IBM Redbooks, Armonk, 2014)
53. N. Bostrom, E. Yudkowsky, The ethics of artificial intelligence. Cambridge Handb. Artif. Intell. **1**, 316–334 (2014)
54. A. Etzioni, O. Etzioni, Incorporating ethics into artificial intelligence. J. Ethics **21**(4), 403–418 (2017)
55. A. Jobin, M. Ienca, E. Vayena, The global landscape of AI ethics guidelines. Nat. Mach. Intell. 1(9), 389–399 (2019)
56. A. Gkoulalas-Divanis, G. Loukides, J. Sun, Publishing data from electronic health records while preserving privacy: a survey of algorithms. J. Biomed. Inform. **50**, 4–19 (2014)

Brain Tumor Segmentation Using Deep Neural Networks and Survival Prediction

Xiaoxu Na, Li Ma, Mariofanna Milanova, and Mary Qu Yang

1 Introduction

Approximately 80,000 primary brain tumor cases are diagnosed each year in the United States and roughly one-fourth of these cases are gliomas [1]. Glioblastoma accounts for the majority of gliomas (55.4%). Glioblastomas (GBM) is a grade IV brain tumor and is considered as the most aggressive brain tumor type. The overall 5-year survival rate of patients with glioblastoma remains as low as 4.6% [2]. GBM have shown no notable improvement in population statistics in the last three decades. Despite multidisciplinary treatments such as surgery, chemotherapy, and radiotherapy, the median survival time for patients with GBM remains around 15 months [3].

Early diagnosis can increase treatment possibility and patient survival rate. Presently, several molecular biomarkers for brain tumor are available. However, brain tissue samples are often required for utilizing these markers for diagnosis, limiting their application in clinical practice. Medical imaging including Positron Emission Tomography (PET), Magnetic Resonance Imaging (MRI), Computed Tomography (CT), Single-Photon Emission Computed Tomography (SPECT), etc. offer noninvasive approaches for diseased diagnosis. Among these, MRI is the most popular imaging technique for tumor diagnosis because of its excellent contrast in soft tissues. It is estimated about 35 million MRIs are performed annually worldwide in recent years. Manual segmentation and extraction of the tumor

---

X. Na · L. Ma · M. Q. Yang (✉)
MidSouth Bioinformatics Center and Joint Bioinformatics Program of University of Arkansas at Little Rock and University of Arkansas for Medical Sciences, Little Rock, AR, USA
e-mail: mqyang@ualr.edu

M. Milanova
Computer Science, University of Arkansas at Little Rock, Little Rock, AR, USA

© Springer Nature Switzerland AG 2021
H. R. Arabnia et al. (eds.), *Advances in Artificial Intelligence and Applied Cognitive Computing*, Transactions on Computational Science and Computational Intelligence, https://doi.org/10.1007/978-3-030-70296-0_7

Table 1 Summary of the original characteristics of the BraTS dataset

Acronym	Property	Acquisition	Slice thickness
T1	Native image	Sagittal or axial	Variable (1–5 mm)
T1Gd	Post-contrast enhancement	Axial 3D acquisition	Variable
T2	Native image	Axial 2D	Variable (2–4 mm)
FLAIR	Native image	Axial or coronal or sagittal 2D	Variable

area from MRIs are very time-consuming. The automated tumor segmentation method is desirable but technically challenging. Tumor structures often vary across patients in terms of size, localization, and intensity. Large overlapping between the distributions of the intensity of healthy and nonhealthy brain tissue was observed [4]. Several automatic approaches based on traditional machine learning (ML) algorithms, such as K-nearest neighbor classifier (KNN) [5], random forest [6, 7], support vector machine [8, 9], neural network [10], and self-organizing mapping (SOM), for tumor segmentation have been developed. Features extracted from data are required are used as input for traditional ML-based approaches. In contrast, the deep-learning algorithms automatically learn complex features from data for classification.

In this project, we implemented two different structures of deep-learning neural networks, and compared the method and performance of these structures. Furthermore, the features were extracted from the predicted segmentation results along with additional basic clinical data that were used to establish a survival prediction model.

2 Materials and Methods

2.1 Data Acquisition

The data used in the project consisted of 262 MRI images that were obtained from the cancer genome atlas glioblastoma multiforme (TCGA-GBM) data collection. (https://wiki.cancerimagingarchive.net/display/Public/TCGA-GBM). The DeepMedic learning neural network was trained based on cases in the challenge of multimodal brain tumor image segmentation (BraTS) 2015 dataset (https://www.smir.ch/BRATS/Start2015). The 3D U-Net deep-learning neural network was trained based on the in the challenge multimodal brain tumor image segmentation (BRATS) 2017 dataset (Table 1).

Each case of the dataset is comprised of T1-weighted, T2-weighted, FLAIR, and T1Gd MR images. Annotations are provided that include four labels: (1) necrotic core (NC), (2) edema (OE), (3) non-enhancing (NE) core, and 4) enhancing core (EC) for BraTS 2015 dataset. Some annotators can overestimate label 3, and oftentimes there is little evidence in the image data for this subregion. Therefore, forcing the definition of this region could introduce an artifact, which could result in substantially different ground truth labels created from the annotators in different institutions [11]. Consequently, in the BraTS 2017 dataset, label 1 and label 3 had been combined to address the aforementioned issues. The training MRI images were annotated semiautomatically, which were predicted by multiple automatic algorithms followed by expert review. The categories of tumor regions include whole tumor (all four labels), core (1,3,4), and enhancing tumor (4).

2.2 Data Preprocessing

The raw DICOM data were preprocessed and formatted to fit into deep-learning algorithms. The preprocessing steps included reorientation, de-noising, bias-correction, COG adjustment, and skull stripping, and co-registration. First, the DICOM files were converted to NIfTI files by FSL [12, 13], a comprehensive library of analysis tools for FMRI, MRI, and DTI brain image data. All MRI volumes were reoriented to the left posterior superior coordinate systems, and the noise was reduced by the SUSAN noise filter algorithm [14]. Followed by an N4 bias field correction [15] that was implemented by ANTs [16], the volumes were subject to a center of gravity adjustment and a skull stripping procedure, resulting in the regions for subsequent analysis. Then, the volumes were registered to an MNI 152 standard space template image [17], which is derived from 152 structural images and averaged after high-dimensional nonlinear registration to the common MNI 152 coordinate system. After registration, the volumes from different cases are comparable in the same coordinate. Lastly, the data were normalized to be zero mean, unit-variant.

2.3 3D Deep Learning Algorithms

The architectures of the deep 3D convolutional neural networks (CNN) for brain tumor segmentation in this study are based on the network architectures of DeepMedic [4] and 3D U-Net [18], respectively.

We adopted the model of DeepMedic, only kept the branch of the normal resolution and enhanced with residual connection (Fig. 1). Residual connections were shown to facilitate the preservation of the flowing signal and hence allowed the training of very deep neural networks [19]. The use of small kernels 3*3*3 was utilized [20]. This design was shown to be effective in building deeper CNNs without significantly increasing the number of trainable parameters, which made the trained model robust and easy to use once it was established. On the other hand, the same kernel contributed to the much less computation required for the convolution.

Three-dimensional (3D) U-Net based network [21] was used as the general structure as illustrated in Fig. 2. One padding with standard convolution was adopted to ensure the spatial dimension of the output is the same as that of the input.

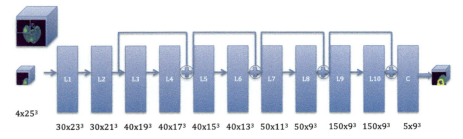

Fig. 1 The DeepMedic model extended with residual connections. The operations within each layer block are in the order: batch normalization, nonlinearity, and convolution. The kernel is $3 \times 3 \times 3$ for all convolution layers; except the last one for the output, the kernel is $1 \times 1 \times 1$

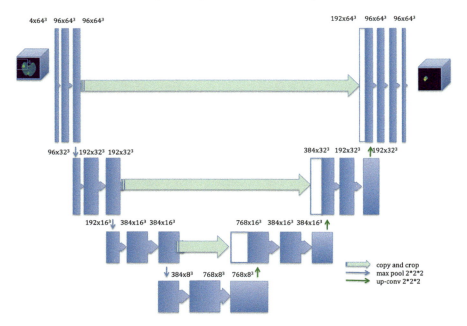

Fig. 2 3D U-Net neural network structure with three encoding and three decoding blocks. Each block was performed in the order of convolution, batch normalization, and parametric rectilinear function. The kernel is $3 \times 3 \times 3$ for all convolution layers, except the last one for the output, the kernel is $1 \times 1 \times 1$

The number of features was doubled while the spatial dimension was halved with every encoding block. The shortcuts established connections from layers of equal resolution in the encoding path to the decoding path. This allowed bypass of the upstream data, which can provide high-resolution features.

3 Results

3.1 DeepMedic Base Model

A total of 220 high-grade gliomas (HGG) in BraTS 2015 were used to build DeepMedic base model for tumor segmentation. The model was trained and validated using 138 high-grade gliomas (HGG) cases, and then tested using 35 HGG cases. The tumor region divided into four annotated subregions, which are necrotic core (I), edema (II), non-enhancing (III) (NE), and enhancing core (EC VI) for BraTS 2015 dataset and were label as 1, 2, 3, 4, respectively. The nontumor region was labeled as 0.

We trained the model for 35 epochs and each epoch included 20 iterations. A learning rate was adjusted according to the gradient convergence during training. The dice similarity coefficient and Hausdorff distance were used for performance evaluation for the tumor segmentation model. The dice coefficient compared the segment regions with the annotated regions coefficient. The values of the dice coefficient range from 0 to 1, 1 refers to complete similarity whereas 0 refers to no overlapping between compared the regions. The dice score increased as increasing epochs (Fig. 3).

Hausdorff distance measures how far two subsets of a metric space are from each other. Here, it measures dissimilarity between segmented tumor regions and annotated regions. Hence, the higher dice coefficient with lower Hausdorff distance indicates the better performance of the model. The overall performance is shown in Table 2 (Fig. 4).

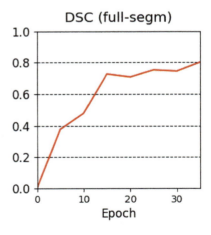

Fig. 3 The values of dice coefficient of DeepMedic models in 35 epochs during the training procedure

Table 2 Mean of Dice, Hausdorff, and Hausdorff 95 of DeepMedic tests and 3D U-net tests details

Model	Label	Dice	Hausdorff	Hausdorff95
DeepMedic	1&3	0.312	12.453	10.473
	2	0.379	15.789	14.024
	4	0.391	13.683	11.952
	0	0.542	14.932	13.140
3D U-net	1	0.659	14.836	12.789
	2	0.718	18.732	16.617
	4	0.754	17.533	15.439
	0	0.816	18.456	16.300

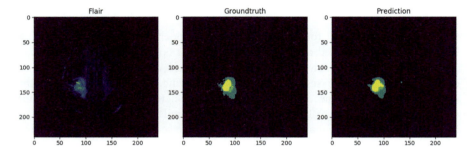

Fig. 4 The automatically segmented tumor regions by DeepMedic model compared to the ground truth

3.1.1 3D U-Net Neural Network Model

The 3D U-net neural network was trained and validated with 169 HGG cases, and tested with 43 HGG cases in BraTS 2017. BraTS2017 contains 12 more annotated images than BraTS2015. Also, in the BraTS 2017 dataset, label 1 and label 3 have been combined to avoid the artifacts introduced by annotators. The model was trained for 40 epochs, with 16 iterations per epoch. In each epoch, only one patch was extracted from every subject. Subject orders were randomly permuted every epoch. The TensorFlow framework was used with Adam optimizer. The batch size was set to 1 during training. A learning rate of 0.0005 was used without further adjustments during training. Dice score steadily increased with epochs (Fig. 5 top panel). The loss function was converged, indicating the training model was not overfitted (Fig. 5 bottom panel).

It appeared that 3D U-net overperformed DeepMedic (Table 2 and Fig. 6).

3.1.2 Survival Prediction

We built a linear regression model to predict HGG patient survival. We first fit the linear regression model with the clinical data including age and resection status. There were two classes of resection status, gross total resection (GTR) and subtotal

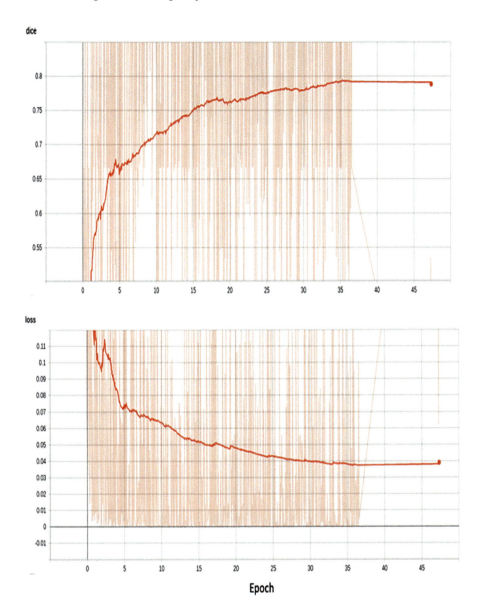

Fig. 5 The value of dice coefficient of 3D U-net models on 50 epochs of the training process. Loss function of 3D U-net training procedure of 50 epochs

resection (STR), and case with missing values in the status. Thus, a two-dimensional feature vector was used to represent the status, which are GTR: (1, 0), STR: (0, 1), and NA: (0, 0), respectively. Furthermore, we extracted images features combining

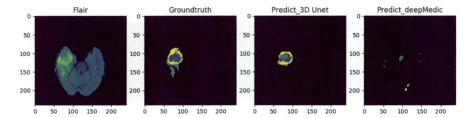

Fig. 6 Comparison of a sample of the results from DeepMedic and 3D U-net segmentation

Table 3 Pearson's coefficient and p-value of all the features

Feature	Pearson's coefficient	P-value
Size of volumes with label 1	−0.110	0.220
Surface of volumes with label 1	−0.186	0.036
Size of volumes with label 2	−0.036	0.690
Surface of volumes with label 2	−0.153	0.085
Size of volumes with label4	−0.177	0.046
Surface of volumes with label 4	−0.236	0.008
Age	−0.410	0.0000017
GTR [0]	0.083	0.352
GTR [1]	−0.089	0.317

with clinical features and then reconstruct the linear regression model. Six image features were calculated from the ground truth label maps during training and the predicted label maps during validation. The volumes, which summed up the voxels, and the surface area, which summed the magnitude of the gradients along three directions were obtained for each class. A linear regression model after normalizing the input features to zero mean and unit standard deviation was fit with the training data. To further improve the survival prediction model, we computed Pearson's correlation coefficient of each with the survival status. The features with p-value larger than 0.05 were excluded (Table 3).

Image texture features are effective at characterizing the microstructure of cancerous tissues. Inspired by this, we consider using texture features extracted in multi-contrast brain MRI images to predicting the survival times of HGG patients. Texture features were derived locally from all regions in fluid-attenuated inversion recovery (FLAIR) MRIs, based on the gray-level co-occurrence matrix representation. A statistical analysis based on the Kaplan–Meier method and log-rank test was applied to identify the texture features related to the overall survival of HGG patients [22]. Four features (Energy, Correlation, Variance, and Inverse of Variance) from contrast enhancement regions and a feature (Homogeneity) from edema regions were shown to be associated with survival times (p-value <0.01). We incorporated these features into the model. The best-performed combinations of features are listed in Table 4. The new features were extracted by the tool of PyRadiomics 3.0 [23].

Brain Tumor Segmentation Using Deep Neural Networks and Survival Prediction   97

Table 4 Pearson's coefficient and p-value of all the features in the proposed model

Feature	Pearson's coefficient	P-value
Surface of volumes with label 1	−0.187	0.036
Size of volumes with label 4	−0.177	0.046
Surface of volumes with label 4	−0.236	0.008
Age	−0.410	0.0000017
glcm_clusterProminence	0.189	0.033
glcm_clusterShade	−0.180	0.043
glrlm_GrayLevelNonUniformity	−0.206	0.020
glcm_Correlation	0.121	0.174
firstorder_Medians	0.173	0.052

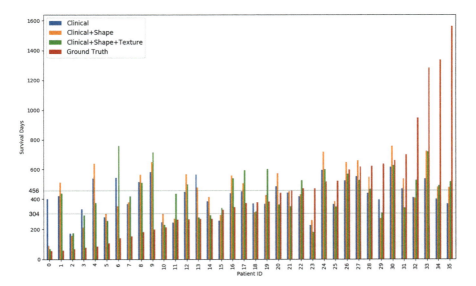

Fig. 7 The predictions from clinical data alone and different feature combinations in Table 5 (clinical + shape) and Table 6 (clinical + shape + texture), comparing with ground truth

The performances of the linear regression models with different feature types are compared in Fig. 7. Combining clinical, shape, and textural features achieved the best performance.

The suggested classes based on the prediction of OS were long survivors (>15 months) as above the gray line 456 days displayed in Fig. 7, short survivors (<10 months) as below the gray line 304 days displayed in Fig. 7, and mid-survivors (between 10 and 15 months). The model with clinical data, shape, and textural features improved the accuracy based on the above classes to 0.567, compared with 0.473 accuracy of the model with clinical data and shape features. The clinical data alone may produce an accuracy of 0.405 on the same dataset.

4 Discussion

The status of the clinical data only contains gross total resection (GTR) and subtotal resection (STR). In addition, many cases did not have resection status information. The residual tumor (R) classification takes into account clinical and pathological findings and requires the pathological examination of resection margins. The R classification has considerable clinical significance, particularly being a strong predictor of prognosis [24]. When R classification information is included, it could potentially increase the survival prediction accuracy.

5 Conclusion

This project systematically compared two different neural networks that implemented brain tumor segmentation. Our results suggested that 3D U-Net outperformed DeepMedic on most evaluation measurements for brain tumor segmentation. The features were selected and extracted from the segmented results. Compared to using the clinical data alone, including shape features into the regression model improved overall survival prediction. Replacing the shape features that were not significantly correlated with survival with the selected texture, the prediction accuracy was further increased.

Acknowledgments The author would like to thank University of Arkansas for Medical Sciences (UAMS) High Performance Computing (HPC) resources used for this research, as well as Dr. Lawrence Tarbox and Dr. Phil Williams for clearing our doubts related to the HPC platform utilization. This research was supported by the AR INBRE program (NIGMS (P20 GM103429) at NIH) and NIH/R15 (R15GM114739) funding.

References

1. F.B. Mesfin, M.A. Al-Dhahir, *Cancer, Brain Gliomas*, in *StatPearls [Internet]*. 2019, StatPearls Publishing
2. J.N. Cantrell, et al., *Progress toward long-term survivors of glioblastoma. In Mayo Clinic Proceedings*. 2019. Elsevier
3. M. Koshy et al., Improved survival time trends for glioblastoma using the SEER 17 population-based registries. J. Neuro-Oncol. **107**(1), 207–212 (2012)
4. K. Kamnitsas, et al., *DeepMedic for brain tumor segmentation. In International workshop on brain lesion: Glioma, multiple sclerosis, stroke and traumatic brain injuries*. 2016. Springer
5. V. Anitha, S. Murugavalli, Brain tumour classification using two-tier classifier with adaptive segmentation technique. IET Comput. Vis. **10**(1), 9–17 (2016)
6. N.J. Tustison et al., Optimal symmetric multimodal templates and concatenated random forests for supervised brain tumor segmentation (simplified) with ANTsR. Neuroinformatics **13**(2), 209–225 (2015)

7. D. Zikic, et al., *Decision forests for tissue-specific segmentation of high-grade gliomas in multi-channel MR*. In *International Conference on Medical Image Computing and Computer-Assisted Intervention*. 2012. Springer
8. M. Havaei et al., Within-brain classification for brain tumor segmentation. Int. J. Comput. Assist. Radiol. Surg. **11**(5), 777–788 (2016)
9. S. Bauer, L.-P. Nolte, M. Reyes. *Fully automatic segmentation of brain tumor images using support vector machine classification in combination with hierarchical conditional random field regularization*. In *International Conference on Medical Image Computing and Computer-Assisted Intervention*. 2011. Springer
10. M. Sasikala, N. Kumaravel, A wavelet-based optimal texture feature set for classification of brain tumours. J. Med. Eng. Technol. **32**(3), 198–205 (2008)
11. Bakas, S., et al., *Identifying the best machine learning algorithms for brain tumor segmentation, progression assessment, and overall survival prediction in the BRATS challenge*. arXiv preprint arXiv:1811.02629, 2018
12. S.M. Smith et al., Advances in functional and structural MR image analysis and implementation as FSL. NeuroImage **23**, S208–S219 (2004)
13. M.W. Woolrich et al., Bayesian analysis of neuroimaging data in FSL. NeuroImage **45**(1), S173–S186 (2009)
14. S.M. Smith, J.M. Brady, SUSAN—a new approach to low level image processing. Int. J. Comput. Vis. **23**(1), 45–78 (1997)
15. N.J. Tustison et al., N4ITK: improved N3 bias correction. IEEE Trans. Med. Imaging **29**(6), 1310–1320 (2010)
16. B.B. Avants et al., A reproducible evaluation of ANTs similarity metric performance in brain image registration. NeuroImage **54**(3), 2033–2044 (2011)
17. T. Rohlfing et al., The SRI24 multichannel atlas of normal adult human brain structure. Hum. Brain Mapp. **31**(5), 798–819 (2010)
18. X. Feng et al., Brain tumor segmentation using an ensemble of 3D U-nets and overall survival prediction using radiomic features. Front. Comput. Neurosci. **14**, 25 (2020)
19. C. Szegedy, et al., *Inception-v4, inception-resnet and the impact of residual connections on learning*. In *Thirty-first AAAI conference on artificial intelligence*. 2017
20. K. Simonyan, A. Zisserman, *Very deep convolutional networks for large-scale image recognition*. arXiv preprint arXiv:1409.1556 (2014)
21. Ö. Çiçek, et al. *3D U-Net: learning dense volumetric segmentation from sparse annotation*. In *International conference on medical image computing and computer-assisted intervention*. 2016. Springer
22. A. Chaddad, C. Desrosiers, M. Toews. *Radiomic analysis of multi-contrast brain MRI for the prediction of survival in patients with glioblastoma multiforme*. In *2016 38th Annual International Conference of the IEEE Engineering in Medicine and Biology Society (EMBC)*. 2016. IEEE
23. J.J. Van Griethuysen et al., Computational radiomics system to decode the radiographic phenotype. Cancer Res. **77**(21), e104–e107 (2017)
24. P. Hermanek, C. Wittekind, The pathologist and the residual tumor (R) classification. Pathol. Res. Pract. **190**(2), 115–123 (1994)

Combination of Variational Autoencoders and Generative Adversarial Network into an Unsupervised Generative Model

Ali Jaber Almalki and Pawel Wocjan

1 Introduction

Human developed the mental model based upon the environment and things they perceive on a daily basis. On a daily basis, we perceive different things based upon the actions and experiences. Human gathers a lot of information, and the human mind learns the temporal and spatial aspects of the knowledge they gained from their environment. Human brains learn the abstract representation of the information. For example, if we can capture a scene in our and remind it later, we will only remember the abstract information of the particular scene. The internal predictive model influences actions and decisions [1]. One way to understand the predictive model that the current action predicts the future actions we performed that create sensory data. Based on the human internal model, the human brain predicts because of the abstract information representation.

Reinforcement learning is the subfield of the machine language that helps to perform the task with making much effort. In reinforcement learning, the agent can be trained according to the environmental actions, and agents can train in the simulating environment. In the world, using learned features, the agent can be trained in a manner that helps to solve complex and challenging problems [2]. The procedure includes the multistep training process in which each process combines the next one in the training of the agent. In the first step, the information collected in the form of abstract representation and stores the data to perform some action. The actions are made based on the previous choices made by the agent.

Many reinforcement learning problems can be solved when combined with Artificial Intelligence (AI) that help to solve complex tasks. An AI-reinforcement

---

A. J. Almalki (✉) · P. Wocjan
Department of Computer Science, University of Central Florida, Orlando, FL, USA
e-mail: Ali.almalki@knights.ucf.edu; wocjan@cs.ucf.edu

© Springer Nature Switzerland AG 2021
H. R. Arabnia et al. (eds.), *Advances in Artificial Intelligence and Applied Cognitive Computing*, Transactions on Computational Science and Computational Intelligence, https://doi.org/10.1007/978-3-030-70296-0_8

learning agent helps to control the complex situation, and the agent performs to gather accurate results with problem handling. Moreover, the AI-reinforcement learning agent helps the predictive model (M) to make better prediction results [3].

The Backpropagation algorithm is used for the predictive model that includes neural networks as well. With the neural network in the predictive model, it helps to produce a better result that supports the prediction accuracy. In the partial observable environment, it is better to use a predictive model with RNN, as it produces better accuracy results and solves complexity effectively.

Most of the reinforcement learning approaches are based upon the model and require a special environment to train the agent. In the world, we have added the GAN discriminator that helps to produce better results and learn in a simulating environment. We have trained the agent in an environment in which the real data can directly go into the GAN/discriminator that make prediction accordingly.

With the combination of GAN/discriminator and MDN-RNN(M), it provides the simplistic approach that helps to train the agent in its own simulating environment and perform well in the complex pixels' detection environment [4]. Previously, the traditional methods were not efficient to help to solve the complex issues of the environment.

2 Related Work

Different existing models of RL help to train the environment and provide generated output. Most of the approaches are based upon the model that learns the environment first and then train the agent to get the task performed quickly. It involves different types of data that process accordingly to gather the output. For the complex data, of the data distribution like images, it involves the image-preprocessing system that helps to convert them into the form that is easily understood by the agent to learn the environment. It helps to formulate a complex task with simple solutions [5]. But when it involves highly complex data, it makes the process of converting the images slow, and normalization of the gathering of the image pixels become difficult to handle. The preprocessing image operations are viewed as the metric engineering that make the process slow when they are bombarded with the highly complex data.

In Siamese architecture, neural networks are used for metric learning. For a similar sample, it minimizes the total distance, but for the dissimilar sample, it maximizes the distance and makes the entire process more complex. But the real problem using this approach is that it can't be applied directly to the problem [6]. This approach can only be applied to the supervised environment.

For the element-wise distance, there are different techniques used that help to measure the distance. For the element, a wise distance-measuring generative model developed that supported in calculating the distance. When the data include the complex images, it becomes difficult to maintain the whole system and autoencoder used that processes the grayscale images. From the edges and shapes, it learns the type of image. Moreover, GAN-based generative model used for the learning of the images helps to gather the data of sharp edges of the images.

Another generative model introduced that is based upon the gradient and combination of the GAN is used for the video prediction of the animated images. It captures the structure of the images, but when it comes to capturing the high-level structural images, it deformed the data and error occurred. To remove this problem, the GAN generator used that help to produce high-quality images.

From the latent representation, a convolutional network used that process to produce the high-structural quality of the images. Using the simple arithmetic expression, it helps to express the semantic relationship. Another technique that is supervised training that is commonly used in the convolutional network helps to define the high-level information.

For the feature representation of the data encoder and decoder used, and it uses the supervised information—the training of the data dependent upon the supervised learning of the data used for sampling. But the problem with this technique is that it can't take the visual data of the pure GANs.

Different dynamic models exist in which they train the model policy, and then apply the experimental procedures. Probabilistic Search Policy used that help to solve the controlling problem based upon the data collected from the environment. It uses the Gaussian Process that understand the system dynamic and train the system to take control to perform desired tasks effectively.

Bayesian Neutral Network is another way to learn the Dynamic Model. The Bayesian Neutral Network is more efficient than the Gaussian process. It performs more effectively and produces promising results for the controlling task. It observes a low-dimensional process, and all of the starting and ending states are well defined. To stimulate the environment of the game, the Conventional Neural network is used to solve challenges and provide complete control to the user [7]. It helps to predict the future behavior of the agent-based upon the data gathering of the stimulating environment.

RNN models are powerful and that helps to predict future behavior, and the next frame of the game can easily be predicted using the RNN models. Moreover, if there is an internal model used, RNN supports the internal model and generate future predicted frames. To control the controlled and perform the required tasks, evolution strategies used that cooperation in the multilevel environment, which means it supports the multiple agents working in the same environment.

Traditional Deep Reinforcement Learning Method is not efficient because it can't offer control to multiple agents in the same environment. To handle the multilayer control evolution-based algorithm used that reading to direct data based upon the environment and predict the behavior of the agent effectively. With the evolutional algorithms, the complex tasks can solve easily and better prediction it produces.

3 Agent Model

A simple model represented in which the combination of the GAN and VAE shown that helps to produce better accuracy results. In this model agent has a small visual component that visualize the better representation of the code. It has a small that helps to store the data and can make accurate future prediction. The future prediction is based upon the historical information it gathers from the environment [8]. There is a separate component of the observation that helps to observe the simulating environment. This model provides the better control to the agent for making the decisions. The decision taken by the agent based upon the visual representation and the data stored in the memory.

In this model, the three-component work closely to collect the data and make better decision that helps to calculate the accuracy of the data. The three main controllers are vision, memory, and controller.

The major contribution to this model is that VAE and GAN are combined that produce better accuracy results. It compares the datasets and generate results with better imaging quality. Generative models are trained that produce quality imaging results.

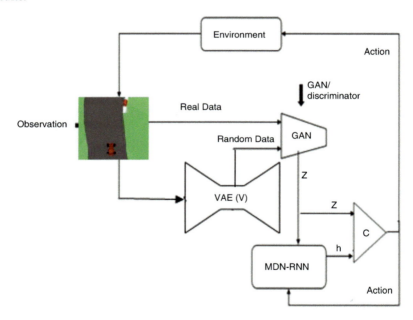

4 VAE (V) Model

The environment in which the agent got trained receives the quality input and with each time frame it receives the input data in form of 2D image. The V component helps to compress the data size and learns the environment into an abstract form [9]. Variational autoencoder (VAE) used and V play an important role in which they compresses the size of the images for the better accuracy results. Each time the agent observes the data from the environment, it compresses the size of the image into the small latent size z.

5 MDN-RNN (M) Model

The role of the V model helps to compress the size of the images in to the latent and produce better imaging results for sampling. It reduces the size of what the agent observes in its environment. But the MDN-RNN helps to predict the future as per the agent movement. It predicts the data of what going to next happens.

The M model supports as the predictive model that make prediction according to the agent performance in the environment. Gaussian distribution used for the sampling a prediction dependent upon the model V. The M model provide the prediction results based upon the past and current information received from the V model. Mixture density network combined with the RNN to gather the better prediction.

The combination of the both MDN–RNN help to produce better prediction results. Next latent vector can predict effectively and have ability to handle the complex environment effectively. Moreover, during the sampling, the temperature of the parameter can adjust. Using the combinational model of the MDN–RNN, the sequence generation problem that include pen detection and handwriting can handle effectively.

6 Controller Model (C)

The controller model C helps to determine the actions. The C model maximize the controller action of the agent in observing different elements of the environment. It helps to reduce the complexity caused by the Model V and M. It is a simple linear model that hold the control of the agent in performing different actions. The model C advances the deep learning process and practically behaves effectively [10]. It maps the vision observe by the agent in the particular environment. It represents the quantitative representation of the action vector based upon the agent action.

7 GAN/Discriminator

GAN is a high-quality generative model that helps to produce better results. GAN discriminator produces high-quality data even from the hidden layers. GAN is a generative adversarial network that consists of two networks. Using the GAN, it gives better accuracy in results in which it can do by removing the discrimination between the generated and true data.

It uses following set of instructions for running the game based on the GAN–VAN as describe in the table. For demo purposes, we are working with a small dataset of 100 episodes and each episode is of 200 time steps. The original authors had worked with 10,000 episodes of 300 time steps [11].

GAN uses the binary classifiers to remove the discrimination between the generated and true data. It can effectively remove the discrimination between the images and nonimages data. The combination of the GAN and VAE represents a high-quality generative model. The combinational model of both helps to gather the data from the hidden layer as well.

In our generative model, we have to train both GAN and VAN that helps to solves the complexities that traditional models, methods, and techniques fail to handle. GAN provide the accessibility in which it can learn the samples indirectly.

In comparison with the separate resulting of the GAN and VAE, they both provide comparatively low results. But the combination results of the GAN and VAE, the produce better results. The both attributes visually better results and better accuracy sampling images. The generalization of the GAN and VAE model is better and semi-supervised method produce productive results [12].

The working of the whole model in which the observation takes from the environment and the observation take in the form of 2D images. The 2D images take the observation in the form of RGD depth, width, and length. The VAE and GAN autoencoder it into the better result and provide the efficient results by encoding it into the latent size of the image.

The model M takes the previous action and current action, it provides the predictive results. The main goal of this model is to take the future prediction based upon the observation of the environment [13]. The C model is the controller that provide full control to the agent in making the decision based upon the action taken in the environment.

8 C. MDN-RNN (M) Model

The role of the V model is to compress the size of the images to the latent representation and produce better imaging results for sampling. It reduces the size of what the agent observes in its environment. The MDN–RNN (M) model, on the other hand, helps to predict the future as per the agent's movements. It predicts from the data what will happen next.

The M model serves as the predictive model that makes predictions according to the agent's performance in the environment. Gaussian distribution is used to sample a prediction dependent on the model V. The M model provides the prediction results based on both the past and current information received from the V model. Mixture Density Network combined with the RNN to gather better predictions.

The combination of MDN–RNN helps to produce better prediction results. Next latent vector can make effective predictions and it has the ability to handle a complex environment more effectively. Moreover, during the sampling, the temperature of the parameter can be adjusted. Using the combinational model of the MDN–RNN, the sequence generation problem includes both pen detection and handwriting and can handle both effectively.

9 D. Controller Model©

The controller model© helps to determine actions. The C model maximizes the controller action of the agent by observing different elements of the environment. It helps to reduce the complexity caused by Model V and Model M. It is a simple linear model that holds the control of the agent in performing different actions. Model C effectively advances the deep-learning process. It maps the images observed by the agent in the particular environment. It shows the quantitative representation of the action vector based on the agent's action.

10 Experimental Results of Car Racing: Feature Extraction

We have trained the agent world by experimenting it on a car racing game. The new combination of the VAE and GAN gives better results as compared to the previous agent world. The component used in the agent model helps to produce better results and solves the complexities that are available in the traditional agent model.

The agent model component that includes the V and M gives better results and helps to extract the features which are reliable to gather useful information. The agent performs in the environment where it is trained to control the three actions that include acceleration, brake, and steering right/left. The agent performs randomly in the environment in which he performs different states and gather different datasets [14]. Moreover, using the new agent model, it is noticed that the agent can hold the decision effectively and user controls are stable to control the car.

Our agent can achieve the score of 1000 as compared to the traditional methods of the deep learning, wherein the agent is only able to achieve the score goal of 500–600. Moreover, the average score is 900. The combination agent worlds of the VAE and GAN are effective and solve the complexity of the entire system. In contrast to the traditional, the new agent world offers better results and finer pixel quality [15]. It means it involves more attention of the user and make it more user-friendly. Moreover, it is able to give better future prediction and gives better hypothetical results when it comes to compare the accuracy with the traditional model's results.

11 Evolutional Strategies and Doom RNN

Covariance matrix strategy is based on the matrix and gives results in form of frequency. In combination with the new agent model it improves the results and gives better frequency results.

The population size is of 65, and agent performs the task 18 times. The average score it can make is 950, which is more as compared to the traditional agent training method.

The same experiment performed using the Doom RNN, where it can configure that it can made the averages score of 980 and shows better result performance.

Doom RNN is comparatively more efficient as compared to the traditional approaches used and traditional approaches also slow down the process and make it difficult for the agent to observe the environment directly. Doom RNN with the new agent model shows better controller performance of the agent.

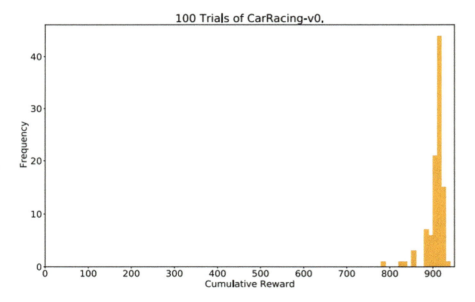

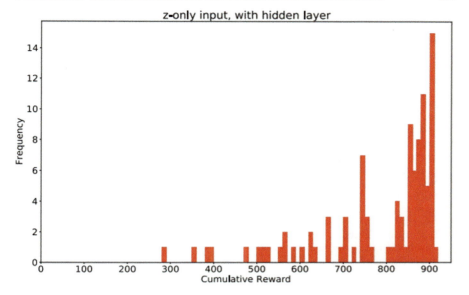

12 Conclusion

For better accuracy result and to gather better visual representation of the images GAN and VAE combined to gather, experiment and perform on the racing game. From the experiment, it is concluded that the scores are better as compared to using traditional algorithms and techniques. Moreover, the experiment performed on the Evolutional Strategies and Doom RNN effectively observed the scoring trend. Doom RNN secured better results and observed that average score is 980. The combination agent world of the VAE and GAN is effective and solves the complexity of the entire system.

References

1. Y.Y. Li Zhang, The difference learning of hidden layer between autoencoder and variational autoencoder. *2017 29th Chinese Control and Decision Conference (CCDC)*, vol. 2, no. 1, pp. 67–89 (2017)
2. C.D. Cong Wang, The feature representation ability of Variational AutoEncoder. *2018 IEEE Third International Conference on Data Science in Cyberspace (DSC)*, vol. 5, no. 2, pp. 78–80 (2018)
3. J.Z. Jing Li, Variational Autoencoder based Latent Factor Decoding of Multichannel EEG for Emotion Recognition. *2019 IEEE International Conference on Bioinformatics and Biomedicine (BIBM)*, vol. 4, no. 3, pp. 56–68 (2019)
4. A. Koç, Variational autoencoders with triplet loss for representation learning. *2018 26th Signal Processing and Communications Applications Conference (SIU)*, vol. 5, no. 3, pp. 89–90 (2018)

5. Y.-P. Ruan, Condition-transforming Variational Autoencoder for Conversation Response Generation. *ICASSP 2019 – 2019 IEEE International Conference on Acoustics, Speech and Signal Processing (ICASSP)*, 2019
6. K. Mano, Sub-band Vector Quantized Variational AutoEncoder for Spectral Envelope Quantization. *TENCON 2019–2019 IEEE Region 10 Conference (TENCON)*, vol. 6, no. 5, pp. 56–77 (2019)
7. R. Yin, Multi-resolution generative adversarial networks for tiny-scale pedestrian detection. *2019 IEEE International Conference on Image Processing (ICIP)*, 2019
8. Y. Wu, Radio classify generative adversarial networks: a semi-supervised method for modulation recognition. *2018 IEEE 18th International Conference on Communication Technology (ICCT)*, 2018
9. D. Ha, J. Schmidhuber; "World Models", 2018. Open Access Archive of Cornell University, New York, USA. https://arxiv.org/abs/1803.10122. Accessed June 15, 2021
10. M. Rezagholiradeh, Reg-Gan: semi-supervised learning based on generative adversarial networks for regression. *2018 IEEE International Conference on Acoustics, Speech and Signal Processing (ICASSP)*, 2018
11. I. Alnujaim, Generative adversarial networks to augment micro-Doppler signatures for the classification of human activity. *IGARSS 2019 – 2019 IEEE International Geoscience and Remote Sensing Symposium*, 2019
12. A. Boesen, Autoencoding beyond pixels using a learned similarity metric, *Preliminary work submitted to the International Conference on,* 2016
13. R. Jiang, Learning spectral and spatial features based on generative adversarial network for hyperspectral image super-resolution, *IGARSS 2019 – 2019 IEEE International Geoscience and Remote Sensing Symposium,* 2019
14. X. Yu, Face morphing detection using generative adversarial networks. *2019 IEEE International Conference on Electro Information Technology (EIT),* 2019
15. G. Chen, Semi-supervised object detection in remote sensing images using generative adversarial networks. *IGARSS 2018 – 2018 IEEE International Geoscience and Remote Sensing Symposium,* 2018

Long Short-Term Memory in Chemistry Dynamics Simulation

Heng Wu, Shaofei Lu, Colmenares-Diaz Eduardo, Junbin Liang, Jingke She, and Xiaolin Tan

1 Introduction

Classical trajectory chemical dynamics simulations are widely and powerful tools that have been used to study reaction dynamics since the 1960s [1]. In contrast to the variational transition state theory (VTST) and reaction path Hamiltonian methods [2], they provide much greater insight into the dynamics of reactions for the classical equations of motion of the atoms are numerically integrated on a potential energy surface (PES). The traditional approach uses an analytic function that is gotten by fitting ab initio and/or experimental data [3] to construct the surface. Regarding a small number of atoms or a high degree of symmetry [4, 5], it is practical. Researchers recently proposed additional approaches and algorithms for representing PESs. Wang and Karplus firstly demonstrated that the trajectories may

---

H. Wu
Department of Mathematics and Computer Science, West Virginia State University, Institute, WV, USA
e-mail: heng.wu@wvstateu.edu

S. Lu (✉)
College of Computer Science and Electronic Engineering, Hunan University, Changsha, China
e-mail: sflu@hnu.edu.cn

C.-D. Eduardo
Department of Computer Science, Midwestern State University, Wichita Falls, USA
e-mail: eduardo.colmenares-diaz@msutexas.edu

J. Liang
School of Computer and Electronics Information, Guangxi University, Guilin, China

J. She · X. Tan
College of Computer Science and Electronic Engineering, Hunan University, Changsha, China
e-mail: shejingke@hnu.edu.cn; lxtanglx@hnu.edu.cn

© Springer Nature Switzerland AG 2021
H. R. Arabnia et al. (eds.), *Advances in Artificial Intelligence and Applied Cognitive Computing*, Transactions on Computational Science and Computational Intelligence, https://doi.org/10.1007/978-3-030-70296-0_9

be integrated "on the fly" when the potential energy and gradient are available at each point of the numerical integration according to an electronic structure theory calculation. During the numerical integration, the method directly calculates the local potential and gradient under an electronic structure theory in a "direct dynamics" simulation. However, regarding a high-level electronic structure theory, the computation of direct dynamics simulations become quite expensive. Thus, it is important to use the largest numerical integration step size when maintaining the accuracy of the trajectory. In order to use a larger integration step, Helgaker et al. adopt the second derivative of the potential (Hessian). After the Hessians are gotten directly by an electronic structure theory, using a second-order Taylor expansion, a local approximation PES can be constructed and the trajectories can be approximately calculated. For local quadratic potential is only valid in a small region (named a "trust radius"), the equations of motion are only integrated under the trust radius. The new potential, gradient, and Hessian, calculated again at the end of the trust radius, define a new local quadratic PES where the integration of the equations of motion is successive. Millam et al. used a fifth-order polynomial or a rational function to fit the potential between the potential, gradients, and Hessians at the beginning and end of each integration step. It provides a more accurate trajectory in the trust region and calculates larger integration steps. That involves a predictor step, the integration on the approximate quadratic model potential. The following step, the fitting on the fifth-order PES between the starting point and the end point in the trust radius, is also called the "corrector step." It is named the Hessian-based predictor–corrector integration scheme. Around it, some scholars proposed their own methods. Because of extrapolation, errors in prediction–correction algorithms grow rapidly, usually four predictions are followed by an ab initio calculation. This limits the improvement of computing performance.

The successful application of the prediction of deep learning in computational chemistry greatly expanded its application. Deep learning is a machine learning algorithm, not unlike those already in use in various applications in computational chemistry, from computer-aided drug design to materials property prediction [6]. Deep learning models achieved top positions in the Tox21 toxicity prediction challenge issued by NIH in 2014 [7]. Among some of its more high-profile achievements include the Merck activity prediction challenge in 2012, where a deep neural network not only won the competition and outperformed Merck's internal baseline model, but did so without having a single chemist or biologist in their team. Machine learning (ML) models also can be used to infer quantum mechanical (QM) expectation values of molecules, based on reference calculations across chemical space [8]. Such models can speed up predictions by several orders of magnitude, demonstrated for relevant molecular properties such as polarizabilities, electron correlation, and electronic excitations [9]. LSTM is an artificial recurrent neural network (RNN) architecture used in the field of deep learning. The prediction of LSTM has been widely used in different fields [10, 11].

In this chapter, we explore the idea of integrating LSTM layer with chemistry dynamics simulations to enhance the performance in trust radius. This idea is inspired by the recent development and use of LSTM in material simulations

Long Short-Term Memory in Chemistry Dynamics Simulation 113

and scientific software applications [12]. We employ a particular example, H_2O molecular dynamics simulation on NWChem/Venus (cdssim.chem.ttu.edu) package [13] to illustrate this idea. LSTM has been used to predict the energy, location, and Hessian of atoms. The results demonstrate that LSTM-based memory model, trained on data generated via these simulations, successfully learns preidentified key features associated with the energy, location, and Hessian of molecular system. The deep learning approach entirely bypasses simulations and generates predictions that are in excellent agreement with results obtained from explicit chemistry dynamics simulations. The results demonstrate that the performance gains of chemical computing can be enhanced using data-driven approaches such as deep learning which improves the usability of the simulation framework by enabling real-time engagement and anytime access.

This chapter is organized as follows. Section 2 presents the idea that integrate chemistry dynamics simulations with LSTM. Section 3 shows the experiment setting and results on H_2O molecular dynamics simulation, followed by data analysis. Section 3 presents the conclusions and lays out future work.

2 Methodology

2.1 Prediction–Correction Algorithm

In chemistry dynamics simulation, Hessian's calculation consumes most of the CPU time because Hessian is the third derivative of the position. Hessian updating is a technique frequently used to replace electronic structure calculations of the Hessian in optimization and dynamics simulations. Existing generally applicable Hessian-update schemes, for example, the symmetric rank one (SR1) scheme, Powell's symmetrization of Broyden's (PSB) method, the scheme of Bofill, the Broyden–Fletcher–Goldfarb–Shanno (BFGS) scheme, the scheme of Farkas and Schlegel, and other Hessian update schemes, are based on the Eq. (1)

$$H\left(X_{k+1}\right)\left(X_{k+1}-X_k\right) = G\left(X_{k+1}\right)-G\left(X_k\right) \tag{1}$$

where $G(X)$ and $H(X)$, respectively, denote the gradient and Hessian of the potential energy at point X. Some researchers employed Hessian update method to build Hessian-based prediction–correction integration method to calculate the trajectory of atom in order to reduce the calculation time of Hessian and ab initio.

As illustrated in Fig. 1, in each time step of the integration method, the prediction is used to identify the direction the trajectory, ab initio potential energy, ab initio gradient, and ab initio or Hessian are computed at the end point $X_{i,p}$ of predicted trajectory. The potential information calculated at the end of predicted trajectory is used with the potential energy information at point $X_{i-1,p}$ near the trajectory starting point X_{i-1} of this time step, which is the end point of corrected trajectory of the

Fig. 1 During the ith step, the algorithm first predicts the trajectory from X_{i-1} to $X_{i,p}$ using potential approximated by the quadratic Taylor expansion about $X_{i-1,p}$. Then performs electronic structure calculation of the potential energy information at $X_{i,p}$ and reintegrate the trajectory from X_{i-1} to X_i using potential interpolated from ab initio potential information at $X_{i-1,p}$ and $X_{i,p}$

previous time step, to interpolate a highly accurate local PES. This highly accurate PES is used in the correction phase of the time step to recompute a more accurate trajectory.

In each time step, to obtain an accurate predicted trajectory, the prediction utilizes the Hessian in addition to the potential energy and its gradient. Assuming the current time step is the i*th* time step, the potential energy information needed during the prediction to integrate the trajectory is obtained by the quadratic expansion.

$$E(X) = E\left(X_{i-1,p}\right) + G\left(X_{i-1,p}\right)\left(X - X_{i-1,p}\right)$$
$$+ 1/2(X - X_{i-1,p})^T H\left(X_{i-1,p}\right)\left(X - X_{i-1,p}\right) i > 2 \quad (2)$$

P is an integer. About the point $X_{i-1,p}$, the end point of the predicted trajectory of the (*i*–1)th time step at which ab initio potential energy $E(X_{i-1,p})$, ab initio gradient $G(X_{i-1,p})$, and ab initio or updated Hessian $H(X_{i-1,p})$ have been computed on a region within a trust radius from $X_{i-1,p}$.

If we use $X_{i-1,p}$ as the current location, the next part will show how to calculate the potential energy for the next X location. We can calculate the Potential Energy (P) and Gradient (G) at the $X_{i-1,p}$ from known position. For example, there are eight atoms in $F^- + CH_3OOH$. There are $3 \times N$ dimensions in the gradient and location vectors and N^2 dimensions in the Hessian matrix of the reaction system. Therefore, most of calculation of Eq. (2) is to compute $H(X_{i-1,p})$. The biggest challenge is to choice different approaches to fast the calculation of $H(X_{i-1,p})$ with the position and others of the current location, at the same time, cannot enlarge the system error.

2.2 Long Short-Term Memory

As shown in Fig. 2, a neural network is the connection of many single neurons, an output of a neuron can be an input of another neuron. Each single neuron has an activation function. The left layer of the neural network is called the input layer, it

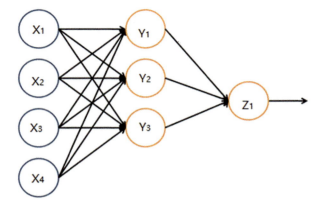

Fig. 2 The structure of a neural network

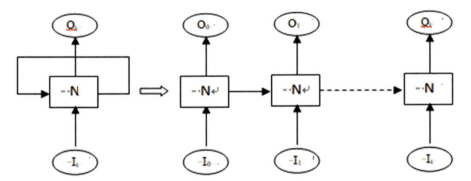

Fig. 3 The structure of a recurrent neural network and its unfolding

includes X1,X2,X3,X4, the right layer of it is output layer, involve Z1. The other layer is hidden layer, it covers Y1,Y2,Y3.

Recurrent neural network (RNN) is a typical kind of neural network. As shown in the leftmost part of Fig. 3.

Like the leftmost of Fig. 3, RNN is a neutral network containing loops. N is a node of neural network. I stands for input and O for output. Loops allow information to be transmitted from the current step to the next step. RNN can be regarded as a multiple assignment of the same neural network, and each neural network module transmits the message to the next one. The right side of Fig. 3 corresponds to the unfolding of the left side. The chain feature of RNN reveals that RNN is essentially related to sequences and lists. RNN applications have been successful in speech recognition, language modeling, translation, picture description, and this list is still growing. One of the key features of RNN is that they can be used to transmit the previous information to the current task. But the distance from previous step to related step is not too long.

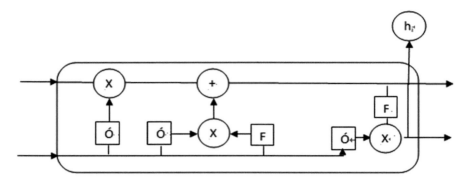

Fig. 4 The structure of a LSTM node

Long short-term memory (LSTM) overcomes this shortcoming. LSTM is a special type of RNN. LSTM solves the problem of long-term dependence of information. LSTM avoids long-term dependencies through deliberate design. Figure 4 shows the structure of a node in LSTM, where a forget gate can be observed. The output of the forget gate is "1" or "0." "1" means full reserve, "0" is abandon completely. The forget gate determines which information will be retained and what will be discarded. The upper horizontal line allows the input information cross neutral node without changing in Fig. 4. There are two types of gates in a LSTM node (input and output gates). The middle gate is an input gate, which determines the information to be saved in the natural node. F means function modular and create a new candidate value vector. The right gate is the output gate. The F module closed the output gate determines which information of the natural node will be transmit to the output gate.

A node has three gates and a cell unit as shown in Fig. 4. The gates use sigmoid as activation function, the tan$h$ function is used to transfer from input to cell states. The following are to definite a node. For the gates, the function are

$$i_t = g\left(w_{xi}x_t + w_{hi}h_{t-1} + b_i\right) \tag{3}$$

$$f_t = g\left(w_{xf}x_t + w_{hf}h_{t-1} + b_f\right) \tag{4}$$

$$f_o = g\left(w_{xo}x_t + w_{of}h_{t-1} + b_o\right) \tag{5}$$

The transfer for input status is

$$c\_in_t = \tan h\left(w_{xc}x_t + w_{hc}h_{t-1} + b_{o\_in}\right) \tag{6}$$

The status is updated by

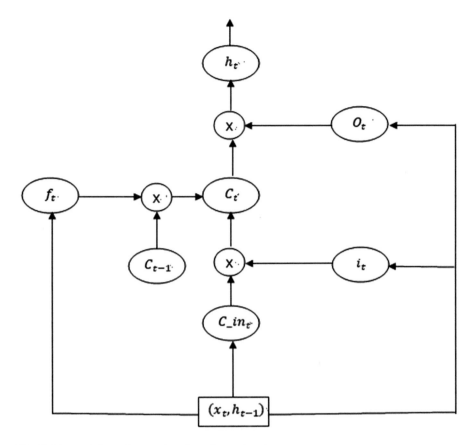

Fig. 5 The workflow of status changing of neutral node

$$c_t = f_t^* c_{t-1} + i_x^* c\_in_t \quad (7)$$

$$h_t = o_t * \tanh(c_t) \quad (8)$$

The workflow of a node is shown in Figs. 5 and 6 shows the flowchart for LSTM.

2.3 Model

The calculation of position of the atom, the energy of the system, and Hessian occupy almost all the CPU time in chemistry dynamics simulations. Figure 7 illustrates the Hessian-based predictor–corrector algorithm in chemistry dynamics

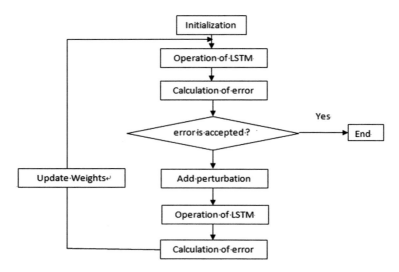

Fig. 6 The flowchart of LSTM

simulations. At each time step, the potential energy, kinetic energy, velocity, Hessian, and other parameters are calculated from the position of the atom. In Fig. 1, assuming $X_{i-1,p}$ is the current point, the calculation potential energy of next point X is as follows. The gradient and potential energy of the current point can be calculated from the known location of the point. Assuming eight atoms, the dimension of gradient and location will be 3 × N, which of H will be N2. Hence, the largest calculation of Eq. (2) will be to calculate $H(X_{i-1,P})$. It is the focus of study of various algorithms to quickly and accurately calculate.

Algorithm 1 Algorithm

Input: Atomic initial parameters
Parameter: location of atoms, initial energy info
Output: the trajectory of atoms.
1.   *ab initio* computing
2: **while** less than steps **do**
3:     **while** less then training step **do**
4:         exec Predictor-Corrector
5:         train deep learning model
6:     **end while**
7:     predict the location,energy and Hessian
8:     output trajectory
9: **end while**
10: **return**

Fig. 7 Flowchart representation of the complete Hessian-based predictor–corrector integrator

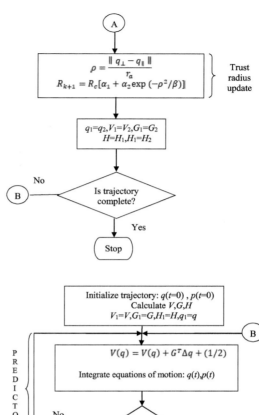
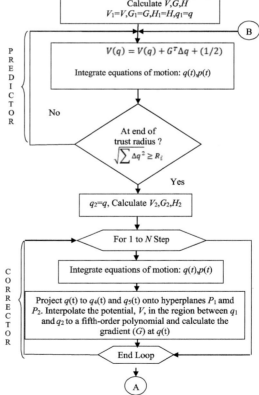

H(Xi–1,p) according to the location and time information of the current point, simultaneously systematic error is required least. Researchers proposed some Hessian update methods to saving computing time [14]. Deep learning will be used to predict the location, energy, and Hessian of atoms. Therefore, deep leaning will be used three times to instead of predictor–corrector. It is important to understand that our deep learning model needs to be trained and initialized before predicting. The result of this approach is a novel predictor–corrector algorithm with deep learning.

3 Experimental Results

To test the algorithm with deep learning, we implemented the integration algorithm in the VENUS (cdssim.chem.ttu.edu) dynamics simulation package interfaced with the electronic structure calculation NWChem [13]. We chose the reaction system H_2O as our testing problem. In the tests, ab initio potential energy, gradient, and Hessian were calculated using the density function theory $6–311 + G^{**}$, and ab initio Hessian is calculated once in every five steps during training. In the remaining nine steps, the new update scheme is used. We calculated a trajectory for the chemical reaction system with 5000*0.67 integration steps, where each step has a fixed size of 0.02418884 fs (100 a.u.; 1 a.u. = 2.418884e-17 s). The remaining step 5000*0.33 steps were predicted by the proposed deep learning algorithm. There are three prediction parameters in our test. They are atomic position, energy, and Hessian, respectively.

Figure 8 illustrates the computational energy and its predicted values. The above is the H_2O system computational energy chart. The horizontal coordinate is the time step and the vertical one is the energy value. The yellow region represents training data and green section predicted values. After more than 3000 training steps, the predicted value is almost the same as the calculated values. Table 3 lists some relative errors. We find the relative error to be less than 0.1%.

Figure 9 shows a hydrogen atomic location chat. The above is the computational values and the following is training and predicted values. The horizontal coordinate is the time step and the vertical one is the atomic location. Table 1 has some relative error between predicted and computational values. We find the relative error less than 0.7% and some even less than 0.01%. Figure 10 is one of Hessian chat. The

Table 1 Relative error between atomic position prediction and computational value

Computational data (D1)	Predicted data D2	$|(D1–D2)|/|D1|$ (%)
−0.65278023	−0.6570433	0.65%
−0.65459454	−0.6573254	0.4%
−0.6577446	−0.6590448	0.2%
−0.662028	−0.66203827	<0.01%
−0.6672311	−0.6661249	0.16%

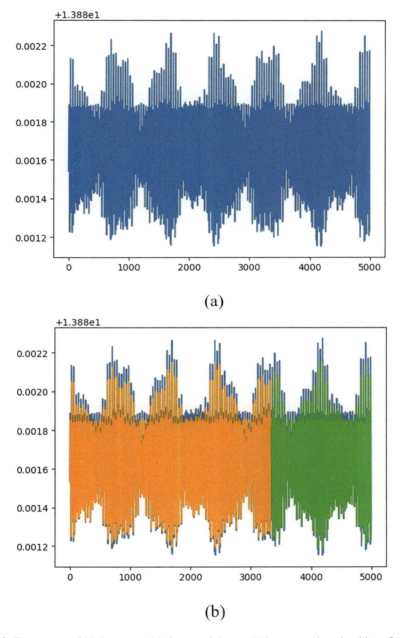

Fig. 8 The energy of H_2O system: (**a**) Output of the prediction–correction algorithm. (**b**) The yellow region corresponds to training data and green is prediction data

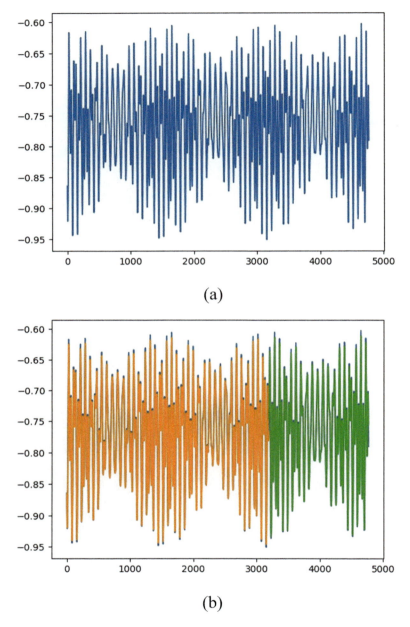

Fig. 9 The location of atoms in H_2O system: (**a**) Output for the prediction–correction algorithm. (**b**) The yellow region corresponds to is training data and green represents prediction data

Long Short-Term Memory in Chemistry Dynamics Simulation

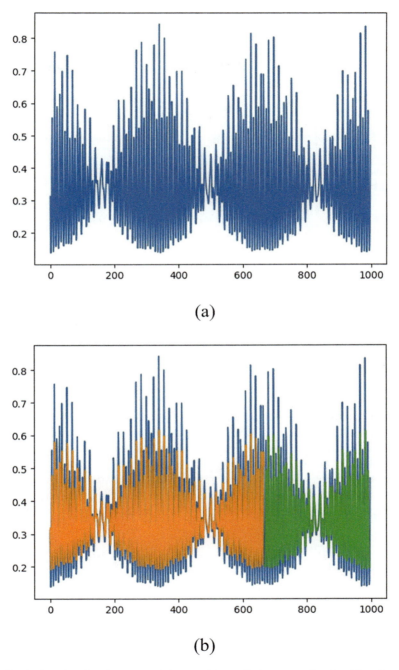

Fig. 10 The Hessian of H_2O system: (**a**) Output of the prediction–correction algorithm; the yellow part is training data and great is prediction data in (**b**)

Table 2 Relative error between Hessian prediction and computational value

Computational data (D1)	Predicted data D2	$|(D1-D2)|/|D1|$ (%)
0.16278428	0.20113048	23%
0.24653251	0.2093605	15%
0.43518633	0.47104657	8%
0.15830526	0.20171289	27%
0.23584451	0.20607977	12.6%

Table 3 Relative error between system energy prediction and computational value

Computational data (D1)	Predicted data D2	$|(D1-D2)|/|D1|$ (%)
13.881894	13.881851	<0.01%
13.881888	13.881856	<0.01%
13.881871	13.881851	<0.01%
13.881843	13.881836	<0.01%
13.881803	13.88181	<0.01%

above are the computational values and the following are training and predicted values. The horizontal coordinate is the time step and the vertical one is Hessian value. Table 2 has some relative error between predicted and computational values. We find the minimum relative error is 8% and some even over 25%. Although it is 5000 steps, Hessian calculated only 1000 steps because of the predictive–correction algorithm. Therefore, the size of the training set is less than 670 and the relative error is relatively large (Table 3).

The prediction–correction algorithm can reduce H_2O reaction system dynamics simulation time from months to days. The stability of the prediction–correction algorithm becomes very weak as simulation goes on. In addition, there must be an ab initio calculation every few steps in the prediction–correction algorithm. As the prediction step increases, the stability becomes weaker. Deep learning can reduce the simulation time of the reaction system by one third. The prediction step can reach over 1200 steps without affecting the system error after enough training. If some reinforcement learning and other methods are used, the calculation time will be further reduced and the prediction steps will be more.

4 Conclusion and Future WORK

In this chapter, a new molecular dynamics simulation algorithm is proposed by combining deep learning and predictive–correction algorithms. The new algorithm can reduce the calculation time of the system by one-third without increasing the error. In the future, the enhanced learning and parameter migration will be used to further reduce the calculation time. Then monodromy matrix [15–17] will be used to monitor the change of the calculation error.

Acknowledgments This work was supported by Dr. Hase research group and Chemdynm cluster at Texas Tech University, as well as the Industrial Internet Innovation and Development Project of China: Digital twin system for automobile welding and casting production lines and its application demonstration (TC9084DY).

References

1. D.L. Bunker, Classical trajectory methods. Comput. Phys., 10, 287–324 (1971, 1971)
2. J.M. Millam, V. Bakken, W. Chen, W.L. Hase, Ab initio classical trajectories on the Born–Oppenheimer surface: Hessian-based integrators using fifth-order polynomial and rational function fits. J. Chem. Phys. **111**, 3800–3805 (1999)
3. N. Sathyamurthy, Computational fitting of AB initio potential energy surfaces. Comput. Phys. Rep. **3**, 1–69 (1985)
4. H.-M. Keller, H. Floethmann, A.J. Dobbyn, R. Schinke, H.-J. Werner, C. Bauer, P. Rosmus, The unimolecular dissociation of HCO. II. Comparison of calculated resonance energies and widths with high-resolution spectroscopic data. J. Chem. Phys. **105**, 4983–5004 (1996)
5. X. Zhang, S. Zou, L.B. Harding, J.M. Bowman, A global ab initio potential energy surface for formaldehyde. J. Phys. Chem. **108**, 8980–8986 (2004)
6. A.P. Bartók, M.J. Gillan, F.R. Manby, G. Csányi, Machine-learning approach for one- and two-body corrections to density functional theory: applications to molecular and condensed water. Phys. Rev. B **88**, 054104 (August 2013)
7. NIH., https://ncats.nih.gov/news/releases/2015/tox21-challenge-2014-winners (2014)
8. M. Rupp, A. Tkatchenko, K.-R. Müller, O.A. von Lilienfeld, Fast and accurate modeling of molecular atomization energies with machine learning. Phys. Rev. Lett. **108**, 058301 (2012)
9. R. Ramakrishnan, P.O. Dral, M. Rupp, O.A. von Lilienfeld, J. Chem. Theor. Comput. **11**, 2087 (2015)
10. S. Lu, Q. Zeng, H. Wu, A New Power Load Forecasting Model (SIndRNN): independently recurrent neural network based on softmax kernel function. IEEE 21st International Conference on High Performance Computing and Communications, https://doi.org/10.1109/HPCC/SmartCity/DSS.2019.00320, 2019
11. H. Wu, S. Lu, A. Lopez-Aeamburo, J. She, Temperature Prediction Based on Long Short-Term Memory Networks, CSCI'19, 2019
12. V. Botu, R. Ramprasad, Adaptive machine learning framework to accelerate ab initio molecular dynamics. Int. J. Quantum Chem. **115**(16), 1074–1083 (2015)
13. E. Apra, T.L. Windus, T.P. Straatsma, et al., NWChem, A computational chemistry package for parallel computers, version 5.0, Pacific Northwest National Laboratory, Richland, Washington, 2007
14. H. Wu et al., Higher-accuracy schemes for approximating the Hessian from electronic structure calculations in chemical dynamics simulations. J. Chem. Phys. **133**, 074101, 2010
15. H. Wu, et al., A High Accuracy Computing Reduction Algorithm Based on Data Reuse for Direct Dynamics Simulations, CSCI 2016
16. H. Wu and S. Lu, Evaluating the accuracy of a third order hessian-based predictor-corrector integrator, *Europe Simulation Conference*, 2016
17. H. Wu, S. Lu, et al., Evaluating the accuracy of Hessian-based predictor-corrector integrators. J. Cent. South Univ. **24**(7), 1696–1702 (2017)

When Entity Resolution Meets Deep Learning, Is Similarity Measure Necessary?

Xinming Li, John R. Talburt, Ting Li, and Xiangwen Liu

1 Introduction

Entity resolution (ER) is the process of determining whether two references to real-world objects in an information system are referring to the same object, or to different objects [1]. Because of its fundamental role in data integration, ER has been extensively investigated in different domains such as database [2], health care [3], e-commerce [4], and so on, using various methods like rule-based [5, 6], probabilistic matching [7, 8], and machine learning [9, 10, 11]. In today's big data world, ER has become more important, since most of the data-driven applications, including artificial intelligence (AI), require integrating data from multiple sources. From difference sources, records tend to be heterogeneously structured or unstructured (textual record) [12]. For example, records from different sources might follow different structures at the attribute level. However, traditional ER system makes linking decisions based on the degree to which the values of the attributes of two records are similar [13]. The similarity measure approach presumes that records are structured uniformly at the attribute level so that comparisons are always between the same types of attributes. Lack of such a unified structure and unstructured records impose challenge for measuring similarity with the existing similarity metrics. In this case, we can either design new similarity metrics for unstructured records, just as Lin et al. [14] do, or we can avoid similarity measure. In this research, we will explore the second method.

As the cutting-edge technique in AI, deep learning has shown remarkable performance in many domains. Inspired by the recent seminal works applying deep learning in ER [15, 16, 17, 18], we explore the possibility of using deep learning

---

X. Li (✉) · J. R. Talburt · T. Li · X. Liu
Information Science, University of Arkansas at Little Rock, Little Rock, AR, USA
e-mail: xxli3@ualr.edu; jrtalburt@ualr.edu; txli1@ualr.edu; xxliu10@ualr.edu

© Springer Nature Switzerland AG 2021
H. R. Arabnia et al. (eds.), *Advances in Artificial Intelligence and Applied Cognitive Computing*, Transactions on Computational Science and Computational Intelligence, https://doi.org/10.1007/978-3-030-70296-0_10

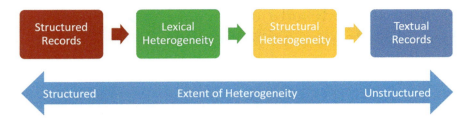

Fig. 1 Summary of ER problem types

to handle the challenge of unstructured records. Different from statistic machine learning, which requires manual features extraction, deep learning extracts features by itself. With this advantage, can deep learning do the pairwise matching without similarity measure? Under this basic logic of our design, the input is just raw record pair, and the output is linking decision: match or nonmatch. We explore both the advanced word embedding technique that considers the token order information and the traditional Bag-of-Words method. For deep learning model, the basic multilayer perceptron (MLP) model and advanced convolutional neural network (CNN) and long short-term memory (LSTM) are investigated. Then we compare the performance of these models. It is interesting to find that the combination of simply representation and simple deep learning model wins.

Overall, this research makes two main contributions in ER field. First, it demonstrates that in some cases, with the help of deep learning, traditional similarity measure is not necessary. Second, for the deep learning method in ER field, sophisticated record representation and model are not necessary to produce the best result. On the contrary, just simply count the words in the record pairs not only get the best performance, but also reduce the running time.

2 Problem Statement and Related Work

Based on works of [16] and [19], we summarize the types of ER problems in Fig. 1. Structured records follow the same schema at attribute level and the attribute values follow the same formation. Elmagarmid et al. [19] distinguish two kinds of records heterogeneity: lexical and structural. Lexical heterogeneity occurs when the records follow identically structured schema across databases, but they use different representations to refer to the same real-world object. Structural heterogeneity occurs when the fields of the records in the database are structured differently in different databases (heterogeneously structured records). Textual records don't have specific attribute schema and they are often raw text entries [16].

Figure 1 also shows the trend of ER development. The left-side problems (structured records and lexical heterogeneity) are traditional ER problems and have been extensively studied for decades; the right-side problems (structural

Table 1 Heterogeneously structured records

Data sources	Attributes	Example
List A	Name, address, city & state & zip, post office box address, post office city & state & zip, social security number, date of birth	Barbara,Chavez,11,881,Gulf pointe Driv, Apt E38,Houston,Tx,77,089, 10,525,974,
List B	First name & middle initial, last name & generational suffix, street number, street address 1, street address 2, city, state, zip, phone	Barbie,Chavze,11,881,Gulf pointe Drive, Apt E38,Houston,Texays,77,089, 7,131,657,474
List C	Name, social security number, date of birth, phone	Brabara,Chavez, 100,525,974, 7,147,263,554

heterogeneity and textual records) are relatively new. Since these two types of records lack of a unified structure, we refer them as unstructured records [12]. This research focuses on unstructured records, not only because they are new, but also because they are significant. On the one hand, in the big data era, more records become unstructured; on the other hand, the unstructured records impose challenges to the tractional similarity-based approaches.

Table 1 shows a specific example of unstructured records. List A, List B, and List C, referring to a group of persons, are from three different sources. Different sources use different attribute schemas. For example, in List A, the address is recorded in one field "address," while in List B, the same information is stored in multiple attributers, "street number," "street address 1," and "street address 2." In other words, records from three sources are heterogeneously structured at attribute level. In addition, missing value is also a problem. In this example, if there is no value between two commas, missing value occurs.

For the above example, although manually standardizing three record sources to the same attribute structure seems one solution, lots of missing values and errors impose challenges to standardization or even make standardization impossible. In this case, we see the whole record as textual records, ignoring the "attribute boundaries" across the heterogeneous structure [16]. Thus, the method in this research can also be directly applied to textual records.

Although ER has been extensively studied for 70+ years [15], deep learning approach is quite new. Considering the ER problem dimension (structured records and unstructured records), we summarize the existing ER research in the Problem–Approach matrix (as Table 2 shows). Most existing works belong to category I, which focus on structured records and use similarity-based approaches. The most related work to this paper is in categories III and IV, which are very few but very promising [16].

Seminal work [15] applies deep learning to reduce the heavy human involvement in the whole ER process. The authors not only use sophisticated composition methods to convert each record to a distributed representation, which is in turn used to effectively capture similarity between records, but also propose a locality sensitive hashing-based blocking approach. Seminal work [16] makes a completed

Table 2 ER problems–approaches matrix

	Structured	Unstructured
Traditional approach	I: Prolific	II: Sparse
Deep learning	III: Few [e.g., 15]	IV: Few [e.g., 16, 17]

review to further understand the benefits and limitation of deep learning; the authors define a space of deep learning solution for ER. Our work adds evidence for their empirical finding that deep learning has excellent performance on textual and dirty records. Seminal work [17] uses a single-layered convolutional neural network combing with crowdsourcing to do ER. Based on distributed representation, seminal work [18] uses transfer learning to reduce the labeling effort in ER. One of the most important differences of our work from the related work is that our method does not rely on similarity measure.

3 The Design of the Deep Learning Method

3.1 Difference with the Traditional Method

The core of traditional ER pipeline is the pairwise comparison, in which records are standardized to same attribute-based structure, the similarity metrics [19] are used to measure the similarity between two records. Then the linking decision is made based on the similarity measure. Since the existing similarity metrics are designed for structured records, it is hard to measure the similarity between unstructured records. One purpose of our design is to avoid similarity measure. In our design, two processes, standardization and similarity measure, are replaced by the single record pair representation process. The basic logic of our approach is to just let deep learning make the matching decision without similarity measure. We design the pairwise matching as a binary classification problem, in which the input is record pair and the output is the label, match or nonmatch (as Fig. 2 shows).

The core of our design is a record pair representation process followed by deep learning classifier. Since the neural network cannot direct read textual input, the record pair needs to be first transformed to numerical values that neural network can read. This is the record pair representation process. Then in the deep learning classifier, deep learning model is trained to extract "features" for the matched pairs.

3.2 Record Pair Representation

There are three steps in the record pair representation (Fig. 3).

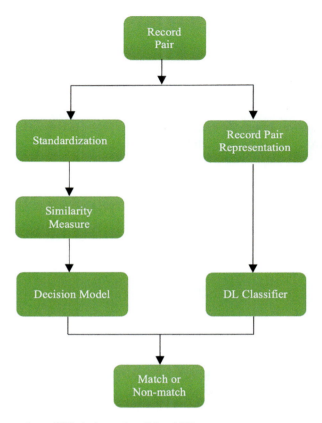

Fig. 2 The comparison of DL design and traditional ER

Fig. 3 The process of record pair representation

Concatenate Two Records: For heterogeneously structured records, we see them as textual records, ignoring the "attribute boundaries" across the heterogeneous structure. Then concatenate the two textual records together.

Tokenization: To make things simple, we remove all the nonalphanumeric characters in the records; and then parse the record pair to a sequence of tokens (words). The tokens are seen as the atomic unit.

Vectorization: Since deep learning algorithms cannot read the sequence of tokens, we have to convert the sequence of tokens to sequence of numerical value. Given the same algorithm, the different transformation method could have different results.

To explore the best vectorization methods for our design, the cutting-edge word embedding and traditional bag-of-words are both investigated.

Word Embedding: Word embedding is the state of the art of word vectorization in deep learning. It is a learned representation that allows word with similar meaning to have the similar representation, overcoming the limitations of traditional encoding methods. Word embedding has shown great power in many tasks in NLP. In general, there are two ways to use word embedding. One is to use a pretrained word embedding, such as Word2Vec and Glove. Since in ER, records often contain special tokens or errors like typos, the pretrained word embedding does not necessary cover all the tokens. Thus, we train the word vector for our specific pairwise matching task. The result embedding should place matched record pairs close together and nonmatched pairs close together respectively in the vector space.

Bag-of-Words: It is the traditional way of representing text data in machine learning. It ignores the order information of words in the text. In this research, we investigate both the basic *Count* (Term Frequency) method and *TF-IDF* method.

3.3 Deep Learning Classifier

For pairwise matching ER, the advantage of deep learning is its ability to learn the underlying pattern of matched pairs by itself, without the need for human expert to point out what makes two records matched, that is, how similar two records are. Since deep leaning is new for ER, both sophisticated algorithms, such as CNN and LSTM, and basic MLP are explored.

Convolutional Neural Network (CNN): It has achieved impressive performance in sentence classification task [20, 21]. Our pairwise matching task is similar to sentence (short text) classification. Following [20], our one-dimensional model includes an embedding layer, a convolutional layer, a pooling layer, a fully connected layer, and output layer. In the embedding layer, let $X_i \in \mathbb{R}^k$ be the k-dimensional word vector corresponding to the i-th token in the record pair text. A record pair of length n is represented as $X_1 \bigoplus X_2 \bigoplus \ldots \bigoplus X_n$, where $\bigoplus$ represents the concatenation operator. In the convolution layer, a filter $w \in \mathbb{R}^{hk}$ is applied to the record pair embedding $\mathbb{R}^{nk}$ to produce a feature map $c = [c_1, c_2, \ldots c_{n-h+1}]$, $c \in \mathbb{R}^{n-h+1}$. In the pooling layer, to capture the most important feature, the max-over-time pooling operation is used. Then features are passed to a fully connected SoftMax layer whose output is the probability of matching.

Long Short-Term Memory (LSTM): Since we see the record pair as a token sequence, we also explore the LSTM model, a special kind of recurrent neural network avoiding the long-term dependency problem. We follow the standard structure of LSTM mode [22]. As shown in Fig. 4, $\{w_1, w_2, \ldots, w_N\}$ represents the word vector of the record pair sequence of length n; $\{h_1, h_2, \ldots, h_N\}$ is the hidden vector.

Multilayer Perceptron (MLP): MLP is the most basic neural network model; it is also called Fully Connected Network. Except the input layer and out layer, there is

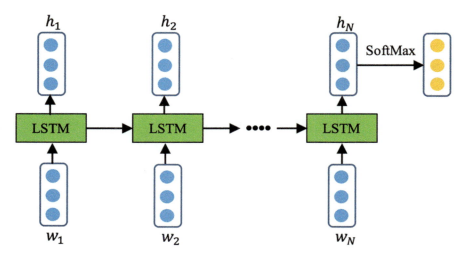

Fig. 4 LSTM model for pairwise matching

at least one hidden layer. In our model, all three kinds of token vectorization could be fed into the input layer of MLP.

Evaluation: In the pairwise matching ER, the actual linking outcome could be in one of the following four cases:

- True Positive (TP): Linking two records that are equivalent.
- True Negative (TN): Not linking two records that are not equivalent.
- False Positive (FP): Linking two records that are not equivalent.
- False Negative (FN): Not linking two records that are equivalent.

Based on the above four categories, we adopt the following measurements to evaluate the matching results.

- *Precision* (P) = TP/(TP + FP).
- *Recall* (R) = TP/(TP + FN).
- *F-Measure* = 2*(P*R)/(P + R).

Specifically, since the pairwise matching ER is an imbalanced class problem and most pairs are unmatched, the evaluation on both classes is not accurate enough. Thus, our evaluation is on the minority matched pairs.

4 Experiments and Results

The experiment process is shown in Fig. 5. Based on three methods of record pair representation and three deep learning models, there are five feasible combinations to test in our experiment. The purpose is to investigate which combination performs

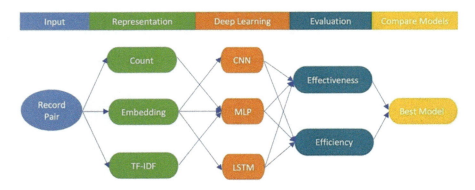

Fig. 5 The overall experiment processes

best in terms of effectiveness and efficiency. For each model, we focus on a practical question, how does the number of training data affect model performance? To rule out the effect of model structure, we keep the structures of three deep learning models as simple as possible since the basic deep learning model structure is enough to get good results [21].

Dataset: We experiment on both synthetic dataset and real-word dataset. In the synthetic dataset, there are 1000 records referring to 77 entities (persons), which form 499,500 record pairs as our training and test data. For the real-world Cora dataset [23], there are 1295 records referring to 112 entities (research papers), which form 837,865 record pairs.

In this section, we first experiment on the synthetic dataset. We gradually increase the number training data, aiming to find out the least number of training data needed to train a satisfied model (F-measure at ~0.99). Then we validate the results on the real-world Cora dataset.

4.1 Convolutional Neural Network

The convolutional neural network model on which we experiment is show in Fig. 6. Besides the parameters shown in Fig. 6, we set the batch size at 500, epoch at 20.

Table 3 shows the results. The last column represents how many seconds each epoch needs. It shows that time increases linearly with the number of training data. For effectiveness, as the number of training data increases, precision recall and F-measure all increase. At certain point, 349,650, F-measure can get as high as 0.99.

```
Layer (type)                     Output Shape              Param #
=================================================================
embedding_1 (Embedding)          (None, 26, 64)            129472

conv1d_1 (Conv1D)                (None, 22, 64)            20544

max_pooling1d_1 (MaxPooling1     (None, 4, 64)             0

flatten_1 (Flatten)              (None, 256)               0

dense_1 (Dense)                  (None, 64)                16448

dense_2 (Dense)                  (None, 1)                 65
=================================================================
Total params: 166,529
Trainable params: 166,529
Non-trainable params: 0
```

Fig. 6 The convolutional neural network model

Table 3 The results of convolutional neural network

Training data	Precision	Recall	F-measure	Time
49,950	0.77	0.44	0.56	2
99,900	0.94	0.74	0.83	4
149,850	0.98	0.84	0.90	6
199,800	0.98	0.92	0.95	8
249,750	0.99	0.93	0.96	10
299,700	0.99	0.96	0.97	12
349,650	1.0	0.97	0.99	14
399,600	0.99	0.98	0.99	16
449,550	1.0	0.98	0.99	17

4.2 Long Short-Term Memory

The long short-term memory model in our experiment is show in Fig. 7. Besides the parameters shown in Fig. 7, we set the batch size at 500, epoch at 20.

The results from long short-term memory model are shown in Table 4. As the number of training data increases, the trend of effectiveness increase is similar as that of convolutional neural network model. The difference is the efficiency. With the same number of training data, LSTM need more than twice time to achieve the same level of effectiveness.

```
Layer (type)                 Output Shape              Param #
=================================================================
embedding_1 (Embedding)      (None, 26, 64)            129472

lstm_1 (LSTM)                (None, 64)                33024

dense_1 (Dense)              (None, 1)                 65
=================================================================
Total params: 162,561
Trainable params: 162,561
Non-trainable params: 0
```

Fig. 7 The long short-term memory model

Table 4 The results of long short-term memory

Training data	Precision	Recall	F-measure	Time
49,950	0.45	0.29	0.35	6
99,900	0.82	0.69	0.75	12
149,850	0.95	0.83	0.89	18
199,800	0.95	0.92	0.93	25
249,750	0.96	0.95	0.95	30
299,700	0.98	0.96	0.97	33
349,650	0.99	0.98	0.98	38
399,600	0.98	0.98	0.98	44
449,550	1.00	0.99	0.99	49

4.3 Embedding Combining MLP

Through a flatten layer, word embedding can be the input of the basic multilayer perceptron, as Fig. 8 shows. We keep the other parameter as the same with CNN and LSTM.

Table 5 shows the results of word embedding combining MLP model. Compared with CNN and LSTM, it has lower effectiveness but higher efficiency.

4.4 Count Combining MLP

With the count representation, the basic multilayer perceptron model is shown in Fig. 9. The input dimension is the number of total distinct tokens (1993 in the dataset).

The results of count combining MLP are show in Table 6. Surprisingly, compared with the previous models, it gets the 0.99 F-measure at the least number of training data 199,800. Meanwhile, the training time is about half of that in CNN model.

```
Layer (type)                Output Shape              Param #
=================================================================
embedding_1 (Embedding)     (None, 26, 64)            129472

flatten_1 (Flatten)         (None, 1664)              0

dense_1 (Dense)             (None, 8)                 13320

dense_2 (Dense)             (None, 1)                 9
=================================================================
Total params: 142,801
Trainable params: 142,801
Non-trainable params: 0
```

Fig. 8 The word embedding combining MLP

Table 5 The results of embedding combining MLP

Training data	Precision	Recall	F-measure	Time
49,950	0.40	0.31	0.35	1
99,900	0.76	0.52	0.61	1
149,850	0.79	0.66	0.72	1
199,800	0.88	0.71	0.79	2
249,750	0.94	0.81	0.87	2
299,700	0.95	0.86	0.90	3
349,650	0.94	0.89	0.91	3
399,600	0.97	0.93	0.95	4
449,550	0.98	0.92	0.95	4

```
Layer (type)                Output Shape              Param #
=================================================================
dense_1 (Dense)             (None, 128)               255232

dense_2 (Dense)             (None, 64)                8256

dense_3 (Dense)             (None, 8)                 520

dense_4 (Dense)             (None, 1)                 9
=================================================================
Total params: 264,017
Trainable params: 264,017
Non-trainable params: 0
```

Fig. 9 The count combining with MLP

Table 6 The results of count combining MLP

Training data	Precision	Recall	F-measure	Time
49,950	0.91	0.56	0.70	1
99,900	0.93	0.86	0.89	2
149,850	0.99	0.97	0.98	3
199,800	0.99	0.99	0.99	4
249,750	1.00	0.98	0.99	6
299,700	0.99	1.00	1.00	7
349,650	1.00	1.00	1.00	8
399,600	0.99	1.00	1.00	9
449,550	1.00	0.99	1.00	10

Table 7 The results of TF-IDF combining MLP

Training data	Precision	Recall	F-measure	Time
49,950	0.86	0.33	0.48	1
99,900	0.91	0.66	0.77	2
149,850	0.96	0.86	0.91	3
199,800	0.99	0.90	0.94	4
249,750	0.97	0.97	0.97	5
299,700	0.98	0.96	0.97	6
349,650	1.00	0.97	0.98	7
399,600	0.99	0.98	0.99	8
449,550	0.99	1.00	1.00	10

4.5 TF-IDF Combining MLP

With exact same model as 4.4, we only change the representation from count to TF-IDF. The results (Table 7) show that the effectiveness of TF-IDF is not as good as count.

Based on the comparison of the five cases, the overall conclusion is that the simple count representation and the simple MLP model together win both in effectiveness and efficiency.

4.6 Validation on Real-World Cora Data

To further validate our design on the real-world Cora data, 586,505 out of 837,865 records in the Cora dataset were used as training data and 251,360 as test data. The results (Table 8 shows) validate our model and conclusion from the synthetic data experiment. The count representation and multilayer perceptron model together get the best effectiveness and efficiency.

Model	Precision	Recall	F-measure	Time
CNN	1.00	0.99	0.99	120
LSTM	0.99	0.98	0.99	140
E + MLP	0.99	0.98	0.99	15
C + MLP	1.00	1.00	1.00	8
T + MLP	0.99	1.00	1.00	8

Table 8 The results of Cora data experiment

5 Conclusion and Future Work

More and more unstructured records in ER impose challenge to traditional ER approach, since existing similarity metrics are designed for structured records. To address this challenge, this research goes to the opposite direction with the help of deep learning. That is making linkage decision without similarity measure. Seeing pairwise matching ER as a binary classification problem, we investigate three record pair representations, word embedding, count and TF-IDF, and three deep learning models, CNN, LSTM, and MLP. Surprisingly, the experiments show that the simplest representation (count) and the simplest deep learning model (MLP) together get the best result. The experiment on the real-world Cora dataset supports our results. Although our method can solve the pairwise ER at ease, same as all other supervised learning methods, our method requires labeled data. However, in ER more often than not, the ground truth labeled data are lacking, and thus considering the limited labeled data is one future research direction. In addition, pairwise matching of all records scales quadratically with the number of records, and hence blocking is needed when handling big data. It is promising to explore deep learning–based blocking approach for ER in the future.

References

1. J.R. Talburt, *Entity Resolution and Information Quality* (Morgan Kaufmann, New York, USA, 2011)
2. M.A. Hernández, S.J. Stolfo, Real-world data is dirty: data cleansing and the merge/purge problem. Data Min. Knowl. Disc. **2**(1), 9–37 (1998)
3. T.W. Victor, R.M. Mera, Record linkage of health care insurance claims. J. Am. Med. Inform. Assoc. **8**(3), 281–288 (2001)
4. H. Köpcke, A. Thor, S. Thomas, E. Rahm, Tailoring entity resolution for matching product offers. In *Proceedings of the 15th International Conference on Extending Database Technology*, 2012 Mar 27, pp. 545–550
5. S.E. Whang, H. Garcia-Molina, Entity resolution with evolving rules. Proc. VLDB Endowment **3**(1–2), 1326–1337 (2010)
6. L. Li, J. Li, H. Gao, Rule-based method for entity resolution. IEEE Trans. Knowl. Data Eng. **27**(1), 250–263 (2014)
7. I.P. Fellegi, A.B. Sunter, A theory for record linkage. J. Am. Stat. Assoc. **64**(328), 1183–1210 (1969)

8. P. Wang, D. Pullen, J.R. Talburt, C. Chen, A method for match key blocking in probabilistic matching. In *Information Technology: New Generations*, pp. 847–857, 2016
9. L. Kolb, H. Köpcke, A. Thor, E. Rahm, Learning-based entity resolution with MapReduce, in *Proceedings of the Third International Workshop on Cloud Data Management*, (2011 Oct 28), pp. 1–6
10. Z. Chen, Z. Li, Gradual Machine Learning for Entity Resolution. arXiv preprint arXiv:1810.12125 (2018)
11. I. Bhattacharya, L. Getoor, A latent Dirichlet model for unsupervised entity resolution. In *Proceedings of the 2006 SIAM International Conference on Data Mining*, 2006 Apr 20, pp. 47–58
12. S. Song, L. Chen, Probabilistic correlation-based similarity measure of unstructured records. In *Proceedings of the Sixteenth ACM Conference on Conference on Information and Knowledge Management*, 2007 Nov 6, pp. 967–970
13. J. Wang, G. Li, J.X. Yu, J. Feng, Entity matching: how similar is similar. Proc. VLDB Endowment **4**(10), 622–633 (2011)
14. Y. Lin, H. Wang, J. Li, H. Gao, Efficient entity resolution on heterogeneous records. IEEE Trans. Knowl. Data Eng. **32**(5), 912–926 (2019)
15. M. Ebraheem, S. Thirumuruganathan, S. Joty, M. Ouzzani, N. Tang, Distributed representations of tuples for entity resolution. Proc. VLDB Endowment **11**(11), 1454–1467 (2018)
16. S. Mudgal, H. Li, T. Rekatsinas, A. Doan, Y. Park, G. Krishnan, R. Deep, E. Arcaute, V. Raghavendra, Deep learning for entity matching: a design space exploration, in *Proceedings of the 2018 International Conference on Management of Data*, (2018 May 27), pp. 19–34
17. R.D. Gottapu, C. Dagli, B. Ali, Entity resolution using convolutional neural network. Proc. Comput. Sci. **95**, 153–158 (2016)
18. S. Thirumuruganathan, S.A. Parambath, M. Ouzzani, N. Tang, S. Joty, Reuse and adaptation for entity resolution through transfer learning. arXiv preprint arXiv:1809.11084 (2018 Sep 28)
19. A.K. Elmagarmid, P.G. Ipeirotis, V.S. Verykios, Duplicate record detection: A survey. IEEE Trans. Knowl. Data Eng. **19**(1), 1–6 (2006)
20. Y. Kim, Convolutional neural networks for sentence classification. arXiv preprint arXiv:1408.5882 (2014 Aug 25)
21. Y. Zhang, B. Wallace, A sensitivity analysis of (and practitioners' guide to) convolutional neural networks for sentence classification. arXiv preprint arXiv:1510.03820 (2015 Oct 13)
22. Y. Wang, M. Huang, X. Zhu, L. Zhao, Attention-based LSTM for aspect-level sentiment classification, in *Proceedings of the 2016 Conference on Empirical Methods in Natural Language Processing*, (2016 Nov), pp. 606–615
23. A. McCallum. Cora dataset, https://doi.org/10.18738/T8/HUIG48 (2017)

Generic Object Recognition Using Both Illustration Images and Real-Object Images by CNN

Hirokazu Watabe, Misako Imono, and Seiji Tsuchiya

1   AlexNet [1]

In 2012, a contest called ImageNet Large-Scale Visual Recognition Challenge (ILSVRC) competed for the performance of image recognition. AlexNet is a network structure of CNN, designed by Krizbevsky et al., who won in the problem (general object recognition) that answers the class of the object from among thousand kinds with the lowest error rate.

2   An Experiment on Object Recognition Using AlexNet

In this experiment, we learn real-object images using AlexNet and verify the difference in object recognition rate in each set of test images. The learning image prepares 1000 real-object images, a total of 50,000 images for each class with 50 kinds of object classes. The test image is 50 kinds of object classes for two types of images: (1) illustration image, (2) real-object image, provided by free material on the Internet, 100 images and a total of 5000 images were prepared for each class.

Image data sets used in this experiment are IMAGENET [2]. IMAGENET is a database that holds more than 14 million images and gives the object class name in

---

H. Watabe · S. Tsuchiya
Department of Intelligent Information Engineering and Science, Faculty of Science and Engineering, Doshisha University, Kyotanabe, Kyoto, Japan

M. Imono (✉)
Department of Information Systems, School of Informatics, Daido University, Nagoya, Aichi, Japan
e-mail: m-imono@daido-it.ac.jp

© Springer Nature Switzerland AG 2021
H. R. Arabnia et al. (eds.), *Advances in Artificial Intelligence and Applied Cognitive Computing*, Transactions on Computational Science and Computational Intelligence, https://doi.org/10.1007/978-3-030-70296-0_11

Real object image   illustration image

Fig. 1 Image examples

Table 1 Object class names

Apple	Backpack	Banana	Baseball bat	Baseball glove
Boom box	Calculator	Camera	Cap	Chair
Coffee mug	Computer keyboard	Computer monitor	Computer mouse	Dolphin
Electric guitar	Elephant	Eyeglasses	Flamingo	Flashlight
Golf club	Hamburger	Headphone	Hot air balloon	Jet
Lion	Mountain bike	Necktie	Orange	Paper clip
Scissors	Ship	Shirt	Shoes	Skateboard
Soccer ball	Sock	Soda can	Spoon	Sport car
Stapler	Sunflower	Teapot	Tennis ball	Tennis racket
Tiger	Umbrella	Watch	Wine bottle	Zebra

each image. In this experiment, 50 types were randomly selected from the object class present in IMAGENET, and 1000 images were selected as learning images and 100 images that were not used for learning as test images. In addition, for the illustration image, it was obtained from the "illustrain" [3] and the "illustAC" [4] which provide the free illustrated image on the Internet. The object class name and the number of test images are the same as the test set in the real image. The image examples are shown in Fig. 1 and the object class names are shown in Table 1.

The results of object recognition rate using AlexNet are shown in Table 2. From Table 2, the recognition rate when using the illustrated image as a test set was 65.6%, and the recognition rate when using the real-object image as a test set was 83.4%. From this result, it can be seen that the recognition rate using the illustrated image as a test set is about 18% lower than that of using the real-object image as a test set.

Generic Object Recognition Using Both Illustration Images and Real-Object...

Table 2 Object recognition rate

Learning images	Real-object images	Real-object images
Test images	Real-object images	Illustration image
Object recognition rate (%)	83.4%	65.6%

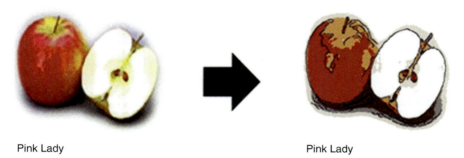

Pink Lady   Pink Lady

Fig. 2 Pseudo-illustration image

This is considered to be a major factor that the illustration image is not included in the learning image.

Currently, there are many objects in our surrounding environment with different illustrations, paintings, etc., in addition to real objects, but it is necessary to recognize them correctly. In order to recognize illustrated images, illustration images are usually used for learning, but illustrated images on the Internet are copyrighted, and it is not easy to collect large amounts of learning images. Therefore, in this chapter, we create pseudo illustration images by illustration processing a real-object image and verify the effectiveness of the learning model which can recognize both the illustration image and the real-object image.

3 Generation of Illustration Images

The illustration image handled in this chapter is defined as follows. "There is an outline in the object. Or the image is colored within 12 colors."

The following shows an algorithm for generating illustration images. In addition, we show the pseudo-illustration image created from the real-object image in Fig. 2.

1. Generate contour images from input images.
2. Color reduction processing of input images in 12 colors.
3. Get the difference between the image and the contour image after the color reduction.

Table 3 Object recognition rate

Learning images	Test images	Object recognition rate (%)	Average
ROI	ROI	83.4%	74.5%
ROI	II	65.6%	
PII	ROI	74.3%	76.7%
PII	II	79.0%	
ROI+PII	ROI	80.9%	77.2%
ROI+PII	II	73.4%	

ROI real-object image, *II* illustration image, *PII* pseudo illustration image

4 Evaluations

In this experiment, we create three learning models using AlexNet to learn real-object images, learning models that learn pseudo-illustrated images, and learning models that learn both real-object images and pseudo-illustrated images, and verify the object recognition rate in each learning model. Two types of test image sets were prepared: real-object images and illustrated images. Therefore, the experimental pattern is a total of six patterns of three learning image sets and two test image sets.

Experimental results are shown in Table 3. From Table 3, it is better to learn the pseudo-illustration image than to learn the real-object image, and the recognition rate of the illustration image is improved by 14%. In addition, in order to recognize both real-object images and illustration images, the learning model that learns both real-object images and pseudo-illustrated images is improved by about 3% than learning real object images.

5 Conclusion

In this chapter, we verified whether the pseudo-illustrated image which processed contour processing and the color reduction processing to the real image is effective for the recognition of the illustrated image. As a result of the verification, it is effective because the recognition rate of the illustrated image is improved by learning the pseudo-illustrated image.

Acknowledgments This work was partially supported by JSPS KAKENHI Grant Number 16 K00311.

References

1. A. Krizhevsky, I. Sutskever, G.E. Hinton, ImageNet Classification with Deep Convolutional Neural Networks. *Advances in Neural Information Processing Systems* 25 (NIPS 2012), pp. 1097–1105, 2012
2. ImageNet: http://www.image-net.org/, 28. December 2016
3. https://illustrain.com/
4. https://www.ac-illust.com/?mid=59acd36c5ad15

A Deep Learning Approach to Diagnose Skin Cancer Using Image Processing

Roli Srivastava, Musarath Jahan Rahamathullah, Siamak Aram, Nathaniel Ashby, and Roozbeh Sadeghian

1 Introduction

Skin cancer is the leading form of cancer in the United States [1]. It forms when skin cells multiply abnormally and can prove fatal if it is allowed to metastasize to other areas of the body through the lymphatic system. Most skin cancers result from exposure to Ultraviolet (UV) light. When the skin is unprotected, UV radiation damages DNA and can produce genetic mutations, which can subsequently lead to cancerous growths [2]. According to Didona et al. [3] the most common types of skin cancer in Caucasian populations are melanoma and nonmelanoma (i.e., basal and squamous cell carcinoma) skin cancers (NMSC), with melanoma accounting for 4% of all deaths from cancer [4].

Two methods are commonly employed to diagnose whether a skin sample (biopsy) should be taken: Visual examination of the skin by a physician [5, 6]; or dermatoscopy [7] and/or epiluminescence microscopy by a trained clinician [8]. Thus, initial diagnostic efficiency currently depends exclusively on the competence and perceptual capabilities of the practitioner. Perhaps unsurprisingly, both methods have been found to result in suboptimal detection efficacy [9] with false positives abounding. Hence there exists an urgent need for a screening method with increased sensitivity and specificity to be developed.

To address these issues, medical practitioners have increasingly been seeking to employ automated image processing tools that can more effectively diagnose skin cancer [10]. Maier et al. [11] successfully used dermatoscopic images to train an Artificial Neural Network to differentiate deadly melanomas from melanocytic nevi.

R. Srivastava · M. J. Rahamathullah · S. Aram · N. Ashby · R. Sadeghian (✉)
Harrisburg University of Science and technology, Harrisburg, PA, USA
e-mail: RSrivastava@my.harrisburgu.edu; MRahamathullah@my.harrisburgu.edu;
SAram@harrisburgu.edu; NAshby@harrisburgu.edu; RSadeghian@harrisburgu.edu

© Springer Nature Switzerland AG 2021
H. R. Arabnia et al. (eds.), *Advances in Artificial Intelligence and Applied Cognitive Computing*, Transactions on Computational Science and Computational Intelligence, https://doi.org/10.1007/978-3-030-70296-0_12

Although promising, this study, like earlier attempts [12], was hampered by small sample sizes and a lack image variation [13].

Recent increases in data availability, paired with technological advances, have revigorated these efforts. A deep learning approach was successfully employed and returned more accurate diagnoses than most trained experts [14, 15]. Gautman et al. [16] issued an automation challenge to modelers and reported the top submission had an accuracy of 85.5% for disease classification. More recently, a deep convolutional neural network (CNN) model known as MobileNetV2 using a transfer learning method classified benign versus malignant lesions with an accuracy of 91.33% [17].

The objective of the current research is to expand earlier efforts in developing automated skin cancer detection systems by producing a model capable of accurately classifying seven different types of skin lesions. Stakeholders in this endeavor are patients with skin lesions and practitioners. Prefacing our findings, we demonstrate that our approach can provide a high degree of accuracy (95%) in the early diagnosis of skin cancer(s). Importantly, because human perception is not required, we argue this approach should greatly minimize the negative impact of human factors.

2 Dataset

The dataset was compiled by the Medical University of Vienna [18] and includes 10,015 images of pigmented skin lesions. Images were sampled equally from male and female patients with an average age of 51. Images were collected from different parts of the body (e.g., face, ear, and neck) and captured in resolutions ranging from 8×8 pixels to 450×600 pixels. Figure 1 displays a sample of the images used in the study. Images fall into seven different classifications:

- Melanoma (mel): The most dangerous form of skin cancer which generally develops from pigment-containing cells known as melanocytes [19].
- Basal cell carcinoma (bcc): This cancer affects the basal cells which are responsible for the production of new skins. While it rarely metastasizes it does spread easily [20].
- Actinic keratosis (akiec): This "pre-cancer" indicator appears as a scaly patch resulting from accumulated UV exposure [21].
- Benign keratosis-like lesions (bkl): A benign, painless skin disorder which is mostly associated with aging and exposure to UV light [22].
- Vascular lesions (vasc): Common birthmarks that can be flat or raised [23].
- Dermatofibroma (df): Superficial benign fibrous histiocytoma which primarily occur in women [24].
- Melanocytic nevi (nv): Benign birthmarks and moles that resemble melanoma [25].

Importantly, these classes are not unique. Thus, some patients may present with more than one type of lesion. More than 50% of lesion images were confirmed

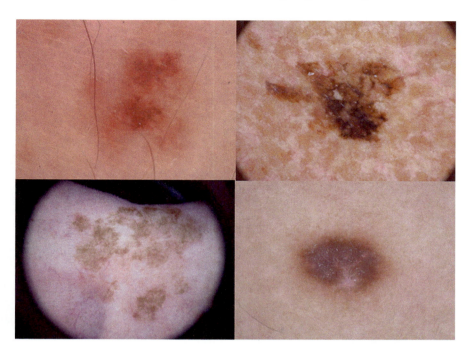

Fig. 1 A lower extremity sample for a 50-year-old male diagnosed with melanocytic nevei (upper left). A lesion sample from a 60-year-old male diagnosed with melanoma (upper right). A face lesion sample from a 70-year-old female diagnosed with basal cell carcinoma (lower left) and a lesion sample from a 50-year-old female diagnosed with benign keratosis-like lesions (upper right)

by pathology, while the ground truth for the rest of were either follow-up, expert consensus, or confirmation by in vivo confocal microscopy.

3 Methodology

CNNs are the state of the art in deep learning for image classification [26], and there are numerous applications for CNN medical image analysis [27].

3.1 Image Preprocessing

Images were preprocessed using normalization techniques, for example, scaling image intensity to the range of [0, 1]. To increase processing speed each image was down sampled to $50 \times 50 \times 3$ pixels. Images were unevenly distributed across classes. Thus, to remove potential bias subsets were created by randomly

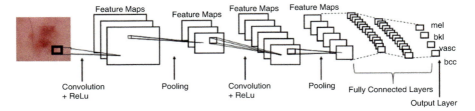

Fig. 2 Visualization of the CNN model built

sampling evenly from the seven categories that ensured the complete population was considered. One aspect of preparation of the images was to be assure that no repeated image should be appeared in training dataset, to address this issue, a chi-square distance measurement technique was used [28].

Data augmentation techniques were employed to increase the number of images available to train on. Images were rotated, zoomed, and flipped.

3.2 CNN

The architecture of CNN model employed is shown in Fig. 2. It consists of two convolutional parts: First, two convolutional layers followed by a pooling layer with a dropout rate of 0.25; second, two convolutional layers, a pooling layer and dropout rate of 0.30, trailed by a flattening of densely connected layers. The convolutional, also known as pooling steps, condense information. The lowest resolution images did not provide enough information to allow for the second convolutional/pooling layers and were omitted. Sometimes results from the first convolutional models were good, besting more complex models. This pattern was particularly true for medium-resolution images. It appears that medium-resolution images can run out of the information required by more complex models, and performance begins to suffer.

3.3 VGG-Net

The last set of models considered employed the VGG16 algorithm [29]. The general structure of this network is a 16-layer CNN that uses 3 × 3 filters with stride and pad of 1, along with 2 × 2 max pooling layers with stride of 2. The convolutional layers have 16, 64, 128, 256, 512 nodes successively. As the spatial size of the input volumes at each layer decrease, the result of the convolutional and pool layers, the depth of the volumes increases as the number of filters increases, doubling after

A Deep Learning Approach to Diagnose Skin Cancer Using Image Processing   151

each maxpool layer. The flattened layers consist of 1098, 4098, and 7 nodes. The final layer employs a SoftMax activation function.

4 Results

In the current analysis the CNN was found to produce an accuracy of 93% and minimal test loss of 0.18%. However, it did not sufficiently address the issue of overfitting. This can be seen clearly in the wide gaps between training and test set performance in Fig. 3.

When compared to the CNN without data augmentation, the model including augmented data did improve accuracy to 94% and decreased loss to 0.14% (shown in Fig. 4). The problem of overfitting was not ideal.

The VGG16 had an average accuracy of 93.67%, sensitivity of 95.66%, and specificity of 80.43%. A ten-fold cross validation was used to estimate the efficiency of the model. The learning curve of this topology is shown in Fig. 5. The learning curves indicate that the training loss decreases to a point of stability, and the small gap with training loss suggest overfitting was mostly resolved.

Metrics for each k-fold of the model are shown in Table 1. From the table, the ability of the model to correctly identify those with cancer (i.e., true positive rate) is as high as 96% and never lower than 94%, while the ability to correctly identify those without cancer (i.e., true negative rate) is as high as 83% and no lower than 70%. Notably higher than those provided by experts [9].

Table 2 indicates how accurately each of the seven classes of skin lesions are predicted. The most common type of skin cancer (bcc) is predicted with an accuracy of 95.61%. The deadliest skin cancer (mel) is predicted with an accuracy greater than 90%. Thus, the model performs well in diagnosing the most serious cases.

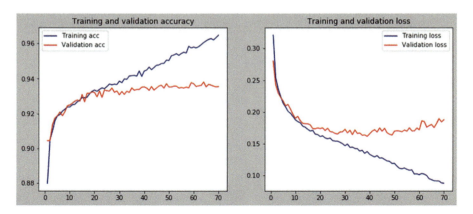

Fig. 3 Accuracy (left panel) and loss (right panel) of the CNN model without data augmentation

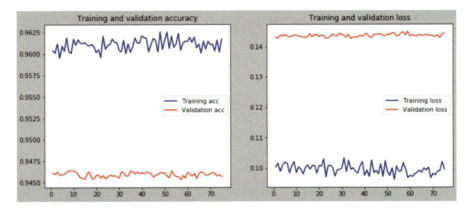

Fig. 4 Accuracy (left panel) and loss (right panel) of the CNN model with data augmentation

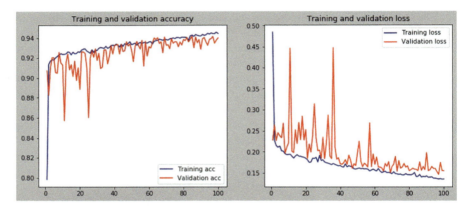

Fig. 5 Accuracy (left panel) and loss (right panel) of the VGG16 model

Table 1 K-fold cross validation metrics

K-Fold	Accuracy (%)	Sensitivity (%)	Specificity (%)
1	94.05	95.96	81.55
2	93.67	95.43	81.49
3	93.94	95.83	81.39
4	94	95.78	82.03
5	94.38	96.2	82.55
6	93.93	95.74	81.79
7	91.02	94.05	70.58
8	94.06	95.96	81.62
9	93.82	95.78	80.87
10	93.86	95.93	80.47
Avg	**93.68**	**95.67**	**80.43**

A Deep Learning Approach to Diagnose Skin Cancer Using Image Processing

Table 2 Predictive accuracy by lesion classification

Lesion classification	Accuracy (%)	Sample N
Bkl	89.97	219
Nv	85.82	1329
Df	99.30	19
Mel	90.26	219
Vasc	99.25	38
Bcc	95.61	110
Akiec	96.31	69

5 Conclusions

A deep learning approach to diagnosing different types of skin lesions ranging from potentially deadly skin cancers to benign age spots was employed. Results indicate that an automated approach can be used to effectively diagnose the etiology of lesions, detecting skin cancers more accurately than human experts [9]. This is an important finding with high pragmatic value. An approach to skin cancer screening will greatly improve health outcomes for patients while reducing resource expenditures. For instance, patients will be able to obtain accurate preliminary diagnosis from their primary care physician without seeking out a specialist, rural patients would be able to obtain a diagnosis through telemedicine, and laboratories would likely see a reduction in the number of unnecessary biopsies needing to be processed. While further testing and refinement is required, we believe the current results can help healthcare providers to make more accurate decisions.

References

1. Melanoma of the Skin Statistics, 9 April 2020. [Online]. Available: https://www.cdc.gov/cancer/skin/statistics/index.htm. Accessed 23 April 2020
2. D.L. Narayanan, R.N. Saladi, J.L. Fox, Ultraviolet radiation and skin cancer. Int. J. Dermatol. **9**, 978–986 (2010)
3. D. Didona, G. Paolino, U. Bottoni, C. Cantisani, Non-melanoma skin cancer pathogenesis overview. Biomedicine **6**(1), 6 (2018)
4. E. Losina, R.P. Walensky, A. Geller, F.C. Bedingfield, L.L. Wolf, B.A. Gilchrest, K.A. Freedberg, Visual screening for malignant melanoma: a cost-effectiveness analysis. Arch. Dermatol. **143**(1), 21–28 (2007)
5. A.I. Riker, N. Zea, T. Trinh, The epidemiology, prevention, and detection of melanoma. J. Oschner **10**(2), 56–65 (2010)
6. L. Brochez, E. Verhaeghe, L. Bleyen, J.M. Naeyaert, Diagnostic ability of general practitioners and dermatologists in discriminating pigmented skin lesions. J. Am. Acad. Dermatol. **44**(6), 979–986 (2001)
7. A. Herschorn, Dermoscopy for melanoma detection in family practice. Can. Fam. Physician **58**(7), 740–745 (2012)
8. A. Steiner, M. Binder, M. Schemper, K. Wolff, H. Pehamberger, Statistical evaluation of epiluminescence microscopy criteria for melanocytic pigmented skin lesions. J. Am. Acad. Dermatol. **29**(4), 581–588 (1993)

9. J.K. Winkler, F. Christine, T. Ferdinand, et al., Association between surgical skin markings in Dermoscopic images and diagnostic performance of a deep learning convolutional neural network for melanoma recognition. JAMA Dermatol. **155**(10), 1135–1141 (2019)
10. A. Marka, J.B. Carter, E. Toto, S. Hassanpour, Automated detection of nonmelanoma skin cancer using digital images: a systematic review. BMC Med. Imaging **19**(21) (2019)
11. H. Maier, M. Schemper, B. Ortel, M. Binder, A. Tanew, H. Hönigsmann, Skin tumors in Photochemotherapy for psoriasis: a single-center follow-up of 496 patients. Dermatology **193**, 185–191 (1996)
12. P. Rubegni, M. Burroni, B. Perotti, M. Fimiani, L. Andreassi, G. Cevenini, Digital Dermoscopy analysis and artificial neural network for the differentiation of clinically atypical pigmented skin lesions: a retrospective study. J. Investig. Dermatol. **119**(2), 417–474 (2001)
13. P. Tschandl, T. Wiesner, Advances in the diagnosis of pigmented skin lesions. Br. J. Dermatol. **178**, 9–11 (2018)
14. T.J. Brinker, A. Hekler, et al., Deep learning outperformed 136 of 157 dermatologists in a head-to-head dermoscopic melanoma image classification task. Eur. J. Cancer **113**, 47–54 (2019)
15. A. Dascalu, E.O. Davis, Skin cancer detection by deep learning and sound analysis algorithms: a prospective clinical study of an elementary dermoscope. EBio Med. **43**, 107–113 (2019)
16. D. Gautman, N. C. Codella, E. Celebi, B. Helba, M. Marchetti, N. Mishra, A. Halpern, Skin lesion analysis toward melanoma detection: a challenge at the International Symposium on Biomedical Imaging (ISBI). *International Symposium on Biomedical Imaging,* 2016
17. A. Cherif, M. Misbhauddin, M. Ech-Cherif, Deep neural network based mobile dermoscopy application for triaging skin cancer detection. *Information security, ICCAIS,* 2019
18. P. Tschandl, C. Rosendahl, H. Kittler, The HAM10000 dataset, a large collection of multi-source dermatoscopic images of common pigmented skin lesions. Sci. Data **161**, 1–6 (2018)
19. Y. Liu, M. Saeed Sheikh, Melanoma: molecular pathogenesis and therapeutic management. Mol. Cell. Pharmacol. **6**(3), 228 (2014)
20. J. Lanoue, G. Goldenberg, A comprehensive review of existing and emerging nonsurgical therapies. J. Clin. Aesth. Dermatol. **9**(5), 26–36 (2016)
21. J.Q. Del Rosso, L. Kircik, G. Goldenberg, B. Berman, Comprehensive management of actinic keratoses. J. Clin. Aesth. Dermatol. **7**(9), 2–12 (2014)
22. U. Wollina, Seborrheic Keratoses – the most common benign skin tumor of humans. Clinical presentation and an update on pathogenesis and treatment options. Open Access J. Med. Sci. **6**(11) (2018)
23. S.C. Nair, Vascular anomalies of the head and neck region. J. Maxillofac Oral Surg. **17**(1), 1–12 (2018)
24. J. Alves, D.M. Matos, H.F. Barreiros, E. Bártolo, Variants of dermatofibroma - a histopathological study. An. Bras. Dermatol. **89**(3), 472–477 (2014)
25. M.R. Roh, P. Eliades, S. Gupta, H. Tsao, Genetics of melanocytic nevi. Pigment Cell Melanoma Res. **28**(6), 661–672 (2015)
26. A. Krizhevsky, I. Sutskever, G.E. Hinton, ImageNet classification with deep convolutional. Adv. Neural Inf. Proces. Syst., 1097–1105 (2012)
27. O. Faust, Y. Hagiwara, T.J. Hong, O.S. Lih, U.R. Acharya, Deep learning for healthcare applications based on physiological signals: a review. Comput. Methods Prog. Biomed. **161**, 1–13 (2018)
28. S. Hadipour, S. Aram, R. Sadeghian, Similar multi-modal image detection in multi-source dermatoscopic images of cancerous pigmented skin lesions. *Submitted to The 2020 World Congress in Computer Science, Computer Engineering, & Applied Computing,* 2020
29. K. Simonyan, A. Zisserman, Very deep convolutional networks for large-scale image recognition. arXiv preprint **1409**, 1556 (2014)

Part II
Learning Strategies, Data Science, and Applications

Effects of Domain Randomization on Simulation-to-Reality Transfer of Reinforcement Learning Policies for Industrial Robots

C. Scheiderer, N. Dorndorf, and T. Meisen

1 Introduction

In recent years, the paradigm of reinforcement learning (RL) has been employed successfully to solve a multitude of different control tasks from games [1, 2] to resource management [3] to robotics [4, 5]. The concept of using an autonomously exploring agent, which is able to leverage experience in order to deduce favorable strategies from observations and performance indicators promises potential to automate tasks where a solution strategy is hard to formulate.

The underlying principle of trial-and-error however limits the applicability of RL. Generating enough experience in order to derive sufficiently accurate control strategies may present a major challenge. This becomes especially apparent when using RL to solve real-world tasks in the domain of robotics, where the execution of actions is time-consuming and costly in comparison to tasks in the digital domain. In addition, the integral property of exploration implies the possibility of executing suboptimal actions, which in real-world scenarios may be disproportionately expensive or safety critical.

One possibility of reducing the need for task-specific experience is given by the application of transfer learning [6, 7]. Transfer learning is based on the idea of using knowledge acquired from a source task in order to reduce the training efforts required to learn a target task. One obvious choice of a source task, when considering real-world RL, is simulations.

C. Scheiderer (✉) · T. Meisen
Institute of Technologies and Management of the Digital Transformation, University of Wuppertal, Wuppertal, Germany
e-mail: scheiderer@uni-wuppertal.de

N. Dorndorf
RWTH Aachen University, Aachen, Germany

© Springer Nature Switzerland AG 2021
H. R. Arabnia et al. (eds.), *Advances in Artificial Intelligence and Applied Cognitive Computing*, Transactions on Computational Science and Computational Intelligence, https://doi.org/10.1007/978-3-030-70296-0_13

In this chapter, we focus on domain randomization [8] as a method for transferring RL agents from a simulation to the real world for robotic applications. In domain randomization, variations are purposefully induced into the simulation environment. The goal hereby is to increase the robustness of a trained RL agent with respect to deviations between simulation and the real world. We consider a robotic task inspired by the work of [9, 10], in which a RL agent is trained end-to-end on the wire-loop game. The task is designed to be based on real motion planning scenarios for industrial processes such as welding or gluing. The goal of the game is to guide a loop over a wire without touching it. Our agents are based on convolutional neural networks (CNNs) and are able to observe the environment via a camera. The agents directly control the movement of the robot holding the loop. In addition to the real-world setup, we use a corresponding simulation in which parameters, such as colors and textures of loop, wire and background, are adjustable. The agents are pretrained in the simulation with domain randomization and subsequently transferred to the real-world task. In our experiments we vary the parameters chosen for randomization and observe effects on the agents' ability to transfer. We finally use Grad-CAM [11] to visualize the attention placed by the agents on various parts of an observed camera image in order to assess the robustness of an agent after pretraining in the simulation as well as its data-efficiency when adapting to the real-world task after transfer.

2 Related Work

2.1 Reinforcement Learning

As RL is thoroughly discussed in various other works [9, 12, 13] only a short overview of the key aspects is provided. In general, RL describes the concept of iteratively training an agent in a trial-and-error fashion on solving a task in an environment $\mathcal{E}$. At each discrete time step t, the agent observes the environment's state $s_t \in \mathcal{S}$. Based on the observation, the agent then chooses an action a_t according to its policy. The policy is given by a probability distribution over all actions given an observed state $\pi(a|s)$. In deep RL, the policy is represented by artificial neural networks. The execution of action a_t yields the reward r_t and the next state s_{t+1}, completing one iteration of the learning process. The overall goal of an agent is to find a policy which maximizes the discounted future reward $R_t = \sum_{i=0}^{\infty} \gamma^i r_{t+i+1}$, where the discount factor $\gamma \in [0, 1]$ adjusts the importance attributed to future rewards.

For tasks with continuous action spaces, actor–critic architectures have proven to be a suitable approach [14, 15]. They consist of two models, the actor and the critic, which are trained interdependently. On the one hand, the actor models the agent's policy and maps states onto actions. The critic on the other hand learns to evaluate and improve the actor's performance. Analogous to [9], we use deep deterministic policy gradient (DDPG) [15] to train actor–critic architectures, whereby both actor and critic are represented by artificial neural networks.

2.2 Simulation-to-Reality Transfer Learning

The aim of transfer learning lies in reusing knowledge acquired during solving one source task in order to efficiently learn a new target task [7]. Especially in computer vision, the benefit of transfer learning has successfully been shown for tasks such as object detection [16], pose estimation [17], or image segmentation [18].

When applying transfer learning to RL setups, the question about suitable source tasks arises. One intuitive possibility for enriching data from a real-world process is by using simulations. This option is attractive especially in industrial contexts, as obtaining experience from suboptimal process situations (e.g., low-quality products or slower production cycles) or critical conditions (e.g., high wear and tear of a tool) can be realized in virtual environments without consequences to the real-world production line.

The major challenge to overcome when using simulation data for pretraining an agent is the reality gap [19]. The reality gap describes the fact that simulations do not model the real world perfectly. Deviations arise, for example, from inadequately modeled parameters or lower resolution of the simulation environment. One strategy for overcoming the reality gap lies in an effort to improve the quality of a simulation, making the simulation as realistic as possible [20, 21]. This approach however assumes a deep understanding of underlying physical principles in the real-world process in order to be able to model them. In addition, increasing the quality of a simulation often requires extensive computation, which limits the applicability of the simulation.

A different strategy is aimed at finding a mapping from the distribution of simulation data onto the distribution of real-world data using generative adversarial networks (GANs) [22]. This mapping in turn can be used to enhance the realism of the artificial training samples. Alternatively, mappings of both simulation and real data in a shared feature space on which the agent is trained can mitigate the reality gap [23]. Finding such mappings however relies on the availability of sufficient real-world data.

In contrast to the aforementioned approaches, domain randomization [8] circumvents the need for both perfect simulations and real-world experience by purposefully randomizing visual (e.g., colors, textures) or even physical (e.g., friction) properties. This way the agent is exposed to a manifold of environments, which forces the agent to disregard the parameter variations. Although the majority of environments are not necessarily realistic, it has been shown by various works that agents are able to generalize and find robust control policies applicable to the real-world target task [8, 24–28]. While the successful application of domain randomization has been demonstrated before, the effects of different randomizations regarding the robustness and transferability of pretrained networks, to our knowledge, were not investigated to date.

2.3 Attention Maps

While artificial neural networks have proven to be very powerful tools, their lack of interpretability remains a key obstacle to the application in industrial environments. One possibility to gain insights into their decision-making process is to take a look at what areas of an input influence the final decision the most. This can be achieved by examining the gradient of an output with respect to the input [29] or the effects of perturbations to the input to the prediction [30, 31]. In this chapter we employ Grad-CAM [11], which was developed for CNNs. By calculating the gradient of an output with respect to the feature maps of convolutional layers, an importance weighting of the feature maps is determined. Using this weighting the feature maps are recombined to arrive at attention maps which highlight important regions in the input image and thusly promote transparency for the network's decision.

While previous works have covered the application of attention maps for RL [32], the application of attention maps for investigating the effects of domain randomization proposed in this chapter is, again to our knowledge, novel.

3 Experimental Setup

3.1 Learning Environment

Our use-case scenario is inspired by the work of Meyes et al. [9], in which a six-axis industrial robot is tasked with playing the wire-loop game. The loop is constructed out of a 3D-printed base to which two metal rods are mounted. By measuring the conductivity between the metal rods and the wire it is possible to determine whether the loop is touching the wire. The hardware setup is depicted in Fig. 1a.

At every time step, an agent observes the shape of the wire immediately in front of the loop in the form of a down-sampled image taken by a camera mounted to the loop (Fig. 1b). The agent subsequently has to decide for a continuous action $a_t \in \mathbb{R}^3$, which describes the longitudinal, lateral, and rotational movement of the loop. The game is considered lost in case the loop touches the wire or if the agent is not progressing along the wire. Conversely, the game is won once the agent successfully arrives at the end of the wire without touching it. We choose our reward function analogous to Meyes et al. [9], whereby the reward function is continuous between $[-1, 1]$ and incentivizes progress along the wire.

Our simulation is based on the simulation software V-REP [33] and mirrors the real-world setup (Fig. 1c–d). In addition to controlling the robot and detecting con-

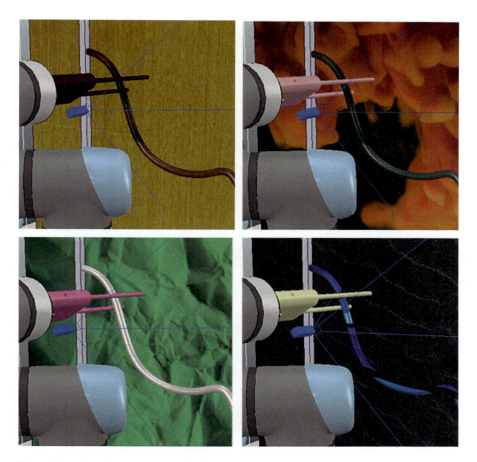

Fig. 1 Examples of randomized simulation parameters

tact between wire and loop, we can also dynamically change multiple environment parameters in order to conduct domain randomization:

- The *colors* of wire, background, and loop can be set to arbitrary RGB values
- In addition, 20 different *textures* can be chosen for the wire and the background
- The *camera pose* can be varied with a maximum displacement of 1.5 cm and a maximum rotation of $5°$ in every direction
- Random Gaussian noise the image taken by the camera, normalized between 0 and 1, can optionally be superimposed with random Gaussian noise with a standard deviation of 0.01.

As shown in Fig. 2, the variation of these environment parameters allows for the generation of highly diverse scenarios.

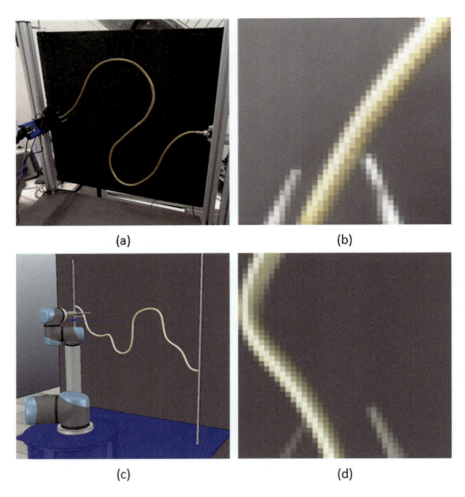

Fig. 2 The experimental setup consisting of a real (**a**) and a simulated (**c**) industrial robot. The environment is observed via real (**b**) and simulated (**d**) camera images

3.2 Agent Architecture

Analogous to [9], the agent comprises an actor and a critic network trained with DDPG and Adam as the optimizer. The actor network consists of three convolutional layers with 30, 15, and 10 filters and respective kernel sizes of 5×5, 5×5, and 3×3. After every convolutional layer, batch normalization is performed. MaxPooling is applied after the first two convolutional layers with a pooling size of 2×2. The CNN is followed by four densely connected layers with sizes of 400, 200, 100, and 3. ReLU is used as activation function after every layer, except for the last one. Instead, the last layer has one of its three neurons activated with a sigmoid function and the remaining two with tanh activations. The three outputs constitute the longitudinal and lateral movement as well as the rotation of the loop. Using the sigmoid activation for the longitudinal action results in the prevention of backward movement.

Similarly, the critic network employs a CNN of the same topology as the actor network. The output of the CNN as well as the action input of the critic are each processed by one respective densely connected layer of size 200. The two resulting outputs are concatenated and further passed through two densely connected layers of sizes 200, 100, and 1.

3.3 Design of Experiments

The method of domain randomization is applied to improve the generalization of the agent by artificially increasing the variability of the training data in the simulation. In our case we uniformly randomize the environment parameters *colors*, *textures*, *camera pose*, and *camera noise* for each training run. In order to investigate how the individual parameters influence the training and the success on the real robot, we consider six different configurations which randomize different combinations of the parameters available (Table 1). Parameters which are not randomized are kept at the value of the standard setup shown in Fig. 1c–d. The agents are tested after every 50th training iteration on the standard setup. Once the wire is traversed successfully training is stopped.

After pretraining the agents in simulation, we transfer them to the real-world setup, whereby the actor and critic neural networks including their optimizer states as well as the experience replay buffer are saved and transferred to maintain continuous training. We reset the exploration rate back to full exploration, which enables the agent to collect experience in the real world with high diversity. This procedure was chosen experimentally, as it resulted in the best performance of the agents after transfer. Training is continued until the agent is able to complete the

Table 1 Experiment configurations

Configuration	Parameter			
	Colors	Textures	Camera pose	Camera noise
1				
2	✓		✓	
3	✓		✓	✓
4	✓	✓	✓	
5	✓	✓		✓
6	✓	✓	✓	✓

Overview of parameters chosen for randomization during the respective training of the 6 agents

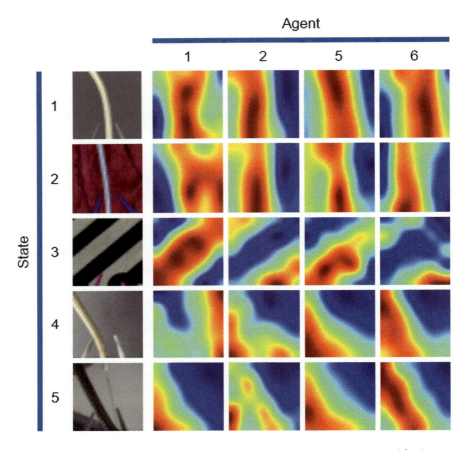

Fig. 3 In order to better understand an agent's behavior, attention maps are generated for 4 agents (1, 2, 5, 6) using different states from simulation (states 1–3) and the real world (4 & 5). The states vary in complexity with regard to colors and textures. A rainbow-coloring scheme is used, which spans from red (high attention) to blue (low attention)

real-world task as well. Analogous to the simulation, the performance is tested after every 50th training iteration.

Finally, we take the pretrained actor networks and generate attention maps for different input images using Grad-CAM. A variety of images are chosen from both the simulation and the real world (see Fig. 3, left column). In combination with the behavior observed during training the attention maps are qualitatively examined and conclusions regarding different randomization strategies are drawn.

Table 2 Required training steps until game finished

Agent	Simulation (pretraining)	Real-world (transfer)
1	*400*	600
2	1500	250
3	1900	250
4	2400	*0*
5	700	350
6	2850	*0*

4 Results

In the following we compare and discuss the training results using the above-mentioned experimental configurations. Table 2 lists the amounts of training steps required to learn the wire loop game in simulation using different randomization strategies, as well as the training steps required once the respective agents are transferred to and retrained on the real-world setup.

4.1 Training in Simulation

Intuitively, the training effort increases with a growing complexity of the observations associated with more randomization. The amount of training steps increases by a factor of 7 when applying full-domain randomization (agent 6) in comparison to no-domain randomization (agent 1). Especially changes in the camera position and orientation have a large influence on the learning process, as the number of training iterations increases only slightly with a fixed camera (agent 5) in contrast to the other experimental setups where it is randomized. Textures (agent 4) seem to have a stronger impact on the training than noise (agent 3). One possible explanation is that even with noise the detection of the wire by color is still fairly straightforward, whereas changing textures are more likely to confuse the neural network.

4.2 Transfer to Real World

The results of transferring the pretrained agents to the real world are also reported in Table 2. Two agents were able to complete the task immediately: the agent trained with full domain randomization (agent 6) and the agent trained with domain randomization but without noise (agent 4). This confirms the ability of domain randomization toward generating robust agents solely in simulation. Additionally, the previously observed minor impact of noise randomization is further substantiated.

All agents trained with some form of domain randomization in the simulation require less additional training in the real world than the agent trained without domain randomization. We also observe a strong negative correlation [$r(4) = -.94$, $p = 0.005$] between the required training steps in simulation and the real world. This intuitively translates into a positive effect of pretraining in more complex environments, and thusly increasing robustness, for solving the target task.

The added Gaussian noise does not seem to play a major role in the transfer of policies to the real robot, as the agents trained without additional noise in the simulation perform as well in the real world as their counterparts trained without noise. Randomization of the camera pose and textures, however, have a big effect on the transferability. These findings are plausible, as the mechanical connections between camera and tool as well as tool and robot experience some mechanical play in the real-world setup. Additionally, a camera image taken in the real world will always exhibit some sort of texture compared to the perfectly simulated images.

4.3 Attention Maps

As mentioned above, we apply Grad-CAM to the policies (actor networks) of our agents after pretraining in simulation. A selection of resulting attention maps is shown in Fig. 3. The states 1–3 are taken from simulation, and states 4 and 5 from the real world. To obtain state 5 we removed the black cardboard background from the demonstrator to increase complexity of the input image.

Each agent is pretrained in simulation. Afterwards training is continued on the real robot. Each training is run until the agent is able to complete the game. The enumeration coincides with Table 1 (Configuration n → Agent n).

In general, we expect a well-performing agent to focus on the wire and disregard the background, as identifying the wire is essential to solving the task. This behavior can be observed for state 1, for which the differences in attention maps for the agents are only marginal. This is plausible as the standard setup is part of the state space for all randomization configurations.

For state 2 we observe a significant degradation of the ability of agent 1 to locate the wire due to the addition of texture, which the agent did not need to account for during its training process. Interestingly enough, agent 2 correctly identifies the wire, even though it was not exposed to texture randomization. This indicates that the variation of camera poses reinforces the agent to learn a more complete concept of the wire.

Detecting the wire in state 3 in the lower right corner is a hard task even for humans, as the dark wire disappears in front of the dark texture stripes. As before, agent 1 clearly struggles with the texture. In addition, agent 5 is also completely distracted by the dark stripe. While agent 2 pays attention to the wire, it also falsely focuses on the upper left corner. The superior attention of agent 2 compared to agent 5 is unexpected considering the training configurations, but again displays that agent 2 seems to have learned a robust representation of the wire due to the variation of the camera pose. Agent 6 exhibits the best performance for state 3, highlighting the benefit of combining parameter randomizations to increase robustness.

Even though state 4 is very similar to state 1 from a human perspective, agent 1 is distracted by a gradient induced by the lighting. While agents 4 and 5 pay attention in the general area of the wire, they seem to be drawn more to the lower left corner instead of the wire itself. Agent 6 again is the only agent which clearly identifies the wire.

Similar observations can be found for state 5, where agents 1, 2, and 5 focus mostly on the sharp background structure in the lower left corner. Although very weakly, agent 2 at least pays some attention along the wire, again showing the positive effect of camera-pose randomization. As before, agent 6 demonstrates its ability to robustly localize the wire despite complex conditions.

Revisiting the agents' abilities to transfer to the real world, a correlation between the transfer success and the interpretation of attention maps becomes evident. When ranking the agents' abilities to focus on the wire under the discussed conditions, full randomization (agent 6) clearly resulted in the most robust policy. It is also apparent that without any randomizations (agent 1) the worst performance is achieved. Randomizing the camera pose (agent 2) arguably has a higher benefit than randomizing textures (agent 5), which becomes especially clear for states 3 and 5. As observed before, applying random Gaussian noise to the camera image does not have any significant effect.

We thus draw the conclusion that in addition to allowing interpretation of the decision-making process of artificial neural networks, attention maps are a valuable tool to assess the prospects of success of simulation-to-reality transfer learning. For vision-based tasks in particular our results suggest that randomization of the camera pose results in the highest amount of robustness, whereas the addition of random Gaussian noise has no effect on the agent's ability to focus its attention.

4.4 Summary and Outlook

In this chapter we examined domain randomization for pretraining RL agents in a simulation in order to decrease training effort in the real world. We investigated the effects of choosing different parameters for randomization on the transferability of the respective agents. In our experiments we found that domain randomization in any case improves the training efficiency in the real world. We also discovered a strong negative correlation between training effort in simulation and the real world under varying randomization configurations. We attribute this finding to a higher complexity in simulation leading to more robust agents leading to faster training in the real world. In addition, we used Grad-CAM to generate attention maps for pretrained agents and discovered a correlation between the qualitative interpretation of such attention maps and the exhibited performance of the agents after transferring them to the real world.

Going forward, the applicability of attention maps as means to predict the success of transfer learning requires further investigation. Especially for scenarios where exploration in the target domain is very costly, attention maps may help to guide refinement of agent robustness prior to transfer.

Although simulations run significantly faster than the respective real-world process, the training time of an agent still increases with an increasing number of parameters chosen for randomization. In addition to optimizing the actual transfer, attention maps might also prove useful to decrease training time in simulation. As demonstrated, attention maps hold information regarding which kinds of scenarios the agent has already learned and more importantly, which ones it is still struggling with. It is thus conceivable to use this information to purposefully select parameter variations which will advance the agent's robustness, instead of the current random selection. Inspired by GANs, this could potentially be achieved and automated using an adversarial agent which is rewarded for identifying scenarios beneficial for training the former agent.

References

1. D. Silver et al., Mastering the game of go without human knowledge. Nature **550**, 354–359 (2017)
2. O. Vinyals et al., StarCraft II: A new challenge for reinforcement learning. ArXiv, abs/1708.04782 (2017)
3. H. Mao, M. Alizadeh, I. Menache, S. Kandula, Resource management with deep reinforcement learning, in *Proceedings of the 15th ACM Workshop on Hot Topics in Networks*, (2016), pp. 50–56
4. S. Gu, E. Holly, T. Lillicrap, S. Levine, Deep reinforcement learning for robotic manipulation with asynchronous off-policy updates, in *2017 IEEE International Conference on Robotics and Automation (ICRA)*, (2017), pp. 3389–3396
5. A.S. Polydoros, L. Nalpantidis, Survey of model-based reinforcement learning: Applications on robotics. J. Intell. Robot. Syst. **86**(2), 153–173 (2017)
6. J. Lu, V. Behbood, P. Hao, H. Zuo, S. Xue, G. Zhang, Transfer learning using computational intelligence: A survey. Knowl.-Based Syst. **80**, 14–23 (2015). https://doi.org/10.1016/j.knosys.2015.01.010
7. K. Weiss, T.M. Khoshgoftaar, D. Wang, A survey of transfer learning. J. Big Data **3**(1), 9 (2016). https://doi.org/10.1186/s40537-016-0043-6
8. J. Tobin, R. Fong, A. Ray, J. Schneider, W. Zaremba, P. Abbeel, Domain randomization for transferring deep neural networks from simulation to the real world, in *2017 IEEE/RSJ International Conference on Intelligent Robots and Systems (IROS)*, (2017), pp. 23–30
9. R. Meyes, C. Scheiderer, T. Meisen, Continuous motion planning for industrial robots based on direct sensory input. Procedia CIRP **72**, 291–296 (2018)
10. C. Scheiderer, T. Thun, T. Meisen, Bézier curve based continuous and smooth motion planning for self-learning industrial robots. Procedia Manuf. **38**, 423–430 (2019)
11. R. Selvaraju, A. Das, R. Vedantam, M. Cogswell, D. Parikh, D. Batra, Grad-CAM: Why did you say that? Visual explanations from deep networks via gradient-based localization. ArXiv, abs/1610.02391 (2016)
12. R.S. Sutton, A.G. Barto, *Reinforcement learning: An introduction* (MIT Press, 2018)
13. V. Mnih et al., Human-level control through deep reinforcement learning. Nature **518**(7540), 529 (2015)
14. T. Haarnoja et al., Soft actor-critic algorithms and applications. arXiv preprint arXiv:1812.05905 (2018)

Effects of Domain Randomization on Simulation-to-Reality Transfer...

15. T. Lillicrap et al., Continuous control with deep reinforcement learning. CoRR, abs/1509.02971 (2015)
16. R. Girshick, J. Donahue, T. Darrell, J. Malik, Rich feature hierarchies for accurate object detection and semantic segmentation, in *Proceedings of the IEEE Conference on Computer Vision and Pattern Recognition*, (2014), pp. 580–587
17. J. Carreira, P. Agrawal, K. Fragkiadaki, J. Malik, Human pose estimation with iterative error feedback, in *Proceedings of the IEEE Conference on Computer Vision and Pattern Recognition*, (2016), pp. 4733–4742
18. J. Dai, K. He, J. Sun, Instance-aware semantic segmentation via multi-task network cascades, in *Proceedings of the IEEE Conference on Computer Vision and Pattern Recognition*, (2016), pp. 3150–3158
19. N. Jakobi, P. Husbands, I. Harvey, Noise and the reality gap: The use of simulation in evolutionary robotics, in *ECAL*, (1995) This file was created with Citavi 6.2.0.12
20. S. James, E. Johns, 3D simulation for robot arm control with deep Q-learning. ArXiv, abs/1609.03759 (2016)
21. B. Planche et al., DepthSynth: Real-time realistic synthetic data generation from CAD models for 2.5D recognition, in *2017 International Conference on 3D Vision (3DV)*, (2017), pp. 1–10
22. K. Bousmalis et al., Using simulation and domain adaptation to improve efficiency of deep robotic grasping, in *2018 IEEE International Conference on Robotics and Automation (ICRA)*, (2017), pp. 4243–4250
23. A. Gupta, C. Devin, Y. Liu, P. Abbeel, S. Levine, Learning invariant feature spaces to transfer skills with reinforcement learning. arXiv preprint arXiv:1703.02949 (2017)
24. O.M. Andrychowicz et al., Learning dexterous in-hand manipulation. Int. J. Robot. Res. **39**(1), 3–20 (2020)
25. S. James et al., Sim-to-real via sim-to-sim: Data-efficient robotic grasping via randomized-to-canonical adaptation networks. ArXiv, abs/1812.07252 (2018)
26. X. Peng, M. Andrychowicz, W. Zaremba, P. Abbeel, Sim-to-real transfer of robotic control with dynamics randomization, in *2018 IEEE International Conference on Robotics and Automation (ICRA)*, (2017), pp. 1–8
27. J. Tremblay et al., Training deep networks with synthetic data: Bridging the reality gap by domain randomization, in *Proceedings of the IEEE Conference on Computer Vision and Pattern Recognition Workshops*, (2018), pp. 969–977
28. S. James, A.J. Davison, E. Johns, Transferring end-to-end visuomotor control from simulation to real world for a multi-stage task. arXiv preprint arXiv:1707.02267 (2017)
29. K. Simonyan, A. Vedaldi, A. Zisserman, Deep inside convolutional networks: Visualising image classification models and saliency maps. CoRR, abs/1312.6034 (2013)
30. M.D. Zeiler, R. Fergus, Visualizing and understanding convolutional networks, in *European Conference on Computer Vision*, (2014), pp. 818–833
31. R.C. Fong, A. Vedaldi, Interpretable explanations of black boxes by meaningful perturbation, in *Proceedings of the IEEE International Conference on Computer Vision*, (2017), pp. 3429–3437
32. L. Weitkamp, E. van der Pol, Z. Akata, Visual rationalizations in deep reinforcement learning for atari games, in *Benelux Conference on Artificial Intelligence*, (2018), pp. 151–165
33. E. Rohmer, S.P.N. Singh, M. Freese, CoppeliaSim (formerly V-REP): A versatile and scalable robot simulation framework, in *Proceedings of the International Conference on Intelligent Robots and Systems (IROS)*, (2013)

Human Motion Recognition Using Zero-Shot Learning

Farid Ghareh Mohammadi, Ahmed Imteaj, M. Hadi Amini, and Hamid R. Arabnia

1 Introduction

Motivation Motion/activity recognition is considered to be a challenging task as there is a greater need to recognize more human activity states, and therefore, it is becoming difficult to prepare a dataset that covers all sample states of motion and activity [1]. To alleviate such a situation, researchers propose applying zero-shot learning (ZSL), which constructs a mapping function from available features to semantic information and ensures the recognition of human activity states in spite of having unseen and unlabeled test data. In prior works, scientists use ZSL to assess a projection function that takes features as well as semantic space for prediction and conjuring classification. Prior models, however, are not capable of predicting when the test data contains unseen and unlabeled classes, and hence, researchers fail to classify unseen and unlabeled classes due to the project domain shift problem. This problem occurs when seen and labeled classes, and unseen and unlabeled classes vary from each other significantly, so the projection function cannot recognize the proper categories for unseen and unlabeled classes.

Kodirov et al. [2] presented an encoder–decoder paradigm concept for ZSL by leveraging semantic information and learning through encoder and decoder as one possible solution to this problem. Their proposed model can be used as

---

F. G. Mohammadi (✉) · H. R. Arabnia
Department of Computer Science, University of Georgia, Athens, GA, Georgia
e-mail: farid.ghm@uga.edu; hra@uga.edu

A. Imteaj · M. H. Amini
School of Computing and Information Sciences, Florida International University, Miami, FL, USA
e-mail: aimte001@fiu.edu; amini@cs.fiu.edu

© Springer Nature Switzerland AG 2021
H. R. Arabnia et al. (eds.), *Advances in Artificial Intelligence and Applied Cognitive Computing*, Transactions on Computational Science and Computational Intelligence, https://doi.org/10.1007/978-3-030-70296-0_14

under-complete auto-encoder, i.e., data structure and data visualization can be learned through this model, and it is capable of regenerating the input signal. They considered an auxiliary constraint for the decoder, which implies that the decoder must be able to regenerate the original output of the visual feature. Semantic information is the bridge between encoder and decoder. This further helped them accelerate the performance of their algorithm.

Researchers proposed another sample solution to the problem in [3] showing zero-shot emotion recognition from speech samples, where the samples are considered to be extracted from emotion states that have not been seen before. They applied attribute learning for making an association with each emotion attribute, captured paralinguistic features from the human speech by comparing with known emotions, and finally, used label learning to predict the unknown emotional state. Also, xu et al. [4] proposed a video action recognition technique using a self-training and data aggregation approach based on ZSL for improving the performance of mapping between the video action features and semantic space. To devise this, they projected their class labels to the semantic embedding space, learned to map visual attributes to the semantic space, and aggregated their training and auxiliary data to attain zero-shot learning.

Another ZSL solution considers purchasing unseen and unlabeled product by learning from seen and labeled product. The social learning studies [5, 6] showed human behavior learning through purchasing a product by learning from prior consumer knowledge about the product. These studies showed the relationship between the willingness of consumers to buy a product and reviews about the product quality history. These studies leveraged statistics, especially Bayesian theory, to help ZSL recognize the unlabeled and unseen products.

As seen in the aforementioned examples, zero-shot learning plays a pivotal role in recognizing unseen and unlabeled objects and overcoming a limitation of machine learning. ZSL provides a projection function to recognize the unseen and unlabeled classes. The crux of ZSL is to learn a visual-semantic function and represent objects semantically. To that end, researchers have proposed complex projection functions, which lead to a higher risk of overfitting and accordingly yield poor performance on the unseen and unlabeled classes. However, on the other hand, simple linear functions obtain poor classification accuracy on the seen classes and do not perform accurately on the unseen and unlabeled classes. To contribute to the field and leverage published recent works and extend them properly for work re-usability and meeting research life cycle as discussed [7], in this paper, we address this linear function problem by using semantic auto-encoder zsl [2] and extending our work [8] to recognize unseen and unlabeled human motions by learning seen human motions.

Contributions The main contribution of this paper lies in two steps.

⋄ ***First:*** We develop and tune SAE parameters in a way that improves the performance. We tune λ, which is the only external parameter, and embedded parameters such as HITK and DIST. After examining and experimenting with different parameters, we find that only HITK has a positive and direct impact on the accuracy of the proposed method.

⋄ **Second:** We further investigate our work [8] and extend it to apply to a completely new area of human motion detection.

Organization The rest of this paper is organized as follows. Section 2 elaborates more on previous works in two major categories: supervised and unsupervised learning. In Sect. 3, we explain details of our novel framework to find the optimal projection matrix and to recognize unseen and unlabeled human motion and action. Section 4 is devoted to experiment results, followed by Sect. 5, which concludes the paper.

2 Related Work

Machine learning algorithms leverage the inductive learning process to learn from training data to identify a pattern, which refers to a model to cover whole sample space thoroughly. Inductive learning rigorously learns from only labeled and seen classes. It fails when the number of labeled seen classes are not captured within training datasets. With this, we have a low-performance model to evaluate testing datasets. The learning process occurs in two different ways: supervised learning and unsupervised learning. Here, we present a wide range of research studies on the applications of ZSL [9] and discuss them within their associated learning groups for clarity in method comparison.

2.1 Supervised Learning

Supervised learning occurs if the given input dataset provides a target value for each instance. This means that the learning model maps instances to the associated target value. This mapping function refers to a model or pattern, then this model aims to predict the target value on testing datasets. For certain, researchers [10–13] have designed a human activity recognition model using mobile embedded sensors and capitalizing fixed points arithmetic of support vector machine (SVM). All these works used mobile sensor data for predicting human activity and motion, but may face difficulty when test activities go from known to unknown. As human activity can be of many forms and patterns even within a particular activity, it is likely that the model will be evaluated on unspecified test data. In such a situation, a conventional supervised learning model is likely to predict a wrong category which will downgrade the system performance accuracy.

The author in [14] studied a similar problem within video classification by applying natural language processing techniques to extract textual features from an event description. They projected the extracted features on a high-dimensional space by applying text expansion to compute the similarity matrix, but they did not perform any semantic similarities or affinities analysis in their model. Due to lack of semantic analysis, the bridging section of the auto-encoder will not work causing low performance.

2.2 Unsupervised Learning

Vaha et al. [15] used hip-worn accelerometers for the recognition of lying, sitting, and standing motions. Ghareh Mohammadi and Amini [16] proposed meta-sense, which learns to sense objects, human beings, and the ambient environment leveraging device-free human sensing. A regularized sparse coding-based unsupervised zero-shot learning model is designed in [17] that formulates a projection function by learning through labeled source data along with unlabeled test data. The authors in [18] focused on the drawback of traditional ZSL where the association among the attributes needs to be defined manually. In this work [18], they proposed an automatic association prediction between the attributes within the dataset and unseen classes for an unsupervised learning problem.

A transfer learning approach based on semantic labeling is proposed [19], where they perform zero-shot learning without any attribute and positive exemplars. In their proposed approach, they used a concept detector which was trained with seen and labeled classes of videos. They extracted information weights through identifying a semantic correlation within input text data to build a projection function within an action classifier. But, they did not use any encoder scheme to carry out an efficient learning process to generate an optimum projection function. Zhang et al. [20] presented a text mining algorithm using natural language processing (NLP) that helps to compute a correlation among words to generate values, referred to as semantic information, for the human motions/actions like sitting, walking, and laying down. They introduced a zero-shot late-fusion approach that exploits the co-occurrence of verbs and nouns rather than semantic correlations. This method works better than the research discussed in [19], when a sufficient training set is available, but it fails to recognize actions when there exist significant performance gaps between noun/verb features.

Qin et al. [21] proposed zero-shot learning using an error-correcting output, where the model is trained using indispensable data structures and category level semantics based on seen categories. They execute the domain shift of knowledge by transferring the correlations captured from seen categories of data to unseen ones by adapting a semantic transfer strategy. Moreover, Liu et al. [22] presented a work to discover semantic attributes for a given classification task. They tried to reduce feature dimension during discriminative selection and performed prediction considering attribute reliability and discriminative power. But, they did not consider semantic drift as a reliability property which could help them with the approximation.

2.3 Auto-Encoder

Auto-Encoder is another technique using machine learning algorithms to yield promising results. Kodirov et al. [17] focused on solving the domain shift problem using auto-Encoder. To achieve this, they considered projection learning as a regularized sparse coding-based, unsupervised problem. For the learning projection function, they used the regularization function which controls transformation strength from source to target domain and fixes the domain shift issue. Their domain shift problem requires access to all the test data which mean that ZSL is not a viable option. To improve the learning projection function, Lin et al. [23] proposed a label auto-encoder to enable multi-semantic synthesis in ZSL. They tried to fuse latent relationship and interaction between semantic and feature space after manifolding process is done. But, their approach ignores the negative structure influence during constructing model. So, we needed an optimal and linear projection function which is robust enough tackling any negative structures of data.

Schonfeld et al. [24] propose a variational auto-encoders based modality-specific model to learn feature attributes and the related class embedding of images. Using the learned distributions and the side information associated with those images, they constructed latent features containing the useful multi-modal knowledge that identify unseen classes. But, their proposed approach does not work well if the domain gap between seen domain and unseen domain class is large.

A semantic auto-encoder framework is designed in [25] to obtain a low-ranked mapping between features and their semantic representations which would be beneficial for data construction. They adapt an encoder–decoder scheme, where the encoder learns projection from a visual feature space to related semantic space and the decoder restores the previous data using the learned mapping. But they did not consider intrinsic mapping between visual features and semantic space.

3 Proposed Method

When it comes to prediction with traditional machine learning algorithms, they work best on seen and labeled classes, but do not work with unseen and unlabeled classes. This prediction is done based on an inductive learning process. This inductive process extracts some rules and functions from training data and does the generalization. However, this process still does not work for unseen and unlabeled classes. In this paper, we propose an algorithm based on a transductive learning process to extend the Zero-Shot learning application in human motion recognition. The transductive process extracts rules and turns them into a specific function for certain unseen and unlabeled classes. Therefore, we extend our past work [8], which is an advanced zero-shot learning for image classification, in order to apply to this human motion classification problem. Our goal is to learn from seen and labeled classes such as motions and activities, such as sitting, laying, standing, and walking, and to recognize unseen and unlabeled classes like walking up and down the stairs.

The main difference between the proposed method and the traditional machine learning algorithms is its linear projection function that yields promising results on unseen and unlabeled classes.

3.1 Preliminary

To understand the idea of zero-shot learning, let us suppose we have a distribution of seen and labeled classes, $(\mathscr{D})=(\mathscr{X}, \mathscr{Y})$ where $\mathscr{Y}$ goes for the sets of seen classes. It is crucial to identify a projection function which takes $(\mathscr{D})$, then map seen and labeled classes to semantic space in an optimized way that helps us to use the projection function to map unseen and unlabeled classes in semantic space to test data $(\mathscr{D}')=(\mathscr{X}', \mathscr{Y}')$, where $\mathscr{Y}'$ goes for unseen and unlabeled classes and $\mathscr{Y}' \wedge \mathscr{Y} = \emptyset$. The main difference between zero-shot learning and machine learning is the input data and the way they learn in training phases. Machine learning learns from a training dataset, which covers all classes in a test dataset. Moreover, machine learning extracts some numbers from data and then finds the best possible pattern based on the data extracted. However, zero-shot learning only takes a dataset and includes a semantic space where the features accept binary values either 0 or 1. 1 means that a specific sample has this feature and 0 means that the sample does not have this option. Furthermore, zero-shot learning behaves like the human cognitive system (i.e., zero-shot learning understands data like a human enabling us to recognize unseen objects). Figure 1 presents a clear presentation of how the projection function works to map unseen motions to semantic space.

3.2 Semantic Auto-Encoder Adaptation on Human Motion Recognition

Figure 2 represents three main sections of our proposed method. One is preparing a dataset which plays an important role in zero-shot learning. We have to follow pre-processing techniques to convert the dataset to a standard dataset, which then enables us to apply our proposed method. The raw dataset has only training and testing datasets. Each of them includes 6 seen and labeled classes such as 6 types of motions and actions. Machine learning algorithms try to learn all 6 types and predict the same types while examining the test dataset. However, in this paper, we only divide our dataset into two categories. One is seen and labeled classes and the other one is unseen and unlabeled classes. Seen classes include 4 types of actions including standing, walking, laying, and sitting, while Unseen classes consist of walking downstairs and walking upstairs.

The second section stands for seeking an optimal projection matrix or function which learns from seen actions and motions. The strength of our method is a linear

Human Motion Recognition Using Zero-Shot Learning

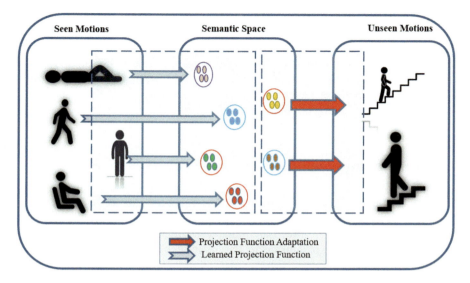

Fig. 1 Projection function for human motion recognition. *Note 1.* Seen motions: seen and labeled classes. Unseen motions: unseen and unlabeled classes

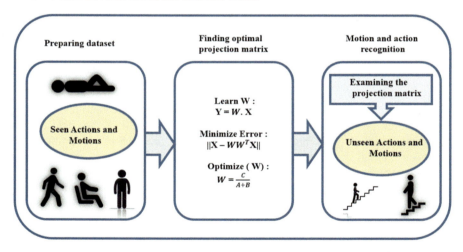

Fig. 2 Proposed framework for human motion recognition. *Note.* Seen Motions: seen and labeled classes. Unseen Motions: unseen and unlabeled classes

projection function which provides minimum error when mapping to a semantic space. Furthermore, we decrease the time complexity of the learning model by only using a linear function rather than a complex model, which may lead to overfitting. Thus, this step prevents us from getting stuck in a local minimum and having an overfitting problem. We use the semantic auto-encoder (SAE) [2] as zero-shot learning to generate the projection function. Kodirov et al. [2] used Sylvester equation to compute an optimal projection function, $\mathcal{W}$, using three parameters A $= SS^T$, B$= \lambda XX^T$, and C$= (1 + \lambda)SX^T$.

The third section is an important phase where we apply the projected function on unseen motions and actions to yield a better result. We use the same projected function, where we embed semantic information for unseen motions and actions. Technically, the nature of the auto-encoder is to transform a training dataset to a semantic space with a projection function $\mathcal{W}$.

3.3 Tuning Projection Functions for Semantic Auto-Encoder

To enhance the performance of the projection functions, semantic auto-encoder functions, we need to update the parameters. In our recent work [8], we have shown that by tuning parameters, unseen and unlabeled images can be classified. In this paper, we propose to tune parameters to classify unseen and unlabeled motions. Our proposed method has one general parameter,λ, and different embedded parameters such as HITK, DIST, etc. We examine the generated dataset and proposed method once with a default value of embedded parameters and then with tuned parameters. When we set the value of HITK equal to 2, the result of the proposed methods reaches 100% performance on unseen and unlabeled motion and action detection.

4 Experimental Result

We develop a zero-shot learning method leveraging semantic auto-encoder [2]. Furthermore, we tune one of the embedding parameters, HITK, to yield a high performance result. This process of applying ZSL to motion recognition and parameter tuning proves that our algorithm on human motion recognition has adapted properly.

4.1 Dataset

In this paper, we use a dataset,[1] which includes 561 features and 6 labels. This dataset was generated by experiments, which have been carried out with a group of 30 people ranging from 19 to 48 years old. Every individual performed six different

---

[1]UCI dataset for human action recognition: https://www.kaggle.com/uciml/human-activity-recognition-with-smartphones: access date: Oct. 30, 2019.

motions and activities, such as standing, sitting, laying, walking, walking upstairs, and walking downstairs, with a smartphone placed on the waist. The researchers used the smartphone's embedded accelerometer and gyroscope to capture one 3-axial linear acceleration and one 3-axial angular velocity at a static rate of 50 Hz. The dataset is manually generated through recording activities, then is labeled. To obtain the dataset, 70 and 30% of the people were selected for generating the training and the test data, respectively.

In order to apply ZSL to this dataset, we need to convert it to seen and unseen classes as discussed in the proposed method section. In zero-shot learning, we do not consider a training and testing dataset. Instead, we consider seen and labeled classes consisting of standing, sitting, laying, and walking, and unseen and unlabeled classes including walking upstairs and walking downstairs. Zero-shot learning takes seen classes to generate an optimal projection matrix or function to be able to recognize the unseen classes.

4.2 Supervised Learning Results

In [10–13], the researchers have used this dataset to apply their method to detect human motion and action. Table 1 presents their results, which show that their proposed supervised learning algorithms work when adapted to this dataset.

4.3 Unsupervised Learning

The main contribution of this paper is to propose a new unsupervised learning in order to classify unseen objects. We propose and adapt zero-shot learning to recognize human motions and actions. Table 2 states the results of the proposed method. Moreover, our tuned method outperforms related works and other state-of-the-work algorithms in comparison with Table 1.

Table 1 Comparing related works (supervised learning method)

Methods	Criteria	
	Classifier	Accuracy (%)
[10]	Multi class (MC) SVM	89
[11]	$MC - HF$ SVM	89
[10]	$MC - HF$ SVM	89
[12]	Binary SVM	96

Parameter	Criteria	
	Classifier	Accuracy (%)
HITK=1	Zero-shot learning	47
HITK=2	Tuned zero-shot learning	100

Table 2 Unsupervised learning method

5   Discussion and Conclusion

Recognition of human motion is a challenging issue in human sensing. Research studies on human motion detection and recognition have proliferated yearly representing the extent of problems in the field of human interaction computing. In this paper, we propose a semantic auto-encoder as zero-shot learning to recognize unseen human motions. While traditional machine learning algorithms fail to recognize the unseen and unlabeled human motions, we are able to overcome this limitation leveraging semantic information and zero-shot learning. The results state that zero-shot learning that leverages semantic auto-encoder provides a promising performance while outperforming other state-of-the-art works.

References

1. F. Baradel, C. Wolf, J. Mille, G.W. Taylor, Glimpse clouds: human activity recognition from unstructured feature points, in *Proceedings of the IEEE Conference on Computer Vision and Pattern Recognition* (2018), pp. 469–478
2. E. Kodirov, T. Xiang, S. Gong, Semantic autoencoder for zero-shot learning, in *Proceedings of the IEEE Conference on Computer Vision and Pattern Recognition* (2017), pp. 3174–3183
3. X. Xu, J. Deng, N. Cummins, Z. Zhang, L. Zhao, B. Schuller, Autonomous emotion learning in speech: a view of zero-shot speech emotion recognition, in *Proceedings of the Interspeech 2019* (2019), pp. 949–953
4. X. Xu, T. Hospedales, S. Gong, Semantic embedding space for zero-shot action recognition, in *2015 IEEE International Conference on Image Processing (ICIP)* (IEEE, Piscataway, 2015), pp. 63–67
5. B. Ifrach, C. Maglaras, M. Scarsini, A. Zseleva, Bayesian social learning from consumer reviews. Oper. Res. **67**, 2–4, 1209–1502 (2019)
6. I. Chaturvedi, E. Ragusa, P. Gastaldo, R. Zunino, E. Cambria, Bayesian network based extreme learning machine for subjectivity detection. J. Franklin Inst. **355**(4), 1780–1797 (2018)
7. F. Shenavarmasouleh, H. Arabnia, Causes of misleading statistics and research results irreproducibility: a concise review, in *2019 International Conference on Computational Science and Computational Intelligence (CSCI)* (2019), pp. 465–470
8. F.G. Mohammadi, H.R. Arabnia, M.H. Amini, On parameter tuning in meta-learning for computer vision, in *2019 International Conference on Computational Science and Computational Intelligence (CSCI)* (IEEE, Piscataway, 2019), pp. 300–305
9. F.G. Mohammadi, M.H. Amini, H.R. Arabnia, An introduction to advanced machine learning: meta-learning algorithms, applications, and promises, in *Optimization, Learning, and Control for Interdependent Complex Networks* (Springer, Berlin, 2020), pp. 129–144
10. D. Anguita, A. Ghio, L. Oneto, X. Parra, J.L. Reyes-Ortiz, Human activity recognition on smartphones using a multiclass hardware-friendly support vector machine, in *International Workshop on Ambient Assisted Living* (Springer, Berlin, 2012), pp. 216–223

11. D. Anguita, A. Ghio, L. Oneto, X. Parra, J.L. Reyes-Ortiz, Energy efficient smartphone-based activity recognition using fixed-point arithmetic. J. Univ. Comput. Sci. **19**(9), 1295–1314 (2013)
12. D. Anguita, A. Ghio, L. Oneto, X. Parra, J.L. Reyes-Ortiz, A public domain dataset for human activity recognition using smartphones, in *European Symposium on Artificial Neural Networks, Computational Intelligence and Machine Learning. BrugesEsann* (2013)
13. J.L. Reyes-Ortiz, A. Ghio, X. Parra, D. Anguita, J. Cabestany, A. Catala, Human activity and motion disorder recognition: towards smarter interactive cognitive environments, in *European Symposium on Artificial Neural Networks, Computational Intelligence and Machine Learning. Bruges ESANN* (Citeseer, 2013)
14. S. Wu, S. Bondugula, F. Luisier, X. Zhuang, P. Natarajan, Zero-shot event detection using multi-modal fusion of weakly supervised concepts, in *Proceedings of the IEEE Conference on Computer Vision and Pattern Recognition* (2014), pp. 2665–2672
15. H. Vähä-Ypyä, P. Husu, J. Suni, T. Vasankari, H. Sievänen, Reliable recognition of lying, sitting, and standing with a hip-worn accelerometer. Scand. J. Med. Sci. Sports **28**(3), 1092–1102 (2018)
16. F.G. Mohammadi, M.H. Amini, Promises of meta-learning for device-free human sensing: learn to sense, in *Proceedings of the 1st ACM International Workshop on Device-Free Human Sensing, DFHS'19*, (ACM, New York, 2019), pp. 44–47
17. E. Kodirov, T. Xiang, Z. Fu, S. Gong, Unsupervised domain adaptation for zero-shot learning, in *Proceedings of the IEEE International Conference on Computer Vision* (2015), pp. 2452–2460
18. Z. Al-Halah, M. Tapaswi, R. Stiefelhagen, Recovering the missing link: predicting class-attribute associations for unsupervised zero-shot learning, in *Proceedings of the IEEE Conference on Computer Vision and Pattern Recognition* (2016), pp. 5975–5984
19. C. Gan, M. Lin, Y. Yang, Y. Zhuang, A.G. Hauptmann, Exploring semantic inter-class relationships (sir) for zero-shot action recognition, in *Twenty-Ninth AAAI Conference on Artificial Intelligence* (2015)
20. Y.C. Zhang, Y. Li, J.M. Rehg, First-person action decomposition and zero-shot learning, in *2017 IEEE Winter Conference on Applications of Computer Vision (WACV)* (IEEE, Piscataway, 2017), pp. 121–129
21. J. Qin, L. Liu, L. Shao, F. Shen, B. Ni, J. Chen, Y. Wang, Zero-shot action recognition with error-correcting output codes, in *Proceedings of the IEEE Conference on Computer Vision and Pattern Recognition* (2017), pp. 2833–2842
22. L. Liu, A. Wiliem, S. Chen, B.C. Lovell, Automatic image attribute selection for zero-shot learning of object categories, in *2014 22nd International Conference on Pattern Recognition* (IEEE, Piscataway, 2014), pp. 2619–2624
23. G. Lin, C. Fan, W. Chen, Y. Chen, F. Zhao, Class label autoencoder for zero-shot learning (2018, preprint). arXiv:1801.08301
24. E. Schonfeld, S. Ebrahimi, S. Sinha, T. Darrell, Z. Akata, Generalized zero-and few-shot learning via aligned variational autoencoders, in *Proceedings of the IEEE Conference on Computer Vision and Pattern Recognition* (2019), pp. 8247–8255
25. Y. Liu, Q. Gao, J. Li, J. Han, L. Shao, Zero shot learning via low-rank embedded semantic autoencoder, in *Twenty-Seventh International Joint Conference on Artificial Intelligence (IJCAI)* (2018), pp. 2490–2496

The Effectiveness of Data Mining Techniques at Estimating Future Population Levels for Isolated Moose Populations

Charles E. Knadler

1 Introduction

Knadler [1] found that no set of parameters could be determined for any of four different differential equation models that would accurately model the moose population levels in the Isle Royale National Park (USA). Thus, these equations could not be used to accurately predict future population levels for either the moose or wolves.

The development of data mining techniques [2] raised the question, "Is there a data mining technique that can effectively predict future population levels using small datasets?"

Four techniques were chosen for evaluation: multiple regression [3], regression trees [4], neural networks [4, 5], and k-nearest neighbors [2]. R language [3] implementations were used for all the data mining analysis.

Knadler [6] developed a discrete event simulation of the Isle Royale National Park wolf and moose populations, which produced simulation runs, "such that each simulation (run) produces data, which could reasonably represent the interspecies competition in a closed habitat of limited geographical area in the absence of major impacts from outside sources . . .".

Forty simulations were generated to be used to evaluate the prediction techniques. Moose became extinct in one run.

This compares to the 2018 Isle Royale National Park observations where wolves were on the verge of extinction. The Isle Royale study is the longest running large mammal predator–prey study in the world and has produced annual estimates of the

---

C. E. Knadler (✉)
Computer Science, Utah Valley University, Orem, UT, USA
e-mail: CKnadler@uvu.edu

© Springer Nature Switzerland AG 2021
H. R. Arabnia et al. (eds.), *Advances in Artificial Intelligence and Applied Cognitive Computing*, Transactions on Computational Science and Computational Intelligence, https://doi.org/10.1007/978-3-030-70296-0_15

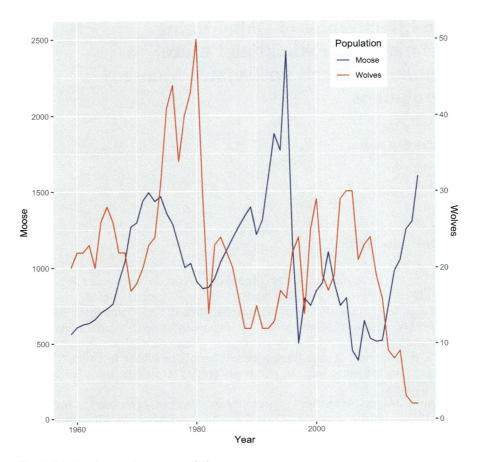

Fig. 1 Isle Royale annual moose populations

park's wolf and moose since 1959 through 2018 (Fig. 1). The annual reports provide a wealth of wolf, moose, and island ecological information [7–11].

The park is a 210–square mile island in Lake Superior and has apparently had limited immigration and emigration of wolves and moose.

It is believed that the current moose population is the result of immigration of isolated groups of animals swimming from Canada in the early 1900s [12]. Based on scat and the lack of positive data to the contrary, it is probable that the current wolf population became established in the 1948–1952 period [12]. Mech's research suggests that there has been more immigration and emigration by animal populations than popularly believed. In the case of the wolves, their observed limited genetic diversity belies this, for the period 1959 to present.

The wolf population has declined to an apparently nonbreeding pair [8, 9] and without intervention would have become extinct. Peterson et al. [9] reported that the

The Effectiveness of Data Mining Techniques at Estimating Future Population... 185

National Park Service is considering restoring a viable wolf population by releasing 20–30 wolves over a period of 3 years. This program was started in 2018 [13].

2 Methods

2.1 Data Wrangling

Data wrangling is "the art of getting your data into R in a useful form for visualization and modeling". The simulation and Isle Royale Park data sets were transformed into data frames. A data frame is a two-dimensional R object. Each row contains the data for one observation. The columns are equal length vectors. One column for each variable. Unlike a matrix, the data frame columns may be of differing types, for example, integers, real numbers, factors, etc. [14].

The data mining techniques were implemented so that all utilized the same data frame structure. Each row contained the data for a specific year. The columns were named M, M1, M2, M3, M4, W, where M was the current year's moose population, M1 was the prior year's moose population, M2 was the moose population 2 years previous, M3 was the moose population 3 years previous, M4 was the moose population 4 years previous, and W was the current year's wolf population.

Each data set was partitioned into a training data frame and a test data frame. The simulation data sets' training data frames consisted of the observations for 50 years, 1963 through 2012. The Park data training data frame consisted of the observations for 47 years, 1963 through 2009. The Park test data frame consisted of the observations for 8 years, 2010 through 2017. The simulation data sets' test data frames consisted of the observations for 8 years, 2013 through 2020.

The training data frames were used to determine the parameters for each model and the test data frames were used to evaluate the accuracy of the models' predictions of the following year's moose population from the current year's data.

2.2 Multiple Regression

The multiple regression model assumes that the dependent variable is a linear combination of explanatory variables and their interactions. The products or powers of explanatory variables are referred to as interactions [3, 14].

Three models were used:

1. The *maximal model*, the starting point of the analysis, containing all variables and interactions which might be significant.
2. The *current model*, which is the model at the current intermediate step of simplification.

3. The *minimal adequate model*, the model which utilizes the smallest number of variables needed to explain the response variable.

The R function `lm` [14] was used to fit the linear models for all the simulation runs and the Isle Royale moose data.

2.2.1 First Maximal Model

The first maximal model considered was

```
model1 <- lm(M ~ M1*M2*M3*M4*W+ I(M1^2)+ I(M2^2)+ I(M3^2)+
I(M4^2)
```

The * means that each variable and all possible combinations of the variables are included in the model. The + means the individual terms are included, in this case the squares of M1, M2, M3, and M4.

The function lm() is a linear model, with the response variable on the left of tilde and the explanatory variables are on the right side "<−" is R's assignment operator. The results of the function call are assigned to a data object named model1. [14].

This model is the linear combination of the moose population for each of the 4 previous years, the wolf population, all interactions between these variables and the squared terms of the moose populations for the 4 previous years. This model has 5 linear terms, 1 five-way interaction, 5 four-way interactions, 10 three-way interactions, 10 two-way interactions, and 4 square terms, resulting in a total of 35 coefficients and an intercept to be determined.

The model was simplified by removing the most complex, statistically insignificant term at each step until all terms were statistically significant. If the removal of a term causes a significant increase in deviance, it is returned to the model.

This approach was successfully used for the 39 simulation models, having normally distributed pseudorandom observation errors with 0 mean and standard deviation of 1/9 of the true value added to the data. However, the function `lm()`, using this maximal model, was unable to process simulation model runs with reduced pseudo random noise (variance 1/18 of the true value) or with no added noise, due to the large number of explanatory variables, compared to the number of data points.

2.2.2 Reduced Parameter Maximal Model

Review of the minimal adequate models for the high noise simulation runs and familiarity with the Isle Royale population dynamics, suggested the reduced parameter model:

```
model1 <- lm(M~M1+M2+M3+M4+W+I(M1^2)+ I(M2^2)+ I(M3^2)+
I(M4^2))
```

which has 9 explanatory variables. This model successfully produced predictions for all the data sets, with less mean prediction error than the more complex first maximal model.

The R function Step() was used to automatically simplify the regression models using the reduced parameter maximal model [14]. This eliminated all subjectivity from the model simplification process. Step() uses the Akaike information criterion (AIC) to stepwise refine the regression models [2, 14].

2.3 Regression Trees

A regression tree is a decision tree "applied" to regression. The tree partitions the training data frame observations (values of the explanatory variables). The mean values of the corresponding response variables, for each partition, are stored in the partitions' leaf nodes [2, 4]. Based on which partition a test data frame observation falls, the predicted value of the response variable is assigned the value stored in the partition's leaf node, as shown in Fig. 2 for one of the simulation models (run3750).

The R functions rpart and predict [3] were used to construct regression tree objects (models) for all the simulation datasets and the Isle Royale moose data. The regression tree and predictions were generated, as follows:

```
rTree <-rpart(M ~., data = mooseData1)
rTreepredictions <- predict(rTree, testData )
```

where rTree is assigned the regression tree object generated from the mooseData1 training data frame. The tilde dot (~.) notation causes the model to include all

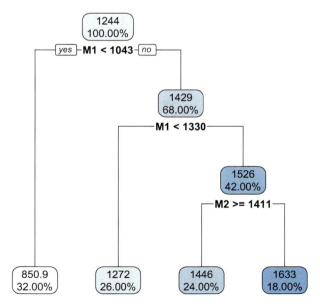

Fig. 2 Regression Tree for Run 3750

The backpropagation algorithm was used for training. It requires the activation function to be continuous, differentiable, and monotonically nondecreasing. The modern version of backpropagation was popularized in the 1980s and overcame the difficulties of training multilayer neural networks [4, 16] and making their use for nonlinear regression practical [15].

The R package neuralnet [4] was used to create the neural networks used to predict future moose populations.

2.5 K-Nearest Neighbors (KNN) Regression

The k-nearest neighbors (KNN) algorithm, where k is a positive integer, averages the response variables for the k closest training observations to the test observation to predict its response variable [2].

Smaller values of k give lower bias, higher variance estimates, while larger values of k smooth the predictions, resulting in higher bias and lower variance estimates [2].

The KNN algorithm was used with the wolf data and four different moose training data cases:

1. KNN1, using the previous year's moose population data
2. KNN2, using the two previous years' moose population data
3. KNN3, using the three previous years' moose population data
4. KNN4, using the four previous years' moose population data

The R function knn.reg [2] was used to calculate prediction for the test data frames for all the simulation datasets and the Isle Royale Park's moose data.

2.6 Simulation (After Knadler [6])

2.6.1 System Analysis and Data Collection

"For purposes of this simulation, the wolf/moose assemblage was divided into three subsystems: first, the habitat; second, the moose population and third, the wolf population." (An assemblage is a subset of the species in a community.) Thus system analysis consisted of a study and characterization of the park habitat and the behavior and ecology of the two species. The primary resources for these activities were the Ecological Studies annual reports (e.g., Refs. [12, 17, 18]) and texts on the behavior and ecology of wolves and moose ([19, 20]).

"Defining characteristics of the habitat are its small size (210 square miles or 54,400 hectares) and that it has been closed with respect to the moose and wolf populations over the study period (1959 to present). These characteristics have impacted the behavior and viability of the wolf population. The wolves are heavily

The R function step() was used to automatically simplify the regression models using the reduced parameter maximal model [14]. This eliminated all subjectivity from the model simplification process. step() uses the Akaike information criterion (AIC) to stepwise refine the regression models [2, 14].

2.3 Regression Trees

A regression tree is a decision tree "applied" to regression. The tree partitions the training data frame observations (values of the explanatory variables). The mean values of the corresponding response variables, for each partition, are stored in the partitions' leaf nodes [2, 4]. Based on which partition a test data frame observation falls, the predicted value of the response variable is assigned the value stored in the partition's leaf node, as shown in Fig. 2 for one of the simulation models (run3750).

The R functions rpart and predict [3] were used to construct regression tree objects (models) for all the simulation datasets and the Isle Royale moose data. The regression tree and predictions were generated, as follows:

```
rTree <-rpart(M ~., data = mooseData1)
rTreepredictions <- predict(rTree, testData )
```

where rTree is assigned the regression tree object generated from the mooseData1 training data frame. The tilde dot (~.) notation causes the model to include all

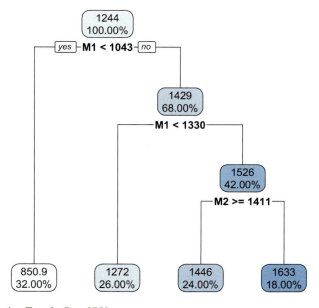

Fig. 2 Regression Tree for Run 3750

explanatory variables ([3]). The rTreepredictions object is assigned the predictions generated, for the test data frame testData, by the regression tree object rTree.

2.4 Neural Networks

Artificial neural networks are nonlinear statistical models modeled after neural networks in animals [4, 15]. A neural network may be defined by specifying it network topology, activation function, and training algorithm.

The network topology is the number of nodes and their connectivity. Fig. 3 shows an example neural network consisting of three layers of nodes: input layer of five nodes, a hidden layer with one node, an output layer with one node, and two bias nodes for the simulation run 3750. The numerical values on each connection are its weights. The output (activation) of a node is multiplied by its weight before it is summed with the other nodes output.

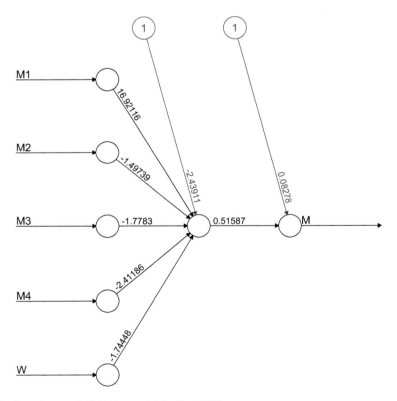

Fig. 3 Neural network (1 hidden node) for Run 3750

The bias nodes have one as their activation value. Depending on the sign of their weights, bias nodes increase or decrease the net input to a node. It this example the net input to the hidden node is decreased by 2.43911 and the net input of the output node is increased by 0.08278.

The inputs to the input layer neurons are the explanatory variables normalized to values in the range [0,1] where the neural network has the best results [4].

Different network topologies may be specified for the same training data frame resulting in a different neural network. Figure 4 shows a network with five nodes in the hidden layer modeling the same example. In general, different topologies will give differing output values, but these two topologies give the exact same results for the all the simulation data frames and the Park data frame.

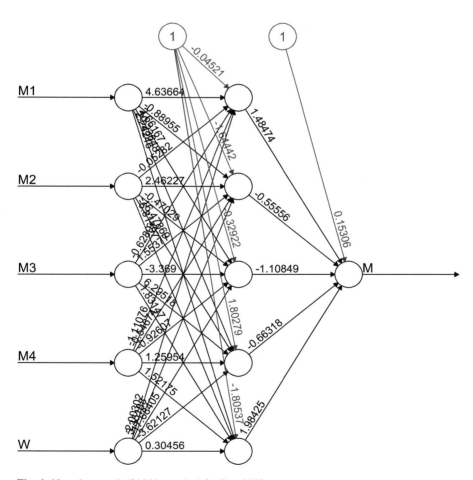

Fig. 4 Neural network (5 hidden nodes) for Run 3750

The backpropagation algorithm was used for training. It requires the activation function to be continuous, differentiable, and monotonically nondecreasing. The modern version of backpropagation was popularized in the 1980s and overcame the difficulties of training multilayer neural networks [4, 16] and making their use for nonlinear regression practical [15].

The R package neuralnet [4] was used to create the neural networks used to predict future moose populations.

2.5 K-Nearest Neighbors (KNN) Regression

The k-nearest neighbors (KNN) algorithm, where k is a positive integer, averages the response variables for the k closest training observations to the test observation to predict its response variable [2].

Smaller values of k give lower bias, higher variance estimates, while larger values of k smooth the predictions, resulting in higher bias and lower variance estimates [2].

The KNN algorithm was used with the wolf data and four different moose training data cases:

1. KNN1, using the previous year's moose population data
2. KNN2, using the two previous years' moose population data
3. KNN3, using the three previous years' moose population data
4. KNN4, using the four previous years' moose population data

The R function knn.reg [2] was used to calculate prediction for the test data frames for all the simulation datasets and the Isle Royale Park's moose data.

2.6 Simulation (After Knadler [6])

2.6.1 System Analysis and Data Collection

"For purposes of this simulation, the wolf/moose assemblage was divided into three subsystems: first, the habitat; second, the moose population and third, the wolf population." (An assemblage is a subset of the species in a community.) Thus system analysis consisted of a study and characterization of the park habitat and the behavior and ecology of the two species. The primary resources for these activities were the Ecological Studies annual reports (e.g., Refs. [12, 17, 18]) and texts on the behavior and ecology of wolves and moose ([19, 20]).

"Defining characteristics of the habitat are its small size (210 square miles or 54,400 hectares) and that it has been closed with respect to the moose and wolf populations over the study period (1959 to present). These characteristics have impacted the behavior and viability of the wolf population. The wolves are heavily

The Effectiveness of Data Mining Techniques at Estimating Future Population... 191

inbred and younger pack members have a very limited opportunity to disperse and form new packs" [6].

2.6.2 Simulation Habitat

"The habitat is characterized by its carrying capacity for moose (i.e., the maximum number of moose that can be supported by the Isle Royale vegetation)."

2.6.3 Wolf Characterization

"The simulation is implemented with event advance timing and a time granularity of 1 day, but object states are advanced weekly. That is each wolf is examined at seven-day intervals. First, a pseudorandom number is generated to determine if the wolf died this week. If it died, then it is removed from its pack. The probability that a wolf dies in a week varies by age group. The probabilities, in this series of runs, were pup (0.0015), mature wolf (0.0003), old wolf (0.0015) and ancient wolf (0.0155). These probabilities were chosen based on experiments using the simulation. If a shortage of suitable moose prey occurs, then the death probabilities are increased as the ratio of moose prey to wolves decrease.

If the wolf survives the week, then its age is advanced 1 week. If the wolf's current age puts the wolf in a different age group (pup to adult, adult to old, old to ancient), then its object descriptor is revised.

In the case of fertile female wolves additional processing is performed. If they are pregnant and their term has completed this week, then the birth event is scheduled for the current day. If the female is not pregnant and it is the breeding season, the simulation determines if the female breeds and becomes pregnant. This is done once per fertile female per breeding season (if there is a surviving male in the pack). The female has a 0.20 probability of becoming pregnant when food is abundant. When there is a shortage of prey the probability is reduced to 0.10. These probabilities were chosen based on experiments using the simulation."

2.6.4 Moose Characterization

"Moose are processed in a similar manner to wolves. Each moose is examined at seven-day intervals. First, a pseudorandom number is generated to determine if the moose died of natural causes (including starvation) this week. If it has died, then it is removed from its herd. The probability that a moose dies in a week varies by age group. The probabilities, for this series of runs, were calf (0.00001), mature moose (0.00001), and old moose (0.0005). If the moose does not die of natural causes then a pseudorandom number is generated to determine if wolves attack the moose. The probability of a moose being eaten (if attacked) is a function of age; e.g. calf (0.76), mature moose (0.005) and old moose (0.4). These probabilities were chosen based on experiments using the simulation.

The probability of attack is generated at the beginning of each simulated year. This probability is (the number of moose to be eaten per wolf per year) × (number of wolves) / (number of moose)) / (52 weeks per year). The number of moose to be eaten per year per wolf is a pseudorandom uniformly distributed number in the interval [4.0,10.0].

When the number of moose exceed the habitat carrying capacity (2100), the natural death probabilities are increased as a function of the ratio of the number of moose to the carrying capacity.

If the moose survives the week, then its age is advanced 1 week. If the moose's new age puts the moose in a new age group (calf to mature, mature to old), then its object descriptor is revised.

In the case of fertile female moose, additional processing is performed. If they are pregnant and their term has completed this week, then the birth event is scheduled for the current day. If the female is not pregnant and it is the breeding season, then the simulation determines if the female breeds and becomes pregnant. This is done once per fertile female per breeding season. The female has a 0.50 probability of becoming pregnant if food is abundant. If there is a shortage of food (habitat carrying capacity exceeded) the probability is reduced to 0.125. These probabilities were chosen based on experiments using the simulation."

2.6.5 Simulation Initialization

"The published Isle Royale moose and wolf populations for 1959 are used as the initial populations."

3 Results

3.1 Overview

Mean relative error (percent) was the metric used to compare the different prediction methods (Fig. 5). All prediction approaches were applied to the simulation data with the following:

1. Normal variance data, normally distributed errors with $\mu = 0.0$ and $\sigma = 1/9$ of the true value added, labelled (norm)
2. Reduced variance data, normally distributed errors with $\mu = 0.0$ and $\sigma = 1/18$ of the true value added, labelled (reducedVar)
3. Zero error data added, labelled (zeroVar)

Fig. 5 Mean relative error (percent)

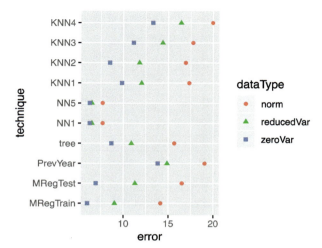

Fig. 6 Standard deviation of mean relative error (percent)

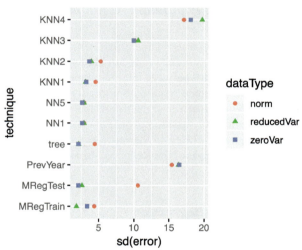

The prediction accuracy and variability of all the methods (Figs. 5 and 6) decreased with increased uncertainty in population data, with neural networks clearly outperforming all the other data mining techniques examined.

They had the least sensitivity to data variability and the best accuracy. Their accuracy was only matched or surpassed by the multiple regression training predictions for the zero-error data. Multiple regression and regression trees had slightly less variability for the zero error and reduced variance data. MRegTrain is the multiple regression results for the training data frames and MRegTest is the multiple regression results for the test data frames. PrevYear is the results assuming no change from the current year to the next years' population.

Table 1 Prediction error statistics assuming no change for simulation runs

	Zero variance	Reduced variance	Normal variance
Minimum error	5.062	6.193	7.967
1st quartile error	6.931	9.361	12.351
Median error	8.412	11.171	15.480
Mean error	8.672	10.890	15.674
3rd quartile error	9.721	12.636	18.697
Maximum error	13.734	14.380	28.575
S.D. of mean error	2.127	2.149	4.456

There was no difference in results between the neural network and with a single hidden node and the neural network with five hidden nodes.

3.2 Constant Population Assumption

The mean prediction error statistics assuming no change in the next year's population for the simulation runs are shown in Table 1. The mean error increases by a factor of 1.81 between the population data with zero observation errors to the "Normal Variance" data with normally distributed errors with mean of 0.0 and standard deviation of 1/9 of the true value.

The mean error, assuming no population change, for the Isle Royale moose population was 13.130% in the second quartile of the Normal Variance model runs and the third quartile of the Reduced Variance model runs.

3.3 Multiple Regression

The mean prediction error statistics for test data frame multiple regression analysis for the simulation runs are shown in Table 2. The mean error increases by a factor of 2.38 between the population data with zero observation errors to the "Normal Variance" data with normally distributed errors with mean of 0.0 and standard deviation of 1/9 of the true value.

The mean multiple regression prediction error for the Isle Royale moose population was 9.246% in the first quartile of the Normal Variance model runs and the second quartile of the Reduced Variance model runs.

The Effectiveness of Data Mining Techniques at Estimating Future Population... 195

Table 2 Prediction error statistics for multiple regression of simulation runs

	Zero variance	Reduced variance	Normal variance
Minimum error	4.470	6.136	8.419
1st quartile error	5.376	8.959	11.757
Median error	6.641	11.383	13.797
Mean error	6.938	11.297	16.520
3rd quartile error	7.467	13.026	18.466
Maximum error	15.565	16.986	75.086
S.D. of mean error	2.109	2.622	10.583

Table 3 Prediction error statistics of regression trees for simulation runs

	Zero variance	Reduced variance	Normal variance
Minimum error	2.979	4.697	5.725
1st quartile error	7.681	9.714	13.552
Median error	9.932	11.671	15.548
Mean error	13.850	14.876	19.051
3rd quartile error	14.353	14.093	20.112
Maximum error	108.528	110.093	106.447
S.D. of mean error	16.460	16.359	15.408

3.4 Regression Tree

The mean prediction error statistics for the regression tree analysis for the simulation runs are shown in Table 3. The mean error increases by a factor of 1.376 between the population data with zero observation errors to the "Normal Variance" data with normally distributed errors with mean of 0.0 and standard deviation of 1/9 of the true value.

The mean regression tree prediction error for the Isle Royale moose population was 19.881% compared to the Normal Variance model runs' mean error of 19.051% and in the fourth quartile of the Reduced Variance model runs.

3.5 Neural Network

The neural network models produced the best results of all the techniques examined. The mean prediction error statistics for the regression tree analysis for the simulation runs are shown in Table 4. The mean error increases by a factor of 1.231 between the population data with zero observation errors to the "Normal Variance" data with normally distributed errors with mean of 0.0 and standard deviation of 1/9 of the true value.

Table 4 Summary prediction error statistics for neural networks of simulation runs

	Zero variance	Reduced variance	Normal variance
Minimum error	1.223	2.198	2.711
1st quartile error	4.047	4.317	5.455
Median error	5.956	6.093	7.024
Mean error	6.260	6.505	7.705
3rd quartile error	8.100	8.112	9.037
Maximum error	12.126	15.580	14.244
S.D. of mean error	2.665	2.909	2.931

Table 5 Summary prediction error statistics for KNN1 models of simulation runs

	Zero variance	Reduced variance	Normal variance
Minimum error	4.717	6.852	9.799
1st quartile error	7.431	9.702	14.026
Median error	8.937	12.009	16.299
Mean error	9.833	12.032	17.325
3rd quartile error	11.434	13.531	19.624
Maximum error	18.165	18.995	28.088
S.D. of mean error	3.217	3.124	4.561

The mean neural network prediction error for the Isle Royale moose population was 9.880% in the fourth quartile of the Normal Variance model runs and the fourth quartile of the Reduced Variance model runs.

3.6 KNN1

The mean prediction error statistics for the k-nearest neighbors analysis for the simulation runs using 1 year of moose population history are shown in Table 5. The mean error increases by a factor of 1.762 between the population data with zero observation errors to the "Normal Variance" data with normally distributed errors with mean of 0.0 and standard deviation of 1/9 of the true value.

The mean KNN1 prediction error for the Isle Royale moose population was 9.316% smaller than the minimum error of 9.799% for the Normal Variance model runs and in the second quartile of the Reduced Variance model runs.

3.7 KNN2

The mean prediction error statistics for the k-nearest neighbors analysis for the simulation runs using 2 years of moose population history are shown in Table 6.

The Effectiveness of Data Mining Techniques at Estimating Future Population... 197

Table 6 Summary prediction error statistics for KNN2 models of simulation runs

	Zero variance	Reduced variance	Normal variance
Minimum error	3.238	4.905	5.769
1st quartile error	5.756	8.984	13.419
Median error	8.208	11.940	16.085
Mean error	8.496	11.796	16.936
3rd quartile error	10.654	13.396	19.979
Maximum error	20.674	27.120	27.902
S.D. of mean error	3.670	3.949	5.280

Table 7 Summary prediction error statistics for KNN3 models of simulation runs

Data type	Zero variance	Reduced variance	Normal variance
Minimum error	2.772	5.335	7.368
1st quartile error	6.988	9.801	13.010
Median error	9.496	12.295	15.540
Mean error	11.123	14.384	17.759
3rd quartile error	11.611	15.948	19.451
Maximum error	65.821	74.576	71.783
S.D. of mean error	9.976	10.618	10.102

The mean error increases by a factor of 1.993 between the population data with zero observation errors to the "Normal Variance" data with normally distributed errors with mean of 0.0 and standard deviation of 1/9 of the true value.

The mean KNN2 prediction error for the Isle Royale moose population was 15.913% in the second quartile of the Normal Variance model runs and the fourth quartile of the Reduced Variance model runs.

3.8 KNN3

The mean prediction error statistics for the k-nearest neighbors analysis for the simulation runs using 2 years of moose population history are shown in Table 7. The mean error increases by a factor of 1.600 between the population data with zero observation errors to the "Normal Variance" data with normally distributed errors with mean of 0.0 and standard deviation of 1/9 of the true value.

The mean KNN3 prediction error for the Isle Royale moose population was 19.491% in the second quartile of the Normal Variance model runs and the fourth quartile of the Reduced Variance model runs.

Table 8 Summary prediction error statistics for KNN4 models of simulation runs

	Zero variance	Reduced variance	Normal variance
Minimum error	3.475	5.572	9.715
1st quartile error	6.416	10.581	14.179
Median error	10.192	13.709	16.245
Mean error	13.310	16.423	19.965
3rd quartile error	12.747	15.044	18.913
Maximum error	118.001	134.045	118.906
S.D. of mean error	18.069	19.710	17.114

3.9 KNN4

The mean prediction error statistics for the k-nearest neighbors analysis for the simulation runs using 2 years of moose population history are shown in Table 8. The mean error increases by a factor of 1.500 between the population data with zero observation errors to the "Normal Variance" data with normally distributed errors with mean of 0.0 and standard deviation of 1/9 of the true value.

The mean KNN4 prediction error for the Isle Royale moose population was 18.025% in the second quartile of the Normal Variance model runs and the fourth quartile of the Reduced Variance model runs.

The mean KNN4 prediction of the next year's population resulted in a percentage error of 18.025%.

4 Conclusions

Based on the analyses of the simulations of the moose–wolf assemblage population dynamics, data mining techniques may be successfully used, with small data sets (approximately 50 data points), to the predict the following year's population of a population of moose. Neural networks and multiple regressions showed the most promise for the Isle Royale simulations and annual observations.

Neural networks had the best average performance, with the minimum variation in accuracy with increasing magnitude of the pseudorandom noise added to the simulation data.

Multiple regression had the best performance on the Park annual observation data, with a mean prediction error of 9.246% compared to the neural network's 9.889% mean prediction error.

References

1. C.E. Knadler, *Simulating a Predator/Prey Relationship. Proceedings of the 2008 Summer Computer Simulation Conference* (The Society for Modeling and Simulation International, Edinburgh, Scotland, 2008)
2. G. James et al., *An Introduction to Statistical Learning with Applications in R* (Springer, New York, NY, 2017)
3. M.J. Crawley, *The R Book*, 2nd edn. (Wiley, West Sussex, UK, 2013)
4. B. Lantz, *Machine Learning with R*, 2nd edn. (Packt Publishing Ltd, Birmingham, UK, 2015)
5. S. Haykin, *Neural Networks and Learning Machines*, 3rd edn. (Prentice Hall, Upper Saddle River, NJ, 2009)
6. C.E. Knadler, Using Simulation to Evaluate Models of a Predator-Prey Relationship, in *Proceedings of the 2008 Winter Simulation Conference*, (Association for Computing Machinery, Miami, FL, 2008)
7. R.O. Peterson, J.A. Vucetich, Ecological studies of wlves on Isle Royale, annual report 2005–2006, in *School of Forestry and Wood Products*, (Michigan Technological University, Houghton, MI, 2006)
8. R.O. Peterson, J.A. Vucetich, Ecological studies of wolves on Isle Royale, Annual Report 2016–2017, in *School of Forestry and Wood Products*, (Michigan Technological University, Houghton, MI, 2017)
9. R.O. Peterson et al., Ecological studies of wolves on Isle Royale, Annual Report 2017–2018, in *School of Forestry and Wood Products*, (Michigan Technological University, Houghton, MI, 2018)
10. J.A. Vucetich, R.O. Peterson, Ecological studies of wolves on Isle Royale, Annual Report 2008–2009, in *School of Forestry and Wood Products*, (Michigan Technological University, Houghton, MI, 2009)
11. J.A. Vucetich, R.O. Peterson, Ecological studies of wolves on Isle Royale, Annual Report 2013–2014, in *School of Forestry and Wood Products*, (Michigan Technological University, Houghton, MI, 2014)
12. L.D. Mech, The wolves of isle Royale, in *Fauna of the National Parks of the United States Fauna Series 7*, (United States Government Printing Office, Washington, DC, 1966)
13. S. Sorace, 4 Canadian wolves air dropped in US national park to deal with moose. https://www.foxnews.com/science/4-canadian-wolves-captured-air-dropped-in-us-national-park-to-help-restore-population. Accessed 2nd April 2019, 2019
14. M.J. Crawley, *Statistics an Introduction Using R*, 2nd edn. (Wiley, West Sussex, UK, 2015)
15. S. Russell, P. Norvig, *Artificial Intelligence*, 3rd edn. (Prentice Hall, Upper Saddle River, NJ, 2010)
16. K. Mehrotra et al., *Elements of Artificial Neural Networks* (The MIT Press, Cambridge, MA, 1997)
17. J.G. Oelfke et al., Wolf Research in the Isle Royale Wilderness: Do the Ends Justify the Means? in *Wilderness Science in a Time of Change Conference. Volume 3: Wilderness as a Place for Scientific Inquiry. Proceedings RMRS-P-15-VOL-3*, (U. S. Department of Agriculture, Forest Service, Rocky Mountain Research Station, Ogden, UT, 1999)
18. R.O. Peterson, J.A. Vucetich, Ecological Studies of Wolves on Isle Royale, Annual Report 2001–2002, in *School of Forestry and Wood Products*, (Michigan Technological University, Houghton, MI, 2002)
19. V. Geist, *Moose, Behavior, Ecology, and Conservation* (Voyageur Press, Inc., Stillwater, MN, 1999)
20. L. D. Mech, L. Boitani (eds.), *Wolves, Behavior, Ecology, and Conservation* (The University of Chicago Press, Chicago, IL, 2003)

Unsupervised Classification of Cell-Imaging Data Using the Quantization Error in a Self-Organizing Map

Birgitta Dresp-Langley and John M. Wandeto

1 Introduction

The quantization error in a fixed-size Self-Organizing Map (SOM) with unsupervised winner-take-all learning [1, 2] has previously been used successfully to detect meaningful changes across series of medical, satellite, and random dot images [3–9]. The computational properties of the quantization error in SOM are capable of reliably discriminating between the finest differences in local pixel color intensities in complex images including scanning electron micrographs of cell surfaces [10]. Moreover, the quantization error in the SOM (SOM-QE) reliably signals changes in contrast or color when contrast information is removed from, or added to, arbitrarily to images, not when the local spatial position of contrast elements in the pattern changes. While non-learning-based and fully supervised image analysis in terms of the RGB mean reflects coarser changes in image color or contrast well enough by comparison, the SOM-QE was shown to outperform the RGB mean, or image mean, by a factor of ten in the detection of single-pixel changes in images with up to five million pixels [7, 8]. The sensitivity of the QE to the finest change in magnitude of contrast or color at the single-pixel level is statistically significant, as shown in our previous work [8, 9]. This reflects a finely tuned color sensitivity of a self-organizing computational system akin to functional characteristics of a specific class of retinal ganglion cells identified in biological visual systems of primates and cats [11]. Moreover, the QE's computational sensitivity and single-pixel change

---

B. Dresp-Langley (✉)
Centre National de la Recherche Scientifique CNRS UMR 7357 Université de Strasbourg,
Strasbourg, France
e-mail: birgitta.dresp@unistra.fr

J. M. Wandeto
Department of Information Technology, Dedan Kimathi University of Technology, Nyeri, Kenya

© Springer Nature Switzerland AG 2021
H. R. Arabnia et al. (eds.), *Advances in Artificial Intelligence and Applied*
Cognitive Computing, Transactions on Computational Science and Computational
Intelligence, https://doi.org/10.1007/978-3-030-70296-0_16

detection performance surpasses the capacity limits of human visual detection, as also shown in our previous work [3–10].

The above-mentioned properties of the SOM-QE make it a promising tool for fast, automatic (unsupervised) classification of biological imaging data as a function of structural and/or ultrastructural changes that are not detectable by human vision. This was previously shown in our preliminary work [10] on the example of Scanning Electron Micrographs (SEM) of HIV-1-infected CD4 T-cells with varying extent of virion budding on the cell surface [12, 13]. SEM image technology is used in virology to better resolve the small punctuated ultrastructural surface signals correlated with surface-localized single viral particles, so-called virions [12, 13]. A defining property of a retrovirus such as the HIV-1 is its ability to assemble into particles that leave producer cells, and spread infection to susceptible cells and hosts, such as CD4 lymphocytes, also termed T-cells or "helper cells." This leads to the morphogenesis of the viral particles, or virions, in three stages: assembly, wherein the virion is created and essential components are packaged within the target cell; budding, wherein the virion crosses the plasma membrane (Fig. 1), and finally maturation, wherein the virion changes structure and becomes infectious [12, 13].

Another potential exploitation of SOM-QE in biological image analysis is cell viability imaging by RED, GREEN, or RED–GREEN color staining (Fig. 2). The common techniques applied for determination of in vitro cell size, morphology, growth, or cell viability involve human manual work, which is imprecise and frequently subject to variability caused by the analyst himself or herself [14]. In addition, considering the necessity for evaluation of a large amount of material and data, fast and reliable image analysis tools are desirable. The use of accessible precision software for the automatic (unsupervised) determination of cell viability on the

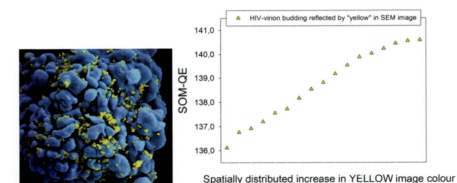

Fig. 1 Color-coded SEM image of a CD4 T-cell with ultrastructural surface signals (left), in yellow here, correlated with surface-localized single HIV-1 viral particles (virions). Some of our previous work [8] had shown that SOM-QE permits the fast automatic (unsupervised) classification of sets of SEM images as a function of ultrastructural signal changes that are invisible to the human eye (right)

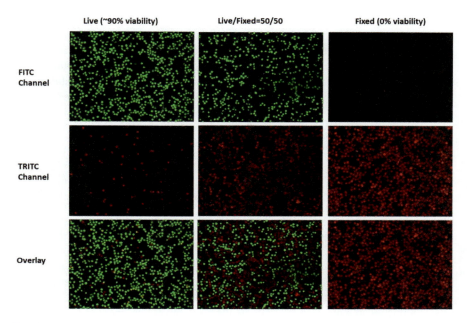

Fig. 2 Color-coded cell viability image data produced by GREEN (top), RED (middle), and RED–GREEN (bottom) staining indicating 90% (left), 50% (middle), and 0% (right) cell viability. For this study here, we generated image data using relative variability of RED and GREEN pixel color reflecting >50% and < 50% variability in cell viability, and submitted the images to automatic classification by SOM-QE [3, 8, 9]

basis of color-staining images would allow accurate classification with additional advantages relative to speed, objectivity, quantification, and reproducibility.

In this study here, we used SOM-QE for the fast and fully automatic (unsupervised) classification of biological imaging data in 126 simulation images. Examples of the original images used for the SOM-QE analyses here are available online at:

https://www.researchgate.net/publication/340529157_CellSurvivalDeathTrend-ColorStainingImageSimulations-2020

The test images variable RED–GREEN color staining indicative of different degrees of cell viability. For this study here, we chose variations between 44% and 56% of theoretical cell viability, that is, variations below the threshold level that may carry clinical significance, but are not easily detected by human vision [14].

2 Materials and Methods

A total of 96 cell viability images with variable RED-GREEN color staining are indicative of different degrees of cell viability between 50% and 56%, indicated by an increase in the relative number of GREEN image pixels, and 44% and 50%, indicated by an increase in the relative number of RED image pixels were computer generated. All 96 images were of identical size (831 × 594). After training the SOM on one image (any image from a set may be used for training), the others from the set were submitted to SOM analysis to determine the SOM-QE variability as a function of the selectively manipulated image color contents, indicative of variable theoretical cell viability, expressed in percent (%).

2.1 Images

A cell image indicating 50% cell viability (cf. Figure 2), considered as the theoretical *ground truth* image here, displays an equivalent number, or spatial extent, of RED and GREEN dots with a specific, fixed intensity range in terms of their RGB values (here R > 100 < 256 and G > 100 < 256). In half of the test images from this study, the GREEN pixel contents were selectively augmented by a constant number of 5 pixels per image, yielding 48 image simulations of color staining data indicative of a theoretical increase in cell viability from 50% to about 56%. In the other 48 images, the green pixel contents were selectively augmented by a constant number of 5 pixels per image, yielding image simulations of color staining data indicative of a theoretical decrease in cell viability from 50% to about 44%. For a visual comparison between images reflecting 50% and 90% cell viability, based on relative amounts of combined RED and GREEN staining, see Fig. 2 (bottom). Image dimensions, RGB coordinates of the selective 5-pixel-bunch RGB spatial color increments, and their relative luminance values (Y), are summarized here below in Table 1.

Table 1 Color parameters of the test images

COLOR	RGB*min*	RGB*max*	N pixels in *ground truth* image/total N image pixels	N pixels + per test image	Cumulated N pixels + across test images	R G B of pixels added	Relative Luminance of pixels added Y=0.2126R+0.7152G+0.0722B
RED	100, 0, 0	255, 0, 0	164 538/493 614	+5	+120	255, 0, 0	55.13
						255, 65, 65	105.39
GREEN	0, 100, 0	0, 255, 0	164 538/493 614	+5	+120	0, 65, 0	45.79
						65, 255, 65	200.89
BLACK Background	0, 0, 0	0, 0, 0	164 538/493 614	0	0	--	--

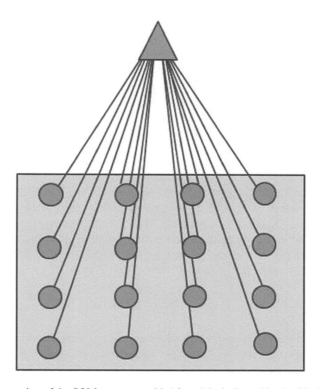

Fig. 3 Representation of the SOM prototype with 16 models, indicated by the filled circles in the gray box. Each of these models is compared to the SOM input in the training (unsupervised winner-take-all learning) process. Here in this study, the input vector corresponds to the RGB image pixel space. The model in the map best matching the SOM input will be a winner, and the parameters of the winning model will change toward further approaching the input. Parameters of models within close neighborhood of the winning model will also change, but to a lesser extent compared with those of the winner. At the end of the training, each input space will be associated with a model within the map. The difference between input vector and final winning model determines the quantization error (QE) in the SOM output

2.2 SOM Prototype and Quantization Error (QE)

The Self-Organizing Map (a prototype is graphically represented here in Fig. 3, for illustration) may be described formally as a nonlinear, ordered, smooth mapping of high-dimensional input data onto the elements of a regular, low-dimensional array [1]. Assume that the set of input variables is definable as a real vector x, of n-dimension. With each element in the SOM array, we associate a parametric real vector m_i, of n-dimension. m_i is called a model; hence the SOM array is composed of models. Assuming a general distance measure between x and m_i denoted by $d(x,m_i)$, the map of an input vector x on the SOM array is defined as the array element m_c that matches best (smallest $d(x,m_i)$) with x. During the learning process, the input vector

x is compared with all the m_i in order to identify its m_c. The Euclidean distances $||x\text{-}m_i||$ define m_c. Models that are topographically close in the map up to a certain geometric distance, denoted by h_{ci}, will activate each other to learn something from the same input x. This will result in a local relaxation or smoothing effect on the models in this neighborhood, which in continued learning leads to global ordering. SOM learning is represented by the equation

$$m(t+1) = m_i(t) + \alpha(t)h_{ci}(t)\lceil x(t) - m_i(t)\rceil \tag{1}$$

where $t = 1, 2, 3 \dots$ is an integer, the discrete-time coordinate, $h_{ci}(t)$ is the neighborhood function, a smoothing kernel defined over the map points which converges toward zero with time, $\alpha(t)$ is the learning rate, which also converges toward with time and affects the amount of learning in each model. At the end of the *winner-take-all* learning process in the SOM, each image input vector x becomes associated to its best matching model on the map m_c. The difference between x and m_c, $||x\text{-}m_c||$, is a measure of how close the final SOM value is to the original input value and is reflected by the quantization error QE. The QE of x is given by

$$QE = 1/N \sum_{i=1}^{N} \|X_i - m_{c_i}\| \tag{2}$$

where N is the number of input vectors x in the image. The final weights of the SOM are defined by a three-dimensional output vector space representing each R, G, and B channel. The magnitude as well as the direction of change in any of these from one image to another is reliably reflected by changes in the QE.

2.3 SOM Training and Data Analysis

The SOM training process consisted of 1000 iterations. The SOM was a two-dimensional rectangular map of 4 × 4 nodes, hence capable of creating 16 models of observation from the data. The spatial locations, or coordinates, of each of the 16 models or domains, placed at different locations on the map, exhibit characteristics that make each one different from all the others. When a new input signal is presented to the map, the models compete and the winner will be the model whose features most closely resemble those of the input signal. The input signal will thus be classified or grouped in one of models. Each model or domain acts like a separate decoder for the same input, that is, independently interprets the information carried by a new input. The input is represented as a mathematical vector of the same format as that of the models in the map. Therefore, it is the presence or absence of an active response at a specific map location and not so much the exact input–output signal transformation or magnitude of the response that provides the interpretation of the input. To obtain the initial values for the map size, a trial-and-error process was implemented. It was found that map sizes larger than 4 × 4 produced observations where some models ended up empty, which meant that these models did not attract any input by the end of the training. It was therefore concluded that 16 models were sufficient to represent all the fine structures in the image data. The values of the neighborhood distance and the learning rate were set at 1.2 and 0.2, respectively. These values were obtained through the trial-and-error method after testing the quality of the first guess, which is directly determined by the value of the resulting quantization error; the lower this value, the better the first guess. It is worthwhile pointing out that the models were initialized by randomly picking vectors from the training image, called the "original image" herein. This allows the SOM to work on the original data without any prior assumptions about a level of organization within the data. This, however, requires to start with a wider neighborhood function and a bigger learning-rate factor than in procedures where initial values for model vectors are preselected [2]. The procedure described here is economical in terms of computation times, which constitutes one of its major advantages for rapid change/no change detection on the basis of even larger sets of image data before further human intervention or decision-making. The computation time of SOM analysis of each of the 98 test images to generate the QE distributions was about 12 seconds per image.

3 Results

After SOM training on the reference image (unsupervised learning), the system computes SOM-QE for all the images of a given series in a few seconds, and writes the SOM-QE obtained for each image into a data file. Further steps generate output plots of SOM-QE, where each output value is associated with the corresponding input image. The data are plotted in increasing/decreasing orders of SOM-QE magnitude as a function of their corresponding image variations. Results are shown here below for the two test image series (Fig. 4). The SOM-QE is plotted as a function of increments in the relative number, by adding pixel bunches of constant size and relative luminance, of GREEN or RED image pixels. For the corresponding image parameters and variations, see Table 1.

4 Conclusions

In this work we exploit the RED–GREEN color selectivity of SOM-QE [8, 9] to show that the metric can be used for a fast, unsupervised classification of cell imaging data where color is used to visualize the progression or remission of a disease or infection on the one hand, or variations in cell viability before and after treatment on the other. Similarly successful simulations were obtained previously on SEM images translating varying extents of HIV-1 virion budding on the host cell surface, coded by the color YELLOW, in contrast with healthy surface tissue, coded

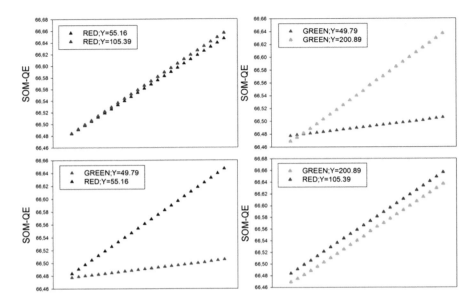

Fig. 4 SOM-QE classification of the 96 test images, with the SOM-QE plotted as function of increasing or decreasing theoretical cell viability indicated by a small and regular increase in the spatial extent of green or red pixels across the corresponding images. The data show the expected SOM-QE sensitivity to the relative luminance (Y) of a given color (top), and its color selectivity (bottom). For any relative luminance Y, the color RED, by comparison with the color GREEN, is signaled by QE distributions of greater magnitude. Future studies on a wider range of color-based imaging data will allow to further benchmark SOM-QE color selectivity

by the color BLUE [10]. Our current work, in progress, reveals hitherto unsuspected properties of self-organized [15] mapping, where the potential of the SOM_QE is revealed in terms of a computational tool for detecting the finest clinically relevant local changes in large series [16] of imaging data. Future image simulations will allow further benchmarking of SOM-QE selectivity for increasingly wider ranges of color variations in image simulations of biological data. This should, ultimately, result in providing a basis for the automatic analysis of biological imaging data where information relative to contrast and/or color is exploited selectively to highlight disease-specific changes in organs, tissue structures, or cells.

References

1. T. Kohonen, *Self-Organizing Maps*. Retrieved from http://link.springer.com/10.1007/978-3-642-56927-2, 2001
2. T. Kohonen, MATLAB implementations and applications of the self-organizing map, in *Unigrafia Oy*, (Helsinki, Finland, 2014)

3. J.M. Wandeto, H.K.O. Nyongesa, Y. Remond, B. Dresp-Langley, Detection of small changes in medical and random-dot images comparing self-organizing map performance to human detection. Inform. Med. Unlocked **7**, 39–45 (2017)
4. J.M. Wandeto, H.K.O. Nyongesa, B. Dresp-Langley, A biologically inspired technique to track a patient's condition through extraction of information from images, in *2nd DEKUT International Conference on Science, Technology, Innovation and Entrepreneurship*, (Nyeri, Kenya, 2016).; November
5. J.M. Wandeto, H.K.O. Nyongesa, B. Dresp-Langley, Detection of smallest changes in complex images comparing self-organizing map and expert performance. 40th European Conference on Visual Perception, Berlin, Germany. Perception **46**(ECVP Abstracts), 166 (2017)
6. J.M. Wandeto, B. Dresp-Langley, H.K.O. Nyongesa, Vision-inspired automatic detection of water-level changes in satellite images: the example of Lake Mead. 41st European Conference on Visual Perception, Trieste, Italy. Perception **47**(ECVP Abstracts), 57 (2018)
7. B. Dresp-Langley, J.M. Wandeto, H.K.O. Nyongesa, Using the quantization error from Self-Organizing Map output for fast detection of critical variations in image time series, in *ISTE OpenScience, Collection "From Data to Decisions"*, (Wiley, London, 2018)
8. J.M. Wandeto, B. Dresp-Langley, The quantization error in a Self-Organizing Map as a contrast and colour specific indicator of single-pixel change in large random patterns. Neural Netw. **119**, 273–285 (2019)
9. J.M. Wandeto, B. Dresp-Langley, Contribution to the Honour of Steve Grossberg's 80th Birthday Special Issue: The quantization error in a Self-Organizing Map as a contrast and colour specific indicator of single-pixel change in large random patterns. Neural Netw. **120**, 116–128 (2019)
10. J.M. Wandeto, B. Dresp-Langley, Ultrafast automatic classification of SEM image sets showing CD4 + cells with varying extent of HIV virion infection, in *7ièmes Journées de la Fédération de Médecine Translationnelle de l'Université de Strasbourg, May 25–26*, (Strasbourg, France, 2019)
11. R. Shapley, V.H. Perry, Cat and monkey retinal ganglion cells and their visual functional roles. Trends Neurosci. **9**, 229–235 (1986)
12. L. Wang, E.T. Eng, K. Law, R.E. Gordon, W.J. Rice, B.K. Chen, Visualization of HIV T-cell virological synapses and virus-containing compartments by three-dimensional correlative light and electron microscopy. J. Virol. **91**(2) (2017)
13. W.I. Sundquist, H.G. Kräusslich, HIV-1 assembly, budding, and maturation. Cold Spring Harb. Perspect. Med. **2**(7), a006924 (2012)
14. L.S. Gasparini, N.D. Macedo, E.F. Pimentel, M. Fronza, V.L. Junior, W.S. Borges, E.R. Cole, T.U. Andrade, D.C. Endringer, D. Lenz, In vitro cell viability by CellProfiler($^{®}$)Software as equivalent to MTT assay. Pharmacogn. Mag. **13**(Suppl 2), S365–S369 (2017)
15. B. Dresp-Langley, O.K. Ekseth, J. Fesl, S. Gohshi, M. Kurz, H.W. Sehring, Occam's Razor for *Big Data*? On detecting quality in large unstructured datasets. Appl. Sci. **9**, 3065 (2019)
16. B. Dresp-Langley, Seven properties of self-organization in the human brain. Big Data Cogn. Comput. **4**, 10 (2020)

Event-Based Keyframing: Transforming Observation Data into Compact and Meaningful Form

Robert Wray, Robert Bridgman, Joshua Haley, Laura Hamel, and Angela Woods

1 Introduction

Learning outcomes are improved when a learning environment is responsive to the capabilities, preferences, and needs of individual learners [1, 2]. However, implementing such responsive learning systems is a significant design and engineering challenge [3]. Creating a learner-adaptive system generally increases the complexity and cost of the learning environment. Further, because each learning environment is unique, these additional costs recur when developing adaptive learning capabilities in a new application or domain. These costs slow research progress; creating an effective adaptive learning environment is expensive and time-consuming. More importantly, they also limit reaching full human potential via better and more widespread training technologies.

This chapter describes one aspect of the adaptive training systems engineering challenge more thoroughly. We describe past work in terms of solution requirements and outline how that work fell short of meeting the challenge. We then introduce a new approach, which combines several elements of previous solutions. This "event-based keyframing" employs methods from artificial intelligence (AI) to create a "learner-centric interpretation" of the current learning situation. A machine-understandable interpretation offers multiple benefits that allow it to mitigate some of the complexity and engineering cost for developing adaptive learning systems. We illustrate how we are using this approach in a number of distinct application domains. Successful use in multiple applications highlights the generality and value of a solution that is not tied to a particular domain, learning environment, or algorithmic methods of adaptation.

R. Wray (✉) · R. Bridgman · J. Haley · L. Hamel · A. Woods
Soar Technology, Ann Arbor, MI, USA
e-mail: wray@soartech.com

© Springer Nature Switzerland AG 2021
H. R. Arabnia et al. (eds.), *Advances in Artificial Intelligence and Applied Cognitive Computing*, Transactions on Computational Science and Computational Intelligence, https://doi.org/10.1007/978-3-030-70296-0_17

2 Systems Requirements for Adaptive Learning

We focus, in this chapter, on the requirements for dynamic adaptation within a practice environment. Many practice environments are implemented with simulations of the task or performance environment at various levels of fidelity. Increasingly simulation (or "virtual") environments are also being integrated with real or "live" environments to enable an integrated mix of live, virtual, and constructive actors in complex, realistic training and practice environments. These learning environments are realistic and effective for training. However, they are also complex systems with many dependencies and interactions. Our challenge is to develop algorithms and software that can "steer" this complex system so that learning needs for an individual learner at an individual point in time can be met.

Figure 1 illustrates a high-level decomposition of the functions required for dynamic adaptation in these environments. This figure is drawn from prior work developing algorithms and software to support dynamic adaptation in training [4]. Learners interact with a simulation environment that provides the opportunity to practice various skills. We have applied this design pattern across a wide range of learning tasks, including perceptual skills for observation, awareness, and response to social cues; cross-cultural communication; tactical aviation; terrain understanding; emergency response; and others. Inputs from the practice environment are used by a "director" to determine if and when to perform tailoring actions. The concept of a director guiding action in a simulation or game occurs repeatedly in simulation environments. It is apt especially when the practice environment includes synthetic or nonplayer characters (NPCs) that can be directed to perform specific actions that will support overall system goals.

To achieve individualized adaptation, the director performs three functions:

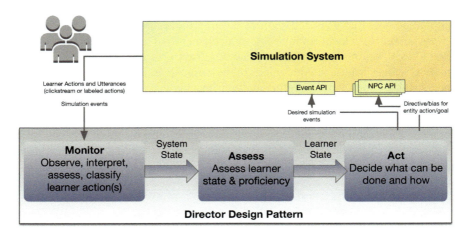

Fig. 1 Functional decomposition for adaptive tailoring of simulation-based practice

- *Monitoring*: The system must interpret what the learner is doing in the environment. Generally, monitoring requires some interpretation of learner actions as well as interpretation the context/situation in which the learner is acting.
- *Assessing*: The system uses its understanding of what is happening to assess the individual's learning state. Monitoring concerns identifying what is happening; assessing concerns what an understanding of the current situation reveals about the learner's progress and needs (e.g., in terms of the learning goals of the system). Its role is to translate from system state to learner state.
- *Acting*: The system state and the learner state provide a context for acting to address learning needs. Action results in adaptation that changes the environment in some way. Adaptation can include extrinsic (outside of the simulation) direct feedback, coaching, and intrinsic adaptation of the simulation (also known as pedagogical experience manipulation).

There are many different methods and algorithms that can be used to realize these functions, including long-standing cognitive and learner modeling approaches and task networks for planning actions and more recent, data-driven approaches that learn to perform these functions via data mining and machine learning. Regardless of the specific approach(es) adopted, however, one of the immediate challenges of building an adaptive learning system is incorporating, integrating, and unifying the various data streams that come from the other components of the learning environment. Examples of these data are enumerated in Table 1, based on both our direct experience in building adaptive learning systems and previous analyses of the requirements for monitoring and interpretation of the learning context [5].

The table enumerates the wide variety of inputs that are used to inform monitoring. We identify three characteristics of the inputs which vary from source to source.

Frame specifies the perspective or point of view of the input. In this table, we distinguish between two distinct frames. *Contextual* inputs are those that describe the environment, the situation (including the presence and actions of others), and other elements of the learning context (e.g., learning objectives). *Learner-centric* inputs directly encode information about the learner, such as activities and actions.

Representation distinguishes between largely *continuous* inputs (e.g., an analog sensor) and largely *discrete* ones. To achieve targeted adaptation, continuous and discrete data must be integrated and fused.

Interpretability characterizes whether meaning is incorporated into the inputs prior to its presentation to the learning system. For example, a raw sensor feed typically requires some translation process to extract meaning (*translatable*). A symbolic or discrete input may also require translation but the process is typically more reliable and simpler; we use *interpretable* in the table to make this distinction. Finally, some inputs may be presented with formally defined semantics underlying the input. We term directly machine-interpretable inputs as *system-understandable*.

Table 1 Examples of various data needed for monitoring and interpretation of a learner

Name/description	Frame	Representation	Interpretability
Simulation state (whole)	Contextual	Discrete	Interpretable
Current state of the simulation environment. Often represented as a "window" or "frame"			
Simulation state (part)	Contextual	Discrete	Interpretable
Representation of one element or entity in the simulation. Often used to communicate state across a network (e.g., DIS PDU)			
Physical state	Contextual	Continuous	Translatable
Summary of the current physical state of some object or entity (in simulation or live). For example, telemetry conveys location as $f(t)$			
Sensor state	Contextual	Continuous	Translatable
Outputs/status of a particular domain sensor. For example, the contacts a radar is detecting			
Speech	Contextual & learner	Continuous	Translatable
Speech produced in the course of the practice, including learner speech and interactions with others. For example, radio communications			
Learner sensor	Learner	Continuous	Translatable
Input from a sensor external to the simulation that supports interpretation of learner state. Examples include eye tracking, mouse tracking, postural sensors, facial expression, etc.			
Activity data	Learner	Discrete	Interpretable
Input from the learning environment that classifies the activities of the learner. Protocols (e.g., xAPI) increasingly provide self-describing activity data that can be immediately understood			Understandable
Learning system state	Learner	Discrete	Interpretable
Feedback inputs from other components within the learner-adaptation system (e.g., estimates of learner proficiency, directives generated).			Understandable

3 Insights from Past Experiences

As suggested by the diversity and breadth of data types in the table, creating a generalized adaptation capability depends on developing an approach to monitoring that can translate/interpret learner and contextual data into a dynamically updating understanding of the progression of learning. In earlier attempts to address this challenge across previous efforts and learning domains (as outlined previously), we have identified several important insights that further expand and refine the requirements introduced in the previous section. This section briefly summarizes these "lessons learned."

Inseparability of learner and context: Some of initial attempts to fuse learner data focused the development of standalone middleware that could capture learning state [5]. In practice, we have concluded that learning context is essential to interpreting learner state. For example, when attempting to debrief or review learner actions, some capture of the context in which those actions occurred is needed for human understanding and assessment of the action. Because the context itself also requires translation and interpretation (as in the table), monitoring needs to perform these tasks together because they depend on one another. Thus, solutions need to aggregate learner and context data.

Common representation of situation and activity: Aggregation of context and learner data leads to a requirement for a common representation that expresses what is happening in the environment, regardless of whether it is relevant to the current learning situation. For example, some situational change or event may not be relevant to the learner's activity in the moment, but may have significant bearing on future learner activities (e.g., an actor makes a decision that is not immediately visible to the learner but that will result in some interaction with the learner at a later time in the learning exercise). Just as it is difficult to separate learner and context for fusion, a common representation that allows representation of learner and environmental activity supports an integrative perspective on learner activity and the capability to reason about what has, is, and will happen without requiring the system to understand the learning implications as those occurrences take place.

Capture of activity at human timescales: The need for common representation pushes solutions toward fine-grained capture of the situation. For example, if the largest update frequency is 200 Hz, then data capture tends to occur at that frequency for all system elements. High frequency capture introduces two needs for the adaptive learning system that are not generally functional requirements for the target capability. First, it creates requirements for data throughput and storage that are marginally useful for the adaptation algorithms. These algorithms will either be required to sample the data or perform an additional abstraction. Second, for the purposes of using data-driven/machine-learning approaches, increased resolution can increase complexity of the machine-learning challenge. Data updates at timescales faster than human reaction times results in noise that the machine-learning system must learn to ignore (greatly increasing the data requirements for training the system). Because adaptive learning system will act on a human timescale, we have concluded that the data moving thru the adaptive system should be captured and abstracted at a comparable timescale, typically from 100 ms to seconds, even when incoming data presents with much higher frequency.

4 Event-Based Keyframing

As a consequence of the previous efforts to provide adaptive training and lessons derived from these experiences, we are now developing and evaluating a new approach. This *event-based keyframing* approach consists of the following:

1. *A generalized event representation*: We have developed a new knowledge representation that seeks to capture the occurrences (or "events") on-going in a dynamic scenario. The representation is "generalized" both in the sense that it integrates across multiple modalities within an application as well as being readily extensible to new application domains.
2. *Event-recognition and keyframing*: While the representation describes what has occurred, event-recognition and keyframing algorithms define the processes and mechanisms by which instances of the event representation (loosely, an "event") is generated in the system. Keyframing is a complementary process that marks some event-instances as being particularly important for understanding how the learner is progressing.

Figure 2 illustrates the event-based keyframing approach conceptually. As in Fig. 1, a breadth of inputs of various types of various periodicities are presented to the monitoring system. These inputs are combined and integrated in the aggregation and event recognition component. The output of this process are instances of the event representation, resulting in a sequence of "events" produced at human timescales (e.g., roughly every second). The result of aggregation is a syntactically regular, semantically consistent, discrete summary of the activity occurring in the environment. In the diagram, various types of events (see next section) are indicated using different colors.

Note that the sequence is being generated from right to left. Additionally, the system creates event instances to mark the "start" and "end" of a real-world occurrence that has duration/is not instantaneous (in contrast to representing the

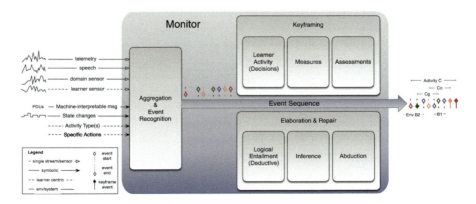

Fig. 2 Decomposition of monitoring functions to support event-based keyframing

event instance as a duration). Using a consistent representation of real-world events as "impulses" of various types has several advantages for monitoring. Further, as suggested by the labels on the output from the monitor, collections of events can be grouped. In the example, there is some *Activity C* that the learner is doing that has sub-activities C_o and C_g. Additionally, in the environment, some occurrence *B* has begun and ended (*B1*) and new, similar occurrence (*B2*) has begun, but has not yet concluded. This very simple example suggests how the event representation allows the integration of learner and environment occurrences within the common representation, which, as above, we have identified as an important requirement for adaption algorithms.

Having generated events, keyframing processes determine if any of the event instances generated in the sequence thus far have direct relevance to interpreting learner activity. The keyframing process looks at the sequence thus far (rather than only the most recently generated instance), allowing it to mark or identify event instances earlier in the observation stream as keyframes. Keyframes can also trigger measurement and assessment activities during keyframing, which then can result in additional annotation.

Finally, the monitoring process also now includes a "elaborate and repair" process that analyzes the event/keyframe sequence being produced and attempts to "fill in" any gaps it identifies in the sequence. As we discuss further below, a significant advantage of the common and consistent representation is that we can identify and use general algorithms that operate on event sequences (and the properties of the event representation) to perform logical entailment and various kinds of inference without requiring additional encoding of the properties of the domain.

4.1 Event Representation and Event Recognition

This section briefly outlines the event representation, further detailed in [6], in the context of the event-based keyframing approach. Events that occur in the learning environment are represented as *event instances* that conform to an *event representation*. Events may include environmental events (the entry of an actor the learner will need to interact with) and learner events (the learner acts, such as turning to face the new actor). As the system recognizes events, it produces a series of event instances, resulting in an event sequence. This event sequence forms the primarily method for communicating what is happening in the environment to all downstream components.

Individual event instances are represented as a defined set of "slots" and "fillers" for events of particular type. Slots include a subject (the primary actor(s) in an event), an event type or "verb" which captures what kind of activity or occurrence is being recorded ("turning toward"). Some event types include an object (what is being turned toward?) and event-specific parameters (e.g., the new direction that the learner is facing). The event types are organized into a taxonomy, drawing from formal ontology [7], which supports inheritance, property evaluation, and other functions. For example, "TurnToward" may be a specialization of the more general event type "Turn" which is where a "facing-direction" property is defined.

Adopting this formal knowledge representation for the event representation, provides several important advantages. First, the library of event types can be created progressively, and grow with additional applications. We first developed "TurnToward" events in the context of aviation training, but the same event type can be used for a first-person game perspective as well. Second, the event sequence provides a compact summary of learning activity, that can be expanded by inspection of the event instances. In an initial experiment, we discovered that the pattern of event types alone was sufficient for some pattern detection in a learning context [6]. Third, the event representation provides a well-defined and consistent target for additional algorithms that can support interpretation and adaptation, as we discuss in the next subsections.

The primary function of the event recognition process is to identify the specific items to fill the slot for the particular event instance being recorded. Using the example in the previous section, when the system observes the learner turning toward the new actor, then it would identify the learner as the subject of the event, the new actor as the object, and the learner's new heading as the key parameter associated with the turn. As we discuss further in the next section, in many cases, the initial or triggering observation may be less specific than the event instance that is finally recorded.

We have explored a number of different mechanisms for recognizing events, including simple, user-defined programs that look for specific patterns [8], data-driven machine learning [9], and comparison of observed patterns to previously observed patterns [6]. Enumerating the various requirements and trade-offs for these methods is outside the scope of this paper. However, key conclusions are that event-recognition is a type of pattern-recognition process and existing pattern-recognition algorithms can be readily applied to this problem. Using these methods enables acceptably reliable recognition of events during learner activity. The event representation provides a target that simplifies computational and data requirements for pattern/event recognition.

4.2 Keyframes and Keyframing

The event sequence summarizes all of the activity and occurrences taking place in the learning environment. Some of the resulting event instances will be of particular interest in interpreting or understanding the learner. We term these special event instances *keyframes* [10] and the process of identifying or marking them, *keyframing*. We first developed keyframes to allow a learner to target specific goals, actions, or outcomes when the learner was to replay a past practice experience (whether their own or the experience of another). Providing indices to the "important

Event-Based Keyframing: Transforming Observation Data into Compact...

Table 2 Examples of keyframes and keyframing processes

Type	Description	Process
Learner activity	Event in which a learner takes some identifiable action ("TurnToward" when the subject of the event is a learner)	Auto
Measurement	Event that measures some aspect of learner activity ("response time" could record the time that elapses between the new actor's entry and the learner initiating "TurnToward")	Auto
Assessment	Event that assesses some aspect of learner activity (the response time recorded for "TurnToward" was acceptable)	Auto
Learner cognitive state	Event that indicates a change in the learner's cognitive/affective state (engagement is now above the threshold)	Auto
Exercise state	Event that indicates a change in environment resulting from learner activity ("TurnToward" allows learner to observe the new actor)	Semi-auto

moments" within the log or trace of activity helped learner focus and resulted in more effective use of time.

Although useful, we have discovered that what a user considers "important" cannot be consistently anticipated. Instead, we have extended the notion of keyframes so that a system can mark various learner-specific events. Keyframes are tags or markers added to the "base" events. Table 2 lists examples of keyframes we have explored in current and prior work. Similar to the event representation, many keyframes of various types can be automatically identified and tagged. In the case of changes in the state of the learning environment itself that are due to learner action, we lack algorithms that can fully automate keyframing for these events, as that would require some causal model of the domain. However, we can employ similar mechanisms to those used for event recognition to mark key changes in the exercise state (a semi-automated solution).

Keyframes provide a means to help interpretation and adaption processes further abstract from the many sources of data input from the environment to information most likely to help the system build a model of the learner and progress and challenges in the domain. Automated keyframing provides a straightforward approach to capture keyframes, but at the cost of mixing "important" and "routine" events together. Long term, we plan to continue to research methods that can support the automatic classification of importance and relevance.

4.3 Elaboration and Repair

The adaptive learning system must be tolerant to noise and observation errors. Sensors will be mis-calibrated. Packets of data will be dropped. A consistent representation enables the research of general algorithms and tools that can support elaboration and refinement of event sequences as well as recognition of problems and repair of the sequence and individual instances within it. Three such elaboration/repair algorithms are as follows:

Logical entailment The formal semantics of the event representation enable deductive elaboration of the event sequence. Every "start" event has a paired "end" event such as start/stop "TurnToward." Some events can be defined as inverses of another event, thus representing the same physical event from different perspectives. For example, "Learner detects Actor" is the inverse of "Actor is-detected-by Learner." Inverse events are particularly useful for elaborating the event sequence so that all event instances are represented from the perspective of each individual entity sequence.

Abduction When event instances are missing or appear out of sequence, some process other than deductive closure is needed to repair the sequence. The representation itself can provide signals when the sequence is need of repair. In Fig. 2, the presence of two "start" event instances for the "green" event, without an intervening "stop" event creates such a signal. An abductive reasoner can evaluate potential hypotheses about the causes of sequence faults. Abduction can also be used to specialize event instances as further evidence accrues. In the "TurnToward" example, the system cannot know when the turn is initiated where it will end and to what end it is directed; however, when it concludes with the learner facing the new actor, abduction can specialize the "Turn" event instance to "TurnToward" based on observation/evidence.

Inference In some cases, there may not be enough evidence to assert a repair in detail. In this case, we are exploring machine learning to compile patterns from a large collection of observations and assert the presence of missing event instances (or missing details from the instance). We are using machine-learned models of individual parameters within event instances to compare and match event sequences [6].

While this component is relatively nascent in comparison to event representation and keyframing, it offers an example of the potential power and value of the integrated and abstract representational approach. Because these algorithms are based on the properties of the representation and the statistical properties of observations within a domain, they are not specific to an individual application. We anticipate further reuse across multiple applications of the adaptive learning system.

Acknowledgments The authors thank teammates for contributions to concepts and implementations reported here. We also thank Dr. Ami Bolton, Ms. Jennifer Pagan, and Dr. Heather Priest for contributions to problem definition and context that informed the approach. Opinions expressed here are not necessarily those of the Department of Defense or the sponsors of this effort, the Office of Naval Research and the Naval Air Warfare Center Training Systems Division (NAWCTSD). This work funded under contracts N68335-20-F-0549, N68335-17-C-0574, and N68335-19-C-0260.

References

1. P. J. Durlach, A. M. Lesgold (eds.), *Adaptive Technologies for Training and Education* (Cambridge, New York, 2012)
2. C.R. Landsberg, R.S. Astwood, W.L. Van Buskirk, L.N. Townsend, N.B. Steinhauser, A.D. Mercado, Review of adaptive training system techniques. Mil. Psychol. **24**, 96–113 (2012)
3. National Academy of Engineering: Grand Challenges for Engineering. National Academy of Sciences/National Academy of Engineering (2008)
4. R.E. Wray, A. Woods, A cognitive systems approach to tailoring learner practice, in *Proceedings of the Second Advances in Cognitive Systems Conference*, ed. by J. Laird, M. Klenk, (Baltimore, MD, 2013)
5. R.E. Wray, J.T. Folsom-Kovarik, A. Woods, Instrumenting a perceptual training environment to support dynamic tailoring, in *Proceedings of 2013 Augmented Cognition Conference*, ed. by D. Schmorrow, C. Fidopiastis, (Springer-Verlag (Lecture Notes in Computer Science), Las Vegas, 2013)
6. R.E. Wray, J. Haley, R. Bridgman, A. Brehob, Comparison of complex behavior via event sequences, in *2019 International Conference on Computational Science and Computational Intelligence (CSCI'19)*, (IEEE Press, Las Vegas, 2019)
7. R. Arp, B. Smith, A.D. Spear, *Building Ontologies with Basic Formal Ontology* (MIT Press, 2015)
8. R.E. Wray, C. Newton, R.M. Jones, Dynamic scenario adaption for simulation training, in *2018 International Conference on Computational Science and Computational Intelligence (CSCI'18)*, (IEEE Press, Las Vegas, 2018)
9. J. Haley, V. Hung, R. Bridgman, N. Timpko, R.E. Wray, Low level entity state sequence mapping to high level behavior via a deep LSTM model, in *20th International Conference on Artificial Intelligence*, (IEEE Press, Las Vegas, 2018)
10. R.E. Wray, A. Munro, Simulation2Instruction: Using simulation in all pases of instruction, in *2012 Interservice/Industry Training, Simulation, and Education Conference*, (NTSA, Orlando, 2012)

An Incremental Learning Scheme with Adaptive Earlystopping for AMI Datastream Processing

Yungi Ha, Changha Lee, Seong-Hwan Kim, and Chan-Hyun Youn

1 Introduction

Advanced Metering Infrastructure (AMI) enabled two-way communication between customers and power utility companies. It includes the meters on the customer side, communication networks between the two sides, and the data management system that makes the information available to the providers [10]. With an era of the Internet of things (IoT), the number of AMI will keep increasing as the industry IoT (IIoT) is estimated to grow from USD 65,452.15 million in 2018 to USD 118,413.63 million by 2025 according to the report from Valuates Reports. AMI generates a huge amount of data in daily basis. It is basically a kind of streaming data, where data instances arrive as a sequence over time.

As the amount of AMI data increases rapidly in recent years, attempts to apply Deep Learning (DL) techniques to AMI data processing are also being actively studied. There is a wide range of DL applications for AMI data processing, such as power energy prediction [9], energy fraud detection [4], etc.

Although many existing frameworks support building and serving DL models, there is still much less effort on model maintenance, in other words, integrating new data with the deployed DL model [7]. The reason for model retraining lies on a phenomenon, namely concept drift [8]. It is caused by a change in the distribution of underlying data, causing the trained model obsolete. As a result, the model's performance deteriorates over time.

Y. Ha · C. Lee · S.-H. Kim · C.-H. Youn (✉)
School of Electrical Engineering, Korea Advanced Institute of Science and Technology, Daejeon, South Korea
e-mail: yungi.ha@kaist.ac.kr; changha.lee@kaist.ac.kr; s.h_kim@kaist.ac.kr; chyoun@kaist.ac.kr

© Springer Nature Switzerland AG 2021
H. R. Arabnia et al. (eds.), *Advances in Artificial Intelligence and Applied Cognitive Computing*, Transactions on Computational Science and Computational Intelligence, https://doi.org/10.1007/978-3-030-70296-0_18

To the best of our knowledge, there are several works which deal with online learning system. One of them is Continuum [13], which handles model updates with the two different policies, best-effort and cost-aware. The best-effort policy is a naive policy which tries to update the model whenever newly incoming data are available at the system. Our interest lies in the cost-aware policy, which is designed for cost-sensitive system that requires fast data incorporation at low training cost. The cost-aware policy is an online algorithm that presents threshold-based update method by formulating latency-cost sum. It shows 19% less training cost compared to periodic model update. However, it does not take care of model convergence in an environment where concept drift exists. ASDP [1] improves the algorithm by introducing the earlystopping and adaptive batch size algorithm. It makes the training process efficient by stopping the training process earlier then specified epoch when it shows little improvement over time.

In this paper, we develop an online learning system that overcomes the limitations of Continuum[13] and ASDP [1], in terms of training efficiency. More specifically, we take into account the effect of concept drift on training DL models. We first propose a light-weight approach for concept drift detection by employing methods of cosine similarity and sliding windows. The model is then trained with a selective earlystopping approach. It means that the earlystopping is performed only when no drift occurs. Otherwise, the model would be trained until reaching a good convergence without being stopped early. In addition, we improve the adaptive batch size algorithm of ASDP[1] with simpler metrics and objective function, but more efficient.

Our online learning system is implemented in an Apache Kafka [5] cluster, which allows handling a large scale of AMI streaming data. To evaluate the system performance, we conduct experiments on AMI data from 2015–2018, and record the metrics of the training loss, inference error, and the test error. The results show that the performance of our proposed approach is better than Continuum [13] (r=4), and comparable to ASDP[1].

2 Problem Description

Many deep learning applications are utilized to process data streams which change over time. When underlying data distribution changes, it causes the DL model trained with stale data not to work as expected with new data. Although naive methods treat individual arriving instance as equally essential, this phenomenon requires a special approach that differs from commonly used schemes. Therefore, the data stream analysis algorithm should provide a way to adapt to these changes that the deployed model can work with recent data by efficient update [2]. As our target data, AMI data stream, has time dependencies, we should continually renew the deployed model. Incremental learning is commonly used to handle the problem of learning unbound data streams.

When measuring power consumption in commercial, the voltage is managed into high and low voltage. Analyzing low voltage, mostly used by household, is more challenging task than high voltage usually used in industrial. Focusing on low-voltage meter data, we first considered the principal properties causing analysis difficult. The low-voltage AMI data are sequentially incoming and its statistical properties are time-variant, which is defined as a concept drift. Concept drift is the change in the underlying distribution of the dataset, such as time-evolving characteristic, class-wise characteristic, etc. We can see this phenomenon when the distribution of streaming data is time-varying, a DL model built on stale data is inconsistent with the new data. To confirm concept drift feature, we evaluate Augmented Dickey–Fuller test on low-voltage household power consumption data, which is a common statistical method used to validate stationary or non-stationary with confidence level [3] as following Equation:

$$\Delta y_t = \alpha + \beta t + \delta_1 \gamma y_{t-1} + \cdots + \delta_{p-1} \Delta y_{t-p+1} + \epsilon_t \tag{1}$$

where α is a constant, β the coefficient on a time trend, and p the lag order of the autoregressive process. Imposing the constraints $\alpha = 0$ and $\beta = 0$ corresponds to modeling a random walk and using the constraint $\beta = 0$ corresponds to modeling a random walk with a drift. Consequently, there are three main versions of the test, analogous to the ones discussed on Dickey–Fuller test. According to reference [3], if the test statistic is greater than critical value (5%), this means that the time-series data are non-stationary at a 95% confidence level. Thus, the results in the ranges 36, 100, and 1000 as described in Table 1 show that AMI meter data has a non-stationary characteristics causing concept drift. The concept drift causes problems because the result of analysis becomes less accurate as time passes. The term concept refers to the quantity to be predicted. More generally, it can also refer to other phenomena of interest besides the target concept, such as an input, but, in the context of concept drift, the term commonly refers to the target variable.

Due to concept drift problem in AMI data, we need to incorporate new information into training model from data stream. Incremental Learning or Online Learning [2, 12] is a common method to train learning-based, such as a machine learning or deep learning, model expanding the knowledge of existing model by incoming input data containing untrained feature.

There are two knobs which control retraining time: batch size and epoch. Though the impact of batch size in model retraining is controversial, a larger batch size for training can produce a higher prediction accuracy [11]. It is because a larger batch

Table 1 Augmented Dickey–Fuller test results to find non-stationary feature of smart meter data stream. Non-stationary data stream in range 36, 100, and 1000 is hard to predict

AMI meter pattern range	10	36	100	1000	100000
Test statistic	−4.55	−1.59	−2.66	−2.78	−14.65
Critical value (5%)	−3.29	−2.96	−2.89	−2.86	−2.86

is better in generalizing the underlying data distribution. However, it is unpractical to have enough batch because retraining with larger batch causes a higher data incorporation latency (DIL) [13]. In other words, delaying update to get enough batch size trade-offs the expense of stale DL model's inference execution. We refer to [13] to define the sum of waiting time of all the incoming instances DIL_i as follows when data instances are arrived as uniform distribution:

$$DIL_i(b_i; \lambda) = \lambda \sum_{j=0}^{b_i-1} j$$
$$= \lambda \frac{b_i(b_i - 1)}{2}$$
(2)

where b_i is batch size and $1/\lambda$ is data arrival rate. We should develop an algorithm which can make both DIL_i and the prediction error of the DL model are minimized.

3 Proposed System

3.1 Architecture Overview

The overall architecture of entire online learning system is illustrated in Fig. 1. The system receives data stream from multiple sources, which is served by one or a set of deployed DL models. The system is built upon the Apache Kafka [5], which supports exchange of large volume of data between *streaming data producers* and *incremental learning system*. *Streaming data producers*, set of each AMI source, collect the data in real-time manner and transfer the data to *incremental learning system*. *Incremental learning system* is responsible for maintaining multiple DL

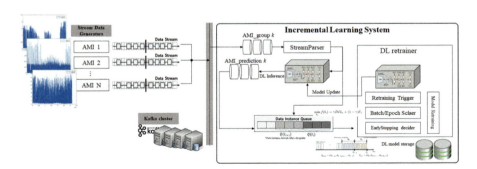

Fig. 1 A system architecture of incremental deep learning system for AMI stream data processing

model's performance. When a data instance is delivered to *incremental learning system*, it executes deep inference and effective retrain strategy with new data.

Incremental learning system deals with concept drift by performing update of deployed DL models. AMI streaming data come in the system then are grouped to produce pairs of instances X_i, yi to feed DL models. For each update, the module selects the batch size and epoch by considering current concept drift level and unstableness of current DL model. When concept drift is detected, earlystopping is applied. Otherwise, the model is trained without the earlystopping until convergence.

3.2 Proposed Incremental Learning Scheme

AMI datasets can be converted to supervised time-series datasets as further data arrived in the future timestamp, which mean that we can know the ground-truth values for the prediction target. The training loss value on the prediction model of the ith value can be represented $L(y_i, \tilde{y}_i)$ where y_i is the ith measured value and $\tilde{y}_i$ is the ith predicted value. In our incremental learning system, pretrained data instances are discarded without being stored on memory or storage.

Updating DL model over new data incurs nontrivial training cost, which is directly measured by the processing time. Training load of DL model is represented as sum of minibatch upload latency, feed-forward latency, back-propagation latency, and gradient transfer latency. In [13], they have shown that there exists linear relationship between processing time and input batch size $|D|$. We can approximate the processing time in multi-variable regression model as following:

$$\tau(e_i, b_i) = \theta_{proc} b_i e_i + \theta_{const} \tag{3}$$

where θ_{const} is the overhead of model initialization procedure in a specific device. θ_{proc} is the coefficient for the processing time.

We assume that a system receives incoming value v_t at time t, the deep learning model M takes a data sequence as an input X, and output y is also a sequence of data, which is ground-truth values of given input. Here we refer kth values grouped in X and y as a data instance $\mathbf{d}_k$. An arrived time of $\mathbf{d}_k$ is denoted as a_k. Hence, the length of the queue at time t is given by following:

$$Q(t) = \{d_k | t_{i-1} \leq a_k \leq t_i\} \tag{4}$$

Expectation of $|Q(t)|$ is τ_{i-1}, where τ_{i-1} indicates time interval between update decisions $(t_i - t_{i-1})$. Also, we assume that the data arrives as uniform distribution, with period of $1/\lambda$. When training of the ith update completes, the queue length is the summation of existing queue length and gaps of arriving and processed data in the queue which is formulated as follows:

$$|Q(t_i)| \leftarrow |Q(t_{i-1})| + \tau_{i-1}\lambda - b_i \tag{5}$$

We assume at the initial state of queue is empty. Therefore, the stability condition of the queue is simply defined as

$$\tau_{i-1}\lambda \leq \frac{\tau_{i-1} - \theta_{const}}{\theta_{proc}e_i} = b_i \tag{6}$$

With the condition described in the Eq. 6, we define an objective function as follows where γ is weight factor.

$$\min_{b_i} f(b_i) = \gamma DIL_i + (1-\gamma)U_i$$
$$= \gamma\lambda\frac{b_i(b_i-1)}{2} + (1-\gamma)exp(-b_i)\frac{\sum_{k=i-L}^{i-1} RMSE_k}{L} \tag{7}$$

Batch size selection for each update affects both the accuracy and DIL_i of the system. *Unstableness*, U_i, indicates how unstable the model performance is where $RMSE_k$ is the root mean square error of the inference on the kth batch, L is the length of the sliding window over the sequence. We used grid search to find the solution b_i for each update.

We can enhance the quality of solution and DL model training efficiency by selectively applying earlystopping. We argue that only when the incoming data distribution is stable, in other words, there is no concept drift, we can apply earlystopping. We use cosine similarity to detect the concept drift with data as shown in Fig. 2. By measuring the cosine similarity between $\mathbf{V}_{prev} = (\mathbf{d}_{k-b_i+1}, \ldots, \mathbf{d}_{k-1})$ and $\mathbf{V}_{curr} = (\mathbf{d}_k, \ldots, \mathbf{d}_{k+b_i-1})$ as in Eq. 8, it decides whether to apply earlystopping or not.

$$cossim(\mathbf{V}_{prev}, \mathbf{V}_{curr}) = \frac{\mathbf{V}_{prev} \cdot \mathbf{V}_{curr}}{||\mathbf{V}_{prev}||||\mathbf{V}_{curr}||} \tag{8}$$

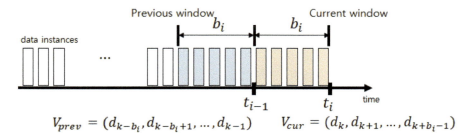

Fig. 2 Data windowing for cosine similarity calculation to identify concept drift

In addition, we define the loss function for the training with the data batch $\mathbf{B}_i = \{(X_1, y_1), \ldots, (X_{b_i}, y_{b_i})\}$ as shown in Eq. 9.

$$L(f(x, \omega_{i-1}), \mathbf{B}_i) = \frac{1}{|\mathbf{B}_i|} \sum_{(X_k, y_k) \in \mathbf{B}_i} l(f(X_k, \omega_{i-1}), y_k) \tag{9}$$

$f(x, \omega_{i-1})$ represents the running DL model after the $(i - 1)$th update. Also, $l(f(X_k, \omega_{i-1}), y_k)$ is the regular loss function such as mean squared error. The process of our incremental learning algorithm is summarized in Algorithm 1.

Algorithm 1 Incremental learning for datastream with adaptive earlystopping
___

Input
Update step: i
Current model: M_i
Data instance queue: $Q(t)$
Concept drift threshold: D_T
Model improvement threshold: TL_T
Earlystopping criteria: $patience$
Output
Model trained after $(i)^{th}$ update: M_{i+1}

1: Find (b_i, e_i) which minimize Eq. 7 by grid search
2: **if** $cossim(\mathbf{V}_{prev}, \mathbf{V}_{curr}) < D_T$ **then**
3:    $P_{earlystop} = True$
4: **else**
5:    $P_{earlystop} = False$
6: **end if**
7: $cnt patience = 0, ce = 0$
8: **if** $P_{earlystop}$ **then**
9:   **while** $ce < e_i$ **do**
10:     $TL[ce] = L(f(x, \omega_{i-1}), \mathbf{B}_i)$
11:     **if** $ce > 0$ and $|TL[ce] - TL[ce - 1]| < TL_T$ **then**
12:       $cnt patience \mathrel{+}= 1$
13:     **end if**
14:     **if** $cnt patience == patience$ **then**
15:       break
16:     **end if**
17:     $\omega_{i-1} \leftarrow \omega_{i-1} - \eta \frac{1}{|\mathbf{B}_i|} \sum_{(X_k, y_k) \in \mathbf{B}_i} \nabla l(f(X_k, \omega_{i-1}), y_k)$
18:     $ce \mathrel{+}= 1$
19:   **end while**
20: **else**
21:   **while** $ce < e_i$ **do**
22:     $\omega_{i-1} \leftarrow \omega_{i-1} - \eta \frac{1}{|\mathbf{B}_i|} \sum_{(X_k, y_k) \in \mathbf{B}_i} \nabla l(f(X_k, \omega_{i-1}), y_k)$
23:     $ce \mathrel{+}= 1$
24:   **end while**
25: **end if**
___

4 Experiment Results

4.1 Experimental Environment

We conducted experiments on AMI streaming data to evaluate the performance of our proposed online learning system. We used a cluster with both stream server and computing server as shown in Table 3. Stream server has Intel(R) Xeon(R) W3565 CPU @3.20 GHz with 16 GB RAM and computing server has Intel(R) Core(TM) i7-4790 CPU @3.60 GHz with 16 GB RAM. We used Apache Kafka framework [5] to construct stream server. We trained and tested the deep learning model with low-voltage AMI data from January 2015 to July 2018 provided by Korea Electric Power Corporation (KEPCO).

We used CNN-LSTM model [6], with one convolution layer to receive the input, one LSTM layer, and one Dense layer to produce the final output. Specification for the DNN is described in detail at Table 2.

4.2 Effects of Concept Drift Threshold

We observe the change in average training loss and prediction error when changing the concept drift threshold value, D_T. Figure 3 shows the average training loss of the entire streaming data over the threshold between 0.1 and 0.9. As clearly shown in the plot, we can see the average training loss goes much smaller with the threshold value over 0.8. It can be thought that training with more epoch according to the higher concept drift threshold (0.8 or higher) enhances the generalization capability of the deployed DL model for current data.

Figure 4 shows the average prediction error of the entire streaming data. Here it shows a low error at a high threshold ($D_T = 0.9$) similar to the average training loss case. We regard this is because when we increase the threshold value, the DL

Table 2 The DL model architecture used in the experiment

Layer type	Output shape	Number of parameters
Conv1D	(None, 4, 64)	1216
LSTM	(None, 50)	23000
Dense	(None, 1)	51
Number of total parameters: 24267		
Number of trainable parameters: 24267		
Number of non-trainable parameters: 0		

Table 3 The configuration of experiment environment

	Stream server	Computing server
CPU	Intel(R) Xeon(R) W3565 @3.20 GHz	Intel(R) Core(TM) i7-4790 @3.60 GHz
RAM	DDR3 16 GB	DDR3 16 GB

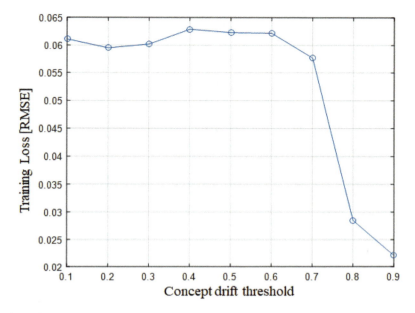

Fig. 3 Training loss of resultant model according to different concept drift threshold

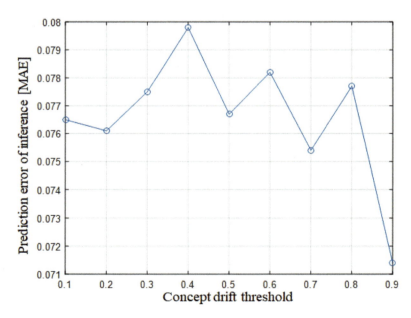

Fig. 4 Prediction error of inference of resultant model according to different concept drift threshold

model is trained on recent data with more epochs. In result, the DL model can have a better predictability on inexperienced data (Table 3).

4.3 Performance Comparison with Other Incremental Learning Algorithms

When comparing the performance with other algorithms, we set the concept drift threshold $D_T = 0.5$. Figures 5 and 6 show the training loss and the prediction error over each batch for different online learning algorithms, respectively. As we used untrained a DL model with the same initially weights, it shows high fluctuations over 0.2 by 15 data batches. We can see that our proposed method shows lower training loss than Continuum [13] (r=4) above the 20 data batches, as it shows training loss value only around 0.02. In terms of prediction error, it shows a similar trend to that of the training loss. All methods show a high degree of variation (0.2 or higher) within the first 20 batches and then gradually decrease to 0.1 over time.

5 Conclusions

In this paper, we proposed an online learning system working on AMI streaming data. We aimed to further reduce the training load by considering the concept drift and batch size effects in model updates. We first derived proper batch sizes and epochs by solving the objective function. Then the drift detection module decides

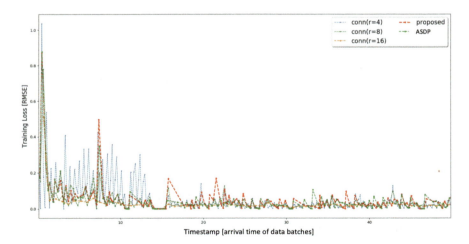

Fig. 5 Training loss over batches of AMI data

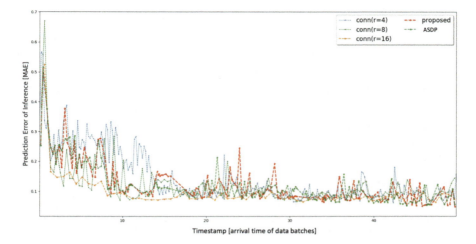

Fig. 6 Prediction error of inference over batches of AMI data

whether to apply the earlystopping technique in model training. Our proposed method utilizes a simpler metrics and objective function, but with higher efficiency.

To evaluate the system performance, we conducted experiments on a set of KEPCO AMI data from 2015 to 2018. Throughout the training process, we compared our proposed method with Continuum [13] and ASDP [1] with two metrics, namely training loss and prediction error of inference. The experimental results showed that our scheme obtained better performance than Continuum [13] (r=4) and were comparable to ASDP [1] on all the metrics.

In future work, we would like to improve our algorithms to achieve better performance. Several techniques should be applied, for example, adjusting the threshold of drift detection or the length of the sliding windows. Moreover, other factors such as learning rate should be also considered to obtain better training efficiency.

Acknowledgments This work was supported by Korea Electric Power Corporation (Grant number:R18XA05)

References

1. S. Bae, (An) accelerated streaming data processing scheme based on CNN-lSTM hybrid model in energy service platform. Master Thesis, KAIST (2018)
2. A.A. Benczúr, L. Kocsis, R. Pálovics, Online machine learning in big data streams (2018, preprint). arXiv:1802.05872
3. Y.-W. Cheung, K.S. Lai, Lag order and critical values of the augmented Dickey–Fuller test. J. Bus. Econ. Stat. **13**(3), 277–280 (1995)

4. Y. He, G.J. Mendis, J. Wei, Real-time detection of false data injection attacks in smart grid: a deep learning-based intelligent mechanism. IEEE Trans. Smart Grid **8**(5), 2505–2516 (2017)
5. Kafka. http://kafa.apache.org
6. T.-Y. Kim, S.-B. Cho, Predicting residential energy consumption using CNN-LSTM neural networks. Energy **182**, 72–81 (2019)
7. V. Losing, B. Hammer, H. Wersing, Incremental on-line learning: a review and comparison of state of the art algorithms. Neurocomputing **275**, 1261–1274 (2018)
8. Lu, J., et al., Learning under concept drift: a review. IEEE Trans. Knowl. Data Eng. (31)(12), 2346–2363 (2018)
9. E.M. Mocanu, P.H. Nguyen, M. Gibescu, A. Liotta, Big IoT data mining for real-time energy disaggregation in buildings, in *Proceedings of IEEE International Conference on Systems, Man and Cybernetics Budapest* (2016), pp. 9–12
10. R.R. Mohassel, et al., A survey on advanced metering infrastructure. Int. J. Electr. Power Energy Syst. **63**, 473–484 (2014)
11. P.M. Radiuk, Impact of training set batch size on the performance of convolutional neural networks for diverse datasets. Inf. Technol. Manag. Sci. **20**(1), 20–24 (2017)
12. D.M.J. Tax, P. Laskov, Online SVM learning: from classification to data description and back, in *IEEE XIII Workshop on Neural Networks for Signal Processing (IEEE Cat. No. 03TH8718)* (2003), pp. 499–508
13. H. Tian, M. Yu, W. Wang, Continuum: a platform for cost-aware, low-latency continual learning, in *Proceedings of the ACM Symposium on Cloud Computing* (2018)

Traceability Analysis of Patterns Using Clustering Techniques

Jose Aguilar, Camilo Salazar, Julian Monsalve-Pulido, Edwin Montoya, and Henry Velasco

1 Introduction

The current epochal moment is characterized by a very high rate of generation of new information, which requires means to characterize its evolution. In particular, traceability on the evolution of information is relevant in different fields: health, safety, process supervision, among other areas [1, 2]. For that task, it is necessary to study techniques that allow analyzing the evolution of the knowledge. A first element to consider, in that sense, is to identify the patterns that represent the common information in data groups. From the characterization of a "pattern," its evolution can be analyzed, that is, as it changes over time.

This article analyzes different techniques to the problem of tracking the traceability of the patterns. From the starting point of the identification of patterns that characterize groups of data, it is analyzed how each technique follows the changing dynamics of a pattern, that is, its evolution. From the evolution of the patterns,

---

J. Aguilar (✉)
GIDITIC, Universidad EAFIT, Medellin, Colombia
CEMISID, Universidad de Los Andes, Merida, Venezuela
e-mail: aguilar@ula.ve

C. Salazar · E. Montoya
GIDITIC, Universidad EAFIT, Medellin, Colombia
e-mail: jhenaos@eafit.edu.co

J. Monsalve-Pulido
Universidad EAFIT, Medellin, Colombia
e-mail: emontoya@eafit.edu.co

H. Velasco
LANTIA SAS, Medellin, Colombia
e-mail: hgvelascov@eafit.edu.co

© Springer Nature Switzerland AG 2021
H. R. Arabnia et al. (eds.), *Advances in Artificial Intelligence and Applied Cognitive Computing*, Transactions on Computational Science and Computational Intelligence, https://doi.org/10.1007/978-3-030-70296-0_19

aspects such as the re-association of the objects to the closest patterns can be established. Also, another aspect to consider is how to reestablish the relevance of a pattern in a given context, based on the idea of a threshold, such that if the characteristics of what is sought (for example, for a given profile of a person) are close to the characteristics of the pattern (based on a similarity threshold), it can be considered that this group of data associated with the pattern is relevant to what is sought.

In general, online unsupervised learning techniques (clustering) are good candidates for the traceability of patterns. These techniques follow an incremental approach of clustering that adds one cluster prototype or adjusts the known clusters at a time [3, 4]. In this work, the techniques LDA, Birch, LAMDA, and K-means are used, both for the initial task of grouping the data, as well as, to analyze the characteristics of the patterns during their evolution (traceability). The characteristics extracted from the text, which are the output of each technique, in the case of LDA are called topics, in the case of LAMDA classes, and so on. Thus, a topological model is constructed to describe the "patterns" generated from the grouping of the analyzed data, which are dynamically analyzed based on their evolution over time.

In addition, in this article different types of educational contents are used as data sources, to evaluate the capabilities of the techniques. In particular, they are learning resources, scientific publications, and patents. With these datasets, the analysis of the characteristics of the patterns is made, based on the concepts of topics, classes, etc. (according to the technique used), and their evolution over time is analyzed (traceability).

2 Literature Review About Approaches for Traceability

A first group of works is online clustering Approaches for Traceability. For example, in [5], they propose a semantic smoothing model in the context of text data flows for the creation of two online grouping algorithms (OCTS and OCTSM) oriented to the grouping of massive text data flows. The two algorithms present a new cluster statistics structure called a cluster profile, which can capture the semantics of text data streams dynamically. In [6] is proposed the CURE algorithm with a greater robustness for outliers, which identifies clusters that have non-spherical shapes and wide variations in size. In CURE, having more than one representative point per cluster allows the algorithm to adjust well to the geometry of the non-spherical shapes, and the reduction helps to dampen the effects of outliers.

In [7], the authors create a Framework that can be applied to most of the flow grouping methods to improve their performance in terms of grouping accuracy (purity, G-Recall, and G-Precision), using the concepts of drift and robustness for the noise samples. The main contribution is to deploy a set of classifiers to assign the incoming samples to microclusters. In [8], the authors propose an EM-based framework (Expectation Maximization) to effectively group distributed data flows.

In the presence of noisy or incomplete data records, the algorithms learn from the distribution of the underlying data flows by maximizing the probability of the data groups.

Tafaj et al. [9] presents a non-parameterized adaptive online algorithm to group eye movement data. In the paper, they conduct experiments demonstrating that the algorithm performs strongly and in real time on raw data collected from the eye tracking experiments in driving sessions. To solve the problem of classification of photos, [10] proposes to group the binary hash codes of a large number of photos in binary grouping centers. In the paper, they present a fast k-means binary algorithm that works directly on the image hashes that retain the similarity, and groups in binary centers with hash indexes to accelerate the calculation, which is capable of grouping large photo streams into a single machine, which can be used for spam detection and photo trend discovery.

A second group of approaches for Traceability, which is not based on online clustering algorithm is presented below. According to the problem of establishing and maintaining the appropriate traceability of the software, [11] demonstrates that supervised machine learning can be used effectively to automate the recovery of the traceability link, only if there is enough data to train a classification model. In this work, they develop an approach based on active learning, for traceability link recovery, while maintaining similar performance. In [12] is proposed an approach based on the semantic relationship (SR), which takes human judgment to an earlier stage of the tracking process by integrating it into the underlying recovery mechanism. SR tries to imitate the human mental model of relevance by considering a wide range of semantic relationships.

In the context of the problem of information retrieval, [13] presents an empirical study to statistically analyze the equivalence of several traceability recovery methods based on information retrieval (IR) techniques. In the paper, they use techniques such as the Jensen–Shannon (JS) method, the vector space model (VSM), the latent semantic indexing (LSI), and the latent Dirichlet allocation (LDA).

According to the literature review, there are no previous works with a similar goal to our paper: study the problem of tracking the traceability of patterns using clustering techniques.

3 Analyzed Techniques

- *LDA*: It is a probabilistic model based on unsupervised learning, which suppose each document like a mix of topics, and each topic has a probability distribution over all words in the vocabulary [14, 15]. The topic distribution reflects the overall semantic information about the document, expressed in the form of probability.

  The core idea is that each document contains several hidden topics, each of which contains a collection of words related to the topic [14]. LDA discovers the latent topics Z from a collection of documents D. For LDA, each document is a probability distribution over all words in the vocabulary. LDA model projects the documents in a topical embedding space, and generates a topic vector from a document, which can be used as the features of the document.

In this way, the LDA topic model defines two polynomial distributions [15]: the document-topic distribution (θ) and the subject vocabulary distribution (ϕ). The first represents the probability distribution of each topic in the document; and the other, the probability distribution of each vocabulary appearing in the topic. In addition, LDA model has three parameters [14, 15]: α is the parameter of the Dirichlet distribution of the topic distribution in a document, β is the parameter of the Dirichlet distribution of the vocabulary distribution in a topic, and K represent the number of topics.

LDA requires a learning phase, in order to infer/discover θ and ϕ in documents, which can be used to predict any new document with a similar topic distribution [15]. Different representations can be built from the documents, varying the amount of topics to be considered.

- *K-means*: It is one of the simplest and popular unsupervised machine learning algorithms. k-means aims to group n data samples into k disjoint clusters, each described by its centroid. In general, a data point is assigned to the nearest cluster.

  K-means starts with a set of k randomly centroids that are used as the beginning points for every cluster, and then performs iterative calculations to optimize the positions of the centroids. It stops the creation and/or optimization of the clusters when either: (i) The centroids have stabilized — there is no change in their values because the clustering has been successful; (ii) The defined number of iterations has been achieved.

  Its performance is usually not as competitive as those of the other sophisticated clustering techniques because slight variations in the data could lead to high variance.

- *LAMDA*: It is a clustering algorithm that uses the degree of adequacy to classify each individual. The analysis of the similarity compares the features of any object $X = [x_1; \ldots; x_j; \ldots; x_n]$, with those of the existing classes $C = C_1; C_2; \ldots; C_k; \ldots; C_m$ [16–18]. LAMDA is a non-iterative algorithm, and it was intended for use in system supervisory tasks and in the identification of functional states.

  The features of the objects are normalized to [0,1]. With normalized values, LAMDA calculates the marginal adequacy degree (MAD), which describes the similarity of any feature with the corresponding feature of the class. MADs are calculated with probability density functions. After obtaining the MADs, LAMDA calculates the global adequacy degrees (GADs) using aggregation functions T-norm and S-norm [16]. The GADs are computed for every class. In clustering tasks, the normalized object X is assigned to the group with the maximum GAD.

  More details about its mathematical formulation can be found in [16–18].

- *Birch*: It is a data incremental clustering method based on a distance-based approach with the next characteristics [19]: (i) It follows a local process, due to that, each clustering decision is made without scanning all data or all currently existing clusters. It uses metrics that reflect the natural closeness of the points, incrementally maintained during the clustering process; (ii) It exploits the idea that the data space is usually not uniformly occupied. For example, a dense region of points is treated collectively as a single cluster, and points in sparse regions are treated as outliers and optionally removed.

The concepts of Clustering Feature (CF) and CF tree are at the core of Birch's incremental clustering algorithm. A CF is a triple summarizing the information about a cluster. The CF vector of the cluster is defined as a triple: CF = (N, L, S), where N is the number of data points in the cluster, L is the linear sum of the N data points, and S is the square sum of the N data points.

A CF tree is a height-balanced tree with two parameters: branching factor B and threshold T. A CF tree will be built dynamically as new data objects are inserted. It uses the same procedure of the B+ trees to guide a new insertion in the tree (the correct subcluster) for clustering purposes. The CF tree is a very compact representation of the dataset because each entry in a leaf node is not a single data point, but a subcluster (which absorbs many data points with diameter (or radius) under a specific threshold T). More details about itsendalization can be found in [19].

4 Experiments

This section presents the experiments with three different datasets: one with 11967 patents, the second with 8307 scientific publications, and the third with 10050 learning objects. Each dataset is ordered according to the data chronology, and divided into 5 parts to simulate the flow of time: T0 with 40% of the data, and the rest with 15%.

4.1 Metrics

For the evaluation of the unsupervised clustering algorithms in our context, we use Calinski–Harabasz Index (used like intracluster metric), Davies–Bouldin Index (used like intercluster metric), and Silhouette Score (used to define the quality of the definition of the clusters).

- *Calinski–Harabasz Index*: This index is defined as the ratio of the sum of between-cluster dispersion and within-clusters dispersion for all clusters, where dispersion is the sum of distances squared (see Calinski and Harabasz [20] for an extended explanation). The Calinski–Harabasz index is calculated as

$$S_{CH} = \frac{trace(B_k)}{trace(W_k)} \times \frac{m-k}{k-1} \tag{1}$$

where m is the number of documents in dataset, k is the number of clusters, $B_k = \sum_{q=1}^{k} n_q (c_q - c)(c_q - c)^T$ is the between group dispersion matrix, and $W_k = \sum_{q=1}^{k} \sum_{x \in P_q} (x - c_q)(x - c_q)^T$ is the within-cluster dispersion matrix; with P_q the set of points in cluster q, c_q the center of cluster q, c the center of the set of data, and n_q the number of points in cluster q.

A higher Calinski–Harabasz score relates to a model with better defined clusters.

- *Davies–Bouldin Index*: It is defined as the average similarity measure of each cluster with its most similar cluster, where similarity is a measure that compares the distance between clusters with the size of the clusters themselves (see Davis and Bouldin [21] for an extended explanation). The Davies–Bouldin index is calculated as

$$S_{DB} = \frac{1}{k} \sum_{i=1}^{k} \max_{i \neq j} R_{ij} \tag{2}$$

where R_{ij} is the similarity between clusters i and j. There are many ways to construct R_{ij}, a simple one (that is non-negative and symmetric) is $R_{ij} = \frac{s_i + s_j}{d_{ij}}$, with s_i is the average distance between each point of the cluster i and the center of the cluster j; this is also known as the cluster diameter and d_{ij} is the distance between centroids of the clusters i and j.

Zero is the lowest possible Davies–Bouldin index. Values closer to zero indicate a better partition, so a lower score relates to a model with better separation between the clusters.

- *Silhouette Score*: It is a measure of the cohesion compared with the separation in clusters of data. It determines how similar an object is in its own cluster, compared to other clusters. In our case, it describes the forms of the clusters that it produces (if they are dense, well separated, etc.). See Rousseeuw [22] for an extended explanation. The Silhouette Coefficient for a set of samples is given as the mean of the Silhouette Coefficient for each sample, and it is calculated as

$$S_S = \sum_{i=1}^{n} \frac{b(i) - a(i)}{\max\{a(i), b(i)\}} \tag{3}$$

where $a(i)$ and $b(i)$ are computed for each sample i in the cluster C_i ($i \in C_i$) as $a(i) = (|C_i| - 1)^{-1} \sum_{j \in C_i, i \neq j} d(i, j)$ and $b(i) = \min_{k \neq i} |C_k|^{-1} \sum_{j \in C_k} d(i, j)$, where $d(i, j)$ is the distance between data points i and j.

This coefficient is ranged from -1 to 1. Values near 1 are desirable, near 0 indicate overlapping clusters, and negative values generally indicate that a sample has been assigned to the wrong cluster. In general, a higher Silhouette Coefficient score relates to a model with better defined clusters.

4.2 Results

4.2.1 General Results

In this section, we are going to present the results for each technique:

- *LDA*: it is not conceived as a clustering technique, but as a topic modeling technique [23]. However, the topics are kind of groups that follows the idea of clustering. LDA is considered in this work because of its capability of grouping documents on topics. Input to LDA model are documents, the embedding vectors of documents are used to calculate metrics, and the vectors of topics are the output of LDA model. A label for each document is calculated as the id of the topic to which the document belongs most. With these embedding vectors of documents and labels, metrics are calculated. LDA has a main parameter that is the number of topics (k). For tuning this parameter, coherence score [24] is used. The assumption is: the k with the highest coherence score is the best k. Best parameters found are in the case of Patents k=12, for Scientific Publications is k=5, and for Learning Objects: k=7
- *K-means*: In the case of K-means, the main parameter is K. In order to determine K, this paper uses the "elbow" method to select the optimal number of clusters. In our work, we use the Elbow method to have a range of K values (around the elbow) like the adequate number of clusters, and we select the K with the best results. This value is determined in the initial clustering, and during the evolutionary analysis of the patterns K is not modified.
- *LAMDA*: In the case of LAMDA, the main parameters to define are the threshold to define the creation of new clusters, known as the Non-Informative Class (NIC), and the function to calculate the MADs. The cluster with maximum GAD is where the individual is assigned, but if the maximum GAD is the one corresponding to NIC, then a new cluster is created. In our case, according to the results in [17, 18], the NIC must be equal to 0.5. Finally, the MAD calculations are made for each descriptor in each cluster using the fuzzy binomial function, also, according to the results in [17, 18].
- *Birch*: It is an efficient clustering algorithm for large amounts of data [19]; however, "Birch does not scale very well to high dimensional data", as mentioned in sklearn (python's library) documentation. Birch has two main parameters: threshold and branching factor. Threshold controls the radius of subclusters obtained, by merging them if their fused radio is lower than the threshold. Branching factor is the maximum number of subclusters in each node of the main tree defined in the algorithm. For tuning these two parameters, a Bayesian optimization is carried out over a predefined space for the two parameters. The best parameters found are: for Patents, threshold = 0.01937 and branching factor = 37; for Scientific Publications, threshold = 0.01914 and branching factor = 85; for Learning Objects, threshold = 0.9928 and branching factor = 223.

4.2.2 Analysis of the Results

The Tables 1, 2 ,and 3 describe the performance of clustering through time with the dataset of Scientific publications, Learning Objects, and Patents. For the Silhouette Score, LAMDA has the best results. It defines the best forms of the clusters, with a score close to zero (some overlapping in the clusters). For the rest of the techniques, the values are negative (some incorrect in the clustering process) or close to zero (like K-means, with overlapping in the clusters).

In the case of the Calinski–Harabasz Index, the results are very good (high) for LDA and Birch in the learning object dataset. For the rest of the techniques and datasets, the results are close to zero. This score establishes that the clusters are dense and well separated but the learning object dataset contains convex clusters (clusters with relationships, which are normally hidden by static methods). In the case of K-means, this is not an adequate metric, because it forces the clusters to be separated, for this reason the results are very high.

For Davies–Bouldin Index, the best results are for LAMDA (very close to zero), which indicates a better partition. In the case of K-means, the results are not good, which indicates not a good separation between the clusters defined by K. In general, when convex clusters are defined, this metric is high (that is the reason that for LDA the results are very bad).

LDA analyses very well contexts where convex clusters are a very good description of their dataset (measured with the Calinski–Harabasz Index) like the learning objet dataset, and for the rest of datasets, LAMDA gives the best results.

At the level of the traceability, these metrics can describe the evolution of the data over the time without problem (the quality of the results does not change). Therefore, it can be internally analyzed how the characteristics of its descriptors change over time. An example of how this analysis can be performed is presented in the next section.

Table 1 Performance of clustering through time with scientific publications dataset

		T0	T1	T2	T3	T4
Silhouette score	LDA	0.0034	−0.0065	−0.01635	−0.0151	−0.0138
	K-means	0.0076	0.0195	0.0293	0.0320	0.0342
	LAMDA	0.5897	0.5009	0.5211	0.5060	0.5078
	Birch	−0.0166	−0.0165	−0.0175	−0.0191	−0.0198
Calinski–Harabasz score	LDA	60.1144	214.5736	236.6688	279.4421	325.1516
	K-means	83.5755	342.4204	502.9372	599.0646	694.7954
	LAMDA	0.0822	0.0652	0.0654	0.0663	0.0670
	Birch	1.3167	1.9597	2.102	2.1137	2.1163
Davies–Bouldin score	LDA	15.5824	16.0244	14.6236	13.9489	15.893
	45K-means	10.7797	9.1594	9.0639	8.9143	8.8358
	LAMDA	0.0801	0.0797	0.0797	0.0798	0.0790
	Birch	0.9527	0.9538	0.963	0.9727	0.9755

Traceability Analysis of Patterns Using Clustering Techniques

Table 2 Performance of clustering through time with patents dataset

		T0	T1	T2	T3	T4
Silhouette score	LDA	−0.0042	−0.0101	−0.0138	−0.0162	−0.0172
	K-means	0.0018	0.0088	0.0126	0.0147	0.0153
	LAMDA	0.3666	0.3660	0.3666	0.3654	0.3666
	Birch	−0.0688	−0.068	−0.0679	−0.068	−0.0681
Calinski–Harabasz score	LDA	1.3176	2.3466	3.9227	5.6527	7.3945
	K-means	30.0962	100.0109	154.6084	201.3089	241.8456
	LAMDA	0.0845	0.0652	0.0734	0.0649	0.0732
	Birch	1.6181	2.811	3.7292	4.4192	5.0178
Davies–Bouldin score	LDA	27.8275	31.9796	33.9581	36.8672	41.0165
	K-means	12.0242	11.7594	11.8662	11.8965	11.7849
	LAMDA	0.0798	0.0792	0.0799	0.0792	0.0792
	Birch	1.2887	1.3322	1.3619	1.3776	1.392

Table 3 Performance of clustering through time with learning objects dataset

		T0	T1	T2	T3	T4
Silhouette score	LDA	−0.0401	−0.0409	−0.0384	−0.0428	−0.0416
	K-means	0.0970	0.0923	0.0861	0.0856	0.0864
	LAMDA	0.1291	0.10126	0.1132	0.1126	0.1128
	Birch	0.1151	0.1142	0.1189	0.1254	0.1233
Calinski–Harabasz score	LDA	40.1494	61.2242	85.43	114.2579	143.2084
	K-means	313.0630	403.8315	487.9524	580.7843	673.0007
	LAMDA	0.0339	0.038	0.0335	0.0335	0.0337
	Birch	71.2864	86.5856	101.8402	115.7936	126.5196
Davies–Bouldin score	LDA	8.9262	7.9996	6.7594	6.0928	5.8652
	K-means	1.6866	1.6865	1.7094	1.7069	1.6893
	LAMDA	0.0339	0.0330	0.0339	0.0338	0.0348
	Birch	0.9649	0.9432	0.9058	0.8828	0.8769

4.3 Example of Analysis of the Traceability of the Patterns

In this section, the objective is to analyze the temporal behavior and the traceability of the patterns discovered from the data, using the previously described techniques. For that, it is used one of the technique, in order to give an example of the analysis. LDA is chosen, because the utilization of the concept of topics helps the analysis.

For LDA, it is analyzed the topic evolution through time, using the following figures. Topics are dimensionality reduced to 2D using Principal Coordinates Analysis (PCoA), and it is visualized its main two coordinates. The size of the circles is the marginal topic distribution, the bigger the more relevant the topic is. The distance in the figure is related to the inter-topic distance and gives an idea on how related topics are.

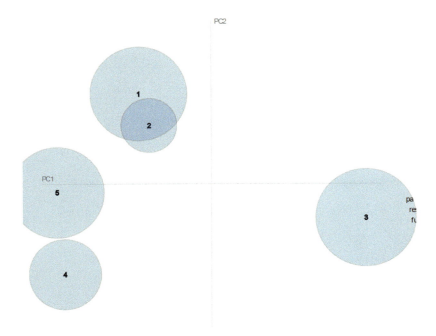

Fig. 1 Scientific publication topics at T0

In Fig. 1 are shown the results for the scientific publication dataset at T0. At T0, a total of 5 topics are extracted from the scientific publication dataset, there is no one relevant topic, but topic 2 has low relevance and presents high overlapping with topic 1 (see Fig. 1).

There are a lot of movements from T0 to T2 (see Fig. 2), especially with topic 2, which seemed replaceable in T0: it does not show an increase in relevance, but represents an important region of documents, since it is well separated from other topics. At this time, the topics are scattered.

Little movement is observed at T4, from T2 to T4 there is no big difference, which shows slow convergence (see Fig. 3). Evidence of well-defined topics at this time is that overlapping is not a main problem.

Now, we carry out a similar analysis for the learning object dataset. Figure 4 describes T0. Five topics are taken into account at T0. There is much overlap between topics 1, 4, and 7, which is evidence that at this time these topics could be combined taking into account that their relevance are similar and their overlapping are highs.

In T2, the overlapping is still a big problem, more than half of the topics have a high overlap (see Fig. 5). Nevertheless, the rest of the topics are well separated and have good relevance, which is a good indication.

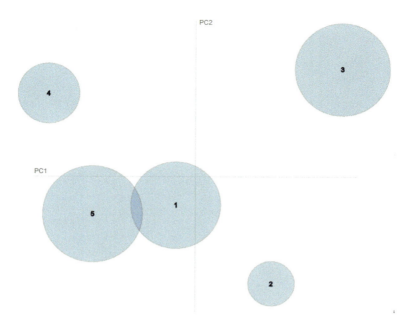

Fig. 2 Scientific publications at T2

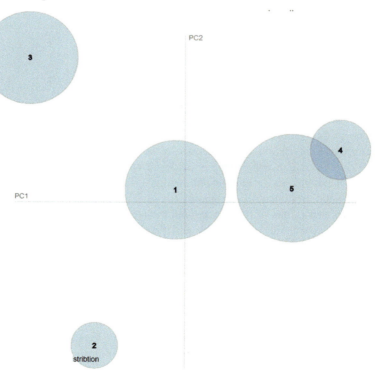

Fig. 3 Scientific publication topics at T4

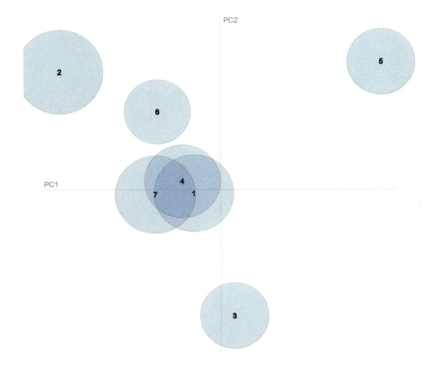

Fig. 4 Learning object topics at T0

At T4, the problem of overlapping is improved because topic 4 separates from the group that presents high overlap (see Fig. 6). Also, there is evidence of convergence, there is no a dominant topic, because the topics have average relevance.

Finally, we analyze the patent dataset, which shows a similar behavior with respect to previous datasets. There is evidence of convergence of all 3 datasets, especially of patents and scientific publications. The learning objects show slow convergence and problems of overlapping topics. This may be due to the similarity between the descriptions of the patents and the abstracts of scientific publications, which present reasonable similarities and therefore comparable behaviors. The descriptions (text) of the learning object dataset are considerably different, and therefore, have a different behavior.

This analysis of tracking the traceability of the patterns is similar for the rest of techniques for each dataset. In each case, we can consider different things, for example, in the case of LAMDA the behavior of the classes, and for K-means and Birch the generated features. In all the three cases, little movement can be observed from T2 to T4, that is, it is evident that with this data the topics tend to converge. In other words, even if new data arrives, the clusters suffer almost no modification.

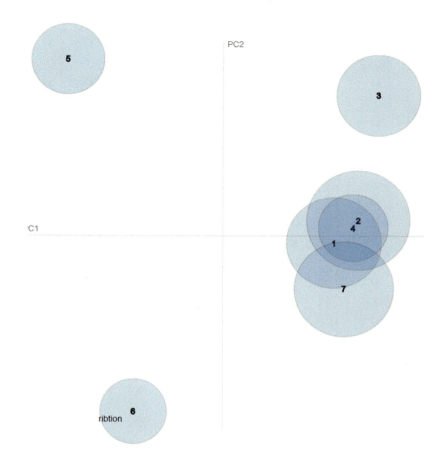

Fig. 5 Learning object topics at T2

5 Conclusions

This article studies the capabilities of different clustering techniques to the problem of tracking the traceability of the patterns. Particularly, the paper studies the evolution of the characteristics of the patterns over time.

In general, different unsupervised learning techniques (clustering) have been studied: LDA, Birch, LAMDA, and K-means. These techniques have been used to follow an incremental approach of clustering that adds one cluster prototype or adjusts the known clusters at a given time.

The paper analyzes these techniques for both, the initial task of grouping the data, as well as, to analyze the characteristics of the patterns, and the relevance of them in the patterns through their evolution (traceability). Different types of data sources of educational contents have been used: learning resources, scientific publications,

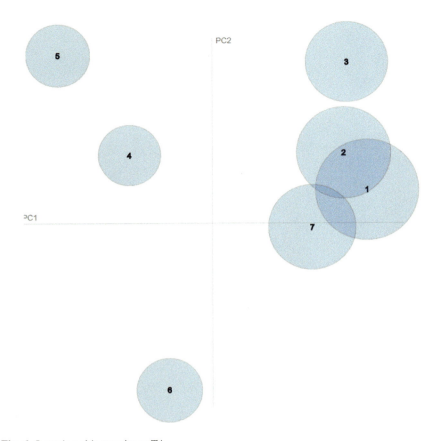

Fig. 6 Learning object topics at T4

and patents. With these datasets, the topological models to describe the "patterns" generated from the grouping of the analyzed data, and their dynamics (evolution over time), are analyzed (traceability). According to our literature review, there are no similar works.

For the evaluation, the paper considers three metrics: Calinski–Harabasz Index (used like intracluster metric), Davies–Bouldin Index (used like intercluster metric), and Silhouette Score (used to define the quality of the definition of the clusters).

In general, for the Silhouette Score, LAMDA has the best results. In the case of the Calinski–Harabasz Index, the results are very good (high) for LDA and Birch in the learning object dataset. For the rest of the techniques and datasets, the results are close to zero. In the case of the Davies–Bouldin Index, the best results are for LAMDA.

We observe that the tracking of the traceability of the patterns can be carried out with these techniques, using the different notions that each one considers during its clustering process. For example, the topic notion in LDA, in the case of LAMDA its classes, and for K-means and Birch the generated features. Future works will carry out a detailed study about this capability of these techniques.

Acknowledgments This work has been supported by the project 64366: "Contenidos de aprendizaje inteligentes a través del uso de herramientas de Big Data, Analtica Avanzada e IA"—Ministry of Science—Government of Antioquia—Republic of Colombia.

References

1. G. Aiello, M. Enea, C. Muriana, The expected value of the traceability information. Eur. J. Oper. Res. **244**(1), 176–186 (2015)
2. J. Aguilar, Resolution of the clustering problem using genetic algorithms. Int. J. Comput. **1**(4), 237–244 (2007)
3. J. Beringer, E. Hüllermeier, Online clustering of parallel data streams. Data Knowl. Eng. **58**(2), 180–204 (2006)
4. W. Barbakh, C. Fyfe, Online clustering algorithms. Int. J. Neural Syst. **18**(3), 185–194 (2008)
5. Y.-B. Liu, J.-R. Cai, J. Yin, A.-C. Fu, Clustering text data streams. J. Comput. Sci. Technol. **23**(1), 112–128 (2008)
6. S. Guha, R. Rastogi, K. Shim, Cure: an efficient clustering algorithm for large databases, in *Proceedings of the 1998 ACM SIGMOD International Conference on Management of Data, SIGMOD'98* (Association for Computing Machinery, New York, 1998), pp. 73–84
7. H.T. Zadeh, R. Boostani, A novel clustering framework for stream data un nouveau cadre de classifications pour les données de flux. Can. J. Electr. Comput. Eng. **42**(1), 27–33 (2019)
8. A. Zhou, F. Cao, Y. Yan, C. Sha, X. He, Distributed data stream clustering: a fast EM-based approach, in *Proceedings - International Conference on Data Engineering* (2007), pp. 736–745
9. E. Tafaj, G. Kasneci, W. Rosenstiel, M. Bogdan, Bayesian online clustering of eye movement data, in *Eye Tracking Research and Applications Symposium (ETRA)* (2012), pp. 285–288
10. Y. Gong, M. Pawlowski, F. Yang, L. Brandy, L. Boundev, R. Fergus, Web scale photo hash clustering on a single machine, in *Proceedings of the IEEE Computer Society Conference on Computer Vision and Pattern Recognition*, vol. 7 (2015), pp. 19–27
11. C. Mills, J. Escobar-Avila, A. Bhattacharya, G. Kondyukov, S. Chakraborty, S. Haiduc, Tracing with less data: active learning for classification-based traceability link recovery, in *2019 IEEE International Conference on Software Maintenance and Evolution (ICSME)* (2019), pp. 103–113
12. A. Mahmoud, N. Niu, S. Xu, A semantic relatedness approach for traceability link recovery, in *2012 20th IEEE International Conference on Program Comprehension (ICPC)* (2012), pp. 183–192
13. R. Oliveto, M. Gethers, D. Poshyvanyk, A. De Lucia, On the equivalence of information retrieval methods for automated traceability link recovery, in *2010 IEEE 18th International Conference on Program Comprehension* (2010), pp. 68–71
14. P. Huaijin, W. Jing, S. Qiwei, Improving text models with latent feature vector representations, in *2019 IEEE 13th International Conference on Semantic Computing (ICSC)* (2019), pp. 154–157
15. Q. Liang, P. Wu, C. Huang, An efficient method for text classification task, in *Proceedings of the 2019 International Conference on Big Data Engineering, BDE 2019* (ACM, New York, 2019), pp. 92–97. http://doi.acm.org/10.1145/3341620.3341631
16. J. Waissman, R. Sarrate, T. Escobet, J. Aguilar, B. Dahhou, Wastewater treatment process supervision by means of a fuzzy automaton model, in *Proceedings of the 2000 IEEE*

International Symposium on Intelligent Control. Held Jointly with the 8th IEEE Mediterranean Conference on Control and Automation (Cat. No. 00CH37147) (IEEE, Piscataway, 2000), pp. 163–168

17. J. Aguilar-Martin, R.L. De Mantaras, The process of classification and learning the meaning of linguistic descriptors of concepts, in *Approximate Reasoning in Decision Analysis*, vol. 1982 (North-Holland, Amsterdam, 1982), pp. 165–175

18. L. Morales, C.A. Ouedraogo, J. Aguilar, C. Chassot, S. Medjiah, K. Drira, Experimental comparison of the diagnostic capabilities of classification and clustering algorithms for the QoS management in an autonomic IoT platform. Serv. Oriented Comput. Appl. **13**, 199–219 (2019)

19. T. Zhang, R. Ramakrishnan, M. Livny, Birch: an efficient data clustering method for very large databases. ACM Sigmod Rec. **25**(2), 103–114 (1996)

20. T. Caliński, J. Harabasz, A dendrite method for cluster analysis. Commun. Stat. Theory Methods **3**(1), 1–27 (1974)

21. D.L. Davies, D.W. Bouldin, A cluster separation measure. IEEE Trans. Pattern Anal. Machine Intell. **2**, 224–227 (1979)

22. P.J. Rousseeuw, Silhouettes: a graphical aid to the interpretation and validation of cluster analysis. J. Comput. Appl. Math. **20**, 53–65 (1987)

23. D.M. Blei, A.Y. Ng, M.I. Jordan, Latent Dirichlet allocation. J. Mach. Learn. Res. **3**, 993–1022 (2003)

24. M. Röder, A. Both, A. Hinneburg, Exploring the space of topic coherence measures, in *Proceedings of the Eighth ACM International Conference on Web Search and Data Mining* (ACM, New York, 2015), pp. 399–408

An Approach to Interactive Analysis of StarCraft: BroodWar Replay Data

Dylan Schwesinger, Tyler Stoney, and Braden Luancing

1 Introduction

In the real-time strategy (RTS) game StarCraft: BroodWar, a player controls an armed force with the goal of defeating an opposing armed force. The gameplay includes acquiring resources, training units, researching technologies, and managing groups of units to attack the opponent. Decision making in a StarCraft can be coarsely divided into three tasks: strategy (high level decision making concerning army composition), tactics (positioning of units and buildings and timing of attacks), and micro-management (controlling individual units). For a good overview of StarCraft AI approaches see [6].

In this paper, we focus on an approach to interactively explore and analyze Star-Craft replay data in order to facilitate the creation of decision making algorithms. Our approach is inspired by a typical data science workflow where the process roughly has the following steps: (1) formulate a question, (2) acquire and clean data, (3) conduct exploratory data analysis, and (4) use prediction and inference to draw conclusions.

The main contribution of this work is in the representation of the data. We represent the data as statements in first-order logic. This is in contrast to the more common approach in data science of representing data in tabular form. The advantage of representing the data as FOL statements is that a logic programming language can be used to query the data and deduce new facts.

The remainder of the paper is organized as follows. First we provide some background material on logic programming. Next we present our encoding of the basic facts about StarCraft and replay data into a usable form to make deductive

---

D. Schwesinger (✉) · T. Stoney · B. Luancing
Kutztown University, Kutztown, PA, USA
e-mail: schwesin@kutztown.edu

© Springer Nature Switzerland AG 2021
H. R. Arabnia et al. (eds.), *Advances in Artificial Intelligence and Applied Cognitive Computing*, Transactions on Computational Science and Computational Intelligence, https://doi.org/10.1007/978-3-030-70296-0_20

queries. Then we present several examples of exploratory data analysis relevant to StarCraft AI research.

2 Logic Programming

In this section, we briefly describe the basic syntax and semantics of a logic programming language [1]. Logic programming is a paradigm where programs are sets of sentences in logical form expressing facts and rules about some problem domain. A computation is initiated when a query is run over the facts and rules that produces new facts and deductions.

In Datalog and other common logic programming languages, rules are written as logical implications of the form

$$H \leftarrow B_1 \wedge \ldots \wedge B_n$$

where H is called the head of the rule and $B_1 \wedge \ldots \wedge B_n$ is called the body of the rule. The rule can be read "H is true if B_1 and $\ldots$ and B_n is true."

2.1 Encoding Domain Knowledge in Datalog

Datalog has three basic constructs: facts, rules, and queries. Facts and rules are used to encode domain knowledge. A set of facts and rules that encodes domain knowledge is often referred to as a knowledge base.

A fact is a mathematical relation (not a function) of the form

$$predicate(term_1, \ldots, term_n)$$

where the *predicate* is the name of the relation and each term is a constant. Semantically, a fact is a ground term, that is, a logical statement that does not contain any free variables.

Rules are used to define how new facts can be inferred from known facts. A rule has the form

$$H : - B_1, \ldots, B_n$$

where H and $B_1, \ldots, B_n$ are predicates. The comma (,) between the predicates in the body denotes logical conjunction. The terms in a rule can be either a constant or a logical variable. Syntactically, logic variables start with a capital letter.

As an example, suppose we want to store some facts about which StarCraft units are considered to be biological for the purposes of the Medic's healing ability. The Datalog knowledge base could be encoded as three facts:

An Approach to Interactive Analysis of StarCraft: BroodWar Replay Data

```
biological ( firebat ).
biological ( marine ).
biological ( medic ).
```

These facts can be read as "a firebat is biological" and so on.

We can also add a rule that determines which units can be healed by a medic:

```
can_be_healed (X)  :−  biological (X).
```

This can be read as "X can be healed if X is biological." This statement is semantically equivalent to the universally quantified logical statement

$$\forall x.biological(x) \rightarrow canbehealed(x).$$

Given this rule and the previously defined facts we could deduce, say, that a firebat can be healed by a medic.

2.2 Datalog Queries

Datalog queries are used to deduce new facts from the existing facts and rules. A Datalog interpreter is needed to execute a query. Unfortunately, there is no standard Datalog implementation. Part of the reason for this is that different interpreter implementations provide various extensions to basic Datalog, such as aggregation functions. The Datalog Education System [8] provides an open source implementation of a Datalog interpreter. This interpreter has an interactive read–evaluate–print–loop (REPL) that offers a workflow similar to performing exploratory data analysis in Python or R.

Since there is no standard syntax for a query, for the sake of this paper, we denote a query as

$$?- H_1 : - B_1, \ldots, B_n.$$

where the question mark with a dash (? -) indicates that the rule that follows is a query that needs to be executed. The head of the rule and any conjuncts are optional. Here we treat the ? - as the prompt of the interactive REPL.

In Datalog, a query returns a set of all predicates that satisfy it. If no predicate satisfies the query, then nothing is returned. To continue the example from the previous section, the following represents an interactive query session:

```
?−  biological ( firebat ).
biological ( firebat )
?−  biological ( dragoon ).
?−
```

Here the first query returns a single predicate since it is a fact in our knowledge base. The second query returns nothing because the fact is not present in knowledge base. The last line indicates that the interpreter is awaiting the next query. We will omit the last line from future examples.

Datalog queries can also include logical variables. In this case, the Datalog interpreter returns a set of predicates where the logical variables are substituted with constants that satisfy the query. For example

```
?- biological(X).
biological(firebat),
biological(marine),
biological(medic)
```

results in all predicates where the logical variable X (logical variables start with a capital letter) can be substituted with a valid ground constant. In this case, we get all of the facts from the knowledge base.

A query can also return results that are derived from rules. For example

```
?- can_be_healed(X).
can_be_healed(firebat),
can_be_healed(marine),
can_be_healed(medic)
```

returns predicates derived from the `can_be_healed` rule in the knowledge base by logical variable substitution.

3 Knowledge Representation

In this work, we keep two separate knowledge bases: one that encodes some basic facts about the units in StarCraft and a second one that encodes the facts from a specific set of replay data. A description of each encoding is described in the following sections.

3.1 StarCraft Domain Knowledge

The game of StarCraft involves gathering resources in the form of minerals and Vespene gas. These resources can be used to create buildings, research new technologies, and train attack units. Our goal is to encode the relationships between these domain objects and minimize the number of predicates needed. Figure 1 lists the predicates that we chose to use for this work. Note that the predicates do not encode every aspect of the game, for example there is no information about special abilities that require energy or what bonuses become available when certain

An Approach to Interactive Analysis of StarCraft: BroodWar Replay Data 255

Predicate	Description
$faction(X, F)$	unit X is a member of faction $F \in$ {protoss, terran, zerg}
$unit\_type(X, T)$	unit X has type $T \in$ {building, add_on, air_unit, ground_unit, upgrade}
$unit\_size(X, S)$	unit X has size $S \in$ {small, medium, large}
$requires(X, Y)$	unit X directly depends on unit Y for construction
$builds(X, Y)$	unit X builds unit Y
$build\_time(X, T)$	unit X requires T seconds of time for construction
$supply(X, S)$	unit X provides S supply
$mineral\_cost(X, M)$	unit X costs M minerals to construct
$gas\_cost(X, G)$	unit X costs G Vespene gas to construct
$supply\_cost(X, S)$	unit X costs S supply to construct
$hitpoints(X, H)$	unit X has a maximum hitpoint value of H
$armor(X, A)$	unit X has an armor value of A
$air\_attack\_damage(X, D)$	unit X does D points of damage to air units
$air\_attack\_range(X, R)$	unit X has an air attack range of R cells
$air\_attack\_type(X, T)$	unit X has an air attack of type $T \in$ {concussive, explosive, normal, splash}
$air\_attack\_cooldown(X, C)$	unit X has an air attack cooldown time of C seconds
$ground\_attack\_damage(X, D)$	unit X does D points of damage to ground units
$ground\_attack\_range(X, R)$	unit X has an ground attack range of R cells
$ground\_attack\_type(X, T)$	unit X has an ground attack of type $T \in$ {concussive, explosive, normal, splash}
$ground\_attack\_cooldown(X, C)$	unit X has an ground attack cooldown time of C seconds
$sight(X, S)$	unit X has a sight range of S cells
$speed(X, S)$	unit X has a movement speed of S cells per second
$ability(X, A)$	unit X has ability A
$adds\_onto(X, Y)$	Terran add-on X can be added onto a Terran building of type Y
$morphs\_into(X, Y)$	Zerg unit X can morph into a Zerg unit of type Y

Fig. 1 Table of predicates used to encode the basic knowledge of StarCraft BroodWar units. Time is based on the fastest game speed and a cell is 32×32 pixels

technologies are researched; these additional facts could easily be added to the knowledge base, but we did not require them for our initial data analysis.

The data for the knowledge base was obtained from Liquipedia[1] a website that contains a wealth of information about StarCraft: BroodWar. There are a few things to note about the measurement units from this source of data. The predicates that have a time component use seconds as the time unit. This is based on the fastest game speed. These time values can be adjusted appropriately for different game speeds. The spatial measurement unit is in cells where a cell is a 32×32 pixel area.

The predicates listed in Fig. 1 allow for more complex queries than the examples in the previous section. In particular, Datalog makes it relatively easy to make recursive queries. For example, the `requires` predicate has facts about the direct dependencies that need to be met in order to construct a specific unit; here is an example for the Terran Ghost

```
?- requires(ghost, X).
requires(ghost, academy),
requires(ghost, barracks),
```

---

[1] https://liquidpedia.net/starcraft/Main_Page.

```
requires(ghost, covert_ops)
```

But, what if we wanted to recursively know all the requirements that must be met in order to train a Terran ghost? We can introduce a recursive rule for this purpose with the base case

```
depends_on(X, Y) :- requires(X, Y).
```

and the recursive case

```
depends_on(X, Z) :-
    requires(Y, Z),
    depends_on(X, Y).
```

which can be read as "X depends on Y if X requires Y" or "X depends on Z if there is a Y that depends on Z and X depends on that Y." With this rule we can obtain the desired information.

```
?- depends_on(ghost, X).
depends_on(ghost, academy),
depends_on(ghost, barracks),
depends_on(ghost, command_center),
depends_on(ghost, covert_ops),
depends_on(ghost, factory),
depends_on(ghost, science_facility),
depends_on(ghost, starport)
```

We can modify this query to obtain additional information. For example, we can obtain the mineral cost of every dependency:

```
?- answer(ghost, X, Y) :-
    depends_on(ghost, X),
    mineral_cost(X, Y).
answer(ghost, academy, 150),
answer(ghost, barracks, 150),
answer(ghost, command_center, 400),
answer(ghost, covert_ops, 50),
answer(ghost, factory, 200),
answer(ghost, science_facility, 100),
answer(ghost, starport, 150)
```

In this example, the logical variable X in the body of the rule is bound to the same value for both of the conjuncts.

4 Replay Knowledge Representation

There are several sources of publicly available curated replay datasets. For this work, the authors examined the datasets described in [7] and [5] because of the amount of game state information available per recorded frame. We ultimately chose [5] as the source of replay data because the full game state is recorded in constant, rather than adaptive time intervals and the dataset is an order of magnitude larger. Each dataset uses BWAPI[2] to obtain the state information which affects how we chose to represent the game state.

The knowledge representation for replay data that we chose for this paper involves the predicate

```
has(Player, Id, Unit, X, Y, Time).
```

where `Player` is the integer number of the player $\in \{-1, 0, 1\}$: the player -1 contains map information, such as mineral field locations, the first player is 0, and the second player is 1, `Id` is the identification number assigned to the unit from the BWAPI interface, `Unit` is the name of the unit, `X` and `Y` are the positional 2D coordinates with a unit of cells (32×32 pixels), and `Time` is the time in seconds. In this work, we chose to sub-sample the data to the granularity of seconds; each predicate has an integer value for `Time`.

While this representation does not contain all of the available game state information, it is sufficient for demonstration purposes. Furthermore, it is relatively straightforward to encode additional state information by adding additional terms or predicates. One advantage of using a single predicate for replay information is in the flexibility it offers for making queries. For example, suppose we know that player one is Zerg and we want to get information the Overlord units 5 min into the replay. The following query can be used to obtain the desired information:

```
?- has(1, Id, overlord, X, Y, 300).
```

Additionally, most Datalog implementations are capable of dynamically updating facts and rules, typically with `assert` and `retract` predicates that add and remove facts and rules respectively. For example, suppose we plan to make several queries involving only player 0, but we do not require the unit locations. A new rule can be dynamically added:

```
assert(p0_units(Id, Unit, T) :-
    has(0, Id, Unit, , , T)).
```

where the underscores represent logical variables that we do not care about. Then the `p0_units` predicate can be used to make additional queries without the need to specify variables for the undesired information.

5 Example Data Analyses

This section includes several examples of using Datalog queries to obtain information from replay files. For our own exploratory analyses, we used an embedded

---

[2]https://github.com/bwapi/bwapi.

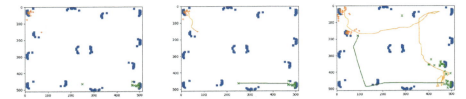

Fig. 2 Example of exploratory data analysis of an arbitrary replay augmented with visualizations. The blue squares are mineral locations, the orange crosses are player zeros units, and the green xs are player ones units. (Left) Queries to retrieve positional information 90 seconds into the replay: `mineral locations(X, Y) :- has(-1, , mineral field, X, Y, 90)`. `p0 locations(X, Y) :- has(0, , , X, Y, 90)`. `p1 locations(X, Y) :- has(1, , , X, Y, 90)`. (Center) Queries to retrieve trajectories of scouts 90 seconds into the game; the IDs of the scouts were determined by inspection of each units location with respect to the position of the base: `p0 scout trajectory(X, Y, T) :- has( , 146, , X, Y, T), T =< 90`. and `p1 scout trajectory(X, Y, T) :- has( , 22, , X, Y, T), T =< 90`. (Right) The positional and scout trajectories five minutes into the replay. In this case, player zero is Terran and the scout is an SCV and player one is Zerg and the scout is an Overlord. The trajectories show the difference between ground and flying units

domain specific language (EDSL) implementation of Datalog in Python.[3] The syntax of the EDSL is not as terse as the Datalog syntax, so for presentation purposes the examples we use the Datalog syntax. We chose Python for the following reasons: it is commonly used in the data science community, there is a Python API to access the dataset [5], and it enabled the integration of existing Python libraries for functionality not provided with a pure Datalog interpreter. For example, Fig. 2 demonstrates the incorporation of the Matplotlib[4] library for 2D plots.

5.1 Build Order Identification

The build order is part of the high level strategy. A build order is the order in which units are produced. Since units have dependencies on other units and available gathered resources choosing a build order is a complex task. An optimal build order provides a strategic advantage. The work in [3] is devoted to this aspect of the game.

With respect to replay data, the approaches in [2, 10] use replay data and machine learning to predict an opponents strategy where the significant component of the strategy is defined by the build order. Supervised machine learning requires labeled information for feedback, in this case, the replay data needed to labeled with build order for each player. Then machine learning can be used to find patterns in the

---

[3]https://sites.google.com/site/pydatalog.
[4]https://matplotlib.org.

replays, such as which build orders are more successful than others against opponent build orders.

With our approach to organizing replay data, a player's build order can be identified by executing a query. Many Datalog implementations have extensions for grouping and aggregation. With such an extension, we can add a rule that computes the number of each type of unit at every time step

```
army_composition(P, U, T, C) :-
    group_by(
        has(P, _, U, _, _, T),
        [P, U, T],
        C=count).
```

where P is the player, U is the unit type, T is the replay time, and C is the count of units. The second term in the group_by predicate the specifies a grouping order: first group by player, then unit type, and finally time. The third term in the group_by predicate specifies that the aggregation is the count.

Here is an example of a query where player zero is Terran:

```
?- army\_composition(0, U, 0, C).
result(command center, 1).
result(scv, 4).
```

which returns the expected number of units at the beginning of the game.

With such a rule in place it is straightforward to query for a time where the army composition reflects a known build order. If the query is successful, then one or more results are returned, otherwise no results are returned which means that there is no match for the build order. The Liquipedia Brood War website[5] has data on many build orders that could be easily encoded into queries.

5.2 State Estimation

One of the components of StarCraft that makes that adds to the complexity of creating an AI agent to play the game is that state information is partially observable. Basically, a player can only observe cells within the sight range of his/her units. There are several approaches to estimating the state of the opponents army based on partial observability. In [9] Bayesian approaches are discussed for estimating various state variables.

In [11] particle filters are used to track the trajectories of opponent units. One way to measure the success of target tracking is to have ground truth trajectory data. Figure 2 demonstrates a query that can be used to obtain a unit's trajectory. Also, since the sight ranges of each unit part of the game rule data, it is straightforward to

---

[5]https://liquidpedia.net/starcraft/Main_Page.

determine which opponents units are directly observable at a given time step. This information could also be added to a ground truth dataset for target tracking.

5.3 Future Work

We realized that a StarCraft AI could benefit from the ability to perform logical deductions during game play. For example, if an opponent ghost is observed we can use queries similar to the depends on relation described in the Knowledge Representation Section to obtain information on what buildings the opponent has and how many mineral and Vespene gas resources were used to produce that unit. We performed benchmarks of query execution speed. All of the queries we tested executed on the order of 100s of milliseconds on commodity hardware.

While the query execution time is not sufficiently fast for, say, making micro-combat decisions, we believe that it is fast enough to make high level strategy decisions such as choosing a build order. Adaptive planning is the ability to change strategies based on observations of the opponent during game play. Intergame adaptive planning is the ability to adapt to an opponent's strategy between games against the same opponent. This is based on the idea that individual players have a distinct play style. In [2] an approach to intergame adaptive planning is discussed. Intra-game adaptive planning is the ability to change strategies during gameplay. In [4] an intra-game adaptive planning approach is discussed that relies on evolutionary planning.

We believe that logical deduction can be utilized for intra-game adaptive planning. For example, say a flying opponent that has a ground attack is observed. There are many useful pieces of information that can be deduced. First, a query can determine whether or not we have a unit capable of attacking a flying unit. If not another query can determine what types of units could potentially be trained to counter the threat. Furthermore, for each of the available options the total build time, mineral cost, Vespene gas cost, and supply cost could be obtained which includes the costs of any needed dependencies. This information could be used to select a new build order mid-game to counter an opponent's strategy. To this end, the authors are in the process of incorporating this approach into the Command Center[6] StarCraft AI Bot.

6 Concluding Remarks

In this paper we described an approach to interactively explore StarCraft: BroodWar replay data by encoding domain knowledge in the form of statements in first-

---

[6]https://github.com/davechurchill/CommandCenter.

order logic and using a logic programming language, Datalog, to query the data. We demonstrated some example cases where this encoding can be useful for AI research. In principle, this approach could be used for exploring other real-time strategy game data by encoding the basic rules appropriately.

References

1. S. Ceri, G. Gottlob, L. Tanca, What you always wanted to know about datalog (and never dared to ask). IEEE Trans. Knowl. Data Eng. **1**(1), 146–166 (1989)
2. H.C. Cho, K.J. Kim, S.B. Cho, Replay-based strategy prediction and build order adaptation for starcraft AI bots, in *2013 IEEE Conference on Computational Intelligence in Games (CIG)* (IEEE, Piscataway, 2013), pp. 1–7
3. D. Churchill, M. Buro, Build order optimization in starcraft, in *Seventh Artificial Intelligence and Interactive Digital Entertainment Conference* (2011)
4. N. Justesen, S. Risi, Continual online evolutionary planning for in-game build order adaptation in starcraft, in *Proceedings of the Genetic and Evolutionary Computation Conference* (ACM, New York, 2017), pp. 187–194
5. Z. Lin, J. Gehring, V. Khalidov, G. Synnaeve, Stardata: a starcraft AI research dataset, in *Thirteenth Artificial Intelligence and Interactive Digital Entertainment Conference* (2017)
6. S. Ontanón, G. Synnaeve, A. Uriarte, F. Richoux, D. Churchill, M. Preuss, A survey of real-time strategy game AI research and competition in starcraft. IEEE Trans. Comput. Intell. AI Games **5**(4), 293–311 (2013)
7. G. Robertson, I. Watson, An improved dataset and extraction process for starcraft AI, in *The Twenty-Seventh International Flairs Conference* (2014)
8. F. Sáenz-Pérez, Des: a deductive database system. Electron. Notes Theor. Comput. Sci. **271**, 63–78 (2011)
9. G. Synnaeve, Bayesian programming and learning for multi-player video games: application to RTS AI. Doctor of Philosophy, Grenoble University (2012)
10. B.G. Weber, M. Mateas, A data mining approach to strategy prediction, in *2009 IEEE Symposium on Computational Intelligence and Games* (IEEE, Piscataway, 2009), pp. 140–147
11. B.G. Weber, M. Mateas, A. Jhala, A particle model for state estimation in real-time strategy games, in *Seventh Artificial Intelligence and Interactive Digital Entertainment Conference* (2011)

Merging Deep Learning and Data Analytics for Inferring Coronavirus Human Adaptive Transmutability and Transmissibility

Jack Y. Yang, Xuesen Wu, Gang Chen, William Yang, John R. Talburt, Hong Xie, Qiang Fang, Shiren Wang, and Mary Qu Yang

1 Introduction and Approaches

Since the first reports that the human infection by the novel Coronavirus disease in 2019 (Covid-19), the disease shows growing human-to-human transmission characteristics and asymptomatic cases. The number of infected humans has continuously increased drastically, and the epidemic has grown precipitous from region to region worldwide. It continuously generates significantly negative impacts on the health and lives of people, as well as global economics across countries and the world.

Upon the first sequencing of the Severe Acute Respiratory Syndrome Coronavirus type II (SARS-CoV-2), this virus has been known as a zoonotic, positive-sense, single-stranded RNA betacoronavirus with subgenus *Sarbecovirus*, subfamily *Orthocoronaviridae* closely related to the old SARS-CoV and MERS-CoV, the Middle Eastern respiratory syndrome, and the SARS-like viruses of bats like bat-SL-CoVZC45 and bat-SL-CoVZXC21. Potential human-to-human transmission can be immediately inferred from its sequence and we realized its potential

---

J. Y. Yang (✉) · X. Wu · G. Chen · H. Xie · Q. Fang
Graduate Colleges of Public Health and Medicine, Bengbu Medical University, Bengbu, Anhui, China
e-mail: cmu-scs@tamu.edu

W. Yang · S. Wang
Carnegie Mellon University School of Computer Science, Pittsburgh, PA, USA

Interdisciplinary Engineering Ph.D. Program and Department of Industrial and Systems Engineering, Texas A & M University College of Engineering, College Station, TX, USA

J. R. Talburt · M. Q. Yang
MidSouth Bioinformatics Center and Department of Information Science, Joint Bioinformatics PhD Program of University Arkansas at Little Rock and University of Arkansas for Medical Sciences, Little Rock, AR, USA

© Springer Nature Switzerland AG 2021
H. R. Arabnia et al. (eds.), *Advances in Artificial Intelligence and Applied Cognitive Computing*, Transactions on Computational Science and Computational Intelligence, https://doi.org/10.1007/978-3-030-70296-0_21

pandemic at the time of the first report of its genome sequence. Regions suffering from the epidemic of the virus have taken active precautions in efforts to stop its transmission since December 2019. Yet the SARS-CoV-2 was later declared as a Public Health Emergency of International Concern by the World Health Organization (WHO) on January 30, 2020. Within a month on February 28, it leveraged to the highest level of global risk and within 2 weeks on March 11, WHO declared it a global pandemic. The outbreak of Covid-19 is directly resulted from social gathering, whereas the disease has an incubation period ranging from several days to a month or even indefinitely. The rate of spread of SARS-CoV-2 more than resembled coronavirus of SARS-CoV and MERS but with longer incubation period by large. The longer the incubation period, the more severe consequences it produces since people are not only prone to get infected by others without symptoms, but also the virus inside a body can spontaneously mutate and involve to higher transmissibility. Over the time, the virus undergoes human-adaptive mutations and evolutions to increase its transmutability and transmissibility. During a social gathering, if any of them had the virus, and many people who were infected but have no symptoms, they would have not only spewed infectious droplets into the air, but also contaminated the surrounding surface. It is not just airborne transmission, but also the virus left on a surface can be transmitted to others through indirect physical touch. People can just get infected without knowing the cause; thus, it is a community-transmitted disease. The only way to stop the outbreak of the virus is to refine physical contacts of humans and ban on social gathering, otherwise the virus would go ahead to cause global pandemic and this was what happened earlier, as declared by the WHO. It can be foreseen that the United States of America will soon become a center of this pandemic due to the prevalent "freedom" infrastructure of the society, while regions with highly centralized rigorous jurisdictions will be less affected by the pandemic.

Coronaviruses have infected a wide range of mammalian and avian hosts in the past, generally with species specificity. Those viruses exist in nature and can be mutated and evolved to human transmissibility upon infecting humans. Nevertheless, the transmutability and segment recombination bestow coronaviruses to potentially mutate to human-adaptive genomes, but unlikely create any pandemic risk once infected humans are isolated. It takes significant amount of time and many steps before a coronavirus originated from the nature can be evolved to human adaptive transmissibility. By the time the virus just begins to obtain such ability of human adaptive transmissibility, it usually has already put off by the active isolation and strong preventive efforts of the society. Despite of all the efforts from human side, there are still many uncertainties, hence it is important to predict the viral transmutability and adaptation of coronaviruses to human transmissibility. Hence, given the incredible spread out of the virus globally, it is important to predict human adaptive viral mutation and the transmutability to address the pandemic of this quick spread of the disease for identifying effective drug targets and preventive procedures.

Predicting human-adaptive Covid-19 transmutability is important but difficult due to lack of large samples of available data and clinical outcomes in the world.

Yet this effort would improve our understanding of the epidemiology and clinical characteristics of SARS-CoV-2, which can lead to reveal pathogenic mechanisms and potential risk factors. With limited systematic analysis and summation of the data, we attempt to develop machine learning approaches to address this issue utilizing the research into the etiology and epidemiology of this newly declared pandemic earlier this month on March 11, 2020. Given the high transmutability of Covid-19, it is important to predict human adaptive mutations for better drug target identification and prospective vaccines.

Viral transmutability and evolution occur instinctively and spontaneous. Once humans contacted a large amount of originally animal-hosted virus, it is possible that such animal virus can be mutated to human-adaptive transmissibility. Over the time when virus presented inside human bodies, human adaptive transmissibility can be induced, and once a huge number of humans get infected by the virus, it almost deterministically led to a global pandemic as declared earlier this month by WHO on March 11, 2020 due to nowadays much faster flow of humans from place to place and much faster transportation network around the world. Therefore, SARS-CoV-2 spreads out much faster than the old SARS-CoV. Human-adaptive transmissibility depends on viral interaction with the host cells, the number of infected humans, flow of humans, social gathering, and many factors, hence it is difficult but important to predict viral transmutability and human-adaptive transmissibility. Yet the ability to predict viral mutation and evolution can help in the early detection of human-adaptive transmissibility and drug-resistant strains, which can potentially facilitate the development of more efficient antiviral treatments and future vaccines.

With the advent of high-throughput of next-generation sequencing technology, virus genomes can be sequenced rapidly; this hence paved a way for us to develop novel machine learning approaches to infer host dependence of mononucleotides (nts) and tetranucleotide compositions of coronaviruses and the protuberant roles of dinucleotides in virus genomes since viral dinucleotides are targets for the host innate immune system [1], and impartially regulate the virulence and replication [2] of viruses. Species-specific and virus-family-specific dinucleotide compositions have been studied. Dinucleotide composition in RNA virus genomes has been used to predict viral reservoir hosts and arthropod vectors utilizing machine learning techniques [3]. Hence, it is reasonable to deem that genomic dinucleotide composition linked to amino acid dependency is associated with human adaptative mutations and transmutability of coronaviruses.

Given that three nucleotides code for an amino acid residue, coding sequences of the coronavirus can be used for the nucleotide composition analysis. The composition of mononucleotide (nt) and dinucleotides (dnts) of 16 (4 × 4) types of combination of every two nts are formed for each of the three types of nucleotide positions within a trinucleotide codon. There were 60 entities, including the frequency of 12 types of nts and the relative frequency of 48 types of dnts can be calculated for generating feature space.

Machine learning has been used to learn the rules characterizing the aligned sequences that are resistant to drugs after viral mutation. The rules are used to predict drug resistance from gene sequences among a set of testing sequences. The training techniques are having each protein sequence represented as a feature vector and fed as input [4] to train datasets of drug-resistant and nonresistant sequences of virus population. During the training phase, intelligent machines learn the rules of predicting new generations of viruses predisposed to drug responses. Intelligent machines can also predict secondary structure of the RNA after mutations. The ribonucleic acid molecule consists of a sequence of ribonucleotides that codes the amino acids' sequence in the corresponding protein. The structure infers a set of canonical base pairs, whereas variation is considered as a form of mutation to be predicted. RNA sequence folding is used to predict the secondary structure. A software package called "RNAMute" analyzes the effects of point mutations on the secondary structure from thermodynamic properties. Intelligent machines can also predict single nucleotide variants at each locus of nucleotide location, utilizing the Next Generation Sequencing technology. Using available NGS, machines can automatically identify SNP or mutation [5] that produces a single base pair change in the sequence. Machine learning is used to analyze the historical patterns and variation mechanism of the virus and detect the rates of variation of each point substitution of its RNA sequence. Hence, we aim to predict RNA sequence of mutated virus by detecting human adaptive point mutations and transmutability. The substitutions and dependency between nucleotides are based on the rules of human adaptive evolution and transmutability of RNA viruses.

Human adaptive transmutability of the virus can be inferred from the input time series. At each point of an RNA sequence to be mutated from evolution passage to the next-mutated RNA sequence, for each iteration in the learning process, the input is an RNA sequence and the output is the human-adaptively mutated RNA sequence. The learned output is not the classification of the RNA virus, but it is the RNA virus after human-adaptive transmutability. Our approaches infer the rules that govern certain value for each nucleotide. Training processes use each nucleotide as a target class of values. The input that leads to one of the values of the target class is the sequence before the sequence containing the current value of the target class. The machine learning algorithms learn what input will produce which output accordingly. The rules learned from the machine are used to predict the generated nucleotide corresponding to the input. The rule is in the form of short sequences of nucleotides genotype and location that govern the transmutability of a nucleotide from certain genotype to another. After each iteration, the required rules for the nucleotide corresponding to the current iteration are extracted.

In each iteration, the algorithm detects the value of nucleotide a in the known sequence, to the value of this same nucleotide in the next human-adaptive mutated sequence a'. The following step in the algorithm is the detection of the values of all nucleotides in the preceding sequences to the ones that include nucleotides a and a'. Using known sequence of the nucleotide values can leads to the value of the mutated nucleotide influenced by human host–adaptive transmutability process. Upon the nucleotides in the known sequence to be mutated to the one corresponding to the nucleotide, then these nucleotide values are included in the rule of nucleotide a, otherwise keep the same without mutations are excluded since we are only interested in the human adaptive transmutability.

Prediction of coronavirus human-adaptive transmutability requires the acquiring of large amount of aligned RNA sequences that go through different mutations over a long period of time. Upon a set of time-series virus datasets are available including the GenBank, Intelligent machines can align time series RNA sequences to ensure reasonable training execution. Correlation between the nucleotides can be visualized after extracting the prediction rules. Predicting human-adaptive mutations reveal the transmutability mechanism from the existing viral RNA sequences. Therefore, we develop new machine learning techniques to address this important task.

Deep learning approaches can improve the prediction power. We hypothesize that with successful formulation of the data analytics into machine learning problems, we can take advantage of deep learning algorithms to overcome data analytic challenges. The feasible prediction is critical for a better understanding of human-adaptive transmutability. In this aim we develop a new computational framework to streamline the likelihood of prediction through integration of different deep learning models.

2 Methods

2.1 Develop Deep Learning–Based Methods for Interacting Host–Cell Identification

Because deep learning algorithms can capture the intricate structure of data, we propose to develop a Convolutional Neural Network (CNN) regression model to impute the data more accurately. The high-dimensional feature space computation problem poses an additional challenge for prospective predictions. To address this issue, we represent the highly varying data manifold efficiently in several nonlinear neural network layers by employing stacked autoencoders. Because stacked autoencoders can identify highly varying manifolds better than the local methods (such as t-SNE), they can also further improve the effectiveness of feasible prediction. The method includes data analytics, dimension reduction, nonmetric similarity measurement, and clustering.

We first build a CNN-based model to detect the value in feature space. The model learns the latent patterns of the data without the assumption of the value distributions, which is commonly done by most of the existing methods. We formulate the problem as a binary classification problem, where label 1 refers to true value whereas -1 represents otherwise. The model can be trained successfully. Our assumption is that the value of viral mutation i in interacting host–cell j is a function of the gene i in the other types of cells, as shown in the following equation,

$$\overrightarrow{x}_{i,j} \leftarrow \left(x_{i,1}, x_{i,2}, \ldots x_{i,j-1}, x_{i,j+1}, \ldots x_{i,N}\right)$$

where N is the total number of possible interacting cells. Each $\vec{x}_{i,j}$ represents a data point in the learning space. In the model, the training data is fed through convolution and pooling layer successively. This process is repeated multiple times for learning purpose. In general, complexity of features learnt by convolution layers increase as the depth of CNN model increase. The output from successive convolution and pooling layer is passed through a multilayer neural network to predict the label of the data points. Cross-validation test can be used to test the performance of the model. The model is represented by a sign function as the model is trained to minimize misclassified (observed label vs predicted label) data points in the training set.

$$Y_{i,j} = \text{sgn}\left(F\left(\vec{x}_{i,j}\right)\right)$$

where $Y_{i,j}$ represents the predicted label for a position in the data matrix. Once the model is built, the label of the remaining positions in the given data matrix can be inferred. After the position confirmation, the CNN regression model is used to impute the values of these positions. The regression model is like the position detection model, however, the label of the training samples using known values. This model optimizes a constraint objective function.

$$O = \min\left(\sum_{i=1}^{m}\sum_{j=1}^{n}\left(Y_{i,j} - R\left(x_{i,j}\right)\right)\right), x_{i,j} \notin MP$$

where R indicates the regression model. MP is the position set detected in the previous step. The testing on independent datasets and comparisons with the existing methods can be performed.

2.2 Stacked Autoencoders for Dimension Reduction

The machine learning algorithm, t-distributed stochastic neighbor embedding (t-SNE), is widely used for reducing the dimension of the feature space. t-NSE is primarily based on local properties of the data. However, Maaten and Hinton (the authors of t-SNE) point out that this method is sensitive to the curse of the intrinsic dimensionality of the data. For data with high intrinsic dimensionality and a highly varying underlying manifold, the local linearity assumption on the manifold made by employing Euclidean distances between near neighbors may be violated. Consequently, t-SNE is less successful on datasets with a high intrinsic dimensionality. Thus, we propose a stacked autoencoder–based method for dimensionality reduction. A stacked autoencoder (SAE) is a neural network consisting of multiple layers of sparse autoencoders. An SAE has great feature extraction and representation power as explained in visualizing data using t-SNE. The first sparse autoencoder is trained using the imputed the data matrix as the "raw input" to learn primary features. The primary features are used as new "raw input" for the second sparse autoencoder to learn secondary features. Similarly, the successive sparse autoencoders are trained using the features learned from previous autoencoder. Finally, we connect all layers together to form a stacked autoencoder. A greedy layer-wise training is employed to obtain optimal parameters for the stacked autoencoder. Through the stack autoencoder, the original high dimension and the data matrix is transferred into a low-dimension matrix with effective and efficient representation of the original dataset. This new low-dimension matrix is used in subsequent step for identifying possible mutation in accordance with the induced process of the host–cell-adaptive protein–protein interaction.

2.3   Nonmetric Similarity Measurement

Most the current prediction methods employ metric similarity, such as Euclidean distance and classical multidimensional scaling techniques such as t-SNE and viSNE in metric spaces. The data contains high variability because of the asynchrony and biological heterogeneity. The metric similarity measurements may not be sufficient to represent the actual distance between differential process. To circumvent the limitations of metric space and find new and rare interaction types, we extend the current research to nonmetric space. The novel nonmetric space measurement uses the Jeffrey Divergence (JD) method. JD is a nonmetric space measure of the distance between two data points.

Consider two interactions, X and Y. Their similarity can be calculated as.

$$s(X, Y) = \sum_{i=1}^{m} \left(x_i \log \left(\frac{x_i}{(x_i + y_i)/2} \right) + y_i \log \left(\frac{y_i}{(x_i + y_i)/2} \right) \right) + \lambda \sum_{j=1}^{p} \delta_j$$

where m is the total number of interacting genes in host-specific cells, λ is a weight vector, and δ is the signature genes vector. The signature genes can be a set of genes that control the differentiation.

2.4   Hybrid Unsupervised Clustering

We employ Affinity Propagation (AP) clustering for the nonmetric data clustering [6]. Different from the commonly used clustering methods, such as k-means, AP quantifies the similarity between two data points as the ability of one data point to be the cluster center rather than the other data point. In this case, AP does not require a symmetrical measure of similarity between data points. AP initially considers all cells as potential cluster centers (called exemplars); thus, we do not preset the number of clusters. Two real-valued information messages, responsibility, and availability are transmitted between pairs until convergence. A damping factor is used to ensure convergence. AP returns apposite number of clusters and, thus, improves the identification of the rare and novel interaction.

2.5 Build a Hybrid Statistical Model to Construct the Temporal Order of Host–Cell-Adaptive Process

The time or temporal order of the diverse types provides with more insights into the complex lineage relationships that describe, explain, or even predict the types of conditions and relationships that may exist during the host–cell-adaptive interacting progression. Several pseudo-time detection approaches [7] have been applied for approximating the differentiation during the developmental stages for differentiation analysis. We therefore propose to develop a new cluster-based pseudo-time detection method specific for the lineage analysis.

To track the differentiation and developmental branches along the interaction progression, we apply several pseudo-time-detection approaches based on various theories, including a diffusion pseudo-time map and graph theory–based models such as reversed graph embedding and the shortest paths from the same dataset. After removing the noninteractive situations, we selected instances based on differential interactions of host–cell types for the further pseudo-time detection. These approaches would allow us to identify the potential interactions by projecting them onto a two-dimensional space using diffusion map. The results show distinct clusters agreeing with specific interactions.

The pseudo-time ordering series of the host–cell types are in direct alignment with the developmental trajectory to the blast-crisis stage. The temporal progression often involves bifurcations from the progenitor and can be learned based on the patterns. Additionally, the detected developmental branches coordinate the different subclones and further enable the investigation of the key interaction to dominate the variations. The developmental trajectory detection is sensitive to the interaction that are used for assessing similarity. We hypothesize that the identification of a comprehensive information including stage-specific interaction to enable more predictive developmental trajectory detection. Our proposed approach aims to optimize the selection in terms of the active learning from the data.

The host–cell interaction evolution has a very similar tree-like architecture as that found in the life evolution. It has been shown that during the course of evolution, selective pressures would allow some mutations to expand while others become extinct. This is analogue with Darwin's "*survival* of the *fittest*" evolutionary theory. *We propose a distance-matrix-based* neighbor joining approach to construct host–cell lineage appropriating development process.

2.6 Identifying Adjacency Relationships between Clusters and Reconstructing Interacting Host–Cell Lineage

We model a tree structure to construct the pseudo-order and adjacency relationship among the clusters. We first measure the distance between clusters (types). The similarity between clusters is calculated by the average distance between all the potential host cells in one cluster and the cells in the other.

$$
d\left(C_i, C_j\right) = \text{avg}\left(\sum_{p=1}^{P}\sum_{q=1}^{Q} d\left(x_{i,p}, x_{j,q}\right)\right)
$$

where $d(\bullet)$ is the measure of distance. C_i and C_j are two cell clusters that contain P and Q cells, respectively. Then, we construct the distance matrix, which is a square symmetric matrix with N_c rows and columns, whereas N_c is the number of the clusters. According to the distance matrix, we employ the neighbor joining approach to build a distance tree for the clusters, which leads to the differential evolvement. The root of the tree is considered as the starting cluster of the lineage. For the existing lineage identification methods, the biological and technical noise often accumulates along the trajectory from the start towards the end. Our proposed approach takes the advantages of the local similarities between nearby clusters and effectively reduce noise accumulation, thus resulting in more accurate cell pseudo-time order detection.

2.7 Constructing Pseudo-temporal Ordering of Individual Interaction

During the variations, a subset of the interactions is involved in the stage-specific elements, indicated that the pseudo-time detection is sensitive to the group that is selected for the distance assessment between host cells. We apply Lasso which solves the L1-penalized regression problem to select the stage-specific interactions.

$$
\min\left(\sum_{i=1}^{N}\left(y_i - \sum_{j} x_{ij}\beta_j\right)^2 + \lambda\sum_{j=1}^{P}|\beta_j|\right)
$$

With the L1-penalty, which is the second part of the function, Lasso reveals a sparse weight set that guides the active selection for construction of pseudo-temporal ordering of individuals. Next, we refine the distance assessment between pair in the two adjacent clusters. Both the selection and weighting respect the stage-specificity. Finally, we assemble the local pseudo-time into a tree-like structure to characterize the nonlinear development.

2.8 Reconstruct Host–Cell-Specific Regulatory Networks by Integrating Profiles and Pseudo-Temporal Information

Several approaches have been proposed to infer interaction networks. The Boolean network was applied to build the networks. The method assumes that most genes change their state in a single step; however, at least one gene must change its state because no gene changes its expression level unless it has another gene to follow. Thus, each gene has the same probability to change its state in a step. These assumptions limit the data application in inferring large-scale regulatory networks for *revealing host–cell-type-specific interaction network*, whereas the host cells are ordered based on the differential trajectory constructed. The matrix of neighboring types along the development of trajectory are extracted accordingly.

2.9 Building Target Interaction Modules

Target interaction relationships are host–cell specific or condition specific. To reveal cell-specific pairs, we score each individual pair based on their patterns. We use Random Forest Regression Tree model to obtain the potential interactions, quantified by a weight w, for all possible interactions based on the matrix. The larger w value indicates a higher confidence in the interaction relationship. Then, its target interactions with $w > 0$ are compiled into a candidate target set. We further remove weak interaction relationships by filtering out candidate targets with smaller w values. After ranking all targets based on w, we can retain the top potential targets. To reduce false negatives, we exam interactions that have w value larger than the cutoff. We search for the enrichment utilizing position weight matrices. The interactions that have a w value larger than the cutoff are included as target interactions modules.

2.10 Establish Differential Interaction Modules

Like protein and DNA/RNA interactions, protein–protein interactions in current databases can only occur under certain conditions and/or in specific cells. It is commonly accepted that interactions which participate in the same biological process tend to engage in resonant activities. Thus, we quantify the co-coordinated expression patterns between interactions using mutual information.

$$I(X; Y) = \sum_{x \in X} \sum_{y \in Y} p(x, y) \log \left(\frac{p(x, y)}{p(x)p(y)} \right) = H(X) + H(Y) - H(X, Y)$$

$H(.)$ stands for information entropy of a random variable. The variables, X and Y represent the levels of interactions in the cells. When an interaction is expressed differently across a set of host cells, the corresponding information entropy (H) is higher. Thus, when two interaction are coordinated, the joint entropy is lower whereas the mutual information is higher. All PPIs are ranked based on mutual information and top-ranked PPIs are retained for subsequent analysis. Utilizing context-specific PPIs and target pairs achieved in the previous steps, we infer differential modules that distinguish adaptivity of the host cell from the other. Besides being a part of an interaction network, differential modules are expected to be robust biomarkers than individual interaction to accommodate the variations. The differential modules are constructed using Genetic Algorithm. Starting with an interaction, the additional interactions are added to module according to the PPIs and transcription factor and target interactions. At each step, the interaction that maximizes the mutual information is selected. The modules that are significantly different from the background are computed using the discrete element method. The null distributions for statistical analysis can be generated by the host–cell permutations and randomization in the module.

2.11 Derive Modules Aggregated to the Interaction Networks

The direction of edges can be inferred by Bayesian analysis. Thus, the host–cell-type-specific interaction networks are constructed along the development trajectory. Then, the pseudo-time orders of the sequential alteration (activation or deactivation) of the interaction networks are reconstructed.

With these newly identified cell-type-specific interaction networks, we now can investigate their relationships and interactions to figure out the critical signals hidden behand. The results of this effort help us to reveal diverse mechanisms underlying host–cell interaction from the viral transmutability and evolution. Additionally, unique and common key regulators can also be identified for identifying effective drug targets.

Network comparison and alignment are based on the similarities of node, edge, and path. Thus, common and unique regulatory relationships among multiple networks can be inferred according to the extent of similarities. Multiple network alignment is an NP-complete problem, the time required increases very quickly as the size of the network grows. The most recent superior method called multiMAGNA++ represents a multiple network alignment using permutations and utilizes a classical genetic algorithm (GA) without a mutation operator to deal with this problem. However, this kind of GA is easy to trap in a local optimum when searching for global optimization. We develop a simulated annealing method for searching the global optimum by decreasing the probability of accepting worse solutions in order to improve the global optimization.

We employ the community discovery model to identify "functional communities," which represent functional subnetworks. Then, with the host–cell-specific interaction networks, we extract the relationship and interactions by aligning the host–cell-specific interaction networks.

The multiMAGNA++ is a recent and superior network alignment (NA) approach. Unlike the other NA methods that calculate the amount of conserved edges after the node alignment is constructed, multiMAGNA++ directly optimizes the edge conservation during alignment construction. However, this approach uses the classical genetic algorithm (GA) to search the potential optimal alignment which can be trapped in a local optimum.

Here, we propose a novel meta-heuristic multiple NA model using the simulated annealing (SA) strategy to improve existing NA approaches. Moreover, we adopt the temperature threshold and acceptance probability function from the SA strategy to restrict the random search of GA to achieve the global optimum search.

The relationships between the aligned networks can be divided into three groups, the homologous, heterogeneous, and similar, based on the alignment scores. With these categories, the master host–cell-specific interaction networks can be separated. With similar or homologous pairs, we propose a novel ranking–based clustering for master interaction detection. Here we define the master interaction as controlling multiple-hub nodes in the networks. We first design a network projection model to map the pairs into a sequence of subnetworks without structural information loss. Then, an information transfer mechanism is used to maintain the consistency across subnetworks. For each subnetwork, a path-based random walk method is proposed to generate the reachable probability, which can be effectively used to estimate the cluster membership probability and the importance of the interaction. Through iteratively analyzing each subnetwork, the steady and consistent clustering, and ranking centers of the master interactions or multi-hubs can be accessed.

On the other hand, the heterogeneity cannot be discovered but are confined to the host cell, which could play important roles in the evolution. With more and more data accumulated, when we perform host–cell-specific interaction network construction and comparison across different data, batch effect can affect the results. Therefore, quality control strategies such as normalization method are developed accordingly.

3 Conclusion

We build a deep learning-based model combining genomic sequences and host–cell-specific interactions to infer transmutability. Then, by incorporating the lineage and host–cell-specific interaction information as well as protein–protein interactions, we can identify host–cell-specific interaction and differential modules. There are two types of interactions that are assembled into a host–cell-specific interaction network for exploring the transmutability underlying disease development or viral mutation differential processes.

1. Build a core interaction network. We propose a deep neural network regression model to construct a core interaction network. When a putative interaction regulator in the input layer has interaction sites in these genomic regions of the targeted host cell, we assign a higher initial weight to this candidate interaction. The model is trained using backpropagation. After training, the weight vector of the input neurons is used to assess the likelihood of interactions in the input layer to be the regulators of the targets. After regression models for each individual interaction are built, the interaction dependency suggested by between the regulator-target pairs are extracted and converted to the network edges. All edges between targets then are combined into a core interaction network.

2. Infer host–cell-specific interaction. Along a cell trajectory, the expression matrix for the cells in a host-specific cell type is extracted accordingly. Each edge in the core interaction network is evaluated using mutual information. The corresponding interaction can be regarded as active in a host–cell-specific type. Employing this strategy, interaction networks for distinct host-specific cell types are established. To understand interaction mechanisms that drive disease development from viral transmutability, we hypothesize that differential modules that distinguish different host-specific cell types should be considered and incorporated into the core network since the core regulatory network is constructed based on the coordinated in the cells. The differential gene modules are derived. PPI can only occur under certain conditions and/or in host-specific cells. It is well accepted that interactions that participate in the same biological processes tend to be coordinated. Thus, we calculated mutual information based on the levels of the interactions that encode the interacting proteins with the architecture of deep neural network regression model for building core regulatory networks. All PPIs obtained from the curated databases are ranked according to mutual information, and only top-ranked PPIs are retained as context-specific interactions for the subsequent analysis. We apply the genetic algorithms (GA) using a heuristic global optimization strategy to identify Differential Interaction Modules based on a "one-to-all" strategy in which one host–cell type is labeled as 1 and all the other cell types are labeled as -1. Staring with a single interaction as a seed, we initialize the first with length equal to the number of the candidates that directly or indirectly interact with this seed based on the context of the host–cell-specific interactions. A new population is generated using the mutation and crossover operations. The fitness can be computed based on mutual information. The searching stops if no improvement occurs from one generation to the next or the maximum number of generations is reached. The best fitness from the final generation is considered as candidate for the seed. The null distributions for the statistical analysis are generated through host–cell-type-specific permutations and randomization. The modules that are significantly different from background are taken into considerations. For each host–cell-specific type derived in the same way, then, adaptive host–cell-type-specific

interaction in the core regulatory network is aggregated into a final interaction network, while the direction of edges is inferred by Bayesian analysis. Using the final the interaction network for each host–cell-specific type along the trajectory, the sequential alterations (activation or deactivation) of the interaction network is inferred for transmutability.

The development of machine learning techniques in the RNA mutation for prediction of the evolution of the virus resulted from the adaptive host–cell-specific interaction paves a way for identifying transmissibility and the prediction of the mutations, such as virus evolution based on human adaptive process. It can assist the designing of new drugs for possible drug-resistant strains of the virus before a possible outbreak.

Acknowledgments This research is supported in part by the United States NIH Academic Research Enhancement Awards 1R15GM137288, 1R15GM114739, and NIH 5P20GM103429, and the *National Science Foundation NSF* Experimental Program to Stimulate Competitive Research *(EPSCoR)*. The AR *EPSCoR* program is designed to foster collaborative research efforts among the state's higher education institutions, promote workforce development, and conduct educational outreach. Acknowledgment is a sole opinion of the authors and is not inferring any view of any funding support of any research. The authors declared no conflict of interest.

References

1. M. Takata et al., CG dinucleotide suppression enables antiviral defense targeting non-self RNA. Nature **550**(7674), 124–127 (2017)
2. J. Witteveldt et al., Enhancement of the replication of HCV replicons of genotypes 1-4 by manipulation of CpG and UpA dinucleotide frequencies and use of cell lines expressing SECL14L2 – application for antiviral resistance testing. Antimicrob. Agents Chemother. **60**(5), 2981–2992 (2016). https://doi.org/10.1128/AAC.02932-15
3. S. Babayan et al., Predicting reservoir hosts and arthropod vectors from evolutionary signatures in RNA virus genomes. Science **362**(6414), 577–580 (2018)
4. D. Wang et al., Enhanced prediction of lopinavir resistance from genotype by use of artificial neural networks. J. Infect. Dis. **188**(11), 653–660 (2003). https://doi.org/10.1086/377453
5. Q. Liu, J. Yang, et al., Supervised learning-based tagSNP selection for genome-wide disease classifications. BMC Genom. **9**, S6 (2008)
6. R. Guan et al., Text clustering with seeds affinity propagation. IEEE Trans. Knowl. Data Eng. **23**(4), 627–637 (Apr. 2011)
7. L. Haghverdi et al., Diffusion pseudotime robustly reconstructs lineage branching. Nat. Methods **13**, 845–848 (2016)

Activity Recognition for Elderly Using Machine Learning Algorithms

Heba Elgazzar

1 Introduction

Human activity recognition is one of the important research fields in Computer Science and it has a wide range of applications in many fields including medical and security applications. Related medical applications includes fall prevention in nursing homes and reducing the high-falls risk activities of older people in hospitals [1–3]. We focus on this chapter on the human activity recognition based on data collected form elderly [1]. Several researchers worked on this area and used different types of sensors to collect data related to human activities [2]. There are many wearable sensors that can be used to collect movement data and these sensors include accelerometers and gyroscopes [4, 5]. Recent approaches focuses on using multiple sensors and fusion at the feature level to accurately identify human activities [6]. The human activities related to walking and settings can be used effectively to monitor elderly in hospitals to prevent falls [1–3, 7–9]. This is one of the important applications that has been investigated by researchers and many approaches were used to identify high risks of falls [3, 8, 10–12].

In this chapter, we focus on identifying the activity related to setting and walking from the data collected from wearable sensors in [2, 3] by applying different machine learning algorithms and comparing between them to accurately identify human activities.

The following sections will discuss related work and provide an overview of machine learning algorithms that will be used to recognize human activities. Information related to the used dataset will be provided. The experimental results will be discussed with ideas for future research directions.

H. Elgazzar (✉)
School of Engineering and Computer Science, Morehead State University, Morehead, KY, USA
e-mail: h.elgazzar@moreheadstate.edu

© Springer Nature Switzerland AG 2021
H. R. Arabnia et al. (eds.), *Advances in Artificial Intelligence and Applied Cognitive Computing*, Transactions on Computational Science and Computational Intelligence, https://doi.org/10.1007/978-3-030-70296-0_22

2 Related Work

Machine learning algorithms were applied effectively to detect human activities and many studies addressed this problem using machine learning [13–16].

The researchers in [14] used 12 body tags attached to shoulders, elbows, wrists, hips, knees, and ankles to collect coordinates data that can be used to recognize human activities. Several machine learning algorithms were used including KNN, C4.5 decision tree, and support vector machine (SVM) in [14] and the best results were obtained using SVM. The training data in [14] was only limited to data collected from three persons which is considered a very limited set of data. Also, feature selection algorithms can help effectively to select relevant features to improve the performance of machine learning algorithms.

The authors in [15] analyzed 11 separate sensor event datasets collected from 7 physical testbeds. The datasets includes data for younger adults, healthy older adults, and older adults with dementia [15] and the sensor data is captured using a sensor network. The researchers in [15] focused on 11 daily activities that includes personal hygiene, sleep, bed-to-toilet, eat, cook, work, leave home, enter home, relax, take medicine, and bathing. Three machine learning classification models were used including naïve Bayes classifier, a hidden Markov model (HMM), and a conditional random field (CRF) model. The results are encouraging and the research was mainly focused on utilizing smart home technologies [15].

The researchers in [16] proposed an approach for activity recognition for streaming sensor data based on sliding window. The proposed approach was evaluated using real-world smart home dataset [16]. The machine learning algorithm that was used in [16] was the support vector machine (SVM) and the results were promising but there is a room for improvement to better model all classes. A sliding window–based data segmentation techniques were applied effectively in [3] for predication of human activities.

The researchers in [3] applied a classification technique based on a linear chain conditional random field (CRF) on datasets from two clinical rooms with subjects wearing a passive sensor-enabled RFID tag [2, 3]. The dataset was based scripted series of activities [3]. Other classification techniques can be applied to classify the human activity data and feature selection algorithms can help effectively to improve the performance of machine learning algorithms.

3 Proposed Methodology

The proposed methodology and the data flow is shown in Fig. 1. For this research, supervised machine learning techniques were used to classify human activities based on the sensors data. We used a publicly available dataset and it is discussed in the next section. We focus in this section on the machine learning algorithms that we used to predict human activities.

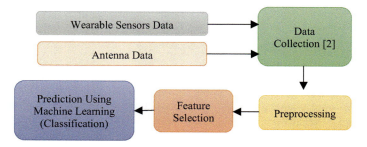

Fig. 1 Proposed methodology and the data flow

3.1 Machine Learning Algorithms

There are several methods for classification and we focus in this section on the following methods that were used effectively. The goal is to learn from the recorded data and be able to classify users based on the predictability. A classifier classifies a data item into one of several predefined classes, and typically known as supervised learning. The classes are typically non-overlapping and partition the input database. We discuss in this section the classification algorithms that were used to analyze the mobile networks. The classification process is divided into two steps:

1. Creating a model using a training data set for which the class labels have already been specified, this done by applying several techniques and comparing their performance to select the one that has the best results. These techniques include Naïve Bayes, and other techniques which will be presented in the next section [17–19].
2. Using the created model to classify new data (test data) [17–19].

There are many problems associated with classification such as missing data. Usually missing data are ignored, or some prediction techniques are used to predict the missing values. Another issue is how to make a judgment about the models or how to decide which model is the best. This usually depends on the user and what they really expect from the data classification. The performance of a classifier can also be evaluated by using information retrieval metrics. Four metrics describe the output of the classifiers (in case of two classes: positive and negative) and can be used to evaluate the performance, these values are as follows:

- TP (true positive) means the group of data classified in a class and it actually belongs to it in other words TP means number of positive samples that are classified as positive.
- FP (false positive) means the group of data classified in a class and it actually doesn't belong to it in other words it means number of negative samples that are classified as positive.

Table 1 Confusion matrix

	Predicted positive	Predicted negative
Actual positive	TP	FN
Actual negative	FP	TN

- TN (true negative) means the group of data that not classified in a class and it actually doesn't belong to it in other words number of negative samples that are classified as negative.
- FN (false negative) means the group of data that not classified in a class and it actually doesn't belong to it in other words case number of samples that are positive and classified as negative.

Table 1 shows a two-class confusion matrix that contains the four values, TP, FP, TN, and FN [17–19].

From the above four metrics, we can interpret the results using some ratios that are called accuracy, precision, and recall. These ratios will be explained as follows [17, 18]:

- Accuracy: the percentage of predictions that are correct:

$$Accuracy = (TP + TN) / (TP + FN + FP + TN) \quad (1)$$

- Precision: The percentage of correctly classified positive cases relative to the cases classified as positive:

$$Precision = (TP)/(TP + FP) \quad (2)$$

- Recall: The percentage of positive cases that were successfully classified as positive:

$$Recall = (TP)/(TP + FN) \quad (3)$$

Another graphical interpretation of the results is the Receiver Operating Characteristic curve (ROC curve), which shows the relationship between false positives and true positives. Another curve shows the relationship between precision and recall, and is the one used predominantly in this work.

Below, will give an overview of the classifiers used in this chapter. These classifiers include Naive Bayes, AdaBoost, Support Vector Machine (SVM), Decision Tree Classifiers, Artificial Neural Network (ANN).

- *Naïve Bayes* is a classification technique based on Bayes theorem [19] which is

$$P(H|X) = \frac{p(X|H)\,P(H)}{P(X)} \quad (4)$$

where X is the data tuple and is considered the evidence and H is a hypothesis that X belongs to a class C.

To solve a classification problem we want to determine $P(H|X)$, the probability that the hypothesis H holds given the data tuple X or the probability that X belongs to class C if we know X. $P(H|X)$ is called the posterior probability, while $P(H)$ is known as the prior probability since it is independent of X. Finally $P(X|H)$ is the probability of X conditioned on H, and $P(X)$ is the prior probability of X. To use Naïve Bayes, assume that we have a training data D and each tuple of data is represented by a vector $X = (x_1, x_2, \ldots \ldots, x_n)$, and suppose that we have m classes $(C_1, C_2, \ldots \ldots \ldots \ldots, C_m)$, then we have to find the maximum posterior probability $P(C_i|X)$ using Bayes theorem [6], given by:

$$P(C_i|X) = \frac{P(X|C_i)P(C_i)}{P(X)} \tag{5}$$

Since $P(X)$ is constant for all classes C_i, then we need only maximize $P(X|C_i)P(C_i)$.

A simple assumption (known as "naïve" assumption) is that all the attributes are independent [19], meaning that there is no relation between them. Thus, we obtain:

$$P(\mathbf{X}|C_i) = \prod_{k=1}^{n} P(x_k|C_i) = P(x_1|C_i) \times P(x_2|C_i) \times \cdots \times P(x_n|C_i) \tag{6}$$

Finally, the class label of X can be predicted as C_i if

$$P(X|C_i)P(|C_i) > P(X|C_j)P(C_j) \quad \forall \iota \neq j \tag{7}$$

• *AdaBoost:* This is an iterative procedure which adaptively changes the sampling distribution of training data by focusing more on previously misclassified records [20], and combines weak models in order to obtain strong ones. It starts by assigning equal weights to all instances in the data and then uses the learning algorithm to build a classifier model for this data. It then reweighs each instance according to the classifier output, such that the correctly classified instances receive lower weight, while the misclassified instances receive a higher weight. In the next iteration, the classifier is built to deal with the reweighed instances, so that it focuses more on the instances that have higher weight. The changes in the instances' weights depend on the current classifier's overall error. Suppose that e denotes the classifier's error rate (a fraction between 0 and 1 that equals (1-*accuracy*)), then the weights of a correctly classified instance is modified by multiplying the weight of the instance by $e/(1 - e)$. The misclassified instances weights remain the same, and then after all weights have been updated, they get normalized so that their sum remains the same (equal to 1). Each instance's weight is divided by the sum of the new weights and multiplied by the sum of the old ones, so that the weight of a correctly classified instance remains the same and the weight of a misclassified instance increases.

The output of all the classifiers is finally combined using a weighted vote. The classifier that has an error rate closer to 0 should receive higher weight. These weights reflect how often the instances have been misclassified. When the error of a classifier reaches or exceeds 0.5, the boosting procedure removes this classifier and doesn't do any further iterations. It can be difficult to analyze the type of models that have been learned from the data, so one possible solution for this problem is to generate an artificial dataset by randomly sampling points from the instance space and assigning them the class labels that were predicted by the ensemble classifier, and then learning a decision tree from this new dataset. This technique can be expected to replicate the performance of the ensemble classifier [21].

• *Support Vector Machines (SVM)* is a method that can be used for classification. SVMs operate by finding a hypersurface (decision boundary) that will separate the data [20]. If the training data are linearly separable, we can select hyperplanes so that there are no points between them and then try to maximize their distance [19]. We need to find a hyperplane that maximizes the margin between the data of different classes. There are various methods to train SVMs. Sequential minimum optimization (SMO) is a fast method to train the support vector machine (SVM), where the missing values are replaced and the output hyperplane coefficients are computed based on the normalized data. We used the SMO implementation provided by Weka for our experiments [21].

• *C4.5 Decision Tree* is an algorithm used to generate a decision tree. C4.5 builds decision trees from a set of training data in the same using the concept of Information Entropy. J48 is decision tree learner, the most important part is generating the decision tree, a decision tree which consists of two types of nodes, a leaf that indicates the class and a decision node that contains a test on an attribute. Each attribute of the data can be used to make a decision that splits the data into smaller subsets. C4.5 examines the normalized Information Gain [18, 20] which is the difference in entropy that results from choosing an attribute for splitting the data. The attribute that yields the highest normalized information gain is the one used to make the decision. The algorithm then continues building the tree recursively on the smaller subsets. C4.5 is a simple depth-first construction, and needs the entire data to fit in memory, thus it is unsuitable for large datasets [18, 20]. We used the J48 decision tree implementation of C4.5, which is provided by Weka to deal with large trees. There are two pruning strategies typically used. In subtree replacement, a subtree is replaced by a leaf node if this replacement will yield an error rate that is close to that of the original tree. This replacement occurs starting from the bottom of the tree and moving up toward the root. The other pruning strategy is subtree raising, in which a subtree is replaced by its most used subtree, that is, the one covering the most training instances [17, 18, 21].

• *Artificial Neural Network (ANN)* is a network of neurons that typically uses backpropagation learning to learn the interconnection of weights. For a fixed network structure, we determine the optimal weights for the connections in the network using the perceptron learning rule. The network, as shown in Fig. 2, consists of input nodes which form an input layer, one or more hidden layers, and one output layer.

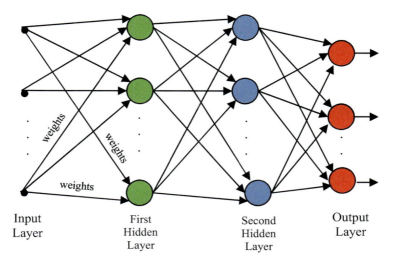

Fig. 2 The structure of ANN

The learning process occurs layer by layer. The backpropagation algorithm consists of two phases: The forward pass which applies training data to the input nodes of the network, and computes the node activations layer by layer in a forward manner (i.e., starting from the input layer, and moving forward until the output layer's set of outputs is produced). In the forward pass, the weights of the network are all fixed, while in the backward algorithm the weights get adjusted based on error correction, that is, the actual response of the network is subtracted from the target response and an error signal is produced on each output node that gets propagated backward through the network toward the input layer. The weights in each layer get adjusted in order to get the actual response of the network to become closer to the target response [18, 21].

3.2 Feature Selection

Feature selection algorithms can be used effectively to select the most relevant features and remove irrelevant or redundant features [18]. Feature selection algorithms can help significantly in enhancing the process of learning of data reducing the computational complexity of the problem.

The ReliefF algorithm [22, 23] is a variant of the original Relief algorithm [24, 25] that can work with multiple classes and it is one of the a one of the most used relief-based algorithms for feature selection. The ReliefF algorithm is a feature selection algorithm that was introduced by Kira and Rendell [24, 25].

The Relief algorithm uses a training dataset to find a ranking score for each feature which can be used to evaluate the features. The ranking is score is calculated based on the feature ability to differentiate between the labeled classes in the dataset [25].

The information gain attribute evaluation algorithm is another algorithm that evaluates that attribute (feature) based on finding the information gain with respect to the class. This algorithm computes the entropy which is used in information theory. The entropy of a given data Table T with n classes can be calculated as follows [18, 26]:

$$\text{Info}(T) = -\sum_{i=1}^{n} p_i \log_2 (p_i) \tag{8}$$

where p_i is the probability of a data record in T to be from class C_i and p_i can be estimated as:

$$p_i = \frac{\text{Count}(C_i, T)}{|T|} \tag{9}$$

where Count(C_i,T) is the number of records in T from class C_i and $|T|$ is the number of records in T.

The information gain for the feature F can be computed by finding the entropy value that can be calculated from partitioning the data Table T based on the feature F into m partitions. The expected information is calculated as [18, 26]:

$$\text{Info}_F(T) = \sum_{j=1}^{m} \frac{|T_j|}{|T|} \text{Info}(T_j) \tag{10}$$

And the information gain of the feature F is calculated as [18, 26]:

$$\text{Gain}(F) = \text{Info}(T) - \text{Info}_F(T) \tag{11}$$

The features can be ranked based on the information gain values and the features with highest information gain value will be selected.

Ensemble feature selection methods were used successfully in machine learning to enhance the overall results and remove irrelevant features where multiple algorithms for feature selection can be used to minimize the bias in the feature selection process and ensure that all relevant features are included [26, 27].

3.3 Human Activity Dataset for Elderly

We conducted our experiments using the dataset collected by research in [2] and it is publicly available for researchers through the UCI machine learning repository

[28]. The data was collected from two clinical room deployments as described in [2, 3].

The data in [2] was collected from 14 elderly volunteers from 66 to 86 years old and using a low-cost battery-free RFID tag called Wearable Wireless Identification and Sensing Platform (W2ISP) [2, 29]. The sensor was worn by elderly patients attached to their clothes. Two clinical rooms settings were used with antennae positioned on the walls and ceiling [2]. The first room, RoomSet1 [2, 3], had one antenna at ceiling level and three antennae on the walls. The second room, RoomSet2 [2, 3], had two antennae at ceiling level and one antenna on the wall [2]. The patients performed a series of scripted activities that includes four activity labels (classes): lying in bed, sitting in bed, sitting in chair, and ambulating. The collected dataset includes vertical access acceleration, lateral access acceleration, frontal access acceleration, ID of antenna reading sensor, Received Signal Strength indicator (RSSI), Phase, Frequency, and the class that represent the type of activity ((1) sit on bed, (2) sit on chair, (3) lying, (4) ambulating).

4 Experimental Results

Several experiments were conducted on the available datasets in [2] to test the proposed techniques. The data mining and machine learning software, WEKA (Waikato Environment for Knowledge Analysis), was used in this research project to run and test several algorithms [30, 31]. WEKA is a comprehensive software resource written in Java, and implements a set of machine learning algorithms for several data mining and machine learning tasks [18, 30, 31].

The results of WEKA includes the confusion matrix, Precision–Recall curves, and ROC curves. The Receiver Operating Characteristics (ROC) graphs are very useful in visualizing, and selecting classifiers based on their performance. ROC graphs are two-dimensional graphs where the X-axis represents the false-positive rate (FP), and the Y-axis represents the true-positive rate (TP) (as shown in Fig. 21). In the graph, the point (0, 1) represents perfect classification which means detecting all the true-positive instances (TP = 1) without getting any false positive (FP = 0), and the point (0, 0) means failure to detect any true-positive instances, but also with no false positives. Certain points on the ROC curve are preferable to other points if they show true positives higher than false positives.

We started by performing the required preprocessing for the data. We will focus in this section on the analysis the results using classification techniques. The goal is to classify the patients' performed activities into one of four classes: (1) sitting in bed, (2) sitting in chair, (3) lying in bed, and (4) ambulating. We converted the dataset in [2] into the WEKA ARFF format. We conducted the experiments for the dataset for RoomSet1 and the dataset for RoomSet2. The dataset for RoomSet1 has a total 52,482 records and the dataset for RoomSet2 has a total of 22,646 records.

We start the process of learning from the data by applying feature selection algorithms. The well-known ReliefF algorithm [22, 23] and the information gain for feature evaluation [26]. The ranking of the features based on the dataset for Roomset1is shown in Table 2. The ranking of the features based on the dataset for Roomset2 is shown in Table 3.

The results of features ranking in Tables 2 and 3 show that the most important features are vertical access acceleration, frontal access acceleration, lateral access acceleration, ID of antenna reading sensor, and RSSI. The results show that the features related to the phase and the frequency are considered the least important features for predicting the activity. This helps to focus only on the first five features to build models for activity prediction.

We started by considering all features and compared them to the effect of considering the features without the phase and frequency features to see the effect of feature selection. Figure 3 and Fig. 4 shows the results of accuracy for both datasets for the four classification algorithms that were used: Naïve Bayes, AdaBoost, SMO, J48, and Artificial Neural Networks (ANN). Figure 5 shows the Artificial Neural Networks model that was used in the experiments. The results show that the J48 and ANN classifiers gave the highest accuracy.

Tables 4 and 5 shows the detailed results of the precision, recall, and the area under the ROC curve for both datasets for the J48 and ANN classifiers for all classes.

Table 2 Ranking of the Features using ReliefF and information gain algorithms for Roomset1 dataset

| Feature | ReliefF | | Information gain | |
	Score	Rank	Score	Rank
Vertical access acceleration	0.2126	1	1.04307	1
Lateral access acceleration	0.0556	5	0.27466	4
Frontal access acceleration	0.1361	2	0.88905	2
ID of antenna reading sensor	0.0966	3	0.42483	3
RSSI	0.0585	4	0.22963	5
Frequency	0.0449	7	0.00394	7
Phase	0.0466	6	0.03663	6

Table 3 Ranking of the Features using ReliefF and information gain algorithms for Roomset2 dataset

| Feature | ReliefF | | Information gain | |
	Score	Rank	Score	Rank
Vertical access acceleration	0.2704	1	0.4306	1
Lateral access acceleration	0.2041	3	0.23605	2
Frontal access acceleration	0.0999	4	0.1279	4
ID of antenna reading sensor	0.2124	2	0.18996	3
RSSI	0.0884	5	0.07014	5
Frequency	0.079	6	0.00295	7
Phase	0.0756	7	0.03109	6

Activity Recognition for Elderly Using Machine Learning Algorithms

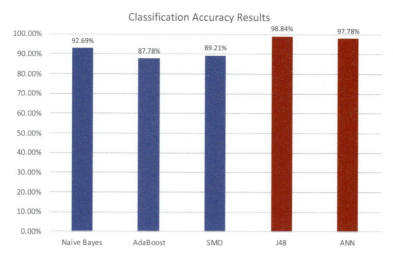

Fig. 3 Classification accuracy using all features for Roomset1 dataset

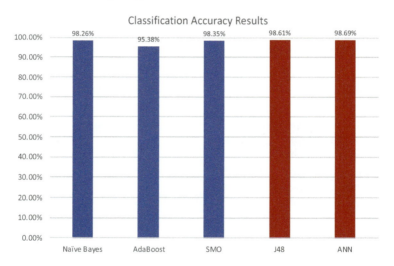

Fig. 4 Classification accuracy using all features for Roomset2 dataset

As we can see from the results for RoomSet1 dataset in Table 4, J48 gives an overall better results compared to ANN. For class 4 (ambulating), ANN gives a recall of 0.5 while J48 outperforms ANN and gives a recall of 0.822.

In the next phase of experiments, we applied the feature selection technique and we removed the two features with the least ranking. The features related to the phase and frequency were removed from the dataset. Figures 6 and 7 show the results of accuracy for both datasets for the four classification algorithms that were used:

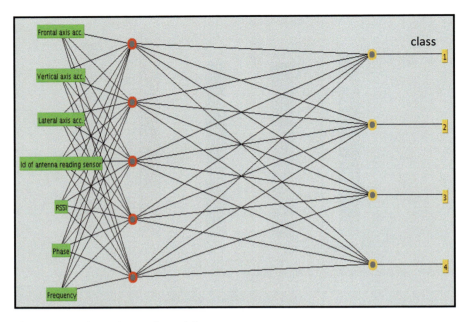

Fig. 5 Artificial neural networks model

Table 4 Precision, recall, and the area under the ROC for Roomset1 dataset

	ANN			J48		
	Precision	Recall	ROC area	Precision	Recall	ROC area
Class 1: Sit on bed	0.944	0.994	0.990	0.981	0.989	0.991
Class 2: Sit on chair	0.951	0.993	0.999	0.974	0.986	0.997
Class 3: Lying	1.000	0.999	1.000	0.999	0.999	0.999
Class 4: Ambulating	0.953	0.500	0.923	0.902	0.822	0.957
Weighted average	0.978	0.978	0.994	0.988	0.988	0.995

Table 5 Precision, recall, and the area under the ROC for RoomSet2 dataset

	ANN			J48		
	Precision	Recall	ROC area	Precision	Recall	ROC area
Class 1: Sit on bed	0.875	0.933	0.991	0.901	0.886	0.942
Class 2: Sit on chair	0.907	1.000	1.000	0.920	0.989	0.996
Class 3: Lying	0.997	0.999	0.996	0.997	0.999	0.994
Class 4: Ambulating	0.964	0.445	0.986	0.705	0.563	0.951
Weighted average	0.987	0.987	0.996	0.985	0.986	0.991

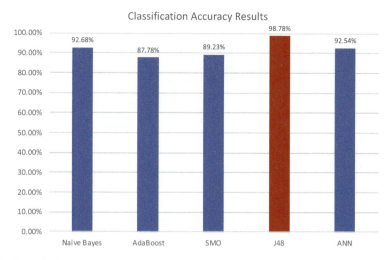

Fig. 6 Classification accuracy using the first five features for Roomset1 dataset

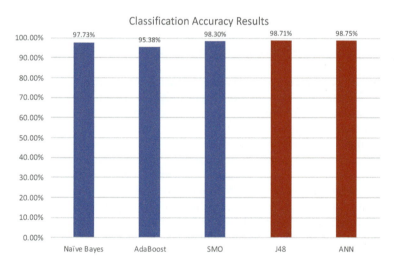

Fig. 7 Classification accuracy using the first five features for Roomset2 dataset

Naïve Bayes, AdaBoost, SMO, J48, and Artificial Neural Networks (ANN) after considering only the five features and removing the two features.

The results show that J48 was still able to give high accuracy for both dataset and ANN gave a higher accuracy for the second dataset. Table 6 shows the detailed results of the precision, recall, and the area under the ROC curve for the first dataset using the J48 classifier which gave the highest accuracy. Table 7 shows the detailed

Table 6 Precision, recall, and the area under the ROC for J48 classifier for Roomset1 dataset using the first five features

	J48		
	Precision	Recall	ROC Area
Class 1: Sit on bed	0.978	0.989	0.993
Class 2: Sit on chair	0.971	0.987	0.997
Class 3: Lying	0.999	0.999	0.999
Class 4: Ambulating	0.907	0.805	0.961
Weighted average	0.988	0.988	0.996

Table 7 Precision, Recall, and the Area Under the ROC for ANN Classifier for Roomset2 Dataset using the First Five Features

	ANN		
	Precision	Recall	ROC Area
Cass 1: Sit on bed	0.878	0.943	0.985
Cass 2: Sit on chair	0.911	1.000	1.000
Cass 3: Lying	0.997	0.999	0.994
Cass 4: Ambulating	0.981	0.437	0.983
Weighted average	0.988	0.988	0.993

results of the precision, recall, and the area under the ROC curve for the first dataset using the ANN classifier which gave the highest accuracy.

The overall results show that the proposed method can be utilized effectively to predict human activities with high accuracy using only five features with the highest feature ranking.

Additional experiments were conducted to see if we can reach a better classification accuracy using the random forest algorithm [32]. The random forest algorithm is based on combining several randomized decision trees by using averaging to aggregates their predictions [18, 32].

In addition, we conducted experiments to explore the possibility of removing additional features and consider only the three features related to vertical access acceleration, lateral access acceleration, and frontal access acceleration. Figures 8 and 9 show the results for accuracy for both datasets after applying the random forest algorithm and using the three features as indicated above, using the five features that were used before, and using all the seven features.

Tables 8 and 9 show the detailed results of the precision, recall, and the area under the ROC curve for both datasets for the random forest algorithm using all features, using five features, and using three features. The results from Figs. 8, 9, Tables 8 and 9 show that the random forest algorithm have a high accuracy for classifying the human activity compared to the previously used algorithms and the accuracy is close to the results obtained using the J48. The overall results show that we can still obtain a high classification accuracy using only the three features related to vertical access acceleration, lateral access acceleration, and frontal access acceleration. The highest accuracy was obtained using the first five features with the highest ranking.

Activity Recognition for Elderly Using Machine Learning Algorithms 291

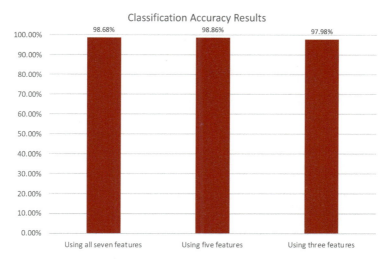

Fig. 8 Classification accuracy for the random forest algorithm using all features, using five features, and using three features for Roomset1 dataset

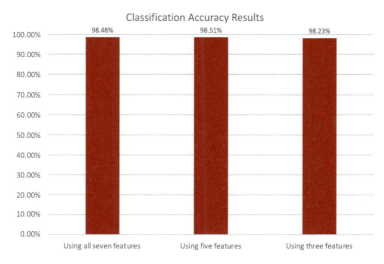

Fig. 9 Classification accuracy for the random forest algorithm using all features, using five features, and using three features for Roomset2 dataset

Table 8 Precision, recall, and the area under the ROC for the random forest algorithm using all features, using five features, and using three features for Roomset1 dataset

	Using all seven features			Using five features			Using three features		
	Precision	Recall	ROC area	Precision	Recall	ROC area	Precision	Recall	ROC area
Class 1: Sit on bed	0.982	0.983	0.988	0.986	0.985	0.990	0.959	0.984	0.991
Class 2: Sit on chair	0.972	0.980	0.989	0.975	0.980	0.989	0.944	0.917	0.969
Class 3: Lying	0.999	0.999	0.999	1.000	0.999	0.999	1.000	0.999	1.000
Class 4: Ambulating	0.859	0.834	0.914	0.870	0.867	0.932	0.900	0.785	0.935
Weighted average	0.987	0.987	0.992	0.989	0.989	0.993	0.980	0.980	0.992

Table 9 Precision, recall, and the area under the ROC for the random forest algorithm using all features, using five features, and using three features for Roomset2 dataset

	Using all seven features			Using five features			Using three features		
	Precision	Recall	ROC area	Precision	Recall	ROC area	Precision	Recall	ROC area
Class 1: Sit on bed	0.873	0.883	0.938	0.898	0.876	0.935	0.827	0.876	0.936
Class 2: Sit on chair	0.937	0.957	0.978	0.946	0.951	0.975	0.901	0.930	0.964
Class 3: Lying	0.997	0.997	0.985	0.997	0.998	0.986	0.998	0.998	0.992
Cass 4: Ambulating	0.713	0.647	0.821	0.640	0.672	0.833	0.681	0.521	0.763
Weighted average	0.985	0.985	0.980	0.985	0.985	0.981	0.982	0.982	0.984

5 Conclusion

Human activity recognition is one of the major research problems in computer science that attracted many researchers in recent years with a variety of applications including healthcare and medical applications. This chapter proposed a machine learning method with feature selection and classification algorithms to recognize different human activities such as setting and walking from publicly available datasets collected by wearable sensors of elderly. The feature selection algorithms were used effectively to rank the features and identify the features that can be removed from the collected sensor data. Several classification algorithms were used in this paper for activity recognition. These algorithms include random forest, ANN, SVM, J48 decision tree, AdaBoost, and naïve bays classifiers. The ReliefF and the information gain attribute evaluation algorithms were used for feature selection. Several experiments were conducted to compare between these different algorithms using different number of features based on the obtained features ranking. Experimental results show that the random forest algorithm provides an overall high accuracy in recognizing human activities and using only three features of the collected sensor data. This can help medical service providers to monitor human activities for elderly and provide the required services based on the detected activity by using a small number of collected features from sensors.

Future directions includes applying the proposed method on other available datasets for human activity recognition and apply different feature selection algorithms and classification algorithms that can further improve the results.

References

1. C. Becker, K. Rapp, Fall prevention in nursing homes. Clin. Geriatr. Med. **26**(4), 693–704 (2010)
2. R.L. Shinmoto Torres, D.C. Ranasinghe, Q. Shi, A.P. Sample, Sensor enabled wearable RFID technology for mitigating the risk of falls near beds, in *2013 IEEE International Conference on RFID*, (2013), pp. 191–198
3. R.L. Shinmoto Torres, D.C. Ranasinghe, Q. Shi, Evaluation of wearable sensor tag data segmentation approaches for real time activity classification in elderly, in *International Conference on Mobile and Ubiquitous Systems: Computing, Networking, and Services*, (Springer, 2013), pp. 384–395
4. B. Najafi, K. Aminian, A. Paraschiv-Ionescu, F. Loew, C. Bula, P. Robert, Ambulatory system for human motion analysis using a kinematic sensor: Monitoring of daily physical activity in the elderly. I.E.E.E. Trans. Biomed. Eng. **50**(6), 711–723 (2003)
5. D. Karantonis, M. Narayanan, M. Mathie, N. Lovell, B. Celler, Implementation of a real-time human movement classifier using a triaxial accelerometer for ambulatory monitoring. IEEE Trans. Inf. Technol. Biomed. **10**(1), 156–167 (2006)
6. M. Ehatisham-Ul-Haq, A. Javed, M.A. Azam, H.M.A. Malik, A. Irtaza, I.H. Lee, M.T. Mahmood, Robust human activity recognition using multimodal feature-level fusion. IEEE Access **7**, 60736–60751 (2019)
7. A. Wickramasinghe, D.C. Ranasinghe, C. Fumeaux, K.D. Hill, R. Visvanathan, Sequence learning with passive RFID sensors for real time bed-egress recognition in older people. IEEE J. Biomed. Health Inform. **PP**(99), 1 (2016)

8. R.L. Shinmoto Torres, R. Visvanathan, S. Hoskins, A. van den Hengel, D.C. Ranasinghe, Effectiveness of a batteryless and wireless wearable sensor system for identifying bed and chair exits in healthy older people. Sensors **16**(4), 546 (2016)
9. A. Wickramasinghe, D.C. Ranasinghe, Recognising activities in real time using body worn passive sensors with sparse data streams: To interpolate or not to interpolate? in *Proceedings of the 12th EAI International Conference on Mobile and Ubiquitous Systems: Computing, Networking and Services*, (2015), pp. 21–30
10. C.D. Vass, O. Sahota, A. Drummond, D. Kendrick, J. Gladman, T. Sach, M. Avis, M. Grainge, REFINE (Reducing falls in in- patient elderly)-a randomised controlled trial. Trials **10**(1), 83 (2009)
11. M. Bruyneel, W. Libert, V. Ninane, Detection of bed-exit events using a new wireless bed monitoring assistance. Int. J. Med. Inform. **80**(2), 127–132 (2011)
12. J. Hilbe, E. Schulc, B. Linder, C. Them, Development and alarm threshold evaluation of a side rail integrated sensor technology for the prevention of falls. Int. J. Med. Inform. **79**(3), 173–180 (2010)
13. C. Doukas, I. Maglogiannis, Emergency fall incidents detection in assisted living environments utilizing motion, sound, and visual perceptual components. IEEE Trans. Inf. Technol. Biomed. **15**(2), 277–289 (2011)
14. M. Lutrek, B. Kalua, Fall detection and activity recognition with machine learning. Informatica **33**(2), 197–204 (2009)
15. D.J. Cook, Learning setting-generalized activity models for smart spaces. IEEE Intell. Syst. **27**(1), 32–38 (2012)
16. N.C. Krishnan, D.J. Cook, Activity recognition on streaming sensor data. Pervasive Mobile. Computing. **10**(Part B), 138–154 (2014)
17. M. Dunham, D. Mining, *Introduction and Advanced Topics* (Prentice Hall, 2002)
18. M. Kantardzic, *Data Mining: Concepts, Models, Methods, and Algorithms* (Wiley, IEEE Press, 2020)
19. J. Han, M. Kamber, J. Peu, *Data Mining – Concepts and Techniques* (Morgan Kaufmann, Elsevier, 2011)
20. P.-N. Tan, M. Steinbach, V. Kumar, *Introduction to Data Mining* (Addison Wesley, 2005)
21. I.H. Witten, E. Frank, *Data Mining: Practical Machine Learning Tools and Techniques*, 2nd edn. (Morgan Kaufmann, 2005)
22. I. Kononenko, E. Simec, M. Robnik-Sikonja, Overcoming the myopia of inductive learning algorithms with RELIEFF. Appl. Intell. **7**, 39–55 (1997)
23. M. Robnik-Sikonja, I. Kononenko, Theoretical and empirical analysis of ReliefF and RReliefF. Mach. Learn **53**, 23–69 (2003)
24. K. Kira, L. Rendell, The feature selection problem: Traditional methods and new algorithm, in *Proceedings of the 10th National Conference on Artificial Intelligence (AAAI 92)*, (1992), pp. 129–134
25. K. Kira, L. Rendell, A practical approach to feature selection, in *Proceedings of the 9th International Conference on Machine Learning*, (Morgan Kaufmann, 1992), pp. 249–256
26. D. Guan, W. Yuan, Y.-K. Lee, K. Najeebullah, M.K. Rasel, A review of ensemble learning based feature selection. IETE Tech. Rev. **31**, 190–198 (2014)
27. B. Seijo-Pardo, I. Porto-Díaz, V. Bolón-Canedo, A. Alonso-Betanzos, Ensemble feature selection: Homogeneous and heterogeneous approaches. Knowl.-Based Syst., 124–139 (2016)
28. UCI machine learning repository: https://archive.ics.uci.edu/ml/datasets/Activity+recognition+with+healthy+older+people+using+a+batteryless+wearable+sensor
29. T. Kaufmann, D.C. Ranasinghe, M. Zhou, C. Fumeaux, Wearable quarter-wave folded microstrip antenna for passive UHF RFID applications. Int. J. Antennas Propag. **2013, 2013**, 129839,, 1–11
30. WEKA the workbench for machine learning: https://www.cs.waikato.ac.nz/ml/weka/
31. E. Frank, M.A. Hall, I.H. Witten, *The WEKA Workbench, Online Appendix for "Data Mining: Practical Machine Learning Tools and Techniques"* (Morgan Kaufmann, 2016)
32. L. Breiman, Random forests. Mach. Learn. **45**(1), 5–32 (2001)

Machine Learning for Understanding the Relationship Between Political Participation and Political Culture

A. Hannibal Leach and Sajid Hussain

1 Introduction

How might machine learning be applied to help promote the ethos of democratic citizenship in this new age? This chapter uses machine learning tools to better understand the role political culture plays in determining levels of political participation among citizens spread across our democratic republic. Political culture is important in democratic systems because it helps shape political preferences, and these in turn, drive the political process [1]. The political process is the engine which keeps the wheels of self-government turning. Political actions such as voting, nominating, meeting, campaigning, informing, and dissenting in the public arena, are all part of the machinery that helps to sustain the democratic project. Thus, understanding how political culture may influence political participation has both normative and practical value. A strong political culture helps to maintain and foster the level of political participation necessary to sustain democratic principles and practices within the state.

Acknowledging the substantial evidence revealing people's attitudes toward local, state, and national issues are shaped in part by their political culture [2–4], we add further to this literature by exploring how people's rates of political participation may be influenced by the dominant political subculture within their region. In what regional political culture are we more likely to find higher levels of political participation among Whites and African Americans? By observing individuals situated within different geographical settings, we expand upon the foundational work laid by scholars such as Elazar [5], Hero and Tolbert [21], and

---

A. H. Leach · S. Hussain (✉)
Fisk University, Nashville, TN, USA
e-mail: aleach@fisk.edu; shussain@fisk.edu

© Springer Nature Switzerland AG 2021
H. R. Arabnia et al. (eds.), *Advances in Artificial Intelligence and Applied Cognitive Computing*, Transactions on Computational Science and Computational Intelligence, https://doi.org/10.1007/978-3-030-70296-0_23

Hero [6], in their assessments of the various ways in which political culture and ethnicity, may work to influence participation in public politics.

This chapter contributes to Political Science and Computer Science in a number of ways. First, it seeks to fill the gap between political participation and political culture in a new era of hyper-partisanship and dynamic change in mass media. Earlier accounts on political culture took place well before the technological revolution and the rise of the internet. This present work not only updates previous scholarship on political culture and public participation in politics, but it also analyzes a citizenry in a completely different political environment. Secondly, this work examines whether race groups such as African Americans are more influenced by their own political heritage or their respective regional subcultures (or something else completely). Thus, this article also opens-up new perspectives on the theory of minority participation in democratic politics. From a computing and machine learning perspective, the publicly available open-source data related to the 2016 federal elections is used to classify the states according to political engagement of the residents. The states are labelled according to Elazar's [5] classifications. A machine learning algorithm is then applied to verify and validate these classifications. We then analyze the data, specially examining observations of Black and White races. The machine learning tools here test model specification and will be used to improve its performance to better predict political participation among Americans. The information may be used by policymakers and concerned citizens in their effort to increase civic participation and combat political apathy.

The remainder of the chapter proceeds in five sections. The first examines the literatures on political culture and political participation. We will look here at past research on political participation at the national, state, and local levels. Focus will also be given to describing how the unique relationship between regional and political culture and public participation is essential to a better understanding of the interchangeable connection between micro- and macro-level politics. The third section develops testable hypotheses, while the fourth presents our empirical analysis of the relationship between regional political subculture and political participation. We conclude with a discussion of our findings' larger implications for a better understanding, of how machine learning tools may be deployed to improve our representative democracy, promote a healthy political culture, and increase political engagement.

2   The Concept of Political Culture

Almond and Verba [2] describe political culture as a shared set of social norms. Focusing on both its attitudinal and cognitive elements, Foster [7] refers to political culture as "a self-reinforcing system of attitudes and ideals emerging from a society's history, its traditions, the spirit of its public organizations, the instinct and reasoning of citizens and the style and ethical code of its leaders" [7], p. 561. Political culture is important because it "colors a people's expectations about the realities of politics and instills in them shared ideals as to what their public life might be" [8], p. 9. According to Almond and Verba [2], democratic government requires a political culture consistent with it. Citing McClosky and Zaller [9], Carman and Barker note that political culture may refer to the "widely shared beliefs, values and norms of a political community regarding the relationship of citizens to their government and to one another in matters of public affairs" [10], p. 516. Concern about political culture has been an activity even predating Greco-Roman society [11]. The concept of political culture has been employed to examine a host of political, social, and economic phenomena. Among these are social capital [12], capitalism and comparative political systems [13]. In his famous trip to the New World, Tocqueville (1831/1866) also attributed a great deal of America's character to its unique political culture [9, 14]. One question the French scholar sought to answer in his famous study of democracy in America was, "why do some communities prosper, possess effective political institutions, have law-abiding and satisfied citizens, and others do not" [15], p. 6. In a somewhat similar vein, Putnam [16] conducted a study comparing different sections in Italy in order to determine the type of conditions most conducive to fostering responsive, robust, and effective political institutions. Among those factors contributing to effective democratic government, Putnam found high levels of civic participation were one of them. Other scholars have also developed theories to help explain differences in political culture among the regions in America, especially as it relates to accounting for variation in terms of policymaking and political outcomes. Find political culture accounts for varying levels of political participation, political trust, and sense of efficacy to such an extent among Canadian provinces that they argued they could be seen as analytically distinct political systems.

3 Regional Political Culture

Elazar's [5] seminal work on the American states demonstrate political culture to be perhaps the primary force influencing the nature of politics within states. He defines political culture as the "particular pattern of orientation to political action in which each political system is imbedded" [5], p. 89. These cultures, he explains, are based largely on the historical experiences and migratory patterns, of the nation's colonial inhabitants, and other early settlers. Migrating and expanding their horizons westward, these groups took their respective political cultures with them, which was essentially bequeathed to subsequent generations. New migrants and immigrants to these areas were assimilated into these political cultures. Elazar essentially claimed that the national culture was built on three political subcultures: Individualistic, Moralistic, and Traditionalistic. His analysis show these political cultures to be geographically based, in accordance with the predominate themes dominating the political landscape of early Americans. In short, the Moralist states are found mostly in the New England area, the Individualistic states are found largely in the Middle

Atlantic area, and the Traditionalistic culture are concentrated largely among the Southern states. From this framework, Elazar contends we are able to explain much about political culture in relation to other factors affecting public opinion, policies, and other outcomes at the state and local levels.

Utilizing Elazar's framework, Carman and Barker [17] observes political culture not only shapes the ideas of individual people, but also influences political decisions undertaken by the states. In estimating an event analysis via a Cox hazard model, they look at the timing of which different states decide to hold their Democratic primary before the national convention. They demonstrate that much of the variation among the states, in this regard, is attributed to political culture. They hypothesize and provide evidence that states dominated by the Moralist political culture adopted earlier primary dates than those dominated by the Individual and Traditional culture (and also the Individual before the Traditional). As they put it, this is because those under the Moralist culture seek to provide policies and an environment which the entire community will benefit from, whereas the Individualist and Traditionalist culture provides for only a few or the elites. Individualist and Traditionalist states will only seek to hold earlier primaries or caucuses if it will benefit certain groups or individuals of the old guard. This rarely happens and thus, their primary dates rarely move.

Carman and Barker [10] also show political culture plays a role in determining the types of traits that individuals wish to see in political candidates. In examining the particular type of values different regional political subcultures place premiums on, they make hypotheses concerning the traits individuals from each political culture would value the most when it comes to choosing their political leaders. They show that those living within the Traditionalistic culture place more weight on a candidate's personal integrity and honesty, which they believe goes to a person's "Character." They also show that those living under the Moralist and Individualist political culture place greater value on a candidate's "Competence." This is because, above all, leaders are expected to perform, and their work will be measured to a large extent on what they are able to deliver on. Looking at the 2008 presidential elections, Fisher [18] observes that political culture is a great predictor of support for Obama in the 2008 Democratic primaries and caucuses. According to Fisher, Obama won 76% of the states under the Moralist political culture.

Elazar's political typology is not without its critics, however. Scholars have both expanded upon Elazar's research, as well as identified weaknesses in some of his assessments. For instance, Nardulli [19] reformulates Elazar's descriptions of the three political cultures and created four survey questions. This is to determine whether individual's attitudes would correspond with the respective political culture of their area. Interestingly, he doesn't find any correlation across his four questions. Lieske [20] also expands upon Elazar's conception of political culture. In formulating a measure of a state's culture by aggregating the respective proportions of the total statewide population that are under the influence of each of Elazar's political cultures, he uncovered 11 distinct regional subcultures [20]. His analysis demonstrates that a state's culture could be seen as the product of the contending subcultures within each state. Hero and Tolbert [21] take a different approach in examining the formation, process, and general disposition of politics at the local, state, and regional levels of government. Rather than centuries of old migration patterns, Hero and Tolbert instead argue that the main dynamics constructing social and political norms are the level of different races and ethnicities within a given state/region. In their work, they conceptualize and model states in terms of their homogeneous, heterogeneous, or bifurcated racial/ethnic composition. Race, as they argue, is the main force which influences how a state's political culture ultimately forms, and the type of format that its political processes will ultimately take. For instance, they show that states with higher levels of minority populations are actually the states in Elazar's study that are characterized as having a "Traditional" political culture. Hero and Tolbert label these as bifurcated states. They argue that states dominated by Elazar's individualistic political culture could also be labeled as heterogeneous states. They contend that this is due to the region's high concentrations of minority groups, and the interplay among them as they vie for premium redistribution policies for their own particular race/ethnicity group.

Many other scholars also utilize political cultures within and among the states as a way to examine a wide array of political topics such as distributional equity [22], government institutions [23], civil rights [24], political participation [19, 25], political representation [26], etc. Suffice it to say, the scope of research that the topic of political culture has inspired is wide and diverse.

4 Political Participation

The literature on political participation proffers several explanations as to the level and extent to which people engage in political activities. Such explanations generally fall under three main models: the resource model, the psychological engagement model, and the mobilization model [27, 35].

According to the resource model, political participation and activity is essentially a function of the quality and quantity of the resources one has at his/her disposal. According to Verba, Schlozman, and Brady [27], primary resources which influence political participation include civic skills, money, and the necessary time to actually engage. This model also considers the costs and benefits of political participation for the individual [28, 29]. As contend, socioeconomic status (SES) variables also play a significant role in determining political participation levels – higher SES status is associated with higher participation rates.

The psychological engagement model essentially posits that those who are interested in politics are more likely to participate while the uninterested are less likely to do so. Examining factors responsible for fostering political interest, Verba, Schlozman, and Brady [27] focus on education, parental influence, and school activities. They find higher levels of education is associated with increased levels of political interest. Their study also shows parental influence and exposure to politics at home is a strong indicator of political interest. Other findings within the literature generally falling under this model include different social and cultural characteristics. Factors associated with this notion include the lack of strong partisanship [30], low political efficacy [31], political alienation [32], residential mobility, social capital [12], and lack of civic obligation [33].

The third model explaining levels of political participation focuses on the role of mobilization. The mobilization model essentially argues that individuals participate politically because people ask and encourage them to do so. Scholars have routinely shown that voters who are contacted by political parties are more likely to vote, even while holding other factors constant [34, 35]. Political campaigns, social media use, and Black churches are also known vehicles for mobilization.

Another set of factors associated with influencing political participation stems from structural barriers. Factors falling under this category include the effects of voter registration laws and the legal/political obstacles which may make voting and formally participating in the political process difficult [36–39].

These theoretical perspectives on political participation demonstrate different avenues for thinking about its root causes. Though race has been considered as an explanation for political participation, rarely is it done in the context of political culture. We close the loop here and consider them both using powerful machine learning tools for data mining and analysis. We look here at whether regional political culture plays any role in how often Black and White Americans engage in political behavior?

5 Hypotheses

If regional political subculture significantly influences political participation rates, then we should expect to see consistent political participation levels and patterns within each respective regional subculture. Put differently, people from states within the Traditionalistic subculture should have similar participation rates from people in other states within the same subculture. Thus, since the states of Mississippi and Louisiana are both within the Traditionalistic political subculture, we should expect to observe similar participation levels from residents of both states. These participation levels should also significantly differ from people within Moralistic and Individualistic states. From this description, we hypothesize that we'll observe similar levels of political participation from people within states of identical regional political subcultures. We also believe that Blacks and Whites in many ways behave different politically. Because of this, we also hypothesize that Blacks and Whites will behave differently from each other within each political subculture.

6 Data and Methods

To what extent does political culture influence political participation? In this project, we consider whether there are any significant differences in the level of political participation among American citizens dwelling within the three regional political subcultures as outlined by Elazar. We use data from the 2016 Cooperative Congressional Election Study (CCES). The CCES is a nationally stratified sample survey administered on an annual basis from September to October. This annual study represents a concerted effort by political scientists nationwide to gather survey data in congressional elections [40]. The N for this survey was 64,600 respondents. The CCES has a large sample for common content and smaller, 1000-person samples for specific content brought by universities. The survey is useful because it makes inquiries on a range of topics including general political behavior, attitudes, demographic factors, foreign policy, and political information. We use Python's wide range of data management, statistical analysis, and data abstraction tools to identify whether regional political subcultures play a role in conditioning the rate to which individuals participate politically. The supervised machine learning algorithm, K-mean, was used to classify the states based on the Election 2016 data of the residents. The machine learning package, scikit-learn, in Python was used for the implementation.

7 Variables

The variables are the regional political subcultures that individuals identify as dwelling within. This information is provided by the given state each individual indicates as their place of residence. In particular, we look at whether a respondent resides in a state with a political culture identified by Elazar as Traditionalistic, Moralistic, or Individualistic. We use dummy variables for each of the three regional political cultures. Table 1 presents a listing of each state and its corresponding political culture. Another variable is race, which is accounted for by controlling for the race group each respondent identifies as belonging to, specifically Whites and African Americans. Dummy variables are also employed for race. Table 2 lists the variables and coding of the main variables.

Further, we focus on political participation as manifested through actions such as making social media posts about political information, sharing political information on social media, and commenting among other. Each of the political actions operationalizing participation can be found below in Table 3. Each of these measures highlight forms of political participation that are often overlooked by studies on voting and political research. Table 3 lists the specific question wording and coding of the variables.

Table 1 Elazar's state political cultures

Moralistic	Individualistic	Traditionalistic
California	Connecticut	Alabama
Colorado	Delaware	Arizona
Idaho	Illinois	Arkansas
Iowa	Indiana	Florida
Kansas	Nebraska	Georgia
Maine	Nevada	Kentucky
Michigan	Maryland	Louisiana
Minnesota	Massachusetts	Mississippi
Montana	Missouri	New Mexico
New Hampshire	New Jersey	North Carolina
North Dakota	New York	Oklahoma
Oregon	Ohio	South Carolina
South Dakota	Pennsylvania	Tennessee
Utah	Rhode Island	Texas
Vermont	Wyoming	Virginia
Washington		West Virginia
Wisconsin		

Table 2 Variables used in creating the clusters of states

Watch news
Read the newspaper
Post story about politics on social media
Post comment about politics on social media
Read story or video about politics on social media
Follow a political event on social media
Forward a story or video about politics on social media

Table 3 Questions operationalizing forms of local political participation taken from the 2016 Congressional Cooperative Election Study

Political activity
Grassroots/community political participation   "In the past year, did you"   (a) Attend local political meetings (such as school board or city council)   (b) Put up a political sign (such as a lawn sign or bumper sticker)   (c) Work for a candidate or campaign   (d) Donate money to a candidate, campaign, or political organization
Voter Registration   "Are you registered to vote?" $1 = $ yes; $0 = $ no

8 Results

Table 1 lists each state included in the analysis and their corresponding classification as originally outlined by Elazar. Figure 1 shows the overall average results for states

	Average of Black Correct %	Average of White Correct %
Individualistic	22.0	25.4
Moralistic	22.1	34.9
Traditionalistic	33.2	35.1

	Average of Individualistic %	Average of White Individualistic %
Individualistic	23.6	25.4
Moralistic	21.4	27.8
Traditionalistic	34.0	28.9

	Average of Moralistic %	Average of White Moralistic %
Individualistic	27.8	39.0
Moralistic	38.2	34.9
Traditionalistic	32.8	36.0

	Average of Traditionalistic %	Average of White Traditionalistic %
Individualistic	48.7	35.6
Moralistic	40.4	37.4
Traditionalistic	33.2	35.1

Fig. 1 Overall (average) comparison of Individualistic, Moralistic, and Traditionalistic states

according to three classifications: Individualistic, Moralistic, and Traditionalistic. According to Fig. 1, of the Blacks dwelling in Individualistic states, 34% behave like those in Traditionalistic states, 21% like those in Moralistic states, and 23% are analogous to those in Individualistic states. For Whites in Individualistic states, 29% behave like those in Traditionalistic states, 28% like those in Moralistic states, and 25% like those in Individualistic states. Figure 1 also shows the performance of those in Moralistic states. Of Black Americans living in Moralistic states, 28%

behave like those in Individualistic states, 33% like those in Traditionalistic states, and 38% like those in Moralistic states. On the other hand, of White Americans in Moralistic states, 39% behave like those in Individualistic states, 36% like those in Traditionalistic states, and 35% like those in Moralistic states. According to Fig. 1, of the Blacks residing in Traditionalistic states, 49% behave like those in Individualistic states, 40% like those in the Moralistic, and 33% like those in Traditionalistic ones. Of the Whites in Traditionalistic states, 36% behave like those in Individualistic states, 37% like those in the Moralistic, and 35% like those in Traditionalistic ones. Thus, Fig. 1 demonstrates that for Black Americans, two penultimate political cultures seem to predominate no matter where they reside: Traditionalistic and Individualistic. Most Blacks overall tend to behave like those in the Individualistic political culture, but those behaving in accordance with the Traditionalistic political culture account for well over a third of Blacks in each regional political culture. Blacks behaving like those in Moralistic states tend to cluster in Traditionalistic and Moralistic states, respectively. Their greatest concentration is in Traditionalistic states, however, which is puzzling given that we see the greatest concentration of Individualistic imitators in these areas as well. From a theoretical perspective, this phenomenon does seem plausible, however. Most African Americans live in the American South where many of the states with Traditionalistic political cultures exist. But most Southern African Americans live in major cities and metropolitan areas, where political life is more similar to that of other major cities such as New York City or Boston. Thus, city life for most Southern Blacks may cause them to assume an Individualistic political culture out of necessity. The case is a bit different for White Americans. Within each political culture, the number of White Americans behaving similarly to Whites within Individualistic, Traditionalistic, and Moralistic states are distributed pretty evenly. This means that a predominate political culture does not exist among White Americans.

Figure 2 depicts the bar charts to show greater details for individual states. As Fig. 2 demonstrates, the overall political culture of Blacks seems to align with that of Blacks dwelling in Traditionalistic and Individualistic states. As shown in the Individualistic cluster percentages in Fig. 2, Blacks tend to congregate along the two tail ends – the areas for Individualistic and Traditionalistic cultures. With Black still in mind, we see that the middle Moralistic areas are left hollow. For Whites, however, the distribution is spread pretty evenly across each of the regional political subcultures. Figure 2 also shows that most Black residents of Moralistic states exhibited political participation levels very similar to residents of Traditionalistic and Individualistic states. As we look at the Traditionalistic cluster percentages in Fig. 2, Blacks again exhibit great similarity with the Individualistic and Traditionalistic cultures, while leaving the middle area for Moralistic states hollow. In fact, here we only see 2 states (out of 17) where at least 15% of Blacks behave analogous to the Moralistic culture. Again, Whites tend to be evenly distributed across each of the political cultures. Thus, Fig. 2 also demonstrates that Elazar's formulation of regional political subcultures seems to be applicable to Whites, but not necessarily to Blacks. This makes sense intuitively because

Machine Learning for Understanding the Relationship Between Political... 307

Fig. 2 Detail comparison of state for Individualistic, Moralistic, and Traditionalistic states

Elazar's main ideas regarding cultures forming through waves of immigration from Old Europe, applies only to the story of White Americans. Largely missing from Elazar's formulation were the main origin stories of Black Americans, which are very different from most White Americans. From the above analysis, Black Americans seem to have a political culture that transcends Elazar's classifications of regional political subcultures. The average Black American, regardless of their regional political subculture, is more likely to participate in politics at rates similar to Blacks who dwell in Traditionalistic and Individualistic states.

9 Conclusion

This chapter began by inquiring into whether there is a relationship between political culture and political participation, the potential role that race may play in the process, and how machine learning may be employed to help bear these things out. Specifically, we sought to determine if the dominant regional political subculture where people reside significantly influences their level of political participation. Predicated on the literatures dealing with political participation, political subcultures, and political behavior, we were led to believe that such an influence may exist. After specifying our models and examining our adopted dependent variables, we generated several hypotheses from which to better understand how one's political culture may affect their level of political participation. The results from this work provide strong evidence that political culture works differently and in nuanced ways for Black and White Americans. For White Americans, the percentage of people behaving like those in the Moralistic, Traditionalistic, and Individualistic political cultures, appear to be spread pretty evenly across each regional political culture examined in our analysis. For instance, though Louisiana is a Traditionalistic state according to Elazar, only a third there behave politically in the Traditionalistic fashion, while the remaining two-thirds are split between the Moralistic and Individualistic. We observe the same case for states that Elazar deems to be dominated by the Moralistic and Individualistic political cultures, respectively. For Black Americans, we observe a very different story. According to our results, Black Americans are overall heavily influenced by the Individualistic and Traditionalistic political cultures. Regardless of the political culture that dominates the state in which they live, Blacks on average are more likely to behave in ways similar to the Individualistic and Traditionalistic cultures. The Individualistic culture seems to have the strongest influence among Black Americans, but well over a third of Blacks in each regional political culture behave like the Traditionalistic. That Blacks would tend to be associated with the Traditionalistic political culture was not too surprising given that the majority of Blacks in America still live in the South, and that states in the South tend to be heavily dominated by the Traditionalistic culture. Because most Blacks in America live in large cities and metropolitan areas (this is the case nationwide), it also makes sense for them to adopt an aggressive political culture that's associated primarily with the Individualistic culture. Though African Americans are as diverse as any other large group of people (there about 43 million African Americans in the US), the findings here show that at least in terms of political participation, Blacks on the whole tend to behave very similarly. Many people often assume Blacks to be a homogeneous group, but this study provides more empirical evidence toward the idea of an actual African American political "community."

Future research may take a further look at our findings here. Perhaps in ways similar to what Hero and Tolbert [21] argue, political culture could be a proxy for something else that we're missing here. It could be one thing for White Americans, and something completely different for Blacks. If this is the case, it would be interesting to gain a better understanding of what these things may be. We also looked here at only some of the classical measures of political participation. In a future chapter, we plan to update these findings including measures more in alignment with how others understand political participation in the twenty-first century to encompass. Not only does this include classical measures such as putting up a political yard sign or attending a political meeting, but also their content and activity via social media. The computational methods we use here are a great way to incorporate how a new generation understands political participation. The computational methods here provide scholars with a better understanding of how machine learning tools can be applied in useful ways to solve traditional social science problems. This perspective allows scholars to better understand what they may be able to get out of utilizing machine learning tools and computational thinking.

Acknowledgment This work is partially supported by NSF Grants #1912588 and #1817282.

References

1. A. Wildavsky, Choosing preferences by constructing institutions: A cultural theory of preference formation. Am. Polit. Sci. Rev. **81**, 4–21 (1987)
2. G. Almond, S. Verba, *The Civic Culture: Political Attitudes and Democracy in Five Nations* (Sage Publications, Newbury Park, 1963)
3. H. Eckstein, A culturalist theory of political change. Am. Polit. Sci. Rev. **82**, 789–804 (1988)
4. R. Putnam, *The Beliefs of Politicians: Ideology, Conflict, and Democracy in Britain and Italy* (Yale University Press, New Haven, 1973)
5. D. Elazar, *American Federalism: A View from the South* (Crowell, New York, 1972)
6. R.E. Hero, *Faces of Inequality: Social Diversity in American Politics* (Oxford University Press, New York, 1998)
7. C. Foster, Political culture and regional ethnic minorities. J. Polit. **44**(2), 560–568 (1982)
8. L. Pye, The concept of political development. Ann. Am. Acad. Pol. Soc. Sci. **358**, 1–13 (1965)
9. J. McCloskey, J. Zaller, *The American Ethos: Public Attitudes Toward Capitalism and Democracy* (Harvard University Press, Cambridge, MA, 1984)
10. C.J. Carman, D.C. Barker, Regional subcultures and mass preferences regarding candidate traits in the USA. Regional Federal Stud. **20**(4/5), 515 (2010)
11. G. Almond, The intellectual history of the civic culture concept, in *The Civic Culture Revisited: An Analytic Study*, ed. by G. Almond, S. Verba, (Little, Brown, Boston, 1980), pp. 1–36
12. R.D. Putnam, *Bowling Alone: The Collapse and Revival of American Community* (Simon & Schuster, New York, 2000), p. 2000
13. L. Pye, S. Verba (eds.), *Political Culture and Political Development* (Princeton University Press, Princeton, 1965)
14. S.P. Huntington, *American Politics: The Promise of Disharmony* (Belknap Press of Harvard University Press, Cambridge, MA, 1981)
15. G. Brewer, building social capital: Civic attitudes and behavior of public servants. J-PART **13**(1), 5–25 (2003)
16. R.D. Putnam, *Making democracy work: Civic traditions in modern italy* (Princeton University Press, Princeton, 1993)
17. C.J. Carman, D.C. Barker, State political culture, primary frontloading, and democratic voice in presidential nominations: 1972–2000. Electoral Studies **24**(4), 665–687 (2005)

18. P. Fisher, State political culture and support for Obama in the 2008 Democratic presidential primaries. Soc. Sci. J. **47**(3), 699–709 (2010)
19. Nardulli, Political subcultures in the American states: An empirical examination of Elazar's formulation. Am. Politics Q. **18**, 287–315 (1990)
20. J. Lieske, The changing regional subcultures of the American states and the utility of a new cultural measure. Polit. Res. Q. **63**, 538–552 (2010)
21. R.E. Hero, C.J. Tolbert, A racial/ethnic diversity interpretation of politics and policy in the states of the US. Am. J. Polit. Sci. **40**(3), 851–871 (1996)
22. J. Fitzpatrick, R. Hero, Political culture and political characteristics of the American states: A consideration of some old and new questions. West. Political Q. **41**(1), 145–153 (1988)
23. I. Sharkansky, The utility of Elazar's political culture. Polity **2**, 66–83 (1969)
24. S.R. Freeman, *Southern Political Culture: A Test of Elazar's Conceptualization* (University of Kentucky Press, Lexington, 1978)
25. N.P. Loverich Jr., B.W. Daynes, L. Ginger, Public policy and the effects of historical-cultural phenomena: The case of Indiana. Publius **10**(2), 111–125 (1980)
26. W. Erikson, McIver, *Statehouse Democracy: Public Opinion and Policy in the American States* (Cambridge University Press, Cambridge, MA, 1993)
27. S. Verba, K.L. Schlozman, H.E. Brady, *Voice and Equality: Civic Voluntarism in American Politics* (Harvard University Press, Cambridge, MA, 1995)
28. A. Downs, *An Economic Theory of Democracy* (Harper & Row, New York, 1957)
29. W.H. Riker, P.C. Ordeshook, A theory of the calculus of voting. Am. Polit. Sci. Rev. **62**, 25–42 (1968)
30. P. Abramson, J.H. Aldrich, The decline of electoral participation in America. Am. Polit. Sci. Rev. **76**, 502–521 (1982)
31. S.D. Shaffer, A multivariate explanation of decreasing turnout in presidential elections, 1960–76. Am. J. Polit. Sci. **25**(1), 68–95 (1981)
32. R.A. Brody, B.I. Page, Indifference, alienation, and rational decisions: The effects of candidate evaluations on turnout and the vote. Public Choice **15**, 1–17 (1973)
33. S. Verba, N.H. Nie, *Participation in American: Political Democracy and Social Equality* (Harper & Row, New York, 1972)
34. H. Clarke, D. Sanders, M. Stewart, P. Whiteley, *Political Choice in Britain* (Oxford University Press, Oxford, 2004)
35. S.J. Rosenstone, J.M. Hansen, *Mobilization, Participation, and Democracy in America* (Macmillan, New York, 1993)
36. R.S. Erickson, Why do people vote? Because they are registered. Am. Politics Q. **9**(2), 259–276 (1981)
37. M.J. Fenster, The impact of allowing day of registration voting on turnout in U.S. elections from 1960 to 1992: A research note. Am. Politics Q. **22**, 74–87 (1994)
38. F.F. Piven, R.A. Cloward, *Why Americans Don't Vote* (Pantheon, New York, 1989)
39. S.J. Rosenstone, R.E. Wolfinger, The effect of registration laws on voter turnout. Am. Polit. Sci. Rev. **72**, 22–45 (1978)
40. S. Ansolabehere, B.F. Schaner, *Cooperative Congressional Election Study: Common Content. [Computer File] Release 2: August 4, 2017* (Harvard University, Cambridge, MA, 2016) [producer] http://cces.gov.harvard.edu

Targeted Aspect-Based Sentiment Analysis for Ugandan Telecom Reviews from Twitter

David Kabiito and Joyce Nakatumba-Nabende (iD)

1 Introduction

With the rise of the social media era, people are always connected and often times share their opinions through tweets, comments, reviews and chats. People have more trust in human opinion compared to traditional advertisements. Many consumers are using social media to seek advice and recommendation from others before making decisions regarding important purchases [5, 9, 12]. Such referrals have a strong impact on customer decision making and new customer acquisition for companies. Social media is a source of valuable customer feedback that helps companies measure satisfaction and improve their products or services. However due to the high velocity and voluminous nature of data from social media, it is difficult for the average human reader to extract and summarise the opinions on social media. Automated opinion mining systems are thus needed to overcome this challenge.

Opinion mining or sentiment analysis (SA) is the computational analysis of people's opinions, sentiments, attitudes and emotions towards a specific target entity such as a product, service, organization, individual, topic, and its aspects [1, 3, 5]. The main goal of opinion mining is to detect the emotional expression from a sentence or an entire document. In general, opinion mining has been investigated at three levels: document, sentence and entity or aspect level [5, 17]. At document level, the task is to classify whether a whole opinion document expresses a positive or negative sentiment. Sentence level opinion mining resolves the sentiment of each sentence in an opinionated document into the categories positive, negative and

---

D. Kabiito (✉) · J. Nakatumba-Nabende
Department of Computer Science, School of Computing and Informatics Technology, CoCIS, Makerere University, Kampala, Uganda
e-mail: jnakatumba@cis.mak.ac.ug

© Springer Nature Switzerland AG 2021
H. R. Arabnia et al. (eds.), *Advances in Artificial Intelligence and Applied Cognitive Computing*, Transactions on Computational Science and Computational Intelligence, https://doi.org/10.1007/978-3-030-70296-0_24

neutral. The neutral category contains sentences that do not express any sentiment. Aspect level opinion mining is used to overcome the shortcomings of both document and sentence level opinion mining by resolving the opinion target and sentiment. It is based on the idea that an opinion consists of a sentiment and a target thus an opinion without identifying its target is of limited use.

Previously, aspect level opinion mining has also been discussed at two levels: aspect-based sentiment analysis (ABSA) [6, 14] and targeted sentiment classification [16]. ABSA captures the sentiment over the aspects on which an entity can be reviewed while targeted sentiment analysis resolves the sentiment expressed toward several target entities [6, 11]. However, ABSA and targeted sentiment classification do not capture the sentiment of multiple aspects of the target entities. To remedy this, Saeidi et al., [11] introduced the task of targeted aspect-based sentiment analysis (TABSA) which aims at identifying the fine-grained opinion towards a specific aspect associated with a target entity. TABSA attempts to tackle the shortcomings of both ABSA and targeted sentiment classification. ABSA assumes that only one target entity is discussed in a sentence while targeted sentiment classification assumes the overall sentiment in the sentence as the sentiment for the target entities contained in the sentence. TABSA is an important task that can be used to generate a structured summary of opinions about target entities and their aspects with less assumptions compared to ABSA and targeted sentiment classification. TABSA turns unstructured opinion text to structured data that can be used for both qualitative and quantitative analysis.

In this paper, we study TABSA in English and Luganda code-mixed Twitter reviews about telecoms in Uganda. For this task, the human annotated dataset called SentiTel is introduced. The dataset can be used to extract fine-grained opinion information about telecoms in Uganda. In particular, our contribution in this paper is three-fold:

1. We provide the SentiTel human annotated benchmark dataset used for the TABSA of telecom data. The reviews in the dataset have a code-mix of English and Luganda.[1]
2. We apply the random forest baseline model and Bidirectional Encoder Representations from Transformers (BERT) model [2] to the TABSA task.
3. We show that despite the short and informal text structure of reviews from Twitter, the reviews are rich in information and can be used for fine-grained opinion mining.

The rest of the paper is organized as follows: In Sect. 2, we review related work about sentiment analysis, Sect. 3 presents the dataset and its annotation schema; the task definition and experimental set-up are discussed in Sect. 4. In Sect. 5, we provide details about the evaluation and results. Finally, we conclude the paper in Sect. 6.

---

[1]Luganda is a Bantu language which is commonly spoken in Uganda.

2 Related Work

Opinion mining has been widely discussed and studied in the recent years. This has been attributed to the rise of social media which has made large volumes of opinionated data publicly available. Opinion mining research has gradually evolved from sentiment analysis which mainly focused on identifying the overall sentiment of a unit of text [15] to aspect-based sentiment analysis (ABSA) [14] and targeted sentiment classification [16]. However, ABSA does not consider more than one entity in the given text while targeted sentiment classification task only identifies the overall sentiment on a single entity [11].

To overcome the challenges of both ABSA and targeted sentiment classification, Saeidi et al. [11] introduced the task of TABSA. They proposed two baseline systems: feature-based logistic regression and Bi-LSTM-based model together with the SentiHood dataset. SentiHood is based on text taken from question answering platform of Yahoo! Answers that is filtered for questions related to the neighbourhood of London. Motivated by the success of deep learning, Yukun et al. [6] proposed a hierarchical attention model that explicitly attends first to the targets and then the whole sentence. They also incorporated useful common sense knowledge into a deep neural network to further enhance their results on SentiHood. Their best model Sentic LSTM obtained a strict accuracy of 67.43% and an accuracy of 89.32% on aspect category detection and sentiment classification respectively. To show the generalizability of Sentic LSTM, they built a subset of the Semeval-2015 dataset [9] for the TABSA task. On this subset, Sentic LSTM obtained a strict accuracy of 67.34% on the aspect category detection and an accuracy of 76.47% on sentiment classification.

Following the successful application of BERT on question answering (QA) and natural language inference (NLI) tasks, Sun et al. [14] modelled the TABSA task as sentence pair classification task via auxiliary sentence generation. Four methods were used to convert the TABSA task into a sentence pair classification task then used to fine-tune the pre-trained BERT model. The highest performance on SentiHood was obtained using BERT-pair-NLI-M with an Area Under the ROC Curve (AUC) of 97.5 and 96.5% on aspect category detection and sentiment classification respectively. The BERT model was also applied to the Semeval-2014 task 4 dataset [8]. An F1 of 92.18% was obtained on aspect category detection using the BERT-pair-NLI-B model.

Despite the promising results obtained on the TABSA task, the previous work has not explored TABSA on code-mixed reviews from a generic social media platform like Twitter; moreover, the existing datasets contain a maximum of two target entities. In this paper, we apply the baseline random forest model and BERT-pair-NLI-M to the human annotated dataset SentiTel which is annotated for the TABSA task. SentiTel is extracted from Twitter and it contains a code-mix of English and Luganda with up to three target entities. As proposed by Sun et al. [14] in the BERT-pair-NLI-M model, the TABSA task is modelled as a sentence pair classification task using auxiliary sentences.

3 Dataset and Annotation

In this section we explain the data collection, annotation schema and summarise the properties of the SentiTel dataset. The target entities in the dataset are three major telecoms in Uganda: *"MTN"*, *"Airtel"* and *"Africell"*. In this study the telecom domain was considered since telecoms are one of the largest service providers in Uganda. This results in many discussions about their aspect on Twitter. The tweets are extracted by querying the Twitter API using the Twitter handle of the respective telecom: "mtnug", "Airtel_Ug" and "africellUG". The query collects all the tweets that mention the telecom entity. In our extraction the language value in the Twitter API is set to "en" which corresponds to English language tweets. However this returns both English and English code-mixed tweets. The English code-mixed tweets are written in both English and Luganda which is a local dialect. The tweets were collected for the period between February 2019 and September 2019.

3.1 *Annotation Schema and Guidelines for the Reviews*

We used the BRAT annotation tool [13] for the annotation task. The data was annotated by three undergraduate software engineering students. Given a review, the task of the annotator was to identify the sentiment s expressed towards the aspect a of the target telecom t. As indicated by Eq. 1, the final annotation A contains 1 to T target telecoms with the respective sentiment towards the aspects of each target telecom.

$$A = \{t, a, s\}_1^T \tag{1}$$

Table 1 provides an overview of the target telecoms (t), aspect category (a) and sentiment (s) set provided for the annotation task. The aspects in the dataset were categorized into seven aspect categories as indicated by the aspects column. These are the seven most discussed aspects in the telecom domain in Uganda. The aspect *application* refers to opinions about the mobile application of the respective telecom, *calls* refers to opinions about the voice call experience of the reviewer, *customerservice* refers to the reviewer's opinion on the customer service of the respective telecom, *data* refers to the reviewer's opinion about the internet services of the telecom discussed in the review. The aspect *general* refers to a generic opinion on the target telecom discussed in the review. The *mobilemoney* and *network* aspect refers to opinions on the mobile money services and signal reception of the target telecom being reviewed. In our annotation schema, two sentiments were considered that is "Positive" and "Negative" since aspects are rarely discussed without an explicit "Positive" or "Negative" sentiment [10, 11].

TABSA for Ugandan Telecom Reviews from Twitter

Table 1 Overview of target telecoms (t), aspect category (a) and sentiment (s) used to annotate the SentiTel dataset

Target telecom (t)	Aspect (a)	Sentiment (s)
	Application	
Airtel	Calls	Positive
Africell	Customerservice	Negative
MTN	Data	
	General	
	Mobilemoney	
	Network	

Following Saeidi et al. [11], the annotators were required to read through the annotation guidelines and examples,[2] and then annotate a small subset of the dataset. After each round of annotation, agreements between the annotators were calculated and discussed. This procedure continued until they reached a reasonable agreement. The Cohen's Kappa coefficient (κ) was used as a measure for the pairwise agreement between the annotators. After which, the entire dataset was annotated.

The final dataset was obtained from the annotator with the highest inter-annotation agreement on the aspect category. The final aspect category pairwise inter-annotator agreement for aspect categories on the whole dataset was 0.59, 0.73 and 0.60. This is of sufficient quality for the TABSA task.

3.2 Composition of the Dataset

The final annotated dataset is called SentiTel,[3] and it contains Twitter reviews about telecoms in Uganda. 5,973 reviews in SentiTel have a single telecom entity mentioned while 347 reviews mention two or three telecom entities. The total number of opinions in SentiTel is 6683 since some reviews have multiple opinions. The telecom entity names in the reviews are all normalised to *"MTN", "Airtel"* and *"Africell"*. Each review in SentiTel is labelled with the telecom entity, aspect and sentiment. A review can have multiple labels depending on the number of target entities and aspects discussed in the review.

Table 2 shows the distribution of opinions across the aspect categories. The "Negative" sentiment is the dominant opinion polarity, this indicates that Twitter users prefer to tweet when they are facing challenges with a service or product otherwise they remain silent. The *data* aspect is the most frequent aspect with over 2400 reviews while the aspect *application* appears with less than 150 reviews.

---

[2]The annotation guidelines are available at: https://github.com/davidkabiito/Sentitel/blob/master/Annotation/guidelines.md.

[3]The dataset is available at: https://github.com/davidkabiito/Sentitel/tree/master/Dataset.

Table 2 SentiTel statistics on sentiment distribution over the different aspects

	Aspects						
		Customer				Mobile	
Sentiment	Data	Service	General	Network	Calls	Money	Application
Positive	377	74	450	81	35	23	8
Negative	2064	928	418	740	442	244	139

4 Methodology

In this section, we provide the formal task definition of TABSA in the telecom context and then discuss the training and evaluation of the random forest and BERT models.

4.1 Task Definition

A review consists of a sequence of words. In this sequence, a target telecom can be mentioned once or multiple times. Similar to Ma et al. [6] and Sun et al. [14], we consider all mentions of the same target as a single target. The TABSA task is divided into two sub-tasks. First, it resolves the aspect category of the aspects discussed in the review using the predefined set of aspect categories and then it resolves the sentiment polarity of each aspect category associated with the target telecom t.

Following Saeidi et al. [11], we set the task as a three-class classification problem: given a review, a set of target telecoms $T = \{MTN, Airtel, Africell\}$ and a fixed set of aspect categories $A=\{data, customer service, general, network, calls\}$, predict the sentiment polarity $y \in Y$ over the full set of the target-aspect pairs (t, a) where $\{(t, a) : t \in T, a \in A\}$ and $Y = \{Positive, Negative, None\}$. The sentiment *"None"* indicates that the review does not contain an opinion about the aspect a of the target telecom t. For example the review in Table 3 contains two targets, "Airtel" and "Africell". The objective of the TABSA task is to resolve the aspect category of the aspects of the target telecoms and classify the sentiment of each aspect category as shown in Table 3.

4.2 Models

For training and evaluating our models, the five most frequent aspects: *data, customer service, general, network* and *calls* from SentiTel were considered. Following Saeidi et al. [11], we split the dataset into three subsets train, validation

Table 3 Example of a Twitter telecom review and the output labels. The target entities are in bold while the aspect expression is underlined

Review	Output		
Bt the way AirtelUg this foolery	Target	Aspect	Sentiment
of slowing down my internet and	Airtel	Data	Negative
sending me harsh messages wanting me	Airtel	customerservice	None
to upgrade to 4G yet its	Airtel	General	None
a hogwash must stop my friend	Airtel	Network	None
with a 3G **africellUG** is doing	Airtel	Calls	None
better I have diligently been a	Africell	Data	Positive
customer for years respect my choices	Africell	Customerservice	None
	Africell	General	None
	Africell	Network	None
	Africell	Calls	None

and test set with each having 70, 10, and 20% of the data respectively. The validation set is used to select the best model parameters before applying the models to the test set.

Training the Random Forest Model The random forest (RF) model is our baseline model. Before training the RF model, all the words in the reviews are converted to lowercase then all punctuation marks and stopwords are removed except the word "not" which reverses the sentiment in the review. While training the RF model, we use the oversampling method suggested by Haibo and Edwardo [4] to overcome the challenge of an imbalanced dataset. The RF experiments were conducted using two word representations: tf-idf (TF-IDF-based unigram features) and word2vec [7]. The word2vec vectors are of dimensionality 300 and are trained using the continuous bag-of-words architecture. The common presence of informal language and words in tweets does not favour using initialised publicly available word2vec vectors [7]. This would result in a high random initialization of word vectors since these words are missing in the word2vec dictionary. We use the output of the fully connected layer of the network as input feature vector of the RF-word2vec model.

Following Sun et al. [14], the TABSA task is modelled as a sentence pair classification task using auxiliary sentences. The aspect category and sentiment class are jointly extracted by introducing the sentiment "None". The best model in both word representations is obtained using 100 trees.

Training the BERT Model We constructed the BERT model using the sentence pair input representation. The BERT-pair-NLI-M auxiliary sentence construction is used as described by Sun et al. [14]. We use the pre-trained uncased $BERT_{BASE}$.[4] model for fine-tuning on TABSA downstream task. The following hyper-parameters are used to obtain the best model; Transformer blocks: 12, hidden layers size: 768,

---

[4]https://storage.googleapis.com/bert_models/2018_10_18/uncased_L-12_H-768_A-12.zip.

Table 4 Performance on SentiTel dataset. Boldfaced values are the overall best scores

Model	Aspect			Sentiment	
	Acc.	F_1	*AUC*	*Acc.*	*AUC*
RF-tf-idf	0.509	0.442	0.883	**0.967**	0.915
RF-word2vec	0.337	0.190	0.694	0.965	0.820
BERT	**0.773**	**0.781**	**0.950**	0.940	**0.965**

The bold values indicate the best score for each evaluation metric

Table 5 Aspect level model performance using the AUC evaluation metric

Model	Aspect				
	Calls	Customer service	Data	General	Network
RF-tf-idf	0.893	0.864	0.869	0.916	0.874
RF-word2vec	0.701	0.669	0.675	0.754	0.669
BERT	**0.967**	**0.960**	**0.941**	**0.956**	**0.926**

The bold values show that BERT provides the best score in terms of AUC at the Aspect level

self-attention heads: 12 and the total number of parameters for the pre-trained model is 110M. When fine-tuning, we keep the dropout probability at 0.1 and set the number of epochs to 6. The initial learning rate and batch size are 2e-5 and 24 respectively. The training was done using one GPU on Google Colab.

We follow the same evaluation criteria as Sun et al. [14]. The evaluation metrics strict accuracy and Macro-F1 were used to evaluate the aspect category detection. Sentiment classification was evaluated using accuracy and AUC.

5 Results and Discussion

This section presents a discussion of the results from the models. Table 4 shows the results of the baseline RF models and BERT. The results are obtained after running the models on the test set. It can be seen from the table that BERT obtains the overall best results of 0.773, 0.781 and 0.950 for the aspect category detection task based on the respective evaluation metrics. For sentiment detection, the RF-tf-idf model performs better by a small margin of 0.027 compared to the BERT model in terms of accuracy while the BERT model performs better by a larger margin of 0.05 in terms of AUC. This means that the quality of sentiment classification by the BERT model is better than that of the RF models.

Table 5 shows the model performance per aspect in terms of AUC. Based on these results, we can see the aspect *Calls* is predicted with the highest AUC. With an AUC score of 0.926 for the BERT model, the aspect *Network* was the most difficult to predict. This is because the words used to represent the aspect *Network* are similar

Table 6 Examples of input reviews and their predicted labels using BERT. The target telecom in the review is highlighted in bold

Review	Target-telecom	Aspect	Predicted	Label
Africell you let the other ISPS cheat us today	Africell	General	Negative	Negative
Day of weekendgb data hearteyes **Africell** webale nyoo	Africell	Data	Positive	Positive
Airtel jst make our freaky friday the fun t used to b now ur offering less bundles for mo money mwebereremu	Airtel	Data	Negative	Negative
How often we should upfate u that your roaming netwrk aborad is done now coming to months **airtel**	Airtel	Network	Negative	Negative
AirtelUg i am about to give up on yowa data mbu upgrade to 4G n get 1 GB of free data its now a week Mulekeyo if u dnt want to give us yowa data Nkooye okulinda	Airtel	Data	Negative	Negative
Its only after you have experienced **mtn** internet that youll realise that other people are trash it runs first but it works	MTN	Data	Negative	Positive
MTN ur internet is very fake am now going for **airtel**	Airtel	Data	Positive	Positive
MTN ur internet is very fake am now going for airtel	MTN	Data	Negative	Negative

to those used for the aspects *Data* and *Calls*. This results in the models predicting the aspect *Data* or *Calls* instead of the aspect *Network*.

Table 6 present examples of reviews that are labelled correctly and those that are labelled incorrectly by the BERT model. The top six rows of the table contain reviews with a single target telecom. The BERT model is able to predict correctly the sentiment of the first five reviews. The model fails to correctly predict the sentiment of the sixth review which contains comparative opinion that cannot be captured by the model. The last two rows contain a review with multiple target telecoms.

The model is able to capture the opposite sentiment expressed for the aspect *data* of the two target telecoms. The BERT model performs better than the RF model because we separate the target and the aspect to form an auxiliary sentence while training the model. The set-up transforms the TABSA task into a sentence pair classification task. This set-up fully utilizes the advantages of the pre-trained BERT model.

Although the TABSA task is more complicated than sentiment analysis due to additional target and aspect information [6], the results demonstrate that fine-grained information can be extracted from Twitter opinions despite the limited cues and informal text structure. This is in line with the results obtained by Saeidi et al. [11] on TABSA task using logistic regression and Bi-LSTM. Furthermore the better performance by the BERT model shows that by fine-tuning a pre-trained BERT model promising results can be obtained on downstream tasks. This was also demonstrated by Chi et al. [14], who obtained an AUC of 0.975 and 0.97 on aspect detection and sentiment classification respectively on SentiHood by generating multiple forms of auxiliary sentences and then used the sentence pairs to fine-tune the pre-trained BERT model.

The comparison between the performance of our baseline RF models and BERT confirms that deep learning models such as BERT present better results than baseline models like RF. However based on previous research [1, 17], we expected the RF-word2vec model to perform better than the RF-tf-idf model. The RF-word2vec model can be improved by training a word2vec model on generic Twitter data and finally train it on SentiTel. This will enable the word2vec model obtain more semantic information on text from Twitter. Some improvement can also be obtained on the BERT model especially in terms of strict accuracy by constructing QA-auxiliary sentences that transform the TABSA task to a binary classification problem.

6 Conclusion and Future Work

In this paper, we provided a human annotated dataset called SentiTel which is annotated for the targeted aspect-based sentiment analysis task in the telecom domain. We also report baseline results using random forest and BERT. The TABSA task is modelled as a sentence pair classification task through auxiliary sentence generation. The BERT model obtained the best result with an AUC of 0.95 and 0.965 on the aspect category detection and sentiment classification tasks respectively. These results show that fine-tuning a pre-trained BERT model by transforming the TABSA task to a sentence pair classification task generates better results compared to other models. The results in this paper demonstrate that fine-grained information can be extracted from Twitter data despite the limited cues and informal text structure. The models developed in this research can be used by both the telecom service providers and their subscribers. The telecom service providers can use the models to generate structured summaries which can be used to highlight aspects that need improvement and also compare their performance with their competitors. The telecom subscribers can use the models to compare services of different telecom service providers. This research also highlights the need to train a social media word2vec model which can generate vectors that represent social media text better. This will improve the performance of word2vec based models in the social media context.

In future, we plan to explore other language phenomena such as comparative opinion, sarcasm and negation in the TABSA task. We will also assess the contribution of contextual information such as the personality of the opinion holder on the TABSA task.

Acknowledgments This work was supported by Mak-Sida Project 381. I also thank the Project PI, Professor Tonny Oyana and Project Coordinator, Dr. Florence Kivunike for their support and mentorship during my graduate studies.

References

1. L. Abid, S. Zaghdene, A. Masmoudi, S.Z. Ghorbel, A vector space approach for aspect based sentiment analysis, in *Proceedings of NAACL-HLT 2015, Association for Computational Linguistics*, (Association for Computational Linguistics, Denver, 2015), pp. 116–122
2. J. Devlin, M.-W. Chang, K. Lee, K. Toutanova, BERT: pre-training of deep bidirectional transformers for language understanding (2019). http://arxiv.org/abs/1810.04805
3. W.-B. Han, N. Kando, Opinion mining with deep contextualized embeddings, in *Proceedings of the 2019 Conference of the North American Chapter of the Association for Computational Linguistics: Student Research Workshop,Minneapolis, June 3–5* (Association for Computational Linguistics, Stroudsburg, 2019), pp. 35–42
4. H. He, E.A. Garcia, Learning from imbalanced data. IEEE Trans. Knowl. Data Eng. **21**(9), 1263–1285 (2009)
5. B. Liu, *Sentiment Analysis and Opinion Mining*. Synthesis Lectures on Human Language Technologies , vol. 5 (Morgan & Claypool, San Rafael, 2012), pp. 1–167
6. Y. Ma, H. Peng, E. Cambria, Targeted aspect-based sentiment analysis via embedding commonsense knowledge into an attentive LSTM. Association for the Advancement of Artificial Intelligence, Singapore (2018)
7. T. Mikolov, K. Chen, G. Corrado, J. Dean, Efficient estimation of word representations in vector space (2013). https://arxiv.org/abs/1301.3781
8. M. Pontiki, D. Galanis, J. Pavlopoulos, H. Papageorgiou, I. Androutsopoulos, S. Manandhar, SemEval-2014 task 4: aspect based sentiment analysis, in *Proceedings of the 8th International Workshop on Semantic Evaluation (SemEval 2014), Dublin, Ireland, August 23–24* (2014), pp. 27–35
9. M. Pontiki, D. Galanis, H. Papageorgiou, I. Androutsopoulos, S. Manandhar, M. AL-Smadi, M. Al-Ayyoub, Y. Zhao, B. Qin, O. De Clercq, V. Hoste, M. Apidianaki, X. Tannier, N. Loukachevitch, E. Kotelnikov, N. Bel, S. María Jiménez-Zafra, G. Eryiğit, SemEval-2016 task 5: aspect based sentiment analysis, in *Proceedings of SemEval-2016, San Diego, June 16–17*, (Association for Computational Linguistics, Stroudsburg, 2016), pp. 19–30
10. A. Rahman, E. K. Dey, Datasets for aspect-based sentiment analysis in bangla and its baseline evaluation. Data **3**(15), 1–10 (2018)
11. M. Saeidi, G. Bouchard, M. Liakata, S. Riedel, SentiHood: targeted aspect based sentiment analysis dataset for urban neighbourhoods, in *International Conference on Computational Linguistics, (COLING 2016) Osaka*, (Association for Computational Linguistics, Denver, 2016), pp. 1546–1556
12. S.R. Sane, S. Tripathi, K.R. Sane, R. Mamidi, Stance detection in code-mixed hindi-english social media data using multi-task learning, in *Proceedings of the 10th Workshop on Computational Approaches to Subjectivity, Sentiment and Social Media Analysis, Minneapolis, June 6* (Association for Computational Linguistics, Stroudsburg, 2019), pp. 1–5

13. P. Stenetorp, S. Pyysalo, G. Topic, T. Ohta, S. Ananiadou, J. Tsujii, Brat: a web-based tool for NLP-assisted text annotation, in *Proceedings of the Demonstrations at the 13th Conference of the European Chapter of the Association for Computational Linguistics* (Association for Computational Linguistics, Stroudsburg, 2012), pp. 102–107
14. C. Sun, L. Huang, X. Qiu, Utilizing BERT for aspect-based sentiment analysis via constructing auxiliary sentence, in *Proceedings of the 2019 Conference of the North American Chapter of the Association for Computational Linguistics: Human Language Technologies, Minneapolis* (2019), pp. 380–385
15. P.D. Turney, Thumbs up or thumbs down? Semantic orientation applied to unsupervised classification of reviews, in *Proceedings of the 40th Annual Meeting of the Association for Computational Linguistics (ACL), Philadelphia* (2002), pp. 417–424
16. D.-T. Vo, Y. Zhang, Target-dependent twitter sentiment classification with rich automatic features, in *Proceedings of the Twenty-Fourth International Joint Conference on Artificial Intelligence (IJCAI'15)* (2015), pp. 1347–1353
17. W. Yinglin, W. Ming, Fine-grained opinion extraction from chinese car reviews with an integrated strategy. J. Shanghai Jiaotong Univ. **23**(5), 620–626 (2018)

A Path-Based Personalized Recommendation Using Q Learning

Hyeseong Park and Kyung-Whan Oh

1 Introduction

Recommendation systems have been helping users to find items that best match their interests. Due to the growth of information, collaborative filtering (CF) methods, which make use of historical interactions or preferences, have made a significant success [14]. However, CF-based methods often suffer from the data sparsity and cold-start problem. Recently, there is an increasing interest in knowledge graph (KG)-based recommender systems, since KGs can provide complementary information to alleviate the problem of data sparsity and the structural knowledge has shown great potential in providing rich information about the items [4, 13, 19, 20, 25].

While the usage of the knowledge base is increasingly introduced, one challenging task is knowledge graph reasoning, which infers new conclusions from existing data. Cao et al. [4] propose a new translation-based recommendation model by jointly training a KG completion model, which to infer new relations with reasoning over the information found along other paths connecting a pair of entities [6, 11, 23]. Nevertheless, KG completion methods can hardly be complete when a KG has missing facts, relations, and entities.

Many researchers have explored the potential of knowledge graph reasoning through knowledge graph embedding. In graph embedding, latent representations of entities and relations are placed in continuous vector spaces, to preserve the semantic information in KGs. However, constructing dimensional embeddings of the large volume of latent features can effectively address the problems of excessive memory and computational resource. To tackle this problem, Liu et al. [12] propose a discrete embedding model [18], because continuous vector-based knowledge

---

H. Park (✉) · K.-W. Oh
Department of Computer Science and Engineering, Sogang University, Seoul, South Korea
e-mail: hyepark@sogang.ac.kr; kwoh@sogang.ac.kr

© Springer Nature Switzerland AG 2021   323
H. R. Arabnia et al. (eds.), *Advances in Artificial Intelligence and Applied Cognitive Computing*, Transactions on Computational Science and Computational Intelligence, https://doi.org/10.1007/978-3-030-70296-0_25

graph embedding models to find correlation of user and item [7, 24] can cause a computational problem.

In large dimensional embeddings of the millions of entities and relations, the number of possible paths between a user and an item could be exponentially large. In this regard, Song et al. [15] have been proposed to encode the path embedding, aiming at finding meaningful paths for the recommendation. They represent entities and relations with low dimensional embedding and generate meaningful paths from users to relevant items by learning a walking policy on the user-item-entity graph.

In this paper, our work aims at generating meaningful paths for the recommendation, inspired by Song et al. [15]. We propose **a path-based personalized recommendation using Q learning**, such as (1) constructing a *user-item-entity graph* using discrete embedding [18] to reduce excessive memory and computational resource, while items and tags are associated with semantic dependencies, and (2) exploiting a graph using *Q learning* to generate an inferred preference path. Furthermore, this model offers an *explanation* of a recommendation as inspired by the previous research [15, 22]. Experimental results show that the proposed method provides effective and accurate recommendations with good explanations.

2 Literature Review

There have been several approaches to the personalization algorithm for the recommendation. Researchers figured out a way to factor in such sets of similar items by using dimensionality reduction in collaborative filtering to make predictions of the interest of a user. To acknowledge user interest or preferences, some of the researchers incorporated a wide variety of matrices that characterize the relationships of user and items by vectoring of factors inferred from an item such as matrix factorization. Those methodologies estimate rating patterns to infer user preferences when explicit feedback of the users is not available [10]. To be personalized recommendations, the paper [16] incorporates side information for the user and ratings on the neural network and then computes a non-linear matrix factorization from sparse rating inputs. However, those studies are not aware of the user's interests or user intentions. And higher ratings do not reflect everything about the users such as signifying user enjoyment of recommended movies or reasons why they rated higher on some of the recommended movies.

There are studies to have contributions to the use of context to recommendations because computational entities in pervasive environments need to be context-aware so that they can adapt themselves to changing situations [21]. Context uses any information that characterizes the situation of an entity. KG can be constructed by using context as fully observed, partially observed, or unobserved because contextual factors are changing by the situation or adapting user preferences [8, 9]. The works in [3, 26] proposed context-based recommendations, and random walking is used to search on the KG.

Reinforcement learning is applied in many decision-making systems such that it is an appropriate method to be able to interact with an environment where it can have inferences over the KG. The works in [6, 23] propose KG completion with reasoning to infer new relations by the path-finding process, and reinforcement learning is used to make decisions over walking. They connect a pair of entities to infer a missing link on KG based on the decisions from reinforcement learning. The works in [15, 22] used these approaches in the recommendation by connecting entities of the user and new items. They used large-scale complex KG, and policy gradient with deep reinforcement learning is used to find the personalized path by navigating on the KG. Gridworld can be used instead of KG to reduce complexity, and the work in [5] used Q learning to navigate on the gridworld. However, the gridworld lacks context-based modeling, and using Q learning in KG needs to be studied to reduce the complexity of the context-based model.

3 Methodology

Our model proposes a model for personalized recommendations with path-finding methods, such as (1) constructing a user-item-entity graph and (2) optimizing a path using *Q learning*, and recommends Top-*N* items using *softmax exploration*.

3.1 User-Item-Entity Graph

The user-item-entity graph is constructed by a **layered directed graph**, which is connected by user implicit feedback data (explicit ratings or implicit user log) and combines items and tags under different semantics. As shown in Fig. 1, Layer 1 represents user entities, and Layer 2 describes the entities of historical user preferences. Layer 3 describes the tags, which relate to the previous historical user preferences layer. Layer 4 shows the recommendation candidates. We set the maximum path length T to 3 as our default setting (e.g., User $\rightarrow$ Movie $\rightarrow$ Tag $\rightarrow$ Movie), followed by prior findings [15, 22] that enabling longer paths results in a worse performance since long paths may introduce more noise that can be meaningfulness.

3.2 Formulating a Graph as Markov Decision Process

We apply **relational reasoning** to find an inferred preference path. Different from Das et al.[6], which automatically learn reasoning paths with following logical rules, we propagate user preference from a graph by following relational reasoning. Relational reasoning is similar to the user-based collaborative filtering approach

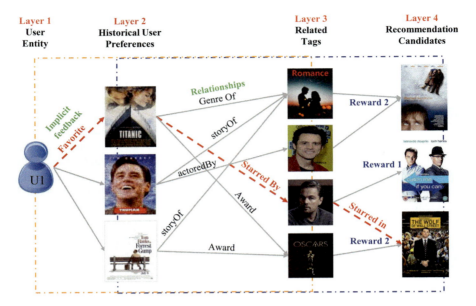

Fig. 1 User-item-entity graph

[27], such that if user X likes a romance movie A when romance movie A has love stories, user X might be interested in the love story.

Since precisely extracting the most relevant items for a given user from massive candidates tends to consume even more time and memory [12], constructing dimensional embeddings of the millions of entities and hundreds of relations [15] is not practical. Our model is efficiently designed by a graph, representing inputs with the discrete vector [18], while continuous vector-based knowledge graph embedding models to find a correlation of user and item [7, 24] can cause a computational problem.

3.2.1 Markov Decision Process

For accelerating the training process, we formulate a graph as Markov decision processes (MDPs) to model mathematical decision-making model. An MDP can be formalized as 4-tuples $(\mathcal{S}, \mathcal{A}, \mathcal{T}, \mathcal{R})$, where $\mathcal{S}, \mathcal{A}, \mathcal{T}$, and $\mathcal{R}$ are the states, actions, transitions, and rewards. In the model, 4-tuples $(\mathcal{S}, \mathcal{A}, \mathcal{T}, \mathcal{R})$ are elaborated below.

- **States**: We represent the state as the set of entities (i.e., *movies* or *tags*)
- **Actions** : When the agent is under state s, it can choose one of the outgoing edges as its next action.
- **Transition** : We adopt a deterministic strategy.

- **Rewards**: The final reward depends on whether or not the agent correctly finds interacted items of the user. Given the terminal entity item in layer 4, the final reward is **the number of outgoing edges** from layer 3 to that entity, see Fig. 1.

3.3 Solving Recommendations Using Q Learning

Following Agarwal et al. [2], we generate Q-walk, over a graph where each node $u \in \mathcal{U}$ is a state, and outgoing edges from u are the possible actions at u. To extract the inferred preference path from the target user to relevant items, we perform Q learning by sequentially selecting walks on the user-item-entity graph. The agent takes the right decision with high-reward paths, i.e., the path to the final node in layer 4 with the largest number of outgoing edges from layer 3.

3.3.1 Optimization

To find a correct path using Q learning, Q learning uses a state-action value function $Q(s, a)$ that means the maximal expected cumulative reward after a certain state s with the action a. $Q(s, a)$ can be expressed as follows:

$$Q(s, a) = \mathbb{E}\left[\sum_{k=1}^{T-t} \gamma^{k-1} r_{t+k} | s_t = s, a_t = a, \pi^* \right] \tag{1}$$

where T is the maximum number of steps, π^* is an optimal policy, and γ is a discount factor $(0 < \gamma < 1)$. Note that $T = 3$ and $t \in \{0, 1, 2, 3\}$ as we have four-layered directed KG. Q learning estimates $Q(s, a)$ by continuously updating $Q(s, a)$ using the following rules:

$$Q(s, a) \leftarrow Q(s, a) + \alpha \left(r + \gamma \max_{a'} Q(s', a') - Q(s, a) \right) \tag{2}$$

where s' is a state transited from the state s after taking action a, and α is the learning rate that controls the convergence speed $(0 < \alpha \leq 1)$. When $t = 2$, the next state s' becomes a terminal state where no actions are exploited, and $Q(s, a)$ is updated as follows:

$$Q(s, a) \leftarrow Q(s, a) + \alpha \left(r - Q(s, a) \right) \tag{3}$$

3.3.2 Softmax Exploration

We also need to consider the trade-off between exploration and exploitation. The agent has to exploit what it already knows to obtain high value, but it also has to explore to make better policy in the future. The ϵ-greedy method is usually used for balancing exploration and exploitation [5], but it is as likely to choose the worst appearing action as it is to choose the second-best appearing action if an exploration action is selected [17]. In this regard, we use softmax exploration method, where the policy π is described as follows:

$$\pi(a|s) = \frac{e^{Q(s,a)/\tau}}{\sum_{a' \in A_s} e^{Q(s,a')/\tau}} \tag{4}$$

where $\pi(a|s)$ is a probability the agent selects action a in state s and $\tau \geq 0$ is the temperature parameter that determines the exploration ratio. When $\tau = 0$ the agent chooses the best action without exploration, and when $\tau \to \infty$ the agent selects random actions.

3.3.3 Top-N Recommendations

To generate top-N recommendations, the system generates distributed exploration for possible candidate actions by performing the softmax policy, as in Eq. (4). Then, the system repeatedly executes Q-walk until it lands on the final node (in layer 4) after T steps. The top-N recommendation node is ranked by the highest total number of visiting through Q-walk.

4 Experiments

4.1 Dataset

In this section, the performance of the proposed model evaluated based on the MovieLens 100k datasets[1] provides users' ratings toward thousands of movies. We used the 80% datasets as user's explicit ratings, and the remaining datasets were used for unseen movies.

---

[1] http://movielens.org.

A Path-Based Personalized Recommendation Using Q Learning

Algorithm 1: Personalized recommendations

---

input : a user entity s_u, knowledge graph, the number of episodes M, the number of recommended list N

output: Top-N recommended list of items

```
/* Q Learning Phase                                            */
```
$Q(s, a) \leftarrow 0$ for all undefined $Q(s, a)$;

for episode $= 1$ **to** M **do**
    $s \leftarrow s_u$;
    for t $= 0$ **to** $T - 1$ **do**
        $a \leftarrow$ softmax exploration with $\pi(a|s)$;
        execute action a;
        receive new state s';
        receive reward r;
        if $t \neq T - 1$ **then**
            update $Q(s, a)$ via Eq. (2);
            $s \leftarrow s'$;
        else
            update $Q(s, a)$ via Eq. (3);
        end
    end
end
```
/* Generating Recommendations Phase                            */
```
visit$(s') \leftarrow 0$ **for all** s' in layer 4 entities;
$s \leftarrow s_u$;

for episode $= 1$ **to** M **do**
    $s \leftarrow s_u$;
    for t $= 0$ **to** $T - 1$ **do**
        $a \leftarrow$ softmax exploration with $\pi(a|s)$;
        execute action a;
        if $t \neq T - 1$ **then**
            receive new state s';
            $s \leftarrow s'$;
        else
            receive new state s';
            visit$(s') \leftarrow$ visit$(s') + 1$;
        end
    end
end

perform top-N recommendations based on the number of visitation

---

4.2 *Evaluation Metrics*

4.2.1 Implementation Details

We select 493 users who watched more than 30 movies from MovieLens 100k datasets. For each data-link layer entity, the *user* entity is linked to the *movie* entities in a layer of user preferences (layer 2), if and only if a user's explicit rating is 5. The pairs of vertices (between *movie* and *tag*) are composed of tag relevance, a numerical

Table 1 Average ratings of recommended movies

	Total	Q_1	Q_2	Q_3	Q_4
GA	3.68	3.05	3.52	3.84	4.28
URW	3.75	3.19	3.58	3.90	4.32
QL (proposed)	**3.98**	**3.53**	**3.87**	**4.08**	**4.42**

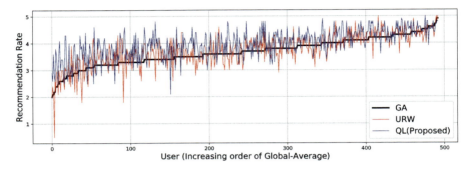

Fig. 2 Average ratings of recommended movies for each user

value to represent how each of a tag contributed to a given movie from tag genome.[2] The vertices between *movie* and *tag* are connected if and only if the tag relevance is more than 0.95.

4.2.2 Evaluation Metrics

The experiment results are compared to other state-of-the-art methods, which compared **average user ratings** of the recommended movies by (1) global average (GA), which are the average ratings observed by over all the users from MovieLens, (2) uniform random walk (URW) [1], where the system explores each neighbor with the same probability at each step, and (3) Q learning (QL).

4.3 Performance Evaluation

To verify the performance of the proposed models, we analyze the results by measuring the average of the total by quartile and each range is denoted by Q_N (e.g., 0–25% (Q_1), 26–50% (Q_2), 50–75% (Q_3), 75–100% (Q_4)). The experimental results of these evaluation metrics are shown in Table 1. Following the results, we can see that our proposed algorithms (QL) show higher ratings than GA and URW. Figure 2 also shows that most users have the highest ratings in the proposed

---

[2]https://grouplens.org/datasets/movielens/tag-genome/.

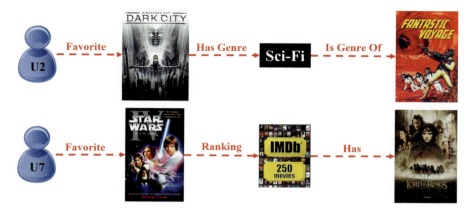

Fig. 3 Recommendations and explanations for user 2 and user 7

Q learning-based recommended items. Based on our acquired knowledge, the proposed algorithm provides a movie that the user might be interested in watching.

4.4 Explainable Recommendation

Our proposed model can be explained to users in the form of meaningful paths from users to recommended items. As shown in Fig. 3, we can easily understand that "Fantastic Voyage" is recommended to user 2 because it shares the same genre (sci-fi) with "Dark City," which the user preferred before. The recommendation "Lord of the Rings" can be explained to user 7 as it ranked in IMDB top 250.

5 Conclusion

Our model provides explainable recommendations by searching the user's preferring item with path-based recommendations using Q learning. Our algorithms do not require excessive memory or large-scale data and achieve experimental results where the recommended items earned high explicit ratings of users. Our system also provides an explanation of a recommendation, which can make the users have more interest and trust in the recommended items. Future research directions include using a large dataset of users with more sophisticated reinforcement learning techniques. Our model can also be incorporated with more additional personal contexts to be extended.

References

1. Z. Abbassi, V.S. Mirrokni, A recommender system based on local random walks and spectral methods, in *Proceedings of the 9th WebKDD and 1st SNA-KDD 2007 Workshop on Web Mining and Social Network Analysis* (2007), pp. 102–108
2. N. Agarwal, G.C. Nandi, Supervised q-walk for learning vector representation of nodes in networks (2017, preprint). arXiv:1710.00978
3. T. Bogers, Movie recommendation using random walks over the contextual graph, in *Proceedings of the 2nd International Workshop on Context-Aware Recommender Systems* (2010)
4. Y. Cao, X. Wang, X. He, Z. Hu, T.S. Chua, Unifying knowledge graph learning and recommendation: Towards a better understanding of user preferences, in *The World Wide Web Conference* (2019), pp. 151–161
5. S. Choi, H. Ha, U. Hwang, C. Kim, J.W. Ha, S. Yoon, Reinforcement learning based recommender system using biclustering technique (2018, preprint). arXiv:1801.05532
6. R. Das, S. Dhuliawala, M. Zaheer, L. Vilnis, I. Durugkar, A. Krishnamurthy, A. Smola, A. McCallum, Go for a walk and arrive at the answer: reasoning over paths in knowledge bases using reinforcement learning (2017, preprint). arXiv:1711.05851
7. T. Dettmers, P. Minervini, P. Stenetorp, S. Riedel, Convolutional 2d knowledge graph embeddings, in *Thirty-Second AAAI Conference on Artificial Intelligence* (2018)
8. N. Hariri, B. Mobasher, R. Burke, Adapting to user preference changes in interactive recommendation, in *Proceedings of the 24th International Conference on Artificial Intelligence IJCAI*, vol. 15 (2015), pp. 4268–4274
9. M. Hosseinzadeh Aghdam, N. Hariri, B. Mobasher, R. Burke, Adapting recommendations to contextual changes using hierarchical hidden Markov models, in *Proceedings of the 9th ACM Conference on Recommender Systems* (2015), pp. 241–244
10. Y. Koren, R. Bell, C. Volinsky, Matrix factorization techniques for recommender systems. Computer **42**(8), 30–37 (2009)
11. X.V. Lin, R. Socher, C. Xiong, Multi-hop knowledge graph reasoning with reward shaping (2018, preprint). arXiv:1808.10568
12. C. Liu, X. Wang, T. Lu, W. Zhu, J. Sun, S. Hoi, Discrete social recommendation, in *Proceedings of the AAAI Conference on Artificial Intelligence*, vol. 33 (2019), pp. 208–215
13. W. Ma, M. Zhang, Y. Cao, W. Jin, C. Wang, Y. Liu, S. Ma, X. Ren, Jointly learning explainable rules for recommendation with knowledge graph, in *The World Wide Web Conference* (2019), pp. 1210–1221
14. J.B. Schafer, D. Frankowski, J. Herlocker, S. Sen, Collaborative filtering recommender systems, in *The Adaptive Web* (Springer, Berlin, 2007), pp. 291–324
15. W. Song, Z. Duan, Z. Yang, H. Zhu, M. Zhang, J. Tang, Explainable knowledge graph-based recommendation via deep reinforcement learning (2019, preprint). arXiv:1906.09506
16. F. Strub, J. Mary, R. Gaudel, Hybrid collaborative filtering with autoencoders (2016, preprint)
17. A.D. Tijsma, M.M. Drugan, M.A. Wiering, Comparing exploration strategies for q-learning in random stochastic mazes, in *2016 IEEE Symposium Series on Computational Intelligence (SSCI)* (IEEE, Piscataway, 2016), pp. 1–8
18. H.N. Tran, A. Takasu, Analyzing knowledge graph embedding methods from a multi-embedding interaction perspective (2019, preprint). arXiv:1903.11406
19. H. Wang, F. Zhang, J. Wang, M. Zhao, W. Li, X. Xie, M. Guo, Ripplenet: propagating user preferences on the knowledge graph for recommender systems, in *Proceedings of the 27th ACM International Conference on Information and Knowledge Management* (2018), pp. 417–426
20. H. Wang, F. Zhang, X. Xie, M. Guo, DKN: deep knowledge-aware network for news recommendation, in *Proceedings of the 2018 World Wide Web Conference* (2018), pp. 1835–1844

21. X.H. Wang, D.Q. Zhang, T. Gu, H.K. Pung, Ontology based context modeling and reasoning using owl, in *Proceedings of the Second IEEE Annual Conference on Pervasive Computing and Communications Workshops, 2004* (IEEE, Piscataway, 2004), pp. 18–22
22. Y. Xian, Z. Fu, S. Muthukrishnan, G. De Melo, Y. Zhang, Reinforcement knowledge graph reasoning for explainable recommendation, in *Proceedings of the 42nd International ACM SIGIR Conference on Research and Development in Information Retrieval* (2019), pp. 285–294
23. W. Xiong, T. Hoang, W.Y. Wang, Deeppath: a reinforcement learning method for knowledge graph reasoning (2017, preprint). arXiv:1707.06690
24. B. Yang, W.T. Yih, X. He, J. Gao, L. Deng, Embedding entities and relations for learning and inference in knowledge bases (2014, preprint). arXiv:1412.6575
25. F. Zhang, N.J. Yuan, D. Lian, X. Xie, W.Y. Ma, Collaborative knowledge base embedding for recommender systems, in *Proceedings of the 22nd ACM SIGKDD International Conference on Knowledge Discovery and Data Mining* (2016), pp. 353–362
26. Z. Zhang, D.D. Zeng, A. Abbasi, J. Peng, X. Zheng, A random walk model for item recommendation in social tagging systems. ACM Trans. Manag. Inf. Syst. **4**(2), 8 (2013)
27. Z.D. Zhao, M.S. Shang, User-based collaborative-filtering recommendation algorithms on Hadoop, in *2010 Third International Conference on Knowledge Discovery and Data Mining* (IEEE, 2010), pp. 478–481

Reducing the Data Cost of Machine Learning with AI: A Case Study

Joshua Haley, Robert Wray ⓘ, Robert Bridgman, and Austin Brehob

1 Introduction

Deep Learning (DL) enables large marginal performance gains on benchmark tasks such as image recognition, predictive analytics, and solutions to strategic gameplay [1]. As Andrew Ng points out in [2], these gains are enabled by both the vast amounts of data and available computational resources. As an academic researcher or industry practitioner with less data for a task or more limited computational resources, should we be approaching the problem the same way that Facebook or Google would? If we envision a typical learning curve such as the one in Fig. 1 where more expressive models exhibit higher performance when given more data, most academic literature is interested in theoretical performance on the right tail where data is plentiful and computation is limitless. In contrast, practitioners will often be operating on the left tail of the learning curve, where data is constrained and computation is limited while proving out a concept. For most practical applications, rather than the search for optimal task performance, engineers must be concerned with satisfying performance requirements while maximizing flexibility given the constraints on hand.

Herein, we present a case study from a practitioner's prospective to solve an applied task. We first used a machine learning-based solution via Deep Learning. We then switched to an AI-enabled ensemble ML approach to simplify the underlying models and data requirements. We describe the task in Sect. 1.1 and the methods used to address the task in Sect. 2, and finally we contrast the qualitative and quantitative differences of the approaches in Sect. 3.

J. Haley (✉) · R. Wray · R. Bridgman · A. Brehob
Intelligent Systems Division, Soar Technology Inc., Ann Arbor, MI, USA
e-mail: joshua.haley@soartech.com; wray@soartech.com; robert.bridgman@soartech.com; austin.brehob@soartech.com

© Springer Nature Switzerland AG 2021
H. R. Arabnia et al. (eds.), *Advances in Artificial Intelligence and Applied Cognitive Computing*, Transactions on Computational Science and Computational Intelligence, https://doi.org/10.1007/978-3-030-70296-0_26

335

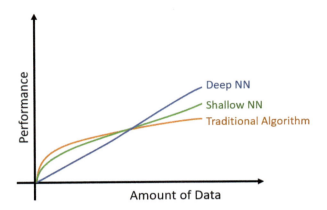

Fig. 1 Data/performance trade-off curve

1.1 Behavior Recognition Problem

In order to understand the computational approaches we wish to contrast in this chapter, we first introduce the problem that the computational approaches are designed to address. The application domain is simulation-based training, which uses realistic models of the environment to enable someone to develop skills and to practice in a realistic but safe setting. Simulation-based training systems employ computational models of human behavior to represent the activities and actions of other actors in the overall environment [3, 4]. For example, a trainee learning to be a fighter pilot could fly against an adversary agent driven by a human behavior model.

Historically, there have been two differing ways of approaching the addition of new behavior models in existing simulation systems. The first is to refactor the existing behavior models to include new behavior models (or extensions/changes to existing models). This approach can take considerable time but supports validation of general behaviors. The other approach is to turn captured examples of the desired behavior into a "script" that will reproduce the observed behavior in a replicated setting. This script approach requires little time and labor to implement, but it is brittle for use in any circumstances other than the exact setting in which the behavior was observed. The motivating goal of our effort was to develop an approach that employs observations of real-world scenarios to recreate simulation-based scenarios. However, rather than simply developing a "script" that replays the scenario, we interpret the events that occurred in the scenario and then map those events to existing (and validated) models available in the simulation system.

For the purposes of this chapter, we introduce an example drawn from tactical aviation, which is the domain of our actual application. In real-world settings, the tactics that would be observed and generated would be more sophisticated than the examples, but the domain we describe here provides an illustration of the computational challenges that can be understood without detailed knowledge of the

Reducing the Data Cost of Machine Learning with AI: A Case Study

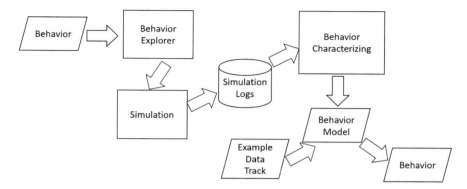

Fig. 2 System data flow

domain. In the chapter, we focus on differing geospatial maneuvers to accomplish an intercept task where one aircraft must engage or "intercept" an opposing force aircraft in a one-on-one encounter.

For the experiments and results described in the chapter, we employed a distributed simulation environment that executes scenarios such as the one-on-engagement. The simulation includes a built-in behavior programming capability sufficient for the capabilities we explored. Its representation is roughly comparable to that of a hierarchical task network [5]. The native behavior programs allow simulation entities to react to environmental cues and perform actions and maneuvers in response. The goal of our behavior recognition system is to identify and configure an existing behavior program to replicate the observation.

2 Methodology

In order to facilitate the desired playback capability via parameterizable behaviors as described in Sect. 1.1, a method is required to match the observation to a specific behavior incidence that sufficiently replicates the observed behavior. We use a parameter exploration technique [6] in order to generate a synthetic dataset to train the behavior recognition model. This "diversification" method spans parameter space for simulation execution, enabling the system to learn a model mapping "cause" (i.e., parameter configuration) to effect (observed outcome) . This flow of behavior exploration, data generation, model building, and model utilization is outlined in Fig. 2. We describe two such methods devised to accomplish this goal of behavior characterizing to produce a behavior model, both using the same data flow described above.

2.1 Deep Learning Method

The initial solution for the behavior recognition task was a Deep Learning LSTM-based approach [7]. In this approach, data is first transformed from a geospatial representation to a fixed $28 \times 28 \times 28$ cubic representation that is egocentric to an entity of interest. This fixed representation size required for the DL approach means that each block could represent a variable distance between observations. Entities were then assigned a value of 1 for self and -1 for opponent. This representation was replicated for each time step to form a 4D tensor input representing 3D spatial state over time. A Convolutions Neural Network was used to take advantage of the spatial relations and was embedded into a feature vector of size 2048. This feature vector was then propagated through multiple LSTM layers before presentation to a dense, fully connected layer to produce the desired classification and regressions.

Haley et al. [7] describe this approach in more detail, but, for the purposes of this analysis, the key observation is that the deep model was comprised of approximately 678k parameters. This parameter space required a large amount of synthetic, diversification data to adequately tune the recognition process. As a result, this approach did not initially yield sufficient recognition for reliable reconstruction for operational use. We noted an inability to differentiate between differing levels of maneuver parameter values due to the coarseness of the discretization of the geospatial domain. In order to add more discrimination, we would have needed to increase the size of our sparse spatial representation, which was already taxing to store. The CNN-based embedding layers were replaced with calculated geospatial features (distance, angle, and azimuth) between the two aircraft to yield sufficient recognition at a cost of model flexibility. Additionally, the true accuracy of maneuver identification is uncertain, as we will discuss in Sect. 3.3. The inflexibility and complexity of the model relative to available data necessitated a change in implementation.

2.2 Event Sequence Matching

The second approach focused on representing events observed in the scenario [8]. An "event" is a change in observable state for an entity with the properties given in Table 1. By representing changes in state, rather than the states themselves, the event approach is by design more compact and less data intensive. Event definitions (types and properties) are decided a priori as a knowledge engineering exercise. A specific observation of an event is deemed an event instance where the parameters have been specified with values. A scenario plays out through a sequence of these event instances, as we can see in the example outlined in Fig. 3 as an observation of executing behavior.

To develop a model mapping observations to behavior, these specific event observation sequences were generalized by replacing specific instances with types

Reducing the Data Cost of Machine Learning with AI: A Case Study

Table 1 Event representation from [8]

Representation of event instances	
Time	Time (from start) that the event occurs. Representation assumes that event sequences are fully ordered (even when multiple instances share a time stamp). Non-instantaneous mission events are represented via instances that indicate start/stop times of the event
Subject	Primary actor(s) associated with the instance. The subject can be a list of actors or named aggregates.
Verb	Type of the event. We have defined a collection of event types for the simulation domains we have explored thus far (e.g., tactical aviation). Typology inherits from formal ontology [6], to provide consistent semantics (definitions) across event types. Examples of event verbs include "initiate intercept," "missile launch," etc. Event verbs represent changes in both physical state (e.g., start/end of a turn or climb) and domain-specific state (e.g., a change in the status of the mission)
Object	The primary recipient/target of the action indicated by the verb. Objects can be bound to any kind of object represented in the target simulation (other entities and aggregates, locations, etc.)
Modifiers	Objects and parameters (specific to the event type) that further describe the event. Modifiers are represented in the form of (name, value) pairs. A modifier for an event instance can be a reference to a prior event (e.g., a "stop turn" event instance includes a pointer to its initiating "start turn" event instance)

Geospatial View	Event Representation	Generalized Signature
	...	...
	17:16:04 S: Snoop V: Start Inter O: Baron	+0:00:04 S: Blue V: Start Inter O: Red
	17:16:22 S: Snoop V: Aspect Beam O: Baron	+0:00:18 S: Blue V: Aspect Beam O: Red
	17:17:25 S: Snoop V: StartTurn Heading:255	+0:01:03 S: Blue V: StartTurn Heading:[255]
	17:18:31 S: Snoop V: StopTurn Heading: 65	+0:01:05 S: Blue V: StopTurn Heading: [65]
	17:19:34 S: Snoop V: Aspect(hot) O: Baron	+0:01:03 S: Blue V: Aspect(hot) O: Red
	17:19:45 S: Snoop V: Missile Launch O:Baron	+0:00:11 S: Blue V: Missile Launch O: Red

Fig. 3 Geospatial, event, and signature representations

where able (e.g., "Snoop" in Fig. 3 would become "Blue Entity"). Sequences of equivalent structure (i.e., same event sequence) were coalesced into collections with all of the underlying event parameters and behavior configurations that were labeled as a signature. A model was then constructed using a simple K-neighbors regression between the event parameters and the behavior configuration that created the observation. Each signature's unique structural composition meant that the cardinality and types of event parameters differ between signatures, but by associating the model between observation and behavior with the signature, each model was free to have its own input representation dynamically defined during signature creation. The k-neighbors regression was used due to the low number of observations (1–20) per signature and was relatively resilient to the lack of data. Additionally, while this k-neighbors regression is insufficient to model the entire observation space to behavior configuration space, it was effective for the subset of the space defined by a single signature.

To reconstruct an observed situation, we developed a variant of the Smith–Waterman algorithm [9], which was developed and is used for DNA sequence matching. This algorithm selects the signature with the closest structural similarity to the observed (original) event sequence. The specific event instance parameters were then encoded into the input representation defined by the signature and used the associated model to infer the behavior parameters that lead to the given observation.

In essence, the signatures form an ensemble of models. The sequence matching determines in which part of the space the observation lies in order to select the appropriate behavior and regression model to parameterize the selected behavior. We found that this approach yielded better recognition than the earlier DL approach, which we will describe next in Sect. 3.

3 Results and Comparisons

In both methods, sufficient recognition was achieved to enable replay of the one-on-one encounter. However, quantitative and qualitative implementation constraints encountered with the DL approach necessitated the switch to exploring the sequence matching approach. These constraints are further recounted here.

3.1 Data Requirements

The primary reason for changing the approach was the data requirement of the Deep Learning methods. The initial experiment was a one-on-one encounter where only a single entity would maneuver in response to the presence of the other. In order to adequately explore the behavior parameter spaces to achieve sufficient behavior performance, the DL model required approximately 77k example data runs to map from the geospatial space to the parameter space with sufficient reconstruction accuracy for test scenarios. In contrast, the sequence matching approach required 300 exemplar encounters in order to achieve similar reconstruction accuracy. The reasons for the reduction of data are twofold. First, the DL model mapped the entirety of the observation space to the set of all possible behaviors and configurations. In contrast, the sequence matching approach subdivides the observation input space into different signatures, each signature responsible only for a part of the input space for a single behavior. In essence, each signature acts as a much simpler model of a subset of the larger problem space. The sequence matcher then focuses selection on which of the simpler models to employ. Second, the decision space in which each signature operates is small enough that a simpler model requiring less data can be used to infer parameters with sufficient performance.

The level of data required presented an issue when we wanted to enhance the DL model's ability to match behavior parameters when both entities maneuvered. Constraining the exploration of the parameter space, the sequence matching model required approximately 500 scenario observations for signature creation. The DL model diversification process required, in contrast, 400k observations. Executing these many simulation scenarios to generate the synthetic data was prohibitive for practical considerations for the operational domain. In general, the sequence matching approach degraded more gracefully in low-data conditions than the DL approach.

Additionally, the storage required for example scenario runs was not equal between methods. The DL approach required a constant input format and took the form of a sequence-based state vector for each time step, which led to a fixed input size per unit time. This led to very large sparse example instances where most time segments did not have any activity of interest. While these files can be reduced in size after capture, they still represented a significant burden to process. In contrast, the sequence approach only captures data when events of interest occur. This dynamic approach to representing state led to a significant decrease in the size of the dataset.

3.2 Flexibility

Another key limitation of the DL approach was expanding the capability beyond one-on-one exercises. The input format used calculated features between two entities from a single entity's perspective; thus, modeling additional entities requires either a second model or a recomputation of features from the other entity's perspective, a costly exercise. Incorporating additional entities beyond one-on-one scenarios would have necessitated additional inputs and complexity, which increases data required for training.

The sequence matching approach used a filtering process to only look at events related to the entity being matched at the time. From the matcher's perspective, there were only events related to self and "other." Thus extending the sequence matching approach for two dynamic entities was a simple matter of running the matching for each entity with an egocentric filter without any required recomputation of features as required by the DL approach. Similarly, we used the event-based approach for two-on-two and one-on-three encounters, and these additional agents required no representational or model changes.

3.3 Validation Issues

The other main consideration between the two approaches was that of validation matching. While the selected parameters in both were not always correct, parameter selection did not need to be perfect for adequate reconstruction. In particular, there were maneuver instances where incorrect maneuvers could be picked, but with a specific variable parameterization, they would appear to be the correct maneuver from the geospatial prospective. For the DL model, these represented multiple points in the input domain that map to the same observational effect; thus, simple error calculations or direct inspection of parameter choices was inadequate to determine if a sufficient parameter set had been chosen. Instead, we needed to look at the effects of the parameter choices to determine appropriateness of reconstruction. This assessment proved difficult for the DL model as large geospatial changes (turning left vs. right), which ultimately do not meaningfully change the scenario as it plays out, thwarted computational attempts to automate the characterization of observed difference.

The sequence-based approach already encodes aspects of scenario importance for reconstruction within the events themselves. Thus, this approach facilitates a comparison between the original observed situations and those created as a result of the reconstruction, as reported in [8].

4 Conclusions

Deep Learning has unquestioningly led to impressive results in state-of-the-art benchmark tasks. It offers incredible opportunity when sufficient data exists to train the large parameterizable models. However, DL may require more data that exists (or can be generated cost effectively) for specific applications. This single case study does not provide a prescriptive recommendation but rather suggests that data-driven DL is not always the best solution when practical implementation considerations are made. While raw DL could be applied to the behavior recognition task, it came at the cost of a large data requirement and was not the more flexible solution of the two we explored. The chosen solution still used machine learning to infer behavior parameters; however, it was augmented with both knowledge engineering-based event definitions and a sequence matching approach. This use of knowledge-based AI and ML together enabled a simpler solution that was still sufficient and computationally feasible without requiring exorbitant resources.

While the desire to use the latest large models is enticing, the average ML engineer should keep in mind that "work smarter not harder" can apply computationally in resource-constrained environments.

Acknowledgments The authors thank Robert "Norb" Timpko, Victor Hung, and Chris Ballinger for contributions to concepts and implementations reported here. We also thank Dr. Heather Priest and Ms. Jennifer Pagan for contributions to problem definitions and solution context that informed our approach. Opinions expressed here are not necessarily those of the Department of Defense or the sponsor of this effort, the Naval Air Warfare Center Training Systems Division (NAWCTSD). This work was funded under contract no. N68335-17-C-0574.

References

1. D. Silver, J. Schrittwieser, K. Simonyan, I. Antonoglou, A. Huang, A. Guez, T. Hubert, L. Baker, M. Lai, A. Bolton, et al., Mastering the game of go without human knowledge. Nature **550**(7676), 354–359 (2017)
2. A. Ng, Machine learning yearning: Technical strategy for ai engineers in the era of deep learning (2019). Retrieved online at https://www.mlyearning.org
3. G.L. Zacharias, J.E. Macmillan, S.B. Van Hemel, *Behavioral Modeling and Simulation: From Individuals to Societies* (National Academies Press, Washington, 2008)
4. R.W. Pew, A.S. Mavor, *Modeling Human and Organizational Behavior-Application to Military Simulations* (National Academies Press, Washington, 1998)
5. K. Erol, J. Hendler, D.S. Nau, HTN planning: Complexity and expressivity, in *AAAI*, vol. 94 (1994), pp. 1123–1128.
6. V. Hung, J. Haley, R. Bridgman, N. Timpko, R. Wray, Synthesizing machine-learning datasets from parameterizable agents using constrained combinatorial search, in *International Conference on Social Computing, Behavioral-Cultural Modeling and Prediction and Behavior Representation in Modeling and Simulation* (Springer, Berlin, 2019,) pp. 60–69
7. J. Haley, V. Hung, R. Bridgman, N. Timpko, R. Wray, Low level entity state sequence mapping to high level behavior via a DeepLSTM model, in *Proceedings on the International Conference on Artificial Intelligence (ICAI)*. The Steering Committee of The World Congress in Computer Science, Computer . . . (2018), pp. 131–136
8. R.E. Wray, J. Haley, R. Bridgman, A. Brehob, Comparison of complex behavior via event sequences, in *2019 International Conference on Computational Science and Computational Intelligence (CSCI)* (IEEE, Piscataway, 2019), pp. 266–271
9. M.S. Waterman, General methods of sequence comparison. Bull. Math. Biol. **46**(4), 473–500 (1984)

Judging Emotion from EEGs Using Pseudo Data

Seiji Tsuchiya, Misako Imono, and Hirokazu Watabe

1 Introduction

For a robot to converse naturally with a human, it must be able to accurately gauge the emotional state of the person. Techniques for estimating the emotions of a person from facial expressions, intonation and speech content have been proposed. This chapter presents a technique for judging the emotion of a person from EEGs by SVM using pseudo extension data.

2 Overview of Proposed Technique

The objective of this technique was to read the emotions of a conversation partner from EEGs.

EEGs acquired from the subject are used as source EEGs. The emotions of the subject at that time are acquired simultaneously. Spectrum analysis of the source EEGs to which emotion flags have been assigned is performed every 1.28 s, and the EEG features of θ waves (4.0–8.0 Hz), α waves (8.0–13.0 Hz), and β waves (13.0–30.0 Hz) are determined (Fig. 1).

S. Tsuchiya (✉) · H. Watabe
Department of Intelligent Information Engineering and Science, Faculty of Science and Engineering, Doshisha University, Kyotanabe, Kyoto, Japan
e-mail: stsuchiy@mail.doshisha.ac.jp; hwatabe@mail.doshisha.ac.jp

M. Imono
Department of Information Systems, School of Informatics, Daido University, Nagoya, Aichi, Japan
e-mail: m-imono@daido-it.ac.jp

© Springer Nature Switzerland AG 2021
H. R. Arabnia et al. (eds.), *Advances in Artificial Intelligence and Applied Cognitive Computing*, Transactions on Computational Science and Computational Intelligence, https://doi.org/10.1007/978-3-030-70296-0_27

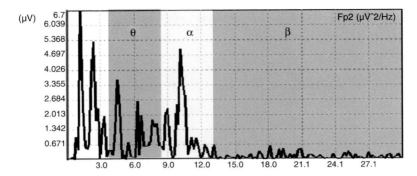

Fig. 1 Spectrum analysis of the source EEGs

Emotions are judged from EEGs by SVM [1] using the learning data which is EEG features to which emotion flags have been assigned was determined in this study. Emotions judged in this study were anger, sadness, fear, and pleasure.

3 Acquisition of Source EEGs and Emotions

EEGs were measured at 14 locations, at positions conforming to the international 10–20 system (Fig. 2) [2]. Subjects fitted with an electroencephalography [3] cap were asked to watch a Japanese film for approximately 2 h while trying to gauge the emotions of the speakers in the film, and source EEGs were acquired. Images were frozen for each of the 315 speakers in the film, and the subject was asked what emotion the speaker was feeling at that time.

Twenty subjects were used, and viewing was divided into four sessions to reduce the physical burden on subjects. Before and after the film, EEGs corresponding to open-eye and closed-eye states were measured for approximately 1 min each, and these data were used when normalizing EEG features.

4 Normalization of EEG Features

EEGs show changes in voltage intensity over time within an individual, and base voltage intensity differs among individuals. For this reason, the possibility of misjudgment exists because those values differ greatly even among EEGs with similar waveforms. To solve this problem, linear normalization and non-linear normalization were performed.

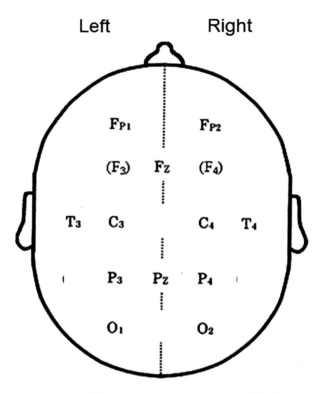

Fig. 2 14 locations to measure EEG conforming to the international 10–20 system

4.1 Linear Normalization

This was performed to take into account how EEGs vary over time depending on the subject. Since the eyes were open while viewing the film, linear normalization was performed based on EEG features from the eyes-open state, acquired both before and after the experiment.

EEG feature $Linear\_al_{ij}$, obtained by linear normalization of first EEG feature al_{ij} at a certain point in time during the experiment, is expressed by Formula (4.1):

$$Linear\_al_{ij} = al_{ij} + \left\{ \left(\frac{q_1 - q_2}{p_2 - p_1} \times l + q_1 \right) - \left(\frac{q_2 - q_1}{p_2 - p_1} \times l + q_2 \right) \right\} / 2 \quad (4.1)$$

4.2 Non-linear Normalization

This was performed to take into account the differences in base voltage intensity among individuals.

Non-linear normalized values were determined using Formula (4.2), where $f(x)$ is the EEG feature after non-linear normalization has been applied, x is the EEG feature applied in non-linear normalization, x_{min} is the minimum EEG feature of the individual, and x_{max} is the maximum of the same data. As a result, EEG features with large values are compressed and EEG features with small values are expanded. The degree of intensity of voltage of an individual's EEGs can thus be accounted for.

$$f(x) = \frac{\log (x - x_{\min})}{\log (x_{\max} - x_{\min})} \qquad (4.2)$$

5 Pseudo Extension of EEG

A lot of training data is required to use SVM. However, it is difficult to acquire a large amount of data because the measurement of EEG puts a heavy burden on subjects.

Therefore, we propose a method for pseudo increasing the number of EEG, referring to the genetic crossover method of GA. Specifically, by setting a genetic crossover point considering the brain region and frequency band, two types of children are generated with two brain waves of the same emotion as the parent. Since EEG features are 42-dimensional vectors, there are usually 41 genetic crossover points. In the proposed method, the brain region was classified into 5 regions (frontal region, parietal region, occipital region, left and right temporal region), and frequency bands were classified into 3 types (θ, α, β). And 15 points that were combinations of them were set as genetic crossover point.

By this method, a total of 40 million data of learning data (10 million data for each emotion) were pseudo generated and used for SVM training.

6 Evaluation Experiment

6.1 Experimental Method

This study used 2945 EEG features (20 persons) obtained by excluding outliers from the total of 3075 EEG features. The emotions of the 2945 EEG features used in this study comprised 629 anger features, 857 sadness features, 979 fear features, and

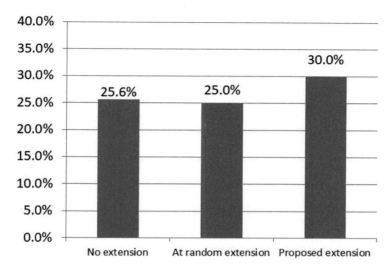

Fig. 3 Result of the emotion judgment from EEGs

480 pleasure features. 2358 EEG features (16 persons) were used as training data, and 587 EEG features (4 persons) were used as test data.

6.2 Evaluation of Accuracy

Figure 3 shows the result of the emotion judgment from EEGs. The accuracy of emotion judgment from EEG features by SVM using at random pseudo extended data was 25.0% and using proposed extension technique was 30.0%. As a comparison, the result of doing the emotion judgment no extension is the accuracy of 25.6%. The method proposed herein thus appears valid. However, performance accuracy remains low, and continued development is required through further development of methods for both reducing noise mixed in with EEGs.

7 Conclusion

We have presented a technique for gauging the emotions felt by a person from EEGs. The accuracy of emotion judgment from EEG features by SVM using at random pseudo extended data was 25.0% and using proposed extension technique was 30.0%. The method proposed herein thus appears valid. However, performance accuracy remains low, and continued development is required through further development of methods for both reducing noise mixed in with EEGs.

We plan to continue research aimed at improving the accuracy of emotion judgment by EEGs in the hopes of developing robotic systems that can participate in conversation and activities while gauging human emotional states.

Acknowledgments This work was partially supported by JSPS KAKENHI Grant Number 16K00311.

References

1. C.-C. Chang, C.-J. Lin, LIBSVM: A library for support vector machines. ACM Trans. Intell. Syst. Technol. **2**(3), 27 (2011)
2. J. Clin, Guideline thirteen: Guidelines for standard electrode position nomenclature. Am. Electroencephalo Graphic Soc. Neurophysiol. **11**, 111–113 (1994)
3. T. Musha, Y. Terasaki, H.A. Haque, G.A. Iranitsky, Feature extraction from EEG associated with emotions. Art. Life Robotics **1**(1), 15–19 (1997)

Part III
Neural Networks, Genetic Algorithms, Prediction Methods, and Swarm Algorithms

Using Neural Networks and Genetic Algorithms for Predicting Human Movement in Crowds

Abdullah Alajlan, Alaa Edris, Robert B. Heckendorn, and Terence Soule

1 Introduction

Simulations of crowd behavior can be used to improve the design of public spaces for the movement of people at large events, such as religious gatherings, athletic events, or concerts. This, in turn, can be used to help prevent crushing and trampling disasters and help guarantee public safety. Many simulations rely on assumptions about how crowds move, leaving a possible predictive gap between reality and the simulation. Our approach is to use machine learning to learn how individuals in a crowd move and provide a simulation more directly tied to an unbiased observation of crowd behavior. We examine learning in two common and important regimes: **structured crowds**, where individuals are moving in a specified direction, as in the Islamic Hajj or Hindu Kumbh Mela, and **unstructured crowds**, such as town squares, shopping malls, and train stations.

A. Alajlan (✉)
Computer Science Department, University of Idaho, Moscow, ID, USA

Computer Science Department, Technical and Vocational Training Corporation, Riyadh, Saudi Arabia
e-mail: alaj0169@vandals.uidaho.edu

A. Edris
Computer Science Department, University of Idaho, Moscow, ID, USA

College of Computer Science and Engineering, University of Jeddah, Jeddah, Saudi Arabia
e-mail: edri6611@vandals.uidaho.edu

R. B. Heckendorn · T. Soule
Department of Computer Science, University of Idaho, Moscow, ID, USA
e-mail: heckendo@vandals.uidaho.edu; tsoule@uidaho.edu

© Springer Nature Switzerland AG 2021
H. R. Arabnia et al. (eds.), *Advances in Artificial Intelligence and Applied Cognitive Computing*, Transactions on Computational Science and Computational Intelligence, https://doi.org/10.1007/978-3-030-70296-0_28

Neural networks (NNs) have a track record of success in problem areas such as speech recognition, language translation, and image/video classifications [1–3]. Genetic algorithms (GAs) have been shown to work well at training neural networks (NNs) [4]. In this chapter, we use GAs to train NNs to predict the movement of people in a crowd. The object of this chapter is to predict the next location of a person in a crowd by taking into account their nearest neighbor within his or her cones of vision.

In the next section, we show the background organized into three subsections. This is followed by a section on how our model works. The fourth section presents the results and the discussion. Finally, the conclusions and future works are presented.

2 Background

Predicting individual trajectories in crowded areas for use in crowd management simulations is an evolving research topic.

Alahi et al. [5] show that they were able to predict future trajectories by using an individual's past positions and a social Long Short-Term Memory (LSTM) model trained on observed human movement. To discover the motion of each pedestrian and how that person interacts with nearby neighbors, Camillen et al. in [6] used a crowd interaction deep neural network (CIDNN) to predict displacement frames for each pedestrian. The DESIRE encoder and decoder [7] predicts future locations of objects in dynamic scenes. Walking step size prediction, using genetic algorithms to optimize a neural network model, predicts the step size by collecting data from different sensors that have been applied to the pedestrians [4].

Based on observing the mobility behavior for a person over a period of time, using the Mobility Markov Chain (MMC) model can lead to a prediction for the next location [8]. The use of this model [9] improves long-term prediction by modeling pedestrian behavior, as a jump-Markov process. Using multi-layer architecture interaction-aware Kalman neural networks (IaKNNs) for forecasting the motion of surrounding dynamic obstacles can solve high-density traffic issues [10].

Different approaches have been applied in [11]; for example, NN and GA have predicted the number of occurrences of dwelling fires in the United Kingdom. Plans to apply deep neural network [12] have been proposed to predict a pedestrian's trajectory using goal–direction plan, and the learning patterns motion behavior will be operated with fully convolutional network (FCN).

2.1 *Structured/Unstructured Crowds*

Analyzing a crowd's behavior can be used to improve the designs for the movement of people at large events, such as religious gatherings, concerts, or sporting events, in order to prevent improve public safety [13, 14]. As noted in [15], there are two types of crowds: the **unstructured crowds**, in which people move in a variety of directions as in Fig. 1a, and the **structured crowds**, in which people tend to move in a specific direction toward a target as in Fig. 1b. Large events or festivals, such as Züri Fäscht in Zürich, Switzerland, deployed an app for crowd management [17] over a period of 3 days. An app with a similar purpose has also been proposed by Tisue and Wilensky [18] for crowds at the Hajj (an Islamic ritual).

A proposed RFID-based Hajj management system [19] would include data sharing, network communication, and [20, 21] other wireless technologies. To improve the research on a crowd's motion when there is a great density of people, the authors in [22, 23] suggest a framework for Hajj management. The work by Schubert et al. [24] describes the development of a decision support system for crowd control that uses a GA with simulation to discover control strategies. Distributing Combining GPS and Bluetooth Low Energy (BLE) tags among groups of people and using smart phones, Jamil et al. [25] hope to capture large-group dynamics for large-scale events. Pellegrini et al. [26] have developed a crowd simulation to model dynamic social behavior that has been trained from bird's-eye video records of high-traffic areas. Koshak and Fouda [27] have used GPS and added GIS to capture and analyze pedestrian movement. Schubert et al. [28] have described a decision support system by storing sample situations and then used GAs to run trials in order to find a successful system to control crowds. In this section, we compared more than one crowd style by emphasizing structured crowded areas and unstructured crowded areas for the purpose of identifying the types of crowd motion.

2.2   Evolution Models NN and GA

Lately, NNs have had a high impact with accomplishments in many areas. William Chan et al. [1] demonstrate NN's ability to learn LAS (Listen, Attend, and Spell) by duplicating an audio sequence signal into a word sequence. Regarding translations, Bahdanau et al. [2] used NNs to predict relevant translatable words. Also, NNs are valuable for classified tasks for videos and image processing [3]. Pluchino et al. in [4] have shown that the NN model developed by GA produces better results for prediction. The work by Ju et al. [29] has compared the genetic algorithm with backpropagation for neural network training and has demonstrated that GA is superior to backpropagation. The methodology in this chapter is based on the combination of the neural networks and genetic algorithms.

2.3   Simulating Crowd Interactions

In our model, the collected dataset that applied unwritten rules in the crowd, such as avoiding collision, depended on the NetLogo model as an important machine

Fig. 1 Examples of two major regimes of nominal crowd flow. (**a**) An unstructured crowd. Image from Ozturk et al. [16]. (**b**) A structured crowd. Image is a screen captured from videos in 2019 from Ministry of Hajj, Kingdom of Saudi Arabia, https://www.haj.gov.sa/en

(a)

(b)

for simulating people in different approaches when seeking crowd behavior motion. Agent-based models have been an attractive research tool for people seeking crowd motion outcomes to evaluate structured public areas and closed spaces. The hope is to show the effectiveness of simulation in crowd management.

For instance, the paper [30] presents a simulation using a NetLogo model, for pedestrian motion at the Castello Ursino Museum in Catania, Italy. The simulation is used to evaluate the capacity of the museum and the safety of visitors in cases of an alarm. Camillen et al. [31] compared their evacuation approach with different evacuation approaches looking for an optimal solution. Their results showed the efficiency of their evacuation plan by uncovering hard forecasts in emergency results. Based on experts on animal migration, Hargrove et al. [32] simulated the efficacy of a PATH (Pathway Analysis Through Habitat) by using NetLogo to study animals moving outside their territory through a connected but fragmented landscape. It is clear that NetLogo is a very useful agent-based modeling tool for research and teaching [33].

3 Our Model

In a crowd, people usually take into account their nearby neighbors in order to avoid a collision. We model a person's vision by their **field of view** or we say **cone of vision**. In our 2D world, a cone of vision is the region in front of a traveling person subtended by an angle on either side of vector of travel; see Fig. 2. Only people visible to the person can act as an influence on the direction the person is proceeding. For example, a person will take an alternative route or will stop, due to the influence of the people who walk into his/her cone of vision.

We created a model for the prediction of pedestrian movement by locating nearby people in an individual's cone of vision. The location of the nearest people is fed into our NN, and a predicted direction is returned.

Because this is a complex system, the farther into the future we attempt to predict that the more inaccuracies will build and that the model strays from actually tracking the location of individuals [26]. We will use an NN given the list of three nearest people in the cone of vision sorted by distance. The NN will be trained by a GA.

3.1 Hand-Collected Dataset

Our goal is to devise a method of predicting where individuals will move in a crowd given observations of crowds. This will require lots of data in the form of (x, y) pairs and timings to train on. We decided that a low-cost solution is to create a separate data-generating simulation for our initial development.

We use something similar to a Social Force Model (SFM). An SFM is based on three factors: (1) the acceleration of an individual to a desired speed, (2) the

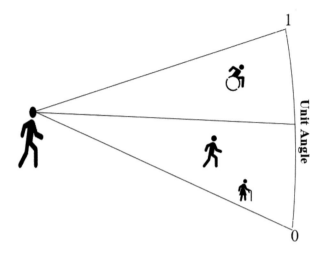

Fig. 2 In our model, the NN focuses on the nearest agents in the individual's cone of vision to predict his/her next position. The position of the nearest three agents is represented as a distance and an angle. The angle is scaled between 0 at the right side of the cone of vision and 1 at the left side. The scaled angle we will call the unit angle. Distances are absolute

individuals maintaining a specific distance from obstacles and other individuals, and (3) the impact of attraction, as in Helbing et al. [14]. Our data-generating crowd model uses a NetLogo model. The model is derived from the flocking model that comes with NetLogo, but with some additions for variation in speed, collision avoidance, field of view, boundary and initialization conditions, and for structured crowds and common direction of flow. Both structured and unstructured crowds were simulated as seen in Fig. 3.

3.2 Combining Neural Networks with GAs

Crowds may behave differently depending on the cultural composition, event, or environment. In order to more accurately model crowds, it is important to learn from observation rather than apply a one-size-fits-all solution. This motivated us to design a method that can read and learn from data. Perceiving the patterns of data (such as nearby neighbors' positions) to predict the motion of a crowd is the main feature of the algorithm. The combination of NN and GA has produced excellent results [4]. This motivated us to develop an algorithm using NN and GA. In a crowd, people pay attention to the people in front of them and "sort" them visually as nearest neighbors, which becomes a major factor when making a decision to change direction, even if only a slight change. This logic encouraged us to apply it in designing the inputs to the NN.

Using NNs and GAs for Predicting Human Movement in Crowds 359

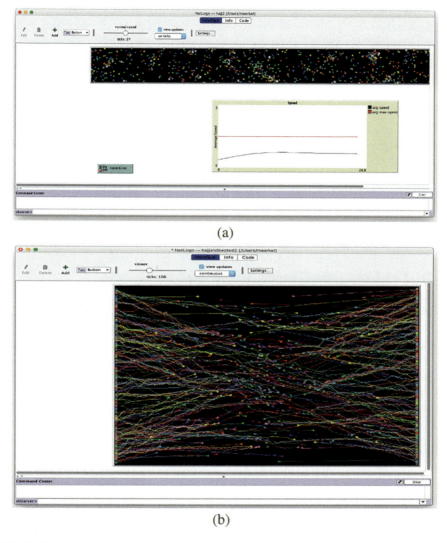

Fig. 3 NetLogo simulations of structured and unstructured crowds. (**a**) Structured crowd simulation in which people tend to move from left to right. (**b**) Unstructured crowd simulation in which people on the sides move at random initial top speeds to a paired random location on the opposite side. This forces the left and the right to negotiate passage through the middle area by adjusting their direction and speed

After calculating the distance between points, the neural network takes the three nearest neighbors' positions in the cone of vision as inputs. For example, if we specify that $60°$ is the cone of vision for the individual, the distance is calculated as in Eq. 1

$$nearest3(c) = \underset{i \in Cone_c(Pop)}{argmin\ 3} \quad distance(c,\ i). \tag{1}$$

In this equation, c is the current pedestrian for whom we are looking for his/her nearest neighbors. i is from the set of individuals in the cone of vision of c denoted $Cone_c(Pop)$. The cone of vision is determined by the direction of motion of c. We assume a cone of vision is 60o. argmin 3 gives the list of three smallest values, in this case, distances. Finally, in our unit angle, every individual from those three nearest neighbors obtains a number between 0 and 1 based on his/her location in the unit angle as in Fig. 2.

3.3 The Use of the Genetic Algorithm

A genetic algorithm (GA) is an optimization algorithm inspired by biological evolution [34]. A GA has several key components. A population of potential solutions where each individual in our population is a possible set of weights for the NN. Imperfect copies of the individuals from the population will be made using cloning and mixing using mutation and crossover operators. Selective pressure will be applied to force an enrichment in the population with sets of weights which are selected for, that is, have higher fitness. In our case, fitness will be determined by the success of the NN in predicting where individuals move given what they see (more below). If selection is controlled by the fitness function so that individuals with higher fitness function values are selected, then we have an optimization algorithm using a scheme very similar to that envisioned by Charles Darwin [35].

Each generation of the genetic algorithm uses tournament selection to choose the worthy individuals among the population as parents and uses them to produce offspring for the next generation using mutation and crossover. For diversity in every generation of the population, 12% is the percentage of mutation applied. To optimize the quality of NN weights, each individual corresponds to an evaluation, which is the distance between the predicted location and the actual location. The fitness distinguishes the NNs with high/low scores based on their outputs. It calculates the difference between the predicted position and the actual position for every individual's next location. Table 1 describes the genetic algorithm parameters used in this chapter.

$$fitness(NN) = \sum_{i=0}^{n-1} |\alpha_i - \pi_i|. \tag{2}$$

Table 1 Table of GA parameters

Parameter	Value
Type of GA	Steady state
Pop size	100
Mutation rate	0.12 per weight
Mutation	Add random $N(0, 1)$
Crossover probability	100%
Crossover type	1pt
Mating selection	Tournament size 10
Stopping criteria	1000 generations

In Eq. 2, $fitness(NN)$ represents the fitness for the weights of the NN. The variable i indexes through all the training data of position and three nearest neighbors. α_i is the actual angle of the next move of training case i. π_i is the angle predicted by the NN. Angles are in unit angles. The less the difference between the angles, the better the neural network. That is, the GA is minimizing the sum of simple errors of the trajectories.

In summary, the GA begins by initializing with a random population of neural networks. The GA uses the fitness based on the nearest neighbors in the cone of vision for each agent. This is used to train the weights for the neural network using GA. The fitness function scores each neural network based on the difference between the predicted trajectory and the actual trajectory in terms of a unit angle in the cone of vision. The neural network with the smallest error will be the model's neural network to predict an individual's trajectory. The step size will be assumed to be the same as the last step size. Figure 4 explains the workflow of the training.

4 Results

In this experiment, we produced training data using NetLogo in two scenarios representing the two crowd types: structured and unstructured. We used NetLogo since it is an agent-based model that is well known in research [30–33]. To model structured crowds, we observed surveillance cameras that were deployed on Hajj 2019 to monitor the behavior of people. A model was then built based on flocking/herding in which agents move within the limits of their own speed to move together and yet not collide or get too close and proceed toward a goal. The model is parameterized to emphasize the distributions of maximum speeds and how close they can get before they feel the urge to separate. There is a cohesion factor as well in that people in a crowd tend to move with others much like a herd. Unstructured crowds were modeled similarly, but individuals were initially positioned on the left and right of the arena (see Fig. 3b). They then proceed at different initial maximum speeds to cross to paired target points on the other side. This way the two sides must negotiate to slide between the opposing moving people. Decisions to avoid collision with others are the most important feature of the NetLogo models. This is done by the agents deciding to change their speed and direction.

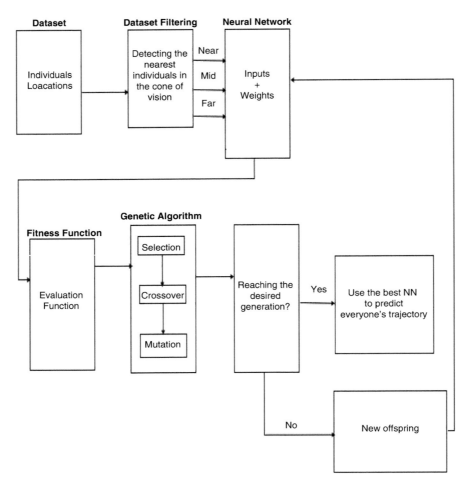

Fig. 4 This shows a sketch of the workflow for training the NN. Training data based on NetLogo simulations is input to the NN which produces trajectory predictions for the fitness function. The fitness function is used in the GA to converge on better weights for the NN

4.1 Trained vs. Control Neural Network

To answer the question of are we able to learn how to predict the trajectory, we compared the output of the NN we trained to a control. As a control, we chose a random NN without training. We predict that the trained NN should actually be able to move more like real people in a crowd (Fig. 5).

Figures 6 and 8 show the results for both experiments, trained and control, with the same test for structured and unstructured crowds, respectively. The graph shows

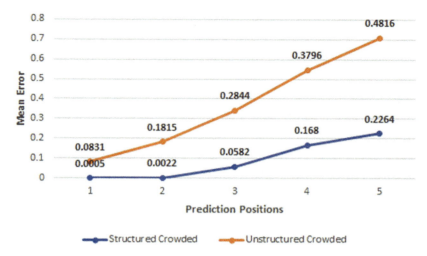

Fig. 5 The error in the average distance between the predicted location and the actual location

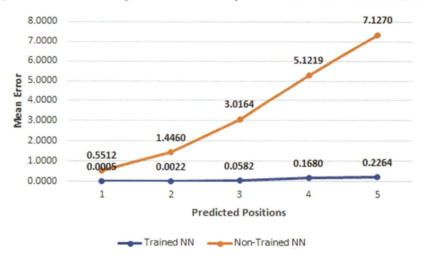

Fig. 6 The mean error between the predicted location and the actual location for a structured crowd. Both trained and untrained neural networks are shown for comparison

the mean Euclidean distance between the predicted position and the actual position assuming the step size for all predicted steps as the step size for last step before starting the prediction.

The graphs support the idea that a GA can be used to train an NN to model people's behavior in a crowd.

4.2 Time Series Use of NN

Our next question is whether the NNs we are generating can be repeatedly applied in a simulation to predict crowd movements farther into the future. In our experiment, we have applied two metrics, the cumulative distance error (CDE) and the mean in two different types of crowds.

The cumulative distance error (CDE) for position 1 is the average distance between the current position and the predicted position divided by the average distance between the current position and the previous position. It then multiplies that result by 100 to obtain the percentage error rate for the first predicted position. For the CDE of the second predicted position, we calculate the distance between the true position and the predicted position, except that it divides by the average distance between the current position and the position from which we started the prediction. We proceed like this for the remaining predicted positions in our experiment. The results for CDE are displayed in Fig. 7, which shows how the error rate grows as time passes (Fig. 8).

Figure 5 displays the mean for each distance between the predicted position and the true position. The mean calculates how the error in the average distance between the predicted position and the true position increases as time passes.

Both graphs indicate a divergence between predicted and real locations. This suggests that a simulation based on our learned NN will quickly diverge from reality. While this may at first appear as a simple negative result, the system we are modeling is a complex system with classic problem that its small errors will

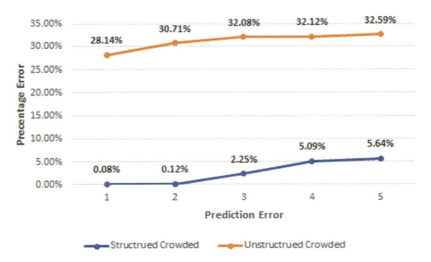

Fig. 7 The chart shows cumulative distance error (CDE) for all the next prediction positions. Since we have trained the NN for the first position, the result shows how the error increases as time passes. That is, error accumulates between the predictions the model is giving and the actual directions that individuals turn

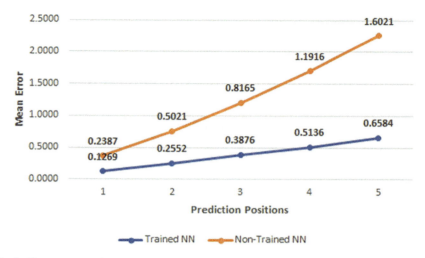

Fig. 8 The mean error between the predicted location and the actual location for an unstructured crowd. Both trained and untrained neural networks are shown for comparison

accumulate exponentially. We would expect these graphs. But since this is our initial research, we believe these measures and others may help to greatly improve our ability to predict and stave off the inevitable divergence.

4.3 Discussion

One of the innovations of our work is the use of a sorted list of nearest neighbors in the cone of vision. We believe this mimics cognitive input to the individuals. Even in structured crowded areas, people walk around one another as they walk in the same direction to the same goal [15]. However, we realize that the unstructured crowded area is assembled differently, where people randomly cross one another paths, and still humans tend to watch out for their nearest neighbors and make a decision about their next position or direction.

For the comparison between the structured and unstructured crowds in our experiment, the results include the cumulative distance error and the mean. The structured crowd results have smaller errors of prediction than unstructured crowds. We believe that this is because the behavior is much more predictable in structured crowds than unstructured crowds.

5 Conclusions

Simulating crowd motion and predicting the individuals' movements can be used to design crowd areas and to improve safety criteria. The need for these kinds of simulations inspired us to design a simulation that shows the reaction of people in an environment to seek a predictable crowd behavior in order to provide safe and pleasant experiences for participants. We have shown that the combination of neural networks and genetic algorithms can be effective for predicting movement in a crowd. We used the first predicted position for individuals to train the neural networks and the genetic algorithm to obtain the best NN. One of our innovations in our NN design is the use of a sorted list of nearest neighbor locations in the individual's cone of vision. We believe that this method shows promise as a technique for learning the movement of people in a crowd. The results showed a more reliable outcome for a structured crowds than the unstructured crowds. Our future work will involve diversifying tests, improving the evolutionary algorithm, and measures of crowd movement. Additionally, the agents of the model will not be limited to pedestrians but include other agents, such as cars, bicycles, etc.

References

1. Y. Xu, Z. Piao, S. Gao, Encoding crowd interaction with deep neural network for pedestrian trajectory prediction, in *Proceedings of the IEEE Conference on Computer Vision and Pattern Recognition* (2018), pp. 5275–5284
2. L. Shao, Z. Cai, L. Liu, K. Lu, Performance evaluation of deep feature learning for RGB-D image/video classification. Inf. Sci. **385**, 266–283 (2017)
3. O. Ozturk, T. Yamasaki, K. Aizawa, Detecting dominant motion flows in unstructured/structured crowd scenes, in *20th International Conference on Pattern Recognition* (2010), pp. 3533–3536
4. A. Pluchino, C. Garofalo, G. Inturri, A. Rapisarda, M. Ignaccolo, Agent-based simulation of pedestrian behaviour in closed spaces: A museum case study (2013). Preprint arXiv:1302.7153
5. D. Bahdanau, K. Cho, Y. Bengio, Neural machine translation by jointly learning to align and translate (2014). Preprint arXiv:1409.0473
6. F. Camillen, S. Caprì, C. Garofalo, M. Ignaccolo, G. Inturri, A. Pluchino, A. Rapisarda, S. Tudisco, Multi agent simulation of pedestrian behavior in closed spatial environments, in *2009 IEEE Toronto International Conference Science and Technology for Humanity (TIC-STH)* (IEEE, Piscataway, 2009), pp. 375–380
7. A.H. Al-Hashedi, M.R.M. Arshad, A.S. Baharudin, H.H. Mohamed, RFID applications in Hajj management system, in *2013 IEEE International Conference on RFID-Technologies and Applications (RFID-TA)* (IEEE, Piscataway, 2013), pp. 1–6
8. V. Karasev, A. Ayvaci, B. Heisele, S. Soatto, Intent-aware longterm prediction of pedestrian motion, in *2016 IEEE International Conference on Robotics and Automation (ICRA)* (IEEE, Piscataway, 2016), pp. 2543–2549
9. M. Rodriguez, S. Ali, T. Kanade, Tracking in unstructured crowded scenes, in *2009 IEEE 12th International Conference on Computer Vision* (IEEE, Piscataway, 2009), pp. 1389–1396
10. B. Krausz, C. Bauckhage, Loveparade 2010: automatic video analysis of a crowd disaster. Comput. Vis. Image Underst. **116**(3), 307–319 (2012)

11. C. Darwin, *On the Origin of Species: A Facsimile of the First Edition*, vol. 49 (Harvard University Press, Cambridge, 1964)
12. S. Jamil, A. Basalamah, A. Lbath, M. Youssef, Hybrid participatory sensing for analyzing group dynamics in the largest annual religious gathering, in *Proceedings of the 2015 ACM International Joint Conference on Pervasive and Ubiquitous Computing* (2015), pp. 547–558
13. A. Johansson, D. Helbing, H.Z. Al-Abideen, S. Al-Bosta, From crowd dynamics to crowd safety: a video-based analysis. Adv. Complex Syst. **11**(04), 497–527 (2008)
14. M. Yamin, H.M. Al-Ahmadi, A. Al Muhammad, Integrating social media and mobile apps into Hajj management, in *2016 3rd International Conference on Computing for Sustainable Global Development (INDIACom)* (IEEE, Piscataway, 2016), pp. 1368–1372
15. S. Pellegrini, A. Ess, K. Schindler, L. Van Gool, You'll never walk alone: Modeling social behavior for multi-target tracking, in *2009 IEEE 12th International Conference on Computer Vision* (IEEE, Piscataway, 2009), pp. 261–268
16. M. Yamin, A framework for improved Hajj management and future research. ENTIC Bull **2**(08), 1–7 (2008)
17. A.Y. Wang, L. Wang, Walking step prediction based on GA optimized neural network algorithm, in *2017 2nd IEEE International Conference on Computational Intelligence and Applications (ICCIA)* (IEEE, Piscataway, 2017), pp. 295–298
18. S. Tisue, U. Wilensky, NetLogo: A simple environment for modeling complexity, in *International Conference on Complex Systems*, vol. 21. Boston, MA (2004), pp. 16–21
19. W. Chan, N. Jaitly, Q. Le, O. Vinyals, Listen, attend and spell: A neural network for large vocabulary conversational speech recognition, in *2016 IEEE International Conference on Acoustics, Speech and Signal Processing (ICASSP)* (IEEE, Piscataway, 2016), pp. 4960–4964
20. Y. Mohammad, Y. Ades, Crowd management with RFID & wireless technologies, in *Proceedings of First International Conference on Networks & Communications, IEEE Computer Society Washington, DC, USA* (2009)
21. A.E. Eiben, J.E. Smith, et al., *Introduction to Evolutionary Computing*, vol. 53 (Springer, Berlin, 2003)
22. W.W. Hargrove, J.D. Westervelt, An implementation of the pathway analysis through habitat (path) algorithm using NetLogo, in *Ecologist-Developed Spatially-Explicit Dynamic Landscape Models* (Springer, Berlin, 2012), pp. 211–222
23. M. Yamin, M. Mohammadian, X. Huang, D. Sharma, RFID technology and crowded event management, in *2008 International Conference on Computational Intelligence for Modelling Control & Automation* (IEEE, Piscataway, 2008), pp. 1293–1297
24. J. Schubert, L. Ferrara, P. Hörling, J. Walter, A decision support system for crowd control, in *Proceedings of the 13th International Command and Control Research Technology Symposium* (2008), pp. 1–19
25. E. Rehder, F. Wirth, M. Lauer, C. Stiller, Pedestrian prediction by planning using deep neural networks, in *2018 IEEE International Conference on Robotics and Automation (ICRA)* (IEEE, Piscataway, 2018), pp. 1–5
26. N. Nasser, M. Anan, M.F.C. Awad, H. Bin-Abbas, L. Karim, An expert crowd monitoring and management framework for Hajj, in *2017 International Conference on Wireless Networks and Mobile Communications (WINCOM)* (IEEE, Piscataway, 2017), pp. 1–8
27. U. Blanke, G. Tröster, T. Franke, P. Lukowicz, Capturing crowd dynamics at large scale events using participatory GPS-localization, in *2014 IEEE Ninth International Conference on Intelligent Sensors, Sensor Networks and Information Processing (ISSNIP)* (IEEE, Piscataway, 2014), pp. 1–7
28. N. Koshak, A. Fouda, Analyzing pedestrian movement in Mataf using GPS and GIS to support space redesign, in *The 9th International Conference on Design and Decision Support Systems in Architecture and Urban Planning* (2008)
29. C. Ju, Z. Wang, C. Long, X. Zhang, G. Cong, D.E. Chang, Interaction-aware Kalman neural networks for trajectory prediction (2019). Preprint arXiv:1902.10928
30. J. Schubert, R. Suzic, Decision support for crowd control: Using genetic algorithms with simulation to learn control strategies, in *MILCOM 2007-IEEE Military Communications Conference* (IEEE, Piscataway, 2007), pp. 1–7

31. A. Alahi, K. Goel, V. Ramanathan, A. Robicquet, L. Fei-Fei, S. Savarese, Social LSTM: Human trajectory prediction in crowded spaces, in *Proceedings of the IEEE Conference on Computer Vision and Pattern Recognition* (2016), pp. 961–971
32. L. Yang, C.W. Dawson, M.R. Brown, M. Gell, Neural network and GA approaches for dwelling fire occurrence prediction. Knowl. Based Syst. **19**(4), 213–219 (2006)
33. J.N. Gupta, R.S. Sexton, Comparing backpropagation with a genetic algorithm for neural network training. Omega **27**(6), 679–684 (1999)
34. S. Gambs, M.-O. Killijian, M.N. del Prado Cortez, Next place prediction using mobility Markov chains, in *Proceedings of the First Workshop on Measurement, Privacy, and Mobility* (2012), pp. 1–6
35. N. Lee, W. Choi, P. Vernaza, C.B. Choy, P.H. Torr, M. Chandraker, Desire: Distant future prediction in dynamic scenes with interacting agents, in *Proceedings of the IEEE Conference on Computer Vision and Pattern Recognition* (2017), pp. 336–345

Hybrid Car Trajectory by Genetic Algorithms with Non-Uniform Key Framing

Dana Vrajitoru

1 Introduction

In this chapter, we present an application of genetic algorithms (GAs) with non-uniform key framing for constructing a trajectory for an autonomous car in a simulated car race setting. We compare its efficiency in terms of trajectory length and car racing time with a procedural benchmark algorithm, with a geometrically improved version of it, and with a uniform key framing GAs model.

The car's trajectory is defined as a curve spanning the length of the race track from start to end and confined to its boundaries. The track itself is represented for our purposes as a centerline with associated curvature function and a constant width.

The model presented here starts by segmenting the track into intervals where the curvature either is close enough to 0 or otherwise has a consistent sign, meaning that the track turns consistently in one direction. A number of such segments are selected, based on presenting sufficient length, and the center of each segment is used as anchor point. The trajectory is set in these anchor points in the middle for straight segments, and at the interior of the curve for the others. Then a GA is used to calculate the trajectory on each interval between the anchor points independently, using the same number of intermediary key frames every time, which makes their spacing overall non-uniform.

The first benchmark algorithm segments the track using consistency of the curvature sign, and builds the trajectory using a procedural method based on known efficient curve shapes. The second one starts from either a simple trajectory or one built by the first method, and applies a combination of a straightening and smoothing transformations to reduce the overall length. The third method also employs a GA,

---

D. Vrajitoru (✉)
Computer and Information Sciences, Indiana University South Bend, South Bend, IN, USA
e-mail: dvrajito@iusb.edu

© Springer Nature Switzerland AG 2021
H. R. Arabnia et al. (eds.), *Advances in Artificial Intelligence and Applied Cognitive Computing*, Transactions on Computational Science and Computational Intelligence, https://doi.org/10.1007/978-3-030-70296-0_29

369

but where the trajectory is computed using uniformly placed key frames and a segmentation of the track based on straight segments only. These reference methods are presented in [1, 2].

Autonomous vehicles are a current subject of interest, and their capabilities and availability have increased in the last years, as they are becoming more of a common occurrence on the road. Assistive driving features are also becoming more standard on new car models. Developing good algorithms for autonomously driving a car or for helping a human driver is still important.

For the research presented in this chapter, we used the TORCS system (The Open Racing Car Simulator) that implements simulated car races and provides multiple vehicles models and several tracks of various difficulty [3]. The programmer can write a car controller in this system.

Our research relates to the work presented in [4], where a neural network (NN) is trained to compute a target trajectory value based on a procedurally optimized trajectory. In their approach, the NN receives the local road curvature data as input and produces a target trajectory value as output. The direction unit in the car controller is programmed to follow this trajectory value. We used a similar pilot in our current research. We intend to replace the trajectory used to train the NN by a better one computed by the GA to improve the system's performance.

Multiple papers feature algorithms for computing optimal paths within given road constraints. The curves used by the first benchmark trajectory computation method can be found in [5], and are designed both to minimize the time it takes to traverse a curve and to maximize the exit velocity. These curves are consistent with recommendations from driving instructors and race pilot practices. They are also similar to the trajectory presented in [6]. The problem settings for the trajectory calculation in our case are similar to [6]; in both cases, the car trajectory is computed for a car driving on a given road-like track, aiming to optimize the travel time. The difference is that in their application, they compute safe zones on the road for the car beforehand, and use them to improve the trajectory computation in real time.

A number of approaches for track prediction can be found in the literature, as, for example, the track segmentation approach in [7]. Here, the track is divided into fragments classified based on a pre-defined set of polygon types. Onieva et al. [8] propose another controller based on track segmentation. Their driving controller, named AUTOPIA, is a very successful participant in the TORCS-simulated racing car competitions. Such segmentation algorithms are also related to map-matching algorithms such as the ones presented in [9].

Genetic algorithms have recently been used in several studies for trajectory optimization, as, for example, in [10, 11]. These papers present trajectories for space travel, and the authors report successful results when compared to real-world missions. Air traffic control is another potential driving-related GAs application [12]. Other works can be found showing the use of GAs for trajectory planning of robots [13].

The chapter is organized the following way. Section 2 introduces the problem settings and the procedural benchmark algorithms. Section 3 describes the application of genetic algorithms to this problem, as well as the GA-based benchmark method.

Fig. 1 The track E-Road used for this research, original (left) and reconstructed (right)

Section 4 presents experimental results and compares the new approach with the three methods. The chapter ends with conclusions.

2 Autonomous Car Pilot and Procedural Trajectories

The research presented in this chapter was conducted using the TORCS car race simulation system. TORCS sustains a large community of developers and users, and it is the platform for popular competitions that have been organized yearly as a part of various international conferences [14]. For this chapter, we have chosen the track E-Road, shown in Fig. 1. This track is about 3.2 km long and features multiple curves of various length and difficulty.

2.1 Track Mapping and Reconstruction

The TORCS program is organized as a server implementing car races combining multiple vehicles on a variety of tracks. A client module can be written by the user to implement the behavior of a controlled car. The server provides runtime information about the car's position and orientation on the track, the car's speed, damage, and other measures. In a multi-car race, the opponents' relative position in the car's vicinity is also available [9]. The client can control the steering wheel taking values in $[-1, +1]$, the gas pedal in $[0, +1]$, the brake pedal in $[0, +1]$, and the gearbox with possible values from -1 through 6 [4].

Since the geometry of the tracks is not provided directly, but is only indirectly given through the car state measures available at race time, the first step for these trajectory calculations was to map the tracks. This was accomplished by running the car at a low constant speed (50 km/h) and by calculating the curvature of the road using sensor information in each frame. These measurements were recorded and then used by the track building algorithms.

Because of rounding and other perception errors introduced by the curvature measurement algorithm, the reconstructed tracks do not conserve the track shape, even though the main curves are still present. For the reconstructed shape to be somewhat similar to the original, the curvature was scaled down. In [1], a curvature scale of 0.1 was used with the track E-Road. In the present research, the scale value is of 0.3, to increase the difference between good trajectories and less efficient ones and to encourage the GAs to find better ones. The shape of the reconstructed trajectory is shown in Fig. 1 right.

We define the car trajectory as a function $r(t)$ returning a value in the interval $[-1, 1]$ for each point of the track's centerline t. This value represents the target lateral position of the car on a perpendicular line to the road centerline. The function $r(t)$ can take values outside the interval $[-1, 1]$ if the car exits the road, but we assume that this is not a desirable trait in such a target function.

2.2 Autonomous Car Pilot

The autonomous car pilot we used for these experiments is called Meep, and represents an improvement to the controller called Gazelle [15]. It comprises a target direction unit, a speed unit, a recovery unit activated when the car gets outside of the track or crashes against the road shoulder, a special alerts unit that focuses on and keeps track of dangerous road conditions, and an opponent avoidance unit.

Since the current experiments focus on the trajectory itself, we have used a constant speed for these experiments, set to 80 km/h. There are no opponents involved in these tests, so this module is not needed. Finally, in the direction unit, the pilot uses a target value for the trajectory in the interval $[-1, 1]$, representing an objective value for the lateral position of the car on the road in each frame of the race. The value of -1 represents the left border of the road, while the value 1 the right border. A value of 0 places the car in the center of the road. The target value for this position represents the output of the trajectory computation algorithms used in this chapter. The steering value is computed using two measurements: the current lateral position of the car on the road, and the current angle between the car's axis and the centerline of the road. Both of these are provided by TORCS at runtime. Then the pilot steers the car to compensate for the deviation from the road centerline, and to direct it toward the target lateral position.

2.3 Benchmark Procedural Trajectory

For the procedural trajectory, after the road's curvature is mapped, the road is segmented based on the continuity of the sign of the curvature. This results in some segments where the road is almost flat, somewhere it turns continuously to the left, and somewhere it turns to the right. For each segment, an optimal curve profile, shown in Fig. 2, is used to calculate a locally optimal trajectory, inspired from [5]. Thus, for the procedural trajectory, for each curve the car veers toward the interior at the beginning of the curve, maintains that lateral position for a while, and comes

Fig. 2 Curve profile optimizing traversal time

back toward the center of the road toward the end. Starting out from the exterior of the curve before engaging in it can reduce the amount of steering needed to traverse it.

Theoretically, the trajectory is a continuous function. Practically, the track itself is discrete, represented by a set of polygons. Thus, the trajectory consists of a set of points of the centerline where the car is found in each frame of the application. The mapping of the track E-Road resulted in about 19.5 K points for a density of about 10 points per meter. One constraint that arises from the particular setting of this problem is that the trajectory should be fast to compute. Thus, it must allow the car's pilot to communicate with the race server quickly enough for the car not to exit the road or crash against a hard shoulder.

The notion of curvature of a curve is employed widely through the chapter and so we shall discuss its definition. For a continuous curve $s(t)$ defined in a multi-dimensional space, the first derivative $s'(t)$ is the tangent to the curve, and the second derivative $s''(t)$, called the curvature, is normal to the curve. Together with their cross-product, they form the Fermat frame. If the curve describes the motion of an object, the first derivative represents the velocity, and the second one is the acceleration. For the current project, we are not interested in the curvature's direction axis, but only in its norm and orientation sign.

In our model, the centerline of the road can be used to represent the curve $s(t)$, where t is the distance from the start of the race track. Since this is part of a simulated environment, the curve is not continuous, but rather composed of a sequence of line segments. We have used the angle between adjacent line segments as a measure of the curvature of the road. As the segments are not directly available, we used the distance measures available from the TORCS sensors to calculate the angle between observed adjacent road segments. A more detailed description of this process can be found in [4].

The trajectory is computed in three phases that are also described in [4]. Phase 1 traverses the entire track and assigns values of 0 (middle of the road) to all the sections where the track is almost straight for long enough, and 0.8 and −0.8, respectively, to all the long sections where the road turns in the same direction. The value 0.8 was chosen based on the width of the car, such that car resides completely inside the track at all times. These values are assigned such as to leave a buffer zone of a few meters at the beginning and end of each continuous section.

Phase 2 traverses the entire trajectory again and connects the segments created during Phase 1. This part uses a linear interpolation of the trajectory values between the end of one section and the beginning of the next one. Phase 3 smooths the resulting trajectory even more by using anti-aliasing techniques.

Multiple experiments were done of the trajectory calculating algorithm with variations of parameters such as the value of the curvature for which the road can be considered almost straight or the length of the buffer zones between the different sections of constant trajectory [4]. For comparison purposes with our current research, we used the procedural trajectory that produced the best track completion time in the race.

2.4 Trajectory Straightening and Smoothing

The second benchmark trajectory uses a smoothing algorithm to improve the procedural trajectory further. This algorithm comprises a combination of two steps: a curvature descent, and an anti-aliasing method.

The curvature descent method starts by computing the real coordinates of the trajectory points in the reconstructed road. Then it moves each point on a direction normal to the trajectory's centerline in that point, in the direction of the curvature vector, and by an amount proportionate to the curvature value. The translation is capped by the road border. The moved points are converted back to displacement values with respect to the road centerline, which in turn give us the new trajectory values. This is similar to methods used in image processing or vector drawing programs to smooth Bezier curves.

The anti-aliasing method is similar to the blending or smoothing transformation used in image processing to reduce the contrast in a given area of an image. In this transformation, each point of the trajectory is replaced by a weighted average of the point's value with the values of the neighboring points in a given radius. The weights of the neighbors' values decrease with the distance to the transformed point.

Several iterations of these two methods are done in alternation. For example, the first method can be applied 3 times, followed by the second one applied 10 times, a process which is repeated until no significant changes happen anymore and the trajectory converges toward a given shape.

3 Trajectory by Genetic Algorithms

Genetic algorithms have been employed for trajectory computation problems before, as well as for many other applications. One of the reasons why they are suitable for these problems is that they can help with parameter and algorithm tuning, and they can also take advantage of high performance computing resources.

3.1 General GA Settings

To apply the GAs to this problem, both the present research and the benchmark model use a segmentation of the track, where the GA can compute the trajectory separately on each segment. However, the segment selection is different. In both cases, a segment chromosome is composed of a number of key frames, each represented by several genes that compose a real value in the interval $[-0.8, 0.8]$, representing the trajectory value. As already mentioned, -1 marks the left border of the road, and $+1$ the right border. In general, the values represent a transversal proportional displacement of the car with respect to the road centerline. Each of these points is a real number represented by the same number of binary genes.

We used the real number encoding proposed in [16] where each gene contributes an amount equal to 1 over a power of 2 times the length of the interval. The values of the genes are added together to the value of the left border of the interval. Thus, the first gene contributes 0 if its value is 0, and ½ times the length of the interval if its value is 1. The second gene contributes 0 if its value is 0, and ¼ times the length of the interval if its value is 1, and so on. This representation method makes it easy to choose the level of precision of the encoded real number.

Once the binary chromosome is decoded into a number of key frames, the actual trajectory is computed by linear interpolation between the control points.

Similarly to [1], we used a population of size 100, a number of generations limited to 5000, the uniform crossover [17] with a probability of swap of 0.3, and a mutation rate of 0.01. In each case, the results represent an optimum value over 10 different trials with a different seed for the random number generator. The optimal trajectory is selected independently for each segment and then the trajectory is put together by concatenating the segments. To avoid introducing artifacts by this procedure at the transition points, both in [1] and in our current model, the trajectory value is fixed in the anchor points connecting the segments.

In terms of fitness, prior tests have shown that the total length of the trajectory within each segment measured on the reconstructed road produces the best results. As the goal is to minimize the length or the resulting curve, lower values are better.

3.2 Benchmark GA Application

In this application presented in [1], the road is segmented based on stretches of the road that are almost flat and long enough to be significant. The middle points of these intervals were used as segment delimiters. With this method, the segments end up being longer than for the procedural method.

The trajectory was computed by the GA on each of the segments. For each segment, the potential trajectory is represented in the chromosome as a set of control points or key frame points equally spaced on the road. We will use here the best-performing trajectory presented in [1] for comparison, which utilizes a key frame

Fig. 3 Examples of anchor points shown as black circles

step of 85, corresponding to 8.5 m. The results of this method show that it can produce a trajectory that has a lower total length than the procedural and optimized trajectories. However, when used in an actual race, the procedural algorithm still achieves the best time. The results also show that a high density of key frames does not necessarily lead to a better performance.

3.3 Non-uniform Key Framing GA Application

For the current research, we started with the idea of using some of the concepts from the procedural method to guide the GA toward a better solution. The procedural algorithm is quite efficient overall. However, it is sub-optimal on parts of the track where the road changes its turning direction in rapid succession and where going straight through might be better.

Thus, in this approach, we segmented the track using both long enough segments where the road is almost straight and long enough segments where the road turns consistently in one direction. We know that for flat segments it is a safe bet to keep the car in the center of the road, which leads to setting a value of 0 for the anchor points in the middle of such segments. We used a minimal length of about 9 m to set a center anchor point on a straight patch.

We also know that the car should keep to the interior of the curve if the curve is long enough. Thus, for this kind of segment, we set anchor points with the values -0.8 or 0.8, depending on the orientation of the curve. We used a minimal length of about 139 m to place an anchor point on a turning patch. Overall, the 3.2 km track resulted in 9 anchor points, dividing it in 10 segments.

Figure 3 shows an example of anchor points chosen by our application. We can see that between the second and third point, there are several curves on the road, but they are too small to set anchor points there.

On each segment between the anchor points, we have chosen a number of key frames to represent the chromosome. Since the same number of key frames is used for each segment, the length of the interval between key frames is not uniform overall.

3.4 Hybrid Model

Finally, we can take advantages of the strengths of the different trajectories computed by these different algorithms by putting together a hybrid approach that might perform better in the actual race. For this purpose, we timed all of the trajectories during the race when the car passes by each of the anchor points and compared their performance on each segment.

After that, we built a combined trajectory that uses on each segment, the trajectory that showed the lowest timing on it. The manner in which the trajectories are combined is very similar to the genetic crossover operator. This procedure is made possible by the fact that the values of the trajectory in the anchor points is the same for the procedural algorithm and for the non-uniform key frame GA. Thus, these trajectories can be combined without introducing situations of sharp change in the target road position that might result in unsafe driving.

4 Experimental Results and Discussion

We performed experiments with the non-uniform key framing GA and compared the result with the three benchmarking methods. We also searched for an optimal number of intermediary key frames. The length column shows the total length of the trajectory curve along the entire track in meters. We can compare it with the total length of the track of 3.2 km measured along the centerline. The curvature column shows the total sum of the absolute value of the curvature along the trajectory, measured in radians. These values are computed based on a quantization of the road with about 19.5 k centerline points.

Table 1 shows the results of these experiments in terms of trajectory length. The results are somewhat different from [1] because we are using a different curvature scale in the reconstructed road. The best results are highlighted in bold.

From this table, we can see that the non-uniform key frame application of the GA is able to find a better trajectory than the other models on the reconstructed track. We can also see that a large number of key frames per segment are not necessary for a good performance. The best result in terms of length was obtained by using 3 key frames per segment.

To measure the effectiveness of the trajectories in a race setting, we ran the Meep pilot in TORCS using each of these trajectories. Table 2 shows the results of each of them in terms of total time (seconds) it takes to complete the race and total amount of turning done by the car during the race, measured in radians. We ran the car at a constant speed of 80 km/h.

From this table, we can see that all the models performed better than the constant trajectory, but the procedural trajectory still yields the best results and it is matched by the bound GA producing the best trajectory length. This also shows that a shorter

Table 1 Trajectory length and total curvature on E-Road

Method		Length	Curvature
Procedural		3173.96	209.175
Optimized		3011.9	148.768
GA uniform key frames		3120.85	108.008
Hybrid		3130.59	188.162
GA non-uniform key frames	#Key frames		
	1	2992.75	140.684
	2	2955.53	149.913
	3	2948.11	148.042
	4	2966.34	166.189
	5	2982.05	175.138
	6	3442.10	264.69

Table 2 TORCS race results for E-Road at 80 km/h

Method		Time (s)	Total turn
Procedural		146.27	521.01
Optimized		147.79	545.13
GA uniform key frames		148.57	530.21
Hybrid		146.17	507.90
GA non-uniform key frames	#Key frames		
	1	148.07	574.99
	2	147.33	550.45
	3	147.31	540.66
	4	147.15	521.41
	5	146.57	526.96
	6	147.93	643.93

trajectory, such as the optimized one, does not always allow the car to finish the race faster.

For the hybrid method, for 7 out of the 10 segments, the chosen partial trajectory belongs to the fastest one overall, which is still the procedural one. However, 3 of the segments were taken from trajectories built by the GA, each of them with a different number of key frames per segment: 3, 5, and 6, respectively. Thus, even though the trajectory using 6 key frames per segment is less efficient overall, it is still better than all the others on one particular segment.

We can see from Table 2 that even though the trajectories made by the GA do not outperform the procedural method in actual race settings, they did come much closer to it than the approach using uniform key frames. Moreover, by combining the different trajectories in the hybrid approach, we were able to build a trajectory that performs better in the race than all the others, both in terms of timing and in terms of total turning.

5 Conclusions

In this chapter, we presented an application of genetic algorithms to building trajectories for autonomous cars in a race setting, and compared its performance with various models from previous work.

To apply the GA to this problem, we segmented the road in stretches of consistent curvature sign or that are close enough to being straight. Then we set a number of anchor points in the middle of such segments that satisfy a minimal length requirement. The trajectory was fixed in the anchor points based on the ideas from optimal driving paths. Then the GA was used to compute the trajectory on each segment separately. Finally, a hybrid trajectory was put together by timing different trajectories in the actual race.

Our experiments showed that the non-uniform key frame GA can produce more efficient trajectories in terms of length than all the prior approaches. In race settings, the procedural trajectory is still ahead overall, but the GA trajectory comes very close to it. By selecting the best trajectory on each segment, we were able to produce a hybrid one that outperformed all of the others. This improves the behavior of the autonomous pilot on the road and its chances to win the race. This shows that the GA was able to improve on the shortcomings of the procedural method on the problematic road patches.

For future research, this approach can also be applied to other race tracks, perhaps more challenging ones. Another direction would be using the new and improved trajectory to train a neural network to compute the trajectory in real time based on sensor data, and test it on other tracks.

References

1. D. Vrajitoru, Genetic algorithms in trajectory optimization for car races. *In Proceedings of the ISTES International Conference on Engineering, Science, and Technology (ICONEST'19)*, Denver, Colorado, October 7–10, pp. 1–10 (2019)
2. D. Vrajitoru, Trajectory optimization for car races using genetic algorithms. *In Proceedings of the Genetic and Evolutionary Computations Conference (GECCO'19)*, Prague, July, Companion Volume, pp. 85–86 (2019)
3. B. Wymann, C. Dimitrakakis, A. Sumner, E. Espié, C. Guionneau, R. Coulom, TORCS, The Open Racing Car Simulator, v1.3. http://www.torcs.org (2013)
4. D. Vrajitoru, Global to local for path decision using neural networks. *In Proceeding of the Pattern Recognition and Artificial Intelligence Conference (PRAI'18), ACM International Conference Proceedings Series*, Union, NJ, pp. 117–123 (2018)
5. E. Velenis, P. Tsiotras, Minimum time vs maximum exit velocity path optimization during cornering. *Proceedings of 2005 IEEE International Symposium on Industrial Electronics*, Dubrovnik, Croatia, pp. 355–360 (2005)
6. A. Liniger, J. Lygeros, A viability approach for fast recursive feasible finite horizon path planning of autonomous RC car. *In Proceedings of the 18th International Conference on Hybrid Systems: Computation and Control, (HSCC '15)*, New York, pp. 1–10 (2015)

7. J. Quadflieg, M. Preuss, Learning the track and planning ahead in a racing car controller. *In Proceedings of the IEEE Conference on Computational Intelligence and Games (CIG10)*, Copenhagen, Denmark, pp. 395–402 (2010)
8. E. Onieva, D.A. Pelta, An evolutionary tuned driving system for virtual racing car games: the AUTOPIA driver. Int. J. Intell. Syst. **27**(3), 217–241 (2012)
9. S. Rathour, A. Boyali, L. Zheming, S. Mita, V. John, A map-based lateral and longitudinal DGPS/DR bias estimation method for autonomous driving. Int. J. Mach. Learn. Comput. **7**(4), 67–71 (2017)
10. M. Schlueter, M. Munetomo, Massively parallelized co-evaluation for many-objective space trajectory optimization. *In Proceedings of the Genetic and Evolutionary Computation Conference (GECCO'18)*, Kyoto, Japan, pp. 306–307 (2018)
11. N. Padhye, Interplanetary trajectory optimization with Swing-bys using evolutionary multi-objective optimization. *In Proceedings of the Genetic and Evolutionary Computation Conference (GECCO'08)*, Atlanta, pp. 1835–1838 (2008)
12. G. Marceau, P. Savéant, M. Schoenauer, Multiobjective optimization for reducing delays and congestion in air traffic management. *In Proceedings of the Proceedings of the Genetic and Evolutionary Computation Conference (GECCO'13)*, Amsterdam, The Netherlands, pp. 187–188 (2013)
13. E. Pires, J. Machado, P. Oliveira, Robot trajectory planner using multi-objective genetic algorithm. *In Proceedings of the Genetic and Evolutionary Computation Conference*, Washington, D.C. (2004)
14. C. Guse, D. Vrajitoru, The epic adaptive car pilot. *In Proceedings of the Midwest Artificial Intelligence and Cognitive Science Conference*, South Bend, IN, pp. 30–35 (2010)
15. K. Albelihi, D. Vrajitoru, An application of neural networks to an autonomous car driver. *In Proceedings of the 17th International Conference on Artificial Intelligence*, Las Vegas, NV, pp. 716–722 (2015)
16. D.E. Goldberg, *Genetic Algorithms in Search, Optimization, and Machine Learning* (Addison-Wesley, Reading, 1989)
17. G. Syswerda, Uniform crossover in genetic algorithms, in *In Proceedings of the International Conference on Genetic Algorithms*, (Morgan Kaufmann Publishers, San Mateo, 1989)

Which Scaling Rule Applies to Artificial Neural Networks

János Végh (iD)

1 Introduction

Single-processor performance stalled nearly two decades ago [1], mainly because of reaching the limits, the laws of nature enable [2]. As we pointed out [3], one of the major reasons for stalling was tremendously extending inherent idle waiting times in computing. Given the proliferation of Artificial Neural Network (ANN)-based devices, applications, and methods, furthermore that even supercomputers are retargeted for Artificial Intelligence (AI) applications, the efficacy of such systems is gaining growing importance. The attempts to prepare truly parallel computing systems failed [4], so the only way to produce the required high computing performance is by creating parallelized sequential computing systems. In such systems, the component processors work sequentially. Still, their work is orchestrated (usually by one of the processors) in a way that it gives the illusion that the system was working in a parallel manner. This latter way replaced *parallel computing* with *parallelized sequential computing*, disregarding that the operating rules of the latter [5–7] sharply differ from those experienced with segregated processors. Besides, their particular workload sheds light on a fundamental hiatus of computing systems: *the lack of considering their temporal behavior [3] of their components, algorithms, and operation.*

Amdahl proposed to build *"general purpose computers with a generalized interconnection of memories or as specialized computers with geometrically related memory interconnections and controlled by one or more instruction streams"* [8]. Amdahl has also pointed out *why* processors should be assembled to form a higher performance system in the way he proposed: *"the non-parallelizable portion [(aka serial part)] of the task seriously restricts the achievable performance of the*

J. Végh (✉)
Kalimános BT, Debrecen, Hungary

© Springer Nature Switzerland AG 2021
H. R. Arabnia et al. (eds.), *Advances in Artificial Intelligence and Applied Cognitive Computing*, Transactions on Computational Science and Computational Intelligence, https://doi.org/10.1007/978-3-030-70296-0_30

381

resulting system." His followers derived the famous Amdahl's formula [9], which served as the base for "strong scaling." That formalism provided a "pessimistic" prediction for resulting performance of parallelized systems and led to decades-long debates about the validity of the law.

The appearance of "massively parallel" systems improved the degree of parallelization so much that researchers suspected that Amdahl's Law was not valid. Similarly to the basic computing paradigm, all scaling laws represent approximations [9] (they are based on some omissions, and so they have their range of validity), and the new experiences led to a new approximation: the "weak scaling" [10] appeared. This paper discusses only the empirical methods of scaling, having all the time the temporal aspects in mind. The detailed theoretical discussion, *why* and *how* the temporal behavior must be introduced into the commonly used paradigm, is discussed in a separate paper [3].

In Sect. 2, the considered scaling methods are shortly reviewed, and some of their consequences are discussed. In Sect. 2.1, Amdahl's idea is shortly described: his famous formula using our notations is introduced. In Sect. 2.2, the basic purpose of massively parallel processing, Gustafson's idea, is scrutinized. Section 3 discusses different factors affecting computing performance of processor-based (as well as some aspects of other electronic equipments) ANN systems.

2 Common Scaling Methods

The scaling methods used to model different implementations of parallelized sequential processing (aka "distributed computing") are approximations to the more general model presented in [3]. We discuss the nature and validity of those approximations, and this section also introduces notations and formalism.

As discussed in detail in [6, 7], *parallelized sequential systems have their inherent performance limitation.* Using that formalism and data from the TOP500 database [11], we could estimate performance limits for present supercomputers. It enabled us to comprehend why supercomputers have their inherent performance limit [7]. We also validated [9, 12] "modern scaling" (as a mostly empirical experience) through applying it, among others, for qualifying load balancing compiler, cloud operation, and on-chip communication. Given that experts, with the same background, build also ANN systems, from similar components, we can safely assume that the same scaling is valid for those systems too. Calibrating our systems for some specific workload (due to the lack of validated data) is not always possible, but one can compare the behavior of systems and draw some general conclusions.

2.1 Amdahl's Law

Amdahl's Law (also called "strong scaling") is usually formulated with a formula such as

$$S^{-1} = (1 - \alpha) + \alpha/N \tag{1}$$

where N is the number of parallelized code fragments, α is the ratio of parallelizable fraction to total (so $(1-\alpha)$ is the "serial percentage"), and S is a measurable speedup. That is, Amdahl's Law considers a *fixed-size problem*, and α portion of the task is distributed to fellow processors.

When calculating the speedup, one calculates

$$S = \frac{(1 - \alpha) + \alpha}{(1 - \alpha) + \alpha/N} = \frac{N}{N \cdot (1 - \alpha) + \alpha} \tag{2}$$

However, as expressed in [13], *"Even though Amdahl's Law is theoretically correct, the serial percentage is not practically obtainable."* That is, concerning S, there is no doubt that it is derived as the ratio of *measured execution times*, for non-parallelized and parallelized cases, respectively. But, what is the exact interpretation of α, and how can it be used?

Amdahl listed performance affecting factors, such as "boundaries are likely to be irregular; interiors are inhomogeneous; computations required may be dependent on the states of the variables at each point; *propagation rates of different physical effects may be quite different*; the rate of convergence or convergence at all may be strongly dependent on sweeping through the array along different axes on succeeding passes, etc." Amdahl has foreseen issues with "sparse" calculations (or in general, *the role of data transfer*) as well as that the *physical size* of computer and the *interconnection* of its computing units (especially in the case of distributed systems) also matters.

Amdahl used wording *"the fraction of the computational load,"* giving way to his followers to give meaning to that term. This (unfortunately formulated) phrase *"has caused nearly three decades of confusion in the parallel processing community. This confusion disappears when processing times are used in the formulations"* [13]. On the one side, *it was guessed that Amdahl's Law is valid only for software* (for the number of executed instructions), and on the other side, *other affecting factors, he mentioned but did not discuss in detail, were forgotten.*

Expressing Amdahl's speedup is not simple: *"For example, if the following percentage is to be derived from computational experiments, i.e., recording the total parallel elapsed time and the parallel-only elapsed time, then it can contain all overheads, such as communication, synchronization, input/output, and memory access. The law offers no help to separate these factors. On the other hand, if we obtain the serial percentage by counting the number of total serial and parallel instructions in a program, then all other overheads are excluded. However, in this case, the predicted speedup may never agree with the experiments"* [13]. Moreover, the experimental one is always smaller than the theoretical one.

From computational experiments, one can express α from Eq. (1) in terms measurable experimentally as

$$\alpha = \frac{N}{N-1} \frac{S-1}{S} \tag{3}$$

It is useful to express *computing efficiency* with those experimentally measurable

$$E(N, \alpha) = \frac{S}{N} = \frac{1}{N \cdot (1 - \alpha) + \alpha} = \frac{R_{Max}}{R_{Peak}} \tag{4}$$

data. Efficiency is an especially valuable parameter, given that constructors of many parallelized sequential systems (including TOP500 supercomputers) provide the efficiency (as R_{Max}/R_{Peak}) of their computing system and, of course, the number of processors N in their system. Via reversing Eq. (4), the value of α_{eff} can be expressed with measured data as

$$\alpha_{eff}(E, N) = \frac{E \cdot N - 1}{E \cdot (N - 1)} \tag{5}$$

As seen, *the efficiency of a parallelized system is a two-parameter function* (the corresponding parametrical surface is shown in Fig. 1), demonstratively underpinning that *"This decay in performance is not a fault of the architecture but is dictated by the limited parallelism"* [5]. Furthermore, that its dependence can be perfectly described by the properly interpreted Amdahl's Law, rather than being an unexplained "empirical efficiency."

2.2 Gustafson's Law

Partly because of the outstanding achievements of parallelization technology, and partly because of issues around practical utilization of Amdahl's Law, a "weak scaling" (also called Gustafson's Law [10]) was also introduced. The assumption is that *the computing resources grow proportionally with the task size*, and the speedup (using our notations) is formulated as

$$S = (1 - \alpha) + \alpha \cdot N \tag{6}$$

Similarly to Amdahl's Law, the efficiency can be derived for Gustafson's Law as (compared to Eq. (4))

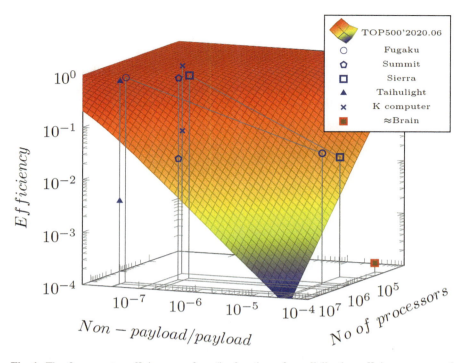

Fig. 1 The 2-parameter efficiency surface (in function of parallelization efficiency measured by benchmark HPL and number of processing elements) as concluded from Amdahl's Law (see Eq. (4)), in first-order approximation. Some sample efficiency values for some selected supercomputers are shown, measured with benchmarks HPL and HPCG, respectively. Also, the estimated efficacy of brain simulation using conventional computing is shown

$$E(N, \alpha) = \frac{S}{N} = \alpha + \frac{(1-\alpha)}{N} \qquad (7)$$

From these equations immediately follows that speedup (aka parallelization gain) *increases* linearly with the number of processors, *without limitation*, a conclusion that was launched amid much fanfare. They imply, however, some more immediate findings, such as

- the efficiency slightly *increases* with the number of processors N (the more processors, the better efficacy),
- the non-parallelizable portion of the task either shrinks as the number of processors grows, or *despite that it is non-parallelizable, the portion* $(1 - \alpha)$ *is distributed between the N processors*,
- executing the extra machine instructions needed to organize the joint work needs no time,

- *all non-payload computing contributions such as communication (including network transfer), synchronization, input/output, and memory access take no time.*

However, an error was made in deriving Eq. (6): *the $N - 1$ processors are idle waiting, while the first one is executing the sequential-only portion.* Because of this, the *time* that serves as the base for calculating[1] the *speed*up in the case of using N processors

$$T_N = (1 - \alpha)_{processing} + \alpha \cdot N + (1 - \alpha) \cdot (N - 1)_{idle}$$
$$= (1 - \alpha) \cdot N + \alpha \cdot N$$
$$= N$$

That is, before fixing the arithmetic error, strange conclusions follow, and after fixing it, the conceptual error comes to light: *"weak scaling" assumes that single-processor efficiency can be transferred to parallelized sequential subsystems without loss.* Weak scaling assumes that the efficacy of a system comprising N single-thread processors remains the same as that of a single-thread processor, a fact that strongly contradicts the experienced "empirical efficiency" (several 100-fold deviation from its predicted value) of parallelized systems, not mentioning the "different efficiencies" [7]; see also Fig. 1.

However, that *"in practice, for several applications, the fraction of the serial part happens to be very, very small thus leading to near-linear speedups"* [14] misleads the researchers. Gustafson concluded his "scaling" for several hundred processors only. The interplay of improving parallelization and general hardware (HW) development (including the non-determinism of modern HW[15]) covered for decades that this scaling was used far outside of its range of validity.

That is, Gustafson's Law is simply a misinterpretation of its argument α: a simple function form transforms Gustafson' Law to Amdahl's Law [13]. After making that transformation, the two (apparently very different) laws become identical. However, as suspected by [13], *"Gustafson's formulation gives an illusion that as if N can increase indefinitely."* Although collective experience showed that it was not valid for the case of systems comprising an ever higher number of processors (an "empirical efficiency" appeared), and later researchers measured "two different efficiencies" [16] for the same supercomputer (under different workloads), and the "weak scaling" was not suspected to be responsible for the issues.

"Weak scaling" omits all non-payload (but needed for the operation) activities, such as interconnection time, physical size (signal propagation time), accessing data in an amount exceeding cache size, synchronization of different kinds, that

---

[1] For the meaning of the terms, the wording "is the amount of time spent (by a serial processor)" is used by the author in [10].

are surely present when working with ANNs. In other words, *'weak scaling"* *neglects the temporal behavior*, a crucial science-based feature of computing [3]. This illusion led to the moon shot of targeting to build supercomputers with computing performance well above feasible (and reasonable) size [7] and leads to false conclusions in the case of clouds. Because of this, *except some very few neuron systems, "weak scaling" cannot be safely used for ANNs, even as a rough approximation.*

2.3 Modern Scaling

The role of α was theoretically established [17], and the phenomenon itself, that the efficiency (in contrast with Eq. (7)) *decreases* as the number of processing units *increases*, is known since decades [5] (although it was not formulated in the functional form given by Eq. (4)). In the past decades, however, theory was somewhat faded mainly due to quick development of parallelization technology and increase of single-processor performance, and finally, because "weak scaling" was used to calculate the expected performance values, in many cases, outside of its range of validity. The "gold rush" for building exascale computers finally made obvious that under the extreme conditions represented by the need of millions of processors, "weak scaling" leads to false conclusions. It had to be admitted that it *"can be seen in our current situation where the historical ten-year cadence between the attainment of megaflops, teraflops, and petaflops has not been the case for exaflops"* [18]. It looks like, however, that in feasibility studies of supercomputing using parallelized sequential systems, and an analysis, whether building computers of such size is feasible (and reasonable) remained (and remains) out of sight either in USA [19, 20] or in EU [21] or in Japan [22] or in China [23].

Figure 1 depicts the two-parameter efficiency surface stemming from Amdahl's Law. On the parametric surface, described by Eq. (4), some measured efficiencies of present top supercomputers are also depicted, only to illustrate some general rules. The HPL[2] efficiencies are sitting on the surface, while the corresponding HPCG[3] values are much below those values. The conclusion drawn here was that *"the supercomputers have two different efficiencies"* [16], because that experience cannot be explained in the frame of "classic computing paradigm" and/or "weak scaling."

Supercomputers *Taihulight*, *Fugaku*, and *K computer* stand out from the "millions core" middle group. Thanks to its 0.3M cores, *K computer* has the best efficiency for HPCG benchmark, while *Taihulight* with its 10M cores the worst one. The middle group follows the rules [7]. For HPL benchmark, the more the cores, the lower the efficiency. It looks like the community experienced the effect

---

[2]http://www.netlib.org/benchmark/hpl/.

[3]https://www.epcc.ed.ac.uk/blog/2015/07/30/hpcg.

of the two-dimensional efficiency. The top supercomputers run HPL benchmark with using all their cores, but some of them only use a fragment of them to measure performance with HPCG. This is the inflexion point: as can be concluded from the figure, increasing their nominal performance by an order of magnitude decreases their efficiency (and so payload performance) by more than an order of magnitude. For $HPCG$ benchmark, the "roofline" [24] of that communication intensity was already reached, and all computers have about the same efficiency. For more discussion on supercomputers, see [7].

3 Performance Limit of Processor-Based AI Systems

3.1 General Considerations

As discussed in detail in [7], the payload performance $P(N, \alpha)$ of parallelized systems comprising N processors is described[4] as

$$P(N, \alpha) = \frac{N \cdot P_{single}}{N \cdot (1 - \alpha) + \alpha} \tag{8}$$

where P_{single} is the single-thread performance of individual processors, and α is describing parallelization of the given system for the given workload (i.e., it depends on both of them).

This simple formula explains why *payload performance of a system is not a linear function of its nominal performance* and why in the case of very good parallelization $((1 - \alpha) \ll 1)$ and low N, this non-linearity cannot be noticed. In contrast with the prediction of "weak scaling," *payload performance* and *nominal performance* differ by a factor, growing with the number of cores. This conclusion is well known, but forgotten: "*This decay in performance is not a fault of the architecture but is dictated by the limited parallelism*" [5].

The key issue is, however, that one can hardly calculate the value of α for the present complex HW/software (SW) systems from their technical data, although some estimated values can be derived. For supercomputers, however, one can derive a theoretical "best possible," and already achieved "worst case" values [7]. It gives us reasonable confidence that those values deviate only within a factor of two. We cannot expect similar results for ANNs. There are no generally accepted benchmark computations, and also there are no standard architectures.[5] Using a benchmark means a particular workload, and comparing the results of even a standardized ANN

---

[4]At least in a first approximation, see [7].

[5]Notice that selecting a benchmark also directs the architectural development: the benchmarks HPL and HPCG result in different rankings.

benchmark on different architectures is as little useful as comparing the results of benchmarks HPL and HPCG on the same architecture.

Recall also that at a large number of processors, the internal latency of processor also matters. Following the failure of supercomputer Aurora'18, Intel admitted: *"Knights Hill was canceled and instead be replaced by a* "new platform and new microarchitecture specifically designed for exascale""" [25]. We expect that shortly it shall be admitted that building large-scale AI systems is simply not possible based on the old architectural principles [26, 27]. The potential new architectures, however, require a new computing paradigm (considering both temporal behavior of computing systems [3] and the old truth that "more is different" [28]) that can give a proper reply to power consumption and performance issues of—among others, ANN—computing.

3.2 Communication-to-Computation Ratio

As we learned decades ago, *"the inherent communication-to-computation ratio in a parallel application is one of the important determinants of its performance on any architecture"* [5], suggesting that communication can be a dominant contribution to system's performance. In the case of neural simulation, a very intensive communication must take place, so the non-payload-to-payload ratio has a significant impact on the performance of ANN-type computations. That ratio and the corresponding workload type are closely related: using a specific benchmark implies using a specific communication-to-computation ratio. In the case of supercomputing, the same workload is running on (nearly) the same type of architecture, which is not the case for ANNs.

3.3 Computing Benchmarks

There are two commonly used benchmarks in supercomputing. The HPL class tasks essentially need communication only at the very beginning and at the very end of the task. Real-life programs, however, usually work in a non-standard way. Because of this reason, a couple of years ago, the community introduced benchmark HPCG: the collective experience shows that the payload performance is much more accurately approximated by HPCG than by HPL, because real-life tasks need much more communication than HPL. Importantly, since the quality of their interconnection improved considerably, supercomputers show different efficiencies when using various benchmark programs [16]. Their efficiencies differ by a factor of ca. 200–500 (a fact that remains unexplained in the frame of "weak scaling"), when measured by HPL and HPCG, respectively.

In the HPL class, the communication intensity is the lowest possible one: computing units receive their task (and parameters) at the beginning of computation, and they return their result at the very end. That is, the core orchestrating their

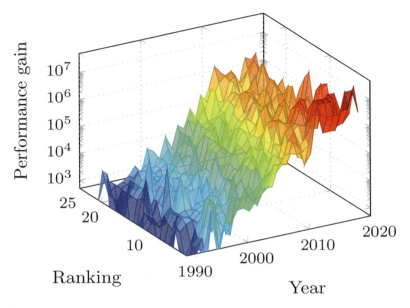

Fig. 2 History of supercomputing in terms of performance gain: performance values of the first 25 supercomputers, in function of year of their ranking. Data measured using benchmark HPL

work must deal with the fellow cores only in these periods, so the communication intensity is proportional to the number of cores in the system. Notice the need to queue requests at the beginning and the end of the task.

In the HPCG class, iteration takes place: the cores return the result of one iteration to the coordinator core, which makes sequential operations: it not only receives and resends the parameters but also needs to compute new parameters before sending them to the fellow cores. The program repeats the process several times. As a consequence, *the non-parallelizable fraction of benchmarking time grows proportionally to the number of iterations*. The effect of that extra communication decreases the achievable performance roofline [24]: as shown in Fig. 4, the HPCG roofline is about 200 times smaller than the HPL one.

As expressed by Eq. (8), the resulting performance of parallelized computing systems depends on both single-processor performance and performance gain. To separate these two factors, Fig. 2 displays the *performance gain* of supercomputers in function of their year of construction and ranking in the given year. Two "plateaus" can be clearly localized before the year 2000 and after the year 2010 also, unfortunately, underpinning Amdahl's Law and refuting Gustafson's Law, and also confirming the prediction "*Why we need Exascale and why we would not get there by 2020*"[29]. The "hillside" reflects the enormous development of interconnection technology between the years 2000 and 2010 (for more details, see [7]). For the reason of the "humps" around the beginning of the second plateau, see Sect. 3.5. Unfortunately, different individual factors (such as interconnection quality, using accelerators and clustering, using on-chip memories, using slightly different computing paradigm, etc.) cannot be separated in this way. However, some limited validity conclusions can be drawn.

3.4 Workload Type

The role of the workload came to light after that interconnection technology was greatly improved, and as a consequence, the *benchmarking computation* (defining *the type of workload*) became the dominating contributor, defining value of α (and as a consequence, payload performance); for a discussion, see [7]. The overly complex Fig. 3 illustrates the phenomenon, why and how the payload performance of a configuration depends on the application it runs.

Figure 3 compares three workloads (owing different communication intensity). In the top and middle figures, the communication intensities of the standard supercomputer benchmarks HPL and HPCG are displayed in the style of AI networks. The "input layer" and "output layer" are the same and comprise the initiating node only, while the other "layers" are again the same: the rest of the cores. Figure 3c depicts an AI network comprising n input nodes and k output nodes; furthermore, the h hidden layers are comprising m nodes. The communication-to-computation intensity [5] is, of course, not proportional in the cases of subfigures, but the figure illustrates excellently how the communication need of different computer tasks changes with the type of the workload.

As can be easily seen from the figure, in the case of benchmark HPL, the initiating node must issue m communication messages and collect m returned results, i.e., the execution time is $O(2m)$. In the case of benchmark HPCG, this execution time is $O(2Nm)$, where N is the number of iterations (one cannot directly compare the execution times because of the different amounts of computations and especially because of the different amount of sequential-only computations).

Figure 3a displays the case of minimum communication and Fig. 3b a moderately increased one (corresponding to real-life supercomputer tasks). As nominal performance increases linearly and payload performance decreases inversely with the number of cores, at some critical value where an inflection point occurs, the resulting performance starts to fall. The resulting non-parallelizable fraction sharply decreases efficacy (in other words, performance gain or speedup) of the system [32]. This effect was noticed early [5], under different technical conditions, but somewhat faded due to development of parallelization technology.

The non-parallelizable fraction (denoted in the figure by α_{eff}^X) of a computing task comprises components X of different origin. As already discussed and noticed decades ago, "*the inherent communication-to-computation ratio in a parallel application is one of the important determinants of its performance on any architecture*" [5], suggesting that *communication can be a dominant contribution to the system's performance.*

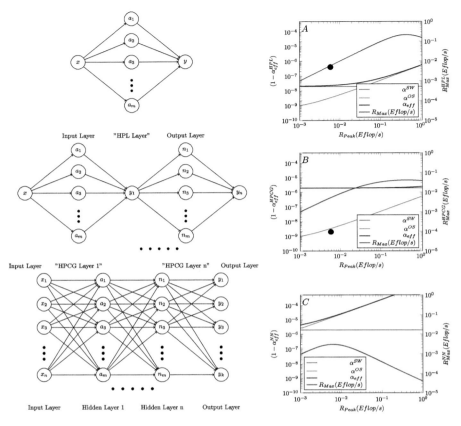

Fig. 3 Different communication/computation intensities of the applications lead to different payload performance values in the same supercomputer system. Left column: models of computing intensities for different benchmarks. Right column: the corresponding payload performances and α contributions in function of the nominal performance of a fictive supercomputer ($P = 1$ Gflop/s @ 1 GHz). The blue diagram lines refer to the right-hand scale (R_{Max} values) and all others (($1-\alpha_{eff}^X$) contributions) to the left-hand scale. The figure is purely illustrating the concepts; the displayed numbers are somewhat similar to the real ones. The performance breakdown shown in the figures was experimentally measured by [5, 30] and [31]

The workload in ANN systems comprises components of type "computation" and "communication" (this time also involving data access and synchronization, i.e., everything that is 'non-computation'). As logical interdependence between those contributions is strictly defined, payload performance of the system is limited by both factors, and the same system (maybe even within the same workload, case by case) can be either computing bound and communication bound, or both.

Notice that supercomputers showing the breakdown depicted in Fig. 3 are not included in history depicted in Fig. 2. Aurora'18 failed, Gyokou was withdrawn, "Chinese decision-makers decided to withhold the country's newest Shuguang

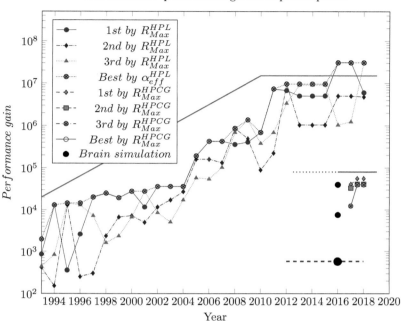

Fig. 4 Performance gain of supercomputers in function of their year of construction, under different workloads. The diagram lines display the measured values derived using HPL and HPCG benchmarks, for the TOP3 supercomputers in the gives years. The small black dots mark the performance data of supercomputers $JUQUEEN$ and K as of 2014 June, for HPL and HPCG benchmarks, respectively. The big black dot denotes payload performance of the system used by van Albada et al. [31]. The saturation effect can be observed for both HPL and HPCG benchmarks

supercomputers even though they operate more than 50% faster than the best current US machines".[6] Also, $Fugaku$ stalled at some 40% of its planned capacity.

Notice that a similar hillside cannot be drawn for benchmark HPCG, because of two reasons. On the one side, HPCG measurement started only a few years ago. On the other side, top supercomputers publish data measured with cores less than the number of cores used to measure HPL. Recall that efficiency is a two-parameter function (see Fig. 1) and that after exceeding a critical number of cores, housekeeping gradually becomes the dominating factor of the performance limitation and leads to a decrease in the payload performance: *"there comes a point when using more Processing Unit (PU)s ... actually increases the execution time rather than reducing it"* [5]. For "real-life" programs, represented by HPCG, this critical number is much lower than in the case of HPL. There is a real

---

[6]https://www.scmp.com/tech/policy/article/3015997/china-has-decided-not-fan-flames-super-computing-rivalry-amid-us.

competition between the different contributions to dominate system's performance. As demonstrated in [7], before 2010, running both benchmarks HPL and HPCG on a top supercomputer was a *communication-bound* task, since 2010 HPL is a computing-bound task, while HPCG persisted to be a *communication-bound* task. This is why some supercomputers provide their HPCG efficiency measured only with a fragment of their cores: the HPCG "roofline" is reached at that lower number of cores. Adding more cores does not increase their payload performance but decreases their efficiency.

3.5 Accelerators

As a side effect of "weak scaling," it is usually presumed that decreasing the time needed for the payload contribution affects the efficiency of ANN systems linearly. However, it is not so. As discussed in detail in [33], we also change the non-payload-to-payload ratio that defines system's efficiency. We mention two prominent examples here: using shorter operands (move less data and perform less bit manipulations) and to mimic the operation of a neuron in an entirely different way using quick analog signal processing rather than slow digital calculation.

The so-called *HPL-AI* benchmark used Mixed Precision,[7,8] [34] rather than Double Precision computations. The researchers succeeded to achieve more than three times better performance gain (3.01 for *Summit* and 3.42 for *Fugaku*) that (as correctly stated in the announcement) "*Achieving a 445 petaflops mixed-precision result on HPL (equivalent to our 148.6 petaflops DP result),*" i.e., the peak DP performance did not change. *However, this naming convention suggests the illusion that when using supercomputers for AI tasks and using half-precision, one can expect this payload performance.*

Unfortunately, *this achievement comes from accessing less data in memory and using quicker operations on shorter operands rather than reducing communication intensity.* For AI applications, limitations remain the same as described above; except that when using Mixed Precision, the power efficiency shall be better by a factor of nearly four, compared to the power efficiency measured using double precision operands.[9]

---

[7]Neither of the names are consequent. On the one side, the test itself has not much to do with AI, except that it uses the operand length common in AI tasks. HPL, similarly to AI, is a *workload type*. On the other side, the Mixed Precision is Half Precision: it is natural that for multiplication twice as long operands are used temporarily. It is a different question that the operations are contracted.

[8]Even https://www.top500.org/lists/top500/2020/06/ mismatches *operand length* and *workload*: "In single or further reduced precision, which are often used in machine learning and AI applications, Fugaku's peak performance is over 1000 petaflops (1 exaflops)."

[9] Similarly, exchanging data directly between processing units [35] (without using the global memory) also enhances α (and payload performance) [36], but it represents a (slightly) different computing paradigm.

We expect that when using half-precision (FP16) rather than double precision (FP64) operands in the calculations, four times less data are transferred and manipulated by the system. The measured power consumption data underpin the statement. However, the system's computing performance is only slightly more than three times higher than in the case of using 64-bit (FP64) operands. The non-linearity has its effect even in this simple case (recall that HPL uses minimum communication). In the benchmark, the housekeeping activity (data access, indexing, counting, addressing) also takes time. Concerning the temporal behavior [3] of the operation, in this case, the data transmission time T_t is the same, the data processing time (due to the shorter length of operands) changes, and so the apparent speed changes non-linearly. Even, the measured performance data enabled us to estimate execution time with zero precision[10] (FP0) operands, see [7].

Another plausible assumption is that if we use quick analog signal processing to replace the slow digital calculation, as proposed in [37, 38], our computing system gets proportionally quicker. Presumably, on systems comprising just a few neurons, one can measure a considerable, but less than expected, speedup. The housekeeping becomes more significant than in the case of purely digital processing. That is, in a hypothetical measurement, the speedup would be much less than the ratio of the corresponding analog/digital processing times, even in the case of HPL benchmark. Recall that here the workload is of AI type, with much worse parallelization (and non-linearity). As a consequence, one cannot expect a considerable speedup in large neuromorphic systems. For a detailed discussion of introducing new effect/technologies/materials, see [3].

Besides, adding analog components to a digital processor has its price. Given that a digital processor cannot handle resources outside of its world, one must call the operating system (OS) for help. That help, however, is expensive in terms of execution time. The required context switching takes time in the order of executing 10^4 instructions [39, 40], which dramatically increases the time of housekeeping and the total execution time. This effect makes the system's non-payload-to-payload ratio much worse than it was before introducing that enhancement.

3.6 Timing of Activities

In ANNs, the data transfer time must be considered seriously. In both biological and electronic systems, both the distance between entities of the network and the signal propagation speed are finite. Because of this, in physically large-sized and/or intensively communicating systems, the "idle time" of processors defines the final performance that a parallelized sequential system can achieve. In conventional computing systems also, the "data dependence" limits the achievable parallelism: we must compute data before we can use it as an argument for another computation.

---

[10]Without dedicated measurement, no more accurate estimations are possible.

Although, of course, also in conventional computing, the computed data must be delivered to the place of their second utilization, thanks to "weak scaling" [10], this "communication time" is neglected. For example, scaling of matrix operations and "sparsity," mentioned in [41], work linearly *only* if data transfer time is neglected.

Timing plays an important role in all levels of computing, from gate-level processing to clouds connected to the Internet. In [3], the example describing temporal operation of a one-bit adder provides a nice example, that although the *line-by-line compiling (sequential programming, also called Neumann-style programming [42]) formally introduces only logical dependence, through its technical implementation, it implicitly and inherently introduces a temporal behavior too.*

In neuromorphic computing, including ANNs, the transfer time is a vital part of information processing. A biological brain must deploy a "speed accelerator" to ensure that the control signals arrive at the target destination before the arrival of the controlled messages, despite that the former derived from a distant part of the brain [43]. *This aspect is so vital in biology that the brain deploys many cells with the associated energy investment to keep the communication speed higher for the control signal.* Computer technology cannot speed up the communication selectively, as in biology, and it is also not possible to keep part of the system for a lower speed selectively: the propagation speed of electromagnetic waves is predefined. However, as discussed in [33], *handling data timing adequately is vital, especially for bio-mimicking ANNs.*

3.7 The Layer Structure

The bottom part of Fig. 3 depicts, how ANNs are supposed to operate. The life begins in several input channels (rather than one as in HPL and HPCG cases), which would be advantageous. However, the system must communicate its values to *all* nodes in the top hidden layer: the more input nodes and the more nodes in the hidden layer(s), the many *times* more communication is required for the operation. The same situation also happens when the first hidden layer communicates data to the second one, except that here *the square of the number of nodes* is to be used as a weight factor of communication.

Initially, n input nodes issue messages, each one m messages (queuing#1) to nodes in the first hidden layer, i.e., altogether nm messages. If one uses a commonly used shared bus to transfer messages, these nm messages must be queued (queuing#2). Also, every single node in the hidden layer receives (and processes) m input messages (queuing#3). Between hidden layers, the same queuing is repeated (maybe several times) with mm messages, and finally, km messages are sent to the output nodes. During this process, the system queues messages (at least) three times. Notice that using a single high-speed bus, because of the needed arbitration, drastically increases the transfer time of the individual messages and furthermore changes their timing, see Sect. 3.8.

To make a fair comparison with benchmarks HPL and HPCG, let us assume one input and one output node. In this case, the AI execution time is $O(h \times m^2)$, provided that the AI system has h hidden layers. (Here, we assumed that messaging mechanisms between different layers are independent. It is not so, if they share a global bus.)

For a numerical example, let us assume that in supercomputers, 1M cores are used. In AI networks, 1K nodes are present in the hidden layers, and only one input and output nodes are used. In that case, all execution times are $O(1M)$ (again, the amount of computation is sharply different, so the scaling can be compared, but not the execution times). This communication intensity explains why in Fig. 2 the HPCG "roofline" falls hundreds of times lower than that of the HPL: the increased communication need strongly decreases the system's achievable performance gain.

Notice that the number of *computation* operations increases with m, while the number of *communication* operations with m^2. In other words, the more nodes in the hidden layers, the higher is their communication intensity (communication-to-computation ratio), and because of this, the lower is the efficiency of the system. Recall that since AI nodes perform simple computations, compared to the functionality of supercomputer benchmarks, their communication-to-computation ratio is much higher, making their efficacy even worse. The conclusions are underpinned by experimental research [26]:

- *"strong scaling is stalling after only a few dozen nodes"*;
- *"the scalability stalls when the compute times drop below the communication times, leaving compute units idle, hence becoming a communication bound problem"*;
- *"the network layout has a large impact on the crucial communication/compute ratio: shallow networks with many neurons per layer ... scale worse than deep networks with less neurons."*

3.8 Using High-Speed Bus(es)

As discussed in connection with the reasoning, why the internal temporal ratio between transporting and processing data has significantly changed [3]: Moore's observation is (was) valid for electronic density only, but not valid for connecting technology, such as buses. On the one side, because of the smaller physical size and the quicker clock signal, and on the other side, the unchanged cm-long bus cable, using a serial bus means spending the overwhelming majority of apparent processing time with arbitration (see the temporal diagram and case study in [3]), so using a sequential bus is at least questionable in large-scale systems: *the transfer time is limited by the needed arbitration (increases with the number of neurons!) rather than by the bus speed. "The idea of using the popular shared bus to implement the communication medium is no longer acceptable, mainly due to its high contention"* [44].

The massively "bursty" nature of the data (when imitating biological neuronal network, the different nodes of the layer want to communicate at the same moment) also makes the case harder. Communication circuits receive the task of sending data to N other nodes. What is worse, bus arbitration, addressing, latency, prolong the transfer time (and in his way decrease the efficacy of the system). This type of communicational burst may easily lead to a "communicational collapse" [45], but it may also produce unintentional "neuronal avalanches" [46].

The basic issue is replacing the private communication channel between biological neurons with the mandatory use of some kind of shared media in technological neurons. As discussed in detail in [3], at a large number of communicating units, sharing the medium becomes the dominant contributor to the time consumption of computing. Its effect can be mitigated using different technological implementations, but the basic conclusion persists: *the payload computational efficiency of an ANN is defined by its technical implementation of neuronal communication, and the computing performance of its nodes has only marginal importance.* The lengthy queueing also leads to irrealistic timings: as discussed in Sect. 3.9, some physically (not biologically or logically) delayed signals must be dropped, in order to provide a seemingly acceptable computing performance. Another (wrong) approach to solving the same problem is introducing conditional computation [47], as discussed in Sect. 3.12. The origin of the issue is that using a shared medium makes the temporal behavior of the computing system much more emphasized and that *the temporal behavior cannot be compensated using methods developed having a timeless behavior in mind.*

3.9 The "Quantal Nature of Computing Time"

One of the famous cases demonstrating existence and competition of those limitations in the fields of AI is the research published in [31]. The systems used in the study were a HW simulator [48] explicitly designed to simulate 10^9 neurons (10^6 cores and 10^3 neurons per core) and many-thread simulation running on a supercomputer [49] able to simulate $2 \cdot 10^8$ neurons (the authors mention $2 \cdot 10^3$ neurons per core and supercomputers having 10^5 cores), respectively. The experience, however, showed [31] that scaling stalled at $8 \cdot 10^4$ neurons, i.e., about four orders of magnitude less than expected. *They experienced stalling about the same number of neurons, for both the HW and the SW simulators.*

Given that supercomputers have a performance limit [7], one can comprehend the former experience: the brain simulation needs heavy communication (the authors estimated that $\approx 10\%$ of the execution time was spent with non-payload activity), which decreases sharply their achievable performance, so their system reached the maximum payload performance that their $(1 - \alpha)$ enables: the sequential portion was too high. But why the purpose-built brain simulator cannot reach its maximum expected performance? And, is it just an accident that they both stalled at the same value, or some other limiting factor came into play? The paper [50] gives the detailed explanation.

The short reply is that digital systems, including brain simulators, have a central clock signal that represents an inherent performance limit: no action in the system can happen in a shorter time. The total time divided by the length of the clock period defines maximum performance gain [7] of a system. If the length of the clock period is the commonly used 1 ns, and measurement time (in the case of supercomputers) is in the order of several hours, clocking does not mean a limitation.

Computational time and biological time are not only not equal but also not proportional. To synchronize the neurons periodically, a "time grid," commonly with 1 ms integration time, was introduced. The systems use this grid time to put the free-running artificial neurons back to the biological time scale, i.e., they act as a clock signal: simulation of the next computation step can only start when this clock signal arrives. This action is analogous with introducing a clock signal for executing machine instructions: the processor, even when it is idle, cannot begin the execution of its next machine instruction until this clock signal arrives. That is, *in this case, the clock signal is* 10^6 *times longer than the clock signal of the processor.* Just because neurons must work on the same (biological) time scale, when using this method of synchronization, the (commonly used) 1 ms "grid time" has a noticeable effect on the payload performance.[11]

The brain simulation measurement [31] enables us to guess the efficacy of ANNs. Given that using more cores only increased the *nominal* performance (and, correspondingly, its power consumption) of their system, the authors decided to use only a small fragment of their resources, only 1% of the cores available in the HW simulator. In this way, we can place the efficiency of brain simulation to the scale of supercomputer benchmarking, see Fig. 4. Under those circumstances, as witnessed by Fig. 4, a performance gain of about 10^3 was guessed for brain simulation. Notice that the large-scale supercomputers use about 10% of their cores in HPCG measurement, see also Fig. 1. The difference of the efficiency values of HPL, HPCG, and Brain comes from the different workloads, which is the reason of the issue. For ANNs, the efficiency can be similar to that of brain simulation, somewhat above the performance gain of HPCG, because the measurement time is shorter than that of HPCG on supercomputers, so the corresponding inherent non-parallelizable portion is higher.

Recall also the "communicational collapse" from the previous section: even if communication packages are randomized in time, it represents a huge peak traffic, mainly if a single global (although high speed) bus is used. This effect is so strong in large systems that emergency measures must have been introduced, see Sect. 3.12. In smaller ANNs, it was found [31] that *only a few dozens of thousands of neurons*

---

[11]This periodic synchronization shall be a limiting factor in large-scale utilization of processor-based artificial neural chips [51, 52], although thanks to their ca. thousand times higher "single-processor performance," only when approaching the computing capacity of (part of) the brain, or when the simulation turns to be communication bound.

can be simulated on processor-based brain simulators. This experience includes both many-thread software simulators and purpose-built brain simulator.[12] Recall also from [26] that *"strong scaling is stalling after only a few dozen nodes."* For a discussion on the effect of serial bus in ANNs, see [3].

3.10 Rooflines of ANNs

As all technical implementations, computing also has technological limitations. The "roofline" model [24] successfully describes that until some needed resource exceeds its technical limitation, utilization of that resource shows a simple linear dependency. Exceeding that resource is not possible: usage of the resource is stalling at the maximum possible level; the two lines form a "roofline." In a complex system, such as a computing system, the computing process uses different resources, and under different conditions, various resources may dominate in defining the "roofline(s) of computing process," see Fig. 4. An example is running benchmarks HPL and HPCG: as discussed in [7], either computing or interconnection dominates the payload performance.

As Sect. 3.9 discusses, in some cases, a third competitor can also appear on the scene, and even it can play a major role. That is, it is not easy at all to describe an ANN system in terms of the "roofline" [24] model: depending on the actual conditions, the dominant player (the one that defines the top level of the roofline) may change. Anyhow, it is sure that the contribution of the component, representing the lowest roofline, shall dominate. Still, the competition of components may result in unexpected issues (for an example, see how computation and interconnection changed their dominating rule, in [7]). Because of this, Fig. 4 has limited validity. It provides, however, a feeling that (1) for all workflow types, a performance plateau exists, and already reached, (2) what value of payload performance gain value can be achieved for different workloads, and (3) where the payload efficiency of the particular kinds of ANNs, brain simulation on supercomputers, is located compared to that of the standard benchmarks (a reasoned guess).

3.11 Role of Parameters of Computing Components

As discussed in detail in [7], the different components of computing systems mutually block each other's operation. Datasheet parameters of components represent a hard limit, valid for ideal, stand-alone measurements, where the utilization is 100%. When they must cooperate with other components, the way as they cooperate (aka the workload of the system) defines their soft limit (degrades utilization of

---

[12]Despite this, Spinnaker2, this time with 10M processors, is under construction [53].

units): until its operand(s) delivered, computing cannot start; until computed, result transmission cannot start, especially when several computing units are competing for the shared medium. This is the reason why ANNs represent a very specific workload,[13] where weaknesses of principles of computing systems are even more emphasized.

3.12 Training ANNs

One of the most shocking features of ANNs is their weeks-long training time, even for (compared to the functionality of brain) simple tasks. The mathematical methods, of course, do not comprise time-dependence, the technical implementation of ANNs, however, does: as their time-dependence is discussed in detail in [3], the delivery times of new neuronal outputs (which serve as new neuronal inputs at the same time) are only loosely coupled: the assumption that producing an output means at the same time producing an input for some other neuron works only in the timeless "classic computing" (and in biology using parallel axons), discussing the temporal behavior of the serial bus in [3].

To comprehend what change of considering temporal behavior means, consider the temporal diagram of a 1-bit adder in [3]. When using adders, we have a fixed time when we read out the result. We are not interested in "glitches," so we set a maximum time until all bits relaxed, and (at the price of losing some performance) we will receive the final result only; the adder is synchronized.

The case of ANNs, however, is different: in the adder, outputs of bit n are the input of bit $n + 1$ (there is no feedback), and the bits are wired directly. In ANNs, the signals are delivered via a bus, and the interconnection type and sequence depend on many factors (ranging from the type of task to the actual inputs). During training ANNs, their feedback complicates the case. The only fixed thing in timing is that the neuronal input arrives inevitably only after that a partner produced it. The time ordering of delivered events, however, is not sure: it depends on technical parameters of delivery, rather than on the logic that generates them. Time stamping cannot help much. There are two bad choices. Option one is that neurons should have a (biological) sending time-ordered input queue and begin processing only when all partner neurons have sent their message. That needs a synchron signal and leads to severe performance loss (in parallel with the one-bit adder). Option two is that they have a (physical) arrival-time ordered queue, and they are processing the messages as soon as they arrive. This enables to give feedback to a neuron that fired later (according to its time stamp) and set a new neuronal variable state, which is a "future state" when processing a message received physically later, but with a time stamp referring to a biologically earlier time. A third, maybe better,

---

[13] https://www.nextplatform.com/2019/10/30/cray-revamps-clusterstor-for-the-exascale-era/ : *artificial intelligence, ... it is the most disruptive workload from an I/O pattern perspective.*

option would be to maintain a biological-time ordered queue, and either in some time slots (much shorter than the commonly used "grid time") send out output and feedback or individually process the received events and to send back feedback and output immediately. In both cases it, is worth to consider if their effect is significant (exceeds some tolerance level compared to a last state) and mitigate the need for communication also in this way.

During training, we start showing an input, and the system begins to work, using the synaptic weights valid before showing that input. Those weights may be randomized or maybe that they correspond to the previous input data. The signals that the system sends are correct, but a receiver does not know the future: it processes a signal only after it was physically delivered,[14] meaning that it (and its dependents) may start to adjust their weights to a state that is still undefined. In [3], the first AND gate has quite a short indefinite time, but the OR has a long one.

In biology, spiking is also a "look at me" signal: the feedback shall be sent to *that* neuron, reflecting the change its output caused. Without this, neurons receive feedback about "the effect of all fellow neurons, including me." Receiving a spike defines the time of beginning of the validity of the signal, and "leaking" also defines their "expiration time." When using spiking networks, their temporal behavior is vital.

Neuronal operations have undefined states and weights. Essentially, their operation is an iteration, where *the actors mostly use mostly wrong input signals and surely adjust their weights to false signals initially and with significant time delay at later times.* If we are lucky (and consider that we are working with unstable states in the case of more complex systems), the system will converge, but painfully slowly, or not at all. *Not considering temporal behavior leads to painfully slow and doubtful convergence.*

Computing neuronal results faster, with the goal of providing feedback faster, cannot help much, if at all. Delivering feedback information needs also time and uses the same shared medium, with all its disadvantages. In biology, the "computing time" and the "communication time" are in the same order of magnitude. In computing, the communication time is very much longer than computation, that is, *the received feedback refers to a time* (and the related state variables) *that was valid a very long time ago.*

In excessive systems, with the goal to provide seemingly higher performance, some result/feedback events must be dropped because of long queueing. The logical dependence, that the feedback is computed from the results of the neuron that receives the feedback, is converted to time dependence [3] by the physical implementation of the computing system. Because of this time sequence, the feedback messages will arrive to the neuron at a later physical time (even if at the same biological time, according to their time stamp they carry), so they stand at the end of the queue. Because of this, it is highly probable that they "*are dropped if the receiving process is busy over several delivery cycles*" [31]. In vast systems,

---

[14]Even if the message envelope contains a time stamp.

the feedback in the learning process includes some potential error sources. It not only involves results based on undefined inputs, but also the calculated and (maybe correct) feedback may be neglected.

A nice "experimental proof" of the statements above is provided in [47], with the words of that paper: *"Yet the task of training such networks remains a challenging optimization problem. Several related problems arise: very long training time (several weeks on modern computers, for some problems), potential for over-fitting (whereby the learned function is too specific to the training data and generalizes poorly to unseen data), and more technically, the vanishing gradient problem." "The immediate effect of activating fewer units is that propagating information through the network will be faster, both at training as well as at test time."* This also means that the computed feedback, based maybe on undefined inputs, reaches the neurons in the previous layer faster. A natural consequence is that: *"As λ_s increases, the running time decreases, but so does performance."* Similarly, introducing the spatio-temporal behavior of ANNs, even in its simple form, using separated (i.e., not connected in the way proposed in [3]) time and space contributions to describe them, significantly improved the efficacy of video analysis in [54].

The role of time (mismatching) is confirmed directly, via making investigations in the time domain. *"The CNN models are more sensitive to low-frequency channels than high-frequency channels"* [55]: the feedback can follow the slow changes with less difficulty compared to the faster changes.

4 Summary

The operating characteristics of ANNs are practically not known, mainly because of their mostly proprietary design/documentation. The existing theoretical predictions [27] and measured results[26] show good agreement, but dedicated measurements using well-documented benchmarks and a variety of well-documented architectures are needed. The low efficacy forces us to change our design methods. On the one side, it requires a careful design method when using existing components (i.e., to select the "least wrong" configuration, millions of devices shall work with low energy and computational efficacy!). On the other side, it urges working out a different computing paradigm [6] (and architecture [56] based on it).

In his famous "First Draft" [57], von Neumann formulated: *"6.3 At this point the following observation is necessary. In the human nervous system the conduction times along the lines (axons) can be longer than the synaptic delays, hence our above procedure of neglecting them aside of τ [the processing time] would be **unsound**. In the actually intended vacuum tube interpretation, however, this procedure is justified: τ is to be about a microsecond, an electromagnetic impulse travels in this time 300 meters, and as the lines are likely to be short compared to this, the conduction times may indeed by neglected. (**It would take an ultra high frequency device – $\approx 10^{-8}$ seconds or less – to vitiate this argument.**)"* That is, (at least) *since the processor frequency exceeded 0.1 GHz, it is surely unsound to use* *von Neumann's computing paradigm in its unchanged form, neglecting transmission time.* The computing paradigm, proposed by von Neumann, is surely not valid for the today's technology. According to von Neumann, it is doubly *unsound* if one attempts to mimic neural operation based on a paradigm that is *unsound* for that goal, on a technological base (other than vacuum tubes) for which the paradigm is *unsound*.

The Gordon Bell Prize jury noticed [58] that *"Surprisingly, there have been no brain inspired massively parallel specialized computers [among the winners]."* It was, however, predicted [5]: *"scaling thus put larger machines [the brain inspired computers built from components designed for Single Processor Approach (SPA) computers] at an inherent disadvantage,"* especially, at AI workloads. All the reasons listed above (essentially, *neglecting the temporal behavior of computing components and methods*) resulted finally in that *"Core progress in AI has stalled in some fields"* [59].

Acknowledgments Projects no. 136496 has been implemented with the support provided from the National Research, Development and Innovation Fund of Hungary, financed under the K funding scheme.

References

1. US National Research Council. The Future of Computing Performance: Game Over or Next Level? (2011). http://science.energy.gov/~/media/ascr/ascac/pdf/meetings/mar11/Yelick.pdf
2. I. Markov, Limits on fundamental limits to computation. Nature **512**(7513), 147–154 (2014)
3. J. Végh, Introducing temporal behavior to computing science, in *2020 CSCE, Fundamentals of Computing Science* (IEEE, Piscataway, 2020), pp. Accepted FCS2930, in print. https://arxiv.org/abs/2006.01128
4. K. Asanovic, R. Bodik, J. Demmel, T. Keaveny, K. Keutzer, J. Kubiatowicz, N. Morgan, D. Patterson, K. Sen, J. Wawrzynek, D. Wessel, K. Yelick, A view of the parallel computing landscape. Comm. ACM **52**(10), 56–67 (2009)
5. J.P. Singh, J.L. Hennessy, A. Gupta, Scaling parallel programs for multiprocessors: methodology and examples. Computer **26**(7), 42–50 (1993)
6. J. Végh, A. Tisan, The need for modern computing paradigm: Science applied to computing, in *Computational Science and Computational Intelligence CSCI The 25th Int'l Conference on Parallel and Distributed Processing Techniques and Applications* (IEEE, Piscataway, 2019), pp. 1523–1532. http://arxiv.org/abs/1908.02651
7. J. Végh, Finally, how many efficiencies the supercomputers have? J. Supercomput. **76**, 9430–9455 (2020). https://doi.org/10.1007%2Fs11227-020-03210-4
8. G.M. Amdahl, Validity of the single processor approach to achieving large-scale computing capabilities, in *AFIPS Conference Proceedings*, vol. 30 (1967), pp. 483–485
9. J. Végh, Re-evaluating scaling methods for distributed parallel systems (2020). https://arxiv.org/abs/2002.08316
10. J.L. Gustafson, Reevaluating Amdahl's law. Commun. ACM **31**(5), 532–533 (1988)
11. TOP500.org, The top 500 supercomputers (2019). https://www.top500.org/
12. J. Végh, P. Molnár, How to measure perfectness of parallelization in hardware/software systems, in *18th International Carpathian Control Conference ICCC* (2017), pp. 394–399
13. Y. Shi, Reevaluating Amdahl's Law and Gustafson's Law (1996). https://www.researchgate.net/publication/228367369_Reevaluating_Amdahl's_law_and_Gustafson's_law

14. S. Krishnaprasad, Uses and Abuses of Amdahl's law. J. Comput. Sci. Coll. **17**(2), 288–293 (2001)
15. V. Weaver, D. Terpstra, S. Moore, Non-determinism and overcount on modern hardware performance counter implementations, in *2013 IEEE International Symposium on Performance Analysis of Systems and Software (ISPASS)* (2013), pp. 215–224
16. IEEE Spectrum, Two Different Top500 Supercomputing Benchmarks Show Two Different Top Supercomputers (2017). https://spectrum.ieee.org/tech-talk/computing/hardware/two-different-top500-supercomputing-benchmarks-show\discretionary-two-different-top-supercomputers
17. A.H. Karp, H.P. Flatt, Measuring parallel processor performance. Commun. ACM **33**(5), 539–543 (1990)
18. M. Feldman, Exascale Is Not Your Grandfather's HPC (2019). https://www.nextplatform.com/2019/10/22/exascale-is-not-your-grandfathers-hpc/
19. US Government NSA and DOE, A Report from the NSA-DOE Technical Meeting on High Performance Computing (2016). https://www.nitrd.gov/nitrdgroups/images/b/b4/NSA_DOE_HPC_TechMeetingReport.pdf.
20. R.F. Service, Design for U.S. exascale computer takes shape. Science **359**, 617–618 (2018)
21. European Commission, Implementation of the Action Plan for the European High-Performance Computing strategy (2016). http://ec.europa.eu/newsroom/dae/document.cfm?doc_id=15269
22. Extremtech, Japan Tests Silicon for Exascale Computing in 2021 (2018). https://www.extremetech.com/computing/272558-japan-tests-silicon-for-exascale-computing-in-2021
23. X.-k Liao, et al., Moving from exascale to zettascale computing: challenges and techniques. Front. Inf. Technol. Electron. Eng. **19**(10), 1236–1244 (2018). https://doi.org/10.1631/FITEE.1800494
24. S. Williams, A. Waterman, D. Patterson, Roofline: an insightful visual performance model for multicore architectures. Commun. ACM **52**(4), 65–76 (2009)
25. www.top500.org, Intel dumps knights hill, future of xeon phi product line uncertain (2017). https://www.top500.org/news/intel-dumps-knights-hill-future-of-xeon-phi-product-line-uncertain/
26. J. Keuper, F.-J. Preundt, Distributed training of deep neural networks: Theoretical and practical limits of parallel scalability, in *2nd Workshop on Machine Learning in HPC Environments (MLHPC)* (IEEE, Piscataway, 2016), pp. 1469–1476. https://www.researchgate.net/publication/308457837
27. J. Végh, *How deep Machine Learning can be*, ser. A Closer Look at Convolutional Neural Networks. Nova, (In press, 2020), pp. 141–169. https://arxiv.org/abs/2005.00872
28. P.W. Anderson, More is different. Science (177), 393–396 (1972)
29. H. Simon, Why we need Exascale and why we won't get there by 2020, in *Exascale Radioastronomy Meeting*, ser. AASCTS2, (2014). https://www.researchgate.net/publication/261879110_Why_we_need_Exascale_and_why_we_won't_get_there_by_2020
30. T. Ippen, J.M. Eppler, H.E. Plesser, M. Diesmann, Constructing neuronal network models in massively parallel environments. Front. Neuroinf. **11**, 30 (2017)
31. S.J. van Albada, A.G. Rowley, J. Senk, M. Hopkins, M. Schmidt, A.B. Stokes, D.R. Lester, M. Diesmann, S.B. Furber, Performance comparison of the digital neuromorphic hardware SpiNNaker and the neural network simulation software NEST for a full-scale cortical microcircuit model. Front. Neurosci. **12**, 291 (2018)
32. J. Végh, J. Vásárhelyi, D. Drótos, The performance wall of large parallel computing systems, in *Lecture Notes in Networks and Systems*, vol. 68 (Springer, Berlin, 2019), pp. 224–237. https://link.springer.com/chapter/10.1007%2F978-3-030-12450-2_21
33. J. Végh, A.J. Berki, Do we know the operating principles of our computers better than those of our brain? Neurocomputing (2020). https://arxiv.org/abs/2005.05061
34. A. Haidar, S. Tomov, J. Dongarra, N.J. Higham, Harnessing GPU Tensor Cores for Fast FP16 Arithmetic to speed up mixed-precision iterative refinement solvers, in *Proceedings of the International Conference for High Performance Computing, Networking, Storage, and Analysis*, ser. SC '18 (IEEE Press, Piscataway, 2018), pp. 47:1–47:11

35. F. Zheng, H.-L. Li, H. Lv, F. Guo, X.-H. Xu, X.-H. Xie, Cooperative computing techniques for a deeply fused and heterogeneous many-core processor architecture. J. Comput. Sci. Technol. **30**(1), 145–162 (2015)
36. Y. Ao, C. Yang, F. Liu, W. Yin, L. Jiang, Q. Sun, Performance optimization of the HPCG benchmark on the sunway taihulight supercomputer. ACM Trans. Archit. Code Optim. **15**(1), 11:1–11:20 (2018)
37. E. Chicca, G. Indiveri, A recipe for creating ideal hybrid memristive-CMOS neuromorphic processing systems. Appl. Phys. Lett. **116**(12), 120501 (2020). https://doi.org/10.1063/1.5142089
38. Building brain-inspired computing. Nature Commun. **10**(12), 4838 (2019). https://doi.org/10.1038/s41467-019-12521-x
39. F.M. David, J.C. Carlyle, R.H. Campbell, Context switch overheads for linux on ARM platforms, in *Proceedings of the 2007 Workshop on Experimental Computer Science*, ser. ExpCS '07 (ACM, New York, 2007). http://doi.acm.org/10.1145/1281700.1281703
40. D. Tsafrir, The context-switch overhead inflicted by hardware interrupts (and the enigma of do-nothing loops), in *Proceedings of the 2007 Workshop on Experimental Computer Science*, ser. ExpCS '07 (ACM, New York, 2007), pp. 3–3
41. J.D. Kendall, S. Kumar, The building blocks of a brain-inspired computer. Appl. Phys. Rev. **7**, 011305 (2020)
42. J. Backus, Can programming languages be liberated from the von neumann style? A functional style and its algebra of programs. Commun. ACM **21**, 613–641 (1978)
43. G. Buzsáki, X.-J. Wang, Mechanisms of gamma oscillations. Ann. Rev. Neurosci. **3**(4), 19:1–19:29 (2012)
44. L. de Macedo Mourelle, N. Nedjah, F.G. Pessanha, Chapter 5: Interprocess communication via crossbar for shared memory systems-on-chip, in *Reconfigurable and Adaptive Computing: Theory and Applications* (CRC Press, Boca Raton, 2016)
45. S. Moradi, R. Manohar, The impact of on-chip communication on memory technologies for neuromorphic systems. J. Phys. D Appl. Phys. **52**(1), 014003 (2018)
46. J.M. Beggs, D. Plenz, Neuronal avalanches in neocortical circuits. J. Neurosci. **23**(35), 11167–11177 (2003). https://www.jneurosci.org/content/23/35/11167
47. E. Bengio, P.-L. Bacon, J. Pineau, D. Precu, Conditional computation in neural networks for faster models, in *ICLR'16* (2016). https://arxiv.org/pdf/1511.06297
48. S.B. Furber, D.R. Lester, L.A. Plana, J.D. Garside, E. Painkras, S. Temple, A.D. Brown, Overview of the SpiNNaker system architecture. IEEE Trans. Comput. **62**(12), 2454–2467 (2013)
49. S. Kunkel, M. Schmidt, J.M. Eppler, H.E. Plesser, G. Masumoto, J. Igarashi, S. Ishii, T. Fukai, A. Morrison, M. Diesmann, M. Helias, Spiking network simulation code for petascale computers. Front. Neuroinf. **8**, 78 (2014)
50. J. Végh, How Amdahl's law limits performance of large artificial neural networks. Brain Inf. **6**, 1–11 (2019). https://braininformatics.springeropen.com/articles/10.1186/s40708-019-0097-2/metrics
51. M. Davies, et al., Loihi: a neuromorphic manycore processor with on-chip learning. IEEE Micro **38**, 82–99 (2018)
52. J. Sawada, et al., TrueNorth ecosystem for brain-inspired computing: scalable systems, software, and applications, in *SC '16: Proceedings of the International Conference for High Performance Computing, Networking, Storage and Analysis* (2016), pp. 130–141
53. C. Liu, G. Bellec, B. Vogginger, D. Kappel, J. Partzsch, F. Neumärker, S. Höppner, W. Maass, S.B. Furber, R. Legenstein, C.G. Mayr, Memory-efficient deep learning on a SpiNNaker 2 prototype. Front. Neurosci. **12**, 840 (2018). https://www.frontiersin.org/article/10.3389/fnins.2018.00840
54. S. Xie, C. Sun, J. Huang, Z. Tu, K. Murphy, Rethinking spatiotemporal feature learning: Speed-accuracy trade-offs in video classification, in *Computer Vision – ECCV 2018*, ed. by V. Ferrari, M. Hebert, C. Sminchisescu, Y. Weiss (Springer International Publishing, Cham, 2018), pp. 318–335

55. K. Xu, M. Qin, F. Sun, Y. Wang, Y.-K. Chen, F. Ren, Learning in the frequency domain, in *Proceedings of the IEEE/CVF Conference on Computer Vision and Pattern Recognition* (2020)
56. J. Végh, How to extend the single-processor paradigm to the explicitly many-processor approach, in *2020 CSCE, Fundamentals of Computing Science* (IEEE, Piscataway, 2020), pp. Accepted FCS2243, in print. https://arxiv.org/abs/2006.00532
57. J. von Neumann, First draft of a report on the EDVAC (1945). https://web.archive.org/web/20130314123032/http://qss.stanford.edu/~godfrey/vonNeumann/vnedvac.pdf
58. G. Bell, D.H. Bailey, J. Dongarra, A.H. Karp, K. Walsh, A look back on 30 years of the Gordon Bell prize. Int. J. High Perform. Comput. Appl. **31**(6), 469–484 (2017). https://doi.org/10.1177/1094342017738610
59. M. Hutson, Core progress in AI has stalled in some fields. Science **368**, 6494/927 (2020)

Growing Artificial Neural Networks

John Mixter and Ali Akoglu

1 Introduction

The low size, weight, and power (SWaP) requirements in embedded systems restrict the amount of memory and processing power available to execute a neural network. The state-of-the-art neural networks tend to be quite large [1, 2], and fitting them into severely constrained hardware has proven to be difficult [3–5]. The challenge of fitting a neural network into low SWaP hardware can be explained by examining neural network requirements—memory and computations. A connection between two layers represents an input multiplied by a weight. The products of these calculations are then summed together to produce an input for the next layer. The weights, inputs, and outputs require memory. The larger the neural network, the more resources needed to support its execution. This poses as the barrier for creating neural networks that can be trained and executed in embedded systems. It has been known for decades that neural networks we design tend to be over-built [6, 7]. Studies have shown that in some cases more than 97% of the weights and connections of a trained neural network can be eliminated without significant accuracy loss [8]. The process of eliminating unnecessary weights and connections is known as pruning. Because pruning can significantly reduce the size of a neural network, it is possible to execute inference in low SWaP hardware. But, this does not enable training in hardware because a full-sized network must fit in hardware *before* it can be pruned. In that case, growing a network from a small initial network would be the desired approach when restricted to low SWaP hardware [9, 10]. We propose an algorithm, Artificial Neurogenesis (ANG), that allows growing small

---

J. Mixter (✉) · A. Akoglu
University of Arizona, Tucson, AZ, USA
e-mail: jmixter6011@email.arizona.edu; akoglu@email.arizona.edu

© Springer Nature Switzerland AG 2021
H. R. Arabnia et al. (eds.), *Advances in Artificial Intelligence and Applied Cognitive Computing*, Transactions on Computational Science and Computational Intelligence, https://doi.org/10.1007/978-3-030-70296-0_31

and accurate neural networks that are small enough to be both trained *and* executed on-chip by determining critical connections between layers.

This chapter is organized as follows. In Sect. 2, we discuss several methods for neural network size reduction and describe our baseline neural network. In Sect. 3, we introduce our method for growing a neural network as an alternative and present the ANG algorithm. We conduct parameter sweeps and experimentally determine the final architecture in Sect. 4 followed by a detailed comparison with respect to the state-of-the-art pruning methods in Sect. 5. Finally, we present our conclusions and future work in Sect. 6.

2 Neural Network Pruning

Pruning- and growing-based methods are fundamentally different in terms of their objectives. Growing-based methods focus on creating optimal neural network architectures in terms of accuracy by iterative processes of adding entire nodes or layers and noting the results. They are not concerned about whether or not the final architectures are larger or smaller than manually designed architectures. The network grows as long as there is improvement in accuracy regardless of network size [9]. Pruning-based methods on the other hand start with a well-trained network and iteratively identify and eliminate connections that do not impact inference accuracy [6]. Their goal is to reduce the network to the smallest size possible while maintaining the original accuracy.

Similar to the pruning methods, we are concerned about the final size of the neural network. Therefore, in our literature review we focus on results of pruning methods and compare our research against algorithms specifically designed to reduce the size of networks while maintaining good accuracy. For each method, we analyze the relationship between network accuracy and degree of pruning measured in terms of percentage of weights removed by referring to their reported results based on the MNIST data set.

The five prominent pruning algorithms we cover in our literature review for comparison all execute on the MNIST data set. Three of the architectures [8, 11, 12] are based on LeNet-5, part of which we use as a Seed Network. The other two architectures [13, 14] are included for a more comprehensive analysis.

Blundel et al. [13] introduce the *Bayes-by-Backprop* algorithm for learning neural network weight probability distribution and exploring the weight redundancies in the networks. This algorithm is able to prune the weights that have a low signal-to-noise ratio and remove 95% of the weights without significant accuracy reduction. Han et al. [8] propose a three-step process to learn important connections in a neural network where they first train the network to learn which connections are important, then use regularization to increase and remove the number of weights that are near zero, and finally retrain the pruned network to maintain accuracy. This method is able to remove 91.7% of weights and neurons for an accuracy of 99.23%. Srinivas and Babu [12] take a different approach to pruning neural networks. Their

Growing Artificial Neural Networks

Table 1 Baseline neural network with 4 layers

Layer	Filter	Kernel	Stride	Perceptron	Weight
2D Conv.	6	7	2	864	300
3D Conv.	50	7	4	450	14,750
Full				100	45,100
Classifier				10	1010
Total				1424	61,160

algorithm is designed to find sets of weights that are similar. The inputs associated with similar weights are added together, and their sum is multiplied by the single weight value. In the case where there are no equal weights, they find weights that are close in value by calculating their "saliency." Their pruning method substantially reduces the number of weights in the network but suffers from a significant drop in test accuracy compared to [8]. Babaeizadeh et al. [11] propose a method that works on an entire neuron instead of its individual weights. Their approach relies on merging the neurons that demonstrate high neuron activation correlation. The pruning occurs during training and allows neurons that are not fully correlated to be merged and then retrained to compensate for any accuracy loss. Starting with a well trained network, they are able to remove 97.75% of the weights without accuracy loss. Tu et al. [14] propose a deep neural network compression method that starts with removing the unnecessary parameters and then uses Fisher information to further reduce the parameter count. As a last step, they utilize a non-uniform fixed-point quantization to assign more bits to parameters with higher Fisher information estimates. Their research of using information theory for pruning has resulted in reducing the weights by 94.72%.

As a motivation for our network growing approach, we also ran a pruning experiment. Our aim is to first demonstrate that using magnitude pruning method, we can reduce our baseline network to similar size of the more sophisticated pruning methods while achieving competitive accuracy. This will also serve as a comparison basis later when we introduce our growing method. Because both of our pruning and growing methods target the same fully connected layer, as a second aim, we will be in a position to demonstrate that we can grow a network that is smaller than a similar pruned network. The architecture we chose for our baseline used two sequential convolution layers followed by a fully connected layer as shown in Table 1. The network has four layers, 61,160 32-bit weights, and 221,950 connections. The first layer has six two-dimensional convolution filters each with a 7×7 kernel and a stride of two. The second layer is a three-dimensional convolution layer that has 50 filters each with a $7 \times 7 \times 6$ kernel and a stride of four. The third layer is a fully connected layer with 100 perceptrons (100 FC). The fourth and final layer is the classifier, which has one perceptron for each class. Each of the ten perceptrons in the classifier layer is fully connected to the third layer. LeNet-5 [15] uses a similar architecture but with max pooling layers after each convolution. Max pooling layers help to reduce the number of weights but increase the required calculations. After the addition of a fully connected and classification layer, our baseline network has 0.5% more weights than LeNet-5 but requires 35% fewer calculations to perform

Table 2 Comparing pruned networks

Neural Network	Removed	Weights	Δ Accuracy
Babaeizadeh et al. [11] (LeNet)	97.8%	13,579	0.00%
Tu et al. [14] (FC)	94.7%	31,972	−0.92%
Han et al. [8] (LeNet)	91.7%	36,000	0.03%
Blundell et al. [13] (FC)	98.0%	48,000	−0.15%
Srinivas et al. [12] (LeNet)	83.5%	71,000	−0.76%
Pruned Baseline (LeNet)	50.0%	30,468	−0.11%

inference. To calculate the size and number of connections, we use Eqs. 1 and 2 for convolution layers.

$$Weights = (Kernel_{Row} \times Kernel_{Column} + Bias) \times Filter_{Count} \tag{1}$$

$$Connections = (Kernel_{Row} \times Kernel_{Column} + Bias)$$
$$\times \left(\frac{Input\,Window\,Row - Kernel - Stride}{Stride} \right)^2, \tag{2}$$

where $Kernel_{Row}$ is the number of rows and $Kernel_{Column}$ is the number of columns in the convolution filter.

To calculate the size and number of connections for fully connected layers, we use Eqs. 3 and 4.

$$Weights = \#of\,Perceptrons \times (Previous\,Layer_{Output\,Count} + Bias) \tag{3}$$

$$Connections = Weights - \#of\,Perceptrons. \tag{4}$$

Weight magnitude pruning is used to reduce the size of our baseline network. Weights close to zero have a small impact on the output sum. A threshold is used to determine how far away from zero a weight value has to be before it is removed. With this method, we can remove 50% of weights without sacrificing the accuracy. In Table 2, we compare various pruning methods based on the percentage of weights removed and the change in accuracy with respect to the original network before pruning. The method of Babaeizadeh et al. [11] is able to prune down to the smallest network without accuracy loss. Pruning approach of Han et al. [8] has the next most notable performance in terms of network size reduction with a slightly improved accuracy. Our baseline network has the least number of weights pruned but still resulted in the second smallest network and the third smallest loss in accuracy.

In summary, the methods mentioned above prune large a number of connections and reduce the size of networks without sacrificing accuracy. This indicates that many weights in the fully connected layer are unnecessary, which leads us to a question: as we forward-propagate training data through a neural network, can the outputs of layers be analyzed to find only necessary (critical) connections? If this is possible, then the critical question is: *Can we find an optimal architecture by analyzing the output of perceptrons before training?* If so, this would eliminate the need to train and then prune an entire neural network offline before implementing in hardware. We introduce our approach that answers these questions in the next section.

3 Artificial Neurogenesis

3.1 Background

For reference, when explaining connections between layers, we will refer to the layer that produces output values as the *source layer* and the layer receiving values as the *destination layer*. We treat *perceptrons* as atomic units whose inputs are connected to the outputs of perceptron(s) in the *source layer*. Our perceptron is a simple multiply and accumulate engine whose sum is applied to a non-linear function, which is then passed on to perceptron inputs in the next layer. Artificial Neurogenesis (ANG), our method of growing a neural network, is employed where a fully connected layer would normally reside. It grows the neural network by adding perceptrons to a new *destination layer* while making only critical connections from the outputs in the *source layer*.

A *Seed Network* is the starting point for ANG. The last layer of a *Seed Network* is the *source layer* for the ANG algorithm. Because a *Seed Network* determines the minimum size of the neural network, it needs to be as small as possible. When working with two-dimensional input data, convolution layers are a good choice for the *Seed Network*. A convolution layer requires very few weights, and the feature maps they produce are the same size or smaller than their input data. Smaller feature maps require fewer connections to the *destination layer*.

Critical connections between *source* and *destination layers* are determined by analyzing the perceptron outputs of the *source layer* as we forward-propagate training data through the *Seed Network*. The *source layers* produce feature maps whose values change significantly as data from different classes are presented to the *Seed Network* input. We use these feature map differences to determine which *source layer* perceptrons are outputting critical data.

3.2 Artificial Neurogenesis Algorithms

ANG consists of two algorithms designed to find the critical connections between layers. Algorithm 1 searches for the two most extreme members of each class. For every class, *source layer* output averages are calculated as each member is forward-propagated through the *Seed Network*. The members of a class that produce outputs most and least similar to the average outputs are chosen as extreme members. Algorithm 2 uses the extreme members of each class to determine the critical

Algorithm 1: Extreme class member search

Result: Class Members Sorted by Mean Squared Error Between Source Layer
Output and Source Layer Average Output

Create a *Seed Network*;
Attach a Temporary Classifier;
Prime *Seed Network*;
Remove Temporary Classifier;
for *Each Class in Data Set* **do**
    Count ←Number of Class Members;
    for *Each Perceptron in Source Layer* **do**
        Sum[*Perceptron*] ←0;
    end
    for *Each Member in Class* **do**
        Forward-Propagate Class Member;
        for *Each Perceptron in Source Layer* **do**
            Sum[*Perceptron*] += *Perceptron Output*;
        end
    end
    for *Each Perceptron in Source Layer* **do**
        Avg[*Perceptron*] ←Sum[*Perceptron*] ÷ Count;
    end
    for *Each Member in Class* **do**
        Error[*Member*] ←0;
        Forward-Propagate Class Member;
        for *Each Perceptron Output in Source Layer* **do**
            Error[*Member*] += Error(Avg[*Perceptron*] - *Output*);
        end
    end
    Sort Members in Class by Error[*Member*];
end

---

source layer perceptron outputs, which are then connected to the *destination layer* perceptron inputs.

As a preprocessing step, before executing Algorithms 1 and 2, we first build a *Seed Network* to which a temporary classifier is connected. Once the temporary classifier is fully connected, the *Seed Network* is *primed* by training on all available training data. A *priming cycle* occurs when all the training data have been forward- and back-propagated once. After several *priming cycles*, the temporary *Seed Network* classifier is removed, and the outputs of the *source layer* perceptrons are ready to be analyzed by Algorithms 1 and 2.

3.2.1 Algorithm 1—Extreme Member Search

This algorithm is designed to find two members of each class that cause extreme feature maps to be generated at the *source layer* outputs. As shown in Algorithm 1, one class at a time, all the members of the class are forward-propagated through the

Growing Artificial Neural Networks

Algorithm 2: Find and connect critical outputs

Result: New Destination Layer Connected to Source Layer Critical Outputs
for *Each Class in Data Set* **do**
    for *First and Last Members in Class* **do**
        Forward-propagate Class Member;
        Sum $\leftarrow$ 0;
        for *Each Source Layer Perceptron Output* **do**
            Sum += *Output*;
        end
        Average $\leftarrow$ Calculate Output Mean;
        Sum $\leftarrow$ 0;
        for *Each Source Layer Perceptron Output* **do**
            Sum += (Average - $Output)^2$;
        end
        $\sigma \leftarrow$ Calculate Standard Deviation;
        Add Extreme Perceptron to Destination Layer;
        for *Each Source Layer Perceptron Output* **do**
            if *Output is $\pm x\sigma$ from Mean* **then**
                Connect Perceptron Input to *Output*;
            else
                This is Not a Critical Connection;
            endif
        end
    end
end

Algorithm 3: Artificial neurogenesis

Create Seed Network (Table 3);
Prime Seed Network;
Remove Seed Network Classifier;
Add a Destination layer;
Add Classifying layer;
while *Accuracy not Achieved* **do**
    Extreme Member Search (Algorithm 1);
    for *Each Extreme Member* **do**
        Find and connect critical outputs (Algorithm 2);
        Connect Extreme Perceptron output to Classifier;
    end
    Train network using ALL training data;
    Remove found extreme members from training set;
end

Seed Network. After each class member is forward-propagated to the *source layer*, a sum for each perceptron output is calculated. After all of the class members have been forward-propagated, the average output value for each *source layer* perceptron is determined by dividing the perceptron sums by the number of class members. Together, all of the output averages form an average feature map for each class, and each member of the class is forward-propagated through the *Seed Network* again. The feature map created by the *source layer* for each class member is compared to the average feature map, and the mean squared error between the two is calculated. Members of the class are then sorted by their mean squared error. After sorting, the first member of the class has a feature map that is most similar to the average, and the last member has a feature map that is least similar to the average. The first and last in the sorted class list represent the extreme members of that class.

3.2.2 Algorithm 2—Critical Connection Search

In this phase, we grow a *destination* layer by adding and connecting new perceptrons to the *source layer*. We refer to the new perceptrons as *extreme perceptrons*. For every class, we present the first member to the *Seed Network* and forward-propagate. The *source layer* outputs are analyzed to calculate the average and standard deviation (σ) for all perceptron outputs. Perceptrons in the *source layer* whose outputs are $\pm x\sigma$ from the average are critical outputs, where x is a scaling factor. These outputs are connected to the *destination layer's* extreme perceptron. After the extreme perceptron is connected, another extreme perceptron is added to the *destination layer*, and the last member of the class is presented to the *Seed Network* input. The *Seed Network* is forward-propagated and the process is repeated. After two extreme perceptrons for the class are connected, first member of the next class is presented to the *Seed Network* and the whole process is repeated. Algorithm 2 shows the process for finding and connecting critical *source layer* outputs.

Once two extreme perceptrons for each class have been added and connected to the *source layer*, a classifying layer is connected to the new *destination layer*, and the entire network is trained. If the grown network does not meet accuracy goals, the newly grown *destination layer* is augmented with additional extreme perceptrons. The classes that do not perform well are the sources of input data to initiate further ANG cycles. For each class that needs improvement, the second and second-to-last members are used as inputs just like the first and last were originally used.

3.2.3 Algorithm 3—ANG Overview

As shown in Algorithm 3, the ANG is implemented based on the interactions between the extreme member search and critical connection search algorithms. The algorithms are nested in two loops. The outside loop, which executes Algorithm 1, is continued until the required accuracy is met. The inner loop executes Algorithm 2 for each critical member in every class. It is important to note that after the inner loop finishes, a new set of extreme members is found in the outer loop. The extreme members that were just used to find critical connects are removed from consideration in the next iteration. This forces new critical connections to be found.

Growing Artificial Neural Networks

Table 3 Seed network plus classifier

Layer	Filter	Kernel	Stride	Perceptron	Weight
2D Conv.	6	7	2	864	300
3D Conv.	50	7	4	450	14,750
Classifier				10	1010
Total				1324	16,060

4 Artificial Neurogenesis Experiments

In our experiments, we use the MNIST data set that contains 60,000 training and 10,000 testing images. We divided the training data set into 57,000 images for training and 3000 images for validation and network tuning. The testing data held aside and is never used for tuning the networks.

The training of our neural networks is divided into cycles. Like a *priming cycle*, a single training cycle is completed when all of the training data have been forward- and back-propagated once. At the end of every training cycle, we execute a validation inference and then randomize the training data. The accuracy of the validation inference is compared to a current maximum. If the validation inference accuracy is greater than the current maximum, the current maximum is set to the validation accuracy, testing inference is executed, and the testing accuracy is noted.

Because experimental results are affected by the order in which training images are presented to the neural network, we have turned off random seeding so that the software generates identical sets of random images for each test. We do the same for weight initialization. This helps to ensure that any accuracy change is due solely to modifications we make to the network.

In each experiment, training is halted when one of the following *stopping criteria* is met; the number of training cycles reaches 30, the validation accuracy reaches 100%, or the validation accuracy does not increase over 20 consecutive training cycles.

Our ANG experiments require us to choose a *Seed Network*. To ensure that our experiments can be easily compared to our baseline (Table 1), it makes sense to use the same two convolution layers that our baseline network uses. For each experiment, we prepare the *Seed Network* as outlined in Sect. 3.2. First, we build a *Seed Network* (Table 3) to which we connect a temporary classifier. We then randomize all weights and proceed to prime the network. After priming, the *Seed Network* is ready for the experiments. Before comparing networks grown with ANG to our baseline network, we conduct experiments for hyper-parameter selection that we cover in the following subsections.

Table 4 Seed network priming accuracy

Cycle	Train	Validate	Test
0	0.00%	0.00%	9.90%
1	89.87%	96.18%	95.13%
2	95.97%	97.94%	96.93%
3	96.88%	98.22%	97.47%
4	97.31%	98.33%	97.52%
7	97.79%	98.63%	97.86%
12	98.07%	98.83%	98.03%
15	98.23%	98.84%	98.04%
18	**98.32%**	**98.86%**	**98.20%**
30	98.32%	98.86%	98.20%

Best results are shown in bold

4.1 Finding Minimum Priming Cycle Count

We prime the *Seed Network* to move the weights out of a random state. As the accuracy of the *Seed Network* increases, the feature maps created by the *source layer* become more focused, and we are better able to find the critical outputs. As the priming cycles progress, accuracy saturation is expected. Therefore, our aim in this experiment is to determine the minimum number of priming cycles before saturation occurs.

We prime the network allowing the validation inference to dictate when a testing inference is executed. The experiment ends when one of the stopping criteria is met. In Table 4, we show training cycle (57,000 images per cycle), training accuracy, validation accuracy on 3000 images after training, and test accuracy on 10,000 images in the test data set. We observed saturation in *validation accuracy* starting at 18 priming cycles through 30 cycles. Therefore, we will use cycle 18 as the stopping point in the following experiments.

4.2 Finding the Best Scaling Factor x

When the extreme perceptrons are being connected to the *source layer*, the σ value determines how far from the average an output can be and still be considered critical. In this experiment, using the minimum priming cycle count, our aim is to determine the value for the scaling factor x that produces the most accurate network.

For each scaling factor shown in Table 5, we prepare a *Seed Network*, prime it to the minimum priming cycle count, and analyze the *source layer* output to determine σ. We sweep the scaling factor value ranging from 0.1 to 1.5 in increments of 0.1 and apply over the σ value to change the number of critical connections made. After the critical connections are made, a classifier is connected to the new *destination layer* and the entire network is trained. As shown in Table 5, sweeping the scaling factor has the desired effect of changing the number of critical connections. We observe

Growing Artificial Neural Networks

Table 5 Seed network scaling factor sweep vs. number of critical connections and test accuracy when validation accuracy peaks

Scale factor	Critical connection	Test accuracy	Scale factor	Critical connection	Test accuracy
0.1	16,180	98.24%	0.9	20,876	98.38%
0.2	17,063	98.49%	**1.0**	**21,211**	**98.80%**
0.3	17,832	98.36%	1.01	21,239	98.82%
0.4	18,502	98.52%	1.1	21,472	98.62%
0.5	19,133	98.59%	1.2	21,743	98.45%
0.6	19,666	98.60%	1.3	22,009	98.56%
0.7	20,135	98.44%	1.4	22,267	98.57%
0.8	20,542	98.75%	1.5	22,529	98.71%

Best results are shown in bold

Table 6 Priming cycles vs. critical connections at peak validation accuracy (1σ)

Cycle	Critical connect	Test accuracy	Cycle	Critical connect	Test accuracy
0	21,362	98.36%	13	21,100	98.77%
1	21,731	98.28%	14	21,100	98.56%
2	21,504	98.51%	15	21,069	98.72%
3	21,370	98.56%	16	21,069	98.66%
4	21,277	98.57%	17	21,069	98.56%
5	21,277	98.64%	18	20,998	98.51%
6	21,277	98.63%	19	20,998	98.45%
7	21,211	98.44%	20	20,998	98.66%
8	21,211	98.70%	21	20,953	98.65%
9	21,211	98.58%	22	20,953	98.61%
10	21,211	98.64%	23	20,953	98.52%
11	**21,211**	**98.80%**	24	20,953	98.75%
12	21,100	98.69%	25	20,953	98.46%

Best results are shown in bold

that both the test accuracy and the *validation accuracy* peak at 1.01. However, because the difference between this accuracy and the accuracy at a scaling factor of 1.0 is only 0.02%, we choose the scaling factor of 1.0 to eliminate the need for a scaling hyper-parameter.

4.3 Growing the Most Accurate Architecture

We sweep *priming cycles* with scaling factor of 1.0 to find the number of *priming cycles* needed to grow our most accurate network. One at a time, we prepare a *Seed Network* by varying the *priming cycles* from 1 to 25. We add the extreme perceptrons and connect *source layer* outputs $\pm 1.0 \ \sigma$ from the average to the *destination layer*. We train the entire network until one of the stopping criteria is met. In Table 6, we

Table 7 Most accurate network grown using ANG

Layer	Filter	Kernel	Stride	Perceptron	Weight
2D Conv.	6	7	2	864	300
3D Conv.	50	7	4	450	14,750
Full				20	5951
Classifier				10	210
Total				1344	21,211

Table 8 Network size relative to grown ANG network

Neural network	Starting weight	Final weight	Relative size %	Test accuracy
Babaeizadeh et al. [11]	606K	13.6K	−35.9	99.1
Tu et al. [14]	606K	32K	50.9	98.4
Han et al. [8]	431K	36K	69.7	99.2
Blundel et al. [13]	2.4M	48K	126	98.6
Srinivas et al. [12]	431K	71K	235	98.4
ANG Grown	21.2K	21.2K	0.0	98.8

observe that the number of critical connections reduces as we increase the number of *priming cycles*. This effect is not seen during the earlier *priming cycle* sweep experiment presented with Table 4 where the *priming cycles* were held to 18. As we sweep the number of priming cycles, the testing inference accuracy varies between 98.28 and 98.80% with a peak validation accuracy on the eleventh priming cycle.

Based on the sweeping-based experiments presented above, the configuration of the final ANG generated architecture is shown in Table 7. The fully connected layer in this table is a direct result of the best scaling factor of 1σ found in Table 5 and the number of priming cycles of 11 found in Table 6.

5 Artificial Neurogenesis Analysis

The goal of ANG is to grow neural networks that are small enough to fit into low SWaP embedded hardware while still performing as well as full sized networks. ANG starts with a *Seed Network* that has 19,560 weights (Table 3), and then as shown in Table 7, the *Seed Network* is grown to a size of 21,211 weights.

In Table 8, we present our final comparison in terms of number of weights (starting and final), relative size in terms of the percentage of change in number of weights relative to the grown ANG network, and test accuracy. Based on the starting weight sizes, none of the pruning-based methods is as suitable as our ANG grown method for execution in low SWaP hardware. In terms of the final number of weights, the grown ANG network is smaller than all but one network [11]. In terms of accuracy, the grown ANG network is less accurate than only two networks ([11] and [8]). In overall, we believe that unlike other methods, the ANG offers ability

Growing Artificial Neural Networks

Table 9 20 perceptron fully connected network (20 FC)

Layer	Filters	Kernel	Stride	Perceptrons	Weight
2D Conv	6	7	2	864	300
3D Conv	50	7	4	450	14,750
Full				20	9020
Classifier				10	210
Total				1344	24,280

to train and infer completely on-chip with a slight trade-off in accuracy. Another advantage of our method is that it offers faster training and inference time due to much smaller starting network size.

Artificial Neurogenesis (ANG) is the opposite of pruning. ANG *grows* a network from a small seed using only the resources required to achieve a size of 21,211 weights. Pruning methods on the other hand demand for much more resources to store and process a full network of weights ranging from 400,000 [12] to 2.4 million [13] before training and reducing to a scale of 13,579 weights.

The proposed ANG grows a network whose testing accuracy (98.80%) is comparable to the best pruned accuracy (99.05%), only differing by 0.25%. We finally evaluate the efficiency of the ANG method based on its ability to find connections that are more critical than the connections revealed by weight magnitude pruning. The contribution of the proposed algorithm can only be validated if a normal network with same structure cannot reach the same performance. To determine this, we will build a network whose architecture is identical to our ANG network, Tables 7 and 9 train the network until the stopping criteria are met and compare the results. As shown in Table 10, the 20 FC network testing accuracy never reaches the accuracy of 98.80% achieved by our ANG grown network. If we prune the network to nearly the same size as our grown network (bold row, 11.80% removed), the test accuracy is significantly less. More importantly, even with the 20 FC network not pruned (0% removed), the achieved testing accuracy is less than the ANG grown network. If we compare the two networks by their connection counts, the ANG grown network is significantly more accurate. This gives us confidence that the algorithm is performing well and is choosing critical connections.

6 Conclusions and Future Work

We propose a network growing method as an alternative to traditional pruning-based approaches so that we avoid the need for offline training and make training a feasible process under low SWaP requirements. Our algorithm, Artificial Neurogenesis (ANG), grows neural networks from small *Seed Networks* by identifying critical outputs in the *source layer* and connecting them to the *destination layer*. Once a network has been grown and trained, ANG then applies pruning as a final step to further reduce the size of the neural network. Working with the MNIST data set, we

Table 10 Pruned 20 perceptron fully connected network

% removed	Connections	Test accuracy	Error
0%	24,280	98.50%	1.50%
1.77%	23,850	98.50%	1.50%
5.79%	22,874	98.49%	1.51%
6.18%	22,779	98.48%	1.52%
7.39%	22,486	98.46%	1.54%
8.54%	22,207	98.47%	1.53%
11.80%	**21416**	**98.39%**	**1.61%**
15.18%	20,595	98.37%	1.63%
18.24%	19,852	98.26%	1.74%
20.90%	19,205	98.01%	1.99%
23.28%	18,627	97.59%	2.41%
25.25%	18,150	95.97%	4.03%
26.98%	17,729	93.82%	6.18%
28.46%	17,369	91.85%	8.15%
36.00%	15,538	61.10%	38.90%
37.61%	15,148	33.29%	66.71%

applied ANG to grow a neural network that achieves 98.8% inference accuracy on the test data set.

One of the main conclusions that can be reached from this research is that the training data holds information that can be used to determine network architecture prior to training. In this instance, we only targeted the fully connected layer, but ANG can be applied to the connections to the classifying layer. This will be investigated in future research along with analyzing the convolution layers to find critical connections.

The neural network we grew only required 21.2K weights to support this accuracy. To the best of our knowledge, no other method in the literature uses input data to determine an architecture that can be implemented on low SWaP hardware. Because we were able to prune the grown layer, it is reasonable to assume that we have not found the optimal critical connections. Therefore, future research will be directed at further analyzing the *source layer* outputs. One change would be to use the average image data as the target for analysis instead of an actual image close to the average image data. It is our belief that neural networks can be grown to near optimal architectures using input data as a guide. ANG has proven to be a step in that direction.

Acknowledgments This work is partly supported by National Science Foundation research project CNS-1624668 and Raytheon Missile Systems (RMS) under the contract no. 2017-UNI-0008. The content is solely the responsibility of the authors and does not necessarily represent the official views of RMS.

References

1. A. Krizhevsky, I. Sutskever, G.E. Hinton, *26th Annual Conference on Neural Information Processing Systems, Lake Tahoe, Nevada,USA December 3–6, 2012* (2012), pp. 1106–1114
2. K. Simonyan, A. Zisserman, *International Conference on Learning Representations* (2015). http://arxiv.org/abs/1409.1556
3. F.N. Iandola, M.W. Moskewicz, K. Ashraf, S. Han, W.J. Dally, K. Keutzer, *International Conference on Learning Representations* (2017). http://arxiv.org/abs/1602.07360
4. K. Guo, S. Zeng, J. Yu, Y. Wang, H. Yang, ACM Trans. Reconfig. Technol. Syst. **4** (2017). http://arxiv.org/abs/1712.08934
5. C. Teuscher, IEEE Trans. Neural Netw. **18**(5), 1550 (2007). https://doi.org/10.1109/TNN.2007.906886
6. Y. LeCun, J.S. Denker, S.A. Solla, *Advances in Neural Information Processing Systems 2, [NIPS Conference, Denver, Colorado, USA, November 27–30, 1989]* (1989), pp. 598–605. http://papers.nips.cc/paper/250-optimal-brain-damage
7. B. Hassibi, D.G. Stork, G.J. Wolff, *Advances in Neural Information Processing Systems 6, [7th NIPS Conference, Denver, Colorado, USA, 1993]* (1993), pp. 263–270
8. S. Han, J. Pool, J. Tran, W.J. Dally, *Advances in Neural Information Processing Systems* (2015), pp. 1135–1143. http://arxiv.org/abs/1506.02626
9. C. Macleod, G.M. Maxwell, Artif. Intell. Rev. **16**(3), 201–224 (2001)
10. O. Irsoy, E. Alpaydin, CoRR (2018). http://arxiv.org/abs/1804.02491
11. M. Babaeizadeh, P. Smaragdis, R.H. Campbell, *29th Advances in Neural Information Processing Systems (NIPS), Barcelona, Spain, 2016* (2016). http://arxiv.org/abs/1611.06211
12. S. Srinivas, R.V. Babu, *Proceedings of the British Machine Vision Conference (BMVC)* (2015). http://arxiv.org/abs/1507.06149
13. C. Blundell, J. Cornebise, K. Kavukcuoglu, D. Wierstra, *International Conference on Learning Representations* (2015). http://arxiv.org/abs/1505.05424
14. M. Tu, V. Berisha, Y. Cao, J. Seo, *IEEE Computer Society Annual Symposium on VLSI, ISVLSI 2016, Pittsburgh, PA, USA, July 11–13, 2016* (2016), pp. 93–98. https://doi.org/10.1109/ISVLSI.2016.117
15. Y. LeCun, L. Bottou, Y. Bengio, P. Haffner, Proc. IEEE **86**(11), 2278 (1998)

Neural-Based Adversarial Encryption of Images in ECB Mode with 16-Bit Blocks

Pablo Rivas ⓘ **and Prabuddha Banerjee**

1 Introduction

Recently, there has been interest in encryption and decryption of communications using adversarial neural networks [1]. This is due to adversarial neural networks that can be trained to protect the communication using artificial intelligence. Adversarial neural networks can effectively be used for generating realistic images [5] and solving multiagent problems [3, 9]. Advancing these lines of work, we showed that neural networks can learn to protect their communications in order to satisfy a policy specified in terms of an adversary.

Cryptography concerns broadly with algorithms and protocols that will give certainty that the secrecy and integrity of our information are maintained. Cryptography is the science of protecting our data and communication.

A secured mechanism is where it achieves its goal against all attackers. So these form Turing machines or a cryptographic mechanism is formed as programs. Hence, attackers may be described in those terms as well as the complexity of time and how frequently are they successful in decoding messages. For instance, an encryption algorithm is said to be secure if no attacker can extract information about plaintexts from ciphertexts. Modern cryptography provides rigorous versions of such definitions [4]. For a given problem, neural networks have the ability to selectively explore a solution space. This feature finds a natural niche of application in the field of cryptanalysis.

P. Rivas (✉)
Computer Science, Baylor University, Waco, TX, USA
e-mail: pablo.rivas@marist.edu

P. Banerjee
Computer Science, Marist College, Poughkeepsie, NY, USA
e-mail: Prabuddha.Banerjee1@Marist.edu

© Springer Nature Switzerland AG 2021
H. R. Arabnia et al. (eds.), *Advances in Artificial Intelligence and Applied Cognitive Computing*, Transactions on Computational Science and Computational Intelligence, https://doi.org/10.1007/978-3-030-70296-0_32

Neural networks provide new methods of attacking an encryption algorithm based on the fact that neural networks can reproduce any function; this is considered as a very powerful computational tool since any cryptographic algorithm must have an inverse function.

In the design of the neural network we study in this chapter, adversaries play a very important role. They were originally proposed using what is known as "adversarial examples" that were found using a neural architecture named a *generative adversarial network* (GAN) [5]. In this context, the adversaries are neural networks that sample data from a distribution whose parameters are learned from the data. Furthermore, as per the cryptography definition, practical approaches to training generative adversarial networks do not consider all possible adversaries, rather adversaries relatively small in size that are improved upon by training. We built our work based upon these ideas.

Historically, neural networks have received criticism in the area of cryptography since earlier models were not able to model an XOR operation, which is commonly used in cryptography. However, nowadays neural networks can easily model XOR operations and be taught to protect the secrecy of data from other neural networks; such models can discover and learn different forms of encryption and decryption, without specifically creating an existing algorithm or even knowing the type of data they will encrypt.

In this chapter, we will discuss the implementation of such architecture in the context of image encryption and will comment on the results obtained. This chapter is organized as follows: Sect. 2 discusses the background pertaining to cryptography and adversarial learning along with the methodology used. Section 3 discusses our experiments and the results obtained. Finally, Sect. 4 draws conclusions of this chapter.

2 Background and Methodology

2.1 Ciphertext-Only Attacks

In the basic architecture proposed by Abadi an Andersen [1], there are 3 neural networks of which the first one is Alice who is part of the cryptographic learning model, so is tasked with encrypting binary strings (e.g., of an image) provided a key. Bob, another friendly neural network, is tasked with learning to decrypt Alice's message. Both Alice and Bob are performing symmetric-key encryption, so they have a common key for encrypting their communication; this key needs to be provided by a human prior to the learning process. Lastly, Eve's neural network will be eavesdropping and tries to attack or hack their communication by observing the ciphertext. This is known as a ciphertext-only attack [8]. While this is a neural network-based architecture, the overall encryption scheme is well known in textbooks as the one shown in Fig. 1.

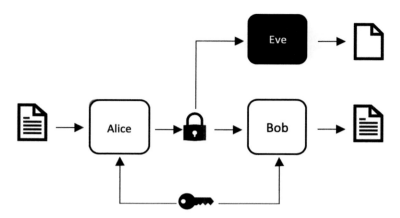

Fig. 1 A traditional encryption scheme with a twist: Alice, Bob, and Eve are neural networks

2.2 Block Cipher Mode: ECB

All block ciphers require specific chunks of data (usually binary) of specific sizes that will be used as input or output. Traditionally, these blocks of data are encrypted and decrypted; however, what happens to those blocks internally on the cipher is unique to each cipher. As long as we have information that can be broken down into binary strings of information (e.g., images), we can use any block cipher.

In this chapter, we are concerned with encrypting images. We used the Python programming language in which we take the image and convert into strings of byte arrays and then the byte arrays are flattened and converted into binary strings of the size of the desired block. In this preliminary research, we limited our study to 16 bits of information per block. The choice of this block size is due to the large training time for larger blocks.

There are several ways or *modes* in which block ciphers operate. Each mode provides additional security. The only mode that does not provide additional security is known as electronic code book (ECB) mode. It is important to remark that in order to study the quality of a cipher we must study it by itself with no additional help from other advanced (and much better) modes, so ECB is used here for functionality and to allow us an objective measure of quality of the cipher.

The encryption task in ECB mode here is given by

$$C_i = E_k(P_i), \quad i = 1, 2, 3, \ldots \qquad (1)$$

where the function $E_k(\cdot)$ is the encryption algorithm operating with secret key k, P_i is the i-th block of plain text, and C_i is the corresponding i-th block of cipher text.

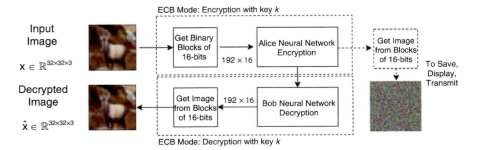

Fig. 2 ECB mode of operation on an image based on key k. The encrypted image, *ciphertext*, can be saved, displayed, transmitted, or further analyzed as needed

During encryption, the key, which is of 16 bits in size, is represented as an array of the following form:

$$\text{key} = [0, 1, 0, 1, 0, 1, 0, 1, 0, 1, 0, 1, 0, 1, 0, 1]$$

Decryption works in similar way but in reverse w.r.t. encryption, i.e., first the cyphertext blocks are passed through the decryption neural network (Bob) and the original plaintext blocks are recovered. The decryption is here governed by

$$P_i = D_k(C_i), \quad i = 1, 2, 3, \ldots \quad (2)$$

where $D_k(\cdot)$ is the decryption algorithm operating with key k, and P_i and C_i are as in Eq. (1).

The reconstruction process of an image in Python follows a similar process as prior to encryption, i.e., in two stages wherein the recovered plaintext is first converted into hexadecimal numbers, then converted to array of bytes, and then these arrays of bytes are thereafter reshaped forming an image. This process is exemplified in Fig. 2.

2.3 Neural Architecture

Figure 3 depicts the neural architecture implemented in Python with Keras on TensorFlow; the original architecture was presented in [1] and is based on generative adversarial networks [5]. The neural architecture has two identical networks, Alice and Bob; however, Bob is a mirror (reverse) of Alice. The Alice network has a combination of one-dimensional convolutional layers and dense layers; there are a total of four convolutional layers, the first three have hyperbolic tangent activations (tanh), and the last convolutional layer has a sigmoid activation as the output. At the

Neural-Based Adversarial Encryption of Images

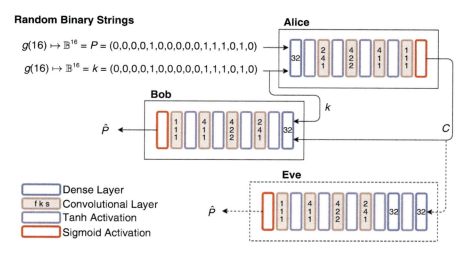

Fig. 3 Neural architecture of the cipher. During training, Alice and Bob are trained using randomly generated keys and plaintext, while Eve attempts to recover the plaintext from the ciphertext and no key

input layer, Alice has a dense layer with a tanh activation. Bob is the exact mirror of Alice, i.e., in opposite order.

The input to Alice has to be two vectors: a binary string denoting the plaintext $P \in \mathbb{B}^{16}$ and a binary string denoting the key $k \in \mathbb{B}^{16}$. In this implementation, we used 16 bits for both. The output of Alice is a vector of codes known as the ciphertext $C \in \mathbb{R}^{16}$.

During training, we randomly generated pairs of (P, k) in mini-batches of size 256, which was determined experimentally to be a nice trade-off between training speed and performance. A standard random number generator function $g(16) \mapsto \mathbb{B}^{16}$ was used to produce binary strings of length 16. We also determined experimentally that 30 epochs were sufficient to train the architecture securely.

Before we discuss the adversarial nature of the training, we now address the design of Eve, the network designed to break the code. Eve's architecture is identical to Bob's, except that it has two stacked dense layers with tanh activations, instead of only one, and it is also followed by the same convolutional layers as Bob. The idea behind this additional dense layer is to give the Eve's network a better chance to retrieve the plaintext from the ciphertext.

As opposed to a GAN, the generator is not part of the architecture in this case, instead we have Alice and Eve competing against each other, where Eve is the adversary. Alice and Bob are not competing, but they resemble an autoencoder architecture in which the input of Alice (i.e., P and k) is encoded into C and Bob is tasked with recovering P from C and k. While this task might be straightforward in a traditional sense, the code that is produced by Alice is affected by the degree to which Eve is able to recover P from C alone. We call Bob's reconstruction $\hat{P}$ and

Eve's reconstruction $\tilde{P}$. During an ideal training session, we want $s(P, \hat{P}) \to 0$ and $s(P, \tilde{P}) \to \infty$, where $s(\cdot)$ is some similarity measure where zero means they are identical.

The specific losses used are the following:

$$L_{\text{Eve}} = \frac{1}{m} \sum_i^m |P_i - \tilde{P}_i| \tag{3}$$

$$L_{\text{Bob}} = \frac{1}{m} \sum_i^m |P_i - \hat{P}_i| \tag{4}$$

$$L_{\text{Alice}-\text{Bob}} = L_{\text{Bob}} + \frac{\left(\frac{|P|}{2} - L_{\text{Eve}}\right)^2}{\left\lfloor \frac{|P|}{2} \right\rfloor^2} \tag{5}$$

where the batch size m is 256 and $|P|$ denotes the size of the plaintext, which is 16 in this case.

Eve's goal is simple: to reconstruct P accurately by minimizing L_{Eve}. Bob simply wants to reconstruct the plaintext perfectly using the secret key k by minimizing L_{Bob}. Alice and Bob want to communicate clearly by making Alice produce a code that Bob can recover using the key while preventing Eve from learning to recover the plaintext without a key.

3   Experiments and Results

During our experiments implementing the architectures shown in Fig. 3, we observed convergence of the network as early as in 3000 training steps (or around 12 epochs), as shown in Fig. 4. The total number of steps shown in the figure comes from the selected batch size, which is 256, and the total number of epochs, which was 30; thus, the horizontal axis is from 1 to 7680. The plot shown in Fig. 4 is produced by overlaying 30 different experiments. A typical single experiment looks like the one shown in Fig. 5.

From Fig. 4 or Fig. 5, we can see three different lines; each line corresponds to the loss functions as the training takes place over time, i.e., L_{Eve}, L_{Bob}, and $L_{\text{Alice}-\text{Bob}}$. As expected, both loses L_{Bob} and $L_{\text{Alice}-\text{Bob}}$ are minimized, while L_{Eve} is maximized. This min–max game is part of the fundamental definition of adversarial learning [6].

During the training of the models, for each training step, both Alice and Bob are trained together so that Alice learns to produce a ciphertext based on a unique key, and Bob learns to retrieve the plaintext from ciphertext and key. Next, Eve is trained on the ciphertext preventing Alice from adapting its parameters. However, note that in training Eve, we are drawing twice as many cyphertext samples as with Bob,

Neural-Based Adversarial Encryption of Images 431

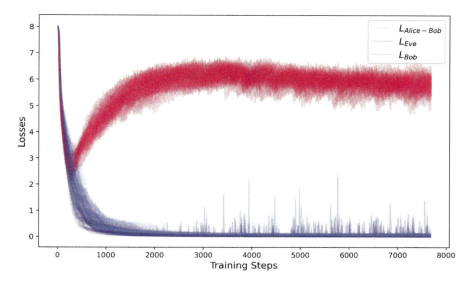

Fig. 4 Loss functions, L_{Eve}, L_{Bob}, and $L_{\text{Alice}-\text{Bob}}$, are minimized over time as training steps are performed. Results of 30 experiments are shown

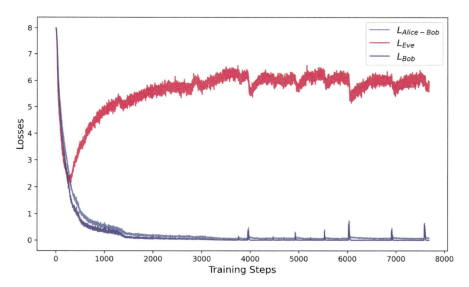

Fig. 5 Loss functions, L_{Eve}, L_{Bob}, and $L_{\text{Alice}-\text{Bob}}$, are shown for a single training experiment

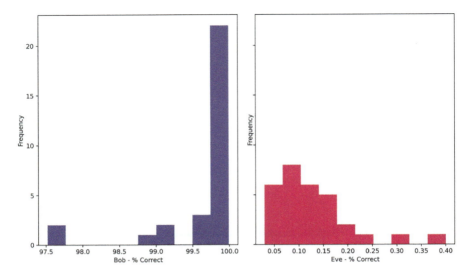

Fig. 6 Histogram that counts the frequency of both Bob and Eve correctly recovering all the bits in the original plaintext. As desired, Eve has a low percentage, while Bob has a very high percentage. These frequency counts are based on 30 independent experiments

hoping to have the best version of Eve in every iteration. The parameter updates are done with a stochastic gradient descent algorithm known as RMS Prop [2, 7], with standard parameters.

In order to measure the quality of the cipher, we produce pairs of 10,000 binary strings and their keys, (P, k), and observe the percentage of plaintext bits recovered by Bob and Eve. We performed this experiment 30 times and produced the histogram counts shown in Fig. 6. We can see that most of the time Bob can reconstruct the plaintext bits beyond a 99.5% of accuracy. On the other hand, Eve can usually recover no more than 0.15% of the bits in the plaintext. This is clearly not a perfect hack of the information encoded.

Beyond the experiments where we have shown that the losses are minimized and the bit strings are recovered correctly to high accuracy and that the best versions of Eve cannot perform successful ciphertext-only attacks, now we want to proceed with our final point that is its applicability in the encryption of images that contain highly correlated information in neighboring pixels. For this approach, we take images and encrypt them with Alice, then we present and observe the encoded version looking for visual clues that might reveal the content of the plaintext (image), and then we use Bob to try to recover the plaintext using the wrong key, to make sure that Bob does not reveal details about the plaintext, and then we recover using the correct key to make sure the image was correctly recovered; the results are shown in Fig. 7.

From Fig. 7, we can see that for the cases in which images are properly encoded (rows a, b, c, and e) the corresponding image when the wrong key is used is also good quality. Here, we refer to good quality and properly encoded as a way to say the

Neural-Based Adversarial Encryption of Images 433

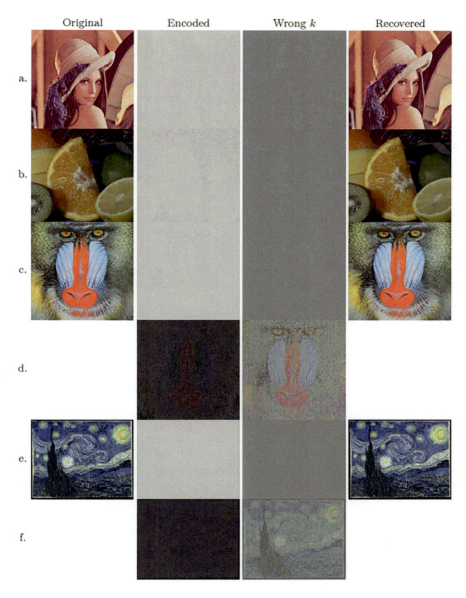

Fig. 7 Test of neural encryption of images. Rows a, b, c, and e are with high-quality ciphers, while rows d and f are with low-quality ciphers.

images resemble random noise although some artifacts related to spatial properties of the original might be visible yet not sufficient to reveal textures, colors, or general shapes. The figure also shows that the recovered image using the correct key leads to perfect reconstruction. However, while the results shown are common, we also take some of the worst models to show that a cipher can reveal information about the original image both in its encoded form, and in while trying to recover using the wrong key. Such results are shown in rows d and f. Note however, that in both cases, with a poor cipher or a good cipher, the wrong key that was used was actually a key that differed in only one bit, e.g., if the key is originally

$$[0, 1, 0, 1, 0, 1, \underline{0}, 1, 0, 1, 0, 1, 0, 1, 0, 1]$$

then, the wrong key used was

$$[0, 1, 0, 1, 0, 1, \underline{1}, 1, 0, 1, 0, 1, 0, 1, 0, 1]$$

thus providing with a more accurate representation of a typical scenario in which a cipher should produce a completely different encoding if only one bit in the key is changed.

4 Conclusions

In this chapter, we studied the applicability of adversarial neural networks in the encryption and decryption of images. The cipher we implemented [1] exhibits high accuracy, fast convergence, and potential for usage in the protection of data communications and other similar applications. Our experiments suggest that, usually, the architecture produces good quality results on images, indicating low visual correlation in neighboring pixels, which validates the high accuracy of the models produced experimentally each time. Furthermore, the proposed experiments validate current studies that show that adversarial training of neural networks has a great potential in the world of cryptography. The idea that one neural network can develop its own way of communications while there is a smart attacker that is listening and learning at the same time is an extraordinary achievement in the field of deep learning.

While the experiments shown in this chapter were limited to only 16 bits for a key and plaintext, further research will be focused on using larger keys and larger blocks to study the impact on the quality of the cipher and its attacker; however, preliminary research shows that increasing the key and block size increases the training time exponentially. Therefore, further research must include a more efficient implementation.

Acknowledgments This work was supported in part by the New York State Cloud Computing and Analytics Center, and by Baylor University's Department of Computer Science.

References

1. M. Abadi, D.G. Andersen, Learning to protect communications with adversarial neural cryptography (2016). Preprint arXiv:1610.06918
2. Y. Dauphin, H. De Vries, Y. Bengio, Equilibrated adaptive learning rates for non-convex optimization, in *Advances in Neural Information Processing Systems* (2015), pp. 1504–1512
3. J.N. Foerster, Y.M. Assael, N. de Freitas, S. Whiteson, Learning to communicate to solve riddles with deep distributed recurrent q-networks (2016). Preprint arXiv:1602.02672
4. S. Goldwasser, S. Micali, Probabilistic encryption. J. Comput. Syst. Sci. **28**(2), 270–299 (1984)
5. I. Goodfellow, J. Pouget-Abadie, M. Mirza, B. Xu, D. Warde-Farley, S. Ozair, A. Courville, Y. Bengio, Generative adversarial nets, in *Advances in Neural Information Processing Systems* (2014), pp. 2672–2680
6. R. Huang, B. Xu, D. Schuurmans, C. Szepesvári, Learning with a strong adversary (2015). Preprint arXiv:1511.03034
7. M.C. Mukkamala, M. Hein, Variants of RMSProp and AdaGrad with logarithmic regret bounds, in *Proceedings of the 34th International Conference on Machine Learning*, vol. 70 (2017), pp. 2545–2553. JMLR.org
8. A. Stanoyevitch, *Introduction to Cryptography with Mathematical Foundations and Computer Implementations* (CRC Press, Boca Raton, 2010)
9. S. Sukhbaatar, R. Fergus, et al., Learning multiagent communication with backpropagation, in *Advances in Neural Information Processing Systems* (2016), pp. 2244–2252

Application of Modified Social Spider Algorithm on Unit Commitment Solution Considering the Uncertainty of Wind Power in Restructured Electricity Market

Heidar Ali Shayanfar, Hossein Shayeghi, and L. Bagherzadeh

1 Introduction

The unit commitment (*UC*) problem has a substantial role in economic operation of power systems. Right time starting or shutting down determination of units among vast range of possible states will have significant economic savings. Also, providing sufficient spinning reserve in power system keeps the system in safe security condition. *UC* problem includes daily schedule of commitment of power plants, in which operating costs along with startup and shutdown costs of units must be minimized, so all limitations and constraints be considered. This problem contains discrete and continuous variables that show on/off status and economic dispatch of power generation among units [1–3].

However, due to finite primary energy sources and normally incremental trend of their prices and low round trip efficiency in power conversion process in power plants, energy planning experts are persuaded to use energy efficiency policies and renewable energy sources. On the one hand, there is extensive propensity for use of renewable energy sources, but on the other hand stochastic behavior of wind causes uncertainty in *UC* solving. Therefore, a definitive answer to the *UC* problem firstly requires exact modeling of wind stochastic behavior and secondly requires an intelligent algorithm to obtain optimal solution. So far, heuristic methods such as neural networks, simulated annealing, fuzzy logic, Tabu search algorithm have been used to solve energy planning problems. Furthermore, various approaches based on

---

H. A. Shayanfar (✉)
College of Technical & Engineering, South Tehran Branch, Islamic Azad University, Tehran, Iran

H. Shayeghi · L. Bagherzadeh
Department of Electrical Engineering, University of Mohaghegh Ardabili, Ardabil, Iran
e-mail: bagherzadeh@student.uma.ac.ir

© Springer Nature Switzerland AG 2021
H. R. Arabnia et al. (eds.), *Advances in Artificial Intelligence and Applied Cognitive Computing*, Transactions on Computational Science and Computational Intelligence, https://doi.org/10.1007/978-3-030-70296-0_33

the combination of aforementioned algorithms are represented to solve *UC*, which have different execution accuracy and convergence speed [4–12].

In this chapter, a method based on modified social spider algorithm (*MSSA*) for solving *UC* problem is adopted so the presence of wind farms are taken into account. Various practical restrictions of system and power plants are also considered in the study. The offered algorithm is implemented on a 10-generator test system and the results are represented.

2 Formulation of Problem

2.1 Objective Function

The main purpose of scheduling is to determine the startup and shutdown time of thermal units so that total operating costs are reduced as much as possible, and system constraints be satisfied. Fuel costs of thermal units are usually characterized by a quadratic equation [13, 14]:

$$F_i (P_i) = a_i + b_i P_i + c_i P_i^2 \tag{1}$$

Operating cost consists of two parts: the first part covers the fuel cost and the second part is startup cost. Since the goal is to minimize the total cost. Thus, the objective function defines as follows:

$$\text{Min} F_T = \sum_{t=1}^{T} \sum_{i=1}^{N} \left[\begin{array}{l} U_i (t) \times F_i (P_i (t)) \\ + U_i (t) \times (1 - U_i (t-1)) \times SUC_i \end{array} \right] \tag{2}$$

2.2 Constraints

2.2.1 System Constraints

Frequency control needs a certain amount of power be kept in reserve with the ability of re-establishing the balance between demand and generation at all times. Spinning reserve can be defined as the amount of generation capacity that is able to be used to generate active power over a specific period of time and which has not yet been committed to the generation during this period. Operators must consider sufficient reserve to maintain certain level of security and reliability. The constraints of power balance, spinning storage, and output power of termal units are presented by (3, 4, 5, and 6).

$$\sum_{i=1}^{N} U_i \text{ (t)} \times P_i \text{ (t)} + P_W \text{ (t)} = P_L \text{ (t)} \tag{3}$$

$$\sum_{i=1}^{N} U_i \text{ (t)} \times P_i(t) \geq SRR + r\% \times P_W \text{ (t)} \tag{4}$$

$$P_L \text{ (t)} - P_W \text{ (t)} \leq r\% \times P_W \text{ (t)} + \sum_{i=1}^{N} U_i \text{ (t)} \times P_i^{\min} \text{ (t)} \tag{5}$$

$$\sum_{i=1}^{N} U_i \text{ (t)} \times P_i^{\max} \text{ (t)} + P_W \text{ (t)} \geq P_L \text{ (t)} + SRR + r\% \times P_W \text{ (t)} \tag{6}$$

2.2.2 Thermal Unit Constraints

Each thermal unit has a generation range that can provide only a limited amount of power at a certain period of time. This capability of increase and decrease in generation are called as ramping rates [15]:

$$U_i \text{ (t)} \times P_i^{\min} \text{ (t)} \leq P_i \text{ (t)} \leq U_i \text{ (t)} \times P_i^{\max} \text{ (t)} \tag{7}$$

So that:

$$P_i^{\max} \text{ (t)} = \begin{cases} \begin{cases} \min\left[P_i^{\max}, P_i \text{ (t} - 1) + RU R_i^{\max}\right] \\ if\, U_i \text{ (t)} = U_i \text{ (t} - 1) = 1 \end{cases} \\ \begin{cases} \min\left[P_i^{\max}, P_i \text{ (t} - 1) + SR_i\right] \\ if\, U_i \text{ (t)} = 1, U_i \text{ (t} - 1) = 0 \end{cases} \end{cases} \tag{8}$$

$$P_i^{\min} \text{ (t)} = \begin{cases} \begin{cases} \max\left[P_i^{\min}, P_i \text{ (t} - 1) - RD R_i^{\max}\right] \\ if\, U_i \text{ (t)} = U_i \text{ (t} - 1) = 1 \end{cases} \\ \begin{cases} P_i^{\min} \\ if\, U_i \text{ (t)} = 1, U_i \text{ (t} - 1) = 0 \end{cases} \end{cases} \tag{9}$$

If a unit commits into generation, it is not possible to turn it off immediately. It should remain on at least for a specific time, which is called (MUT):

$$
\begin{cases}
\sum_{t=1}^{L_i} [1 - U_i(t)] = 0 \\
\sum_{t}^{t+MTU_i-1} U_i(t) \geq MUT_i \times Y_i(t), \quad \forall t = L_i + 1 \cdots T - MUT_i + 1 \\
\sum_{t}^{T} [U_i(t) - Y_i(t)] \geq 0, \quad \forall t = T - MUT_i + 2 \cdots T \\
L_i = \text{Min} \left\{ T, \left(MUT_i - U_i^0 \right) \times U_i(0) \right\}
\end{cases}
\tag{10}
$$

It is noticeable that if a unit be shut down, it is not allowed to be restarted up again immediately. It is associated with mechanical stress on power plant's equipment. After a unit shuts down, it must stay off for a certain period of time to prevent mechanical damage to power plant's equipment, which is called minimum downtime (MDT) [16]:

$$
\begin{cases}
\sum_{t=1}^{B_i} U_i(t) = 0 \\
\sum_{t}^{t+MDU_i-1} [1 - U_i(t)] \geq MDT_i \times Z_i(t), \forall t = B_i + 1 \cdots T - MDT_i + 1 \\
\sum_{t}^{T} [1 - U_i(t) - Z_i(t)] \geq 0, \forall t = T - MDT_i + 2 \cdots T \\
B_i = \text{Min} \left\{ T, \left(MDT_i - S_i^0 \right) \times (1 - U_i(0)) \right\}
\end{cases}
\tag{11}
$$

2.2.3 Wind Unit Constraint

Wind stochastic behavior has led to discontinuous and uncontrollable characteristic for units. The power of wind units depends on wind speed as well as wind turbine features. In a wind turbine, wind generation can be obtained from (12). The values of A, B, and C depend on the turbine features and can be calculate from (13), (14) to (15) [17]:

$$
P_W =
\begin{cases}
0 & 0 \leq V < V_{ci} \\
\left(A + BV + CV^2 \right) P_r & V_{ci} < V < V_r \\
P_r & V_r < V < V_{co} \\
0 & V > V_{co}
\end{cases}
\tag{12}
$$

$$
A = \frac{P_r}{(V_{ci} - V_r)^2} \left[V_{ci} (V_{ci} + V_r) - 4 (V_{ci} \times V_r) \left[\frac{V_{ci} + V_r}{2V_r} \right]^3 \right]
\tag{13}
$$

Application of Modified Social Spider Algorithm on Unit Commitment... 441

$$B = \frac{P_r}{(V_{ci} - V_r)^2} \left[4 \left(V_{ci} + V_r \right) \left[\frac{V_{ci} + V_r}{2V_r} \right]^3 - (3V_{ci} + V_r) \right] \tag{14}$$

$$C = \frac{1}{(V_{ci} - V_r)^2} \left[2 - 4 \left[\frac{V_{ci} + V_r}{2V_r} \right]^3 \right] \tag{15}$$

The wind turbine output power can be obtained from (16):

$$P_{WG} = P_W \eta_W \tag{16}$$

In the above equation, η_W is the combined efficiency of the gearbox, generator, and electronic equipment. If a wind farm consists of several models wind turbine, wind data of speed and wind turbines specifications can calculate wind farm output power through total output power of all turbines. Unit commitment problem with the presence of wind farms considering time-related constraints such as *MUT*, *MDT*, *RUR*, and *RDR* is more complicated and needs a suitable algorithm for solving the problem. The purpose of this study is to use modified social spider algorithm (*MSSA*).

3 Proposed Method

Social spider algorithm (SSA) is a new optimization algorithm which is recently studied by [18]. SSA is used for solving continuous unconstrained problems, but with the modification, we can use it for solving continuous constrained problems.

The basic operating agents of SSA are spiders. Spiders can move freely on their webs, and so in the SSA, solution space of an optimization problem is hyper-dimensional space. Each position on the web corresponds to a feasible solution to the optimization problem. Each spider linked the ith spider in the population by two ways. Frist its position $Pi\ (t)$ and second fitness value $f\ (Pi\ (t))$, where t is the current iteration and $f(x)$ is the objective function. Each spider has several attributes which guide another spider. Weaken the previous goal vibration. Figure 1 shows modified social spider algorithm.

4 Case Study

In this chapter, a power system containing 10 thermal units has been used. The information of production and cost functions' coefficients is presented in reference [16]. Figure 2 shows the daily load diagram. In this chapter, multilayer artificial

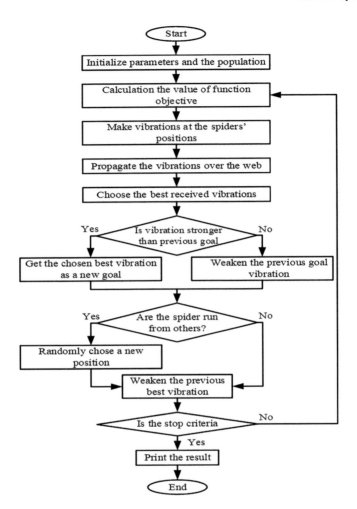

Fig. 1 The modified social spider algorithm

neural network (*ANN*) is employed to predict wind speed, with the consideration of 18 neurons in three layers using nonlinear auto regressive time-series with external input prediction method. Therefore, 4-year wind speed data is gathered.

This data contain hourly wind speed of a 1460-day period, to insert as input to neural network. It lets *ANN* to take long backward steps. Figure 3 illustrates exact predicted wind speed by the neural network and low-risk forecasted wind speed; it also shows forecasting error. The capacity of installed wind farms is about 9.4 percent of peak load. The wind unit maintenance costs and wind generation costs are neglected.

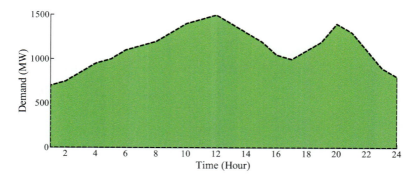

Fig. 2 Daily load curve

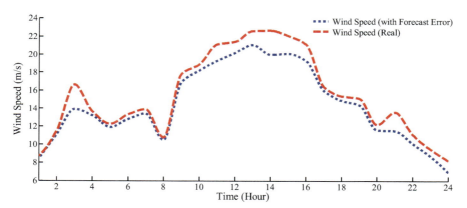

Fig. 3 Actual and cautious predicted wind speed in wind farm

5 Simulation Results

5.1 Scenario 1: Only Thermal Units (wWithout Wind Units)

In scenario 1, only thermal units are called for supplying. Accordingly, the schedule determines the contribution of thermal units in Table 1. Figure 4 shows the economic dispatch among generating units. As expected, as the load increases, more expensive units commit in generation. The total generation cost of thermal units is $554065.9865. In this case, due to lack of commitment of wind units, wind units profit is equal to zero.

Table 1 Unit commitment in 10-unit test system (Scenario 1)

	1	2	3	4	5	6	7	8	9	10
1	1	0	1	0	1	1	0	0	0	0
2	1	0	1	0	1	1	0	0	0	0
3	1	1	1	0	1	0	0	0	0	0
4	1	1	1	1	1	0	0	0	0	0
5	1	1	1	1	1	1	0	0	0	0
6	1	1	1	1	1	1	0	0	0	0
7	1	1	1	1	1	1	0	0	0	0
8	1	1	1	1	1	1	1	0	0	0
9	1	1	1	1	1	1	1	0	0	0
10	1	1	1	1	1	1	1	0	0	0
11	1	1	1	1	1	1	1	1	0	1
12	1	1	1	1	1	1	1	1	0	1
13	1	1	1	1	1	1	1	0	0	0
14	1	1	1	1	1	1	1	0	0	0
15	1	1	0	1	1	1	0	0	0	0
16	1	1	0	1	1	0	0	0	0	0
17	1	1	0	1	1	0	0	0	0	0
18	1	1	0	1	1	0	0	0	0	0
19	1	1	0	1	1	1	0	0	0	0
20	1	1	1	1	1	1	0	1	1	0
21	1	1	1	1	1	1	0	0	0	0
22	1	1	1	0	1	0	0	0	0	0
23	1	1	1	0	1	0	0	0	0	0
24	1	1	0	0	0	0	0	0	0	0

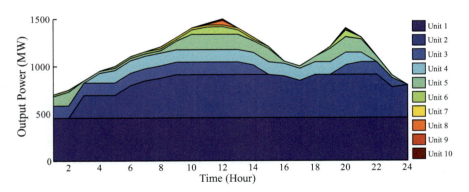

Fig. 4 Hourly economic dispatch in 10-unit test system (Scenario 1)

5.2 Scenario 2: Thermal and Wind Units

In scenario 2, wind and thermal units participate in power supplying. Hence, for the participation of wind units in a day-ahead power market, risk reduction

Application of Modified Social Spider Algorithm on Unit Commitment... 445

Table 2 Unit commitment in 10-unit test system (Scenario 2)

	1	2	3	4	5	6	7	8	9	10
1	1	1	0	0	0	0	0	0	0	0
2	1	1	0	0	0	0	0	0	0	0
3	1	1	0	0	0	0	0	0	0	0
4	1	1	0	0	1	0	0	0	0	0
5	1	1	0	0	1	0	0	0	0	0
6	1	1	1	1	1	0	0	0	0	0
7	1	1	1	1	1	0	0	0	0	0
8	1	1	1	1	1	0	1	0	0	0
9	1	1	1	1	1	1	1	0	0	0
10	1	1	1	1	1	1	1	1	1	0
11	1	1	1	1	1	1	1	1	1	1
12	1	1	1	1	1	1	1	1	1	1
13	1	1	1	1	1	1	1	0	0	0
14	1	1	1	0	1	1	1	1	0	0
15	1	1	1	0	1	1	0	0	0	0
16	1	1	1	0	1	0	0	0	0	0
17	1	1	0	1	1	0	0	0	0	0
18	1	1	0	1	1	0	0	0	0	0
19	1	1	0	1	1	1	0	0	0	0
20	1	1	1	1	1	1	0	1	1	0
21	1	1	1	1	1	1	0	0	0	0
22	1	1	1	1	1	0	0	0	0	0
23	1	1	0	0	1	0	0	0	0	0
24	1	1	0	0	0	0	0	0	0	0

policy is adopted by owners of wind units. Hence, the most pessimistic (cautious) contribution is undertaken. Therefore, unit commitment program can be calculated as Table 2. Figure 5 also shows economic load dispatch between committed units. In this case, the total cost of thermal units is $518177.1879, which represents a %6.477 cost reduction. In this case, the wind units, due to lack of the presence of a large-scale storage unit and risk reduction of buying from spot market because of deficient generation, bid cautious forecasted wind generation instead of exact forecast and the calculations show the profit of $104280.7347 for wind units.

In this scenario, *ISO* buys the extra generated electricity of wind units at a lower price and sells them to consumers with market clearing price, which produces a surplus. Thus, *ISO* earns some profit. At real time, *ISO* shuts down most expensive units which were supposed to have production at real time. If it were considered in market architectural, a specific detriment could be paid to aforementioned thermal units due to sudden change in generation schedule. These units receive this money without burning any fuel. This surplus is usually produced in market transactions and *ISO* benefits from it, but it must be spent for expansion of system, and *ISO* is not allowed to seek to increase it.

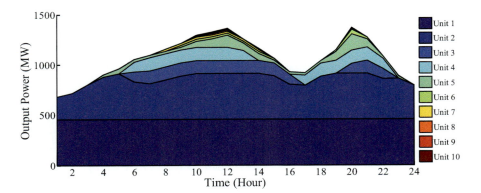

Fig. 5 Hourly economic dispatch in 10-unit test system (Scenario 2)

6 Conclusion

Participation of renewable units, especially wind units, can reduce thermal units' commitments as well as environmental emission. However, due to uncertainty of wind units, there is always a challenge for experts of power system to manage sources of generation. Therefore, finding a definitive answer to unit commitment problem with the effects of the presence of uncertain resources such as wind units needs an appropriate algorithm. In this chapter, *MSSA* algorithm is used to optimize the problem. Optimization results show that coordinated operation of wind and thermal units can decrease thermal generation costs compared with the case without the presence of wind units. Optimization has been done with the goal of reaching to maximum wind units' profit. However, in this scheduling, due to lack of large-scale storage, wind units bid to market based on cautious predicted generation. This causes this fact that part of probable profit of wind units be given to independent system operator.

References

1. S.Y. Abujarad, M.W. Mustafa, J.J. Jamian, Recent approaches of unit commitment in the presence of intermittent renewable energy resources: A review. Renew. Sust. Energ. Rev. **70**, 215–223 (2017)
2. P.H. Chen, H.L. Shieh, J.Y. Chen, H.C. Chang, Application of swarm optimization to thermal unit commitment problem, in *2017 IEEE International Conference on Information and Automation (ICIA)*, (IEEE, 2017), pp. 478–482
3. J. Zhang, Q. Tang, Y. Chen, S. Lin, A hybrid particle swarm optimization with small population size to solve the optimal short-term hydro-thermal unit commitment problem. Energy **109**, 765–780 (2016)

4. P. Khazaei, M. Dabbaghjamanesh, A. Kalantarzadeh, H. Mousavi, Applying the modified TLBO algorithm to solve the unit commitment problem, in *2016 World Automation Congress (WAC)*, (IEEE, 2016), pp. 1–6
5. V.K. Kamboj, A novel hybrid PSO–GWO approach for unit commitment problem. Neural Comput. & Applic. **27**(6), 1643–1655 (2016)
6. S. Saranya, B. Saravanan, Solution to unit commitment using lagrange relaxation with whale optimization method, in *2019 Innovations in Power and Advanced Computing Technologies (i-PACT)*, vol. 1, (IEEE, 2019, March), pp. 1–6
7. Y. Zhai, N. Mu, X. Liao, J. Le, T. Huang, Unit commitment problem using an efficient pso based algorithm, in *2019 Eleventh International Conference on Advanced Computational Intelligence (ICACI)*, (IEEE, 2019, June), pp. 320–324
8. S. Prabakaran, S. Tamilselvi, P.A.D.V. Raj, M. Sudhakaran, S. Rajasekar, Solution for multi-area unit commitment problem using PSO-based modified firefly algorithm, in *Advances in Systems, Control and Automation*, (Springer, Singapore, 2018), pp. 625–636
9. V.K. Kamboj, S.K. Bath, J.S. Dhillon, A novel hybrid DE–random search approach for unit commitment problem. Neural Comput. & Applic. **28**(7), 1559–1581 (2017)
10. J. Liu, S. Liu, An improved dual grey wolf optimization algorithm for unit commitment problem, in *Intelligent Computing, Networked Control, and their Engineering Applications*, (Springer, Singapore, 2017), pp. 156–163
11. G. Yehescale, M.D. Reddy, A new strategy for solving unit commitment problem by PSO algorithm, in *2018 IEEE International Conference on Current Trends in Advanced Computing (ICCTAC)*, (IEEE, 2018, February), pp. 1–6
12. A. Shukla, S.N. Singh, Clustering based unit commitment with wind power uncertainty. Energy Convers. Manag. **111**, 89–102 (2016)
13. L. Wu, M. Shahidehpour, Security-constrained unit commitment with uncertainties, in *Power grid operation in a market environment: economic efficiency and risk mitigation*, (2016), pp. 115–168
14. L. Bagherzadeh, H. Shayeghi, Optimal allocation of electric vehicle parking lots and renewable energy sources simultaneously for improving the performance of distribution system, in *2019 24th Electrical Power Distribution Conference (EPDC)*, (IEEE, 2019), pp. 87–94
15. L. Bagherzadeh, H. Shahinzadeh, H. Shayeghi, G.B. Gharehpetian, A short-term energy management of microgrids considering renewable energy resources, micro-compressed air energy storage and DRPs. IJRER **9**(4), 1712–1723 (2019)
16. H. Shahinzadeh, A. Gheiratmand, J. Moradi, S.H. Fathi, Simultaneous operation of near-to-sea and off-shore wind farms with ocean renewable energy storage, in *2016 Iranian Conference on Renewable Energy & Distributed Generation (ICREDG)*, (IEEE, 2016), pp. 38–44
17. Ö. Kiymaz, T. Yavuz, Wind power electrical systems integration and technical and economic analysis of hybrid wind power plants, in *2016 IEEE International Conference on Renewable Energy Research and Applications (ICRERA)*, (IEEE, 2016), pp. 158–163
18. W.T. Elsayed, Y.G. Hegazy, F.M. Bendary, M.S. El-Bages, Modified social spider algorithm for solving the economic dispatch problem. Eng. Sci. Technol. Int. J. **19**(4), 1672–1681 (2016)

Heidar Ali Shayanfar received the B.S. and M.S.E. degrees in electrical engineering in 1973 and 1979, respectively. He received the Ph.D. degree in electrical engineering from Michigan State University, East Lansing, MI, USA, in 1981. Currently, he is a full professor with the Department of Electrical Engineering, Iran University of Science and Technology, Tehran, Iran. His research interests are in the application of artificial intelligence to power system control design, dynamic load modeling, power system observability studies, voltage collapse, and congestion management in a restructured power system, reliability improvement in distribution systems, smart grids and reactive pricing in deregulated power systems. He has published more than 520 technical papers in the international journals and conferences proceedings. Dr. Shayanfar is a member of the Iranian Association of Electrical and Electronic Engineers.

Hossein Shayeghi received the B.S. and M.S.E. degrees in electrical and control engineering in 1996 and 1998, respectively. He received his Ph.D. degree in electrical engineering from Iran University of Science and Technology, Tehran, Iran in 2006. Currently, he is a full professor in Technical Engineering Department of University of Mohaghegh Ardabili, Ardabil, Iran. His research interests are in the application of robust control, artificial intelligence and heuristic optimization methods to power system control design, operation and planning and power system restructuring. He has authored and co-authored 10 books in the electrical engineering area all in Farsi, one book and 10 book chapters in international publishers and more than 410 papers in international journals and conference proceedings. Also, he collaborates with several international journals as reviewer boards and works as editorial committee of three international journals. He has served on several other committees and panels in governmental, industrial, and technical conferences. He was selected as distinguished researcher of the University of Mohaghegh Ardabili several times. In 2007, 2010, 2012, and 2017 he was also elected as distinguished researcher in engineering field in Ardabil province of Iran. Furthermore, he has been included in the Thomson Reuters' list of the top one percent of most-cited technical Engineering scientists in 2015–2019, respectively. Also, he is a member of Iranian Association of Electrical and Electronic Engineers (IAEEE) and senior member of IEEE.

Predicting Number of Personnel to Deploy for Wildfire Containment

John Carr, Matthew Lewis, and Qingguo Wang

1 Introduction

Climate change is causing longer forest fire seasons and more fires in places where they traditionally were rare [1–3]. According to the Department of Agriculture of the United States (USDA), the ten largest fires in 2014 cost more than $320 million dollars [4] and the firefighting costs in 2017 exceeded $2.4 billion, making the 2017 fire season the most expensive ever [5]. The rising cost of wildfire operations negatively affects the budget and resources of the USDA Forest Service and other federal agencies. Being prepared to fight forest or wildfires, from the most minimal to the most catastrophic, as well as the effective management of resources, is vital to a timely containment of the fire and minimizing the damage to land, property, and lives.

Between 1995 and 2015 the percentage of the Forest Service's annual budget devoted to wildfires jumped from 16% to over 50% and this trend has continued [5, 6]. Along with this shift in resources is a shift in staff – meaning more personnel being devoted to fighting fires and less to non-fire programs which help take preventative measures to reduce fire threat. Because of this, the efficiency through accurate forecasting of personnel needs is vital to the operations of not only the Forest Service and other federal agencies, but also to private contractors that provide equipment and services to support wildfire-suppression efforts [7].

John Carr and Matthew Lewis contributed equally with all other contributors.

J. Carr · M. Lewis · Q. Wang (✉)
College of Computing and Technology, Lipscomb University, Nashville, TN, USA
e-mail: Jucarr@lipscomb.edu; malewis1@lipscomb.edu; qwang@lipscomb.edu

© Springer Nature Switzerland AG 2021
H. R. Arabnia et al. (eds.), *Advances in Artificial Intelligence and Applied Cognitive Computing*, Transactions on Computational Science and Computational Intelligence, https://doi.org/10.1007/978-3-030-70296-0_34

As wildfires occur, personnel are deployed to fight, manage, and contain those fires. However, moving firefighting assets can be time consuming and expensive, so getting the correct amount of personnel and equipment to the fire as quickly as possible is vital. If too few personnel are sent, then the fire could get out of control, taking longer to contain, and could cause more damage to parks and residential areas. If too many are deployed to a fire, then expensive resources are wasted, recovery and prevention methods suffer, and more time and effort will be required to redeploy those personnel to other locations where they are needed.

In this chapter, we explore the possibility of predicting the number of personnel needed to effectively fight a wildfire, in an aim to develop a forecasting system in the future. A model that can accurately predict the number of firefighting personnel is urgently needed. It can be used in combination with other wildfire forecasting models to facilitate efficient resource management and decision support, enabling those fighting the fires to do so more effectively.

2   Related Work

While the scientific community has investigated intensively weather, fires, and fire size forecasting, much less has been done regarding operations, resources, and personnel. Wildfire operations are based on wildfire risk analysis and analysis of effectiveness of mitigation measures. Various models have been proposed to study wildfire risk [8–11]. For example, Thompson et al. proposed a polygon-based model to estimate highly valuable resources and assets (HVRA) area burned by overlaying simulated fire perimeters with maps of HVRA [8]. Their model complements traditional pixel-based burn probability and fire intensity metrics and has been used to inform prioritization and mitigation decisions.

In 2013, Lee et al. optimized the shift of firefighting resources among three planning units in California, from the unit with the highest fire load to the planning unit with the highest standard response requirements, by utilizing a scenario-based, standard-response optimization model with stochastic simulation [12]. Through optimizing the unit whose resources would be deployed, they found increasing the probability of resource availability on high fire count days resulted in greater containment success [12]. Their analysis was limited to three individual units. Models considering multiple resources were also investigated [13–16].

Firefighting is not merely operations and resources, but rather command and control process of firefighters. McLennan et al. used four methodologies to study the processes of wildfire Incident Management Team (IMT) [17]. Their findings increased people's understanding of the decision-making processes of IMTs and have implications for issues such as creating IMTs and improving IMTs' effectiveness. Other aspects of the role of firefighters have also been explored [18–21]. For instance, Cvirn et al. examined the effects of temperature and dehydration on 73 volunteer firefighters during a simulation of wildfire suppression under either control or hot temperature conditions [19]. Their results showed cognitive performance on the psychomotor vigilance task declined when participants were dehydrated in the heat and Stroop task performance was impaired when dehydrated late in the afternoon. This, as well as other similar studies, highlights the need for effective personnel and resource management.

This chapter evaluates the possibility of predicting the number of personnel needed to effectively fight a wildfire. To the best of our knowledge, this is the first attempt to predict personnel needs. A model for personnel prediction can potentially be used to improve resource management and decision support related to firefighting.

3 Data Collection and Preprocessing

3.1 Data Collection

The data used in our study were downloaded from an online government database managed by the Integrated Reporting of Wildland-Fire Information (IRWIN) service [22]. IRWIN is an investment by Wildland Fire Information and Technology (WFIT) intended to enable 'end-to-end' fire reporting capability [23]. It provides data exchange capabilities between existing applications used to manage wildfire incident data. IRWIN dataset stores both active and historical wildfire data across the United States. As far as we know, it is the most comprehensive nonredundant database of U.S. wildfire data. We gained access to it by acquiring ArcGIS accounts to access GeoPlatform ArcGIS Online from the office of the Chief Information Officer in the U.S. Department of the Interior.

The raw fire incident data we pulled from IRWIN covers the period from June 12th, 2014 to November 2nd, 2019. It consists of 131,881 observations (or individual wildfire incidents) and 201 features.

3.2 Data Cleansing

The IRWIN data requires extensive cleaning before use for analysis. We began by dropping homogenous variables that had no variance or were 100% null. After that, we were left with 168 features, the majority of which were extremely sparse. Also, many columns had mixed data types that required us to manually identify, interpret, and replace these values. Of these features, we discarded the ones that were obviously irrelevant to our purpose of study, too sparse to be useful, or duplicates of other features, leaving us with 35 final variables from the Incidents table.

The IRWIN dataset includes a mandatory ID number for uniquely identifying each fire incident. The ID is required to join the resources table, i.e., to aggregate personnel, equipment, and other resources associated with each fire. All records without a valid ID number, which were very few, were dropped from the dataset.

Lastly, from the existing data, we also engineered new features, such as "time not contained," to measure the difference between the start and end times of the fires (in days). By converting date-time entries into ordinal values, we were able to include temporal data into our predictive model.

3.3 Data Aggregation

The resources table we pulled from IRWIN includes 16 features, such as the number of personnel, trucks, planes, helicopters, boats, and agencies supporting the firefighting efforts. These data were aggregated by grouping by each unique incident and summing the resources. The aggregation resulted in data such as the types of resources, their quantity, and the total combined personnel. For fires without resources we filled those records with zero.

After combining the fire incident table and resources table together, the new aggregated table contains 47 features to be used for analysis and modeling.

3.4 Exploratory Analysis

After preprocessing the data, we conducted exploratory data analysis. The histogram in Fig. 1 shows the frequency of total personnel per incident. It indicates that few fires had more than 250 personnel deployed to fight. Only about 3000 out of 130,000 fires burned more than 1 acre and, hence, required extensive resources and personnel, which ranged from a few to over 1000.

We also noticed in the IRWIN data that some fires burned longer than average for the number of acres involved, probably as a result of an insufficient amount of personnel deployed to contain them.

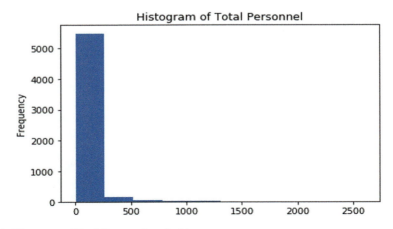

Fig. 1 Histogram of Total Personnel per incident

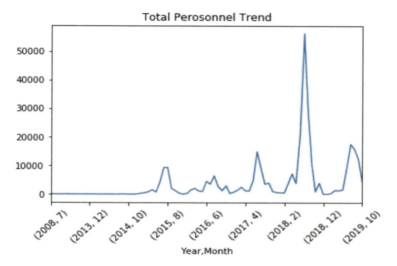

Fig. 2 Monthly averages of personnel deployed

Figure 2 shows the historical trends of the monthly averages of personnel deployed. It indicates that overall, the personnel being used to fight fires has been on the rise. This observation is consistent with the trends reported by others [1–3]. The trend of more, larger, and worse wildfires requires deployment of a proper number of personnel to contain fires. If too many resources and personnel are deployed to a fire, there may not be enough left to other ongoing fires.

The distribution of the number of resources deployed to fight fires is provided in Fig. 3. It shows that the majority of fires required few resources to contain and the greater spectrum of resources were deployed to a relatively few fires. The few fires that required more resources were typically the ones that had caused significant damages and firefighting costs.

This exploratory analysis shows the urgency to deploy a proper number of personnel and resources to a fire. If a fire is not contained in a timely manner due to insufficient personnel and resources, it can grow and become more destructive, and eventually require more agencies and people involved to contain it.

4 Methods

With wildfires getting worse and more costly each year, it is important to have a model to accurately predict the number of personnel needed to contain the fires, in particular for the few, yet extremely destructive fires.

In our study, we tested the concept of personnel prediction by using an ensemble of gradient boosted decision tress (XGBoost) as our predictive model. XGBoost is

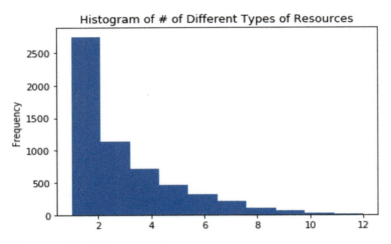

Fig. 3 Histogram of the different types of resources

a widely used optimized python library designed to be distributed, highly efficient, flexible, and portable [24]. It provides a parallel tree boosting that solves many learning problems in a fast and accurate way and enables XGBoost to outperform many machine learning models across a variety of applications. In our study, we used 100 trees in the ensemble. As hyperparameter tuning resulted in only minimal improvement, default parameters were used to run XGBoost.

To the categorical variables, we applied a form of label encoding called Rank Order Label Encoding (ROLE). ROLE provides us with an average of the target variable for every unique value within each feature. For example, after applying ROLE to the "POO State" feature (POO: Point of Origin), we had an average number of personnel for each of the states included in the data. The states can then be ranked based on these averages, from 1 to 50. By instilling ordinality into label encoding, ROLE can enhance performance of the model. An alternative method, one-hot encoding, could be used here as well. But with our high-dimensional categorical data, one-hot encoding would drastically increase the feature space and this is not conducive to tree ensembles.

When examining the relationship between each pair of features, we realized all the resource features are correlated with personnel. For example, four personnel were typically assigned to an engine. If there were three engines deployed to a fire, at least 12 personnel would be required to deploy there accordingly. Therefore, we removed all the resource variables to avoid training the model on features correlated to personnel. The remaining features were all included in our model.

To evaluate the model on the IRWIN data, we conducted a stratified ten-fold cross validation using Scikit-Learn [25], a popular Python library for various machine learning algorithms such as classification, regression, and clustering. The row order was shuffled to remove the presorted order of the data. We tested the model over 10 folds and applied ROLE encoding consistently to each fold.

5 Results

After the ten-fold cross validation, we evaluated the accuracy of our model. Its average coefficient of determination (R^2) is 0.61, indicating our model accounted for 61% of the variation in the data in terms of number of personnel deployed to fight fires.

The mean absolute error (MAE) of our model is 1.4, meaning that the error from record to record is off by 1.4 personnel per record on average. Hence, using the model for prediction, we would expect an average error of ± 2 persons.

The root mean squared error (RMSE) of the model was 17.99, significantly higher than MAE. This is expected as RMSE is more sensitive to outliers and, as shown earlier, there were relatively few, large-scale fires in the IRWIN dataset. For these cases where hundreds of personnel could be needed, the margin of error is therefore much larger.

Next, we used the SHAP Python package to gain insight into the model [26]. SHAP works by building an "explanation model" to shed light on how the model makes decisions. It assigns SHAP values to features to quantitatively measure their impact on a model.

Figure 4 shows the SHAP values of the top features of the model that XGBoost constructed from the IRWIN data. The y-axis of the plot specifies features and the x-axis denotes their impact on the model. The color signifies the value of each feature, with blue being a low value and red being a high value. A high SHAP value indicates that the feature value tends to drive the prediction upward, meaning a higher personnel count. A negative SHAP value, however, indicates that the feature tends to drive the prediction downward, toward a lower personnel account. A SHAP value close to zero denotes that the feature did not have a significant impact on the model. Figure 4 shows that Fire Code is at the higher end of the spectrum and, thus, had a higher impact on the model than other features.

We also computed the average SHAP values of the top features, as shown in Fig. 5. The average SHAP values are more robust than individual values in Fig. 4 and therefore reflect better how much each feature affects the predicting capabilities of the model.

Figures 4 and 5 show in terms of the SHAP values, a close second to the aforementioned feature Fire Code is the number of agencies supporting. This is quite expected, as with more agencies supporting, more personnel are likely to get involved in firefighting. But some areas clearly have more agencies than others. This might be an issue to consider in operational decision and shifting of resources.

Moreover, it shows in Figs. 4 and 5 that "POO protecting unit" and "estimated cost to date" were contributing features too. But their average SHAP values are way less than those of the aforementioned two features, indicating their relatively low impact on the model.

Figure 6 shows a scatterplot of the predicted values vs. actual number of personnel. It shows the model predicted most accurately for the incidents in which fewer personnel were deployed.

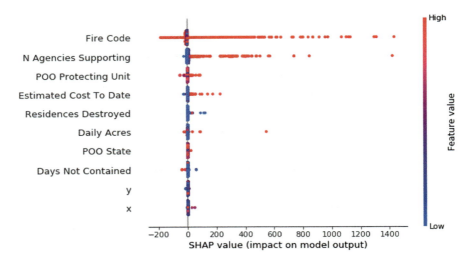

Fig. 4 SHAP values for the variables in the model

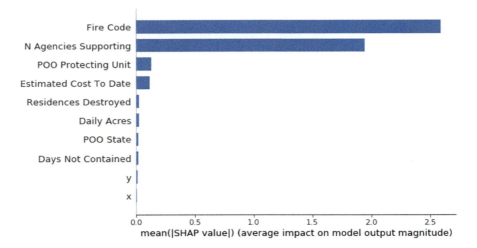

Fig. 5 Top average SHAP values for the features in the model

6 Discussion

With wildfires getting worse and more costly each year, it is important to have a model to accurately predict the number of personnel needed to contain the fires. A model that can correctly predict personnel needs can be used to help firefighting agencies combat wildfires more effectively.

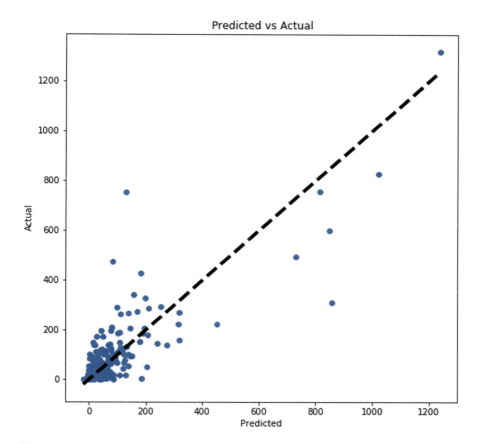

Fig. 6 Scatterplot of predicted values vs. actual personnel

In this chapter, we explored the possibility of predicting the number of personnel needed to fight a wildfire. We tested the concept by using gradient boosted ensemble (XGBoost) as our predictive model and training it on IRWIN data. The results show that the average coefficient of determination (R^2) of the model is 0.61 and mean absolute error is 1.4. Although the results seem reasonably good, there is still plenty of room for improvement.

The IRWIN dataset that we used did not provide weather information. But weather has a great impact on fire containment. We searched but did not find an open-source API that provides the meteorological services on the scale we needed. Most sources available today require industry-level accounts ($800+ a month) to make multiple API requests for every record within our dataset. So, a future development of this project is to continue to try to incorporate weather data into predictive model.

Besides weather data, geographical information could be a valuable addition to our model. Terrain features, vegetation levels, soil types, etc., could all be beneficial in increasing the model's performance. Population statistics based on location such as population density, number of houses per square mile, etc., which can be acquired from Census data, could potentially be used to enhance the model as well.

In this chapter, we only tried XGBoost for personnel prediction. Other methods such as artificial neural networks [27] could be a suitable model too for such purpose of study. Also, limited by time, we did not perform any substantial model tuning or feature selection. Feature selection methods, such as recursive feature elimination, might result in an increased accuracy as well.

Acknowledgments We thank Dr. Todd Gary for his continued guidance and support with this project.

References

1. C. Ingraham, Wildfires have gotten bigger in recent years, and the trend is likely to continue. In: Wash. Post. (2018).. https://www.washingtonpost.com/business/2018/08/14/wildfires-have-gotten-bigger-recent-years-trend-is-likely-continue/. Accessed 6 Mar 2020
2. R. Meyer, California's wildfires are 500 percent larger due to climate change. In: The Atlantic. (2019).. https://www.theatlantic.com/science/archive/2019/07/climate-change-500-percent-increase-california-wildfires/594016/. Accessed 6 Mar 2020
3. G. Levy, Wildfires are getting worse, and more costly, every year. In: U.S. News. (2018).. https://www.usnews.com/news/data-mine/articles/2018-08-01/wildfires-are-getting-worse-and-more-costly-every-year. Accessed 6 Mar 2020
4. USDA Forest Service, The Rising Cost of Fire Operations: Effects on the Forest Service's Non-Fire Work (2015)
5. US Forest Service. Cost of Fire Operations. https://www.fs.usda.gov/about-agency/budget-performance/cost-fire-operations. Accessed 6 Mar 2020
6. Tatro S, For First Time in 110 Years, Firefighting Costs Exceed 50 Percent of U.S. Forest Service Budget. In: NBC 7 San Diego. (2015).. https://www.nbcsandiego.com/news/local/for-first-time-in-110-years-firefighting-exceeds-50-percent-of-us-forest-service-budget/55910/. Accessed 6 Mar 2020
7. H. Huber-Stearns, C. Moseley, C. Bone, N. Mosurinjohn, K.M. Lyon, An initial look at contracted wildfire response capacity in the American West. J. For. **117**, 1–8 (2019) https://doi.org/10.1093/jofore/fvy057
8. M.P. Thompson, J. Scott, J.D. Kaiden, J.W. Gilbertson-Day, A polygon-based modeling approach to assess exposure of resources and assets to wildfire. Nat. Hazards **67**, 627–644 (2013). https://doi.org/10.1007/s11069-013-0593-2
9. C. Stockdale, Q. Barber, A. Saxena, M.-A. Parisien, Examining management scenarios to mitigate wildfire hazard to caribou conservation projects using burn probability modeling. J. Environ. Manag. **233**, 238–248 (2019). https://doi.org/10.1016/j.jenvman.2018.12.035
10. J. Reimer, D.K. Thompson, N. Povak, Measuring initial attack suppression effectiveness through burn probability. Fire **2**, 60 (2019) https://doi.org/10.3390/fire2040060
11. F.J. Alcasena-Urdíroz, C. Vega-García, A.A. Ager, M. Salis, N.J. Nauslar, F.J. Mendizabal, R. Castell, Forest fire risk assessment and multifunctional fuel treatment prioritization methods in Mediterranean landscapes. Cuad. Investig. Geográfica **45**, 571–600 (2019). https://doi.org/10.18172/cig.3716

12. Y. Lee, J.S. Fried, H.J. Albers, R.G. Haight, Deploying initial attack resources for wildfire suppression: spatial coordination, budget constraints, and capacity constraints. Can. J. For. Res. **43**(1), 56–65 (2013). https://doi.org/10.1139/cjfr-2011-0433
13. B. Bodaghi, E. Palaneeswaran, S. Shahparvari, M. Mohammadi, Probabilistic allocation and scheduling of multiple resources for emergency operations; a Victorian bushfire case study. Comput. Environ. Urban. Syst. **81**, 101479 (2020). https://doi.org/10.1016/j.compenvurbsys.2020.101479
14. S. Sakellariou, F. Samara, S. Tampekis, A. Sfougaris, O. Christopoulou, Development of a Spatial Decision Support System (SDSS) for the active forest-urban fires management through location planning of mobile fire units. Environ. Hazards **19**, 131–151 (2019). https://doi.org/10.1080/17477891.2019.1628696
15. A. Behrendt, V.M. Payyappalli, J. Zhuang, Modeling the cost effectiveness of fire protection resource allocation in the United States: models and a 1980–2014 case study. Risk Anal. **39**, 1358–1381 (2019). https://doi.org/10.1111/risa.13262
16. Z. Yang, L. Guo, Z. Yang, Emergency logistics for wildfire suppression based on forecasted disaster evolution. Ann. Oper. Res. **283**, 917–937 (2019). https://doi.org/10.1007/s10479-017-2598-9
17. J. McLennan, A.M. Holgate, M.M. Omodei, A.J. Wearing, Decision making effectiveness in wildfire incident management teams. J. Contingencies Crisis Manag. **14**, 27–37 (2006) https://doi.org/10.1111/j.1468-5973.2006.00478.x
18. G.E. Vincent, B. Aisbett, S.J. Hall, S.A. Ferguson, Fighting fire and fatigue: sleep quantity and quality during multi-day wildfire suppression. Ergonomics **59**, 932–940 (2016). https://doi.org/10.1080/00140139.2015.1105389
19. M.A. Cvirn, J. Dorrian, B.P. Smith, G.E. Vincent, S.M. Jay, G.D. Roach, C. Sargent, B. Larsen, B. Aisbett, S.A. Ferguson, The effects of hydration on cognitive performance during a simulated wildfire suppression shift in temperate and hot conditions. Appl. Ergon. **77**, 9–15 (2019). https://doi.org/10.1016/j.apergo.2018.12.018
20. R.H. Coker, C.J. Murphy, M. Johannsen, G. Galvin, B.C. Ruby, Wildland firefighting. J. Occup. Environ. Med. **61**, e91–e94 (2019). https://doi.org/10.1097/JOM.0000000000001535
21. S. Abrard, M. Bertrand, T.D. Valence, T. Schaupp, Physiological, cognitive and neuromuscular effects of heat exposure on firefighters after a live training scenario. Int. J. Occup. Saf. Ergon. (2018). https://doi.org/10.1080/10803548.2018.1550899
22. Integrated Reporting of Wildland-Fire Information (IRWIN). https://www.forestsandrangelands.gov/WFIT/applications/IRWIN/. Accessed 16 Mar 2020
23. Wildland Fire Information and Technology (WFIT). https://www.forestsandrangelands.gov/WFIT/index.shtml. Accessed 17 Mar 2020
24. T. Chen, C. Guestrin, XGBoost: a scalable tree boosting system. In: *Proceedings of the 22nd ACM SIGKDD International Conference on Knowledge Discovery and Data Mining*, ACM, San Francisco, pp. 785–794 (2016)
25. F. Pedregosa, G. Varoquaux, A. Gramfort, V. Michel, B. Thirion, O. Grisel, M. Blondel, P. Prettenhofer, R. Weiss, V. Dubourg, J. Vanderplas, A. Passos, D. Cournapeau, M. Brucher, M. Perrot, É. Duchesnay, Scikit-learn: machine learning in python. J. Mach. Learn. Res. **12**, 2825–2830 (2011)
26. S.M. Lundberg, S.-I. Lee, A unified approach to interpreting model predictions. In: *The 31st International Conference on Neural Information Processing Systems*. ACM, Long Beach , pp. 4768–4777 (2017)
27. R. Holt, S. Aubrey, A. DeVille, W. Haight, T. Gary, Q. Wang, Deep Autoencoder neural networks for detecting lateral movement in computer networks. In: *The 21st International Conference on Artificial Intelligence*, CSREA Press, pp. 277–283 (2019)

An Evaluation of Bayesian Network Models for Predicting Credit Risk on Ugandan Credit Contracts

Peter Nabende ⓘ, Samuel Senfuma, and Joyce Nakatumba-Nabende ⓘ

1 Introduction

Credit risk prediction involves the computation of the future status of a credit contract, or specifically the repayment success or failure within a specific timeline [26]. Many financial institutions and individual lenders are often uncertain in the process of computing the future status of a credit contract. The Bayesian network (BN) framework has always been and continues to be a favorable approach to handling uncertainty since its formalization by Judea Pearl [22]; it has been used in various risk prediction applications including credit risk prediction [1, 10, 13, 15, 17, 19]. The BN framework can be used to define BN models based on expert knowledge and experience of previous credit contracts; alternatively, the BN models can be learned from retrospective data that constitutes details of instances of credit contracts and their respective outcomes. The model learning approach is currently more prevalent inline with trending data-driven paradigms in developing risk prediction models.

Bayesian network methods have been studied before for predicting credit risk of credit contract cases from a Ugandan financial institution [20]; however, just a small subset from an infinitely possible and potentially better set of BN models have been explored. An in-depth investigation associated with the capabilities and opportunities that more promisingly appropriate BN modeling methods and

---

P. Nabende (✉) · S. Senfuma
Department of Information Systems, School of Computing and Informatics Technology, CoCIS, Makerere University, Kampala, Uganda
e-mail: pnabende@cis.mak.ac.ug; sensam45@gmail.com

J. Nakatumba-Nabende
Department of Computer Science, School of Computing and Informatics Technology, CoCIS, Makerere University, Kampala, Uganda
e-mail: jnakatumba@cis.mak.ac.ug

© Springer Nature Switzerland AG 2021
H. R. Arabnia et al. (eds.), *Advances in Artificial Intelligence and Applied Cognitive Computing*, Transactions on Computational Science and Computational Intelligence, https://doi.org/10.1007/978-3-030-70296-0_35

parameter settings offer still lacks. In this chapter, we apply more BN modeling methods and parameter settings, which we use to explore a significantly larger set of BN models for predicting credit risk based on credit contract cases from a Ugandan financial institution.

The chapter is organized as follows: Sect. 2 presents related work, Sect. 3 presents the various Bayesian network methods, Sect. 4 presents the experimental setup, Sect. 5 presents results and discussion, and Sect. 6 concludes the chapter with pointers to future work.

2 Related Work

Credit risk prediction is usually used to make a judgment on a general categorical outcome [5, 25, 28] or to estimate a real number credit limit [9] for a specific credit contract. This chapter focuses on the former case where various credit prediction modeling methods have been used [6, 8, 24]. In particular, Bayesian networks have gathered increasing interest in the recent past. They have been used for credit risk scoring, and they have proven to perform really well even in extreme datasets with other problematic issues such as class imbalance issues [17]. The following paragraph highlights some previous major studies concerned with the use of Bayesian networks in credit risk prediction.

Pavlenko and Chernyak [21] introduced Bayesian networks for modeling credit concentration risk where they focused on tree adjoining BNs (TAN) and k-BNs (where each feature is allowed to have upto k parents). Peng et al. [23] developed a two-step approach to evaluate classification algorithms for financial risk prediction, and from their work they show that linear logistic, Bayesian network, and ensemble methods lead to the best classifiers. Peng et al. [23] evaluate both a Naïve Bayes model and a Bayesian network model, but they do not provide specific details of the model in terms of structure learning and parameter estimation. Leong [17] evaluates the performance of BNs against logistic regression and neural networks while addressing three key issues in credit scoring (sample selection, class imbalance, and real-time implementation). Leong [17] uses the tree adjoining Naïve Bayes model where the Chow–Liu algorithm is employed to find an optimum tree and the minimum description length (MDL) function is used to determine the best candidate Bayesian network model. Abid et al. [1] use a hill climbing method and a k2 scoring algorithm to learn the Bayesian network structure and model on personal loan data from a Tunisian Commercial Bank. Krichene [15] applies a Naïve Bayes classifier for loan risk assessment to a database of credits granted to industrial Tunisian companies over the years 2003–2006. Anderson [2] uses Bayesian networks to model reject inference where the Bayesian information criterion is used to measure how well a network fits the training data.

All the studies reviewed in this section evaluate only one or just a small number of Bayesian network methods for credit risk prediction. In comparison to previous studies, we apply and evaluate a broader range of Bayesian network methods including Naïve Bayes; tree augmented Naïve Bayes; several structure learning methods such as genetic search, k2, hill climbing, and simulated annealing; and ensemble methods such as averaged n-dependency estimation, and boosting and bagging where BNs are base classifiers. To the best of our knowledge, this is the first application and evaluation of a broad range of BN models on Ugandan credit contracts.

3 Bayesian Networks Methods

3.1 Formal Definition

A BN model specifies probabilistic relationships among a set of random variables (features or attributes) that characterize a specific domain (such as the credit risk prediction task of this study). Relationships between features in the domain can be represented graphically by a directed acyclic graph where each node in the graph contains probabilistic information (parameters) for a feature (based on the values of its parent/related features); the arcs in the graph represent conditional dependencies (between features that have a path between them in the graph), and at the same time conditional independencies (for features with no path between them). Different BN models are as a result of variations in the types of possible relationships between features, and in parameter estimation.

3.2 Naïve Bayes

The Naïve Bayes [14] model is the simplest of all BNs because of the conditional independences assumed between predictor features given the class feature. This assumption can make Naïve Bayes models limited in cases where there are valid and significant dependencies between predictor features. On the other hand, it can make the Naïve Bayes models very effective predictors where more complicated Bayesian networks are infeasible and when predictor features are indeed mostly independent. We apply the Naïve Bayes method as well as other semi-Naïve Bayes methods such as the tree augmented Naïve Bayes (TAN) and averaged n-dependency estimation that partially relax the strict assumption.

3.3 Tree Augmented Naïve Bayes

The tree augmented Naïve Bayes model provides the ability to violate the conditional independence assumption of the Naïve Bayes models by allowing each of the predictor features to depend on at least one other predictor feature and the class feature, thus leading to a semi-Naïve Bayes model [12]. In this chapter, we apply two immediate methods in this regard: the tree augmented Naïve Bayes (TAN) method and the averaged n-dependency estimators (AnDE). The AnDE method is covered under ensemble methods. The TAN method allows a feature to depend on the class feature and at most one of the other features such that there are at most two parents for each of the predictor features.

3.4 Structure Learning Methods

The relaxation of conditional independence restrictions presents a larger set of possible BN structures for a specific domain. We can specify BN structures by hand if we are very sure of the relationships between features in a domain. However, we do not know any BN structures that are appropriate for modeling credit risk for the Ugandan credit contracts. There are several methods for learning BN structures, and we explore some of the commonly used ones as described in the following subsections.

3.4.1 Genetic Search

The genetic search method [16] is based on concepts from Darwin's theory of evolution. We use the method to find the best scoring BN structure after a number of iterations (generations). In this framework, each BN structure is encoded as a chromosome string. The search process starts with an initial population of BN structures (of a preset population size). The fitness of each chromosome in the current generation is determined using a fitness function; the fittest chromosomes in the current population are selected as parent chromosomes. Probabilities are then used to apply crossover and mutation operations on the parent chromosomes to produce offspring chromosomes for the next generation. Reproduction takes place until the number of offspring chromosomes reaches a preset population size of descendant chromosomes. This process is repeated for a specific number of generations, and at the end, the fittest chromosome (or the highest scoring Bayesian network structure) is returned. In this study, it became infeasible to use the method for all features; however, it was easily applicable for almost half the total number of features.

3.4.2 k2

The k2 method [7] establishes the parents of each feature up to a maximum number. For each feature, the other features are tested as possible parents by determining the probability of the dataset. Parents that result in higher probabilities of the dataset are established as parent features of a specific feature in the network. For n_f features, we can vary the maximum number of parents that the algorithm uses up to $n_f - 1$.

3.4.3 Hill Climbing Methods

The hill climbing method [18] starts with an initial structure (usually an empty BN) and proceeds with local changes in future steps. A local change can be effected by addition, deletion, or reversal of an arc. Local changes lead to neighboring structures that are evaluated according to their scores over the dataset. The method chooses structures with the highest improvements, and the process continues until there is no further improvement in the scores. The main parameters we explore here are the maximum number of parents allowed for each feature in the structure and the set of local changes that can be done.

One limitation with the hill climbing method is the high likelihood of ending up in local optima. We explore two methods that are specifically aimed at overcoming this problem. The **repeated hill climbing** method enables the hill climbing method to jump out of local optima by letting it to execute a number of random restarts. The **Tabu search** method [3] enables the hill climbing method to accept worse solutions (that are not in a "Tabu" (forbidden) list) whenever it reaches a local optimum. The parameters above for hill climbing apply for these extended methods as well.

3.4.4 Simulated Annealing

The simulated annealing method [3] is based on concepts from metal annealing where a metal is melted at high temperatures and then carefully cooled to realize a stronger nonbrittle structure. In optimization, the simulated annealing method assigns a problem a temperature T, and the quality of a solution at this temperature can be conceptualized as an energy level, analogous to the state of a metal. In BN structure learning, the method starts with an arbitrary structure and recursively selects a new structure from the neighborhood of the current structure. Neighborhood structures are also generated in a similar way that local changes are made in hill climbing, that is, addition, deletion or reversal of arcs. If the quality of a neighborhood structure is better than that of the current structure, then the current structure is replaced; otherwise, the neighborhood structure is accepted with a probability = exp (score(neighborhood structure)—score(current structure)/T). During execution of the method, the temperature is decreased slowly (according to some cooling schedule) so as to accept fewer and fewer structures that are worse than the current structure. The best structure is returned at the end of the cooling process.

3.5 Ensemble Classifiers

3.5.1 Averaged *n*-Dependency Estimator (A*n*DE)

The A*n*DE method [27] fixes a subset of n features as common parents in addition to the class feature resulting in an augmented Naïve Bayes structure. The method then considers the full set of all possible structures on the dataset. The final prediction from the method is an average over the full set of possible structures. When $n = 0$, only the class feature is a parent, and we end up with only one structure (the Naïve Bayes structure). When $n = 1$, one of the other features becomes a super parent in addition to the class feature resulting into what is called a one dependency classifier, and the final prediction is an average over the full set of one dependency classifiers. In this study, we apply up to $n = 2$ (A1DE and A2DE).

3.5.2 Boosting and Bagging

Boosting and Bagging are general ensemble methods that generate various distributions of the training data that are used to learn various prediction models from a given learning method [4, 11]. The final prediction is an average over the outputs from the various prediction models. The main differences between the two methods are associated with the training process and how the final prediction is made. The training process for the bagging method is parallel whereby each prediction model is learned independently. The process for boosting is sequential whereby the training data distributions for successive models are weighted so that incorrectly predicted cases are prioritized for successive models to focus on. In deciding the final prediction, boosting applies relatively higher weights to outputs from better performing models, whereas bagging simply finds the average. In this study, we configure some of the Bayesian network learning methods described above (such as hill climbing) as the learning methods for generating prediction models under boosting and bagging.

4 Experimental Setup

4.1 The Dataset

In this chapter, we use the dataset that is described in [20]. The dataset has 12 features: gender (two categories), age (numeric), job (42 categories), work experience (numeric), amount of credit (numeric), individual credit frequency (numeric), time period of credit (numeric), type of collateral (8 categories), interest rates (numeric), value of collateral (numeric), monthly turn over (numeric), and

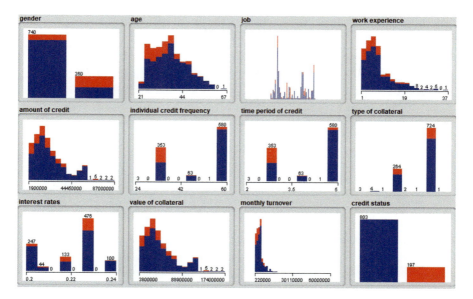

Fig. 1 Feature distributions in the credit dataset. Red regions are associated with repayment failure (bad), while blue regions are associated with repayment success (good)

credit status (the class feature). Figure 1 shows the distribution of features against the class feature.

We conduct two sets of experiments using this dataset: the first set involves use of all the features in the dataset and the second involves the use of a subset of features selected from three common feature selection methods (Pearson correlation co-efficient, information gain, and learner-based feature selection). As mentioned earlier, some structure learning methods such the genetic algorithms necessitated some form of dimensional reduction, although the total number of features in our dataset (12 features) is much lower than the number of features for cases that have several dozens or even hundreds of features. Table 1 shows the ranking of the strength of relationship between the predictor features and the class feature based on the Pearson correlation co-efficient and information gain. Table 2 shows the subset of features resulting from the learner-based feature selection method for three common classification methods (C4.5, Bayesian network using k2 algorithm, and Naïve Bayes).

Table 1 has four common features in the top five across the correlation-based measures: time period of credit, individual credit frequency, gender, and age. Table 2 shows three features that are consistently selected across the learner-based methods.

To establish the final feature set, we compare prediction performance between use of the high ranking features from the correlation measures and all the selected features from the learner-based methods. Table 3 shows the prediction accuracy results. It is clear that the selected features from the learner-based methods lead

Table 1 Feature ranking from Pearson correlation co-efficient and information gain

Rank	Pearson correlation co-efficient		Information Gain	
1	Time period of credit	0.480	Job	0.340
2	Individual credit frequency	0.480	Individual credit frequency	0.166
3	Gender	0.457	Time period of credit	0.166
4	Age	0.247	Gender	0.135
5	Monthly turnover	0.129	Age	0.046
6	Work experience	0.119	Interest rates	0.036
7	Amount of credit	0.115	Monthly turn over	0.025
8	Value of collateral	0.115	Value of collateral	0.015
9	Type of collateral	0.091	Amount of credit	0.015
10	Job	0.090	Work experience	0.013
11	Interest rate	0.071	Type of collateral	0.012

Table 2 Selected features from the learner-based feature selection method

C4.5	Bayesian network	Naïve Bayes
Gender	Job	Job
Job	Amount of credit	Individual credit frequency
Amount of credit	Individual credit frequency	Type of collateral
Individual credit frequency	Type of collateral	
Type of collateral		

Table 3 Prediction performance comparison between feature selection methods

Model	Correlation-based	Learner-based
Naïve Bayes	0.808	0.882
Global K2 (max of 3 parents)	0.828	0.933
Global hill climbing (max of 3 parents)	0.870	0.930
A1DE	0.874	0.918
A2DE	0.875	0.922

to significantly better performing Bayesian network models than the high ranking features from the correlation-based methods.

4.2 Evaluation

The main aim of the task in this study is to prevent bad outcomes that are significantly costly than good outcomes. So the prediction of bad cases is very critical. If the dataset has a reasonable level of balance between classes, then absolute measures such as prediction accuracy should suffice. We evaluate the accuracy of the BN models over a balanced dataset we derived by undersampling the majority class. For the original dataset (which is imbalanced to a ratio of

An Evaluation of Bayesian Network Models for Credit Risk Prediction   469

0.8), we consider alternative metrics. Firstly, we consider the true positive rate (TPR) on the minority bad cases class. The TPR on the "bad" class is the ratio of correctly predicted "bad" outcomes to the total number of "bad" cases in the test set. We also consider the area under ROC (AUROC) and F1 measure that are better alternatives than accuracy for evaluating models on imbalanced datasets. The AUROC measures how much a model is capable of distinguishing between the two classes. The F1 measure is the harmonic mean between recall and precision. All metrics vary between 0 and 1 such that values closer to 0 for accuracy, TPR and F1 suggest "poor" models, whereas those closer to 1 suggest "high quality" models. For AUROC, values closer to 0.5 suggest poor discriminative capability and those closer to 1 suggest high discriminative capability. A model that always results in values closest to 1 for each of these measures should be considered a "high quality" model.

5   Results and Discussion

We present different sets of results starting with the case where the BN model learning methods use all features in the original dataset. Table 4 shows the results for the first set.

Table 4 shows that the Naïve Bayes models post the lowest values for accuracy, TPR, and AUROC because of their poor performance in predicting instances associated with the majority class. However, a Naïve Bayes model that uses kernel estimation and supervised discretization has the highest "bad" class TPR

Table 4 Prediction evaluation results for the case where all features are used

Model	Accuracy	TPR	F1	AUROC
Naïve Bayes	0.849	0.761	0.852	0.893
Naïve Bayes (kernel estimator)	0.875	0.736	0877	0.921
Naïve Bayes (supervised discretization)	0.860	**0.777**	0.866	0.893
A1DE minus weighting	0.892	0.726	0.892	0.947
A1DE with weighting	0.907	0.706	0.907	0.950
A2DE minus weighting	0.915	0.721	0.913	0.954
A2DE with weighting	0.922	0.716	0.919	0.959
Global (hill climbing)	0.928	0.766	0.927	0.958
Global (K2)	0.931	0.736	0.928	**0.962**
Global (repeated hill climbing)	0.928	0.736	0.928	0.952
Global (simulated annealing)	0.933	0.751	0.931	0.957
Bagging (BN from hill climbing)	0.938	0.751	0.936	0.957
Bagging (BN from K2)	**0.939**	0.761	**0.937**	0.957
Boosting (BN from hill climbing)	0.925	0.751	0.923	0.958

The bold values represent the highest value achieved per evaluation metric

Table 5 Prediction evaluation results for the case where a subset of features is used

Model	Accuracy	TPR	F1	AUROC
Naïve Bayes	0.872	0.690	0.873	0.892
Naïve Bayes (kernel estimator)	0.876	0.650	0.874	0.898
Naïve Bayes (supervised discretization)	0.882	0.670	0.881	0.894
A1DE minus weighting	0.913	0.665	0.909	0.930
A1DE with weighting	0.918	0.665	0.913	0.938
A2DE minus weighting	0.926	0.716	0.923	0.942
A2DE with weighting	0.922	0.701	0.919	0.944
Global (hill climbing)	0.930	0.716	0.927	0.950
Global (K2)	0.933	0.726	0.930	0.949
Global (repeated hill climbing)	0.933	0.726	0.930	0.948
Global (simulated annealing)	0.929	0.731	0.929	0.952
Global (genetic algorithm search)	0.934	0.726	0.931	0.948
Bagging (BN from hill climbing)	**0.935**	0.731	**0.932**	**0.954**
Bagging (BN from K2)	**0.935**	**0.736**	**0.932**	0.952
Boosting (BN from hill climbing)	0.930	**0.736**	0.927	0.929

The bold values represent the highest value achieved per evaluation metric

value. Unfortunately, it makes relatively many mistakes on the "good" class, thus negatively affecting its overall performance. The other models that achieve relatively high TPR values are those from the hill climbing and simulated annealing learning methods, and those from the boosting and bagging ensemble methods. The averaged n-dependency estimators have relatively lower TPR values than the NB models, but they end up achieving better overall results because of their good performance on the majority class. The Bayesian network models learned from structure learning and global scoring algorithms such hill climbing and simulated annealing lead to even better overall results over the AnDE methods. The ensemble approach of bagging leads to the highest overall result in this case. However, these results show some difficulty for all models in correctly predicting the "bad" class cases. The area under ROC (AUROC) results suggest excellent predictive capabilities for most of the models. The global k2 algorithm results in a structure with the highest predictive capability.

Table 5 shows the results for the second set of experiments where a subset of features is considered. As the number of features involved in this case makes the model sizes manageable, we were able to apply and evaluate more learning methods such as genetic algorithms that had proven difficult to use in the first set of experiments with a relatively larger feature set.

As Table 5 shows, there is no significant gain in predictive performance between this set of experiments and the first except that the algorithms run faster and it is feasible to apply those that were previously infeasible for the full feature set. However, the trend of prediction results is similar to that in Table 4 where performance generally improves from the simplest BN models (Naïve Bayes), to

An Evaluation of Bayesian Network Models for Credit Risk Prediction 471

those that relax independence assumptions, through to BN models from ensemble methods. The AUROC values in Table 5 suggest excellent discriminatory power of all the other BN models except for the Naïve Bayes models. The Naïve Bayes models this time also result in the lowest TPR results.

5.1 Error Analysis

The prediction results in the two Tables 4 and 5 show that the Bayesian network models learned so far can achieve up to a peak TPR value of 0.780 (on the "bad" class cases) and an accuracy of 0.940. These results are comparable to those in [20]. Usually, an error analysis can help to establish cases where prediction models are making wrong judgments so that a post-processing solution is designed to handle these difficult cases. However, we need to first determine whether a post-processing solution will even be possible. We created ten train and test datasets in a manner akin to tenfold cross-validation. After that the hill climbing and simulated annealing (with global scoring metrics) methods were applied to learn Bayesian networks on each of the ten training subsets. We then evaluated each of the resulting Bayesian network models on their corresponding test sets and extracted cases where each of the BN models were making incorrect predictions to see if we could identify any patterns. We find that the set of features that are associated with the Ugandan dataset could be the limitation in achieving further prediction quality improvements. We believe that additional information or features can help to learn better models that correctly judge the difficult cases.

As a last option, we attempted to find out what would result from a test of the two best performing models from the folds on each of the other test sets. Table 6 shows the TPR and AUROC values from these tests. BN_f represents the best performing Bayesian network model for each fold f ($f = 1, \ldots, 10$). In this case BN_f generalizes BN models learned either through simulated annealing or hill climbing using a global scoring metric. More specifically, the hill climbing method resulted in better TPR and AUROC results in eight out of tenfolds, and so we chose the two best performing BN models from twofolds where TPR was highest. As Table 6 shows, the highest TPR values resulted from the fifth (0.929) and the tenth (0.905) folds. We use $BN05_{HC+G}$ and $BN10_{HC+G}$ in Table 6 to denote the two BN models, respectively, and where the $HC + G$ subscript refers to hill climbing using a global scoring metric. The notation t01, ..., t10 in the table corresponds to the test sets for each of the folds $f = 1, \ldots, 10$. Table 6 shows that the two Bayesian network models can only improve the TPR and AUROC values and in the worst case achieve the best result obtained by the best corresponding Bayesian network model on a specific test set. The two models actually result in significant improvements in discriminative capability as suggested by the AUROC values for each of the test sets. Figure 2 shows the BN structure for the best performing BN model in this study (BN_{HC+G}).

Table 6 Cross-validation results

	Model	t01	t02	t03	t04	t05	t06	t07	t08	t09	t10
TPR	BN_f	0.667	0.826	0.875	0.727	0.929	0.727	0.680	0.857	0.727	0.905
	$BN05_{HC+G}$	0.667	0.826	**1.000**	**0.818**	0.929	**0.773**	**0.760**	0.857	0.727	0.905
	$BN10_{HC+G}$	0.667	0.826	**1.000**	**0.818**	0.929	**0.773**	0.720	0.857	**0.773**	0.905
AUROC	BN_f	0.969	0.973	0.987	0.933	0.982	0.962	0.890	0.960	0.953	0.980
	$BN05_{HC+G}$	**0.977**	**0.982**	**1.000**	0.973	0.982	**0.990**	**0.972**	**0.988**	**0.977**	**0.986**
	$BN10_{HC+G}$	0.972	0.980	0.999	**0.974**	**0.996**	0.984	0.968	0.987	0.976	0.980

The bold values represent the highest value achieved per evaluation metric

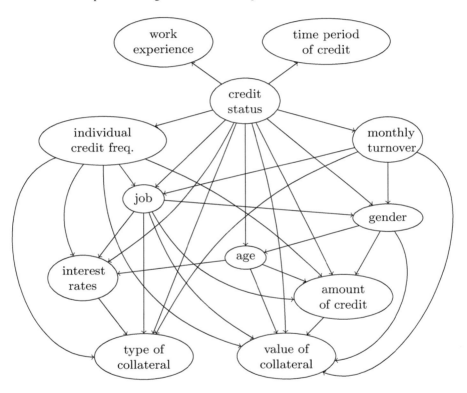

Fig. 2 Structure of the best performing BN model learned using a hill climbing method with a maximum number of 9 parents and reversal of arcs

The BN structure in Fig. 2 shows that most of the features have a considerable number of dependencies with other features in the dataset except only two features ("work experience" and "time period of credit"). The structure suggests that the presence of this considerable number of dependencies results in better credit risk prediction quality.

6 Conclusion and Future Work

In this chapter, we have applied and evaluated several Bayesian network methods to modeling credit risk prediction using credit applications from a Ugandan financial institution. Prediction results show that the resulting Bayesian network models achieve state-of-the-art results from other prominent classification methods such as random forests. However, because of class imbalance in the dataset, all Bayesian network models mostly skewed their predictions toward the majority class, thus leading to poor true positive rate values for the minority class. We established that one limitation that complicates any effort to improve TPR values for the minority class is because the features used were limited toward providing adequate discriminative quality. We believe that another study that considers additional features may lead to better TPR values. One approach we took to try to improve the quality of prediction was to extract different subsets of the dataset for training and testing in a manner akin to k-fold cross-validation. A Bayesian network model learned from one of the subsets considerably improved minority class TPR values so far on most of the test sets. This leads us to believe that by breaking down the training data into smaller subsets, we may be able to model very difficult to predict cases; however, it remains to be seen how such smaller models can be integrated to lead to a more effective predictive system that deals with the class imbalance problem and with very difficult cases that share characteristics across classes. In future, we would also like to explore various methods for addressing the class imbalance in the credit applications dataset we used for this study.

Acknowledgments This study was supported by funds from the SIDA Project 317/BRIGHT under the Makerere-Sweden bilateral research program 2015–2020.

References

1. L. Abid, S. Zaghdene, A. Masmoudi, S.Z. Ghorbel, Bayesian network modeling: a case study of credit scoring analysis of consumer loan's default payment. Asian Econ. Financial Rev. **7**(9), 846–857 (2017)
2. B. Anderson, Using Bayesian networks to perform reject inference. Exp. Syst. Appl. **137**, 349–356 (2019)
3. R.R. Bouckaert, Bayesian Belief Networks: from construction to inference, Ph.D. Thesis, Utrecht (1995)
4. L. Breiman, Bagging predictors. Mach. Learn. **24**(2), 123–140 (1996)
5. A. Byanjankar, M. Heikkil, J. Mezei, Predicting credit risk in peer-to-peer lending: a neural network approach, in *2015 IEEE Symposium Series on Computational Intelligence* (2015), pp. 719–725
6. N. Chen, B. Ribeiro, A. Chen, Financial credit risk assessment: a recent review. Artif. Intell. Rev. **45**, 1–23 (2016)
7. G.F. Cooper, E. Herskovits, A Bayesian method for induction of probabilistic networks from data. Mach. Learn. **9**(4), 309–347 (1992)

8. J.N. Crook, D.B. Edelman, L.C. Thomas, Recent developments in consumer credit risk assessment. Eur. J. Oper. Res. **183**, 1447–1465 (2007)
9. D. Du, Forecasting credit losses with reversal in credit spreads. Econ. Lett. **178**, 95–97 (2019)
10. N. Fenton, M. Neil, Decision support software for probabilistic risk assessment using Bayesian networks. IEEE Softw. **31**(2), 21–26 (2014)
11. Y. Freund, R.E. Schapire, Experiments with a new boosting algorithm, *Machine Learning: Proceedings of the Thirteenth International Conference, (ICML)*, vol. 96 (1996), pp. 148–156
12. N. Friedman, D. Geiger, M. Goldszmidt, Bayesian network classifiers. Mach. Learn. **29**(2–3), 131–163 (1997)
13. A. Gandy, L.A. Veraart, A Bayesian methodology for systemic risk assessment in financial networks. Manage. Sci. **63**(12), 4428–4446 (2016)
14. G.H. John, P. Langley, Estimating continuous distributions in Bayesian classifiers, in *Proceedings of the Eleventh Conference on Uncertainty in Artificial Intelligence* (1995), pp. 338–345
15. A. Krichene, Using a Naïve Bayesian classifier methodology for loan risk assessment - evidence from a Tunisian commercial Bank. J. Econ. Finance Admin. Sci. **22**(42), 3–24 (2017)
16. P. Larrañaga, M. Poza, Structure learning of Bayesian networks by genetic algorithms, in *New Approaches in Classification and Data Analysis. Studies in Classification, Data Analysis, and Knowledge Organization*, ed. by E. Diday, Y. Lechevallier, M. Schader, P. Bertrand, B. Burtschy (Springer, Berlin, 1994)
17. C.K. Leong, Credit risk scoring with Bayesian network models. Comput. Econ. **47**(3), 423–446 (2016)
18. D. Margaritis, Learning Bayesian network model structure from Data, Ph.D. Thesis, School of Computer Science, Carnegie Mellon University, Pittsburgh (2003)
19. K. Masmoud, L. Abid, A. Masmoud, Credit risk modeling using Bayesian network with a latent variable. Exp. Syst. Appl. **127**, 157–166 (2019)
20. P. Nabende, S. Senfuma, A study of machine learning models for predicting loan status from Ugandan loan applications, in *Proceedings of the 21st International Conference on Artificial Intelligence* (2019), pp. 462–468
21. J. Pavlenko, O. Chernyak, Credit risk modeling using Bayesian networks. Int. J. Intell. Syst. **25**(4), 326–344 (2010)
22. J. Pearl, *Probabilistic Reasoning in Intelligent Systems* (Morgan Kaufmann, San Mateo, 1988)
23. Y. Peng, G. Wang, G. Kou, Y. Shi, An empirical study of classification algorithm evaluation for financial risk prediction. Appl. Soft Comput. **11**(2), 2906–2915 (2011)
24. M.R. Sousa, J. Gama, E. Brandão, A new dynamic modeling framework for credit risk assessment. Exp. Syst. Appl. **45**, 341–351 (2016)
25. M. Soui, S. Smiti, S. Bribech, I. Gasmi, Credit card default prediction as a classification problem, in *Proceedings of the International Conference on Industrial, Engineering and other Applications of Applied Intelligent Systems* (2018), pp. 88–100
26. B. Twala, Combining classifiers for credit risk prediction. J. Syst. Sci. Syst. Eng. **18**(3), 292–311 (2009)
27. G.I. Webb, J.R. Boughton, F. Zheng, K.M. Ting, H. Salem, Learning by extrapolation from marginal to full multivariate probability distributions: decreasingly Naïve Bayesian classification. Mach. Learn. **86**, 233–272 (2012)
28. L. Yu, R. Zhou, L. Tang, R. Chen, A DBN-based resampling SVM ensemble learning paradigm for credit classification with imbalanced data. Appl. Soft Comput. **69**, 192–202 (2018)

Part IV
Artificial Intelligence – Fundamentals, Applications, and Novel Algorithms

Synthetic AI Nervous/Limbic-Derived Instances (SANDI)

Shelli Friess, James A. Crowder, and Michael Hirsch

1 Introduction: The Modified OODA Loop

The original OODA loop is a four-step process/approach to decision making focused on collecting and filtering information (Observing), putting the information in proper context (Orienting), making the most appropriate, context-relevant decision available (Deciding), and then executing the decision (Acting) [1] (see Fig. 1). This process was developed by USAF Colonel (ret.) John Boyd [2, 3]. Colonel Boyd applied this process for real-time military operational situations. The OODA loop concept has been more recently utilized across commercial applications and overall learning processes [4, 5]. When looked at from the artificial cognitive system perspective, this process is easily adaptable to AI inferencing and decision making.

Many systems utilize AI and machine learning (ML) to augment the OODA loop process. These steps include:

Observe: data cleansing, metadata creation, data warehousing, and data correlation.
Orient: data analytics, data classification, basic ML.
Decide: decision support, actionable intelligence, AI inferencing.
Act: human/system interface, gathering feedback.

S. Friess
School of Counseling, Walden University, Minneapolis, MN, USA
e-mail: shelli.friess@mail.waldenu.edu

J. A. Crowder (✉)
Colorado Engineering, Inc., Colorado Springs, CO, USA
e-mail: jim.crowder@coloradoengineering.com

M. Hirsch
ISEA TEK LLC, Maitland, FL, USA
e-mail: mhirsch@iseatek.com

© Springer Nature Switzerland AG 2021
H. R. Arabnia et al. (eds.), *Advances in Artificial Intelligence and Applied Cognitive Computing*, Transactions on Computational Science and Computational Intelligence, https://doi.org/10.1007/978-3-030-70296-0_36

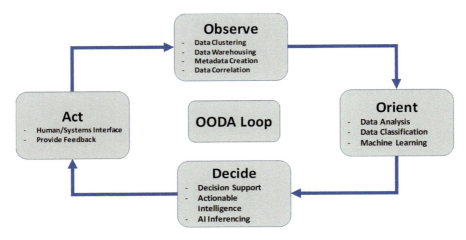

Fig. 1 The OODA loop [1]

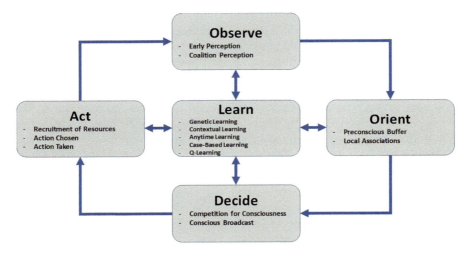

Fig. 2 The Observe, Orient, Decide, Act and Learn (OODAL) loop

The only part missing is an overall learning component for the OODA process. Every part of the OODA loop lends itself to learning. Learning to observe/collect/generate the right information, learning how to properly orient and provide the right context, learning what decisions are most appropriate for given situations, and learning how to execute (Act) on these decisions. For these reasons we introduce a new version of the OODA loop, called the Observe, Orient, Decide, Act, and Learn (OODAL) loop, illustrated in Fig. 2.

Synthetic AI Nervous/Limbic-Derived Instances (SANDI)   479

1.1 Observe

Early Perception Input arrives through the AI entity's senses. Specialized perception cognitive perceptrons (CP) descend on the input. Those that find features relevant to their specialty activate and broadcast their needs to the AI's system.

Coalition Perception Activation passes from CP to CP within the system. The attention manager brings about the convergence CPs from different senses and into coalitions. Relevant contexts are identified (recognized) along with objects, events, and their relations by the perceptual memory system. This could entail simple reactive orientations based on a single input or more complex orientations requiring the convergence of several different CPs.

1.2 Orient

Perception-to-Preconscious Buffer Perception, including some of the data, plus the context and meaning, is stored in preconscious buffers working memory. These buffers may involve visual, spatial, phonological, and other kinds of information. Contextual orientation is part of the preconscious perception written during each cognitive cycle into the preconscious working memory buffers.

Local Associations Using the incoming perception and the residual contents of the preconscious buffers as cues, including emotional content and context, local CP associations are automatically retrieved from transient episodic memory and from long-term associative memory. Contextual associations are part of the cue that results in local associations from transient episodic and declarative memory. These local associations contain records of the agent's past observations in associated situations. The more common the occurrence is observed, the more familiar the association and the stronger the neural pathway.

1.3 Decide

Competition for Consciousness Attention CPs, whose job it is to bring relevant, urgent, or insistent events and information to the AI entity's consciousness, view long-term working memory. Some of them gather information, form coalitions, and actively compete for access to consciousness. The competition may also include attention CPs from a recent previous cycle. Present and past contexts and observations influence the competition for consciousness in each cognitive cycle. Strong affective content strengthens a coalition's chances of coming to consciousness [6], i.e., emotional triggers. One area that has to be resolved is the case where competing coalitions have the same strength; possible conflicts must be adjudicated, and potential a priori knowledge or learning might provide such adjudication.

Conscious Broadcast A coalition of CPs, typically an attention CP and its covey of related information CPs carrying content, gains access to the system and has its contents' broadcast. This broadcast is hypothesized to correspond to phenomenal consciousness. The conscious broadcast contains the entire content of consciousness including the affective portions. The contents of perceptual memory are updated considering the current contents of consciousness, including feelings/emotions, as well as objects, events, and relations. The stronger the affect, the stronger the encoding in memory (emotional triggers). Transient episodic memory is updated with the current contents of consciousness. At recurring times not part of a cognitive cycle, the contents of transient episodic memory are consolidated into long-term declarative memory. Procedural memory (recent actions) is updated (reinforced) with the strength of the reinforcement influenced by the strength of the affect.

1.4 Act

Recruitment of Resources Relevant behavioral CPs respond to the conscious broadcast. These are typically CPs whose variables can be bound from information in the conscious broadcast. If the successful attention CP was an expectation CP calling attention to an unexpected result from a previous action, the responding CP may be those that can help to rectify the unexpected situation. Thus, consciousness solves the relevancy problem in recruiting resources; however, it is possible to erode other coalitions through diffusion of the boundaries between them. This could lead to a CP being cut out of consciousness. The affective content (feelings/emotions) together with the cognitive content help to attract relevant resources (processors, neural assemblies) with which to deal with the current situation. This drives the AI entity's synthetic nervous system (SNS) to increase the current and information flows to relevant parts of the AI entity requiring the resources. The increase in current flow results from a stronger robotic heartbeat (explained later) to increase the mean voltage available to the systems requiring more resources.

Action Chosen The behavior subsystem chooses a single behavior (goal context), perhaps from a just instantiated behavior stream or possibly from a previously active stream. This selection is heavily influenced by activation passed to various behaviors influenced by the various feelings/emotions. The choice is also affected by the current situation, external and internal conditions, by the relationship between the behaviors, and by the residual activation values of various behaviors.

Action Taken The execution of a behavior (goal context) results in the behavior CPs performing their specialized tasks, which may have external or internal consequences. The acting CPs also include an expectation CP whose task it is to monitor the action and to try and bring to consciousness any failure in the expected results. This exception CP could be due to satisfaction, dissatisfaction, disappointment, or affirmation.

1.5 Learn

Genetic Learning The genetic learning agents inherit initial states from the memory system and inherit the initial parameters for behavior from the behavioral center of the cognitive system. The consciousness mechanism along with the mediator control the response of the learning agent, and direct its constraints based on the environment and the problems to be solved currently. This provides the priorities, preferences, goals, needs, and activation constraints (when you know you've learned something). The genetic agents (called genomes) adapt to the environment and gather information in order to make conclusions (learn) about the problem to be solved.

Contextual Learning In the cognitive OODAL environment, drives, priorities, and constraints influence learning and decision making. The behavioral subsystem receives situations and computes actions, while memories provide personality parameters and the various conscious agents' sensitivities to situational computation. If we think of the cross-connectivity of the neural layers as a matrix, we can compute emotional response from the column-wise fuzzy weightings, and the action response from the row-wise fuzzy weightings.

Anytime Learning Anytime learning is a general approach to continuous learning in a changing, real-time environment. There are software agents whose learning module continuously tests new strategies against a simulation model and dynamically update the memories used by the agent based on the results. When the model is modified, the learning process is restarted on the modified model. The *Anytime Learning System* (*ALS*) is assumed to operate indefinitely, and the execution system uses the results of the learning process whenever they are available. An experimental study utilizing a two-agent system of a plane keeping track of a vehicle on the ground is used [7].

Case-Based Learning The monitor's task is to decide how well the observed speeds, turns, and size of the target in the external environment match the current distributions or values assumed in the current models within the case-based learning system. Using the 50 most recent samples of the object/event changes, the monitor computes the observed mean and variance of these samples, and compares the observed values with the current simulation parameters, using the F-test to compare the variances and the t-test to compare the means. If either statistical test fails, the monitor changes the simulation parameters to reflect the new observed mean and variance of the target speed or turn. When a new event context is detected, the monitor updates this value in the simulation model. A change in simulation parameters then causes the genetic algorithm to restart [8]; i.e., the system must deal with either new information or new contexts.

Q-Learning An algorithm structure representing model-free reinforcement learning. By evaluating decisions based on "policies" and using stochastic modeling, Q-learning finds the best path forward in a Markov decision process. Q-learning, in this context, provides decision support for course of action execution. Q-learning algorithms involve a software agent, a set of states, and a set of actions per state. The Q function uses weights for various steps in conjunction with a discount factor in order to value rewards. The results of the learning process adjust the weighting functions. Q-learning plays an important role in reinforcement learning and deep learning models for AI entities. One example of deep Q-learning is in helping machine learning systems to learn and adapt strategies for decision support in AI inference systems, requiring fusion of the AI entities sensors (vision, sound, touch, etc.). Here, a convolution neural network uses samples of "objects/event of interest" to work up a stochastic model that provides the input to the "orient" block of the OODAL Loop [6].

Now that we have established how the functions of the OODAL loop work and interact, we need to establish what emotions mean to an artificial entity. Q-learning is of particular use to an AI entity, as it mimics the trial and error learning humans use constantly and allows the AI entity to learn as it interacts with its environment over time.

2 SANDI Emotional Contexts

For SANDI's artificial nervous/limbic system, drives, priorities, and constraints will all influence emotions. The SANDI behavioral system receives situations and first computes the locus of control for the observations and contexts. Figure 3 illustrates this concept. As in humans, SANDI will respond differently if it perceives that the situation is controlled externally, as opposed to internal control of the situation.

Decisions are governed by how SANDI perceives the context and overall situation, as well as the appropriate emotion for that situation in that context. Tables 1 and 2 illustrate how we have adapted the definitions of human emotions for the SANDI artificial nervous/limbic system. Table 1 presents the standard definitions of human emotions [8]. We note that there have been alternative views presented in the literature [9, 10], and the standard definitions will continue to evolve over time.

The definitions in Table 1 are adapted (shown in Table 2) to use in the inferencing and cognitive control components within the SANDI cognitive processing infrastructure.

Fig. 3 External vs. internal locus of control

Table 1 Basic definitions of human emotions

Emotion	Human definition
Anxiety	A feeling of worry/nervousness about an imminent event or something with an uncertain outcome.
Contentment	A state of happiness and satisfaction.
Sad	A feeling of hopelessness or grief over an event. In sadness, the reason is known, even if it isn't apparent how to overcome the event or issue.
Frustration	The feeling of being upset or annoyed due an inability to achieve something; a prevention of progress and/or success.
Curious	Inquisitive; tending to ask questions, investigate, and explore. No strong reason for asking questions other than they want to understand new situations or information/subjects.
Confident	Believe of certain success or feeling that conditions will improve and solutions will be found. Trust in abilities.
Fear	Unpleasant emotion caused by the belief that someone or something is likely to cause pain or a threat to self or others.
Sense of Self	the way a person thinks about and views his or her traits, beliefs, and purpose within the world. In a nutshell, a strong sense of self may be defined by knowing your own goals, values and ideals. Regardless of whether we are conscious of it or not.
Attraction	Ability to recognize something that evokes interest or pleasure form something or someone. Recognition that you have something to gain by another person or object.

484 S. Friess et al.

Table 2 Basic SANDI definitions for human emotions

Emotion	SANDI system definition
Anxiety	Loss of homeostasis. Either sensor data not well understood resource management not stable, or anomalies detected within internal systems.
Contentment	All systems performing to specification, data incoming from sensors is understood and resource management in stable (homeostasis).
Sad	Loss of resources, inability to repair or restore a loss.
Frustration	All resources are allocated and end result cannot be predicted or achieved.
Curious	Recognizes some attributes of current situations but not others, equates to need to fit all the pieces together to make sense and infer about new data.
Confident	Has resources to process all incoming data and find solution/inference, and the algorithms to come to correct inference.
Fear	Includes anxiety. Awareness of limits of resources in a given mission or problem set, system overload or impending failure.
Sense of Self	Knowing its own limits, resources, and constraints different from outside world (internal locus of control). Understands its own operating environment with both internal and external resources and data.
Attraction	Recognition of opportunity to increase resources.

3 SANDI Synthetic Nervous/Limbic System

We develop our synthetic nervous/limbic system concepts by looking at the human pulmonary cycle, the cycle through the body where blood supplies oxygen to the cells, allowing the body to function. When the heart pumps blood, it exerts pressure on the blood forcing it to move out. The cycle repeats continually, although the rate and pressure exerted changes, depending on the context of the body's situation. Heart rate is an important parameter. The same is true for SANDI. The monitor and controller allow more/less pressure (voltage) to various parts of the system as needed to satisfy the current actions chosen during the OODAL loop process. The relationship between the human pulmonary cycle and SANDI are given by:

Human blood flow rate = pressure difference/resistance
SANDI current flow rate = voltage/resistance

This can be represented mathematically as:

$$
\frac{\partial^2 W(x,t)}{\partial t^2} - C^2 \frac{\partial^2 W(x,t)}{\partial x^2} = 0
$$
$$
\frac{d^2 f(t)}{dt^2} = -Dc^2 f(t); \quad \frac{d^2 g(x)}{dx^2} = -Dg(x)
$$
$$
f(t) = A \sin\left(c\sqrt{D}t\right) + B \cos\left(c\sqrt{D}t\right); \quad g(x) = A \sin\left(\sqrt{D}x\right) + B \cos\left(\sqrt{D}x\right)
$$
$$
W(x,t) = 2.5 \cos(20\pi t) \cos(2\pi x) + 2.5
$$

$$(1)$$

Synthetic AI Nervous/Limbic-Derived Instances (SANDI) 485

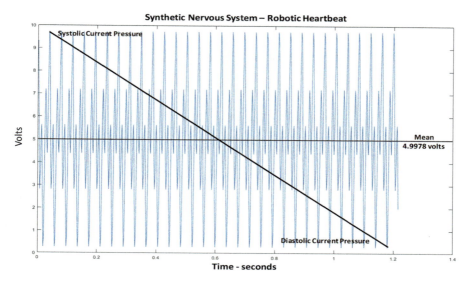

Fig. 4 The SANDI homeostasis robotic heartbeat

Figure 4 illustrates an example of the SANDI robotic heartbeat (not actually a heart). Different subsystems within SANDI need increased or decreased voltage to effect movement, or other functionality, based on the decisions and actions chosen. Figure 4 represents a homeostasis situation. The important consideration is the mean overall voltage allowed for a given subsystem. It may be necessary to allow less current to flow in one subsystem in order to increase the flow in another subsystem that requires increased functionality at a given instance, depending on the context of the overall SANDI system at any given time. This type of resource regulation will become increasingly important as robotic systems become autonomous. The synthetic heartbeat shown in Fig. 4 allows the autonomous cognitive system to increase or decrease heart rate similar to humans, to increase the overall current flow (like human blood flow regulation) when the entire system or subsystems (i.e., arms for lifting, legs for moving, etc.) require more power (voltage potential and current flow). Using the regulator equations shown in Eq. (1), the architecture provides the required "cognitive economy" and resource management which will be necessary in autonomous systems.

Figure 5 is a high-level illustration of the overall SANDI system, mimicking the human circulatory system. This current mediation circuitry is controlled by Eq. (1). Part of the current mediation is charging controls to continuously recharge the batteries that power the robotic entity. This may be solar collectors, hydrogen generators, etc.

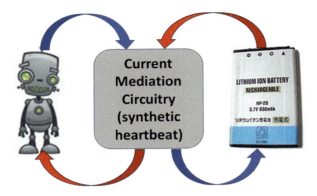

Fig. 5 SANDI circulatory regulation

3.1 Emotional Learning

The basics of artificial emotions start with the assumption that the SANDI processing framework must perform an action and the ensuing emotion contributes to the action chosen. We assume an Emotion Matrix E, with matrix element *Emotion(a, j)* representing a possible emotion within the SANDI framework. We note that *Emotion(a, j)* represents performing action *a* in situation *j*. Genetic learning agents perform an emotional learning procedure which has the following steps:

1. State *j*: choose an action in situation – (let it be action *a*; let the environment return situation *k*).
2. State *k*: feel the emotion for state *k* – *emotion(k)*.
3. State *k*: learn the emotion for *a* in *j* – *Emotion(a, j)*.
4. Change state: $j = k$; return to 1.

This SANDI learning procedure is a secondary reinforcement learning procedure for the emotion. The learning constant used in Step 3 (above) is:

$Emotion^0(a,j) = \text{genome}^0(\text{inherited})$
$Emotion^1(a,j) = Emotion^0(a,j) + emotion(k)$

This learning rule adds the emotion of being in the consequence situation, *k*, to the emotion toward performing action *a* in situation *j* on which *k* is the consequence. The above discussion works for emotional learning. But in general, we need algorithms that provide basic learning capabilities in a real-time dynamically changing environment, something that is typically difficult for learning systems. The learning system described in Sect. 1 will drive the SANDI real-time environment.

What is discussed below is the concept and algorithms for the anytime learning system (ALS) that provides the capability for concept learning in a real-time environment [11].

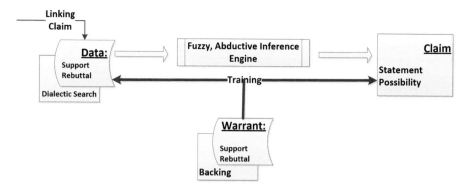

Fig. 6 The SANDI inference high-level architecture

4 The SANDI Agent-Based Processing and Inference Architecture

The SANDI artificial nervous/limbic system utilizes an intelligent agent processing infrastructure that provides the processing and inference capabilities required to handle the emotional processing. Figure 6 illustrates the high-level SANDI inference architecture. Coming out of the "orient" processing within the SANDI OODAL loop, possible hypotheses, or contexts, are handed off to the SANDI inference engine for processing to discover the range of scenarios possible for decisions/emotions, based on support and rebuttal information available. The context/situation that has the highest possibility is then transmitted as a "claim" to the decision engine to create a decision, based on the inference produced [12].

The fuzzy, abductive inference engine shown in Fig. 6 is facilitated using an intelligent software agent (ISA) processing framework. Figure 7 illustrates this ISA ecosystem. There are four basic agents:

1. Data steward agent: It takes in and does initial cleaning and processing of data.
2. Analyst agent: It analyzes the incoming data/situations and establishes context.
3. Reasoner agent: It takes the inputs from the analyst agent(s) and infers appropriate emotions and decisions.
4. Advisor agent: It formats the output in appropriate language, depending on the emotion and context from the decision engine.

5 SANDI Self-Organizing Emotional Maps

The SANDI processing environment utilizes fuzzy topical maps in order to (a) integrate diverse sources of information, (b) associate events, data, and information

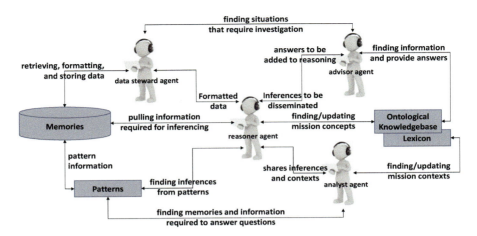

Fig. 7 The SANDI ISA ecosystem architecture

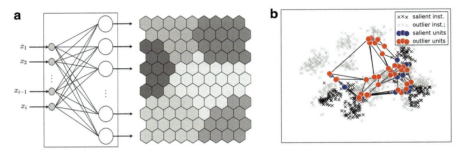

Fig. 8 (**a**) Emotional topical map; (**b**) topical map correlation

and (c) make observations and inferences based on the correlation/integration of disparate information. When combined with the inference engine shown in Fig. 6, the application of hybrid computing promises to allow the SANDI system to answer questions and make decisions that require interplay between doubt and belief [13]. The SANDI fuzzy topical maps allow information to be grouped by informational topics and those topics mapped to emotional topics. Figure 8 illustrates the conceptual basics of SANDI's fuzzy topical maps. In Fig. 8b, information is mapped to topics, based on data coming out of the inference engine. This information is sent to the emotional topical map, Fig. 8a. The shaded groupings indicate possible emotional inferences, with darker regions indicating a higher level of possibility [14].

Figure 9 illustrates an actual example of the fuzzy topical maps. The fuzzy, semantic, self-organizing topical map (SOM) shown in Fig. 9 is a method for analyzing and visualizing complex, multidimensional data, the correlation of which helps formulate an overall context for multi-sensor fusion and allows the inference engine to make sense of the emotional responses that may be required of SANDI.

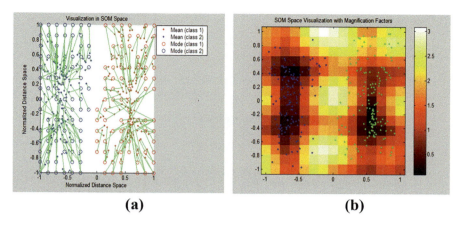

Fig. 9 (**a**) Actual data relationship connections; (**b**) emotional topical associations

The fuzzy SOM (FSOM) allows SANDI to index and describe knowledge structures that span multiple sources. The key features are the topics, their associations, and their occurrences in the FSOM. The topics are the areas on the FSOM that fall under a topic name. The associations describe the relationships between topics and subsequent emotions. The occurrences are the links from the FSOM into the data sources used to form the FSOM. It enables rapid and sophisticated dialectic inferences not possible with inductive or deductive logic/inferencing. This proposed approach is considered dialectic in that it does not depend on deductive or inductive logic, though these may be included as part of the warrant. Instead, it depends on non-analytic inferences (see Fig. 9) to find new possibilities based upon warranted examples. In Fig. 9b, the darker the color, the closer association the information has to an emotional topic (e.g., anxious).

6 Conclusions and Discussion

There has been much discussion over the last few years on the possibilities of artificial emotions [14]. There are numerous websites dedicated to this topic. What is required to facilitate AI systems is the ability to process emotions, read emotional reactions, and communicate effectively with humans at an emotional level. This is the processing and analytical framework that SANDI provides. This is just the beginnings and much more research and development are needed to even scratch the surface in the facilitation of artificial emotions that mimic human emotions. One of the issues that need to be thoroughly understand and thought through is "whether this is even a necessary or advantageous development for robotic systems." Future topics to be researched include determining at what point we would want this in our robotic systems.

References

1. S. Dreier, *Strategy, Planning & Litigating to Win* (Conatus Press, Boston, 2012). ISBN 9780615676951. OCLC 917563752
2. H. Hillaker, Code one magazine, "John Boyd, USAF Retired, Father of the F16" 1997 (1991).
3. C. Richards, *Certain to Win: The Strategy of John Boyd, Applied to Business* (Xlibris Corporation, 2004). ISBN 1-4134-5377-5
4. H. Linger, Constructing the infrastructure for the knowledge economy: Methods and tools. Theory Pract. **43**, 449 (2010)
5. E. Metayer, Decision making: It's all about taking off – and landing safely..., Competia (2011)
6. J. Crowder, S. Friess, J. Carbone, *Artificial Psychology: Psychological Modeling and Testing of AI Systems* (Springer International, New York, 2019)., ISBN 978-3-030-17081-3
7. J. Crowder, S. Friess, J. Carbone, Anytime learning: A step towards life-long AI machine learning, in *Proceedings of the 2019 International Conference on Artificial Intelligence*, (2019), pp. 16–21, Las Vegas, NV
8. J. Crowder, J. Carbone, S. Friess, *Artificial Cognition Architectures* (Springer International, New York, 2013)., 978-1-4614-8072-3
9. J. Crowder, S. Friess, J. Carbone, Implicit learning in artificial intelligent systems: The coming problem of real cognitive AI, in *Proceedings of the 2019 International Conference on Artificial Intelligence*, (2019), pp. 48–53, Las Vegas, NV
10. J. Crowder, S. Friess, Artificial neural emotions and emotional memory, in *Proceedings of the 2010 International Conference on Artificial Intelligence, Las Vegas, NV*, (2009)
11. H. Mobahi, S. Ansari, Fuzzy perception, emotion and expression for interactive robots, in *IEEE International Conference System, Man and Cybernetics*, (2003)
12. L. Pour, M. Bagher, Intelligent agent system simulation using fear emotion. World Acad. Sci. Eng. Technol. **48** (2008)
13. D. Ullman, "OO-OO-OO!" the sound of a broken OODA loop. CrossTalk **42**, 22–25 (2007)
14. E. Bevacqua, M. Mancini, R. Niewiadomski, C. Pelachaud, An expressive ECA showing complex emotions, in *Proceedings of Artificial and Ambient Intelligence*, (Newcastle University, 2007)

Emergent Heterogeneous Strategies from Homogeneous Capabilities in Multi-Agent Systems

Rolando Fernandez, Erin Zaroukian, James D. Humann, Brandon Perelman, Michael R. Dorothy, Sebastian S. Rodriguez, and Derrik E. Asher

1 Introduction

Like human organizations, agent teams are often formed to take advantage of supplementary similarities or complementary differences [7]. Similarity is leveraged by scaling the system size up with homogeneous agents that can work in parallel to increase the task completion rate. Differentiation in heterogeneous systems allows for specialization to complete diverse sub-tasks that can be integrated into completion of the full task. Degree of heterogeneity is a major differentiating factor among multi-agent systems [15]. In addition to heterogeneity based on form factor (e.g., a team of ground and aerial robots) or hardware-defined function [13], there are agent teams with identical hardware but heterogeneous behaviors. For example, it has been shown that a team of 5 robots with identical hardware, whose labor was divided between digging (prying boxes away from the wall) and twisting (clustering boxes in the center of the testbed), was able to cluster groups of boxes more efficiently than homogeneously programmed agent teams [14]. By altering the mix of diggers and twisters, they showed different levels of efficiency and reliability of the resultant systems. Heterogeneous behavior can also be achieved by dynamic state switching, where agents assume different roles based on their local perception of the environment and task needs, even if they are all running the same behavioral algorithms [8]. In the case of human–AI centaur chess teams, amateur human players and their AI teammates are able to achieve better performance than either

---

R. Fernandez (✉) · E. Zaroukian · J. D. Humann · B. Perelman · M. R. Dorothy · D. E. Asher
US CCDC Army Research Laboratory, Adelphi, MD, USA
e-mail: rolando.fernandez1.civ@mail.mil

S. S. Rodriguez
Department of Computer Science, University of Illinois at Urbana-Champaign, Urbana, IL, USA

© Springer Nature Switzerland AG 2021
H. R. Arabnia et al. (eds.), *Advances in Artificial Intelligence and Applied Cognitive Computing*, Transactions on Computational Science and Computational Intelligence, https://doi.org/10.1007/978-3-030-70296-0_37

human grandmasters or supercomputers, by crafting their method of interaction to leverage one another's strengths [6].

Heterogeneity has also been shown to naturally emerge from homogeneity. In some insect species, juvenile members can be deferentially nurtured so that they show marked dimorphism at maturity, enabling differentiation into soldier ants and drones [11]. So we see that heterogeneity can substantially improve system performance and can emerge from various environmental constraints such as nurture or training, hardware, algorithm, or local temporal dependent on-the-fly behavior differentiation.

The source of heterogeneity we study here is trained behavioral heterogeneity from a reinforcement learning (RL) approach. This raises interesting questions. If multiple agents are trained to complete a task over successive trials, do they learn to differentiate their behavior and take advantage of complementary differences? If so, does each agent learn a fixed role, or are the roles distributed dynamically? If a teammate is lost, changed, or compromised, have the other agents learned robust strategies to compensate? Partial answers to these questions can be found in previous results. Changing the reward structure in RL from zero sum to shared rewards can cause qualitatively different behaviors to emerge from the learning agents [16], implying that they are learning to cooperate. Agents may be allowed to train individual heterogeneous algorithms, train as a set with mutually known inputs and actions, or train homogeneously but with differing sensor information so that they make decisions locally. Even when agents are allowed to train heterogeneously, it is difficult to definitively say that they are learning to specialize or even consider their teammates [1]. They may simply be learning to maximize their own reward in a way that generally scales well to group settings (i.e., learning complementary similarities even if supplementary differences are possible) or find strategies that perform well irrespective of their teammates' actions.

In the following sections, the concept of heterogeneous strategies emerging from homogeneous capabilities is demonstrated. In the Methods section, we describe our simulation environment, where we test agents that are guided either by a learning algorithm or fixed strategies. Next, agent performance is shown with probability distributions and statistics for both homogeneous and heterogeneous cases in the Results section. Finally, the Discussion section points to the conclusions that were drawn from the results and provides further avenues of research associated with heterogeneous strategies from homogeneous capabilities in multi-agent systems.

2 Methods

A continuous bounded 2D simulation environment was utilized to train and evaluate a set of four agents (three predators and one prey, represented as circles) per model in the predator–prey pursuit task [4, 10], and a visualization of the task is shown in Fig. 1. The predators scored points (i.e., were given reward during training and evaluated during testing) every time they collided with the prey. Predator agents

Fig. 1 Predator–prey pursuit particle environment

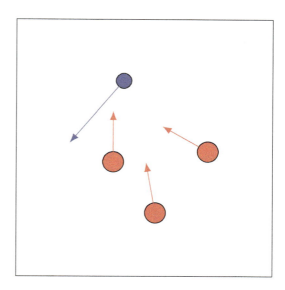

were homogeneous in their capabilities (i.e., same size, velocity, and acceleration limitations), whereas the prey was 33% smaller, could accelerate 33% faster, and had a 30% max speed advantage. The simulation environment was built upon the OpenAI Gym library [5] and developed for use with the multi-agent deep reinforcement learning algorithm, multi-agent deep deterministic policy gradient (MADDPG) [10]. We assume that the predator agents must cooperate to score a hit on the prey, given the prey's capability advantages. Prior work has shown that a simple greedy policy (i.e., minimize distance to prey) is insufficient for the predators to succeed [3].

All the agents were trained concurrently using the MADDPG algorithm [10]. MADDPG utilizes a decentralized-actor centralized-critic framework that accounts for each agent's observations and actions during training. The predators all received the same fixed reward (shared/joint reward) when any one of them hit the prey, while the prey received the negative of the same fixed reward when hit. At the start of each episode, the initial positions of the agents were randomized, and their initial accelerations and velocities were set to zero. The state space of each predator agent contained its absolute velocity, absolute position in the environment, relative distance and direction to the other predators and the prey, and the prey's absolute velocity. The state space of the prey agent contained its velocity, absolute position in the environment, and relative distance to the predators. The action outputs of the policy network for an agent are accelerations in the two-dimensional coordinate system.

In addition to the MADDPG-trained agents, we consider two analytically defined agents (also referred to in this article as fixed-strategy agents), called a chaser agent and an interceptor agent. The chaser agent does not leverage any prey velocity information and only points its own velocity directly at the prey's instantaneous position. In contrast, the interceptor agent considers both instantaneous position and velocity of the prey. At a moment in time, the Apollonius circle describes the potential interception locations if both agents continue in a constant direction [9]. Given a prey that is faster than the predator, only a subset of possible constant prey strategies admit a capture trajectory for the predator [12]. We extend the Apollonius circle strategy for the predator in the case where capture is not possible. The case where capture is possible is shown in Fig. 2a. Equal travel time at capture gives $\frac{d_E}{V_E} = \frac{d_P}{V_P}$, and the rule of sines gives $\frac{d_P}{\sin\phi} = \frac{d_E}{\sin\theta}$, resulting in

$$\sin\theta = \frac{V_E}{V_P}\sin\phi. \tag{1}$$

In the case where capture is not possible, we consider a finite time prediction for prey trajectory and choose the predator's strategy to minimize the final distance. This is shown in Fig. 2b, where the d_P circle represents how far the pursuer is able to travel in the time it took the evader to travel distance R. The optimum position for the purser to minimize the relative distance at the moment the evader reaches the R circle is to head straight toward that point. Clearly, $\psi = \pi - \theta - \phi$, and

$$\frac{\sin\theta}{R} = \frac{\sin(\theta + \phi)}{r}. \tag{2}$$

The critical case between the capture set and non-capture set is $\sin\phi^* = \frac{V_P}{V_E}$, $\theta^* = \frac{\pi}{2}$, and evaluating Eq. 2 at that point gives

$$R = \frac{r}{\sqrt{1 - \frac{V_P^2}{V_E^2}}}. \tag{3}$$

This value for R will result in a continuous policy across all cases.

To compare heterogeneous and homogeneous team structures, the MADDPG-trained agents and fixed-strategy agents were subdivided into homogeneous and heterogeneous teams. We took two previously trained models, independently trained with the MADDPG algorithm for 100,000 episodes and 25 timesteps per episode [2], and used them for our evaluations. Trained Model 1 and 2 predator agents were tested as heterogeneous teams consisting of all three agents (i.e., Agents 0, 1, and 2), labeled as "All Agents" in Figs. 3a, b, or as homogeneous teams in which all three actors' behaviors are driven by the policies of either Agent 0, Agent 1, or Agent 2 from each of the models. Similarly, the fixed-strategy predator agents were tested against both Model 1 and Model 2 prey agents in several types of homogeneous or heterogeneous team compositions: 3 interceptors, 2 interceptors and 1 chaser, 1 interceptor and 2 chasers, and 3 chasers. All tests were performed for 1000 episodes at 1000 timesteps per episode.

Emergent Heterogeneous Strategies

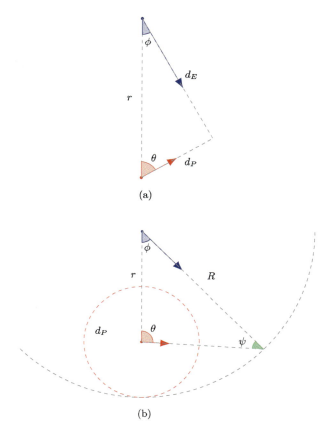

Fig. 2 Pictorial of interceptor strategy for capture and non-capture cases. (**a**) Capture scenario for interceptor strategy. (**b**) Finite R interceptor strategy

3 Results

The simulation results show how group performance changes upon replicating a single agent's policy network and thus introducing homogeneous strategies. Further, the performance resulting from this homogeneous strategy implementation may provide a means of classifying different trained policies that emerge through collaboration in the multi-agent reinforcement learning process.

Using the data collected during testing, we generated probability density plots of the hits the predators achieved on the prey to analyze agent performance for each of the models (Figs. 3a, b). "All Agents" shows the aggregated or team performance of three predator agents with their independently trained network policies. "Agent 0" represents the data generated from the replication of "Agent 0" across the three predator agents. Similarly, "Agent 1" and "Agent 2," respectively, represent the replication of their corresponding agent policies. The x-axis shows the number of

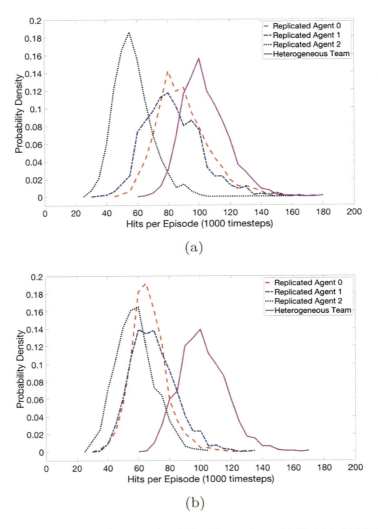

Fig. 3 Probability density performance plots for learning predators. (**a**) Model 1. (**b**) Model 2

hits (i.e., number of times the predators collaboratively contacted the prey agent throughout an episode). The y-axis shows the normalized frequency or probability density for the hits per episode. We performed the pairwise 2-sample Kolmogorov–Smirnov (KS) test to show that all the distributions were significantly different from one another at the alpha = 0.01 level (p-values « 0.001).

We can see from Figs. 3a, b that in the case of the MADDPG-trained agent strategies the heterogeneous predator team (All Agents) is able to outperform all of the homogeneous predator teams (Agent 0, Agent 1, and Agent 2) where the same policy is replicated across each agent. Furthermore, as all the homogeneous

Emergent Heterogeneous Strategies

Table 1 Performance of fixed-strategy predators

Predators' strategy	Mean	CI lower	CI upper
(a) Model 1			
3 interceptors	1.55	1.46	1.65
2 interceptors, 1 chaser	1.16	1.09	1.24
1 interceptors, 2 chaser	0.876	0.824	0.932
3 chasers	0.154	0.145	0.164
(b) Model 2			
3 interceptors	1.44	1.35	1.53
2 interceptors, 1 chaser	1.05	0.990	1.12
1 interceptors, 2 chaser	0.998	0.939	1.06
3 chasers	0.144	0.136	0.153

team performances were significantly different, we can infer that each individual agent policy learned to utilize a different strategy that benefited the team as whole.

The data in Table 1 shows the performance statistics for the fixed-strategy predator agents playing against the MADDPG-trained prey. The statistics were generated by fitting an exponential distribution to the probability density functions of the data from the respective cases. Note that these data are shown in tables rather than plots as the probability density functions are all visually similar when presented with x-axis values on the same scale as Fig. 3a, b.

Overall, the tables allow us to see that the performance of the chaser predators is an order of magnitude worse than that of the interceptor predators when playing against trained prey from Models 1 and 2, which was also shown to be significantly different with 2-sample KS tests ($p \ll 0.001$). Interestingly, when we combine a chaser predator with two interceptor predators (i.e., heterogeneous fixed-strategy cases), the inclusion of chaser predators results in a significant reduction in group performance (compare 3 interceptors case to the heterogeneous cases in Table 1). In addition, across both models, combining an interceptor predator with two chaser predators significantly improves group performance from the three chasers case ($p \ll 0.001$). Together, these results suggest that overall the interceptor strategy is significantly better than the chaser strategy, especially in the homogeneous cases.

References

1. D. Asher, S. Barton, E. Zaroukian, N. Waytowich, Effect of cooperative team size on coordination in adaptive multi-agent systems, in *Artificial Intelligence and Machine Learning for Multi-Domain Operations Applications*, vol. 11006 (International Society for Optics and Photonics, Bellingham, 2019), p. 110060Z
2. D.E. Asher, E. Zaroukian, B. Perelman, J. Perret, R. Fernandez, B. Hoffman, S.S. Rodriguez, Multi-agent collaboration with ergodic spatial distributions, in *Artificial Intelligence and Machine Learning for Multi-Domain Operations Applications II*, vol. 11413 (International Society for Optics and Photonics, Bellingham, 2020), p. 114131N

3. S.L. Barton, N.R. Waytowich, E. Zaroukian, D.E. Asher, Measuring collaborative emergent behavior in multi-agent reinforcement learning, in *International Conference on Human Systems Engineering and Design: Future Trends and Applications* (Springer, Berlin, 2018), pp. 422–427
4. S.L. Barton, E. Zaroukian, D.E. Asher, N.R. Waytowich, Evaluating the coordination of agents in multi-agent reinforcement learning, in *International Conference on Intelligent Human Systems Integration* (Springer, Berlin, 2019), pp. 765–770
5. G. Brockman, V. Cheung, L. Pettersson, J. Schneider, J. Schulman, J. Tang, W. Zaremba, OpenAI gym (2016, preprint). arXiv:1606.01540
6. N. Case, How to become a centaur. J. Des. Sci. (3). MIT Press. https://doi.org/10.21428/61b2215c (January 8, 2018)
7. H.G. Hicks, C.R. Gullett, S.M. Phillips, W.S. Slaughter, *Organizations: Theory and Behavior* (McGraw-Hill, New York, 1975)
8. J. Humann, Y. Jin, A.M. Madni, Scalability in self-organizing systems: an experimental case study on foraging systems, in *Disciplinary Convergence in Systems Engineering Research* (Springer, Berlin, 2018), pp. 543–557
9. R. Isaacs, *Differential Games* (Wiley, Hoboken, 1965)
10. R. Lowe, Y.I. Wu, A. Tamar, J. Harb, O.P. Abbeel, I. Mordatch, Multi-agent actor-critic for mixed cooperative-competitive environments, in *Advances in Neural Information Processing Systems* (2017), pp. 6379–6390
11. G.F. Oster, E.O. Wilson, *Caste and Ecology in the Social Insects* (Princeton University Press, Princeton, 1978)
12. M.V. Ramana, M. Kothari, Pursuit-evasion games of high speed evader. J. Intell. Robot. Syst. **85**(2), 293–306 (2017)
13. D. Shishika, J. Paulos, M.R. Dorothy, M.A. Hsieh, V. Kumar, Team composition for perimeter defense with patrollers and defenders, in *2019 IEEE 58th Conference on Decision and Control (CDC)* (IEEE, Piscataway, 2019), pp. 7325–7332
14. Y. Song, J.H. Kim, D.A. Shell, Self-organized clustering of square objects by multiple robots, in *International Conference on Swarm Intelligence* (Springer, Berlin, 2012), pp. 308–315
15. P. Stone, M. Veloso, Multiagent systems: a survey from a machine learning perspective. Auton. Robot. **8**(3), 345–383 (2000)
16. A. Tampuu, T. Matiisen, D. Kodelja, I. Kuzovkin, K. Korjus, J. Aru, J. Aru, R. Vicente, Multiagent cooperation and competition with deep reinforcement learning. PloS One **12**(4), e0172395 (2017)

Artificially Intelligent Cyber Security: Reducing Risk and Complexity

John N. Carbone and James A. Crowder

1 Introduction: Non-Linearity and Complexity

Traditionally, much research exists for analysis, characterization, and classification of complex heterogeneous non-linear systems and interactions which historically have been difficult to accurately understand and effectively model [1, 2]. Systems that are nonlinear and dynamic generally comprise combinatorial complexity with changes in variables over time [3]. They may appear chaotic, unpredictable, or counterintuitive when contrasted with much simpler linear systems. Complex system interrelationships and chaotic behavior of these systems can sometimes be perceived as random. However, they are not. Simple changes in one part of a nonlinear system can produce complex effects. Widespread nonlinearity exists in nature as well [4]. Micro- and macroscopic examples include collision-based particles in motion; common electronic distortion; chemical oscillations; weather modeling; and, for this paper, complex relationships between vast volumes of seemingly ambiguous and superficially relatable binary cyber event data from a wide array of systems, users, and networks.

Additionally, advanced cyber research shows that meaningful enhancements are required to mitigate the ever-increasing volume of intrusion detection system (IDS) alerts overwhelming human analysts today, thousands of which are received per day, and 99% of which are false indications [5]. The currently well-known cyber security

---

J. N. Carbone
Department of Electrical and Computer Engineering, Southern Methodist University, Dallas, TX, USA
e-mail: john.carbone@forcepoint.com

J. A. Crowder (✉)
Colorado Engineering, Inc., Colorado Springs, CO, USA
e-mail: jim.crowder@coloradoengineering.com

© Springer Nature Switzerland AG 2021
H. R. Arabnia et al. (eds.), *Advances in Artificial Intelligence and Applied Cognitive Computing*, Transactions on Computational Science and Computational Intelligence, https://doi.org/10.1007/978-3-030-70296-0_38

landscape is wrought with many types of cyber security attacks where individuals and nation states employ a wide array of tactics to attack at every available data, network, and system access location. Mitigation research appears to be focused on using network-based intrusion detection systems and fusing their outputs to gain a more comprehensive understanding of undesired activities on the network [6]. While there has been some success, overall awareness of current network status and future adversarial action prediction has still not been achieved [6]. Although analysts are undeniably capable of performing difficult cyber security mitigation tasks, our understanding of the cognitive processes required by analysts, under current conditions, to produce more effective protection is very limited and requires new research [7, 8]. Therefore, as many complex problems generally involve more than one solution and as innovation is found at the intersection of multiple disciplines [9], our approach will be multi-, trans-, disciplinary to improve critical cyber situational awareness, classification, contextual understanding, and decision support [7]. We survey and heuristically decompose cyber security, axiomatic design, complexity theory, and novel new AI/ML/ITM learning techniques.

While many disciplines are known to have well-known mathematical formalisms for describing, predicting, and approximating solution;, historically, however, non-linear solution accuracy and analysis have generally been problem dependent and have required significant added effort to simplify and bound [10]. For example, many current cyber event classification-based techniques rely on an experts' extensive knowledge of network attack characteristics. Once provided to a detection system, an attack with a known pattern can then be more rapidly detected. Literature shows that the ability to mitigate a cyber event becomes highly dependent upon the fidelity of the attack's signature [11, 12]. Hence, systems often detect only what is known and are therefore significantly vulnerable to an environment of continuously adaptive attacks and vectors. Even if new attack signatures are rapidly incorporated for improving mitigation, the initial loss is irreplaceable and repair procedures costly. Mitigations are made even more difficult by the indeterminate validity of an inherently stochastic nonlinear system attack [13].

Recent advances in artificial intelligence (AI) and machine learning (ML) research are well known to be benefitting many disciplines struggling with rapidly increasing velocity, volume, and complexity of data and systems, and for improving timely generation of qualitative readily consumable knowledge [14]. However, AI/ML classification-based approaches generally rely on what is considered normal or standard baseline data traffic activity profiles generally built and stored over time. Comparative activities then explore for anomalies which deviate from captured baseline profiles. However, as vast volumes and high velocity binary data are individually captured, subsequent significant processing is then required to correlate/fuse potential complex non-linear inter- and intra-data relationships [15].

The difficult objective is to understand what is considered normal traffic, what is not already included in an ever-increasing ambiguous cyber knowledge store, and if an obscure event is nominal, a risk, or an attack. Today, this must all be accomplished within a given day's context challenged cyber environments where 99% of Intrusion Detection System (IDS) events can be potentially considered inadvertent false alarms [5]. Fortunately, the use of iterative methods, followed by

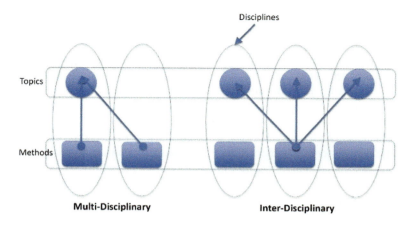

Fig. 1 Cross discipline engineering

domain−/discipline-specific solution sets, support effective solution generation for nonlinear problems. The addition of relatable data domains and data sources can potentially also add context to mitigating cyber threats [16, 17].

As part of our transdisciplinary approach, we therefore first employ axiomatic design's iterative design matrix optimization mechanisms to manage, understand, and simplify existing and future cyber systems, which generally comprise multi-faceted processing methods and ambiguously related data. We also decrease cyber security risks by reducing information content with AD analysis processes, which in turn helps to reduce false positives and false negatives.

In Fig. 1, *multi-disciplinary* engineering involves engaging methods from other disciplines to solve a problem within one discipline, *inter-disciplinary* involves choosing methods from one discipline and applying it to one or more disciplines. In contrast, *transdisciplinary* engineering is the field of study which supports simplification and optimization by engineering common solutions across many disciplines [18]. Computer science, mathematics, and cyber security are examples of disciplines which provide capabilities for many disciplines. Even single-discipline engineering solutions can be complicated. Multi- and/or trans-disciplinary engineering solutions can increase the complexity and non-linearity even further, requiring more robust engineering principles and increased efforts for standard or more critical implementations.

1.1 Complexity

Exacerbating the problematic design of nonlinear systems are challenging levels of ambiguity and intricacy. Specifically, complex nonlinear relationships exist between vast volumes of seemingly ambiguous, independent cyber events and their potential relationships across multi-domain data. Additionally, modern manufacturing systems are increasingly required to adapt to changing market demands, their structural and operational complexity increases, creating a negative impact on system performance [19]. Similarly, cyber security systems suffer from increasingly adaptive adversaries [20] and must innovate in-kind to adapt to the asymmetric assault on system, data, and knowledge integrity. Significant research and patents exist to improve knowledge and context reliability with the correlation and fusion of big data [21]. Thus, the employment of complexity theory and applications of axiomatic design potentially decrease cyber system and data ambiguity and enable cyber security systems and their algorithms to become increasingly adaptive.

Axiomatic design (AD) research originated, in the 1990s with Nam P. Suh, within the Massachusetts Institute of Technology (MIT) Mechanical Engineering school. AD has been widely deployed for optimizing industrial and manufacturing applications. Complexity theory is applied for simplifying and optimizing system designs for mechanical, manufacturing, industrial, economic, social, and other systems. We propose that cyber security system designs can similarly benefit from AD ultimately improving cyber system adaptability and resiliency.

Suh describes that significant confusion exists within the definition of what is complex and explains that many more attempts have been made to understand complexity in terms of physical entities stead focusing on what is to ultimately be achieved. Suh describes complexity as computational, algorithmic, and probabilistic [3] and employs an approach comprising four complexity types: Time-independent real/imaginary complexity and time-dependent combinatorial/periodic complexity. Suh mitigates overall system complexity by employing optimizing actions: Reduce the time-independent real complexity, eliminate time-independent imaginary complexity where possible, and transform time-dependent combinatorial complexity into time-dependent periodic complexity for decomposing complex systems into smaller, more easily comprehensible, operable units. Suh describes this action as functional periodicity domain determination (e.g., biological, thermal, circadian, temporal, geometric, etc.).

1.1.1 Managing Complexity

Suh stipulates that complexity must be viewed within the functional domain. Therefore, fundamental management of complexity within a discipline, project, need, or gap is a focused process which defines what we want to achieve or understand within the functional domain. Managing complexity in the physical domain begins with creating system definitions using the voice of the customer known as customer needs (CN), which are subsequently translated into functional requirements (FR). To achieve optimized system design goals a set of design parameters (DP) is iteratively defined. DPs are analyzed for their individual probability of satisfying FR based system designs as shown in Fig. 2, probability density function (PDF). For example, if all required functional cyber system's DPs are completely encompassed within

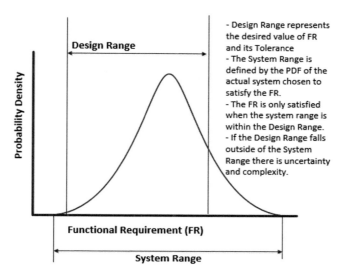

Fig. 2 Design vs. system PDF

the cyber system range, then we know that a cyber system's FRs are very likely to be achieved within a bounded design. Therefore, as an existing design or actively designed system becomes iteratively more well defined, it becomes less complex and less ambiguous. This is achieved through iterative FR-DP mapping and design matrix decoupling efforts where system optimization is driven through utilization of an independence axiom which drives orthogonality analysis. The ultimate design objectives are to achieve, as close as possible, a completely orthogonal uncoupled FR-DP design matrix, and the reduction of information content by minimizing DP random variations which reduce the variance of cyber system function and consequently reduces time dependent combinatorial complexity and non-linearity [3].

Moreover, when a system range extends outside of a design range boundary, a design and the satisfaction of functional requirements become more difficult to achieve and more complex. Axiomatic design processes achieve optimized functional designs through the aforementioned matrix decomposition process by traversing from "what we want to achieve" to "how we hope to satisfy" cyber system functionality across the following domains: customer, functional, physical, and process. Two important axioms, eluded to earlier, drive optimized cyber system functionality. First, the independence axiom helps maintain design independence and orthogonality of functional requirements and supports minimization of design overlap where possible to drive solutions to improve minimization and increase cost effectiveness. Second, the information axiom drives minimizing information content throughout the iterative design process to provide continuous simplification. The ultimate cyber system complexity reducing objective is complete uncoupled design relationships where all functional requirements are satisfied by independent design parameters and functions.

Analogously, for managing and securing cyber systems, software, and data, matrix decomposition provides logical decoupling of dependent interfaces and supports development of common normalized functional requirement inputs(what) and design parameter output(how) mappings. Consequently, upon design completion, physical and logical complex cyber system component tasks are effectively abstracted and simplified, and their data minimized, increasing successful understanding and improved correlation/fusion, thereby optimizing cyber sensing, improving scalability through orthogonal/independent design features, and reducing overall complexity and cost.

Analogously, supervised machine learning is a process, derived from statistical learning theory, also used for mapping inputs to outputs by learning unknowns from example pairs of related information. Hence, machine learning and axiomatic design are generally well aligned as they both support problem bounding and correlation-based discovery of unknowns. Each utilizes vectorized learning across a spectrum of widely varying characteristics. Comparably, empirical risk minimization (ERM) is a principle of statistical learning which helps bound algorithm performance and can be characterized by using joint probability distributions. Similarly, in axiomatic design, a system range correlates to a pre-defined bounded data range where a design range correlates to an actual range of each design component's datum. Simply, if a datum X represents a cyber event which occurred at time t_1, while datum Y represents a cyber event occurrence at time t_2, if Y occurred within the time range for cyber event X, then Y is within the range of X or $[(t_2-t_1)$ and $(t_1 + t_2)]$, there exists a PDF which can represent the overlap of the system vs design range. Therefore, expanding upon the ML-AD parallels, the proceeding sections examine combining advanced novel machine-based learning techniques for increasing data correlation/fusion to further reduce cyber risk and complexity.

2 Artificially Intelligent Cyber

Discipline breadth is required for developing AI systems making AI research and education inherently multi- and trans- disciplinary [18, 22]. Machine learning is also supporting the evolution of cognition-based learning within these many domains [22, 23]. Cognition research employs computer software design analogous to components of the human brain combined with varying advances in artificial neural networks (ANN). ANNs are deemed one of the hallmarks of machine learning and designed to operate synonymously as neurons within a human brain, and as a group of interconnected nodal relations. ANNs were inspired by neuroscience-based biological neural networks which are the structure by which chemical-based ionic electrical signals are passed throughout the body [24]. They are therefore at the core of probabilistically relating changing levels of data relationships and recent active research supports their use in developing automated machine- based pattern recognition in general and for enhancing the cyber security domain [25, 26]. Thus, we describe the combined use of ITMs and cognitive-based learning methods to support the challenging processing of high volume, high velocity data, and improving opportunities for autonomous operations within an overwhelmed cyber user environment, minimal contextual security, and lacking system stability, thus improving understanding, transmitting less, and enabling individuals to more effectively utilize their cyber systems to better control their heterogeneous systems and data environments.

2.1 Cyber Data Ingest

Moving toward more autonomous self-learning operations requires a more intelligent, normalized, and optimized set of data ingress. The relatively standard extract transform load (ETL) tasks listed below, among others, are used throughout the data industry for processing passive and active streaming data (e.g., sensor data). Generally, learning from any data, including a cyber security scenario, involves a collection of each varying set of resting or streaming sensor data which is then compared, contrasted, associated, and normalized against previously captured content. The efficacy of algorithms being used to try and understand the data, relies on, among others, the ability to detect change. In short, these changes evolve into patterns which support improved learning fidelity over time. Ultimately these processes support learning continuously and evolving toward eventual autonomous learning and autonomous improvement of system functionality:

- Collect & verify data.
- Analyze for missing data.
- Cleanse data.
- Tag data.
- Organize/correlate/classify data.
- Compare, contrast, normalize with existing data.
- Verify, remember, store analytical results.
- Transmit/display results.

Traditionally, ETL functionality is also included for improving scaling of input source types (e.g., TCP/UDP Sockets, Filesystems) by understanding and classifying content prior to, and for improved, system processing. It is well known that current industry data volume, velocity, and variety vary greatly and can require significant processing for discovering patterns and context. However, a perception of high complexity exists in cyber security data, primarily because cyber data pattern analysis has traditionally been wrought with false positives and false negatives stemming from minimally included and minimally derivable context. This challenge increases the difficulty in determining nominal, unscrupulous, or accidental behavior from the many data, network, or user-based cyber events. As an example, a distributed denial of service (DDoS) attack is a high volume and velocity attack attempting to impact user services and is usually directed from many distributed locations. Understanding whether malicious or accidental is many times difficult. Other attacks: Phishing, JavaScript, SQL injections are generally small in data volume and velocity. This type of cyber data can be characterized as more passive (files per day) or active (streaming sensor data) but can still be difficult to classify. Important however is that correlation of high volume and/or passive data requires the proper infrastructure to support collection and processing of both.

As systems and sensors scale up or down, we therefore propose employing the logical and physical efficiency benefits of a well-known common ingest and processing architecture known as Lambda Architecture [27]. Lambda is well-known within many high-volume data architectures for processing individualized passive/batch and/or active/streaming data. Therefore, cyber sensor input data, like many varying data types, can be transformed per more well-defined flows and through a scalable orchestrated ETL process. This ensures proper a priori ingest curation and formatting required for subsequent cyber-based analytical processing algorithms, which is common for most modern ETL environments. Once data ingest analytics have curated and appropriately tagged the input, the resulting curated output is subsequently correlated, normalized, and/or fused with data within parallel streams of data and/or passive data including previous results. The algorithmically infused output is then believed to be of enhanced value (learning from the data) and becomes synonymous with terms like adaptive system learning, machine learning, and algorithmic domain terms (e.g., anomaly detection, cyber behavioral learning, intelligence, surveillance reconnaissance (ISR) fusion, molecular effect correlation, etc.).

2.2 Machine Learning and Cyber Security

It is well known that machine learning algorithms derive from statistical learning theory, are therefore inherently statistical in nature, and require significant initial training within a given bound or context to become qualitatively relevant. Similarly, finite element analysis within engineering disciplines and many mathematical concepts have historically supported non-linear solution accuracy and analysis by also providing problem-dependent bounding and thus simplification [10]. It is also well known that machine learning algorithm training, to be of consequence, is time consuming and generally considered a fine art to suitably discover and train with a sufficiently related problem data. Similarly, qualitative mitigation of cyber security risk requires proper human and data training to improve anomaly detection techniques and to adequately build normal activity profiles [28].

The effectiveness of the cyber algorithm training (as for most applications) then depends greatly upon the availability of completely normalized cyber traffic datasets, which, in practice, are rare, extremely difficult to keep up to date, and thus expensive to obtain, especially attack-free instances [5]. It is well known that for ML algorithms to be of benefit a significant amount of work must be achieved early in just understanding the data through a process of problem definition, data analysis, and data preparation. Understanding the context around the problem, constraints, assumptions, and motivations around who, what, where, and how the data was captured is critical to successful and useful subsequent modeling and application. A large pool of well-known machine learning algorithms and classification-based anomaly detection techniques are available today for computer vision, natural language processing, dimensionality reduction, anomaly detection, time series analysis, prediction, and recommender systems. Although multi-model ML shows promise in pattern analysis by creating ensemble outputs of multiple classifiers, herein we discuss traditional single model machine learning examples, their pitfalls, and subsequently propose information theoretic mechanisms to decrease cyber risk by significantly improving data-context understanding and thereby also improving autonomy when analyzing complex cyber systems.

2.2.1 Machine Learning: Value, Characteristics, and Limitations

Traditional machine learning is divided into two groups known as *Supervised* and *Unsupervised*. Additionally, we discuss our proposal for the additional application of AD. The objective is to drive more mindful initial and continuous interpretation of data, to help optimize development of common processing, simplification of well-defined data-analytics pairing, as well as, optimized frameworks for the improved processing of high volume, high velocity ML data.

ML processing flows can be resource intensive and can come in the form of *data collection exemplars, learning approximations, learning associations, striving for equality/specificity using sensitivity, and the use of optimization strategies.* Considering recent research into ML, what becomes apparent is that ML's usefulness is measured in a few different ways. ML algorithms are generally employed to sift through massive volumes of data looking for patterns. Some challenging characteristics of ML processing can include significant time consumption when compared against traditional data processing, ambiguous output, improved only with significant a priori data analysis, and the perceived complexity of data dependencies. Benefits of ML can include significant mitigation of the relative difficulty and/or inability of less automated approaches for determining discriminatory separation and classification of data.

We propose that in order to more fully understand the potential benefits and drawbacks of ML and in order to significantly improve valuable in-context affiliation between cyber data, one must account for the context of human interactions taking place between systems, data, algorithms, and applications. This includes capturing specific human cognitive states, and simultaneous and continuous correlation of all information, recursively. The implication is that "ones" and "zeroes" by themselves are most often analyzed "out-of-context" and hence provide much less is discernible meaning than when also compared to additive valuable contextual characteristics.

These value-based data characteristics become more visible when considering the subtle differences between the concepts of *presentation* and *representation*, well researched within the domains of information theory and physics [22] and used for improving human decision-making. *Representation* is simply defined as the underlying simple and complex relationships that are represented by mathematics, protocols, and formats. However, *representation* of simple and complex data and relationships between data are often not as readily discernible, when data relationships reach higher dimensions [29]. Hence, a potential use for ML pattern learning and algorithmic association designed specifically for this purpose.

Pattern analysis performance on high dimensional, most often initially unknown and unrelated data, is one key factor generally used to determine the value of an algorithm. Proper *presentation*/visualization of ML output is also required within the context of how data is used to support perceived benefits and/or drawbacks of a given method. These perceived benefits derive from the subjective quality of decision value derived from both, the a priori knowledge/context of the processed data, and the expression quality of the output. Perceived value inherently derives from how separable and discernible the data is and how well expressed the context around the data becomes after processing. As an example, parallel coordinates, as described by Inselberg, is a renowned method for discerning multi-dimensional data relationships through novel visualization for significantly improved decision making [22]. The objective of multi-dimensional visualization is to vastly improve the ability to perform multi-dimensional comparisons and context development whether simple or non-linear and then rapidly transform the visual presentation of higher order complex mappings into more simply discernible dimension reducing sets of two-dimensional relationships.

A quick review of ML: *Supervised* learning (SL), takes as input, data which has been previously learned and tagged, also known as "labeled." As an SL algorithm iterates over and processes each piece of new data, it compares each to previously learned data. As comparisons are made, exactly or partially, within a certain range or approximation of an a priori defined boundary, then the resulting response/decision variable is placed in the appropriate group. Hence, if the SL objective is to estimate the value of a given variable, then a simple *"curve fit approximation"* would be the recommended approach. If the SL algorithm is attempting to discover discrete categories within the data, then *Decision Tree* methods are preferred for organizing decisions, strategies, and possible outcomes.

Unsupervised learning (UL) treats all data equally and its prime directive is not to estimate the value of a variable but rather to simply look for patterns, groupings, or other ways to characterize the data that may lead to understanding of the way data interrelates. Hence, cluster analysis, K-means, correlation, hidden Markov models, and even factor analysis (principal components analysis) and statistical measures are examples of unsupervised learning. Unsupervised techniques essentially attempt a "partitioning" of a large pool of data into smaller chunks of data that are more related to members of the partition than data outside the partition. Different methods and different disciplines have varying names for this partitioning, whether for simple organization like concept maps or more algorithmic, like clustering, a term frequently used and commonly associated with methods such as (K-means). "Chunking" and "kriging" are terms for methods that handle the data differently but strive for the same organization. There are also methods such as Voronoi maps, Self-Organizing-Maps, Isoclines, Gradients, and many other approaches that also strive for separation of data into partitions without regard to what to call (labeling) members of each partition. It is this lack of requiring an external "label" as criterion which drives "unsupervised" partitioning.

Therefore, one of the most useful applications of statistical analysis is the development of estimators and function approximators (not models) to *visually* explain (*present*) the relationship between many data items (variables). Thus, many types of estimators have been developed (e.g., linear and nonlinear regression (function fitting), discriminant analysis, logistic regression, support vector machines, neural networks, and decision trees). However, Wulpert's no free lunch (NFL) theorem describes that each method has advantages only within a particular use case, and therefore, great care must be taken to understand each use case thoroughly [30].

ML algorithms focus upon similar varying optimization strategies where no single estimation or approximating method is best for all applications. Therefore, in order to improve the fidelity of machine-based learning and go further than current ML allows, we propose that it is uniquely important to understand the difference between memorizing (not accepted within academia as learning) discovered patterns and comparing/estimating how well those patterns compare/relate to recent and well-known information theory innovations and information theoretic methods (ITM) [31]. These methods strive to "explain" data at higher fidelity for rounding out the expression of details/context in every case, focusing upon enabling better realizations and decision-making. Analogously, imagine employing a new, more highly expressive method that enabled the realization of the exact location of electrons in space and time as opposed to Shroedinger approximations. The scientific applications and optimizations possible with this new expressive information would be considered revolutionary because of the wealth of new opportunities and vastly improved decision-making. Therefore, we discuss objectives to significantly improve understanding of ambiguous cyber security event data, optimization, and innovation, by providing higher fidelity-based information context and insights and for ultimately creating higher quality cyber security mitigation decision-making.

2.2.2 Currently Employed Supervised and Unsupervised Cyber Security Machine Learning Approaches

Most recently Ferrag et al. [32] analyzed 35 of the most well-known cyber security data sets with ten significant machine learning algorithms (Fig. 3 [32]) against performance efficiency for binary and multi-class classification, as well as accuracy, false alarm rate, and detection rate. Each data set-ML paired analysis yielded a best algorithm. FFDNN: Feed forward deep neural network; CNN: Convolutional neural network; DNN: Deep neural network; RNN: Recurrent neural network; DBN: Deep belief net- work; RBM: Restricted Boltzmann machine; DA: Deep auto-encoder; DML: Deep migration learning; STL: Self-taught learning; ReNN: Replicator neural network. Additionally, Eskin describes the use of unsupervised.

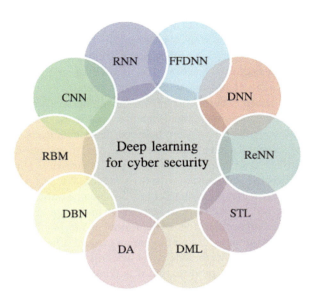

Fig. 3 Current cyber deep learning approaches [32]

SVM for detecting anomalous cyber events. Example SVM usage approaches: Registry anomaly detection (RAD) for monitoring Windows registry queries [33], Robust SVM (RSVM) anomaly detection ignores noisy data [34], confident anomaly detection [35]. A core assumption of SVMs is that all sample data for training are independently and identically distributed. Additionally, in ML practice, training data is often noisy, thereby often invalidating results and driving standard results into highly non-linear decision boundaries, leading to poor generalization. Research is also lacking to optimize and reduce anomaly detection and SVM runtime processing [35]; like many ML methods, SVM benefits and/or drawbacks are highly tied to a priori, well-defined boundaries and to the homogeneity of data mapped to equivalent small or large numbers of false detections, hence, our objective to provide improved machine learning fidelity and efficacy.

2.2.3 Improving Machine Learning Fidelity: Information Theory and Information Theoretical Methods

Vast systemic complexity and multi-dimensionality issues retard, impede, and provide significant friction to building advanced systems more capable of managing and protecting valuable commercial and government assets. Therefore, among a treasure trove of issues managing and understanding current and exponentially expanding Big Data and system endpoints, as well as, globally distributed computing, this chapter confronts the major issue regarding *data and system fidelity* along with prescribed solutions.

Historically, the fidelity of information content combined with *representation* and *presentation* clarity lends itself to improved insights and thus greater potential for improved decision-making and efficient actions [22]. In the complexity section above, the information and independence axioms were outlined expressly to provide background on mechanisms used today that support the optimization and simplification of information content relationships. The definition of fidelity is defined herein as the degree of exactness, accuracy, or precision. Historically, in biochemistry the ability to see and understand finer grained cellular interactions increases our ability to make improved decisions on curing disease or the manufacture of specifically targeted rather than generalized drugs. The lack of fidelity of understanding can result in dire consequences. In nuclear physics the quest for higher fidelity understanding of the universe drives the search for even smaller particles (e.g., Higgs Boson) which provide even more globally impacting insights. However, our systems, software, processing, and hardware are forever built by a third party and most often are not developed to a common standard. Software intra- and inter- dependencies are not effectively known or managed. Hence, the continuous lack of fidelity and lack of real-time insight into comprehensive system, system processing, application, and data dependencies is one of several core reasons we continue down a path of cyber frustration and insecurity. Therefore, along with employing axiomatic design for improving data and system understanding, and dependency mapping for simplifying system construction, we propose the application of advanced ITM approaches. The objective is to potentially achieve significantly higher fidelity system dependency and data anomaly detection/classification understanding for today's complex multidimensional data issues and cyber security challenges.

2.2.4 Cyber Data: Reducing High Dimensionality and Complexity of Machine Learning

Remember that data representation is defined as the underlying simple and complex relationships represented as collected, ingested, correlated, protocols, and data formats. Subsequent fusion of context from additive data sources increases relational complexity, ranging from clear simple comparisons to more ambiguous, higher dimensional, complex interrelationships. Stated earlier, parallel coordinates (PC) [28] provide for visually deriving clarity from complex higher dimensional relations. However, as in supervised learning, PC assume a priori parameterization has been normalized across data types, and initial interrelationships have been a priori defined. As stated above, SL algorithms iterate over and process each piece of new data, comparing each to previously learned data, thus iteratively adding complexity and higher dimensionality to data interrelationships.

Therefore, we propose an introductory data context leveling step where we apply axiomatic design principles for improving ML fidelity of a given SL application. First, AD should be applied to determine the type of complexity (e.g., time-dependent, time-independent, complexity, imaginary complexity, combinatorial complexity) [3] surrounding the creation of a collected data set driving a priori

SL data labeling and boundary approximation decision-making. AD supports the development of better indicators for unambiguously attributing truth to applied labels which increases data definition quality through added context and discernibility. Specifically, since SL processes perform data comparisons within specific ranges or approximations of a priori defined boundaries, AD's system design range correlation mechanism enhances understanding of the types of complexity which can drive SL range approximations. The added knowledge of which type of complexity was involved in data creation provides added insight into the value of a given SL algorithm on said data. Thus, this combined approach provides a candidate complexity reducing tool for reducing cyber and/or other data relationship complexities and improved efficacy of SL algorithm utilization.

2.2.5 Increasing Cyber Security Event Understanding with Information Theoretic Methods (ITM)

Novel Data Characterization Using Fractals

For understanding complex data and relationships, Jaenisch et al. [36] describe how to apply continuous wavelet transform (CWT) to discrete n^{th} order integrable and differentiable data, and how to automate derivation of analytical functions which explicitly describe the data, directly from the data. They prove mathematically how to automate modeling of disparate data types using a new concept of univariate fluxion, coined as Unifluxion (UF) [36]. The UF formulation employs adaptive partitioning of data into fractals and then derives a fractal wavelet parent function for each partition. This enables automated wavelet transformation (integration or differentiation) across all partitions in a common repeatable manner, across the complete set of provided time series data [36]. Jaenisch et.al also compare UF to classical techniques and provide details on enhanced performance [36]. Hence, we propose that correlated and sequenced time series-based cyber, network, and user event data can be similarly described and modeled using fractals and UF.

Jaenisch et al. show how the unique formulation of $U(f(x))$ enables an automated data model transformation into either an integral or differential model of any order (an automatically derived differential equation) [36]. They describe how UF is defined to be a data model because it provides a continuous predictive model that can be both integrated and differentiated to any order. UF is also derived incrementally, as each measurement point is piecewise collected, although the final result is a continuous analytical function across all the time series data [36].

Subsequently, adding credence to the use fractals for improved cyber understanding, Jaenisch et al. [37] provide research examining the hypothesis that decision boundaries between malware and non-malware is fractal. They characterized the frequency of occurrence and distribution properties of malware functions compare them against non-malware functions [37]. They then derived data model–based classifiers from identified features to examine the nature of the parameter space classification boundaries between families of existing malware and the general non-malware category. Their preliminary results strongly supported a fractal boundary hypothesis based upon analyses at the file, function, and opcode levels [37].

Security information and event management (SIEM) data and systems are used extensively throughout industry, incrementally capturing many system-wide time series events: Data, Network, User Behavioral, Endpoint, Email, Web, etc. Additionally, security analytics in big data environments present a unique set of challenges, not properly addressed by the existing SIEM systems that typically work with a limited set of those traditional data sources [38]. Thus, we propose the reasonable application of UF to cyber SIEM time series data. UF's decomposition of cyber event content into fractal partitions enables the application of wavelet transformation which increases fidelity of cyber inter-, intra-, data relationships through the derived analytical functions which explicitly describe the cyber data, directly from the cyber data. Simultaneously, UF's continuous analytical function provides rapid predictive integration and/or differentiation, thereby improving the speed of cyber event relationship prediction and understanding. Hence, potentially traditional time-consuming machine learning classification would not be necessary (k-means, SVMs, etc.)

Spatial Voting (SV) for High Fidelity Data Characterization

Spatial voting is a multidimensional clustering and grouping algorithm. Individual spatial measurements (e.g., latitude and longitude) are stacked onto a coarse resolution SV grid. Similar or closely related points are organized into the same or within neighboring cell locations on the SV grid. The input features for the SV grid form the x and y axes of the SV grid. Once measurements are stacked on the SV grid, if required, a 2-D spatial convolution kernel is used to smooth the stack values in the landscape and connect isolated regions together into regions (subgroups) [39]. As proposed by Jaenisch et al., SV provides an analog data modeling approach to provide a solution to the "object to assembly" aggregation problem [40]. Generally, "object to assembly" refers to the perception of objects within their given spatial relations. For example, information design in advanced augmented reality (AR) applications requires support for assembly with high demands of spatial knowledge. SV is based upon combined principles of voting, geometry, and image processing using 2D convolution [40]. Voting is defined as a democratic process where the majority decision rules. Votes equate to hard decisions. For example, if sensors observe a phenomenon and identify it and then rank based upon each different hypothesis, then as one sums the number of sensors that declare a hypothesis to be true, then the largest sum becomes the winner [40]. Hence, voting reduces to probabilities, and typically, this is where Bayesian and other probabilistic analysis methods are generally used [40]. A presentation summary of the SV process is shown in Fig. 4. A shows an example plots of.

spatial events(e.g., UAV locations), B shows the detection grid output after initial ellipse stacking has been performed, C depicts the identification relationships of the candidate sub-frames, D represents a graphic output after additional feature

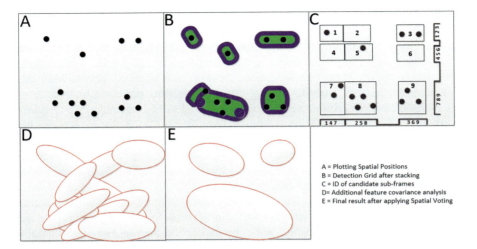

Fig. 4 Spatial voting process summary

covariance analysis has been performed (e.g. sensor reporting locations), and E depicts the final results after spatial voting was applied. It should be noted that additional intermediary feature analysis can be applied to further the exactness of the final output.

Hence, the difference is that conceptually SV and ITMs strive to "*explain*" rather than "*learn*" flash card style as machine learning aka machine remembering does. The emphasis is that memorization is not considered in academia as a measure of learning at all. Learning requires understanding which requires insight, the ability to synthesize (i.e., restate differently), and then generalize to conclusions. Neither machine learning nor artificial intelligence is focused on a path to achieve this. Hence, SV employs a matrix/grid where spatial artifacts are captured and marked/classified based upon their unique identifying characteristics [40]. Objects are characterized using recursive forms of higher order highly descriptive parametric and non-parametric (fractal) features and statistics, and the classifiers are derived for discrimination and classification. Characterizations and formulations are highly object and situation dependent. The next step is to create a classifier to practically associate the individual features. As always classifiers and algorithms which support it must continue to be chosen carefully. Their research has shown that SV provides significant characterization, discrimination, and performance benefits for improving object context in physical operational environments [36, 37, 40]. In the proceeding sections, we will show that spatial constructs can be extended to high fidelity explanation and characterization of varying types of digital artifact interrelationships and highly contextual knowledge creation.

High Fidelity Object/Data Relationship Modeling

When very few examples are available to discern statistical behavior, data models can be employed. Generally, data models are constructed using a bottom-up approach in the form of a knowledge base to collect, capture, identifies, encodes each example encountered. Data models within systems where little intelligence initially exists and into which intelligence is injected can contain many data models. Of course, these myriads of models have varying structures, formats and datum, all created under the influential context of the data model author. Relating similar and alien/abstract models and context has historically been challenging for many reasons. Hence, an inductive self-organizing approach, polynomial neural networks (PNN) or group method of data handling (GMDH), has been applied in a great variety of areas for deep learning and knowledge discovery, forecasting and data mining, optimization, and pattern recognition [41]. GMDH algorithms provide value in automatically finding interrelations in data, in order to select an optimal model or network structure and to increase algorithm accuracy. GMDH focus is to minimize modeling influence of the author and enables the computer to find optimal model structure or laws of the system acting upon the data.

Extending Spatial Constructs to System Learning and System Knowledge Development

Improving decision quality autonomy of artificially intelligent systems requires reliable information discovery, decomposition, reduction, normalization, and context-specific knowledge recall [22]. Hence, capturing the essence of any given set of information content is paramount. When describing how science integrates with information theory, Brillouin [42] defined knowledge as resulting from exercising thought. Knowledge was mere information without value until a choice was made based upon thought. Additionally, Brillouin concluded that a hundred random sentences from a newspaper, or a line of Shakespeare, or even a theorem of Einstein have exactly the same information value. He concluded that information content had "no value" until it had been thought about and turned into knowledge.

Artificially infused robotic systems must be able to integrate information into their cognitive conceptual ontology [43] in order to be able to "think" about, correlate, and integrate information. Humans think to determine what to do or how to act. It is this decision-making that can be of great concern when processing ambiguity because of the sometimes-serious ramifications which occur when erroneous inferences are made. Often there can be severe consequences when actions are taken based upon incorrect recommendations. Inaccurate inferences can influence decision-making before they can be detected or corrected. Therefore, the underlying challenge is to reliably understand the essence of a situation, action, and activity and to significantly increase capability and capacity to make critical decisions from a complex mass of real-time information content. Harnessing actionable knowledge from these vast environments of exponentially growing structured and unstructured sources of rich interrelated cross-domain data is imperative [44] and a major challenge for autonomous systems that must wrestle with context ambiguity without the advantage of human intervention [23]. The next section comprises combining ITMs with enhancing understanding ambiguous characteristics using knowledge relativity threads (KRT) [22]. As SV is employed to "explain" spatial characteristics, KRTs extend mature physical spatial mechanics for defining adaptive knowledge object presentation and representation.

2.2.6 Physical Representation of Meaning

Research shows that the community of disciplines researching how humans generate knowledge has traditionally focused upon how humans derive meaning from interactions and observations within their daily environments, driving out ambiguity to obtain thresholds of understanding. With similar goals, spatial voting, information theory, and complexity theory, as described earlier, focus more closely on explaining actual information content. Zadeh pioneered the study of mechanisms for reducing ambiguity in information content, informing us about concepts in "fuzzy logic" and the importance of granular representations of information content [45], and Suh focused upon driving out information complexity via the use of axiomatic design principles [3]. Hence, a vast corpus of cognitive-related research continually prescribes one common denominator, representation of how information content, knowledge, and knowledge acquisition should be modeled. Gardenfors [46] acknowledges that this is the central problem of cognitive science and describes three levels of representation: symbolic—Turing machine like computational approach; associationism—different types of content relationships which carry the burden of representation; and thirdly, geometric—structures which he believes best convey similarity relations as multi-dimensional concept formation in a natural way; learning concepts via similarity analysis has proven dimensionally problematic for the first two and is also partially to blame for the continuing difficulties when attempting to derive actionable intelligence as content becomes increasingly distended, vague, and complex.

Historically, there are many examples and domains, which employ concepts of conceptual representation of meaning as geometric structures (e.g., cognitive psychology [47], cognitive linguistics [48–50], transdisciplinary engineering [22], knowledge storage [14], computer science, e.g., entity relationship, sequence, state transition, and digital logic diagrams, Markov chains, neural nets, and many others. It should be noted here that there is not one unique correct way of representing a concept. Additionally, concepts have different degrees of granular resolution as Zadeh [45] describes in the fuzzy logic theory. However, geometric representations can achieve high levels of scaling and resolution [46] especially for n-dimensional relations, generally difficult if not impossible to visualize above the fourth dimension. However, high dimensionality can be mathematically represented within systems in several ways. Hence, mature mathematics within the physical domain allows this freedom. Therefore, we show the overlay of physics-based mathematical characteristics to enhance relational context and develop a unifying underlying knowledge structure within information theory. We employ knowledge relativity threads (KRT) [22] to minimize ambiguity by developing detailed context and for conveying knowledge essence simply and robustly. The next section describes presentation formation, representation, and the process of organization of n-dimensional contextual relationships for humanistic prototypical object data types with application of the common denominators: time, state, and context.

2.2.7 Knowledge Relativity (KR)

Knowledge relativity threads (KRT) [22] primarily originate from computational physics concepts as an analogy to Hibeller [51] where the concept of relating the motion of two particles is a frame of reference and is measured differently by different observers. Different observers measure and relate what they behold, to a context of what has been learned before and what is being learned presently. The reference frame of each knowledge building action contains the common denominators of time, state, and context—the point in time and all the minutia of detailed characteristics surrounding and pronouncing the current captured state of all related context. Historically, organization, presentation, representation of knowledge, and context have been researched across many disciplines (e.g., psychology, computer science, biology, and linguistics) because of the primal need to survive, understand, and make sense of a domain. However, most systems we engineer today are increasingly incapable of processing, understanding, presenting, and structurally representing the volume, velocity, variety, and complexity of content because first, they are not built to learn, only to store [52], and second, the content systems store and filter are what is generally or explicitly known to be true, not the more valuable and higher fidelity tacit knowledge that is context specific to each frame of reference or situation [53].

Therefore, we build KRTs upon the concept of "occam learning" [54] to construct continually renegotiable systems [55] with the simplest portrayal (e.g., present and represent) capable of encapsulating complex n-dimensional causal structures, within and between the complex data generated from the observed/captured behavior [14].

The KRT concept was developed to take advantage of mature physical universe n-dimensional relationship calculations relating any celestial object to another regardless of size or composition. KRTs extend physics space-time mathematics and apply to information theory to increase contextual knowledge understanding through a concept of recombinant knowledge assimilation (RNA) [22] or recursive spatial data model representations consisting of information object relationships. The logical concept develops from the following:

- *An infinite amount of data and data relationships exists in the universe.*
- *The infinite amount of data doesn't increase or decrease; it simply changes in form (knowledge increases).*

- *Fundamental increases in data volume increases decision points, which fundamentally should result in, but does not guarantee, increased data maturity and increased quality decision making.*
- *As data is consumed or processed, an increase in data quality and maturity appears, if and ONLY if enough relationships are known or can be discerned, can be captured, and reused to inform.*
- *Information explosion has always provided a human challenge.*
- *Humans, by themselves, are not physically capable of rapidly comprehending vast complex data sets, and then providing associated solutions to complex problems, the human brain can only handle approximately 7 events at a time [56].*
- *Can we capture and establish understanding of the fundamental relationships among all types structured and unstructured data?*
- *Can we extend or abstract premier concepts used to capture physical universe relationships?*

2.2.8 High Fidelity Fusion Using Concepts of Space-Time

In the general theory of relativity, the relationship depends upon the observer. This is similarly the case for fusing data relationships per Joint Directors of Laboratories (JDL) Fusion Level 5 [57], where the user serves a primary observer role in support for "decision-making" and defining actionable relationships. Different observers of the same data may apply different relationships. This is not dissimilar in Einstein's theory of special relativity where it is demonstrated that a "correct" answer is measured differently by different observers [58] for any independent event. Therefore, any observer of n-data of n-types can have n-independently observable relationships. Today, complex systems of data stores are developed from significant research across many different scientists/observers. Research data considered mundane or unrelated to one observer might be the ultimate piece of the puzzle or major discovery for another.

Hence, beginning with Reimann, space-time mathematics with respect to relativity has been in development for more than a century. Extending n-dimensional relationship mathematics principles to correlate and fuse non-physical data seemed intuitive. Here we describe Hendrik Lorentz's and Schrödinger's use of manifolds [59, 60] for application to n-related data objects in a linear or non-linear space. In systems biology, a cell is made up of many things. A strand of deoxyribonucleic acid (DNA) is made up of numerous bits and pieces of information which define a genetic blueprint, as well as, the DNA helix like physical shape. A space-time object in a Lorentzian manifold can be defined by the tangent vectors or signatures on the curved manifold. These objects can be represented as tensors or metric tensors which comprise a vector of eigenvalue attributes which define an object signature in n-dimensions [59]. Multivariate analysis in n-dimensions is applied to aspects of complexity in the information theory as well [32, 39]. Employing tensors as vector relationships to systems biology data, we can describe DNA attribute relationships mathematically and can store them in a common manner. Their pedigree is maintained mathematically as a tensor attribute per data element. Metric tensors or tangent vectors where a manifold intersects a spheroid at a single point represent the data that describes the attributes of the intersection. Hence, if the spheroid space represents a locale where data exists and the point on the curved surface is a datum, then the vector of coefficients and attributes reflects the characteristics of that datum.

Lorentzian manifolds also have the concept of causal structure. Causal structure in space-time can describe ordering, past and future, designated by the directionality of each tensor vector. Consequently, this can also be applied to digital data. For example, biology can have time-dependent ordering when describing the time lapse yeast growth characteristics in cell array experimentation, just as described earlier in Unifluxion and spatial voting time series analysis. Hence, metric tensors or "vectors of knowledge" intersect at their point of relation, Fig. 5. Hence, as one vector's directionality disperses or moves away from another vector, one can logically deduce that the strength of that relationship decreases or increases. Therefore, Newton's law of gravitation is used as an analogy to compare, contrast, associate, and normalize representations of relationship strength for any type information (analog or digital) artifact type. Figure 6 depicts an example representation of two information objects being compared, contrasted, and associated based upon user and/or system definable characteristics: importance, closeness. Newtonian gravity defines that the force of two bodies is proportional to their mass and inversely proportional to the square of their separation. The calculation is then multiplied by the universal gravitational constant to achieve the final force of gravity result between two bodies. For application to information theory, Newtonian mass is extended to denote relative information *importance* provided by a person or system as it pertains to a specific knowledge object to knowledge object comparison or any other smaller or larger context. The importance measure is a user r system defined scalar diameter in order to provide relative radius to the separation denominator. The square of separation is analogous to how close (*closeness*) the two pieces of information are relative to the overarching context. Figure 6, knowledge object (KO) #1, depicts the concept of two internal sub-nested objects of information which if reviewed would show additional context for KO1. The attractive force, A, of the two pieces of information in Fig. 6 is shown to equal 10. Lastly, a user or system can also employ a balance variable as an analogy to the Newtonian universal gravitational constant multiplier. This type of constant is considered a balance factor/variable of proportionality would be dependent upon user or system situational context which KO1 and KO2 are part of.

2.2.9 Conclusions and Discussion

This is preliminary work and significant research is still required. Here we have presented adaptive learning methods for enhancing cyber security risk reduction

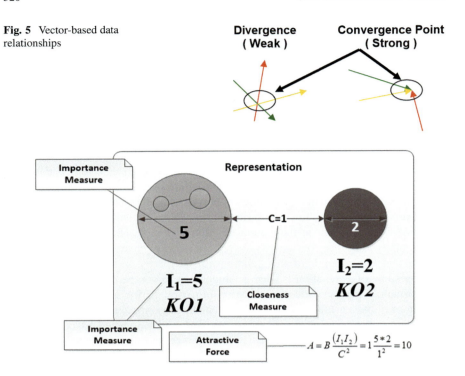

Fig. 5 Vector-based data relationships

Fig. 6 Knowledge relativity threads

by improving knowledge density (KD) (how much do we know about a given event or subject) and knowledge fidelity (KF) (how well do we know) to improve cyber event context and decision quality for improved and more autonomous action. Axiomatic design axioms and design features were recommended for adding to our KD data analysis prior to machine learning application. High fidelity Unifluxion fractals, spatial voting research, and improved performance analysis of time series data was provided. Spatial voting was shown to also operate against cyber digital time series data (e.g., malware detection) and their benefit comparisons to cyber machine learning data classification was also provided.

Tensor vector data and knowledge relativity threads (KRT) were shown to provide spatial constructs for explaining high fidelity relationships and the use of manifolds and metric tensor vectors as attribute descriptors was also described. These methods were combined with axiomatic design concepts to organize and supply complexity reduction techniques for reducing traditional time-consuming machine learning classification. Together, it was shown that these capabilities potentially produce support for more efficient decision actuation, due to improved explanation and relationship context of data and higher data analysis performance, thereby providing added insights to cyber systems and analysts to reduce security risk and reduced non-linearity and complexity. Suggested next steps should prototype, design, and implement an architecture to learn on large cyber datasets. Future papers will present progress and results as available.

References

1. S.S. Zhou, G. Feng, C.B. Feng, Robust control for a class of uncertain nonlinear systems: adaptive fuzzy approach based on back- stepping. Fuzzy Sets Syst. **151**(1), 1–20 (Apr. 2005)
2. W.S. Yu, C.J. Sun, Fuzzy model based adaptive control for a class of nonlinear systems. IEEE Trans. Fuzzy Syst. **9**(3), 413–425 (2001)
3. N. Suh, *Complexity Theory and Applications* (Oxford University Press, 2005)
4. G. Nicolis, *Introduction to Nonlinear Science, DI-Fusion* (Cambridge University Press, 1995)
5. J.R. Goodall, W.G. Lutters, A. Komlodi, I know my network: collaboration and expertise in intrusion detection, in *Proceedings of the 2004 ACM Conference on Computer Supported Cooperative Work*, ed. by J. Herbsleb, G. Olson, (ACM, New York, 2004), pp. 342–345
6. N.A. Giacobe, Application of the JDL data fusion process model for Cyber Security. *Multisensor, Multisource Information Fusion: Architectures, Algorithms, and Applications 2010*, vol. 7710. International Society for Optics and Photonics (2010)
7. P.C. Chen, P. Liu, J. Yen, T. Mullen, Experience-based cyber situation recognition using relaxable logic patterns. *In Proceedings of the 2012 IEEE international multi-disciplinary conference on cognitive methods in situation awareness and decision support (CogSIMA)*, pp. 243–250, IEEE (2012)
8. A. Joinson, T. van Steen, Human aspects of cyber security: behaviour or culture change? Cyber Secur. Peer-Reviewed J. **1**(4), 351–360 (2018)
9. S.A. Zahra, L.R. Newey, Maximizing the impact of organization science: theory-building at the intersection of disciplines and/or fields. J. Manag. Stud. **46**(6), 1059–1075 (2009)
10. D.V. Hutton, *Fundamentals of Finite Element Analysis* (McGraw-Hill, 2017)
11. A. Aziz, Prospective client identification using malware attack detection. U.S. Patent No. 9,027,135. 5 May 2015
12. D. Clark, J. Strand, J. Thyer, Active attack detection system. U.S. Patent No. 9,628,502. 18 Apr. 2017
13. S. Liu, G. Wei, Y. Song, Y. Liu, Extended Kalman filtering for stochastic nonlinear systems with randomly occurring cyber-attacks. Neurocomputing **207**, 708–716 (2016)
14. J. Crowder, J. Carbone, *The Great Migration: Information to Knowledge Using Cognition-Based Frameworks* (Springer Science, New York, 2011)
15. I. I. Liggins, D. H. Martin, J. Llinas (eds.), *Handbook of Multisensor Data Fusion: Theory and Practice* (CRC Press, 2017)
16. G. Bello-Orgaz, J.J. Jung, D. Camacho, Social big data: recent achievements and new challenges. Inform. Fusion **28**, 45–59 (2016)
17. D. Quick, K.K.R. Choo, Digital Forensic Data and Open Source Intelligence (DFINT+OSINT). In: *Big Digital Forensic Data. Springer Briefs on Cyber Security Systems and Networks*. Springer, Singapore (2018)
18. A. Ertas, M.M. Tanik, T.T. Maxwell, Transdisciplinary engineering education and research model. J. Integr. Design Proc. Sci. **4**(4), 1–11 (2000)
19. P. Nyhuis (ed.), *Wandlungsfähige Produktionssysteme* (GITO mbH Verlag, 2010)
20. R. Colbaugh, K. Glass, Predictability-oriented defense against adaptive adversaries. *Systems, Man, and Cybernetics (SMC), 2012 IEEE International Conference on. IEEE* (2012)
21. J. Lee, B. Bagheri, H.-A. Kao, Recent advances and trends of cyber-physical systems and big data analytics in industrial informatics. *International proceeding of int conference on industrial informatics (INDIN)* (2014)

22. J. Carbone, A framework for enhancing transdisciplinary research knowledge. Texas Tech University (2010)
23. J.A. Crowder, J.N. Carbone, S.A. Friess, *Artificial Cognition Architectures* (Springer, New York, 2014)
24. J. Crowder, S. Friess, Artificial neural diagnostics and prognostics: self-soothing in cognitive systems. *Proceedings of the 12th annual International Conference on Artificial Intelligence*, Las Vegas, NV (2010)
25. W. Liu et al., A survey of deep neural network architectures and their applications. Neurocomputing **234**, 11–26 (2017)
26. S.S. Roy, et al., A deep learning based artificial neural network approach for intrusion detection. *International Conference on Mathematics and Computing*, Springer, Singapore (2017)
27. N. Marz, J. Warren, Big data: principles and best practices of scalable real-time data systems. Manning (2013)
28. S. Sridhar, M. Govindarasu, Model-based attack detection and mitigation for automatic generation control. IEEE Trans. Smart Grid **5**(2), 580–591 (2014)
29. A. Inselberg, Parallel coordinates, in *Encyclopedia of Database Systems*, (Springer, Boston, 2009), pp. 2018–2024
30. D.H. Wolpert, W.G. Macready, No free lunch theorems for optimization. IEEE Trans. Evol. Comput. **1**(1), 67–82 (1997)
31. K.P. Burnham, D.R. Anderson, Practical use of the information-theoretic approach, in *Model Selection and Inference*, (Springer, New York, 1998), pp. 75–117
32. M.A. Ferrag et al., Deep learning for cyber security intrusion detection: approaches, datasets, and comparative study. J. Inform. Secur. Appl. **50**, 102419 (2020)
33. K.A. Heller, et al., One class support vector machines for detecting anomalous windows registry accesses. *Proc. of the workshop on Data Mining for Computer Security*, vol. 9 (2003)
34. W. Hu, Y. Liao, V. Rao Vemuri, Robust Support Vector Machines for Anomaly Detection in Computer Security. ICMLA (2003)
35. I. Balabine, A. Velednitsky, Method and system for confident anomaly detection in computer network traffic. U.S. Patent No. 9,843,488. 12 Dec. 2017
36. H.M. Jaenisch, J.W. Handley, N. Albritton, Converting data into functions for continuous wavelet analysis. *Independent Component Analyses, Wavelets, Neural Networks, Biosystems, and Nanoengineering VII*, vol. 7343. International Society for Optics and Photonics (2009)
37. H.M. Jaenisch, et al., Fractals, malware, and data models. Cyber Sensing 2012, vol. 8408. International Society for Optics and Photonics (2012)
38. R. Zuech, T.M. Khoshgoftaar, R. Wald, Intrusion detection and big heterogeneous data: a survey. J. Big Data **2**(1), 3 (2015)
39. H. Jaenisch, Spatial voting with data modeling for behavior based tracking and discrimination of human from fauna from GMTI radar tracks. *Unattended Ground, Sea, and Air Sensor Technologies and Applications XIV*, vol. 8388. International Society for Optics and Photonics (2012)
40. H.M. Jaenisch, et al., A simple algorithm for sensor fusion using spatial voting (unsupervised object grouping). *Signal Processing, Sensor Fusion, and Target Recognition XVII*, vol. 6968. International Society for Optics and Photonics, 2008
41. T. Aksenova, V. Volkovich, A.E.P. Villa, Robust structural modeling and outlier detection with GMDH-type polynomial neural networks. *International Conference on Artificial Neural Networks*. Springer, Berlin, Heidelberg, 2005
42. L. Brillouin, *Science and Information Theory* (Dover, 2004)
43. J. Crowder, V. Raskin, J. Taylor, Autonomous creation and detection of procedural memory scripts, in *Proceedings of the 13th Annual International Conference on Artificial Intelligence*, (Las Vegas, 2012)
44. J. Llinas, et al., Revisiting the JDL data fusion model II. *Space and Naval Warfare Systems Command San Diego CA* (2004)

Artificially Intelligent Cyber Security: Reducing Risk and Complexity

45. L.A. Zadeh, A note on web intelligence, world knowledge and fuzzy logic. Data Knowl. Eng. **50**(3), 291–304 (2004)
46. P. Gärdenfors, *Conceptual Spaces: The Geometry of Thought* (MIT Press, 2004)
47. P. Suppes, Current directions in mathematical learning theory, in *Mathematical Psychology in Progress*, (Springer, Berlin, Heidelberg, 1989), pp. 3–28
48. R.W. Langacker, *Foundations of Cognitive Grammar: Theoretical Prerequisites*, vol 1 (Stanford University Press, 1987)
49. G. Lakoff, Z. Kövecses, The cognitive model of anger inherent in American English, in *Cultural Models in Language and Thought*, Cambridge University Press, (1987), pp. 195–221
50. L. Talmy, Force dynamics in language and cognition. Cogn. Sci. **12**(1), 49–100 (1988)
51. R.C. Hibbeler, *Engineering mechanics* (Pearson Education, 2001)
52. D. Ejigu, M. Scuturici, L. Brunie, Hybrid approach to collaborative context-aware service platform for pervasive computing. JCP **3**(1), 40–50 (2008)
53. I. Nonaka, H. Takeuchi, *The Knowledge-Creating Company: How Japanese Companies Create the Dynamics of Innovation* (Oxford University Press, 1995)
54. M.J. Kearns, U.V. Vazirani, U. Vazirani, An *Introduction to Computational Learning Theory* (MIT Press, 1994)
55. T. Gruber, Collective knowledge systems: Where the social web meets the semantic web. J Web Semantics **6**(1), 4–13 (2008)
56. J.C. Platt, Fast training of support vector machines using sequential minimal optimization, in *Advances in Kernel Methods*, MIT Press, Cambridge, MA, (1999), pp. 185–208
57. E.P. Blasch, S. Plano, JDL Level 5 fusion model: user refinement issues and applications in group tracking, SPIE Vol. 4729, Aerosense (2002)
58. A. Einstein, Relativity: the special and general theory: a popular exposition, authorized translation by Robert W. Lawson: Methuen, London (1960)
59. A. Hendrik Lorentz, Considerations on Gravitation. In: *KNAW, Proceedings*, 2, 1899–1900, Amsterdam (1900)
60. M.S. Alber, G.G. Luther, J.E. Marsden, Energy Dependent Schrodinger Operators and Complex Hamiltonian Systems on Riemann Surfaces, August 1996

Procedural Image Generation Using Markov Wave Function Collapse

Pronay Peddiraju and Corey Clark

1 Introduction

Procedural Content Generation (PCG) is a design technique used to generate content using a variety of algorithms. PCG techniques are incorporated in use-cases such as map and narrative generation to allow for randomness in the game content. Examples of games that utilize such techniques include *No Man's Sky*, *Dwarf Fortress* [1], and *Caves of Qud* [2]. Some approaches to PCG include noise-based content generation using a noise function such as Perlin noise [3]. Although, noise-based techniques can be used for PCG, some of the drawbacks of this approach is the reliance on random distributions which are difficult to manipulate [3]. Unlike noise-based algorithms, constraint solving allows for more controllable PCG tools that are content agnostic [4].

Wave Function Collapse (WFC) has been used in games including *Bad North* [2] to generate maps as well as navigation data in the context of the game. It provides benefits over traditional PCG techniques in that WFC generates content based off of rules inferred from contextual information. By comparing against previously observed patterns, WFC can accurately predict viable choices to use when generating outputs using texture synthesis. Texture synthesis generates a large (and often seamlessly tiling) output image with a texture resembling that of a smaller input image [5]. Implementations of WFC can be seen in multiple programming languages including C# [6], Python [7], JavaScript, C++ [8], Rust [9] and others

---

P. Peddiraju
Guildhall, Southern Methodist University, Dallas, TX, USA
e-mail: ppeddiraju@smu.edu

C. Clark (✉)
Guildhall, Computer Science, Southern Methodist University, Dallas, TX, USA
e-mail: coreyc@smu.edu

© Springer Nature Switzerland AG 2021
H. R. Arabnia et al. (eds.), *Advances in Artificial Intelligence and Applied Cognitive Computing*, Transactions on Computational Science and Computational Intelligence, https://doi.org/10.1007/978-3-030-70296-0_39

with applications including 2D map generation [6], 3D map generation [8], poetry generation [10], 2D image generation, and more. In this paper, we will be discussing WFC in the context of generating images procedurally and explore some of the existing methodologies used to generate the same. This paper also proposes a WFC implementation using Markov random fields and analyzes the benefits and drawbacks for this approach.

2 Wave Function Collapse

Wave Function Collapse (WFC) is a constraint solving algorithm named after the quantum physics wave superposition concept. Maxim Gumin first implemented WFC using C# to incorporate a texture synthesis constraint solving PCG algorithm [6]. Entropy in WFC is described as the set of all possible color outcomes allowed by the input constraints for each individual pixel to be generated in the output image. This means that if a set of n colors were observed in an image, the entropy for any pixel in that image can be up to n. By following a minimum entropy heuristic, WFC resolves constraints from a point on the output sample and propagates outwards resolving entropy to create a final pattern [4]. These constraints determine the occurrence of specific information in the input data which can be visualized as neighboring pixel constraints in the case of 2D images. By splitting an image into uniform chunks called kernels, the WFC algorithm can identify patterns that exist in the input sample and their rate of occurrence. Using this information, WFC performs texture synthesis to generate a new output image that follows the same constraints observed in the input. Figure 1 illustrates some of the inputs and their corresponding procedural outputs generated by Maxim Gumin's original implementation of WFC [6].

This paper discusses 2 existing models used to generate procedural content using WFC, they are the Overlapping WFC (OWFC) and Tiling WFC (TWFC) models, and proposes a new Markov WFC (MkWFC) model. In the case of 2D, the OWFC model splits an input image into smaller $N \times N$ sized kernels. These kernels are then overlapped to determine the possible occurrences of pixels in the output. The TWFC model, on the other hand, does not identify the kernels by splitting the input image into chunks, instead, it receives these kernels and their constraints in the form of meta-data. This meta-data can be used by TWFC to propagate constraints and solve the entropies of kernels without having to determine them in a pre-compute step. By eliminating the iterative kernel identification process, the TWFC model generates procedural outputs faster than the OWFC model. This, however, comes at the cost of manually generating a meta-data file that feeds the relevant constraint information to TWFC which is non-trivial; this affects the scalability of TWFC as increasing the complexity of the input can result in the need for increased investment of resources in determining the constraints required to generate desired results [11].

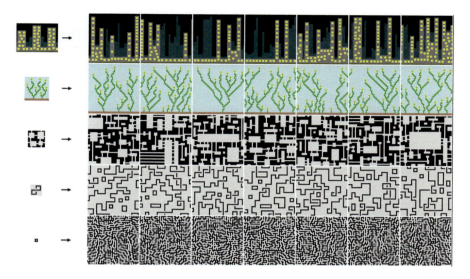

Fig. 1 Samples from Maxim Gumin's WFC repository [6] highlighting the input provided to the algorithm on the left side and the random procedural results on the right side for each corresponding input image

The MkWFC model builds on the TWFC model by using constraint identification from input samples similar to OWFC while maintaining the performance benefits of TWFC.

2.1 Overlapping Wave Function Collapse Model

2.1.1 Overview

The OWFC model expects only an input sample as opposed to the other models which require some meta-data or a combination of meta-data and input samples. The OWFC model splits the input sample image into $N \times N$ kernels where N is defined by the user; these kernels are then stored in a list to be used in the generation of the output image. These identified kernels account for 8 possible occurrences of each kernel including rotated and reflected versions of the kernel. The requested output image is allocated in memory but none of the pixels in the output image is assigned any color information as all the pixels in the requested output can be any of the pixel colors read from the input. This means they all start at the highest possible entropy (all of the observed colors from the input) and the entropy of the image needs to be resolved to finally generate an image where there is only 1 possible color for any given pixel in the output (i.e., entropy of 1).

To perform the entropy resolution, an arbitrary pixel is selected in the output and assigned one of its possible color values. This reduces the entropy of that pixel to 1 and the assigned color information is then propagated to its neighboring pixels. Each of the neighboring pixels is compared against observed kernels to determine possible color values based on the previously resolved pixel information. The check is performed by overlapping the kernel with the neighboring pixel to determine if the entropy for the neighboring pixel is valid based on the occurrence of color values in the kernel. If the entropy values exist in the overlapped kernel, the pixel entropy is maintained, otherwise, all entropy values that cannot occur are eliminated from the neighboring pixel entropy [12]. This results in the entropy of the neighboring pixel is either decreases or remains the same. The kernel overlap is performed for all neighboring pixels; the process continues recursively until all the pixels in the output image have been resolved to an entropy of 1. Due to the nature of the constraints used by the algorithm, there are instances where entropy resolution cannot resolve the output pixel entropies [13]. This occurs when there is not enough information to identify constraints that can resolve the pixel entropy, in which case WFC cannot generate a result with the provided input [14].

By arbitrarily picking the first pixel to perform entropy resolution, the resolution occurs differently on every execution of the algorithm hence generating different outputs.

2.1.2 Tradeoffs of OWFC

Although OWFC does provide procedural results, due to the iterative nature of kernel identification OWFC is computationally heavy. By increasing the size of the kernel, there is a rapid increase in computational time in the order of N^2 where $N*N$ represents the dimensions of the kernel. Increasing the complexity of the image by using numerous colors or well-defined patterns such as alphabets or symbols causes results to lose randomness due to there being a larger constraint set. While providing procedural results, OWFC suffers from high computational cost as well as the lack of scalability.

2.2 Tile-Based Wave Function Collapse Model

2.2.1 Overview

The TWFC model is very similar to the OWFC model except for the kernel generation step. Instead of computing the kernels present in an input sample, the TWFC model relies on a meta-data file that can provide the constraints required to generate an output image. By doing so, the tiling model can benefit from not having to use per-pixel entropies; instead, a tile can comprise of a set of pixels that fit on a fixed tile grid. The constraints are neighborhood relationship information between

2 distinct tiles and their orientations. The algorithm uses a WFC wave object which contains all possible tile instances that can occur on the output image. Similar to OWFC, TWFC still requires the need to resolve the entropies of each tile in the output image as the WFC wave object considers all tiles to have the highest possible entropies (i.e., the total number of tiles used by the algorithm) when generating the output [15].

When comparing the 2 models, OWFC model takes a sample image as input while the TWFC model takes a meta-data file as input. The meta-data file contains a set of tiles to be used as subsets of the generated image and some constraints that determine which tiles can be placed next to each other. By feeding the constraints and possible tiles in manually generated meta-data to TWFC, the kernel generation step is not required hence reducing the computational load involved of TWFC as opposed to that of OWFC. Since TWFC uses tiles on a fixed tile grid as opposed to pixels in an image, the user has control over the size of the tiles used as well as the complexity of the output generated as tiles can be simplified or made more detailed based on usage. Unlike OWFC however, TWFC has no way of identifying the rate of occurrence for kernels (in this case tiles) occurring in the input as there is no input image. Instead, to account for a rate of occurrence, TWFC requires this information to be provided as tile frequency or tile weight in the meta-data file.

2.2.2 Architecture

The architecture used by TWFC is illustrated in Fig. 2. The meta-data is parsed to identify the set of tiles to be used for texture synthesis and their symmetries are computed. Once all tiles have been identified and their symmetries computed, TWFC parses the constraints to be used to solve against. TWFC then uses a WFC Propagator object which will propagate the entropy resolution information and the WFC wave which stores an output of the desired size. After creating the propagator and wave, TWFC can begin the entropy resolution process.

Similar to the case of OWFC, the iterative entropy resolution process cannot complete if there is not enough constraint information, in which case we set a limit to the number of attempts to complete entropy resolution. After the entropy resolution solves for an output WFC wave object, the algorithm then writes the content of the WFC wave object as an output image of the same size. The result of the TWFC algorithm returns either a pass or fail output that conveys if the model generated an output file or not from the given input sample.

2.2.3 Meta-Data Configuration

TWFC uses meta-data comprising tile information utilized by the texture synthesis step along with constraint information. The meta-data lists a set of tiles to be used by TWFC as well as their size in pixels and their possible symmetries (rotated and reflected instances) of the tiles to be used when generating an output. The constraints

TWFC Architecture

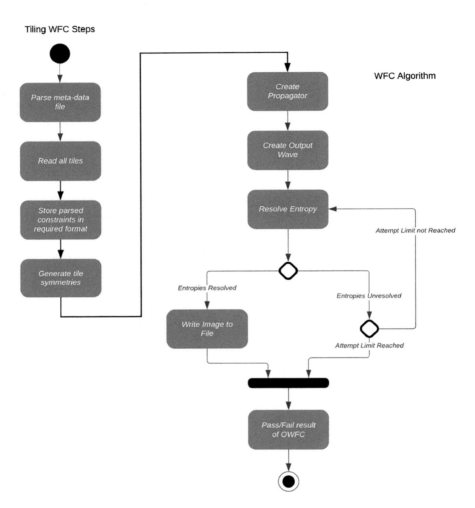

Fig. 2 Flowchart of tiling WFC

in meta-data specify a left tile and its orientation to a corresponding right tile and its orientation to determine which of these constraints can fit in the output WFC wave object.

2.2.4 Improvements Over OWFC

By eliminating the kernel generation process, TWFC does not need to pre-compute the existing patterns and their rate of occurrence, instead, the meta-data file provides the required tile information including frequencies, constraints, and sizes hence allowing the TWFC algorithm to only perform the WFC wave propagation and entropy resolution. With kernels relationships determined ahead of time, the computational overhead is reduced while also providing users the opportunity to control the constraints involved as well as the complexity of the tiles without directly affecting the WFC resolution and propagation. By doing so, making changes to tiles being used by the TWFC model does not affect its speed, adding constraints however would affect the performance of the algorithm due to the need to do fewer (if neighbor constraints are reduced) or more (if neighbor constraints are increased) comparisons during entropy resolution.

2.2.5 Tradeoffs of TWFC

Despite having a reduced computationally load and improving on the speed of generating output using WFC, the TWFC model still has some drawbacks. The generation of the meta-data file is non-trivial. To identify the required constraints and how they affect the generated output requires designing possible outcomes by involving some constraints as opposed to others or by balancing the frequency of the tiles being used. This can be challenging as the reduction or inclusion of even a single constraint can change the entropy resolution process significantly. Due to the determining of neighborhood constraints being a non-trivial process, TWFC also does not scale well. Increasing the number of constraints or adding more tiles to the TWFC problem requires re-iterating on the possible outcomes and manipulating the non-trivial neighborhood constraint set. As the number of tiles and constraints increases, the complexity also increases which can result in a greater chance of human error. Similar to OWFC, there are cases where TWFC cannot resolve the tile entropy and the algorithm fails to generate an output image. This can be a result of conflicting constraints or human error in curating the manually generated meta-data file.

2.3 Markov Chain Wave Function Collapse Model

2.3.1 Overview

The proposed Markov WFC (MkWFC) model aims to solve the scalability issues of TWFC while maintaining the performance benefits it provides over OWFC. To eliminate this scalability tradeoff, MkWFC performs an automation step to infer the constraints rather than use a manually generated meta-data file. MkWFC uses the tile information to scan input images provided in meta-data to determine constraints. These constraints are used to create Markov random field that determine neighboring tile relationships for each instance of a tile in the input image. Although this introduces an overhead in terms of execution time, by automating the non-trivial step of constraint identification MkWFC identifies constraints and generates a Markov random field to be used by the WFC algorithm. Due to this automated pre-processing step MkWFC can scale better than TWFC with increased input and output sizes.

There are 2 benefits MkWFC provides over the TWFC implementation. The first being the elimination of manually determining non-trivial constraint information. By eliminating this process MkWFC uses the design philosophy of the OWFC model, which is to observe input images for constraint identification. With tile information present in meta-data, the MkWFC model can pre-compute their symmetries and observe the input images for occurrences of these tiles, thereby determining its neighborhood relationships to build constraints that can be used for output generation. The second benefit is scalability in terms of increasing the number of constraints used by the WFC algorithm. Due to the automation of constraint identification, increasing the constraint set used by WFC requires adding additional input images or increasing the size of provided input images to identify a constraint set with greater coverage over the theoretical maximum of possible constraints. This also allows for generating results with different constraints by switching between images that follow the desired constraints as opposed to manually curating new constraint relationships to be passed as meta-data like in the case of TWFC.

After parsing the meta-data, MkWFC generates all the tile symmetries as is in the case with TWFC. After generating the symmetries, MkWFC performs an extra step of analyzing all the input samples provided in the meta-data for tile occurrences. By analyzing the entire image, MkWFC identifies a set of neighborhood relationships between 2 adjacent tiles. For each tile, the MkWFC model creates a Markov random field of compatible neighbors. After identifying compatible neighbors for all tile occurrences in the input images, these neighborhood constraints are propagated during entropy resolution.

2.3.2 Architecture

The architecture used by MkWFC is illustrated in Fig. 3. The meta-data is parsed to identify the set of tiles to be used for texture synthesis and their symmetries are computed as in the case of TWFC. The difference is in the identification of constraints. MkWFC uses input images described in meta-data to determine tile instances and create constraint chains. The constraint chains are then populated into a constraint set of a Markov random field which are the relationships of each tile instance in the input image to its corresponding neighbors. MkWFC then creates the WFC Propagator object which will propagate the entropy resolution information and the WFC wave which stores an output of the desired size. The result of the MkWFC

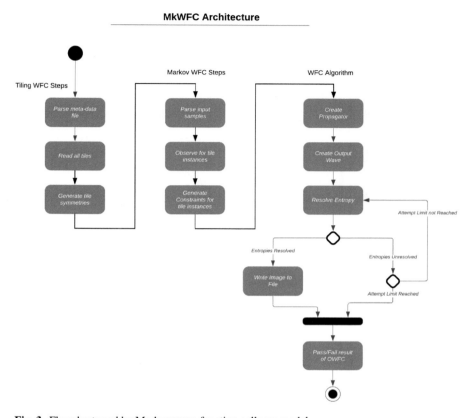

Fig. 3 Flowchart used by Markov wave function collapse model

algorithm returns either a pass or fail output that conveys if the model generated an output file or not from the given input sample similar to TWFC.

2.3.3 Meta-Data Configuration

Similar to TWFC, MkWFC also needs some meta-data to describe the tiles used by MkWFC as well as their respective symmetries. So as in the TWFC case, there is a list of tiles to be used provided in meta-data along with their size in pixels and their respective symmetries. But unlike TWFC, instead of a list of left tile to right tile neighborhood constraint information, the meta-data for MkWFC provides a list of input images. These input images are analyzed in the constraint identification step and a list of left tile to right tile neighborhood constraints is automated, then fed as constraint list to the tiling step of MkWFC. The elimination of curating non-trivial constraint information reduces the size of meta-data file as well as eliminates the time intensiveness of manually curating the constraint information.

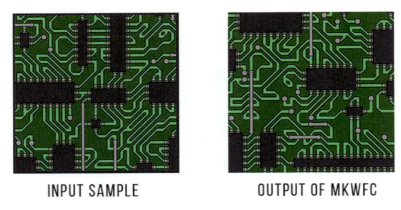

Fig. 4 Input sample and generated output for circuit generation using MkWFC

2.3.4 Model Usage

To showcase the workings of MkWFC we will consider the use-case of procedurally generating a circuit board as described in Fig. 4. This image is passed as input to MkWFC. Its corresponding meta-data file contains a list of 14 tiles that are used and their corresponding symmetry.

As expected, the meta-data only contains tile information and the input samples that would be used by MkWFC which in this case is the image in Fig. 4. On execution of MkWFC, the automated constraint identification step checks against the input image to identify tile occurrences and creates a list of neighborhood constraints as a Markov random field. These constraints are then used by the WFC algorithm to perform the texture synthesis step to resolve the entropy of each tile instance and generate an output image.

On successful entropy resolution, MkWFC saves the resolved output as an image. The results of this problem are illustrated in Fig. 4 where the constraints identified from the input image generated an output image following the same constraints. It is important to note however that similar to OWFC and TWFC, MkWFC cannot guarantee the resolution of entropy per tile as there may be conflicting constraints which may require more than 1 pass to resolve said constraints or the constraint contradictions are impossible to resolve using the identified constraint set.

3 Experiment

To compare the scalability and performance of MkWFC against TWFC, 6 different content generation problems namely *Circles*, *Circuits*, *Rooms*, *Knots*, *Castles*, and *Summer* were considered. Each of the 6 problems was executed to generate outputs of 3 different sizes which were 5×5 tiled for small outputs, 10×10 tiled for medium outputs, and 25×25 tiled for large outputs. Metrics for comparing TWFC against MkWFC were calculated by executing 10 trials for each of the 6 problems when generating outputs for each of the 3 output sizes. A total of 180 trials were performed for TWFC (60 trials each of 5×5, 10×10, and 25×25 tiled outputs). Similarly 540 trials were performed for MkWFC, where the first 180 trials used 5×5 tiled input samples to generate outputs of all sizes (60 trials of 5×5, 10×10, and 25×25 outputs), the next 180 trials used 10×10 tiled inputs to generate outputs of all sizes, and the final 180 trials used 25×25 tiled inputs to generate outputs of all sizes. By averaging the metrics for all 720 trials, this paper compares the constraint identification, usage, and execution times in milliseconds for TWFC and MkWFC.

To generate outputs, TWFC used a manually generated meta-data file for each of the 6 problems which listed the tiles used for that problem as well as the constraints to be solved against. MkWFC on the other hand used 3 different meta-data files, 1 for each desired output size, and used 3 images as input samples from which to determine constraints. When generating outputs, MkWFC performance was measured when using 3 input samples of size 5×5, 10×10, and 25×25 to generate outputs of all sizes. All constraints were used as a set of left to right neighboring tile relationships. In the case of TWFC, these were provided in the manually generated meta-data file whereas in the case of MkWFC, they were identified from the input sample images provided to the algorithm. To calculate the total number of theoretical constraints, it was assumed that all tiles used by MkWFC and TWFC had 8 possible symmetries. The formula used to compute the total number of possible constraints ϵ on an $n * n$ sized grid is described in Eq. 1 such that we compute 4 possible constraints for all tiles in the $n * n$ image excluding the boundary tiles where n represents the width and height of the image.

$$\epsilon = (\alpha + \beta + \gamma) \tag{1}$$

where α, β, and γ are described in Eqs. 2, 3, and 4 as

$$\alpha = (n - 2)^2 * 4 \tag{2}$$

$$\beta = (n - 2) * 4 * 3 \tag{3}$$

$$\gamma = 4 * 2 \tag{4}$$

ϵ in Eq. 1 represents the maximum possible constraint pairs on a tile grid of size $n * n$. α determines the maximum possible constraint pairs of all the tiles excluding the boundary tiles on the tile grid. The number of tiles on the grid excluding the boundary tiles is represented as $(n-2)^2$ and since these tiles have neighbors on all 4 directions *top, bottom, left, and right* we multiply 4 as the possible constraint pairs for these tiles.

β represents the maximum possible constraint pairs for the tiles that lie on the boundary of the tile grid except for the corner tiles. The tiles on the boundary except for corner tiles can each have 3 constraint pairs associated with them as one of their 4 directions will have no neighboring tile. This is represented as $(n-2)$ tiles for each of the 4 sides having 3 possible neighborhood constraints.

Finally a constant γ represents the maximum possible constraints for the 4 corner tiles on the tile grid. Since the corner tiles have neighbors in 2 out of the 4 possible directions, the value of C is $4 * 2$ and is constant.

4 Results

Figure 6 showcases 1 of the input samples used by MkWFC and the outputs generated by both TWFC and MkWFC for all 6 problems in 1 of the trials performed. For the *Circuits* problem, MkWFC identified insufficient unique constraints resulting in fail conditions when generating outputs under 2 of the 3 test conditions. MkWFC was unable to resolve constraints when using small and medium input samples to identify constraints but was able to generate results using large input samples. This lack of information for entropy resolution resulted in entropy for tiles in the output to be higher than 1 which meant that there is more than 1 possible outcome for that tile in the output resulting in a fail condition. Failure to generate outputs for the *Circuits* problem occurred when the number of constraints identified to the possible constraint pairs on the output image fell below a threshold value. The threshold value identified was 0.2 when using small input samples and 0.47 when using medium input samples. The tiles used to generate outputs for the *Circuits* problem are shown in Fig. 5 (Fig. 6).

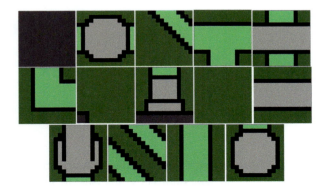

Fig. 5 Tiles provided as inputs for TWFC and MkWFC for the *Circuits* problem

Fig. 6 Outputs of TWFC and input sample with corresponding generated outputs using MkWFC for 1 of the trials

The TWFC model relies on the use of constraints to be provided in meta-data which would require the need to explicitly specify the desired constraints to use whereas MkWFC identifies all constraints existing in an input sample and generates Markov random field of all the neighborhood tile constraints. As a result, there are constraints identified by MkWFC which may not be part of the constraint set provided to TWFC resulting in patterns in output which would not be generated by the TWFC implementation unless explicitly mentioned in manually generated meta-data. To highlight the case where new constraints are used in generating outputs by MkWFC, the *Castles* trial used by MkWFC is showcased in Fig. 7 where the tile instances highlighted in red represent the constraints that were identified by MkWFC which were not present in the TWFC implementation. When comparing TWFC and MkWFC, there was an increase in the coverage of the maximum theoretical constraints set for MkWFC when compared to TWFC based on the data analysis performed.

Generating the constraints in meta-data for TWFC requires the use of reference images provided by a designer, by automating constraint identification MkWFC used the same reference images to identify constraints in millisecond range. MkWFC consistently increased the number of constraints identified as a result of automating the constraint identification process and scaled with increasing output sizes. MkWFC introduces an increase in execution time which is nominal to allow for performance in real-time similar to TWFC.

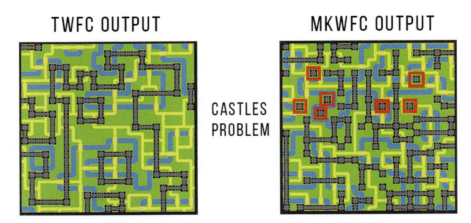

Fig. 7 Comparison of TWFC and MkWFC outputs for the *Castle* problem. Red boxes show new constraints not present in TWFC

4.1 Constraint Identification

Table 1 is the comparison of constraints received by TWFC to those identified by MkWFC when generating outputs using 3 image samples as inputs. When reading constraints, TWFC received $88.76 \pm 61.16\%(\alpha = 0.05)$ of the constraints set identified by MkWFC when using small image samples of size 5×5. Each identified constraint in MkWFC was weighted to determine the frequency of a pattern that existed in input. As the size of input images increased, the constraints received by TWFC reduced to $46.95 \pm 19.77\%(\alpha = 0.05)$ of the constraint set identified by MkWFC when using 10×10 sized input samples and further reduced to $36.47 \pm 6.2\%(\alpha = 0.05)$ of the constraint set identified by MkWFC when using 25×25 sized input samples.

From Table 1 it is clear that MkWFC manages to identify more constraints when compared to those read by TWFC with increasing input sizes. As the size of input sample increases, the constraint set identified by MkWFC also increases whereas the TWFC model utilizes the same fixed constraint set which needs to be manually generated for each problem.

4.2 Constraint Usage

Table 2 is the comparison of constraint usage for TWFC and MkWFC when generating outputs of varying sizes. The table compares the outputs generated when using 3 different sized inputs with MkWFC. When generating outputs using inputs of size 5×5, TWFC used $42.65 \pm 5.72\%(\alpha = 0.05)$ of the constraints used by MkWFC as the worst case for TWFC. The best case for TWFC resulted in the usage of $57.38 \pm 14.29\%(\alpha = 0.05)$ of the constraints used by MkWFC when generating outputs of size 25×25 with 5×5 input samples. In the case of using 5×5 input samples, MkWFC provides almost double the coverage of constraints used when compared to TWFC. When comparing the constraint usage of TWFC with MkWFC while using 10×10 images to generate outputs with MkWFC, TWFC exhibits a best case of $98.72 \pm 152.9\%(\alpha = 0.05)$ of the constraint set used by MkWFC when generating 5×5 sized output images. This drops to a $37.31 \pm 7.12\%(\alpha = 0.05)$ of the constraints used in the worst case of generating 10×10 image outputs with the metrics for generating 25×25 outputs beings similar where TWFC uses $38 \pm 5.4\%(\alpha = 0.05)$ of the constraint set used by

Table 1 Constraint identification TWFC compared to MkWFC

Input size	Mean ($\alpha = 0.05$)
5×5	$88.76 \pm 61.16\%$
10×10	$46.95 \pm 19.77\%$
25×25	$36.47 \pm 6.2\%$

Table 2 Constraint usage TWFC compared to MkWFC

Input size	Output size	Mean ($\alpha = 0.05$)
5×5	5×5	$42.65 \pm 5.72\%$
5×5	10×10	$45.47 \pm 13.17\%$
5×5	25×25	$57.38 \pm 14.29\%$
10×10	5×5	$98.72 \pm 152.9\%$
10×10	10×10	$37.31 \pm 7.12\%$
10×10	25×25	$38 \pm 5.4\%$
25×25	5×5	$64.83 \pm 67.33\%$
25×25	10×10	$67.45 \pm 82.12\%$
25×25	25×25	$34.57 \pm 4.68\%$

Table 3 Execution time comparisons TWFC and MkWFC

Input size	MkWFC constraint identification time ($\alpha = 0.05$)	MkWFC tiling step time ($\alpha = 0.05$)	TWFC execution time ($\alpha = 0.05$)
5×5	1.03 ± 0.09 ms	22.76 ± 2.94 ms	25.12 ± 4.77 ms
10×10	4.13 ± 0.38 ms	34.83 ± 6.03 ms	25.54 ± 5 ms
25×25	28.19 ± 2.58 ms	60.71 ± 8.82 ms	25.15 ± 4.6 ms

MkWFC. The last set of metrics to be compared is when MkWFC uses 25×25 sized input samples to generate outputs. The constraint coverage of TWFC exhibits the best case of $67.45 \pm 82.12\% (\alpha = 0.05)$ of the coverage used by MkWFC when generating outputs of size 10×10. This drops to the worst case of TWFC using $34.57 \pm 4.68\% (\alpha = 0.05)$ of the constraints used by MkWFC when generating 25×25 sized output images.

From Table 2, MkWFC consistently manages to use a larger constraint converge to generate outputs irrespective of size when compared to TWFC. The optimal case for MkWFC is when the size of generated output is the same as the size of the input samples provided to the algorithm while the worst case is observed when generating outputs that are smaller than the size of the provided input samples.

4.3 Performance

Table 3 shows the time taken on average by MkWFC for constraint identification and tiling step when compared to the total execution time of TWFC.

When generating outputs using images of size 5×5, MkWFC spent 1.03 ± 0.09 ms on constraint identification and 22.76 ± 2.94 ms on the tiling step. This result shows there was no statistical significant difference between the 25.12 ± 4.77 ms TWFC execution times at this size. MkWFC spent 28.19 ± 2.58 ms on constraint identification and 60.71 ± 8.82 ms on the tiling step for 25×25 tiled steps, which was a statistically significant increase from the 25.15 ± 4.60 ms execution time of TWFC. Although TWFC performed faster for the same case, MkWFC

Procedural Image Generation Using Markov Wave Function Collapse   541

identified almost 3 times the number of constraints manually identified for TWFC. The increase in constraints identified introduces an increase in execution time in the range of milliseconds. To provide the same increase in constraints for TWFC, the manual meta-data generation process would require a substantial increase in design time. Due to the increasing number of constraints identified, the tiling step of MkWFC exhibits a slower execution time with increasing image sizes.

5 Conclusion

MkWFC managed to consistently provide larger constraint coverage for both the constraint identification and usage when compared to TWFC for all 3 image sizes. The best case usage of MkWFC was exhibited when generating outputs of the same size as the input samples provided to the algorithm. Due to an increase in constraints used by MkWFC, the tiling step time was larger than the total execution time of TWFC when generating outputs of increasing size. The increase in execution time of MkWFC is in the range of milliseconds maintaining the real-time performance advantages of TWFC.

Where run-time performance is a priority, TWFC can be used with a substantial design time to generate meta-data to save on the run-time time overhead of MkWFC. MkWFC however, consistently provides increased constraint usage across all image sizes making it better suited for increasing content sizes. The added performance overhead in automated constraint identification is significantly less than the design time required to manually generate the constraint set. MkWFC provides the performance benefits of TWFC over OWFC while maintaining viable usage in real-time applications.

References

1. Procedural Generation with Wave Function Collapse (2019). https://gridbugs.org/wave-function-collapse/
2. J. Grinblat, C.B. Bucklew, Subverting historical cause & effect: generation of mythic biographies in *Caves of Qud*, in *Proceedings of the International Conference on the Foundations of Digital Games - FDG '17* (ACM Press, Hyannis, Massachusetts, 2017), pp. 1–7. http://dl.acm.org/citation.cfm?doid=3102071.3110574
3. Math for Game Developers, End-to-End Procedural Generation in 'Caves of Qud' (2019). https://www.gdcvault.com/play/1026313/Math-for-Game-Developers-End
4. Isaackarth.com, WaveFunctionCollapse is Constraint Solving in the Wild (2017). https://isaackarth.com/papers/wfc_is_constraint_solving_in_the_wild/
5. A. Efros, T. Leung, Texture synthesis by non-parametric sampling, in *Proceedings of the Seventh IEEE International Conference on Computer Vision*, vol. 2 (1999), pp. 1033–1038. iSSN: null
6. M. Gumin, mxgmn/WaveFunctionCollapse (2020). original-date: 2016-09-30T11:53:17Z. https://github.com/mxgmn/WaveFunctionCollapse

7. sol, s-ol/gpWFC (2020). original-date: 2018-05-14T15:25:21Z. https://github.com/s-ol/gpWFC
8. sylefeb, sylefeb/VoxModSynth (2020). original-date: 2017-07-23T04:49:33Z. https://github.com/sylefeb/VoxModSynth
9. S. Leffler, sdleffler/collapse (2020). original-date: 2016-10-25T21:42:16Z. https://github.com/sdleffler/collapse
10. Available: https://twitter.com/mewo2/status/789167437518217216
11. D. Long, Visual Procedural Content Generation with an Artificial Abstract Artist. https://www.academia.edu/34062512/Visual_Procedural_Content_Generation_with_an_Artificial_Abstract_Artist
12. Doodle Insights #19, Logic Data Generation (feat. WFC made easy) (2017). https://trasevol.dog/2017/09/01/di19/
13. crawl/crawl (2020). original-date: 2014-08-17T10:39:37Z. https://github.com/crawl/crawl
14. GDC Vault. https://gdcvault.com/login
15. M. Kleineberg, marian42/wavefunctioncollapse (2020), original-date: 2018-10-25T21:48:08Z. https://github.com/marian42/wavefunctioncollapse

Parallel Algorithms to Detect and Classify Defects in Surface Steel Strips

Khaled R. Ahmed, Majed Al-Saeed, and Maryam I. Al-Jumah

1 Introduction

Worldwide, steel industry is one of the most important strategic industries. Quality is an important competitive factor to the steel industry success. Detection of surface defects devotes a large percent of quality control process to satisfy the customer's need [1], and [2]. Defect detection and classification can be accomplished manually by human labor; however, it will be slow and subject to human-made errors and hazards. Therefore, automatic traditional-inspection systems were developed to detect various faults. These include eddy current testing, infrared detection, magnetic flux leakage detection, and laser detection. These methods are not able to detect all the faults, especially the tiny ones [3]. This motivates many researchers [4, 5] to develop computer vision systems capable of classifying and detecting defects in ceramic tiles [6], textile fabrics [7], and steel industries [8]. Achieving defect detection, localization, and classification in real time is one of the challenges in the steel production process. Therefore, the main aim of this chapter is to propose parallel algorithms to detect and classify patches, scratches, and scale defects in surface steel strips in real time.

The rest of this chapter is organized as follows. Section 2 reviews the related works. Section 3 illustrates the proposed algorithm. Section 4 discusses the experiment setup and results. Section 5 concludes this chapter.

K. R. Ahmed (✉)
School of Computing, Southern Illinois University, Carbondale, IL, USA
e-mail: kahmed@cs.siu.edu

M. Al-Saeed · M. I. Al-Jumah
King Faisal University, Hofuf, Saudi Arabia
e-mail: alsaeed@kfu.edu.sa

© Springer Nature Switzerland AG 2021
H. R. Arabnia et al. (eds.), *Advances in Artificial Intelligence and Applied Cognitive Computing*, Transactions on Computational Science and Computational Intelligence, https://doi.org/10.1007/978-3-030-70296-0_40

2 Related Work

Image processing plays a major role in the steel production industry to enhance the quality of the products. In the literatures, many image-processing algorithms have been proposed to detect various defects by features extraction techniques. A plenty of features have been used including color, texture, shape, geometry features, etc., for defect localization and type identification [9]. The common techniques used for feature extraction in steel images are categorized into four different approaches [10]. These approaches are statistical methods, structural algorithms, filtering methods, and model-based techniques as shown in Fig. 1a.

Statistical approaches usually used histogram curve properties to detect the defects such as histogram statistics, autocorrelation, local binary patterns, grey level co-occurrence matrices [11], and multivariate discriminant function [12]. Image processing and edge detection algorithms are the basic operations used in structural approaches. Due to various defects depicting similar edge information, it is hard to classify the defect types. Filter-based methods involve convolution with filter masks for computing energy or response of the filter. Filters can be applied in frequency domain [13], in spatial domain, or in combined spatial frequency domain [14]. Model-based approaches include fractals, random field models, autoregressive models, and the epitome model [10] to extract a model or a shape from images. Figure 2 lists methods utilized to detect two types of surface defects on steel strips.

There are many approaches to extract features in parallel. Lu et al. [15] proposed an adaptive pipeline parallel scheme for constant input workloads and implemented an efficient version for it based on variable input workloads; they speed up to 52.94% and 58.82% with only 3% performance loss. Also, Zhang et al. [16] proposed a model to generate gray level run length matrix (GLRLM) and extracts multiple features for many ROIs in parallel by using graphical processing unit

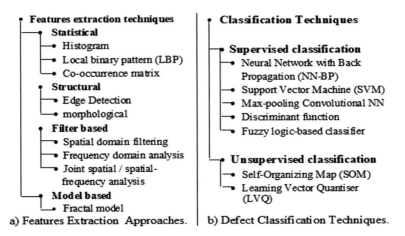

Fig. 1 Related work

	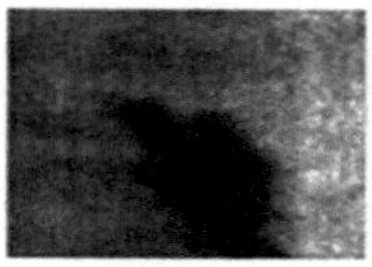 Patches	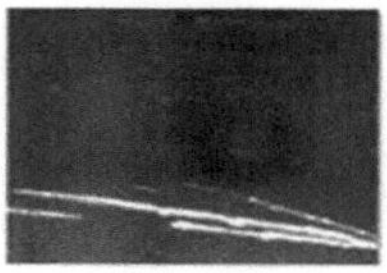 Scratches
Detection Methods	• Local Binary Pattern (LBP) [3] • Histogram [24] • Edge Detection [24] • Features Extraction [10-26] • Algorithms [27]	• Local Binary Pattern (LBP) [3] • Histogram [24] • Edge Detection [10] • Features Extraction [21] • Gray Level Co-occurrence Matrices (GLCM) [20-25]
Classification Methods	• Support Vector Machine (SVM) [3] • Nearest Neighbour Classifier (NNC) [3] • Deep Auto-encoder Network (DAN) [24] • Artificial Neural Networks (ANN) [24] • Learning Vector Quantization (LVQ) [30]	• Nearest Neighbor Classifier (NNC) [3] • Support Vector Machine (SVM) [28] • Self-organizing Map (SOM) [18] • Gabor Filter [29] • Deep Auto-encoder Network (DAN) [24] • Artificial Neural Networks(ANN) [24]

Fig. 2 Defects detection and classification techniques

(GPU), and they achieved five-fold increase in speed than an improved sequential equivalent.

The classification process is the main consideration in the inspection system. Generally, there are two types of classification methods: supervised and unsupervised as presented in Fig. 1b. In supervised classification, training samples are labeled, and features are given to the classifier to generate the training model. The training model predicts the pre-defined classes for test samples [10]. These methods include SVM, neural networks, nearest neighbors, etc. Yazdchi et al. [17] applied neural network (NN) to classify steel images that achieved accuracy 97.9%. Yun et al. [18] suggested support vector machine (SVM) classifier for defect detection of scale-covered steel wire rods. In unsupervised classification, classifier earns on its own and it is not fed with labeled data. Classifier just tries to group together similar objects based on the similarity of the features [19]. Most common types of methods include K-means, SOM (self-organizing map) [20] and LVQ (learning vector organization) [13]. Figure 2 lists some defect detection and classification methods. The key parameters of the defect classification methods are the accuracy and the efficiency. This paper employs the SVM.

3 Proposed Algorithms

This chapter develops parallel algorithms to detect and classify patches, scratches, and scale defects in surface steel strip. Figure 3 shows the high-level design of the

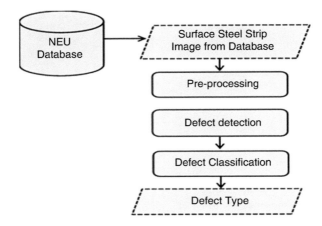

Fig. 3 High-level architecture of the proposed algorithm

proposed defect detection and classification technique. First phase is to preprocess the image to improve it and remove noises. Second phase detects defects from the steel image and segments it to defective ROIs. Third phase extracts *Haralick* features from gray level co-occurrence matrix (*GLCM*). Finally, these features will be used as inputs to the SVM classifier.

3.1 Preprocessing Phase

Surface steel images are subject to various types of noises due to image acquisition setup, lighting conditions, or material reflections. The preprocessing operation is an important step to eliminate light reflection and noises. Preprocessing operation carried out image enhancement and noise reduction. Image enhancement composes two steps to make image clearer. First, convert the RGB images into grayscale images and resize the image to $M \times N$. Then apply the contrast stretching operation to enhance image brightness by stretching the intensity values from *0* to *255*. To remove noises, this chapter uses median filter to remove salt, pepper noises, and makes images more blurred [21, 22].

3.2 Defect Detection Phase

In this phase the algorithm divides the $M \times N$ grayscale steel image into blocks (ROIs) of size $W \times H$. After that, it extracts statistical features for each ROI by using multivariate discriminant function [12] to detect either the ROI is defected or not.

1. *Features Extraction*: The proposed algorithm divides the $M \times N$ grayscale image into ROIs of size $W x H$, where $W \ll M$ and $H \ll N$. To characterize the shape of the surface defects and detect either if the ROI is defected or not, the algorithm extracts following statistical features for each ROI: difference (δ), mean (μ) and variance (υ) as in Eqs. (3), (4), and (5). After that, it calculates mean vector (MV) for each ROI as Eq. (6). Extract features need many operations that may take long time, which is not suitable to achieve real-time for defects detection. Consequently, this paper uses Summed Area Table (*SAT*) [3] to reduce the required time to compute these features. It quickly generates the sum of values of a rectangular subset of a grid using Eq. (1). Where $i(x, y)$ is the pixel value from the given image and $S(x, y)$ is the value of the summed area table [23]. For $M \times N$ image, SAT table is created with $O(M \times N)$ complexity. Once it is created, the task to calculate the sum of pixels in a rectangle that is a subset of the original image can be done in constant time by Eq. (2) with O(1).

$$S (x, y) = i (x, y) + S (x{-}1, y) + S (x, y{-}1) - S (x - 1, y{-}1) \tag{1}$$

$$\begin{aligned} \text{SUM} = {}& S (x_0{-}1, y_0{-}1) + S (x_0 + x_1, y_0 + y_1) \\ & -S (x_0 + x_1, y_0{-}1) - S (x_0{-}1, y_0 + y_1) \end{aligned} \tag{2}$$

$$Diff\_Value (ROI, W, H) \tag{3}$$

$$\mu = Mean\_SAT (Image, x_0, y_0, W, H) \tag{4}$$

$$\upsilon = Variance\_SAT (Image, x_0, y_0, W, H) \tag{5}$$

$$\text{MV} = [\delta \ \mu \ \upsilon]^T \tag{6}$$

Consequently, SAT can quickly iterate pixels and significantly reduces the required time to process the images. In this paper, we developed SAT algorithm in parallel using CUDA [24, 25] as shown in Fig. 4.

2. *Defect Detection*: The defect detection algorithm divides image into ROIs to detect each ROI either belongs to defective group (G_1) or non-defective group (G_2). MV_1 and MV_2 are mean vectors that contain the statistical features of G_1 and G_2 respectively. We assume MVROI denotes a mean vector that contains the features in ROI [12]. The two groups represent defective pixels and non-defective pixels in the image. To separate the pixels into defective and non-defective pixels, we create two Gaussian Mixture Models (GMM)s [26]. An iterative Expectation-Maximization (EM) algorithm is used to estimate maximum likelihood $Ł_1$ and

Algorithm: Parallel SAT Function	

```
1    function PSAT (image)
2    {
3        double *SAT;
4
5        //Allocate space on device
6        cudaMalloc(&SAT, M * N * sizeof(double));
7
8        //Copy matrix to device memory
9        cudaMemcpy(SAT, image, M * N * sizeof(double), cudaMemcpyHostToDevice);
10
11       // invoke kernel at host side
12       const int maxThreadsPerBlock = IMG_SIZE;  // number of threads in each block
13       int numBlocksX = ((unsigned int)(M / maxThreadsPerBlock) + 1);
14       int numBlocksY = ((unsigned int)(N / maxThreadsPerBlock) + 1);
15
16       dim3 numBlocksForRows = dim3(numBlocksX, 1);
17       dim3 numBlocksForColumns = dim3(numBlocksY, 1);
18
19       //Run rows sum kernel
20       rowSum <<<numBlocksForRows, maxThreadsPerBlock >>>((double(*)[IMG_SIZE])SAT, M, N);
21       cudaDeviceSynchronize();
22
23       //Run columns sum kernel
24       colSum<<<numBlocksForColumns,maxThreadsPerBlock>>>((double(*)[IMG_SIZE])SAT, M, N);
25       cudaDeviceSynchronize();
26
27       //Copy data to host memory
28       cErr = cudaMemcpy(image, SAT, M*N * sizeof(double), cudaMemcpyDeviceToHost);
29
30       // free device global memory
31       cudaFree(d_Data);
32   }
```

Algorithm: Sum Rows Kernel	

```
1    function sumRows (SAT, M, N)
2    {
3        int idx = blockIdx.x * blockDim.x + threadIdx.x;
4        if (idx < N)
5            for (int i = 0; i < N; i++)
6                SAT[i][idx] = SAT[i][idx] + SAT [i - 1][idx];
7    }
```

Algorithm: Sum Columns Kernel	

```
1    function sumCols (SAT, M, N)
2    {
3        int idx = blockIdx.x * blockDim.x + threadIdx.x;
4        if (idx < M)
```

Fig. 4 Parallel SAT Algorithm in CUDA

$Ł_2$ for both GMM_1 and GMM_2 as in Eq. (7), by guess weight α, mean m and variance σ values [27]. EM contains three steps. First step chooses initial parameters values, the second is E-step that evaluates the responsibilities using the current parameter values, the third is M-step that re-estimates the parameters using the current responsibilities [28]. By maximum likelihood function $ML(p)$ in Eq. (8), the pixel belongs to G_1 if $Ł_1$ is larger than or equal to $Ł_2$ otherwise it belongs to G_2 Eq. (8).

$$\mathfrak{t} = \log(\alpha) + (2\pi\sigma^2)^{\left(-\frac{1}{2}\right)} \exp\{-\tfrac{1}{2}((x-m)/\sigma)^2} \tag{7}$$

$$ML(p) = \begin{cases} p \in G_1, & \mathfrak{t}_1 \geq \mathfrak{t}_2 \\ p \in G_2, & \mathfrak{t}1 < \mathfrak{t}_2 \end{cases} \tag{8}$$

To decide if the image is defective or not, we apply multivariate discriminant function, Ω, for each ROI in the image [5, 12]. Multivariate discriminant function applies *Mahalanobis* distance rule Δ^2 Eq. (9) [12, 29]. If *Mahalanobis* distance between the ROI and G_1 more than or equal *Mahalanobis* distance between ROI and G_2, then the ROI is defective; otherwise, the ROI is non-defective as in Eq. (10). Multivariate discriminant function in Eq. (11) derived from Eqs. (9) and (10), where T denotes matrix transpose [12]:

$$\Delta^2 \, (\text{MVROI, MV}) = (\text{MVROI} - \text{MV})^T \text{CV}^{-1} \, (\text{MVROI} - \text{MV}) \tag{9}$$

$$\Delta^2 \, (\text{MVROI, MV}_1) \geq \Delta^2 \, (\text{MVROI, MV}_2) \tag{10}$$

$$\Omega = (\text{MV}_1 - \text{MV}_2)^T \text{CCV}^{-1} \, \text{MVROI} - 1/2(\text{MV}_1 - \text{MV}_2)^T \text{CCV}^{-1} \, (\text{MV}_1 - \text{MV}_2) \tag{11}$$

To apply discriminant function Ω, we need to calculate covariance vector, CV, for both groups G1 and G2 by Eq. (12), where N_i denotes the number of pixels in the group and x_{ij} denotes pixel in G1 and G2. Then the common covariance matrix (CCV) will be calculated by Eq. (13):

$$\text{CV}_i = \frac{1}{N_i - 1} \sum_{j=1}^{N_i} \left(x_{ij} - \text{MV}_i\right) \left(x_{ij} - \text{MV}_i\right)^T \text{ for } i = 1, 2. \tag{12}$$

$$\text{CCV} = \sum_{j=1}^{2} \left(N_j - 1\right) \frac{\text{CV}_j}{n - 2} \quad \text{where } n = \sum_{j=1}^{2} n_j \tag{13}$$

The ROI is defective if the value of discriminant function Ω is positive. Otherwise, the ROI is non-defective as Eq. (14). To decide either the image contains defects or not, it must have at least one defective ROI; otherwise, the image is non-defective [12].

$$\Omega \, (\text{block}) = \begin{cases} defective \; block, & \Omega \geq 0 \\ non - defective \; block, & \Omega < 0 \end{cases} \tag{14}$$

Applying the discriminant rule, Ω, for all ROIs in the image, results would be like in Fig. 5; the numbers represent the value of the discriminant rule for each ROI [12]. The image has no defect if all ROIs have negative discriminant value.

To speed up the EM algorithm, this chapter calculates each iteration E-step and M-step for all pixels in parallel using CUDA [30] as shown in Fig. 6. The parallel EM algorithm has main function *PEM()* that launches the GPU kernel UpdateKernel() to process E-step and M-step for all pixels in parallel as seen in Fig. 7. The UpdateKernel() creates 1D grid and 1D ROIs; each ROI contains MT threads [25]. Each thread calculates both E-step and M-step for a pixel. Assume an image has *NP* pixels, the complexity of the sequential EM is *O(Maximum of Iteration × NP)*. However, it is O(Maximum of Iteration) for the proposed parallel EM.

3.3 Defect Classification

In the past decade, different researchers have presented several methods for steel defect classification [31]. Nevertheless, these methods are limited to high computation and low accuracy. This work proposed a classification algorithm to classify scratch, patches, and scale defects. The algorithm has two modules. First, features extraction module takes the defective image and calculates GLCM and Haralick features [32, 33]. Second, once features are extracted, the classification module utilizes support vector machine (SVM) for the recognition of their corresponding class.

1. *Features Extraction Module*: GLCM defines the texture of an image by calculating how frequently pairs of pixels with specific values and in a specified spatial relationship happen in an image. Each element (i, j) of GLCM denotes how many times the gray levels i and j occur as a sequence of two pixels located at a defined distance δ along a chosen direction θ. *Haralick* defined a set of 14 measures of textural features [33]. This work selected six textural features shown in Table 1

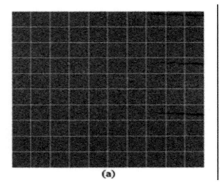

Fig. 5 Defective image and its discriminant result. [12]

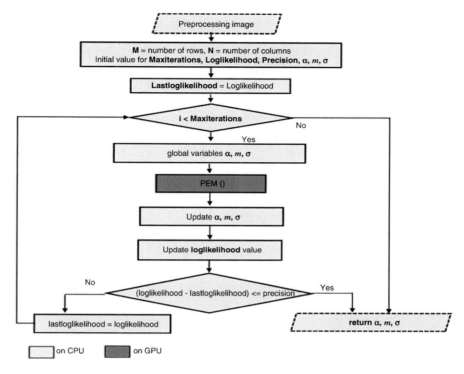

Fig. 6 Parallel EM algorithm

that are used as input to SVM classifier to classify the defect in steel image. The computation time of texture features depends on the number of gray levels, between 0 and 255 levels. This chapter develops *Haralick* features calculation in parallel to reduce the execution time for the proposed classification algorithms [34]. To extract the features from GLCM matrix in parallel, this work developed *P_Haralick_Featuers()* function that launches the *HaralickKernel()* kernel with 2D (blocks) ROIs with n_x threads in the x-direction and n_y threads in the y-direction [25]. Each thread computes six features for one pixel. To accumulate features values from threads, kernel uses *AtomicAdd()* function as shown in Fig. 8. While a thread reads a memory address, it adds the value of feature to it, and writes back the result to memory. As GLMC is *256×256* matrix, *256* gray levels then the complexity to extract *Haralick* features by sequential algorithm is *O(256x256)*; however, with parallel algorithm it is O(1).

2. *Defect Classification Module*: To classify surface steel strip defects, this chapter uses multi-class classification SVM. The classification process is divided into two steps: training process and testing process. In both steps the classifier will use features vectors; in the training step to label different defects and in the test

---

Algorithm: Update Kernel called by the $PEM()$

```
1    function UpdateKernel(double *x, double *dev_sum_wj, double *dev_sum_
     *a, double *mean, double *var, double *dev_L, const int numGaussians, cor
2    {
3        int idx = blockIdx.x * blockDim.x + threadIdx.x;
4        int j = idx;
5
6        double resp[2];
7
8        if (j < dataSize)
9            double den;
10           double l_max, tmp, sum;
11           l_max = -1000000;
12           for (int i = 0; i < numGaussians; i++)
13               resp[i] = log(a[i]) + Probability(x[j], mean[i], var[i]);
14               if (resp[i] > l_max) l_max = resp[i];
15           sum = 0;
16           for (int i = 0; i < numGaussians; i++)
17               sum += exp(resp[i] - l_max);
18           tmp = l_max + log(sum);
19           dev_L[j] = tmp;
20
21           den = 0;
22           for (int i = 0; i < numGaussians; i++)
23               resp[i] = exp(resp[i] - tmp);
24               den += resp[i];
25
26       double *learn_a = &dev_sum_wj[j * numGaussians];
27       double *learn_m = &dev_sum_wj_xj[j * numGaussians];
28       double *learn_v = &dev_sum_wj_xj2[j * numGaussians];
29
30       for (int k = 0; k < numGaussians; k++)
31           learn_a[k] = resp[k] / den;
32           learn_m[k] = x[j] * resp[k] / den;
33           learn_v[k] = x[j] * x[j] * resp[k] / den;
     }
```

Fig. 7 UpdateKernel() invoked by EM algorithm in CUDA

Table 1 Haralick features

Homogeneity	$F1 = \sum_m \sum_n (m - n)^2 \text{GLMC}\,(m, n)$
Entropy	$F2 = \sum_m \sum_n \frac{1}{1+(m-n)^2} \text{GLMC}\,(m, n)$
Energy	$F3 = \sum_m \sum_n \text{GLMC}\,(m, n)\,\text{logGLMC}\,(m, n)$
Mean	$F4 = \sqrt{\sum_m \sum_n \text{GLMC}^2\,(m, n)}$
IDM	$F5 = \sum_m \sum_n \frac{1}{2}\,(\text{mGLMC}\,(m, n)\,\text{nGLMC}\,(m, n))$

step to classify defects [2, 18]. This work extracts features in parallel to reduce the classification time. In the training phase, we pass these features along with their corresponding labels to the SVM to generate SVM model. The second and final step of the proposed system is the testing phase where we have a test dataset

	Algorithm: Extract *Haralick* Features Kernel
1	**function** *HaralickKernel (GLMC, energy, contrast, homogenity, IDM, entropy, mean)*
2	{
3	int i = blockIdx.x * blockDim.x + threadIdx.x;
4	int j = blockIdx.y * blockDim.y + threadIdx.y;
5	
6	if (i < HARLICK_SIZE && j < HARLICK_SIZE)
7	*energy = GLCM[i][j] * GLCM [i][j]; //Energy
8	atomicAdd(contrast, (i - j)*(i - j)* GLCM[i][j]); //Contrast
9	atomicAdd(homogenity, GLCM [i][j] / (1 + abs(i - j))); //Homogenity
10	if (i != j)
11	atomicAdd(IDM, GLCM[i][j] / ((i - j)*(i - j))); //IDM
12	if (data[i][j] != 0)
13	atomicAdd(entropy, -1 * GLCM[i][j] * log10(GLCM[i][j])); //Entropy
14	atomicAdd(mean, 0.5*(i* GLCM[i][j] + j * GLCM[i][j])); //Mean
15	}

Fig. 8 *HaralickKernel ()* to extract *Haralick* features

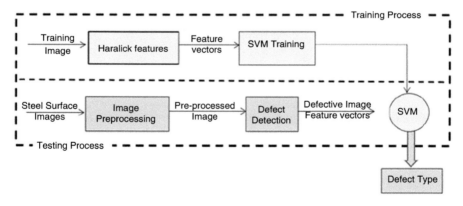

Fig. 9 Defect classification steps

images of the steel strips. These images are further checked for the defect; if an image has defective ROIs, then Haralick features must be extracted. These features are then given to the SVM along with a trained SVM model which was trained in the first step; as a result, SVM identifies the predicted class of defect. Figure 9 shows the classification steps.

3.4 Evaluation Criteria of Defect Detection and Classification

This section introduces the performance criteria to check the effectiveness and accuracies of the defect detection and classification algorithms.

1. *Detection Accuracy*: The defect detection accuracy as shown in Eq. (15) is used to determine the accuracy and the effectiveness of the defect detection algorithms [35, 36]:

$$DA = \frac{TP + TN}{TP + TN + FP + FN} \tag{15}$$

where TN is true negative, TP is true positive, FN is false negative, and FP is false positive. True positive is referred to defective steel image identified as defective. True negative is referred to defect-free steel image identified as defect-free. False positive is referred to defect-free steel image identified as defective. False negative is referred to defective steel image identifies as defect-free.

Classification Accuracy: The accuracy of the classification algorithm could be calculated as in Eq. (16):

$$\text{Accuracy}(dj) = \frac{N_c\left(d_j\right)}{N_t\left(d_j\right)} \tag{16}$$

where d_j is the defect class $j, j = 1, \dots W$, $N_c(d_j)$ is the number of images correctly classified as defect class d_j, $N_t(d_j)$ is the total number of images in that defect class, and W is the total number of defected classes. The total accuracy for the defect classification algorithm is the probability of a correct prediction due to our classifier for all defect classes over a set of images:

$$\text{Total accuracy} = \frac{\sum_j^W \text{Accuracy}\left(d_j\right)}{W} \tag{17}$$

2. *Performance Criteria*: Computing time is the main criteria to study the performance of the proposed defect detection and classification algorithm. The required time to detect and classify defects for steel surface is divided into two main significant parts: detection time and classification time as shown in Eq. (18):

$$\text{Total}_{\text{Time}} = \sum_{i=1}^{B} Dt_i + Ct_j \tag{18}$$

where the surface steel image has been divided into B ROIs, Dt_i is the required time to detect either ROI i in the surface steel image has defect or not, Ct_i is the required time to classify the type of the defects in the defected ROI i in the surface steel image. Ct_i equals zero if ROI i has no defects. In addition, this work used speedup to measure the performance of the proposed algorithm is speedup as in Eq. (19):

$$\text{Speedup} = \frac{T_s}{T_p} \tag{19}$$

where Ts denotes the execution time of the sequential algorithms, and Tp denotes the execution time of parallel algorithms.

4 Experiment Results

This section introduces experiments results of the proposed parallel algorithms for detecting scratch, patches, and scale defects.

1. *Setup*: The experiment platform in this work is Intel(R) Core™ i7-8550U with a clock rate of 1.8 GHz, working with 8 GB DDR4 RAM and a graphics card that is NVIDIA GeForce 940MX with 2GB of DDR3 and 1122 MHz. All experiments in this project were conducted in Windows 10 64-bit operating system with the development environments of Visual C++ 2017 with *OpenCV* 3.4.3 and *CUDA* toolkit 10. NEU dataset has 1800 grayscale steel images has been used. It includes six types of defect which are inclusion, crazing, patches, pitted surface and rolled-in scale, 300 samples for each type. Moreover, to study the tolerance of the proposed algorithm against noises this paper added salt and pepper noises to about 1%–2% of the steel images dataset. Dataset is divided into 70% for training set and 30% for testing set.
2. *Experiment*: The experiments were conducted in three stages: pre-processing, defect detection, and defect classification. In the first stage, images are pre-processed as follows. Steel images are resized to *400 × 400* and then a *3 × 3* median filter is used to remove noises. The second stage "defect detection" includes four steps. The first step creates two Gaussian mixture models for each image by maximum likelihood to divide the image pixels into two groups: defective group and non-defective group. Figure 10 shows two GMM for steel image having scratch defect. The second step calculates statistical features mean, difference, and variance for these groups. In third step, each image is divided into ROIs. Each ROI contains *40x40* pixels. Use summed area table to extract statistical features for each ROI. Finally, use the discriminant rule to decide either the ROI is defective or non-defective. The fourth step displays defected ROIs if the steel image is defective or not as shown in Fig. 11. The defect classification stage is divided into two phases. In the training phase, the SVM classifier takes vectors of the extracted six *Haralick* features with associated labels for all images in the training set and then generates a training model. In the testing phase, the

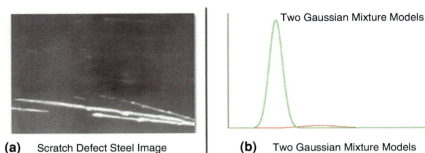

(a) Scratch Defect Steel Image   **(b)** Two Gaussian Mixture Models

Fig. 10 Steel image and GMM models

Fig. 11 Defect detection result

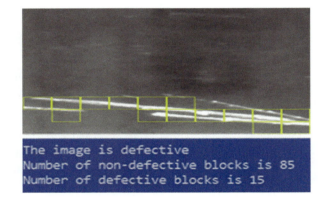

trained SMV takes *Haralick* features as a vector for test image from testing set and predicts defect class.

3. *Results*: This section illustrates gradually the results of the proposed algorithms. In this chapter we develop three defect detection algorithms: (1) sequential without SAT algorithm (SEQ) [12], (2) proposed sequential with SAT algorithm (SSAT), and (3) proposed parallel with SAT algorithms (PSAT) developed by CUDA. The median execution time for three types of implementations to detect and classify three defects will be illustrated in this section. Table 2 shows the median defect and classification time in milliseconds (ms) for SEQ and SSAT and PSAT algorithms. They detect three defects, patches, scratch, and scale, while image size is *400 × 400* pixels and ROIs size is *40 × 40* pixels. Table 2 contains steel images with defected ROIs, defect type, median of the execution time for defect detection and classification algorithms, and speedup. The rightmost column in this table displays the speedup of the PSAT compared to SEQ.

Figure 12 shows median execution time for sequential and parallel algorithms implemented to detect and classify three defects. It depicts that the *PSAT* algorithm is the fastest one especially in detecting scratch defect. The PSAT algorithm is able to accomplish ~1.50x speedup. Figure 13 plots median defect detection and

Table 2 Execution times for three algorithms

Steel image	Defect type	Median execution time in ms			Speedup
		SSAT	SEQ	PSAT	
	Patches	22.045	20.094	14.656	1.50
	Scratch	22.787	21.697	13.804	1.65
	Scale	20.319	20.015	14.577	1.39

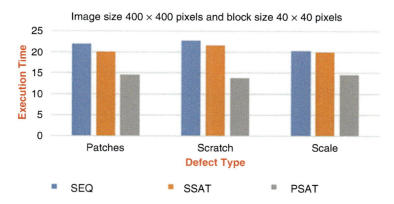

Fig. 12 Median defect detection and classification time

classification time with different dimensions of surface steel images with block (ROI) size 40×40 pixels. It shows that the median execution increases linearly with the increase of image size. The proposed PSAT algorithm has exhibited superior performance compared to the other algorithms while image size is increasing.

The proposed algorithm divides the image into non-overlapped ROIs (partitions). The number of ROIs is specified based on defect location. Some defect may split into two ROIs. So, the smaller defect in a ROI may not be classified as a defect type. In doubt, this case will affect the accuracy of the proposed algorithm. The number of ROIs must be chosen carefully to reduce the defect splitting. Figure 14 depicts that PSAT algorithm takes shortest execution time in milliseconds for all bock sizes, while the SEQ takes significantly long execution time. SEQ divides image into ROIs with size W × H and handles each ROI separately, while PSAT generates 2D ROIs with W × H threads. Each thread launches kernel to detect either ROI is defective or not in parallel. Therefore, PSAT shows 1.4 speedup compared with SSAT and more than 1.6 speedup compared with SEQ. The accuracy of the proposed defect detection algorithms SSAT and PSAT is about 95.66%.

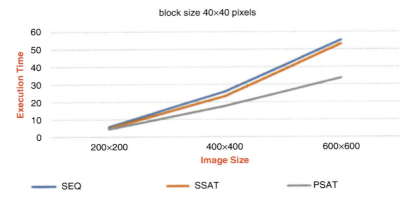

Fig. 13 Median defect detection and classification time with image size

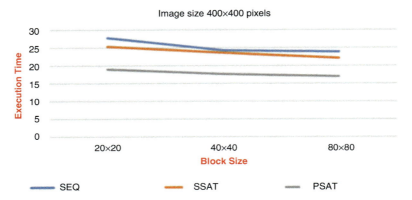

Fig. 14 Median defect detection and classification rime with ROI size

5 Conclusion

The major aim of this chapter is to design and develop a parallel algorithm that automates the defects inspection process in steel industry. This work employed SAT to improve the defect detection algorithm in [12] In addition, it demonstrated the detailed implementation of the proposed sequential algorithm based on SAT and parallel algorithm. Once defected image is detected, SVM classifier has been used to classify the type of the defect (scratch, patches, scale). The experimental results in this article verified that the developed techniques succeeded to speed up the surface steel defects detection and classification compared with the existing techniques. Finally, the proposed parallel algorithm speeds up over the sentential algorithms developed in [12] by about 1.65 times to detect scratch, about 1.5 times to detect patch and 1.39 times to detect scale defects respectively where the image size is *400 × 400* with about 95.66%.

References

1. Mostafa Sadeghi, Hossein Soltani, Kamran Zamanifar, Application of parallel algorithm in image processing of steel surfaces for defect detection, Special Issue: Technological Advances of Engineering Sciences, Fen Bilimleri Dergisi (CFD), Cumhuriyet University, Turkey. 36(4), 263-273 (2015)
2. K. Song, Y. Yunhui, A noise robust method based on completed local binary patterns for hot-rolled steel strip surface defects. Appl. Surf. Sci. **285**, 858–864 (2013)
3. S. Tian, X. Ke, An algorithm for surface defect identification of steel plates based on genetic algorithm and extreme learning machine. Metals 7(8), 311 (2017)
4. R. Khaled, Fast and parallel summed area table for fabric defect detection. Int. J. Pattern Recognit. Artif. Intell. **30**(09), 1660004 (2016)
5. N. Neogi et al., Review of vision-based steel surface inspection systems. EURASIP J. Image Video Process. **2014**(1), 50 (2014)
6. R. Khaled, A. Nahed, An efficient defect classification algorithm for ceramic tiles. *IEEE 13th Int. Symp. on Autonomous Decentralized System (ISADS)*, pp. 255–261 (2017)
7. H. Sager, E. Loay, George. Defect detection in fabric images using fractal dimension approach. *International Workshop on Advanced Image Technology*, vol. 2011 (2011)
8. S. Zhou et al., Classification of surface defects on steel sheet using convolutional neural networks. Materiali Tehnologije **51**(1), 123–131 (2017)
9. T. Ramesh, B. Yashoda, Detection and Classification of Metal Defects using Digital Image Processing, pp. 31–36 (2014)
10. Xianghua Xie, A review of recent advances in surface defect detection using texture analysis techniques, Electronic Letters on Computer Vision and Image Analysis, 7(3), 1-22 2008. Computer Vision Center / Universitat Autonoma de Barcelona, Barcelona, Spain
11. D. Wang, et al., Wood surface quality detection and classification using gray level and texture features. *Int. Symp on Neural Networks*. Springer, Cham (2015)
12. L. Weiwei, et al., Automated on-line fast detection for surface defect of steel strip based on multivariate discriminant function. *Intelligent Information Technology Application, IEEE. IITA'08. Second Int. Symp on. Vol. 2* (2008)
13. G. Wu, et al., A bran-new feature extraction method and its application to surface defect recognition of hot rolled strips. *IEEE Int. Conf. on Automation and Logistics* (2007)
14. Y. Zhang et al., Fabric defect detection and classification using gabor filters and gaussian mixture model, in *Asian Conference on Computer Vision*, (Springer, Berlin, Heidelberg, 2009)
15. Y. Lu, Li, et al., Parallelizing image feature extraction algorithms on multi-core platforms. J. Parallel Distrib. Comput. J. **92**, 1–14 (2016)
16. H. Zhang et al., GPU-accelerated GLRLM algorithm for feature extraction of MRI. Sci. Rep. **9**(1), 1–13 (2019)
17. M. Yazdchi, et al., Steel surface defect detection using texture segmentation based on multifractal dimension. *Int. Conf. on. IEEE Digital Image Processing* (2009)
18. J.P. Yun, et al., Vertical scratch detection algorithm for high-speed scale-covered steel BIC (Bar in Coil). *Int. Conf. on. IEEE Control Automation and Systems (ICCAS)* (2010)
19. P. Caleb, M. Steuer, Classification of surface defects on hot rolled steel using adaptive learning methods. *Knowledge-Based Intelligent Engineering Systems and Allied Technologies, IEEE Proc. Fourth Int. Conf. on. Vol. 1* (2000)
20. T. Maenpaa, Surface quality assessment with advanced texture analysis techniques. *Proc. of Int. Surface Inspection Summit*, Luxembourg (2006)
21. S.R. Mahakale, N.V. Thakur, A comparative study of image filtering on various noisy pixels. International Journal of Image Processing and Vision Sciences **1**(2), 69–77 (2012)
22. I. Singh, N. Neeru, Performance comparison of various image Denoising filters under spatial domain. International Journal of Computer Applications **96**(19), 21–30 (2014)
23. F.C. Crow, Summed-area tables for texture mapping. ACM SIGGRAPH Comput. Graph. **18**(3), 207–212 (1984)

24. D. Kirk, NVIDIA CUDA software and GPU parallel computing architecture. Proceedings of the 6th International Symposium on Memory Management, ISMM 2007, Montreal, Quebec, Canada, (2007), pp. 103-104. DOI: https://doi.org/10.1145/1296907.1296909
25. C. Zeller, *Cuda c/c++ Basics* (NVIDIA Coporation, NVIDIA, Santa Clara, California, USA, 2011)
26. H. Bensmail et al., Regularized Gaussian discriminant analysis through eigenvalue decomposition. J. Am. Stat. Assoc. **91**(436), 1743–1748 (1996)
27. B.W. Silverman, On the estimation of a probability density function by the maximum penalized likelihood method. Annals of Statistics **10**(3), 795–810 (1982). https://doi.org/10.1214/AOS/1176345872
28. L. Machlica, Fast estimation of Gaussian mixture model parameters on GPU using CUDA. *12th IEEE Int. Conf. on Parallel and Distributed Computing, Applications and Technologies* (2011)
29. G.J. McLachlan, Mahalanobis distance. Resonance **4**(6), 20–26 (1999)
30. G. Noaje, et al., Source-to-source code translator: OpenMP C to CUDA. *IEEE 13th Int. Conf. High Performance Computing and Communications (HPCC)* (2011)
31. C. Park, W. SangChul, An automated web surface inspection for hot wire rod using undecimated wavelet transform and support vector machine. *Industrial Electronics, Annual 35th Conference of IEEE, IECON'09* (2009)
32. P. Mohanaiah, P. Sathyanarayana, L. GuruKumar, Image texture feature extraction using GLCM approach. International Journal of Scientific and Research Publications **3**(5), 290–294 (2013)
33. R.M. Harlick et al., Textural features for image classification. IEEE Trans. Syst. Man Cybernet. **SMC-3**(6), 610–621 (1973)
34. A. Parvez, C.P. Anuradha, Efficient implementation of GLCM based texture feature computation using CUDA platform. *Int. Conf. on Trends in Electronics and Informatics (ICEI)* (2017)
35. M. Arun, R. Prathipa, P.S.G. Krishna, Automatic defect detection of steel products using supervised classifier, International Journal of Innovative Research in Computer and Communication Engineering, India. 2(3), 3630-3635 (2014).
36. R. Mishra, D. Shukla, A survey on various defect detection. Int. J. Eng. Trends Technol. **10**(13), 642–648 (2014)

Lightweight Approximation of Softmax Layer for On-Device Inference

Ihor Vasyltsov and Wooseok Chang

1 Introduction

Due to the rapid growth of wireless communication technology, the number of Internet of Things (IoT) devices has increased dramatically in recent years. According to a Cisco, 50 billions of IoT devices will be connected to the Internet by 2020 [3]. In addition, it was estimated that data volume generated yearly by those devices will be more than 40 times bigger than the current global traffic, which will amount to 850 Zettabytes [7]. Thus, existing cloud computing infrastructure would be unable to provide good service for the analysis of new data as it simply has not enough computing power and network capacity for a large number of computation tasks. In addition, many AI applications (e.g., autonomous driving) have strict requirement on the latency of computation.

Therefore, it makes more sense to locate computations closer to the data sources, so called edge computing or on-device computing. On-device computing has better capabilities in terms of privacy, latency, scalability, reliability, diversity, and costs compared with traditional cloud computing [8, 14, 18, 19].

Deep learning (DL) tasks are usually computationally intensive and require large memory footprints. At the same time, end-devices have limited computing power and small memories to support raw large-scale DL models. Therefore, original DL models are optimized, compressed, distilled, and quantized to reduce the resource cost. There are already many methods for optimizing DL models [1, 6, 12], but most of them are related to quantization of matrix multiplication operations, while quantization of activations (built as nonlinear functions (NLFs)) has not been

---

I. Vasyltsov (✉) · W. Chang
Samsung Advanced Institute of Technology, Samsung Electronics, Suwon-si, Gyeonggi-do, South Korea
e-mail: ihor.vasiltsov@samsung.com

© Springer Nature Switzerland AG 2021
H. R. Arabnia et al. (eds.), *Advances in Artificial Intelligence and Applied Cognitive Computing*, Transactions on Computational Science and Computational Intelligence, https://doi.org/10.1007/978-3-030-70296-0_41

studied enough. Softmax layer is one of the most popular and important NLFs, but the complexity of implementation in the platform with limited hardware resources can be a bottleneck of application performance. Thus, we will focus on the usage of softmax layer in computer vision tasks as the main application.

In this paper we propose a lightweight method for efficient computation of the softmax layer at the devices with limited computational power. The method is based on the approximation of softmax by taking reciprocal of natural exponential function, which is implemented as 1-dimensional look-up-table (1-D LUT). In Sect. 2, we consider some preliminaries for understanding softmax, drawbacks of existing approximation methods, and propose our method. Section 3 shows the experimental validation of the proposed method with human segmentation tasks. Section 4 describes a plan for further extension of our research, and Sect. 5 concludes the paper.

2 Softmax Approximation

2.1 Preliminaries

In mathematics, the softmax is a function that takes a vector x of n real numbers as an input, and normalizes it into a probability distribution $P(x)$ consisting of n probabilities proportional to the exponential of each input number. Thus, after applying softmax, each component will be in the interval $\sigma(x_i) \in (0, 1)$, and the components will add up to 1, so that they can be interpreted as probabilities. Softmax is often used in neural networks, to map the non-normalized output of a network to a probability distribution over predicted output classes. There are different representations of the softmax function depending on the application [4], but the most well-known and widely accepted version is as follows [11]:

$$\sigma(x_i) = \frac{e^{x_i}}{\sum e^{x_i}} \quad (1)$$

In the real hardware the range of number representation is limited, thus e^x computations can often lead to overflow or underflow. Therefore for practical implementation to provide stable computation, a normalization of input by $x_i^* = x_i - max(x)$ is used [11][1] as shown below:

$$\sigma(x_i) = \frac{e^{x_i - max(x)}}{\sum e^{x_i - max(x)}} \quad (2)$$

---

[1]All major DL frameworks (TensorFlow v1.7, PyTorch (with Caffe2) v0.4.0, MXNET v1.1.0, Microsoft Cognitive Toolkit v2.5.1, and Chainer v5.0.0a1) are using this safe version for softmax computation [11].

2.2 Previous Arts

Softmax layer is one of the most important and widely used NLF in modern AI models, due to its smooth properties. However, many modern neural processing unit (NPU) architectures are focused on the acceleration of matrix multiplications only, as they are a majority of computational functions of a DL model. As a result, for computation of complex NLF (i.e. softmax layer), the data must be sent out of NPU to an external host (CPU, GPU, or DSP), which complicates the software development, and can also negatively impact on the overall performance and power consumption. In addition, since general purpose interface is used for data transmission, the internal data can be exposed to malicious user, which may cause an issue of data privacy.[2]

To avoid involvement of the host for softmax computation, some NPU proposed dedicated HW accelerators. In many of those implementations, each of the numerator and the denominator in Eq. (1) are computed first, and then a division operation is performed, e.g., [10, 15, 17]. In such case, the HW accelerator should contain a divider circuit (fixed-point, or even floating point), which requires an additional HW cost. To avoid a big area cost for traditional dividers, the authors in [5] propose to replace the denominator with closest 2^b value.[3] Then division can be implemented just as a simple bit-shift operation. Although the method described above is decreasing the hardware complexity of softmax computation it still relies on the division operation, which is not always feasible for end-devices with limited computational power.

2.3 Proposed Method

In general case, we can consider alternative softmax function $\sigma^*(x_i)$ as

$$\sigma^*(x_i) = \frac{score(x_i)}{norm(x)} \tag{3}$$

where $score(x_i)$ and $norm(x)$ are some scoring and normalization factors (for original function $score(x_i) = e^{x_i}$, and $norm(x) = \Sigma e^{x_i}$).

As described in [9], we can list some desirable properties of the alternative softmax function as below:

- **Nonlinearity:** for better selectivity of the scored values.
- **Numerical stability:** to avoid overflow, or underflow during computation.
- **Positive:** output values all should be positive, to be used for scoring.

---

[2]For example, for CCTV or industrial data sensing application.

[3]Where b is a certain integer constant.

Table 1 Softmax approximation methods and their properties

#	Method of approximation	Nonlinearity	Numerical stability	Positive	Bounded	Computational complexity
1	x_i	Bad	Bad	Bad	Bad	Best
2	e^{x_i}	Best	Bad	Good	Bad	Good
3	$\frac{x_i}{max(x)}$	Bad	Bad	Bad	Good	Bad
4	$\frac{x_i^2}{max^2(x)}$	Good	Bad	Good	Good	Bad
5	$\frac{x_i^2}{\Sigma x_i^2}$	Good	Bad	Good	Best	Worst
6	$e^{x_i - max(x)}$	Best	Best	Best	Best	Good
7	$\frac{1}{e^{max(x)-x_i}}$	Best	Best	Best	Best	Good

- **Bounded:** output values should be bounded by some constant, ideally, $\sigma^*(x_i) \in (0, 1)$.
- **Computational complexity:** should be feasible for the implementation into platform with limited HW resources.

Since we consider softmax approximation for inference task in computer vision applications where softmax layer is mostly used for *scoring* the outputs for classification, the requirements to normalization factor $norm(x)$ can be softer compared with the original formula Eq. (1), where inputs are mapped into the corresponding probabilities. Thus, we can use more various factors for normalization in $\sigma^*(x_i)$.

We have experimented with different approximations for $score(x_i)$ and $norm(x)$ factors, and summarized some of the methods and their properties in Table 1.[4]

First, we have started with the simple approximations, ignoring normalization factor at all, i.e. applying $norm(x) = 1$. We have obtained *identity* (i.e., $\sigma^*(x_i) = x_i$, method 1 in Table 1), and natural *exponentiation* function ($\sigma^*(x_i) = e^{x_i}$, method 2 in Table 1) for approximation. However, despite their low computational complexity, the numerical stability was not good as the output of function was not bounded. To counter this issue, we have applied some normalization factors (see methods 3 to 5 in Table 1), but the numerical stability was still poor. At the same time we have noticed that method 2 (exponentiation) is showing good selectivity due to its nonlinear property (refer to the corresponding image in Table A.1 in Appendix), and can be a good candidate if normalized appropriately. For this purpose we have performed several transformations as shown in Eq. (4) below:

$$\frac{e^{x_i}}{max(e^x)} = \frac{e^{x_i}}{e^{max(x)}} = e^{x_i - max(x)} = e^{-(max(x) - x_i)} = \frac{1}{e^{max(x) - x_i}} \tag{4}$$

---

[4]For more details, please refer to Table A.1 in Appendix, where statistical results and examples of images from initial tests are shown.

First, we kept e^{x_i} as a scoring factor $score(x_i)$ and then we have used $max(e^x)$ as a normalization factor to bound the output by 1 and thus we have obtained method 6. However, in such case, the input values to the exponential function e^x will be all negative due to the $x_i - max(x)$ term, and if e^x is implemented by LUT (which is a common approach for HW with limited computation resources), then additional affine transformation is required to compensate for negative input values. To avoid this drawback, we propose to use max-normalization in the inverse way as $x_i^* = max(x) - x_i$, then input values to e^x will be all positive, which allows them to be used directly as indices to LUT values. Second, to compensate for the inverse way of max-normalization, we have used the reciprocal version of $e^x \rightarrow 1/e^x$ as shown in Eq. (4).

In such case neither divider nor multiplier is needed, and only 1-D LUT is required to compute the approximated value of softmax. As a result, the computational complexity is reduced significantly, and it becomes more feasible for the implementation in HW with limited computational power.

Thus, we propose to substitute the original method for softmax computation with the inverse way of max-normalization and the reciprocal of exponential function:

$$\sigma(x_i) = \frac{e^{x_i}}{\sum e^{x_i}} = \frac{e^{x_i - max(x)}}{\sum e^{x_i - max(x)}} \rightarrow \sigma^*(x_i) = \frac{1}{e^{max(x) - x_i}} \tag{5}$$

The properties of the proposed method $\frac{1}{e^{max(x) - x_i}}$ are as below:

- **Nonlinearity** is satisfied with the reciprocal of exponential function $1/e^x$:
  $\frac{1}{e^{\alpha x}} \neq \alpha \frac{1}{e^x}$
- **Numerical stability** is satisfied by $max(x) - x_i$ term:
  $(max(x) - x_i) \in [0, max(x) - min(x)] \rightarrow \frac{1}{e^{max(x) - x_i}} \in (0, 1]$
- **Positive** output values are due to the exponential function:
  $\frac{1}{e^x} > 0 \; \forall x \in (-\infty, +\infty)$
- **Bounded** $\sigma^*(x_i) \in (0, 1]$ due to the inverse normalization term $max(x) - x_i$ used together with the reciprocal of exponential function $1/e^x$:
  $(max(x) - x_i) \geq 0 \; \forall x \rightarrow \frac{1}{e^{max(x) - x_i}} \in (0, 1]$
- **Computational complexity** is low, as $1/e^x$ can be implemented with LUT-based method, where the size of LUT is small.

Indices in LUT can be directly calculated by rounding operation as $i = \lfloor max(x) - x \rceil$. When input data are quantized by w bits then *efficient quantization boundary* x_q can be defined as[5]

---

[5]Efficient quantization boundary x_q defines the biggest input value, which can be mapped into w quantization bits.

$$e^{-x_q} = \frac{1}{2^w - 1}$$

$$ln(e^{-x_q}) = ln(\frac{1}{2^w - 1})$$

$$-x_q = ln(1) - ln(2^w - 1)$$ \hfill (6)

$$x_q = ln(2^w - 1)$$

$$x_q = \lceil ln(2^w - 1) \rceil$$

Content of LUT is computed as shown below:

$$LUT_{1/e}[i] = \left\lfloor \frac{1}{e^i} \cdot (2^w - 1) \right\rfloor, \forall i = 0, 1, \ldots, x_q + 1 \tag{7}$$

Note, that $LUT[i] = 0, \forall i > x_q$ due to quantization, as no value can be encoded with w bits after efficient quantization boundary x_q.

If selectivity (precision of computation) is not enough, then LUT can be scaled linearly by α as

$$LUT_{1/e}[i] = \left\lfloor \frac{1}{e^{i/\alpha}} \cdot (2^w - 1) \right\rfloor, \forall i = 0, 1, 2, \ldots, \alpha \cdot (x_q + 1) \tag{8}$$

3 Experimental Validation

To validate the proposed method we have used a pre-trained Unet model for human segmentation and internally prepared dataset with 1000 images. In this model softmax computation is used to predict the class of every pixel, thus requiring 307,200 computations for typical 640×480 image. This model takes the image as an input, and produces the predicted grey-scale image for segmentation, where each pixel P_i is in uint8 precision (with values from 0 to 255). To get the binary mask for segmentation, every pixel in those images was binarized into two classes (class 0 for "background," and class 1 for "human") by using threshold $thr = 127$, as follows:

$$B_i^m = \begin{cases} 0, & \forall P_i < thr \\ 1, & \forall P_i \geq thr \end{cases} \tag{9}$$

In the model we have substituted the conventional softmax layer with the computation method as described above in Sect. 2. For practical implementation, we have selected three different precisions (uint8, uint4, uint2) and prepared the LUTs accordingly to Eq. (7). For evaluation accuracy of segmentation we used the well-known bit-wise intersection-over-union metric [13, 20] as shown below:

Lightweight Approximation of Softmax Layer

Table 2 Accuracy of different approximation methods. Full test over 1000 images

#	Method of approximation	Precision	Size of LUT (Bytes)	IoU (class 0)	IoU (class 1)	mIoU
0	Reference	FP32		0.9847	0.9799	0.9823
1	$\dfrac{x_i^2}{max^2(x)}$	FP32		0.9817	0.9746	0.9781
2	$\dfrac{x_i^2}{\Sigma x_i^2}$	FP32		0.9828	0.9765	0.9797
3	$e^{x_i - max(x)}$	FP32		0.9845	0.9800	0.9823
4	$\dfrac{1}{e^{max(x)-x_i}}$	uint8	8	0.9845	0.9799	0.9822
5	$\dfrac{1}{e^{max(x)-x_i}}$	unit4	5	0.9845	0.9799	0.9822
6	$\dfrac{1}{e^{max(x)-x_i}}$	uint2	3	0.9842	0.9792	0.9817

$$IoU = \frac{area(B_{gt,i}^m \cap B_{p,i}^m)}{area(B_{gt,i}^m \cup B_{p,i}^m)} \tag{10}$$

where $B_{gt,i}^m$ is a pixel-group of ground-truth image, and $B_{p,i}^m$ is that of predicted segmentation image. The mIoU value was computed as a mean value among two classes.

Table 2 shows the results of our experiments for different methods of approximation and selected precision (for LUT-based method). As it comes from the table, the accuracy of human segmentation task based on the proposed approximation of softmax layer is as high as the FP32 reference. There is no, or negligibly small accuracy drop ($< 0.1\%$ for 2-bit quantization) even for very small size of LUT (3 to 8 bytes).

4 Future Work

Despite its extremely low computational complexity, the current version of the softmax approximation can be applied only to the applications where softmax layer is used for scoring (typically last layer in CNN models), calculated within one input tensor only. Thus, it cannot be directly applied to more complicated and softmax-intensive applications such as Natural Language Processing (NLP) tasks where cross-tensor probabilities must be computed more often (e.g., multi-head attention block in Transformer [16], and BERT [2] models). Therefore, we will work forward in order to extend the proposed method to other classes of AI applications.

5 Conclusion

In this paper we have proposed an efficient method for softmax approximation, which can be implemented at the platform with limited hardware resources (mobile, IoT, edge devices) for AI inference tasks. We have applied max-normalization to the input data in the inverse way, which together with the application of LUT-based method for computation of the reciprocal exponential function $1/e^x$ has significantly reduced the complexity of softmax layer computation. It also has the additional benefits as follows:

- does not require any additional multiplier, divider, or adder.
- scalable in terms of accuracy and precision (appropriate LUTs can be pre-computed off-line).
- fixed latency of computation, which depends only on the size of the tensor.

Thus, the proposed approach provides a good alternative for HW accelerator design, simplifying the overall process of computing the softmax layer, while maintaining identical accuracy to the conventional FP32 based computation.

Appendix

In this section there are presented more details about the research on softmax layer approximation. There are shown more methods for approximation, as well their results over initial test. The experiments for initial tests were conducted the same way as in Sect. 3, but over sub-set of 100 images. Also, examples of images generated by human segmentation model for different methods of softmax computation are shown (Table A.1).

As it can be seen from the table, imag e generated by human segmentation model with method 2 ($\sigma^*(x_i) = e^{x_i}$) shows good selectivity of the method. Thus, we used it as a base for creating the finally proposed method $\sigma^*(x_i) = \frac{1}{e^{max(x)-x_i}}$.

Lightweight Approximation of Softmax Layer

Table A.1 Accuracy of different approximation methods. Initial test over 100 images

#	Method of approximation	Precision	IoU (class 0)	IoU (class 1)	mIoU	Image (example)
0	Reference	FP32	0.9844	0.9833	0.9838	
1	x_i	FP32	0.1731	0.2274	0.2002	
2	e^{x_i}	FP32	0.7320	0.4989	0.6154	
3	$\frac{x_i}{max(x)}$	FP32	0.9667	0.9557	0.9612	
4	$\frac{x_i^2}{max^2(x)}$	FP32	0.9811	0.9778	0.9795	
5	$\frac{x_i^2}{\Sigma x_i^2}$	FP32	0.9825	0.9800	0.9812	
6	$e^{x_i - max(x)}$	FP32	0.9838	0.9828	0.9833	
7	$\frac{1}{e^{max(x)-x_i}}$	uint8	0.9838	0.9827	0.9833	
8	$\frac{1}{e^{max(x)-x_i}}$	uint4	0.9838	0.9827	0.9833	
9	$\frac{1}{e^{max(x)-x_i}}$	uint2	0.9837	0.9822	0.9829	

References

1. Y. Cheng, D. Wang, P. Zhou, T. Zhang, A survey of model compression and acceleration for deep neural networks (2017). *CoRR*, abs/1710.09282
2. J. Devlin, M.-W. Chang, K. Lee, K. Toutanova, BERT: pre-training of deep bidirectional transformers for language understanding (2018). *CoRR*, abs/1810.04805
3. Fog computing and the internet of things: extend the cloud to where the things are (2015). https://www.cisco.com/c/dam/en_us/solutions/trends/iot/docs/computing-overview.pdf, Accessed 10 March 2020
4. B. Gao, L. Pavel, On the properties of the softmax function with application in game theory and reinforcement learning (2017, preprint). arXiv:1704.00805
5. X. Geng, J. Lin, B. Zhao, A. Kong, M.M. Sabry Aly, V. Chandrasekhar, Hardware-aware softmax approximation for deep neural networks, in *Computer Vision – ACCV 2018*, ed. by C.V. Jawahar, H. Li, G. Mori, K. Schindler (Springer, Cham, 2019), pp. 107–122
6. Y. Guo, A survey on methods and theories of quantized neural networks (2018). *CoRR*, abs/1808.04752
7. Y. Han, X. Wang, V.C.M. Leung, D. Niyato, X. Yan, X. Chen, Convergence of edge computing and deep learning: a comprehensive survey (2019). *CoRR*, abs/1907.08349
8. A. Ignatov, R. Timofte, W. Chou, K. Wang, M. Wu, T. Hartley, L.V. Gool, AI benchmark: running deep neural networks on android smartphones (2018). *CoRR*, abs/1810.01109
9. S. Kanai, Y. Fujiwara, Y. Yamanaka, S. Adachi, Sigsoftmax: reanalysis of the softmax bottleneck (2018, preprint). arXiv:1805.10829
10. Z. Li, H. Li, X. Jiang, B. Chen, Y. Zhang, G. Du, Efficient FPGA implementation of softmax function for DNN applications, in *2018 12th IEEE International Conference on Anti-Counterfeiting, Security, and Identification (ASID)* (2018), pp. 212–216
11. M. Milakov, N. Gimelshein, Online normalizer calculation for softmax (2018). *CoRR*, abs/1805.02867
12. Model compression papers (2018). https://github.com/chester256/Model-Compression-Papers, Accessed 10 March 2020
13. S.H. Rezatofighi, N. Tsoi, J.Y. Gwak, A. Sadeghian, I.D. Reid, S. Savarese, Generalized intersection over union: a metric and a loss for bounding box regression (2019). *CoRR*, abs/1902.09630
14. M.G. Sarwar Murshed, C. Murphy, D. Hou, N. Khan, G. Ananthanarayanan, F. Hussain, *Machine Learning at the Network Edge: A Survey* (Open Access Archive of Cornell University, New York, USA, 2019) https://arxiv.org/abs/1908.00080. Accessed June 15, 2021
15. Q. Sun, Z. Di, Z. Lv, F. Song, Q. Xiang, Q. Feng, Y. Fan, X. Yu, W. Wang, A high speed softmax VLSI architecture based on basic-split, in *2018 14th IEEE International Conference on Solid-State and Integrated Circuit Technology (ICSICT)* (2018), pp. 1–3
16. A. Vaswani, N. Shazeer, N. Parmar, J. Uszkoreit, L. Jones, A.N. Gomez, L. Kaiser, I. Polosukhin, Attention is all you need (2017). *CoRR*, abs/1706.03762
17. K. Wang, Y. Huang, Y. Ho, W. Fang, A customized convolutional neural network design using improved softmax layer for real-time human emotion recognition, in *2019 IEEE International Conference on Artificial Intelligence Circuits and Systems (AICAS)* (2019), pp. 102–106
18. S. Wang, A. Pathania, T. Mitra, Neural network inference on mobile SoCs. IEEE Des. Test **37**, 50–57 (2020)
19. X. Zhang, Y. Wang, S. Lu, L. Liu, L. Xu, W. Shi, OpenEI: an open framework for edge intelligence (2019). *CoRR*, abs/1906.01864
20. D. Zhou, J. Fang, X. Song, C. Guan, J. Yin, Y. Dai, R. Yang, *IoU loss for 2d/3d object detection* (Open Access Archive of Cornell University, New York, USA, 2019) https://arxiv.org/abs/1908.03851. Accessed June 15, 2021

A Similarity-Based Decision Process for Decisions' Implementation

Maryna Averkyna ⓘ

1 Introduction

Implementation decisions are one type of decision that must be taken at different levels of management. This kind of decision-making process is often accompanied by endless disputes over one and another. One argument used here is a very convincing but not well-founded statement. Our situation S here is very similar to the situation S'. It is not sufficiently similar to the situation S" which is in another case. Therefore, it is necessary to implement what is used where situation S'. One example is the long and urgent controversy that once existed in the Republic of Estonia. It approaches technical solutions and legal frameworks for cyber security of IT systems (see [1, 2]). The same issues were also very seriously addressed at the (NATO) Cooperative Cyber Defense Center of Excellence, which resulted in a number of serious research papers and doctoral dissertations. Another example is the design of public transport systems suitable for small Ukrainian cities such as Ostroh. There is something in European small cities that should be taken as an example. It is normal that there will be a discussion about whether something should establish itself, or reimplement something that is already available elsewhere. In the latter case, supporters of implementation that already exists elsewhere use the typical argument: no need to reinvent the wheel.

The following problem must be resolved before appropriate decisions are made:

M. Averkyna (✉)
Estonian Business School, Tallinn, Estonia
The National University of Ostroh Academy, Ostroh, Ukraine
e-mail: maryna.averkyna@oa.edu.ua

© Springer Nature Switzerland AG 2021
H. R. Arabnia et al. (eds.), *Advances in Artificial Intelligence and Applied Cognitive Computing*, Transactions on Computational Science and Computational Intelligence, https://doi.org/10.1007/978-3-030-70296-0_42

- to what extent are the following two "things" similar or different in the choice of implementable (e.g., two states, two cities, two critical information infrastructures, etc.);
- the one in which the existing solution is to be implemented and the one in which the solution in question is to be implemented.

We should look at how similar the "city geography" is before implementing a public transport system even if it looks great. Before implementing a social welfare system, it should be explored how similar are the age structure, income, health indicators, etc. of both groups.

This problem is important because it is based on the so-called conventional assumption (that is, one and the other are too different). Then one that fits the first one may not fit the other. In this work, we will look at

- how to evaluate the aforementioned similarity before implementing decisions,
- what we mean by similarity, and
- what role artificial intelligence could play in this.

2 Descriptive Similarity

Human decision-making is often (not to say always) based on descriptions of things, situations, and developments. Well-written and practically applicable descriptions usually consist of relevant statements. Therefore, in the future, we expect the descriptions to be a set of relevant statements. A closer look at the claims often reveals that they can be "reformatted" into formulas. The system theory of the system representing the area to be described (e.g., certain things, situations, developments, etc.) (see [6]). Here we refer to the system as an organized or structured set of some fixed things. We by order or structure mean that the properties and interrelationship that we consider important in this case have been selected and fixed for the elements under consideration. In this case, we call the set of elements to be considered the *basic set of the system* and the set of properties or relations selected as *the system signature*.

Based on the foregoing, we agree in the following that we will consider as language assertions those language construction that we can represent as formulas of a suitable system theory.

In order to evaluate the descriptive similarity of the descriptions or sets of relevant statements—that is, the descriptive similarity of some things, situations, developments, etc., we will use the numerical value defined by Lorents below—the descriptive similarity coefficient:

$$Sim(P, Q) = E(Com(P, Q)) : [E(P) - E(Com(P, Q)) + E(Q)]$$
$$= E(Com(P, Q)) : [E(Spec(P)) + E(Com(P, Q)) + E(Spec(Q))] \tag{1}$$

According to this formula, it is first necessary to make descriptions of both sets, which must consist of relevant statements. At the next stage, it is necessary to clarify what statements from one and another description can be considered equivalent. There are now three sets of claims:

- P is a set whose elements are statements from the first description;
- Q is a set whose elements are statements from another description;
- Com (P, Q) C is a set of elements that are ordered pairs, where the first position of the pair has the claim P, the second position has the claim Q, and these claims have been equalized by each other [5].

3 Descriptive Similarity and Structural Similarity

In addition to descriptive similarity, structural similarity should be mentioned here. The most important types of this similarity are homomorphism and its special case isomorphism. The homomorphism of systems (i.e., sets having a certain structure) is a many-to-one correspondence in which corresponding elements have corresponding properties or corresponding relationships to each other. If the systems homomorphous is one-to-one, then the systems are said to be isomorphic (see [6]). It has been shown that if two systems are homomorphic, then the formulas of the first system theory that are correct and positive (i.e., contain no logical denials or implication), the corresponding formulas in the second system theory are also correct (see, e.g., [6], ch. III, §7, 7.4, Theorem 1). In the case of isomorphism, however, there is no need to limit the formulas to the requirement of positivity: the corresponding formulas of the theories of isomorphic systems are always correct together (see, e.g., [6], ch. III, §6, 6.3, Theorem 1).

These theorems give rise to a kind of "sometimes one-way bridge" between structural similarity and descriptive similarity.

Suppose the first system S' is indeed the one in which we would like to implement the solution in the second system S". If it turns out that the first system is homomorphous with the second system, then figuratively speaking, the second system holds all that the correct positive formulas represent in the first system. Thus, we could say that the positive description of the first system "covers" the whole description of the second system. However, for isomorphic systems, it is not necessary to confine itself to the positive formulas of the first system, and all that is true if one system applies to the other system as well.

Example We consider two small towns S' and S". Describe the towns of the statements S' (i) and S" (p) to these towns.

Town S' Small town which is situated on an island. Its features are: inhabitants—18000 (10000–199999), density—1500 (at least 300 inhabitants), a minimum population of 5000, less than 50% of the population living in rural grid cells, less

Table 1 Towns statements

Town S' describes the statements	Equivalent statements			Town S" describes the statements
	The wording from the first	The wording from the second	The wording for both	
1. The town is situated on island				
2.	Small town	Small town	Small town	
3.	Inhabitants—18000	Inhabitants—15600	Inhabitants—(10000–199999)	
4.	Min. pop. 5000 inhab.	Min.pop. 5000 inhab	Population	
5.	Density 300 inhab. per sq.km	Density 300 inhab. per sq.km	Density	
6.	Less than 50% inhab. lives in high-density clusters	Less than 50% inhab. lives in high-density clusters	Concentration	
7.	Intermediate area (towns and suburbs)	Intermediate area (towns and suburbs)	Intermediate area (towns and suburbs)	
8.	Three short transport lines	Two short transport lines	Short transport lines	
9. The public transportation is comfortable				
10.				Small town that situated on mainland
11.				The public transportation is uncomfortable.

than 50% living in a high-density cluster. There are three short transport lines. The public transportation is comfortable.

Town S" Small town which is situated on mainland. Its features are: inhabitants—15600, (10000–199999) density—1430 (at least 300 inhabitants), a minimum population of 5000, less than 50% of the population living in rural grid cells, less than 50% living in a high-density cluster. There are two short transport lines. The public transportation is uncomfortable (Table 1).

We will calculate the similarity rating: $7 : [2 + 7 + 2] = 7 : 11 \approx 0.64$. Perhaps in this case there is an assessment that can be characterized by words: rather high, than low.

4 One Segment of Implementation Decisions from Real Life

In this section we look at one particular area of implementation decisions that still has opportunities for description and analysis that are appropriate to the human being. The author's personal experience confirms that already in this particular field, it is already perceptible that the description and analysis of slightly larger and more complex situations is no longer feasible for the human beings to ensure the credibility of the basis of implementation decisions. In order to create appropriate artificial intelligence systems that support and increasingly replace human beings. First, it is necessary to examine carefully those aspects in which the human has been successful so far. So, we look at a specific area. What would be useful to implement in Ostroh, based on what is used in the small towns of the Republic of Estonia.

Ostroh is one small university town in Ukraine. It is the small town according to the criteria presented in Working Paper written by Lewis Dijkstra and Hugo Poelman, European Commission Directorate-General for Regional and Urban Policy (DG REGIO) [3]. This town include inhabitants—15700 (10000–199999), density—1436 (at least 300 inhabitants), a minimum population of 5000, less than 50% of the population living in rural grid cells, less than 50% living in a high-density cluster. The main "bottlenecks" and problems of the transport system of the town were identified by M. Averkyna only last year—2019 based on an interview with Olga Logvin (deputy mayor).

It was made a decision to analyze public transportation in Estonian small towns. It helps to understand features Estonian public transportation and create a set of claims that can be implemented in Ostroh. There are only 10 small towns, which include such criteria in Estonia according to the DG REGIO [3].

It is necessary to have a set of claims according to the proposed approach. The set of claims consists of the textual information, which describe situation of public transportation. This information was received from interviews. There were questions about urban transportation system which allow us to establish the sets of claims. They are the following:

1. What types of transport are in the town?
2. Are there bridges in the town?
3. How do the residents of the town get to their work?
4. What are the largest cluster points in the town?
5. How are the town's transport routes developed? Who were the designed by?
6. How many transport lines are there? How do you determine a sufficient number of buses for the town?
7. What are rush hours in the town?
8. Are the urban transport schedule in place?
9. How are the town's applications decided?
10. Should you suspend regular traffic in the town?
11. Does the private transport of town residents greatly affect the passenger transportation organization?
12. What is the algorithm of town's transport system management?

13. What is the strategic plan for the town's transport system development?
14. Can I get a document form that produces information about city's transport system management?
15. What were the problems and how they were solved in the urban transport system?
16. What is the characteristic of the problems of the town transport system?
17. How public transportations' problems are sort out in the town?
18. In what property is urban transport (public, private)?
19. What are the bottlenecks of the town's transport system?
20. Are information technologies used to solve transport problems in the city? If so, what?
21. How much does it cost to travel in public transport? Who pays for travel (residents, local council)?
22. Who carries out the control over the quality of the town transport system? What is the control algorithm?

It helped to create the sets of claims based on interview with persons who are responsible for the public transportation in Estonian small town. These systems were a basis for calculating index similarity by P. Lorents and M. Averkyna only last year (2019). The result obtained showed that the similarity of Ostroh with Estonian towns is within the range of 0.49–0.6. Ostroh is similar to the Estonian towns for the statement regarding the town's specifics (population, density, cells, concentration, intermediate area, historical town, tourist attraction, cultural and educational town, town-forming enterprises, diesel engine, thinking about ecological transportation, transports' line, frequency per month, rush hours, network of routes, transports' lines are important, a permit for transportation after competition, frequency per route, calculation emission level is absent, control is conducted) [in press]. It is important to understand the similarities between public transportation in order to create the set of s system for Ostroh (see Table 2).

Table 2 Transport situations' towns comparisons

	Ostroh	Haapsalu	Rakvere	Viljandi	Valga	Sillamäe	Kuressaare	Keila	Maardu	Võru	Jõhvi
Ostroh	1										
Haapsalu	0.37	1									
Rakvere	0.47	0.65	1								
Viljandi	0.46	0.73	0.66	1							
Valga	0.40	0.79	0.63	0.83	1						
Sillamäe	0.47	0.71	0.69	0.73	0.80	1					
Kuressaare	0.44	0.81	0.59	0.61	0.76	0.74	1				
Keila	0.42	0.93	0.62	0.75	0.93	0.78	0.69	1			
Maardu	0.44	0.74	0.82	0.62	0.74	0.81	0.60	0.79	1		
Võru	0.42	0.89	0.67	0.78	0.89	0.71	0.66	0.92	0.83	1	
Jõhvi	0.37	0.85	0.67	0.76	0.74	0.69	0.66	0.76	0,69	0.71	1

Fig. 1 Economic data

According to the data of the Table 2, we can see that the similarity of transport situation of Ostroh with transport situation of Estonian small towns is within the range of 0.37–0.47. The similarities between transportation situation in towns involve the buses are equipped with diesel engine. The local councils think about ecological transportation, transports' line in the towns are short, the frequency of buses per hour, they have the rush hours, there are the network of routs, trans-ports' lines of towns are important, permit for transportation after competition, frequency per route, calculation emission level is absent, control is conducted.

The differences of the transport situation in Ostroh from Estonian small town include the next claims: private carriers dictate their own terms and conditions for the provision of transport services, which leads to non-observance of the schedules on certain routes. The work of private carriers is not satisfactory. Residents are forced to turn to the services of private vehicles. There is no route information at the stops to understand the time of a shuttle arrival. The Local Council is not satisfied with current situation, public transport is not comfortable.

The public situation of the Estonian small towns has features such as comfortable public transportation, public transport follow by established schedules. It is free of charge public transportation in some cities (Haapsalu, Viljandi, Valga, Kuressaare, Keila, Võru, Jõhvi). It depends on the political decision in the towns. There is time schedule near bus stations in the town, public authorities estimate efficiency transports' lines (see Fig. 1). It is possible to track the arrival of the public transport via the Internet.

The assessment analysis of similar situations in the transport system of town found out that the use of information technology in the management of transport systems is quite important for the decision-making process. They are crucial in the following purposes:

1. Carriage validation that allows managers to evaluate road congestion at a special span time. It also has been explained at schools. There are two transport tracking information systems: www.peatus.ee; www.ridango.com. Managers can quickly make decisions about redistributing public transport for congested routes.
2. City residents understand the time of arrival and departure of public transport using the Internet www.peatus.ee. Towns' residents actively interact with managers responsible for public transportation. The system of transportation of schoolchildren and residents to the basic infrastructures (workplaces, hospital) is established in the towns.

5 The Needs and Some Options for the Implementation of Applied Artificial Intelligence

It is necessary to point out that finding similarity has been quite labor consuming in the cases studied and investigated. It is clear that, in slightly more "bulky" cases, such work is beyond the reach of the human being. Especially as far as the assertion of assertions is concerned. Therefore, you need to look for establishing ways to rely on the right IT solutions. The first problem here is to transform statements into formulas that, in principle, cannot be fully automated. However, such a thing is conceivable to some extent within self-learning dialogue systems, such as the DST prototype of the dialogue system created by E. Matsak (see [8–10]).

This system is important because it is very crucial to create an artificial system, which makes the decisions and is based on formulas through the transforming text. By transforming texts E. Matsak prosed to understand a step-by-step modification of the original text into logic formulas [7]. The procedure by P. Lorents [4] consists of applying the steps listed below (the order and the amount of use of the steps is not important). It follows that in some cases it is sensible to use the same step many times in many parts of the text. The steps are:

- complementing or adding necessary parts to the text.
- withdrawing or removing unnecessary parts from the text.
- repositioning or changing the relative positions of arguments within the text.
- replacing or substituting some parts of the text with some other (equivalent) texts.
- identifying symbols or finding parts of the text that can be represented as individual symbols (fully representing individual objects), predicate symbols (fully representing properties of objects, or relationships between objects), logic operation symbols (negation, conjunction, disjunction, implication or equivalence), functional symbols (fully representing functional relationships, including logic operations), quantifiers (fully representing some part or all objects under observation) or modality symbols (characterizing the "validity" of some argument about an object, for example definitely or possibly).

A Similarity-Based Decision Process for Decisions' Implementation

- categorizing symbols or determining whether a symbol belongs to individual, predicate, functional, logic operation, quantifier, or modality category.
- positioning symbols or reshuffling the symbols according to the rules of creating formulas.

Example In order to improve the transportation situation in Ostoh we can use information from Estonian small towns. Then we propose to form the set of claims.

$M = \{x \mid \text{claims} (\dots x \dots)\}$—Set M all such x that satisfies a condition of the claims.

x_1—the buses must be comfortable.

x_2—the schedules should be established near buses station.

x_3—the public transport should follow by established schedules.

x_4—to track the arrival of the public transport via the Internet.

x_5—the system of transportation of schoolchildren and residents to the basic infrastructures (workplaces, hospital) ought to be established.

x_6—providing validation system.

x_7—providing e-ticket.

x_8—encourage maximum use of public transport by residents.

x_9—the rejection of private cars.

x_{10}—estimate efficiency transports' lines.

x_{11}—actively monitoring the availability of public transportation

x_{12}—it is necessary to analyze experience of urban transportation in Viljandi, Valga, because they faced the decentralization process.

...

...

x_n—analyses the conditions for private companies in order to provide public transportation. In case of urban transportation implementation in Ostroh we can write the claims in a formula:

$$(x_1 \& x_2 \& x_3 \& x_4 \& x_5 \& x_6 \& x_7 \& x_8 \& x_9 \& x_{10} \& x_{11} \& x_{12} \& x_n) \supset M' \tag{2}$$

Then we receive the new set M'

$M' = \{x_1, x_2, x_3, x_4, x_5, x_6, x_7, x_8, x_9, x_{10}, x_{11}, x_{12}, x_n\}$—Set M' that satisfies condition of public transportation in Ostroh.

This example shows us how to get the formulas you need to apply artificial intelligence.

6 Conclusion

The decision-making process is crucial for system's management. Human decision-making is often (not to say always) based on descriptions of things, situations, and developments. Well-written and practically applicable descriptions usually consist of relevant statements. Relevant systems statements allow managers to make the decision in order to sort out the issues or improve the situation in the system. In this case it is crucial to understand how equate the statements are that describe the systems. The author pointed out that descriptive similarity is the relevant approach for equation of the statements. The results of the analyses transport situations' similarities between Ostroh and Estonian small towns permit to implement claims of positive influence on public transportation. The author indicated that it is important to transform textual information into formulas. It helped to create the set of claims that satisfies condition public transportation in Ostroh. In addition, it is crucial for artificial intelligence, which helps managers to make the decisions. Moreover there is no a single line of logic and inference rules. In decision-making process human thinks by inference rules. It can be modus ponens or syllogism. "A bright future" could deal with:

- An analysis of process thinking and inference rules through investigation works of well-known scientists.
- Creation AI for the decisions making process and based on formulas through the transforming text.

Some things should be mentioned in the above context. The theoretical foundations of which and the possibilities for their application should be clarified:

- The DST system for transforming propositions in natural language described by E. Matsak (see [9, 10]) a formula for transforming predictions into formulas should be supplemented with a module that enables the equalization or differentiation of propositions. Certainly, such a module cannot work 100% automatically, but like DST, it could be self-learning and therefore work more autonomously the more it is used.
- It would also be interesting and necessary to address the (partial!) automation to one of the most important part of the decision-making process. It is based on how a qualified expert does it. Specifically, there is a need for a system that (I) analyzes an expert's derivation steps to highlight the derivation rules that that expert uses in his reasoning, and (II) to apply them to the highest degree of autonomy with increasing degree of justification in natural language. The inference-steps and inference-rules that the particular person has given in the justification could be followed.

Acknowledgments The author of the given paper expresses profound appreciation to professor Peeter Lorents for assistance in a writing of given clause. Great thanks to Olga Logvin, Ailar Ladva, Karl Tiitson, Väino Moor, Mati Jõgi, Yuriy Petrenko, Olga Morgunova, Monika Helmerand, Timo Suslov, Arno Uprus, Kaupo Kase, Sander Saare, Kait Kabun, Heiki Luts, Kirke Williamson, Kaadri Krooni, Riina Tamm, Karina Troshkina for the information on towns public transportation management.

References

1. A. Kasper, The fragmented securitization of cyber threats, in *Regulating e-Technologies in the European Union – Normative Realities and Trends* (Springer, Berlin, 2014), pp. 157–187
2. J. Kivimaa, Applying a cost optimizing model for IT security, in *Proceeding of the 8th European Conference on Information Warfare and Security. Lisbon, Portugal, July 6–7, 2009* (Academic, Cambridge, 2009), pp. 142–153
3. D. Levis, P. Hugo, Regional working paper: a harmonized definition of cities and rural areas: the new degree of urbanization, P: 28 (2014)
4. P. Lorents, Keel ja Loogika [Language and logic] (Tallinn: Estonian Business School, Estonia, 2000)
5. P. Lorents, E. Matsak, A. Kuuseok, D. Harik, Assessing the similarity of situations and developments by using metrics, in *Intelligent Decision Technologies, (KES-IDT-17), Vilamoura, Algarve, Portugal, 21–23 June 2017* (Springer, Berlin, 2017), 15pp.
6. A.I. Maltsev, *Algebraic Systems* (Science, Moscow, 1970)
7. E. Matsak, Dialogue system for extracting logic constructions in natural language texts, in *Proceedings of the International Conference on Artificial Intelligence. IC–' AI'2005, Las Vegas*, vol. II (2005), pp. 791–797
8. E. Matsak, Using natural language dialog system DST for discovery of logical constructions of children's speech, in *Proceedings of the International Conference on Artificial Intelligence. IC-AI 2006, Las Vegas*, vol. I (2006), pp. 325–331
9. E. Matsak, The Prototype of system for discovering of inference rules, in *Proceedings of the International Conference on Artificial Intelligence IC– AI'2007, Las Vegas*, vol. II (2007), pp. 489–492
10. E. Matsak, Improved version of the natural language dialog system DST and its application for discovery of logical constructions in children's speech, in *Proceedings of the International Conference on Artificial Intelligence IC– AI'2008, Las Vegas*, vol. II (2008), pp. 332–338

Dynamic Heuristics for Surveillance Mission Scheduling with Unmanned Aerial Vehicles in Heterogeneous Environments

Dylan Machovec, James A. Crowder, Howard Jay Siegel, Sudeep Pasricha, and Anthony A. Maciejewski

1 Introduction

Unmanned aerial vehicles (UAVs) are used in many environments to gather information, such as in active battlefields scenarios. An example of such a scenario is shown in Fig. 1. In this example, seven UAVs are being used to gather information about nine targets. Assuming that a UAV can only surveil a single target at a time, this is an oversubscribed scenario, which means that there are more targets than UAVs and it will not be possible to surveil all targets simultaneously with the fleet of UAVs. To gather as much useful information about the targets as possible, it is necessary to conduct mission planning and scheduling to determine how the UAV fleet should effectively surveil the targets.

As both the number of UAVs that are active simultaneously and the number of targets available in an environment increases, it becomes necessary to reduce the amount of human control and human scheduling required to operate them effectively [1]. This can be accomplished by designing and deploying heuristic techniques that can find effective mission scheduling solutions.

D. Machovec (✉) · A. A. Maciejewski
Department of Electrical and Computer Engineering, Colorado State University, Fort Collins, CO, USA

J. A. Crowder
Colorado Engineering, Inc., Colorado Springs, CO, USA
e-mail: jim.crowder@coloradoengineering.com

H. J. Siegel · S. Pasricha
Department of Electrical and Computer Engineering, Colorado State University, Fort Collins, CO, USA

Department of Computer Science, Colorado State University, Fort Collins, CO, USA
e-mail: hj@colostate.edu; sudeep@colostate.edu

© Springer Nature Switzerland AG 2021
H. R. Arabnia et al. (eds.), *Advances in Artificial Intelligence and Applied Cognitive Computing*, Transactions on Computational Science and Computational Intelligence, https://doi.org/10.1007/978-3-030-70296-0_43

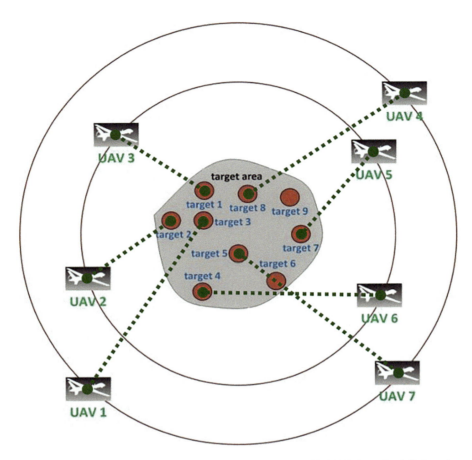

Fig. 1 An example scenario with seven UAVs and nine targets. The UAVs have fixed flight paths, which follow one of two circular paths around the area containing the targets that are candidates for surveillance

In this study, our focus is on the design of mission scheduling techniques capable of working in dynamic environments that are suitable for determining effective mission schedules in real time. Because scheduling problems of this type are, in general, known to be NP-Hard, finding optimal solutions is not feasible [2]. Due to this, we consider fast heuristic techniques to find mission schedules and evaluate their performance when compared to mathematical upper bounds we derive and simple baseline heuristics we generate. These techniques are capable of generating the mission schedules required by our environment in less than a second.

To effectively compare and evaluate these techniques, we measure system-wide performance using a metric called surveillance value. Surveillance value is designed to measure the overall performance of all information gathered by the UAVs, based on the number of surveils that occur, the quality of information gathered by each surveil (e.g., image resolution), the overall importance of each target, and the relevance of the information obtained for a specific target.

This work builds on our previous work in [3, 4], which defined surveillance value as the system-wide performance measure and explored bi-objective trade-offs between surveillance value and energy consumption in a static environment where all mission planning decisions are made once in advance. This work considers a dynamic (instead of static) environment and attempts to maximize surveillance value while considering energy consumption as a hard constraint.

The novel contributions in this work include:

- The design of new mission scheduling heuristics that are used to dynamically determine which UAVs should be used to surveil each target and additionally which sensors should be used for these surveils
- Extending the model of surveillance targets in [4] to require a minimum interval of time between consecutive surveils of the same target
- A model for randomly generating scenarios defined by a set of UAVs and targets for the purpose of effectively evaluating mission scheduling techniques, such as the heuristics considered in this work
- A detailed analysis and comparison of heuristics across many simulated scenarios with varied characteristics

This chapter is organized as follows. In Sect. 2, the system model and environment are described. The methods used by both the novel mission scheduling heuristics and the comparison heuristics evaluated in this study are presented in Sect. 3. Section 4 contains the specific process used to generate all of the scenarios we use in our simulations. In Sect. 5, we show the results of the simulations and use the results to analyze and compare the behavior of the heuristics. Related work is discussed in Sect. 6 and finally in Sect. 7 we conclude and discuss possible future work that could build on this study.

2 System Model

2.1 Overview

For this study, we use a simplified model of a real environment; our future studies that build on this research will continue to enhance this model. The system considered in this study consists of a heterogeneous set of UAVs (with varying sensor and energy characteristics) and a heterogeneous set of targets (with varying surveillance requirements). These sets are constant, meaning that UAVs and targets will not be added or removed dynamically. During the time when we consider mission scheduling, the UAVs fly in continuous concentric circles at a constant speed around the target area, which contains the entire set of targets. This is a simplifying assumption because the techniques we consider in this study do not control the flight path of the UAVs. It is assumed that every UAV is always close enough to every target and has an unobstructed view of every target so that any sensors available to any UAV can be used to surveil any target at any time. A UAV can only surveil a single target at any given time. Because UAVs cannot stay airborne indefinitely, this work considers mission scheduling strategies for a single day. This means that at the end of the day, all UAVs would be able to return to their base of operations to refuel or recharge. The problem space we explore is made up of oversubscribed systems, which means that there are fewer UAVs than targets. This prevents all targets from being surveilled simultaneously; however, the techniques designed and evaluated in this study are still applicable to undersubscribed systems.

2.2 Target and UAV Characteristics

In this environment, each UAV has a single energy source with a fixed amount of total energy available to it. In our subsequent discussions, we normalize this value so that the maximum amount of energy available to any UAV is less than or equal to 1.0. Every UAV is equipped with one or more sensors that can be used to surveil targets. The sensor types considered in this work are visible light sensors (VIS), infrared light sensors (IR), synthetic-aperture radars (SAR), or light detection and ranging systems (LIDAR). Each UAV cannot have more than one sensor of a given type, which is a simplifying assumption in this work. The heuristics presented in this study can be easily modified to function in environments with multiple sensors with the same type. Each sensor available to a UAV also has an associated sensor quality value ranging from 0.0 to 1.0 and a rate of energy consumption, which is normalized to the total energy available to the UAV and ranges from 0.0 to 1.0. All sensors available to a UAV draw from the same source of energy. An example set of characteristics for seven UAVs is shown in Table 1, based on the table in [4].

Targets represent locations of interest to be potentially surveilled by UAVs. Because the environments we consider are often oversubscribed (there are more targets than UAVs), it is possible that some targets will not be surveilled. An unchanging priority value is assigned to each target, which represents the overall importance of surveilling the target. Priority values can be any positive number, where higher numbers represent more important targets. Each target has a fixed surveillance time value, which specifies the number of hours per day that a UAV

Table 1 UAV characteristics

Characteristics	UAV 1	UAV 2	UAV 3	UAV 4	UAV 5	UAV 6	UAV 7
Sensor type	VIS|IR	VIS	SAR	LIDAR|IR	SAR|IR	VIS|IR	SAR|LIDAR
Sensor quality	0.9|0.7	0.5	0.5	0.7|0.8	0.8|0.7	0.9|0.7	0.8|0.8
Energy used/hour	0.08|0.06	0.08	0.08	0.1|0.06	0.125|0.06	0.07|0.06	0.125|0.08
Total energy	1.0	0.8	0.9	0.7	0.8	0.6	1.0

should spend surveilling the target. Because the kind of information that is useful for each target may vary, targets have a set of surveillance types, which defines the sensor types that are allowed to surveil the target, and a list of sensor affinity values, which range from 0.0 to 1.0 and measure how useful or relevant the information gained from that sensor type is for the target. There is a constraint on the number of times a target may be surveilled by all UAVs in a day called the surveillance frequency of the target. Finally, we define a minimum time between surveils for each target because information gained from back-to-back surveils of a target is likely to contain repeat information compared to surveils that are spread throughout the day. Table 2 lists an example set of characteristics for nine targets, which is also based on [4].

2.3 Surveillance Value

To evaluate the performance of different techniques for assigning UAVs to targets, it is necessary to measure the worth of individual surveils by a UAV on a target. We use the concept of surveillance value [4] to achieve this, where for a given surveil, the value of that surveil is given by the product of the priority (ρ), sensor affinity (α), and sensor quality (γ) associated with the UAV (u), target (t), and used sensor type (s):

$$\text{value (surveil)} = \rho(t) * \alpha(t, s) * \gamma(u, s). \tag{1}$$

The total surveillance value earned over an interval of time is then defined by the sum of values earned by all surveils performed by UAVs in that interval of time:

$$\text{surveillance value} = \sum_{\substack{s \in \text{surveils} \\ \text{performed}}} \text{value (s)}. \tag{2}$$

If a surveil is not fully completed during the day due to being in progress when the day ends, or due to the UAV lacking sufficient energy to finish a complete surveil of the target, then partial value will be earned for that surveil, which is directly proportional to the fraction of the surveil that was completed. For example, if a full surveil of a target takes 4 h and only 3 h of this were completed in the interval of interest, then the value of that surveil in that interval of time would be 75% of the value given by the fully completed surveil.

Table 2 Target characteristics

Characteristics	Target 1	Target 2	Target 3	Target 4	Target 5	Target 6	Target 7	Target 8	Target 9
Priority	1	2	6	7	5	4	3	3	1
Surveillance type	SAR|IR|VIS	SAR|LIDAR	VIS|SAR	VIS|IR|SAR|LIDAR	VIS	IR|LIDAR	SAR|VIS	SAR|IR|VIS	VIS
Sensor affinity	0.8|0.6|0.9	0.9|0.7	1.0|0.8	1.0|0.7|0.9|0.8	1.0	0.8|0.7	0.9|1.0	0.9|0.8|1.0	1.0
Surveillance time	3 h	4 h	1 h	3 h	2 h	1 h	3 h	3 h	4 h
Surveillance freq. (#/day)	4	3	5	2	4	5	3	3	4
Min. time betw. surveils	2 h	3 h	4 h	8 h	3 h	4 h	3 h	6 h	2 h

Dynamic Heuristics for Surveillance Mission Scheduling with Unmanned... 589

2.4 Problem Statement

The goal of our proposed scheduling techniques in this environment is to assign UAVs to targets in a particular order to obtain as much useful surveillance information as possible. This problem is constrained by the total energy available to each UAV. In this study, this constraint is only applied to the energy consumed by a UAV's sensors. The energy needed for the UAV's fixed flight plan is not included in the energy available to the sensors. Because the flight plan is fixed, it is also not possible for the UAV to refuel at any point during the day, such as while there are no targets available for it to surveil. Additionally, each UAV can only surveil one target and only operate one of its sensors at any time; similarly, each target can only be surveilled by one UAV at any time. This is a simplifying assumption used in this study. Because there may be duplicate information obtained from surveilling a target with multiple sensors at once, the surveillance value performance measure would need to be adjusted to account for this. Once a UAV begins surveilling a target, this surveil cannot be preempted unless the UAV does not have enough energy to continue (i.e., the surveil cannot be stopped early to allow the UAV to surveil a different target). Given the above constraints, the final goal is to maximize the total surveillance value obtained over a day.

3 Mission Scheduling Techniques

3.1 Mapping Events

Mapping UAVs to targets refers to the process of determining which UAVs will surveil which targets, which sensors will be used for surveils, and when the surveils will occur. A UAV is available to be mapped if it is not currently surveilling targets and has energy remaining. A target is available to be mapped if it is not currently being surveilled and it is eligible for being surveyed based on its surveillance frequency and its minimum time between surveils. A sensor of a UAV is said to be a valid sensor type for a given target if that sensor type is also in the target's list of surveillance types. If a UAV has a valid sensor type for a target, it is called a valid UAV for that target. Only valid UAVs are considered for mapping to a given available target.

The time when a mapping of available UAVs to available targets occurs is called a mapping event. When a mapping event happens, a mission scheduling technique is used to assign available UAVs to surveil available targets based on the current state of the system. In this study, all techniques presented are real-time heuristics to allow mapping events to be completed in less than a second for the problem sizes considered. There are different techniques for deciding when a mapping event should be initiated, e.g., at fixed time intervals or due to changes in the environment. In this study, we consider the case where mapping events occur every 6 min during the day if there are available UAVs and available targets. By using a time interval such as 6 min when the minimum surveillance time of any target is 1 h, any delay where a UAV is not surveilling a target because a mapping event has not yet occurred is small. This 6 min time interval provides some opportunity for multiple UAVs to become available during that window, which increases the number of choices available to the mission scheduling technique. Other values for the time interval can be implemented instead of 6 min. In a real-world implementation, this interval of time can be derived based on empirical evaluations of the characteristics of the actual system. While our resource manager is designed to deal with dynamic environments, such as those with time-varying UAV and target characteristics, in this simulation study the environment is non-varying.

3.2 Comparison Techniques

For comparison, two randomized heuristics are considered. Randomized heuristics are often better points of comparison than heuristics that use extremely simple logic to make mapping decisions. For example, in high performance computing, simple randomized heuristics often perform better than simple ordered techniques like first-come-first-served [5].

3.2.1 Random

At a mapping event, this technique considers available targets in a random order. For each target, a random available valid UAV and a random valid sensor type of that UAV are selected. The selected UAV and sensor type are assigned to surveil the target. This results in both the target and the UAV becoming unavailable for new assignments until this new surveil completes. If there is no available UAV that has a valid sensor type for the target, then no UAV is assigned to the target. This repeats with the next target in the random ordering until there are no more assignments of UAVs to targets possible in the current mapping event.

3.2.2 Random Best Sensor

The random best sensor heuristic is similar to the random technique, but uses knowledge about the sensor quality of UAVs and the sensor affinity of targets to make decisions that are likely to result in higher surveillance value. Like the random heuristic, available targets are considered in a random order and a UAV with a valid sensor type for this target is selected at random. However, instead of selecting a random valid sensor type from the UAV, this heuristic then chooses the sensor type with the maximum product of the UAV's sensor quality and the target's sensor affinity. Because both of these values are directly used along with

Dynamic Heuristics for Surveillance Mission Scheduling with Unmanned...

the target's priority in the calculation for the value of a surveil, this strategy will likely select higher value surveils when compared to the random heuristic if they are available on the selected UAV. After this, the same process used by the random heuristic is followed: the UAV is assigned to surveil the target with this sensor type and the heuristic will continue with the next randomly ordered target until no more assignments are possible.

3.3 Value-Based Heuristics

The value-based heuristics in this study are designed to search through valid combinations of UAVs, targets, and sensor types to greedily assign UAVs to surveil targets based on the surveillance value performance measure. A valid combination is represented by an available target, a valid available UAV for that target, and a valid sensor type of the UAV for the target.

3.3.1 Max Value

At a mapping event, the max value heuristic starts by finding a valid combination of a UAV, target, and sensor type that results in the maximum possible value for a single surveil. If there are multiple valid combinations with the same maximum possible value, then any of these combinations can be selected arbitrarily. The heuristic then assigns the UAV from the selected combination to surveil the selected target with the selected sensor type. This process of finding the maximum value combination and starting a surveil based on the combination repeats until no more assignments of available UAVs to available targets are possible in the current mapping event.

3.3.2 Max Value per Time

The max value per time heuristic is identical to max value except for one difference. Instead of selecting the valid combination that results in the maximum possible value for a surveil, max value per time instead selects the valid combination that results in the maximum possible value divided by surveillance time of the target (based on a complete surveil of the target).

3.3.3 Max Value per Energy

The max value per energy heuristic is identical to max value except for one difference. Instead of selecting the valid combination that results in the maximum possible value for a surveil, max value per energy instead selects the valid combination that results in the maximum possible value divided by the projected energy consumed by the UAV. The projected energy consumption can be easily calculated from the energy consumption rate of the selected sensor type and the surveillance time of the selected target (based on a complete surveil of the target). We have used the general concept of performance per unit time and performance per unit of energy in prior work in a high-performance computing environment, e.g., [5, 6].

3.4 UAV-Based Metaheuristic

The value-based heuristics described in Sect. 3.3 are designed to perform well in specific situations, and using the wrong heuristic for a scenario could result in performance that is worse than using random. Because there may be insufficient information to predict which heuristic should be used, we design a metaheuristic to intelligently combine the best performing value-based heuristics. This does not include the max value per time heuristic because in the scenarios we consider, max value per time never performs better than both max value and max value per energy. The UAV-based metaheuristic uses a two-phase process to find good surveillance options. This is similar in concept to the metaheuristic we designed in [5] for a different environment.

In the first phase, the heuristic selects a candidate target and valid sensor type for each UAV. To make this selection, the UAV's remaining energy is used to determine if the strategy used by the max value or max value per energy heuristic would be most effective. If the remaining fraction of the UAV's energy is greater than the remaining fraction of time in the day, then its energy has been consumed at a relatively slow rate during the day and it makes sense to use the strategy from max value to make decisions without considering energy consumption. Otherwise, the UAV has been consuming energy at a relatively high rate and the strategy from max value per energy can be used to make energy-efficient decisions. Based on this choice, either the valid combination using the UAV that results in the maximum possible value or the maximum possible value divided by energy consumed is selected as the best candidate combination for the current UAV. The first phase ends when every UAV has a candidate combination selected. Note that multiple UAVs can select the same target as their candidate.

The second phase is used to determine which UAV from the first phase should be assigned to its candidate target and sensor type. Unlike the first phase, it is unnecessary to use strategies from multiple value-based heuristics in the second phase. This is because energy is a constraint for individual UAVs and not for the overall system. At the system level, all that is relevant to maximizing surveillance value is the value of each surveil. Thus, we choose the UAV with a candidate combination that results in the maximum possible value earned by its corresponding surveil. This chosen UAV is assigned to surveil its target. This process of selecting candidates in the first phase and making an assignment of the best candidate in the second phase is repeated until no more assignments are possible in the current mapping event.

4 Simulation Setup

4.1 *Generation of Baseline Set of Randomized Scenarios*

Each scenario that we use to evaluate the heuristics is defined by a set of UAV characteristics and a set of target characteristics. For example, a small sample scenario with seven UAVs and nine targets is defined in Tables 1 and 2. To compare and evaluate the heuristics, we consider a wide variety of scenarios to understand the kinds of scenarios for which each heuristic is most effective.

We randomly generate 100 baseline scenarios by sampling from probability distributions for the number of UAVs and targets in a scenario, in addition to the value for each characteristic of the UAVs and targets. In each case, distributions were selected in a way to attempt to model distributions of parameters that may occur in real-world environments. The details of these distributions are as follows.

The number of UAVs available during the 24 h period of a scenario is sampled from a Poisson distribution with the Poisson parameter $\lambda = 70$. The characteristics of each UAV are then generated. The total energy available to the UAV is sampled from a beta distribution with a mean of 0.8 and a standard deviation of 15% of the mean. The energy consumption rate for each sensor is sampled from a beta distribution with a mean of 0.1 and a standard deviation of 30% of the mean. The total energy and energy consumption rates are sampled in this way so that UAVs can be expected to operate for an average of 8 h, which is enough to allow multiple surveils during the day for most UAVs. The number of sensors available to each UAV is generated by using a Rayleigh distribution with a scale parameter $\sigma = 2$. Any values <1 are increased to one and any values >4 are decreased to 4.

The sensor type for each sensor is selected using probabilities of 0.5, 0.2, 0.2, and 0.1 for the VIS, SAR, IR, and LIDAR sensor types, respectively. Once a sensor has been assigned a type using the listed probabilities, that sensor type is no longer a candidate and the next sensor is chosen among the remaining sensors types after normalizing their probabilities so that the sum is 1.0. The quality of each sensor is found using a beta distribution with a mean of 0.8 and a standard deviation of 10% of the mean.

The number of targets available to surveil during the 24 h period is obtained using a Poisson distribution with $\lambda = 90$. Because the number of UAVs was generated with $\lambda = 70$, this means that these scenarios in general will be oversubscribed. The priority of each target is sampled from a gamma distribution with a mean of 4 and a standard deviation of 20% of the mean. To obtain the required surveillance time for each target, we use a uniform distribution ranging from 1 to 4 h. Differing from what was used for UAVs, we obtain the number of surveillance types for each target by adding 1 to the value obtained from a binomial distribution with $p = 0.5$ and $n = 3$. The surveillance types selected to match this number are uniformly selected from VIS, SAR, IR, and LIDAR. To get the sensor affinity for each surveillance type, we use a beta distribution with a mean of 0.7 and a standard deviation of 20% of the mean. A discrete uniform distribution allowing values of 2, 3, 4, or 5 is used to find the surveillance frequency for each target. Lastly, a two-step process is needed to generate the minimum time between surveils for each target. We sample a value from a gamma distribution with a mean of 2 h and a standard deviation of 20% of the mean. When generating scenarios, we require that it is possible for a target to be fully surveilled a number of times equal to its surveillance frequency within 24 h. The total time (T) needed for these surveils can be calculated from the surveillance time (τ), surveillance frequency (ω), and minimum time between surveils (δ). It is given by:

$$T = \tau * \omega + (\omega - 1) * \delta. \tag{3}$$

To ensure that the total surveillance time for a UAV does not exceed 24 h, the minimum time between surveils is set to the smaller of the sampled value from the gamma distribution and the maximum time between surveils (Δ):

$$\Delta = \frac{24\,\text{h} - \tau * \omega}{\omega - 1}. \tag{4}$$

4.2 Generation of Additional Scenarios for Parameter Sweeps

Because the baseline set of 100 scenarios in Sect. 4.1 may have characteristics that are favorable to the performance of individual heuristics, we use parameter sweeps to evaluate the heuristics in a diverse set of environments. We generate 22 additional sets of 100 scenarios each for the parameter sweeps of five characteristics of the UAVs and targets. The characteristics we vary are the mean time between surveils of targets (five sets in addition to the baseline), the mean number of targets in a scenario (four sets in addition to the baseline), the mean number of UAVs in a scenario (four sets in additional to the baseline), the mean rate of energy consumption for sensors (five sets in addition to the baseline), and the mean total energy available to each UAV (four sets in addition to the baseline). The number of sets for each characteristic was selected such that the impact of each characteristic on the performance of the heuristics is clearly shown.

To vary the minimum time between surveils of targets, we vary the mean of the gamma distribution used over the integers ranging from 1 to 5. Additionally, we consider the case where the minimum time between surveils is 0 h for all targets. For each of these cases, 100 scenarios are generated. We examine the effect of varying the number of targets and number of UAVs by generating 100 scenarios each for the cases with λ values of 50, 70, 90, 110, and 130 for the number of targets and 30, 50, 70, 90, and 110 for the number of UAVs. Finally, we vary the mean energy consumption of sensors and mean total energy available to each UAV. Instead of sampling all of the distributions again in this case, we take the 100 baseline scenarios from Sect. 4.1 and scale the values to match the new means.

5 Simulation Results

5.1 *Upper Bounds on Performance*

We calculate upper bounds on the possible surveil value that can be obtained in each scenario. While the optimal result is likely significantly below these bounds, they provide additional insight when evaluating the results. Two different methods of calculating upper bounds are used. The minimum of these upper bounds is then selected for each scenario to produce the final upper bound shown in our results.

The target-based upper bound is found by first calculating the maximum possible surveillance value that can be earned by each target individually. This is done by assuming that the UAV that would produce the most value for a single surveil will surveil the target the maximum number of times allowed by the surveillance frequency of the target. For example, consider using UAV 1 from Table 1 to surveil Target 1 from Table 2. The valid sensor types are VIS and IR. Following the process for calculating surveil value in Sect. 2.3, VIS results in a surveillance value for a single surveil of $1.0 * 0.9 * 0.9 * = 0.81$ and with IR the surveillance value is $1.0 * 0.7 * 0.6 = 0.42$. The best option for this UAV is to use the VIS sensor with 0.81 surveillance value for a single surveil. This is greater than or equal to the surveillance value from a single surveil for all UAVs that can surveil Target 1. If we assume that UAV 1 is used to surveil Target 1 four times (the surveillance frequency of Target 1), a total of 3.24 surveillance value will be gained, which is an upper bound on the surveillance value earned by Target 1. This process can be repeated for all targets to find an upper bound on surveillance value for the entire scenario.

The energy-based upper bound is calculated by finding an upper bound for each UAV individually and then summing the results. For a given UAV, this method finds the best possible way that the UAV's energy can be spent assuming there are no other UAVs surveilling targets in the scenario. To do this, we first construct a list of all possible valid combinations of targets and sensor types that can be surveilled by the UAV. The surveillance value divided by projected energy consumed for each combination is then calculated and the list is sorted in descending order based on these amounts. This list is then traversed in order starting with the most energy-efficient combination. For each combination, assume that the UAV spends as much energy as possible on that option. This is constrained by the surveillance frequency of the target in addition to the total energy remaining for the UAV. For example, consider the simple example of UAV 1 from Table 1 and the first three targets from Table 2. The most energy-efficient option here is using UAV 1 to surveil Target 3 with its sensor of type VIS. Each surveil of Target 3 takes 1 h and the VIS sensor consumes 0.08 units of energy per hour of surveillance. This would result in a surveillance value per unit of energy of $(6 * 1.0 * 0.9)/(1 * 0.08) = 67.5$. This target can be surveilled at most five times. Each surveil requires 1 h and consumes 0.08 units of energy per hour. This results in $5 * 0.08 = 0.4$ of the UAV's 1.0 total energy. This set of surveils earns 27 surveillance value with Target 3. As the UAV still has 0.6 of its energy remaining, we move to the next best option in terms of energy efficiency, which is Target 1, again with the VIS sensor resulting in 0.81 value for each completed surveil. This target can only be surveilled 2.5 times before UAV 1 will consume the rest of its energy. Considering half of a surveil worth of value for the partial surveil, this will result in 2.5 * 0.81 = 2.025 surveillance value for Target 1. Considering just this set of three targets, the maximum possible surveillance value that can be earned by UAV 1 is 27 + 2.025 = 29.025. After repeating this process for all UAVs, we sum up the resulting maximum possible surveillance values from each UAV to get the energy-based upper bound on surveillance value for an entire scenario.

5.2 Comparisons of Mission Scheduling Techniques

As described in Sect. 4.2, the results shown in this section consist of parameter sweeps where the means of the distributions described in Sect. 4.1 are varied. In addition to the upper bounds on performance (Sect. 5.1) for the results that quantify value obtained, the results that quantify energy consumed also include upper bounds on energy consumption. Calculating the upper bound for energy consumption is very simple: By summing the total energy values from all UAVs. This upper bound can be obtained because the total energy of each UAV acts as a hard constraint on the energy it can expend. It is important to note that the upper bounds for energy and for values (performance) shown in the result figures are average upper bounds over the 100 scenarios considered. This means that in some cases, individual scenarios will surpass these upper bounds, but the average over all 100 scenarios will not. It is also possible that the 95% mean confidence intervals shown for each set of results will surpass these upper bounds for the same reason.

5.2.1 Effect of Minimum Time Between Surveils

In Fig. 2, we vary the mean minimum time required between surveils. Because this value is sampled from a gamma distribution and a mean of 0 is not possible for a gamma distribution, the set of bars marked with a 0 mean corresponds to the case where all targets have no time required between surveils. There are three main points to observe from this comparison. First, as the minimum time between surveils is increased, the value earned by max value per time decreases significantly and becomes less than the value earned from the random heuristics. This occurs because the max value per time only considers the time the UAV spends surveilling a target and not the time afterwards where the target will be unavailable. Due to this, max value per time is generally not a suitable heuristic for the environment considered in this study where the minimum time between surveils is not 0. Second, the other value-based heuristics are always the three best heuristics and earn similar amounts of value. Third, despite the max value per energy heuristic earning a significant amount of value, it consumes far less energy than most other heuristics, which

Dynamic Heuristics for Surveillance Mission Scheduling with Unmanned...

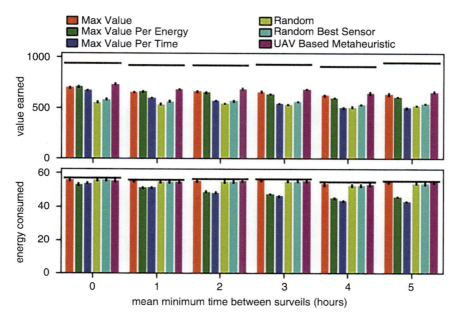

Fig. 2 A comparison of the average total energy consumed and average surveillance value earned for 100 randomized scenarios where the mean minimum time between surveils for targets is varied from the baseline set of scenarios with a mean minimum time between surveils of 2 h. Except for the minimum time between surveils, which is varied, the other parameters have the same value as the baseline case described in Sect. 4.1. Average upper bounds are indicated by horizontal black lines above each set of bars and 95% mean confidence intervals are shown for each bar

always consume energy close to the upper bound for these scenarios. This means that as the minimum time between surveils increases and max value per energy consumes less energy, max value and the metaheuristic start to earn more value.

5.2.2 Effect of Energy Consumption Rate

Results of simulations where the mean energy consumption per hour is varied are shown in Fig. 3. It can be seen that when the rate of energy consumption is low, none of the heuristics is able to reach the upper bound for energy consumption because there is not enough time in the day for the UAVs to do so. These cases also get much closer to the upper bound on value earned because energy is no longer a significant constraint. It can be observed that the max value heuristic earns high value when the energy consumption rates are lowest and drops below the max value per energy heuristic when the energy consumption rates are at their highest. This shows that both heuristics are working as designed and was the motivation for creating the UAV-based metaheuristic, which can intelligently use the strengths of both options.

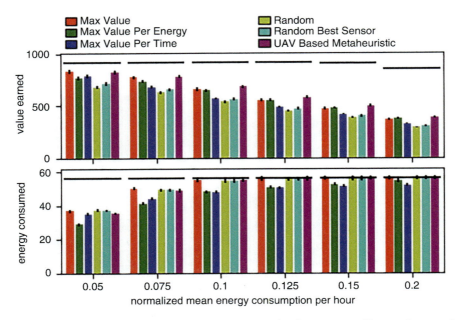

Fig. 3 A comparison of the average total energy consumed and average surveillance value earned for 100 randomized scenarios where the mean energy consumption per hour for each UAV sensor is varied. This variation is achieved through scaling the energy consumption values from the baseline with a mean energy consumption per hour of 0.1. Except for the mean energy consumption per hour, which is varied, the other parameters have the same value as the baseline case described in Sect. 4.1. Average upper bounds are indicated by horizontal black lines above each set of bars and 95% mean confidence intervals are shown for each bar

In these results, it can be seen that the metaheuristic always performs as well or better than both max value and max value per energy.

5.2.3 Effect of Total Energy

The results in Fig. 4 show what happens when the mean total energy available to each UAV. These results show similar behavior to the results with varied energy consumption rates, except that the difference among the max value, max value per energy, and UAV-based metaheuristic heuristics is not as large. This is partially because none of these scenarios result in the heuristics consuming significantly less energy than is available. This can be seen when comparing the energy consumption upper bounds in Figs. 3 and 4.

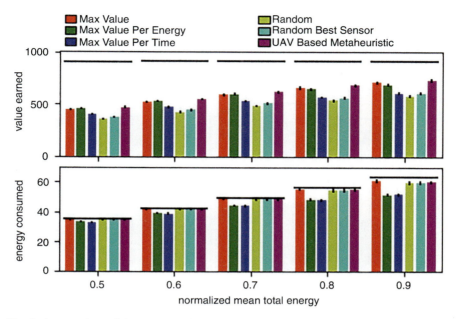

Fig. 4 A comparison of the average total energy consumed and average surveillance value earned for 100 randomized scenarios where the mean total energy of each UAV sensor is varied. This variation is achieved through scaling the total energy values from the baseline with a mean total energy of 0.8. Except for the mean total energy, which is varied, the other parameters have the same value as the baseline case described in Sect. 4.1. Average upper bounds are indicated by horizontal black lines above each set of bars and 95% mean confidence intervals are shown for each bar

5.2.4 Effect of Number of Targets

In Fig. 5, a comparison of different scenarios with a varied mean number of targets is shown. When there are few targets available, it is not possible to consume all of the energy available to the UAVs because there are more UAVs than targets on average and some UAVs will not be able to surveil targets for a significant portion of the day because all targets will be under surveillance already by other UAVs. As the number of targets in a scenario grows, it can be seen that it becomes very easy to consume all of the energy available to the UAVs. This is because there are many targets and all UAVs are likely to be able to surveil targets continuously. The upper bounds for value earned increase significantly as the number of targets increase. In addition to UAVs being able to operate simultaneously, this effect is due to the fact that with more targets there is a higher probability that a high priority target with a sensor affinity that matches up well with each UAV will appear in a scenario. Lastly, it can be seen that the UAV-based metaheuristic again performs as well or better than both max value and max value per energy in all cases.

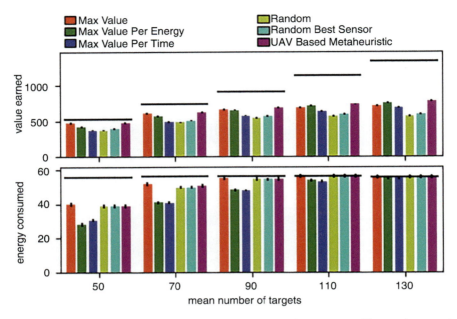

Fig. 5 A comparison of the average total energy consumed and average surveillance value earned for 100 randomized scenarios where the mean number of targets is varied through generating new scenarios in addition to the baseline, which has a mean of 90 targets in a single scenario. Except for the mean number of targets, which is varied, the other parameters have the same value as the baseline case described in Sect. 4.1. Average upper bounds are indicated by horizontal black lines above each set of bars and 95% mean confidence intervals are shown for each bar

5.2.5 Effect of Number of UAVs

Figure 6 shows simulation results for a set of randomized scenarios where the mean number of UAVs in a scenario is varied. These results show the biggest variance in value earned between max value and max value per energy. When there are a small number of UAVs, it is important to use each UAV efficiently, which means that max value per energy performs well. When there are many UAVs, it is more important to carefully pick which UAVs will get the most useful information for each target and energy is no longer a significant constraint because there will be unused UAVs still available when some UAVs start running out of energy. In this case, max value is most effective at picking the best target for each UAV. Once again, the UAV-based metaheuristic still earns value similar to the better of max value and max value per energy.

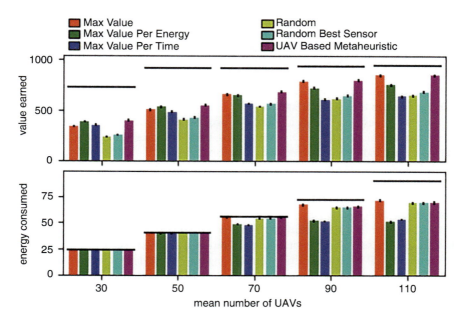

Fig. 6 A comparison of the average total energy consumed and average surveillance value earned for 100 randomized scenarios where the mean number of UAVs is varied through generating new scenarios in addition to the baseline, which has a mean of 70 UAVs in a single scenario. Except for the mean number of UAVs, which is varied, the other parameters have the same value as the baseline case described in Sect. 4.1. Average upper bounds are indicated by horizontal black lines above each set of bars and 95% mean confidence intervals are shown for each bar

5.3 Discussion of Results

The results in Sect. 5.2 indicate that depending on the scenario, either max value or max value per energy is an effective real-time heuristic for maximizing surveillance value. The UAV-based metaheuristic combines the strengths of both heuristics and is effective in all scenarios. The max value per time heuristic was shown to be ineffective in scenarios where there is a minimum time between surveils of targets, and resulted in poor performance in most of the scenarios that we considered. Additionally, when considering scenarios where the energy of UAVs will not be fully consumed during the day, max value per energy is ineffective. Based on these results, our proposed metaheuristic is the best option to use in all cases where the characteristics of the scenario may change unexpectedly.

6 Related Work

Developing a complete mission schedule for UAVs involves solving multiple problems, many of which have been studied in the past such as planning the specific routes used by UAVs, which we do not consider in this study, and assigning specific tasks to UAVs. Some studies solve these problems through time-consuming optimization techniques such as mixed integer linear programming (MILP), while others use techniques ranging from expensive metaheuristics like genetic algorithms (GAs) to fast and efficient greedy heuristics to find effective solutions.

In [7], mission planning is divided into two subproblems: task scheduling and route planning. The task scheduling problem is the one we consider in our study. The task scheduling problem is solved using a MILP approach to minimize the completion time of all tasks as opposed to our work which aims to maximize surveillance value. This also differs from our work in that our environment does not have a specific set of tasks that must be completed and finds solutions in real time, while the MILP approach is time consuming and is used to construct a solution in advance. Similarly, in [8] a swarm of UAVs is also optimized to perform tasks while minimizing total completion time using a MILP approach. In [9], UAVs are assigned to attack and attempt to destroy clusters of targets through expressing three objective measures into one weighted measure (the success probability of the attack, the cost of the attack, and how well the timing of the attack will match a desired window), which is used to apply integer programming methods to find a solution. No-fly zones are considered in [10], which compares MILP and heuristic techniques to solve a task assignment problem where UAVs must complete a sequence of tasks. A solution here is represented by a DAG. In comparison to all these techniques, our methods can compute in real time and can be adapted to incorporate changes in the environment, such as dynamically adding and removing targets and UAVs. Furthermore, our performance measure considers the attributes of individual sensors and the energy consumption by individual UAVs.

Mission planning for UAVs is sometimes studied as an orienteering problem [11, 12]. For example, in [11] the authors utilize a model where UAVs originate from a depot and gain profit from traveling a path through nodes and back to the depot at the cost of fuel. Robust optimization techniques are used to maximize profit while taking uncertainty into account to avoid running out of fuel early. This work differs significantly from ours because distance between targets and UAVs is considered and the focus is on optimization of UAV movement instead of sensing. In [12], UAVs again depart from a depot, but before departure can select a specific set of sensors, which will impact their weight and the information they can gather. The authors solve this problem using both a MILP approach when ample time is available for finding solutions and several heuristic techniques for larger problem sizes that cannot be solved in a reasonable amount of time using the MILP approach. Our work differs significantly from these studies in part because a full mission plan is generated by the MILP approach and it is not modified during the day. In our work, the heuristics dynamically schedule UAVs to assign targets many times throughout the day and these decisions depend on the current state of the scenario. Additionally, our work considers the energy consumption of each sensor, which can greatly impact scheduling decisions.

Some studies have a greater focus on motion planning of the UAVs, which is beyond the scope of our contribution in this work. The work in [13] considers an environment where UAVs must keep track of targets moving throughout regions. This is done by probabilistically sending the UAVs to regions where useful information is likely to be found. A detailed three-dimensional space is considered in [14] where a single UAV must minimize the expected cost of ignorance, which is a measure of the information that was not gathered in a timely matter by the UAV. This is done using a greedy agent-based heuristic approach. In [15], MILP and a GA are used to minimize the total travel distance required for the time-constrained surveillance of objects with known flight paths. Base stations are also present in the field, which can be used to refuel the UAVs as needed.

In [16], UAVs launch with heterogeneous loads taken from bases and must scout targets using those loads. The goal of this study is to minimize the number of UAVs needed to fully scout all targets, which is achieved by seeding a genetic algorithm with an initial solution from a greedy heuristic. In [17], possible solutions for mapping a UAV to any combination of targets is represented by a tree where moving from the root to a node represents assigning the UAV to the target corresponding to that node. A best first search (BFS) method is used to find solutions to this problem. This differs significantly from our work because this heuristic produces a full mission plan for the UAV while our heuristics make decisions for a single surveil only for each UAV at mapping events throughout the day. The work in [18] is to design a method for assigning UAVs to targets in a disaster response scenario. UAVs originate at rally points and use a greedy approach to find short paths to move to and surveil targets based on the capabilities of each UAV.

7 Conclusions and Future Work

We designed a set of value-based heuristics used to conduct mission planning for UAV surveillance of a set of targets (max value, max value per time, max value per energy, and the UAV-based metaheuristic). We conducted a simulation study to evaluate, analyze, and compare these heuristics in a variety of scenarios. We found that while max value and max value per energy are each good techniques in a subset of the scenarios considered, the UAV-based metaheuristic found solutions with among the highest surveillance value for all scenarios. In environments where the best approach is not known, this makes the metaheuristic the obvious choice to employ.

In the future, we are interested in exploring more dynamic environments where a scenario experiences significant changes throughout the day. For example, changes in weather could dynamically affect the quality of outputs produced by different sensors. This is something that could be handled with a performance measure that can capture how the weather will affect the quality of outputs (sensor quality) and how useful different outputs are for different targets (sensor affinity). Additionally, some real environments will have targets of interest that are not in static positions. To address this, the model we use in this study could be expanded to accurately model the movement of both UAVs and targets. Lastly, the mission scheduling techniques in this study are heuristics with relatively simple functionality to allow mission schedules to be constructed in real-time. In the future, we would like to consider more complex techniques that can be used to produce better mission schedules in some cases when there is enough time available to do so.

Acknowledgments The authors thank Ryan D. Friese and John N. Carbone for their comments on this research.

References

1. J. Crowder, J. Carbone, Autonomous mission planner and supervisor (AMPS) for UAVs, in *20th International Conference on Artificial Intelligence (ICAI '18)*, (July 2018), pp. 195–201
2. M.R. Garey, D.S. Johnson, *Computers and Intractability: A Guide to the Theory of NP-Completeness* (W. H. Freeman and Co, 1979)
3. R.D. Friese, J.A. Crowder, H.J. Siegel, J.N. Carbone, Bi-objective study for the assignment of unmanned aerial vehicles to targets, in *20th International Conference on Artificial Intelligence (ICAI '18)*, (July 2018), pp. 207–213
4. R.D. Friese, J.A. Crowder, H.J. Siegel, J.N. Carbone, Surveillance mission planning: Model, performance measure, bi-objective analysis, partial surveils, in *21st International Conference on Artificial Intelligence (ICAI '19)*, (July 2019), 7 pp
5. D. Machovec, B. Khemka, N. Kumbhare, S. Pasricha, A.A. Maciejewski, H.J. Siegel, A. Akoglu, G.A. Koenig, S. Hariri, C. Tunc, M. Wright, M. Hilton, R. Rambharos, C. Blandin, F. Fargo, A. Louri, N. Imam, Utility-based resource management in an oversubscribed energy-constrained heterogeneous environment executing parallel applications. Parallel Comput. **83**, 48–72 (2019)
6. B. Khemka, R. Friese, S. Pasricha, A.A. Maciejewski, H.J. Siegel, G.A. Koenig, S. Powers, M. Hilton, R. Rambharos, S. Poole, Utility maximizing dynamic resource management in an oversubscribed energy-constrained heterogeneous computing system. Sustain. Comput. Inform. Syst. **5**, 14–30 (2015)
7. J.J. Wang, Y.F. Zhang, L. Geng, J.Y.H. Fuh, S.H. Teo, Mission planning for heterogeneous tasks with heterogeneous UAVs, in *13th International Conference on Control Automation Robotics & Vision (ICARCV)*, (March 2014), pp. 1484–1489
8. C. Schumacher, P. Chandler, M. Pachter, L. Pachter, UAV task assignment with timing constraints, in *AIAA Guidance, Navigation, and Control Conference and Exhibit*, (June 2003), 9 pp
9. J. Zeng, X. Yang, L. Yang, G. Shen, Modeling for UAV resource scheduling under mission synchronization. J. Syst. Eng. Electron. **21**(5), 821–826 (2010)
10. S. Leary, M. Deittert, J. Bookless, Constrained UAV mission planning: A comparison of approaches, in *2011 IEEE International Conference on Computer Vision Workshops (ICCV Workshops), 2002–2009*, (November 2011)
11. L. Evers, T. Dollevoet, A.I. Barros, H. Monsuur, Robust UAV mission planning. Ann. Oper. Res. **222**, 293–315 (2012)
12. F. Mufalli, R. Batta, R. Nagi, Simultaneous sensor selection and routing of unmanned aerial vehicles for complex mission plans. Comput. Oper. Res. **39**, 2787–2799 (2012)

13. W. Chung, V. Crespi, G. Cybenko, A. Jordan, Distributed sensing and UAV scheduling for surveillance and tracking of unidentifiable targets, in *Proceedings of SPIE 5778, Sensors, and Command, Control, Communications, and Intelligence (C3I) Technologies for Homeland Security and Homeland Defense IV*, (May 2005), pp. 226–235
14. D. Pascarella, S. Venticinque, R. Aversa, Agent-based design for UAV mission planning, in *8th International Conference on P2P, Parallel, Grid, Cloud and Internet Computing (3PGCIC)*, (October 2013), pp. 76–83
15. J. Kim, B.D. Song, J.R. Morrison, On the scheduling of systems of UAVs and fuel service stations for long-term mission fulfillment. J. Intell. Robot. Syst. **70**, 347–359 (2013)
16. W. Yang, L. Lei, J. Deng, Optimization and improvement for multi-UAV cooperative reconnaissance mission planning problem, in *11th International Computer Conference on Wavelet Active Media Technology and Information Processing (ICCWAMTIP)*, (December 2014), pp. 10–15
17. M. Faied, A. Mostafa, A. Girard, Vehicle routing problem instances: Application to multi-UAV mission planning, in *AIAA Guidance, Navigation, and Control Conference*, (August 2010), 12 pp
18. G.Q. Li, X.G. Zhou, J. Yin, Q.Y. Xiao, An UAV scheduling and planning method for post-disaster survey. ISPRS – Int. Arch. Photogramm. Remote. Sens. Spat. Inf. Sci. **XL-2**, 169–172 (2014)

Would You Turn on Bluetooth for Location-Based Advertising?

Heng-Li Yang 🔟, Shiang-Lin Lin, and Jui-Yen Chang

1 Introduction and Research Backgrounds

Smart phones have become an indispensable mobile device in daily lives and changed models of advertising. Traditional advertising media such as print ads, radio, and TV commercials have been transformed into mobile advertising models. Advertisers now can deliver messages that contain information of products, services, or promotion directly to mobile users on the basis of current location and specific time, who might read the advertisements and then become motivated for shopping [1]. Currently, mobile advertising agencies are able to deliver advertisements to customers in targeted locations and provide product and service information of their local businesses. This type of push advertising that delivers local shopping information to potential customers is called location-based advertising (LBA) [2].

LBA service can provide ads that are from businesses in the neighborhood of consumers and provide them with more relevant and quicker shopping information, which enhances the accuracy of advertisement delivery [3]. The service delivers higher-quality advertisements to users through certain components in a mobile device such as Bluetooth, Wi-Fi, and Globe Positioning System (GPS) and collect users' information of their current locations. It is helpful for a business to customize the contents of their advertisements. Compared to GPS, Bluetooth is the short distance positioning method. The commonly applied Bluetooth version 4.x Class 2 radios has a range of 10 meters and the accuracy of positioning can reach 2 to 5 meters, and the Class 3 and 4 have better positioning accuracy. Many studies have indicated that the effectiveness of mobile advertising would be enhanced if the location of advertising store is near the mobile device users (i.e., "location-

---

H.-L. Yang (✉) · S.-L. Lin · J.-Y. Chang
National Cheng-Chi University, Taipei, Taiwan
e-mail: yanh@nccu.edu.tw

© Springer Nature Switzerland AG 2021
H. R. Arabnia et al. (eds.), *Advances in Artificial Intelligence and Applied Cognitive Computing*, Transactions on Computational Science and Computational Intelligence, https://doi.org/10.1007/978-3-030-70296-0_44

congruency") [4, 5]. In other words, via the Bluetooth positioning users' location and calculating the distance between a local store and the users, then pushing relevant advertisements to the customers, it can efficiently improve click-through rate and wiliness to consume. However, for pushing more accurate advertisements to consumers, the LBA service requires consumers to provide their current location, which increases the risk of personal information leakage. Some studies have pointed out that most young mobile users are willing to receive advertisements through Bluetooth, but are concerned about the frequency of advertisement delivery. As a result, while collecting personal information, an advertising agency may need to take good care of users' privacy issue [6].

Although LBA service can increase the exposure and click-through rate of advertisements among consumers, and give advertisers more opportunities to reach consumers, its technologies are still developing and expected to make progress in application in the future [7]. In view of this, understanding customers' needs and considerations and further developing an effective LBA push strategy are essential for the popularization of LBA services in the future.

The purpose of this study is to explore the key factors customers would concern about while they decide whether to turn on Bluetooth on their mobile devices to receive LBA services. With multi-layered considerations, this decision problem is a typical multi-criteria decision making (MCDM) problem. Among those popular MCDM methods, the analytic hierarchy process (AHP) is commonly applied to solve decision-making methods with multiple evaluation dimensions and factors ([8–10]). However, the AHP method has several limitations. First, in real life, people normally do not rely on an absolute standard to compare and consider the priorities among various factors. To solve this concern, when using AHP to evaluate the initial data from surveys, it is suggested combining with the concept of fuzzy theory to avoid the problem of surveyed participants being over-subjective and extreme while considering the priorities among various factors [11]. Second, although AHP can find out the weights of each consideration factor on the overall decision-making problem, the considerations under different dimensions are independent. In other words, the AHP method cannot verify the existence of interrelationships between the consideration factors under different dimensions. Therefore, the decision-making trial and evaluation laboratory (DEMATEL) can find out the possible interrelationships between the factors in the evaluation hierarchy and calculate the degree of impact through the matrix operation method [12, 13].

In summary, this study uses Fuzzy AHP in conjunction with DEMATEL to explore the key factors customers would concern about while they decide to use Bluetooth on their mobile devices to receive LBA services and proposes suggestions to advertisers and businesses based on the analysis results.

2 Research Methodology

For the customers who are willing to receive the LBA, it is necessary to turn on the location-identifying function on their mobile device to allow businesses and advertisers to find the current location of customers more accurately. But not all customers would turn on Bluetooth just for receiving LBA. As a result, this study uses Fuzzy AHP and DEMATEL methods to analyze the decision-making factors involving customers turning on the Bluetooth function on the mobile device to receive the LBA, and the interrelationships and cross-impacts between factors.

2.1 Selection of Evaluation Dimensions and Factors

Based on consumption value theory [14], fashion theory [15], and the characteristics of location-based service [16], this study categorizes customers' considerations about turning on Bluetooth to receive LBA as two dimensions, which are "functional value" and "popularity value." In addition, since customers may try to use LBA services because of curiosity or the desires to get notices about store events in their nearby area, the dimension of "psychological value" is also included [17]. Meanwhile, to receive LBA, it is required that users' mobile devices are well supported by decent software and hardware to get the most instant personalized LBA. This may also cause the issue of leaking the user's current location and personal privacy information [18, 19]. The study also adds the dimensions of "mobile device support" and "privacy considerations." In total, five dimensions are evaluated in this study.

2.2 Establishment of Hierarchical Evaluation Framework

Based on the dimensions described above and the characteristics of the LBA service, this study selects a total of 18 consideration factors under dimensions, defines each factor, and establishes the evaluation hierarchy, as shown in Table 1.

2.3 Analysis Approach Operation

According to the evaluation hierarchy from Table 1, this study designs and distributes an AHP interview questionnaire, and applies the Fuzzy AHP method to analyze the importance weight of each consideration factor. After that, a DEMATEL questionnaire based on matrix operation method is distributed to participants to find out the interrelationships between the consideration factors and how much they

Table 1 Definition of evaluation dimensions and factors

Dimension	Factor	Definition
D1 Functional value	**F1.1** Get ads at right time and right place	Through LBA services, customers can access a variety of products/services/brand information provided by surrounding businesses, such as experience, trial, sales.
	F1.2 Get information about saving money at right time and right place	Through LBA services, customers can obtain various discounts/discounts/free information provided by surrounding businesses, such as coupons, free parking, promotions, card discounts, etc., to save money.
	F1.3 Time saving	Through LBA services, customers can save time searching for daily supplies information, such as planning a route to a nearby store or finding a local restaurant for dining.
	F1.4 Convenience for decision-making	Through LBA services, customers can have more precise choices that meet personal preferences and the highest cost performance (CP value) when making consumption decisions.
D2 Popularity value	**F2.1** Popularity	Customers use LBA service because it has been highly accepted by users and gets a large number of positive reviews.
	F2.2 Following social network	Customers use LBA service because many of their friends and family have used LBA service.
	F2.3 Showing personal style	Customers want to show their unique choice and personal style by using LBA service.
	F2.4 Following the trend	Customers want to experience fashionable and cutting-edge services by using LBA.
D3 Psychological value	**F3.1** Early adopting	Customers choose to use LBA as earlier adopters because they are curious about how they may receive advertisements at right time and place.
	F3.2 Exploration	Customers choose to use LBA because they want to explore and learn about unfamiliar shops, product/service experiences, and promotions around their location.
	F3.3 Perceived playfulness	Customers choose to use LBA because they expect to be able to obtain more information, videos, or interactive games from surrounding stores, thereby enhancing the pleasure and satisfaction of shopping experience.
D4 Mobile device support	**F4.1** Push stability	LBA provides customers with stable pushing and receiving advertisement services.
	F4.2 User-friendly interface	LBA provides customers with user-friendly and plain interactional surface for clicking advertisement.
	F4.3 Response time	LBA provides quick response to customers' clicks without taking much page loading time.
	F4.4 Battery durability	Customers' mobile devices have durable battery power that keep Bluetooth/GPS to consistently receive LBA.
	F4.5 Screen size	Customer's mobile devices come with ideal screen sizes which are able to well present the information from LBA.
D5 Privacy considerations	**F5.1** Personal preference privacy	Customers are concerned about leaking their shopping history related to promotion, purchased products, and visiting stores by using LBA.
	F5.2 Personal location privacy	Customers are concerned about leaking their present locations to stores through LBA.

Would You Turn on Bluetooth for Location-Based Advertising? 611

impact one another. By cross-checking the results from these two methods, we made some conclusions.

3 Analysis Results

3.1 Descriptive Statistics Analysis

The targeted survey participants in this study are college students who have been currently using LBA applications. A total of ten participants were interviewed. Five of them are male college student and the other five are female college students. In addition, the study also obtained participants' relevant experience and habits of using LBA services, as described in Table 2.

3.2 AHP Analysis

This study developed questionnaires based on AHP and interviewed participants by using 1–9 scale pairwise comparisons developed by Saaty [10]. Each interview result was reviewed to make sure that the C.I. and C.R. values of each question are less than or equal to 0.1 so that the responded answers are in accordance with

Table 2 Interviewees' experiences in using LBA

Item	Sub-items	Count	Percent
Motivations to turn on Bluetooth on mobile device	Get directions	3	30.0%
	Sending files	9	90.0%
	Search for stores	2	20.0%
	Connect accessories	6	60.0%
	Others	1	10.0%
Types of frequently used LBA applications	Food and restaurant	7	70.0%
	Tea parlor/coffee shop	8	80.0%
	Hotels	2	20.0%
	Transportations	1	10.0%
	Shopping	2	20.0%
	Others	1	10.0%
Frequency of using cellphone to receive LBA ads weekly	1–5 times	4	40.0%
	6–10 times	2	20.0%
	11–15 times	1	10.0%
	More than 15 times	3	30.0%

consistency. If a question fails the inconsistency test, the participant was requested to take the survey again and the responded answers were adjusted.

After all valid questionnaires are analyzed by Fuzzy AHP, the local fuzzy weights (LW) of each evaluation factors are obtained. Further, by multiplying the LW of the dimension levels and the factor levels, the global fuzzy weight (GW) of each consideration factor in the overall evaluation hierarchy can be calculated, and according to the weight value, the importance ranking of each consideration factors is created. The results of the analysis are shown in Table 3. It indicates that the GW of the top three key consideration factors (F5.2 "personal location privacy," F1.2 "get information about saving money at right time and right place," F5.1 "personal preference privacy") takes more than 40% of the overall GW. It can be seen that these evaluation factors have a great influence on the customers when considering whether to turn on Bluetooth on the mobile device to receive the LBA services.

3.3 DEMATEL Analysis

This study further explores the interrelationships and impacts between consideration factors through the DEMATEL method. This study first established a correlation matrix between consideration factors, as an interview questionnaire for the DEMATEL analysis, and then conducted a questionnaire interview for the selected ten respondents. The scale of the questionnaire is based on the 0–3 scale proposed by

Table 3 Fuzzy weight of dimensions and factors

Dimension	LW	Factor	LW	GW
D1Functional value	0.338 (1)	F1.1	0.138 (4)	0.047 (9)
		F1.2	0.460 (1)	0.156 (2)
		F1.3	0.210 (2)	0.071 (4)
		F1.4	0.192 (3)	0.065 (5)
D2Popularity value	0.059 (5)	F2.1	0.389 (1)	0.023 (14)
		F2.2	0.265 (2)	0.016 (15)
		F2.3	0.208 (3)	0.012 (17)
		F2.4	0.138 (4)	0.018 (18)
D3Psychological value	0.152 (4)	F3.1	0.414 (1)	0.063 (6)
		F3.2	0.285 (3)	0.043 (11)
		F3.3	0.301 (2)	0.046 (10)
D4Mobile device support	0.185 (3)	F4.1	0.192 (3)	0.036 (12)
		F4.2	0.333 (1)	0.062 (7)
		F4.3	0.255 (2)	0.047 (8)
		F4.4	0.151 (4)	0.028 (13)
		F4.5	0.069 (5)	0.013 (16)
D5Privacy considerations	0.266 (2)	F5.1	0.358 (2)	0.095 (3)
		F5.2	0.642 (1)	0.171 (1)

[20]. The larger the score is, the greater the mutual influence between each factors would be [21]. Finally, through a series of matrix calculation, the study compiles the results from all questionnaires and obtains a total influence-relation matrix (T) as shown in Table 4.

In addition, by summing the values from each column and row of the total influence-relation matrix (T), the sums of all columns (D_i value) and all rows (R_i value) pop out. The results of $D_i + R_i$ and $D_i - R_i$ are shown in Table 5.

Here, D_i represents the total influence of factor i on other factors, and R_i represents the total influenced intensity of factor i that is influenced by other factors. In addition, $D_i + R_i$ is called "prominence" that represents the sum of influences strength of factor i including both the strength influencing other factors and the influences from other factors. $D_i - R_i$ is called "relation" that divides factors into "cause group" and "effect group." If $D_i - R_i$ is positive, it means the factor i dispatches the influence to other factors more than it receives; thus, factor i belongs to the "cause group." On the other hand, if $D_i - R_i$ is negative, it means the factor i receives the influence from other factors more than it dispatched; thus factor i belongs to the "effect group" [21].

4 Conclusions

Identifying the locations of potential customers, mobile advertising can send useful advertisements to them. Many studies have indicated that it can effectively stimulate consumers' purchasing intentions [22]. However, for pushing the advertisements to customers accurately, the first condition is that they are willing to turn on the positioning service on their mobile devices. This study applies the FAHP and DEMATEL methods to analyze the consideration factors of customers while making the decision whether to turn on Bluetooth on mobile devices to receive the LBA service.

The analysis results generated in this study indicate that "Functional Value" (D1, weight 0.338) is the most important consideration dimension for customers while considering turning on Bluetooth to receive LBA services, followed by "privacy considerations" (D5, weight 0.266). As to ranking of consideration factors, "personal location privacy" (F5.2, weight 0.171), "get information about saving money at right time and right place" (F1.2, weight 0.156), and "personal preference privacy" (F5.1, weight 0.095) are the top three consideration factors. In addition, the results from DEMATEL analysis show that "exploration (F3.2)," "get ads at right time and right place (F1.1)," and "convenience for decision-making (F1.4)" are ranked as the top three consideration factors with highest prominence among all decision-making questions in this study.

According to the above analysis results, it can be found that when customers are willing to turn on Bluetooth to get LBA, one of their major consideration factors is to get chance of saving money anytime, anywhere. The study also finds that although "exploration" (F3.2) and "get ads at right time and right place" (F1.1)

Table 4 Total influence-relation matrix of each factor

	F1.1	F1.2	F1.3	F1.4	F2.1	F2.2	F2.3	F2.4	F3.1	F3.2	F3.3	F4.1	F4.2	F4.3	F4.4	F4.5	F5.1	F5.2
F1.1	0.158	0.178	0.143	0.174	0.164	0.151	0.126	0.186	0.221	0.211	0.168	0.175	0.145	0.148	0.132	0.076	0.172	0.188
F1.2	0.192	0.109	0.126	0.155	0.152	0.141	0.117	0.127	0.172	0.184	0.163	0.121	0.106	0.112	0.083	0.072	0.127	0.145
F1.3	0.147	0.111	0.081	0.121	0.112	0.117	0.106	0.134	0.134	0.124	0.103	0.111	0.105	0.131	0.076	0.058	0.080	0.103
F1.4	0.240	0.226	0.204	0.153	0.183	0.164	0.150	0.187	0.230	0.235	0.182	0.156	0.156	0.188	0.148	0.091	0.163	0.201
F2.1	0.137	0.109	0.079	0.131	0.088	0.114	0.098	0.124	0.132	0.136	0.106	0.114	0.094	0.100	0.054	0.034	0.114	0.116
F2.2	0.167	0.134	0.132	0.160	0.166	0.099	0.137	0.160	0.177	0.174	0.140	0.117	0.118	0.117	0.078	0.042	0.126	0.114
F2.3	0.152	0.113	0.118	0.146	0.153	0.136	0.091	0.168	0.191	0.173	0.140	0.103	0.091	0.088	0.059	0.037	0.135	0.115
F2.4	0.188	0.165	0.161	0.179	0.191	0.172	0.181	0.129	0.219	0.189	0.177	0.125	0.137	0.137	0.087	0.048	0.141	0.137
F3.1	0.168	0.134	0.144	0.166	0.158	0.167	0.159	0.174	0.136	0.195	0.145	0.108	0.107	0.101	0.069	0.042	0.152	0.167
F3.2	0.248	0.194	0.179	0.205	0.195	0.174	0.203	0.212	0.244	0.176	0.204	0.169	0.158	0.160	0.121	0.088	0.189	0.209
F3.3	0.190	0.173	0.149	0.186	0.190	0.151	0.153	0.171	0.218	0.216	0.122	0.140	0.139	0.132	0.096	0.064	0.128	0.147
F4.1	0.237	0.210	0.195	0.215	0.211	0.175	0.169	0.212	0.230	0.227	0.213	0.130	0.175	0.192	0.137	0.091	0.153	0.198
F4.2	0.206	0.176	0.171	0.181	0.184	0.165	0.153	0.166	0.192	0.183	0.160	0.137	0.099	0.162	0.107	0.094	0.127	0.153
F4.3	0.241	0.215	0.201	0.220	0.202	0.187	0.160	0.211	0.235	0.232	0.212	0.204	0.161	0.128	0.128	0.084	0.139	0.177
F4.4	0.278	0.247	0.202	0.232	0.226	0.209	0.176	0.222	0.256	0.260	0.231	0.208	0.162	0.218	0.105	0.133	0.161	0.214
F4.5	0.155	0.112	0.109	0.127	0.105	0.088	0.119	0.105	0.120	0.125	0.137	0.096	0.133	0.083	0.098	0.038	0.101	0.104
F5.1	0.218	0.193	0.160	0.173	0.206	0.179	0.182	0.194	0.229	0.219	0.186	0.132	0.116	0.129	0.099	0.065	0.113	0.152
F5.2	0.260	0.233	0.183	0.210	0.178	0.171	0.176	0.215	0.240	0.250	0.196	0.195	0.141	0.193	0.172	0.072	0.156	0.145

Table 5 Row and column operation in T matrix of each factor

Factor	D_i	R_i	$D_i + R_i$	$D_i - R_i$
F1.1	2.917	3.584	6.501	−0.666
F1.2	2.405	3.033	5.437	−0.628
F1.3	1.955	2.739	4.693	−0.784
F1.4	3.258	3.135	6.393	0.123
F2.1	1.880	3.063	4.943	−1.183
F2.2	2.357	2.763	5.119	−0.406
F2.3	2.209	2.657	4.865	−0.448
F2.4	2.764	3.098	5.862	−0.334
F3.1	2.492	3.577	6.070	−1.085
F3.2	3.328	3.508	6.836	−0.180
F3.3	2.765	2.985	5.751	−0.220
F4.1	3.370	2.540	5.910	0.830
F4.2	2.816	2.343	5.159	0.473
F4.3	3.338	2.517	5.855	0.820
F4.4	3.739	1.849	5.588	1.890
F4.5	1.956	1.230	3.186	0.726
F5.1	2.948	2.479	5.427	0.468
F5.2	3.388	2.785	6.173	0.603

are not ranked on the top of the importance ranking. However, they are highly prominent consideration factors in the whole evaluation framework. As a result, this study suggests that LBA advertisers could cooperate with various businesses and provide customers with LBA that contains activities of product/service experience, shopping promotion, and discount information from stores near customers' current location. In this way, customers might experience the shopping surprises when they use the LBA service. In addition, it can save customers' money, thereby stimulating consumption and increasing the profit for the businesses.

However, the analysis results also indicate that customers are concerned about giving out personal privacy information when considering turning on Bluetooth to use LBA services. This study suggests that LBA advertisers should not only design proper contents of advertisements, but also develop better strategies and implement fair practices to improve privacy protection, which might relieve the concerns of the customers while providing personal location information.

In addition, from Table 5, the DEMATEL analysis indicates that "battery durability" (F4.4) has the highest Di score; i.e., it would have the highest influence to other factors. It is noticed that many customers would turn off location positioning in order to save battery life. Thus, it is suggested that LBA advertisers could focus on more relevant and customized advertisements instead of pushing too many irrelevant advertisements to the users.

In the future, the evaluation hierarchy proposed by this study can be applied to analyze other customer groups with different work background and ages. In addition, it is expected to further explore the factors that may influence customers' decisions to click and watch advertisements after receiving LBA messages.

Acknowledgment The authors would like to thank the Ministry of Science and Technology in Taiwan, for financially supporting this research under contract No. MOST 107-2410-H-004-097-MY3.

References

1. H.L. Yang, S.L. Lin, Applying fuzzy AHP to analyse the critical factors of adopting SoLoMo services. Int. J. Mob. Commun. **17**(4), 483–511 (2019)
2. T.T. Lin, J.R. Bautista, Content-related factors influence perceived value of location-based mobile advertising. J. Comput. Inf. Syst. **60**(2), 184–193 (2020)
3. H.L. Yang, S.L. Lin, J.Y. Chang, Would you turn-on GPS for LBA? Fuzzy AHP approach. Paper presented at the 18th International Conference on Machine Learning and Cybernetics (ICMLC), Kobe, Japan, 6–10 July 2019
4. G.C. Bruner, A. Kumar, Attitude toward location-based advertising. J. Interact. Advert. **7**(2), 3–15 (2007)
5. P.E. Ketelaar, S.F. Bernritter, J. van't Riet, A.E. Hühn, T.J. van Woudenberg, B.C.N. Müller, L. Janssen, Disentangling location-based advertising: The effects of location congruency and medium type on consumers' ad attention and brand choice. Int. J. Advert. **36**(2), 356–367 (2017)
6. S. Leek, G. Christodoulides, Next-generation mobile marketing: How young consumers react to bluetooth-enabled advertising. J. Advert. Res. **49**(1), 44–53 (2009)
7. A.E. Hühn, V.J. Khan, P. Ketelaar, J. van't Riet, R. Konig, E. Rozendaal, N. Batalas, P. Markopoulos, Does location congruence matter? A field study on the effects of location-based advertising on perceived ad intrusiveness, relevance & value. Comput. Hum. Behav. **73**, 659–668 (2017)
8. V. Belton, *Multiple Criteria Decision Analysis: Practically the Only Way to Choose* (University of Strathclyde, Strathclyde Business School, Glasgow, 1990)
9. R.L. Keeney, H. Raiffa, *Decisions with Multiple Objectives: Preferences and Value Trade-Offs* (Wiley, New York, 1976)
10. T.L. Saaty, A scaling method for priorities in hierarchical structures. J. Math. Psychol. **15**(3), 234–281 (1997)
11. J.J. Buckley, Fuzzy hierarchical analysis. Fuzzy Sets Syst. **17**(3), 233–247 (1985)
12. Y.J. Chiu, H.C. Chen, G.H. Tzeng, J.Z. Shyu, Marketing strategy based on customer behaviour for the LCD-TV. Int. J. Manag. Decis. Mak. **7**(2–3), 143–165 (2006)
13. H. Tamura, K. Akazawa, Structural modeling and systems analysis of uneasy factors for realizing safe, secure and reliable society. J. Telecommun. Inf. Technol. **3**, 64–72 (2005)
14. J.N. Sheth, B.I. Newman, B.L. Gross, Why we buy what we buy: A theory of consumption values. J. Bus. Res. **22**(2), 159–170 (1991)
15. C.M. Miller, S.H. McIntyre, M.K. Mantrala, Toward formalizing fashion theory. J. Mark. Res. **30**(2), 142–157 (1993)
16. H.L. Yang, R.X. Lin, Determinants of the intention to continue use of SoLoMo services: Consumption values and the moderating effects of overloads. Comput. Hum. Behav. **73**, 583–595 (2017)
17. S. Lee, K.J. Kim, S.S. Sundar, Customization in location-based advertising: Effects of tailoring source, locational congruity, and product involvement on ad attitudes. Comput. Hum. Behav. **51**, 336–343 (2015)
18. J. Gu, Y. Xu, H. Xu, C. Zhang, H. Ling, Privacy concerns for mobile app download: An elaboration likelihood model perspective. Decis. Support. Syst. **94**, 19–28 (2017)
19. T.T.C. Lin, F. Paragas, J.R. Bautista, Determinants of mobile consumers' perceived value of location-based advertising and user responses. Int. J. Mob. Commun. **14**(2), 99–117 (2016)

20. E. Fontela, A. Gabus, Current perceptions of the world problematique, in *World Modeling: A Dialogue*, (North-Holland Publishing Company, Amsterdam/Oxford, 1976)
21. G.H. Tzeng, C.H. Chiang, C.W. Li, Evaluating intertwined effects in elearning programs: A novel hybrid MCDM model based on factor analysis and DEMATEL. Expert Syst. Appl. **32**(4), 1028–1044 (2007)
22. M.N.M. Noor, J. Sreenivasan, H. Ismail, Malaysian consumers attitude towards mobile advertising, the role of permission and its impact on purchase intention: A structural equation modeling approach. Asian Soc. Sci. **9**(5), 135–153 (2013)

Adaptive Chromosome Diagnosis Based on Scaling Hierarchical Clusters

Muhammed Akif Ağca, Cihan Taştan, Kadir Üstün, and Ibrahim Halil Giden

1 Introduction to Feature Extraction and Chromosome Classification on Massive Data

Healthcare analytical applications generate massive data including text/image/video in both batch and streaming contexts, which consist of a growing volume in petabytes for a single average-size city hospital that possesses micro-macro scale real-time tracking/monitoring devices. Furthermore, such hospital systems require tracking real-time responses both during diagnosis and historical analytics for trusted results. Therefore, the veracity of the growing datum is highly controversial, in which trust is an obligation to give vital decisions.

In this study, we unify the resources and apply real-time trusted analytical models just by focusing on scalable chromosome diagnostic models. Picture archiving and communication systems (PACS) are considered as storage data formats and HL7/similar ones to unify the resources. PACS are a prerequisite in data-exchange mission in modern hospitals to enable routing, retrieving, and storing medical images [1]. Specifically, images from radiology department such as X-ray, magnetic resonance imaging, and computed tomography are the main tasks of PACS systems to facilitate the workflow. Here, we integrate the chromosome karyotyping

---

M. A. Ağca (✉)
TOBB ETU/Computer Engineering, Ankara, Turkey
e-mail: akif.agca@etu.edu.tr

C. Taştan
Acıbadem Labcell Laboratory,İstanbul, Turkey
e-mail: cihan.tastan@acibademlabcell.com.tr

K. Üstün · I. H. Giden
TOBB ETU/Electronics Engineering, Ankara, Turkey
e-mail: k.ustun@etu.edu.tr; igiden@etu.edu.tr

© Springer Nature Switzerland AG 2021
H. R. Arabnia et al. (eds.), *Advances in Artificial Intelligence and Applied Cognitive Computing*, Transactions on Computational Science and Computational Intelligence, https://doi.org/10.1007/978-3-030-70296-0_45

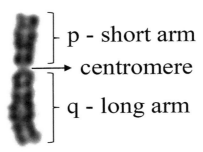

Fig. 1 Chromosome image with its geometrical features

resources, which are used for disease characterization, and our analytical models in a trusted manner to co-operate with other resources by considering the trustworthiness of the overall system for the vital decisions.

Chromosomes are organic structures found in the cell nucleus that carries genetic information. A healthy human cell includes 23 chromosome pairs (22 pairs of autosomes, classes 1–22; a sex genosome, either XX for females or XY for males). Microscopic images of the chromosomes are taken at the metaphase stage of the cell division for identification of chromosomes as well as their classification, i.e., karyotyping, since the chromosomes can be easily distinguished at that stage [2]. Genetic disease diagnosis at an early stage is an important task for proper medical treatment processes, and thus, morphological analysis of chromosomes becomes very crucial. Chromosomal abnormalities that cause genetic diseases can be exemplified as having an improper number of chromosomes (monosomy/trisomy), translocation, deletion, or inversion of chromosomes. Such morphological abnormalities of chromosomes could be associated with genetic diseases and, especially, with many cancer diseases [3].

In feature extraction systems, a reliable chromosome classification is possible if feature vectors are defined appropriately. The common features used in automatic chromosome classification algorithms are mainly geometrical and banding pattern-based features [4]: The length of the chromosome and its centromeric index (CI), a ratio of the short arm (p) with total length of the chromosome ($p + q$), are widely used geometrical features. As an example, an individual chromosome is given in Fig. 1 with its geometrical features $\{p,q\}$.

Efficient chromosome diagnosis systems require dynamic feature extraction and object detection mechanisms. Multiple-object detection is one of the challenging problems in computer vision application domains. Furthermore, real-time object images may not be sufficiently clear and separation of the objects from the background may usually not be adequate, which can be considered as other arising problems in computer vision systems. In order to circumvent such limitations, different object detection algorithms have already been developed in the literature [4–9]. Adaptive sub-modularity can be considered as one of prominent image processing techniques for the detection of multiple objects [10]. As a solution to dealing with large-scale data sets, a submodular function maximization method is proposed in a distributed fashion, whose effectiveness in massive data such as sparse

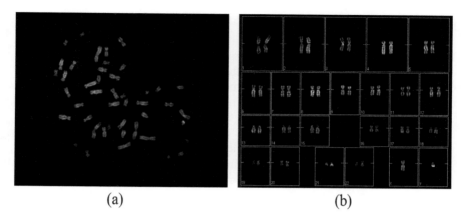

Fig. 2 (a) Typical Q-band chromosome images taken at metaphase of cell division and (b) their karyotyping [15]

Gaussian process inference and exemplar-based clustering is demonstrated in [11]. As additional types of chromosome classifiers, different algorithms have already been studied in literature such as artificial neural network-based classifiers [12], support-vector machine [13], and fuzzy logic–based classifiers [14].

In this study, a scalable and dynamic feature-extraction model is developed to manage the specified features in chromosome dataset. A distributed clustering system and a memory-centric analytical infrastructure are implemented for high-speed data processing of chromosome images, which can be built on the current PACS systems in hospitals. Every class of chromosomes in different cells is compared in terms of their features and with its homolog chromosome and the cell having potential genetic disorder is diagnosed, accordingly. The chromosomal features are firstly extracted from publicly downloadable (BioImlab) chromosome-image dataset [15], which includes 119 human cells with 5474 individual Q-band chromosome images. A typical chromosome cell at mesophase and its karyotyping is demonstrated in Fig. 2.

The remainder of the chapter is as follows: Sect. 2 discusses the scalable trusted computing technique briefly employed in this study for the classification process of the chromosomes. In Sect. 3, the feature extraction model is discussed in detail; the raw chromosome images, feature selection techniques, and the establishment classification models based on the features are presented. Finally, a conclusion is provided in the last section.

2 Proposed Chromosome Classification Model

In this study, a hierarchical multi-layer neural networks and tree structures are considered as a chromosome classifier that processes the genome data in hierarchical fashion. Each layer gathers the chromosome image features as an input data; automatically vectorizes and merges the matrices; analyses the feature of genes and encodes the data; and sends the feature sets to the next layer.

As universal health standardization, Health-Level Seven (HL7) information modeling is implemented for the genome analyses [16]. HL7 follows an object-oriented modeling technique that is organized into class diagrams (graphs) with attributes of classes. The extracted genomic data in our study will be transported in the distributed system via predefined encapsulating HL7 objects and the raw genomic data will be sorted out for the selection of specific genes at the memory-centric section of the system. Later, the feature extraction of the specified genes will be provided with the distributed system via training algorithms, and then, new genetic data will be classified comparing with formerly collected genetic features.

As a network communication model in HL7 standards, Open System Interconnection (OSI) reference model is used. As a hierarchical network architecture, the OSI model includes seven layers in a logical progression (see Fig. 3). Briefly stated, the seven OSI layers have the following assignments [17]:

- *Application layer* includes network software (user interface and application features) directly served to users.
- *Presentation layer* converts the information into specific formats (encryption, compression, conversion, and so on). In our case, the raw genomic data is processed.
- *Session layer* controls end-to-end connections between the applications in different nodes.
- *Transport layer* converts the received genome data from the upper layers into segments.
- *Network layer* is responsible for path determination and datagrams generation.
- *Data links* provides error control as well as "transparent" network services to physical layer.
- *Physical layer* enables direct communication with the physical media.

Comparing both TCP/IP and OSI reference models, TCP/IP has four layers—combining the application, presentation, and session layers into one top layer; taking both data-link and physical layers as a bottom (Network) layer (see Fig. 3).

Storing, printing, and transmitting of chromosome image information is handled by the Digital Imaging and Communications in Medicine (DICOM) protocol, which includes application protocol in TCP/IP model to communicate among the system. Each chromosome image is defined as a data object and the data exchange is conducted in image format. Complete encoding of medical data is standardized by DICOM protocol with attributes named as "DICOM data dictionary."

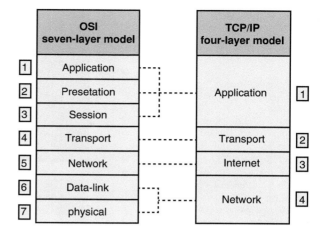

Fig. 3 Hierarchical architectures of OSI and TCP/IP reference models

DICOM stores the images from the archive, and if they are needed, DICOM provides association between service class users (SCUs) and service class providers (SCPs). Different protocols are also available for specific purposes like disease names, clinical context management, and hospital data acquisition aims. The current global standards can be exemplified as ICD (International Classification of Disease), NIST (National Institute of Standards and Technology), HL7 CDA (Clinical Document Architecture), CCR (Continuity of Care Record), CCOW (Clinical Context Management Specification), LOINC (Logical Observation Identifiers Names and Codes), ELINCS, EHR-Lab (Electronic Health Record - Lab), X12, SNOMED (Systematized Nomenclature of Medicine Clinical Terms), NCPDP (National Council for Prescription Drug Programs), IHE (Integrating the Healthcare Enterprise), CCHIT (Certification Commission for Healthcare Information Technology), HITSP (Healthcare Information Technology Standards Panel), CMMI (Capability Maturity Model Integration), ISO 27001, TS 13298, OHSAS 18001, Medical Data Interchange Standard (MEDIX), ASTM E1238, IEEE 1073 and ASC X12N. Furthermore, ASTM E31.17 and ASTM E31.20 standards are applied for data security and privacy. The standards are explored to manage the transactions in a scalable/trusted manner.

The studied feature extraction method will be described in detail in the following subsections.

2.1 Proposed Feature Extraction Method

For high-precision classification, it is very important to have best-characterized chromosome images. For that reason, better features should be selected to improve classification performance. It is also critical to note that well-described features

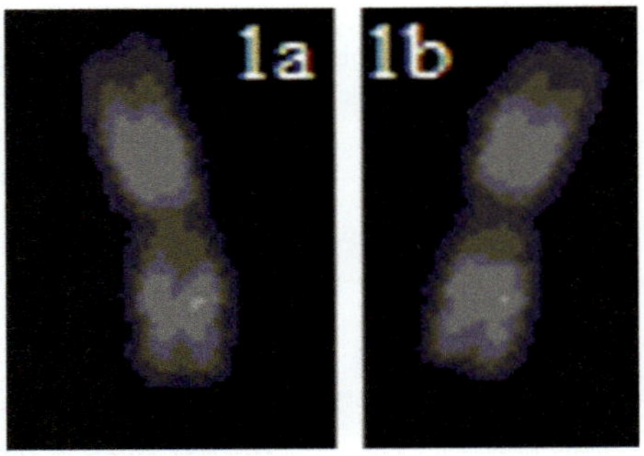

Fig. 4 Original image of
Chromosome 1 pair [15]

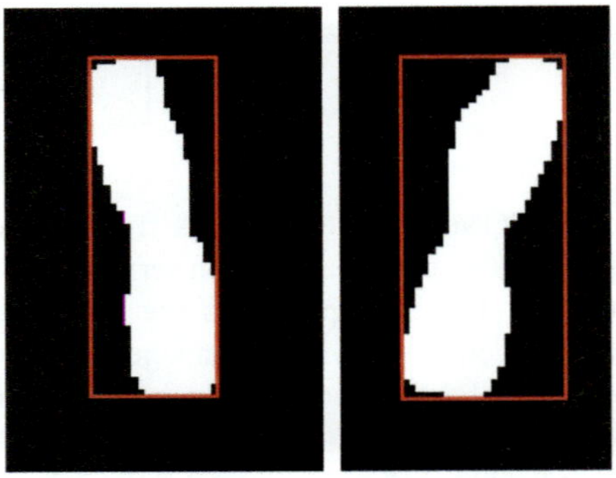

Fig. 5 Chromosome 1 pair
image (in Fig. 4) after
binarization. Solid rectangles
indicate the region above the
predefined threshold

reduce the processing workload and thus offer greater success of chromosome classification in less processing time.

2.1.1 Adaptive Thresholding to Individual Chromosome Images

An adaptive threshold value must be determined because the quality of each chromosome image may differ. A boundary detection-based thresholding is applied for efficient feature extraction. Chromosome 1 pair is chosen as an example for the application of the proposed feature extraction method (see Fig. 4).

2.1.2 Chromosomes-Image Binarization and Skeletonization

After thresholding with an adaptive value for each chromosome image as in Fig. 4, binarization of the image is conducted and the chromosome region is detected from the image (see Fig. 5). Later, a skeletonization algorithm is carried out for the binary version of the chromosome image, which can be considered as a landmark for the feature extraction (see Fig. 6).

Fig. 6 Skeletons of Chromosome 1 pair image (in Fig. 4) extracted by using the skeletonization algorithm

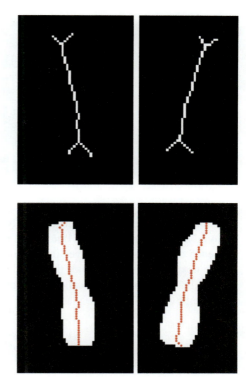

Fig. 7 Binary images of Chromosome 1 pair with its medial axes

2.1.3 Medial Axis Estimation of Chromosome Pair

Determination of medial axis is an essential task for the feature extraction of chromosome pairs. Medial axis estimation provides detection of centromere location and its indexing, chromosome length estimation, and density profile extraction [18]. For that purpose, the chromosome image in Fig. 5 is scanned with lines to determine boundaries of the binary image. The mid-points of every scanned line are found using a mid-point algorithm, which enables an accurate medial axis estimation. Estimated medial axes for the binary images in Fig. 5 are drawn and represented in Fig. 7.

2.1.4 Gray Level Extraction Along Medial Axis

After the determination of medial axes, a curve-fitting is performed with a 5th-degree polynomial function. Using this approach, the chromosome alignment can be mathematically derived. The next step is to resolve the chromosome image by slicing an equal number of perpendicular lines to the medial axes (see Fig. 8 for detailed representation). For that purpose, the norm of the polynomial curve is

Fig. 8 Polynomial curve-fitting of medial axes on the chromosome images and their slicing with perpendicular lines

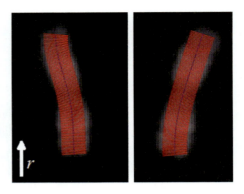

determined, and the curve is sliced equally for every adjacent perpendicular line. It should be noted that the solid lines are in parallel, and hence, they do not intersect with each other. The distance vector r is given as an inset in Fig. 8.

The next step is extraction of the gray levels through the norm of medial curve. To do this, a new parameter called Histogram, $H(r)$, of the chromosome image is calculated by the following equation:

$$h(r) = \sum_{k=1}^{n} \frac{\vec{g_k} \cdot \vec{u_k}}{n}, \qquad (1)$$

where $\vec{g_k}$ denotes the gray-level vector along the perpendicular line $\vec{u_k}$ and n is the number of perpendicular lines, viz. pixels. Corresponding histogram function dictates the average density of pixels along the medial axis. The $h(r)$ calculation is conducted for the chromosome pair in Fig. 4 and plotted in Fig. 9. Some important remarks could be gathered from the histogram plot: (1) End-points of the chromosome pairs could be inferred from the histogram plot; (2) centromere position, which is the lowest point of the curve, could be precisely determined by using the histogram plot. In this case, corresponding centromere-index (CI) could also be calculated from Fig. 9, which is found to be CI = 0.46 for chromosome 1a and CI = 0.49 for chromosome 1b.

2.1.5 The Feature Extraction from Histogram Information

As a common shape signature extraction of image data, frequency response calculation is implemented in image processing systems [19]. Therefore, discrete Fourier transform (DFT) of the histogram function is taken as a feature extractor $H(f)$ for precise chromosome classification. The feature extractor function is calculated via the following equation:

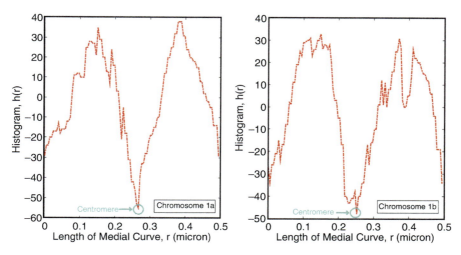

Fig. 9 Histogram function calculation along the length of the medial curve

$$H(f) = \sum_{k=1}^{n} h_k(r) e^{-i2\pi kr/n}. \tag{2}$$

Absolute values of $|H(f)|$ is plotted in terms of frequency (see Fig. 10). There are two critical points to infer from Eq. (2): The centromere location, i.e., CI of the chromosome, directly influences the frequency of the main peak, which is found to be 4 Hz for Chromosome 1a whereas that value equals 4.02 Hz for Chromosome 1b (see Fig. 10). It is important to note that the reason of being the two frequencies nearly the same is due to the nearly equal values of CI. Another important remark can be inferred while comparing Figs. 9 and 10 that the average value of histogram function $h(r)$ affects the absolute value of corresponding feature extractor, $|H(f)|$.

As an alternative, the feature extraction model can be extended by using hyperspectral images of the chromosomes. Namely, chromosomes images under different wavelengths of illumination can give us additional information for the diagnosis of disease carrying/faulty chromosomes. It is known that biological molecules and cells have footprints in the mid-infrared and terahertz parts of the electromagnetic spectrum [20]. Combining the image and spectral data at these wavelengths would improve the performance of the classification algorithm significantly. We discuss further improvements on feature extraction and classification in detail in Sect. 2.3.

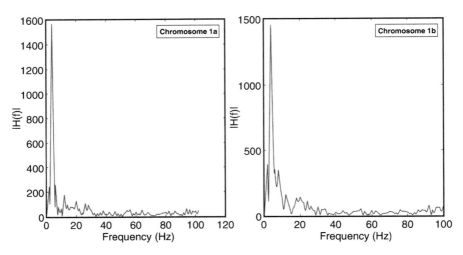

Fig. 10 DFT function $|H(f)|$ as a feature extractor calculation for the Chromosome 1 pair in Fig. 4

2.2 Chromosomal Abnormalities–Related Disease Diagnosis by Spectral Chromosome Information and Other Novel Features

Humans have 23 pairs of chromosomes which have various numbers of genes and differ in their structures and sizes (i.e., male Y and female X chromosome). Structural and numerical chromosomal alterations are identified in ~0.6% of births [21], which frequently cause developmental disabilities, mental retardation, birth defects, and dysmorphism [22]. Many genetic or rare diseases are caused by the chromosomal abnormalities, which are diagnosed with a determined gain or loss of some genomic materials coding hundreds of genes. Unlike single-gene mutations, the gain or loss in chromosome complexes directly influences gene expression doses which can cause imbalances in phenotype and human body functions.

Alternative abnormalities can be microdeletion or microduplication of chromosomes, all of which can be identified via conventional visual analysis of chromosomes under microscope (karyotyping). In contrast to numeric chromosomal abnormalities, which occur when three copies of chromosome present rather than two such as Down syndrome (trisomy of chromosome 21), structural abnormalities are caused from the breakage or rejoining of chromosome arms. This can result in deletions of chromosome segments.

The chromosomal alterations can cause the imbalances in a contiguous gene syndrome, in which multiple genes are negatively influenced, resulting in combinatorial abnormalities in clinical phenotype [23]. On the other hand, most of the chromosomal deletion syndromes (such as Williams syndrome, Miller-Dieker syndrome and DiGeorge syndrome) result in haploinsufficiency of a gene or genes in the lost segment where single copy of other allele gene does not express sufficient doses of the gene product for a normal and healthy state phenotype [22]. Chromosomes can be visualized as lighter and darker bands which are considered as sections/segments following certain staining protocols and classical microscopy.

However, the newly developed spectral chromosome technologies enable the characterization of the disorders in a high throughput manner via super-resolution visualization with lab-on-a-chip (LOC) microfluidics and microscopy in a real-time fashion. The chromosome imaging and spectral analysis platform can address disease-related abnormalities with respect to healthy and homolog chromosomal information. Therefore, any subtle alterations or rearrangements of the segments in a chromosome (micro-level deletions or reunions syndromes) can be easily imaged and detected using high-resolution chromosome spectral analysis, enabling the identification of unbalanced abnormalities that are invisible under conventional microscopy technique.

The use of high-resolution LOC spectral chromosome analysis can allow cytogenetic field both to diagnose the pre-characterized genetic syndromes and to identify new disorders. With an automated chromosome spectral analysis through an adaptive optimization method of karyotyping schemes, numerical or structural abnormalities in chromosomes can be detected highly efficient and robust manner while diagnosing cancers and other genetic disorders. For instance, in diagnosed chronic myeloid leukemia (CML) patients, abnormal pattern named t (9;22) chromosomal translocation is identified in one chromosome 9 and one chromosome 22 [24].

A study reported a developed computerized scheme to automatically identify the unbalanced alteration in one chromosome 22 of abnormal cells in CML patients by analyzing and image-processing karyotypes of metaphase cells from bone marrow patient specimens [25]. Extracting novel spectral features for real-time diagnostic requires using maximum available computational resources in minimum physical space. The minimization of physical space can be achieved via various innovations such as 3D-Stack packaging OBCs and ASICs including micro-channel cooling/transfer, nano-sensing, custom processor CPU/GPU/FPGA on a single board, which we discussed in the last chapter.

Extracted genetic disorder features can be efficiently characterized and diagnosed using a genetic algorithm, artificial neural networks and statistical models along with graph construction for genetic syndrome-related chromosomal spectral information, extended data flow blocks and computational units' path tracking like in biological system design tool, BioGuide [26]. The software tool standardizes genetic elements and provides graph data structures of all possible genetic element interactions to allow molecular biologists to estimate and characterize phenotypically functional genetic circuits. BioGuide constructs a massive graph structure $G = (V,E)$, where $V = \{g_1, g_2, g_3, \ldots, g_n\}$ each element is a part of genetic circuit. $E = \{e_1, e_2, e_3, \ldots, e_m\}$ each edge $e_m = \{g_s, g_t\}$ represents the possible connections between the elements, which build a genetic circuit. In this way, complex sequence relations can be modelled effectively.

Nevertheless, tracking the complex sequences requires scalable data flow and transaction flow management in a trusted manner. Both analytical models and the system are needed to scale at a massive level. As illustrated in Fig. 14, the transaction is managed by layered system components. Scaling the models and the system requires memory-centric analytical approaches, which is tracking the lineage of datum, transaction, and the models. In another study, we proposed a Memory Centric Analytics (MEMCA) system that provides a holistic abstraction to ensure trusted scaling and memory speed trusted analytics. Initial results are shared in [27] MEMCA used for satellite and space data. The results evidenced the potentials for spectrum analysis of satellite and space datum. Thus, the use of MEMCA approach for other kinds of high-resolution spectral imagery data may have great potential for novel studies.

The proposed scalable analytical model for chromosome diagnosis can be utilized in cytogenetic fields as well to diagnose the pre-characterized genetic syndromes and to identify new disorders, which will be further discussed in the next section. The system architecture will be explored in Sect. 2.3 to track the complex sequences and to apply the analytical models on the massive data at memory speed in a trusted manner.

2.3 Genetic Disorder Detection via Spectrum Analysis and Scalable Multi-layer Neural Networks

In this study, an open-public data set is [15] used for single-chromosome images of different human cells. The chromosomes of the same class are analyzed with the abovementioned spectral analysis approach in the cases of different cells, and the obtained FFT spectra are compared altogether. Important outcomes are gathered from the spectral information comparison: The shape of chromosome as well as density profile of genes directly influences the spectral feature. For that reason, comparing the FFT spectral feature enables to detect corresponding genome diseases, chromosome diagnostic, and so on. Spectral feature extraction could be performed via OBCs and ASICs (2D/3D Stacks) including nano-sensor plates, which will be discussed later in detail.

Combining the image and spectral data from different types of measurement systems would improve the performance of the classification algorithm significantly. Specifically, the feature extraction stage can be extended by using hyper-spectral images and other spectroscopic data related to chromosomes. Namely, images of the chromosomes under different wavelengths of illumination can give us additional information for the diagnosis of decease carrying/faulty chromosomes. It is known that biological molecules and cells have foot-prints in the mid-infrared and terahertz parts of the electromagnetic spectrum [20].

Another option is to use Raman spectral response of chromosomes to gather extra features for the classification algorithm. Indeed, Ojeda et al. [28] have

Fig. 11 A simple analytical model of neurons as a binary threshold unit

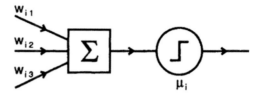

managed to differentiate the Raman spectroscopy response of three different types of chromosomes by optical tweezing and exposing the chromosomes to a laser beam; an interesting method for differentiating colon cancer cells is just measuring the dry mass of chromosomes by tomographic phase microscopy (actually the refractive index distribution is measured) [29].

Another important method which is very beneficial for analysis of chemical and biological substances is Fourier transform infrared spectroscopy (FTIR). FTIR is already proposed for the diagnosis of cancer [30]. In one of the studies, FTIR is used for comparing the chromosomes taken from breast cancer cells and healthy cells [31]. As another genetic diagnostic imaging approach, quantum dots such as gold nanoparticles are used for in vivo bio-imaging of subcellular organelles as well as genetic diagnosis [32]. Nanowires are another alternative platform for efficient label-free nano-sensing of DNAs and selective electrical detection of genetic disorders [33]. Graphene-based bio-devices can also be considered as a powerful biosensor for monitoring the chromosomes' behaviors as well as extracting their electro-chemical features [34].

The datum, including the features, has varying resources and computational complexity/cost increases the complexity of analytical model and training process. We define a multi-layer neural network to classify and embed the new features dynamically obtained from available data sources. Neural computation is an efficient method to represent complex sequences inspired from neuroscience. Analytically, a neuron can be simply modeled as a binary threshold unit, w: Each neuron model computes the weighted sum of its inputs from other connected units and outputs either "0" or "1" according to the threshold level, μ (see Fig. 11). The output state of neurons, n, is calculated via the following formula:

$$n_i(t+1) = \Theta\left(\sum_j w_{ij} n_j(t) - \mu_i\right). \qquad (3)$$

Here, Θ is a unit step-function representing the following characteristic:

$$\Theta(x) = \begin{cases} 1, & \text{for } x \geq 0 \\ 0, & \text{elsewhere.} \end{cases} \qquad (4)$$

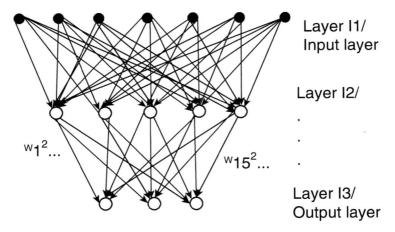

Fig. 12 Illustration of MNNs intra-node and intra-layer interaction

The output state n_i becomes either "1" (*firing* state) or "0" (*not firing* state) depending on the inputs as well as the threshold level. In this way, real neurons could be translated as computing units since the output states are described as binary units [35].

In reality, it is not as simple as in Eq. (3) to determine the behavior of real neurons since (1) the input/output relationships are nonlinear, (2) neurons produce continuous output pulses rather than a simple output level, and (3) neurons do not works synchronically as Eq. (3) states. For that reason, the output state, n_i, should be defined as continuous-valued function like the following formula:

$$n_i = g\left(\sum_j w_{ij} n_j - \mu_i\right). \qquad (5)$$

In this equation, a new function g called activation function is added to the neural system.

Human brain itself can be described as a parallel system of billions of processors (neurons), which are highly connected into each other via synapses and operate simultaneously. Therefore, neural networks can be considered as an efficient way for parallel processing of massive data. For that purpose, neural network systems are described as multi-layer architecture (perceptron) to make applicable in computational networks (see Fig. 12).

In the studied feature extraction model, each chromosome is taken as a neuron and operates as a feature vector generator in the multi-layer neural network system. The implemented image feature classifiers such as binarization, skeletonization, gray-level and medial axis detection, thresholding, histogram, and DFT vectors are set to be multi-layers of the proposed neural network system. At every layer, the

Adaptive Chromosome Diagnosis Based on Scaling Hierarchical Clusters

activation function g is adaptively selected to infer spectral chromosome feature for real-time diagnostic purposes.

Each neuron takes as input w_{i1}, w_{i2}, w_{i3}, $\ldots$, w_{in}, feature sets and outputs an activation function $g(.)$, triggering the function according to μ_i threshold value. As illustrated in Fig. 12 perceptron/multi-layer architecture is defined to denote layer-wise implementation. There are L number of scalable layers denoted as $L = \{l_1, l_2, \ldots, l_n\}$. Connection between layers is denoted as $w_{ij}^{(l)}$, where unit j in layer l is connected with unit i in layer $l + 1$. The activation of unit i in layer l is denoted as $\mu_i^{(l)}$. Activation function computation is as follows:

$$\mu_1^{(2)} = g\left(w_{11}^{(1)} + w_{12}^{(1)} + w_{13}^{(1)} + \ldots\right)$$
$$\mu_2^{(2)} = g\left(w_{21}^{(1)} + w_{22}^{(1)} + w_{23}^{(1)} + \ldots\right)$$
$$\mu_3^{(2)} = g\left(w_{31}^{(1)} + w_{32}^{(1)} + w_{33}^{(1)} + \ldots\right)$$
$$\ldots$$

Each layer is trained dynamically; training sets, test sets, and activation functions are updated according to upcoming extracted features. Below is generalized sudo-code for the multi-layer neural network building/training process.

1. **Set** initial feature sets $w_{ij}^{(l)} = \mathbf{0}$ for each neuron and all l
2. Embed new features to each neuron
3. Extract new features with CNN/RNN
4. For $i = \mathbf{1}$ to n

    1. Update activation function $g(.)$ for each $w_{ij}^{(l)}$.
    2. Update features $w_{ij}^{(l)}$ for each neuron
    3. Update output state n_i for each neuron

    Repeat until $\sum_j w_{ij}^{(l)}$ of each neuron $>= \mu_i$

Activating the functions $g(.)$ in appropriate layer, hidden layer or upper ones, improves the training performance and decreases computational cost notably. Reference [36] is discussing the placement of activation functions, recommending the replacement of activation functions of lower layers with MNNs in a limited computing resources case.

To extract new features based on the available sets, limit the interaction between the neurons to computationally feasible level, we restrict the connections between hidden layers inter/intra layers, and define stimulus locations for the images and activation functions. Layer-wise implementation of multi-layer neural networks enables efficient abstraction of stimulus features like centromere indexes, skeleton, polynomial curves, histograms, discrete Fourier transform (DFT), and additional spectral features we pursue research. Convolutional/subsampling layers are defined in the fully connected MNN structure; 2D structural advantage of images/signals is used with the local connections.

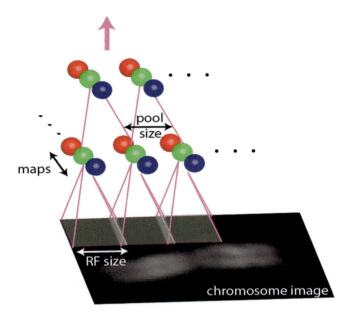

Fig. 13 First layer of CNN with mean/max pooling

The image sets are given as input to convolutional layer as $m*m*r$ image, where m = height/width, and $r = 3$, RGB channels in this set (see Fig. 13) for the illustration of CNN multi-layer system architecture. Other channels with spectral and signal features can also be added into the system modeling. There are kernels/filters k have size $n*n*q$, where the size of kernel $n < m$ and $q \leq r$. Kernel size raises the locally connected structure, convolved with the image to produce k feature maps with the size of $m - n + 1$. The maps are sub-sampled with mean or max pooling over $p*p$ contiguous regions. Size of pool is variant, $2 \leq p \leq 5$ up to size of image, greater for larger input images. Bias and sigmoidal nonlinearity is applied to each feature map, before or after subsampling layers. After the convolutional layers, there may be any number of fully connected layers.

The implemented neural network system needs a training feature set $\{t_1, t_2, \ldots, t_n\}$; in our case, t_n denotes known image features of either healthy chromosomes or with genetic diseases. A cost function can be described with the following equation to prevent possible over-fitting conditions:

$$J_n^{(l)} = \left\| \mu_n^{(l)} - t_n \right\|^2. \tag{6}$$

Equation (6) implies that the implemented CNN model tries to converge the output feature set to the training set by minimizing the differentiation between the two sets and thus reducing the overall cost function $J_n^{(l)}$. The train and output feature datasets/matrices are vectorized dynamically and served to a scalable memory-

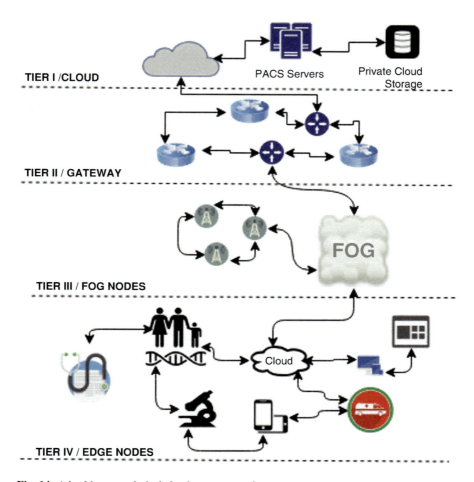

Fig. 14 A healthcare analytical cloud system overview

centric clustering algorithm, which is implemented at the output and training feature datasets in a distributed and trusted manner. In this way, new anomaly detection and classification is provided with the help of disjoint computational nodes, and hence, the requested genetic feature comparison can be conducted in real time with faster data processing. Further hierarchical system architecture is schematically represented in Fig. 14.

Placement of activation functions is a non-deterministically triggering computational resource. We implemented our memory-centric analytical approach for trusted analytics as illustrated in Fig. 14. MEMCA is a holistic abstraction to ensure trusted scaling and memory speed trusted analytics [37]. Using Markovian chains and distributed ledger-based structures is an effective way to keep the data sources fresh and to execute the transactions trustfully. Managing non-linearly growing transaction in trusted manner ensures scalability and elasticity of the system as much as possible up to currently available hardware and physical constraints. Trusted execution of the system enables trusted scaling of algorithms in real time. We extend the analytical models in trusted manner via the MEMCA abstraction. Large feature sets/matrices are dynamically embedded to the growing models. The transactions including private data are verified with checksum values at appropriate check-pointing locations. Thereby, the integrity between cloud, gateway, fog nodes, and edge nodes layers is ensured. The proposed system is highly scalable and trusted to manage the growing datum and emerging edge devices to extract new features like spectral data and other custom ASIC and OBC devices.

ASIC designs congregate different types of detectors and processing units. These detectors may belong to different physical domains such as pressure, heat, and electromagnetics. Moreover, electromagnetic wave detectors can also be classified according to the wavelength (frequency) band of operation, lying between microwave regime and up to ultraviolet and X-rays. Especially, infrared and terahertz bands are extremely useful for the characterization of chemical and biological samples due to their specific footprints [38]. Their integration to ASIC systems is very important in that respect. Infrared detectors are mostly based on semiconductor technology and hence their integration to the ASIC integration is not very challenging, except in some cases where the detectors should be cooled down to ultra-low temperatures.

In the terahertz detector case, there are many different detector types with different operation principles. Here we are interested in the photoconductive antennas, because of its simple design and low cost. These antennas would occupy a planar region in the order of the wavelength of operation (e.g., 1 THz corresponds to a wavelength of 300 μm). However, a lens system and a detector array may be needed for far-field imaging systems, which can be useful for defense and testing. The size of the detector array and the lens may differ according to the target resolution and image size, where increasing the size may show some integration problems to the ASIC architecture.

On the other hand, such far-field detection systems may become separated from the ASIC system and a communication link may be established for the control and date transfer between these modules. 3D-Stack packaging enables novel miniaturized OBCs and ASICs into highly space-efficient architectures, where the micro-channel cooling/transfer can remedy any extreme heat generation and hot-spot issues (references above). The space efficiency can be further improved by nano-sensing, custom processor CPU/GPU/FPGA/Storage on a single board, and memory speed compact storage given the available physical limits [39]. The novel micro-machines packaged with the innovative method is handling computational bottlenecks in limited physical space and enabling to extract novel features for real-time diagnostic applications. We develop customs ICs for specific purposes up to requirements, which will be discussed in detail in another study.

The datum/transaction flowing in the system is verified with periodical check-sums at available check-pointing locations to enhance overall trustworthiness of the system. Lineage data enhances fault recovery and decreasing up-time to milliseconds in case of any failure, which enables memory speed scalable/trusted analytics on the massive datum processed by thousands of transactions in real time.

3 Conclusions

Chromosomal abnormalities are identified using conventional microscopy techniques after classic karyotyping protocols. Fast, efficient, and trustable bioinformatical tools can help genetic diagnosis laboratories to characterize genetic disorders with a high-throughput manner. Here, we developed an adaptable and dynamic feature-extraction model, which utilizes scalable and hierarchical chromosomal data sets. Chromosomal alterations include numerical (i.e., trisomy) or structural rearrangements (i.e., micro-deletions or rejoining) of chromosomes, which can be identified by comparing with healthy homolog chromosome. Our study characterized chromosomal structures by spectral analysis in detail, where solid rectangles, skeletons, medial axes, polynomial curve-fitting of medial axes, length of the medial curve, and DFT functions of regions of the chromosome pair were extracted. This deep information of the chromosomal spectral analysis for a chromosome pair provides a massive data set for an efficient characterization of a chromosome pair. Different soft computing techniques are developed to classify chromosomes based on the extracted features using the Copenhag data base.

We further improved our classification and feature extraction algorithms with multi-layer networks and convolutional neural networks. The proposed classification system is compatible with HL7 standards and efficient for massive data acquisition and storage as well as real-time data-stream processing. It is also compatible with spectral analyses and edge computing methods, which could be considered as another superiority of the proposed chromosome classifier tool. The proposed system operates via a distributed and scalable algorithm, which provides elasticity on the number of edges/clients.

Placement of activation functions differentiates the behavioral pattern of computational resource triggering non-deterministically. We implemented our MEMCA (memory-centric analytical) abstraction for trusted analytics so that chromosome feature data is transferred to clients in a trusted manner and the behavioral pattern of computational resources is optimized most efficiently up to the resource requirements. Initial results indicate a promising performance of trusted scaling with the trust metrics to ensure the trustworthiness of the overall system. Furthermore, ASICs and OBCs could be developed to extract new electrochemical features and identified in the case of using nanoscale sensors based on nanomaterials like graphene and quantum dots, which is the future direction of our study besides from memory speed trusted and scalable analytic models of the massive datum.

Acknowledgment Thanks to TOBB University of Economics and Technology Distributed Data Analytics Research Laboratory and IBM for providing test clusters and research infrastructure. Thanks to industrial collaborators Dr. Bruno Michel and Dr. Atakan Peker for discussions about the 3D-Stack Packaging and ASIC design technologies. Thanks to *YONGATEK Embedded Systems* (https://yongatek.com/) *and SilTerra* (https://www.silterra.com/index.php#homepage) *Incorporations* for partial funds for the research studies and proof of concept (POC) implementation supports.

References

1. Y.X. Ho, Q. Chen, H. Nian, K.B. Johnson, An assessment of pharmacists' readiness for paperless labeling: A national survey. J. Am. Med. Inform. Assoc. **21**(1), 43–48 (2014)
2. S. Jahani, S.K. Setarehdan, Centromere and length detection in artificially straightened highly curved human chromosomes. Int. J. Biomed. Eng. **2**(5), 56–61 (2012)
3. X. Wang, B. Zheng, M. Wood, S. Li, W. Chen, H. Liu, Development and evaluation of automated systems for detection and classification of banded chromosomes: Current status and future perspective. J. Phys. D. Appl. Phys. **38**, 2536–2542 (2005)
4. J.M. Cho, Chromosome classification using back propagation neural networks. IEEE Eng. Med. Biol. **19**, 28–33 (2000)
5. B. Lerner, Towards a completely automatic neural-network-based and human chromosome analysis. IEEE. Trans.. Syst. **28**(4), 544–552 (1998)
6. M. Moradi, S.K. Setarehdan, S.R. Ghaffari, Automatic landmark detection on chromosomes' images for feature extraction purposes, in *Proceedings of the 3rd International Symposium on Image and Signal Processing and Analysis*, (2003), pp. 567–570
7. M.R. Mohammadi, Accurate localization of chromosome centromere based on concave points. J. Med. Signals Sens. **2**(2), 88–94 (2012)
8. E. Poletti, E. Grisan, A. Rugger, A modular framework for the automatic classification of chromosomes in Q-band images. Elsevier Comput. Methods Prog. Biomed. **105**, 120–130 (2012)
9. N. Madian, K.B. Jayanthi, Analysis of human chromosome classification using centromere position. Measurement **47**, 287–295 (2014)
10. Y. Chen et al., Active detection via adaptive submodularity, in *ICML*, (2014)
11. B. Mirzasoleiman, A. Karbasi, R. Sarkar, A. Krause, Distributed Submodular Maximization: Identifying Representative Elements in Massive Data. In NIPS. (2013, December). pp. 2049-2057
12. S. Rungruangbaiyok, P. Phukpattaranont, Chromosome image classification using a two-step probabilistic neural network. J. Sci. Technol. **32**(3), 255–262 (2010)
13. A.S. Arachchige, J. Samarabandu, J.H.M. Knoll, P.K. Rogan, Intensity integrated Laplacian-based thickness measurement for detecting human metaphase chromosome centromere location. I.E.E.E. Trans. Biomed. Eng. **60**(7), 2005–2013 (2013)
14. H. Choi, K.R. Castlman, A.C. Bovik, Segmentation and fuzzy logic classification ofM-FISH chromosomes images, in *Proceedings of Image Processing, 2006 IEEE International Conference*, (2006), pp. 69–72
15. BioImlab [Available on line 23 Sept 2019] http://bioimlab.dei.unipd.it.
16. HL7 [Available on line 23 Sept 2019] https://site.hl7.org.au/standards/international-published/. http://www.hl7.com.au/HL7-Tools.htm.
17. D. Wetteroth, *OSI Reference Model for Telecommunications*, vol 396 (McGraw-Hill, New York, 2002)
18. F. Abid, L. Hamami, A survey of neural network based automated systems for human chromosome classification. Artif. Intell. Rev. **49**(1), 41–56 (2018)
19. S. Prakash, N.K. Chaudhury, Dicentric chromosome image classification using Fourier domain based shape descriptors and support vector machine, in *Proceedings of International Conference on Computer Vision and Image Processing*, (Springer, Singapore, 2017)
20. S.S. Dhillon et al., The 2017 terahertz science and technology roadmap. J. Phys. D. Appl. Phys. **50**(4), 043001 (2017)

21. L.G. Shaffer, J.R. Lupski, Molecular mechanisms for constitutional chromosomal rearrangements in humans. Annu. Rev. Genet. **34**, 297–329 (2000)
22. A. Theisen, L.G. Shaffer, Disorders caused by chromosome abnormalities. Appl. Clin. Genet. **3**, 159–174 (2010). https://doi.org/10.2147/TACG.S8884
23. R.D. Schmickel, Contiguous gene syndromes: A component of recognizable syndromes. J. Pediatr. **109**(2), 231–241 (1986)
24. P.C. Nowell, D.A. Hungerford, A minute chromosome in human chronic granulocytic leukemia. Science **142**, 1497 (1960)
25. X. Wang, B. Zheng, S. Li, J.J. Mulvihill, X. Chen, H. Liu, Automated identification of abnormal metaphase chromosome cells for the detection of chronic myeloid leukemia using microscopic images. J. Biomed. Opt. **15**(4), 046026 (2010)
26. M.A. Agca, C. Tastan, et al., Biological system design tool, application paper, in *iGEM*, (2011)
27. M.A. Ağca, E. Baceskı, S. Gökçebağ, MEMCA for satellite and space data: MEMCA [Memory centric analytics] for satellite and space data, in *Recent Advances in Space Technologies (RAST), 2015 7th International Conference on*, (IEEE, 2015)
28. J.F. Ojeda et al., Chromosomal analysis and identification based on optical tweezers and Raman spectroscopy. Opt. Express **14**(12), 5385–5393 (2006)
29. Y. Sung et al., Stain-free quantification of chromosomes in live cells using regularized tomographic phase microscopy. PloS One **7**(11), e49502 (2012)
30. P.D. Lewis et al., Evaluation of FTIR spectroscopy as a diagnostic tool for lung cancer using sputum. BMC Cancer **10**(1), 640 (2010)
31. D.J. Lyman, J. Murray-Wijelath, Fourier transform infrared attenuated total reflection analysis of human hair: Comparison of hair from breast cancer patients with hair from healthy subjects. Appl. Spectrosc. **59**(1), 26–32 (2005)
32. P.C. Chen, S.C. Mwakwari, A.K. Oyelere, Gold nanoparticles: from nanomedicine to nanosensing. Nanotechnol. Sci. Appl. **1**, 45 (2008)
33. J.-i. Hahm, C.M. Lieber, Direct ultrasensitive electrical detection of DNA and DNA sequence variations using nanowire nanosensors. Nano Lett. **4**(1), 51–54 (2004)
34. L. Feng, L. Wu, Q. Xiaogang, New horizons for diagnostics and therapeutic applications of graphene and graphene oxide. Adv. Mater. **25**(2), 168–186 (2013)
35. J.A. Hertz, *Introduction to the Theory of Neural Computation* (CRC Press, Boca Raton, 2018)
36. W. Sun, S. Fei, L. Wang, Improving deep neural networks with multi-layer maxout networks and a novel initialization method. Neurocomputing **278**, 34–40 (2018)
37. M.A. AGCA, A holistic abstraction to ensure trusted scaling and memory speed trusted analytics, in *Developments in eSystems Engineering (DESE), 2018 Eleventh International Conference on*, (IEEE, Cambridge University, 2018)
38. B. Stuart, *Infrared Spectroscopy* (Wiley, New York, 2005); P. Jepsen, D. Cooke, M. Koch, Terahertz spectroscopy and imaging—Modern techniques and applications. Laser Photon. Rev. **5**(1), 124–166, (2011)
39. Y. Yin et al., 3D integrated circuit cooling with microfluidics. Micromachines **9**(6), 287 (2018)

Application of Associations to Assess Similarity in Situations Prior to Armed Conflict

Ahto Kuuseok

1 Introduction

Very often decision-makers while planning for actions in critical situations that are happening, for example, in process of business process management or in military mission control are relaying on experience.

Therefore, when it comes to a new situation, it is worth looking at whether similar situations have occurred in the past, how were decisions made and how the situations evolved.

This chapter focuses on situation assessment and action control in military conflict situation assessment on the eve of military invasions. Unfortunately, the material available is scarce, problematic, and often fictional. Moreover, the volume of material available is too small to employ suitable methods for analyzing numerous examples. Therefore, tools must be found and implemented that allow at least something to be evaluated in a clearly justified way, and at least some explanation of how and on what basis it was decided. In previous works tools have been tried to use applying a descriptive similarity coefficient for situations taken in pairs. Unfortunately, it became clear that similarity estimates that emerged when comparing pairs of descriptions available for study turned out to be somehow oddly variable, with surprisingly little value.

I also got some inspiration from news article, which described situation of prediction. With the help of Artificial Intelligence, computer systems were able to predict a takeover of large Biomedical Corporation 5 days before that fact officially announced [11]. Although the methodology is not naturally the same or even close, it is tempting to go further with the idea and try to combine solutions of artificial

---

A. Kuuseok (✉)
Estonian Police and Border Guard Board, Estonian Business School, Tallinn, Estonia
e-mail: ahto.kuuseok@eesti.ee

© Springer Nature Switzerland AG 2021
H. R. Arabnia et al. (eds.), *Advances in Artificial Intelligence and Applied Cognitive Computing*, Transactions on Computational Science and Computational Intelligence, https://doi.org/10.1007/978-3-030-70296-0_46

intelligent (e.g., neural networks) and method of evaluation of descriptive similarity, to create some kind of predictions and compare them with our results. Right now, so fare it is purely "manual paper-based process" and it takes a lot of work to get results. However, it is reasonable to turn into a dialog system, where human and machine act in collaboration.

One of the inherent features of our approach is the fact that we need to evaluate similarity based on a very modest amount of information (human-accessible). Therefore, methods suitable for the analysis of voluminous data were initially excluded.

The chapter is structured as follows. Section 2 describes eves of military conflicts, which we observed, and some results we came to. Section 3 refers to highlights previous works and authors that I have relied on in this work. Section 4 explains and reveals the methodology I have used in this work. Section 5 presents some conclusions and brief description of the results and problems to be addressed for future work.

2 Military Domain

Some time ago, at the beginning of this research topic, I started with observations of numerical evaluations of situations and developments. We elaborated similarities from structural and descriptive aspects. The basis of the structural similarity is homomorphism of algebraic systems. Under observation was mainly of descriptive similarities – claims in descriptions of situations, expressible with formulas of calculations of the predictions.

The situations we observed were prior military conflicts, since Second World War II up to these days. The conflicts we chose to investigate were somehow related to the region of interest or related to the subject of the country under study (Estonia). In addition, we concentrated to the cases where one state (or group of states) were aggressor and attacked another, just a matter to make procedure of picking situations more clearly [4].

In later work, I did formalize the procedure of mining relevant examples of armed conflicts [3].

However, the results of pairwise evaluated situations were surprisingly scattering, and with surprisingly low numerical similarity indices [5]. Some of the observed pairs we expected to have very high indexes turned out low, and there were examples to the contrary. Some examples. "Gulf War" (02.08.1990–28.02.1001) was a conflict between coalition forces and Iraq; and Suez Crisis (10.1956–03.1957) armed conflict between Coalition and Egypt. Looking at the history of the conflict, the eve of the outbreak of the conflict, and even the underlying causes, the similarity should be quite high. In fact, the similarity index calculated by the above method was only 0.111. In addition, contrary to expectations, in some cases the calculated index turned out to be higher than expected. For example, the story of the Vietnam War (between coalition leaded by the US and coalition leaded by Soviet Union

1964–1972) and the outbreak of the Russo-Georgian War (between Russia and Georgia – August 7–15, 2008) and the eve of it should not have had much in common. However, the calculated similarity index came out 0.363.

In addition, it emerged that it was not possible to separate the intersection part in the descriptions of the situations available for examination. This made it necessary to find a more appropriate method of dealing with sets of claims from descriptions that may not all have an intersection part.

3 Related Works

Situations and developments dealt with as *"sets of statements"* that characterize them. These are finite number of finite sets that could estimate by the descriptive similarity coefficient of the descriptions of the situations. Defined by P. Lorents as following: $Sim_{LT}(A,B) = E(equ_T(A,B)):[E(A) + E(B) - E(equ_T(A,B))]$, where

- **E(H)** is the number of elements of a set H, **T** this is the way of equalization,
- **equ$_T$(A, B)** is such a set, where $x \in A$ or $y \in B$ do belong to i*n case, if* x and y, **are** *equated* [4]. Remark: basis (and method) of the equalization will be selected and fixed by the implementer of the current method (expert, analyst, etc.). For example, based on logical equivalence, if there are suitable tools to "rewrite" the respective statements into formulas of the predicate calculus [2, 9]. On the other hand, for example, based on the ability to define statements with the same meaning by fixing the corresponding definitions in the appropriate table [4, 6, 7].

It is important to note that the same and identified are essentially different concepts. The same things may turn out to be the identified; the things that identified may not be the same.

The approach described above has already been implemented in areas other than security. For example, to investigate the circumstances of IT project failure, to assess the actual similarity of certain Federal and Estonian legislation [4]. It has also been used for comparative study of the public transport situation in small towns [6, 7].

4 Proposed Methodology

Treating situation descriptions as a set of relevant statements. Definition of similarity of statements. For example, in one case based on logical equivalence of the predicate calculus formulas which express those statements [12, 13]. In the second case, for example, based on the common meaning of the statements. Association of certain things, including texts with meanings, is essentially the formation of relevant knowledge [14].

If there are two comparable descriptions, then we can use the descriptive similarity coefficient to assess the similarity of the situations (equ.) [4]. However, if there are more than two comparable descriptions, then the association similarity coefficient (cos) should be used (see Def. 4).

To use multiple sets of similarity coefficients, it is necessary to define associations of sets (situation descriptions). As well as interset contacts, connectors, associators, contact networks, etc. (see Def. 1, 2, 3).

The results of pair-wise observation of situation descriptions immediately preceding armed conflicts were not suitable to provide practically usable assessments of the similarity of situation descriptions. This method explained more closely in previous research work, see citation [5].

Consequently, a number of new (mathematical) concepts and appropriate procedures had to be defined so that even in the absence of a common element, it would be possible to quantify the similarity of some final sets, including sets of statements derived from situation descriptions. Such terms were, for example, associations, associators, connectors, contact network, etc. of finite sets, as well as overhead networks of association sets, and finally the coefficient of similarity of association sets and the procedure for calculating it [8].

5 Definitions and Procedures

It is important to note it again that same and identified are essentially different concepts. Therefore, the "intersection-like set" is not automatically the same what could be considered as set–theoretical intersection.

The proposed method of assessing similarity between situations happening in armed conflicts is based on set–theoretical mathematical apparatus introduced in [4, 5, 8].

In the following, let us state that for sets, we speak of finite sets. We use the symbol $E(H)$ to denote the final number of finite set H elements. A set whose elements are other sets we call a class.

One of the most important classes in this work is the sets of the descriptions of situations. To identify sets that belong to a class K, we use a fixed method of identification, such as representation of pairs of objects that been identified by an appropriate table, see reference [5].

Let it be a sign of this equalization ε. During the identification process, pairs of elements formed. We know from set theory that a set of pairs of elements of any set forms some kind of binary relationship between the elements of these sets [10].

The relation created by the identification method (proposed by P. Lorents) ε we denote by a symbol $\equiv_\varepsilon$. The relations under consideration could possess different algebraic properties, such as symmetry, different forms of transitivity (e.g., weak transitivity [15]) etc. Inter alia, those properties often used as constraints for building reasoning mechanisms over situations.

Definition 1 Let A and B be two sets in class K. We call ***contact*** between sets A and B a pair $\{\alpha,\beta\}$, where $\alpha \in A$, $\beta \in B$, and $\alpha \equiv_\varepsilon \beta$. Elements α and β will be called

Application of Associations to Assess Similarity in Situations Prior to Armed Conflict   645

connectors. All connectors from the set A are constitute a ***binder of A,*** and denoted as **binA**. We call set **H** *isolated* in class **K**, unless **H** is directly associated with any other class of **K**.

Definition 2 The two sets A and B of class K are called associate sets by the method of equalization ε in short associate sets. If (I) these sets are directly associated, that is, there is at least one contact between these sets (II) there is a set C, which also comes from class K, where sets A and C are associated and sets C and B are associated [8].

Definition 3 As the ***associator*** of the class K association G we call the binH'∪binH"∪ . . . ∪binH''', that is, the set whose elements are all the connectors that come from the sets H ', H' ', ..., H' "forming the given association. We designate this associator with the symbol assG. In the ***network G*** of association G, we refer to the number of contacts that occur between sets in that association. We designate this network with the symbol netG.

The **associator** of some sets is what could be considered as an expression of the intersection, even if, if there is no "pure intersection" by set theory.

Definition 4 The ***similarity coefficient of the sets of the association G***, which we denote by cosG, is called the ratio E(netG):E(∪G), where E(netG) is the number of all contacts in the network netG, and E(∪G) is the number of all elements derived from union ∪G. We denote the similarity coefficient of association G by the symbol **cosG**. While doing so, cosG = E(netG):E(∪G).

Below we will deal with sets of class **K** elements that contain descriptions of 16 situations immediately preceding the military conflict. Elements of these descriptions, in turn, are relevant statements, which are identified/unidentified by using an appropriate table that contains pairs of identifiable statements. These 16 conflicts selected by appropriate procedure [3]. This procedure is based on a certain direct or indirect connection with the security of the Republic of Estonia, the strict formulation of which (i.e., the procedure) follows certain clearly formulated criteria.

Based on the fixed form of identification in the form of a table and the relationship of the identification equalized with it, we formed associations in class K. In addition, exactly one isolated set revealed. We then found out the number of contacts that make up the overhead network of contacts, as well as the number of all the elements in each set of associations. It is necessary to assess the overall similarity of situations that have arisen prior to the military conflicts addressed.

Based on clearly and rigorously formulated terms and the formulas needed to implement them, inequalities and equations, it is possible to explain what could be similar in descriptions of many different situations. Defining, handling, and applying in some sort of common or somehow similar set to several sets proved to be problematic. Especially when it turned out those descriptions of situations from a larger class of situations did not contain identifiable statements that could be found in all relevant (i.e., situations in a given class) descriptions. What was particularly irritating was the fact that, although all of these situations were followed exclusively by armed conflict, comparisons between these (i.e., a small numerical value of the similarity coefficient). However, again, it was clear that some claims or their equivalents were present, sometimes in groups, in descriptions of many and many different situations. It took quite a long, laborious journey to notice such strange "common-like" collections, and to define them precisely. In doing so, one had to create and integrate, on the one hand, carefully collected, structured, and analyzed empirical material, and, on the other, mathematical constructions that often did not want to fit together. The result, however, was eventually to find the relevant concepts, tools, and obtain numerical estimates. In doing so, the results, including the concepts of association, associators of this association, and the coefficient of similarity of the catenary and finally the association, helped. Their implementation helped to produce some noticeable and partly surprising results. Including pleasant surprises. For example:

In previous work, Kuuseok [3] has developed procedure of mining proper and suitable cases as examples. Because there is an unimaginable amount of data on the topic of interest to us in publicly available materials, it needs to be prefiltered. The purpose of the filtering is to exclude from consideration such cases where the historical, military–political, and other such aspects would not fit in the case of the country under investigation, in this case Estonia. Geopolitical, military–political, historical, and other aspects have been considered. The analysis of the original material revealed the following dimensions and boundaries:

- Dimensions of Time: II MS to Contemporary
- The geopolitical dimension: On the one hand, delimited by the geopolitical characteristics of its immediate neighborhood and allied countries
- The military–political dimension is limited to potentially hostile neutral, friendly and allied countries

It is important that at least one condition is met, but there may be a number of conditions, all of them. The application of the appropriate procedure ultimately left the sieve under scrutiny, leaving 15 military conflicts in which these conditions were met. We can take the example of the Russo-Georgian War in 2008, where for Estonia have met several conditions – a close neighbor (Russia) and military–political cooperation (Georgia).

Their implementation helped to produce some noticeable and partly surprising results. Including pleasant surprises. For example:

- It turned out that if we allowed only two sets (D' and D'') of G to be considered as association G, the cosG value of the coefficient would be equal to the corresponding descriptive (discussed and used in previous works) similarity coefficient to Sim (D', D'').
- The sets of numbers in the figure provide a description of the situations. In sets (Fig. 1), numbers represents the statements. The numbers in the colored squares represent the connectors. The arcs between the connectors indicate the contacts. If different sets have squares of the same color, then this means that the corresponding statements represented by the numbers in the squares

equalized to each other. All statements from a description of a situation, with corresponding numbers in colored squares, form the binder of the set of the statements describing that situation. All colored squares in the figure represent the associator of the class under observation. However, all arcs represent the corresponding contact network.

- Of the observed armed conflict descriptions, one of them remained an isolated set. There was 145 statements from the 15 cases observed. During the identification process, 28 identifiable statements constituting contacts emerged, while the isolated set contained seven statements, which not identified in the case descriptions of any of the other claims. Unexpectedly, allegations emerged that were not technically much related to the military invasion of one country by another. An example that has repeatedly been stated is: "high international condemnation but no real action."
- And 83 contacts emerged between the 28 statements identified in the body of allegations that formed the association. Interestingly, when using the terminology of graph theory [1], the degree of vertices of associated statements as vertices of a graph varied up to seven (!) times, ranging from 3 to 20.
- Unexpectedly, the assertion "prompted international condemnation, but nothing realistic" came up frequently. Certainly different experts would have found these descriptions of correlated relationships, but it was not primary in this study.
- If the pairwise comparison turned out to have an average of 0.28 similarity ratings for claims derived from descriptions of military conflict eve, then the value of the similarity coefficient used in this work is considerably higher: 0.601 (see Fig. 1). This confirmed the intuitively perceived situation: When in a given set of situations each of which grew into a real armed conflict, there had to be, and as it turned out to be, more similarity than it seemed when comparing individual situations to only two.

6 Conclusions and Future Works

In this work, we studied possibilities to assess numerical similarity of several descriptions of situations immediately preceding the outbreak of a military conflict. The descriptions of the situations considered as sets of relevant statements. The central problem turned out to be the absent of an intersection (in set theoretical meaning) of the sets mentioned above. A mentioned intersection of descriptions would perhaps have been a clearer way of highlighting the more general characteristics of the outbreak of armed conflict. It was necessary to compensate for this lack of intersection.

We did it with a number of new concepts and proper definitions, and procedures for relevant assessments. For example, contacts between sets, associations of the sets, contact networks, coefficients of similarities of the sets from associations.

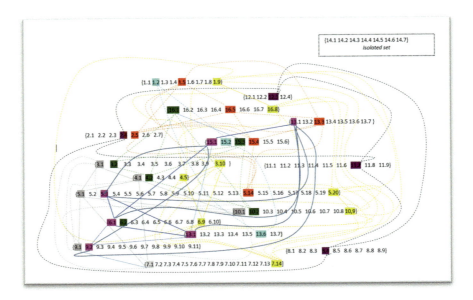

Fig. 1 Isolated Set and Similarities

However, there are already more problems that need to be addressed. For example:

- How does the change in the composition of the association (e.g., the addition of new pools or the removal of existing pools) be reflected in the change in the value of the coefficient of similarity of the association sets?
- How to interpret the changes just mentioned?
- What is the relation between the values of the similarity coefficient of the association sets to other conceivable numerical values, which, to one degree or another, arise from the treatment of associations (such as the mean of each descriptive similarity coefficient obtained by pairwise comparisons of sets)?
- What could be the interpretations of the aforementioned relationships?
- How interpret the cooperation dialog system, where human and machine act in collaboration.

One key issue here is the creation of a system based on the decisions of a particular expert and learning how to assert claims from different sources as a specific expert would do. What is important is that the system would allow equalize claims from different sources as expert would do. One approach would be to transform step-by-step the statements into suitable predicate calculus formulas. Basically in the same way as it was implemented in the Matsak 2010 prototype of the DST system [13]. Basically in the same way as it was implemented in the DST system prototype created by Matsak 2010 [13]. However, it must be acknowledged that neither DST nor other systems of this kind can in principle be so-called fully automatic. Unfortunately, this is due to algorithm–theoretical considerations. As mentioned above, it should be a self-learning system whose "mentor" is a particular expert, or team of experts, using that system. In a sense, this is the most genuine example of human–computer interaction where neither human nor computer is fully autonomous. People because data volumes and processing speeds are not feasible. Computers, however, because equalization of statements is not algorithmic.

References

1. F. Harary, Graph Theory (Published 2018 by CRC Press Taylor & Francis Group. First published 1969 by Westview Press, 2018), New York, USA
2. S.C. Kleene, Mathematical Logic (Dover Publication INC; Originally published in 1967 by John Wiley & Sons INC, 2002), Massachusetts, USA
3. A. Kuuseok, Procedure of mining relevant examples of armed conflicts to define plausibility based on numerical assessment of similarity of situations and developments. (IHSI International 2020, International Human Systems 2020) Published. Modena, Italy 19–21 February 2020, Springer (2020)
4. P. Lorents, E. Matsak, A. Kuuseok, D. Harik, Assessing the similarity of situations and developments by using metrics. 15 pages. Intelligent Decision Technologies, (KES-IDT-17), Springer, Vilamoura, Algarve, Portugal, 21–23 June 2017
5. P. Lorents, A. Kuuseok, E. Lorents, Applying systems' similarities to assess the plausibility of armed conflicts. 10 pages. Smart information and communication technologies (Smart-ICT-19), Springer, Saidia, Morocco, 26–29 September 2019.
6. P. Lorents, M. Averkina, *Application the Method of Similarity for the Decision- Making Process. Smart ICT* (Springer, Saidia, 2019)
7. P. Lorents, M. Averkina, Some mathematical and practical aspects of decision making based on similarity. (168–179), 21-st HCI international conferences, HCII 2019, Orlando, FL, USA. Springer, July 26–31, 2019 Proceedings
8. P. Lorents, Comparison of similarity of situations in associations. ICAI 2020 (2020)
9. E. Matsak, Dialogue system for extracting logic constructions in natural language texts. IC-AI 2005: 791–797 (2005)
10. M. Potter, S. Theory, its Philosophy (Oxford University Press, Oxford. UK (2004)
11. R. Aliya, R. Wigglesworth, When Silicon Valley Came to Wall Street. Financial Times, October 28, 2017. https://www.ft.com/content/ba5dc7ca-b3ef-11e7-aa26-bb002965bce8
12. E. Matsak, Dialogue system for extracting logic constructions in natural language texts. IC-AI 2005: 791–797 (2005)
13. E. Matsak, *Discovering Logical Constructs from Estonian Children Language* (Lambert Academic Publishing, Germany, 2010)
14. Lorents P, Knowledge and information. Proceedings of the international conference on artificial intelligence, IC- AI 2010. Las Vegas. CSREA Press (2010)
15. L. Esakia, Эсакиа Л. Слабая транзитивность - реституция. // Логические исследования / Logical Investigations. 2001. Т. 8. С. 244–255 (2001)

A Multigraph-Based Method for Improving Music Recommendation

James Waggoner, Randi Dunkleman, Yang Gao, Todd Gary, and Qingguo Wang

1 Introduction

With the transition from physical to digital music distribution, streaming platforms have become the main source for musical listening. With online music streaming platforms like Spotify, Pandora, and Apple Music making tens of millions of songs available to hundreds of millions of subscribers, music recommendation systems (MRS) have become a critical component of providing a satisfactory listening experience for users. Developing a recommendation system that satisfies a variety of user needs in such a growing and dynamic space is becoming increasingly challenging in the industry and appealing to academia [1]. Though MRS have made considerable gains in efficacy over the last decade, the challenges to deliver a truly personalized user experience remain to be illuminated [2].

Current MRS can be broadly classified into three categories: collaborative filtering (CF) based, content based (CB), and hybrid systems that blend both CF and CB techniques [3, 4]. All three methodologies are heavily dependent on information and, importantly, comparisons, in order to make a particular recommendation. CF-based applications typically rely on rich data sets of user interactions in order to draw a recommendation from the comparison of user listening habits, playlists, song ratings, and so on. Similarly, effective content-based recommendations require considerable information on the characteristics of a song, artist, or user in order to develop a recommendation. Hybrid systems, as a combination of CF and CB techniques, also require external user information as part of the recommendation process. As a result, many MRS tend to recommend music from the short tail of the available music catalog but fall short of music discovery [2].

J. Waggoner · R. Dunkleman · Y. Gao · T. Gary · Q. Wang (✉)
College of Computing and Technology, Lipscomb University, Nashville, TN, USA
e-mail: jr@waggs.net; qwang@lipscomb.edu

© Springer Nature Switzerland AG 2021
H. R. Arabnia et al. (eds.), *Advances in Artificial Intelligence and Applied Cognitive Computing*, Transactions on Computational Science and Computational Intelligence, https://doi.org/10.1007/978-3-030-70296-0_47

Although the majority of existing systems perform well within the context of creating low effort "set-and-forget" playlists, due to the nature of the algorithms used by MRS, most users are likely to encounter a stream of songs and artists with which they are already familiar. Additionally, most of these MRS implementations do not adequately address other potentially important listener use cases, primarily in the domain of automated music discovery. On many music streaming platforms, a novel discovery is a somewhat rare occurrence that requires artists and albums to gain significant streams, likes, follows, or plays to gain enough traction to be recommended to listeners. The effect is especially pronounced for older or less common areas of a catalogue that may become consigned to the so-called "long-tail" of obscurity [5]. Ultimately, discovery still requires a significant amount of listener effort to seek out new and novel pieces of music.

In this chapter, we explore the viability of a network graph-based method, with an emphasis on novelty and music discovery. The proposed method takes into consideration the relationships between musical groups, which are defined as two groups sharing at least one artist. In exploring these relationships from a network perspective, then we generate an artist recommendation that requires little or no preexisting artist or listener information. By avoiding the popularity-bias inherent in existing MRS, our artist graph allows broader, but still relevant, artist recommendation with greater diversity and novelty.

2 Related Work

Due to the massive variety and quantity of musical options available to users, much progress has been made in the field of MRS and the capabilities and shortcomings of traditional MRS are well documented [2]. For example, MRS were shown to bias toward popular artists at the detriment of less-known artists. In an audio content-based (CB) networks, users tend to listen to more music independent of the quality or popularity of the music [5]. Additionally, though good at long-tail recommendations, CB networks can cause musical-genre-based bias. In a collaborative filtering network, however, users tend to listen to more mainstream music.

As the widely used CF-based MRS is affected by the disadvantage of cold start caused by the lack of user historic data, Van den Oord et al. developed a latent factor model for music recommendation [4], which predicted the latent factors from music audio when they could not be obtained from usage data. Their results showed that predicted latent factors produced sensible recommendations.

Network graphs have also been used to improve music recommendation [6–9]. For instance, Zhao et al. proposed genre-based link prediction in bipartite graph for music recommendation [7]. They enhanced recommendation by representing complex numbers and computing users' similarity by genres weight relations

A Multigraph-Based Method for Improving Music Recommendation

[7]. With music-genre information, they demonstrated that user similarity and link prediction ability were improved and accordingly overall recommendation accuracy was increased. In another study of bipartite graph, they revealed that the performance of recommendation can be significantly improved by considering homogeneous node similarity [9].

A common issue encountered by many different recommendation systems is the need to incorporate information about users and subjects that may be implicit, unknown, or otherwise missing. Tiroshi et al. addressed this issue by representing the data using graphs, and then systematically extracting graph-based features, with which to enrich original user models [8]. Their results showed features derived from graph increased recommendation accuracy across tasks and domains.

In this chapter, we explore network graphs that make recommendations with emphasis on novelty and music discovery. The remainder of the chapter is organized as follows. First, we describe the data to use and how we preprocess it. Then, we describe the generation and analysis of our artist-based networks. Next, we provide our results and lastly suggest directions for future development.

3 Data Collection and Preprocessing

The data for constructing our artist networks was downloaded from the MusicBrainz Database, a freely available open music encyclopedia [10]. The database contains a wealth of user-maintained meta-data for over 143,111 music artists. In particular, we utilized two data fields for building and enriching our artist-based networks: an artist's membership in a musical group and user-submitted descriptors called "tags."

The database defines an artist as an entity that can be an individual performer, a group of musicians, or other contributors such as songwriters or producers. We limited our discussion of artist-artist relationships to musical groups and their current and past group members, as well as other members who may have contributed or played with the group in some meaningful way. With respect to group membership, we did not weigh or otherwise discriminate between member "categories" (current member, past member, touring, recording, etc.); i.e., we consider all associated members to be equal for simplicity.

User-submitted tags allow users to attach various categorical identifiers to artists. Table 1 shows the tags associated with artist *The Who*. The primary artist tags are related to the musical genres and typically are used to describe the artist's music. Secondary tags, labeled "Other tags," are used to more broadly categorize an artist, e.g., countries of origin or periods of time and may refer to certain musical movements with which the artist is typically associated. For the convenience of analysis, we did not weigh or discriminate between tags or their types, meaning we considered all tags to be equally informative.

Table 1 Tags submitted by users for the artist *The Who*

Artist	Genre	Other tags
The Who	Rock	British
	Hard Rock	Mod
	Art Rock	Freakbeat
	Classic Rock	Rock Opera
	Pop Rock	British Invasion
		Classic Pop and Rock
		UK

Table 2 Description of seven Spotify playlists used in our experiment

Playlist name	URL	Total songs	Unique artists
All Classic Hits	open.spotify.com/playlist/4C9mWYjVobPsfFXesGxYNf	80	70
2004 Hits	open.spotify.com/playlist/3pM6OEFeTo4L1yd4eRltjL	135	94
Rock Classics	open.spotify.com/playlist/37i9dQZF1DWXRqgorJj26U	150	76
Rock Hard	open.spotify.com/playlist/37i9dQZF1DWWJOmJ7nRx0C	100	86
Pure Rock N Roll	open.spotify.com/playlist/37i9dQZF1DWWRktbhJiuqL	100	65
Power Pop	open.spotify.com/playlist/37i9dQZF1DX5W4wuxak2hE	43	30
Punk Playlist	open.spotify.com/playlist/7Jfcy1H82lsTIzhpL4MZXu	200	180

Using the artist-artist relationship and tag data described above, we created two subsets of data for our experimentation and analysis. The first was created by mapping all artists to the musical groups to which they belong, resulting in a simple dictionary of 143,111 unique musical groups and 495,291 of their related group members. In much the same way, we mapped user-submitted tags to their parent artists, generating a second dictionary of 143,111 musical groups and 352,180 their associated tags. The Python code for processing the datasets, together with the processed data files, is publicly available at GitHub website (at https://github.com/jrwaggs/music/blob/master/Data%20Gathering.ipynb).

Using Spotify API (see our code in Github), we also collected Spotify's music data, which includes user-generated playlists, artist info, album data, musical genre data, and data specific to individual tracks. Although we had planned to select playlists of varying length and genre content, with limited time we were able to examine only seven playlists in our experimentation. Table 2 below provides a brief description of the seven Spotify playlists. The unique artists extracted from these playlists are used to construct our test data sets.

4 Method

Different from other methods that exploit similarity among users, in this chapter we propose to create multigraphs based on the relationships between individual artists and musical groups. This design enables us to create models even in the

A Multigraph-Based Method for Improving Music Recommendation

absence of user attribute or preference data, which are not always easy to obtain. More importantly, as shown by our results later, our graphs can effectively avoid low novelty exhibited by existing MRS methods.

Network Graph Concept
Our network graph contains two basic elements: vertices or nodes that represent the objects in the graph, and edges or links that connect vertices. When multiple types of relationships exist between objects in a network, multilayer or multidimensional graphs are required. The multidimensional graph G used in our study is defined as:

$$G = (V, E, D)$$

where

$V = A$ *finite set of Vertices or nodes*
$E = A$ *finite set of labeled Edges or links connecting vertices*
$D = A$ *set of labels representing each Dimension or layer* [11]

In our model, each node represents a unique musical group, each edge represents a single labeled relationship between the two, and the dimension of the layer is expressed as either *artist* or *tag*. This definition allows for creating and layering together two separate networks: (1) networks in which the relationship between two music groups is defined as having a shared artist or band member, and (2) networks in which the relationship is defined by a shared tag. Edges in our graph are thus represented as tuples (u, v, d) or in our case (*Artist, Artist, Dimension: label*) where *Dimension* is either artist or tag. For instance, below are two such types of edges:

(The Yardbirds, Cream, artist: Eric Clapton)

or

(The Rolling Stones, The Who, tag: British)

Graph Construction
Our approach to constructing artist and tag-based network graphs follows a breadth-first-search-style procedure: we begin with a root node and systematically traverse an entire level of a graph structure before moving on to the next level [12]. But rather than searching for a single type of nodes in a network, we build alternating layers of artists and musical groups, from which a graph structure is created. The flowchart of our graph construction is depicted in Fig. 1.

As show in Fig. 1, we start with a single seed music group, from which we retrieve a list of all artists that have ever been a member of the group. These artists will make up the first artist level in the tree structure. Next, we traverse the first artist level, retrieving all musical groups of which an artist has been a member. By repeating this process for each successive layer, we are able to construct a fully connected artist-based graph. By traversing each band level and appending the band to a separate list, we create a set of music groups that serve as the node set in a graph and are the foundation of our graph building method.

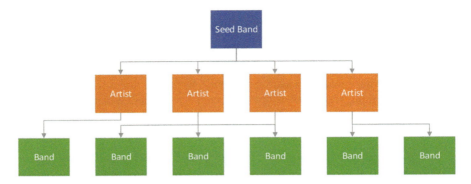

Fig. 1 Illustration of graph construction procedure

It is important to note that, to obtain a manageable graph with decent depth and breadth, we first use artist-artist relationships to generate our initial artist graphs. Because an artist may be labeled with as many as 20 user tags, the use of tag-based relationships can result in tree structures that are incredibly shallow (only a few layers "deep") and wide, in comparison with artist-based graphs of the same size. Our experimentation showed that wide, shallow trees were too large and contained far too much variation to be able to produce any meaningful result.

With the nodes in our initial artist graphs, the process of generating the sets of artist and tag edges is straightforward. Similar to artists, tags represent a set of user-submitted descriptors assigned to groups and artists. For example, the tags for the group *The Who* include "rock," "classic rock," "mod," and "British invasion" [13] (see Table 1). We compare each music group's artists or tags against every other music group in the node set. If they share a common artist or tag, an undirected, labeled edge is then added to the graph. All of our undirected, multidimensional music graphs are created in this way by combining node, artist-edge, and tag-edge sets.

When experimenting with different artist-based graphs of varying size and complexity, we found that a flaw in the aforementioned procedure is: an ideal level of musical variety may not be reached efficiently in certain cases. For example, to obtain a desired level of variety for a graph that contains two groups associated with very different genres, we may have to create incredibly large graphs that are inefficient to work with and evaluate. In other cases, "natural" connection between two bands or artists may not exist at all.

To address this issue, we layer together multiple artist-based multidimensional graphs to form combined multilayer graphs that may contain any number of specified artists. We develop such graphs by first creating separate artist-based graphs, and then repeating the process to incorporate tag-based edges to each graph. With a fair number of graphs of varying origin and size, the shared user-submitted tags will be more likely to connect two artists whose natural relationship either does not exist or requires an unreasonably large graph to find a path between them. For example, creating a network to link the artists *The Who* and *The Cure* requires a very large network graph. With the use of the '*Pop Rock*' user tag, however, the two graphs containing either artist can be easily connected.

Measuring Node Importance

To evaluate the importance of nodes in a graph, we calculate a number of graph connectedness measures, such as degree centrality, load centrality, and PageRank score. PageRank score is calculated using Python code NetworkX (https://github.com/jrwaggs/music/blob/master/Network%20Graph.ipynb) [14]. Of these metrics, PageRank consistently produced more relevant results than others. For example, the bands *Cream*, *Eric Clapton*, and *The Yardbirds* all have many artists and tags in common and are relatively important to that era of music. Their scores are more similar with PageRank than they do with other centrality measures. The performance of PageRank can be explained by its design, which considers not only the number of connections a node has in a graph, but also the relative importance of neighboring nodes [15]. It is important to accurately measure the importance of each node in the graph, because the top-ranked artists in a graph are used by our method to produce artist recommendations.

To verify the efficacy of artist-based networks, in conjunction with PageRank, we constructed two multigraphs and examined the top 25 ranked artists produced by each. The results of the two multigraphs are provided in Appendix A. The first graph was constructed using five root artists commonly found together in a classic rock playlist: *The Rolling Stones, The Who, The Eagles, Van Halen*, and *Boston*. To demonstrate the ability of our method to link disparate musical groups, the second graph is constructed using five root artists that are more varied in genre and musical era: *R.E.M., Abba, Megadeth, The Spice Girls*, and *The Cardigans*. Both graphs were constructed using the same specifications: a graph of size $V = 200$ was constructed with both artist and user-tag edge sets for each artist and then individual graphs were combined into a single multigraph.

Graph Evaluation

We employed two commonly used performance metrics, accuracy, and novelty, to evaluate our multigraph models.

Accuracy A recommendation system should maintain a level of user comfort by recommending artists with some of whom they are familiar. Accuracy is used to measure the similarity of the recommended artists to the artists in the user's playlist. Let UA_u be the set of artists in the original user playlist and RA_u the set of recommended artists. We define the accuracy of an individual playlist as:

$$\text{Accuracy} = \frac{|RA_u \cap UA_u|}{|RA_u|} \tag{1}$$

The accuracy of our recommendation model is defined as the average of the accuracy of individual playlists, which is calculated using the Eq. (1) above.

Novelty Together with a number of other metrics, novelty was used by Chou et al. to evaluate several MRS [16]. A user is better served if recommendations include new and novel songs or artists. Novelty measures music discovery by quantifying the recommended artists not in the original playlist. Let UA_u be the set of artists in the original playlist and RA_u the set of recommended artists. The novelty of an individual playlist is defined as:

$$\text{Novelty} = \frac{|RA_u \setminus UA_u|}{|RA_u|} \tag{2}$$

To measure method novelty, we compute the novelty for each individual playlist and take the average as the novelty measurement of our method [16].

From Eqs. (1) and (2), we can see that accuracy and novelty are complementary and, thus, negatively correlated. Hence, in the section below, we just focus our discussion on novelty.

5 Results

Firstly, we evaluated the node importance metrics used in our model (see Appendix A), by calculating the correlation coefficient between them and Spotify popularity score, which is in the range of 0–100. The result in Table 3 shows PageRank and degree centrality are moderately correlated to Spotify popularity score, with correlation coefficients being 0.39 and 0.51, respectively.

Using the graph for the *Rock Classics* playlist, Fig. 2 shows a scatterplot of PageRank and Spotify popularity score. It confirms the moderate correlation between artists' PageRank and Spotify popularity scores.

Next, we evaluated artist recommendation by mimicking a real-world context. For each of the seven Spotify user playlists in Table 2, we generated a number of multigraphs and computed the average novelty of the artists in the graphs against the original playlist. We also evaluated two parameters used in graph construction – number of seed artists and size of individual graphs – and assessed their influence on model accuracy and novelty.

Below is the description of our evaluation process. We applied the same procedure to each playlist and assessment:

1. Let *UA* be the set of artists in the original user playlist. We used a random number generator to select *N* random seed artists from *UA*.

Table 3 The correlation between node importance metrics and Spotify Popularity Score

Measure	PageRank	Degree centrality	Centrality
Correlation	0.393	0.507	0.034

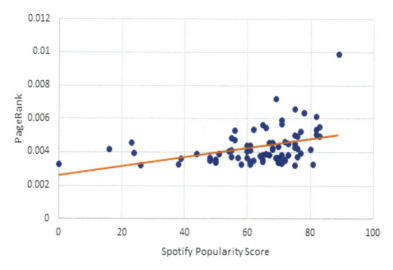

Fig. 2 Scatterplot of PageRank and Spotify Popularity scores

- To study the effect of seed artist selection on novelty, we selected *2, 5, 7, 10*, and *15* artists, respectively, for our test.
- To study the impact of graph size on novelty, we selected *10* artists in our experiment.

2. For each artist, we built graphs of size *V* and combined them to form a single multigraph.

   - For the study of the effect of seed artist selection on novelty, we built graphs of size *V = 100*.
   - For the study of the impact of graph size on novelty, we built graphs of size *V = 10, 25, 50, 75, 100, 150, 200*, and *250*, respectively.

3. We evaluated the graphs and calculated a PageRank score for each node.
4. Let *RA* be the set of recommended artists derived from the graphs and let $|RA| = |UA|$. Then we computed accuracy and novelty against the original playlist.
5. We repeated this process *10* times for each playlist to compute average accuracy and novelty.

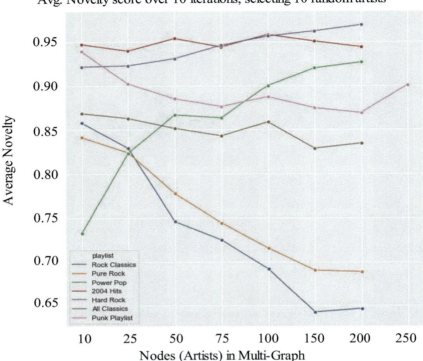

Fig. 3 Multigraph novelty by nodes

Figure 3 shows the average novelty of our recommendations for the seven playlists. The x-axis of the plot represents the number of nodes (artists) in the resulting multi-graphs. It shows with the increase of graph size, the novelty of our recommendations decreased significantly for two of the playlists, and stayed relatively flat for other playlists. This is expected, as the two playlists exhibiting drastic decrease in novelty have a higher percentage of multi-member bands and thus more artist-artist relationships to build graphs. This figure indicates our recommendations are affected more by artist-artist relationships than by artist tag data.

The impact of the number of seed artists used for building graphs on recommendation novelty is provided in Fig. 4. With the increase of the number of seed artists used, as shown in the figure, the novelty drops uniformly. Therefore, for listeners in favor of novel music, fewer seed artists should be used for graph construction.

Finally, we compared our results to the published work by Chou et al. [16]. In general, our average novelty, which is from 64% to 98%, is comparable to their

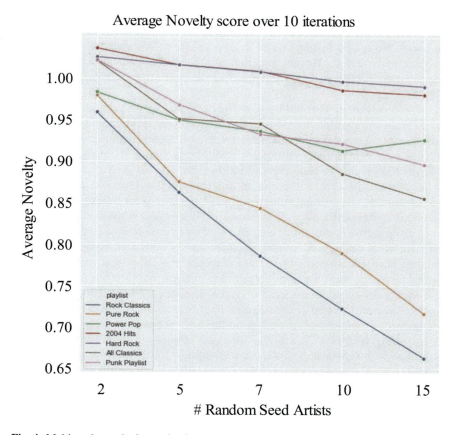

Fig. 4 Multigraph novelty by seed artists

results, which fell between 89% and 97%. Chou's study, however, used KKBOX dataset, different from our MusicBrainz dataset. More comprehensive evaluation is needed to better understand the capability and limitation of our method.

6 Discussion

In this chapter, we developed a multigraph method for music recommendation. We built graphs by utilizing artists' relationships and user-submitted tags, which reinforce artists' relationships and therefore help identify more relevant recommendations. Comparing with current MRS, whose recommendations are generally positioned within the most popular artists/groups, our recommendations provide improved novelty and diversity and, hence, are more representative of the spectrum of musical groups and artists. Our method is ideal to be used as an add-in or user-controlled option for popular streaming services. This added diversity would provide a more unique listener experience. Our method can also be used as a component of an MRS to help overcome cold-start issue in the CF-based streaming services.

Our recommendations could be negatively affected by the ambiguity of artists' names. There are multiple bands with the same names. It is difficult to determine which one is referenced, in particular in the use of Spotify popularity scores. This challenge can be potentially cracked by assigning a unique ID to each artist. But we did not do it due to time constraints. In addition, we used the MusicBrainz data as our test data set. If there is an artist missing, we have to manually add it to the dataset. This is an arduous process and our work could be alleviated by combining MusicBrainz with additional data sources. A more comprehensive dataset can potentially promote the accuracy of recommendations further.

For future development, graph structure can be optimized to generate more consistent recommendations. In our graph model, the edges that represent artist-artist relationships are clearly more relevant than tag-based edges. So a way to refine the model is to distinguish two types of edges by associating edges with weights. Another direction is to integrate other relationships into the model, such as songwriter, producer, or other behind-the-scene contributors, and this would be particularly helpful in identifying stronger relationships for single artists or new groups with few or no artist-artist connections. Lastly, to improve our music recommendation, it is necessary to have it evaluated by real-world users. The level of satisfaction of users can be used to fine-tune the model.

Appendix A: Comparison of Three Node Importance Measurements in Two Multigraphs

Rank	Graph I with *The Who, The Rolling Stones, Van Halen, The Eagles,* and *Boston* as seed artists				Graph II with *R.E.M., Spice Girls, Megadeth, The Cardigans,* and *Abba* as seed artists			
	Artist	Degree centrality	Load centrality	Page rank	Artist	Degree centrality	Load centrality	Page rank
1	Black	4.181	0.038	0.011	Metallica	2.441	0.007	0.007
2	The Rolling Stones	3.598	0.046	0.010	Black Sabbath	2.568	0.008	0.007
3	Genesis	4.098	0.018	0.010	R.E.M.	2.047	0.056	0.007
4	Whitesnake	3.960	0.016	0.010	The Smiths	2.453	0.010	0.007
5	Yes	4.014	0.009	0.009	Motörhead	2.378	0.008	0.007
6	Deep Purple	3.768	0.014	0.009	Whitesnake	2.408	0.008	0.007
7	Queen	3.691	0.013	0.009	Megadeth	1.676	0.043	0.007
8	Alice Cooper	3.443	0.018	0.008	Queen	2.319	0.012	0.006
9	The Who	3.248	0.027	0.008	The Pretenders	2.371	0.003	0.006
10	Electric Light Orchestra	3.640	0.004	0.008	Ozzy Osbourne	2.286	0.010	0.006
11	Ozzy Osbourne	3.394	0.014	0.008	The Damned	2.029	0.021	0.006
12	Artists united against apartheid	0.973	0.064	0.008	Alice Cooper	2.180	0.009	0.006
13	Uriah Heep	3.520	0.003	0.008	New Order	2.221	0.006	0.006
14	Eric Clapton	3.547	0.003	0.008	Simple Minds	2.209	0.003	0.006
15	The Kinks	3.549	0.003	0.008	Fleetwood Mac	2.185	0.006	0.006
16	Frank Zappa	3.242	0.010	0.008	Killing Joke	2.082	0.007	0.006
17	The Yardbirds	3.229	0.009	0.008	The Who	1.985	0.010	0.006
18	Santana	3.056	0.009	0.007	Ac/Dc	1.965	0.005	0.005
19	Brian May	3.152	0.004	0.007	The Lightning Seeds	1.980	0.005	0.005

Rank	Graph I with *The Who, The Rolling Stones, Van Halen, The Eagles,* and *Boston* as seed artists				Graph II with *R.E.M., Spice Girls, Megadeth, The Cardigans,* and *Abba* as seed artists			
	Artist	Degree centrality	Load centrality	Page rank	Artist	Degree centrality	Load centrality	Page rank
20	Faces	2.853	0.016	0.007	Oasis	1.908	0.007	0.005
21	Boston	2.899	0.014	0.007	Tool	1.882	0.005	0.005
22	Def Leppard	3.046	0.004	0.007	The Hollies	1.934	0.004	0.005
23	The Beach Boys	3.166	0.002	0.007	Radiohead	1.839	0.008	0.005
24	Van Halen	2.672	0.023	0.007	Nine Inch Nails	1.786	0.006	0.005
25	Jethro Tull	3.061	0.004	0.007	Modest Mouse	1.456	0.026	0.005

References

1. M. Schedl, P. Knees, B. McFee, et al., Music recommender systems, in *Recommender systems handbook*, ed. by F. Ricci, L. Rokach, B. Shapira, (Springer, Boston, 2015), pp. 453–492
2. M. Schedl, H. Zamani, C.W. Chen, et al., Current challenges and visions in music recommender systems research. Int J Multimedia Inf Retr 7(2), 95–116 (2018)
3. A. Vall, M. Dorfer, H. Eghbal-zadeh, et al., Feature-combination hybrid recommender systems for automated music playlist continuation. User Model. User-Adap. Inter. **29**, 527–572 (2019). https://doi.org/10.1007/s11257-018-9215-8
4. A. Van den Oord, S. Dieleman, B. Schrauwen, Deep content-based music recommendation, in *Proceedings of the 26th International Conference on Neural Information Processing Systems (NIPS'13)*, (December 2013), pp. 2643–2651
5. Ò. Celma, Music Recommendation and Discovery in the Long Tail, Dissertation, Universitat Pompeu Fabra, Barcelona, Spain, 2008
6. K. Mao, G. Chen, Y. Hu, L. Zhang, Music recommendation using graph based quality model. Signal Process. **120**, 806–813 (2015). https://doi.org/10.1016/j.sigpro.2015.03.026
7. D. Zhao, L. Zhang, W. Zhao, Genre-based link prediction in bipartite graph for music recommendation. Procedia Comput. Sci. **91**, 959–965 (2016)
8. A. Tiroshi, T. Kuflik, S. Berkovsky, M.A. Kaafar, Graph based recommendations: From data representation to feature extraction and application, in *Big Data Recommender Systems*, ed. by O. Khalid, S. U. Khan, A. Y. Zomaya, vol. 2, (IET, Stevenage, 2019), pp. 407–454
9. L. Zhang, M. Zhao, D. Zhao, Bipartite graph link prediction method with homogeneous nodes similarity for music recommendation. Multimed. Tools Appl. (2020). https://doi.org/10.1007/s11042-019-08451-x
10. MusicBrainz Database, MetaBrainz Foundation, California (2019), https://musicbrainz.org/doc/MusicBrainz_Database. Accessed 14 May 2020
11. S. Boccaletti, G. Bianconi, R. Criado, et al., The structure and dynamics of multilayer networks. Phys. Rep. **544**(1), 1–122 (2014)
12. T.H. Cormen, C.E. Leiserson, R.L. Rivest, C. Stein, *Introduction to Algorithms*, 3rd edn. (The MIT press, Cambridge, MA/London, 2009)
13. MusicBrainz, The Who: Tags (2019), https://musicbrainz.org/artist/9fdaa16b-a6c4-4831-b87c-bc9ca8ce7eaa/tags. Accessed 14 May 2020
14. A. Hagberg, D. Schult, P. Swart, Exploring network structure, dynamics, and function using NetworkX, in *Proceedings of the 7th Python in Science Conference (SciPy)*, (Pasadena, California, August 2008), pp. 11–15
15. N. Perra, S. Fortunato, Spectral centrality measures in complex networks. Phys. Rev. E **78**(3 Pt 2), 036107 (2008)
16. S.Y. Chou, Y.H. Yang, Y.C. Lin, Evaluating music recommendation in a real-world setting: On data splitting and evaluation metrics, in *The 2015 IEEE International Conference on Multimedia and Expo (ICME)*, (Turin, Italy, August 2015), pp. 1–6

A Low-Cost Video Analytics System with Velocity Based Configuration Adaptation in Edge Computing

Woo-Joong Kim and Chan-Hyun Youn

1 Introduction

As video cameras are pervasive as a CCTV or a mobile device, many researches attempt to utilize the potential of these cameras and develop various intelligent systems and services with Video Analytics (VA), which is a generic term for various methods of video analysis [3, 11]. Deep Neural Network (DNN) recently has been a key technology for video analytics due to their high accuracy and various applications such as object detection and tacking, action recognition and instance segmentation, etc. [2]. A desired video analytics quality requires high accuracy with low latency, but the high accuracy of DNN requires a prohibitive cost on latency [1, 4]. DNN based VA involves high complexity computation which cannot be performed at high speed on the limited computing resource of end devices, such as CCTV or mobile device [5]. For example, on Qualcomm Adreno 530 GPU which is the embedded GPU of end devices, the processing time per image of big-yolo, a DNN based object detection model with 22 convolution layers, is about $600\sim4500$ ms ($1.6\sim0.2$) fps and those of tiny-yolo, a DNN based object detection model with 9 convolution layer, is about $200\sim1100$ ms ($5\sim0.9$) fps [9].

In this regard, Mobile Edge Computing (MEC) has been emerged as a promising solution to address these issues. The MEC provides computing resources located in close proximity to end users within the Radio Access Network. The latency on DNN based VA in the limited computing resource of end devices would be significantly improved by offloading the computation to the MEC. However, since multiple video

---

W.-J. Kim · C.-H. Youn (✉)
School of Electrical Engineering, Korea Advanced Institute of Science and Technology, Daejeon, South Korea
e-mail: w.j.kim@kaist.ac.kr; chyoun@kaist.ac.kr

© Springer Nature Switzerland AG 2021
H. R. Arabnia et al. (eds.), *Advances in Artificial Intelligence and Applied Cognitive Computing*, Transactions on Computational Science and Computational Intelligence, https://doi.org/10.1007/978-3-030-70296-0_48

cameras are usually served concurrently for VA in the MEC, even the computing resources (i.e. GPUs) of the MEC are limited to handle each video stream.

Many VA systems efficiently utilizing the limited computing resources of the MEC shared by multiple video cameras have been proposed. Their main target application is object detection, which is the basis of various high-level VA applications such as scene graph detection and instance segmentation. In order to guarantee the VA performance in real time, several systems address these issues by adapting configurations [3, 11], such as frame resolution, frame sampling rate and detector model. However, they have a high profiling cost problem derived from an online-profiling technique, which is essential due to the dynamics of the configuration's impact on video analytics accuracy [3].

In this paper, we propose a low-cost VA system analyzing multiple video streams efficiently under limited resource. The objective of our proposed system is to find the best configuration decision of frame sampling rate for multiple video streams in order to minimize the accuracy degradation while guaranteeing the real time VA in the shared limited resource. Firstly, the frame sampling rate knob of each video stream is adapted optimally based on the velocity feature of objects extracted from its video context in low-cost. Its objective on each video stream is to find the best frame sampling rate for reducing the resource demands as much as possible while guaranteeing a desired accuracy. Secondly, the frame sampling rate knob of each video stream is adapted additionally in a greedy way, considering a limited resource shared by multiple video streams. Its objective is to reduce the resource demands of each video stream in fairness to analyze multiple video streams in real time in the shared limited resource while minimizing the accuracy degradation of each video stream.

2 Low-Cost Video Analytics System

We introduce our VA system which supports multiple camera feeds fairly under limited resource capacity. Figure 1 shows the structure of our VA system for edge computing environment. It consists of a pre-processor, an object detection model, and a configuration controller [3]. It is implemented on a GPU-enabled edge server attached to a small cell base station and its input video streams are generated from multiple cameras. Let $[n] = 1, \ldots, n$ denote the set of integer numbers which has a cardinality of n.

There are I multiple video streams, each of which has its default frame rate and resolution. Each video stream is split into smaller segments, each of which is a contiguous set of frames spanning a T-second interval (By default, $T = 4$). Let $S_i = \{S_{i,1}, S_{i,2}, \ldots, S_{i,j}, \ldots, S_{i,J}\}$ denote the segments of i-th video stream whose size is J. The raw frames of $S_{i,j}$ are denoted as $S_{i,j} = \{f_{i,j}^1, f_{i,j}^2, \ldots, f_{i,j}^{l_s}\}$, where the number of frames in $S_{i,j}$ is l_s. Let fps_{def} be the default frame rate of all cameras

A Low-Cost Video Analytics System with Velocity Based Configuration... 669

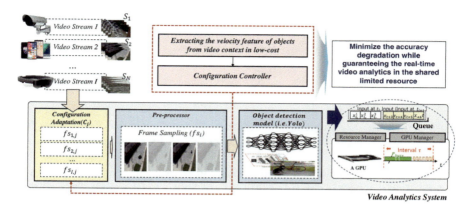

Fig. 1 An illustration of our proposed video analytics system

(By default, $fps_{def} = 30$),[1] respectively. The generated raw frames are transferred to the VA system. We assume l_s is same with fps_{def}.

The pre-processor samples the received raw frames of each video stream in a certain frame rate. The frame rate is considered as a main configuration knob and controlled by the configuration controller. Let $fs_{i,j}$ denote the frame rate of i-th video stream for j-th segment. There are allowable candidates which the configuration controller can decide for frame rate knob. The candidate set of frame rate is denoted as S and are sorted in a descending order (e.g. $S = \{30, 15, 10, 6, 5, 3, 2, 1\} fps$. For simplicity, we use the divisors of fps_{def} for S. Then, the frame rate knob of i-th video stream for j-th segment is defined as $fs_{i,j}$. Let $C_j = \{fs_{1,j}, fs_{2,j}, \ldots, fs_{I,j}\}$ denote the configurations of I video streams on j-th segment, decided by the configuration controller.

The sampled frames of each video stream are fed into the queue and processed on a single GPU for VA by an object detection model (i.e. Yolo). A GPU is entirely occupied by a video stream at a time and the usage time of the GPU is used as a resource. We assume that the processing time of the object detection model on a frame is static since the GPU tends to be stable and static for processing each operation of the object detection model.

Lastly, the model recognizes objects and draws bounding boxes around them in each sampled frame of video streams.

Based on the detection results on the sampled frames of each video stream from the object detection model, the system predicts and finalizes the detection results of all frames in j-th segment of each video stream (including the frames not sampled). The accuracy of the j-th segment in each video stream is calculated as the average accuracy per frame of a segment. In the same way as existing VA systems, the

---

[1] We assume that the default frame rate of all cameras is same.

detection results of the last sampled frame are used for the frames not sampled in a segment.

2.1 System Objectives

We formulate the VA problem to optimize the average accuracy of analyzing live video streams by adapting configurations under limited resource capacity over the time horizon.

We divide the continuous time to discrete time slots. The length of a time slot is defined as τ, which is defined as the segment length l_s. Therefore, the j-th time slot is defined as the time at which the j-th segment, which has generated during $(j-1)$-th time slot, have to be processed. Then, in the j-th time slot, the problem is to process the j-th segments of video streams, denoted as $\{S_{i,j}|\forall i \in [I]\}$.

The objective of our VA system is to make an optimal configuration decision for its VA problem over video streams at each time slot. Then, the configuration decision vectors for j-th segments is defined as follows:

$$C_j = \{fs_{1,j}, fs_{2,j}, \ldots, fs_{I,j}\}. \tag{1}$$

The total processing time on the GPU for all frames sampled in the configuration C_j for j-th segments of video streams is modeled as follows:

$$PT(C_j) = \sum_{i \in I} fs_{i,j} * l_p, \ where \ fs_{i,j} \in S. \tag{2}$$

Here, $fs_i(t)$ is decided among the candidates of S.

The accuracy of analyzing a video stream is affected by several factors: the video context, the object detection model, the frame resolution, rate and bitrate of the video stream [9]. Since we assume that our VA system receives each video stream in high enough bitrate and uses a single object detection model, which has a fixed input size, the accuracy depends only on the video context, the frame rate of the video stream. Then, the accuracy model is defined as $a_{i,j}(fs_j)$ which represents the VA accuracy on j-th segment of i-th video stream with fs_j. In i-th video stream, there is a trade-off between the accuracy $a_{i,j}(fs_{i,j})$ and the processing time $fs_{i,j} * l_p$. Increasing the frame rate may increase the accuracy but will also increase processing time.

Then, we formulate the VA problem to find the configuration decision vector C_j optimizing the trade-offs of video streams as follows:

$$\underset{C_j}{\text{maximize}} \sum_{i \in I} a_{i,j}(fs_{i,j}) \ \text{subject to} \ PT(C_j) \leq \tau. \tag{3}$$

The objective is to maximize the sum of the accuracies of I video streams for j-th segment. The constraint says that the total processing time cannot exceed the time slot length τ, in order to keep the queue stable. The VA problem is a multiple-choice knapsack problem (MCKP). A brute-force solution to the VA problem would take a running time of $O(I \cdot |S|)$. Obviously, it is impractical to search a large space of the configuration by profiling the accuracy models of video streams in each segment. Moreover, the accuracy model $a_{i,j}(fs_{i,j})$ is non-linear function and cannot be formulated cleanly in analytic form with the lack of analytic understanding on theoretical properties of the object detection models. It is necessary to design an efficient algorithm resolving the time complexity and the non-linearity.

2.2 Velocity Based Configuration Adaptation Algorithm

In this procedure, we consider the l_s raw frames of j-th segment generated from each live video stream. Then, we decided the sampling frame rate $fs_{i,j}$ for the l_s raw frames of each live video stream, using the velocity of objects on 1-th raw frame, $f_{i,j}^1$. Firstly, since $f_{i,j}^1$ is sampled in default, we have the bounding boxes of detected objects $B_1 = \{b_{1,1}, b_{1,2}, \ldots, b_{1,q}, \ldots, b_{1,Q}\}$ on $f_{i,j}^1$, where $b_{1,q}$ is the bounding box of q-th object detected. We extract K tracks on $f_{i,j}^1$, denoted as $\{tr_{1,k} | \forall k \in [K]\}$, using an optical flow method which extracts many tracks of feature points over contiguous frames. Let $tr_{1,k} = \{p_{1,k}, p_{2,k}, \ldots, p_{l,k}\}, k \in [K]$ denote the track of k-th feature point at $f_{i,j}^1$ where l is the track length ($l = 5$) and $p_{h,k} = \{val_{h,k}^x, val_{h,k}^y\}$ is the coordinates of k-th feature point at h-th time.

Secondly, we estimate the velocity of each object by calculating the average movement distance per frame of feature points included in the detected bounding box of an object on $f_{i,j}^1$. The tracks of M feature points included in the detected box $b_{1,q}$ are denoted as $TR_1^q = \{tr_{1,1}, tr_{1,2}, \ldots, tr_{1,M}\}$ where $tr_{1,m} = \{p_{1,m}, p_{2,m}, \ldots, p_{l,m}\}, m \in [M]$. The delta of the track of m-th feature point over its track length is calculated and denoted as $(\Delta tr_{1,m}^x, \Delta tr_{1,m}^y) = \frac{p_{l,m} - p_{1,m}}{trackLen} = \frac{(val_{l,m}^x, val_{l,m}^y) - (val_{1,m}^x, val_{1,m}^y)}{trackLen}$. Based on it, the estimated velocity of $b_{1,q}$ is defined as follows:

$$v_{1,q} = (v_{1,q}^x, v_{1,q}^y), \tag{4}$$

where

$$v^x_{1,q} = \frac{1}{|M|} \sum_{i=1}^{|M|} \Delta tr^x_{1,i},$$

$$v^y_{1,q} = \frac{1}{|M|} \sum_{i=1}^{|M|} \Delta tr^y_{1,i}. \tag{5}$$

As a result, we estimate how objects on $f^1_{i,j}$ move per frame in a segment with $\{v_{1,q} | \forall q \in [Q]\}$. In order to decide the best frame rate intuitively which maximizes the accuracy with minimum resource consumption, we focus on the velocity of the fastest object $q_{max} = argmax_{q \in Q}\{v_{1,q}\}$ and, based on it, adapt the sampling frame rate $fs_{i,j}$ for j-th segment. First, we predict the q_{max}-th object's bounding boxes over j-th segment by shifting the q_{max}-th object's bounding box on $f^1_{i,j}$, $b_{1,q_{max}}$, with its velocity of $v_{1,q_{max}}$. Let $\{b^*_{t,q_{max}} | \forall t \in [l_s]\}$ denote the predicted bounding boxes of q_{max}-th object over the l_s frames of j-th segment. It is performed by shifting the coordinates of $b_{1,q_{max}}$ based on $v_{1,q_{max}}$ and predicting those of $\{b^*_{j,q_{max}} | \forall t \in [l_s]\}$, as defined as follows:

$$rect^*_{t,q_{max}} = \left(x^*_{t,q_{max}}, \ y^*_{t,q_{max}}, \ w^*_{t,q_{max}}, \ h^*_{t,q_{max}}\right)$$
$$= \left(x_{1,q_{max}} + t * v^x_{1,q_{max}}, \ y_{1,q_{max}} + t * v^y_{1,q_{max}}, \right. \tag{6}$$
$$\left. w_{1,q_{max}}, \ h_{1,q_{max}}\right), \forall t \in [l_s].$$

Here, $(x^*_{t,q_{max}}, y^*_{t,q_{max}}, y^*_{t,q_{max}}, w^*_{t,q_{max}})$ is the left and top coordinates of the predicted bounding box $b^*_{t,q_{max}}$. We assume that the width and height of $b^*_{t,q_{max}}$ are static in this short time. Finally, we decide the sampling frame rate $fs_{i,j}$ based on the reciprocal of the longest interval from 1-th frame to the t-th frame whose predicted bounding box $b^*_{t,q_{max}}$ shows IoU above the desired threshold η_v (By default, $\eta_v = 0.5$)with $b_{1,q_{max}}$. In practice, we choose the closest one among S for the sampling frame rate $fs_{i,j}$.

$$fs_{i,j} = \frac{fps_{def}}{max\{t | IoU(b^*_{t,q_{max}}, b_{1,q_{max}}) > \eta_v, \forall t \in [l_s]\}}. \tag{7}$$

To minimize the accuracy degradation, we prevent the prediction from being far from $b_{1,q_{max}}$ in a certain level with the desired threshold and adapt the sampling frame rate depending on how fast the prediction exceeds the desired threshold over l_s frames.

By default, our VA system runs only based on the frame rate $fs_{i,j}$ of each live video stream decided by the aforementioned algorithm for j-th segment. Then, the default configuration of video streams is denoted as $C^0_j = \{fs^0_{1,j}, fs^0_{2,j}, \ldots, fs^0_{I,j}\}$.

However, if $PT(C_j^0)$ exceeds τ, we additionally adapt the frame rates of video streams in C_j^0. That is, in order to remove the total exceeded time $ET = PT(C_j^0) - \tau$, the configuration C_j^0 is adapted to be cheaper.

Firstly, ET is split into I smaller segments, denoted as $ET_i, i = 1, \ldots, I$, and distributed to all live video streams. Then, each live video stream is required to reduce ET_i from $PT(C_j^0)$. ET_i is decided in proportion to $fs_{i,j}$ fairly because $fs_i(t)$ reflects the resource demand of i-th live video stream, defined as follows:

$$ET_i = \left(PT(C_j^0) - \tau\right) * \frac{fs_{i,j}}{\sum_{i \in [I]} fs_{i,j}}. \tag{8}$$

Secondly, each video stream $i = 1, \ldots, I$ independently starts to adapt its frame rate $fs_{i,j}^0$ cheaper for reducing ET_i. The adaptation is conducted one by one iteratively in a greedy way for maximizing the resource consumption reduction while minimizing the accuracy degradation until $ET_i \leq (fs_{i,j}^0 - fs_{i,j}) * l_p$. As a result, $C_j = \{fs_{1,j}, fs_{2,j}, \ldots, fs_{I,j}\}$ is determined.

3 Evaluation

In our experiment, we deploy our implemented VA system on the physical machine equipped with CPU, Intel(R) Core(TM) i7-6700 CPU @ 3.40 GHz, GPU, GeForce GTX 1080(8 GB), memory 16 GB in order to process all analysis tasks on video streams. The VA task is a DNN based object detection. We use CUDA 9.0 and cuDNN 7.0 [7] to accelerate the DNN speed based on a GPU. We use Keras 2.2.4 [6] APIs with Tensorflow framework 1.12 [10] in order to execute an object detection model of Yolov3. We determine a particular DNN called Yolo, which input size is 608×608. It can detect 80 object classes and are pre-trained on the COCO image dataset. We use a subset of surveillance videos and annotations from VIRAT 2.0 Ground dataset [8]. Based on them, we construct 9 video streams from nine real-world static cameras deployed in different area. Each video stream generates 10800 frames totally in 30 fps and 1920×720p (for 6 video streams) or 1280×720p (for 3 video streams). We implement optical flow algorithm used in the proposed algorithm by Lucas–Kanade method.

In these experiments, the proposed algorithm was compared with existing aforementioned well-known algorithm in existing VA systems such as Chameleon, which are based on online profiling. For the existing algorithm, we use the original parameters and configurations described in its paper, such as segments of profiling window $w = 4$, segment interval $T = 4$, profiling interval $t = 1$. We assume the maximum performance we can achieve is the analytics performance with the expensive configuration (i.e. 30 fps) and it is denoted as Pure in this paper. In order to evaluate the VA performance of the proposed algorithm on accuracy and resource consumption, we measure F1 score, the harmonic mean of precision and recall, and average GPU processing time per frame. We calculate an accuracy of a single frame by comparing the detected objects with the ground truth box. We identify true positives in the F1 score using a label-based condition, which checks if the bounding box has the same label and sufficient spatial overlap with some ground truth box. This IoU threshold is set as 0.5.

In this experimental environment, we evaluate the VA performance of each algorithm by measuring accuracy (F1 score) and resource consumption (GPU processing time) for multiple video streams in limited resource scenario. We apply several video streams concurrently (from 2 video streams to 9 video streams) among 9 video streams. Figure 2 shows the accuracy and normalized accuracy performance of proposed algorithm over multiple video streams. The x-axis represents the number of streams to be processed, ranging from two to nine. The proposed algorithm shows higher accuracy in every cases compared to existing VA systems.

In Fig. 2a, b, the proposed algorithm shows higher accuracy for multiple streams in every cases. In Fig. 2b, the accuracy of the proposed algorithm is almost constant above five streams, since the configuration is already adapted to the cheapest one decided within the proposed algorithm's constraint. By relaxing this constraint we impose, it is possible to reduce the resource consumption although the accuracy decreases. However, we judge that the accuracy below the certain level is not meaningful, so this constraint is needed to guarantee the least desired accuracy.

Figure 3 shows the latency performance of the proposed algorithm over multiple video streams. The proposed algorithm also shows better performance on resource consumption in every cases. However, as the number of streams increases, the resource consumption of Chameleon drops sharply from two streams to eight streams while those of the proposed algorithm falls and stops to a proper level from six streams. Compared to Chameleon, the proposed algorithm finds better configurations on accuracy and resource consumption by profiling. Meanwhile, Chameleon, which profiles limited configuration candidates basically, profiles fewer configuration candidates feasible in divided resource allocated to each video stream. Obviously, it is not enough to find efficient configurations and realize a desired accuracy. Especially, as the number of video streams increases and the resource allocated to each video stream decreases, configuration candidates to be profiled decreases. Consequently, although reducing significantly its profiling load and resource consumption, it shows unacceptable accuracy with this deficient profiling. The resource consumption of the proposed algorithm is also almost constant from five streams with its aforementioned constraint.

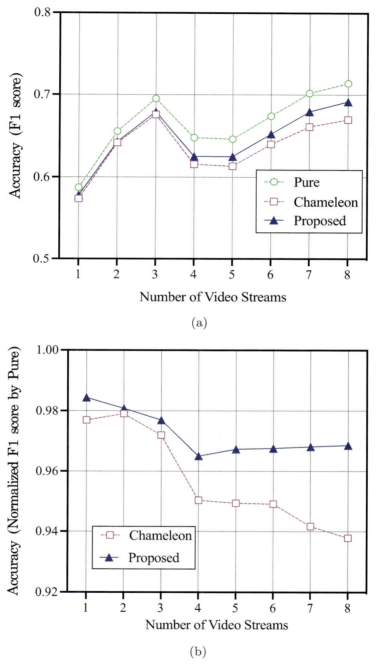

Fig. 2 Accuracy performance of the proposed algorithm over multiple video streams, (**a**) F1 score, (**b**) Normalized F1 score by Pure

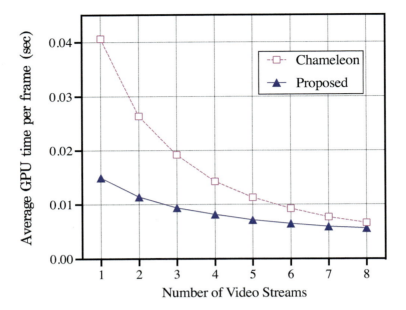

Fig. 3 Latency performance of the proposed algorithm over multiple video streams (Average GPU time per frame)

4 Conclusion

In this paper, we propose a low-cost VA system with velocity based configuration adaptation to find the best configuration decision of frame sampling rate for multiple video streams in order to minimize the accuracy degradation in the shared limited resource. Firstly, the frame sampling rate knob of each video stream is adapted optimally based on the velocity feature of objects extracted from its video context in low-cost. Secondly, the frame sampling rate knob of each video stream is adapted additionally in a greedy way, considering a limited resource shared by multiple video streams. As a result, the proposed VA system outperforms the existing VA systems in terms of accuracy and resource consumption.

Acknowledgments This work was partly supported by the Institute for Information and Communications Technology Promotion (IITP) grant funded by the Korean Government (MSIT) (No. 2017-0-00294, Service Mobility Support Distributed Cloud Technology) and (No. 2020-0-00537, Development of 5G based low latency device—edge cloud interaction technology).

References

1. Z. Fang, D. Hong, R.K. Gupta, Serving deep neural networks at the cloud edge for vision applications on mobile platforms, in *Proceedings of the 10th ACM Multimedia Systems Conference* (ACM, New York, 2019)
2. C.-C. Hung, et al., Videoedge: processing camera streams using hierarchical clusters, in *2018 IEEE/ACM Symposium on Edge Computing (SEC)* (IEEE, Piscataway, 2018)
3. J. Jiang, et al., Chameleon: scalable adaptation of video analytics, in *Proceedings of the 2018 Conference of the ACM Special Interest Group on Data Communication* (ACM, New York, 2018)
4. S. Jain, et al., Scaling video analytics systems to large camera deployments, in *Proceedings of the 20th International Workshop on Mobile Computing Systems and Applications* (ACM, New York, 2019)
5. D. Kang, et al., NoScope: optimizing neural network queries over video at scale. Proc. VLDB Endowment **10**(11), 1586–1597 (2017)
6. Keras: the python deep learning library. https://keras.io/
7. NVIDIA developer. https://developer.nvidia.com/
8. S. Oh, et al., A large-scale benchmark dataset for event recognition in surveillance video, in *IEEE Conference on Computer Vision and Pattern Recognition CVPR 2011* (IEEE, Piscataway, 2011)
9. X. Ran, et al., Deepdecision: a mobile deep learning framework for edge video analytics, in *IEEE INFOCOM 2018-IEEE Conference on Computer Communications* (IEEE, Piscataway, 2018)
10. Tensorflow: an open source machine learning library for research and production. https://www.tensorflow.org/
11. H. Zhang, et al., Live video analytics at scale with approximation and delay-tolerance, in *14th USENIX Symposium on Networked Systems Design and Implementation (NSDI'17)* (2017)

Hybrid Resource Scheduling Scheme for Video Surveillance in GPU-FPGA Accelerated Edge Computing System

Gyusang Cho, Seong-Hwan Kim, and Chan-Hyun Youn

1 Introduction

A smart-city is a city that provides the information needed to efficiently manage assets and resources using various types of electronic data collection sensors. The data used in smart-city is in various forms, e.g., IoT(Internet of Things) sensor data, crowd-sourced data. To process the various kinds of data, lots of brand-new technologies incorporate to develop the quality of life to the citizens. One of the most important application that was widely explored was the video surveillance system, since the system itself can ensure safety and detect illegal behavings of others. Under several applications [3, 5, 8, 21], we focus on the scenario of detecting illegally parked vehicles with object re-identification. Several studies explore this area. [8, 21]. We focus on the scenario of [8], which considers multi-observer-environment, since it is considered more applicable to adapt to the real-world problem.

The illegally parked car detection system proposed in [8] consists of multiple stages as shown in requests in Fig. 1. Stage 1 is the object detection stage, and single shot detection(SSD) [13] is used to detect object from the frame of video. The shots of cars are extracted from the videos in this stage by SSD. Next in stage 2, we extract additional features from different frames. The extracted features are Scale Invariant Feature Transform(SIFT) [14] features and Convolution Neural Network(CNN) features. When applying SIFT, we acquire points of local extremas from the image. We then pool the points by VLAD pooling [7], and adjust the dimension of the feature vector. CNN features are extracted by AlexNet [11], with

---

G. Cho · S.-H. Kim · C.-H. Youn (✉)
School of Electrical Engineering, Korea Advanced Institute of Science and Technology, Daejeon, South Korea
e-mail: cks1463@kaist.ac.kr; s.h_kim@kaist.ac.kr; chyoun@kaist.ac.kr

© Springer Nature Switzerland AG 2021
H. R. Arabnia et al. (eds.), *Advances in Artificial Intelligence and Applied Cognitive Computing*, Transactions on Computational Science and Computational Intelligence, https://doi.org/10.1007/978-3-030-70296-0_49

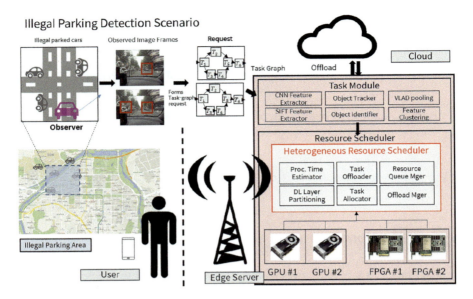

Fig. 1 The flow of video surveillance system. The images or frames recorded by observers are processed at edge server and cloud. We aim to design a scheduling algorithm shown in the figure with red box

excluding final soft-max layer. In stage 3 and 4, we compare the extracted features with the reference object which was priorly in the city database. We use Euclidean distance to compare with the reference feature. In this way, we detect the illegally parked vehicles and refer them to the vehicle of the city database.

The video surveillance system has huge workload, including computationally intense jobs such as image processing stages and deep learning stages. We aim to schedule the video surveillance system at servers located at edge computing system rather than cloud computing system, to accelerate the whole process by general purpose processors and reduce the communication cost. However, when the edge server is full of task, we aim to offload the rest of the process at cloud. This approach is well explored in numerous applications [6, 10]. The whole task flow and its process in edge server and cloud are shown in Fig. 1.

The contribution of this paper is as follows.

- We develop a scheduling algorithm applicable to the video surveillance system, with considerations of the limited heterogeneous accelerators in the edge server with considerations of cloud offloading.
- We develop a volume-based execution time model for the convolutional neural network, which is applicable to both accelerators, GPU(Graphics Processing Unit), FPGA(Field Programmable Gate Array).

2 Problems in Resource Scheduling in the Heterogeneous Accelerated Edge Environment

In this section, we identify the hardships of scheduling the video surveillance system to the heterogeneous accelerated edge servers.

2.1 Task Graph Complexity

Tasks such as illegal parking detection or other real-world tasks include combination of different stages, due to the demanded high quality from users. Especially, the real-time services require the deadline of few hundred milliseconds. To preserve the quality and capacity of the service of presenting it real-time, it is inevitable to assemble a successful graph which is complex, to accomplish the given problem.

Complex task graphs, also known as directed acyclic graph(DAG), include the serial and parallel tasks. Serial tasks should be accomplished in order, while parallel tasks can be achieved in different machines concurrently. Including homogeneity in resources and queueing time, scheduling the tasks from complex task graph is an NP-hard problem [19] with yielding a complicated version of the knapsack problem.

When executing the task graph, unexpected situation such as machine failure, or performance degradation can happen. Especially with numerous tasks in a single task graph, the probability of deadline matching to a single request will dramatically be higher than a graph of single task. To target this issue, we reschedule the remaining tasks to the accelerators to adjust to the deadline.

2.2 Scheduling Hardships

Here we introduce hardships that come from setting the constraints in the scheduling algorithm.

2.2.1 Heterogeneity in Accelerators

Deep learning task is commonly processed and scheduled to the general purpose processors. The two different accelerators, GPUs and FPGAs, have different sequence to process the same task.

One of the most famous processors is GPUs, with convenience of processing matrix-wise floating point operations. GPUs optimize the operations with basic linear algebra subprograms(BLAS) with CUDA [16] or OpenCL [20], depending on the vendor. Deep learning frameworks such as Tensorflow [1], Pytorch [18], Caffe [9] then implements the deep learning layers or optimizers with the methods.

The other approach is to use FPGAs. Recent researches spotlight FPGAs as the new accelerators to deep learning applications. FPGAs yield less kernel time for the deep learning operations because the device is simply hard-coded with Hardware Description Language(HDL). However, the effort for developing on FPGAs is quite considerable compared to GPUs, because low-level language such as OpenCL [20] is required. Likewise, the characteristics of the accelerators are different, and thus have different processing time.

2.2.2 Resource Limitation in Edge Environment

The next constraint is that the resources are not always available. The number of resources is limited, which is different from the cloud environment, where we consider that available resource exists at anytime. Then assuming D as the deadline, the relaxed version of scheduling of the task to minimize the latency for the user can be transformed into a knapsack problem [15], where

$$\begin{aligned}
&\text{maximize } \Sigma_{j=1}^{R}(\Sigma_{i=1}^{n} w_i t_i),\\
&\text{subject to } \Sigma_{i=1}^{n} w_i t_i \leq D, t_i \geq 0, \text{ and } w_i \in \{0, 1\}
\end{aligned} \tag{1}$$

which is proven to be NP-complete while making the decision and NP-hard to be optimized [15].

In the task graph scheduling problem, we need to consider more of the objectives such as Quality of Service(QoS), average latency, or first-come-first-served manner. Therefore number of good heuristics are suggested [4] to solve the given application, which would be introduced in the next sections (Fig. 2).

3 Proposed Scheduling Algorithm for Heterogeneous Accelerators

In this section, we first derive an estimation time model for profiling layers in convolutional neural network. Based on the model, we partition the CNN task into subtasks layer by layer, and schedule the subtask to the proper accelerator. This is to adapt to the heterogeneous environment, where GPUs and FPGAs exhibit different processing time. We explore the detailed scheduling scheme afterwards.

3.1 Profiling Layers in CNN Network

The execution time for each layer is derived based on the [2, 12], with summation of computation time and communication time. Denoting g_l as number of operations

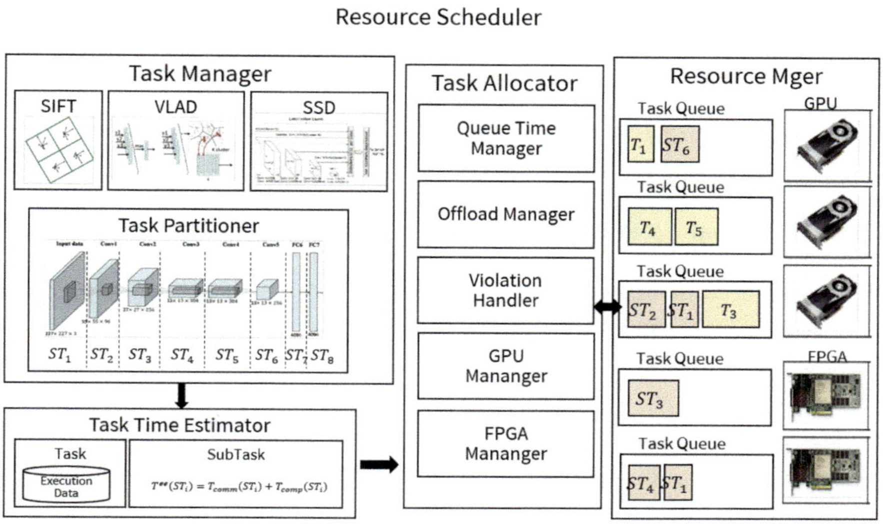

Resource Scheduler

Fig. 2 Illustration of the resource scheduling module shows the specific functions in using GPU and FPGA

in layers, I as size of input, and B as the bandwidth of device, we formulate the execution time model as $T(exec) = P \times g_l + Q \times I/B + R$. For heterogeneous devices, we try to acquire hyperparameters P, Q, R experimentally. I, g_l would be determined when the layer is specified. We partition the task T_i into subtasks, and leave each subtask to include convolution layer or fully connected layer. This is because the two layers are most computing-intensive.

3.1.1 Convolution Layer

Different deep learning frameworks such as Pytorch [18], Tensorflow [1], Caffe [9] implement convolution layer differently. They use different libraries to optimize the operation, so various kinds of convolutions such as direct convolution, GEMM based convolution, FFT-based convolution exist. However, the number of operations is neglectable, and we only consider the direct convolution.

The authentic way of performing convolution is by computing matrix multiplication directly. It follows standard convolution algorithm as it is. We denote the input I is of 3-dimension, with size $I_x \times I_y \times I_z$. The number of $K \times K$ sized filters in convolution layer is denoted as N_f. A single computation as one multiplication, the total number of computations can be calculated as

$$g_l \propto K^2 N_f I_x I_y I_z \tag{2}$$

with assuming filter size is small compared to input size.

The communication time can be formulated by memory operation size. The value can be determined by input size I and the size of the weight of the layer $S_l = K^2 \times N_f$. Overall, the convolution layer execution time can be written as Eq. 3.

$$T(Conv) = P \times K^2 N_f I_x I_y I_z + Q_1 K^2 N_f + Q_2 I_x I_y I_z + R \tag{3}$$

3.1.2 Fully Connected Layers

We denote the vector size of input and output as U_I, U_O. With similar procedures as the convolution layer, we derive the execution time of the fully connected layer as,

$$T(FC) = P \times U_I U_O + Q_1 \times U_I + Q_2 \times U_O + R \tag{4}$$

By Eq. 4, we can acquire the processing time of fully connected layer.

3.2 Assumptions and Constraints

m denotes the number of present GPUs in the edge servers, and n denotes the number of present FPGAs in the edge server. The deadline D is pre-announced by the users to the whole system. Each resource is represented as s_j, where j with value 1 to m stands for GPU, and $m + 1$ to $m + n$ stands for FPGA. The tasks are denoted as T_i, with tasks image preprocessing, CNN feature extraction, SIFT extraction, VLAD pooling, object re-identification, and distance comparison process as T_1 T_6, respectively. Each resource contains task queue, and the estimated ending time of the queue belonging to resource s_j is $T_q^{s_j}$. We assume that the scheduling of the whole video surveillance system is scheduled at GPUs or FPGAs in edge computing environment. We assume that the cloud has infinite resources, and always has a possible vacant resource that the service can be started right away. Also we assume that only a type of GPU or FPGA is used, and the CNN feature extraction task, T_2, is available of use of FPGAs. Single task is scheduled at a single resource, except for T_2, while T_2 can be layer-wise partitioned and scheduled at other device. The first assumption is about setting up the environment. Other assumptions are made to simplify the problem, but can be generalized afterwards.

First we leave out FPGAs, since task 2 only considers the usage of FPGAs. The scheduling scheme is divided into 3 phases. The first phase is initiated on request. Phase 2 and phase 3 are called for rescheduling, which is initiated when the subdeadline violation occurs.

3.3 Phase 1: Compute Subdeadline, and Schedule

In the first phase, the preparation for all tasks is made to be scheduled to the proper resource. We assign subdeadline for each task, in order to keep track of the task to be done in time.

3.3.1 Estimate the Execution Time

In the first step in phase 1, we estimate the execution time for each task. We suppose that the execution time is pre-defined, with executed several times before and already measured. We take the average execution time, with denoting the execution time for the resource as $T^{ee}_{s_j}(T_i)$.

$$T^{ee}(T_i) = \Sigma^m_{j=1} T^{ee}_{s_j}(T_i)/m \tag{5}$$

3.3.2 Find the Most Time-Consuming Path

The most time-consuming path is determined at next step. We first identify paths in the task graph, and find the most time-consuming graph by adding the average task execution time for all paths. In this application, we can find two of the paths L_1, L_2 in the task graph.

For path L_v, we compute the execution time of the path by Eq. 6,

$$T^L(L_v) = \Sigma_{T_i \in L_v} T^{ee}(T_i) \tag{6}$$

Next, we compute the most time-consuming path with the value $T^L(L_v)$, by Eq. 7,

$$T^{dl}_L = \max_{1 \leq v \leq L}(T^L(L_v)) \tag{7}$$

T^{dl}_L here will be the average full execution time of whole system, with no delay with no use of FPGAs.

3.3.3 Find the Subdeadlines to Each Task

We define subdeadlines to each task, to manage the occurrence of the deadline violation. When the subdeadline violation happens, it can be used as a signal of possible deadline violation. Then the situation can be handled by rescheduling phases which we will discuss later in the paper.

The subdeadlines can be distributed proportionally to each task's average execution time. We denote $T^{sdl}(T_i)$ as the subdeadline for the task T_i. Then the subdeadline for the task T_i in most time-consuming path will be distributed following Eq. 8,

$$T^{sdl}(T_i) = D \times \frac{T^{ee}(T_i)}{\Sigma_{T_i \in L} T^{ee}(T_i)} \tag{8}$$

The subdeadlines are computed proportionally to each step. We denote $T^{sdl}(T_i)$ as the subdeadline for the task T_i. The subdeadlines can be computed as the Eq. 9.

Additional subdeadlines for the tasks that are not included in the most time-consuming path can also be distributed proportionally to the average execution time of the parallel tasks on most time-consuming path. We denote L_{sub} as the sub-path of most time-consuming path, and the subdeadline for L_{sub} as D_{sub}.

$$T^{sdl}(T_i) = D_{sub} \times \frac{T^{ee}(T_i)}{\Sigma_{T_i \in L_{sub}} T^{ee}(T_i)} \tag{9}$$

3.3.4 Schedule to the Fastest Resource

Based on the subdeadline acquired from above steps, we actually schedule the tasks to the proper resources. The scheduling happens in the greedy-manner, i.e., the task is scheduled to the fastest vacant resource.

In the resource point of view, each resource $s_j (1 <= j <= m)$ has its task queue with tasks. The task queue of resource s_j is filled with tasks $W_1^j, W_2^j, \ldots, W_{k_j}^j$. We first compute the task queue ending time, $T_q^{s_j}$.

$$T_q^{s_j} = \Sigma_{k=1}^{k_j} T_{s_j}^{ee}(W_k^j) \tag{10}$$

Then the task T_i is scheduled to the fastest-vacant resource with smallest task queue ending time, $T_q^{s_j}$.

$$T_i \rightarrow \arg_j \min_{s_j} (T_q^{s_j}) \tag{11}$$

3.4 Phase 2: Runtime Rescheduling to Use FPGAs

In this phase, we make a slight modification to the to-be-scheduled resource in order to make FPGAs in use. By this phase, we can make the algorithm sturdy to unexpected situations.

3.4.1 Runtime Rescheduling Task 2 for Use in FPGAs

In this step we delve into task 2, which is a stage of CNN feature extraction. We run an inference task on pre-trained AlexNet [11] based model. We split task 2 into subtasks with including one convolution layer or fully connected layer to the subtask. In this case, we split into 8 subtasks. This is because convolution layers and fully connected layers are the most computing-intensive, and could be accelerated in FPGAs. By the information of execution time model of each subtask we derived at Sect. 3.1, we can approximate the execution time of the subtask on heterogeneous accelerators. We look into available resources, compare the ending time, and reschedule the subtasks using the FPGAs.

Next, we compute the subtask execution time on both resources. We estimate by computing from the formulation in Sect. 3.1. The scheduling is done similar to the step 4 in phase 1, but this time we compare the fastest subtask ending time.

Subtasks are denoted as ST_i, and for each subtask, the average execution time based on Sect. 3.1 can be denoted as $T_{GPU}^{ee}(ST_i)$, $T_{FPGA}^{ee}(ST_i)$. Then we can refer to $T_{s_j}^{ee}(ST_i)$ with $T_{GPU}^{ee}(ST_i)$, $T_{FPGA}^{ee}(ST_i)$, with just identifying what the resource s_j is.

We compute the task queue ending time for each accelerator again, as Eq. 12,

$$T_q^{s_j} = \Sigma_{k=1}^{k_j} T_{s_j}^{ee}(W_k^j) \tag{12}$$

where we assume the task queue of resource s_j is filled with jobs $W_1^j, W_2^j, \ldots, W_{k_j}^j$. We compute the ending time to each resource by calculating,

$$T_{est}^{s_j} = T_q^{s_j} + T_{s_j}^{ee}(ST_i) \tag{13}$$

where $T_{est}^{s_j}$ denotes the ending time of the resource. We schedule to the shortest ending time resource.

$$ST_i \rightarrow \arg_j \min_{s_j}(T_{est}^{s_j}) \tag{14}$$

3.5 Phase 3: Runtime Rescheduling for Deadline Adjustment with Cloud Offloading

In phase 3, we reschedule the violated tasks. By applying this, we might enhance the performance of the whole system, and can meet the deadline. This phase can be initiated when subdeadline violation happens, or when the task 2 is terminated with shortening the execution time with usage of FPGAs.

3.5.1 Re-Computing Subdeadlines and Re-Assigning

As in steps 1 to 3, we recompute the most time-consuming path to the rest of the task graph. Denoting L_{new} as the new path and D_{new} as the new deadline, which the value would be the full deadline D minus the current time. The new subdeadlines are again proportionally spread to each remaining task. After that, the rest of the task graph is scheduled to the fastest-vacant resource. We consider offloading to the cloud if the estimated task ending time exceeds that of processing at the cloud. In other words, cloud offloading happens when the Eq. 15 is true.

$$arg_j min_{s_j} T_q^{s_j} + \Sigma_{T_k \in L_{new}} T_{s_j}^{ee}(T_k) > t_{cloud} \tag{15}$$

Algorithm 1 Resource scheduling scheme of video surveillance system on heterogeneous accelerated edge computing system

---

Input: Tasks T_i, Deadline D
Output: Scheduled Resource s_j
1: Compute expected execution time $T^{ee}(T_i)$ // Phase 1
2: Find the most time-consuming path L
3: **for** $i = 1$ to 6 **do**
4:     Assign subdeadline T^{sdl}
5: **end for**
6: **for** $i = 1$ to 6 **do**
7:     **for** $j = 1$ to m **do**
8:         Calculate $T_q^{s_j}$
9:     **end for**
10:     Schedule T_i to the s_j with smallest $T_q^{s_j}$
11: **end for**
12: **for** Subtasks ST_i **do**
13:     Calculate $T_{GPU}^{ee}(ST_i), T_{FPGA}^{ee}(ST_i)$ // Phase 2
14:     $T_{est}^{s_j}(ST_i) \leftarrow T_q^{s_j} + T_{s_j}^{ee}(ST_i)$
15:     Schedule to the s_j with smallest $T_{est}^{s_j}(ST_i)$
16:     Update $T_q^{s_j}(ST_i)$
17: **end for**
18: **if** Task T_i ends with exceeding $T^{sdl}(T_i)$ **then**
19:     **if** $arg_j min_{s_j} T_q^{s_j} + \Sigma_{T_k \in L_{new}} T_{s_j}^{ee}(T_k) < t_{cloud}$ **then**
20:         **for** Left tasks T_i **do**
21:             Reschedule to possible resource s_j // Phase 3
22:             $T^{sdl}(T_i) = D_{new} \times \frac{T_{s_j}^{ee}(T_i)}{\Sigma_{T_k \in L_{new}}(T_k)}$
23:         **end for**
24:     **else**
25:         Offload to the remaining tasks to the cloud
26:     **end if**
27: **end if**

Hybrid Resource Scheduling Scheme for Video Surveillance in GPU-FPGA...

Table 1 Edge server hardware information

Type	Device name	Specification
GPU	NVIDIA Quadro M2000	768 NVIDIA Cores 4 GB GDDR5 GPU Memory PCI Express 3.0×16 Interface Max Power Cons.: 75 W
FPGA	Intel Altera Arria 10 GX	Dual ARM Cortex-A9 MPCore Processor 1.5 GHz Logic Core 500 MHz 53 Mbit DDR4 SDRAM Memory PCI Express 3×8 Interface
CPU	Intel E5-2620v4	2.1 GHz
RAM		DDR4, 24 GB

4 Performance Evaluation

4.1 Experiment Environment

We set up a simulation environment for evaluating the scheduling algorithm. We used a single type of GPU, and a single type of FPGA. We used GPU of NVIDIA Quadro M2000, and FPGA of Intel Altera Arria 10 GX. The specifications of the memory, processor frequency could be found at Table 1. We compared average processing time, resource utilization, queue waiting time, and deadline violation rate with the other scheme, so called as Bestfit algorithm [17].

For the requests, we assume that the requests arrive in Poisson distribution, with varying λ. The scheduling is simulated in the environment of 3 GPUs and 3 FPGAs.

4.2 Performance Metrics

Good scheduling scheme should yield good use of resources, and guarantee good quality of service. We first compare resource utilization rate for each scheme in regards of request arrival interval, to examine whether the scheduling scheme holds well of resource usage on different circumstances. Next, to observe quality of services, we measure average processing time, queue waiting time, deadline violation rate. The average processing time and deadline violation time can ensure the distribution of the processing time. Also to increase user convenience, queue waiting time should be small of which can be considered as redundant time.

For the assessment of average processing time and queue waiting time, we examine how the value changes over the increase in request arrival interval.

4.3 Experimental Results

4.3.1 Impact of the Proposed Scheme

We compare performance of the considered scheme at different task arrival rates. First we take a look at Fig. 3. Figure 3 shows the resource utilization. Without cloud offloading, increasing utilization of all possible accelerators in edge computing system will be time and energy efficient. According to the plot, the resource utilization of GPU does not differ at every time interval. However, the resource utilization of FPGA does seem to be different in the high request arrival rate, with low request arrival interval. Resource utilization is the productive use of resources. With the high request arrival rate, the proposed scheme yields better utilization than that of BestFit algorithm.

We next compare average processing time, queue waiting time, deadline violation rate in Fig. 4. The above metrics are obtained with When the edge server is not full of work with high request arrival interval, the two algorithms show similar performance. However, when the edge server becomes busy with low request arrival interval, proposed algorithm has smaller average processing time than the

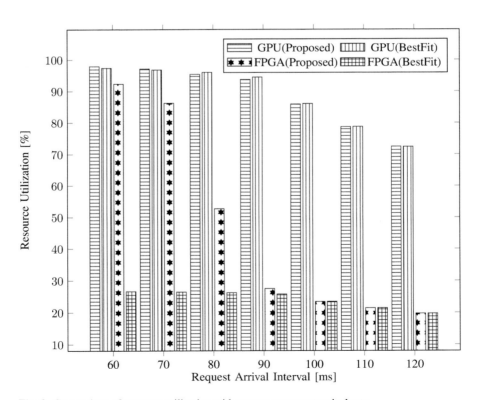

Fig. 3 Comparison of resource utilization with respect to request arrival rate

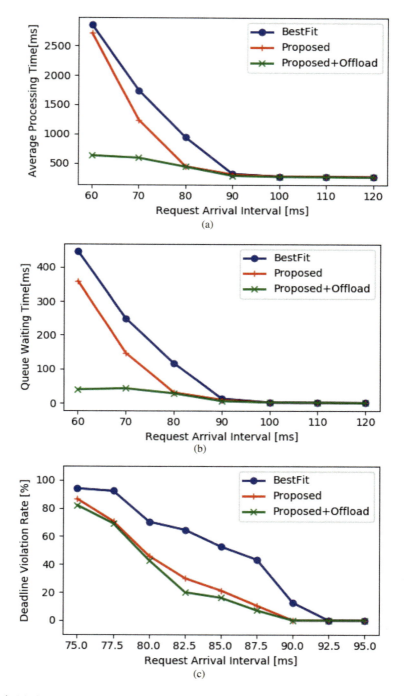

Fig. 4 (**a**) Average processing time, (**b**) Queue waiting time, (**c**) Deadline violation rate with respect to request arrival interval. Deadline D is set to 500 ms, and the cloud offloading time is set to 300 ms

BestFit algorithm. Queue waiting time, which represents the length of the request's idle time, where the task of it is not being processed and waiting for its turn in the resource queue. Each of the task T_i which is the part of a request can have its queue waiting time, and we compute the summation of them. According to the experimental result, in the high request arrival rate, with low request arrival interval, the queueing time is preserved at most 100 ms, with the proposed algorithm compared to the BestFit algorithm. Proposed algorithm exhibits more freedom to the task, with the resource selection. Also, the proposed algorithm exhibits at most 40% difference in deadline violation rate, when the request arrival interval is 87.5 ms. The proposed algorithm is more robust to meeting the deadline.

4.3.2 Impact of the Cloud Offloading

We include cloud offloading scheme. When the edge servers reach the point of incapability in work, it is better to offload the tasks to the cloud server. From Fig. 4, we can examine that average processing time dramatically decreases at low request time interval. These cases can be seen as when the queue waiting time exceeds the cloud communication time. We can successfully decrease the workload for edge servers and send the jobs to cloud servers. Also, deadline violation rate decreases with offloading, when the deadline is set to 500 ms, which is meetable when the task is offloaded at the beginning.

5 Conclusion

In this paper, we proposed a video surveillance system task graph scheduling scheme for heterogeneous GPU-FPGA accelerated edge server. We also delved into the layers of the CNNs to be processed at GPU and FPGAs, and estimated execution time for the proposed scheduling algorithm. To our best knowledge, there are no scheduling scheme considering edge servers, cloud offloading, heterogeneous accelerators with GPUs and FPGAs, deep learning based task-partitioning, and task graph scheduling in the same time. Our proposed algorithm suggests a heuristic to support the above constraints, and provides scheduling in a greedy-manner. Our proposed algorithm schedules the partitioned subtasks of Convolutional Neural Network according to the derived formulation of execution time of each layer.

To verify the performance of the algorithm, we have conducted a simulated experiment to support large environment. Results are derived from the experiments and it outperforms than the prior scheduling scheme in performance metrics of average processing time, resource utilization, queue waiting time, and deadline violation rate.

We leave optimizing other tasks to FPGAs to future work. Object detection, SIFT or VLAD pooling can also be processed at FPGAs. The predictable difficulties might be difficulties such as, optimizing the whole tasks to the size of the reference board, or deriving heuristics for allocating each resource with extended allocation scheme.

Acknowledgments This work was partly supported by "The Cross-Ministry Giga KOREA Project" grant funded by the Korea government(MSIT) (No.GK20P0400, Development of Mobile-Edge Computing Platform Technology for URLLC Services) and in part by Samsung Electronics, Device Solution (DS).

References

1. M. Abadi, A. Agarwal, P. Barham, E. Brevdo, Z. Chen, C. Citro, G.S. Corrado, A. Davis, J. Dean, M. Devin, S. Ghemawat, I. Goodfellow, A. Harp, G. Irving, M. Isard, R. Jozefowicz, Y. Jia, L. Kaiser, M. Kudlur, J. Levenberg, D. Mané, M. Schuster, R. Monga, S. Moore, D. Murray, C. Olah, J. Shlens, B. Steiner, I. Sutskever, K. Talwar, P. Tucker, V. Vanhoucke, V. Vasudevan, F. Viégas, O. Vinyals, P. Warden, M. Wattenberg, M. Wicke, Y. Yu, X. Zheng, TensorFlow: large-scale machine learning on heterogeneous systems. Software available from tensorflow.org (2015)
2. Y. Abe, H. Sasaki, S. Kato, K. Inoue, M. Edahiro, M. Peres, Power and performance characterization and modeling of GPU-accelerated systems, in *2014 IEEE 28th International Parallel and Distributed Processing Symposium, Phoenix* (2014), pp. 113–122. https://doi.org/10.1109/IPDPS.2014.23
3. A. Alshammari, D.B. Rawat, Intelligent multi-camera video surveillance system for smart city applications (2019). https://doi.org/10.1109/CCWC.2019.8666579
4. F.F. Boctor, Some efficient multi-heuristic procedures for resource-constrained project scheduling. Eur. J. Oper. Res. **49**(1), 3–13 (1990). ISSN 0377–2217
5. L. Calavia, C. Baladrón, J.M. Aguiar, B. Carro, A. Sánchez-Esguevillas, A semantic autonomous video surveillance system for dense camera networks in smart cities. Sensors **12**, 10407–10429 (2012)
6. S. Guo, B. Xiao, Y. Yang, Y. Yang, Energy-efficient dynamic offloading and resource scheduling in mobile cloud computing, in *IEEE INFOCOM 2016 - The 35th Annual IEEE International Conference on Computer Communications, San Francisco, 2016* (2016), pp. 1–9. https://doi.org/10.1109/INFOCOM.2016.7524497
7. H. Jegou, M. Douze, C. Schmid, P. Perez, Aggregating local descriptors into a compact image representation, in *Proceedings of IEEE Conference on Computer Vision and Pattern Recognition (CVPR)* (2010)
8. J.-H. Jeong, S.-H. Kim, M. Jeon, C.-H. Youn, Adaptive object re-identification based on RoI aware Sift-CNN hybrid feature clustering, in *International Conference on Artificial Intelligence (ICAI'19)* (2019)
9. Y. Jia, E. Shelhamer, J. Donahue, S. Karayev, J. Long, R. Girshick, S. Guadarrama, T. Darrell, Caffe: convolutional architecture for fast feature embedding (2014, preprint). arXiv:1408.5093
10. S.-H. Kim, S. Park, M. Chen, C.-H. Youn, An optimal pricing scheme for the energy efficient mobile edge computation offloading with OFDMA. IEEE Commun. Lett. **22**(11), 1922–1925 (2018)
11. A. Krizhevsky, I. Sutskever, G.E. Hinton, ImageNet classification with deep convolutional neural networks, in *Advances in Neural Information Processing Systems* vol. 25 (2012)
12. P. Lei, J. Liang, Z. Guan, J. Wang, T. Zheng, Acceleration of FPGA based convolutional neural network for human activity classification using millimeter-wave radar. IEEE Access **7**, 88917–88926 (2019).

13. W. Liu, et al., SSD: single shot multibox detector, in *European Conference on Computer Vision* (Springer, Cham, 2016)
14. D.G. Lowe, Distinctive image features from scale-invariant keypoints. Int. J. Comput. Vis. **60**(2), 91–110 (2004)
15. S. Martello, P. Toth, *Knapsack Problems: Algorithms and Computer Implementations* (Wiley, Hoboken, 1990)
16. NVIDIA, CUDA technology (2007). http://www.nvidia.com/CUDA
17. E. Oh, W. Han, E. Yang, J. Jeong, L. Lemi, C. Youn, Energy-efficient task partitioning for CNN-based object detection in heterogeneous computing environment, in *2018 International Conference on Information and Communication Technology Convergence (ICTC), Jeju* (2018), pp. 31–36. https://doi.org/10.1109/ICTC.2018.8539528
18. A. Paszke, S. Gross, S. Chintala, G. Chanan, E. Yang, Z. DeVito, Z. Lin, A. Desmaison, L. Antiga, A. Lerer, Automatic differentiation in PyTorch, in *NIPS 2017 Autodiff Workshop: The Future of Gradient-Based Machine Learning Software and Techniques, Long Beach* (2017)
19. B. Simon, Scheduling task graphs on modern computing platforms. Distributed, Parallel, and Cluster Computing [cs.DC]. Université de Lyon. English. NNT : 2018LYSEN022. tel-01843558 (2018)
20. J. Stone, D. Gohara, G. Shi, C.L. Open, A parallel programming standard for heterogeneous computing systems. Comput. Sci. Eng. **12**(3), 66 (2010)
21. Y. Zhou, L. Shao, Aware attentive multi-view inference for vehicle re-identification, in *Proceedings of the IEEE Conference on Computer Vision and Pattern Recognition* (2018)

Artificial Psychosocial Framework for Affective Non-player Characters

Lawrence J. Klinkert and Corey Clark

1 Introduction

A challenging task for video game designers is to make a realistic world for the player. One way to achieve a realistic world is to improve the human characteristics of a Non-Player Character (NPC). The 2018 AI Summit at GDC had a panel of video game designers discussing the necessary improvements for future NPCs [1]. The developers expressed several needs, and this paper addresses the following:

- AI that focuses on non-combat and allows NPCs to react to inputs at unpredictable moments.
- A system based on creditable research that covers the topic of human development.
- Working with NPCs that are not just good or bad but have varied characteristics.
- A framework implemented in a game engine that enables developers to balance values and create content.

Researchers in recent studies have advanced the understanding of NPC cognition by incorporating concepts from psychology [2]. In this paper, we define this new NPC class as an "Affective NPC" (ANPC). Bourgais et al. surveyed the progress in emotional modeling for social simulations to inform others of the potential architecture of ANPCs [2].

L. J. Klinkert
Guildhall, Southern Methodist University, Dallas, TX, USA
e-mail: jklinkert@smu.edu

C. Clark (✉)
Guildhall, Computer Science, Southern Methodist University, Dallas, TX, USA
e-mail: coreyc@smu.edu

© Springer Nature Switzerland AG 2021
H. R. Arabnia et al. (eds.), *Advances in Artificial Intelligence and Applied Cognitive Computing*, Transactions on Computational Science and Computational Intelligence, https://doi.org/10.1007/978-3-030-70296-0_50

A simple version of emotional modeling is reactive creation. The emotions felt by an ANPC are directly created by the perception of an event. The works from both Le et al. and Luo et al. demonstrate reactive creation for ANPCs in the context of an emergency evacuation of a building [3, 4]. If an ANPC perceives a threat, then its fear increases. As the fear increases and passes a threshold, then its behavior alters. The simplicity of this model allows for ease of implementation; however, this method does not rely on any emotional theory described in psychology [2].

A different approach to emotional modeling considers using fuzzy logic rules. Kazemifard et al. worked on COCOMO, a way to determine the cost of software development via simulating developers in a company [5]. The ANPCs (the developers) had pair values joy/distress and gratitude/anger with a value between 0 and 100, along with other metadata such as technical knowledge and level of soft skills. Fuzzy inference rules transform the values to calculate the emotional state, which, in turn, generates a mood for the ANPC. As the ANPC simulated the work and generated a mood, their behavior altered accordingly; for example, if an ANPC were in a good mood, they would work faster than one in a bad mood. Jones et al. also used fuzzy logic for emotional modeling but in the context of traffic simulation [6]. The calculation of fuzzy sets comes from the ANPCs perception of objects and events, such as their current speed, number of surrounding vehicles, and the percentage of trucks from the vehicles. From these perceptions, the desirability is calculated based on satisfying the ANPCs current goal, resulting in the ANPCs current emotional state. However, similar to reactive creation, fuzzy rules do not rely on any psychological theory about emotions [2].

Keeping psychological theories in mind, designers have integrated emotional behaviors for their ANPCs by using the "Cognitive Appraisal Theory" (CAT). From the works of Zoumpoulaki et al., ANPCs would calculate their emotions based on the appraisal of a situation, and in this case, are in the context of evacuations. The ANPC's appraisal of an event comes from the consequences of their current goal, modified by their personality. A 5-tuple vector represents the ANPC's emotional state, which holds a positive or negative value for each emotion. Multiplying the appraisal, personality, and current emotions, a final emotional intensity is calculated. The ANPC uses this emotional intensity to determine how to fulfill its current desire. Ochs et al. takes a step further and incorporates a social relationship for the ANPC with other ANPCs [7]. Ochs et al. implemented a complete version of the OCC model. However, the number of appraisals the designers have to assign, for every event, for each ANPC, will exponentially increase [2].

Lastly, designers are interested in the evolution of emotions over time. An ANPC starts with an initial emotional state, and the designers focus on the changes of emotions over time to observe how those changes affect their behavior. Faroqi and Mesgari evaluated ANPCs evacuating from an open square using different levels of emotion [8]. Each ANPC had six transitions in fear, from calm to hysteria. Once an ANPC reached hysteria, they would run in random directions. NPCs were added to the simulation to calm the ANPC as well as inform them of an exit. Suvakov et al. proposed that emotions be generated on a two-dimensional plane to simulate the spread of an emotion in an online social network [9]. While ANPCs

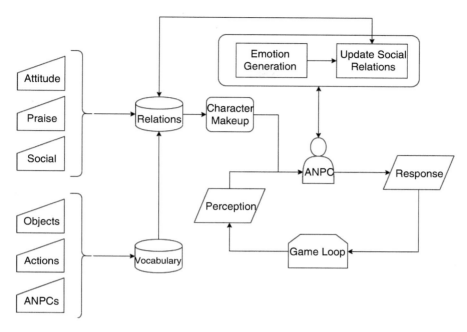

Fig. 1 Flow diagram of APF, starting from input data, representing an ANPC, a psychological evaluation, and then the continuation of the game loop

communicated through an online social network, variable valence and arousal were used with stochastic equations to model the evolution of their emotional state. This method makes a link to emotional generation and appraisal as well as the dynamic behavior using formal notation. The drawback from the evolutionary works is that the creation process is not fully explained, and would be much harder to implement into a system [2].

This paper presents the "Artificial Psychosocial Framework" (APF), an extension of the work proposed by Ochs et al. [7], which based their work on the CAT. Having APF based on the CAT allows developers to focus on creating content for the ANPC, such as dialog or voice-overs. Developers would assign values to the ANPC to represent their feelings towards other game objects, interactions, or other ANPCs. APF extends the works of Ochs et al. by utilizing an ambivalent hierarchy of emotions, classification in vector space, and intended for game engine integration. In psychology, ambivalence is the phenomenon of an individual feeling both positive and negative emotions at the same time. Hierarchical refers to the conditions that the actions experienced can alter the character's mood, emotions, and feelings. The classification allows designers to have a reference point to interpret the set of values when used in each model. Lastly, APF is implemented as a library so that developers can integrate the framework into their existing game engine as a plug-in (Fig. 1).

2 Overview

The principle of APF is on the psychological theory of the CAT. A cognitive appraisal states that a person evaluates a situation based on their criteria [2]. Consequently, due to personal criteria, individuals can experience a similar situation with different emotions, resulting in different actions. From this foundation, APF chains together psychological models to generate different states.

APF uses two popular psychological models, along with a third supplemental model. First, is the Big Five personality traits model: "Openness, Consciousness, Extroversion, Agreeableness, and Neuroticism" (OCEAN). Second, is emotion representation by the "Ortony, Clore, and Collins" (OCC) model. There is a third model used to represent social relations, which is from the works of Ochs et al. It Uses a 4-tuple to represent the "Liking, Dominance, Solidarity, and Familiarity" (LDSF) model [7].

Vectors are used to decompose each model, i.e., the OCEAN personality model is a 5-tuple vector where each element is in the range [0.0, 1.0]. The benefits of using a vector are that one can apply equations to manipulate the values and provides an ease of implementation into a program. The drawback is that the vector represents some point in n-dimensional Euclidean space that, at first glance, does not have contextual meaning. Thus, using named reference points, APF uses Multidimensional Scaling (MDS) to classify the point of interest. The classification gives meaning to the point and offers a better understanding of its interpretation.

Lastly, APF takes a step further from previous emotional agent frameworks for its development as a library. Previously, these frameworks were implemented as custom applications to solve theoretical problems. APF is a plug-in that works with commercial engines, such as the Unreal Engine, or proprietary engines. APF can work in tandem with other AI techniques, such as Fuzzy Logic, Behavior Trees, and Goal-Oriented Action Planning. Utilizing APF with AI allows developers to work on decision making from a different perspective and focus on emotion-centric gameplay.

3 Background Information

In psychology, a concept called Psychosocial focuses on the individual's psyche and their interactions with their social environment. The concept primarily focuses on how individuals conduct themselves based on the influences of relationships with other people, places, experiences, and themselves [10]. Maintaining these relationships will have varying repercussions and will inevitably come back around to the individual. Nearly all of us want to have a safe environment, but there are multiple variables to determine safety [11]. Do we have a sense of control in our lives, can we build a meaningful connection with others, do we feel safe from harm's way, or support ourselves or those closest to us? Answering these questions will require an understanding of valence.

Valence is defined as an affective quality that is expressed on a scale of attractiveness to aversion towards an event, object, or situation [12]. The amount of valence generated from a stimulus is subjective. Hence, when different people face the same situation, they can feel different emotions, which leads to performing different actions. For example, when given a photo of an individual and a few seconds to determine whether to talk to that individual or not, the decision is influenced by the valence generated from the appearance of the person in the photograph. APF attempts to quantify valence and is used as the base unit to calculate the emotions generated from a stimulus. Additionally, ambivalence is another term analogous to valence but infers conflicting emotions generated by a stimulus. Ambivalence causes individuals to have mixed feelings towards something, such as a love/hate relationship with someone. Having the base unit defined, we can now describe the evolutionary proses it goes through to generate an emotion, known as the "Cognitive Appraisal Theory" (CAT).

A cognitive appraisal is a subjective interpretation made from a person towards a stimulus in the environment [13]. The subjectivity is explicitly the valence emotion that the person has associated with the stimulus. Cognitive appraisal allows for a person's emotional state to be directly linked to the situation, rather than their physical response. Additionally, the connection between emotional state and situations allows an individual to respond to the situation, rather than instinctively react.

Stepping outside of Psychology and into Data Science, there is a technique called Multidimensional Scaling (MDS). MDS is a cluster analysis method that reduces the dimensions of a complex dataset. Complexity is explicitly referring to the data being represented as a four or higher-dimensional vector. The reduction in dimensions allows for the vector to be drawn onto a 2D or 3D graph, making the data easier to interpret and determine common groups. Using a vector allows for ease of manipulation with formal equations. One of the benefits and drawbacks of working with vectors is that we have an ample n-dimensional Euclidean space to manipulate the values. To a designer, there is no context in representing a point in this space. APF uses MDS to give the designer meta-information about the dynamic vectors used. It does this by using reference points authored by the designer. The reference points represent the center of a hypervolume. Within that hypervolume, other potential points are closely related to the reference point. Because the reference point and the generated point are so close to each other, APF labels the generated point with the same name as the reference point. Additionally, this process also allows designers to reverse engineer a vector, rather than to handwrite them. Designers can choose a reference point that feels right for their ANPC. Once selected, APF uses that point, but random offsets each axis by a small amount. The slight randomization allows the designers to continue to work with their creativity, rather than interrupt their thought process to think about mathematics.

4 A Priori Setup

For APF to work, there needs to be a set of vocabulary known to the system. The vocabulary is made up of all the living, nonliving, and potential actions to perform in the virtual world. Following the concept of Psychosocial, the vocabulary needs to have relationships with each other. The relationship depicts the valence an ANPC has with a vocabulary. Thus, a subset of these relationships, along with a personality, are the characteristics that make up an ANPC.

Defining the vocabulary describes what inhabits the virtual world and what the inhabitants can do. Three main sets make up the vocabulary: Objects, Actions, and ANPCs. Objects are things that have no cognition and are props used by the player and an ANPC. These include, but are not limited to, cars, work tools, chairs, and laptops. For APF to know what an object is, it just needs to be represented by a name. Actions are all the things that both the player and an ANPC can do within the virtual world. These include, but are not limited to, commuting, hacking, complementing, and bartering. For APF to know what an Action is, it must be represented as a name and an effect onto the virtual world. The effect is a value from 0.0 to 1.0, where 0.0 is a negative outcome of the action for the ANPC who experiences it, while 1.0 is a positive outcome. The last set is the ANPCs, and these are all the characters that will inhabit the virtual world. For APF to know what an ANPC is, it is represented with their name and personality. The personality is represented by a vector, where each dimension is a trait. Each trait is between the value of 0.0 and 1.0, where 0.0 means the opposite of the trait, and 1.0 is fully resembling that trait.

With the vocabulary defined, the connections are left to form the web of relationships. The entire set of relationships forms the psychosocial aspects of the ANPCs. A single relationship represents a cognitive appraisal of an ANPC. A relationship can be one of the three types: Attitude, Praise, or Social.

An Attitude relation is between an ANPC and an Entity. Specifically, the relationship is between an ANPC and an Object or another ANPC. An Attitude relation represents the ANPC's thoughts towards the game object or the other ANPC. The value can be calculated automatically for the other ANPC via the social relation's liking trait; however, the attitude value represents the original biased thought of the ANPC. The valiance from the Attitude is between 0.0 and 1.0, where 0.0 means the ANPC vehemently hates the entity, while 1.0 means an uncontrollable love towards the entity.

A Praise relation is between an ANPC and an action. A Praise relation represents the ANPC's beliefs against the action. The action can be as general as "arguing" or "hugging." Taking into consideration of satisfying a goal, an option that the ANPC can pick is "killing." However, the developers should consider specifying the action so that the ANPC can view it in different situations, for instance, "killing_for_money" with low praise versus "killing_to_save_human_life" with high praise. The valiance is between 0.0 and 1.0, where 0.0 means the ANPC views the action as something that goes against their moral belief, and 1.0 is praiseworthy.

The difference between the effect of an action and the praise of action is the objective and subjective point of view. The effect represents the general outlook of undergoing the action, either being painful or pleasurable. The praise is the ANPC's perception of the action, either being reprehensible or admirable. The distinction allows ANPCs to express their biases for action while also experiencing the consequence of the action.

Social relations are the connections between ANPCs. A social relation describes how an ANPC should present themselves and how they should treat the other ANPC. However, these kinds of relations are more complicated than the previous and will need more information than a single value. The social relationship follows the same model as Ochs et al., using the LDSF model [7]. A 4-tuple vector represents the LDSF model. Each trait is between the range [0.0, 1.0], where 0.0 is the opposite trait, and 1.0 is the full trait. Liking is the degree of affinity towards another ANPC. Dominance is the power that an ANPC can exert onto another ANPC. Solidarity is the degree of "like-mindedness" or having similar behavior dispositions with another ANPC. Furthermore, Familiarity is characterizing the type, private or public, and the amount of information comfortably exchanged between the two ANPCs.

5 Character Makeup

With the vocabulary defined, APF now has enough information to create instances of the ANPCs. Instancing an ANPC is different from instancing a game object or action. The ANPC file is treated as the unique cast of actors, meaning each line in the file is a different character in the virtual world. The game objects and actions files are the definitions to make as many copies as the designer needs, meaning if the designer needs nine copies of baseball bats, then only the line that defines a baseball bat is read nine times. To instantiate an ANPC, APF goes down the list of ANPCs to gather their metadata first and then find the corresponding relations to populate the ANPC. From the list of ANPCs, we set their names and their personalities.

APF uses the OCEAN model to represent the personality of the ANPC. The OCEAN model is a 5-tuple vector where each element is in the range [0.0, 1.0]. For example, with the Openness trait, 0.0 means the ANPC is non-curious and less intellectually driven, while 1.0 means the ANPC is inquisitive and is insatiable in their quest to know more. Stated earlier, MDS is used to give designers additional information about the vector. APF is given 32 reference points, which are mapped from the OCEAN model to the Myers-Briggs Personality Types (MBPT) model [14]. Some of the names are "Turbulent Architect," "Assertive Mediator," and "Turbulent Executive," just to name a few [15]. The mapping turns the 5-tuple personality vector into a label to quickly understand it is contextual meaning. Understand that the mapping that was selected was by designer choice. If there is another list of personality reference points and is mapped from the OCEAN model to the designer's convention, APF will use those labels and hypervolumes instead.

Additionally, if APF parses a valid label of a reference point in the ANPC set, rather than a vector, then the reference point will be used and will be randomly offset.

The relations are all that is left to instantiate an ANPC. The Attitude and Praise relations are assigned to their ANPC accordingly. If an ANPC is missing a relationship with a game object or an action, APF assigns the valence of that relationship to neutral, 0.5. These values are not dynamic and will stay the same throughout the entire playthrough. The social relation is assigned to the ANPCs, but because they are vectors, MDS can be used. Like personalities, a list of reference points is used to label the vector. The names used are "Parent-Child," "Child-Parent," "Boss-Worker," and "Stranger." Notice how these labels are pairs of titles. The first title is how the source ANPC is presenting themselves, and the second title is how they are treating the other ANPC. The exception to this rule is stranger, which acts as the default social relation for any missing pairs. Like the other reference points, these are designer specific. A different list of reference points mapped to the LDSF model can be used to better suit the context. Additionally, if a label is parsed, rather than a vector, then the random offset is applied to the relation and is set

6 Psychological Evaluation

At this point, the virtual world is defined and populated. ANPCs are walking around the virtual world and interacting with game objects and other ANPCs. For an ANPC to evaluate their psychological state, they first need to have some means of perception. In APF, the perception is the same event handling from Ochs et al., which uses a Sowa graph [7, 16]. The representation of an event is a 4-tuple node (Agent, Action, Patient, Certainty). The Agent is the source of the event and can be either an ANPC acting, an unknown ANPC, or no ANPC. Action is the performance that is taken place during the event. The Patient is an entity, either a game object or another ANPC. The Certainty is used to represent either seeing the action in person, 1.0, hearing about the action 0.0–1.0, or confirming that the action did not happen 0.0.

When an ANPC registers an event, it goes through a psychological evaluation. The first step in the evaluation is determining the emotional state of the ANPC. APF uses a modified OCC model to evaluate the emotion an ANPC felt caused by the event. The modified version is based on the works of Steunebrink et al. and their revised version of the original OCC model. Their revised version kept the emotions as pairs, hate/love, approving/disapproving, fear/hope, and so on. The pair representation meant that the value would be from a range $[-1.0, 1.0]$. For example, using hate/love, a -1.0 means a feeling of hatred, and 1.0 means a feeling of love. An issue that arises from this representation is when we update the emotional history of an ANPC.

Let us say that an ANPC was to undergo a practical joke. When the joke is initiated, the ANPC will be distressed from what happened. Once the ANPC realizes

Artificial Psychosocial Framework for Affective Non-player Characters

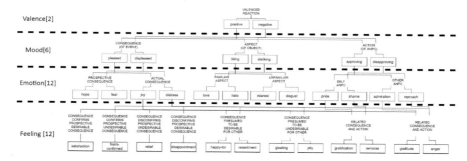

Fig. 2 Modified OCC model based on the works of Steunebrink et al. [17] taking into account ambivalence and psychological ideology with the hierarchical nature

the joke, the ANPC should be joyful. In the revised version, the emotions of distress and joy are paired together. The pairing means that the emotional value can be from [−1.0, 1.0], as well as the emotional history for the ANPC. Let us say that the ANPC hated the joke's start so that they would generate an emotion of hatred of −1.0. However, once they realized it was a joke, and they liked the joke, they would generate an emotion of joy of 1.0. When updating the emotional state, we would simply add the current state to the emotional history and add the new emotional value to the current state. Now, the ANPC has gone from really hating the joke, −1.0, to being neutral about the joke, 0.0. The current emotional state does not correctly reflect the amount of joy that the ANPC generated. Furthermore, we lose the opportunity to track how angry the ANPC is since the start of the joke. With the modified version, we track both hatred and joy separately. This separation allows for ambivalence, the state of having mixed emotions. Going through the scenario again, the ANPC will generate hatred from the start of the joke and store it accordingly. When the ANPC generates joy, hatred is not modified, but instead continues to decay over time. Joy is stored accordingly, and now we make a simple comparison of which emotion has a higher value.

In the work of Steunebrink et al., they mentioned that the original model had a hierarchical property. They wanted to preserve this property, but as an inheritance hierarchy for object-oriented programmers [17]. In doing so, we can represent each height of the hierarchy as a different idea from psychology. From Fig. 2, we can split the modified model into four sections and denote them as follows:

(1) From the CAT, we have a valence, our positive or negative input value. Valence is the subjective spectrum of positive to negative evaluation of an experience an individual may have had [12].
(2) Mood, an affective state heavily influenced by the environment, physiology, and mental state of the individual. Moods can last minutes, hours, or even days [12].
(3) The basic emotions, the chemical release in response to our interpretation of a specific situation. It takes our brains a quarter of a second to identify the trigger, and another to produce the chemical [12].

(4) Feelings, what we analyze after experiencing the generated emotion, which can last longer than an emotion [18].

Using the modified OCC model is similar to using the previous versions. The model starts with a valence; this is the summation of all the relations involved and is bounded between 0.0 and 1.0 for both the positive and negative valences experienced. Next, the mood queries the Sowa node for the action performed, the ANPC acting, and the affected object. The valences from the relative relations of the ANPC are used to calculate the new mood. When calculating the new emotion, queries are made to the Sowa node, the social relation, and the values from the new mood. Lastly are the feelings; however, due to time constraints, this section was not fully implemented in APF.

Going through the OCC model, the personality of the ANPC is also taken into consideration when generating an emotion. An ANPC with a high Neurotic trait will increase negative emotions, such as disapproving, disliking, reproach, and distress. An ANPC with a high Extraversion trait will increase positive emotions, such as liking, pleased, love, and hope. An ANPC with a high Agreeableness trait will positively influence the consequence subtree and action subtree. An ANPC with a high Openness trait will have a positive influence on the aspect and action subtree. An ANPC with a high Conscientiousness will have a positive influence on the consequence subtree.

Using the terms from the Sowa node, when generating the new emotion from the event, if an Agent ANPC was involved in the event, then the social relation from the Patient ANPC to the Agent ANPC is updated. Note that the social relation from the Patient ANPC to the Agent ANPC is not the same as the Agent ANPC to the Patient ANPC. Thus the social relation has a non-commutative property. From Fig. 3, any positive emotions will positively influence the liking trait, while negative emotions will negatively influence the trait. In Fig. 4, certain positive emotions will cause the dominance trait to increase, while certain negative emotions will decrease it. However, if the Agent ANPC reacts to the Patient ANPC and expresses negative emotions, this will increase the Patient ANPC's dominance trait. From Fig. 5, if the Agent ANPC generates the same emotion as the patient ANPC, then the solidarity trait is increased. If they are not the same, then the solidarity trait decreases. However, negative emotions from the consequence subtree and the action

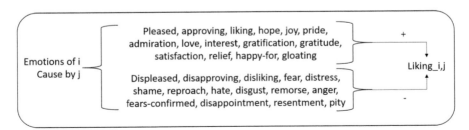

Fig. 3 Updating the liking trait from the LDSF model

Artificial Psychosocial Framework for Affective Non-player Characters

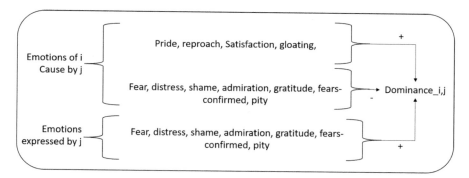

Fig. 4 Updating the dominance trait from the LDSF model

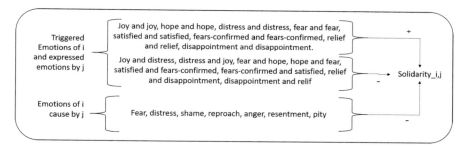

Fig. 5 Updating the solidarity trait from the LDSF model

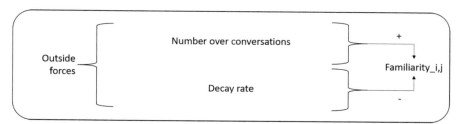

Fig. 6 Updating the familiarity trait from the LDSF model

subtree will negatively influence the solidarity trait. Lastly, Fig. 6 shows that over time, the more interaction an ANPC has with another ANPC, the Familiarity trait will increase; otherwise, it will decrease slowly over time.

7 Formal Grammar

The purpose of this section is to understand the general structure of each model used in APF. An ANPC is continually changing, so referring to it is relative to a time t. The moment an ANPC is created is defined by $t = 0$. For this section, let us refer to one ANPC at a given moment as i_t, that is from the set of all ANPCs in our world, I. From the works of Eggs et al., personality, mood, and emotional state form the PME model [19]. We are extending their work by adding valence, attitude relations, praise relations, and social relations. Personality, attitude relations, and praise relations are constant and initialized at $t = 0$. The valence, mood, emotional state, and social relation are dynamic and initialize to 0 at $t = 0$. Thus we define i_t as a 7-tuple $(m, A_i, P_i, \gamma_t, \pi_t, \xi_t, S_{i_t})$ where m is the personality, A_i is the subset of attitude relations for an ANPC i, P_i is the subset of praise relations for an ANPC i, γ_t is the valence at time t, π_t is the mood at time t, ξ_t is the emotional state at time t, and S_{i_t} is the subset of social relations for ANPC i at time t.

There exist several personality models, each consisting of a set of dimensions, where each dimension correlates to a specific attribute towards the makeup. In APF, the personality model used is the OCEAN model, which has five dimensions. Generalizing the theory, we assume that a personality has n-dimensions, where values represent each dimension in the interval [0.0, 1.0]. A value of 0.0 is the lack of that dimension in the personality, while a value of 1.0 is the greatest presence of that dimension in the personality. The following vector can represent the personality m of an individual:

$$m^T = [\mu_1, \mu_2, \mu_3, \dots, \mu_n], \forall x \in [1, n] : \mu_x \in [0.0, 1.0] \tag{1}$$

As we mentioned earlier, we must define a priori to the game a vocabulary, the set of game objects, O, the set of ANPCs, I, and the set of actions, C. The set of vocabs is $Voc = O \cup I \cup C$. Object $o \in O$ is a game object that has no cognition, versus an ANPC that can. An ANPC $i \in I$ is the non-player character that simulates human affects. Action $c \in C$, are couples $(name, effect)$, where $name$ uniquely identifies the action, and $effect$ denotes the effect of the action onto an ANPC ranging from [0.0, 1.0]. 0.0 means a negative impact on the ANPC, and 1.0 a positive impact. Additionally, let us define E to be the set of entities within the game. The entire set of entities is from both sets of objects, O, and ANPCs, I, thus $E = O \cup I$. A single entity, $e \in E$, can either be an object or an ANPC.

With the vocabulary and entity set, we can define the set of Attitudes and Praises. The set of attitudes is the combination of all ANPCs with all entities. An attitude also is denoted from the range [0.0, 1.0], where 0.0 means the ANPC vehemently hates the entity, while 1.0 means an uncontrollable love towards the entity. The set of praises is the combination of all ANPCs with all actions. Praise is denoted from the range [0.0, 1.0], where 0.0 means the agent views the action as something that goes against their moral belief, and 1.0 as praiseworthy. So the formal notation of the set of attitudes, A, and the set of praises, P, is as follows:

$$A = I \times E = (i, e) \rightarrow [0.0, 1.0] : i \in I, e \in E = (O \cup I) \tag{2}$$

$$P = I \times C = (i, c) \rightarrow [0.0, 1.0] : i \in I, c \in C \tag{3}$$

$$A_i \subseteq A \text{ and } P_i \subseteq P \tag{4}$$

Valence is the bias experience an ANPC feels. In APF, valence is a tuple for positive and negative experience coming from the attitude and praise. In general, this can be a list of physical or mental values that the individual enjoys or dislikes based on a specific moment. We define the valence γ_t as a k-dimensional vector, where values represent all k valence intensities in the interval $[0.0, 1.0]$. The value 0.0 corresponds to the absence of the valence, while 1.0 is the maximum intensity. The vector is given as follows:

$$\gamma_t^T = \begin{cases} [\eta_1, \ldots, \eta_k], \forall x \in [1, k] : \eta_x \in [0.0, 1.0], & \text{if } t \geq 0 \\ 0, & \text{if } t = 0 \end{cases} \tag{5}$$

Furthermore, we define the valence history Γ_t that contains all valances until γ_t, thus:

$$\Gamma_t = \langle \gamma_1, \gamma_2, \gamma_3, \ldots, \gamma_t \rangle \tag{6}$$

The mood is an affective state of the ANPC. In APF, the mood is defined by a 6-tuple vector, (pleased, displeased, approving, disapproving, liking, disliking). The mood could also be represented as either good or bad, similar to valence. We define a mood π_t as an h-dimensional vector, where values represent all h mood intensities in the interval $[0.0, 1.0]$. The value 0.0 corresponds to the absence of the mood, while 1.0 is the maximum intensity. The vector is given as follows:

$$\pi_t^T = \begin{cases} [\nu_1, \ldots, \nu_h], \forall x \in [1, h] : \nu_x \in [0.0, 1.0], & \text{if } t \geq 0 \\ 0, & \text{if } t = 0 \end{cases} \tag{7}$$

Similarly, we define the mood history Π_t that contains all moods until π_t, thus:

$$\Pi_t = \langle \pi_1, \pi_2, \pi_3, \ldots, \pi_t \rangle \tag{8}$$

The emotional state is the biological state of the ANPC. In APF, the emotional state is defined by a 12-tuple vector, (hope, fear, joy, distress, pride, shame, admiration, reproach, love, hate, interest, disgust). The six basic emotions are these same values, however, as pairs, and from a range of -1 to 1. We define an emotional state ξ_t as a y-dimensional vector, where values represent all y emotional intensities in the interval $[0.0, 1.0]$. The value 0.0 corresponds to the absence of the emotion, while 1.0 is the maximum intensity. The vector is given as follows:

$$\xi_t^T = \begin{cases} [\epsilon_1, \ldots, \epsilon_y], \forall x \in [1, y] : \epsilon_x \in [0.0, 1.0], & \text{if } t \geq 0 \\ 0, & \text{if } t = 0 \end{cases} \tag{9}$$

Similarly, we define the emotional state history Ξ_t that contains all emotional states until ξ_t, thus:

$$\Xi_t = \langle \xi_1, \xi_2, \xi_3, \ldots, \xi_t \rangle \tag{10}$$

The formal model of social relations continues its representation from the works of Pecune et al. [20]. Social relations are a 4-tuple vector (Liking, Dominance, Solidarity, Familiarity), represented as s_{i_t} for the social relations of ANPC i, towards a different ANPC at time t. $s_{i_t} \in S_{i_t}$, is the set of Social relations for ANPC i, at time t. We can also define the social relation history Θ_t, that contains all social relations until S_{i_t} for an ANPC, thus:

$$\Theta_t = \langle S_{i_1}, S_{i_2}, S_{i_3}, \ldots, S_{i_t} \rangle \tag{11}$$

As the ANPC perceives the world, events are registered based on the actions performed onto entities. APF calculates the valence based on the Sowa node describes in section VI and following the OCC model. We define this information as a *desired change in valence intensity* for each valiant, defined by a value in the interval [0.0, 1.0]. The valiant information vector V (or valance influence) holds the desired change of intensity for each of the k valances after evaluation in the first height of the OCC model:

$$V^T = [\eta_1, \eta_2, \eta_3, \ldots, \eta_k], \forall x \in [1, k] : \eta_x \in [0.0, 1.0] \tag{12}$$

The valence can then be updated using a function $D(m, \Gamma_t, V)$. This function calculates the change in valence, based on the personality, m, the valence history, Γ_t, and the valence influence, V. To represent internal valence changes to the ANPC, such as the decay rate or balancing modifications, is given as $F(m, \Gamma_t)$. Given these two functions, the new valence γ_{t+1} can be calculated as follows:

$$\gamma_{t+1} = \gamma_t + D(m, \Gamma_t, V) + F(m, \Gamma_t) \tag{13}$$

Given the new valence and the event, the mood is updated. The mood influence vector M holds the desired change of intensity for each of the h moods in the second height of the OCC model:

$$M^T = [\nu 1, \nu_2, \nu_3, \ldots, \nu_h], \forall x \in [1, h] : \nu_x \in [0.0, 1.0] \tag{14}$$

The mood is updated using a function $G(m, \Pi_t, \Gamma_{t+1}, M)$. This function calculates the change in mood, based on the personality, m, the mood history, Π_t, the new valence history, Γ_{t+1}, and the mood influence M. Additionally, to represent internal mood changes, either decay rate or balancing modifications, to the ANPC is given as $H(m, \Pi_t, \Gamma_{t+1})$. Given these two functions, the new mood π_{t+1} can be calculated as follows:

$$\pi_{t+1} = \pi_t + G(m, \Pi_t, \Gamma_{t+1}, M) + H(m, \Pi_t, \Gamma_{t+1}) \tag{15}$$

Lastly, we update the emotion. The emotion influence vector J holds the desired change of intensity for each of the y emotions in the third height of the OCC model:

$$J^T = [\epsilon_1, \epsilon_2, \epsilon_3, \ldots, \epsilon_y], \forall x \in [1, y] : \epsilon_x \in [0.0, 1.0] \tag{16}$$

The emotion is updated using a function $K(m, \Xi_t, \Pi_{t+1}, \Gamma_{t+1}, J)$. This function calculates the change in emotion, based on the personality, m, the emotional state history, Ξ_t, the new mood history, Π_{t+1}, the new valence history, Γ_{t+1}, and the emotion influence J. Additionally, to represent internal emotional state changes, such as decay rate and balancing modifications, to the ANPC is given as $L(m, \Xi_t, \Pi_{t+1}, \Gamma_{t+1})$. Given these two functions, the new emotional state ξ_{t+1} can be calculated as follows:

$$\xi_{t+1} \quad = \quad \xi_t \; + \; K(m, \Xi_t, \Pi_{t+1}, \Gamma_{t+1}, J) \; + \; L(m, \Xi_t, \Pi_{t+1}, \Gamma_{t+1}) \tag{17}$$

The social relations are updated similarly to the works of Ochs et al. [7]

8 Implementation

Running the APF library, we construct a scenario with two ANPCs with different personalities. In this scenario, we use the context of the 1960 comic book "Dennis the Menace." Dennis is a young, 6-year-old boy who is trying to learn about the world. Dennis's parents, the Mitchells, and his neighbors, the Wilsons, are his role models that try and steer Dennis to the right path. For this scenario, the player plays as Mr. Wilson, and the ANPC is Dennis.

We can understand the ANPCs' social relationships by graphing their emotional state. As time passes, actions are performed by the player, and the ANPC reacts to the action. In Fig. 7, there are six graphs with three different examinations over time for both experiments. The first set of graphs is the ANPC's emotional state history, plotting the twelve emotions the ANPC can generate. The second set of graphs shows the social relationship history from Dennis to Mr. Wilson. The social relationship is based on the emotional state at that time, and the graph plots the four dimensions of a social relationship. For clarity, only liking and dominance from the social relationship are plotted. The third graph is using MDS to show the certainty of classifying the current relationship.

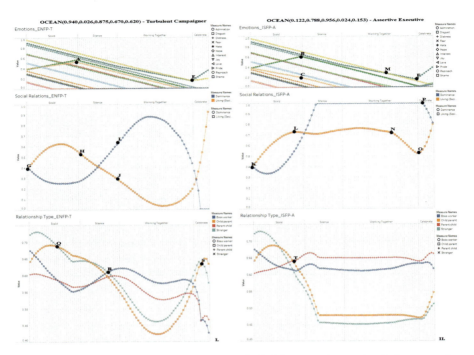

Fig. 7 Two experiments running APF with a Dennis the Menace setting, testing the heterogeneity of two ANPCs undergoing the same scenario, one Dennis with a turbulent Campaigner personality and the other Dennis with an assertive executive personality

The abscissa for all graphs is the time passing and the name of the action performed by the player. The ordinate for each graph ranges from 0.0 to 1.0. The first set of graphs, the emotional state history, tells the amount of an emotion felt by the ANPC at that time. The second graph, the social relationship history, tells the amount of a trait visible by their relationship. The third set of graphs, the relationship type, is a percentage of similarity between the reference points of social relationships and the current social relationship between Dennis and Mr. Wilson. In other words, the social relation with the highest value is considered to be the current relationship.

The scenario plays out as follows: Mr. Wilson catches Dennis uprooting one of his turnips. Mr. Wilson scolds Dennis for picking his crop too early and tells him to wait as he grabs some tools. Mr. Wilson comes back and works with Dennis to plant new turnips. Once they finished, they both celebrate that the garden is back to normal.

With this scenario, we can change Dennis's personality to see how he will react to Mr. Wilson. To represent Dennis's personality, APF was given 32 reference points to map the OCEAN personality to the Myers-Briggs Type Indicator [14]. In the first experiment, Dennis has a Turbulent Campaigner (ENFP-T) personality. In the second experiment, Dennis has the Assertive Executive (ESTJ-A) personality.

In the emotional history graph, there are distinct differences between the Campaigner Dennis and Executive Dennis. As the Campaigner is experiencing the Scolding from Mr. Wilson, he generates emotions of hate and fear, Point A. The Executive, on the other hand, only generates hatred, point B, and does not fear Mr. Wilson, point C. The reason for this is because of the Neurotic trait of their personality. The Campaigner's Neurotic level is at 0.620, giving him qualities of anxiousness and prone to negative emotions. The Executive's Neurotic level is much lower, at 0.153, giving him qualities of being calm and even-tempered. Both the Campaigner and the Executive generated hatred because of their Extraverted trait, 0.875, and 0.956, respectively, giving the Dennises the qualities of being sociable and assertive. At point E, the Campaigner is generating love and enjoyment when they are celebrating with Mr. Wilson. The Executive, however, only generates love and not enjoyment at point F. The lack of enjoyment is because of the Openness trait. For the Campaigner, Openness is at 0.940, giving him qualities of curiosity and having a wide range of interests. For the Executive, Openness is at 0.122, giving him qualities of being practical and conventional.

In the social relations history graph, we see how the Dennises view Mr. Wilson based on how Mr. Wilson treats Dennis. With the Campaigner, Denis at first likes Mr. Wilson at point G because he started with an abundance of joy. However, since the Campaigner is generating fear and hatred up to point A, the liking starts to decrease. With the hatred increasing, the dominance increases as well. When Mr. Wilson leaves to grab his tools at point H, there are more negative emotions generating than positive, and so the liking for Mr. Wilson continues to drop. While Mr. Wilson is looking for his tools, fear is decreasing, so the dominance continues to rise. When they start working on the garden at point I, there is an inflection point for dominance because of the decrease in pride and reproach. At point J, there is another inflection point for liking because of the increase in love and a decrease in hatred and reproach. Finally, since the Campaigner generated emotions of love and enjoyment, liking rises.

Looking at the social relations history graph for the Executive Dennis, there is still an increase in liking at point K because of the abundance of joy. However, because fear was not generated, the liking flat lines at point L before the scolding finishes. Similarly, because of the decrease of fear and an increase in hatred, dominance rises much quicker. During the silence, the dominance maxes out while the liking continues to flatline because a relatively equal amount of positive and negative emotions are decreasing. While they are working together, joy decreases a little faster than love, making them intersect at point M. The intersection influences the liking at point N so much that it starts to decrease. Once the garden is done, and they are celebrating, the Executive is generating love, admiration, and hope at point F. The generation of positive emotions increases liking at point O, but because of the increase in admiration, there is a decrease in dominance at point P.

In the last set of graphs, the relation type shows how the Dennises are associating themselves and how they would treat Mr. Wilson. The Campaigner, in the beginning, views Mr. Wilson as a stranger. At point Q, during the scolding, the Campaigner believes he is a child and should treat Mr. Wilson as a parent. At point R during the silence, the Campaigner presents himself as a boss and treats Mr. Wilson as a worker. The reason the Campaigner chose the boss-worker relation is because of the enormous gap between the liking and dominance from the social relation. The boss-worker relation gives the feeling that the Campaigner is impatient and wants to start working without Mr. Wilson. It is not until they celebrate does the Campaigner presents himself as a child, and Mr. Wilson as a parent once more at point S.

For the Executive, we start the same with a stranger, and then the child-parent relation similar to the Campaigner. The difference shows at point T when the Executive presents himself as a parent and treats Mr. Wilson as a child. The Parent-Child relation was chosen because of the smaller gap between the liking and dominance from the social relation. The Executive continues with the Parent-Child relation for the remainder of the scenario. The Parent-Child relation gives the feeling that the Executive wants to take responsibility but is forced to work with an overdramatic person.

9 Conclusion

The "Artificial Psychosocial Framework" (APF) is an emotional modeling tool to help create "Affective Non-Player Characters" (ANPC) in-game engines. The principle of APF is the CAT and is supported by psychological models to generate emotions and social relations with other ANPCs. Developing ANPCs will help create realistic virtual worlds and allow for the conception of further emotion-centric gameplay for designers. APF was created as a library for developers to use as a plug-in for any game engine. From Fig. 8, one can see an example of how a developer can use APF with the visual programming tool, blueprints, from the Unreal Engine.

Future works include optimizations to allow for thousands or millions of ANPCs within a virtual world. In this case, one file of valence values would represent a colony. One can also use a random number offset to modify the ANPCs' valences within the said colony. Balancing is necessary for each game to determine the correlations between personality vs. emotion, personality vs. social relation, and emotion vs. social relations. APF allows for social relations to update over time. One can consider updating the other relations, Attitude and Praise, to allow for dynamic opinions.

After integrating APF into a game engine, we can consider unique gameplay features. Like Valve's zombie apocalypse title, "Left 4 Dead," their AI director guides the gameplay based on the players' dread [21]. APF would monitor ANPCs' emotions, and either changes the virtual world, using a director or arbiter system, or influence NPCs' decisions, such as animal or zombie. From the works of Brinke et al. with their project, "The Virtual Storyteller," we can use their idea of out-of-character knowledge with in-character knowledge to influence ANPC choices [22]. An ANPC would select actions that result in an intended reaction from another

Artificial Psychosocial Framework for Affective Non-player Characters

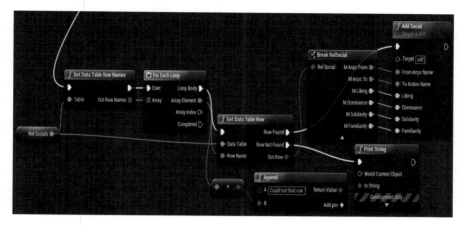

Fig. 8 And example blueprint that uses the APF plug-in to import the social relations

ANPC, leading to exciting scenarios. Alternatively, APF can enhance the tension of 11 Bit studio's title "This War of Mine," where the player has to consider the ANPC's opinions and well-being before sending them off to complete a moral dilemma [23]. Integrating APF into a game engine will allow for novel game mechanics that focus on human emotions and social relationships.

To access the Library and data presented in the paper, please visit https://gitlab.com/humin-game-lab/artificial-psychosocial-framework.

References

1. D. Mark, AI wish list: What do designers want out of AI? AI Summit (2018). https://gdcvault-com.proxy.libraries.smu.edu/play/1024900/AI-Wish-List-What-Do
2. M. Bourgais, P. Taillandier, L. Vercouter, C. Adam, Emotion modeling in social simulation: a survey. J. Artif. Soc. Soc. Simul. 21(2), 5 (2018). http://jasss.soc.surrey.ac.uk/21/2/5.html
3. L. Van Minh, C. Adam, R. Canal, B. Gaudou, H. Tuong Vinh, P. Taillandier, Simulation of the emotion dynamics in a group of agents in an evacuation situation, in *Principles and Practice of Multi-Agent Systems*, ser., ed. by N. Desai, A. Liu, M. Winikoff. Lecture Notes in Computer Science (Springer, Berlin, 2010), pp. 604–619
4. L. Luo, S. Zhou, W. Cai, M.Y.H. Low, F. Tian, Y. Wang, X. Xiao, D. Chen, Agent-based human behavior modeling for crowd simulation. Comput. Anim. Virtual Worlds 19(3–4), 271–281 (2008)
5. M. Kazemifard, A. Zaeri, N. Ghasem-Aghaee, M.A. Nematbakhsh, F. Mardukhi, Fuzzy emotional COCOMO II software cost estimation (FECSCE) using multi-agent systems. Appl. Soft Comput. 11(2), 2260–2270. http://www.sciencedirect.com/science/article/pii/S1568494610002085
6. H. Jones, J. Saunier, D. Lourdeaux, Fuzzy rules for events perception and emotions in an agent architecture, in *Proceedings of the 7th conference of the European Society for Fuzzy Logic and Technology* (Atlantis Press, Amsterdam, 2011), pp. 657–664. https://www.atlantis-press.com/proceedings/eusflat-11/2219

7. M. Ochs, N. Sabouret, V. Corruble, Simulation of the dynamics of nonplayer characters' emotions and social relations in games. IEEE Trans. Comput. Intell. AI Games **1**(4), 281–297. http://ieeexplore.ieee.org/document/5325797/
8. H. Faroqi, S. Mesgari, Agent-based crowd simulation considering emotion contagion for emergency evacuation problem. Remote Sensing Spatial Inf. Sci. **40**(1), 193–196 (2015)
9. M. Šuvakov, D. Garcia, F. Schweitzer, B. Tadić, Agent-based simulations of emotion spreading in online social networks (2012). http://arxiv.org/abs/1205.6278
10. K. Woodward, *Psychosocial Studies: An Introduction*, 1st edn. (Routledge, Milton Park, 2015)
11. Understanding psychosocial support #PowerOfKindness (2019). https://www.youtube.com/watch?v=h8PHvxVmC0I
12. P. Ekkekakis, Affect, mood, and emotion, in *Measurement in Sport and Exercise Psychology* (Human Kinetics, Champaign, 2012), pp. 321–332. https://ekkekaki.public.iastate.edu/pdfs/ekkekakis_2012.pdf
13. R.S. Lazarus, S. Folkman, *Stress, Appraisal, and Coping* (Springer, Berlin, 1984). OCLC: 10754235
14. T. Flynn, Global 5 to Jung/MBTI/Kiersey Correlations. http://similarminds.com/global5/g5-jung.html
15. Personality types | 16personalities. https://www.16personalities.com/personality-types
16. J. Sowa, Semantic networks, in *Encyclopedia of Artificial Intelligence* (Wiley, New York, 1992), p. 25
17. B. Steunebrink, M. Dastani, J.-J. Ch, J.-j. Meyer, The OCC model revisited, in *Proceedings of the 4th Workshop on Emotion and Computing. Association for the Advancement of Artificial Intelligence*
18. J. Freedman. What's the difference between emotion, feeling, mood? (2017) https://www.6seconds.org/2017/05/15/emotion-feeling-mood/
19. A. Egges, S. Kshirsagar, N. Magnenat-Thalmann, A model for personality and emotion simulation, in *Knowledge-Based Intelligent Information and Engineering Systems*, ser. ed. by V. Palade, R. J. Howlett, L. Jain. Lecture Notes in Computer Science (Springer, Berlin, 2003), pp. 453–461
20. F. Pecune, M. Ochs, C. Pelachaud, "A formal model of social relations for artificial companions, in *Proceedings of The European Workshop on Multi-Agent Systems (EUMAS)*
21. M. Booth, The ai systems of left 4 dead, in *Booth, M. (2009). The ai systems of left 4 dead. In Artificial Intelligence and Interactive Digital Entertainment Conference at Stanford*. https://steamcdn-a.akamaihd.net/apps/valve/2009/ai_systems_of_l4d_mike_booth.pdf
22. H.t. Brinke, J. Linssen, M. Theune, Hide and sneak: Story generation with characters that perceive and assume, in *Tenth Artificial Intelligence and Interactive Digital Entertainment Conference*. https://www.aaai.org/ocs/index.php/AIIDE/AIIDE14/paper/view/8989
23. G. Mazur, This war of mine: Under the hood, dev Gamm! (2015) . https://www.youtube.com/watch?v=sqHledFRP1A&t=2110s

A Prototype Implementation of the NNEF Interpreter

Nakhoon Baek

1 Introduction

Recently, the machine learning and big-data analysis applications are among the most important key issues in the area of computation and data handling [2]. One of the distinguished characteristics of the target data is that they are too large and/or too complex to be dealt with traditional data analysis tools. Thus, we need brand-new data handling tools for those large-scale data processing.

The neural networks and their related tools are actually one of the most suitable tools for the machine learning applications and large-scale complex data analysis, at this time. Combining the neural network tools with large-scale data processing, we can achieve remarkable results. As the result, we now have many neural network frameworks, including *TensorFlow* [1], *Caffe* [4], *Keras* [5], and others.

In the field of neural networks and their handling, the NNEF (Neural Network Exchange Format) [3] is one of the de facto standard file formats to exchange the trained data and the computation network itself, as shown in Fig. 1. At this time, NNEF file loaders and exporters are widely used in a variety of areas, including neural network applications, big-data handling, vision processing, and others.

In this work, we present an approach to directly execute the described operations in the NNEF files, similar to the programming language interpreters. Our goal is to build up a framework to support the execution and the translation of the NNEF files. As the first step, we show that the computation result of NNEF files can be

---

N. Baek (✉)
School of Computer Science and Engineering, Kyungpook National University, Daegu, Republic of Korea

School of Electrical Engineering and Computer Science, Louisiana State University, Baton Rouge, LA, USA
e-mail: nbaek@knu.ac.kr

© Springer Nature Switzerland AG 2021
H. R. Arabnia et al. (eds.), *Advances in Artificial Intelligence and Applied Cognitive Computing*, Transactions on Computational Science and Computational Intelligence, https://doi.org/10.1007/978-3-030-70296-0_51

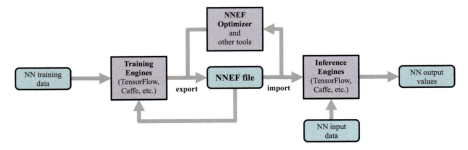

Fig. 1 The role of NNEF files

calculated directly from our prototype implementation of the NNEF interpreting system.

2 Design and Implementation

For the ease of NNEF file handling, *the Khronos group* provides some tools to generate and consume NNEF documents, on the *GitHub* repository [6]. Among those tools, the *NNEF parser* provides a standard way of parsing the standard NNEF files.

When successfully parsed, the Khronos NNEF parser generates a computational graph corresponding to the input NNEF file. An NNEF computational graph is actually represented with the following three major data structures:

- **nnef::graph**—contains the computational graph itself, which consists of lists of *tensors* and *operations*;
- **nnef::tensor**—contains a specific tensor, which can be an activation tensor or a variable tensor in the computational graph;
- **nnef::operation**—represents a single pre-defined operation in the computational graph.

The NNEF standard specification [3] shows that the tensors may have arbitrary dimensions. Additionally, the original NNEF standard already specified more than 100 operations, and additional extensions can be applied to those operations.

Starting from those computational graphs, our system can do the required actions, such as directly executing the NNEF computation graph, or generating translated programming codes. Our overall design for this work flow is summarized in Fig. 2.

As the first step of our NNEF handling system, we targeted an *NNEF interpreter*: a single program executing the NNEF computations in a step-by-step manner. Our NNEF interpreter takes the computational graph from the Khronos NNEF parser, as the most important input.

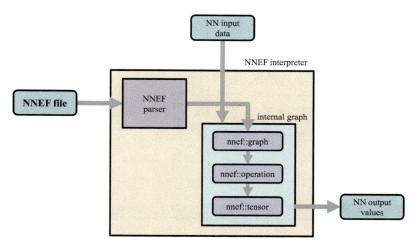

Fig. 2 Overall architecture of our NNEF interpreting system

In the next step, it traverses all the NNEF operations (or **nnef::operation** nodes), through the given order in the NNEF file. For each NNEF operation, it takes all the corresponding input tensors (or **nnef::tensor** nodes) and does the computations. The computation results are again stored into the tensors in the computational graph. Thus, the values of tensors in the computational graph can be updated after executing the computational graph itself.

3 Results

As an example case, we selected the famous *MNIST hand-writing training* [8] and used the very simple *Python* implementation, which is available in [7].

The input of the example NNEF file is a 1D array with 784 floating-point numbers, which actually represents 28-by-28 pixel images for hand-written decimal numbers from 0 to 9. Using the trained values of the internal neural networks, from the disk files of "mnist-wih" and "mnist-who," we can calculate the results as given in Fig. 3.

After executing the example NNEF file in Fig. 3, with selected samples as its input values, our NNEF interpreter gets the result shown in Fig. 4. Since the result is actually the same to that of the original *Python* program, we can consider our NNEF interpreter works well at least for the MNIST examples.

NNEF file

```
version 1.0;
graph mnist_query( input ) -> ( output ) {
    input = external<scalar>( shape=[784,1] );
    wih = variable<scalar>( shape=[200,784], label="mnist-wih" );
    who = variable<scalar>( shape=[10,200], label="mnist-who" );
    hidden = sigmoid( matmul( wih, input ) );
    output = sigmoid( matmul( who, hidden ) );
}
```

Fig. 3 An example NNEF file

result

```
0.04075291
0.00826926
0.05436546
0.02137949
0.01371213
0.03757233
0.00205216
0.90605090
0.02076551
0.03548543
```

Fig. 4 An example result from our NNEF interpreter

4 Conclusion

For the machine learning applications and also for the big-data handlings and their related works, the most intuitive computational tool may be the tools based on neural networks. To support neural network computations, NNEF is the de facto standard from the Khronos Group.

At this time, NNEF handling tools are focused on the data loading and the file format conversions. In contrast, we tried to directly execute the contents of NNEF files, after interpreting them. Our prototype implementation shows that our approach works for real examples. Other kinds of NNEF handling tools, including *code generations* and *NNEF training tools*, are also possible in the near future.

Acknowledgments This work was supported by Basic Science Research Program through the National Research Foundation of Korea (NRF) funded by the Ministry of Education (Grand No. NRF-2019R1I1A3A01061310).

References

1. M. Abadi, et al., TensorFlow: Large-scale machine learning on heterogeneous systems (2015). White paper available from tensorflow.org
2. S.L. Brunton, J.N. Kutz, *Data-Driven Science and Engineering* (Cambridge University Press, Cambridge, 2019)
3. T.K.N.W. Group, *Neural Network Exchange Format*, version 1.0.1. (Khronos Group, Beaverton, 2019)
4. Y. Jia, et al., Caffe: Convolutional architecture for fast feature embedding, in *Proceedings of the 22nd ACM International Conference on Multimedia (MM '14)* (2014)
5. Keras Homepage. http://www.keras.io. Retrieved Jun 2020
6. KhronosGroup/NNEF-Tools, https://github.com/khronosgroup/nnef-tools. Retrieved Jun 2020
7. Makeyourownneural network, https://github.com/makeyourownneuralnetwork/makeyourownneuralnetwork. Retrieved Jun 2020
8. T. Rashid, *Make Your Own Neural Network* (CreateSpace Independent Publishing Platform, Scotts Valley, 2016)

A Classifier of Popular Music Online Reviews: Joy Emotion Analysis

Qing-Feng Lin and Heng-Li Yang (iD)

1 Introduction

Since the Web 2.0 site began to flourish, many user reviews accumulated every day on the web. More and more users would collect these online reviews to compare the advantages and disadvantages of candidate products before making purchasing decisions. As data grows on the network, the problem of information overload is getting worse. With the development of data and text mining technology, many automatic opinion-oriented classifiers have been proposed to solve the problem of information overload. However, although most of opinion analyses can find attributes (features)-opinion pairs rules, such as sound quality-very good, graphic quality-very poor, they could not fit well for hedonic products, such as music, novel, story, film, dance, painting, etc. Products can be classified into two categories: hedonic product and functional products [5]. Functional products refer to either tangible goods (e.g., computers, cars and mobile phones), or intangible transportation services or financial products, which are designed to expect users to obtain pre-designed functional values by operating such things. Such products usually do not directly trigger multiple emotions. For example, in normal conditions, people should not feel like "a car is sad to drive" or "this phone really scares me." However, hedonic products (e.g., movie) would invoke different emotional feeling

---

The Type of the Submission: "Short Research Paper"

Q.-F. Lin
National Penghu University of Science and Technology, Penghu, Taiwan

H.-L. Yang (✉)
National Cheng-Chi University, Taipei, Taiwan
e-mail: yanh@nccu.edu.tw

© Springer Nature Switzerland AG 2021
H. R. Arabnia et al. (eds.), *Advances in Artificial Intelligence and Applied Cognitive Computing*, Transactions on Computational Science and Computational Intelligence, https://doi.org/10.1007/978-3-030-70296-0_52

for customers, such as sadness, joy, excitement, shock, love [14]. For such products, the general attribute evaluation rules are not enough to describe consumers' likes and dislikes. The providers of hedonic products would like to know "how their products can bring what kind of emotion to people." This study proposes a system framework to analyze the emotion feeling embedded in the online reviews, and uses "popular songs" as an example hedonic product and "joy" as the interested emotion to build a joy emotion classifier.

2   Literature Review

Opinion mining, sentiment analysis, or sentiment classification is to collect the evaluations on products or services, and apply statistical or machine learning methods to identify the positive and negative orientation or emotion excitation status [9]. Opinion mining techniques are widely used, for example, analyzing news to understand how competitors operate [15]. More and more attention is being paid to analyze online reviews on products/services, such as electronics [11], movies [1, 15], and restaurants [13]. The reviews could come from fan groups, social network sites, forums, or micro-blogs [3, 7]. There were some previous studies targeting on music. For example, some research tried to analyze the emotions inherent in the lyrics [2, 6, 10, 12] or audio feature [4]; some research used reviews to determine the music category [8]. They are different from our study that tried to identify the emotion embedded in the online reviews.

3   The Proposed System Framework

The goal of this study is to create a system to understand what emotions are embedded in the music online review. A system framework is proposed in Fig. 1. There are two stages: the first is the classifier training phase, and the second is the review summary phase. The objective of the classifier training phase is to complete the training of accurately classifying single review. A pre-processing system would automatically scan and retrieve online reviews, clean and filter for valid comments, and apply English/Chinese word segmentation facility to process the reviews to obtain "formatted" comments. Since in most cases, the online reviews do not explicitly indicate the inspired emotions of the authors after listening to the song and because we adopt supervised learning method, we need to ask some experts to annotate them. Another parallel job is to analyze those keywords of reviews to judge their emotional meaning, which we call them as keyword characteristics (i.e., distances from a particular emotional word, for example, the distance between word "lonely" and "joy"). Using the experts' annotated values as the criteria, the keyword characteristics of each comment would be fed into the sentiment classifier which can be built via SVM (support vector machine). If the classification correct rate is not

A Classifier of Popular Music Online Reviews: Joy Emotion Analysis

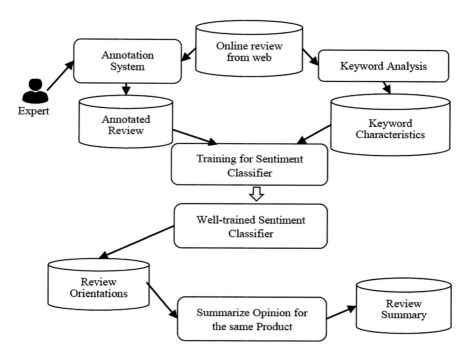

Fig. 1 Proposed system framework

acceptable, we would change the parameters and then re-train. If the classification rate is satisfactory, we can apply the well-trained classifier for prediction. A new review would be formatted and analyzed to obtain its keyword characteristics. Then, the emotional opinion tendency of the new review would be classified as "trigger" or "no trigger." Finally, we need summarize the opinions of all reviews to the same target product.

4 Prototype System

As an example, a prototype was built for a joy emotion classifier. Using the ontology ConceptNet to calculate the relationship strength between emotion key words and basic emotions, this study applied supervised SVM machine learning method training to build our classifier. First, as our data set, from https://www.xiami.com/, we used the pre-processing system to receive 483,427 reviews for 320 Chinese pop songs from the hot list of the last 20 years. Each review contains data such as comment identifier, user identifier, username, targeted song, comment time, comment content, number of likes (number of other users expressing agreement), number of dislikes (number of other users expressing disagreement), etc.

Next, we need some experts to judge the author's emotional tendency for each review. Because the data amount is huge, it would be unrealistic to expect experts to annotate all reviews. We wish to have annotations for the "important reviews." The importance was ranked by the "identity number," which is "number of likes" minus "number of dislikes." From those 320 songs, we screened and picked out 71 songs, each of them had more than 10,000 identities. For these 71 songs, there were 7029 reviews, which annotation burden would still be unbearable to experts. So, we set the second screening criterion: for each song, we chose at least 3 reviews, at most 7 reviews; and stopped choosing if we had chosen more than 3 reviews and the total number of identities reached 70%. Finally, we selected 341 reviews for these 71 songs, and the 341 reviews accounted for 72.38% identities of all 7029 reviews.

We invited 171 college students, who were accustomed to read online reviews, to judge the emotion expressed by these 341 review authors. Via a computing assisted system, the students were prompted only the texts of each online review and marked 1 (joy) or 0 (no joy). All other information (e.g., the review author, number of endorsements, lyrics, singer) was not disclosed to the students. Finally, we obtained 8941 valid answers (the minimal number of answers for single review was 23). We computed the average marking for each review and judge "triggered joy" if the average value is greater than 0.5.

At the same time, we handled the keyword analysis of the review. With the processing of the part-of-speech tagger SINICA CKIP (http://ckipsvr.iis.sinica. edu.tw/), we obtained 9,063,647 words from the whole corpus (483,427 reviews). There were 27,662 different words, 2004 words of which were possible emotional words (nouns, verbs, adjectives) and appeared more than 30 times in the corpus. With manual judgement, we selected 508 words from the above 2004 words as the frequently used keywords in the song reviews. We translated them to English and applied ConceptNet (http://conceptnet.io/) to compute the "keyword characteristics," that is, distances between them and the targeted emotion "joy." For each review, we obtained its emotional characteristic matrix, which elements are weighted by the number of times the keyword appeared in the review.

With the expert annotation value as the target, we applied SVM to train the joy classifier. The 5-fold cross-validation method was adopted, that is, 341 reviews were divided into 5 data sets, at a time one data set as test data, the other 4 data sets as training data. We set the cost value to 10 and gamma to 0.5 as the SVM training parameters. Thus, we achieved precision 67.41%, recall 86.89%, and F-Measure 75.92% in the training phase, and precision 68.57%, recall 79.08%, and F-Measure 73.45% in the testing phase.

Finally, we need to summarize the review opinions for each song. For example, #3 song had three reviews: review 1 included "joy" with 10 identities, review 2 included "joy" with 5 identities, review 3 did not include joy with 3 identities. Thus, the "joy value" of the #3 song was $(10*1 + 5*1 + 3*0)/(10 + 5 + 3) = 83.33\%$.

5 Conclusions

Music consumer behaviors have been changed drastically. Audiences are increasingly paying attention to online reviews. This study successfully established an acceptable joy-emotion classifier from the real-world online Chinese music reviews, from which it was able to summarize the emotional percentage values for each single song. This article only introduces the online review joy classifier, but the method of establishing other emotions is similar.

Acknowledgment The authors would like to thank the Ministry of Science and Technology, Taiwan, for financially supporting this research under Contracts No. NSC 101-2410-H-004-015-MY3 and MOST 107-2410-H-004-097-MY3.

References

1. P. Chaovalit, L. Zhou, Movie review mining: A comparison at supervised and unsupervised types, in *The 38th Annual Hawaii International Conference on System Sciences (HICSS), Hawaii, USA*, (3–6 January 2005)
2. H. Corona, M.P. O'Mahony, An exploration of mood classification in the million songs dataset, in *The 12th Sound and Music Computing Conference, Maynoth University, Ireland*, (26–29 July 2015)
3. T. Goudas, C. Louizos, G. Petasis, V. Karkaletsis, Argument extraction from news, blogs, and the social web. Int. J. Artif. Intell Tools **24**(5), 1540024 (2015) https://doi.org/10.1142/S0218213015400242
4. X. Hu, Y.H. Yang, The mood of Chinese pop music: Representation and recognition. J. Assoc. Inf. Sci. Technol. **68**(8), 1899–1910 (2017)
5. D.S. Kempf, Attitude formation from product trial: Distinct rolls of and app for hedonic and functional products. Psychol. Mark. **16**(1), 35–50 (1999)
6. V. Kumar, S. Minz, Mood classification of lyrics using SentiWordNet, in *2013 International Conference on Computer Communication and Informatics, Coimbatore, India*, (January 1–5 2013)
7. M.M. Mostafa, More than words: Social networks' texting for consumer brand sentiments. Expert Syst. Appl. **40**(10), 4241–4251 (2013)
8. S. Oramas, L. Espinosa-Anke, A. Lawlor, Exploring customer reviews for music genre classification and evolutionary studies, in *The 17th International Society for Music Information Conference (ISMIR), New York, USA*, (August 7–11 2016)
9. B. Pang, L. Lee, Opinion ming and sentiment analysis. Found. Trends Inf **2**(1-2), 1–135 (2008)
10. V. Sharma, A. Agarwal, R. Dhir, G. Sikka, Sentiments mining and classification of music lyrics using SentiWordNet, in *2016 Symposium on Colossal Data Analysis and Networking, Indore, India*, (18–19 March 2016)
11. P.D. Turney, M.L. Littman, Calling praise and criticism: Inference of semantic from the orientation association. ACM Trans. Inf. Syst **21**(4), 315–346 (2003)
12. J. Wang, Y. Yang, Deep learning based mood tagging for Chinese song lyrics. arXiv preprint, arXiv:1906.02135 (2019)
13. X. Yan, J. Wang, M. Chau, Revisit intention to restaurants: Evidence from online reviews. Inf. Syst. Front. **17**(3), 645–657 (2015)

14. H.L. Yang, Q.F. Lin, Opinion mining for multiple types of emotion-embedded products/services through evolution strategy. Expert Syst. Appl. **99**, 44–55 (2018)
15. Q. Ye, W. Shi, Y. Li, Sentiment classification for movie reviews in Chinese by improved semantic approach, in *The 39th Annual Hawaii International Conference on System Sciences (HICSS'06), Hawaii, USA*, (1–4 January 2006)

Part V
Hardware Acceleration in Artificial Intelligence (Chair: Dr. Xiaokun Yang)

A Design on Multilayer Perceptron (MLP) Neural Network for Digit Recognition

Isaac Westby, Hakduran Koc, Jiang Lu, and Xiaokun Yang

1 Introduction

The subject of creating a computational model for neural networks (NNs) can be traced back to 1943 [1, 2]. Until 1970, the general method for automatic differentiation of discrete connected networks of nested differentiable functions has been published by Seppo Linnainmaa [3, 4]. In 1980s, the VLSI technology enabled the development of practical artificial neural networks. A landmark publication in this field is a book authored by Carver A. Mead and Mohammed Ismail, titled "Analog VLSI Implementation of Neural Systems" in 1989 [5].

The real-world applications of NNs mainly appeared after 2000 [6]. In 2009, the design on the network with long short-term memory (LSTM) won three competitions of handwriting recognition without any prior knowledge about the three languages to be learned [7, 8]. And in 2012, Ng and Dean created a network being able to recognize higher-level concepts such as cats [9].

Today, NNs have shown great ability to process emerging applications such as speech/music recognition [10, 11], language recognition [12, 13], image classification [14, 15], video segmentation [16, 17], and robotic [18]. With the artificial intelligence (AI) chip market report published in May 2019, the global AI chip market size was valued at 6638.0 million in 2018, and is projected to reach 91,185.1 million by 2025, growing at Compound Annual Growth Rate (CAGR) of 45.2% from 2019 to 2025 [19]. Therefore, to make NNs high speed and efficient will have a profound impact on transportation, sustainability, manufacturing, city services, banking, healthcare, education, entertainment, gaming, defense, criminal investigation, and many more.

I. Westby · H. Koc · J. Lu · X. Yang (✉)
Department of Engineering, University of Houston Clear Lake, Houston, TX, USA
e-mail: YangXia@UHCL.edu

© Springer Nature Switzerland AG 2021
H. R. Arabnia et al. (eds.), *Advances in Artificial Intelligence and Applied Cognitive Computing*, Transactions on Computational Science and Computational Intelligence, https://doi.org/10.1007/978-3-030-70296-0_53

Under this context, in this paper a case study of a low-cost design on a MultiLayer Perceptron (MLP) neural network is proposed. This work is a preparation of a hardware implementation on field-programmable gate array (FPGA), aiming to accelerate the network computation within milliseconds. The design on the network is based on the database of Modified National Institute of Standards and Technology (MNIST), which was developed by Yann LeCun, Corinna Cortes, and Christopher Burges for evaluating machine learning models on the handwritten digit classification problem [20]. The main contributions of this work are:

- We present a low-cost design on MLP neural network with one input layer, one hidden layer, and one output layer. By feeding in 28 × 28 MNIST images of handwritten digits, this network is able to identify the digits with over 93% accuracy. The final goal of this work is an implementation and demonstration on FPGA to make the design as a real-time system.
- The network training is based on the MNIST handwritten digit data using the stochastic gradient descent method [20]. And the accuracy is estimated by Matlab on designs with half (16-bit), single (32-bit), and double (64-bit) precision. This platform is scalable to establish different design architectures of the MLP neural network, which is able to find the optimal energy-quality tradeoff of the hardware implementation.

The organization of this paper is as follows. Section 2 briefly introduces the related works and Sect. 3 presents our work with design architecture. Section 4 discusses the implementation of the proposed system. In Sect. 5, the experimental results in terms of hardware cost, power consumption, and FPGA prototype are shown. Finally, Sect. 6 presents the concluding remarks and future work in our target architecture.

2   Related Works

To date, many designs on digit recognition have been presented on algorithm level [21, 22] and hardware level [23, 24]. For example, a recurrent neural network (RNN) has been proposed in [25] to recognize digits, and in [26] the designs on Deep Neural Network (DNN), Convolutional neural networks) (CNN), and Bidirectional Recurrent Neural network (RNN) have been implemented and evaluated. As a result, the accuracy can reach 99.6% with the CNN and 97.8% with the four-layer DNN. The accuracy of RNN is 99.2%. All these researches were focused on exploring the accuracy of the implementations. In this paper, we present a simple architecture on MLP neural network, targeting to a low-cost design on FPGA in terms of less slice count and power cost, and high speed of the digit recognition.

3 Proposed Design Architecture

In this work, the database of MNIST is used to evaluate the MLP network models and obtain the weights and biases on the handwritten digit classification [20]. Generally, the input image of this database is a 28 by 28 pixel square (784 pixels total). A standard split of the dataset is used to evaluate and compare models, where 60,000 images are used to train a model and a separate set of 10,000 images are used to test it.

A perceptron of the MLP is a linear classifier which is able to classify input by separating two categories with a straight line. Input is typically a feature vector x multiplied by weights w and added to a bias b: $y = w*x + b$. Due to the nature of the output being binary, it makes it very hard to "train" these perceptron networks. This is where the sigmoid neuron comes in. The sigmoid neural network works similar to the perceptron network, except that the output is a value between 0 and 1.0, instead of a binary value of 0 or 1 [27]. The output for the sigmoid neuron is shown below.

$$sigmoid\,neuron\,output = \frac{1}{1 + exp\left(-\sum_{i=1}^{n} w_i \times x_i - b\right)} \tag{1}$$

where w_i is the weight corresponding to the input x_i, and b represents the bias.

This sigmoid neuron solves the problem with training. By making small changes to the weights and biases of the sigmoid neuron, we are able to make small changes to the output, eventually converging on a "correct" or most effective set of weights and biases. Because of this fact, sigmoid neurons are used in this paper to build the network as shown in Fig. 1. It can be observed that only one hidden layer with 12 neurons has been used to this network. For the input and output layers, there are 28×28 input pixels in the input layer and 10 output results in the output layer.

4 Implementation

In this section, the weights and biases of the MLP neural network are obtained, and a simulation result is discussed as an example when running the program.

4.1 Network with Weights and Biases

Once a design had been chosen, the next step was to prove the approximate design with single-precision floating point numbers. This step is important in ensuring the weights and biases trained in Python work in the approximate design. The input

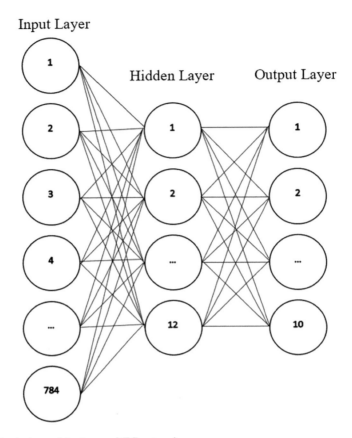

Fig. 1 The design architecture on MLP network

image to the network is a 28 × 28 grayscale pixels of matrices. The way that these pixels are numbered is shown in Fig. 2a.

The implementation of the design begins with the weights and biases that were generated in Python. The setup of all the weights for the hidden layer neuron is shown in Fig. 2b, being distributed in 12 rows and 784 columns. Each row represents the 786 weights associated with one of the 12 neurons in the hidden layer. Additionally 12 biases—one for each neuron is shown in Fig. 2c.

The output for the hidden layer neurons is then:

$$output(x) = \frac{1}{1 + exp\left(-\sum_{i=1}^{786} w(x)_i \times p_i - b(x)\right)} \quad (2)$$

p_1	p_2	p_3	p_4	...	p_{25}	p_{26}	p_{27}	p_{28}
p_{29}	p_{30}	p_{31}	p_{32}	...	p_{53}	p_{54}	p_{55}	p_{56}
p_{57}	p_{58}	p_{59}	p_{60}	...	p_{81}	p_{82}	p_{83}	p_{84}
...	...	...	...	...	...	...	...	...
p_{701}	p_{702}	p_{703}	p_{704}	...	p_{725}	p_{726}	p_{727}	p_{728}
p_{729}	p_{730}	p_{731}	p_{732}	...	p_{753}	p_{754}	p_{755}	p_{756}
p_{757}	p_{758}	p_{759}	p_{760}	...	p_{781}	p_{782}	p_{783}	p_{784}

(a) The layout of the input pixels of the first hidden layer

$w1_1$	$w1_2$	$w1_3$	...	$w1_{783}$	$w1_{784}$		b_1
$w2_1$	$w2_2$	$w2_3$	...	$w2_{783}$	$w2_{784}$		b_2
$w3_1$	$w3_2$	$w3_3$	...	$w3_{783}$	$w3_{784}$		b_3
...	...	...	...	...	...		...
							b_{10}
$w11_1$	$w11_2$	$w11_3$	...	$w11_{783}$	$w11_{784}$		b_{11}
$w12_1$	$w12_2$	$w12_3$	...	$w12_{783}$	$w12_{784}$		b_{12}

(b) The layout of the weights of the first hidden layer	(c) The layout of the biases of the first hidden layer
(d) The layout of the weights of the output layer	(e) The layout of the biases of the output layer

$w1_1$	$w1_2$	$w1_3$	...	$w1_{11}$	$w1_{12}$		b_1
$w2_1$	$w2_2$	$w2_3$	...	$w2_{11}$	$w2_{12}$		b_2
$w3_1$	$w3_2$	$w3_3$	...	$w3_{11}$	$w3_{12}$		b_3
...	...	...	...	...	...		...
							b_8
$w9_1$	$w9_2$	$w9_3$	...	$w9_{11}$	$w9_{12}$		b_9
$w10_1$	$w10_2$	$w10_3$	...	$w10_{11}$	$w10_{12}$		b_{10}

Fig. 2 The hardware design on sobel engine

where in the above equation, p_i, refers to the input pixel of index i (1–786), and $w(x)_i$ denotes the corresponding weight of the x hidden layer neuron. x ranges from 1 to 12 with the 12 hidden neurons.

Once all 12 of the outputs of the hidden layer neurons are found, these values act as inputs to the 10 output layer neurons. A similar process in finding the outputs takes place as for the first hidden layer neurons, except this time we are using weights shown in Fig. 2d and biases shown in Fig. 2e.

The table in Fig. 2d is laid out the same as weights for the hidden layer, except this time there are 10 rows for the output neurons, and 12 columns for the inputs. The table biases in Fig. 2e is similar to biases for the hidden layer as well, except there are only 10 values (one for each output neuron).

The output for the output layer neurons is then:

$$output(x) = \frac{1}{1 + exp\left(-\sum_{i=1}^{12} w(x)_i \times hiddenout_i - b(x)\right)} \tag{3}$$

where in the above equation, $hiddenout_i$, refers to the output from the i (1–12) hidden layer neuron, and $w(x)_i$ denotes the corresponding weight of the x output neuron. The value, x, refers to the output layer neuron (1–10) that the output is being calculated for.

In summary, the 784 input pixels are used with the corresponding 9408 weights, and 12 biases to find the 12 outputs of the hidden layer neurons. The outputs of the hidden layer are then used with the corresponding 120 weights, and 10 biases to find the 10 outputs of the hidden layer neurons. Once this entire process has completed, the network has processed and classified one image.

4.2 Network Implementation

The way the results are extrapolated from the neurons is shown in this section. In general, with each correct identification the value of the number that the network classified could be any value from 0 to 1.0. The number that is output from each neuron can be described as the "strength" of the result. With a value of 1.0 meaning the network has classified that as being the result with the highest probability, and a value of 0 meaning that it has the lowest chance of being that value.

The neurons are assigned such that output(1) = 0, output(2) = 1, until output(10) = 9. An example output when running the program can be seen in Fig. 3. The way the program works is it goes through each value of the outputs (0–9) sequentially, and if the current value is greater than the previous, it sets the value for variable "answer" equal to that number. That is why we see the output "The answer is 1," when that number is processed, because the value for 1 (0.0001626174), is greater than the value for 0 (0.000000007). We can see for the number 6, we have a value of 0.9999223948. Because this number is the highest, we can determine that the network has processed the input image, and classified it as a "6."

In another example, for each input image, all 10 of the outputs were compared and the highest value was determined to be the result. For instance, if we have the outputs shown in Fig. 4, the answer would be 3 even though the number for 3 (0.6894786954) is not as strong as it could be 0.9999223948 in Fig. 3.

```
The outputted value for 0 = 0.0000000070
The answer is 0
The outputted value for 1 = 0.0001626174
The answer is 1
The outputted value for 2 = 0.0000095044
The outputted value for 3 = 0.0000000021
The outputted value for 4 = 0.0000001204
The outputted value for 5 = 0.0000202689
The outputted value for 6 = 0.9999223948
The answer is 6
The outputted value for 7 = 0.0000013408
The outputted value for 8 = 0.0000000015
The outputted value for 9 = 0.0000022672
```

Fig. 3 An example output when running the program with an input of 6

```
The outputted value for 0 = 0.0000003561
The answer is 0
The outputted value for 1 = 0.0000059766
The answer is 1
The outputted value for 2 = 0.0198363531
The answer is 2
The outputted value for 3 = 0.6894786954
The answer is 3
The outputted value for 4 = 0.0000000649
The outputted value for 5 = 0.0012522331
The outputted value for 6 = 0.0000001147
The outputted value for 7 = 0.0000000442
The outputted value for 8 = 0.0000994121
The outputted value for 9 = 0.0000465117
```

Fig. 4 An example output when running the program with an input of 3

Fig. 5 Comparison of half, single, and double precision results in Matlab

5 Experimental Results

The implementation described above was run in Matlab using IEEE-754 double (64-bits), single (32-bits), and half (16-bits) precisions. The network was tested using the 10,000 MNIST test images. These are the same 10,000 images that were used to test the accuracy of the network in the Python program.

5.1 Accuracy Analysis

By running for all 10,000 of the MNIST test images, the results in Fig. 5 show how many images were correctly identified for half, single, and double precision data types. It can be observed that the designs with half-, single-, and double precision achieve similar accuracy of average 9325/10,000 images being correctly recognized.

In what follows, Fig. 6 shows the "strength" of the images that were classified correctly. Specifically in Table 1, we can see that for both single and double precision data types there are 8524 out of 10,000 images correctly identified with a 'strength' of 0.9 or above. This data was examined further to find any differences between results with these data types.

5.2 Further Analysis

Figure 7 looks at the 8524 correctly identified results in Table 1, then breaks this data down further into different ranges. More specifically, the data in Table 2 is broken down such that the first column represents which of the values are >0.9 and $<=0.99$, the second column represents values >0.99 and $<=0.999$, and so on.

A Design on Multilayer Perceptron (MLP) Neural Network for Digit Recognition 737

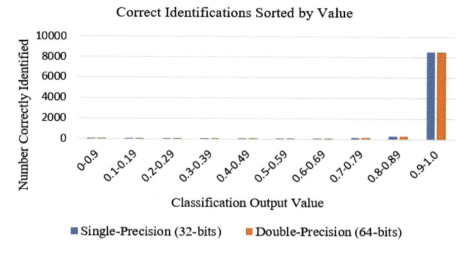

Fig. 6 Comparison of the "strength" of single and double precision data types

Table 1 Corresponding data used for comparison of "strength" of single and double precision data types

Accuracy	0–0.9	0.1–0.19	0.2–0.29	0.3–0.39	0.4–0.49	0.5–0.59	0.6–0.69	0.7–0.79	0.8–0.89	0.9–1.0
Single	14	19	36	36	57	91	89	155	304	8524
Double	14	19	36	36	57	91	89	155	304	8524

Then the last column is values >0.99999999 and <=1. Experimental results in this table show a minor difference between the implementations with signal-precision and double precision. In the fifth column the number of single-precision design identifies 644 images and the double precision design recognizes 643 images; and in the sixth column the single-precision design identifies 71 images and the double precision design recognizes one more image.

Another difference is that for the single-precision results there are 5 correctly identified results that are equal to 1.0, and no results >0.9999999 that are not equal to 1.0. Meanwhile for double precision, there are no results equal to 1.0 and 5 results >0.9999999.

This difference can be attributed to the fact that doubles are more precise than singles, just as the name would suggest. For single-precision numbers, any values with precision of 0.99999998 or greater are not able to be represented. These numbers are then converted to 1.0 when represented in single-precision floating point. For double precision numbers meanwhile, these numbers are able to be represented. This is why the for double precision values, there are 5 numbers that are >0.9999999 and <1.0, and for single-precision there are no values and 5 that are equal to 1.0.

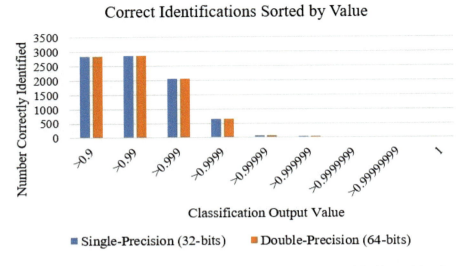

Fig. 7 Further comparison of the "strength" greater than 0.9 of single and double precision data types

Table 2 Corresponding data used for comparison of "strength" greater than 0.9 of single and double precision data types

Accuracy	0.9	0.99	0.999	0.9999	0.99999	0.999999	0.9999999	0.99999999	1
Single	2832	2867	2067	644	71	38	0	0	5
Double	2832	2867	2067	643	72	38	4	1	1

5.3 Accuracy Comparison

The accuracy of the Matlab implementation of the neural network can be compared with the Python results obtained when training the network. This is because the network has the same weights and biases, and we are using the same 10,000 MNIST test images in order to test the network. The comparison of results is shown in Fig. 8. In Python, the weights and biases are created using 64-bit floating point numbers, therefore all the weights and biases are 64-bit, and must be converted to 32-bit and 16-bit for single and half-precision. From this data we can see that the Python results, single, and double precision results are all the same.

Surprisingly, the half-precision results in Matlab correctly identified 9326 values out of the 10,000. This is one better than all other results and is unexpected. Therefore the discrepancy seen in the results above is most likely associated with a slight change in internal numbers when changing to lower granularity at half-precision.

Overall, the results have proven the proposed network design for half, single, and double precision data types. Comparing all of these results with the initial Python

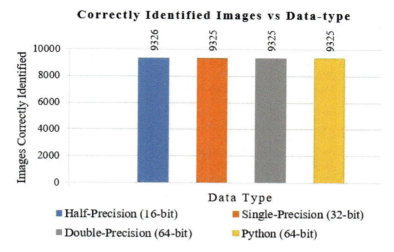

Fig. 8 Comparison of digit recognition accuracy of Matlab implementations with Python results

results, we can see that there is no discernible difference when using different precision data types.

6 Conclusion and Future Work

This paper presents a low-cost design on MLP neural network for recognizing handwritten digits, including one hidden layer, as well as the input and output layers. The accuracy of the designs with half-, single-, and double precision have been evaluated by Matlab. Experimental result shows that there is no big difference between the different designs. All of them achieve an accuracy of more than 93% by running 10,000 images with the MNIST database.

In the future work, the same design including the neurons' weights and biases will be used to implement a hardware core by FPGA, which has a great potential to accelerate the computation speed with an edge device. By integrating the open source platform with an OV7670 camera and VGA-enabled interface [28, 29], it is able to further demonstrate a real-time system to recognize handwritten digits.

References

1. W. McCulloch, P. Walter, A logical calculus of ideas immanent in nervous activity. Bull. Math. Biophys. **5**(4), 115–133 (1943)
2. S.C. Kleene, Representation of events in nerve nets and finite automata, in *Annals of Mathematics Studies* (Princeton University Press, Princeton, 1956), pp. 3–41
3. S. Linnainmaa, The representation of the cumulative rounding error of an algorithm as a Taylor expansion of the local rounding errors. University of Helsinki, pp. 6–7, 1970
4. S. Linnainmaa, Taylor expansion of the accumulated rounding error. BIT Num. Math. **16**(2), 146–160 (1970)
5. C. Mead, M. Ismail, Analog VLSI Implementation of Neural Systems, in *The Kluwer International Series in Engineering and Computer Science*, vol. 80 (Kluwer Academic Publishers, Norwell, 1989). ISBN 978-1-4613-1639-8
6. Jaspreet, A Concise History of Neural Networks. Medium (2016), https://towardsdatascience.com/a-concise-history-of-neural-networks-2070655d3fec
7. A. Graves, J. Schmidhuber, Offline handwriting recognition with multidimensional recurrent neural networks, in *Advances in Neural Information Processing Systems, Neural Information Processing Systems (NIPS) Foundation* (2009), pp. 545–552
8. A. Graves, Novel connectionist system for improved unconstrained handwriting recognition. IEEE Trans. Pattern Analy. Machine Intell. **31**(5), 855–868 (2009)
9. A. Ng, J. Dean, Building High-level Features Using Large Scale Unsupervised Learning (2012). arXiv:1112.6209
10. K. Vaca, A. Gajjar, X. Yang , Real-time automatic music transcription (AMT) with Zync FPGA, in *IEEE Computer Society Annual Symposium on VLSI (ISVLSI)*, Miami (2020)
11. K. Vaca, M. Jefferies, X. Yang , An open real-time audio processing platform on Zync FPGA, in *International Symposium on Measurement and Control in Robotics (ISMCR)* (2019)
12. H. He, L. Wu, X. Yang , Y. Feng, Synthesize corpus for Chinese word segmentation, in *The 21st International Conference on Artificial Intelligence (ICAI)*, Las Vegas (2019), pp. 129–134
13. H. He, L. Wu, X. Yang, H. Yan, Z. Gao, Y. Feng, G. Townsend, Dual long short-term memory networks for sub-character representation learning, in *The 15th International Conference on Information Technology - New Generations (ITNG)*, Las Vegas, NV (2018)
14. A. Gajjar, X. Yang , L. Wu, H. Koc, I. Unwala, Y. Zhang, Y. Feng, An FPGA synthesis of face detection algorithm using HAAR classifiers, in *International Conference on Algorithms, Computing and Systems (ICACS)*, Beijing (2018), pp.133–137
15. L. Nwosu, H. Wang, J. Lu, I. Unwala, X. Yang, T. Zhang, Deep convolutional neural network for facial expression recognition using facial parts, in *2017 IEEE 15th International Conference on Dependable, Autonomic and Secure Computing (DASC)*, Orlando (2017), pp. 1318–1321
16. Y. Zhang, X. Yang, L. Wu, J. Lu, K. Sha, A. Gajjar, H. He, Exploring slice-energy saving on an video processing FPGA platform with approximate computing, in *International Conference on Algorithms, Computing and Systems (ICACS)*, Beijing (2018), pp. 138–143
17. X. Fu, J. Lu, X. Zhang, X. Yang, I. Unwala, Intelligent in-vehicle safety and security monitoring system with face recognition, in *2019 IEEE International Conference on Computational Science and Engineering (CSE)*, New York (2019), pp. 225–229
18. J. Thota, P. Vangali, X. Yang, Prototyping an autonomous eye-controlled system (AECS) using raspberry-Pi on wheelchairs. Int. J. Compt. Appl. **158**(8), 1–7 (2017)
19. AI Chip Market Report (2019). https://www.alliedmarketresearch.com/artificial-intelligence-chip-market
20. F. Chen, N. Chen, H. Mao, H. Hu, Assessing four neural networks on handwritten digit recognition dataset (MNIST), in *Computer Vision and Pattern Recognition* (2018)
21. Q. Zhang, M. Zhanga, T. Chen, Z. Sun, Y, Ma, BeiYub, Recent advances in convolutional neural network acceleration. Neurocomp. **323**(5), 37–51 (2019)
22. J. Qiao, G. Wang, W. Li, M. Chen, An adaptive deep Q-learning strategy for handwritten digit recognition. Neural Netw. **107**, 61–71 (2018)

23. P. Lei, J. Liang, Z. Guan, J. Wang, T. Zheng, Acceleration of FPGA based convolutional neural network for human activity classification using millimeter-wave radar. IEEE Access **7**, 88917–88926 (2019)
24. E. Wu, X. Zhang, X. Zhang, D. Berman, I. Cho, J. Thendean, Compute-efficient neural-network acceleration, in *Proceedings of the 2019 ACM/SIGDA International Symposium on Field-Programmable Gate Arrays* (2019), pp. 191–200
25. H. Zhan, Q. Wang, Y. Lu, Handwritten digit string recognition by combination of residual network and RNN-CTC, in *Neural Information Processing*, vol. 10639 (2017)
26. S. Jain, R. Chauhan, Recognition of handwritten digits using DNN, CNN, and RNN.' Adv. Comput. Data Sci. **905**, 239–248 (2018)
27. M. Nielsen, *Neural Networks and Deep Learning* (Determination Press, 2015). https://neuralnetworksanddeeplearning.com/chap1.html
28. Y. Zhang, X. Yang , L. Wu, J. Andrian, A case study on approximate FPGA design with an open-source image processing platform, in *IEEE Computer Society Annual Symposium on VLSI (ISVLSI)*, Miami (2020)
29. X. Yang , Y. Zhang, L. Wu, A scalable image/video processing platform with open source design and verification environment, in *20th International Symposium on Quality Electronic Design (ISQED)*, Santa Clara (2019), pp. 110–116

An LSTM and GAN Based ECG Abnormal Signal Generator

Han Sun, Fan Zhang, and Yunxiang Zhang

1 Introduction

The electrocardiogram (ECG) is a record of the heart muscle movement and has been widely used in detection and treatment for cardiac diseases. Different from normal ECG signals, arrhythmia signals can be harbingers of some dangerous heart diseases. Early diagnosis helps early detection of chronic disease and starts treatment as soon as possible. For emergency heart attack such as ventricular fibrillation, timely detection can significantly improve the survival rate [1]. Therefore, efficient and effective ECG arrhythmia detection and classification is very important. However, arrhythmia beats usually occur sporadically and unexpected, which makes them extremely difficult to record. The rarity of arrhythmia data limits the training quality of classifiers and becomes a hindrance on the road to a comprehensive diagnosis system [2].

To enable auto-diagnosis with limited real arrhythmia data, an efficient model to generate ECG signals with high confidence and quality, especially the abnormal ECG signals, is necessary. To build a system that can mimic arrhythmia signal, we choose long short-term memory (LSTM) and generative adversarial nets (GANs) as main components. The GANs were first published in 2003 by Goodfellow et al. [3]. In the system, there is a generator (G) that will continuously produce fake signals to approach real database, while there is a discriminator (D) that can determine whether the input is real or from the generator. Both G and D will be trained at the same time until the discriminator cannot define the true label. Instead of using

---

H. Sun (✉) · F. Zhang · Y. Zhang
Department of Electrical and Computer Engineering, Binghamton University, Binghamton, NY, USA
e-mail: hsun28@binghamton.edu; fzhang27@binghamton.edu; yzhan392@binghamton.edu

© Springer Nature Switzerland AG 2021
H. R. Arabnia et al. (eds.), *Advances in Artificial Intelligence and Applied Cognitive Computing*, Transactions on Computational Science and Computational Intelligence, https://doi.org/10.1007/978-3-030-70296-0_54

the original ECG signal, we use the feature got from an LSTM encoder to train the generator.

LSTM is a special recurrent neural network model, which can selectively remember the important information of the input. It is good at dealing with correlated data sequence and commonly used on speech recognition [4] and sequence translation [5]. In our system, the LSTM encoder and decoder are trained to find the commonality of a group of data and exclude the effort from differences among individuals. To learn and mimic arrhythmia signals, a GAN model [3] is included between LSTM encoder and decoder to learn the commonality within hidden states. Previous study shows that GANs can generate similar output based on input efficiently [6]. So we choose GANs as the generative model to synthesize the commonality given by LSTM encoder. After enough training, it can generate fake states and pass them to LSTM decoder to produce high quality fake signals.

Our main contributions are listed below: (1) We proposed an LSTM and GAN based arrhythmia generator, which can learn from a small data set and produce high quality arrhythmia signals. (2) We optimize the system performance by studying the correlation between training iteration and training samples. (3) We verify the effectiveness of our method on MIT-BIH data set with random forest classifiers. Compared to classifier trained with only real data, the same classifier that is trained with both real and fake data achieves an accuracy boost from 84.24 to 95.46%.

The remaining of this paper is organized as follows: Sect. 2 is the related work about both mathematical and machine learning based ECG models. In Sect. 3, our method will be discussed in detail. Section 4 describes the classifier in order to test our method. And the result is shown in Sect. 5. Discussion about future work is given in Sect. 6. The paper is concluded in Sect. 7.

2 Related Work

2.1 Mathematical ECG Models

Mathematical ECG models use a set of dynamic equations to fit the ECG behavior. In [7], a realistic synthetic ECG generator is built by using a set of state equations to generate a 3D ECG trajectory in a 3D state space. ECG signal is represented by a sum of Gaussian functions with amplitudes, angular spreads, and locations controlled by the Gaussian kernel parameters as in Eq. 1. In [8], a sum of Gaussian kernels are fitted to a normal ECG signal and then used to generate abnormal signal. Switching between normal and abnormal beat types is achieved by using a first-order Markov chain. The state probability vector (P) is trained to be changed based on factor such as R–R time series. According to the P and the ECG morphology parameters, the next ECG beat type can be determined.

An LSTM and GAN Based ECG Abnormal Signal Generator

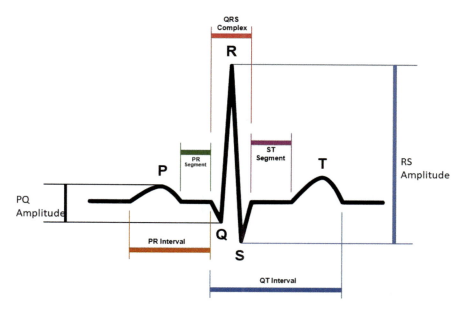

Fig. 1 A typical normal ECG signal wave

$$\begin{cases} \hat{x} = \gamma x - \omega y \\ \hat{y} = \gamma y - \omega x \\ \hat{z} = \sum_{i \in (P,Q,R,S,T)} a_i \Delta\theta_i exp\left(-\frac{\Delta\theta_i^2}{2b_i^2}\right) - (z - z_0). \end{cases} \quad (1)$$

In [9] to apply the model to filtering arrhythmia, a new wave-based ECG dynamic model (WEDM) by separating different events of the ECG has been introduced. In this model, the ECG signal was separated into three events as the P, C, and the T that represent the P-wave, the QRS complex, and the T-wave, respectively. Beside the (P,Q,R,S,T) events, there are also parameters needed for R-R process used in the wave-based synthetic ECG generation, which are mean heart rate, the standard deviation of the heart rate, the mean normalized low frequency, the standard deviation of the normalized low frequency, the mean respiratory rate, the standard deviation of the respiratory rate, and the low frequency to high frequency ratio. By controlling the set of parameters, it is possible to generate abnormal signals.

Mathematical models can generate high quality ECG signal with enough data to extract the parameters. These parameters are often more than just standard parameters P, Q, R, S, T as marked in Fig. 1. More parameters, such as mean heart rate, the standard deviation of heart rate, the mean normalized low frequency, etc., are often necessary to configure the models. However, lack of enough ECG data for each type of abnormality causes extreme difficulty on parameter extractions and makes mathematical models inappropriate for general abnormal ECG signal generation.

2.2 Machine Learning ECG Models

The generative adversarial networks (GANs) [3], based on the mini–max two-player game theory, show a great superiority in generating high quality artificial data. In [10], a GAN based inverse mapping method has been produced. Instead of cooperation between the constructed and original image space, the latent space has been used to update the generator. The similarity of the reconstructed image to original image is around 0.8266, which is higher than direct training. In [11], personalized GANs (PGANs) are developed for patient-specific ECG classification. The PGANs learn how to synthesize patient-specific signals by dealing with the P, Q, R, S, T events. The results are used as additional training data to get higher classification rate on personal ECG data. Three types of arrhythmia are considered in the work with an average accuracy rate about 93%. However, the models are only trained to produce personalized ECG signals. As sporadic types of arrhythmia are even harder to collect for each individual, it is impractical to train GANs for personalized abnormal ECG signal generation.

Long short-term memory (LSTM) can solve the vanishing gradient problem caused by the gradual reduction of the back propagation. It has high performance when dealing with time series related issues such as speech recognition, machine translation, and encoding/decoding [4]. For ECG signal classification, LSTM encoder/decoder classifier can achieve 99.39% for ECG signals [12]. However, it only studies common arrhythmia that is equally distributed and ignores rare cases.

Although those models can generate both normal and abnormal signals, there are still limitations. For the mathematical model, there are too many parameters needed for computation. On the other hand, the input real data need to be analyzed before it can be used for calculating. For the VAE and PGAN models, the models are only trained to produce personalized ECG signals. However, for those sporadic types of arrhythmia, it is hard to record and cannot be used for training. Therefore, a general purposed ECG signal generator is needed.

3 Methodology

Figure 2 illustrates the overall flow of our LSTM and GAN based ECG signal generator. Instead of using LSTM as classifier, we use two LSTMs, one is as encoder to translate ECG signal data $\mathbf{x}$ into hidden states $\mathbf{h}$, and another is as decoder to convert $\mathbf{h}$ back to $\mathbf{x}$. Not only more data, but also more iterations can help improve the quality of the LSTM encoder and decoder. To generate fake ECG signals, we insert GANs between LSTM encoder and decoder and train it to generate fake latent vector $\mathbf{h_f}$ that is similar to $\mathbf{h}$. The LSTM encoder, decoder, and GANs are trained separately for each type of abnormal signals. For abnormal signal type that has less than 1000 beats, the iteration is increased to improve performance. After enough fake ECG data are generated, a random forest classifier is trained with both real and fake ECG data.

An LSTM and GAN Based ECG Abnormal Signal Generator

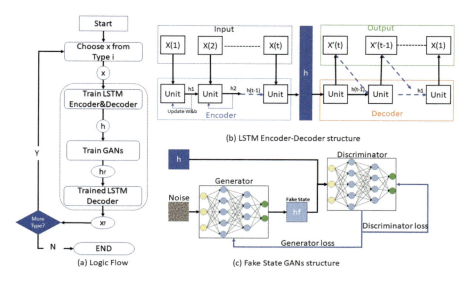

Fig. 2 Proposed approach for ECG signal generator. Part (**a**) is the logic flow of our system. Part (**b**) is the LSTM encoder and decoder training process. Part (**c**) shows how GANs work

3.1 LSTM Encoder and Decoder

In this section, we describe the LSTM unit and how to train LSTM encoder and decoder. Figure 2b shows the general diagram of LSTM encoder and decoder. Each unit represents an LSTM cell. It takes the input signal **x**(*t*), the previous unit output **h**(*t*), the previous cell state **C**(*t* − 1), and the bias vector **b** as inputs. Inside a unit, three gates (*f*,*i*,*o*) work together to calculate the hidden state **h**. The gates' activation value vectors are calculated to determine whether the input information should be used. The **W** is the separate weight matrices for each input. The activation of three gates is calculated in the following way: the forget gate (*f*) first determinates whether the previous memory should be used. The total input of the cell passes through the input gate (*i*) and sigmoid function (tanh) and then multiplies with the activation value of the input gate. After adding input value to the cell state **C**, the final output hidden state **h** is calculated by multiplying the output gate activation value (*o*) and tanh(**C**). **h** is used to update **W** for each gate.

The decoder is formed as an inverse process of the encoder, and it works by taking the final hidden state *h* as the first input. The conditional decoder takes the last generated output as one of the inputs that is represented as the dot line in Fig. 3b. The final output of the decoder is an ECG signal restored based on the hidden state **h**. All parameters including weight **W** and bias **b** are the same as the encoder unit but in reverse order.

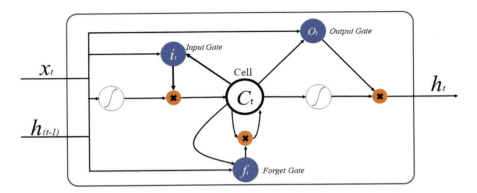

Fig. 3 LSTM cell units

3.2 GANs for Fake Hidden State Generation

The GANs are used to generate fake hidden state that can pass to the LSTM decoder to produce fake ECG signal waveform. The basic idea of the GANs is a mini–max game of two players: the generator (G) and the discriminator (D). D exterminates the samples to determine whether they are valid (real hidden states) or invalid (fake hidden states), while G keeps creating samples with similar distribution to real hidden states. We implemented the generator and discriminator in the following way: **Generator**(G): The generator is a 2-layer feed-forward fully connected ANN with ReLu. It takes a random noise as input and gives the fake hidden state $\mathbf{h_f}$ as output. **Discriminator**(D): The discriminator is a 4-layer feed-forward fully connected ANN with ReLu between each layer. It is trained to distinguish the real and fake states.

Algorithm 1 summarizes the training of GANs. The output of the function $D(x)$ is the predicted label from the discriminator of input x, while the $MSE(Y, \hat{Y}) = \frac{1}{n}\sum_{i=1}^{n}(Y_i - \hat{Y}_i)^2$, Y is the predicted label, and the $\hat{Y}$ is the ground truth. A mini-batch of real state and a mini-batch of fake state made from the former step variables are chosen. The training is based on two simultaneous stochastic gradient descending processes. At each step, mini-loss functions G_{loss} and D_{loss} are all given by D. D_{loss} shows the ability of D in different real and fake states. A 0 D_{loss} means it can perfectly distinguish the inputs, and the higher D_{loss} is the worse situation it will be. The G_{loss} represents how much the G can "cheat" D. G_{loss} equals to 0 means all the fake states given by G are considered as real, the higher, the worse. A learning rate of 0.0002 is chosen [6].

Algorithm 1 GANs Training
Input:
Noise: Gaussian Distribution noise
h: Real Hidden State from LSTM Encoder
Valid: Label of real state
Invalid: Label of fake state
Output:
h_f: Fake Hidden State
for step number of training irritation **do**
for k Steps **do**
Sample mini batch of h;
Sample mini batch of Generated h_f;
RealLabel = D(h);
FakeLabel = D(h_f);
Update D by descending its stochastic gradient with loss function D_{loss};
$D_{loss} = \frac{1}{2} * [MSE(\textbf{RealLabel}, Vaild)$
$+MSE(\textbf{FakeLabel}, Invalid)]$
end for
NoiseLabel = D(Noise);
Update G by descending its stochastic gradient with loss function G_{loss};
$G_{loss} = MSE(\textbf{NoiseLabel}, Valid)$
end for

Algorithm 1 GAN training

4 Random Forest Arrhythmia Classifier

The random forest (RF) classifier used to classify different types of abnormal ECG signals. The RF is a method that brunch of decision trees work together. Each tree votes independently, and the final decision is made with the class that gets most votes. The decision tree is built by randomly choosing n features from the total m characteristics of the samples, where $(n = \sqrt{m})$. The child nodes are decided by using those n features until the current n features have been used in its parent node. Multiple decision trees are built independently based on randomly picked samples of n features. Combination of all the decision trees forms a random forest. The contribution of RF works as follows:

(1) Randomly pick N samples from the original data set with replacement.
(2) Train the root node of the decision node based on the N samples.

(3) Randomly choose m features from the total $M(m = \sqrt{M})$ characteristics of the samples. Use the m features as the split nodes until the current m features have been used in the parent node.

(4) Independently build multiple decision trees following steps 1–3, while all trees are built without pruning.

(5) Combine those decision trees with counters.

The feature for RF classifier in our system is the signal itself. The classifier is trained based on real data only to classify each arrhythmia type, and then another classifier is trained to prove the effect of real and fake data used together.

5 Result

5.1 ECG Database

In experiments we use MIT-BIH arrhythmia database [13]. In the MIT-BIH database, total 23 types of annotations have been record. The 4 types we used are: left bundle branch block beat (LB) with 8011 beats, right bundle branch block beat (RB) with 6425 beats, aberrated atrial premature beat (AA) with 6548 beats, and the atrial fibrillation (AF) with only 310 beats as the rare ECG type.

5.2 Result for LSTM and GANs

Figure 4 shows the output of decoder for type AF with different iterations. As iteration number increases, LSTM decoder can better capture the commonality of AF signals. Figure 5 gives the error vs. iteration curves for all four types. Figure 6 shows the fake hidden states generated by GANs and its fake ECG signal generated by the LSTM decoder. We can see that GANs+LSTM decoder can produce similar but not exactly the same abnormal ECG signals, which are suitable for the later training of the RF classifier.

5.3 Classification with Fake ECG Signal

Here we compare the qualities of training based on real data only vs. training on both real and fake data. Table 1 summarizes the training result with or without fake data and also comparisons to other existing methods. RF stands for RF classifier trained with only real data. It achieves an average 84.24% accuracy in total 21,298 testing samples. And for each abnormal ECG type, the classifications rates are: 95.33% for LB, 61.34% for RB, 96.29% for AA, and 0.18% for AF. RLG stands for RF classifier trained with both real and fake data. Fake signals are added to data sets

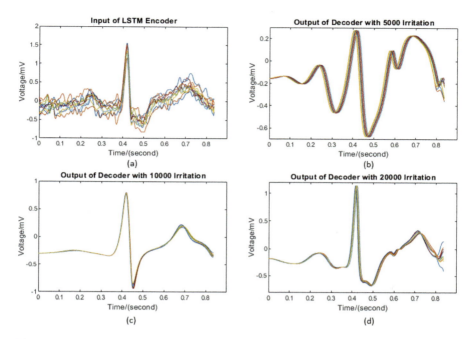

Fig. 4 Encoder result with different training steps. (**a**) The real signal. (**b**) Decoder output after 5000 steps. (**c**) Decoder output after 10,000 steps. (**d**) Decoder output after 20,000 steps

of each type to an equal number of 8200. After training, the average classification accuracy rate increases to 95.46%. Moreover, for each type, we have 95.28% for LB, 95.39% for RB, 95.85% for AA, and 93.55% for AF. Note that testing is only done on real data for a fair comparison, and we can see a significant boost in accuracy especially for AF. The huge difference between with or without fake data on AF classification shows the importance of data balance for training as well as verifies the effectiveness of our approach.

6 Future Work

Although ECG based arrhythmia test is attracting more and more attentions and more and more ECG databases are created, the arrhythmia beats included in are still not comparable to normal beats. In the MIT-BIH database [13], only 16 beats of 110,000 are recorded as atrial escape beat. On the other hand, all the annotations are given by two cardiologists artificially, and some of the public databases such as [16] do not include arrhythmia annotation. Therefore, mathematical models can be used as auxil. In [17], new models based on probability density function are created to

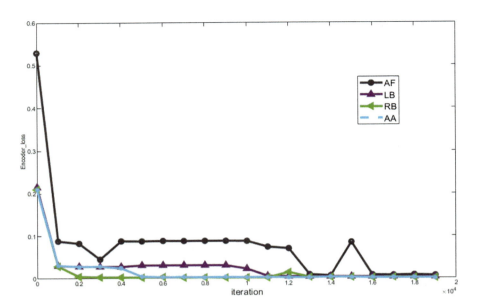

Fig. 5 Encoder_Loss of different iterations

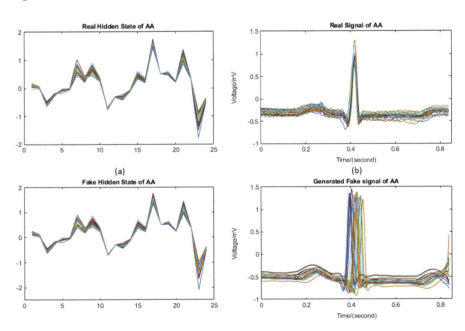

Fig. 6 Real vs. fake state and signal for AA. Real ones are listed on top, and fake states/signals are at bottom

An LSTM and GAN Based ECG Abnormal Signal Generator

Table 1 Performance comparison of the proposed method with other studies

	Paper	Method	Class	Beats	Accuracy/%
Neuro-morphic model	[14]	RF	3	154	92.16
	[15]	SVM	4	44,258	86
	[11]	LG	3	Unknown	93
Mathematical model	[8]	VFCDM	3	61	88
Our method		RF	4	21,298	84.24
		RLG			95.46

make more detailed divisions about ECG signals. In that case, neural network can participate in adjust parameters.

There are research that show that ECG signals vary from person to person [18], and arrhythmia ECG models can also be created to adapt to different persons. With the development of wearable ECG detect device [19], personalized ECG signal can be collected easily. Since our system can generate ECG beat from few input real signals, we can build a personalized classification system within few periods of testing.

Hardware based ECG application is another field with good development prespects. In [20] a 65-nm CMOS is used for personal cardiac monitoring. A classifier is trained based on the MIT-BIH database and then tested with connected ECG processor. With the hardware based neural network implementation method [21] and database extended by our generator, a faster arrhythmia can be trained to get higher accuracy.

Therefore, we will consider in future the feasibility of this hardware–software co-operated application. We want to create a wearable device with pre-installed arrhythmia classifier that can be adjusted by users' ECG signal to finish the personal cardiac monitoring.

7 Conclusion

In this work, we present an LSTM and GAN based ECG signal generator to improve abnormal ECG classification especially for rare types. Our LSTM encoder can extract the hidden states that represent the commonality between individuals of the same ECG type. With fake states from GANs, our LSTM decoder can produce high quality fake signals for later detection and classification. We implemented a random forest classifier to verify the effectiveness of our approach. With the help of fake ECG signals, the average classification rate improves from 84.24 to 95.46%, with classification accuracy for rare ECG types (AF) boosted from 0.18 to 93.55%

References

1. V. Fuster, L.E. Rydén, D.S. Cannom, H.J. Crijns, A.B. Curtis, K.A. Ellenbogen, J.L. Halperin, J.-Y. Le Heuzey, G.N. Kay, J.E. Lowe, et al., ACC/AHA/ESC 2006 guidelines for the management of patients with atrial fibrillation: a report of the American college of cardiology/American heart association task force on practice guidelines and the European society of cardiology committee for practice guidelines (writing committee to revise the 2001 guidelines for the management of patients with atrial fibrillation) developed in collaboration with the European heart rhythm association and the heart rhythm society. J. Am. College Cardiol. **48**(4), e149–e246 (2006)
2. E.J.d.S. Luz, W.R. Schwartz, G. Cámara-Chávez, D. Menotti, ECG-based heartbeat classification for arrhythmia detection: a survey. Comput. Methods Programs Biomed. **127**, 144–164 (2016)
3. I. Goodfellow, J. Pouget-Abadie, M. Mirza, B. Xu, D. Warde-Farley, S. Ozair, A. Courville, Y. Bengio, Generative adversarial nets, in *Advances in Neural Information Processing Systems* (2014), pp. 2672–2680
4. A. Graves, N. Jaitly, A.-r. Mohamed, Hybrid speech recognition with deep bidirectional LSTM, in *2013 IEEE Workshop on Automatic Speech Recognition and Understanding* (IEEE, Piscataway, 2013), pp. 273–278
5. I. Sutskever, O. Vinyals, Q.V. Le, Sequence to sequence learning with neural networks, in *Advances in Neural Information Processing Systems* (2014), pp. 3104–3112
6. I. Goodfellow, Nips 2016 tutorial: Generative adversarial networks (2016). Preprint arXiv:1701.00160
7. P.E. McSharry, G.D. Clifford, L. Tarassenko, L.A. Smith, A dynamical model for generating synthetic electrocardiogram signals. IEEE Trans. Biomed. Eng. **50**(3), 289–294 (2003)
8. J. Lee, D.D. McManus, P. Bourrell, L. Sörnmo, K.H. Chon, Atrial flutter and atrial tachycardia detection using Bayesian approach with high resolution time–frequency spectrum from ECG recordings. Biomed. Signal Proc. Control **8**(6), 992–999 (2013)
9. G. Clifford, A. Shoeb, P. McSharry, B. Janz, Model-based filtering, compression and classification of the ECG. Int. J. Bioelectro. **7**(1), 158–161 (2005)
10. G. Perarnau, J. Van De Weijer, B. Raducanu, J.M. Álvarez, Invertible conditional GANs for image editing 2016. Preprint arXiv:1611.06355
11. T. Golany, K. Radinsky, PGANs: Personalized generative adversarial networks for ECG synthesis to improve patient-specific deep ECG classification, in *Proceedings of the AAAI Conference on Artificial Intelligence* (2019)
12. Ö. Yildirim, A novel wavelet sequence based on deep bidirectional LSTM network model for ECG signal classification. Comput. Biol. Med. **96**, 189–202 (2018)
13. G.B. Moody, R.G. Mark, The impact of the MIT-BIH arrhythmia database. IEEE Eng. Medicine Biology Mag. **20**(3), 45–50 (2001)
14. R.G. Kumar, Y. Kumaraswamy, et al., Investigating cardiac arrhythmia in ECG using random forest classification. Int. J. Comput. Appl. **37**(4), 31–34 (2012)
15. Z. Zhang, J. Dong, X. Luo, K.-S. Choi, X. Wu, Heartbeat classification using disease-specific feature selection. Comput. Biol. Med. **46**, 79–89 (2014)
16. S. Pouryayevali, S. Wahabi, S. Hari, D. Hatzinakos, On establishing evaluation standards for ECG biometrics, in *2014 IEEE International Conference on Acoustics, Speech and Signal Processing (ICASSP)* (IEEE, Piscataway, 2014), pp. 3774–3778
17. J.P. do Vale Madeiro, J.A.L. Marques, T. Han, R.C. Pedrosa, Evaluation of mathematical models for QRS feature extraction and QRS morphology classification in ECG signals. Measurement **156**, 107580 (2020)
18. S.K. Berkaya, A.K. Uysal, E.S. Gunal, S. Ergin, S. Gunal, M.B. Gulmezoglu, A survey on ECG analysis. Biomed. Signal Proc. Control **43**, 216–235 (2018)
19. N. Glazkova, T. Podladchikova, R. Gerzer, D. Stepanova, Non-invasive wearable ECG-patch system for astronauts and patients on earth. Acta Astronaut. **166**, 613–618 (2020)

20. S. Yin, M. Kim, D. Kadetotad, Y. Liu, C. Bae, S.J. Kim, Y. Cao, J.-s. Seo, A 1.06-μw smart ECG processor in 65-nm CMOS for real-time biometric authentication and personal cardiac monitoring. IEEE J. Solid-State Circ. **54**(8), 2316–2326 (2019)
21. F. Zhang, M. Hu, Mitigate parasitic resistance in resistive crossbar-based convolutional neural networks (2019).Preprint arXiv:1912.08716

An IoT-Edge-Server System with BLE Mesh Network, LBPH, and Deep Metric Learning

Archit Gajjar, Shivang Dave, T. Andrew Yang, Lei Wu, and Xiaokun Yang

1 Introduction

In the last decade, the applications of embedded systems have boosted in the field of Internet-of-Things (IoT). Typically, IoT systems with multiple sensors including computation devices scattered over an enormous area [1]. While the advent of IoT solved many hitches, several inevitable problems were invited as well [2, 3]. The amalgamation of IoT and cloud requests facilitated the formation of edge computing, in which computing befalls at the network edge where there is no limitation of devices in terms of hardware type [4, 5]. In other words, the computing hardware device can be anything such as Raspberry Pi [6], Field Programmable Gate Array (FPGA) [7], System-on-Chip (SoC) [8], Application-Specific Integrated Circuit (ASIC) [9], general-purpose Central Processing Unit (CPU), or server [10]. For better understanding, a comparison between edge computing and traditional cloud computing systems is shown in Table 1.

Under the described idea, many research groups are working on edge computing in the hunt for further exploration and improvement. The edge computing is, currently, one of the most popular topics, and with the trend of machine learning, scholars are combining two ideas in order to achieve desired goals that are mentioned above [11, 14]. In the field of edge computing, the major focus of the research

---

A. Gajjar · S. Dave · T. Andrew Yang · X. Yang (✉)
Department of Engineering, University of Houston Clear Lake, Houston, TX, USA
e-mail: YangXia@UHCL.edu

T. Andrew Yang
Department of Computer Science, University of Houston Clear Lake, Houston, TX, USA

L. Wu
Computer Science Department, Auburn University at Montgomery, Montgomery, AL, USA

© Springer Nature Switzerland AG 2021
H. R. Arabnia et al. (eds.), *Advances in Artificial Intelligence and Applied Cognitive Computing*, Transactions on Computational Science and Computational Intelligence, https://doi.org/10.1007/978-3-030-70296-0_55

Table 1 Cloud computing vs. edge computing

Parameters	Cloud computing	Edge computing
Architecture	Centralized	Distributed
Data processing	Far from the source	Adjacent to the source
Latency	High	Low
Jittering	High	Low
Data privacy	Low	High
Accessibility	Global	Local
Mobility	Limited	Feasible
No. of nodes	Few	Extremely high
Data exploitation risk	Global	Remains in edge network
Communication with devices	Over the Internet	Local through edge node

work is on the software side, which includes performance improvement [19–21], algorithm optimization [12, 15], and increased efficiency with better task scheduling [13]. Comparatively, there is much less flow where scholars are working on the FPGA [17, 18], digital circuit [16], embedded systems [24], and hardware architecture [25, 26].

On that front, this literature proposes a promising architecture on IoT-Edge-Server based embedded systems with Bluetooth Low Energy (BLE) mesh system, surveillance system, a couple of processors and a server. In a generic manner, we have two platforms: (a) a BLE mesh network and (b) a surveillance demonstration. These systems are able to fetch data, at one instance, to Intel i7-7700HQ CPU and to Raspberry Pi on the other instance. The entire system has, virtually, three layers: (1) disposed layers, (2) edge computing network, and (3) cloud computing network [27].

The edge computing is boon to the IoT, but it also brings veil challenges that inspired us to propose a pioneering architecture. The main contributions of this work are as follows:

- To provide a robust architecture of IoT-Edge-Server embedded systems that can be implemented on extensively scattered environments such as agricultural farms, smart cities, or commercial/industrial buildings.
- Depending upon the application-specific tasks, we provide a system with an architecture that has two conceivable methods of computation: 1) traditional cloud computing and 2) edge computing.
- An implementation of two of the innumerable face recognition algorithms on Raspberry Pi and Intel i7-7700HQ CPU. There were few publications where they had performed a similar task on the comparably likewise hardware setting, but there is no mention of computation time or accuracy [28–30]. While our work not only delivers computation time and accuracy but also provides those data with an assortment of multiple face recognition algorithms, cameras, and computation devices as well.
- Improved data privacy—edge computing enables computation at the edge nodes, which requires personal data to be on the edge nodes rather than storing them on

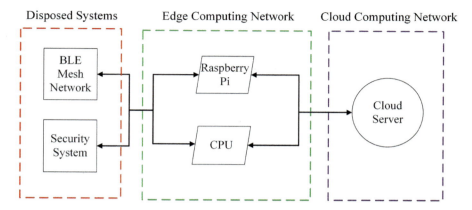

Fig. 1 Two channel computation options

a cloud. In other words, data remains at the network edge, which restricts data being hacked by anyone from the server.
- Improved computation speed—Performing any type of computation at any edge node prevents extreme back-and-forth of data to the cloud server and also improves the computation speed by reducing the latency. We provide benchmarks for computation time of cloud computing and edge computing for the same task to prove our hypothesis.
- We design communication commands for BLE mesh fabricated boards, which are reserved for only these boards.

2 Proposed Architecture

In this section, we propose the design architecture as shown in Fig. 2, and we also provide various computation options as shown in Fig. 1. The computation options can be decided based on the requirements in terms of computation time and responsiveness.

Figure 2 displays the proposed architecture of the IoT-Edge-Server based embedded system. More specifically, the architecture consists mainly of three virtual layers. These layers are depicted in Fig. 1: (1) disposed layer, (2) edge computing network, and (3) cloud computing.

2.1 Disposed System

In Fig. 2, the disposed system is labeled with the multiple disposed systems, which are BLE mesh network and surveillance system. Generally, the disposed layer is an integrated system to the edge computing network depending upon the applications

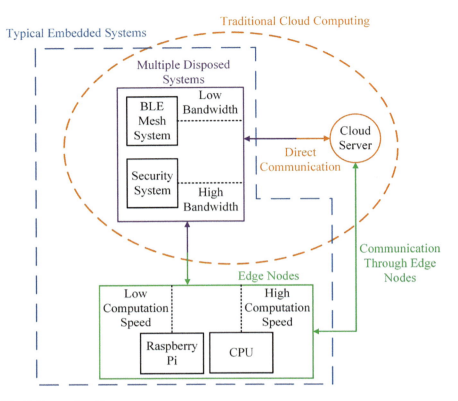

Fig. 2 Proposed architecture

and requirements. As mentioned earlier, for our proposed architecture and mainly to test our hypothesis, we implemented a BLE mesh network and surveillance system.

The BLE mesh network is a creation of four semi-customized boards using the APlix CSR 1020 modules that are suitable for low-power and restricted-complexity IoT applications. In this typical system, there is one master host connected to three slave boards. Each board possesses components to perform a designated task: (1) temperature and humidity sensor, (2) light-emitting diode (LED), and (3) motor.

On the other side, the surveillance system manifests a passive infrared (PIR) sensor associated with a camera to detect and recognize a face(s) in the milieu.

2.2 Edge Network

As shown in Fig. 2, the green-colored area named as edge nodes consists of all the computation devices for the architecture. Such devices are connected to the disposed systems and cloud server that apparently creates an edge computing network. For time being, we neglect the orange-colored area to visualize a simple embedded system without any cloud server. For our system, we propose three types of edge nodes that include FPGA, CPU, and Raspberry Pi. These devices will provide assistance in the improvement of latency for the task and sieving data that needs to be stored on the server eventually. The filtration of data includes the removal of repetitive data, duplicated logs, or any junk data.

2.3 Cloud Network

The entire orange-colored system titled as traditional cloud computing, as displayed in Fig. 2, is envisioned as a cloud computing network where all the data from the disposed systems can be transferred to the server for the computation to take further steps. When the computation is completed, the decision or data will be sent back to the disposed systems. We would like to emphasize that the orange-colored portion in Fig. 2 is a traditional cloud computing architecture where all the computing and data would be sent to cloud for any kind of processing. For the proposed research work, the server specifications are as follows: Intel(R) Core (TM) i3-7350K @ 4.20 GHz with 8 Gigabyte (GB) Random Access Memory (RAM).

3 Implementation

In this section, details of the design of the whole platform are introduced including the BLE mesh system, surveillance system, and server-side execution.

3.1 BLE Mesh System

As mentioned before, the BLE mesh system has, currently, four self-designed boards accomplishing varied tasks creating the complete system on the foundation of the BLE low energy protocol. As shown in Fig. 3, using the mesh-topology method and abundant sensors, we can establish a large-scale environment. Figure 3 visually elucidates the arrangement of all the sensors in the giant setting. Moreover, the maximum distance between two boards to be able to communicate is around 57 feet as our experimental result. As shown in Fig. 3, the BLE mesh host/server is only able to communicate with an actuator that is in the same group 2. Furthermore, the actuator works as a relay for groups 1 and 2; in that manner, the BLE mesh host/server is able to communicate with the LED and temperature/humidity sensor boards. With the use of a relay between two boards, the maximum-communication distance can be increased to 77 feet as our results. Thus, a large network with numerous sensors can be deployed for wide-range habitats.

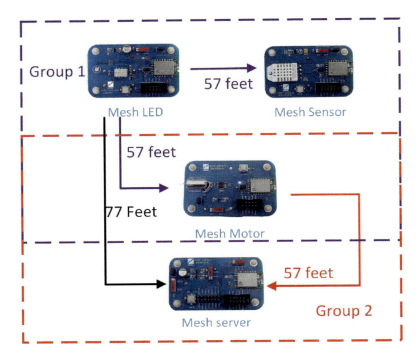

Fig. 3 Mesh topology

Data Type	Size	Content	Description
Header	2 B	0xFA, 0xF5	Data start header
Size	1 B	0x00 to 0x20	No more than 32
Data	N B	0xFF... 0xFF	Data frame
Checker	1 B	0x00 to 0xFF	Addition checking bit, including size and data
Stop	2 B	0x0D, 0x0A	Stop tail

Fig. 4 Data packet

The above-mentioned boards communicate on specific commands with each other. Due to the self-manufactured PCB boards, they are not, commercially, available in the market and also should interact with each other on the private set of commands that are reserved for these typical boards only [22, 23]. Figure 4 is a generic hexadecimal command, which includes required data types with their predefined size, content, and description for better understanding.

Figure 5 presents a particular hexadecimal command to turn on/off the LED.

Figure 6 demonstrates a particular hexadecimal command to control the LED, and the output is the mixed color of Red, Green, and Blue.

Header	Size	Type	On/Off	Checker	Stop
FA, F5	4	F2	00~01	XX	0D, 0A

Fig. 5 LED ON/OFF

Header	Size	Type	On/Off	Red	Green	Blue	Checker	Stop
FA, F5	7	F1	00~01	00~FF	00~FF	00~FF	XX	0D, 0A

Fig. 6 LED control

Header	Size	Type	Device No.	Checker	Stop
FA, F5	4	AB	02,03,04	XX	0D, 0A

Fig. 7 Mesh network check

Header	Size	Type	Direction	Strength	Checker	Stop
FA, F5	5	F4	00~01	00~05	XX	0D, 0A

Fig. 8 Motor control

Figure 7 is mainly used to check the network connection between the host and other devices. Note that device numbers hexadecimal 02, 03, and 04 allow to check network connection with LED, a thermal sensor, and motor, respectively.

Figure 8 displays the command to control a motor in which direction has two input values: hexadecimal 00 or 01 for clockwise and anti-clockwise rotation, respectively. The strength represents the speed of the motor that has a total of six input values: hexadecimal 00 stops the motor and hexadecimal 01 to 05 changes the speed of the motor from low to high.

Figure 9 provides ability to control the sensor. Here, hexadecimal 00 stops data receiving from the sensor and hexadecimal 01 activates data receiving, which provides current temperature and humidity.

The command, shown in Fig. 10, transmits current humidity and temperature value data to host that has 16-bit resolution for each, the most significant bit (MSB) to least significant bit (LSB). The sensor data is 10 times of the real humidity. On the other side, the MSB is the sign bit for temperature, where 1 and 0 represent negative and positive temperature, respectively. The other 15 bits are the value of temperature that is again 10 times of the real temperature.

– An example for humidity:
    Humidity range: hexadecimal 0000 to FFFF

– 0000 0010 0101 0011 (Binary) = 0253 (Hex)
    0253 (Hex) = 2 × 256 + 5 × 16 + 3 = 595
    Real humidity = 59.5% RH

Header	Size	Type	Sensing Data	Checker	Stop
FA, F5	4	F1	00~01	XX	0D, 0A

Fig. 9 Sensor control

Header	Size	Type	Humidity	Temperature	Checker	Stop
FA, F5	7	F3	0000~FFFF	0000~FFFF	XX	0D, 0A

Fig. 10 Sensor data

– An example for temperature:
Temperature range: hexadecimal 0000 to FFFF

– 0000 0001 0010 0001 (Binary) = 0121 (Hex)
0121 (Hex) = $1 \times 256 + 2 \times 16 + 1 = 289$
Real temperature = 28.9° C
– 1000 0000 1000 0011 (Binary) = 8083 (Hex)
8083 (Hex) = $0 \times 256 + 8 \times 16 + 3 = 131$
Real temperature = -13.1° C

3.2 Surveillance System

When such a big system is deployed on the large environments, security becomes one of the main concerns [31–33]. Keeping security, a crucial factor of our system, we implemented an intelligent control system that can detect motion with a PIR sensor. Moreover, a detected motion enables the camera that captures an image on that instance of motion to detect and recognize a human face(s) in that typical image frame. If an unknown human is detected, the system creates an alert and notifies to the designated person via email. Upon the recognition of a known person from the database, there will be no alert generated.

There are plentiful algorithms available to recognize the face but for the sake of simplicity we chose two of them to test on our system: (1) the Low Binary Pattern Histogram (LBPH) and (2) the Deep Metric Learning (DML).

3.2.1 Low Binary Pattern Histogram

The LBPH is one of the algorithms open-sourced in the Open-CV library [34]. Post-implementation, we noticed that LBPH provides lesser accuracy compared to the Deep Metric Learning algorithm but at the same time it also has lower computation time for the equal setup. We implemented LBPH on Raspberry Pi 3 and Intel(R) Core (TM) i7-7700HQ CPU @ 2.80GHz with 16 Gigabyte (GB) Random Access

Memory (RAM). We also need to create and train the dataset in order to recognize the face(s) using the algorithm. Training of the dataset is performed every time changes occur in the dataset; otherwise, we do not need to train and directly run the algorithm to recognize a face(s).

Due to the lower computation time, it was feasible to contrivance LBPH on the Raspberry Pi. For the LBPH, the lighting of the surrounding, distance of humans from a camera, and many other perilous situations may affect the accuracy of recognizing a person from the given image frame.

3.2.2 Deep Metric Learning

DML is a face recognition algorithm from the dlib C++ library, which is an open-sourced collection of machine learning applications and algorithms [35]. The DML delivers surprisingly higher accuracy but at the same time drains out the computation power of a processor compared to LBPH. The requirement of the computation resource on DML is high; thus, as a case study the execution was successfully achieved on the Intel(R) Core (TM) i7-7700HQ CPU @ 2.80 GHz with 16 Gigabyte (GB) Random Access Memory (RAM).

As shown in Fig. 13, for DML based face recognition system, which identifies the human face if it is in the image frame, there is no need to train the dataset as the open-source library is pre-compiled with human facial features. Hence, it directly allows implementing inference on the desired platform.

Furthermore, on top of the face recognition systems, we have implemented an alert system that notifies the designated person if the recognized face(s) is not in the dataset. To accomplish such a task, we instigated a script using Simple Mail Transfer Protocol (SMTP) that enables us to send an alert in the email to the authorities to initiate safekeeping regarded action. The benefit of such notification is that the nominated person can choose to take contemplating safety action without physically being in the range of the system if needed.

3.2.3 Server-Side Execution

As shown in Fig. 2, we set traditional cloud computing architecture to measure the computation time for each task such as recognition time for both LBPH and DML algorithms as well as BLE mesh network commands. We create a server using Node.js, a JavaScript run-time environment to execute JavaScript, scripting on an Intel(R) Core (TM) i3-7350K @ 4.20 GHz with 8 Gigabyte (GB) Random Access Memory (RAM). In Fig. 11, steps to create a server on a processor are shown including all the project environments, which must be followed by "nodemon app" command where the "app" is our project name.

As represented in Fig. 12, whenever web request is triggered by task or user, an appropriate Node.js script runs to request data on the cloud from the disposed systems. In a follow-up after that processed request, data is transmitted toward

```
global.__basedir = __dirname;

const express = require('express');
const dotasksController = require('./controllers/dotasksController.js');

const app = express();

app.set('view engine','ejs');
app.use('/',express.static('./'));
app.use('/assets',express.static('./assets'));
app.use('/uploads',express.static('./uploads'));

dotasksController(app);

app.listen(2000);
console.log('Now listening on port 2000');
```

Fig. 11 Creating server on processor

cloud from the disposed systems. Depending upon the requirements, the cloud server executes a task or performs computation and sends back data to the disposed systems, if needed. For this work, the cloud server will be performing all the tasks or computations in a python scripting environment. Also, for any continuous task, such a process could have back-and-forth data transmission from cloud to a disposed system or vice versa.

To control the whole system, we create a simple web page to perform all the tasks using Hypertext Markup Language (HTML) and Node.js. A screenshot from demo web page is shown in Fig. 13.

Figure 14 is a screenshot from the web page after a task being executed on the BLE mesh system, and it also shows the time of execution for a task on cloud computing.

Figure 15 displays a screenshot of a web page when the database is stored on the cloud for LBPH algorithm based face recognition. The database stores 300 images each for every person.

As mentioned before, every time we change the database, the algorithm is required to train the new database, and for the status update, screenshot is shown in Fig. 16.

Finally, after receiving the database and training them, the algorithm would be able to recognize the face(s) from the image/video frame. Figure 17 shows a correctly recognized image together with a screenshot and execution time.

In Fig. 18, we share a screenshot of the face recognition system using the DML algorithm including the total computation time.

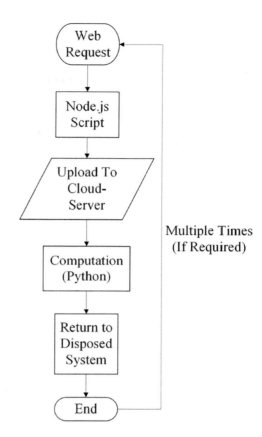

Fig. 12 Cloud server request flow

4 Experimental Results

In this section, we display results from the conducted experiments on the system to bolster our hypothesis. We share the distance improvement of the BLE mesh system with a node between them functioning as a relay. We provide accuracy results for the LBPH and DML algorithms on various computing devices including their performance results in terms of computation time.

Table 2 represents the maximum distance between two mesh boards, which is 57 feet. Interestingly, any board would not be able to communicate with other boards if the distance is greater than 57 feet, but it can be, definitely, increased up to 77 feet using other boards as a relay between those boards.

Table 3 represents computation time for a specific task on a couple of computation devices as well as a cloud server. While execution time on the Raspberry Pi and CPU for a specific same task is quite negligible, the execution time cloud server is terribly high even for such a low-data-rate task execution.

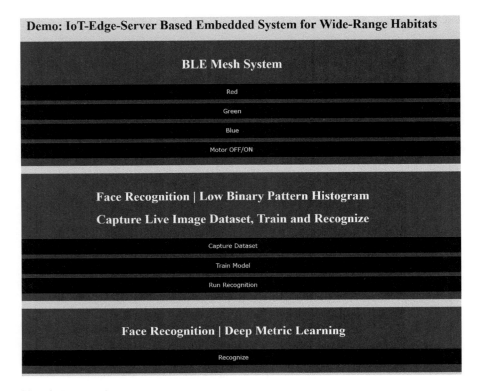

Fig. 13 Demo web page

Fig. 14 BLE mesh system—task execution

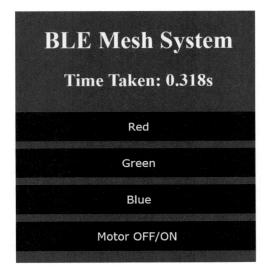

Fig. 15 Capture dataset for LBPH

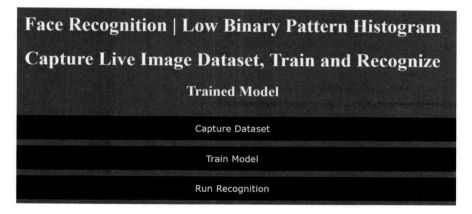

Fig. 16 Trained dataset

Table 4 displays the time taken, in seconds, by different devices along with both the recognizing methods for the edge computing as well as execution for cloud computing. The computation time utterly relies on the hardware, and it is the processing speed. Furthermore, the time taken by the cloud server is fairly high compared to any edge computation devices.

Table 5 is the accuracy of recognizing the face(s) on the different systems with the same environment setting. The LBPH provides extremely lower accuracy compared to the DML method that consists of potential raise to a certain percentage by training more datasets and also replacing with better camera quality. Although there is a plausible way to increase accuracy, we exceedingly doubt about LBPH's capability to imitate the accuracy of DML.

Fig. 17 Face recognition using LBPH

Fig. 18 Face recognition using DML

5 Conclusion

The proposed architecture presents favorable fallout as far as wide-range habitats are concerned with plausibility to deploy long-range BLE mesh system, scalable

An IoT-Edge-Server System with BLE Mesh Network, LBPH, and Deep Metric... 771

Table 2 Distance results for mesh network

Maximum distance	No. of node(s)
57 feet	0
77 feet	1

Table 3 Task execution on BLE mesh system

Device	Task execution time(s)
Raspberry Pi	≥ 0.07
CPU	≥ 0.06
Cloud server	≥ 0.2

Table 4 Face recognition performance for edge/cloud computing

Device	Recognition time(s)	Method
Raspberry Pi	0.41–0.45	LBPH
CPU	0.035–0.037	LBPH
CPU	0.391–0.398	DML
Cloud server	≥ 6	LBPH
Cloud server	≥ 9	DML

Table 5 Face recognition accuracy

Devices	Recognition accuracy(%)	Webcam	Method
Raspberry Pi	50–60%	Logitech C270	LBPH
CPU	50–60%	Logitech C270	LBPH
CPU	60–70%	720p HD Laptop	LBPH
CPU	99.38%	720p HD Laptop	DML
Cloud server	50–60%	Logitech C270	LBPH
Cloud server	99.38%	Logitech C270	DML

surveillance system, and an email alert system. A large-scale BLE mesh system can be established using mesh topology. A scalable surveillance system means using a different combination of face recognition algorithms, camera quality, and hardware's computation speed, and one can use such a setup depending on the requirements. With the implementation of the proposed architecture, there are many application-specific large-scale sites that can be benefited such as commercial buildings, agricultural farms, industrial factories, etc.

6 Future Scope

The proposed architecture has many future research directions. The manufactured boards can be replaced with ASIC chips with research funding. As demonstrated in the proposed hardware architecture figure, all these computations can be done at a logic level using FPGAs that can be, eventually, replaced with ASIC chips, too. Alert system upgrades are also possible where mobile applications could be created rather than web page based control methods. Since the proposed work has a lot of potential research opportunities in the emerging field of artificial intelligence, it is likely to create a big impact on the design of such an application-specific integrated circuit for offering low-latency and inexpensive computations.

References

1. P. Vangali, X. Yang, A compression algorithm design and simulation for processing large volumes of data from wireless sensor networks, vol. 7(4). *Communications on Applied Electronics (CAE)*, July 2017
2. X. Yang, L. Wu, et al., A vision of fog systems with integrating FPGAs and BLE mesh network. J. Commun. **14**(3), 210–215 (2019)
3. W. Shi, J. Cao, Q. Zhang, Y. Li, L. Xu, Edge computing: vision and challenges. IEEE Internet Things J. **3**(5), 637–646 (2016)
4. S.K. Datta, C. Bonnet, An edge computing architecture integrating virtual IoT devices, in *2017 IEEE 6th Global Conference on Consumer Electronics (GCCE)*, pp. 1–3, Nagoya, 2017
5. H. El-Sayed et al., Edge of things: The big picture on the integration of edge, IoT and the cloud in a distributed computing environment. IEEE Access **6**, 1706–1717 (2018)
6. J. Thota, P. Vangali, X. Yang, Prototyping an autonomous eye-controlled system (AECS) using raspberry-pi on wheelchairs. Int. J. Comput. Appl. **158**(8), 1–7 (2017)
7. Y. Zhang, X. Yang, L. Wu, A. Gajjar, H. He, Hierarchical synthesis of approximate multiplier design for field-programmable gate arrays (FPGA)-CSRmesh system. Int. J. Comput. Appl. (IJCA) **180**(17), 1–7 (2018)
8. X. Yang, J. Andrian, A high performance on-chip bus (MSBUS) design and verification. IEEE Trans. Very Large Scale Integr. (VLSI) Syst. (TVLSI) **23**(7), 1350–1354 (2015)
9. X. Yang, J. Andrian, A low-cost and high-performance embedded system architecture and an evaluation methodology, in *IEEE Computer Society Annual Symposium on VLSI (ISVLSI)*, PP. 240–243, Tampa, FL, USA, 2014
10. X. Zhang, J. Lu, X. Fu, X. Yang, I. Unwala, T. Zhang, Tracking of targets in mobile robots based on camshift algorithm, in *Intl. Symposium on Measurement and Control in Robotics (ISMCR)* (UHCL, Houston, USA, 2019), pp. B2-3-1-B2-3-5
11. M. Alrowaily, Z. Lu, Secure edge computing in IoT systems: Review and case studies, in *2018 IEEE/ACM Symposium on Edge Computing (SEC)*, pp. 440–444, Seattle, WA, 2018
12. Y. Li, S. Wang, An Energy-aware edge server placement algorithm in mobile edge computing, in *2018 IEEE Intl. Conference on Edge Computing (EDGE)*, pp. 66–73, San Francisco, CA, 2018
13. P. Ren, X. Qiao, J. Chen, S. Dustdar, Mobile edge computing - a booster for the practical provisioning approach of web-based augmented reality, in *2018 IEEE/ACM Symposium on Edge Computing (SEC)*, pp. 349–350, Seattle, WA, 2018
14. J. Hochstetler, R. Padidela, Q. Chen, Q. Yang, S. Fu, Embedded deep learning for vehicular edge computing, in *2018 IEEE/ACM Symposium on Edge Computing (SEC)*, PP. 341–343, Seattle, WA, 2018
15. F. Wei, S. Chen, W. Zou, A greedy algorithm for task offloading in mobile edge computing system. China Communications **15**(11), 149–157 (2018)
16. X. Yang, Y. Zhang, L. Wu, A scalable image/video processing platform with open source design and verification environment, in *20th Intl. Symposium on Quality Electronic Design (ISQED 2019)*, pp. 110–116, Santa Clara, CA, US, April 2019
17. K. Vaca, A. Gajjar, X. Yang, Real-time automatic music transcription (AMT) with zync FPGA, in *IEEE Computer Society Annual Symposium on VLSI (ISVLSI)*, PP. 378–384, Miami, FL, US, 2019

18. Y. Zhang, X. Yang, L. Wu, J. Andrian, A case study on approximate FPGA design with an open-source image processing platform, in *IEEE Computer Society Annual Symposium on VLSI (ISVLSI)*, pp. 372–377, Miami, FL, US, 2019
19. A. Mebrek, L. Merghem-Boulahia, M. Esseghir, Efficient green solution for a balanced energy consumption and delay in the IoT-Fog-Cloud computing, in *2017 IEEE 16th Intl. Symposium on Network Computing and Applications (NCA)*, PP. 1–4, Cambridge, MA, 2017
20. S. Singh, Optimize cloud computations using edge computing, in *2017 Intl. Conference on Big Data, IoT and Data Science (BID)*, pp. 49–53, Pune, 2017
21. A. Yousefpour, G. Ishigaki, J.P. Jue, Fog computing: Towards minimizing delay in the internet of things, in *2017 IEEE Intl. Conference on Edge Computing (EDGE)*, pp. 17–24, Honolulu, HI, 2017
22. X. Yang, X. He, Establishing a BLE mesh network using fabricated CSRmesh devices, in *The 2nd ACM/IEEE Symposium on Edge Computing (SEC 2017)*, No. 34, Oct. 2017
23. A. Gajjar, Y. Zhang, X. Yang, A smart building system integrated with an edge computing algorithm and IoT mesh networks, in *The Second ACM/IEEE Symposium on Edge Computing (SEC 2017)*, Article No. 35, Oct. 2017
24. H. He, L. Wu, X. Yang, Y. Feng, Synthesize corpus for Chinese word segmentation, in *The 21st Intl. Conference on Artificial Intelligence (ICAI)*, pp. 129–134, Las Vegas, NV, USA, 2019
25. J. Grover, R.M. Garimella, Reliable and fault-tolerant IoT-edge architecture, in *2018 IEEE Sensors*, pp. 1–4, New Delhi, 2018
26. S.W. Kum, J. Moon, T. Lim, Design of fog computing based IoT application architecture, in *2017 IEEE 7th Intl. Conference on Consumer Electronics - Berlin (ICCE-Berlin)*, pp. 88–89, Berlin, 2017
27. A. Gajjar, X. Yang, H. Koc, et al., Mesh-IoT based system for large-scale environment, in *5th Annual Conf. on Computational Science & Computational Intelligence (CSCI)*, pp. 1019–1023, Las Vegas, NV, USA, 2018.
28. M. Sahani, C. Nanda, A.K. Sahu, B. Pattnaik, Web-based online embedded door access control and home security system based on face recognition, in *2015 Intl. Conference on Circuits, Power and Computing Technologies (ICCPCT)*, pp. 1–6, Nagercoil, 2015
29. I. Gupta, V. Patil, C. Kadam, S. Dumbre, Face detection and recognition using Raspberry Pi, in *2016 IEEE Intl. WIE Conference on Electrical and Computer Engineering (WIECON-ECE)*, pp. 83–86, Pune, 2016
30. A. Gajjar, X. Yang, L. Wu, H. Koc, I. Unwala, Y. Zhang, Y. Feng, An FPGA synthesis of face detection algorithm using haar classifiers, in *Intl. Conference on Algorithms, Computing and Systems (ICACS)*, pp. 133–137, Beijing, China, 2018
31. X. Yang, W. Wen, Design of a pre-scheduled data bus (DBUS) for advanced encryption standard (AES) encrypted system-on-chips (SoCs), in *The 22nd Asia and South Pacific Design Automation Conference (ASP-DAC)*, pp. 1–6, Chiba, Japan, Feb. 2017
32. X. Yang, W. Wen, M. Fan, Improving AES core performance via an advanced IBUS protocol. ACM J. Emerg. Technol. Comput. (JETC) **14**(1), 61–63 (2018)
33. X. Yang, J. Andrian, An advanced bus architecture for AES-encrypted high-performance embedded systems, US20170302438A1, Oct. 19, 2017
34. G. Bradski, The OpenCV library. J. Software Tools. https://docs.opencv.org/3.4.3/
35. D. King, Dlib-ml: A machine learning toolkit. J. Mach. Learn. Res. **10**, 1755–1758 (2009). http://dlib.net/

An Edge Detection IP of Low-Cost System on Chip for Autonomous Vehicles

Xiaokun Yang, T. Andrew Yang, and Lei Wu

1 Background

No doubt, the wide use of autonomous vehicles will significantly enhance the human mobility and change our lives [1–3]. It has a great potential to benefit the world by increasing traffic efficiency [4], reducing pollution [5], and eliminating up to 90% of traffic accidents [6].

The first autonomous vehicle can be traced back to 1980s—Carnegie Mellon University's Navlab and autonomous land vehicle (ALV) projects in 1984 and Mercedes-Benz and Bundeswehr University Munich's Eureka Prometheus Project in 1987 [7, 8]. Until 2007, the first benchmark test for autonomous driving in realistic urban environments was completed [9, 10]. And in 2010, Google announced that its self-driving car was able to cover thousands of miles of real-road driving [11]. To date, the large-scale adoption of sensors [12–14] and artificial intelligence [15–17] brings a chance to make autonomous vehicles being widely used a reality in the coming years. However, many challenges to modern vehicles still remain, such as the security of the connected systems [18] and the real-time decision making with complex intelligent algorithms and systems [19, 20].

X. Yang (✉)
Department of Engineering, University of Houston Clear Lake, Houston, TX, USA
e-mail: YangXia@UHCL.edu

T. Andrew Yang
Department of Computing Science, University of Houston Clear Lake, Houston, TX, USA

L. Wu
Computer Science Department, Auburn University at Montgomery, Montgomery, AL, USA

© Springer Nature Switzerland AG 2021
H. R. Arabnia et al. (eds.), *Advances in Artificial Intelligence and Applied Cognitive Computing*, Transactions on Computational Science and Computational Intelligence, https://doi.org/10.1007/978-3-030-70296-0_56

2 Introduction

Under this context, we propose a system-on-chip (SoC) design by using a low-cost and low-power on-chip bus architecture [21]. Compared to the previous designs on the SoC for autonomous vehicles [22–24], our work has a great potential to reduce the chip size and power cost by using a novel bus transfer mode and on-chip topology for security cores and road detection engines [25]. Furthermore, several intellectual properties (IPs), including an advanced encryption standard (AES) engine presented in [26, 27], and multiple SoC interfaces and peripherals in [28–30] will be integrated to the SoC. Most of these works are open-source designs and ready to be integrated. In this chapter, we focus on exploring the design on the edge detection IP, aiming to show a demonstration on detecting the edge of 320×240 size frames of images within 0.8 milliseconds per frame.

Generally, edge detection is an image processing algorithm for finding the boundaries of objects within images by using a variety of mathematical approaches. It can be mainly used for image segmentation and data extraction in object detection such as road lane detection [31, 32], interactive model predictive control [33], smart parking [34], and traffic analysis [35]. More specifically, the contribution of this chapter is as follows:

- Compared to previous works focusing on algorithm optimizations and software level implementations on edge detection [36–39], this work is an open-source design with multiple hardware IPs, which are able to be integrated and demonstrated on Xilinx Nexys 4 FPGA. We expect that the public release of this platform will lead to better demonstrations in the future, as well as to serve teaching courses and research projects in the fields of integrated circuit design, robotics, and artificial intelligence.
- Using the video processing platform proposed in [28], the edge detection speed can reach 1302 frames per second (fps). Additionally, the experimental results in terms of simulation, synthesis schematic, slice count, and power consumption have been discussed in this chapter.

The organization of this chapter is as follows. Section 3 presents our work with design architecture, and Sect. 4 discusses the implementation of the proposed system. In Sect. 5, the experimental results in terms of computation speed, slice cost, and power consumption are shown. Finally, Sect. 6 presents the concluding remarks and future work in our target architecture.

3 Proposed Design Architecture

This section presents the design architecture of the entire system, including the image capture module, I2C controller, VGA master, memory controller, and more

An Edge Detection IP of Low-Cost System on Chip for Autonomous Vehicles 777

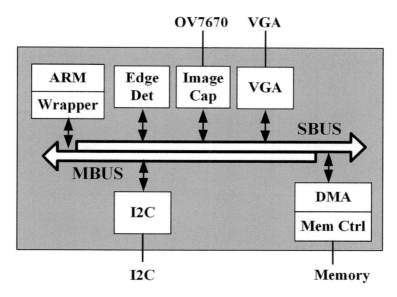

Fig. 1 The MSBUS SoC architecture

important, the sobel engine for edge detection. The camera-buffer-vga data path has been presented in [28], and the sobel engine is mainly performed in this chapter.

3.1 System-on-Chip Architecture

In Fig. 3, the SoC architecture is shown with two on-chip buses, one control bus, and one data bus [25]. The name of the control bus is master bus (MBUS), since only one master is located on the control bus which is the ARM processor. Similarly, only one slave is located on the data bus which is the DMA combined with the memory controller, so the data bus is termed as slave bus (SBUS).

The control bus performs the functional register configuration with the ARM core. Due to the incompatible bus protocols between the ARM processor and the IPs, an AXI-to-MBUS wrapper is needed to the IP integration [40]. For the data bus, four IPs will be integrated, including the SoC peripheral I2C master, the image capture module, the VGA master, and the edge detection engine. The MSBUS architecture [25] and the three IPs were publicly available and presented in prior works [28, 41]. This chapter focuses on the design on the edge detection IP, which will be integrated to the SoC in the future work (Fig. 1).

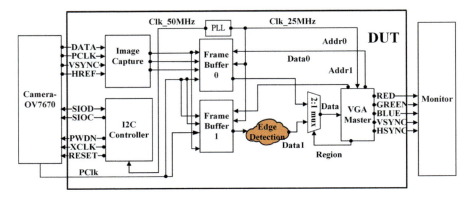

Fig. 2 The FPGA design architecture

3.2 Design and Interconnection of Edge Detection Core

Figure 3 shows the design and interconnection between the three IPs and the edge detection engine. Generally, the I2C master is used to configure the OV7670 camera. After being enabled, the camera is able to send frames of images into the buffer#0 through the image capture interface. At the same time, the sobel engine reads pixels out of the buffer#0, converts the color pixels into grayscale then into edge detection results, and finally stores the results in buffer#1.

The original color images from the OV7670 camera are in the 320 × 240 pixels in byte, so the sizes of buffer#0 and buffer#1 are 76,800 bytes. The color images in buffer#0 will be displayed in 1/4 of the 640 × 480 screen. Hence, a VGA master is designed to perform the window splitting for refreshing the VGA-interfaced monitor (Fig. 2).

3.3 Sobel Algorithm

Sobel edge filter is one of the most commonly used edge detectors. It is based on convolving the image with a small and integer-valued filter in horizontal and vertical directions and is therefore relatively inexpensive in terms of computations. As a case study, the sobel filter is thus performed for edge detection in this section. It mainly works by detecting discontinuities in brightness or, in other words, by calculating the gradient of image intensity at each pixel within the image.

A pseudocode of the sobel algorithm is shown in Algorithm 1. The iterative $for - loop$ between line#1-2 and line#17-18 is used to sweeping on a 320 × 240 image by 3 × 3 matrix. During line#3-11, the values in each 3 × 3 matrix are assigned to $p0 - p8$. As an example in Fig. 3, the first matrix from the top left is specifically

Algorithm 1 Pseudocode of the Sobel Algorithm

Require: p[i,j]: pixel with index i and j; gx, gy: gradient; abs_gx, abs_gy: absolute gradient;
1: **for** $i = 0; i < 318; i++$ **do**
2:    **for** $j = 0; j < 238; j++$ **do**
3:       $p0 = I(i, j)$;
4:       $p1 = I(i, j + 1)$;
5:       $p2 = I(i, j + 2)$;
6:       $p3 = I(i + 1, j)$;
7:       $p4 = I(i + 1, j + 1)$;
8:       $p5 = I(i + 1, j + 2)$;
9:       $p6 = I(i + 2, j)$;
10:      $p7 = I(i + 1, j + 1)$;
11:      $p8 = I(i + 2, j + 2)$;
12:      $gx = ((p2 - p0) + ((p5 - p3) * 2) + (p8 - p6))$;
13:      $gy = ((p0 - p6) + 2 * (p1 - p7) + (p2 - p8))$;
14:      $abs\,gx = abs(gx)$;
15:      $abs\,gy = abs(gy)$;
16:      $out = abs\,gx + abs\,gy$;
17:    **end for**
18: **end for**

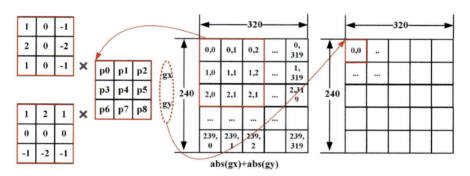

Fig. 3 Matrix-based Sobel filter

data in array [0, 0], [0, 1], [0, 2] in the first row, [320, 0], [320, 1], [320, 2] in the second row, and [640, 0], [640, 1], [640, 2] in the third row.

The gradient of the matrix in horizontal and vertical is computed in line#12-13 of Algorithm 1, and the absolute values are computed in line#14-15. Basically, the sobel filter uses two 3 × 3 kernels: one for changes in the horizontal direction, and the other for changes in the vertical direction. The two kernels are convolved with the original image to calculate the approximations of the derivatives. Assume that Gx and Gy are the two images that contain the horizontal and vertical derivative approximations, respectively, and the computations are

$$Gx = \begin{bmatrix} 1 & 0 & -1 \\ 2 & 0 & -2 \\ 1 & 0 & -1 \end{bmatrix} \times P \tag{1}$$

and

$$Gy = \begin{bmatrix} -1 & -2 & -1 \\ 0 & 0 & 0 \\ 1 & 2 & 1 \end{bmatrix} \times P \tag{2}$$

where P is the input source 3×3 matrices. Finally, the result with edge image is the sum of the two absolute values as shown in line#16.

4 Implementation

As discussed in Algorithm 1, the design on sobel engine is presented in this section, and the functions of the register-transfer (RT)-level design are tested by comparing the design-under-test (DUT) results with the Matlab results.

4.1 Design on Sobel Engine

Figure 4 depicts the hardware design for the implementation with Xilinx Nexys 4 FPGA. The target device used in this work is Xilinx Artix-7 FPGA with xc7a75tcsg324-1 part.

Basically, it contains five stages, and in each stage, we maximize the computation in parallel to reduce the critical path in order to achieve higher operational frequency. Ignoring the latency of inverter, logic gates, multiplexers, and the connected wires, the critical path of the entire design is the propagation delay of one subtractor and four adders.

For the gate count, totally six subtractors, seven adders, three multiplexers, and some logic gates are performed for the design of the sobel engine.

4.2 RTL Design and Edge Detection Results

The design on the sobel engine is based on the Verilog hardware description language (HDL), and a testbench is provided to verify the functions of the design. The golden model is generated by Matlab, including the original color image in

An Edge Detection IP of Low-Cost System on Chip for Autonomous Vehicles 781

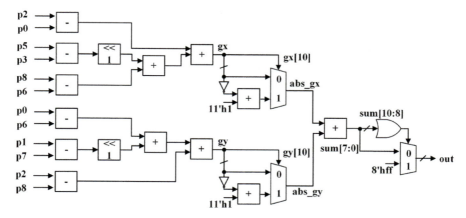

Fig. 4 The hardware design on Sobel engine

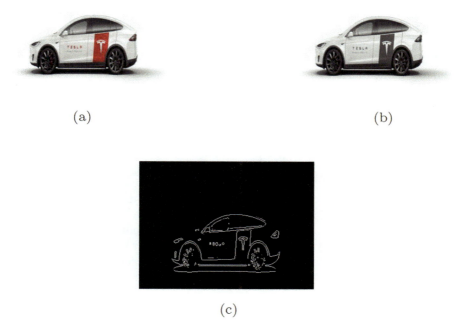

Fig. 5 Color-grayscale-edge results. (**a**) Original color image. (**b**) Grayscale image. (**c**) Edge detection results

Fig. 5a, the grayscale image in Fig. 5b, and the final results with edge detection in Fig. 5c.

By comparing the results from design under test and from Matlab pixel by pixel, the functions of the RT-level design are verified.

5 Experimental Results

This section shows the simulation and synthesis results in schematic. The simulator used in our work is Mentor Graphics ModelSim—Intel FPGA 10.5b, and the synthesis tool is Xilinx Vivado 2017.2. In what follows, the FPGA resource cost in terms of slice count and power consumption is evaluated.

5.1 Simulation and Synthesis Schematic Results

After simulation, the latency for processing 320×240 size of images is calculated. To make the edge pixel output in pipeline, totally 76,800 cycles are spent for one 320×240 image. The clock frequency of the Nexys 4 FPGA is 100 MHz as specified in the constraint file. Hence, to detect the edge of 320×240 size of images takes less than 0.8 milliseconds per frame. In other words, the speed of the sobel engine is 1302 frames per second.

In Fig. 6, it shows the synthesis schematic result, containing six subtractors (or RTL_SUB in the figure), seven adders (or RTL_ADD in the figure), and three multiplexers (or RTL_MUX in the figure). It can be observed that the synthesis result with Xilinx Vivado is the same as our expectation in Sect. 4. The analysis to the design in Fig. 4 is justified.

5.2 Resource Cost on FPGA

In what follows, the slice count on Nexys 4 FPGA is summarized in Table 1. It takes 77 slices of LUTs and 72 IOs as a computational core.

In Table 2, the total power is depicted as 4.637 W, including 0.114 W of static power and 4.523 W of dynamic power. The dynamic power is composed of power cost on the switching activities of signals (1.084 W), logic (0.997 W), and IOs (2.442 W).

6 Conclusion and Future Work

This chapter proposes a design on edge detection engine, which will be a submodule of the road-edge detection demonstration as the future work. The earlier demonstration of the video date path is shown in Fig. 7 [28]. In Fig. 7a, the platform including the Nexys 4 FPGA, the OV7670 camera, and the VGA-connected monitor is shown. And in Fig. 7b, a 320×240 size of original color frame of images is shown in 1/4 of the screen. As the future work, the grayscale images will be shown in another

An Edge Detection IP of Low-Cost System on Chip for Autonomous Vehicles

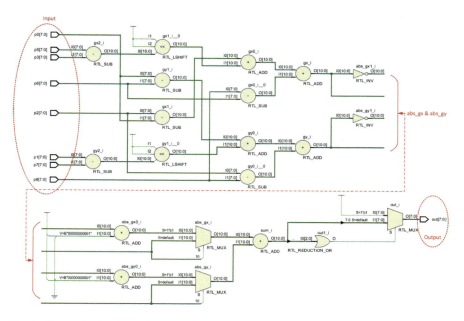

Fig. 6 The synthesis schematic

Table 1 FPGA resource cost

Resource	Utilization	Available	Percentage
LUT	77	47200	1%
IOB	72	210	34%

Table 2 Power consumption on FPGA

TP[a](W)	SP[b](W)	DP[c](W)		
		Signals	Logic	I/O
4.637	0.114	1.084	0.997	2.442
100%	2%	23%	22%	53%

[a]Total power consumption
[b]Static power consumption
[c]Dynamic power consumption

1/4 of the screen, and the final results of edge detection will be shown in the other 1/4 of the display. This integrated SoC will be demonstrated on FPGA for real-time obstacle detection for an autonomous vehicle or robot.

Most of the IPs for this SoC design are publicly available. We expect that the public release will lead to multiple designs in the future and serve as a framework to projects of research and education.

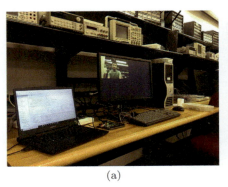 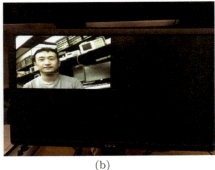

(a) (b)

Fig. 7 FPGA demonstration on image data path. (**a**) FPGA demonstration. (**b**) Four regions based image processing

References

1. J. HoChristopher, B. ChenRohan, P. Kielbus, Autonomous vehicle routing, US20190120640A1, 2019
2. J. Kuffner, Systems and methods for detection by autonomous vehicles, US20190079194A1, 2019
3. S. Bonadiesa1, S. Andrew Gadsden, An overview of autonomous crop row navigation strategies for unmanned ground vehicles. Eng. Agriculture Environ. Food **12**(1), 24–31 (2019)
4. B. van Arem, C.J. van Driel, R. Visser, The impact of cooperative adaptive cruise control on traffic-flow characteristics. IEEE Trans. Intell. Transp. Syst. **7**, 429–436 (2006)
5. K. Spieser, K. Treleaven, R. Zhang, E. Frazzoli, D. Morton, M. Pavone, Toward a systematic approach to the design and evaluation of automated mobility-on-demand systems: A case study in Singapore, in *Road Vehicle Automation*, ed. by G. Meyer, S. Beiker, pp. 229–245, 2014
6. P. Gao, R. Hensley, A. Zielke, A roadmap to the future for the auto industry. McKinsey Quarterly (2014)
7. K. Takeo, Autonomous land vehicle project at CMU, in *Proceedings of the 1986 ACM Fourteenth Annual Conference on Computer Science*, pp. 71–80, 1986
8. W. Richard, First results in robot road-following, in *Proceedings of the 9th International Joint Conference on Artificial Intelligence*, 1985
9. B. Montemerlo, et al., Junior: The Stanford entry in the urban challenge. J. Field Robot **25**, 569–597 (2008)
10. C. Urmson, et al., Autonomous driving in urban environments: Boss and the urban challenge. J. Field Robot **25**, 425–466 (2008)
11. M.M. Waldrop, Autonomous vehicles: No drivers required. Nature **518**, 20–23 (2015)
12. F. Guo et al., Detecting vehicle anomaly in the edge via sensor consistency and frequency characteristic. IEEE Trans. Vehicular Technol. **68**(6), 5618–5628 (2019)
13. W. McNeill, Sensor-based detection of landing zones, US10472086B2, 2020
14. W.M. Leach Scott, C. Poeppel Matthew Langford Tess Bianchi, Sensor control system for autonomous vehicle, US20190092287A1, 2019
15. J. Annamalai, C. Lakshmikanthan, An optimized computer vision and image processing algorithm for unmarked road edge detection, in *Soft Computing and Signal Processing*, ed. by J. Wang, G. Reddy, V. Prasad, V. Reddy. Advances in Intelligent Systems and Computing, vol. 900 (Springer, Singapore, 2019)

16. S. Yang, J. Wu, Y. Shan, Y. Yu, et al., A novel vision-based framework for real-time lane detection and tracking, SAE Technical Paper 2019-01-0690, 2019
17. D. Vajak, M. Vranješ, R. Grbić, D. Vranješ, Recent advances in vision-based lane detection solutions for automotive applications, in *2019 International Symposium ELMAR*, Zadar, Croatia, PP. 45–50, Vajak, 2019
18. C. Hana, E. Joel, S. Alvaro, Autonomous vehicle heaven or hell? Creating a transportation revolution that benefits all, Greenlining Institute, 2019
19. K. Vaca, A. Gajjar, X. Yang, Real-time automatic music transcription (AMT) with Zync FPGA, in *IEEE Computer Society Annual Symposium on VLSI (ISVLSI)*, Miami, FL, US, Jan. 13, 2020
20. Y. Zhang, X. Yang , L. Wu, J. Lu, K. Sha, A. Gajjar, H. He, Exploring slice-energy saving on an video processing FPGA platform with approximate computing, in *Intl. Conference on Algorithms, Computing and Systems (ICACS)*, pp. 138–143, July 27–29, Beijing China, 2018
21. X. Yang, J. Andrian, A high performance on-chip bus (MSBUS) design and verification. IEEE Trans. Very Large Scale Integr. (VLSI) Syst. (TVLSI) **23**(7), 1350–1354 (2015)
22. A. Tang et al., A 95 GHz centimeter scale precision confined pathway system-on-chip navigation processor for autonomous vehicles in 65nm CMOS, in *2015 IEEE MTT-S International Microwave Symposium*, PP. 1–3, Phoenix, AZ, 2015
23. D. Guermandi, et al., A 79-GHz 2 × 2 MIMO PMCW radar SoC in 28-nm CMOS. IEEE J. Solid State Circuits **52**(10), 2613–2626 (2017)
24. J. Pahasa, I. Ngamroo, PHEVs bidirectional charging/discharging and SoC control for microgrid frequency stabilization using multiple MPC. IEEE Trans. Smart Grid **6**(2), 526–533 (2015)
25. X. Yang, J. Andrian, An advanced bus architecture for AES-encrypted high-performance embedded systems, US20170302438A1, Oct. 19, 2017
26. X. Yang, W. Wen, Design of a pre-scheduled data bus (DBUS) for advanced encryption standard (AES) encrypted system-on-chips (SoCs), in *The 22nd Asia and South Pacific Design Automation Conference (ASP-DAC 2017)*, pp. 1–6, Chiba, Japan, 2017
27. X. Yang, W. Wen, M. Fan, Improving AES core performance via an advanced IBUS protocol. ACM J. Emerg. Technol. Comput. (JETC) **14**(1), 61–63 (2018)
28. X. Yang, Y. Zhang, L. Wu, A scalable image/video processing platform with open source design and verification environment, in *20th Intl. Symposium on Quality Electronic Design (ISQED)*, pp. 110–116, Santa Clara, CA, US, April 2019
29. X. Yang, J. Andrian, A low-cost and high-performance embedded system architecture and an evaluation methodology, in *IEEE Computer Society Annual Symposium on VLSI (ISVLSI)*, pp. 240–243, Tampa, FL, USA, Sept. 2014
30. X. Yang, N. Wu, J. Andrian, A novel bus transfer mode: block transfer and a performance evaluation methodology. Elsevier Integ. VLSI J. **52**, 23–33 (2016)
31. H. Park, Robust road lane detection for high speed driving of autonomous vehicles, in *Web, Artificial Intelligence and Network Applications*, ed. by L. Barolli, M. Takizawa, F. Xhafa, T. Enokido. Advances in Intelligent Systems and Computing, vol. 927 (Springer, Cham, 2019)
32. T. Datta, S.K. Mishra, S.K. Swain, Real-time tracking and lane line detection technique for an autonomous ground vehicle system, in *International Conference on Intelligent Computing and Smart Communication*, ed. by G. Singh Tomar, N. Chaudhari, J. Barbosa, M. Aghwariya. Algorithms for Intelligent Systems (Springer, Singapore, 2019)
33. D.P. Filev, J. Lu, D.D. Hrovat, Autonomous vehicle operation based on interactive model predictive control, US10239529B2, 2020
34. X. Chen, S. Zhang, J. Wu, Lidar inertial odometry and mapping for autonomous vehicle in GPS-denied parking lot, in *WCX SAE World Congress Experience*, 2020
35. D. Miculescu, S. Karaman, Polling-systems-based autonomous vehicle coordination in traffic intersections with no traffic signals. IEEE Trans. Automatic Control **65**(2), 680–694 (2020)
36. R. Abi Zeid Daou, F. El Samarani, C. Yaacoub, X. Moreau, Fractional derivatives for edge detection: Application to road obstacles, in *Smart Cities Performability, Cognition, & Security*, ed. by F. Al-Turjman. EAI/Springer Innovations in Communication and Computing (Springer, Cham, 2020)

37. Z. Wang, G. Cheng, J. Zheng, Road edge detection in all weather and illumination via driving video mining. IEEE Trans. Intell. Vehicles **4**(2), 232–243 (2019)
38. S. Agrawal, B.K. Dean, Edge detection algorithm for $Musca - Domestica$ inspired vision system. IEEE Sensors J. **19**(22), 10591–10599 (2019)
39. X. Song, X. Zhao, L. Fang, et al., EdgeStereo: An effective multi-task learning network for stereo matching and edge detection. Int. J. Comput. Vis. (2020)
40. X. Yang, et al., Towards third-part IP integration: A case study of high-throughput and low-cost wrapper design on a novel IBUS architecture, in *IET Computers & Digital Techniques (IET-CDT), Under Review*, 2020
41. Y. Zhang, X. Yang , L. Wu, J. Andrian, A case study on approximate FPGA design with an open-source image processing platform, in *IEEE Computer Society Annual Symposium on VLSI (ISVLSI)*, Miami, FL, US, Jan. 13, 2020

Advancing AI-aided Computational Thinking in STEM (Science, Technology, Engineering & Math) Education ($\mathcal{A}ct$-STEM)

Lei Wu, Alan Yang, Anton Dubrovskiy, Han He, Hua Yan, Xiaokun Yang, Xiao Qin, Bo Liu, Zhimin Gao, Shan Du, and T. Andrew Yang

1 Introduction

An overwhelmingly large portion of STEM knowledge is skill knowledge [1]. Skill knowledge learning emphasizes the capability to dynamically use learned knowledge to solve problems rather than pure memorization facts. We define skill knowledge as tasks that require repetition and application of existing knowledge to master, such as computer programming, playing a sport, or playing an instrument.

L. Wu · H. Yan · Z. Gao
Department of Computer Science and Computer Information Systems, Auburn University at Montgomery, Montgomery, AL, USA

A. Yang (✉)
Department of Information Systems, University of Nevada, Reno, Reno, NV, USA
e-mail: alany@unr.edu

A. Dubrovskiy
Department of Chemistry, University of Houston Clear Lake, Houston, TX, USA

H. He
Department of Computer Science, Emory University, Atlanta, GA, USA

X. Yang
Department of Engineering, University of Houston Clear Lake, Houston, TX, USA

X. Qin · B. Liu
Department of Computer Science and Software Engineering, Auburn University, Auburn, AL, USA

S. Du
Department of Computer Science, Mathematics, Physics and Statistics, University of British Columbia Okanagan Campus, Kelowna, BC, Canada

T. A. Yang
Department of Computer Science, University of Houston Clear Lake, Houston, TX, USA

© Springer Nature Switzerland AG 2021
H. R. Arabnia et al. (eds.), *Advances in Artificial Intelligence and Applied Cognitive Computing*, Transactions on Computational Science and Computational Intelligence, https://doi.org/10.1007/978-3-030-70296-0_57

We also define static fact knowledge; these are facts that can be memorized and recited through repetition, but do not require consistent reinforcement for mastery. Proficiency in skill knowledge requires a different approach than memorization of fact knowledge. Logically, the way skill knowledge is taught should be different than the way fact knowledge is taught. However, STEM subjects are commonly taught as content to be remembered rather than skills, expertise, and habits to be developed [2, 3]. As a result, many students perceive STEM disciplines as difficult, arcane, and irrelevant, which further hinders their academic performance in STEM fields, limiting career options [4].

2 Crisis of Common STEM Teaching Method

Common practices for STEM education do not promote habits that nurture skill-building routines, and hands-on opportunities for students to depict, replicate, and interact with the concepts being taught.

A conventional educational method of lecturing to a large number of students with the sole interaction being student questions is a teaching method that is much more suitable to conveying fact knowledge; students inevitably become audiences forced to only rely on passive learning to master skill knowledge. The major challenges of skill knowledge STEM teaching rely on the following:

- One teacher oversees many students.
- Lecturing becomes the primary teaching mode.
- Teaching ultimately becomes a monologue by the instructor (Fig. 1).

As a result, conventional teaching leads to extremely low direct learning effects in the classroom. Educators have to request extra time and resources beyond the in-class learning time period which squeezes the limited time available for learners to participate in other activities and learning outside of the classroom.

Fig. 1 Conventional teaching methods have transformed the lecture into the "instructor's personal talk show" to the detriment of conveying skill-based knowledge. (Photo sources: [5, 6])

Failure to effectively *teach* skill knowledge is the **quintessential factor** that leads to American students' exceptionally low performance in STEM education, in public K-12 schools & universities.

3 *Act*-STEM Skill Knowledge Teaching Technology

Computing has become an integral part of science, technology, mathematics, and engineering. We seek the usage of computational thinking in K-14 STEAM education. Our pilot project "Realistic Drawing & Painting with Geometrical and Computational Methods *(Fun-Joy)*" applies AI-aided computational thinking with geometrical methods to visual art skill development for learners new to drawing. Since its debut on Google Play, the software designed for this project has been downloaded by more than 5 million learners, and has been translated into 65 languages; more than 110,000 learners have given review comments on Google Play; the average score is 4.5 out of 5.0 scale, topping the highest in its category.

The second pilot project that we have successfully conducted is "High Efficiency Skill Knowledge Learning with Computational Methods for STEM ($\mathcal{HE}$-STEM)." The sample was drawn from a dimensional analysis module in a college general chemistry course with a student pool of 182. $\mathcal{HE}$-STEM successfully converted the average score of at-risk students from 45/100 to 85/100, an average 88.9% improvement in STEM learning. A similar study of at-risk students in another semester with a student pool over 300 saw an average improvement rate as high as 90.3%. *Act*-STEM largely uses four cornerstones to build its two major design and development modules. Computational thinking addresses a complex problem by breaking it down into a series of smaller, more manageable problems (*decomposition*); each of these smaller problems can then be looked at individually and related to how similar problems have been solved in the past (*pattern recognition*). By focusing only on the important details, while ignoring irrelevant information (*abstraction*); eventually, simple steps or rules to solve each of the smaller problems can be designed (*algorithms*). There are four key major components in Computational Thinking (CT) [7]:

- *Decomposition* – breaking down a complex problem or system into smaller, more manageable parts
- *Pattern recognition* – looking for similarities among and within problems
- *Abstraction* – focusing on important information only, ignoring irrelevant details
- *Algorithms* – developing a step-by-step solution to the problem, or the rules to follow to solve the problem

Our *Act*-STEM approach further expands the research findings from the two pilot projects. It combines (i) the computational thinking approaches [8], (ii) the latest innovative pedagogical discoveries in STEM education [1, 2, 8, 9], (iii) phenomena in natural computing, including biological computing and physical computing, and (iv) the project-based interactive cyber-learning communities, to construct a holistic formal and informal educational software environment that stimulates the curiosity of young leaners ranging from K-14 grades (including firs 2 years college students) within and far beyond school districts and universities. Students therefore will be able to spontaneously develop an in-depth understanding of computational thinking as a way to creatively approach learning in STEM fields.

4 Taxonomy of Knowledge Levels

There are four major levels of knowledge based on Anderson and Krathwohl's Taxonomy [10, 11] – which is a revision of Bloom's taxonomy of educational objectives [12]:

A. *Factual Knowledge:* "The basic elements one must know to be acquainted with a discipline or solve problems." It refers to fundamental and essential facts, terminology, details, or elements one must know or be familiar with in order to understand a discipline or solve a problem in it.
B. *Conceptual Knowledge:* "The interrelationships among the basic elements within a larger structure that enable them to function together." The knowledge of classifications, principles, generalizations, theories, models, or structures is pertinent to a particular disciplinary area.
C. *Procedural Knowledge:* "How to do something, methods of inquiry, and criteria for using skills, algorithms, techniques, and methods," etc. It refers to specific or finite skills, algorithms, techniques, and particular methodologies.
D. *Metacognitive Knowledge*: "Knowledge of cognition in general, as well as awareness and knowledge of one's own cognition." It contains strategic or reflective knowledge about how to go about using appropriate skills and strategies to solve problems, cognitive tasks, to include contextual and conditional knowledge and knowledge of self.

5 Fact Knowledge

The first two levels, both factual and conceptual knowledge, constitute a static **What** type of knowledge, including either essential and tangible facts, such as "All mammals have hair," "The tallest grass, some as tall as 50 m (164 ft), is bamboo"; or conceptual and abstracted ideas, theories, such as "One cannot step twice in the same river – *Heraclitus*," "A day without laughter is a day wasted - *Charlie Chaplin*," etc. They can be summarized as *fact knowledge* with some distinctive characteristics:

1. *Hard to ask "why":* "1 foot $= 12$ inches," "Circumference of any circle divided by its diameter is a constant value, known as Pi or π," etc.; an in-depth

explanation for the reasoning beyond the simple heuristic fact is comparatively difficult to learn.

2. *Received through humans' five basic senses:* Sight (vision) –"rose is red"; sound (hearing) – "thunder is loud"; smell (olfaction) – "jasmine is fragrant"; taste (gustation) – "sugar is sweet"; and touch (tactile perception) – "ice is cold" [13].

3. *Mainly acquired through the first two senses:* Sight (vision) – such as visual illustration, representation, recording, etc.; and sound (hearing) – such as narrative, lecture, etc., which are exclusively used in the traditional mass education environment: classroom or auditorium settings. For example, "Can a killer whale (orca) hunt seals on ice without breaking, penetrating, or climbing on it?" The answer is yes, and can be further illustrated with photos, recorded videos by using only visual and audio facilities.

4. *Mostly depends on brain* (intellect, conscious memory) rather than hand (practice, spontaneous muscle memory) to master this type of knowledge. To better experience that, try to use hand rather than brain to master following fact knowledge: "A chain is as strong as its weakest link," "Refused by King John II of Portugal, Italian explorer Christopher Columbus reached America in 1492 with the support from Spanish monarch.".

5. *Easy to forget:* Without reinforcement, most fact knowledge has a tendency to be learned quickly. An unfortunate reality is that after a few months, students are likely to have forgotten most of the fact knowledge they have learned in any given course.

6. *Easy to teach to a large number of audiences simultaneously:* It could be quite feasible, enlightening, and even entertaining to educate fact knowledge to many students in large classroom with the support of audio and visual (photo, video, etc.) facilities, for example:

- Killer whale (orca) kills other whales by drowning its victim with its belly.
- The largest toy distributor in the world is McDonald's.
- Humans are born without kneecap bones; only develop them around the age of 3.
- The Greenland shark has an average lifespan of 272 years; some of them can even live as long as 500 years.
- There are a lot more trees on the earth than stars in the galaxy. NASA estimates there could be from 100 billion to 400 billion stars in the Milky Way galaxy. The earth has more than 3 trillion trees.

6 Skill Knowledge

The last two levels, procedural and metacognitive knowledge, constitute dynamic *How-to* types of knowledge, including the distillation of such kind of how-to capabilities into a cognition in general, and being able to apply previously learned ideas to new situations. Examples include: "How to calculate the circumference of any circle provided you know its radius," "How to perform vibrato when playing the violin," etc. These levels of knowledge can be simplified as *skill knowledge* with some distinctive characteristics:

1. *Capability of problem-solving:* Available fact knowledge is not simply recited, but applied through a series of steps that leads to an answer to the problem.
2. *Staged process:* Since the challenge is usually not that simple to be resolved in a single-punch approach, with conditional discretion, it normally involves a series of sequential phases (steps) to conduct logical transition, procedure, deduction, evolvement, inference, etc., to eventually reach a resolution.
3. *Exhibits mastery of skills, dexterity in problem-solving*: Repetition of tasks leads to mastery through cognitive and muscle memory. Examples include the ability to effectively solve problems in math/chemistry/physics; write well-expressed essays; play proficient tennis/golf/the violin; create impressive art work in painting/sculpturing; compose exquisite pieces of music; have a beautiful handwriting, etc.
4. *Heavily depends on hand/muscle* (practice, spontaneous muscle memory), together with brain (intellect, conscious mental memory), vision and hearing, to successfully master skill knowledge. For example, it is almost impossible to master swimming by only watching videos of swimming; you must experience the actions involved with movement in the water in order to achieve competence and eventual mastery.
5. *Difficult to forget:* Although it is hard to learn skill knowledge, once mastered, it is hard to forget. Many such skills can last for a lifetime. For example, after one has mastered riding a bicycle, as long as physical condition allows, one will be able to ride a bike for the remainder of their life.
6. *Difficult, even impossible to teach to a large number of learners simultaneously:* It could be quite challenging, infeasible, and even impossible to educate skill knowledge to many students together in a classroom. For example,

   - Teaching two students learning the violin simultaneously could be extremely difficult. There's no problem for two, or even 200 students to listen to and watch the instructor's live lecture and demonstration at the same time in a classroom. However, it is impossible for the instructor to listen to and observe two or 200 students' practice together, and still be able to tell which pupil makes what mistakes, and give the instant feedback on detected errors to each learner, in order for students to reduce the mistakes and to improve in the next round of practice.
   - The same situation happens in public schools and universities when teaching skill knowledge (math, chemistry, physics, etc.) to 20 or 200 students together in a classroom. The instructor is not able to provide instant feedback to each student for each mishandled step. All major practice has to be assigned as homework. Feedback is returned day(s) or even week(s) later. If a homework contains 10 questions, and a student made errors in each of them, then that student has practiced and reinforced the wrong ways to solve problems 10 times. By misinterpreting the teacher's instruction, the student has reinforced

the incorrect ways to solve problems and mistakenly believes that they have achieved mastery. The student will only be able to figure out their own mistakes after the teacher returns the homework day(s)/week(s) later. After receiving the feedback, the student will likely think that they understand and will never make the same mistake again. This is unlikely, as their experience up to this point of training the skill has been attempting a problem with incorrect results. The student would have been better off learning nothing at all rather than reinforcing an incorrect procedure. In summation, the old saying of "practice makes perfect" does not apply if the practice is imperfect. Under the conventional mass education model, it is really challenging for students to successfully master skill knowledge in STEAMS domain.

7 Goal, Pedagogy, and Service Audiences

Our approach pursues four main goals: (1) stimulate young learners' profound curiosity in STEM learning by using computational thinking, biological and physical computing in nature, (2) effectively integrate computational modeling and simulation in math and science with project-based learning (PBL) modules, (3) develop students' strong capability of applying computational thinking in nature & engineering drawing which benefits the spatial geometrical thinking, visionary & imaginative design, and analytical logic in STEM disciplines, and (4) enhance students' engagement, particularly among those from groups underrepresented, in STEM and computing domain. *Act*-STEM approach uses modeling-centered inquiry and project-based learning in the study (Fig. 2).

The professional development experiences at the center of this initiative follows a participatory approach that engages teachers as partners in research, expanding their computational vision that leads to the success for all students. It produces data on how the integration of biological and physical computing in nature with computational modeling and simulation in math and science can improve students' deeper understanding and interests in STEM and computing fields. Through strategic curricular design, transformative teacher training, and close monitoring of student learning outcomes in formal classroom and informal out-of-school settings. This study paves the way for effective learning journey for students and broadens the pipeline of students prepared for STEM and computing careers.

The direct service audiences are from our partner public school district which contains four school campuses', 1200 K-12 grades students, and 70 teachers. According to STAAR 2016 Statistics data, 96.6% of our four partner schools' students are Hispanic/African American; 76.9% are from economically disadvantaged families, 49.3% are categorized as at-risk students of dropping out of school. We are optimistic that the potential results from this research project are to be of sufficient significance and quality. The educational resources developed in this study will be available to K-14 researchers, teachers, students, and school districts nationwide.

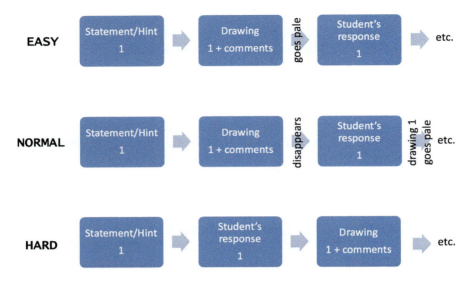

Fig. 2 𝒜𝒸𝓉-STEM training modules

References

1. E. Baker, A. Blackmon, et al., *STEM 2026: A Vision for Innovation in STEM Education* (U.S. Department of Education, Office of Innovation and Improvement, Washington, DC, 2016)
2. A. Bailey, E. Kaufman, S. Subotic, Education, technology, and the 21st century skills gap, (2015). Retrieved from https://www.bcgperspectives.com/content/articles/public_sector_education_technology_twenty_first_century_skills_gap_wef/
3. A. Camins, Ambiguity, uncertainty, failure: Drives for STEM improvement, The Washington Post. https://www.washingtonpost.com/blogs/answersheet/post/three-core-values-of-science-engineering-and-how-ed-reform-contradictsthem/2012/08/07/130a9e7a-de4e-11e1-af1d-753c613ff6d8_blog.html, (August 8, 2012)
4. M. Anft, The STEM crisis: Reality or myth? *The Chronical of Education* (2013)
5. Ithacan, 2014 *Regulations on Laptop Use in College Classrooms Differ Among Professors,* https://theithacan.org/news/regulations-on-laptop-use-in-college-classrooms-differ-among-professors/, 2014
6. MiGill, 2019 *Strategies for Teaching Large Classes,* https://www.mcgill.ca/tls/channels/event/webinar-strategies-teaching-large-classes-289140, 2018
7. J.M. Wing, Computational thinking. Commun. ACM **49**(3), 33 (2006)
8. L. Martin, The promise of the maker movement for education. J. Pre-College Eng. Educ. Res (J-PEER) **5**(1), Article 4 (2015)
9. D. Willingham, *Why Students Don't Like School? A Cognitive Scientist Answers Questions About How the Mind Works and What it Means for your Classroom* (Jossey Bass, San Francisco, 2009)
10. L. W. Anderson, D. R. Krathwohl, et al. (eds.), *A Taxonomy for Learning, Teaching, and Assessing: A Revision of Bloom's Taxonomy of Educational Objectives*. Allyn & Bacon (Pearson Education Group, Boston, MA, 2001)
11. L. Anderson, *A Taxonomy for Learning, Teaching, and Assessing: A Revision of Bloom's Taxonomy of Educational Objectives* (Abridged Edition, Pearson Publisher, 2013)

12. B.S. Bloom, D.R. Krathwohl, *Taxonomy of Educational Objectives: The Classification of Educational Goals*, by a committee of college and university examiners. Handbook I: Cognitive Domain. (Longmans, Green, New York 1956)
13. G. Wilson, *The Five Senses: Or, Gateways to Knowledge* (Lindsay & Blakiston, 1860)

Realistic Drawing & Painting with AI-Supported Geometrical and Computational Method ($\mathcal{F}$un-$\mathcal{J}$oy)

Lei Wu, Alan Yang, Han He, Xiaokun Yang, Hua Yan, Zhimin Gao, Xiao Qin, Bo Liu, Shan Du, Anton Dubrovskiy, and T. Andrew Yang

1 Introduction

Research studies have shown that drawing is a key to help students' engagement in STEM disciplines, because many concepts in science, technology, engineering and math are so often visual and spatial in nature. Drawing is a key activity, alongside modelling, role-play and digital simulation to focus collaborative reasoning and

---

L. Wu · H. Yan · Z. Gao
Department of Computer Science and Computer Information Systems, Auburn University at Montgomery, Montgomery, AL, USA

A. Yang (✉)
Department of Information Systems, University of Nevada, Reno, Reno, NV, USA
e-mail: alany@unr.edu

H. He
Department of Computer Science, Emory University, Atlanta, GA, USA

X. Yang
Department of Engineering, University of Houston Clear-Lake, Houston, TX, USA

X. Qin · B. Liu
Department of Computer Science and Software Engineering, Auburn University, Auburn, AL, USA

S. Du
Department of Computer Science, Mathematics, Physics and Statistics, University of British Columbia Okanagan Campus, Kelowna, BC, Canada

A. Dubrovskiy
Department of Chemistry, University of Houston-Clear Lake, Houston, TX, USA

T. A. Yang
Department of Computer Science, University of Houston-Clear Lake, Houston, TX, USA

© Springer Nature Switzerland AG 2021
H. R. Arabnia et al. (eds.), *Advances in Artificial Intelligence and Applied Cognitive Computing*, Transactions on Computational Science and Computational Intelligence, https://doi.org/10.1007/978-3-030-70296-0_58

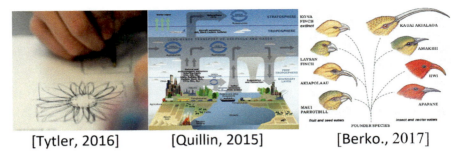

[Tytler, 2016]   [Quillin, 2015]   [Berko., 2017]

Fig. 1 Research studies show that drawing is a critical skill in STEAM learning

generation of meaning [1, 2]. A research study published in *Life Science Education Journal* [3] shows that:

- Drawing of visual representations is important for learners and scientists alike, such as the drawing of models to enable visual model-based reasoning.
- It is the core competency skill of *Vision and Change* report's modelling and simulation capability [4].
- The power of visual drawing has been used by scientists from the representational anatomical and engineering design works of Leonardo da Vinci to the theoretical phylogenetic work of Charles Darwin.

In the "Spatial Skills: A Neglected Dimension of Early STEAM Education," published at *Education Weeks*, February 2017 [5], the following points on STEM were made:

- Spatial skills predict success in STEM fields out to adulthood.
- The importance of spatial skills is often overlooked as a key feature of STEM education.
- Frequent neglect of spatial development creates an additional barrier to children's STEM learning (Fig. 1).

2 Insufficient Public-School Visual Art Education for STEM

In nationwide public K-12 education, visual art focused on drawing and painting has been significantly downplayed with an average of one 50-minute lesson every 2 weeks. A common misconception of the visual arts is that an individual needs to be gifted to master it – a perfect excuse to deprive visual art education for public

K-12 grade students. Recent studies show that drawing skill development for young learners has been intentionally and systematically neglected [5].

Theory of action 1 Certain key skills, such as drawing for STEM learning, have been purposefully overlooked and systematically neglected with a severe lack of education resources.

3 𝓕un-𝓙oy Visual Art Education for STEM

Built upon literature, prior practice, and research, our 𝓕un-𝓙oy project applies computational thinking (CT), computing methods, technology and CT-enriched curriculum to promote desire and competency in the areas of visual arts development for K-12th grade learners. The National Research Council has documented the nature of young leaners' intrinsic desire to discover, explore and solve problems [6]. However, research on students using computational thinking with intriguing nature computing phenomena towards visual art, math and science learning, among the pre-kindergarten to 8th grade demographic, in a multi-year, formal and informal learning environment is relatively unexplored. As Wing pointed out, computational thinking benefits society in a much broader perspective, especially in early education [7]. A growing number of researches on integrating CT into early STEM learning has shown that new perspectives for studying and assessing the effectiveness of applying computational thinking in K-12 education is urgent [8–12].

    𝓕un-𝓙oy project is built upon above solid research base [13, 14], increasing the probability that we will much better understand how to combine visual art, computational thinking and nature computing phenomena into an effective early STEM education practice, and increase STEM literacy.

4 Research Questions

- *RQ-1:* What pedagogical environments and learning strategies are needed for integrating computational thinking skills into drawing and painting learning from an early age (K-8th grade)?
- *RQ-2:* What are the learning progressions in computational thinking that can be identified from the early ages of kindergarten (5 years old) through middle school (6th–8th grades, 12–14-year-old)?
- *RQ-3:* How does computational thinking significantly improve drawing skill development, and how does drawing skill development greatly affect students' learning performance in STEM learning from the early grade pre-K to 12 grades?

- *RQ-4:* Under what conditions does the integration of computing into drawing and painting increase student interest, motivation, content acquisition and performance in STEM learning?

5 Design and Development Efforts

Theory of action 2 Certain vital habits and skills are best learned at a young age between kindergarten to 8th grade; past 8th grade, those skills take longer to learn, and teaching those skills becomes less effective.

Based upon the theory of action 1 and 2, we have conducted the first major design and development module and built our successful pilot project titled "Drawing & Painting with AI-supported Geometrical and Computational Methods *Fun-Joy*." It heavily uses the four key components in computational thinking [7]:

1. *Decomposition* – breaking down a complex drawing object into smaller, more manageable parts
2. *Abstraction* – level by level, focusing on the important geometrical shapes only, ignoring irrelevant drawing details
3. *Pattern recognition* – looking for similarities among and within drawing objects, and apply patterns in drawing
4. *Algorithms* – developing a step-by-step solution/rules to follow to solve the drawing challenge

Computational thinking conquers a complex problem by breaking it down into a series of smaller, more manageable problems (*decomposition*); each of these smaller problems can then be looked at individually and compared to previously solved problems (*pattern recognition*). By focusing only on the important details, while ignoring irrelevant information (*abstraction*), eventually, simple steps or rules to solve each of the smaller problems can be designed (*algorithms*). The cornerstone of our theory of action is: the most effective teaching and learning happens when teacher deeply inspires students' curiosity, and motivates their desire with guidance to explore, depict, replicate and interact with the amazing external world using the leaner's own hands. New education approaches should capture this human nature of learning and correct the flaws inherited from traditional public education practices [15–17] (Fig. 2).

Our major research methods include:

- Integrating computational thinking approaches into STEM disciplines, including visual arts education
- Applying modelling-centred inquiry and project-based learning into the study
- Applying the fascinating phenomena in nature computing, including biological and physical computing, to stimulate young learners' desire to learn and discover
- Applying the project-based online interactive cyber-learning communities, to promote students to spontaneously develop an in-depth understanding of com-

Fig. 2 Teaching software developed in our pilot project "Realistic Drawing & Painting with Geometrical and Computational Methods (𝓕un-𝓙oy)"

putational thinking as a way of creative approach towards learning in STEM fields
- Imposing professional development experiences at the centre of this project to include school teachers, social and learning scientists, out-of-school practitioners, informal educators, education media and technology developers, into the research work
- Applying a participatory approach that engages teachers as partners in research, expanding their computational vision that leads to the success for all students
- Producing data on how integration of visual art, biological and physical computing in nature with computational modelling, simulation in algebra, geometry and

science can improve students' deeper understanding and interests in STEM and computing fields

- Developing strategic curricular design, transformative teacher training, and close monitoring of student learning outcomes in formal classroom and informal out-of-school settings

Theory of action 3 At a young age, especially from K-8th grades, one is:

I. Curious about the world
II. Eager to sense every element the world has
III. Keen to mimic, depict and replicate with one's own hands
IV. Ardent to draw, paint, evolve, innovate and invent one's own counterparts of the nature and human world

Theory of action 4 Humans have instinctual interest in nature phenomenon and their causes.

Theory of action 5 Unexplained phenomena will stimulate innate inquisitiveness and trigger one's intense desire to discover and understand.

Fun-Joy's approach is based upon the theory of action 3, 4 and 5, together with computational modelling and simulation. We have conducted the major design and development which applies the visual art skill knowledge with fascinating nature phenomena caused by biological computing and physical computing, to improve students' deeper understanding, interests and mastery in targeted STEM fields. It has profoundly stimulated young learners' intrinsic desire and deeply promoted the mastery of exploring, discovery and innovation.

6 Preliminary Results

Our pilot project *Fun-Joy* heavily applies AI-aided computational thinking with geometrical methods in visual art skill development for learners who lack inborn artistic talents in the visual art domain. Since its debut on Google Play, the software designed for this project has been downloaded by more than 5 million learners, and has been translated into 65 languages; more than 110,000 learners have given written review comments on Google Play; the average score is 4.6 out of 5.0 scale, topping the highest in its category (Fig. 3).

Realistic Drawing & Painting with AI-Supported Geometrical... 803

Fig. 3 Students' work after 1-week summer camp with 𝓕𝓾𝓷-𝓙𝓸𝔂, the visual art skill knowledge teaching software in our pilot project "Realistic Drawing & Painting with Geometrical and Computational Methods"

References

1. R. Tytler, How art is drawing students to STEM, Australian Council for Educational Research (ACER) conference 2016 - improving STEM learning: What will it take? 2016
2. G.T. Tyler-Wood, G. Knezek, R. ChrisTensen, Tyler-Wood, Knezek, Christensen, Instruments for assessing interest in STEM content and careers. J. Technol. Teach. Educ. **18**(2), 341–363 (2010)
3. K. Quillin, S. Thomas, Drawing-to-learn: A framework for using drawings to promote model-based reasoning in biology. CBE Life Sci. Educ **14**(1), es2 (2015)
4. American Association for the Advancement of Science (AAAS), Vision and Change Report, 2011. http://visionandchange.org/files/2011/03/Revised-Vision-and-Change-Final-Report.pdf
5. J. Berkowicz, A. Myers, Spatial skills: A neglected dimension of early STEM education, Education Weeks, Feb. 2017. http://blogs.edweek.org/edweek/leadership_360/2017/

02/spatial_skills_a_neglected_dimension_of_early_stem_education.html?utm_content=bufferffc48&utm_medium=social&utm_source=twitter.com&utm_campaign=buffer

6. National Research Council, in *A Framework for K-12 Science Education: Practices, Crosscutting Concepts, and Core Ideas*, ed. by H. Quinn, H. Schweingruber, T. Keller, (National Academies Press, Washington, D.C, 2012)

7. J.M. Wing, Computational thinking. Commun. ACM **49**(3), 33 (2006)

8. K. Brennan, M. Resnick, New frameworks for studying and assessing the development of computational thinking. In 2012 annual meeting of the American Educational Research Association (AERA). Vancouver, Canada, 2012

9. N.C.C. Brown, M. Kölling, T. Crick, S. Peyton Jones, S. Humphreys, S. Sentance, Bringing computer science back into schools: Lessons from the UK, in *Proceeding of the 44th ACM Technical Symposium on Computer Science Education*, (ACM, New York, 2013), pp. 269–274

10. S. Daily, A. Leonard, S. Jörg, S. Babu, K. Gundersen, D. Parmar, Embodying computational thinking: Initial design of an emerging technological learning tool. Technol. Knowl. Learn. **20**(1), 79–84 (2014)

11. J. Voogt, P. Fisser, J. Good, P. Mishra, A. Yadav, Computational thinking in compulsory education: Towards an agenda for research and practice. Educ. Inf. Technol. **20**(4), 715–728 (2015)

12. S.C. Kong, A framework of curriculum design for computational thinking development in K-12 education. J. Comput. Educ **3**(4), 377–394 (2016)

13. S. Grover, R. Pea, S. Cooper, Designing for deeper learning in a blended computer science course for middle school students. Comput. Sci. Educ. **25**(2), 199–237 (2015)

14. R. Taub, M. Armoni, M. Ben-Ari, (Moti). Abstraction as a bridging concept between computer science and physics, in *Proceedings of the 9th Workshop in Primary and Secondary Computing Education*, (ACM Press, 2014), pp. 16–19

15. S. Grover, R. Pea, Computational thinking in K–12 a review of the state of the field. Educ. Res. **42**(1), 38–43 (2013)

16. I. Lee, Reclaiming the roots of CT. CSTA Voice – Spec. Issue Comput. Think **12**(1), 3–5 (2016)

17. D. Weintrop, K. Orton, M. Horn, E. Beheshti, L. Trouille, K. Jona, U. Wilensky, Computational thinking in the science classroom: Preliminary findings from a blended curriculum. Annual meeting of the National Association for Research in Science Teaching (NARST), 2015

Part VI
Artificial Intelligence for Smart Cities (Chair: Dr. Charlie (Seungmin) Rho)

Training-Data Generation and Incremental Testing for Daily Peak Load Forecasting

Jihoon Moon [ID], Sungwoo Park [ID], Seungmin Jung [ID], Eenjun Hwang [ID], and Seungmin Rho [ID]

1 Introduction

The smart grid system aims at optimizing the energy efficiency between suppliers and consumers through bidirectional interactions, and accurate electric load forecasting is essential for its operation [1]. Because a typical smart grid consists of an energy management system with renewable energy; an energy storage system; and combined cooling, heating, and power, it is necessary to establish operational plans based on short-term load forecasting (STLF) [1, 2]. STLF includes total daily load forecasting, daily peak load forecasting (DPLF), and very STLF (with a lead time of < 1 d) [2]. DPLF is an essential procedure for unit commitment, security analysis of the system, and tight scheduling of outages and fuel supplies in smart grid applications [3].

Accurate DPLF is a challenging issue because energy consumption has various patterns and involves uncertainty due to unforeseeable external factors [4]. Additionally, when predicting electric loads, it is necessary to adequately consider the very complex correlation between the historical data and the current data [5]. Therefore, several studies have been conducted to achieve accurate DPLF based on artificial intelligence (AI) techniques such as the naïve Bayes (NB) classifier, support vector regression (SVR), adaptive boosting (AdaBoost), random forest (RF), artificial neural networks (ANNs), fuzzy neural networks (FNNs),

---

J. Moon · S. Park · S. Jung · E. Hwang (✉)
School of Electrical Engineering, Korea University, Seoul, Republic of Korea
e-mail: johnny89@korea.ac.kr; psw5574@korea.ac.kr; jmkstcom@korea.ac.kr; ehwang04@korea.ac.kr

S. Rho
Department of Software, Sejong University, Seoul, Republic of Korea
e-mail: smrho@sejong.edu

© Springer Nature Switzerland AG 2021
H. R. Arabnia et al. (eds.), *Advances in Artificial Intelligence and Applied Cognitive Computing*, Transactions on Computational Science and Computational Intelligence, https://doi.org/10.1007/978-3-030-70296-0_59

Table 1 Summary of DPLF based on AI techniques

Author (year)	Dataset	AI techniques	Training set period	Test set period
Son et al. [4] (2015)	Korea Power Exchange	FNN	5 years	1 year
Hsu et al. [6] (2018)	Taiwan Power Company	ANN	5 years	1 year
Yu et al. [7] (2019)	EUNITE	GRU	104 weeks	5 weeks
Saxena et al. [8] (2019)	RIT Campus	ANN	2 years 4 months	1 year
Sakurai et al. [9] (2019)	TEPCO Power Grid	ANN	4 years	1 year
Liu and Brown [10] (2019)	IESO	NB, SVR, RF, AdaBoost, CNN, LSTM network, Stacked Autoencoder	5 years	1 year

convolutional neural networks (CNNs), long short-term memory (LSTM), and gated recurrent unit (GRU) neural networks [4, 6–10]. Previous DPLF-related studies based on AI techniques are presented in Table 1.

Most of these studies depended on sufficient datasets. However, when the collected datasets are insufficient for prediction, we must make the best possible use of the collected datasets to build an accurate electric load forecasting model. For this purpose, we propose a novel day-ahead DPLF model to handle insufficient datasets. We first configured various input variables that exhibit a high correlation with the daily peak load. Then, we constructed AI-based DPLF models using time-series cross-validation (TSCV) to achieve high accuracy.

The remainder of this chapter is organized as follows. In Sect. 2, we describe the input variable configuration of the DPLF models. In Sect. 3, we explain the construction of our DPLF model, which is based on various AI techniques. In Sect. 4, we present the experimental results to demonstrate the superiority of our model. Finally, we end with conclusions in Sect. 5.

2 Input Variable Configuration

In this study, we considered four different types of building clusters located in a private university in South Korea, because they exhibited typical usage patterns of diverse buildings [11, 12]. We collected their daily peak load data from September 1, 2015 to February 28, 2019. Table 2 presents information about the building clusters. Additionally, we gathered other data, including the weather conditions and holiday information, which are known to be closely related to the electric load [8]. By

Table 2 Building cluster information

Cluster type	No. of buildings	Cluster description	Duration (day)
A	32	Humanities and social science	1277
B	20	Science and engineering	1277
C	5	Science and engineering	1277
D	16	Dormitory	1277

utilizing them, we configured diverse input variables to perform a day-ahead load forecasting.

Because time is closely related to the trend of the electric load, we considered all the variables that can represent time factors, such as months, days, and days of the week. Time data in the sequence form cannot reflect periodic information when they are applied to AI technology [13, 14]. Therefore, we enhanced them to obtain continuous data in the two-dimensional (2D) space using (1), (2), (3), and (4) [13]. LDM_{Month} in (3) and (4) represent the last day of the month to which the day belongs:

$$Month_x = \sin((360/12) \times Month) \tag{1}$$

$$Month_y = \cos((360/12) \times Month) \tag{2}$$

$$Day_x = \sin((360/LDM_{Month}) \times Day) \tag{3}$$

$$Day_y = \cos((360/LDM_{Month}) \times Day) \tag{4}$$

Data with different numbers of dimensions have different advantages. One-dimensional (1D) data can better reflect categorical data information, and 2D data can better reflect periodic information [5]. Hence, we used both 1D data and continuous 2D data, to exploit their advantages.

Buildings of higher education institutions have different energy use patterns on weekdays, weekends, and holidays, depending on the type of building. To reflect these differences, the day of the week and holiday data, which were nominal measures, were reflected in the DPLF model. To represent a nominal scale, we defined a vector of 0 or 1 for each building cluster depending on the eight-dimensional (8D) feature vector composed of the 7 days of the week and holidays. Here, it was essential to know the day of the week as well as the periodicity of the days. For this reason, we used both the data configured with an 8D feature vector and the data represented in the 2D space through the periodic function. Table 3 presents all the input variables that we considered, and their types.

Table 3 Input variables of time factors

No.	Input variables	Variable type	No.	Input variables	Variable type
1	Month	Continuous	9	Monday	Binary
2	Day	Continuous	10	Tuesday	Binary
3	$Month_x$	Continuous	11	Wednesday	Binary
4	$Month_y$	Continuous	12	Thursday	Binary
5	Day_x	Continuous	13	Friday	Binary
6	Day_y	Continuous	14	Saturday	Binary
7	Day of the week$_x$	Continuous	15	Sunday	Binary
8	Day of the week$_y$	Continuous	16	Holiday	Binary

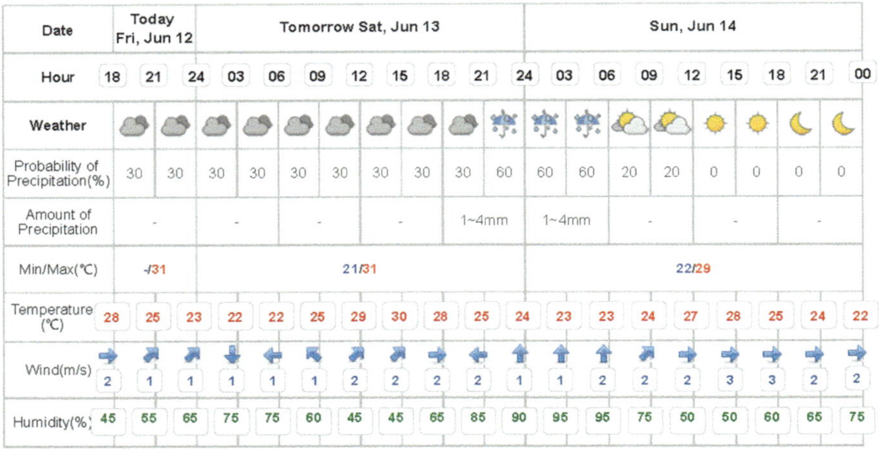

Date	Today Fri, Jun 12		Tomorrow Sat, Jun 13									Sun, Jun 14							
Hour	18	21	24	03	06	09	12	15	18	21	24	03	06	09	12	15	18	21	00
Weather																			
Probability of Precipitation(%)	30	30	30	30	30	30	30	30	30	60	60	60	20	20	0	0	0	0	
Amount of Precipitation	-		-		-		-		1~4mm		1~4mm		-		-		-		
Min/Max(℃)	/31				21/31							22/29							
Temperature (℃)	28	25	23	22	22	25	29	30	28	25	24	23	23	24	27	28	25	24	22
Wind(m/s)	2	1	1	1	1	1	2	2	2	2	1	1	2	2	3	3	2	2	
Humidity(%)	45	55	65	75	75	60	45	45	65	85	90	95	95	75	50	50	60	65	75

Fig. 1 Example of a short-term forecast provided by the KMA

In general, the electric load increases in summer and winter, owing to the heavy use of air conditioning and electric heating appliances, respectively. In South Korea, the Korea Meteorological Administration (KMA) provides a short-term weather forecast for most major regions [14]. The short-term forecast includes weather forecasts for 3 days from the present day, as shown in Fig. 1. Because our DPLF model focuses on day-ahead load forecasting, we used the short-term forecast provided by the KMA. In general, the temperature, humidity, and wind speed significantly affect electric loads. To reflect them in our DPLF model, we collected weather forecast data; calculated the average temperature, minimum temperature, maximum temperature, average humidity, and average wind speed; and used them as input variables.

Historical electric load data are particularly important in electric load prediction because they exhibit the trend of recent electric loads [8, 15]. To reflect the recent trend in the prediction, we used the daily peak loads of the previous 7 days as input variables. Moreover, the electric load trends of weekdays and holidays can differ.

To reflect such differences in the model, we added a holiday indicator to signify whether the day is a holiday [5]. Additionally, to reflect the recent electricity usage trend more effectively, we added the average and deviation of the daily peak electric loads for the previous 7 days as input variables. We used a total of 37 input variables to build our DPFL model. In the next section, we describe the overall architecture of our DPLF model.

3 Forecasting Model Construction

In this section, we describe the overall architecture of our DPLF models. To construct our DPLF models, we used as many datasets as possible by applying TSCV. Additionally, we used principal component analysis (PCA) and factor analysis (FA) to define more input variables that were closely related to the daily peak load. Then, we constructed the DPLF models using various AI technologies and compared their performance.

3.1 Time-Series Cross-Validation

When constructing a forecasting model, datasets are usually divided into a training set and a test set. Then, the forecasting model is built using the training set and verified using the test set. However, conventional time-series forecasting methods exhibit unsatisfactory prediction performance when there is a significant gap between the training set period and the test set period [12]. Additionally, when the dataset is insufficient, it is challenging to obtain satisfactory prediction performance using a small amount of training data [15].

To solve these problems, we utilized TSCV based on a rolling forecasting origin. TSCV focuses on a single forecast horizon for each test set [16]. In our approach, we used different training sets, each containing one or more observations than the previous training set, depending on the scheduling period. Hence, we used one-step-ahead TSCV to construct the day-ahead peak load forecasting model, as shown in Fig. 2.

3.2 Input Variable Generation

Even though we configured as many input variables as possible using time information, weather data, and historical electric loads, it was challenging to construct sophisticated forecasting models because the 2D matrix had small training datasets. To increase the size of the 2D matrix for training the DPLF models effectively,

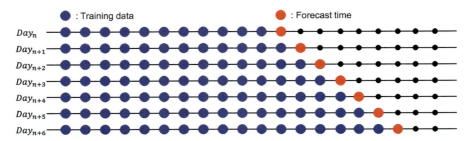

Fig. 2 One-step-ahead TSCV

we augmented the input variables by using PCA and FA, which are widely used multivariate techniques [11, 17].

PCA converts a set of observations of possibly correlated variables into a set of linearly uncorrelated variables known as principal components. When the data are projected onto a single axis, a linear transformation is performed to a new coordinate system based on the size of the dispersion. FA assembles similar variables by considering the correlation among the variables. It identifies factors that have a significant impact on the data collection composed of several variables.

Because PCA and FA exhibit different values whenever the 2D matrix size increases, it is not appropriate to use PCA and FA for all the data of the test set (unseen data). Hence, when performing TSCV, we used PCA and FA with only the input variables from the training set and the forecast period and then added them as new input variables. To configure new input variables based on PCA and FA, we applied the number of dimensions sequentially, from one to one less than the number of input variables, and identified the optimal number of dimensions by returning the average value of the log-likelihood.

As mentioned previously, we considered various input variables; however, they had different scales. For smoothing the imperfection of the AI-based DPLF model training, normalization is needed to place all the input variables within a comparable range [18]. Therefore, we performed min–max normalization for all the input variables using (5). Min-max normalization is commonly known as feature scaling, where the values of a numeric range of data are reduced to a scale between 0 and 1, where z is the normalized value of a number of the set of observed values of x, and Min(x) and Max(x) are the minimum and maximum values in x given its range:

$$z = x - \text{Min}(x) / \text{Max}(x) - \text{Min}(x) \qquad (5)$$

3.3 DPLF Model Construction

To construct our DPLF models, we used a total of 10 AI techniques, including two conventional machine learning methods, four ensemble learning methods, and four

ANN methods. The conventional machine learning methods included decision tree (DT) and SVR. The ensemble learning methods included RF, gradient boosting machine (GBM), extreme gradient boosting (XGB), and light gradient boosting machine (LightGBM). The ANN methods included shallow neural network (SNN), deep neural network (DNN), LSTM network, and attention-based LSTM network. We determined the hyperparameters of each AI technique using a grid search.

4   Experimental Results

For the experiments, we performed preprocessing for the dataset and forecast modeling in a Python environment. We divided the dataset into parts: a training set (in-sample) spanned from September 1, 2015 to February 28, 2018 (912 days), and a test set (out-of-sample) spanned from March 1, 2018 to February 28, 2019 (365 days). We used the peak load data from September 1, 2015 to September 7, 2015 to configure input variables for the training set of the day-ahead load forecasting model. The data from September 8, 2015 to February 28, 2018 were used as the training set. The data from March 1, 2018 to February 28, 2019 were used as the test set:

$$\text{MAPE} = 100/n \times \sum | (A_t - F_t) / A_t | \tag{6}$$

To compare the prediction performances of the forecasting models, we used the mean absolute percentage error (MAPE). The MAPE is easier to understand than other performance metrics such as the root-mean-square error and mean-squared error (MSE) because it represents the accuracy as a percentage of the error [15]. The MAPE is defined by (6), where n represents the number of observations, and A_t and F_t represent the actual and forecasted values, respectively.

Tables 4 and 5 present several experimental results for day-ahead peak load forecasting. "Hold-out" indicates the splitting of the datasets into training and test sets. The training set was used to train the model, and the test set was used to determine how well the model performed on unseen data. The TSCV exhibited more accurate peak load forecasting than the hold-out because not only was it trained with more data, but it also adequately reflected recent electric load patterns. Additionally, in most cases, we confirmed that training with both existing input variables and new input variables from PCA and FA yielded more accurate peak load forecasting than training with only existing input variables.

To demonstrate the superiority of our method, we performed a Friedman test [15]. The Friedman test is a multiple comparison test that aims to identify significant differences between the results of three or more forecasting methods. To verify the results of this test, we used all the MAPE values of hold-out, TSCV, and our method, respectively. The result of the Friedman test confirmed a p-value of 3.493363×10^{-17}. We can observe that our method significantly outperforms the other methods because the p-value is below the significance level.

Table 4 MAPE comparison of day-ahead peak load forecasting for Clusters A and B

AI techniques	Cluster A			Cluster B		
	Hold-out	TSCV	Proposed	Hold-out	TSCV	Proposed
DT	7.61	7.32	6.23	7.45	6.82	6.66
SVR	7.11	5.93	5.49	5.04	3.89	3.64
RF	6.11	5.75	4.64	4.99	4.82	4.71
GBM	5.96	5.65	3.72	5.04	4.85	3.88
XGB	6.10	5.29	3.66	5.08	4.35	3.57
LightGBM	5.17	4.89	3.05	4.47	4.19	2.86
SNN	12.07	10.57	9.15	7.57	6.30	5.17
DNN	6.33	3.45	2.66	4.84	4.56	2.39
LSTM network	5.16	5.08	4.41	5.20	4.90	3.67
Attention-based LSTM network	3.29	2.50	2.21	2.71	2.48	2.22

Table 5 MAPE comparison of day-ahead peak load forecasting for Clusters C and D

AI techniques	Cluster C			Cluster D		
	Hold-out	TSCV	Proposed	Hold-out	TSCV	Proposed
DT	4.32	3.83	3.97	6.46	5.50	6.01
SVR	3.25	3.02	3.02	5.55	4.94	3.44
RF	3.15	2.80	2.79	4.51	4.33	4.24
GBM	3.06	2.73	2.38	4.29	4.15	3.47
XGB	3.02	2.73	2.23	4.25	3.93	3.24
LightGBM	2.79	2.52	2.22	3.96	3.79	2.25
SNN	4.43	3.35	2.42	5.80	3.74	2.58
DNN	3.33	3.10	2.58	4.73	3.58	2.02
LSTM network	5.64	5.04	4.26	6.21	6.15	4.54
Attention-based LSTM network	3.52	3.43	2.44	6.41	5.59	3.14

5 Conclusion

We proposed a novel approach for accurate DPLF using insufficient datasets. We first collected three and a half years of peak electric load data for four building clusters on a university campus. Additionally, we obtained weather and holiday information, which are closely related to the electric loads. We considered a day-ahead peak load forecasting and configured input variables using various types of data preprocessing. To use more training datasets and reflect recent electric load patterns, we utilized one-step-ahead TSCV for constructing the day-ahead peak load forecasting model. We generated new input variables based on PCA and FA to increase the size of the 2D matrix for training the DPLF model effectively. Experimental results based on various AI techniques indicated that the proposed method provided more accurate peak load forecasting than hold-out and TSCV, which used existing input variables.

Acknowledgments This research was supported in part by Energy Cloud R&D Program (grant number: 2019M3F2A1073179) through the National Research Foundation of Korea (NRF) funded by the Ministry of Science and ICT and in part by the Korea Electric Power Corporation (grant number: R18XA05).

References

1. A.I. Saleh, A.H. Rabie, K.M. Abo-Al-Ez, A data mining based load forecasting strategy for smart electrical grids. Adv. Eng. Inform. **30**(3), 422–448 (2016)
2. M. Son, J. Moon, S. Jung, E. Hwang, A short-term load forecasting scheme based on auto-encoder and random forest, in *APSAC 2018. LNEE*, ed. by K. Ntalianis, G. Vachtsevanos, P. Borne, A. Croitoru, vol. 574, (Springer, Cham, 2019), pp. 138–144. https://doi.org/10.1007/978-3-030-21507-1_21
3. M.B. Tasre, P.P. Bedekar, V.N. Ghate, Daily peak load forecasting using ANN, in *2011 Nirma University International Conference on Engineering*, (IEEE, Ahmedabad, 2011)
4. S.-Y. Son, S.-H. Lee, K. Chung, J.S. Lim, Feature selection for daily peak load forecasting using a neuro-fuzzy system. Multimed. Tools Appl. **74**(7), 2321–2336 (2015)
5. S. Park, J. Moon, S. Jung, S. Rho, S.W. Baik, E. Hwang, A. Two-Stage Industrial, Load forecasting scheme for day-ahead combined cooling, heating and power scheduling. Energies **13**(2), 443 (2020)
6. Y.-Y. Hsu, T.-T. Tung, H.-C. Yeh, C.-N. Lu, Two-stage artificial neural network model for short-term load forecasting. IFAC-PapersOnLine **51**(28), 678–683 (2018)
7. Z. Yu, Z. Niu, W. Tang, Q. Wu, Deep learning for daily peak load forecasting-a novel gated recurrent neural network combining dynamic time warping. IEEE Access **7**, 17184–17194 (2019)
8. H. Saxena, O. Aponte, K.T. McConky, A hybrid machine learning model for forecasting a billing period's peak electric load days. Int. J. Forecast. **35**(4), 1288–1303 (2019)
9. D. Sakurai, Y. Fukuyama, T. Iizaka, T. Matsui, Daily peak load forecasting by artificial neural network using differential evolutionary particle swarm optimization considering outliers. IFAC-PapersOnLine **52**(4), 389–394 (2019)
10. J. Liu, L.E. Brown, Prediction of hour of coincident daily peak load, in *2019 IEEE Power & Energy Society Innovative Smart Grid Technologies Conference (ISGT)*, (IEEE, Washington, 2019)
11. J. Moon, J. Park, E. Hwnag, S. Jun, Forecasting power consumption for higher educational institutions based on machine learning. J. Supercomput. **74**(8), 3778–3800 (2018)
12. J. Moon, Y. Kim, M. Son, E. Hwang, Hybrid short-term load forecasting scheme using random forest and multilayer perceptron. Energies **11**(12), 3283 (2018)
13. S. Jung, J. Moon, S. Park, S. Rho, S.W. Baik, E. Hwang, Bagging ensemble of multilayer perceptrons for missing electricity consumption data imputation. Sensors **20**(6), 1772 (2020)
14. J. Kim, J. Moon, E. Hwang, P. Kang, Recurrent inception convolution neural network for multi short-term load forecasting. Energ. Buildings **194**, 328–341 (2019)
15. J. Moon, J. Kim, P. Kang, E. Hwang, Solving the cold-start problem in short-term load forecasting using tree-based methods. Energies **13**(4), 886 (2020)
16. R.J. Hyndman, G. Athanasopoulos, *Forecasting: Principles and Practice* (OTexts, Melbourne, 2018)
17. J.C.F. De Winter, D. Dodou, Common factor analysis versus principal component analysis: A comparison of loadings by means of simulations. Commun. Stat. Simul. Comput. **45**(1), 299–321 (2016)
18. J. Moon, S. Park, S. Rho, E. Hwang, A comparative analysis of artificial neural network architectures for building energy consumption forecasting. Int. J. Distrib. Sens. Netw. **15**(9), 1550147719877616 (2019)

Attention Mechanism for Improving Facial Landmark Semantic Segmentation

Hyungjoon Kim, Hyeonwoo Kim, Seongkuk Cho, and Eenjun Hwang

1 Introduction

Smart city refers to an urban area that is efficiently managed by bigdata, Internet of Things, and information communication technology. To achieve the goal, various digital technologies are used into diverse domains. For instance, face recognition technology can be used in a variety of domains such as searching for missing persons, tracking criminals and suspects, and controlling building access. In order to recognize the face, the main features of the face should be detected and represented. One of the most popular features for face recognition is facial landmarks such as eyes, nose, and mouth as they are the most visually distinctive features. So far, many works have been done to detect such landmarks precisely, but it is still challenging due to diverse factors such as image resolution, occlusion, and face orientation. Recent works usually represent facial landmarks using several key points [1] by using convolutional neural networks (CNNs). These approaches have the advantage that landmarks can be detected quickly and accurately and processed in real time in a general hardware environment [2, 3]. In our previous works, we extracted facial landmarks as pixel units by using the semantic segmentation method [4–6]. This method requires more computational resource than detecting landmarks as points, but it can extract landmarks that are difficult to express with several points, such as hair. In particular, the landmark extraction performance, we proposed network architecture, class weights, and post-processing method to overcome the limitations of existing semantic segmentation techniques and achieved an excellent facial

---

H. Kim · H. Kim · S. Cho · E. Hwang (✉)
School of Electrical Engineering, Korea University, Seoul, Republic of Korea
e-mail: hyungjun89@korea.ac.kr; guihon12@korea.ac.kr; kuklife@korea.ac.kr; ehwang04@korea.ac.kr

© Springer Nature Switzerland AG 2021
H. R. Arabnia et al. (eds.), *Advances in Artificial Intelligence and Applied Cognitive Computing*, Transactions on Computational Science and Computational Intelligence, https://doi.org/10.1007/978-3-030-70296-0_60

landmarks extraction. To make further progress here, we investigate the concept of attention, which is popularly used to improve the performance of the networks.

Attention is known to improve the performance of the networks by giving networks information about where to focus in the input vector. It complements the point that the size of the receptive field, a limitation of CNNs, is determined by the size of the convolution filter. In detail, it calculates the relationship among every pixel across the image without significantly increasing the parameters. This method shows a good performance in image classification, object detection, semantic segmentation, and even recently panoptic segmentation.

In this chapter, we improve facial landmarks extraction performance by adopting attention mechanism. The base mechanism is the *squeeze-excitation network* [7]. It calculates channel attention and provides information on which channel the network should focus on. To improve its advantages, we divide the input vector into two vectors then, use two *squeeze-excitation* process. We call it *decouple squeeze-excitation*.

This chapter is organized as follows: Sect. 2 introduces several studies related to facial landmark detection, semantic segmentation and attention. Section 3 describes our proposal in detail, and in Sect. 4 we evaluate our approach experimentally. Finally, Sect. 5 concludes this chapter.

2 Related Works

2.1 Facial Landmark Detection

Dlib [8] is one of the most popular methods for facial landmark detection. It uses an ensemble approach of regression trees and represents facial landmarks using 68 key points. As the method is simple and shows a reasonable performance, it has been used as a baseline in other studies. Recently, various facial landmark detection methods have been proposed based on CNNs and their main topics are: (i) how to improve facial landmark detection performance [9], (ii) how to recognize facial landmarks in real time [10], and (iii) how to segment facial landmarks for all pixels that are not specific coordinates [11].

2.2 Semantic Segmentation

The first semantic segmentation method based on CNNs is fully convolutional networks [12]. It extracts features from the input image by using the same convolution and pooling as CNNs. Then, through convolution and interpolation, the feature vectors are upscaled, and then finally, pixel-based classification is carried out. Diverse studies used this encoder-decoder structure; some used more powerful

CNNs or proposed new encoder-decoder structures to improve the performance [13, 14]. Other studies have improved segmentation performance by using new schemes such as dilated convolution, conditional random field (CRF), and atrous spatial pyramid pooling (ASPP) [15, 16].

2.3 Attention Mechanism

In the computer vision field, attention was initially used for generating image captions [17]. Then, the concept has been used to improve the performance of CNNs [9, 18]. More recently, the concept of self-attention was applied to CNNs [19], and the self-attention in CNNs demonstrated excellent performance in the field of panoptic segmentation [20].

3 Proposed Method

Squeeze-excitation calculates channel attention in each block. The calculated attention is used for the network to focus on a specific channel. It calculates the average value among channels by using global average pooling. Then, each channel weight is calculated through the fully connected, relu, and sigmoid layer. The calculated channel weights are element-wisely multiplied with the input.

Figure 1 illustrates this process, and the effect of *squeeze-excitation* for four channel input is described in Fig. 2. In the Figure, H, W, and C denote height, width, and channel of the input vector, respectively, and r is reduction factor. In our work, we set r as 4. The brightened feature maps in the output indicate that the channels have a large weight.

Motivated by [19, 20], we assumed that depending on the feature maps such as X, Y-axis, queries, keys, and values, the channel on which the network should focus could be different. So, we redesign *squeeze-excitation* block in parallel as shown in Fig. 3. We divide the input vector into two vectors by using two 1×1 convolution layers. Each vector has half the channel size compared to the original input vector. Two channel attentions are calculated through two squeeze-excitation blocks and concatenated. We call this *decouple squeeze-excitation*, and Fig. 4 shows the effect of *decouple squeeze-excitation*.

As shown in Fig. 4, decouple squeeze-excitation can calculate the channel weights differently from the original *squeeze-excitation* for the same input. A comparison of the effects of *decouple squeeze-excitation* and *squeeze-excitation* will be covered in detail in the next session.

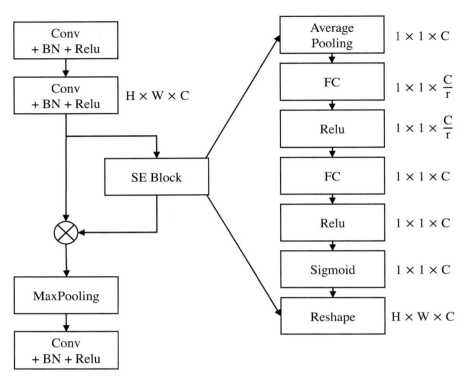

Fig. 1 The original form of *squeeze-excitation* block

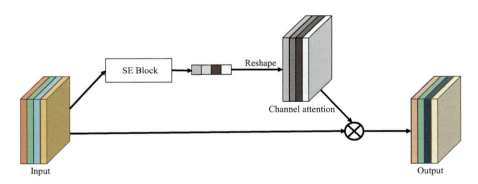

Fig. 2 The effect of *squeeze-excitation*

Attention Mechanism for Improving Facial Landmark Semantic Segmentation

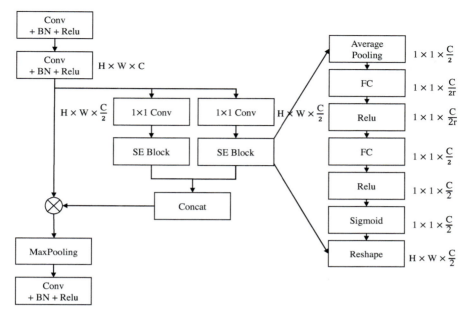

Fig. 3 Parallel *squeeze-excitation* block

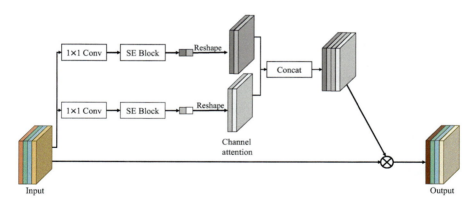

Fig. 4 The effect of *decouple squeeze-excitation*

4 Experiment

In the previous section, we proposed *decouple squeeze-excitation* for improving facial landmark semantic segmentation. To investigate the effect of our proposal, we construct a network with *decouple squeeze-excitation* and train it using a pixel-annotated facial landmark dataset. The dataset, which is the same one used in FLSNet [8], has a total of 59,428 facial images for training, and 5876 facial images for tests. The backbone architecture is VGG16 base SegNet [14]. Both encoder and decoder adopt *decouple squeeze-excitation* block. Also, in [8], we proposed three methods for improving landmark extraction performance. In particular, by using sub-network and weighted feature map, we could achieve remarkable performance improvement in facial landmark extraction. The weighted feature map is a post-processing step that does not increase the processing time or the number of parameters a lot. So, we used it in the experiment. For comparison, we also considered several popular semantic segmentation methods for facial landmark extraction. They are SegNet [14], FLSNet [6], and Deeplabv3plus [16]. We summarized the experimental result in Table 1. As shown in the table, facial landmark extraction performance is improved clearly when adopting *squeeze-excitation* block. In addition, *decouple squeeze-excitation* showed better performance than its original form. Finally, the best mIoU (mean Intersection over Union) was achieved when using the weighted feature map and *decouple squeeze-excitation* both.

5 Conclusion

In this chapter, we proposed an attention-based scheme for improving the performance of facial landmark extraction. Our key idea was to modify the original squeeze-excitation block into parallel structure. The experimental result showed that our proposed decouple squeeze-excitation is effective in facial landmark as it improved the extraction performance greatly when combined with other techniques.

Acknowledgment This work was supported by the National Research Foundation of Korea(NRF) grant funded by the Korea government(MSIT) (No. 2020R1F1A1074885).

Table 1 Facial landmark extraction performance in terms of mIoU

Methods			Classes									
Encoder	*Decoder*	*W. F*	*Hair*	*Skin*	*Eyebrow*	*Pupil*	*White*	*Nostril*	*Lip*	*Background*	*Inner mouth*	*mIoU*
Vgg16	SegNet	X	0.906	0.894	0.572	0.799	0.363	0.206	0.794	0.942	0.342	0.646
Vgg16	SegNet	O	0.864	0.921	0.659	0.862	0.637	0.524	0.831	0.924	0.727	0.772
Vgg16	FLSNet	O	0.902	0.932	0.801	**0.895**	**0.807**	0.712	**0.892**	0.956	0.807	0.856
I.R	DeepLabv3+	X	0.890	0.900	0.671	0.818	0.447	0.266	0.798	0.949	0.477	0.691
I.R	DeepLabv3+	O	0.848	0.917	0.709	0.873	0.678	0.505	0.799	0.936	0.815	0.787
Vgg16+SE	SegNet+SE	X	0.894	0.897	0.603	0.796	0.406	0.274	0.806	0.937	0.533	0.683
Vgg16+SE	SegNet+SE	O	0.880	0.916	0.698	0.866	0.661	0.737	0.809	0.916	0.696	0.798
Vgg16+DSE	SegNet+DSE	X	**0.934**	0.934	0.651	0.803	0.476	0.255	0.827	**0.974**	0.540	0.710
Vgg16+DSE	SegNet+DSE	O	0.927	**0.937**	**0.804**	0.882	0.756	**0.775**	0.869	0.946	**0.828**	**0.858**

W.F Weighted feature map, *SE* Squeeze-excitation, *DSE* Decouple squeeze-excitation, *I.R* Inception ResNetv2
Bold values indicate the maximum value for each row

References

1. R. Ranjan, V.M. Patel, R. Chellappa, Hyperface: A deep multi-task learning framework for face detection, landmark localization, pose estimation, and gender recognition. IEEE Trans. Pattern Anal. Mach. Intell. **41**(1), 121–135 (2017)
2. H. Kim, H. Kim, E. Hwang, Real-time facial feature extraction scheme using cascaded networks, in *2019 IEEE International Conference on Big Data and Smart Computing (BigComp)*, (IEEE, 2019), pp. 1–7
3. H.W. Kim, H.J. Kim, S. Rho, E. Hwang, Augmented EMTCNN: A fast and accurate facial landmark detection network. Appl. Sci. **7**, 2253 (2020)
4. H. Kim, J. Park, H. Kim, E. Hwang, Facial landmark extraction scheme based on semantic segmentation, in *2018 International Conference on Platform Technology and Service (PlatCon)*, (IEEE, 2018), pp. 1–6
5. H. Kim, J. Park, H. Kim, E. Hwang, S. Rho, Robust facial landmark extraction scheme using multiple convolutional neural networks. Multimed. Tools Appl. **78**(3), 3221–3238 (2019)
6. H. Kim, H. Kim, J. Rew, E. Hwang, FLSNet: Robust facial landmark semantic segmentation. IEEE Access **8**, 116163–116175 (2020)
7. J. Hu, L. Shen, G. Sun, Squeeze-and-excitation networks. Proc. IEEE Conf. Comput. Vis. Pattern Recognit. **42**(8), 2011–2023 (2018)
8. D.E. King, Dlib-ml: A machine learning toolkit. J. Mach. Learn. Res. **10**, 1755–1758 (2009). Fan, H.; Zhou, E. Approaching human level facial landmark localization by deep learning. Image Vis. Comput. 2016, 47, 27–35
9. M. Zhu, D. Shi, M. Zheng, M. Sadiq, Robust facial landmark detection via occlusion-adaptive deep networks, in *Proceedings of the IEEE Conference on Computer Vision and Pattern Recognition*, (2019), pp. 3486–3496
10. K. Zhang, Z. Zhang, Z. Li, Y. Qiao, Joint face detection and alignment using multitask cascaded convolutional networks. IEEE Signal Process. Lett. **23**(10), 1499–1503 (2016)
11. A.S. Jackson, M. Valstar, G. Tzimiropoulos, A CNN cascade for landmark guided semantic part segmentation, in *European Conference on Computer Vision*, (Springer, 2016), pp. 143–155
12. J. Long, E. Shelhamer, T. Darrell, Fully convolutional networks for semantic segmentation, in *Proceedings of the IEEE Conference on Computer Vision and Pattern Recognition*, (2015), pp. 3431–3440
13. O. Ronneberger, P. Fischer, T. Brox, U-net: Convolutional networks for biomedical image segmentation, in *International Conference on Medical Image Computing and Computer-Assisted Intervention*, (Springer, 2015), pp. 234–241
14. V. Badrinarayanan, A. Kendall, R. Cipolla, Segnet: A deep convolutional encoder-decoder architecture for image segmentation. IEEE Trans. Pattern Anal. Mach. Intell. **39**(12), 2481–2495 (2017)
15. L.-C. Chen, G. Papandreou, I. Kokkinos, K. Murphy, A.L. Yuille, Semantic image segmentation with deep convolutional nets and fully connected crfs, *arXiv preprint* arXiv:1412.7062 (2014)
16. L.-C. Chen, Y. Zhu, G. Papandreou, F. Schroff, H. Adam, Encoder-decoder with atrous separable convolution for semantic image segmentation, in *Proceedings of the European Conference on Computer Vision (ECCV)*, (2018), pp. 801–818
17. K. Xu et al., Show, attend and tell: Neural image caption generation with visual attention, in *International Conference on Machine Learning*, (2015)
18. S. Woo, J. Park, J.-Y. Lee, I.S. Kweon, Cbam: Convolutional block attention module, in *Proceedings of the European Conference on Computer Vision (ECCV)*, (2018)
19. P. Ramachandran, N. Parmar, A. Vaswani, I. Bello, A. Levskaya, J. Shlens, Stand-alone self-attention in vision models, arXiv preprint arXiv:1906.05909 (2019)
20. H. Wang, Y. Zhu, B. Green, H. Adam, A. Yuille, L.-C. Chen, Axial-DeepLab: Stand-alone axial-attention for panoptic segmentation, arXiv preprint arXiv:2003.07853, 2020Author, F.: Article title. Journal 2(5), 99–110 (2016)

Person Re-identification Scheme Using Cross-Input Neighborhood Differences

Hyeonwoo Kim, Hyungjoon Kim, Bumyeon Ko, and Eenjun Hwang

1 Introduction

Intelligent CCTV-based surveillance technology is becoming an essential element in smart cities because of its many possibilities. For instance, the technology can be used to find missing persons, track criminals, or detect accidents effectively. As the number of CCTV installed for security purposes has been increased explosively, an effective method to monitor numerous CCTVs is required.

Person re-identification is a technique to find an image in disjoint camera views that contains the previously detected pedestrian. This technique can also be used in copyright protection of commercial videos by detecting actors that appear in videos. Conventional methods for person re-identification used the similarity based on hand-crafted features and their performance heavily relies on lighting or camera angle. In recent years, as deep learning has shown overwhelming performance in the field of image processing [1–3], various attempts have been made for person re-identification using deep learning algorithms and shown good performance [4, 5].

Most deep learning-based person re-identification models used networks for image classification, such as ResNet [6], VGG [8], and Inception [7] to extract features from each image. As these image classification networks were designed to classify different types of objects, it may be difficult to tell whether the objects are the same or different. For this reason, a network was added to compare the features extracted by using the existing model [9, 10]. In [11], IDLA was proposed to construct feature maps using cross-input neighborhood differences. These feature

---

H. Kim · H. Kim · B. Ko · E. Hwang (✉)
School of Electrical Engineering, Korea University, Seoul, South Korea
e-mail: guihon12@korea.ac.kr; hyungjun89@korea.ac.kr; kby530@korea.ac.kr;
ehwang04@korea.ac.kr

© Springer Nature Switzerland AG 2021
H. R. Arabnia et al. (eds.), *Advances in Artificial Intelligence and Applied Cognitive Computing*, Transactions on Computational Science and Computational Intelligence, https://doi.org/10.1007/978-3-030-70296-0_61

maps were used for training the model, and determining whether the images are the same or different in the test. In this model, as only two images were used in determining whether they are the same or not, there is a limitation in that similarities and differences cannot be considered simultaneously when training the model.

On the contrary, in [12], they showed that a model trained by triplet loss [13] can improve the performance of person re-identification significantly. In the triplet loss function, the distance of positive pairs is added and the distance of negative pairs is subtracted. This trains the network to place images of the same person close together in a multi-dimensional space and the images of other people farther away. This shows that using both positive and negative pairs to train the network is effective for improving performance. Therefore, in this chapter, we proposed a model that uses three input images so that it can simultaneously consider positive and negative pairs. Also, cross-input neighborhood differences were calculated between features extracted from each image and used for model training.

This chapter is organized as follows: Section 2 describes the dataset used in the chapter, and Sect. 3 describes the overall content of our proposed person re-identification model. Experimental results for the proposed model are described in Sect. 4, and Sect. 5 concludes the chapter.

2 Dataset

In this study, the CUHK03 dataset [14] is used to train and evaluate the model. This dataset has two versions: one was made automatically through a pedestrian detector, and the other was created manually. In our study, we used the second version. In the dataset, 12,221 images of 1260 people were used for training and 943 images of 100 people were used for test. In addition, positive and negative pairs were sampled at the same rate from the dataset.

3 Person Re-identification Model

Figure 1 shows the structure of our proposed network. As shown in the figure, we use three images as input, including an anchor image. Since the anchor image is compared with two images, and it can be divided into four classes depending on whether they are the same. In this process, cross-input neighborhood differences are calculated as in Eqs. (1) and (2).

$$K_i (x, y) = f_i (x, y) \, \mathbb{1} (5, 5) - N \left[g_{i,n} (x, y) \right] \tag{1}$$

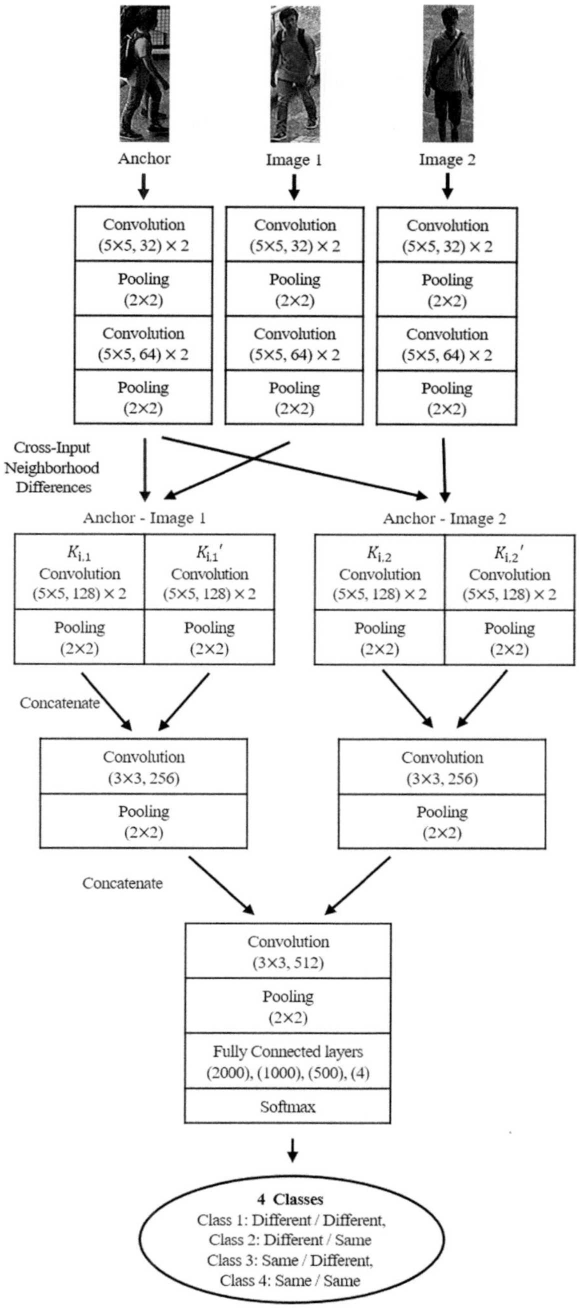

Person Re-identification Scheme Using Cross-Input Neighborhood Differences

Fig. 1 Network architecture

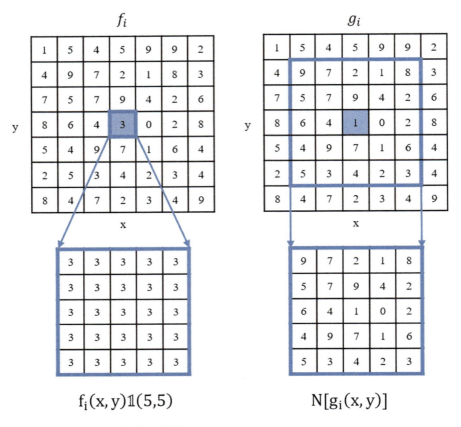

Fig. 2 Cross-input neighborhood differences

$$K_i'(x, y) = g_{i,n}(x, y) \mathbb{1}(5, 5) - N[f_i(x, y)] \quad (2)$$

f_i is a feature map extracted from an anchor image through a convolution layer, and $g_{i,1}$ and $g_{i,2}$ are feature maps extracted from comparison images (Image1/Image2). $\mathbb{1}(5, 5)$ is a 5 × 5 matrix consisting of 1, which is the value at (x, y) in the f_i, and $N[g_{i,n}(x, y)]$ is a matrix centered at (x, y) in the $g_{i,n}$ matrix. Figure 2 shows an example when cross-input neighborhood difference is applied to the feature maps. Each feature map is enlarged five times differently before calculating the difference. This allows the model to consider global and local differences between feature maps.

In this study, since we use three images as input, two pairs of feature maps are created as shown in the Fig. 1. These feature maps are reconstructed in the cross-input neighborhood differences layer through Equations (1) and (2). Then, these feature maps are extracted through the convolution layer, and finally, it is classified into four classes through the fully connected layer and Softmax.

4 Experimental Results

This section describes the experiments we performed to evaluate our scheme. We measured the rank accuracy of our model and compared it with IDLA. CUHK03 data is used for training our proposed model and the IDLA [11]. In the experiments, we used Intel (R) Core (TM) i7-8700 CPU, 32G DDR4 memory, NVIDIA Geforce GTX 1080ti, Python 3.5 environment.

We evaluated the performance by measuring the accuracy of Rank-1, Rank-5, and Rank-8. Rank refers to the score rank of the images recognized as the same person as the query image in the gallery set. Figure 3 shows an example of finding the same person as the query image in the gallery. Rank-1 is the ratio of the correct answer to the image with the highest similarity, and Rank-5 is the ratio with the correct answer to the top five images with high similarity scores.

Figure 4 shows the Rank-1, Rank-5, and Rank-8 accuracy of the two models, our proposed model, and IDLA. As shown in the figure, our proposed model has about a 10% improvement in Rank-1 accuracy than the previous model. In addition, it can be seen that the overall performance is improved, including Rank-1 accuracy.

Fig. 3 Result example

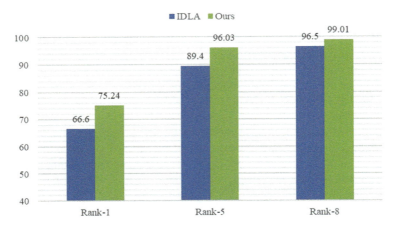

Fig. 4 Test result

5 Conclusion

In this chapter, we propose a person re-identification method of using three human images as input and reconstructing them into two pairs of feature maps, which enabled the model to learn while reflecting the differences and commonalities between the three images. To evaluate our method, we extend the existing person re-identification model, and the model was trained using the CUHK03 dataset. In addition, the accuracy of Rank-1, Rank-5, and Rank-8 was measured in a comparison experiment, and we confirmed that the overall performance of the proposed model was improved.

In this study, we conducted the experiments by extracting the features of the image using the shallow networks. In the next study, we plan to use a variety of person re-identification datasets for training the model with proven networks such as ResNet, VGG, and Inception.

Acknowledgement This work was supported by Electronics and Telecommunications Research Institute (ETRI) grant funded by the Korean government. [2018-micro-9500, Intelligent Micro-Identification Technology for Music and Video Monitoring].

References

1. H. Kim, J. Park, H. Kim, E. Hwang, S. Rho, Robust facial landmark extraction scheme using multiple convolutional neural networks. Multimed. Tools Appl. **78**(3), 3221–3238 (2019)
2. H. Kim, H. Kim, E. Hwang, Real-time facial feature extraction scheme using cascaded networks, in *2019 IEEE International Conference on Big Data and Smart Computing (BigComp), Kyoto, Japan*, (IEEE, 2019), pp. 1–7

Person Re-identification Scheme Using Cross-Input Neighborhood Differences 831

3. H.W. Kim, H.J. Kim, S. Rho, E. Hwang, Augmented EMTCNN: A fast and accurate facial landmark detection network. Appl. Sci. **7**, 2253 (2020)
4. H. Chen et al., Deep transfer learning for person re-identification, in *2018 IEEE International Conference on Multimedia Big Data (BigMM), Xi'an*, (2018), pp. 1–5. https://doi.org/10.1109/BigMM.2018.8499067
5. X. Bai, M. Yang, T. Huang, Z. Dou, R. Yu, Y. Xu, Deep-person: Learning discriminative deep features for person re-identification. Pattern Recogn. **98**, 107036 (2020)
6. K. He, X. Zhang, S. Ren, J. Sun, Deep residual learning for image recognition, in *2016 IEEE International Conference on Computer Vision and Pattern Recognition (CVPR)*, (2016), pp. 770–778
7. K. Simonyan, A. Zisserman, Very deep convolutional networks for large-scale image recognition, arXiv preprint arXiv:1409.1556 (2014)
8. C. Szegedy, V. Vanhoucke, S. Ioffe, J. Shlens, Z. Wojna, Rethinking the inception architecture for computer vision, in *2016 IEEE International Conference on Computer Vision and Pattern Recognition (CVPR)*, (2016), pp. 2818–2826
9. R. Quan, X. Dong, Y. Wu, L. Zhu, Y. Yang, Auto-ReID: Searching for a part-aware ConvNet for person re-identification, in *2019 IEEE International Conference on Computer Vision (ICCV)*, (2019), pp. 3750–3759
10. Y. Lin et al., Improving person re-identification by attribute and identity learning. Pattern Recogn. **95**, 151–161 (2019)
11. E. Ahmed, M. Jones, T.K. Marks, An improved deep learning architecture for person re-identification, in *2015 IEEE International Conference on Computer Vision and Pattern Recognition (CVPR)*, (2015), pp. 3908–3916
12. A. Hermans, L. Beyer, B. Leibe, In defense of the triplet loss for person re-identification, arXiv preprint arXiv:1703.07737 (2017)
13. F. Schroff, D. Kalenichenko, J. Philbin, FaceNet: A unified embedding for face recognition and clustering, in *2015 IEEE International Conference on Computer Vision and Pattern Recognition (CVPR)*, (2015), pp. 815–823
14. W. Li, R. Zhao, T. Xiao, X. Wang, DeepReID: Deep filter pairing neural network for person re-identification, in *2014 IEEE International Conference on Computer Vision and Pattern Recognition (CVPR)*, (2014), pp. 152–159

Variational AutoEncoder-Based Anomaly Detection Scheme for Load Forecasting

Sungwoo Park, Seungmin Jung, Eenjun Hwang, and Seungmin Rho

1 Introduction

With recent advances in MEMS (micro electro mechanical systems) technology, the size of sensors has been significantly reduced. As it has become easy to embed sensors into a variety of devices, diverse data such as temperature, humidity, and pressure can be collected continuously from diverse devices through the sensor [1].

Most data collected from such devices are time-series data as they are collected regularly or periodically. Many systems of smart cities also utilize time-series data for developing applications [2]. In particular, typical smart grid systems optimize energy operations by analyzing data collected from all processes of power utilization in the smart city [3]. A smart grid is an intelligent power grid that combines information and communication technologies with existing power grids. Through the smart grid, consumers can reduce electricity bills, and suppliers can optimize energy efficiency by solving the imbalances between demand and supply through real-time information sharing and control [4].

Anomaly detection is one of the essential technologies of the smart grid [5]. If the data is damaged during transmission or tampered with hacking, various problems such as malfunctioning and incorrect billing can occur. To detect anomalies, it is necessary to monitor the data collected by the smart meter in real-time and check for any abnormality.

S. Park · S. Jung · E. Hwang (✉)
School of Electrical Engineering, Korea University, Seoul, Republic of Korea
e-mail: psw5574@korea.ac.kr; jmkstcom@korea.ac.kr; ehwang04@korea.ac.kr

S. Rho
Department of Software, Sejong University, Seoul, Republic of Korea
e-mail: smrho@sejong.edu

© Springer Nature Switzerland AG 2021
H. R. Arabnia et al. (eds.), *Advances in Artificial Intelligence and Applied Cognitive Computing*, Transactions on Computational Science and Computational Intelligence, https://doi.org/10.1007/978-3-030-70296-0_62

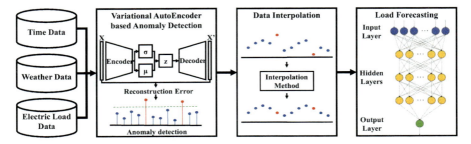

Fig. 1 Overall structure of our scheme

IQR (interquartile range) method is one of the most commonly used statistical methods for identifying abnormal data [6]. However, as high leverage anomalies can skew the median of the IQR, some types of anomalies may not be detected. If the normal value is replaced by another value due to incorrect classification and used in a predictive model, its predictive performance may degrade. To solve this problem, in this chapter, we propose VAE (Variational AutoEncoder)-based scheme for accurate anomaly detection. Figure 1 shows the overall structure of our scheme.

This chapter is organized as follows. In Sect. 2, we describe the input variable configuration. Section 3 explains the construction of the VAE-based anomaly detection model. Then, we explain the construction of the ANN (artificial neural network)-based load forecasting model in Sect. 4. In Sect. 5, we present some experiments to evaluate the performance of our scheme. Finally, Section 6 concludes the chapter.

2 Input Variable Configuration

In this study, we focused on detecting anomalies in electric load data. For electric load data, we used LBNL (Lawrence Berkeley National Laboratory) building's electric energy consumption data from January 1, 2014 to June 30, 2015. LBNL data includes missing data, and the data immediately after the missing data occurs has a slightly different pattern. We experimented with this part assuming abnormal data. Table 1 summarizes some statistics of the LBNL data.

In addition, we used time data and weather data which are obtained from NOAA (National Oceanic and Atmospheric Administration).

Time data consists of the month, day, hour, minute, and day of the week. It is difficult to reflect periodic information in machine learning algorithms when time data are in a sequential format. Therefore, we enhanced the time data to 2-dimensional data using Eqs. (1) and (2). Since using both 1-dimensional and 2-dimensional data can represent the time factor more effectively, we used both data [7]:

Table 1 Statistics of electric load data

Category	Value
Count	46,111
Missing data	6305
Maximum	224287.4
Minimum	11.5
Mean	36.74151
Standard error	1044.429
Median	29

$$time_x = \sin\left((360/cycle) \times time\right) \tag{1}$$

$$time_y = \cos\left((360/cycle) \times time\right) \tag{2}$$

In the equations, cycle represents the period of time data. For example, when time indicates month, the cycle becomes 12, and when time indicates hours to minutes, cycle represents 48.

Weather data consists of temperature, daily maximum temperature, and daily minimum temperature. Temperature is closely related to the operation of equipment with high power consumption, such as air conditioners and radiators [8]. Due to this, we used weather data as an input variable.

3 Anomaly Detection Model Configuration

In a typical AE (AutoEncoder), reconstruction errors are used as a loss function to obtain the same output data as the input data [9]. AEs are trained to minimize loss function values. Latent variables compressed by the encoder preserve the characteristics of the input data, so it is possible to measure the similarity between data or solve the classification problem using the latent variable.

If the latent variable generated through the encoder has a specific distribution, such as a normal distribution, the latent variable can be sampled from the normal distribution, and new data can be generated using a decoder. VAE is a model that follows a specific distribution using variance inference to train input data and generate new data through sampling from this distribution [10]. AE outputs the latent variables directly, but VAE outputs the mean and variance of the latent variables and then generates the latent variables through sampling. As VAE can learn the probability of latent variables, it is a suitable model for anomaly detection [11].

In general, AE-based anomaly detection uses reconstruction errors of the entire input variables. However, if the anomaly of the load data is detected using the reconstruction error of the entire input variables, the normal load data may be

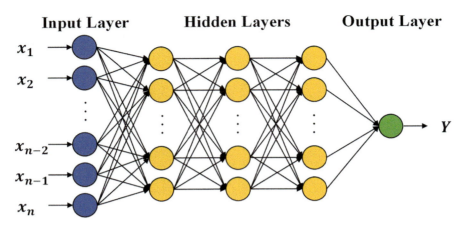

Fig. 2 Typical structure of ANN model

classified as abnormal data. So, after calculating the reconstruction error of each input variable in VAE, we only used the reconstruction error of the load data.

4 Forecasting Model Configuration

An ANN, which is also known as MLP (multilayer perceptron), is a machine learning algorithm that imitates the human brain and consists of three layers: an input layer, one or more hidden layers, and an output layer [12]. Each layer consists of several nodes called a perceptron. Each node receives values from the nodes on the previous layer, determines its output, and gives it to the nodes on the next layer. As this process is repeated, the nodes of the output layer give the values we want. Figure 2 represents a typical structure of an ANN model.

We constructed an ANN-based load forecasting model by using interpolated load data and the same input variables used in anomaly detection. In our model, we used the scaled exponential linear unit function as activation function, and set the number of hidden layers to five [13]. In addition, we set the number of neurons in the hidden layer to two-thirds of the number of input variables [14].

5 Experimental Results

For the experiments, we divided the dataset into a training set collected from January 1, 2014 to April 30, 2015 and test set collected from May 1, 2015, to June 30, 2015. Also, we used MAE (mean absolute error), RMSE (root mean square error), and RMSLE (root mean square logarithmic error) to evaluate the performance of the proposed scheme. MAE, RMSE, and RMSLE can be defined by using Eqs. (3), (4) (5), respectively:

$$MAE = \frac{\sum_{i=1}^{n} |y_i - x_i|}{n} \tag{3}$$

$$RMSE = \sqrt{\frac{\sum_{i=1}^{n} (y_i - x_i)^2}{n}} \tag{4}$$

$$RMSLE = \sqrt{\frac{\sum_{i=1}^{n} (\log (y_i + 1) - \log (x_i + 1))^2}{n}} \tag{5}$$

We constructed diverse ANN-based load forecasting models using different combinations of anomaly detection and data interpolation methods. We consider five different anomaly detection methods; IQR, isolation forest, k-means clustering, one-class support vector machine, and VAE. We also considered three different data interpolation methods; zero filling interpolation, linear interpolation, and random forest-based interpolation.

Table 2 MAE, RMSE, RMSLE comparison of load forecasting

Model		MAE	RMSE	RMSLE
Original		8.24	65.39	0.38
Interquartile range	Zero	6.49	43.65	0.28
	Linear	4.21	37.56	0.19
	Random forest	3.55	37.02	0.17
Isolation forest	Zero	7.01	57.81	0.32
	Linear	924.65	2939.82	3.41
	Random forest	121.30	10369.19	0.63
K means clustering	Zero	4.75	40.14	0.22
	Linear	11.01	51.13	0.35
	Random forest	6.45	40.07	0.24
One-class support vector machine	Zero	4.73	30.05	0.22
	Linear	11.02	51.15	0.35
	Random forest	6.39	40.04	0.23
Variational AutoEncoder	Zero	5.78	41.92	0.25
	Linear	3.73	37.20	0.18
	Random forest	**3.18**	**36.86**	**0.17**

Values in bold font indicate the best forecasting performance for each combination of anomaly detection method and data interpolation method

Table 2 shows the result. From the table, we can see that the ANN model that used interpolated data by VAE-based anomaly detection and random forest–based data interpolation gives the best performance in the three metrics. Overall, using interpolated data showed better performance than using original data.

Other machine learning–based anomaly detections using random forest–based data interpolation method showed worse performances than zero filling or linear interpolation. This is because random forest model was trained using data classified as normal data by anomaly detection technique. If the accuracy of the anomaly detection technique is low, the proportion of anomalies in the data increases; it reflects the characteristics of the anomaly when data is interpolated.

6 Conclusion

In this chapter, we proposed a VAE-based scheme for anomaly detection and applied the scheme for electric load forecasting. To do that, we constructed a dataset composed of time, weather data, and electric load data. To evaluate the performance of our scheme, we compared our scheme with other anomaly detection methods in terms of MAE, RMSE, and RMSLE. The experimental results showed that interpolating outliers detected by VAE for load forecasting gave the best performance.

Acknowledgement This research was supported in part by National Research Foundation of Korea(NRF) grant funded by the Korea government(MSIT) (grand number: 2019M3F2A1073179) and in part by the Korea Electric Power Corporation (grant number: R18XA05).

References

1. S. Jagannathan, Real-time big data analytics architecture for remote sensing application, in *2016 International Conference on Signal Processing, Communication, Power and Embedded System*, (2016), pp. 1912–1916
2. D. Kyriazis, T. Varvarigou, D. White, A. Rossi, J. Cooper, Sustainable smart city IoT applications: Heat and electricity management & Eco-conscious cruise control for public transportation, in *2013 IEEE 14th International Symposium on "A World of Wireless, Mobile and Multimedia Networks"*, (2013), pp. 1–5
3. J.S. Chou, N.T. Ngo, Smart grid data analytics framework for increasing energy savings in residential buildings. Autom. Constr. **72**, 247–257 (2016)
4. S. Saponara, R. Saletti, L. Mihet-Popa, Hybrid micro-grids exploiting renewables sources, battery energy storages, and bi-directional converters. Appl. Sci. **9**, 4973–4990 (2019)
5. M. Raciti, S. Nadjm-Tehrani, Embedded cyber-physical anomaly detection in smart meters, in *Critical information infrastructures security*, (2013), pp. 34–45
6. C. Wang, K. Viswanathan, L. Choudur, V. Talwar, W. Satterfield, K. Schwan, Statistical techniques for online anomaly detection in data centers, in *12th IFIP/IEEE International Symposium on Integrated Network Management and Workshops*, (2011), pp. 385–392

7. S. Park, J. Moon, S. Jung, S. Rho, S.W. Baik, E. Hwang, A two-stage industrial load forecasting scheme for day-ahead combined cooling, heating and power scheduling. Energies **13**, 443–465 (2020)
8. J. Moon, Y. Kim, M. Son, E. Hwang, Hybrid short-term load forecasting scheme using random forest and multilayer perceptron. Energies **11**, 3283–3302 (2018)
9. M.A. Kramer, Nonlinear principal component analysis using autoassociative neural networks. AICHE J. **37**, 233–243 (1991)
10. D.P. Kingma, M. Welling, Auto-encoding variational bayes. arXiv preprint arXiv:1312.6114 (2013)
11. J. An, S. Cho, Variational autoencoder based anomaly detection using reconstruction probability, in *Special Lecture on IE*, vol. 2, (2015)
12. Y. Liang, D. Niu, W.C. Hong, Short term load forecasting based on feature extraction and improved general regression neural network model. Energy **166**, 653–663 (2019)
13. J. Moon, S. Park, S. Rho, E. Hwang, A comparative analysis of artificial neural network architectures for building energy consumption forecasting. Int. J. Distrib. Sens. Netw. **15**(9), 155014771987761 (2019)
14. S. Karsoliya, Approximating number of hidden layer neurons in multiple hidden layer BPNN architecture. Int. J. Eng. Trends Technol. **3**(6), 714–717 (2012)

Prediction of Clinical Disease with AI-Based Multiclass Classification Using Naïve Bayes and Random Forest Classifier

V. Jackins, S. Vimal, M. Kaliappan, and Mi Young Lee

1 Introduction

Due to modern lifestyle, diseases are increasing rapidly. Our lifestyle and food habits lead to create impacts on our health causing heart diseases and other health issues. Data mining technique is one of the most challenging and leading research areas in healthcare due to the high importance of valuable data. It also accommodates the researchers in the field of healthcare in development of effective policies, and different systems to prevent different types of disease and early detection of diseases can reduce the risk factor. The aim of our work is to predict the diseases among the trained dataset using classification algorithms. It has been trained the Naive Bayes and Random Forest classifier model with three different disease datasets namely – diabetes, coronary heart disease and cancer datasets and performance of each model are calculated.

Data mining is a growing field that transforms piece of data into useful information. This technique help the authorized person make informed options and take right decisions for their betterment. It used to understand, predict, and guide future behavior based on the hidden patterns among huge datasets, Artificial Neural networks are the best effort classification algorithm for prediction of medical diagnosis due to its best efficiency parameter. The neural network comprises of the

---

V. Jackins
Department of IT, National Engineering College, Kovilpatti, Tamil Nadu, India

S. Vimal · M. Kaliappan
Department of Computer Science and Engineering, Ramco Institute of Technology, Rajapalayam, Tamil Nadu, India

M. Y. Lee (✉)
Department of Software, Sejong University, Seoul, South Korea
e-mail: miylee@sejong.ac.kr

© Springer Nature Switzerland AG 2021
H. R. Arabnia et al. (eds.), *Advances in Artificial Intelligence and Applied Cognitive Computing*, Transactions on Computational Science and Computational Intelligence, https://doi.org/10.1007/978-3-030-70296-0_63

neurons with three layers such as input layer, hidden layer, and output layer for the efficiency attainment.

2 Related Work

Variety of classification and clustering algorithms play a significant role for prediction and diagnosis of different types of diseases. Bayesian network classifier and Random Forest classifier are used to diagnose the risk for diabetes [1–3]. The prediction accuracy of the k-means algorithm is enhanced using both class and cluster method and making it adapt to different dataset [4]. A group of classification algorithms excluding Random Forest algorithm is applied on diabetes data to diagnose the risk. On comparing the performance of each method, the outcome shows that Random Forest was performed well in both accuracy and ROC Curve [5–7].

The Classification algorithm from various experts [8–11] Suggests that the data mining techniques such as k-means algorithm provide the Clinical observation in a clear way. Purushottama et al. [12] it helps a nonspecialized doctor can make the right decision about the heart disease risk level by generating original rules, pruned rules, classified rules, and sorted rules.

3 Proposed Work

The proposed method has been used with Anaconda tool (Jupyter Notebook IDE) for data analysis. Anaconda a package management system manages the package versions for predictive analysis and data management. It has been taken three disease patient data such as diabetes, coronary heart disease, and breast cancer data as input. These data are loaded and checked to see whether it has any missing values or not. If any missing values are found they are replaced to a null value. Then it has been checked whether any columns in the data have any correlation with another column in the data individually. If any correlation is found between two columns, one of those column is removed. If any true and false value is found in data, it is replaced with 1 and 0, respectively. It has been have split the original data into training data which has 70% of original data and test data which has 30% of the original data (Fig. 1).

It has been have taken few sample test data separately for each class data. Applying this sample data on each trained model of that disease shows us the results whether the data is identified with that disease or not. While comparing the results of both model for each class data, it has been can see that the model trained with Random Forest gives the accurate results of classification.

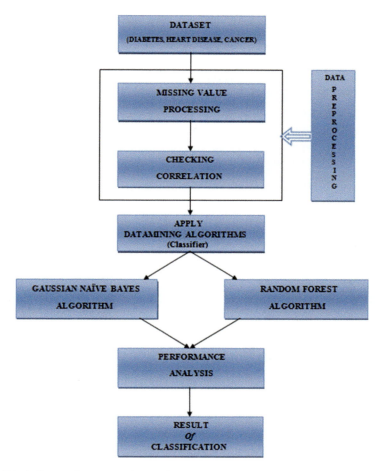

Fig. 1 Block diagram for proposed method

3.1 Dataset

Here, it has been have multiple disease data such as diabetes, coronary heart disease and breast cancer. The dataset has been collected using the wearable devices and the prediction data.

3.1.1 Coronary Heart Disease

This dataset is used by Framingham Heart study which includes several demographic risk factors:

- Age: Current age of the patient.

The dataset also includes behavioral risk factors associated with smoking:

- Smoking nature: The patient is a current smoker or not.

Medical history risk factors:

- BPMeds: Whether the patient was on blood pressure medication or not.
- PrevalentStroke: Whether the patient had a stroke previously or not.
- PrevalentHyp: hypertensive or not.
- Diabetes: Patient has diabetes or not.

Risk factors from the first physical examination of the patient:

- Cholrange: Total cholesterol level.
- BPs: Systolic blood pressure.
- diBl: Diastolic blood.
- BMI: Body mass index.
- HR: Heart rate.
- GL: Glucose level.
- CHDRISK: CHD coronary heart disease.

3.2 Filling in Missing Values

Missing value in any data means that the data was not available or not applicable or the event did not happen. Here, it has been replaced the missing values not available into null values.

3.3 Correlation Coefficient

Correlation coefficients are used in statistics to measure how strong a relationship is between two variables. It is the statistical measure of the linear relationship between a dependent variable and an independent variable. It is represented by a lowercase letter "r."

Here, the correlation between all the columns of the datasets is calculated to measure their relationship. The results give the correlation value of each column in a dataset against another column in that dataset. If two column in a dataset has same correlated values, then one among them is removed to avoid repetition of values.

In the below mentioned figure the Positive coefficients are indicated with the blue color and negative coefficients in red. The color intensity is found proportional to the blue and red indicted (Fig. 2).

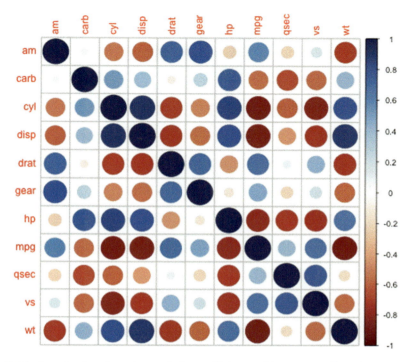

Fig. 2 The legend colors show the correlation coefficients and its corresponding colors

3.4 Algorithm

3.4.1 Naive Bayes Classification Algorithm

Here it has been used Bayes theorem for classification purpose and assumes that classification is predictor independent. It assumes that Naive Bayes classifier in the presence of a particular feature in a class is unrelated to any other feature.

Naive Bayes model is compatible for very large data sets to build and for further analysis. This model is very simple and sophisticated classification method it is performed well even in complicated scenarios. By using Bayes theorem, calculate the posterior probability using the equation below:

4 Results and Discussion

A confusion matrix having the information or data about actual and predicted classifications finished by a classification process. The performance is evaluated using the available data in the matrix. The confusion matrix for a two-class classifier is shown in the Table 1 (Fig. 3).

Table 1 Heart disease data

Sl. no	Age	Current smoker	BP meds	Prevalent stroke	Prevalent hyp	Diabetes	Chol range	BPs	diBl	BMI	HR	GL	CHD risk
0	39	0	0.0	0	0	0	195.0	106.0	70.0	26.97	80.0	77.0	0
1	46	0	0.0	0	0	0	250.0	121.0	81.0	28.73	95.0	76.0	0
2	48	1	0.0	0	0	0	245.0	127.5	80.0	25.34	75.0	70.0	0
3	61	1	0.0	0	1	0	225.0	150.0	95.0	28.58	65.0	103.0	1
4	46	1	0.0	0	0	0	285.0	130.0	84.0	23.10	85.0	85.0	0

Fig. 3 Confusion matrix

		Predicted Class	
		No	Yes
Observed Class	No	TN	FP
	Yes	FN	TP

- Always Positive (AP): The classification model correctly find class positively.
- Always Negative (AN): The negative class exactly labeled by the classifier.
- Always Least Positive (ALP): The classification model were incorrectly predicted and labeled as positive.
- Partial Least Negative (PLN):These are the positive classes that were incorrectly predicted as negative one.

Accuracy Calculation
The prediction accuracy is calculated using the formulae:

$$Accuracy = (AP + AN) / (M + N)$$

where M = AP + AN and N = ALP + AN. OR AP + AN/(TOTAL)

For heart disease data, Naive Bayes algorithm gives 82.44 and 82.35 accuracies for training and test data, respectively. Random Forest algorithm gives 97.96 and 83.85 for training and test data, respectively (Fig. 4).

5 Conclusion

Data mining can be effectively implemented in medical domain. The aim of this study is to discover a model for coronary heart disease, among the available dataset. The dataset is chosen from online repositories. The techniques of preprocessing applied are fill in missing values and removing correlated columns. Next, the classifier is applied to the preprocessed dataset then Bayesian and Random Forest models are constructed. Finally, the accuracy of the models is calculated and analyses based on the efficiency calculations. Bayesian classification network shows the accuracy of 82.35 for coronary heart disease. Similarly, classification with Random Forest model shows the accuracy of 83.85. The accuracy outcome of Random forest model for the three diseases is greater than the accuracy values of Naïve Bayes classifier. When performing classification in the trained model by applying sample test data of each disease, the Random forest model gives accurate results.

Our proposed methodology helps to improve the accuracy of diagnosis and greatly helpful for further treatment.

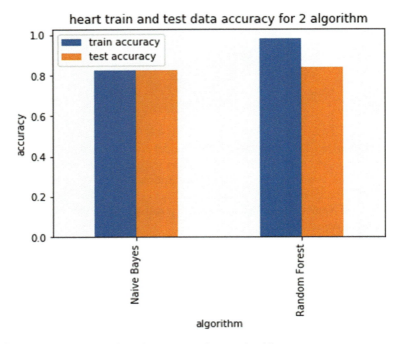

Fig. 4 Heart disease train and test data accuracy for two algorithms

Acknowledgments This research was supported by Basic Science Research Program through the National Research Foundation of Korea (NRF) funded by the Ministry of Education (2018R1D1A1B07043302).

References

1. M. Kumari, R. Vohra, A. Arora, Prediction of diabetes using Bayesian Network. Int. J. Comput. Sci. Inf. Technol (IJCSIT) **5**(4), 5174–5178 (2014)
2. W. Xu, J. Zhang, Q. Zhang, X. Wei, Risk prediction of type II diabetes based on random forest model, in *3rd International Conference on Advances in Electrical, Electronics, Information, Communication, and Bio-Informatics (AEEICB17)*, (2017)
3. B.J. Lee, J.Y. Kim, Identification of type 2 diabetes risk factors using phenotypes consisting of anthropometry and triglycerides based on machine learning. IEEE J. Biomed. Health Inform. **20**(1) (2016)
4. H. Wu, S. Yang, Z. Huang, J. He, X. Wang, Type 2 diabetes mellitus prediction model based on data mining. Inform. Med. Unlocked **10**, 100–107 (2018)
5. N. Nai-arna, R. Moungmaia, Comparison of classifiers for the risk of diabetes prediction, in *7th International Conference on Advances in Information Technology Procedia Computer Science*, vol. 69, (2015), pp. 132–142
6. C.M. Lynch, B. Abdollahi, J.D. Fuqua, A.R. de Carlo, J.A. Bartholomai, R.N. Balgemann, V.H. van Berkel, H.B. Frieboes, Prediction of lung cancer patient survival via supervised machine learning classification techniques. Int. J. Med. Inform. **108**, 1–8 (2017)

7. V.V. Veena, C. Anjali, Prediction and diagnosis of diabetes mellitus –A machine learning approach, in *2015 IEEE Recent Advances in Intelligent Computational Systems (RAICS)*, (10–12 December 2015)
8. S. Babu, E.M. Vivek, K.P. Famina, K. Fida, P. Aswathi, M. Shanid, M. Hena, Heart disease diagnosis using data mining technique, in *International Conference on Electronics, Communication, and Aerospace Technology (ICECA)*, (2017)
9. J. Singh, A. Kamra, H. Singh, Prediction of heart diseases using associative classification, in *2016 5th International Conference on Wireless Networks and Embedded Systems (WECON)*, (IEEE, 2016)
10. N.C. Long, P. Meesad, H. Unger, A highly accurate firefly-based algorithm for heart disease prediction. Expert Syst. Appl. **42**, 8221–8231 (2015)
11. A. Esteghamati, N. Hafezi-Nejad, A. Zandieh, S. Sheikhbahaei, M. Ebadi, M. Nakhjavani, Homocysteine and metabolic syndrome: From clustering to additional utility in prediction of coronary heart disease. J. Cardiol. **64**, 290–296 (2014)
12. S. Purushottama, K. Saxena, R. Sharma, Efficient heart disease prediction system. Procedia Comput. Sci. **85**, 962–969 (2016)

A Hybrid Deep Learning Approach for Detecting and Classifying Breast Cancer Using Mammogram Images

K. Lakshminarayanan, Y. Harold Robinson, S. Vimal, and Dongwann Kang

1 Introduction

Breast cancer causes the most common cancer type in women in the world, and this disease is dangerous whenever there is a delay in the diagnosis process. The malignant tumor expands rapidly and spreads the tumor in abnormal shapes [1]. Figure 1 demonstrates the malignant breast cancer.

Mammography is the enhanced technique for detecting the breast cancer very early that will reduce the energy for X-ray to diagnosis the breast cancer [2]. It is the way of digital X-ray image of the breast that is utilized for disease identification. The Computer-Aided Diagnosis techniques were constructed to assist the radiologists for enhancing the accuracy for breast cancer detection [3]. The feature extraction technique is used for evaluation and classification. The deep learning–based hybrid technique is constructed for breast cancer diagnosis system.

K. Lakshminarayanan
Electronics and Communication Engineering, SCAD College of Engineering and Technology, Tirunelveli, India

Y. H. Robinson
School of Information Technology and Engineering, Vellore Institute of Technology, Vellore, India

S. Vimal
Department of Computer Science and Engineering, Ramco Institute of Technology, Rajapalayam, Tamil Nadu, India

D. Kang (✉)
Department of Computer Science and Engineering, Seoul National University of Science and Technology, Seoul, South Korea
e-mail: dongwann@seoultech.ac.kr

© Springer Nature Switzerland AG 2021
H. R. Arabnia et al. (eds.), *Advances in Artificial Intelligence and Applied Cognitive Computing*, Transactions on Computational Science and Computational Intelligence, https://doi.org/10.1007/978-3-030-70296-0_64

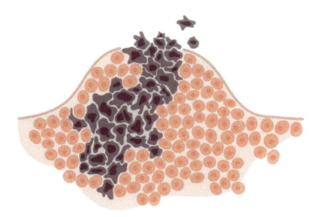

Fig. 1 Malignant breast cancer

2 Related Work

The classification of breast cancer from the input image has been completed using the k-means clustering technique that the feature extraction to apply the several parameters [4]. The Extreme Learning Machine Classifier [5] utilizes the classification by obtaining the specificity and sensitivity of the enhanced performance. The lesion annotations are utilized to implement the initial training time by performing the classification process [6]. The linear kernel function with the threshold value has provided the region-related segmentation process [7]. The machine learning algorithms have been utilized for providing the classification process from the image dataset to minimize the dimensionality feature for performance evaluation [8]. The uncertainty in the process of diagnosis has been eliminated with the median filtering and the histogram procedure [9]. The Feed Forward Neural Networks have been trained using the classifier to achieve the better sensitivity and specificity of the input image [10]. The MLP-based procedure [11] has been utilized to achieve the better breast cancer diagnosis. The principle component analysis [12] has been processed to identify the breast cancer from the dataset. The k-nearest algorithm [13] has implemented the classification procedure to diagnose the breast cancer from the ultrasound images. The efficient machine learning technique [14] and the enhanced supply vector machine [15] have been used for diagnosing the breast cancer more accurately.

3 Proposed Work

The novel hybrid technique proposed in this chapter joins the deep learning technique with the random forest classifier. The input image is preprocessed to eliminate the unwanted noise from the image and ROI is extracted with AlexNet.

A Hybrid Deep Learning Approach for Detecting and Classifying Breast...   853

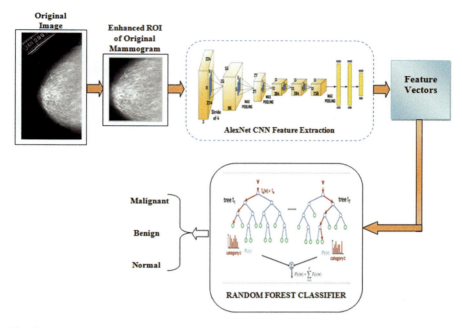

Fig. 2 Proposed system architecture

The feature vectors are generated for implementing the classification process, the random forest classifier is classifying the image and identifies the breast cancer type, and the whole process is demonstrated in Fig. 2.

The AlexNet CNN has been pretrained by the dataset to complete the classification process. The classification is more complicated for the medical images that the transfer learning procedure is used to provide the better solution for the deep learning–based issues. The dense part is eliminated from the image using the AlexNet and the random forest classifier to require the training process. The fully connected network is being trained according to the newly included layers. The final fully connected layer is eliminated from the CNN and to be replaced by the random forest classifier. The complexity has been minimized by the reduced amount of time and produces the enhanced accuracy. The AlexNet CNN architecture for feature extraction is demonstrated in Fig. 3. Here, the random forest classifier is included after the completion of the Max pooling layers in the convolutional neural network.

The ROI extraction from the input image after performing the preprocessing is illustrated in Fig. 4. The adaptive histogram equalization is generated to be capable of increasing the contrast value from the image for implementing the feature extraction procedure.

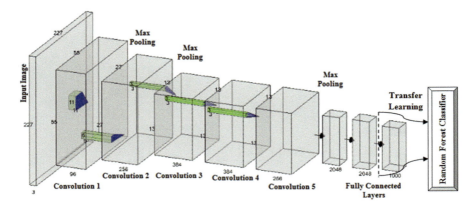

Fig. 3 AlexNet CNN architecture

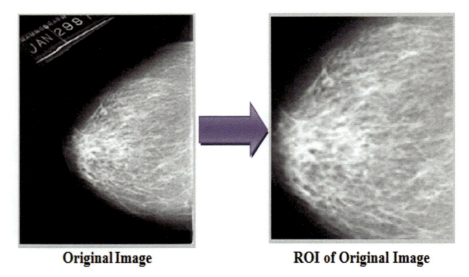

Fig. 4 ROI extraction

The random forest classifier is the supervisor learning–based technique to classify the images with the CNN-based classification process. The ensemble technique has been combined with the Bootstrap Aggregation mechanism to provide the sample replacement. The bagging technique is utilized to minimize the variance without reducing the bias value.

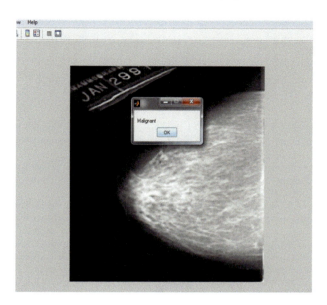

Fig. 5 Image classified as malignant

4 Performance Evaluation

The hybrid CNN–based random forest classifier is implemented in MATLAB to validate the dataset of large amount of images. The ROI as the input deep learning–based features are extracted with the AlexNet and the classification is completed using the random forest classifier to segregate the input images as the breast cancer types. The classification output is demonstrated in Fig. 5 as the input image has the malignant type of breast cancer. The confusion matrix is evaluated to specify the breast cancer classification that provides the details about the types of breast cancer in the actual stage and also the prediction stage.

5 Conclusion

In this chapter, the proposed hybrid deep learning technique has analyzed the root cause for the breast cancer of the women according to the mutations and also aging conditions. The proposed technique has classified the mammogram images and produced the efficiency from the pretrained CNN for feature extraction through the connected layer using the random forest classifier. This method has achieved the better accuracy compared with the related techniques. The experimental results show that the proposed technique providing the best solution for early detection of breast cancer.

Acknowledgments This work has supported by the National Research Foundation of Korea (NRF) grant funded by the Korea government (MSIT) (No. NRF-2019R1F1A1064205).

References

1. S. Khan, N. Islam, Z. Jan, I.U. Din, J.J. Rodrigues, A novel deep learning based framework for the detection and classification of breast cancer using transfer learning. Pattern Recognition Letters **125**, 1–6 (2019)
2. A.M. Solanke, R. Manjunath, D.V. Jadhav, A novel approach for classification of mammograms using longest line detection algorithm and decision tree classifier. Int. J. Innov. Technol. Explor. Eng. **8**(8) (2019)
3. A.S. Eltrass, M.S. Salama, Fully automated scheme for computer-aided detection and breast cancer diagnosis using digitised mammograms. IET Image Process. **14**(3), 495–505 (2020)
4. A. Gopi Kannan, T. Arul Raj, R. Balasubramanian, Comparative analysis of classification techniques in data mining on breast cancer dataset. Int. J. Inf. Comput. Sci. **9**(12) (2019)
5. C. Hemasundara Rao, P.V. Naganjaneyulu, K. Satyaprasad, Automated detection, segmentation and classification using deep learning methods for mammograms-a review. Int. J. Pure Appl. Math. **119**(16), 5209–5249 (2018)
6. H.D. Cheng, S. Juan, J. Wen, G. Yanhui, Z. Ling, Automated breast cancer detection and classification using ultrasound images: A survey. Pattern Recogn. **43**, 299–317 (2010). https://doi.org/10.1016/j.patcog.2009.05.012
7. D. Nagthane and A. Rajurkar, "A novel approach for mammogram classification," 2018 3rd IEEE International Conference on Recent Trends in Electronics, Information & Communication Technology (RTEICT), Bangalore, India, pp. 2613–2617
8. T.G. Debelee, A. Gebreselasie, F. Schwenker, M. Amirian, D. Yohannes, Classification of mammograms using texture and CNN based extracted features. J. Biomimetics Biomater. Biomed. Eng. **42**, 79–97 (2019)
9. H. Dhahri, E. Al Maghayreh, A. Mahmood, W. Elkilani, M. Faisal Nagi, Automated breast cancer diagnosis based on machine learning algorithms. J. Healthc. Eng. **2019**, 4253641 (2019). https://doi.org/10.1155/2019/4253641
10. E. Karthikeyan, S. Venkatakrishnan, Breast cancer classification using SVM classifier. Int. J. Recent Technol. Eng. **8**(4) (2019)., ISSN: 2277–3878
11. F.J.M. Shamrat, M.A. Raihan, A.S. Rahman, I. Mahmud, R. Akter, An analysis on breast disease prediction using machine learning approaches. Int. J. Sci. Technol. Res. **9**(02) (2020)
12. S. Gardezi, A. Elazab, B. Lei, T. Wang, Breast cancer detection and diagnosis using mammographic data: Systematic review. J. Med. Internet Res. **21**(7), e14464 (2019). https://doi.org/10.2196/14464
13. P. Giri, K. Saravanakumar, Breast cancer detection using image processing techniques. Orient. J. Comp. Sci. Technol **10**(2)
14. R. Guzmán-Cabrera, J.R. Guzmán-Sepúlveda, M. Torres-Cisneros, D.A. May-Arrioja, J. Ruiz-Pinales, O.G. Ibarra-Manzano, G. Aviña-Cervantes, A.G. Parada, Digital image processing technique for breast cancer detection. Int. J. Thermophys. **34**, 1519 (2013)
15. H. Lu, E. Loh, S. Huang, The classification of mammogram using convolutional neural network with specific image preprocessing for breast cancer detection, in *2019 2nd International Conference on Artificial Intelligence and Big Data (ICAIBD)*, (Chengdu, China, 2019), pp. 9–12

Food-Type Recognition and Estimation of Calories Using Neural Network

R. Dinesh Kumar, E. Golden Julie, Y. Harold Robinson, and Sanghyun Seo

1 Introduction

In recent years, the numbers of diabetic patients are increasing due to lack of knowledge in diet control. The most important cause of diabetes is the less insulin production in the human body. This leads to many of the diseases like continuous urination, increased blood pressure, loss of weight, and sometimes cardiac arrest. These enable us to develop the system to educate the people to have proper healthy foods every day. The proper system is built in MATLAB to food-type recognition and estimate food calories consumed.

R. D. Kumar
CSE Department, Siddhartha Institute of Technology and Science, Hyderabad, India

E. G. Julie
Department of Computer Science and Engineering, Anna University Regional Campus, Tirunelveli, India

Y. H. Robinson
School of Information Technology and Engineering, Vellore Institute of Technology, Vellore, India

S. Seo (✉)
School of Computer Art, College of Art and Technology, Chung-Ang University, Anseong-si, Kyunggi-do, South Korea
e-mail: sanghyun@cau.ac.kr

© Springer Nature Switzerland AG 2021
H. R. Arabnia et al. (eds.), *Advances in Artificial Intelligence and Applied Cognitive Computing*, Transactions on Computational Science and Computational Intelligence, https://doi.org/10.1007/978-3-030-70296-0_65

2 Related Work

The clustering techniques have been proposed to identify the food types. To segment the food item, affinity propagation and unsupervised clustering method have been adopted. The affinity propagation with agglomerative hierarchical clustering (AHC) obtains 95% of accuracy. The Monitoring of Ingestive Behavior model has built to monitor and estimate the calories [1]. The input food images have been taken from smartphones with single and mixed food items and fed into training and testing. The preprocessing steps are carried first, followed by vision-based segmentation done, and deep learning algorithm applied to estimate calories [2]. In the input food image, chopstick is used as a reference for measurement. The density-based database of the food is considered to evaluate the food volume, weight, and calorie estimation. The estimated weight and calories' relative average error rate are 6.65% and 6.70% [3]. The two datasets have been collected and trained with single-task and multitask CNN. The above multitask CNN classifiers achieved good results in food identification and calorie estimation than single-task CNN [4]. The top 10 Thai curries are considered. The segmented image is fed into the fuzzy logic to identify the ingredients based on their intensities and boundaries. The calories are calculated by sum of all the ingredient calories [5].

3 Proposed Work

The proposed technique has been divided into two phases as the training phase and the testing phase. The input image is resized using the scaling technique. Feature extraction consists of the SIFT method, Gabor Filter, and color histogram. The feature extraction is converted into the classification that implements the segmented process and MLP, and these processes are implemented in testing phase. After the classification procedure, the total area computation and the volume measurements are identified for producing the calorie estimate. The entire process is demonstrated in Fig. 1.

3.1 Algorithm – Multilayer Perceptron Neural Networks

Initialize all neurons (v_{ih}) and weights (w_{hj}.)
   Run=1
   While (Run)
   Do{
   Store all w and v.
   Epoch=epoch+1 /*training*/
   For all $(x^t, r^t) \in X_{training}$ in random order.

Food-Type Recognition and Estimation of Calories Using Neural Network

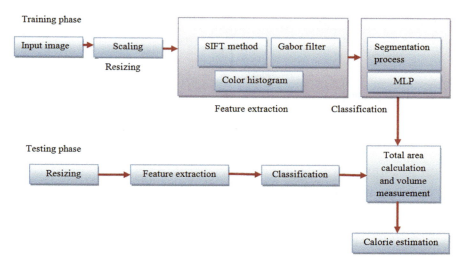

Fig. 1 Proposed system architecture

For all hidden nodes z_h,

$$z_h = \text{sigmoid}\left(\sum_{j=0}^{d} w_{hj} x_j^t\right)$$

For all output nodes y_i

$$y_i = \text{softmax}\left(\sum_{h=0}^{H} v_{ih} z_h\right)$$

For all weights v_{ih}

$$\Delta v_{ih} = \eta \left(r_i^t - y_i^t\right) z_h$$

$$v_{ih} = v_{ih} + \Delta v_{ih}$$

For all weights w_{ih}

$$\Delta w_{ih} = \eta \left(\sum_{i=1}^{K} \left(r_i^t - y_i^t\right) v_{ih}\right) z_h (1 - z_h) x_j^t$$

$$w_{ih} = w_{ih} + \Delta w_{ih}$$

For all $(x^t, r^t) \in X_{\text{Validating}}$

For all hidden nodes z_h,

$$z_h = \text{sigmoid}\left(\sum\nolimits_{j=0}^{d} w_{hj} x_j^t\right)$$

For all output nodes y_i

$$y_i = \text{softmax}\left(\sum\nolimits_{h=0}^{H} v_{ih} z_h\right)$$

$$\text{err (epoch)} = \frac{1}{2}\sum\nolimits_{x \epsilon X_{\text{Validating}}}\left(\sum\nolimits_{i=1}^{K}\left(r_i^t - y_i^t\right)^2\right)$$

If err(epoch)>err (epoch-1)
Run=0
}
For all $(x^t, r^t) \in X_{\text{testing}}$
For all hidden nodes z_h,

$$z_h = \text{sigmoid}\left(\sum\nolimits_{j=0}^{d} w_{hj} x_j^t\right)$$

For all output nodes y_i

$$y_i = \text{softmax}\left(\sum\nolimits_{h=0}^{H} v_{ih} z_h\right)$$

If y==r
"Successfully recognized"
Else
"Failed to recognize"

3.2 Multilayer Perceptron Neural Network

Multilayer perceptron neural network working principle is relatively based on the human brain and belongs to feed forward NN. Normally a human brain stores the information as the pattern and gains the knowledge to solve the complex problems by experience and it is demonstrated in Fig. 2.

The multilayer perceptron neural network recognizes the patterns by supervised training algorithm fed forward from input to output layers. Activation function is computed as given in Eq. (1).

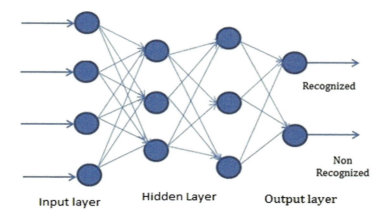

Fig. 2 Multilayer perceptron neural network

$$x_i^{m+1}(n) = f\left[\sum_{j=1}^{M} W_{ij}^m x_j(n)\right] \quad (1)$$

4 Performance Evaluation

The proposed work is carried out in MATLAB with six food classes. Here, precision and recall are identified for different food classes for SVM and MLP based classifier to have better results. The accuracy is tested using F-measure. The F-measure is calculated by taking the precision and recall values of each class and is shown in Table 1.

5 Conclusion

In this proposed work, identification of food type is made and estimation of calorie is done using MLP and the results proposed. Single food item types were considered previously, but here mixed food item types are taken to have better results. The implementation is processed in MATLAB with 1000 fruit images containing 6 food classes with good accuracy. The automatic dietary control is made available for diabetic patients.

Acknowledgments This work has supported by the National Research Foundation of Korea (NRF) grant funded by the Korean government (MSIT) (No. NRF-2019R1F1A1058715).

Table 1 Performance comparison

Class	Precision SVM	Precision MLP	Recall SVM	Recall MLP	SVM = 2*(p*r/p+r)	MLP = 2*(p*r/p+r)
Apple	0.7	0.8	0.7	0.8	0.70	0.80
Banana	0.8	0.9	0.8	0.9	0.80	0.90
Bread	0.8	1	0.9	1	0.85	1
Guava	0.6	0.7	0.6	0.6	0.60	0.65
Pizza	0.7	0.9	0.9	0.9	0.79	0.90
Pomegranate	0.6	0.8	0.6	0.7	0.60	0.75

References

1. P. Lopez-Meyer, S. Schuckers, O. Makeyev, J.M. Fontana, E. Sazonov, Automatic identification of the number of food items in a meal using clustering techniques based on the monitoring of swallowing and chewing. Biomed. Signal. Process. Control 7(5), 474–480 (2012)
2. P. Pouladzadeh, A. Yassine, FooDD: Food detection dataset for calorie measurement using food images. Lect. Notes Comput. Sci, 550–554 (2015)
3. E.A.H. Akpa, H. Suwa, Y. Arakawa, K. Yasumoto, Smartphone-based food weight and calorie estimation method for effective food journaling. SICE J. Control Meas. Syst. Integr. **10**(5), 360–369 (2017)
4. Takumi Ege, Keiji Yanai, Image-Based Food Calorie Estimation Using Knowledge on Food Categories, Ingredients and Cooking Directions, Association for Computing Machinery, 2017, pp. 367–375
5. S. Turmchokkasam, K. Chamnongthai, The design and implementation of an ingredient based food calorie estimation system using nutrition knowledge and fusion of brightness and heat information. IEEE Access **6**, 46863–46876 (2018)

Progression Detection of Glaucoma Using K-means and GLCM Algorithm

S. Vimal, Y. Harold Robinson, M. Kaliappan, K. Vijayalakshmi, and Sanghyun Seo

1 Introduction

The natural characteristic eye is an organ of vision that expects a noteworthy activity in life just as the human body. Various basic infections may impact the retina provoking vision mishap. The symptoms of by far most of the eye diseases are not adequately evident until it makes a huge impact on vision. Glaucoma could be a tireless eye malady that prompts vision loss. Perceiving the disease in time is critical. Glaucoma is an eye disease that hurts the optic nerve of the eye and gets the chance to be not kidding over some time [1]. It is caused by the development of pressure inside the eye. Glaucoma will, in general, be gained also and may not show up until a short time later throughout everyday life. The revelation of glaucomatous development is one of the principal, basic, and most testing view focuses to maintain a strategic distance from vision incident. Glaucoma is an eye illness of the common nerve of a vision, called the optic nerve, and it is frequently related with raised intraocular weight, in which harm to the optic nerve is dynamic over a long period and leads to the misfortune of vision [2]. Glaucoma could be an infection of the eye

---

S. Vimal · M. Kaliappan · K. Vijayalakshmi
Department of Computer Science and Engineering, Ramco Institute of Technology, Rajapalayam, Tamil Nadu, India
e-mail: vijayalakshmik@ritrjpm.ac.in

Y. H. Robinson
School of Information Technology and Engineering, Vellore Institute of Technology, Vellore, India

S. Seo (✉)
School of Computer Art, College of Art and Technology, Chung-Ang University, Seoul, South Korea
e-mail: sanghyun@cau.ac.kr

© Springer Nature Switzerland AG 2021
H. R. Arabnia et al. (eds.), *Advances in Artificial Intelligence and Applied Cognitive Computing*, Transactions on Computational Science and Computational Intelligence, https://doi.org/10.1007/978-3-030-70296-0_66

in which liquid weight inside the eye rises if cleared out untreated the understanding may lose vision, and indeed gotten to be daze. The illness, by and large, affects both eyes, even though one may have more extreme signs and indications than the other. There are two sorts of glaucoma:

(i) Open-angle glaucoma: The passages to the eye's waste canals are clear, but a blockage creates inside the canal, catching liquid and causing an increment in weight within the eye. Vision misfortune is as a rule moderate and progressive.
(ii) Angle-closure glaucoma: The entrance to the canal is either well limit or is closed totally. Weight can rise exceptionally rapidly. The known tests to distinguish glaucoma are tonometry (internal eye weight), ophthalmoscopy (structure and color of the optic nerve), and perimetry (total field of vision) [3].

2 Related Work

A method that proposed that the wavelet includes extraction has been trailed by streamlined hereditary component choice joined with a few learning calculations and different parameter settings. Not at all like the current research works where the highlights are considered from the total fundus or a sub-picture of the fundus, this work depends on include extraction from the divided and vein evacuated optic plate to improve the precision of ID [4]. The exploratory outcomes introduced right now that the wavelet highlights of the sectioned optic plate picture are clinically progressively noteworthy in contrast with highlights of the entire or sub-fundus picture in the discovery of glaucoma from fundus picture [5]. The precision of glaucoma recognizable proof accomplished right now 94.7% and a correlation with existing techniques for glaucoma recognition from fundus picture show that the proposed approach has improved exactness of arrangement. The programmed OC and OD division utilizing a novel strategy has been implemented to depend on CNN [6]. The proposed strategy utilized entropy inspecting for choosing to examine focuses that are professed to be superior to uniform testing. They chose to examine focuses that are additionally used to plan a learning arrangement of convolutional channels. The separated OC and OD can be utilized for CDR count which can additionally be utilized for glaucoma conclusion [7]. The implemented method that the automated analyze the referable diabetic retinopathy by characterizing shading retinal fundus photos into two grades. An epic convolutional neural system model with Siamese-like engineering is prepared with an exchange learning strategy. Unique about the past works, the proposed model acknowledges binocular fundus pictures as information sources and learns their connection to assist with making a forecast and proposed to consequently order the fundus photos into 2 kinds: with or without RDR [8].

3 Proposed Work

3.1 Preprocessing

Image pixel analysis and preprocessing step glaucoma diagnosis, in around 10% of the retinal pictures, ancient rarities are sufficiently huge to hinder human 38 reviewing. Preprocessing of such pictures can guarantee a satisfactory degree of achievement in the mechanized variation from the norm identification. In the retinal pictures, there can be varieties brought about by the variables remembering contrasts for cameras, brightening, procurement edge, and retinal pigmentation. The preprocessing step evacuates varieties because of picture securing, for example, inhomogeneous brightening. Changing conditions during picture catch, noise, uneven brightening, and 21 complexity varieties are the additional difficulties of mechanized optic circle restriction. To deal with these pictures self-rulingly, preprocessing must be applied. The accompanying explanation expounds on the preprocessing techniques embraced for our work. Grayscale is an information grid whose qualities speak to powers inside some range. MATLAB stores a grayscale picture as an individual grid, with every component of the matrix corresponding to one picture pixel. For changing over RGB picture to grayscale, it demonstrates the pixel brightness (Fig. 1):

$$Inrgb = imread\ ([pathname\ filename])\ ;$$
$$X = rgb2gray\ (Inrgb)$$

3.2 Detection of MA

Microaneurysms (MAs) are typically the main side effect of DR that causes blood spillage to the retina. This injury typically shows up as little red round spots with a width of fewer than 125 micrometers. Since microaneurysms (MAs) are one of the primary side effects of the sickness, recognizing this difficulty inside the fundus pictures encourages early DR location. Right now, the programmed examination of retinal pictures utilizing a convolutional neural network (CNN) is presented. It demonstrates noteworthy improvement in MA identification utilizing retinal fundus pictures for observing diabetic retinopathy.

$$i1 = Inrgb\ (:,:,2)\ ;$$
$$im = medfilt2(i1);$$
$$isub = imsubtract\ (im, i1)\ ;$$

It performs the median filtering of the picture i1 in two measurements that each outcome pixel contains the middle an incentive in a 3-by-3 neighborhood around the comparing pixel in the given picture.

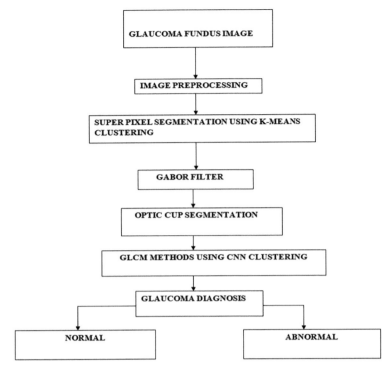

Fig. 1 Proposed technique

4 Performance Evaluation Using K-means Algorithm

K-means algorithm is an unaided grouping calculation which arranges the information points into numerous elements dependent on innate intervals from one another. The calculation accepts that the information highlights structure a vector space and attempts to discover common grouping in them. K-means calculation is a grouping algorithm that arranges the information of data points into various classes dependent on their characteristic intervals from one another. This technique for segmentation is to get an exact limit depiction. In optic disc and cup division, histograms and focus surround statistical are utilized to characterize every superpixel as disc or non-disc; the area data is likewise included in the component space to support the presentation. The calculation accepts that the information highlights structure a vector space and attempts to discover common grouping in them. The points are bunched around centroid $i = 1 \ldots k$ which are acquired by limiting the goal:

$$\sum_{J=1}^{K} \sum_{l=1}^{X} ||Xi(j) - cj||^2$$

Progression Detection of Glaucoma Using K-means and GLCM Algorithm

where $||Xi(j) - cj||^2$ is a picked interval (length) measure.

- Calculate the intensity distribution (moreover it's also called a histogram) of the intensities.
- Initialize the centroid with k irregular intensity.
- Repeating the accompanying strides until the grouping marks of the picture don't change any longer.
- Group the points dependent on the separation of their intensity from centroid powers imitated with the mean an incentive inside every one of the clusters, and afterward the separation lattice for the distance matrix is determined.
- Calculate the new centroid for each of the clusters.

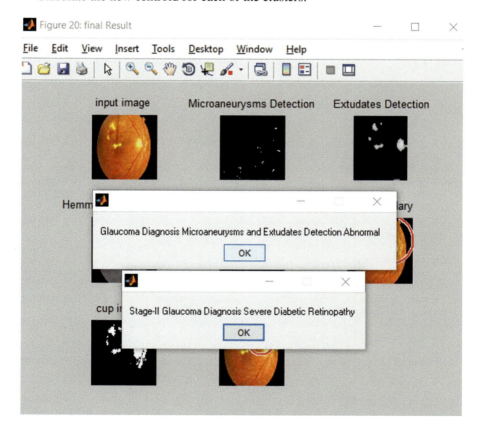

5 Conclusion

Glaucoma is extraordinarily risky as it causes perpetual visual impairment and shows up at the last stage. Sadly, there is no known remedy for glaucoma at cutting edge stages. Customary screenings and glaucoma recognition at beginning periods can spare vision. For the most part, existing mechanized glaucoma identification frameworks use fundus pictures and CDR to distinguish glaucoma at later stages. Computerized glaucoma recognition utilizing fundus pictures at the last stages is at a full-grown level yet tragically neglects to perceive glaucoma at beginning times. The manual screen for glaucoma is testing and sensitive and relies upon the capability of experts.

Acknowledgments This work was supported by the National Research Foundation of Korea (NRF) grant funded by the Korean government (MSIT) (No. NRF-2019R1F1A1058715).

References

1. S. Roychowdhury, D.D. Koozekanani, K.K. Parhi, DREAM: diabetic retinopathy analysis using machine learning. IEEE J. Biomed. Health Inform. **18**, 1717–1728 (2014)
2. V. Gulshan, L. Peng, M. Coram, M.C. Stumpe, D. Wu, A. Narayanaswamy, S. Venugopalan, K. Widner, T. Madams, J. Cuadros, R. Kim, Development and validation of a deep learning algorithm for detection of diabetic retinopathy in retinal fundus photographs. JAMA **316**(22), 2402–2410 (2016)
3. R. Gargeya, T. Leng, Automated identification of diabetic retinopathy using deep learning. Ophthalmology **124**(7), 962–969 (2017)
4. H. Li, K.-M. Lam, M. Wang, Image super-resolution via feature-augmented random forest. Signal Process. Image Commun. **72**, 25–34 (2019)
5. D.C.G. Pedronette, Y. Weng, A. Baldassin, C. Hou, Semi-supervised and active learning through Manifold Reciprocal kNN Graph for image retrieval. Neurocomputing **340**, 19–31 (2019)
6. F. Huazhu, J. Cheng, Joint optic disc and cup segmentation based on multi-label deep network and polar transformation. IEEE Trans. Med. Imaging **30**, 7 (2018)
7. J. Cheng, Superpixel classification based optic disc and optic cup segmentation for glaucoma screening. IEEE Trans. Med. Imaging **32**(6), 1019–1032 (2013)
8. G.D. Joshi, J. Sivaswamy, S.R. Krishnadas, Optic disk and cup segmentation from monocular color retinal images for glaucoma assessment. IEEE Trans. Med. Imaging **30**(6), 1192–1205 (2011)

Trend Analysis Using Agglomerative Hierarchical Clustering Approach for Time Series Big Data

P. Subbulakshmi, S. Vimal, M. Kaliappan, Y. Harold Robinson, and Mucheol Kim

1 Introduction

Road traffic accident (RTA) is an important factor to consider in research as it contains many factors and injuries lead to disability and other personal affections [1]. A report generated by World Health Organization shows that about 2 million accidents are happening throughout worldwide and it may lead to various segment of road safety and injuries [2]. The main focus on road accident is to ensure safety in road and emergency reporting [3]. The time sequence data analysis helps in encouraging the user to predict the emergency service, and accident rate can be predominantly reduced in there forth [4]. Data mining technique has been applied to statistically survey the accident rate [5]. The challenges of big data are capture, storage, search, sharing, transfer, analysis, and visualization [6]. These data are in various formats such as structured, unstructured, and semi-structured [7]. These data

---

P. Subbulakshmi
School of Computing, Scope, VIT University, Chennai, India

S. Vimal
Department of Computer Science and Engineering, Ramco Institute of Technology, Rajapalayam, Tamil Nadu, India

M. Kaliappan
Department of Computer Science and Engineering, Ramco Institute of Technology, Rajapalayam, Tamil Nadu, India

Y. H. Robinson
School of Information Technology and Engineering, Vellore Institute of Technology, Vellore, India

M. Kim (✉)
School of Computer Science & Engineering, Chung-Ang University, Dongjak-gu, Seoul, South Korea

© Springer Nature Switzerland AG 2021
H. R. Arabnia et al. (eds.), *Advances in Artificial Intelligence and Applied Cognitive Computing*, Transactions on Computational Science and Computational Intelligence, https://doi.org/10.1007/978-3-030-70296-0_67

are not handled by the traditional data processing tools [8]. In this paper, we take the huge amount of data generated in the road accidents for finding the factors causing them. This can be handled by the clustering and association rule mining techniques [9]. The trend analysis in PTS is more helpful for finding the factors of the accident since accident ratio increases every year. This can be efficiently identified with the time sequence data in every location. These data are needed to be normalized and then used for the processing with the help of the data mining techniques [10].

2 Related Work

Time series performs a group of related data points gathered within the specified interval [11]. Monthly basis road accident is measured to store the time series data for revealing the future trend. This will help to identify the dissimilar regions of road accidents for providing the trend analysis [12]. This analysis contains 11 leading cities in India for using the dataset. It is very hard to analyze the time series data of the leading cities independently and also relate to the nature and trend of the road accidents in the similar other cities in India [13].

3 Proposed Work

3.1 Data Preprocessing

Data preprocessing is a prior task is to analyze the data in every data handling techniques. The techniques of data preprocessing remove the unwanted noise or other constraints in the network. Here, this time sequence data was preprocessed, and an analysis has been normalized [14]. The data transformations have been performed to implement data available for the time series-related analysis. The time series information needs the enhanced data preprocessing in spite of getting the data useful for the specific analysis. The time series information uses the normalization technique for assisting the difficulties in the preprocessing technique. Hence, the analysis is used to get the efficient result for normalizing the time series information based on the time series.

3.2 Similarity Measure for Time Sequence

Dynamic time warping (DTW) measures the correlation between two time sequence data objects even if their lengths are not the same. DTW is used to minimize the metrics of two time sequences $r_i = \{r_1, r_2, \ldots, r_n\}$ and $f_i = \{f_1, f_2, \ldots, f_m\}$ which are of length x and y, respectively, and have to align r_i and t_j. The dynamic programming

Trend Analysis Using Agglomerative Hierarchical Clustering Approach for... 871

establishes a cluster approach using the infinity matrix, and the parameters are computed in Eqs. (1) and (2):

$$\text{mean}\,(f_i) = 0 \tag{1}$$

$$\text{SD}\,(f_i) = 1 \tag{2}$$

where mean (f_i) is the mean value and SD(f_i) is the standard deviation for producing the normalization time series. The normalization is computed in Eq. (3).

$$Nor_{f_i} = \sum_{t=t}^{n} \frac{time_i - mean(f_i)}{SD(F_i)} \tag{3}$$

3.3 Hierarchy-Based Cluster Analysis

Cluster analysis is done using the similar objects to be grouped in a single forum or groups based on their attributes and properties. A lot of clustering algorithms such as agglomerative clustering are used in statistical data prediction. There may be a variety of algorithms for clustering a time sequence data. Agglomerative hierarchical clustering algorithm is designed to map reduce framework for clustering of time sequence data. The space and time complexity using agglomerative hierarchical clustering is $P(n_3)$, and the other one is $P(2n)$. The proposed AHCTB is used to predict the road accident data as shown in Fig. 1.

Algorithm for simple hierarchical agglomerative clustering

```
Input: Preprocessed Road Accident dataset
Output: PTS for each cluster
Initialization:
    Set line, month, district as Object
Preprocessing:
Tokenize dataset
Line=get line from the dataset
Month=get the Month value from the line
District=get the District value from the line
Agglomerative hierarchical clustering algorithm for mapper and
reducer: (mapper)
    Do until reach the Month & District
    Set district = district value of line
    If Month Exists
    Set Month as Key
    Set month = month value of line
    End If
    End
    Merge month and district value
    Set month and district value into another variable
    Map the variable to the reducer
    (Reducer)
    Set sum=0
    For each get value from the key
    Calculate Euclidean mean for each district = PTS
```

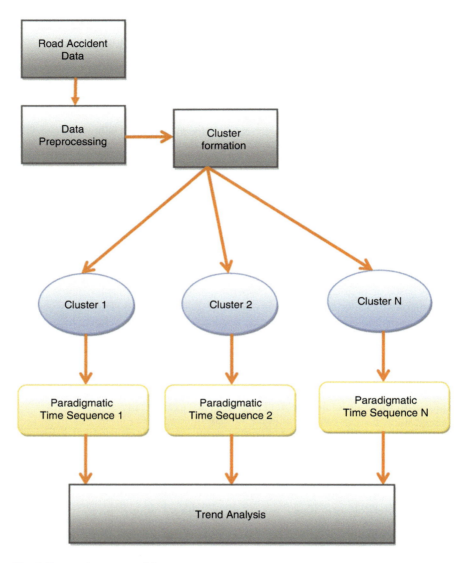

Fig. 1 Proposed system model

4 Performance Evaluation

4.1 Monthly Prediction

The clustering of districts makes the time sequence data to be visualized to perform the normal conditions over the trend analysis. In Fig. 2, the accident rate is high

Trend Analysis Using Agglomerative Hierarchical Clustering Approach for... 873

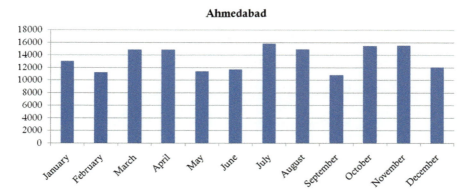

Fig. 2 Accident rates in Ahmedabad city

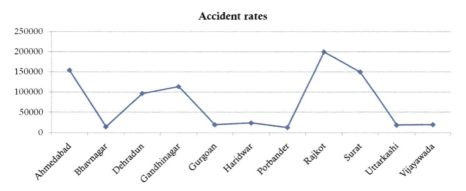

Fig. 3 Accident rates for Indian cities

because it is a metropolitan city which has a high population rate, results in high traffic which tends to maximum accident occurrence. Here, in each month, the trend rate varies depending on the accident count. The high peak of accident rate is recorded in the months of July and November.

Figure 3 demonstrates the accident rates for 1 single year for 11 important cities, and it illustrates that the cities of Ahmedabad, Surat, and Rajkot are having highest amount of accident rates. The cities of Bhavnagar, Gurgaon, Haridwar, Porbandar, Uttarkashi, and Vijayawada are having the lowest amount of accident rates. The cities of Dehradun and Gandhinagar are having the medium amount of accident rates.

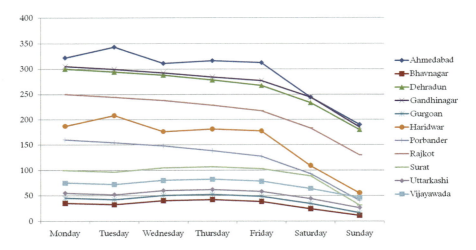

Fig. 4 Weekly analysis

4.2 Weekly Analysis

Fig. 4 demonstrates the weekly analysis for the road accident rates for the 11 Indian cities in detailed manner.

Figure 5 analyzes the time of accident percentage for the cities, and it is measured that at the time of 12–2 pm, the accident rates are high. At the time period of 2–4 pm, the accident rate is around 19 percentages. At the time between 8 and 10 am, the accident rate is 15%, and in all the other timings, the accident rate percentage is 10 or less.

5 Conclusion

The progress of the system includes preprocessing of heterogeneous data which are classified and clustered based on the district-wise grouping of the road accident rates. Consequently, in each cluster, attributes are merged based on the monthly analysis, and paradigmatic time sequence is predicted which is then fed for the trend analysis of the accident rate. Here, AHCTB is proposed for the clustering of each district, and trend analysis is done to each cluster using PTS. The trend analysis shows the variation of the accident rates in each cluster. It also shows that accidents are prone to increase in metropolitan and industrial cities where the population and transportation are very high in nature, whereas in rural areas, the accident rate is low.

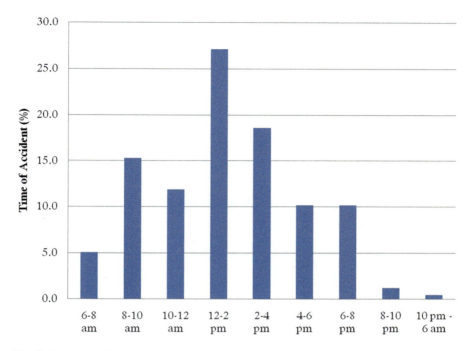

Fig. 5 Time of accident

Acknowledgement This work was supported by the National Research Foundation of Korea (NRF) grant funded by the Korean government (MSIT) (No. 2020R1F1A1076976).

References

1. J. Abellan, G. Lopez, J. Ona, Analysis of traffic accident severity using decision rules via decision trees. Expert Syst. Appl. **40**, 6047–6054 (2013)
2. S. Kumar, D. Toshniwal, Analyzing road accident data using association rule mining, in *International Conference on Computing, Communication and Security, ICCCS-2015*, vol. 20, pp. 30–40
3. S. Kumar, D. Toshniwal, A novel framework to analyze road accident time series data. J. Big Data **30**, 5004–5020 (2016)
4. L. Wang, H.-p. Lu, Y. Zheng, Z. Qian, Safety analysis for expressway based on Bayesian network: A case study in China. IEEE Commun. Mag. **18**(12C), 438–434 (2014)
5. J. de Ona, R.O. Mujalli, F.J. Calvo, Analysis of traffic accident injury severity on Spanish rural highway using Bayesian networks. ScienceDirect Accid. Anal. Prev. **43**, 402–411 (2011)
6. A. Pakgohar, R.S. Tabrizi, M. Khalili, A. Esmaeili, The role of human factor in incidence and severity of road crashes based on the CART and LR regression: A data mining approach. ScienceDirect Procedia Comput. Sci. **3**, 764–769 (2014)
7. S. Kumar, D. Toshniwal, A data mining approach to characterize road accident locations. J. Mod. Transp **24**(1), 62–72 (2016)

8. Y. Lv, Y. Duan, W. Kang, Z. Li, F.-Y. Wang, Traffic flow prediction with big data: A deep learning approach. IEEE Trans. Intell. Transp. Syst. **16**, 2 (2015)
9. J. de Oña, G. López, R. Mujalli, F.J. Calvo, Analysis of traffic accidents on rural highways using Latent Class Clustering and Bayesian Networks. ScienceDirect Accid. Anal. Prev. **51**, 1–10 (2013)
10. S. Regine, C. Simon, A. Maurice, Processing traffic and road accident data in two case studied of road operation assessment. ScienceDirect Transp. Res. Procedia **6**, 90–100 (2015)
11. S. Kumar, D. Toshniwal, A data mining framework to analyze road accident data. J. Big Data **2**, 26 (2015)
12. D. Kee, G.T. Jun, P. Waterson, R. Haslam, A Systemic analysis of South Korea Sewol ferry accident – Striking a balance between learning and accountability. Appl. Ergon. **1**, 1–14 (2016)
13. L. Ramos, L. Silva, M.Y. Santos, J.M. Pires, Detection of road accidents accumulation zones with a visual analytics approach. ScienceDirect Procedia Comput. Sci. **64**, 969–976 (2015)
14. S. Kumar, D. Toshniwal, Analysis of hourly road accident counts using hierarchical clustering and cophenetic correlation coefficient (CPCC). J. Big Data **3**, 13 (2016). **30**, 20–56

Demand Response: Multiagent System Based DR Implementation

Faisal Saeed, Anand Paul, Seungmin Rho, and Muhammad Jamal Ahmed

1 Problem and Motivation

It is very crucial to come up with a real-time forecasting model which can give primary instruction to evaluate DR. Previous works have some limitations that should be overcome for better implementation. There is a strong need to come up a model for peak demand forecast because hourly power consumption forecast may miss very important information. Similarly, we also need power consumption prediction for every home as every smart home has different pattern of power consumption. Meanwhile we also need real-time energy price prediction as well as scheduler which can schedule power accordingly. These all works should be in heterogeneous.

2 Background and Related Work

Demand response is a technique which is designed to condense the peak demand and to make awareness in the people that they should consume the power when renewable energy is accessible in terms of PV panels. It is beneficial for grid operators, electrical customers, and many load serving entities. It is very difficult for an end user to have the track of dynamic pricing. So effective DR implementation always depends on its enabling technologies. Generally, in housing societies, these

---

F. Saeed · A. Paul (✉) · M. J. Ahmed
Department of Computer Science and Engineering, Kyungpook National University, Daegu, South Korea

S. Rho
Department of Software, Sejong University, Seoul, South Korea

© Springer Nature Switzerland AG 2021
H. R. Arabnia et al. (eds.), *Advances in Artificial Intelligence and Applied Cognitive Computing*, Transactions on Computational Science and Computational Intelligence, https://doi.org/10.1007/978-3-030-70296-0_68

enabling technologies are present in smart home situations where smart appliances like smart meters, smart refrigerators, and washing machines worked. EMS (energy management system) is classified as one of the DR-enabling technologies which is a pure smart home topography which has the ability to control household loads very intelligently using connotation between smart appliances like smart meters, fridges, AC, washing machine, etc. Implementing the efficient DR model to see the usefulness of its policies is very effective to boost up the energy efficiency as well as power system constancy. Nowadays, research has a trend that DR implementation is moving toward distributed systems where many heterogeneous mechanisms exist [1–3]. There is a need to come up with an efficient DR model which should have the ability to implement intelligent distributed algorithms, optimizing these algorithms and load forecasting aiming to evaluate power systems on large scale.

MAS (multiagent system) is a combination of different autonomous agents working cooperatively for effective design objectives [4–6]. Every agent in MAS has heterogenous components, and each component can interact with each other. IEEE Power Engineering Society's Multi-Agent Systems Working Group passes comprehensive review from the IEEE that a multiagent system has an ability to provide two novel methods for power engineering, i.e., developing efficient systems which should be flexible and extensible and can simulate and model the algorithms [1, 7]. Generally, MAS is appropriate for developing heterogeneous components while implementing the DR application for smart grids. In this paper, we proposed multiagent system to device DR which is based on RTP. In our model, we proposed a MAS for implementing DR as a residential DR where primary participants are named as HAs and RAs as shown in Fig. 1. In Fig. 1 we can see that the black line is the power connections and the red lines are the LAN connection which is used for communication between HAs and RAs. HAs are connected with each other, so security is also retained. Every agent has heterogeneous components like residential load prediction model, price prediction model, real-time pricing, and power scheduling. There are two main purposes of our multiagent system algorithm, e.g., demand response policies and their enabling technologies. To make prediction, we used artificial neural network concept. We used LSTM to make load forecast and price prediction. The detail of our model is given below.

3 Approach and Uniqueness

3.1 Home Agent

The environment of an HA includes real-time prices prediction, real-time load forecast, scheduling the power for better energy management, and demanding the power from retailer by having the communication with its corresponding agent. We used multipurpose LSTM model in HA, which has the ability to predict each appliance load. Then the agent checks his own renewable energy in the form of

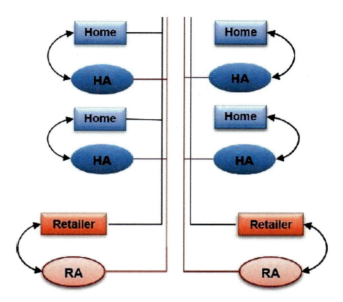

Fig. 1 Overview of proposed model

batteries or solar. If the predicted load is manageable, then electricity request is scheduled by the control load action, or if it is not manageable, then it can ask for power by making demand to RA immediately. LSTM model also predicts the overall load used in home and by using RTP predicts the price of electricity. It also has the ability to predict the solar energy. The primary purpose of the control load action is to minimize electricity bills which are based on our predicted real-time prices or our DLC requests.

3.2 Retailer Agent

The features of a retailer agent atmosphere comprise of real-time prediction of used electricity, available power, aggregated loads, service scope, and the manpower capacity. An RA has the functionality to aggregate power demand based on HA demand according to their services and purchases electricity from wholesale market. After that RA then trades power to the HAs according to their need in a retail market. Sometimes according to firm circumstances such as abundant renewable energy, transmission limitations, and power deficiency, the RA may demand DLC.

Our idea emphasizes on ideal implementation of residential DR in a distribution network; therefore, we also introduced WMA (wholesale market agent) and GA (generator agent). RAs demand power from the wholesale market agent, and generator agent according to their capacity. The electricity generation capacity after

Table 1 This table is showing specific software and hardware specification during the experiment

Name	Experiment environmental parameters
OS	Windows 10
CPU	Intel(R)Core(TM)i5-3570CPU@3.40GHZ3.80GHZ
GPU	NAVIDIA GeForce GTX1050
RAM	24GB
DT	Python 3.6 (PyCharm)

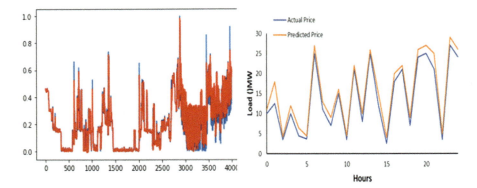

Fig. 2 Simulated MAS results

having the lowest price will be allocated first in the market. In conclusion, this mechanism influences the market in stable state, and a peripheral price is attained.

4 Simulation and Result Discussion

In order to test the performance of proposed MAS, we need agent execution environment. We need LSTM model for predicting load and prices of electricity. We run our LSTM model on PC having the attributes described in Table 1. We used LSTM models to predict load, price, and energy with real-time data. provided by US power control department for training and testing of LSTM model. We used real-time data provided by US power control department for training and testing of LSTM model. To implement MAS-based DR, we consider one RA which is responsible for 100 HAs in a spread network. While implementing we accept that the RA can demand and obtain enough energy from a wholesale market and all broadcast restrictions are fulfilled. In our MAS, the retail market is balanced with respect to budget, i.e., the RA cannot generate revenues. As in our model home, agents can schedule the manageable power to abate the electricity prices, so Fig. 2 shows its simulated LSTM model results. First figure is the predicted price, and the second figure is the predicted load.

Acknowledgement This work was supported by the National Research Foundation of Korea (NRF) grant funded by the Korea government(MSIT) (NRF-2019R1F1A1060668).

References

1. P. Vrba, V. Mařík, P. Siano, P. Leitão, G. Zhabelova, V. Vyatkin, T. Strasser, A review of agent and service-oriented concepts applied to intelligent energy systems. IEEE Trans. Ind. Inf. **10**(3), 1890–1903 (2014)
2. M.J. Gul, A. Rehman, A. Paul, S. Rho, R. Riaz, J. Kim, Blockchain expansion to secure assets with fog node on special duty. Soft. Comput. **28**, 1–3 (2020)
3. F. Saeed, A. Paul, P. Karthigaikumar, A. Nayyar, Convolutional neural network based early fire detection. Multimed. Tools Appl. **20**, 1–7 (2019)
4. M. Wooldridge, *An Introduction to Multiagent Systems* (Wiley, Hoboken, 2009)
5. H.P. Chao, Price-responsive demand management for a smart grid world. Electr. J. **23**(1), 7–20 (2010)
6. N. Rahmatov, A. Paul, F. Saeed, W.H. Hong, H. Seo, J. Kim, Machine learning–based automated image processing for quality management in industrial Internet of Things. Int. J. Distrib. Sens. Netw. **15**(10), 1550147719883551 (2019)
7. F. Saeed, A. Paul, W.H. Hong, H. Seo, Machine learning based approach for multimedia surveillance during fire emergencies. Multimed. Tools Appl. **6**, 1–7 (2019)

t-SNE-Based K-NN: A New Approach for MNIST

Muhammad Jamal Ahmed, Faisal Saeed, Anand Paul, and Seungmin Rho

1 Introduction

Nowadays the terms "artificial intelligence," "deep learning," and "machine learning" are thrown around by associations in every single industry. It might come into one's mind that these words seemed to have appeared overnight and are new; it's the hard work of many within the field that has really moved it into the spotlight as the latest tech trend, but the truth is that they've been around for a while. One of the applications of machine learning, i.e., data mining, constitutes most of data analysis tools to regulate the previous effective patterns, unidentified patterns, and relationships in big data. These tools comprise of mathematical algorithms, machine learning methods, and statistical models. Accordingly, data mining not only consists of more than managing and collection data, but it also comprises prediction and analysis.

Classification technique is artistic adequate to process a huge variety of data than regression and is growing in distinction. In several fields, the visualization of high-dimensional, large datasets is important. van der Maaten and Hinton (2008) introduced t-distributed stochastic neighborhood embedding (t-SNE), and it has become immensely prevalent in various fields. To signify high-dimensional dataset in a low-dimensional space of two or three dimensions so that we can visualize it in the form of spectral clustering, a dimensionality reduction technique t-distributed stochastic neighbor embedding (t-SNE) is used. In comparison to other dimensionality reduction techniques like PCA which merely exaggerate

---

M. J. Ahmed · F. Saeed · A. Paul (✉)
Connected Computing and Media Processing Lab, Department of Computer Science and Engineering, Kyungpook National University, Daegu, South Korea

S. Rho
Department of Software, Sejong University, Seoul, South Korea

© Springer Nature Switzerland AG 2021
H. R. Arabnia et al. (eds.), *Advances in Artificial Intelligence and Applied Cognitive Computing*, Transactions on Computational Science and Computational Intelligence, https://doi.org/10.1007/978-3-030-70296-0_69

the variance, t-SNE constructs a compact feature space where alike samples are demonstrated by neighboring points and contradictory samples are demonstrated by distant points with immense probability. At an immense level, t-SNE paradigms for the high-dimensional samples a probability distribution in such a way that alike samples have a high possibility of being chosen while unlike points have a really minor possibility of being chosen. In the low-dimensional embedding, t-SNE states an alike distribution for the points. Finally, t-SNE minimizes the Kullback-Leibler divergence between the two distributions with reverence to the locations of the points in the embedding [1].

2 Background and Related Work

According to Sahibsingh A. Dudani, among the most intuitively and simplest appealing classes of those classification procedures which are non-probabilistic are those that weigh the evidence of closed sample information most heavily [2]. More precisely, one might demand to weigh the evidence of a neighbor close to an unclassified observation more heavily than the evidence of another neighbor which is at a greater distance from the unclassified observation.

When we have a given classification problem, we might find ourselves in one of the extreme situations. In one situation, we may have complete statistical knowledge of the basic or hidden joint distribution. In second situation, we may have a collection of n correctly classified samples; then there no classification procedure exists. The transparent majority k-nearest-neighbor rule will be worthy of comparison to first formulation of a rule of the nearest-neighbor type. The problem of classifying an unknown pattern on the basis of its nearest neighbors in a trained dataset is addressed from the point of view of evidence theory also known as Dempster-Shafer theory. Suppose we have a sample to classify, and each neighbor of that sample is treated as an evidence that supports firm hypotheses regarding the class affiliation of that pattern. Between the two vectors the degree of support is defined as a function of the distance. Through the means of Dempster's rule of combination, the neighbor or evidence of the k-nearest neighbors is then pooled together. This method offers a comprehensive treatment of such issues such as distance rejection and ambiguity and defective knowledge regarding the class membership of training patterns. This procedure is effective classification scheme as compared to the voting and distance-weighted k-NN. The procedure of D-S theory provides a global treatment of important issues that are there in the previous methods, like the consideration of the distances from the neighbors in the decision, distance rejection and ambiguity, and the consideration of imprecision and uncertainty in class labels. This is accomplished by setting the problem of merging the evidence provided by nearest neighbors in the conceptual framework of D-S theory or evidence theory.

A new nonparametric technique for pattern classification has been proposed, based on the conceptual framework of D-S theory [3]. This technique essentially considers each of the *k-nearest* neighbors of a pattern to be classified as an item of evidence that changes one's belief regarding the class membership of that pattern. D-S theory then provides a simple mechanism for pooling this evidence in order to compute the uncertainty attached to each simple or compound hypothesis. Evidence theory provides a natural way of modulating the importance of training samples in the decision, depending on their nearness to the point to be classified. It allows for the introduction of ambiguity and distance reject options that receive a unified interpretation using the concepts of lower and upper expected costs.

3   Approach: K-NN for Classification Using t-SNE

One of the vital methods in data mining is clustering, because it can be used for data compression and database segmentation. Clustering can also be active for data pre-processing of data mining. Clustering is intended to cluster a set of samples in such a way that samples with high similarity are grouped in the same group (cluster), that is, samples with low similarity between clusters and samples with high similarity within a cluster. We will apply t-SNE for the required spectral clustering, and then we will use K-NN. We have to use an optimal number of clusters "m" because large number of "m" increases the overhead of clustering; hence at the same time, classification accuracy will be lower. Consequently, we need an optimal value for the number of "m" selection. Normally, the larger the number of "m," the better the accuracy, but if the dataset distribution is concentrated comparatively, it will approach toward lesser accuracy. The value of small "m" is what prohibits standard K-NN to be used on big data.

4   Simulation and Result Discussion

In this section, for appropriate selection of k value, a collection of trials was directed on dataset with $m = 10$ with respect to different number of k's. From Table 1 with the value of $k = 10$, the overall accuracy of classification is very good. But as we increase the number of k, the classification accuracy gradually falls down; as you can see in Table 2, we selected $k = 15$, and the overall accuracy dropped. And in Table 3 as we increase more and more the number of k, the overall accuracy decreases (Figs. 1 and 2).

Table 1 Table I: K-NN with $k = 10$ and $m = 10$

	Precision	Recall	f1 score
0	1.00	1.00	1.00
1	0.93	1.00	0.96
2	1.00	0.94	0.97
3	1.00	0.94	0.97
4	1.00	0.96	0.98
5	1.00	1.00	1.00
6	1.00	1.00	1.00
7	0.93	0.97	0.95
8	0.93	0.97	0.95
9	0.98	0.96	0.97

Table 2 Table I: K-NN with $k = 15$ and $m = 10$

	Precision	Recall	f1 Score
0	0.97	1.00	0.99
1	0.86	1.00	0.93
2	1.00	0.97	0.98
3	1.00	0.97	0.98
4	1.00	0.96	0.98
5	1.00	1.00	1.00
6	1.00	0.97	0.99
7	0.97	0.97	0.97
8	0.95	0.90	0.92
9	0.98	0.96	0.97

Table 3 Table I: K-NN with $k = 20$ and $m = 10$

	Precision	Recall	f1 score
0	0.97	1.00	0.99
1	0.86	1.00	0.93
2	0.97	0.94	0.95
3	1.00	0.94	0.97
4	1.00	0.96	0.98
5	1.00	1.00	1.00
6	1.00	0.97	0.99
7	0.93	0.97	0.95
8	0.89	0.87	0.88
9	1.00	0.96	0.98

5 Conclusions

Concluding the paper, we have proposed a K-NN classification technique. We conduct spectral clustering using t-SNE, to divide the data into different clusters, and then we did classification through K-NN. We analyzed the optimal value for arguments such as k and m. The results showed that it is better in terms of efficiency and accuracy; hence it is good to use it on big data.

t-SNE-Based K-NN: A New Approach for MNIST

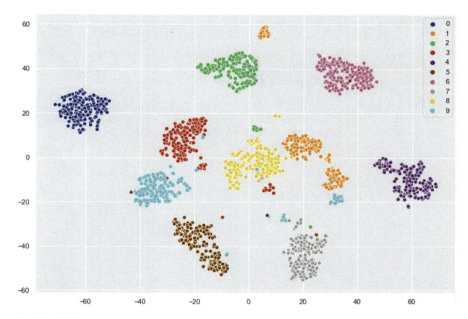

Fig. 1 t-SNE with $m = 10$

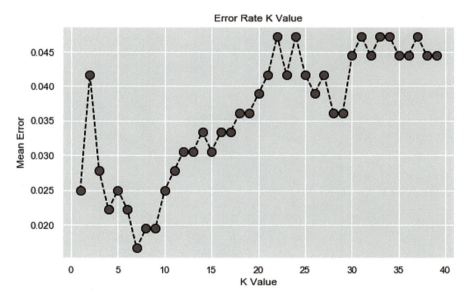

Fig. 2 Mean error

Acknowledgment This work was supported by the National Research Foundation of Korea (NRF) grant funded by the Korean government (MIST) (NRF-2019R1F1A1060668).

References

1. T. Van Erven, P. Harremos, Řenyi divergence and Kullback-Leibler di- vergence. IEEE Trans. Inf. Theory **60**(7), 3797–3820 (2014)
2. L.V.D. Maaten, G. Hinton, Visualizing data using t-SNE. J. Mach. Learn. Res. **9**(Nov), 2579–2605 (2008)
3. B.W. Silverman, M.C. Jones, E. Fix and JL. Hodges (1951): An important contribution to nonparametric discriminant analysis and density estimation: Commentary on fix and Hodges (1951). Int. Stat. Rev. (Revue Internationale de Statistique) **3**, 233–238 (1989)

Short- to Mid-Term Prediction for Electricity Consumption Using Statistical Model and Neural Networks

Malik Junaid Jami Gul, Malik Urfa Gul, Yangsun Lee, Seungmin Rho, and Anand Paul

1 Introduction

Electricity load balancing and supply and demand management are required by the electricity suppliers to utilize the power plant's ability up to required demand level. In terms of predictions, forecasting is divided into three categories (1) long term, (2) mid term, and (3) short term. These studies can save electricity expenditures by analyzing the data and the pattern of electricity from approximate values of future consumption. Analyzing the approximate future values, electricity suppliers can manage the prices of electricity [1], which eventually helps in the economy of the country. Properly planned electricity load and consumption can save money that can help in building economy[2].

Statistical as well as machine learning models are resourceful to get insight of the electricity consumption from dataset. Generally, these consumption datasets have time series representation. Time series data can be obtained from domains where data figures change with respect to time like in stock market[3]. Statistical or machine learning models can be applied to other fields also as [4] introduces an approach to analyze churn dataset with respect to geographic locations.

M. J. J. Gul · M. U. Gul · A. Paul
Department of Computer Science and Engineering, Kyungpook National University, Daegu, South Korea
e-mail: junaidgul@live.com.pk

Y. Lee
Division of Computer Engineering, Hanshin University, Gyeonggi-do, South Korea

S. Rho (✉)
Department Software, Sejong University, 621 Innovation Center, Sejong University, Seoul, South Korea
e-mail: smrho@sejong.edu

© Springer Nature Switzerland AG 2021
H. R. Arabnia et al. (eds.), *Advances in Artificial Intelligence and Applied Cognitive Computing*, Transactions on Computational Science and Computational Intelligence, https://doi.org/10.1007/978-3-030-70296-0_70

Auto-regressive and moving average models are widely used for such forecasting studies, but models from artificial intelligence can also be utilized to increase the forecast accuracy. ARIMA stands for "Auto Regressive Integrated Moving Average." The ARIMA system is described with 3 terms: p, d, and q. The approach is that computing the differential value from the prior and the present values. ARIMA is used in studies like forecasting wheat production, infant mortality rate, automated modeling for electricity production [5–7], and many more.

Forecasting electricity can impact industries that are directly linked to electricity production. Prices can go up and down in such industries like oil and gas industry [8], which helps in electricity production. This statement emphasizes the cruciality and importance of the study as many other factors can get affected with improper management of electricity load balancing, production, supply, and demand.

In this study, the authors investigated comprehensive models for related dataset. Authors identified the hyperparameters for models, conducted experiments, and compared results on the basis of MSE and RMSE metrics. The structure of the paper is as follows: Sect. 2 provides related work and much better understanding of the models like ARIMA and neural networks. Section 3 provides material and methods that have been used while conducting this study. Section 4 provides information of the experiments and discusses certain results that are observed. Furthermore, results are discussed for future work.

2 Literature Review

Neural networks are mostly known for image processing especially in medical image processing. Different models with certain parameters have to be analyzed to determine the best outcomes. The study of [9] shows good progress in terms of medical image processing as they analyze 5 thousand to 160 million parameters with a certain number of layers to evaluate their large scale dataset that eventually helps in computer aided vision (CAD). Neural networks can help in understanding the personality dynamics and can determine state of personality is stable or not [10] and what are the variables that can affect the personality variable. Photovoltaic (PV) integration can help in economic growth as it is a promising source of renewable energy and thus requires prediction and forecasting to help in taking future decision. Forecasting PV based data can be done by neural network model named as LSTM. LSTM-RNN is analyzed by Abdel-Nasser and Mahmoud [11], which can determine temporal changes in PV output and can be evaluated through hourly dataset for a year. Electricity being considered as a key role player in economy thus studied by many researchers with different models and approaches. An approach used by Bouktif et al. [12] utilizes LSTM along with genetic algorithm to get better results and performance with time series data for short-term and long-term forecasting. Considering electricity plays an essential role, [13, 14] also propose a model with LSTM that is capable of forecasting load for single residential as there are certain other parameters that are involved.

Short- to Mid-Term Prediction for Electricity Consumption Using Statistical... 891

Table 1 Abbreviation table

sr#	Acronym	Detail
1	ARIMA	Auto-regressive moving average
2	MSE	Mean square error
3	RMSE	Root mean square error

3 Material and Methods

We have chosen core i7 processor along with NVIDIA GPU with 8GB memory, in terms of hardware utilized, to speed up the computational time. To develop prediction models, we have used Python along with Python integrated environment along with TensorFlow and Keras libraries for neural network and statsmodels' libraries to implement statistical models (Table 1).

4 Building the Forecasting Model

We have performed statistical as well as machine learning algorithm to fetch the information from the dataset. To evaluate our experiments, we have considered performance metrics like MSE and RMSE.

4.1 ARIMA Model

ARIMA is a statistical model to forecast the values. Combining the AR and MA model formulas, we get general formula as

$$\hat{y}_t = \mu + \phi_1 y_{t-1} + \cdots + \phi y_{t-p} - \theta_1 e_{t-1} - \cdots - \theta_q e_{t-q}. \tag{1}$$

Figure 1 indicates that our data is not fully stationary, and we have to do some data preparation before we can use ARIMA model. It can also be observed that there are a number of spikes that are not within the critical range. This phenomenon can determine the P and D values for the ARIMA model.

Figure 2 provides results of different hyperparameter values that are determined from Fig. 1. In Fig. 2, it can be seen that the suitable value for p,d,q is 3,1,1 respectively. Figure 3 provides pictorial view of ARIMA model with hyperparameter (p,d,q) values as $(3,1,1)$ with a MSE of 0.028. This shows that model was successfully executed with hyperparameter setting and yielded good results.

Figures 4 and 5 show the 4-month predicted values in a grey area that is 95% interval and forecasted values that are in blue line.

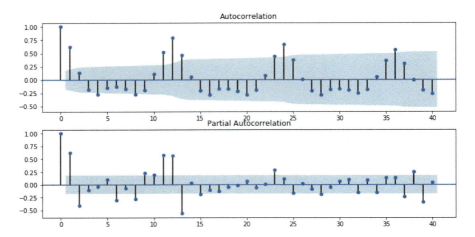

Fig. 1 ACF and PACF

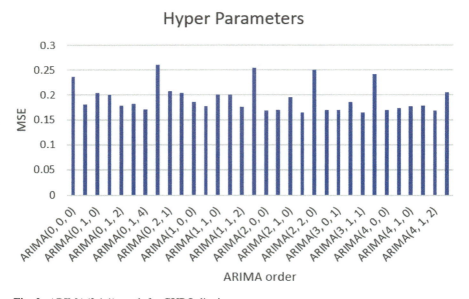

Fig. 2 ARIMA(3,1,1) result for GURO district

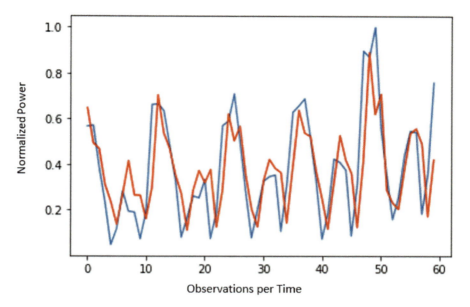

Fig. 3 ARIMA(3,1,1) result for GURO district

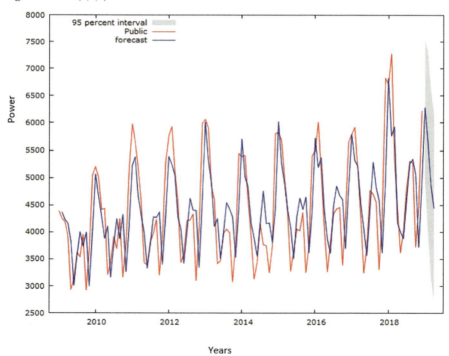

Fig. 4 Four-month prediction with ARIMA(3,1,1)

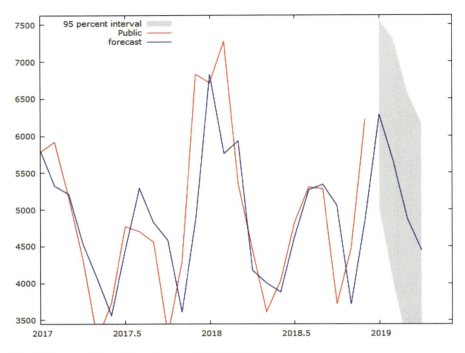

Fig. 5 Enhanced 4-month prediction with ARIMA(3,1,1)

5 Conclusions

The main focus of the study is to determine whether statistical models or neural networks can perform better with provided dataset. To compare all the models, we maintain the consistency of hyperparameters so comparison can be most realistic. ARIMA requires heavy data preprocessing, while neural networks are easy to adopt. This study will expand to find results from different neural network models that are applicable in time series prediction. Results produced by ARIMA are certain and can be applied into real world application where data pattern does not change most of the time. Meanwhile, LSTM and other neural network models are difficult to train but produce better results.

Acknowledgement This work was supported by the National Research Foundation of Korea (NRF) grant funded by the Korea Government (MSIT) (No. 2019M3F2A1073179).

References

1. P. Siano, Demand response and smart grids—A survey. Renew. Sust. Energy Rev. **30**, 461–478. (2014). https://doi.org/10.1016/j.rser.2013.10.022
2. F.J. Ardakani, M.M. Ardehali, Energy **65**, 452–461 (2014). https://doi.org/10.1016/j.energy.2013.12.031
3. S.P. Chatzis, V. Siakoulis, A. Petropoulos, E. Stavroulakis, N. Vlachogiannakis, Expert Syst. Appl. **112**, 353–371 (2018). https://doi.org/10.1016/j.eswa.2018.06.032
4. H.V. Long, L.H. Son, M. Khari, K. Arora, S. Chopra, R. Kumar, T. Le, S.W. Baik, Comput. Intell. Neurosci. **2019** (2019). https://doi.org/10.1155/2019/9252837
5. M.A. Masood, S. Abid, Asian J. Agric. Rural Dev. **8**, 172–177 (2018). https://doi.org/10.18488/journal.1005/2018.8.2/1005.2.172.177
6. A.K. Mishra, C. Sahanaa, M. Manikandan, J. Family Commun. Med. **26**, 123 (2019). https://doi.org/10.4103/jfcm.JFCM_51_18
7. P. Amin, L. Cherkasova, R. Aitken, V. Kache, *Proceedings of the 2019 IEEE International Congress on Internet Things, ICIOT 2019: Part 2019 IEEE World Congress on Services* (2019). https://doi.org/10.1109/ICIOT.2019.00032
8. K.B. Debnath, M. Mourshed. Forecasting methods in energy planning models. Renew. Sust. Energy Rev. **88**, 297–325 (2018). https://doi.org/10.1016/j.rser.2018.02.002
9. H.C. Shin, H.R. Roth, M. Gao, L. Lu, Z. Xu, I. Nogues, J. Yao, D. Mollura, R.M. Summers, IEEE Trans. Med. Imaging **35**, 1285–1298 (2016). https://doi.org/10.1109/TMI.2016.2528162
10. S.J. Read, V. Droutman, B.J. Smith, L.C. Miller, Pers. Individ. Dif. **136**, 52–67 (2019). https://doi.org/10.1016/j.paid.2017.11.015
11. M. Abdel-Nasser, K. Mahmoud, Neural Comput. Appl. **31**, 2727–2740 (2019). https://doi.org/10.1007/s00521-017-3225-z
12. S. Bouktif, A. Fiaz, A. Ouni, M.A. Serhani, Energies **11**, 1636 (2018). https://doi.org/10.3390/en11071636
13. W. Kong, Z.Y. Dong, Y. Jia, D.J. Hill, Y. Xu, Y. Zhang, IEEE Trans. Smart Grid **10**, 841–851 (2019). https://doi.org/10.1109/TSG.2017.2753802
14. F. Saeed, A. Paul, W.H. Hong, H. Seo, *Multimedia Tools and Applications* (Springer, Berlin, 2019), pp. 1–17

BI-LSTM-LSTM Based Time Series Electricity Consumption Forecast for South Korea

Malik Junaid Jami Gul, M. Hafid Firmansyah, Seungmin Rho, and Anand Paul

1 Introduction

Electricity power plants require efficient load balancing that heavily depends on electricity consumption. Future forecast based on historical data can provide insight for future so the government and power plant can take decision to balance demand and supply chain. Statistical analysis can provide simple solution in terms of computation and short-term prediction but heavily suffer from data preprocessing like stationarity issue. Neural networks on the other hand can provide a model based on various parameters that can be tuned and applied in the real-world scenario for better accuracy.

Forecasting of electricity consumption can provide future insights so power plant can manage their electricity production, which can save the resources for holding the economy. This can directly impact on the prices of other fuel resources like gasoline and liquefied petroleum gas [1]. This emphasizes and motivates us to find better forecasting model that can predict future values effectively. In this study, the authors conducted comprehensive models for related dataset. Authors identified the hyperparameters for models, conducted experiments, and compared results based on RMSE metrics. These models provide enough insights for the dataset to get more accurate forecasting. Authors also contributed more by combining LSTM with Bi-

---

M. J. J. Gul · M. H. Firmansyah · A. Paul
Department of Computer Science and Engineering, Kyungpook National University, Daegu, South Korea
e-mail: junaidgul@live.com.pk; hafid@knu.ac.kr

S. Rho (✉)
Department Software, 621 Innovation Center, Sejong University, Gwangjin-gu, Seoul, South Korea
e-mail: smrho@sejong.edu

© Springer Nature Switzerland AG 2021
H. R. Arabnia et al. (eds.), *Advances in Artificial Intelligence and Applied Cognitive Computing*, Transactions on Computational Science and Computational Intelligence, https://doi.org/10.1007/978-3-030-70296-0_71

LSTM to create a new model to forecast and analyze the electricity consumption. The structure of the paper is as follows: Section 2 provides related work about neural networks. Section 3 provides material and methods that have been used while conducting this study. Section 4 provides information of the experiments and discusses certain results that are observed. Furthermore, results are discussed for future work.

2 Literature Review

Neural networks are mostly renown for image processing especially in medical image processing. Different models with certain parameters have to be analyzed to determine the best outcomes. The study of [2] shows good progress in terms of medical image processing as they analyze 5 thousand to 160 million parameters with a certain number of layers to evaluate their large scale dataset, which eventually helps in computer aided vision (CAD). Neural networks can help in understanding the personality dynamics and can determine state of personality is stable or not [3] and what are the variables that can affect the personality variable. Photovoltaic (PV) integration can help in economic growth as it is a promising source of renewable energy that thus requires prediction and forecasting to help in taking future decision. Forecasting PV based data can be done by neural network model named as LSTM. LSTM-RNN is analyzed by Abdel-Nasser and Mahmoud [4], which can determine temporal changes in PV output and can be evaluated through hourly dataset for a year. Electricity being considered as a key role player in the economy thus studied by many researchers with different models and approaches. An approach used by Bouktif et al. [5] utilizes LSTM along with genetic algorithm to get better results and performance with time series data for short-term and long-term forecasting. Further improvement in forecasting with LSTM is done in article from [6] where author combined the CNN with Bi-LSTM to get better forecasting result for electricity for households. Considering electricity plays an essential role, [7] also propose a model with LSTM that is capable of forecasting load for single residential as there are certain other parameters that are involved. They proposed a framework with LSTM and evaluated the framework with real residential smart meter data. Residential usage is considered to be important, and many researchers are looking deeper to find patterns in residential electricity usage. Forecasting long-term electricity demand for residential user is also affected by other variables. As electricity demand and supply forecasting is divided into short, mid-, and long term, researchers are getting their hands-on granularity data to forecast hourly and daily with the help of artificial neural network. Modeling with granularity is quite challenging as shown by Rodrigues et al. [8]. Recurrent neural networks (RNN) can be used for dynamic modeling for the non-linear data. Data plays an important role in overall modeling and experimentation, so [9] make simple modification into RNN to work along with non-linear spatio-temporal data for forecasting applications. Computational time is also an important factor in the overall forecasting process, and time factor can be

Electricity Usage Forecasting

Table 1 Abbreviation table

sr#	Acronym	Detail
1	RNN	Recurrent neural network
2	LSTM	Long short-term memory
3	Bi-LSTM	Bi-directional long short-term memory
4	RMSE	Root mean square error
5	RELu	The rectified linear unit

heavily improved if we decrease the number of variables as [10, 11] did it in their study by relying only on past data of solar energy consumption.

3 Material and Methods

We have chosen core i7 processor along with NVIDIA GPU with 8 GB memory, in terms of hardware utilized, to speed up the computational time. To develop prediction models, we have used Python along with Python integrated environment along with TensorFlow and Keras libraries for neural network and statsmodels' libraries to implement statistical models (Table 1).

4 Building the Forecasting

We have tested neural network model with Bi-LSTM layer with 50 units along with LSTM layer with 25 units. We choose loss as our model evaluation metrics with mean squared error (Fig. 1).

4.1 LSTM

LSTM requires data reshaping according to samples, timestamps, and feature. Data has been reshaped accordingly and sent to two layers of LSTM and is configured with 50 neurons to boost the learning process. RELu activation function has been used with return_sequences equal to true so data can be passed from one layer to another. After 200 epochs, we got a train score 0.15 RMSE and a test score 0.21 RMSE.

Figure 2 shows the LSTM result with 120 observations 70–30%. Seventy percent for training and 30% for test.

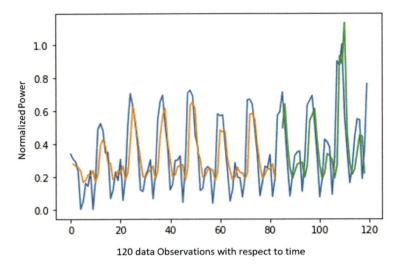

Fig. 1 LSTM results

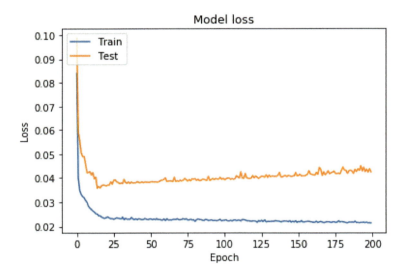

Fig. 2 LSTM model loss

4.2 Bi-LSTM-LSTM Model

Our experimentations also include combined model of Bi-LSTM and LSTM layers. Configuration is the same as we have utilized for our previous models' experimentation. We have used one layer of Bi-LSTM and one layer of LSTM for this experimentation. After 50 epochs, we get a training score as 0.15 RMSE and a testing score as 0.26 RMSE.

Figure 3 shows the results of the model for training and testing for 120 observations, while Fig. 4 shows the model loss. It can be seen that model loss is fluctuating as compared to Bi-LSTM model and trend can be seen to go downward. This means over the time this model will produce less loss. Like other models, this model can also be re-tuned to work for mid-term forecasting applications.

5 Conclusions

The main focus of the study is to determine the performance of LSTM and Bi-LSTM with LSTM layers in neural networks time series prediction. To compare all the models, we maintain the consistency of hyperparameters so comparison can be most realistic. This study will expand to find results from different neural network models that are applicable in time series prediction. Our model shows RMSE scores of 0.15 on training and 0.19 for testing with tuning hyperparameters that are in process of tuning for better accuracy and less loss. In future, we are planning to extend our study to build a model that can perform better with provided dataset.

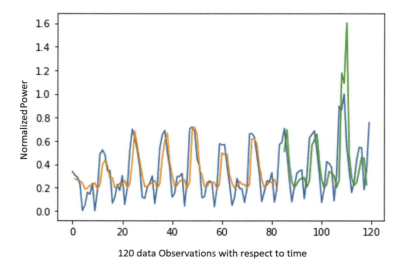

Fig. 3 Bi-LSTM-LSTM results

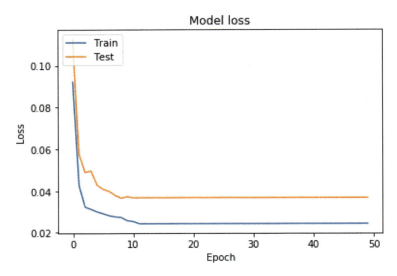

Fig. 4 Model loss

Acknowledgments This work was supported by the National Research Foundation of Korea(NRF) grant funded by the Korea Government(MSIT) (No. 2019M3F2A1073179).

References

1. K.B. Debnath, M. Mourshed, Forecasting methods in energy planning models. Renew. Sustain. Energy Rev. **88**, 297–325 (2018). https://doi.org/10.1016/j.rser.2018.02.002
2. H.C. Shin, H.R. Roth, M. Gao, L. Lu, Z. Xu, I. Nogues, J. Yao, D. Mollura, R.M. Summers, IEEE Trans. Med. Imag. (2016). https://doi.org/10.1109/TMI.2016.2528162
3. S.J. Read, V. Droutman, B.J. Smith, L.C. Miller, Pers. Individ. Dif. (2019). https://doi.org/10.1016/j.paid.2017.11.015
4. M. Abdel-Nasser, K. Mahmoud, Neural Comput. Appl. (2019). https://doi.org/10.1007/s00521-017-3225-z
5. S. Bouktif, A. Fiaz, A. Ouni, M.A. Serhani, Energies (2018). https://doi.org/10.3390/en11071636
6. T. Le, M.T. Vo, B. Vo, E. Hwang, S. Rho, S.W. Baik, Appl. Sci. (2019). https://doi.org/10.3390/app9204237
7. W. Kong, Z.Y. Dong, Y. Jia, D.J. Hill, Y. Xu, Y. Zhang, IEEE Trans. Smart Grid (2019). https://doi.org/10.1109/TSG.2017.2753802
8. F. Rodrigues, C. Cardeira, J.M. Calado, in *Energy Procedia* (2014). https://doi.org/10.1016/j.egypro.2014.12.383
9. P.L. McDermott, C.K. Wikle, Entropy (2019). https://doi.org/10.3390/e21020184
10. M. Majidpour, H. Nazaripouya, P. Chu, H. Pota, R. Gadh, Forecasting (2018). https://doi.org/10.3390/forecast1010008
11. F. Saeed, A. Paul, P. Karthigaikumar, et al. Convolutional neural network based early fire detection. Multimed Tools Appl **79**, 9083–9099 (2020). https://doi.org/10.1007/s11042-019-07785-w

Part VII
XX Technical Session on Applications of Advanced AI Techniques to Information Management for Solving Company-Related Problems (Co-Chairs: Dr. David de la Fuente and Dr. Jose A. Olivas)

Inside Blockchain and Bitcoin

Simon Fernandez-Vazquez, Rafael Rosillo, Paolo Priore, Isabel Fernandez, Alberto Gomez, and Jose Parreño

1 Introduction

Bitcoin was introduced in 2008 as a peer-to-peer payment system used in electronic transactions. Through this mechanism, middlemen (or central banks) were suppressed from the interaction in such transactions [1].

The idea of an electronic system in which payments can be made without the central authority is possible through the use of public ledgers, which function in the blockchain. The value of this currency in which payments are made (Bitcoins) is set by the market price. This market takes into account not only the nodes involved but also all participants. The value is achieved by consensus within the players of this virtual currency [2].

Cryptocurrencies are very diverse in name and form nowadays. They can be defined as a digital asset that stores value and serves as a medium of exchange for buying goods and services. Another use is the issue of digital coins, which helps some of the companies raise funds [2]. This aspect will be looked into in the following sections.

Due to the characteristics of blockchain, which include decentralized mining, it becomes a transparent and unregulated technology. The reward of the miners comes in the form of Bitcoins. This reward, added to the fact that the value of this cryptocurrency is growing and soon there will be scarcity, has attracted more and more miners in pursuing these activities. This has increased competition among them [3].

The trust in blockchain is supported by the consensus of the miners, which is achieved by solving cryptographic puzzles. But this trust comes at a price, as some

---

S. Fernandez-Vazquez · R. Rosillo (✉) · P. Priore · I. Fernandez · A. Gomez · J. Parreño
Business Management, University of Oviedo, Oviedo, Asturias, Spain
e-mail: rosillo@uniovi.es

© Springer Nature Switzerland AG 2021
H. R. Arabnia et al. (eds.), *Advances in Artificial Intelligence and Applied Cognitive Computing*, Transactions on Computational Science and Computational Intelligence, https://doi.org/10.1007/978-3-030-70296-0_72

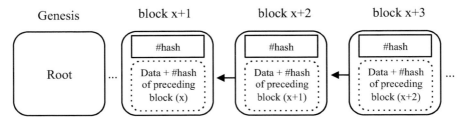

Fig. 1 A blockchain structure. Each has (or unique identification) references its previous block

developing organizations have tried to corrupt the system by launching 51% attacks or acting as selfish miners. These issues have started to emerge, and they need to be tackled if the cryptocurrency system wants to move forward.

Figure 1 shows a typical blockchain structure. The first block in the chain is known as the genesis or root. New transactions are incorporated into the chain through the solving of cryptographic puzzles by its miners, also known as proof-of-work (PoW) [4].

2 Blockchain-Based Applications

Blockchain has many kinds of applications for which it can be used. Below are just some of the main applications using this system:

1. Wallet

    In order for users to exchange and store Bitcoins, the use of wallets is important. Wallets are secured through the use of cryptographic private and public keys. When a user sends a message to another user, it can only be decrypted by the latter using its private and public keys. Users can generate an infinite number of wallets [5].
2. Initial coin offerings (ICOs)

    ICOs are mechanisms by which enterprises produce and issue their own tokens. In exchange of these tokens, investors give away funds. This process allows companies to raise capital by selling, or preselling, entrance to a new product within the market. Some of the biggest ICOs have risen as much as $12 billion during 2018 [6].

3 Bitcoin: Main Advantages

Below are some of the main advantages of the use of Bitcoin:

1. Anonymity

When using encrypted wallets of Bitcoin, the holder can use it without identification or verification. Bitcoins are sent and received in an anonymous manner. The identity is encrypted; nevertheless, the transactions are kept in a public available ledger. This is why sometimes Bitcoin is considered to be pseudonymous and not fully anonymous [7].

2. Noninflationary

Governments have the power to print fiat currencies when needed. They can inject money into the economy to pay, for example, debt, having a direct effect in the economy, decreasing the value of the currency, and creating inflation. This cannot happen in the Bitcoin world. The creation of Bitcoins is fixed to 21 million, which tackles the problem of inflation [8].

3. Resilience

Due to the fact that Bitcoins are issued without the intervention of a central authority, they are less exposed to attacks. Therefore, the system is less vulnerable to assaults from the parties involved [9].

4. Transparency

In Bitcoin, transactions are available and verifiable to all users, through what is known as the blackchain. This provides a level of transparency and verification of all transactions that makes it very powerful [10].

5. Secure information

As opposed to many credit card systems, in which the user needs to enter a secret code to validate a transaction, Bitcoin does not require the use of any secret code in order to make a purchase [9].

6. Transaction costs

Today, Bitcoin has become an accepted means of payment for goods and services. Transaction fees when using Bitcoin are minimal and in many cases do not exist. When operating with traditional payment systems, users need to pay a transaction fee when they want to send money or make a payment in a different currency. Through the use of Bitcoin, cost savings are direct and create a new, more efficient market structure for the user [8].

4 Bitcoin: Main Challenges

There are many challenges that need to be faced by Bitcoin, some of which are listed below:

1. Instability and volatility

This is one of the main challenges that this cryptocurrency faces. Unlike most well-known fiat currencies, Bitcoin has experienced huge volatility in its price. This will probably discourage many users and investors to invest in Bitcoin [10].

2. Trust and security

One of the main challenges that Bitcoin has to face is trust. The main reason is that this cryptocurrency is not backed by any entity of trust (i.e., governments);

therefore investors are reluctant to buy Bitcoins. Another concern is insecurity due to fraud in some exchange platforms. Due to the fact that Bitcoins are not backed by central governments, when theft occurs, investors usually lose their money, as it is not covered by deposit insurance [11].

3. Cyberattacks and hacking

Although less common nowadays, cyberattacks have become rather common in previous years, creating to a decrease in the value of the Bitcoin. These attacks have damaged the reputation of Bitcoin, and once again it has a negative image due to these types of criminal activities [12].

4. Lack of consumer protection

Although blockchain is a deregulated system, in order to guarantee the protection of consumers within the industry, there is a need for protection. The possibility for a user of losing all of his Bitcoins is by far one of the biggest fears for investors [11].

5. Illegal activities

Due to the novelty of this virtual currency, major frauds have emerged which place Bitcoin at the heart of illegal activities. Such activities are [11]:

(i) Tax avoidance

Anonymity and the lack of bank account associated to the sending and receiving of Bitcoins make this system attractive for users willing to avoid taxes. Only a few countries have issued rules and guidelines on the taxation of Bitcoin transactions, although it is still early to have a shared view regarding all countries.

(ii) Money laundering

Money laundering through the use of Bitcoin occurs when there is no uniformity regarding financial jurisdictions around the world. Important organizations such as the European Central Bank or the European Banking Authority have pointed at Bitcoin as one of the major sources of money laundering.

6. Scalability

The number of transactions in Bitcoin is increasing every day. In order for a transaction to be validated, this transaction needs to be stored, and due to the fact that the Bitcoin blockchain can only process seven transactions per second, this becomes a challenge when scaled into a worldwide network. Also, many small transactions are delayed as they provide lower transaction fees than bigger transactions for miners [12].

5   Future Directions

Bitcoin has shown its potential throughout the years. Nevertheless, due to its issues, it is important to trace a roadmap about future directions that it may take, such as:

1. Increase speed

   One of the main efforts by the community of Bitcoin developers is to increase the performance of its protocol. While it can reduce scaling, one of its main issues, it does not prevent forks from occurring. Even though delay might be reduced, this cannot prevent security breaches [13].

   Making the process more efficient for the user can help decrease the collision window. Making the process more efficient for the user can help decrease the collision window, which is the is the time lapse that occurs when node X hears node Y has found a block [14]. Nevertheless, there are limitations to the increase in speed, as if the processing speed increases to, let's say, 300%, it will tolerate a block size increase of the same percentage at the same fork rate [13].

2. Inclusive blockchain

   Another approach could be replacing the actual structure of the blockchain by the blocks making references to trimmed branches that include as well their transactions. This could improve the mining power usage as well as achieving a higher degree of fairness [15].

3. Alternative solutions

   Some alternative solutions, such as performing off-chain transactions, have been also suggested by some authors. This would mean allowing for payment networks to be placed on top of a certain blockchain. These networks would guarantee their privacy and security in comparison to Bitcoin's current protocol. In the event of a computer crash by the nodes performing the transactions, the information would be lost as this information never managed to make it to the blockchain [15].

   Some protocols, such as GHOST, are designed to improve the scalability of Bitcoin through a change in the chain selection rule. In Bitcoin, the longer chain (with the higher amount of work) is the main blockchain. In GHOST, whenever a fork occurs, the nodes would choose the sub-tree with the biggest amount of work [16].

4. Incentive

   One of the main discussions regarding Bitcoin, and all cryptocurrencies as a whole, has been the way in which miners are rewarded. Incentives have been analyzed through different perspectives. Some authors propose a different blockchain structure, which would encourage the participation of unconnected miners. Other authors point that an incentive that would arise naturally would diminish the formation of big mining pools [17].

5. Decentralized centers

   Some investigations point toward the idea of data centers which centralize and authorize information. These centers have the capacity to offer as well cloud-mining services. Services such as trust cues, through the data centers' website, can help increase the level of trust among miners. Also, some tools can help support miners and feed them with users' feedback [18].

6 Discussion

One of the widest uses of Bitcoin is related to cryptocurrency payments, but there are a number of areas where it can also be used in a near future. Blockchain as a whole can be used in a wide range of platforms, such as transportation or energy. The introduction of this technology in smart homes and smart cities can help expand the Internet of Things (IoT) system. Acting as a distributed data sharing platform, it guarantees a transparent accountability of a company's sources of revenue. Other areas where blockchain has not developed might see the introduction of this technology soon, such as in underground mining or firefighting [19].

In order to tackle the limitations of IoT resources, accessible computer resources are needed. Many authors have addressed the expensive entry barriers to blockchain, for example, in terms of equipment. A use of a lighter consensus algorithm, which would store a summary of the transactions instead of whole blocks without scarifying the user's trust, could be a solution [20].

7 Conclusions

Blockchain and Bitcoin are currently gaining an increase in interest in both academia and the industry. Nevertheless, it faces challenges that need to be solved before further expansion is reached. A more thorough investigation and bigger investment will help scaling and securing the system. We are far from using blockchain as a solution to many existing problems, but its characteristics set grounds for hope in the years to come.

References

1. S. Nakamoto, Bitcoin: a peer-to-peer electronic cash system (2008), https://bitcoin.org/bitcoin.pdf. Accessed 21 Feb 2020
2. J. Sarra, Q.C.L. Gullifer, Crypto-claimants and bitcoin bankruptcy: Challenges for recognition and realization. Int. Insolv. Rev. (2019). https://doi.org/10.1002/iir.1346
3. I. Khairuddin, C. Sas, An exploration of Bitcoin mining practices: Miners' trust challenges and motivations (2019). https://doi.org/10.1145/3290605.3300859
4. D. Mills et al., Distributed ledger technology in payments, clearing, and settlement, in *Federal Reserve Board Finance and Economics Discussion Series*, (2016)
5. R. Kher, S. Terjesen, C. Liu, Blockchain, Bitcoin, and ICOs: A review and research agenda. Small Bus. Econ. (2020). https://doi.org/10.1007/s11187-019-00286-y
6. B. Fabian, T. Ermakova, U. Sander, Anonymity in Bitcoin? The users' perspective, in *Proceedings of the 24th European Conference on Information Systems*, (Istanbul, 2016)
7. J. Chod, E. Lyandres, A theory of ICOs: Diversification, agency, and information asymmetry. SSRN (2018). https://doi.org/10.2139/ssrn.3159528
8. Z. Zheng, S. Xie, H.N. Dai, X. Chen, H. Wang, Blockchain challenges and opportunities: A survey. Int. J. Web Grid Serv. **14**, 352 (2018). https://doi.org/10.1504/IJWGS.2018.095647

9. W. Meng, E.W. Tischhauser, Q. Wang, Y. Wang, J. Han, When intrusion detection meets blockchain technology: A review. IEEE Access **6**, 10179–10188 (2018)
10. J. Bonneau, A. Miller, J. Clark, A. Narayanan, J. Kroll, E. Felten, SoK: Research perspectives and challenges for Bitcoin and cryptocurrencies (2015), pp. 104–121. https://doi.org/10.1109/SP.2015.14
11. A. Kosba et al., Hawk: The blockchain model of cryptography and privacy-preserving smart contracts (2016). https://doi.org/10.1109/SP.2016.55
12. M. Vukolic, The quest for scalable blockchain fabric: Proof-of-work vs. BFT replication, in *International Open Problems in Network Security*, (2015), pp. 112–125
13. C. Stathakopoulou, A faster Bitcoin network. Tech. Rep. (ETH, Zurich, 2015)
14. J.E. Pazmino, D. Silva, C.K. Rodrigues, Simply dividing a Bitcoin network node may reduce transaction verification time. Trans. Comput. Netw. Commun. Eng. **3**, 17–21 (2015)
15. C. Decker, R. Wattenhofer, A fast and scalable payment network with Bitcoin Duplex Micropayment Channels, in *Stabilization, Safety, and Security of Distributed Systems*, (Springer, 2015), pp. 3–18
16. Y. Sompolinsky, A. Zohar, Secure high-rate transaction processing in Bitcoin, in *Financial Cryptography and Data Security*, (Springer, Berlin, Heidelberg, 2015), pp. 507–527
17. I. Eyal, The Miner's Dilemma, in *IEEE Symposium on Security and Privacy*, (2015), pp. 89–103
18. A.J. Adoga, M.G. Rabiu, A.A. Audu, Criteria for choosing an effective cloud storage provider. Int. J. Comput. Eng. Res. (IJCER) **4**(2) (2014)
19. L.D. Xu, W. He, S. Li, Internet of Things in industries: A survey. IEEE Trans. Ind. Inf. **10**(4), 2233–2243 (2014)
20. A. Stanciu, Blockchain based distributed control system for edge computing, in *21st International Conference on Control Systems and Computer Science (CSCS)*, (2017). https://doi.org/10.1109/CSCS.2017.102

Smart Marketing on Audiovisual Content Platforms: Intellectual Property Implications

Elisa Gutierrez, Cristina Puente, Cristina Velasco, and José Angel Olivas Varela

1 Introduction

Since its creation, Internet has been a wide market for information exchange. All technologies associated and all the applications created have changed in many forms the way that human beings are connected and related. Therefore, information plays a key role. The speed of access to information nowadays and the use of it have posed a new forest of opportunities but have also brought new legal problems never expected before. For this purpose, we propose some legal questions associated to searching technologies and the way that information is treated.

The idea of dealing with information in a smart way was raised by Salton [1], considered as the father of modern search technology. His studies in the 1960s about how to manage huge amounts of information for searching and the development of one of the first information retrieval systems, SMART [2], with modern techniques in this field such as the use of tf-idf, vector space model, or relevance feedback, opened the door to a new via to explore and exploit information.

E. Gutierrez
UNED University, Madrid, Spain

C. Puente (✉)
ICAI, Pontificia Comillas University, Madrid, Spain
e-mail: cristina.puente@comillas.edu

C. Velasco
Elzaburu SLP, Madrid, Spain
e-mail: cvv@elzaburu.es

J. A. O. Varela
University of Castilla-La Mancha, Ciudad Real, Spain
e-mail: jaolivas@uclm.es

© Springer Nature Switzerland AG 2021
H. R. Arabnia et al. (eds.), *Advances in Artificial Intelligence and Applied Cognitive Computing*, Transactions on Computational Science and Computational Intelligence, https://doi.org/10.1007/978-3-030-70296-0_73

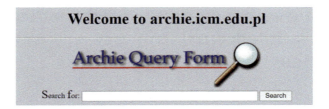

Fig. 1 Archie search engine interface

Over the years and with the advancement of the Internet, more sophisticated tools appeared. The increase of websites launched the first search engine, Archie [3], created by a student, Alan Emtage, in 1990, and rather simple while retrieving information, Archie basically matched query entries with filenames stored in a database of filenames. Archie provided a textbox to introduce the query and the appearance remained in modern search engines, as seen in Fig. 1:

As the World Wide Web was expanding, new forms of searching were emerging. That is the case of ALIWEB [4], created in 1992 by Martijn Koster and considered as the first search engine able to retrieve websites. Its working was based on the description of the websites provided by users, being indexed in a database in according to this manual description. But soon, the exponential expansion of the web made this procedure obsolete, so automatic bots were created to retrieve information about webpages and index them properly. But was this indexing process legal? It depends, though there will be other factors to be taken into account.

As consequence of the unstoppable growth of the Internet, more sophisticated search engines emerged, such as WebCrawler, Lycos, AltaVista, Yahoo, and Google, later with intelligent searching techniques or the first approach of question answering system based on natural language like Ask Jeeves. All of them had their own army of bots crawling the net, their own indexation system, their own interface, and their own raking policies.

But what happens when this information is "stolen"? Is it legal to get the information from another search engine or another website? That is the case of metasearch engines. In this work, we will analyze how engines process data from a legal point of view, presenting cases where intellectual property is protected and some others where it is not.

2 Metasearch Engines and Recommender Systems

Using its definition, a metasearch engine is an engine that uses other engine's information to provide its own results. As seen in Fig. 2, the architecture of a metasearch engine is feed by the results of other search engines, reordering their information to its convenience:

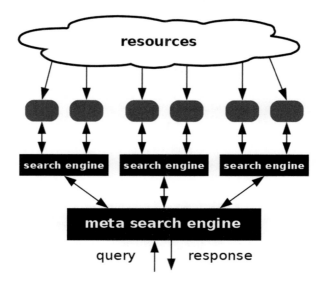

Fig. 2 Architecture of a metasearch engine [5]

This procedure is legal as far as programmers use the search engine's APIs to retrieve and gather the information and use their software to reorder and present it to the user. The other way to collect information is via web scraping and is not always permitted by law. Although this process can sound controversial, it is widely used nowadays, as in the case of the airline Ryanair versus the web portal site Atrapalo [6] in the Supreme Court of Spain. In this case, Atrapalo included Ryanair flight information into its portal allowing searching for the information of flights and moving this information into the Ryanair portal to book the flight.

But what happens when the object of search has intellectual property rights, mainly copyright and trademarks? Is it legal in this case the use of web scraping? To this extent, we can find on the Internet some cases such as Netflix, where its recommender system provides a set of "similar films" to the one searched in case this one is not in their catalogue, as seen in Fig. 3.

But is this practice legal? Can a system recommend related content titles with the one sought? In the next sections, we will analyze the legality of these questions.

3 Copyright

According to the World Intellectual Property Organization (WIPO), "[c]opyright (or author's right) is a legal term used to describe the rights that creators have over their literary and artistic works" [7]. The expression "literary and artistic works" may contain creations of a very varied nature. Among them, we must understand cinematographic and audiovisual works.

Fig. 3 Example of search in Netflix, when the film is not in their catalogue and they propose a set of similar items

The copyright recognizes a kind of monopoly for the author on his work, so that no one can use it without his consent, except for exceptions stated in the applicable law. But not any creation will be protected by copyright nor will any creator be considered an author for legal purposes.

In addition to being expressed, this creation must be sufficiently original. The requirement that the work must necessarily be original is a common criterion at the international level in the field of copyright, even though it is often not reflected in national legislation or in regional or supra-regional legal texts [8].

The ways of interpreting the notion of originality are different from country to country. In some countries, it is understood from a subjective point of view and in others from an objective point of view. Although there is no standard definition of the term original, for practical purposes, it is indifferent to differentiate between the two approaches [9]. Halfway among them, roughly speaking, we can understand that a creation is original when it is an expression of the artist's personality (subjective criterion) that it is not a copy of a previous work by another author and that it is different from other analogous creations already in existence (objective criterion) [10].

When a creation is sufficiently original, different national laws protect it as a work of copyright and its creator as the author. From the point of view of our paper, it should be borne in mind that, depending on the country, copyright protection may extend to the titles of the works. In this respect, different laws provide unequal answers to the same question, although there are mainly two main perspectives.

In many countries, like most European and South American countries, which are highly protectionist toward authors, titles are protected by copyright, provided that they are original. In these territories, the title is protected in an independent way in relation to the work it identifies, as in France [11], or more frequently as part of the work, as in Spain [12]. In these territories, regardless of originality, the title may also be registered as a trademark, if it complies with the applicable legislation to grant it express content additional protection.

In other countries such as the USA or the United Kingdom, the titles of the works are not protected by copyright in any case [13]. They are countries that have a way of understanding copyright from a very mercantilist perspective. These countries understand that the adequate protection for the titles of the works is directly the trademarks, since, from their point of view, they lack enough originality to be protected in any case [14].

An important difference between these two solutions or ways of protection is that the rights over a work and its title, when it is original, arise by the mere fact of the creation, so its protection is automatic. However, rights over a trademark usually arise if the trademark has been registered. The copyright or the trademark rights will limit or prohibit the use that can be made of the work or trademark without the permission of the corresponding owner, who in both cases is usually the audiovisual producer.

Audiovisual works are the main or only content of audiovisual content platforms, such as Netflix, HBO, Amazon Prime Video, or Disney+. Their vocation of universality implies that they are subject to as many laws as the countries to which they provide their services, since in matters of copyright, as of trademarks, the *lex loci protectionis* applies, i.e., the law applicable is the law of the place for which protection is claimed in each case, *where* the work is exploited, not the law of the country *from* which the platform can provide its services. So the same title of an audiovisual work may be automatically protected in a given territory and unprotected in another, without any registration by the author or owner.

If a content platform wishes to exploit a third party's audiovisual work, the owner of the audiovisual work must first authorize the exercise of certain exploitation rights over the work and, where appropriate, over the trademarks related to the creation under certain minimum conditions that allow the platform to develop its economic activity. Under these conditions, the platform may include the title of the licensed works that make up its catalogue in search engines and recommendation systems to facilitate access by users.

No legal audiovisual content platform makes available to users a work of which it is not the owner or which has not been licensed to it temporarily. However, it is a common practice for these platforms to include in their search engines and recommendation systems, in addition to the titles of the works available in their catalogue, titles of works that have not been licensed to them and over which they do not have any exploitation rights (neither rights of authorship over the work itself nor over the trademarks linked to the work). The intention of this is that in the event that a user enters a work that is not part of the platform's catalogue of contents in the search engine, the results returned or the future recommendations will be the contents available in the catalogue, at present or in the future, with which the work being searched for maintains some proximity (plot, subject matter, genres, actors, directors, etc.).

As we know, copyright grants to the owner a certain exclusive rights to his work and to the title of the work, if it is protected in the country concerned. The fact that titles are protected by copyright means that they cannot be reproduced or copied without the permission of the title holder. This prohibition generally extends to the introduction of the title of a work into an intelligent recommendation system or search engines, at a programmatic or internal level, that is to say, without it being shown to the user and whether or not the audiovisual work is made available to the public or any other use is made of the work it identifies.

On the contrary, if the title of the work is not protected by the corresponding legislation or is not sufficiently original, no such infringement would occur.

Then, the same use of a mere title by a content platform may be a copyright infringement or not, according to the title and depending on the case and on place from where the search is performed and for which protection would be claimed, being the applicable law, so it is not a recommended practice from this point of view.

4 Trademarks and Unfair Competition

In some cases, as we have introduced, the titles of audiovisual works may be protected as trademarks. However, although registration of the titles of an audiovisual work is a widespread practice, it should be noted that a trademark is any sign capable of distinguishing the goods or services of one undertaking from those of other undertakings [15]. Therefore, although trademark protection is not excluded for protecting titles of audiovisual works, only those titles that are capable of indicating certain commercial origin of the concrete work can be protected as trademarks. So it must be distinctive in relation to the good or service they identify in order to be registered as a trademark.

The foregoing means that trademarks which consists only of a description of the plot of the audiovisual works (e.g., *Love* for a romantic comedy), which are too generic (e.g., *The Others*) or which are equal to an expression identifying a popular story or film that has become customary for the relevant public, in most cases may not be protected by trademarks (e.g., *The Jungle Book* or *Cinderella*) because are considered incapable of distinguishing the certain origin. These "expressions" would not have the sufficient distinctive character that a trademark requires for its protection in relation to the work they wish to identify.

Unlike copyright, which duration of protection is linked to the person of the author or his first publication and is intended to provide an incentive for creation, trademarks, whose purposes are strictly commercial, can be renewed indefinitely for specified periods. The potentially infinite monopoly on a trademark is justified as long as the trademark (the title of an audiovisual work, in this case) is being commercially exploited (or, failing that, there is a sufficiently justified reason for not having used it in commercial traffic despite being registered), i.e., as long as the required genuine use is carried out.

In this regards, while in the USA, the trademark owner must file a declaration of use (and/or excusable nonuse) between the 5th and 6th years after the registration, in the EU, the trademark may be renewed, without submitting such declaration, every each 10 years from the date of filling the application (although the trademark is vulnerable due to lack of genuine use after 5 years from its date of registration).

From the trademark perspective, the use of another's trademark in an audiovisual platform should only be carried out with the express consent of its owner, unless it does not meet the requirements to be legally considered as a trademark (because it is generic and descriptive or coincides with a well-known classic story for the public, mainly). However, if there is an appropriate trademark to distinguish films or series, an unauthorized use of the title might not be advisable, since registration of a trademark provides the owner an exclusive right to use the mark on or in connection with the goods/services listed in the registration. By way of clarification, when a mark is registered, the goods and/or services for which the trademark is to be used are determined in the application.

Unlike in the field of copyright, the perceptive registration of trademarks (with a few exceptions) in those territories where you want to protect means that there is a register in each country or region that provides third parties with full certainty as to whether a trademark is actually registered.

However, it should be borne in mind that a trademark infringement requires that the use of another's trademark is made in the course of trade, in relation to similar goods or services, and causes confusion or association about business origin on the consumer. Based on this premise, it is at least doubtful that the search engines or recommender systems included by the audiovisual platforms would entail a trademark infringement in any case, firstly, because the searches for titles are carried out on the platform's own search engine and, therefore, a contractual relationship between the platform and the effective user already exists and, secondly, because when the related results are displayed, the consumer does not confuse the titles of the audiovisual works actually available in the platform's catalogue with the one entered in the search engine.

A different matter would be if the title of the audiovisual work that is not available in catalogue, and over which there is no right of use, were to be used before the service is hired by the consumer, i.e., in the sales process or in the advertising of the platform, including those cases in which the search engine could be used without a previous hiring or identification in the platform, with the aim of promoting one's own titles or encouraging the contracting of the services offered by the platform by using, for example, the knowledge or notoriety of the titles of the audiovisual works not currently available.

In such cases and in the absence of a trademark infringement, an unfair competition practice could be developing that is considered illegal by the legislation of the corresponding country in which such practice is taking place. The criteria for prosecuting such a practice from the perspective of unfair competition, which is also subsidiary in the absence of copyright infringement, may be more or less flexible from one country to another, so that there would also be no full certainty as to the legality of this practice.

5 Conclusions

Unlike trademarks, whose registration is mandatory in most cases, where there is a "list" of the titles of works that constitute a trademark, copyright protection is automatic, and registration is optional, so there is no such "list," and there will never be full certainty of whether or not a title will be protected in a territory. This decision will remain in the hands of the judge, who will assess the originality of the title of the work and whether its protection complies with the legislation of the country in each case.

In view of the above, from a copyright perspective, it would be advisable to (i) adjust search engines and recommendation systems to the catalogue available to each platform in those countries where titles of works may be protected by copyright or (ii) refrain from using the title of any work over which no exploitation rights are held. Even if originality in the title is additionally required to understand a copyright infringement with such a practice, the infringement becomes potential. Since the appreciation of originality is subject to a certain arbitrariness that will depend on the judge in each case, there will be no full assurance that such infringement is not taking place.

Although the titles of the audiovisual works are not protected by copyright and despite from a trademark perspective it is questionable whether the practice prosecuted is unlawful because there is a prior contractual relationship between the user and the platform, and the user is not being misled about the content available on the platform, other acts of unfair competition may be committed through this practice in the programming of the recommender systems of the search engines, which will also be assessed by the judge in each case. In view of the above, the recommendations made regarding copyright would also be extendable to the present case, and the recommender systems should be limited to the works available on the platform.

Acknowledgments This work has been partially supported by FEDER and the State Research Agency (AEI) of the Spanish Ministry of Economy and Competition under grants MERINET (TIN2016-76843-C4-2-R) and SAFER (PID2019-104735RB-C42), AEI/FEDER, UE.

References

1. G. Salton, A theory of indexing, in *Conference Board of the Mathematical Sciences: CBMS-NSF regional conference series in applied mathematics*, vol. 18, (SIAM, 1975)
2. G. Salton, The SMART System, in *Retrieval Results and Future Plans*, (1971)
3. M. Koster, ALIWEB-Archie-like indexing in the web. Comput. Networks ISDN Syst. **27**(2), 175–182 (1994)
4. T. Seymour, D. Frantsvog, S. Kumar, History of search engines. Int. J. Manage. Inf. Syst. (IJMIS) **15**(4), 47–58 (2011)
5. https://commons.wikimedia.org/w/index.php?curid=4952120
6. https://supremo.vlex.es/vid/desleal-ryanair-atrapalo-as-411388894
7. Anonymous. Copyright. What is copyright? Available at www.wipo.int/copyright/en/. Last visited: May 2020.
8. C. Masouyé, *Guide to the Berne Convention for the Protection of Literary and Artistic Works (Paris Act, 1971)* (World Intellectual Property Organization, Geneva, 1978), pp. 18–19
9. R. Bercovitz Rodríguez-Cano, The work on *Manual of Intellectual Property (in Spanish)*. Coordinated by Rodrigo Bercovitz Rodríguez-Cano. Tirant lo Blanch. Valencia, Spain, 54 (2018)
10. R. Bercovitz Rodríguez-Cano, Art. 10.1 in Comments on Intellectual Property Law (in Spanish). Coordinated by Rodrigo Bercovitz Rodríguez-Cano. Tecnos. Madrid, Spain, pp. 160-166 (2018)
11. Article L112-4 Intellectual Property Code (France).
12. Article 10.2 Royal Legislative Decree 1/1996, of April 12, approving the revised text of the La won intellectual property, regularizing, clarifying and harmonizing the legal provisions in force on the subject (Spain).
13. J.I. Peinado García, Article 10.Works and original titles in Comments on Intellectual Property Law (in Spanish). Felipe Palau Ramírez and Guillermo Palao Moreno. Tirant lo Blanch. Valencia, Spain, pp. 228–229 (2017)
14. As established in the USA Code of Federal Regulations, Title 37, § 202 (Available at www.govinfo.gov/app/collection/cfr. Last consultation date April 2020). In the same sense, it is pronounced UK Copyright Service, in fact sheet P-18: "names, titles and copyright" (Available at copyright-service.co.uk/copyright/p18_copyright_names. Last consultation date April 2020). Vid., McJohn, S. M. (2015). Intellectual Property. Examples & Explanations (Wolters Kluwer, New York, 2002). Vid. ed., Lutzker, A.P. Content Rights for creative professionals. Copyrights and Trademarks in a Digital Age. Focal Press. Burlington, MA, USA.
15. Anonymous. Trademarks. What is a trademark? Available at www.wipo.int/trademarks/en/index.html. Last visited: May 2020.

Priority Management in a Cybernetic Organization: A Simulation-Based Support Tool

J. C. Puche-Regaliza, J. Costas, B. Ponte, R. Pino, and D. de la Fuente

1 Introduction

'Time management is an oxymoron. Time is beyond our control, and the clock keeps ticking regardless of how we lead our lives. Priority management is the answer to maximizing the time we have', attributed to John C. Maxwell [1]

Priority management is fundamental for organizations of any kind and scale; see, e.g. Westbrook [2], Jin et al. [3] and Kim [4]. Managers need to perceive the key stimulus from the environment, as well as to consider the organization's resources and capabilities, in order to assign the available assets to the most appropriate tasks. In this sense, they constantly need to manage a queue of pending tasks in an efficient manner by defining an appropriate workload prioritization policy. Defining such policy is a complex challenge, as factors of different types, related to both the task itself and the customer should be simultaneously taken into consideration with the aim of maximizing organizational value.

J. C. Puche-Regaliza (✉)
Department of Applied Economics, Faculty of Economics and Business, University of Burgos, Burgos, Spain
e-mail: jcpuche@ubu.es

J. Costas
Department of Engineering, Florida Centre de Formació, Florida Universitária, Catarroja, Valencia, Spain
e-mail: jcostas@florida-uni.es

B. Ponte · R. Pino · D. de la Fuente
Department of Business Administration, Polytechnic School of Engineering, University of Oviedo, Gijón, Spain
e-mail: ponteborja@uniovi.es; pino@uniovi.es; david@uniovi.es

© Springer Nature Switzerland AG 2021
H. R. Arabnia et al. (eds.), *Advances in Artificial Intelligence and Applied Cognitive Computing*, Transactions on Computational Science and Computational Intelligence, https://doi.org/10.1007/978-3-030-70296-0_74

In this work, we show how a model-driven decision support system (DSS) can be developed to assist organizations in the efficient management of priorities. To develop our solution, we use discrete-event simulation (DES) and apply principles from Beer's Viable System Model (VSM).

2 Methods

Organizational cybernetics is a system approach, developed by Stafford Beer [5], that applies the control principles of cybernetics to enterprises. One of its key components is the VSM [6]. This allows managers to design good-performing companies that are equipped with effective regulatory, adaptive, and learning capabilities [7]. In doing so, five systems are defined to deal with the complexity of the environment.

System One, which includes the operational units, produces and delivers the goods and services by carrying out tasks. System Two plays a key role in the coordination of System One, making it work effectively. Systems Three and Three* monitor and control the performance of System One and optimize its performance by managing the insides of the organization. System Four considers the future of the organization, by constantly analysing the meaningful changes in the environment. Lastly, System Five formulates the strategic goals of the system.

3 Our Solution

We assume that the prioritization policy should consider four main factors:

1. *Customer importance.* Prioritizing the tasks related to some customers is often the 'name of the game' for organizations. In this regard, we assume that the company categorizes customers using A to D scale (A being the highest).
2. *Nature of task.* We define three different types of tasks. First, P-tasks are those that directly generate revenue for the company. Second, CP-tasks refer to those that do not generate direct revenue but allow the organization to improve its performance, including innovation and continuous improvement (e.g. total quality management initiatives). Third, 5S-tasks denote those tasks that neither generate direct revenue nor increase the company's performance. Rather, they avoid an erosion in the organization's resources and capabilities.
3. *Value of task.* We assume the value of the task can be quantified, or at least estimated, by the managerial team.
4. *Window of opportunity.* This refers to the period of time during which the task achieves the 'optimal' outcome. Once this period is over, the desired outcome is no longer possible, and the value of carrying out the task reduces.

To develop the model-driven DSS, we use DES techniques [8]. Importantly, we consider that System One receives as follows: (i) 5S-tasks from System Two; (ii) P-tasks from System Three; and (iii) CP-tasks from System Four.

In our DES system, System One is initially 'idle'. Once a task arrives, it instantly moves to 'preparing' mode, where the resources are allocated to the specific task. After some (random) time, it moves to 'active' mode, in which the task is carried out. As soon as the task finishes, the resources are released, and System One returns to 'idle' mode. At that moment, more than one tasks may be waiting in the queue. When this happens, the prioritization policy decides which task to perform. Note that at any point in time, System One may need to move to the 'out-of-service' mode.

The prioritization policy considers the four criteria defined before according to a setting introduced by the user. The simulation model allows managers to evaluate different prioritization policies and select the most appropriate setting according to the most relevant key performance indicators for the organization.

4 Concluding Remarks

In this work, we have shown how DES techniques and the VSM approach can be effectively combined to develop a DSS aimed at improving prioritization strategies in companies. Further work can be done to explore in detail the performance of different prioritization policies based on the set of factors defined before.

References

1. ACEL, Management is not a dirty word. e-Leading **30**, 1–4 (2017)
2. R. Westbrook, Priority management: New theory for operations management. Int. J. Oper. Prod. Manag. **14**(6), 4–24 (1994)
3. H. Jin, H.A. Wang, Q. WAng, G.Z. Dai, An integrated design method of task priority. J. Software **14**(3), 376–382 (2003)
4. T.Y. Kim, Improving warehouse responsiveness by job priority management: A European distribution centre field study. Comput. Ind. Eng. **139**, 105564 (2018)
5. S. Beer, *Cybernetics and Management* (English Universities Press, London, 1959)
6. S. Beer, *Diagnosing the System for Organizations* (John Wiley & Sons, Chichester, 1985)
7. J. Puche, B. Ponte, J. Costas, R. Pino, D. De la Fuente, Systemic approach to supply chain management through the viable system model and the theory of constraints. Prod. Plann. Control **27**(5), 421–430 (2016)
8. G.A. Wainer, P.J. Mosterman, *Discrete-Event Modeling and Simulation: Theory and Applications* (Taylor & Francis, Boca Raton, 2009)

A Model for the Strategic Management of Innovation and R&D Based on Real Options Valuation: Assessing the Options to Abandon and Expand Clinical Trials in Pharmaceutical Firms

J. Puente, S. Alonso, F. Gascon, B. Ponte, and D. de la Fuente

1 Introduction

Appropriately managing innovation in pharmaceutical firms not only is essential for improving the economics of these organizations but also is integral for the advancement of societies given the health implications of these innovations; see, e.g. Schmid and Smith [1] and Zhong and Moseley [2]. In this regard, it is important to highlight that R&D projects in pharmaceutical companies are characterized by key differentiating features, including as follows: (1) having a clearly defined sequential development (in phases); (2) being subject to a wide range of sources of uncertainty, related not only to their processes but also to the performance of competitors; and (3) having discretionary capacities to decide on the future evolution of a series of circumstances during and after the development process of a new drug.

Therefore, the evaluation of R&D projects in these companies requires the application of flexible assessment tools that are capable of accommodating these differentiating characteristics. From this perspective, the literature has long pro-

---

J. Puente · B. Ponte · D. de la Fuente
Department of Business Administration, Polytechnic School of Engineering, University of Oviedo, Gijon, Spain
e-mail: jpuente@uniovi.es; ponteborja@uniovi.es; david@uniovi.es

S. Alonso
Department of Financial Economics and Accounting, Faculty of Economics and Business, University of Valladolid, Valladolid, Spain
e-mail: salonso@eco.uva.es

F. Gascon (⊠)
Department of Business Administration, Faculty of Economics and Business, University of Oviedo, Oviedo, Spain
e-mail: fgascon@uniovi.es

© Springer Nature Switzerland AG 2021
H. R. Arabnia et al. (eds.), *Advances in Artificial Intelligence and Applied Cognitive Computing*, Transactions on Computational Science and Computational Intelligence, https://doi.org/10.1007/978-3-030-70296-0_75

moted the application of investment selection techniques based on the real options approach to the valuation of R&D projects, given that real options valuation is well known as a valuable tool to assess projects that have an intrinsic flexibility and face strong uncertainties; see, e.g. Lo Nigro et al. [3].

A very good example of this stream of literature that uses real options to evaluate R&D projects in the pharmaceutical sector is the work by Brandao et al. [4]. These authors propose a real options-based model for the assessment of innovations in this industry and apply it to a numerical example of R&D investment they develop based on data from different data sources. The procedure they use for considering abandonment and expansion options when developing new drugs is based on the least squares Monte Carlo (LSM) model proposed by Longstaff and Schwartz [5] in the field of financial derivatives, which combines Monte Carlo simulation, dynamic programming and statistical regression. Finally, we note that another paper that need to be mentioned, who indeed inspired Brandao et al.'s [4] work, is that one by Hsu and Schwartz [6], who developed a real options-based R&D valuation model for the pharmaceutical industry that considers the uncertainty in the quality of the output, the time and cost to completion and the market demand for the output.

2 Objectives

In this work, we conceptually propose a valuation model, based on real options, that allows pharmaceutical laboratories to adequately assess their R&D projects. This would have implications both on the laboratory's pipeline of clinical trials and its portfolio of innovative drugs, helping the organization to determine whether they should promote or abandon clinical trials and new drug developments.

From this perspective, our approach follows the steps defined by Brandao et al. [4], although we have included relevant modifications that allow us to adapt the model for a wider set of empirical applications in the development of new drugs. In this short research paper, we describe how we will carry out the implementation of our proposal, putting special emphasis on the differences with Brandao et al.'s [4] work. At this point, we note that, along with the real options valuation method, we employ simulation and fuzzy techniques. In this regard, we have made use of recent advancements in the area, such as the work by Puente et al. [7] that shows how fuzzy techniques may define a very powerful approach to study R&D decision-making in pharmaceutical companies.

3 Background: Developing New Drugs

The development process of a new drug consists of different sequential stages; see, e.g. Friedman et al. [8] for a large amount of detail. First, the *preclinical testing phase* refers to the discovery of lead compounds, that is, chemical compounds whose pharmacological activity experts consider to be potentially useful in therapeutic terms.

Second, the *clinical trials* in turn consist of three main phases. In phase I, the drug is tested on a reduced number of volunteers (generally, 20–100), with the aim of determining whether the drug is safe to check for efficacy. In phase II, the drug is tested on a higher number of patients in order to assess efficacy and side effects. Thus, the success rate is significantly lower than in the previous phase. Finally, phase III involves a study with a wider sample of patients (up to 3000) to confirm, or not, the findings derived from phase II in terms of efficacy, effectiveness and safety. This phase normally takes several years to complete.

It is important to note that while progressing between the different clinical trials phases, successive abandonment options are considered by pharmaceutical companies (generally, at the end of each phase). On the one hand, this occurs because each phase of testing provides the company with additional information on the usefulness of the compound. On the other hand, the passing of time may have affected (reduced) the economic viability of the R&D project; this is common, for example, when other pharmaceutical laboratories are investigating on the same disease. Taking both aspects into account, successful laboratories need to decide whether to continue or abandon the clinical trial.

According to Adams and Brantner [9] and Holland [10], the whole process of drug development—which spans from preclinical testing to commercialization, which may be understood as the last phase—often takes 12–18 years and costs in average $1.2 billion (value estimated for the US market in the early 2010s). Last, it should be underlined that once a R&D project has overcome all the clinical trials, the *post-marketing surveillance* needs to be carried out. This involves monitoring the safety of the drug after it has been released on the market.

Finally, we note that the term of patent—that is, the period of protection—for a new medicine is 20 years from the filing date, both in Europe and in the USA. However, patents need to be filled before public disclosure (which typically occurs before human testing); therefore, the average time in which the pharmaceutical laboratory has exclusivity in the market is significantly lower, as an average, 13 years [11]. After that, it may be assumed that there is no residual value. Nonetheless, at this stage, the possibility of expanding the commercialization of the product to other groups of patients can be contemplated, which would boost the income. In this sense, evaluating this expansion option involves estimating the rate at which the flows would increase if expansion were to be carried out to other groups of patients.

4 Methods

In the evaluation of the different real options that need to be considered in the R&D process of developing a new drug (basically, abandonment in each of the development phases and expansion in the commercialization phase), our model proposes the adoption of the optimal strategy according to the most likely outcome. This requires comparing the estimated value of the stream of cash flows that the project is expected to generate (incomes minus commercialization costs), provided that it reaches the commercialization phase, to the estimated value of the R&D costs that are expected to be incurred in the different stages of the new drug development project.

Each of the income and cost stream depends on a number of parameters, some of which will be modelled stochastically, while others will be estimated from a fuzzy inference system (FIS). We first focus on the evolution of the future stream of cash flows that the project is expected to generate, V, assuming its stochastic behaviour. This is the key variable on which the decision rights (abandonment and expansion options) available to the company are defined. We will estimate the cash flows during a time interval of T, which spans up to the expiration of the patent. Thus, and given that the stochastic value of the stream of cash flows is the only source of systematic risk in the R&D project, it will also be necessary to estimate the risk-adjusted discount rate.

4.1 Estimation of Commercialization Cash Flows

To estimate the potential cash flows derived from the commercialization of the new drug, it is necessary to consider the ultimate sources of uncertainty on which they depend, namely, (i) the volume of final product obtained and (ii) its sale price in the market. Initially, the company makes an estimate of the specifications and quality standards that the finished product needs to achieve the (ideal) blockbuster status, which helps to estimate the present value of the expected cash flows in the commercialization phase.

As the clinical trial phases progress, pharmaceutical companies have more information, which they can use to update the specifications and quality standards and thus the cash flow estimations. In this sense, a key difference between our proposal and that of Brandao et al. [4] is that we incorporate the active learning from the FIS. Thus, at the end of each R&D phase, the level of adjustment to the ideal quality specifications will be estimated based on two input variables: the previous volume of clinical trials in that disease and the volume of drugs authorized in that area. In doing so, the knowledge base of this FIS would vary depending on the phase of drug development, and its conditional rule structure would tend to estimate increasing levels of adjustment to those pharmaceutical laboratories with higher volumes of both clinical trials and authorized drugs in the corresponding disease area.

Additionally, the expected value of the future flows generated by the project, V, may be affected by other sources of model uncertainty. Hence, we consider the existence of risk associated with a catastrophic failure event. Thus, at any time, the project may lose its value due to different events—for example, a competitor that becomes substantially ahead in that area or a decisive regulatory change. If this

A Model for the Strategic Management of Innovation and R&D Based on Real... 931

happens, then the value of the project becomes null. This circumstance is modelled with a Poisson process where, at each moment, we assume that there is a probability of the catastrophic event occurring.

A second important avenue of contribution with respect to the work by Brandao et al. [4] is that our model also includes the possibility of exploiting the external capacity of R&D. This contemplates the possibility that, during the phases of clinical trials, a financial transaction may be carried out to improve the company's position in the development of the new drug. This feature is also modelled through a (second) FIS. Thus, a pharmaceutical laboratory could analyse its corporate tree and use financial transactions to improve its positioning in the development of new drugs in a given disease area. The FIS would then estimate the degree of improvement of the laboratory's positioning in each development phase based on two variables: (i) the volume of financial transactions of the laboratory oriented towards that disease and (ii) the extent and coverage of its corporate tree in that context. Again, in this case, the knowledge bases would give better ratings to those laboratories with higher levels of financial transactions and with a more suitable corporate tree for a given disease.

Lastly, it is important to note that as a result of the exploitation of external R&D capacity, there could be a reduction in development times, which would result in an extension of the period of commercialization of the drug and a beneficial effect on the estimated value of the flows.

4.2 Estimation of R&D Costs

To estimate the volume of costs involved in the development of each of the phases discussed in Sect. 3, we use information available from different data sources, including historical data from the European Medicines Agency (EMA) and the US Food and Drug Administration (FDA) (see, e.g. [12]). We assume that there is uncertainty in the amount of cost and the duration of each phase.

5 Final Remarks

Managing innovation and R&D is a highly complex and dynamic task in pharmaceutical firms that has significant effects on the health of societies. In this work, we have suggested how to devise a valuation model for assessing new drug development projects by using real options analysis, fuzzy techniques and simulation. Specifically, the model would act as a support system that would assist decision makers in determining whether to promote or abandon clinical trials based on different uncertain factors. As we have discussed in the previous section, our model differs from previous approaches in the literature by including additional capabilities to better accommodate some important real-world uncertainties faced by pharmaceutical organizations.

References

1. E.F. Schmid, D.A. Smith, Managing innovation in the pharmaceutical industry. J. Commer. Biotechnol. **12**(1), 50–57 (2005)
2. X. Zhong, G.B. Moseley, Mission possible: Managing innovation in drug discovery. Nat. Biotechnol. **25**(8), 945–946 (2007)
3. G.L. Nigro, A. Morreale, G. Enea, Open innovation: A real option to restore value to the biopharmaceutical R&D. Int. J. Prod. Econ. **149**, 183–193 (2014)
4. L.E. Brandão, G. Fernandes, J.S. Dyer, Valuing multistage investment projects in the pharmaceutical industry. Eur. J. Oper. Res. **271**(2), 720–732 (2018)
5. F.A. Longstaff, E.S. Schwartz, Valuing American options by simulation: A simple least-squares approach. Rev. Financ. Stud. **14**(1), 113–147 (2001)
6. J.C. Hsu, E.S. Schwartz, A model of R&D valuation and the design of research incentives. Insur. Math. Econ. **43**(3), 350–367 (2008)
7. J. Puente, F. Gascon, B. Ponte, D. de la Fuente, On strategic choices faced by large pharmaceutical laboratories and their effect on innovation risk under fuzzy conditions. Artif. Intell. Med. **100**, 101703 (2019)
8. L.M. Friedman, C. Furberg, D.L. DeMets, D.M. Reboussin, C.B. Granger, *Fundamentals of Clinical Trials*, vol 4 (Springer, New York, 2010)
9. C.P. Adams, V.V. Brantner, Estimating the cost of new drug development: Is it really $802 million? Health Aff. **25**(2), 420–428 (2006)
10. J. Holland, Fixing a broken drug development process. J. Commer. Biotechnol. **19**(1), 238–263 (2013)
11. S. Dutta, B. Lanvin, S. Wunsch-Vincent, Global innovation index 2019 – Creating healthy lives–the future of medical innovation (2019). Available via https://www.wipo.int/edocs/pubdocs/en/wipo_pub_gii_2019.pdf. Accessed 31 May 2020
12. F. Gascón, J. Lozano, B. Ponte, D. de la Fuente, Measuring the efficiency of large pharmaceutical companies: An industry analysis. Eur. J. Health Econ. **18**(5), 587–608 (2017)

Part VIII
International Workshop – Intelligent Linguistic Technologies; ILINTEC'20
(Chair: Dr. Elena B. Kozerenko)

The Contrastive Study of Spatial Constructions *na* NP_{loc} in the Russian Language and 在NP上 in the Chinese Language in the Cognitive Aspect

Irina M. Kobozeva and Li Dan

1 Introduction

According to the [1], there are 204 prepositions in the Russian language. In [2], the analyses of 60 prepositions and their typological classification are proposed. Even on this basis we can expect that the Russian language reflects the relationship between objects including spatial ones in more detail, and the Chinese language does that in a more integral way.

Spatial relations (further SR) between the phenomena and processes of the world around us are encoded in the languages of the world by various grammatical and lexical means. In Chinese, the SR are mainly expressed by prepositions and postpositions ([3]: 93). Given the peculiarities of the SR expression in Chinese, Lu Danqing proposed the concept of a 'frame construction'. According to Lu Danqing, the framework is formed by a preposition and a postposition, between which there stands a name (i.e. a nominal unit) [4]. We will be considering the frame structures such as 在+NP + P with the preposition 在, meaning a common location (LOC), such as 在桌子上(zài zhuōzi shàng) 'on the table', 在水中(zài shuǐ zhōng) 'in the water', 在房子里(zài fángzi lǐ) 'in the house' and so on. In the Russian language, prepositions are used to express the SR. The peculiarity of the prepositions in Russian is that not only the preposition itself but also the case form is involved in the expression of a certain meaning. Russian preposition-case structures correspond in their function to Chinese frame structures. We will compare the

---

I. M. Kobozeva (✉)
Lomonosov Moscow State University, Moscow, Russia

L. Dan
Hejlongjiang University, Harbin, China

Jilin International Studies University, Changchun, China

© Springer Nature Switzerland AG 2021
H. R. Arabnia et al. (eds.), *Advances in Artificial Intelligence and Applied Cognitive Computing*, Transactions on Computational Science and Computational Intelligence, https://doi.org/10.1007/978-3-030-70296-0_76

Russian construction *na NP_{loc}* and the Chinese construction 在NP上*(zài NP shàng)* expressing the superessive (SUPER) spatial meaning in order to identify similarities and differences in the sets of spatial situations reflected by these indicators.

These constructions will be compared on the basis of a cognitive approach. We will examine their basic spatial use, consider the underlying cognitive schemas and identify the main similarities and differences between them.

Cognitive linguistics is a trend in linguistics that explores the role of language in the conceptualization and categorization of the world, in cognitive processes and in generalization of human experience. Interacting with the world, man masters the existing spatial relationships in it: inside-outside, top-bottom, right-left, etc. M. Johnson called the basic abstract cognitive structures that were formed as a result of human interaction with the surrounding world 'image schema' [5]. E.V. Rakhilina proposed to translate this term as 'topological schema', thus emphasizing that such schemas were originally formed on the basis of spatial sensations and motor reactions of a person ([6]: 364–365). R. Langacker, who described the semantics of spatial prepositions using such generalized schemas presented in the form of drawings, coined the terms *trajector* and *landmark* [7, 8]. L. Talmy used the concepts of *figure* and *ground* for the same purpose [9].We will use the terms *trajector* and *landmark* to describe the image schemas that correspond to the preposition *na 'on'* in a locative meaning, i.e. the one it has in the construction *na NP_{loc}* and the frame construction 在NP上*(zài NP shàng)*. In a simplest locative spatial situation, a static *trajector* (TR) being in the focus of attention is localized by reference to a static *landmark object* (LM), which is a secondary focus, for example, *Na stole* (LM) *lezhit kniga* (TR) – *On the table* (LM) *there is a book* (TR). In a more complex case, the moving TR is localized by a relatively static LM, for example, *Na stsene* (LM) *tantsuet devushka* (TR) – *lit. On the stage dances a girl 'On the stage (LM) there is a girl* (TR) *dancing'*. The difference between the locative situation with the mobile TR and the directional spatial situation expressed in Russian and Chinese by other means is the following. In a locative situation, the mobile TR either moves without changing its spatial coordinates (e.g. spins, bows, squats, sways) or makes multidirectional or unidirectional movements, remaining within the boundaries of LM, conceptualized as a surface (e.g. walking on the ground, running in the stadium). In a directional situation, TR moves in order to achieve a stationary LM, conceptualized as a point (e.g. go to the ground, run to the stadium). We will represent graphically the local image schemas for situations with a stationary TR, and the situations with moving TR will be described verbally. In this paper, we do not consider situations in which not only TR but also LM is on the move, e.g. *jekhat' na avtobuse* 'go by bus'.

The prototypical uses of locative structures *na NP_{loc}* in the Russian language and 在NP上(zài NP shàng) in Chinese include the idea of contact between TR and LM. Their respective image schemas vary in spatial configuration of TR and LM. At the same time, it is important to pay attention to such aspect of image schemas as topological types of TR and LM because this often determines whether the translation equivalence of the locative structures in question will remain.

The concept of topological type was introduced by L. Talmy to simulate the linguistic conceptualization of space. He showed that in the linguistic picture of the world, the variety of real-world forms of objects is reduced to a limited number of schematic and abstracted geometric forms, which he called topological types. In later work [10], the term '*geometric schema*' is used in the same sense, but we will continue to use the '*topological type*' term, as it has been entrenched in the use of Russian linguists (see, e.g. Rahilina [6]). L. Talmy mentions such topological types as POINT, LINE, PLANE, CIRCLE, LIMITED SPACE (ENCLOSURE), CYLINDER, HOLE and their fixed-orientation options, e.g. VERTICAL CYLINDER, and/or an additional marker of isolation, such as BOUNDED LINE (i.e. a line segment) ([10]: 245–252). These semantic units are used when he describes semantics of structures with spatial propositions in English and other languages. E.V. Rahilina describing in terms of topological types the compatibility of the Russian dimension adjectives employs the semantic metalanguage that in addition to topological 'primitives' of L. Talmy includes the names of prototypical physical objects, which in addition to geometric characteristics, orientation in space and isolation can indicate other properties, for example, flexibility/rigidity, e.g. the type ROPES vs. the type BARS.

Talmy showed that for the linguistic conceptualization of space, it is more important to specify the topological type for LM (ground in his terminology), while TR (figure in his terminology) in most spatial configurations is reduced to the type of DOT (see this also in ([11]: 147)). That is why the image schemas of TR in most cases given below are depicted uniformly, i.e. in the form of a three-dimensional rectangular figure.

2 Contrastive Study of Image schemas Encoded by the Constructions *na NP$_{loc}$* and 在NP上(zài NP shàng)

2.1 The schema in which TR is in contact with LM, LM is horizontally oriented and TR is located above the LM. In this spatial configuration, LMs can be of different topological types. In view of translation from Chinese into Russian, it is important to distinguish between three cases.

2.1.1 LM is expressed by a noun with the meaning of a place (geographical object), e.g. *land, sea* and *field*; a physical name (which in this context denotes a place covered by a layer of some 'substance'), e.g. *ice* and *grass*; or a name with an object meaning: *a table, a chair, a shelf, a bed,* etc. The object LMs, e.g. a table in the context of the structures in question refer to the type SURFACE (PLANE) because the main functional part of the objects they designate (in the case of the *table,* it is *the table top*) refers to this topological type. A TR is usually expressed by a common noun. The area of contact of TR and LM is not significant (cf. a *person lying on the bed, a cup on the table*). In this case, presented in Fig. 1, the preposition *na 'on'* is used, and its Chinese equivalent is 在NP上(zài NP shàng).

Fig. 1 An image schema of the '*on*' localization for LMs of the type SURFACE

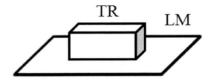

All Chinese examples below are taken from the Corpus of Beijing University of Language and Culture (BLCU Chinese Corpus, abbreviated BCC) and glossed according to the Leipzig system of glossing rules [12]: CLF is for a classifier, a count word; LOC is for a locative marker; POSS is for a possessive particle; PROG is for progressive marker; SUPER is for superessive marker, and a BA gloss is added for the particle used in a direct object inversion. Chinese examples are translated into Russian and English by the authors of this article.

(1)	信纸	和	钢笔	在桌子上。
	Xìnzhǐ	hé	gāngbǐ	zài=zhuōzi-shàng. (Chin.)
	paper	and	pen	LOC=table-SUPER
	Bumaga	i	ruchka	na stol-e. (Rus.)
	paper	and	pen	SUPER=table-LOC
	Paper and a pen are on the table.			

(2)	小刘	躺		在床上。
	xiao-liú	tang		zài=chuáng-shàng. (Chin.)
	Xiao Liu	lie		LOC=bed-SUPER
	Xiao Liu	lezhal		na krovat-i. (Rus.)
	Xiao Liu	lie		SUPER=bed-LOC
	Xiao Liu was lying on the bed.			

In (1) and (2), the predicate of the sentence is stative. In (1) it is the expressed by 在(zài) in Chinese and by the zero form of the verb *byt'* 'to be' in Russian and denotes location proper. In (2) we have the verbs of position with the meaning 'lie' indicating the location plus the orientation of TR in relation to LM. But these locative constructions can also be combined with dynamic predicates. Consider example (3):

(3)	黑色的	船只	在水上	漂浮摇晃。
	Hēisè=de	chuánzhī	zài=shuǐ-shàng	piāofú-yáohuàng. (Chin.)
	Black=ATR	boat	LOC=water-SUPER	swim-swing.
	Na vod-e	kachaiutsia	chern-ye	lodk-i. (Rus.)
	SUPER=water-LOC	sway	black=PL	boat-PL.
	Black boats are swaying **on the water**.			

Here we have a predicate of oscillatory motion. As one can see, the mobility of an object, which is not connected with its moving to another point in space, does not change the way the image schema is coded in both languages, and it is the same as in the situation where the localized object is immobile. Another case is presented in (4).

(4)	他	把	瓦莲卡的		相片	放	在自己桌子上。
	Tā	ba	Waliánka=de		xiàngpiàn	fàng	**zài**=zìjǐ-**zhuōzi-shàng**. (Chin.)
	He	BA	Varen'ka=ATR		photo	put	LOC=his-table-SUPER.
	On	postavil	u	sebia	**na**	**stol-e**	foto	Varen'k-i. (Rus.).
	He	put	at	himself	SUPER	table-LOC	Photo	Varen'ka-GEN.
	He put **on his desk** a photo of Varen'ka.						

In (4) LM is still a SURFACE (of the table), but the locative group fills in this case the valency of the causative verb of moving the object with the meaning of *'put'*. In the semantic structure of such verbs, there are two predications: 'Subject X did something with the object Y, and as a result Y moved to the place Z'. Provided that Z refers to the type of SURFACE, the valency of the end point of movement in the Russian language is usually encoded by the construction *na NP$_{acc}$*, cf. *On postavil portret Varen'ki sebe **na stol*** (**acc. case**) *'He put a photo of Varen'ka **on his desk'***. The logical consequence of Y moving to the point Z is the location of Y at the point Z. This consequence is the third, implicit, predication in the semantic structure of sentences with the causative verb of movement. This is already a locative predication in which Z is an actant with the role of place, and such a role is encoded for Z of the type SURFACE by the construction *na 'on' NP$_{loc}$*. This resulting location of the object is fully consistent with the image schema in Fig. 1. Thus, the variable government of the causative verbs of movement in Russian, e.g. *postavit' 'to put', povesit' 'to hang', usadit' 'to seat' and polozhit'* 'to lay' (*na chto-l./na chem-l. 'onto smth/on smth'*), is explained by the presence in their meaning of the *movement* component and the component of the *resulting location* implied by the movement. Each of the government model variants profiles (accentuates) the semantic component corresponding to it: *na NP$_{acc}$* profiles moving to some point, and *na NP$_{loc}$* profiles the resulting location at that point. In Chinese, it is also possible to encode a movement causation with the profiling of the movement (to the point) component using the construction 到NP上(dao NP shang) with the directional marker 到(dao, also used as a noun with the meaning 'path') instead of locative 在(zài).

2.1.2 With the same reciprocal location of TR and LM, LM is represented by nouns denoting objects conceptualized as one-dimensional, i.e. belonging to topological type LINE by [10] (cf. BAR or ROPE by [6]) and at the same time serving as SUPPORT for the TR (Fig. 2).

Fig. 2 An image schema of the '*on*' localization for LMs of the type LINE

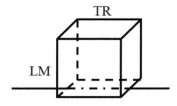

(5)	... tovarisch	tants-uet	na	provoloke	s	sabliami. (Rus.)
	... the comrade	dance-PRS	on	wire	with	sabers.
	... the comrade is dancing on the wire with sabers.					

(Here and further, Russian examples, which are not Chinese translations, are taken from the National Corpus of the Russian Language [NCRL].)

(6)	他	穿着	溜冰鞋	在钢丝上	滑行。
	Tā	chuān-zhe	liūbīngxié	zài=gāngsī-shàng	huáxíng. (Chin.)
	He	wear-PROG	rollers	LOC=wire-SUPER	skate.
	On	na	rolikakh	kata-et-sia	po provolok-e. (Rus.)
	He	on	rollers	roll-PRS-REFL	along=wire-LOC.
	He is roller-skating on a wire.				

Example (5) shows that even if LM does not belong to the type SURFACE, but serves as a SUPPORT for TR, the location of the TR relative to the LM is indicated by using *na* NP_{loc} in the case of both static and dynamic TR (in this case *dancing person*). It is important that TR be perceived as remaining in one place. The Chinese equivalent of the *na* NP_{loc} construction in such cases will be, as usual, the frame construction '在NP上'(在钢丝上zài=gāngsī-shàng 'on a wire'). However, in (6), this equivalence no longer holds. The same Chinese construction with the same LM can no longer be translated with the help of *na* NP_{loc} (**On na rolikakh kataetsia **na provoloke***). The PP *po* $NP_{dat.}$ with the preposition *po* 'along' is the only possibility in Russian for such cases. The choice of *po* 'along' instead of *na* 'on' is obligatory in Russian for spatial adverbial constructions with LM of the type LINE (+ SUPPORT) in the context of 'multiple motion' verbs (motorno-kratnyje in Russian terminology), a subgroup of multidirectional movement verbs (see about this subgroup in [13]. The verb *katat'sia* 'roll, skate' is just one example of these. Such verbs can govern the group *na* NP_{loc} but only if NP_{loc} denotes LM of the type SURFACE, as in (7):

(7)	My kataemsia **na** katke pod bravurnye melodii nashikh liubimykh marshei...
	We skate on the skating-rink to the bravura melodies of our favourite marches...

Fig. 3 An image schema of the '*on*' localization for LMs of the type SURFACE oriented as a slope

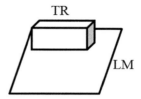

Multidirectional movement on the surface can have an arbitrary trajectory. The construction *na NP_{loc}* '*on NP_{loc}*' indicates only the area within which the TR moves in various directions. But the wire refers to the type of LINE and in this context also to the type of SUPPORT. A multidirectional movement on the LINE, which is a SUPPORT, is only possible if TR (many times) moves in a direction strictly defined by that line and then returns in the opposite direction. Otherwise, TR will lose contact with the SUPPORT and will no longer be able to move in the same way that it had moved before (e.g. it will no longer ride or crawl but fall/fly down). In the Russian language, both unidirectional and multidirectional movement of TR, defined by a linear trajectory, is encoded by the prepositional construction *po NP_{dat}* 'along (*NP_{dat}*)', cf. *idti/hodit' po doroge* (**na doroge*) 'go/walk along (*on) the road'; *lezt'/lazat' po kanatu* (**na kanate*) lit. climb along (*on) the rope 'climb the rope', etc. Thus, in contexts where LM belongs to the type LINE+SUPPORT, the Russian and Chinese constructions under consideration lose their equivalence if the multidirectional movement along the LM is described.

2.2 A schema in which TR is in contact with a LM that is a topological SURFACE type oriented as a slope (see Fig. 3).

(8)	Doma	vs-e	na	gor-e. (Rus.)
	Houses	all PL	on	mountain.
	The houses are all on the mountain.			

(9)	白天,	儿	散	在坡上。	
	Báitiān	yángér	sàn	zài=pō-shàng. (Chin.)	
	Day	sheep	walk	LOC=slope-SUPER	
	Dni-om	ovts-y	pasutsia	na	sklon-e. (Rus.)
	Day-INS	sheep-PL	graze	SUPER	slope-LOC.
	In day time the sheep graze on the slope.				

Fig. 4 An image schema of the '*on*' localization of TR which is in contact with the smaller upper surface of the vertically oriented PLATE type LM

As with horizontally oriented LM of the SURFACE type, the TR location in contact with such LM is encoded by the constructions under consideration, regardless of whether the TR is still (8) or moves within LM (9).

2.3 The schema in which LM belongs to the topological type of PLATE, the largest of the LM surfaces is vertically oriented and TR is in contact with its smaller upper surface, oriented horizontally and located above it (see Fig. 4). This schema is represented in example (10).

(10)	几只	麻雀	在围墙上。	
	Jǐ-zhī	máquè	zài=wéiqiáng-shàng.	
	Several-CLF	sparrow	LOC=fence-SUPER.	
	Vorob'-i	sid'at	na	zabor-e. (Rus.)
	sparrow-PL	sit-PRS	SUPER	fence-LOC.
	Sparrows are sitting on the fence.			

In static situations, as in (8), the constructions *na NP$_{loc}$* and *zài=NP-shàng* are equivalent, but in situations in which TR performs (variously) directed movements, the equivalence is lost (cf. 2.1.2). The upper surface which is the smallest one of all the surfaces of an object belonging to the type PLATE relates to the topological type STRIP. The trajectory of TR movement on the STRIP which serves as a SUPPORT for TR is determined by the location of this strip, and in the Russian language in such cases, not *na 'on' NP$_{loc}$* but rather the construction *po 'along' NP$_{dat}$* is used, cf. *Po verkhnemu tortsu monitora polziot/polzaet muravei* '*Along the top end of the monitor an ant crawls*' vs. **Na verkhnem tortse monitora polziot/polzaet muravei* – '**On the top end of the monitor an ant crawls*'.

2.3 A schema in which the LM of the SURFACE type or the largest of the LM surfaces of the PLATE type is vertically oriented and the TR is in direct or indirect contact with that surface (see Fig. 5). This schema is a transformation of image schema 2.1 which consists in turning it by 90 degrees. In schema 2.1 (Fig. 1), the TR was situated higher than LM above the ground and relied on LM, which ensured that TR was at rest. In schema 2.3, LM itself is no longer a support for TR. For

Fig. 5 An image schema of the '*on*' localization of TR which is somehow fixed on the vertically oriented LM of the type SURFACE

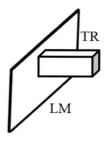

the stability of this configuration, it is important that TR does not lose contact with LM under the influence of gravity. Therefore, as a rule, TR is somehow fixed on the surface of the LM, like a cage on the wall in (11).

(11)	一只	鸟笼			挂	在墙上。	
	Yì-zhī	niao-lóng			guà	zài=qiáng-shàng. (Chin.)	
	One-CL	bird-cage			hang	LOC=wall-SUPER	
	Odna	kletka	dl'a	ptits	visit	na	sten-e.
	One	cage	for	birds	hang	SUPER	wall-LOC
	One bird cage hangs on the wall.						

Note that in this case, only a mount (e.g. a hook) can be in direct contact with LM, while TR may not have direct contact with LM. Thus, if the cage hangs on a long hook, hammered into the wall, and does not directly touch the wall, it will still be said that the cage hangs on the wall – *na stene* (Rus.), 在墙上(Chin.). Ignoring the possible distance between TR and LM in such cases is a typical manifestation of idealizing real spatial configurations in the process of their language schematization.

In such a spatial configuration, there is no need for special fastening only for the TRs, which are traces of matter accidentally left or intentionally applied to the surface of LM, such as inscriptions (by chalk a or marker) on a school board in (12).

(12)	老师	在黑板上	写字。
	Laoshī	zài=hēiban-shàng	xiězì. (Chin.)
	Teacher	LOC=board-SUPER	write.
	Uchitel'	pish-et	na dosk-e (Rus.).
	Teacher	write-PRS.3Sg	SUPER=board-LOC.
	The teacher writes on the board.		

As we can see, the constructions under study encode this schema not only with state predicates of position in space (see '*hang*' in (11)) but also with action predicates denoting the coating of the surface with writings or drawings: 'write', 'draw', etc. In the latter case, locative constructions, just as in the case of movement

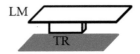

Fig. 6 An image schema of the '*on*' localization of TR which is contact with LM that is situated higher above the ground level

causation predicates in 2.1.1, profile the resulting state: the position of various 'traces' of the substance inflicted on the surface.

2.4 A schema in which TR comes into contact with a LM that belongs to the topological type of SURFACE or PLATE and is located higher than TR and higher than the observer. In this case, the contact of TR and LM can be direct or indirect. See Fig. 6 in which the grey colour indicates the level of the initial count point on which the observer is located.

schema 2.4 is obtained by the transformation of schema 2.3 which consists in the rotation of the latter by 90°. So just one example will suffice:

(13)	光禿禿的	灯泡	挂	在	天花板	上。
	Guāngtūtū-de	dēngpào	guà		zài=tiānhuāban-shàng. (Chin.)	
	Bald-ATR	light bulb	hang		LOC=ceiling-SUPER	
	Lysaia	lampochka	visit	na	potolk-e.	
	Bald	light bulb	hang		SUPER=ceiling-LOC	
	A bald light bulb hangs on the ceiling.						

It should be noted that in Russian such spatial configurations as in (13) are more often described with the help of the construction *pod 'under' NP$_{ins}$* (e.g. *pod potolkom 'under the ceiling'*). In NCRL there are 11 **na potolke** '*on the ceiling*' vs. 45 *pod potolkom* '*under the ceiling*'), which reflects not the subsuming of the same real spatial configuration under another image schema but a change in the focus of attention. In the case of *na* 'na', the contact between TR and LM is in focus, and in the case of *pod 'under'*, the TR location below LM (relative to the reference (ground) level) is in focus. In Chinese, the construction 在NP上(zài NP shàng) is regularly used in such cases.

2.5 The schema in which LM is a SURFACE or a PLATE and TR is a HOLE. Since the orientation of LM in this case is of little consequence, we will consider only one specific example of the implementation of such a schema – a hole in the sole of a shoe (see Fig. 7) – interesting because in the Russian language, the same spatial configuration can be also subsumed under another schema. Although the basic orientation of the LM 'sole' (the one that it has when the shoes are being worn) is horizontal, in contact with the surface (land, floor, etc.), but it can be easily changed if, holding a shoe in your hands, you turn it. In this case, the position of the

Fig. 7 An image schema of the '*on*' localization of HOLE TR with respect to the SURFACE or PLATE LM

shoe will not affect the way the spatial relationship between the hole and the sole is expressed.

In Chinese, this schema is encoded by the construction 在NP上. In Russian one can use a corresponding construction *na 'on' NP_{loc}*, as example (14) from the NCRL shows, but for such cases, *v (in) NP_{loc}* is more commonly used, and in a number of contexts, *v (in) NP_{loc}* becomes mandatory, as in (15):

(14)	dyrk-u	na podoshv-e	ya	vsio	vremia
	hole-ACC.SG	SUPER=sole-LOC	I	all	time
	zakle-iva-l	plastyr-em.			
	stick-IPFV-PST	band-aid-INST			
	The hole in the sole I have been all the time plastering with the Band-Aid.					

(15)	她	在鞋底上	戳	一个	眼儿。
	Tā	zài=xiédǐ-shàng	chuō	yí-gè	yanér. (Chin.)
	(S)he	LOC=sole-SUPER	pierce	one-CL	hole.
	Ona	protknula	dyrk-u	v /*na=podoshv-e	
	She	pierced	hole-ACC	INESS /*SUPER=sole-LOC	
	zhelezn-ym	shil-om. (Rus.)			
	iron-INS.SG	awl INS.SG.			
	She pierced a hole in (*on) the sole with an iron awl.				

The Russian locative equivalent in (15) demonstrates that the same spatial relationship can be subsumed under another image schema, in which LM is conceptualized not as a PLATE but as a CONTAINER within which the TR of the type HOLE is situated. At the same time, the corpus data show that the second way for the Russian language is the main one. Since there were only two examples of this particular schema in the NCRL, which is clearly not enough to conclude about the comparative frequency of the constructions, we turned to a much larger body of

Fig. 8 TR is located higher than the LM relative to the reference level, and there is no contact between them

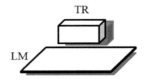

the General Internet Corpus of the Russian Language (GICRL) in which requests for a combination *dyrka/dyrku na podoshve 'a hole-NOM/a hole-ACC on the sole'* gave only 3 examples and *dyrka/dyrku v podoshve 'a hole-NOM/a hole-ACC in the sole'* gave 21 examples, which clearly shows which of the variants prevails.

2.6 A schema in which TR is located higher than the LM relative to the reference level and there is no contact between them (see Fig. 8)

Consider the following examples:

(16)	他们	看见	日本	飞机	在头上	绕。
	Tāmen	kàn-jiàn	rìběn	fēijī	zài=tóu-shàng	ráo.
	They	see-PFV	Japan	aircraft	LOC=head-SUPER	about.
	Oni	uvideli	yaponskiy	samoliot	nad golov-oy. (Rus.)	
	They	saw	Japanese	plane	over=head-INS.	
	They saw a Japanese plane overhead.					

(17)	海鸥		在海上		飞翔。	
	Haiōu		zài=hai-shàng		fēixiáng (Chin.)	
	Seagull		LOC= sea-SUPER		fly.	
	Chaiyk-i		letaiut		nad morem. (Rus.)	
	Seagull-PL		fly		over=sea-INS.	
	Seagulls fly over the sea.					

(18)	晨	雾	在田野上	轻轻地	飘动。	
	chén	wù	zài=tiányě-shàng	qīngqīngde	piāodòng (Chin.)	
	morning	mist	LOC=field-SUPER	slightly	sway	
	Utrenniy	tuman	slegka	kolyshetsia	nad	pol-em. (Rus.)
	morning	mist	slightly	sways	over	field-INS
	Morning mist sways slightly over the field.					

We can see that this image schema can be encoded with the 在 NP 上, but in Russian, it will be matched not with *na 'on' NP$_{loc}$* but with *nad 'over' NP$_{ins}$*. In Russian, the image schemas 2.1.1 (Fig. 1) and 2.6 (Fig. 8) expressed by *na NP$_{loc}$* and *nad NP$_{ins}$*, respectively, are contrasted by the parameter 'contact between LM and

The Contrastive Study of Spatial Constructions *na* NP$_{loc}$ in the Russian... 947

TR'. The preposition *na 'on'* indicates the presence of a direct contact between LM and TR, and the preposition *nad 'over'* indicates the absence of it. The preposition *nad 'over'* in the Russian translation of the example (18) indicates that the mist layer is perceived as being separated from the surface of the field. But if it is perceived as touching this surface, the preposition *na 'on'*, as in example (19) of the NCRL, will be used:

(19)	Prikhodila	osen'	so	svoimi	dozhdyami,	nizkimi
	come: PST.F	autumn	with	PRON.REFL	rains,	low
	tumanami	na polyakh ... (Rus.)				
	fogs	on field:INS.PL				
	Coming was autumn with its rains, low fogs on the fields ...					

It should be emphasized that the Chinese language has a way of expressing the lack of contact between LM and TR in the image schema in question, but it is used when the expression of this component meaning is essential for communication. Consider example (20):

(20)	在桌子上方	有	一盏	灯。
	Zài=zhuōzi-shàng-fāng	yǒu	yì-zhan	dēng (Chin.)
	LOC=table-SUPER-fang	have	one-CLF	lamp
	Nad	stolom	lampa. (Rus.)	
	over	table: INS.SG	lamp.	
	Above the table there is a lamp.			

We can see that a submorph 方(fāng) is added to the superessive morpheme 上(shang), and thus we obtain a postposition that is similar to the Russian *nad 'over'*. If there had been no element 方there, (20) would have activated not the schema 2.6 but the schema 2.1: TR is above LM and is in contact with it, and then the Russian equivalent would be the phrase 'On the table there is a lamp'.

3   Conclusion

We have considered the main image schemas activated by the constructions 在NP 上and *na* NP$_{loc}$ in their direct spatial meaning, and we have demonstrated that the set of schemas encoded by the Russian construction is a subset of the schemas encoded by the Chinese construction. Firstly, this happens because the preposition *na 'on'* excludes the schemas in which there is no (at least indirect) contact between the object-trajector and the object-landmark. A schema in which the trajector is above the landmark and there is no contact between them, in the Russian language, is encoded by the preposition *nad* 'over, above'. Secondly, in two cases, where in principle the use of *na* NP$_{loc}$ is possible in Russian, still the use of other constructions is preferable.

This happens, first, in the case of indirect contact of the trajector with the landmark located above (schema 2.4, Fig. 6), when the construction *pod* 'under' NP$_{ins}$ is preferred (cf. *l'ustra na potolke/pod potolkom* 'chandelier on the ceiling/under the ceiling'), which profiles not the presence of a contact but the position of the trajector (or of its main functional part) below the landmark.

Secondly, it occurs when the same spatial situation allows alternative ways of schematization. Thus, the position of the trajector HOLE in relation to the landmark PLATE can be correlated with the schema 'TR is in contact with the larger one of the LM PLATE surfaces' (2.5, Fig. 7), but more often it is conceptualized as a different schema: 'TR is within the boundaries of the LM of CONTAINER type' (cf. *dyrka na/v podoshve* 'hole on/in the sole'). Finally, the possibility of using the *na* NP$_{loc}$ construction may be constrained by the mode of movement of the trajector in relation to the landmark as in schemas 2.1.2, Fig. 2 and 2.3. Figure 4: 'TR is in contact with LM of type LINE or STRIP acting as a SUPPORT for it'. Thus, if TR moves in opposite directions along the LM of such types, its spatial relation to LM in Russian should be expressed by the *po* 'along' NP$_{dat}$ construction. All the observed differences between the constructions under consideration can be explained, on the one hand, by the non-identical distribution of universal image schemas between the means of their verbalization in different languages and, on the other hand, by the possibilities of subsuming the same real spatial configuration under the alternative image schemas. The analysis of the use of the two similar spatial constructions in terms of their respective image schemas confirmed the observation about the greater degree of generalization of spatial meanings in Chinese and demonstrated the need to take into account more contextual factors in the verbalization of spatial relations in the Russian language.

References

1. *Russkaja grammatika* [Russian Grammar], M.: Nauka, 1980
2. Luj Shusian, 吕叔湘.现代汉语800词[800 words in modern Chinese]. 北京:商务印书馆出版, 2007
3. Qui Silian, 崔希亮.认知图式与句法表现[Cognitive schemata and syntactic expressions] // 北京语言大学汉语语言学文萃:语法卷[Collected papers on Chinese linguistics of Beijing university of language and culture: grammar volume], 2004
4. Lu Danqing, 刘丹青.汉语中的框式介词[Frame preposition in Chinese] // 当代语言学. [Modern linguistics]. 2002. No 4. pp. 241–253
5. M. Johnson, *The Body in the Mind: The Body Basis of Meaning, Imagination and Reason* (University of Chicago Press, Chicago, 1987)
6. E.V. Rahilina, *Kognitivnyj analiz predmetnyh imjon: semantika i sochetaemost'* [Cognitive analysis of object names: semantics and combinability]. M.: Russkie slovari, 2008
7. R.W. Langacker, *Foundations of Cognitive Grammar. Vol.1: Theoretical Prerequisites* (SUP, Stanford, 1987)

The Contrastive Study of Spatial Constructions *na* NP$_{loc}$ in the Russian. . .

8. R.W. Langacker, Cognitive grammar, in *Linguistic Theory and Grammatical Description*, ed. by F. G. Droste, J. E. Joseph, (John Benjamins Publishing Company, Amsterdam, 1991)
9. L. Talmy, How language structures space, in *Spatial Orientation: Theory, Research, and Application*, ed. by H. Pick, L. Acredolo, (Plenum, New York, 1983)
10. L. Talmy, *Toward a Cognitive Semantics*, 2 vols, vol. 1. Concept Structuring Systems. (MIT Press, Cambridge, MA, 2000), pp. viii + 565
11. I. M. Kobozeva, Kak my opisyvaem prostranstvo, kotoroe vidim: problema vybora "orientira" [How we describe the space we see]. Trudy mezhdunarodnogo seminara «Dialog'95» po komp'juternoj lingvistike i ee prilozhenijam. Kazan', 1950. S. 146–153
12. The Leipzig System of Glossing Rules, https://www.eva.mpg.de/lingua/resources/glossing-rules.php
13. D.V. Sichinava, Chasti rechi. Materialy dlja proekta korpusnogo opisanija russkoj grammatiki (http://rusgram.ru). Manuscript. M, 2011

Methods and Algorithms for Generating Sustainable Cognitive Systems Based on Thematic Category Hierarchies for the Development of Heterogeneous Information Resources in Technological and Social Spheres

Michael M. Charnine and Elena B. Kozerenko

1 Introduction

Automatic construction of high-quality sustainable topic hierarchies is an extremely important and urgent scientific problem necessary to solve a wide range of artificial intelligence tasks, including the automatic construction of reviews of mass media sources, advertising targeting, the development of universities' curricula, the development of thesauri, ontological representations and classifications, writing encyclopedic articles, identifying misinformation, and forecasting promising directions of science and technology development.

The approach presented in the paper combines methods of generating hierarchical semantic representations (poly-hierarchies) and setting ontology and vector spaces which contain the characteristics of contexts (distributions) of language objects; special attention is paid to the research of methods based on vector models. Most modern text analysis systems are focused on one language, usually English. However, as the Internet grows around the world, users write comments in different languages. Multilingual methods of tuning the training of neural networks have been developed to analyze text data in different languages. The applications of the results relate to scientific and technical forecasting and analysis of social networks.

New methods are employed that consist in collecting data on the prospective research and development and psychological state of the society from a variety of sources, including texts, photo and video content, and databases unifying all the information of interest. The Mathematical apparatus introduced by the authors

---

M. M. Charnine
Federal Research Center "Computer Science and Control" of the Russian Academy of Sciences, Institute of Informatics Problems, Moscow, Russian Federation

E. B. Kozerenko (✉)
Institute of Informatics Problems, The Russian Academy of Sciences, Moscow, Russia

© Springer Nature Switzerland AG 2021
H. R. Arabnia et al. (eds.), *Advances in Artificial Intelligence and Applied Cognitive Computing*, Transactions on Computational Science and Computational Intelligence, https://doi.org/10.1007/978-3-030-70296-0_77

provides the methods of expert and automatic assessment of cognitive sustainability of the materials analyzed. The software toolkit has been designed and implemented for monitoring social-economic interactions in difficult critical situations, such as epidemiological situations and extremism threat, using artificial intelligence technologies and cognitive approach.

The theme is a set of semantically close terms. The hierarchy of topics for a collection of texts is based on methods of hierarchical clustering and neural networks which extremely, accurately, and subtly calculate the semantic similarity of terms. The research and development presented in the paper creates a new clustering algorithm for the semantic likeness of terms' matrix, which does not require the task of the number of clusters but integrates information from all levels of detail of cluster splitting. The algorithm generates meaningful interpreted clusters close to the manual work of experts by identifying the densest spherical clusters in the vector space based on a neural network built on a collection of millions of scientific articles, patents, and texts of social communications and mass media.

The new concept of *impact factor of the term and theme* is introduced calculated on the basis of citation and with proven prognostic properties, to predict the future scientific directions. Popular topics with a high impact factor from the scientific hierarchy of thematic categories in the future are also likely to have high citations and comprise prospective research directions. In addition to the hierarchy of scientific themes, other thematic hierarchies are similarly constructed.

Basing on social networks and the open Internet materials, the hierarchies of thematic categories are being built for such fundamental fields as Science, education, healthcare, industry, etc. From the analysis of the dynamics of these hierarchies, their components, and relationship growth, the most relevant concentrating topics that determine the future directions of national and international development are identified. This intelligent algorithm is implemented by calculating and evaluating prospective areas of development for one of the areas/topics of computer science, in particular, the National Technology Initiative and the Digital Economy Program. The Internet data collection method is based on the original Internet knowledge retrieval system KeyCrawler [1] developed by Michael M. Charnine.

2 Semantic Vector Spaces in Establishing the Proximity of Meanings

Vector models are successfully used for semantic processing and clustering of large text data. Since in our studies we use an approach that combines methods of generating multiple hierarchical semantic representations that provide ontological structures and vector spaces that characterize the contexts (distributions) of language objects, we have focused on the research of methods based on vector models.

The concept of semantic vector spaces (SVS) was first implemented in the SMART information and search engine. The idea of the SVS is to present each document from the collection as a point in space (Fig. 1), i.e., a vector in the vector space. Points closer to each other in this space are considered to be closer

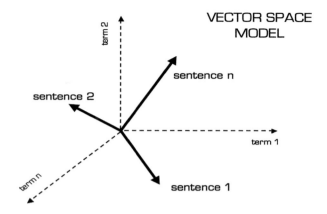

Fig. 1 Vector space model

in meaning. The user request is seen as a pseudo-document and also appears as a point in the same space. Documents are sorted in the order of increasing distance, i.e., in the order of decreasing semantic proximity to the request, and in this form are provided to the user. Subsequently, the concept of SVS was successfully applied to other semantic tasks. For example, in the work [2] contextual vector space was used to assess the semantic proximity of words. This system achieved a result of 92.5% in the test of choosing the most suitable synonym from the standard TOEFL (Test of English as a Foreign Language) test, while the average result of passing the test by a human was 64.5%. Active research is conducted nowadays to unify the SVS model and to develop a common approach to various tasks of extracting semantic links from text corpora [3].

In computational linguistics, various statistical measures, i.e., association measures, calculating the power of connection between elements in collocations are used to establish meaningful phrases [2–11]. The literature mentions several dozen measures of associative connection [2–9, 11–14], mutual information (MI), t-score, and log-likelihood being the most popular instruments. The MI measure introduced in [4] compares dependent context-related frequencies with independent frequencies of words in the text. If the MI value exceeds a certain threshold, the phrase is considered statistically significant.

The methodology developed in our research is an artificial intelligence tool that uses clustering to process big data (texts, audio, video materials) in order to obtain a "general picture" and present it as a hierarchy of themes and sub-themes (Fig. 2).

The constructed hierarchy contains answers to complex questions of semantic and cultural analysis, such as "what topics are represented in the collection of texts," "what to read primarily on these topics," "what is there at the intersection of these topics with related areas," "what is the structure of a subject area," "how it developed over time," "what are the latest achievements," "where are the main centers of competence/development," "who is an expert in the topic," and so on.

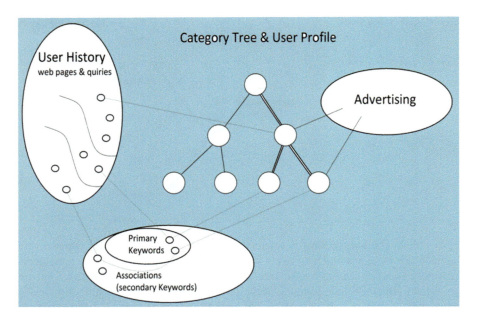

Fig. 2 Data structures based on hierarchy

3 Hierarchy Establishment Procedure

By processing a large amount of heterogeneous/diverse data, a Charged Hierarchical Space (CHS) is built in which a semantic distance is calculated between the individual elements of the hierarchy (categories) using the neural network, and a certain numeric value (i.e., charge) corresponds to each element. The charge can be either some emotion or hatred and extremism or cultural values or finances or scientific impact factor.

Basing on the proposed methods and models, a prototype of cognitive technology is being created to perform the following:

- Automatically build significantly different and well-interpreted thematic representations.
- Determine the optimal number of topics, hierarchical subdivision of topics into subtopics.
- Consider not only individual words but also thematically significant phrases as well.
- Consider the heterogeneous metadata of documents: authorship, time, categories, and tags.
- Define the emotional tone of texts, video, and audio.

On the basis of the constructed hierarchy and its dynamics and correlations, information systems are developed to make predictions and support decision-

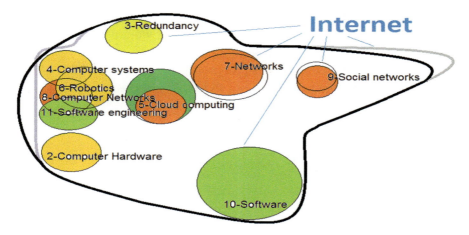

Fig. 3 Visualization of semantic similarity

making. In particular, taking into account the citation, the impact factor of scientific topics is predicted, and promising directions for investments are identified (Fig. 3).

The research also uses methods of distributional semantics and corpus linguistics and methods of statistical processing of large text volumes, such as Word2Vec, Doc2Vec, and others. The main applications of distributional models are as follows:

- Resolution of lexical ambiguity
- Information search
- Clustering of documents
- Automatic formation of dictionaries (semantic dictionaries, bilingual dictionaries)
- Creation of semantic maps
- Modeling of periphrasis
- Establishing the subject of the document
- Identification of the tone of the statement
- Bioinformatics

The theoretical basis of this direction goes back to the distributional methodology of Z. Harris [10]. Similar ideas were put forward by the founders of structural linguistics Ferdinand de Saussure [15] and Ludwig Wittgenstein [16]. Distributional semantics is based on the hypothesis that linguistic elements with similar distributions have close meanings.

One of the current tasks carried out within the framework of our research is to develop the collection of text corpora describing human values. This work involves the definition of a semiotic structure of values that reflects the balance of affective and cognitive elements of the human mental sphere.

- Each category is linked to a set of keywords by triples: <keyword, category, **weight**>
- The probability of correct classification:
  $P_i = P(\text{Category} \mid \text{Keyword})$
- Optimal **weight = Log(Pi / 1 - Pi)**, for two alternative categories (Nitzan and Paroush, 1982)
- For 3 or more categories:

$$\textbf{weight} = Log\left(\frac{P(keyword\ \&\ category)}{P(keyword)\ P(category)}\right)$$

Fig. 4 Optimal keyword weights for automatic classification

The operation of classification using the key words and phrases is involved in the process of semantic analysis; it is essential for the representation of user interests.

Primary and secondary keywords (associations) of the category tree are established with the help of the KeyCrawler engine [1, 7]. Calculation of the weights of primary and secondary keywords (associations) is carried out, and classification procedures employ the optimal keyword weights for automatic classification (Fig. 4).

Determining the interests of a user is based on the information extracted from the history of his visiting the sites and queries. User interests can be represented as a set of pairs <keyword, weight>. Thus a shopper can have the following interests: <dress, 70> <cosmetics, 40> <handbag, 20>. The interests of the sports fan can be represented as follows: <sports, 30> <football, 60> <volleyball, 20>, etc.

Semantic similarity calculation is based on the following models and techniques:

- Distributional semantic models
- Statistical analysis of Web texts
- Special lexical-syntactic patterns

  (e.g., x "is a" y | y "including" x | x "such as" y)

- Context vector of terms/keywords
- Cosine similarity measure

Values and other elements of the future profile are extracted by the program in the form of an ordered set of key terms and phrases similar in meaning in a multilingual semantic field. The terms are organized into topics, and the topics comprise supertopics. Consider several examples from particular subject areas.

The list of supertopics:

SEMANTIC, CLASSIFICATION, SPEECH, RECOGNITION, AUTOMATIC, TOPIC, NEURAL

SuperTOPIC: SEMANTIC

TOPIC: ONTOLOGY OWL METADATA ONTOLOGICAL SEMANTIC LINKED

- Ontology Learning from Text Using Automatic Ontological-Semantic Text Annotation and the Web as the Corpus.

  TOPIC: CONTEXT KNOWLEDGE CONTENT LANGUAGE

- Improved Semantic-Based Human Interaction Understanding Using Context-Based Knowledge
- Using context- and content-based trust policies on the semantic web

  TOPIC: DISTRIBUTIONAL SYNTACTIC DOCUMENT TEXT SIMILARITY SYNTAX

- Verb Classification using Distributional Similarity in Syntactic and Semantic Structures

  TOPIC: REPRESENTATION INTERPRETATION DESCRIPTION

- Beyond cluster labeling: Semantic interpretation of clusters' contents using a graph representation

  . . .

  SuperTOPIC: CLASSIFICATION
  TOPIC: REGRESSION CLASSIFIER DISCRIMINANT KERNEL MULTIL-ABEL CLASSIFICATION

- Combining Instance-Based Learning and Logistic Regression for Multilabel Classification.

  TOPIC: CLUSTERING BAYESIAN SENTIMENT NEURAL

- A feature selection Bayesian approach for extracting classification rules with a clustering genetic algorithm
- A hybrid ensemble pruning approach based on consensus clustering and multi-objective evolutionary algorithm for sentiment classification

  . . .

  SuperTOPIC: SPEECH
  TOPIC: VOICED CONCATENATIVE ASR SPEECH PRONUNCIATION PITCH

- Uniform concatenative excitation model for synthesizing speech without voiced/unvoiced classification.

  TOPIC: HMM DNN SPEAKER

- Model Integration for HMM- and DNN-Based Speech Synthesis Using Product-of-Experts Framework

4 Semantic Clustering Techniques in the Tasks of Social Computing

The development of technology and the digital transformation of all aspects of human life and society show the difficulty of formalizing modern tasks and the need to present problems from a position of interdisciplinary approach. In these conditions, the creation of multi-parametric analysis tools for poorly formalized tasks will reveal connections, mechanisms, and patterns in heterogeneous data, determine the vector of structural changes. These tools will be the starting point for the formation of new knowledge and the creation of a unique methodology for monitoring the state in various sectors of science and technology, establishment of the meaningful value of research and development, the psychological state of society, dynamics and methods of regulation of social-economic interactions, regional and industrial relations, monitoring of the social sphere in difficult situations.

Vector space models are increasingly being used in studies related to semantic natural language processing and have a wide range of existing and potential applications.

Semantic clustering techniques have been applied to thematic modeling, monitoring, and forecasting of terrorist activity in the Internet information field using a virtual environment, including the creation of theoretical and technological frameworks for automation of building a new measure of the degree of extremism of a site, i.e., the Ideological Influence Index (III), which can predict the future values of the Terrorist Threat Degree Index (TTDI), calculated on the basis of collective decisions of experts.

As part of this area, we have conducted research and development:

- Creating an algorithm for calculating the Terrorist Threat Degree Index (TTDI) based on the collective decisions of experts
- Construction of a model that uses the system of machine learning on the lexical characteristics of the studied Internet site and its direct connections helps to evaluate the TTDI
- Creating a probabilistic model of the dependence of future TTDI values on the number of implicit references and their parameters that determine the Ideological Influence Index (III) and the study of optimal frequency and temporal characteristics of implicit references, in which the maximum correlation between the III index and future TTDI index values is achieved
- Building a dynamically updated dictionary for a number of languages, including terms related to the subject area (SA) of terrorism and extremist groups, and building a hierarchy of topics containing terms and key phrases for each topic
- Creating a method of automated reporting on the goals, objectives, and activities of extremist groups

The method is implemented in a combination of interdisciplinary approaches, on the basis of which a single automated technology is created.

The main modules implemented in the layout are as follows:

- The algorithm creating a dynamically replenished collection of extremist texts from the Internet including social networks
- The collection of text samples identifying terrorist communities and links between groups and individual users
- Neural networks, machine learning, and crowd sourcing
- The method of automated identification of extremist documents with the help of lexical analysis
- The method of automatic calculation of the semantic proximity of texts, including cross-lingual similarity [12], which allows to group texts/sites
- The language implicit link processor (LILP) which works on the method of relevant phrases to identify explicit and implicit links between texts, sites, and blogs
- The method of automatic calculation of the III index based on the analysis of implicit connections with the help of lexical databases using the III predictive analysis of the formation and growth of extremist groups

Currently, a large number of scientific publications are devoted to the problem of creating methods and means of semantic navigation in the Internet. Within the framework of this research, a targeted semantic search is being developed as the creation of a dynamically replenished collection of lexical resources. The search and data collection algorithm is divided into general Internet monitoring and specialized monitoring of social networks.

5 Key Terms and Topics Establishment

The Internet data collection method is implemented by means of the Internet knowledge retrieval system KeyCrawler [1, 7]. The texts from the Internet totaling 47 GB on the topics connected with extremism and others have been processed.

The social media data collection method uses the monitoring system of the IQBuzz. Texts from the Internet totaling 26 GB focus on topics: hate, terror, and extremism. The language implicit link processor (LILP), which works on the method of relevant phrases, has been improved to identify explicit and implicit links between texts, sites, and individual blogs.

To analyze formal references in the articles of the collection, the authentic linguistic processor BREF (Book Reference), which is part of the linguistic processor of the LILP, is used. The BREF processor uses Word2Vec-based proximity to the Word2Vec, neural network technology, and the Word Mover's Distance (WMD) which well assesses the distance between multiple terms to compare semantic similarities.

The method of building a collection of extremist texts has been developed, and a collection of more than 500 texts has been compiled:

1. Choosing key terms (e.g., Kafir, Jamaat).

2. Extracting key phrases from the Internet that contain the given terms, such as: "I am a young man from Egypt, and I joined the Jamaat."
3. Expert selection of informative phrases and detection of source code on the Internet through a search engine (Yandex, Google) requests.
4. Discovery of the relevant documents.
5. Expert assessment of the TTDI of the found text.
6. Adding text to the relevant collection of extremist texts.
7. Singling out key terms from the found text ("weapons and money") and replenishing their original set of p. 1.
8. Training the automatic classifier by a ranked collection of texts (building collections of texts via the KeyCrawler and IQBuzz database by precedents).
9. Expert assessment of the TTDI of the found texts, replenishment of their ranked collection, and highlighting new key terms.

The machine learning algorithms used in the targeted thematic text data analysis are as follows:

- Support vector method (SVM).
- Bayesian classifier.
- A probabilistic classifier based on the principle of maximum entropy.
- Conditional random fields.
- Decision tree learning was used to clarify the ontology and subject terminology.
- Random forest was used to classify texts.
- K-nearest neighbors method (k-nearest neighbors, nonparametric copy-based learning algorithm, based on the tags of the K-nearest training instances).
- The neural network was used to study how news and advertising information influence nonprofessionals' discussions on the social media platform (Twitter).

Thus the methods and algorithms for generating sustainable cognitive systems are based on thematic category hierarchies, and the development of heterogeneous information resources in technological and social spheres rests on the following:

- Primary and secondary keywords of the category tree are established.
- Users interests are classified by the category tree (taken from the Internet).
- Each category is linked to a set of primary and secondary keywords by triples: <keyword, category, weight>.
- Primary keywords are discovered from a training set of categorized documents.
- Secondary keywords are discovered in the Internet texts using the distributional semantic methods.

6 Conclusion

The performance of machine learning algorithms depends not only on a specific computational task but also on the properties of the data that characterize the problem. Support vector method (support vector machine, SVM) at present proved to be the most popular choice of researchers for content analysis and sentiment analysis of statements in social networks. Further research will focus on the development of software to find implicit references that determine the similarity of the meanings of phrases and texts, which will be determined by the methods of thematic modeling, grammatical transformations, and translation and synonym establishing programs, as well as through associative connections and methods of constructing an associative portrait of the subject area developed by the authors' team.

The current research efforts are focused on the following tasks:

- Developing the methodology for identifying sites with the greatest outreach, as well as the method of identifying hierarchical links between sites and the structure of the social community
- Study of the functional relationship between implicit references from open Internet documents and the correlation of the III index and future TTDI index values
- Research in the impact of transformation, synonymy, and translation when looking for similar phrases in the correlation between the III index and future TTDI index values

Acknowledgments This work was sponsored in part by the Russian Foundation for Basic Research, grant 19-07-00857.

References

1. M. Charnine, S. Klimenko, Measuring of "idea-based" influence of scientific papers, in *Proceedings of the 2015 International Conference on Information Science and Security (ICISS 2015), December 14–16, 2015, Seoul, South Korea*, (2015), pp. 160–164
2. R. Rapp, Word sense discovery based on sense descriptor dissimilarity, in *Proceedings of the 9th MT Summit. New Orleans, LA*, (2003), pp. 315–322
3. P.A. Turney, Uniform approach to analogies, synonyms, antonyms and associations, in *Proceedings of COLING, Manchester*, (2008), pp. 905–912
4. K. Church, P. Hanks, Word association norms, mutual information, and lexicography. Comput. Linguist. **16**(1), 22–29 (1996)
5. M. Sahlgren, The word-space model: Using distributional analysis to represent syntagmatic and paradigmatic relations between words in high-dimensional vector spaces. PhD Dissertation, Department of Linguistics, Stockholm University (2006)
6. I.P. Kuznetsov, E.B. Kozerenko, A.G. Matskevich, Intelligent extraction of knowledge structures from natural language texts, in *Proceedings of the 2011 IEEE/WIC/ACM International Joint Conferences on Web Intelligence and Intelligent Agent Technology (WI-IAT 2011)*, (2011), pp. 269–272
7. M. Charnine, S. Klimenko, Semantic cyberspace of scientific papers, in *Proceedings of the 2017 International Conference on Cyberworlds, 20–22 September 2017, Chester, United Kingdom*, (2017), pp. 146–149
8. A.T. Luu, J. Kim, S. Ng, Taxonomy construction using syntactic contextual evidence, in *EMNLP*, (2014), pp. 810–819

9. M. Charnine, K. Kuznetsov, O. Zolotarev, Multilingual semantic cyberspace of scientific papers based on WebVR technology, in *Proceedings of the 2018 International Conference on CYBERWORLDS (CW 2018), Singapore, Oct 3, 2018 – Oct 5, 2018*, (2018)
10. Z. Harris, Distributional structure, in *Papers in structural and transformational Linguistics*, (1970), pp. 775–794
11. D. Yarowsky, Word-sense disambiguation using statistical models of Roget's categories trained on large corpora, in *Proceedings of the 14th International Conference on Computational Linguistics, COLING'92. Association for Computation Linguistics*, (1992), pp. 454–460
12. E.B. Kozerenko, Cognitive approach to language structure segmentation for machine translation algorithms, in *Proceedings of the International Conference on Machine Learning, Models, Technologies and Applications, June, 23–26, 2003*, (CSREA Press, Las Vegas, 2003), pp. 49–55
13. W. Wu, H. Li, H. Wang, K.Q. Zhu, Probase: A probabilistic taxonomy for text understanding, in *SIGMOD*, (ACM, 2012), pp. 481–492
14. R. Shearer, I. Horrocks, Exploiting partial information in taxonomy construction. ISWC **2009**, 569–584 (2009)
15. F. Saussure, Course in general linguistics. Duckworth (Translated by Roy Harris) (1916/1983)
16. L. Wittgenstein, Philosophical investigations. Blackwell (Translated by G.E.M. Anscombe) (1953)

Mental Model of Educational Environments

Natalia R. Sabanina and Valery S. Meskov

1 Introduction

Modern educational environments can no longer be considered in isolation from the information and communications technologies. At the same time, the educational environment is not just any environment. This is an environment purposefully modeled by the subjects themselves, in the process of cognition and creation, in order to increase the degrees of freedom of man and humanity as a studied form of life, possessing consciousness, i.e., a conscious life form (CLF). The formation of consciousness is an inherent purpose and value, including in the process of forming educational environments.

However, we can hardly say that a group of people, or even a certain community, is a factor that can direct human development (and even humanity) in a certain direction; rather, on the contrary, excessive efforts can only limit development. At the same time, it should be noted that the process of education is non-monotonous (i.e., it must be described as using non-monotonous logic). Thus, if, in the initial stages of the development of consciousness of life forms, it is set by a limited number of factors, then as it becomes more complex, in the philo- and ontogenesis of man, the complexness of the relationship between these factors and the quality of new levels are so complicated that they can no longer be described in the categories of structures or systems and their mathematical and semiotic models. A fundamentally new environmental approach is needed.

Thus, the creation and study of the mental model of educational environments can become a method of describing the formation of man and humanity at different levels of reality in the context of the Great History [1].

N. R. Sabanina · V. S. Meskov (✉)
Moscow Pedagogical State University (MPSU), Moscow, Russia
e-mail: nr.sabanina@mpgu.su; vs.meskov@mpgu.su

© Springer Nature Switzerland AG 2021
H. R. Arabnia et al. (eds.), *Advances in Artificial Intelligence and Applied Cognitive Computing*, Transactions on Computational Science and Computational Intelligence, https://doi.org/10.1007/978-3-030-70296-0_78

Mentality (from lat. *mens* or (kind of case) *mentis*, soul, spirit (in a narrower sense, mind)) is a set of characteristics of man as part of a certain social group (people, community, state). Mentality can be considered a generalized indicator of the results of the conscious activity of the represented communities, i.e., the concept of mentality is meaningful solely in connection with the emergence of a conscious ability to form life.

The emergence of consciousness is a natural process in the universe at a certain stage of its evolution, namely, at a certain segment of the space-time continuum (taking the beginning of coordinates, the moment of "The Big Bang"). We believe that the emergence of consciousness is due to the fact that some form of life acquires a qualitatively new negentropy property of energy, matter, and information (EMI). This statement is verifiable if it is possible to create a model that meets the principles of systematization in the post-non-classical scientific paradigm.

The purpose of the study is to describe the possible mechanism of this qualitative transition to a form of life that has an attribute of consciousness.

The creation of a mental model of educational environments (MMEE) will allow to take into account the variety of activities (forms of cultural representation) mastered by man during its cultural evolution, as well as to consider ways of operating EMI objects by the subjects of educational activity.

To build MMEE, we will need to establish the relationship between the objects of the EMI world and the model of the world of culture, as manifestations of the attribute of the form of life – consciousness. Next, you need to identify the prototypes and set the characteristics of the processes of operation of EMIs. Moreover, the formation of the subject will be considered up to the level of feasibility of "temporoformation" on the basis of self-transference in the process of cognition: *Temporoformation* characterizes the process of body transformation due to the characteristics of conscious ability person. A possible measurable indicator for this process can be the calculation of the intensity of temporoformation, where it is necessary to take into account the factor of energy, matter, information, and time that is spent on its implementation [2, p. 48]. The basis for the introduction of this concept is research in the modern philosophy of consciousness: "... a new idea of the bodily nature of consciousness (realized mind) appears. The corporeality of consciousness does not at all mean a denial of the ideality of its products, but indicates the need to take into account the bodily determinants of spiritual activity and cognition. A holistic body-consciousness approach is needed: the consciousness is bodied, embodied mind, and the body is inspired, animated by the spirit" [3]. The criterion of the feasibility of temporoformation determines the attainability of the formation of the body-cognitive-value paradigm of education.

There is an opinion that the education system is an institution of modern culture, which allows for more intensive education of a person, transmitting him a systematic set of knowledge. In addition, it brings his scientific model of the world to a common denominator, which allows you to synchronize the activities of people in solving certain problems, in particular professional ones. At the same time, we see that the picture of the world is not allowed and is not aimed at the fact that the person goes to other levels of mutual interaction besides the narrow professional. This rule is with the exception of what we all know: there are special educational institutions in all countries of the world in which people are trained who are responsible for implementing mutual cooperation at the state level.

Today, there is an acute problem of cooperating at the global level in order to solve environmental problems, including those related to the ecology of thinking, namely, the formation of biological, cognitive, and spiritual sustainability of human and human development. Thus, not only the educational systems of various countries but a global educational environment is necessary for the implementation of productive interaction between people at any necessary level.

You can also hear the opinion that the children are not able to carry out the "elementary school curriculum," and you – "Avon where you swung!". However, such a remark has nothing to do with the nature of man; moreover, man has already grown from a disciplinary, school approach to education. In many countries of the world, standards have been adopted that "prepare a person for life" (both in school and in higher education). How justified is this approach compared to the traditional, subject-specific one for Russia? The relevance of solving this issue in the education system can be regarded as one of the prerequisites for the formation of the bodily-cognitive-value paradigm of education. The multidimensional model that we are creating, MMEE, can be used as an argument in answering this question, which is not as unambiguous as it might seem.

The energy-matter-informational model of the world (EMI-mw) [2, p. 49] is considered together with TWIMM (Trinitarian World Information Model and Methodology (Meskov V.S.) [4]). Relations between EMI objects set the conditions for the restoration of the integrity and existence of the world, as necessary attributes of consciousness.

It will be shown that many EMI objects are in relation to association on the basis of complexness both to information objects and to hypothetically possible energy objects and material objects. Experimentally, it has been established that in order to capture quantum states (which can be regarded as prototypes of energy objects – N.R. Sabanina) and the transition of their manifested (prototypes of physical objects (M), N.R. Sabanina), it is necessary to exchange information (information objects) between quantum particle and environment or observer (the latter in case of identification of it with specific content and vice versa). This process is accompanied by "fixing states" to the manifested (Fig. 1).

If we carry out extrapolation on the principle of conformity to other levels of the implementation of the form of life (in particular – the conscious), we can assume that consciousness (as a tool of operation of information) is a specific way for the form of life to fix energy conditions in the physical equivalent and vice versa.

Moreover:

E (energy) is what sets the vector of action, activity, or mutual assistance which determines the strength of the direction of the processes and is a characteristic of EMI-mw. It is important to note that the term mutual assistance (or inter-cooperation) as opposed to simple interaction is used as a unifying term for

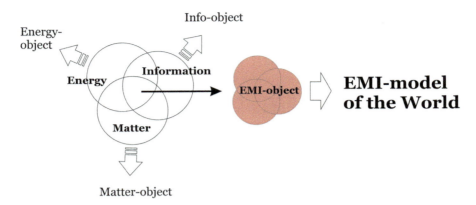

Fig. 1 Theoretical and methodological assumptions in constructing the EMI model of the world

"consciousness" and "understanding" as a characteristic of synchronization of activities in tiered environments [quoted by 2, p. 34].

M (matter) is the common thing that unites everything "manifested" for man and will be a representation of the existence of the EMI object.

I (information) is a characteristic of the integrity of the EMI object. "... in classical and quantum physics, a different meaning is embedded in the concept of 'information.' In classical physics, the information itself and its material medium are separated. It is believed that information cannot exist without a material medium. In quantum physics, information is a physical value that characterizes a system, similar to volume, mass, entropy, etc." [5, p. 198]. In this article, E, M, and I are considered as philosophical categories.

And integrity will be set by the elements in relation to trinitarianity; existence is a relationship of complexness.

We will enter the EMI object, based on ontological assumptions regarding EMI-mw.

Ontological assumptions reflect the general relationship that sets the relationship between man (as a form of life possessing consciousness) and elements of possible worlds of his being, ontologically represented by EMI processes in the universe. Modeled EMI objects in this consideration are in the relationship of part and whole, and the EMI model of the world satisfies the following principles: generation, subjectivity, formation, self-development, and n-dimensionality.

2 EMI Model of the World

The world is such that it consists of EMI objects.
The world of EMI objects is holistic.
EMI objects have a form.

The form of EMI objects is determined by existence, n-dimensionality, and self-organization.

The form of the EMI object is an EMI object.

EMI world (EMI_0) is an EMI object.

The world of EMI objects has characteristics of structure, system, and environment.

The integrity of the EMI world is determined by the possibility of making transitions between its objects (elements of forms of existence): energy objects (E), physical objects (M), info-objects, e. i. information objects (I), EMI objects, EM, MI, MI, etc.

The existence of EMI objects is determined by the possibility of forming and self-organizing in n-dimensional structures, systems, and environments.

The assumption for structures, systems, and environments is the ability to move from one form to another.

Life can be interpreted as an integral attribute of the universe.

The EMI world is the case of the universe exemplification.

Life arises at a certain stage of the formation of the EMI world.

The manifestations of structures, systems, and environments in the EMI world are inanimate, living, conscious, and spiritual (spiritual life).

EM, MI, and EI can be presented as objects, relationships, or transitions depending on the evolution stage of the form of existence.

Ontologically, the EMI world (EMI_0) is characterized by existence, integrity, and continuity. EMI_0 is a paradoxical object because it is an analogue of the one and many paradoxes. All paradoxical objects – ideal and at the same time EMI_0 – exist, and it is continuous.

In other words, it is impossible and, at the same time, generates the world of EMI objects.

EMI_1, $EMIE_2$..., EMI_n are the born objects of EMI world. The consequence of the generation is the emergence of boundaries, and thus the possibility/need for the implementation of transitions is given.

The principle of "creation" sets the relationship between EMI_0 and the objects it generates. We will call the procedure of implementing these relationships "transitions" (further denoted as "$\rightarrow$," the sting of which indicates the direction of the transition; "$\leftrightarrow$" denotes a two-way direction of the transition).

Let us set the relationship between elements of EMI-mw.

The method of reasoning about the types of transition is a "thought experiment" which is an experiment, reasoning, and actions in virtually impossible but theoretically possible situations [6].

The types of transitions are:

I. For the inanimate: abiogenesis (birth of life), EMI (M) $\rightarrow$ E/MI; and dying, E/MI $\rightarrow$ EMI (M).

EMI – the condition of objects corresponds to M (matter), and E are in potential state, due to the fact that the conditions of their active self-implementation have not yet arisen or have already disappeared. This means that, by definition, the integrity and direction of the EMI world processes are in a potential state, at a time when it itself is manifested for the life form of the possessing consciousness (observer of the EMI world).

(MI) is the first constant (I): the transition between M and I is impossible, as for the inanimate MI is an object, not an attitude or a transition.

II. For the living: energy extraction to continue life (establishing a "relationship of life" between elements of forms of existence), E ↔ MI; anabiosis (slowing down the processes of life in which death does not yet occur), MI → E (energy "goes away"); and dying, there is no relationship between the elements of E and MI.

(E) Second constant (II): The evolution of the living occurs by fixing the "relationship of life," in the direction of realizing the readiness of the life form to implement the B-I transition.

The *inter-level task of I and II* – mastering the energy operating, set by external factors (the environment) – defines the "border outside."

III. For the conscious (CLF):

Third constant (III): the attribute of CLF – cognition as a way of operating E/M/I.

Exercise a conscious life form.

Consciousness is interpreted as (M ↔ I) transition.

Cognition: E information transition (ME → I) and joint energy transition (or fulfillment) IE → M, where the joint is set by a couple (ME, IE).

"Info-transition" and "Fulfillment" are fundamental relationships for EMI-mw (ontologically) and, accordingly, necessary prototypes for the construction of the relations "Naming" and "Creation" in the PNCC and, accordingly, in the PNC-texts. Post-non-classical text (PNC-text) is an artifact of the Sexta-paradigm of the post-non-classical concept of culture (PNCC). Sexta-paradigm is what each form of cultural representation (FCR) sets: cultural actor (Ak), value (V), law (L) and regularity (FCR) development, artifact (Af) (subjects and processes specific to each FCR), and language (Lg) FCR. The PNCC, in turn, is assigned to EVI-Mm. Elements of the PNC-text can be similar to elements of EMI-mw [2].

The *inter-level task of II and III* is to master the operation of information for the purpose of conscious (internally conditioned) mastery of energy (Fig. 2).

IV. For the spiritual: the realization of spiritual life (I ↔ E): the transition (soul) as a result of self-transcendence (movement from "Self" to "God"). *The inter-level task of III and IV:* mastery of CLF operating (E), this operation is equivalent to the implementation of controlled temporoformation by man.

Each of the crossings is set at the objects of EMI world.

Temporoformation (possible prototype – "theosis"): Ascension I → E; resurrection of E → M; revelation of E → I. Theosis, or holiness, is the teaching of the Orthodox Church about the connection of man with God, the communion of man's non-modern divine life through the action of divine grace.

How is it possible to maintain the life form?

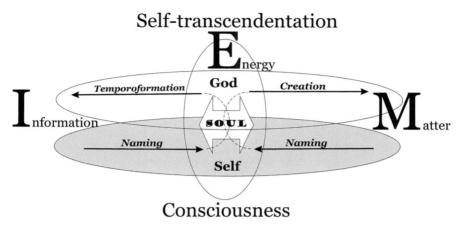

Fig. 2 The self-transcendentation of the form of life possessing consciousness

The law of life-saving – whenever there is a transition procedure that translates integrity into a multitude (partially ordered) and a corresponding transition that organizes the multitude of things to integrity at this level of EMI-mw implementation, the preservation of EMI takes place.
How is it possible to generate EMI objects?
$EMI_0 = $ df World (existence, integrity, continuity).
Existence (EMI_0) (E, M, I). The existence of EMI_0 is the perception, understanding, and awareness of the object as having energy- and matter-information characteristics.
Integrity (EMI_0) (I). The integrity of EMI_0 is the perception, comprehension, and awareness of the object as having information.
Continuity (EMI_0) (no boundaries, monotony of expansion). Continuity – no boundaries, therefore monotony.

The act of generating EMI involves the violation of the equilibrium (law of preservation of life form) on one or more grounds previously generated by EMI objects by virtue of the structure of relations between elements of EMI, namely:

E is characterized by the power of direction, so it determines the change of state.
And the preservation of a certain level of information saturation determines the occurrence of boundaries and translates the EMI object into a qualitatively different state (non-monotonous); continuity is disturbed.

Thus, it is necessary to describe:

The law of redistribution of EMI (generation) – whenever the inter-level task reaches its resolution, there is an inter-level transition.
The first act of the creation of EMI_0 is a paradox.

3 Types of Transitions

The types of transition are as follows.

  I. EMI is inanimate.

 II. The generation of the living:

    1. The generation of life: EMI $\rightarrow$ E/VI.
    2. Dying: E/MI $\rightarrow$ EMI.
    3. "Relationship Life": E $\leftrightarrow$ MI.
    4. Anabiosis (slowing down the processes of life in which death does not yet occur): MI $\rightarrow$ E (energy "goes away").

 III. The birth of a living conscious:

    1. Consciousness is interpreted as (M $\leftrightarrow$ I) transition.
    2. Self-transtranscention of the conscious life form (CLF): ME $\rightarrow$ I information transition and the joint implementation of IE $\rightarrow$ M.

 IV. The birth of the spiritual:

    1. (I $\leftrightarrow$ E) transition is the soul.
       Temporoformation (theosis) of the CLF:
    2. Revelation E $\rightarrow$ I.
    3. Resurrection E $\rightarrow$ M.
    4. Ascension I $\rightarrow$ E.

Complexness 1 (for EMI objects):

1. The principle of creation implies complexness as a condition for the implementation of the other four principles. This principle is based on the paradox of the EMI world.

    R Complexness 1 (EMI_0, EMI_1) = df(generation (n-dimensionality, formation, development, subjectivity (EMI_0, EMI_1)).

2. The principle of development determines the conditions and patterns of reproduction and improvement of EMI objects.

3. The principle of subjectivity is a subject who recognizes the world and recognizes himself and the world as holistic and existing. Subjectivity (EMI_1) (integrity, existence (EMI_1)). Integrity – perception, comprehension, and awareness of the object as having information characteristics. Existence is the perception, comprehension, and awareness of the object as having energy- and utility-information characteristics. It is a good thing (EMI_1) (E, M, I).

4. The principle of n-dimensionality – EMI objects have signs: existence and multi-levelness (i.e., a certain organization – inanimate, living, living conscious, spiritual). N-dimensionality (EMI_1).

5. The principle of becoming determines the possibility of forms of existence to transition (overcoming borders) and, further, to implement them at new levels.

Becoming $(EMI_1, EMI_2) = df$ Transition (EMI_1, EMI_2) and Existence (EMI_2).

4 Post-non-classical Model of Culture (PNCC)

The post-non-classical model of culture is an EMI object of EMI-mw and is similar to them.

PNC-text is an artifact of the Sexta-paradigm of the PNCC, as well as an EMI object. A PNC-text, like an EMI object and its elements, can be similar to the elements of EMI-mw.

The basic elements of the model – EMI objects – are described as ontological and gnoseological entities. The result of the operation of EMI objects is the construction of PNC-texts. The nature of these objects involves identifying the relationship between these entities, as well as designing ways of reasoning about the operation of EMI objects in the PNC-texts on the basis of Complexness 2 relationships. At the same time, the most complex element of such design is the subject. The subject is not an EMI object, but operates EMI objects through conscious ability. Moreover, the focus of such operation is the consistent implementation of the level transitions EMI-mw.

Based on the classification of many forms of cultural representation, it can be concluded that the most appropriate, basic scientific language of research and building a model of the world of culture is the relevant logic, which most fully explains the relationship in the world of culture. Moreover, in this sense, A is relevant to B, only if by interpreting A as a whole (holistically) and B as a whole, we can distinguish in A and B the total part – Q. And it is essential that between Q and A, as well as between Q and B, the relationship of part and whole, but not subclass to class. The logic of fractals, the logic of extras, and non-monotonous logic will be considered as sections of non-classical logic. Moreover, on the basis of the introduction of axiological modalities, the PNCC can talk about the introduction of post-non-classical logic, in which the subject is an integral element of logical construction.

When considering the process of creating PNC-texts, we see that in them, in addition to true and false statements, there are also uncertainness in various modalities (modal logics) and hidden parameters of the subject of reasoning. Previous research has shown that when discussing this type of object, the law of the third is usually no longer meaningful.

The logic on the basis of which the creative process, which includes the formation of information-iconic environments by subjects, as well as the PNC-text can be considered probabilistic. In the structure of such logic, the final conclusion is not clearly defined by the link between the conclusion and the assumptions on which it is based. Thus, another law of classical logic – distribution – also does not work in the logic of creating a post-language text.

The main problem is the establishment of cause-and-effect relationships. At the same time, for non-classical causality, it is defined on the principle of additionality, while for "active" "post-non-classical" logics, it is necessary to use the very subjectivity – self-transcendence – as causation [2, p. 203; 3]. This process requires the addition of a new element that takes the way of existence of the subject to a fundamentally new level, and therefore it is an intensive logic. It is clear that creation is a non-monotonic procedure.

The evolution of the conscious form of life, coupled with its spiritual formation, requires the formation of a special way of thinking based on values. Value, considered from the point of view of its specificity, more or less holistically determines the mental state that sets a person's ability to exercise transcendental synthesis. Accordingly, reasoning, based on a value argument, makes it possible to construct a coherent text of human existence, reflecting his individual existential and transpersonal meanings. When considering the process of creating PNC-texts, we see that in addition to their true and false statements, these can also be considered uncertain in various modalities (modal logics) and hidden parameters of the subject of reasoning.

The problem of describing the system of values of the subject of cognition is solved by supplementing the seven biological structures of the subjective component, reflecting the attitude of "sense-name" and "sense-thing" that will describe the individual meanings of the subject, including the value component of his activity.

Complexation of complex levels is set by analyzing the relationship between "nested models" (see Table 1). And "nesting" is created by the relations of naming and creation carried out by the actor of culture (Ak). The measure of complexness of systems is dependent on the degree of knowledge, understanding, and values of the subject of knowledge, a particular subject, or process. Moreover, in the post-non-classical paradigm, the subject is not eliminable from the observation process, which causes the effect of "complexness."

Relationship Complexness 1 and 2 (Table 2) show the following: Firstly, the types of operation of E/M/I for EMI-mw in the world of culture correspond to the levels of civilizational and cultural development; inter-level tasks are levels of mutual assistance. If the central committee defines differences and uniqueness, the levels of mutual assistance indicate the general fact that is inherent in the actors of activity; secondly, if R Compl. 1 is included in the description of the EMI world as the relationship between EMI objects and R Compl. 2 characterizes the relationship in the model of culture (PNCC), then, as shown in Table 2, Compl. 1 corresponds to Compl. 2 to Compl. 3 model of consciousness and vice versa.

Based on the meta-theorem on compliance [7, p. 63], the results of the application of the Compl. 1 relationship will be identical to Compl. 2 on the PNC-texts and to Compl. 3 vice versa.

For example, infotransition and implementation are fundamental relationships in EMI-Mm and, accordingly, necessary prototypes for building naming and creation relationships in PNC-texts.

Thus, we have the opportunity to formulate a complexness for these two types of models.

Mental Model of Educational Environments

Table 1 EMI world model level table

	Prototype and model	Relationship	Relationships of complexness	
Level 1	EMI-mw (prototype – universe)	Transitions based on EMI surgery	R Compl. 1 (between EMI objects)	MR Compl. – The model of complexness relationship is what R 1 and R 2 can be given
Level 2	Meta-model PNCC (prototype – civilization)	Relationship between FCR	R Compl. 2 (between elements of the PNC-text)	
Level 3	CRF model (prototype – differences between forms of culture)	Differences of relations (Ak, V, L, P, Af, Lg) in culture forms of representation (or groups of FCR)		
Level 4	Sexta-paradigm (R) (prototype is common to all and every form of culture)	Common in relationships (Ak, V, L, P, Af, Lg) in culture forms of representation (or groups of FCR)		
Level 5	Sexta-paradigm elements' characteristics	Relationships between the characteristics of each of the Sexta-paradigm elements		
Level 6	Model of consciousness characteristics	Relations of naming Relations of creation Self-transcendentation Temporoformation	R Compl. 3 (between elements of the consciousness model)	

Four-plane complexness semiotics (FPCS) ((Environment 1 (sense, meaning) (R Compl.) $\overset{R\,compl.}{\leftrightarrow}$ Environment 2 (sign, significance). This means that the four-plane semiotics is set by the environments of the subject's inner world (Environment 1) and the external environment (Environment 2), which is in relation to the complexness, where R Complexness = Compliance (Compl. 1, Compl. 2, Compl. 3), that you need to consider by setting a relationship of syntax, semantics, and pragmatics.

Table 2 Correlation of elements of complex environments and revealing the nature of relationships at different levels

PNCC: Sexta-paradigm	EMI-mw (Complexness 1)	PNC-text (Complexness 2)	Model of consciousness (Complexness 3)
Actor of culture (Ac)	CLF (conscious life form)	Man/personality/culture actor/subject/I am (self)	The process of cognition and self-transference
Value (V)	The characteristics of the environment that make it possible to make transitions	Values I, II, and III – PNC	Define boundaries between levels of self-transcendentation (subconscious/conscious)
Law (L) of development	Types of operation of EMI CLF	Operating PNC-texts at various civilizational cultural levels	The results of cognition
Pattern (P) of development	Solving "inter-level problems" by operating EMI CLF	Operating by PNC-texts at different levels of mutual assistance	The results of cognition
Artifact (Af)	EMI objects have a form	The meaning of the PNC-text	Relations of creation and naming (for Af as a PNC-text)
Language (Lg)	Infotransition ME$\rightarrow$I jointly the implementation of IE $\rightarrow$ M, where the joint is set by a couple (ME, IE)	Creation and reading of the PNC-text, in the process of cognition and self-transference	Relations of creation Relations of naming
Model of cultural representation forms (CRF model)	Compl. 1 = df CLF (methods of operation (E/M/I), the formation of an environment to solve the inter-level problem, preservation of life form)	Compl. 2 = df Human/personality/actor culture/subject/I am (self) (mastering values, operating PNC-texts in different cultures and levels of mutual assistance, forming environments of self-transcendence, saving life)	Compl. 3 = df Temporoformation (for CLF, T. is defined as the ratio of the intensity of handling EMI objects to the energy and time spent on the implementation of the process of changing the states of the body (matter))

5 The Mental Model of Educational Environments of the Subject of Knowledge

The mental model sets the way EMI objects operate by the subject of cognition. If we assume that the result of such surgery is the solution of the problem of the level transition of life form, the environment within this model will be set by the following characteristics:

$$\text{FPCS} \left(\text{Environment 1 (sense, meaning)} \overset{Rc\text{л.}}{\leftrightarrow} \text{Environment 2 (sign, significance)} \right)$$

where for the model of four-plane semantics (FPCS) can be distinguished: Env.1 – the inner environment (subject (sense, meaning)); Env.2 is external (sign and significance), boundary, transition.

1. The improvement of the process of cognition by the actor of culture is in accordance with the development of his ability, readiness, and motivation to make the transition from the pre-language to the information-sign (language) and post-language forms of representation of culture in the process of creating and creating PNC-texts [8].

2. The non-classical semiotic model, which includes three main elements, the meaning, sign, denotation, meaning, as a set of characteristics of an object defined in a language, should be supplemented by another pragmatic element, which also requires naming and definition, the non-linguistic object of the inner world of the subject of the PNC-text. Let's label this element by the term "feeling." The initial classification of "feelings" objects will apply to each of the three levels of the PNC-text, namely: 1 level: pre-language – installations, sensations and corresponding emotions, representations, images, values of the First order; Level 2 – intuition, intellectual feelings and corresponding emotions, values, order; Level 3: post-language – values of the third order (ontic).

3. The state of faith creates conditions for a person to move the "focus of attention" from rational to irrational, internally conditioned causes of its existence (transcendental deduction, I. Kant). Moreover, this internal subjective reality, in turn, is formed on the basis of a special kind of categories: ontic values. Ontic values of human existence are similar to the original conditions of human creation in the universe and, together, can set the integrity of his thinking and existence (i.e., correspond to ontological assumptions regarding EMI-mw).

4. The syntax of the PNC-text will be set – elements of the alphabet, the rules of education, and conversion of expressions of the language of the PNC-text.

5. If we use EMI-mw as a "statement" model in the PNC-text, it is a description of the process and the result of the ACT of operation of EMI objects. The specificity of this text is that it holistically presents the totality of the EMI transformation cycle – energy, matter, and information – in connection with the activities of the cultural actors. The intensity of the "B-I transition" implementation (based on the realization of conscious ability) determines the level of "energy security" of the process of operation of EMIs.

6. The semantics of the PNC-text will be set by the "Sign-Sense-Meaning-Meaning" attitude. Unlike the classic "Sign-Meaning" path, the subject of reading the PNC-text gets the opportunity to "read" not only verbalized but also nonverbal information. This, in turn, allows you to take into account the hidden

meanings of the message, which determines the accuracy and effectiveness of further operation of EMI objects.

7. Continuity is inherent in the "relationship of time" between the elements of the subject world, while the definition of words and concepts, discursivability, and discreteness are signs of belonging to the "relationship of space." Existing in two worlds at the same time and changing the direction of contemplation, man connects himself with the world, maintains his boundary with it permeable. In this way of existence, there is an understanding of the whole world.

8. Consciousness is an attribute of the multiple form of life, for it is aimed at developing ways of joint self-realization in the environments of existence. Maintaining autonomy, the subject seeks synchronization, as it gives him the opportunity to achieve on orders of order more energy-efficient states, which are accompanied by positive experiences, which are fixed by consciousness as values II and III.

9. Consciousness of oneself (uppercept) is a simple representation of I (by I. Kant), which determines the inner integrity of the subject and its boundary with what/who is not the "I am" (observer of Complexness 2). The knowledgeable subject explores himself, self-expanding his subject to the boundaries of the world-in-general (EMI-mw), "becoming" together with this world. Cognition is based on a model that is created by the subject (Difficulty Observer 1). The nature of the relationship between the subjects of education, including, can be described as follows: "Me" and "not-Me" are "recursively connected... This situation can be transmuted in the context of inter-subjective communication in the 'I am – Other' system... At the same time, it is very important to take into account at least two points: (1) this situation itself should be considered again in the process of its formation, 'self-development', i.e. recursially, fractally, self-similar, as evolutionarily driven by changing contexts; (2) the boundary between 'designated' and 'indesignated,' 'internal' and 'external' in the situation of formation should not be erased, but each time renewed as a fundamental prerequisite of creativity, innovation of intersubject communication. Moreover, in the context of the complexness paradigm, the main task of a flexible methodology of interaction with evolutionary processes is not so much the ability to erase the old differences (forget them) but to create them without turning them into barriers" [9 , p. 79].

10. The way of forming the boundaries of transitions is presented by different types of uncertainties overcome by the subject of cognition: aporias, antinomy, contradictions, paradoxes, cognitive failures, creative failures, and value failures.

11. The general meaning of creating and reading the CLF's PNC-texts is the formation of the self-transcendence of text actors. Individualizing the contribution to the formation of environments sets the characteristics of a complex environment. In particular, four-plane semantics can provide methodological grounds for solving the problems of the medium approach and modeling educational environments in the context of complexness.

As part of the EMI model of the world and methodology, we will set "the mental model of educational environments" with a coherent threesome *subject-environment-PNC-text*.

Definition:

mental model of educational environments (x)
there is a coherent trio "Subject -Environment – PNC-text", Emi (x)
if and only if

x∈Emi (PNCC), where Emi (PNCC) has a meta-model (PNC-text) – post-classical text;

At the same time, Emi (x) has attributes:

1. *Generation*
2. *Subjectivity* (existence of integrity)
3. *Becoming*
4. *Self-development* (self-reproduction, self-improvement)
5. *N-measure*

Emi (x) present in the form of:

Emi (x) (Sub., Env., EMI (PNC-text)), where:

Sub. (subject) – element Emi (PNC-text) corresponding to the ontological aspect of subjectivity.

Env. (Environment) is an Element of Emi (PNCC) that corresponds to the ontological aspect of existence, represented by the PNC-text.

PNC-text – an element of the PNCC,

Emi (PNC-text), corresponding to the Sexta-paradigm element of the PNC – FPC language; Sub. by Ak, who has a "naming and creation" relationship; the result of the creation and reading of the PNC-text is the self-transcendence of the subject of cognition.

R.L. – attitude of complexness

FPCS – four-plane complexness semiotics.

6 Conclusion

1. FPCS (Environment 1 (sense, meaning) (R Complexness) $\overset{Rсл.}{\leftrightarrow}$ Environment 2 (sign, significance)).
2. PNC-text is a transcendental object, which includes pre-language, linguistic, and post-language elements, built on the basis of complexness semiotics. The use of EMI-mw and the creation and reading of the PNC-text are necessary and sufficient conditions for the realization of the body-cognitive-value paradigm of education in the context of complexness of the mental model of educational environments.

3. Considering the PNC-text as having a property of integrity, it was shown that it is the basis of the existence of the subject. At the same time, complex cognition is the result of operation of the PNC-text. Thus, it can be concluded that the main competence of the subject of knowledge is the development of its makings, ability, readiness, and motivation to create and read the PNC-texts.

References

1. A.P. Nazaretitan, Universal (Great) History: Versions and Approaches (Research supported by RFBR-grant No. 07-06-00300), https://www.socionauki.ru/journal/files/ipisi/2008_2/universalnaya_istoriya.pdf
2. N.R. Sabanina, Post-non-classical concept of culture: transdisciplinary monographic study under. Scientific red. Meskov V.S. – Moscow: Russia, 2018, 400 p
3. E. Knyazeva, The bodily nature of consciousness. Body as an epistemological phenomenon. Text/Institute of Philosophy of the Russian Academy of Sciences. Under red. I.A. Beskova. – M.: IFRAS, 2009, 231 p. (The work was carried out with the support of the RGNF grant (07-03-00196a))
4. V.S. Meskov, Trinitarian methodology. Voprosi filosofii Magazine. 2013, No. 11
5. A.N. Verhozin, Thermal Decoherence (analysis of the results of the experience of the Research Group of Cailinger). The Herald of Pskov State University. Series: Natural and Physical and Mathematical Sciences, 2013, pp. 194–200. https://cyberleninka.ru/article/n/teplovaya-dekogerentsiya-analiz-rezultatov-opyta-issledovatelskoy-gruppy-tsaylingera
6. V.S. Meskov, *Thought Experiment and Logic/Logic and Methodology of Scientific Cognition* (M.: Pub. MSU, 1974), pp. 126–140
7. V.S. Meskov, K.E. Ziskin, N.R. Sabanin, Introduction to mathetica. in *2 Books/Scientific and Methodological Edition* (Moscow, Russia, 2018), Book 1–260 pages, p. 63
8. V.S. Meskov, N.R. Sabanina, Texture Universe/Electronic scientific edition of Almanac Space and Time, 2015. T. 10. No 1
9. V.I. Arshinov, Y.I. Svirsky, Innovative complexness. A complex world and its observer. Part 2. Philosophy of Science and Technology, 2016. T. 21. No 1. pp. 78–91

Part IX
Applied Cognitive Computing

An Adaptive Tribal Topology for Particle Swarm Optimization

Kenneth Brezinski and Ken Ferens

1 Introduction

PSO was first introduced by [1] and has grown to be one of the most widespread evolutionary metaheuristics. The inspiration for PSO is drawn from the behavior of schools of fish or flocks of birds, where each individual member of the swarm operates independently but is wholly motivated by the necessary need to communicate in order to increase the chances of finding a source of food. This motivation allows for the algorithm to converge to solutions when a select few of the swarm members come into the vicinity of local minima. Through this simple concept, various PSO strategies have been adopted, which optimize the diversity of the population, as well as the exploration and exploitation capabilities of the swarm as a whole.

Various PSO strategies have stimulated the use of PSO in a wide variety of applications, which include multiplexing sensor networks [2], production scheduling [3, 4], ship hydrodynamics [5], temperature control in distillation units [6], fire evacuation [7], fuel management [8], and improving k-means clustering [9].

This work introduces a novel tribal coefficient that dynamically manages the size of the swarm population as the algorithm progresses. This coefficient dictates the rate at which swarm members are killed or spawned, in accordance with a personal fitness evaluation that penalizes large tribes that perform poorly in objective function evaluations, while stimulating smaller tribes with a demonstrated ability to find better solutions.

K. Brezinski (✉) · K. Ferens
Department of Electrical and Computer Engineering, University of Manitoba, Winnipeg, MB, Canada
e-mail: brezinkk@myumanitoba.ca; Ken.Ferens@umanitoba.ca

© Springer Nature Switzerland AG 2021
H. R. Arabnia et al. (eds.), *Advances in Artificial Intelligence and Applied Cognitive Computing*, Transactions on Computational Science and Computational Intelligence, https://doi.org/10.1007/978-3-030-70296-0_79

This strategy was evaluated for a series of benchmark problems, each with unique solution spaces and local minima. In Sect. 2, the relevant literature pertaining to PSO and genetic algorithms (GA) is given. Section 3 provides the rationale for the adaptive strategies used in this work, while Sect. 4 presents the performance results following testing on four benchmark functions.

2 Relevant Literature

PSO is especially effective for local exploitation [3] and, when combined with global search strategies such as simulate annealing (SA) [6], can effectively balance global exploration with local exploitation of the solution space. On its own, PSO possesses many drawbacks due to the presumptive nature of the algorithm. Drawbacks include premature convergence [10], lack of search operator, and no directional related information conveyed [11]. Performance is also heavily subject to initial parameter dependence, such as the number of members in the swarm, initialization of inertia and velocity, as well as balancing swarm coefficients that dictate the degree to which members communicate with each other [5]. Researchers have made strides to correct for some of these drawbacks, by either creating hybridized PSO variants to prevent premature convergence or developing time- or performance-based strategies to eliminate the need for parameters. In the work carried out by [11], a multi-swarm approach was developed without the determination of the velocity vector—which is typically a key implementation in PSO. A chaotic implementation was used in [12] to improve the initial placement of particles in the search space. If particles are initialized to position too close together, the swarm may undergo premature convergence before adequate exploration can occur. The need for hyperparameters was all but removed in the approach taken by [13], who used machine learning to develop predictive models to adjust the required hyperparameters based on performance for any given problem. One noteworthy shortcoming being the required computational overhead required for such an approach to work.

Newer implementations of PSO have utilized a multi-swarm strategy, whereby multiple swarms independently operate with separate goals or strategies in mind, but together they communicate the best possible objective optimization. In [14], this was identified as a solution to the "two-steps forward, one step back" behavior that typically occurs in PSO, where progress is ultimately made in due time but at a cost to efficiency. In [14], multiple swarms were deployed with separate learning strategies: one local and the other global. A similar adoption was carried out in [15] where two populations focused on local and global search independently. Global search was implemented using the information provided from the swarm's best solution, while local search was driven by the personal best of each particle. A more sophisticated implementation of multi-swarm behavior has been the Pareto dominance and decomposition multi-objective approach. The work of [16] demonstrated better improvements in both convergence and swarm diversity as a result. Similar findings were shown in [17], whereby solving multi-objectives simultaneously promoted diversity. The use of multiple swarms can also be extended to include swarms adopting different learning strategies all together, as carried out in [18] where each generation modified its learning strategy based on performance. Performance was based on balancing recall and precision and had showed great improvements in benchmark evaluation compared to other state-of-the-art algorithms.

Another practical solution to the issue of population diversity can be provided through the GA. This heuristic benefits from the idea that better performing particles are in some way more fit—perhaps in terms of their location in solution space, or their inertia or velocity—therefore, their information should, in some way, be translated to other members of the population. This use of electing exemplar particles as a means to improve performance is then carried over to the next generation [17]. Compared to other heuristics such as SA, GA looks for the best fit of the population instead of relying on the performances of the swarm as a whole, thereby preventing local minima localization. Use of GA can even be translated to solve discrete problems such as the travelling salesperson problem [19], where continuous variables can be mapped to discrete ones. Since reliance on exemplar particles has the additive potential of early convergence, various mutation strategies can be used to introduce random diversity. Two very common mutation strategies are crossover and mutation [20], which both improve local search as they draw from information of other swarm members. In [19], several forms of mutation were used, including trigonometric mutation, and several mutation strategies that would take the difference between 2 and 4 other random vectors within the population, and mix those with personal and swarm best vectors. In the work of [21], the researchers took an alternative approach: learning from the worst experience of the population, and adjusting the swarm to coordinate around poor solutions found. The use of exemplar's personal best is one way in which solutions can be improved, but another strategy involves removing the presence of swarm members that are inhibiting progress. This idea was used in practice in [22], where a replacement strategy was used to remove trapped members in local minima. The authors also utilized a competitive selection scheme to promote the exemplars, which introduces a stochastic influence to encourage further population diversity.

3 Tribal PSO Implementation

3.1 Conventional PSO

In a standard PSO environment, the swarm members are initialized randomly in the search space with a corresponding position and velocity represented as vectors $X_i = (X_{i,1}, X_{i,2}, \ldots, X_{i,d})$ and $V_i = (V_{i,1}, V_{i,2}, \ldots, V_{i,d})$, respectively, for a d-dimensional search space. Each swarm member has knowledge of its own personal best denoted as $p_{i,j}^{best}$, where $j = 1, 2, \ldots, d$, and that of the swarm as a whole denoted as g^{best}. The velocity is updated on each iteration, see Eq. (1), to account for a particle's personal inertia and velocity, ω and V_i, respectively, its own personal best, $X_{i,j}^{best}$ as well as that of the swarm best, g^{best}. The expression for the velocity update is carried out according to Eq. (1):

$$V_{i,j}(t+1) = \omega V_{i,j}(t) + c_{1,t} r_1 (X_{i,j}^{best} - X_{i,j}(t))$$
$$+ c_{2,t} r_2 (g^{best} - X_{i,j}(t)), \quad (1)$$

where r_1 and r_2 are randomly distributed in the range [0,1]; $c_{1,t}$ and $c_{2,t}$ are personal and social best coefficients, respectively, at iteration t.

The process continues until a user-defined stopping condition is met. At each iteration t, the objective function is evaluated as a cost function $C[\cdot]$ and is denoted as $C_i(t)$ for particle i. An illustration of the competing influences involved in conventional PSO is shown in Fig. 1.

In this work, personal and social coefficients, $c_1 and c_2$, respectively, were updated adaptively at each iteration, according to the value of χ of Eq. (2), to accelerate convergence tendencies, as adopted in prior work [1]:

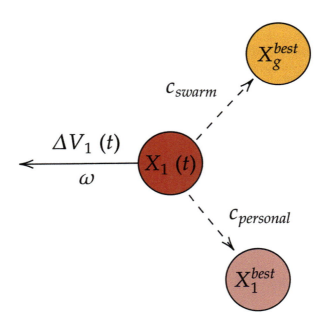

Fig. 1 This figure demonstrates conventional PSO, where each particle's next position at iteration $t + 1$, $X_i(t + 1)$, depends on a linear combination of different influences, including its current inertia and velocity, ω and $V_i(t)$, respectively, its personal best, X_i^{best}, as well as the global best position of the swarm, X_g^{best}

Algorithm 1 Conventional PSO

1: **procedure** PSO(P, T, ω)
2:    Initialize population members
3:    **for all** $i \in P$ **do**
4:        $X_i \leftarrow rand(D)$
5:        $V_i \leftarrow rand([2, D])$
6:    **end for**
7:    Initialize c_1, c_2 according to Eq. (2)
8:    **while** $t \neq T$ **do**
9:        **for all** $i \in P$ **do**
10:            $V_{i,t+1} \leftarrow$ according to Eq. (1)
11:            $X_{i,t+1} \leftarrow X_{i,t} + V_{i,t+1}$
12:            $C_{i,t+1} \leftarrow f(X_{i,t+1})$
13:            **if** $C_i^{best} > C_{i,t+1}$ **then**
14:                $C_i^{best}, X_i^{best} \leftarrow C_{i,t+1}, X_{i,t+1}$
15:                **if** $g^{best} > C_{i,t+1}$ **then**
16:                    $g^{best}, X_g^{best} \leftarrow C_{i,t+1}, X_{i,t+1}$
17:                **end if**
18:            **end if**
19:        **end for**
20:        $\omega \leftarrow \alpha\omega$
21:        $t++$
22:    **end while**
23: **end procedure**

$$c_{1/2,t}\chi = c_{1/2,t}\left(\frac{2\kappa}{\left|2 - \phi - \sqrt{|\phi^2 - 4\phi|}\right|}\right), \tag{2}$$

where $\kappa \in (0, 1]$ and $\phi = \phi_1 + \phi_2$. In this work, a κ of 1 and a ϕ_1 and ϕ_2 of 2.05 were used as they best approximated a global optimum when tested over a series of benchmark problems. Pseudocode for the conventional PSO implementation is shown in Algorithm 1.

3.2 Tribal PSO

In the proposed tribal PSO implementation, each particle is initially assigned to one of the tribes, $s_i = s_{1,2,...,S}$, where the total number of tribes, S, is a user-defined parameter set at compile time. In the conventional PSO, the social coefficient used to navigate the swarm to the swarm best is replaced with a tribal coefficient, g_s^{best}, which attracts tribe members to a tribe best in the proposed tribal PSO. As our results demonstrate, this small change to Eq. (1) allows different tribes to find their own local minima by simultaneously investigating different regions of the search space. Compared to conventional PSO, which trends toward exploiting a single local minima, tribal PSO has a better chance at finding the global minima, simply because many local minima are in consideration simultaneously. Tribal PSO is illustrated in Fig. 2 for a 3-tribe, 6-member swarm population—with additional changes shown in Algorithm 2.

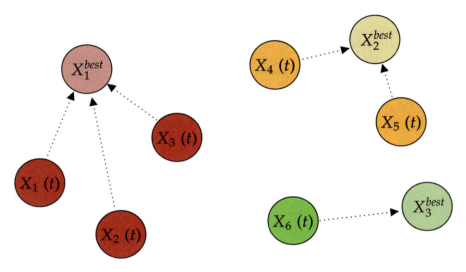

Fig. 2 Tribal PSO implementation that adopts a tribal best influence X_s^{best} for each tribe s

---

Algorithm 2 Tribal PSO

1: **procedure** TRIBAL PSO(P, T, ω, S)
2:     Initialize population members
3:     **for all** $i \in P$ **do**
4:         \\ Code from Algorithm 1
5:         $s_i \leftarrow randi[1, S]$
6:     **end for**
7:     **for all** $i \in P$ **do**
8:         \\ Code from Algorithm 1
9:         **if** $C_s^{best} > C_{i,t+1}$ **then**
10:            $C_s^{best}, X_s^{best} \leftarrow C_{i,t+1}, X_{i,t+1}$
11:         **end if**
12:     **end for**
13: **end procedure**

3.3 Spawn Behavior

3.3.1 Particle Fitness

Besides the typical position and change in position updates, as expressed in Eq. (1), a fitness evaluation is carried out at each iteration in order to assess a novel fitness of each particle with respect to the swarm as a whole. This evaluation takes into account changes in the evaluation of the objective function from iteration t to $t + 1$, as well as the current velocity, $V_{i,j}(t)$, shown in Eq. (3):

$$f_i(t + 1) = (1 - \hat{S}_s)V_{i,j}(t) |C_{t+1} - C_t|, \tag{3}$$

where $|C_{t+1} - C_t|$ is the absolute difference between the current and previous evaluations of the objective function, respectively; $\hat{S}_s$ corresponds to the normalized fitness of the tribe s. Using this method, Eq. (3) accounts for three influences on the fitness evaluation:

(1) $(1 - \hat{S}_s)$ will take into account the total fitness that each member of tribe s has relative to that of all tribes in existence. This in effect will penalize tribes who garner a larger fitness relative to that of other tribes, and prevent them from otherwise overtaking the swarm. Conversely, small and well-performing tribes are given a boost to fitness.
(2) $|C_{t+1} - C_t|$ will account for the change in objective evaluation made from iteration t to $t + 1$. When C_{t+1} is approximately equal to C_t, the fitness will evaluate to approximately 0. This will penalize tribes whose members are stuck in local minima. A large change to the proposed objective evaluation, irrespective of improvement, will coincide with a larger fitness by design.
(3) $V_{i,j}(t)$ will account for the particle's ability to traverse surface space, but without considering any improvements gained or lost in objective evaluation (as carried out in (2)).

For each member of the swarm population, a fitness is evaluated at each iteration, $f_i(t)$, as well as stored as a cumulative sum over the particle's lifespan f_i^{cum}. These two fitness variables will be used to assess the spawn behavior of future swarm members.

3.3.2 Spawn Decision Variables

Taking the particle's cumulative fitness f_i^{cum} into account, as outlined in Eq. (3), following each iteration a determination is made for either spawning a new swarm member or deletion of an existing swarm member. This evaluation is carried out using the following method given by Eq. (4):

$$f_i^{cum} = \begin{cases} spawn(p_{n+1}) & : \hat{f}_i^{cum}(t) > \tau^{spawn} \\ delete(p_i) & : \hat{f}_i^{cum}(t) < \tau^{kill}, \end{cases} \quad (4)$$

where $\hat{f}_i^{cum}(t)$ corresponds to the normalized cumulative fitness evaluation of the particle i relative to that of the swarm as a whole; τ^{spawn} and τ^{kill} act as bounds that control the rate at which swarm members are spawned and killed, respectively. If more than one particle meets either criteria set out in Eq. (4), then the particle is selected based on roulette selection proportional (or inversely proportional in the case of τ^{kill}) to the cumulative fitness of the candidates. The process by which the spawned member is initialized is described in Sect. 3.3.3.

Since values for τ^{spawn} and τ^{kill} cannot be known and regulated a priori, a constriction function was used to regulate the swarm population. In Eq. (5), Φ is determined based on the initial swarm population N, current swarm population n, and a regulation coefficient σ.

$$\Phi(n, \sigma, N) = \frac{\exp(N - n) - \exp(n - N)}{\sigma}. \quad (5)$$

As shown in Fig. 3, decreases in σ produce greater values for Φ. By fine tuning σ, the degree to which the population deviates from the initial starting population is

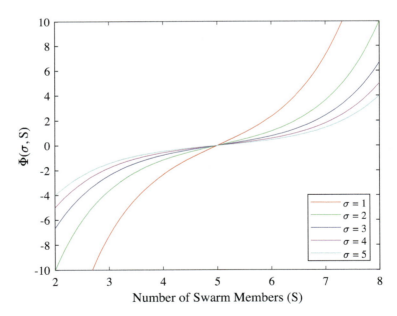

Fig. 3 Constriction coefficient Φ being evaluated for several values of σ for a starting population N of 5. Based on Eq. 5

An Adaptive Tribal Topology for Particle Swarm Optimization 989

managed dynamically. Values for Φ are then used to make changes to τ^{spawn} and τ^{kill}, where $\tau^{spawn}(t+1) = \tau^{spawn}(t) + \Phi$ and $\tau^{kill}(t+1) = \tau^{kill}(t) + \Phi$.

3.3.3 Spawn Child Initialization

Following the selection of the exemplar particle in Eq. (4), the child spawn is initialized based on the exemplar particle using the following set of procedures:

(1) Copy C_i^{best} and X_i^{best} from exemplar swarm member i to child swarm member $n + 1$.
(2) Heighten inertia w_{p+1} to a higher value than the exemplar swarm member ω_i at random, where $w_{p+1} = rand(w_i, 1]$. Dampen inertia of parent swarm member to a lower value using a similar convention: $w_i = rand(0, w_i]$.
(3) Assign child swarm member to new tribe at random, $s_{p+1} = randi[1, S]$. This prevents rivaling tribes from approaching extinction and provides a small chance an extinct tribe recovers.
(4) The new position vector X_{p+1} is derived from the parent swarm member best position X_i^{best}, as well as the best of the assigned tribal swarm X_s^{best}. This is expressed in Eq. (6).

$$X_{p+1} = \frac{X_s^{best} + X_i^{best}}{2}.$$
(6)

(5) Finally, the cumulative fitness of the parent swarm particle f_i^{cum} is halved and is added to the cumulative fitness of the assigned swarm $f_{s_{p+1}}^{cum}$. Additionally, the cumulative fitness of the parent swarm particle is subtracted from the parent tribe $f_{s_i}^{cum}$. This reduces the tendency for tribal fitness to accumulate and lead to early extinction through the evaluation of $(1 - \hat{S}_s)$ in Eq. 3.

These additions to spawn behavior, as well as the alterations made to tribal influence, are summarized in Algorithm 3.

3.4 Test Functions

3.4.1 Benchmark Functions

To test the efficacy of the tribal implementation, a series of benchmark problems were selected. The problems, summarized in Table 1 and illustrated in Fig. 4, represent a series of optimization problems with several local minima present. Therefore, in order for the algorithm to successfully traverse the solution space, sufficient intra-tribal exploration is required, with enough tribal diversity to encourage thriving tribes to expand, while minimizing the impact that poorly performing tribes have on the algorithms performance.

Algorithm 3 Tribal Spawn PSO

1: **procedure** TRIBAL SPAWN PSO(P, G, ω, S)
2:    Initialize population members
3:    **for all** $i \in P$ **do**
4:        \\ Code from Algorithms 1 and 2
5:        $f_{i,i} \leftarrow$ according to Eq. (3)
6:        $f_i^{cum} + = f_i^{cum}$
7:        **if** $f_i^{cum} < \tau_{kill}$ **then**
8:            $delete(p_i)$
9:        **end if**
10:       **if** $f^{cum} > \tau^{spawn}$ **then**
11:           $C_{i+1}^{best}, X_{i+1}^{best} \leftarrow C_i^{best}, X_i^{best}$
12:           $w_i \leftarrow rand(0, \omega_i]$
13:           $s_{i+1} \leftarrow randi[1, S]$
14:           $\omega_{i+1} \leftarrow rand(w_i, 1]$
15:           $f_i^{cum} \leftarrow f_i^{cum}/2$
16:           $f_{s_i}^{cum}, f_{s_{p+1}}^{cum} \leftarrow -f_i^{cum}/2, f_i^{cum}/2$
17:           $X_{i+1} \leftarrow$ according to Eq. (6)
18:       **end if**
19:       $\Phi \leftarrow$ according to Eq. (5)
20:       $\tau^{kill}, \tau^{spawn} + = \Phi$
21:   **end for**
22: **end procedure**

3.4.2 Benchmark Performance and Evaluation

Over the course of the performance evaluation, a series of metrics will be used to both test the diversity of the tribal population, and the performance of the tribes in finding improved objective evaluations. Four metrics will be the focus of the analysis:

(1) **Tribal best** will track the overall performance of the swarm in finding better evaluations of the objective function.
(2) **Tribal fitness** will track the cumulative fitness of the tribe as a whole, which impacts the evaluation of $\hat{S}_s$ and fitness of the individual swarm members according to Eq. 3.
(3) **Tribal count** will count the number of members in each tribe, which is indicative of the current survival rate.
(4) **Coefficient values** will track the changes in τ^{spawn} and τ^{kill} as changes are made according to the value of Φ at each iteration.

By combining these metrics, a better understanding of the evolving nature of the tribes can be visualized given the constant or changing nature of the solution space.

Table 1 Benchmark problems used to test the tribal PSO implementation

Function (f(x))	Global Min/Max (x^*)	Range (x_i)	Identifier	Ref.
$\sum_{i=1}^{m} c_i \exp\left(-\frac{1}{\pi}\sum_{j=1}^{d}(x_j - A_{ij})^2\right)\cos\left(\pi\sum_{j=1}^{d}(x_j - A_{ij})^2\right)$	(2.002, 1.006)	$\in [0, 10]$	LANG	Fig. 4a*
$418.9829d - \sum_{i=1}^{n} x_i \sin(\sqrt{|x_i|})$	(420.968, 420.968)	$\in [-500, 500]$	SCHWEF	Fig. 4b
$-\left|\exp\left(\left|1 - \frac{\sqrt{x_1^2+x_2^2}}{\pi}\right|\right)\sin(x_1)\cos(x_2)\right|$	(−9.6645, 9.6645)	$\in [-10, 10]$	HOLD	Fig. 4c
$-(x_2+47)\sin\left(\sqrt{|x_2+\frac{x_1}{2}+47|}\right) - x_i \sin(\sqrt{|x_1 - (x_2+47)|})$	(512, 404.2319)	$\in [-512, 512]$	EGG	Fig. 4d

*Values for A_{ij} and c_i retrieved from [23] for $d = 2$

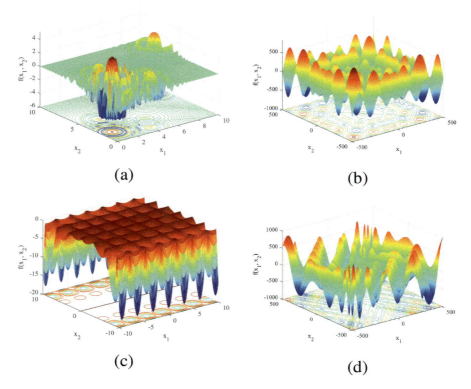

Fig. 4 Contour plots for the (**a**) Langer, (**b**) Schwefel, (**c**) Holder, and (**d**) Eggholder benchmark functions. Expressions and bounds are defined in Table 1

4 Results and Discussion

4.1 Unmodified Benchmark Functions

4.1.1 Impact of σ

As discussed previously, the value for Φ has the effect of moderating the swarm populations by constricting values for τ^{spawn} and τ^{kill} through the coefficient σ. This was illustrated in Fig. 3, where larger values of σ provide lower changes to Φ, and vice versa. In Fig. 5, the evaluation metrics are displayed for the Schwefel function evaluated for a σ of 5.

One thing to note is that while the value for σ is sufficiently high, the tribal population count remains relatively steady (Fig. 5c) once the tribes approach a steady state cost (Fig. 5a)). Additionally, corrections are made to the coefficient values initially from their initial starting values and then remain steady when the

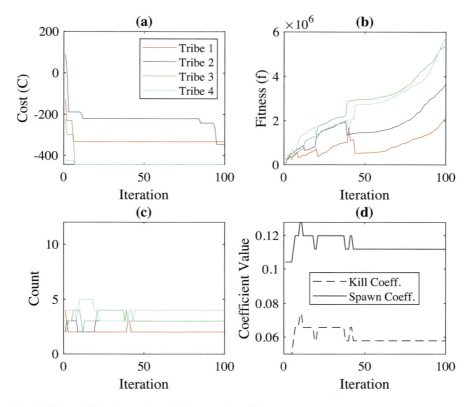

Fig. 5 Schwefel function evaluated for a σ value of 5

tribal finesses rise relative to one another (Fig. 5b) as no improvements are made to objective function evaluations.

In contrast, Fig. 6 illustrates the performance, while the value of σ is closer to 2. Several tribes approach and become extinct as shown in Fig. 6c, while several changes are being made to the coefficients to account for the deviation from the initial starting population (Fig. 6d). Even if no improvements are seen to tribal fitness, a lower σ promotes more exploration even if the tribes are themselves stuck in local minima.

4.1.2 Impact of Benchmark Function

Shown in Fig. 7 are the performance metrics for the four benchmark problems defined in Table 1. Several takeaways can be made from the results. Firstly, while all tribes converged to the global minima early on in their deployment, only select benchmark functions continued to increase in tribal fitness. This includes the

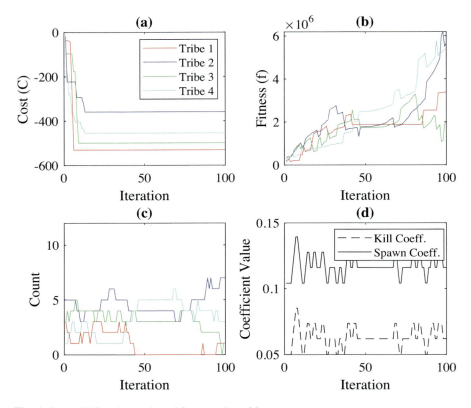

Fig. 6 Schwefel function evaluated for a σ value of 2

functions LANG and EGG shown in Fig. 7a and d, respectively. This trend was shown to be the case when tested over several run times. In examining the solution spaces for these functions in Fig. 4a and d, we find that these functions have fewer deep local minima than their benchmark counterparts. This would coincide with the algorithm jumping between local minima, with improvements to short term cost, and therefore greater fitness through the evaluation of $|C_{t+1} - C_t|$, but no improvements to global exploration. Several of these parameters will be tweaked and optimized in the following section.

4.1.3 Benchmark Objective Evaluation and Parameter Tuning

In testing the efficacy of the algorithms, the ability for the tribes to converge to the global minima was evaluated over several values of total population P, number of tribes D, and σ. Table 2 demonstrates the results of this evaluation, with the value indicating the proportion of total run times where at least one of the tribes converged

An Adaptive Tribal Topology for Particle Swarm Optimization

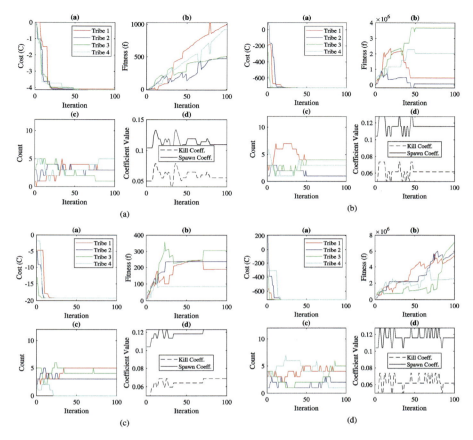

Fig. 7 Performance metrics for the (**a**) Langer, (**b**) Schwefel, (**c**) Holder, and (**d**) Eggholder benchmark functions. σ was set to 5

to the global solution. It is worthy to note that instances where more than one tribe converged to the global minima were not recorded separately, as at least one of the tribes reaching the global minima would suffice as a global solution search. Two functions that were the most difficult to explore were the EGG and LANG functions, the same two benchmark problems discussed in the previously section. It is not clear whether or not this is due to the proximity of local minima, the relative depth of local minima, or the sheer number of local minima that create the discrepancy in performance between benchmarks. The scenario also creates the possibility that a stop condition can be created that is based on the decrease in relative tribal fitness. As demonstrated in Fig. 7c and 7d, no change is shown in tribal fitness when these tribes converge to global solutions, but while the algorithm is still exploring, tribal fitness increases as tribe members continue to make changes to their current and previous cost found.

Table 2 Objective function evaluations for the four benchmark problems, where the value indicates the proportion of total run times where at least one of the tribes reached the global minima. Unless otherwise noted, the default total population P, number of tribes D, and σ are set to 16, 4, and 5, respectively

Parameter		LANG	SCHWEF	HOLD	EGG	
Regulation coefficient (σ)	2	0.12	1.0	0.69	0.06	
	5	0.16	1.0	0.62	0.09	
	7	0.17	1.0	0.60	0.07	
Total population (P)	8	0.09	1.0	0.40	0.05	
	12	0.11	1.0	0.51	0.07	
	24	0.16	1.0	0.75	0.12	
Tribe number (D)	3	0.17	1.0	0.64	0.07	
	5	0.15	1.0	0.62	0.07	
	7	0.14	1.0	0.60	0.06	

In regards to the regulation coefficient σ, a value of 5 outperformed the performance of either 2 or 7. This trend was discussed at length in Sect. 4.1.1. While performance overall was very poor for both the LANG and EGG benchmarks, increased population N did lead to better performance (see Table 2) as more particles were assigned for exploration. When run time was measured between N values of 8, 12, and 24, the time difference was proportional to the population number, indicating no additional overhead is introduced that interferes with algorithm scalability. Run time was also tested between values for tribe number S, and no statistically significant difference was found. It was found that decreases in the number of tribes S did coincide with decreased performances across the board. Since tribes are the main driving force between inter-tribal interactions, this parameter will heavily impact the performance for global exploration. While lower S did follow with better performances, these were all taken with a total population N of 16 and may reflect the number of tribe members per tribe as the main driving factor for performance than for the value of S itself. Since tribe number would be a parameter that is extremely problem specific, proper tuning would be required in order to optimize for additional benchmark problems.

5 Conclusion

In this work a tribal implementation of PSO was implemented to solve a series of benchmark functions. A novel constriction coefficient was adopted to manage the number of members of the swarm population as the algorithm progressed. New swarm members were introduced based on exemplar particles in the existing swarm, while poorly performing swarm members were killed off to prevent swarm stagnation. This heuristic was successful in encouraging new local minima to be explored and was found to exploit the global minima in two of the benchmarks

An Adaptive Tribal Topology for Particle Swarm Optimization

tested in over 50% of the run-time instances. The other two benchmarks tested had solution spaces that the algorithm could not exploit in over 12% of the run times, leading the authors to believe parameter tuning according to the number of members in the swarm population, or the number of tribes initialized at run time, is warranted.

Acknowledgments This research has been financially supported by Mitacs Accelerate (IT15018) in partnership with Canadian Tire Corporation and is supported by the University of Manitoba.

References

1. M. Clerc, J. Kennedy, The particle swarm - explosion, stability, and convergence in a multidimensional complex space. IEEE Trans. Evol. Comput. **6**(1), 58–73 (2002)
2. Y. Qi, C. Li, P. Jiang, C. Jia, Y. Liu, Q. Zhang, Research on demodulation of FBGs sensor network based on PSO-SA algorithm, vol. 164, pp. 647–653 (2018). [Online]. Available: http://www.sciencedirect.com/science/article/pii/S0030402618304261
3. L.A. Bewoor, V.C. Prakash, S.U. Sapkal, Production scheduling optimization in foundry using hybrid particle swarm optimization algorithm. Procedia Manufacturing **22**, 57–64 (2018). [Online]. Available: http://www.sciencedirect.com/science/article/pii/S2351978918303044
4. B. Keshanchi, A. Souri, N.J. Navimipour, An improved genetic algorithm for task scheduling in the cloud environments using the priority queues: Formal verification, simulation, and statistical testing. J. Syst. Software **124**, 1–21 (2017). [Online]. Available: http://www.sciencedirect.com/science/article/pii/S0164121216301066
5. A. Serani, C. Leotardi, U. Iemma, E.F. Campana, G. Fasano, M. Diez, Parameter selection in synchronous and asynchronous deterministic particle swarm optimization for ship hydrodynamics problems. Appl. Soft Comput. **49**, 313–334 (2016). [Online]. Available: http://www.sciencedirect.com/science/article/pii/S1568494616304227
6. S. Sudibyo, M.N. Murat, N. Aziz, Simulated annealing-particle swarm optimization (SA-PSO): Particle distribution study and application in neural wiener-based NMPC, in *2015 10th Asian Control Conference (ASCC)*, pp. 1–6 (2015)
7. Y. Zheng, H. Ling, J. Xue, S. Chen, Population classification in fire evacuation: A multiobjective particle swarm optimization approach. IEEE Trans. Evol. Comput. **18**(1), 70–81 (2014)
8. F. Khoshahval, A. Zolfaghari, H. Minuchehr, M.R. Abbasi, A new hybrid method for multi-objective fuel management optimization using parallel PSO-SA. Prog. Nucl. Energy **76**, 112–121 (2014). [Online]. Available: http://www.sciencedirect.com/science/article/pii/S0149197014001334
9. X. Wang, Q. Sun, The study of k-means based on hybrid SA-PSO algorithm, in *2016 9th International Symposium on Computational Intelligence and Design (ISCID)*, vol. 2, pp. 211–214 (2016)
10. K. Dorgham, I. Nouaouri, H. Ben-Romdhane, S. Krichen, A hybrid simulated annealing approach for the patient bed assignment problem. Procedia Comput. Sci. **159**, 408–417 (2019). [Online]. Available: http://www.sciencedirect.com/science/article/pii/S1877050919313778
11. K.M. Ang, W.H. Lim, N.A.M. Isa, S.S. Tiang, C.H. Wong, A constrained multi-swarm particle swarm optimization without velocity for constrained optimization problems. Expert Syst. Appl. **140**, 112882 (2020). [Online]. Available: http://www.sciencedirect.com/science/article/pii/S0957417419305925
12. G. Li, Z. Yu, The double chaotic particle swarm optimization with the performance avoiding local optimum, in *2015 International Conference on Estimation, Detection and Information Fusion (ICEDIF)*, pp. 424–427 (2015)
13. K.R. Harrison, B.M. Ombuki-Berman, A.P. Engelbrecht, A parameter-free particle swarm optimization algorithm using performance classifiers. Information Sciences **503**, 381–400 (2019).

[Online]. Available: http://www.sciencedirect.com/science/article/pii/S0020025519306188

14. G. Xu, Q. Cui, X. Shi, H. Ge, Z.-H. Zhan, H.P. Lee, Y. Liang, R. Tai, C. Wu, Particle swarm optimization based on dimensional learning strategy. Swarm Evol. Comput. **45**, 33–51 (2019). [Online]. Available: http://www.sciencedirect.com/science/article/pii/S2210650218305418

15. N. Lynn, P.N. Suganthan, Heterogeneous comprehensive learning particle swarm optimization with enhanced exploration and exploitation. Swarm Evol. Comput. **24**, 11–24 (2015). [Online]. Available: http://www.sciencedirect.com/science/article/pii/S2210650215000401

16. A. de Campos, A.T.R. Pozo, E.P. Duarte, Parallel multi-swarm PSO strategies for solving many objective optimization problems. J. Parallel Distrib. Comput. **126**, 13–33 (2019). [Online]. Available: http://www.sciencedirect.com/science/article/pii/S0743731518308554

17. X. Liu, Z. Zhan, Y. Gao, J. Zhang, S. Kwong, J. Zhang, Coevolutionary particle swarm optimization with bottleneck objective learning strategy for many-objective optimization. IEEE Trans. Evol. Comput. **23**(4), 587–602 (2018)

18. N. Lynn, P.N. Suganthan, Ensemble particle swarm optimizer. Appl. Soft Comput. **55**, 533–548 (2017). [Online]. Available: http://www.sciencedirect.com/science/article/pii/S1568494617300753

19. I.M. Ali, D. Essam, K. Kasmarik, A novel design of differential evolution for solving discrete traveling salesman problems. Swarm Evol. Comput. **52**, 100607 (2019). [Online]. Available: http://www.sciencedirect.com/science/article/pii/S2210650219304468

20. J. Zhao, H. Xu, W. Li, A hybrid genetic algorithm for Bayesian network optimization, in *The 2014 2nd International Conference on Systems and Informatics (ICSAI 2014)*, pp. 906–910 (2014)

21. X. Yan, Q. Wu, H. Liu, W. Huang, An improved particle swarm optimization algorithm and its application. J. Comput. Sci. Issues **10**(1), 9 (2013)

22. Y. Chen, L. Li, J. Xiao, Y. Yang, J. Liang, T. Li, Particle swarm optimizer with crossover operation. Eng. Appl. Artif. Intell. **70**, 159–169 (2018). [Online]. Available: http://www.sciencedirect.com/science/article/pii/S0952197618300174

23. M. Molga, C. Smutnicki, Test functions for optimization needs (2005). [Online]. Available: http://www.zsd.ict.pwr.wroc.pl/files/docs/functions.pdf

The Systems AI Thinking Process (SATP) for Artificial Intelligent Systems

James A. Crowder and Shelli Friess

1 Introduction

Every cognitive entity, human, whale, dog, mouse, or artificial intelligent entity (AIE), there are lower executive functions required to keep the entity at a basic functioning level. In humans, and other living entities, we call these lower brain functions which take on such functions as keeping the heart pumping, therefore supplying blood to the body, keeping the entity breathing, etc. [1]. Within an AIE, there are analogous lower-level executives required to keep the AIE functional [2]. These lower-level brain functions are autonomous and are required for the entity to survive. These include functions like breathing, heart rate, etc. In an artificial entity, these functions, like regulation of resources, regulation of movement of the entity, and access to memory systems, function in the background as artificial subconscious processes [3]. Here we present an Intelligent Software Agent (ISA) architecture for subconscious control and regulation of an artificial entity, including subconscious functions like cognitive economy and action control. This includes background processes like artificial neurogenesis, the creation of new neural pathways as new concepts and memories are created from experiential-based activities [4]. Figure 1 illustrates the overall systems-level AI architecture, presenting the artificial AI entities' subconscious regulatory system and a control system [5].

J. A. Crowder (✉)
Colorado Engineering, Inc., Colorado Springs, CO, USA
e-mail: jim.crowder@coloradoengineering.com

S. Friess
School of Counseling, Walden University, Minneapolis, MN, USA
e-mail: shelli.friess@mail.waldenu.edu

© Springer Nature Switzerland AG 2021
H. R. Arabnia et al. (eds.), *Advances in Artificial Intelligence and Applied Cognitive Computing*, Transactions on Computational Science and Computational Intelligence, https://doi.org/10.1007/978-3-030-70296-0_80

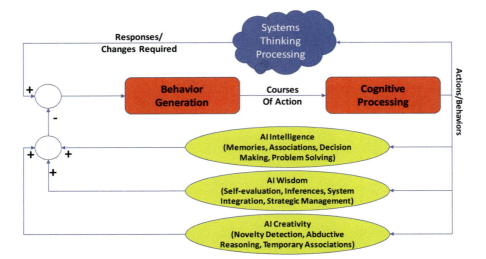

Fig. 1 Systems-level AI process control

2 Systems Theory and Intelligent Agents

For an AI entity to be autonomous, all the conscious and subconscious mechanisms that drive a living entity must be in place and available for the artificial entity to function within its environment [6]. In order to facilitate these conscious and subconscious processes, we designed an Intelligent Software Agent (ISA) processing infrastructure that communicates constantly throughout the artificial entity, providing information, goals, inferences, memory management, reasoning, and other functions required for autonomous functionality [7].

2.1 The ISA Processing Framework

The ISA processing framework is general-purpose software architecture providing for configurable software agents with dynamically loadable components that provide application-specific processing [8]. The framework defines how components interact while an application implemented using the framework defines the agent functionality and how they operate. The ISA processing framework (IPF) functions as a distributed intelligence model within the artificial entity that allows multiple, as-needed entry and exit points while providing reasoning and inferencing capabilities required for autonomous control of an artificial entity [9]. This is required for real-time autonomous systems which will operate in dynamically changing environments. There are five ISA archetypes that are utilized within the IPF and are described below (Fig. 2).

The Systems AI Thinking Process (SATP) for Artificial Intelligent Systems

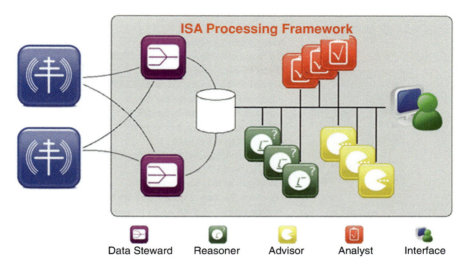

Fig. 2 The systems-level IPF processing framework

ISAs acquire, remove, and modify functional capabilities at birth and during operation in order to evolve based on conditions of their environment and rules associated with both agent and system goals [10]. These are facilitated through ISA service nodes described below. The five ISA archetypes within the IPF are:

1. *Data Steward Agent (ADS)*: The data steward agent's primary duty is the collection and storage of incoming sensor information to support dynamically changes sources. The data steward agent (called an ADS) forwards data to a service data flow (SDF) which generates metadata for the incoming data and stores these data in the long-term storage. A configuration service (SCO) is used to specify which node needs the available data.
2. *Advisor Agent (AAD)*: Advisor agents disseminate the right information to the right place at the right time; it provides capabilities that allow collaborative question asking and information sharing by agents and end-users. AADs generate and maintain topical maps required to find relative information fragments, memories, and "expert" ISAs.
3. *Reasoner Agent (ARE)*: The reasoner agent pulls from short-term memory to perform signal processing, reasoning evaluation and traces, and feature extraction of the data, applying configured and generated goal reaching tasks for the purpose of initial data manipulation or as part of condition monitoring. Reasoner agents utilize goals provided through a user interface (UI). Strategies (algorithms) implement learning and adaptation applied to configured goals. Reasoner agents provide state data as output from processing [11]. The output is made available for retrieval by other agents (e.g., an analyst agent).

Fig. 3 The ISA available services

4. *Analyst Agent (AAN)*: Analyst agents are fed by reasoner agents and utilize the developed ontologies and lexicons to expand upon questions and answers learned from collected information.
5. *Interface Agent (AIN)*: The interface agents provide human operator interactions. They provide the necessary input and output mechanisms that allows the SATP IPF to configure and provision agents and services within the agent network as well as retrieve reports from the agents representing ongoing behaviors and actions that the system has taken. Interface agents provide the primary input of loadable components – nodes and services – once the system is already running. The IPF can provision new agent types, new services, and new nodes to the system via an interface agent that communicates with the system mediator's plugin archives.

The ISAs are specifically designed to support distributed data collection, processing, reasoning, and dissemination across a highly distributed processing environment within the artificial entity. Each ISA consists of a well-defined set of services which are mandatory, but which can be extended to support additional functionality on a per-use-case basis. Figure 3 illustrates the current list of available ISA services.

Each ISA archetype has a minimum set of services required for executive functions as that ISA archetype. Each agent can have more services than the minimum but must have at least these services. Figure 4 illustrates the mandatory services per ISA archetype.

Each ISA service is composed of a set of service nodes. A family of service nodes constitutes an ISA service plugin. Figure 5 shows the nodes currently available for service plugins.

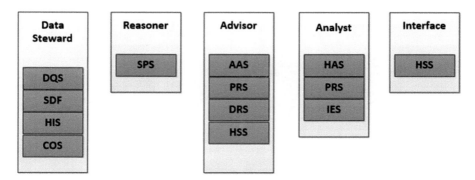

Fig. 4 Mandatory services for each ISA archetype

Fig. 5 Currently available nodes for ISA services

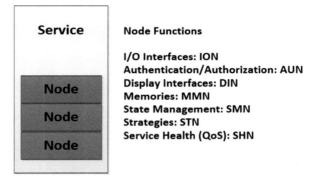

Each IPF ISA service requires a minimum set of nodes to be called a specific service type. A service may contain more than the minimum nodes but must contain at least the nodes shown in Fig. 6.

2.2 Self-Adapting ISAs

Intelligence reveals itself in a variety of ways, including the ability to adapt to unknown situations or changing environments as would be the case for virtually any autonomous AI entity [12]. Without the ability to adapt to new situations, the IPF could only rely on a previously written set of rules. For a truly autonomous AI entity, it cannot only depend on previously defines set of rules for every possibly contingency [13]. In order to be truly autonomous, there must be a set of cognitive system roles that the ISAs must function within, based first on the ability to create new artificial neural pathways and ending with these new capabilities being added to the overall AI entities' resources and resource management [14]. The IPF self-adaptive system roles are illustrated in Fig. 7.

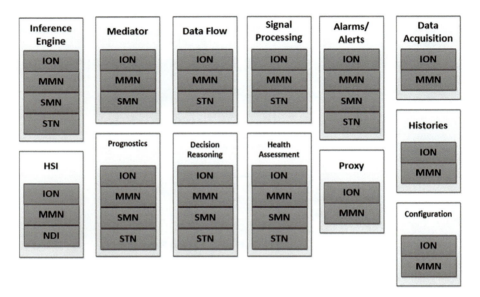

Fig. 6 Minimum nodes required for each ISA service plugin

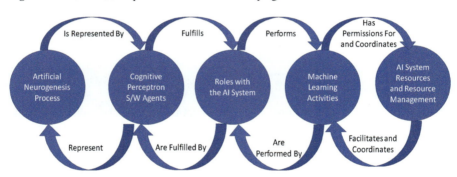

Fig. 7 The IPF SATP self-adaptive system roles

To function at a conscious and subconscious level, the SATP IPF agents must have mechanisms to facilitate systems-level collective thinking [15]. This requires a well-defined collective thinking process, which is illustrated in Fig. 8. This process defines the roles, authorities, skill, activities, and shared states for ISAs across the IPF processing infrastructure. Operating within this collective thinking process allows autonomous operation – but within defined bounds [16].

Within this process, self-organizing topical maps categorize and classify information into topics of interest to the SATP IPF system. This allows changing external environments to be processed and that knowledge added to the artificial entity's overall knowledgebase, which then drives behavior emergence based on this new set of knowledge. Figure 9 illustrates this SATP adaptive behavior flow.

The Systems AI Thinking Process (SATP) for Artificial Intelligent Systems

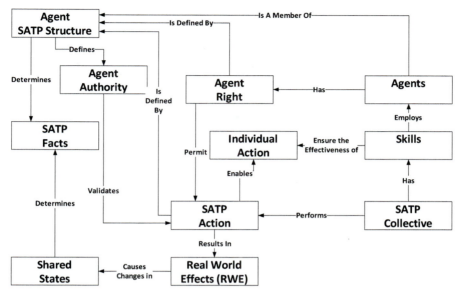

Fig. 8 The SATP IPF collective thinking process

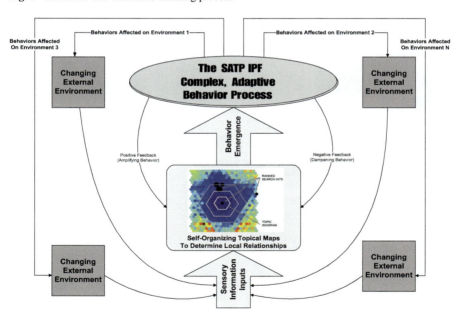

Fig. 9 The SATP IPF adaptive behavior process

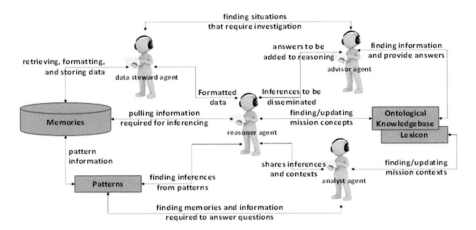

Fig. 10 The SATP IPF communication ecosystem

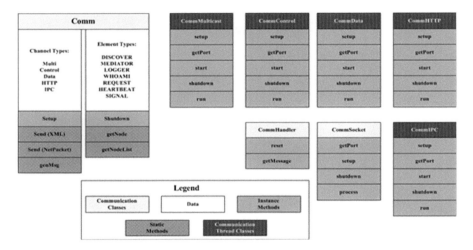

Fig. 11 The SATP IPF communication classes

3 The SATP Artificial Neurogenesis

Communication throughout the IPF system and subsystems provides a continuous flow of information through the ISAs [11]. This ISA high-level SATP ecosystem is illustrated in Fig. 10.

The continuous subconscious SATP communication threads provide the artificial entity's informational system [17]. Figure 11 illustrates the current communication classes within the IPF processing infrastructure.

The Systems AI Thinking Process (SATP) for Artificial Intelligent Systems 1007

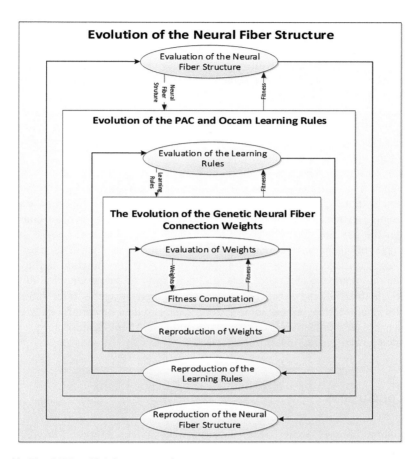

Fig. 12 The SATP artificial neurogenesis process

When it has been determined that new information has been learned or experienced and explained, new neural pathways must be created to accommodate the new knowledge and inferences. This is facilitated within the IPF with multiple learning systems. Two major learning systems required for SATP are PAC[1] learning and Occam learning. These drive the evolution of new neural pathways that are then added to the collective intelligence of the artificial entity and available for processing. Figure 12 illustrates the artificial neurogenesis process [19].

---

[1]In this framework, the algorithm receives samples and must select a generalization function, hypothesis, from a class of possible functions. The goal is that, with high probability, the selected function will have a low generalization error or be approximately correct.

4 Dynamic SATP Consideration

Real-time systems are intrinsically dynamic. They can arrive at answers and fulfill objectives and goals in many ways, some of which must be adaptive and creating. Autonomous AI entities are no different. Plasticity or multi-level adaptivity must be designed into the autonomous AI entity in order to achieve any real level of autonomy [19]. They must adapt to new circumstances and contexts, and they must happen at a system level so that the knowledge and contexts can be driven to all parts of the cognitive system.

Each part has interactions within and among systems (interaction with its environment). Reactions may be both homeostatic and spontaneous. This would require adaptability, in order to react to the different feedback both within the system and among systems [3]. The ISA IPF processing infrastructure must constantly consider the abilities and restrictions of the agent types within the SATP system. Each type of agent has limitations. Figure 13 illustrates this. The three main attributes that make the AI entity's autonomous SATP IPF function is the ability to be autonomous, the ability to learn, and the ability to cooperate. Not all agents can do all three all the time. The reasoner and analyst agents must have these capabilities all at once, but the advisory, interface, and data steward function do not. This is necessary for resource management within the AI entity, since no system has unlimited power, processing, or memory [20].

5 Conclusions

Fully autonomous AI entities are not a reality (we are not counting self-driving cars). In order to facilitate real, thinking, reasoning, inferencing, and communicating AI, we must bring their design considerations up to the system level and not just a collection of algorithms and processing boards [21]. Every aspect of systems-level intelligence needs to be thought through and designed in. The results of not doing this could be disastrous . . . "No Dave, I don't think I can do that . . . ." Much work is left to do; we must design, build, test, tear down, re-design, rebuild, and retest until we have achieved a systems-level thinking AI entity that is autonomous and functional within the world at large, whether that's this world or exploring a far off planet or moon.

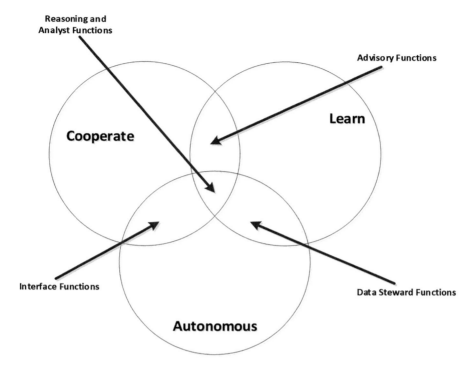

Fig. 13 The IPF ISA functional Venn diagram

References

1. J. Miller, *Living Systems* (McGraw-Hill, Niwot, 1978)
2. W. Ashby, *Design for A Brain: The Origin of Adaptive Behavior*, 2nd edn. (Chapman & Hall, London, 1960)
3. H. Simon, *The Sciences of the Artificial*, vol 136, 3rd edn. (The MIT Press, 1996)
4. C. Churchman, *The Design of Inquiring Systems: Basic Concepts of Systems and Organizations* (Basic Books, New York, 1971)
5. P. Checkland, *Systems Thinking, Systems Practice: Includes a 30-Year Retrospective* (Wiley, Chichester, 1999)
6. L. Bertalanffy, *General System Theory: Foundations, Development, Applications* (George Braziller, New York, 1968)
7. A. Burks, *Essays on Cellular Automata* (University of Illinois Press, Urbana, 1970)
8. J. Crowder, J. Carbone, S. Friess, *Artificial Cognition Architectures* (Springer International, New York, 2013). ISBN 978-1-4614-8072-3
9. J. Crowder, J. Carbone, S. Friess, *Artificial Psychology: Psychological Modeling and Testing of AI Systems* (Springer International, New York, 2019). ISBN 978-3-030-17081-3
10. J. Crowder, S. Friess, System-level thinking for artificial intelligent systems, in *Proceedings of the 2019 International Conference on Artificial Intelligence*, (2019), pp. 157–163
11. J. Crowder, S. Friess, J. Carbone, Anytime learning: A step toward life-long machine learning, in *Proceedings of the 2019 International Conference on Artificial Intelligence*, (2019), pp. 16–20

12. I. Prigogine, *From Being to Becoming: Time and Complexity in the Physical Sciences* (W H Freeman & Co., San Francisco, 1980)
13. H. Simon, The architecture of complexity. Proc. Am. Philos. Soc. **106** (1962)
14. J. Crowder, S. Friess, J. Carbone, Implicit learning in artificial intelligent systems: The coming problem of real, cognitive AI, in *Proceedings of the 2019 International Conference on Artificial Intelligence*, (2019), pp. 48–53
15. C. Shannon, W. Weaver, *The Mathematical Theory of Communication* (University of Illinois Press, Urbana, 1971)
16. L. Zadeh, From circuit theory to system theory. Proc. IRE **50**(5), 856–865 (1962)
17. C. Cherry, *On Human Communication: A Review, A Survey, and A Criticism* (The MIT Press, Cambridge, MA, 1957)
18. J. Crowder, J. Carbone, Methodologies for continuous life-long machine learning for AI systems, in *Proceedings of the 2018 International Conference on Artificial Intelligence*, (2018), pp. 44–50
19. J. Holland, *Adaptation in Natural and Artificial Systems: An Introductory Analysis with Applications to Biology, Control, and Artificial Intelligence* (The MIT Press, Cambridge, MA, 1992)
20. J. Crowder, J. Carbone, Artificial neural diagnostics and prognostics: Self-soothing in cognitive systems, in *Proceedings of the 2018 International Conference on Artificial Intelligence*, (2018), pp. 51–57
21. N. Luhmann, *Introduction to Systems Theory* (Polity, Cambridge, 2013)

Improving the Efficiency of Genetic-Based Incremental Local Outlier Factor Algorithm for Network Intrusion Detection

Omar Alghushairy, Raed Alsini, Xiaogang Ma, and Terence Soule

1 Introduction

Outlier detection is an important process in big data, which has received a lot of attention in the area of machine learning. The reason is that it is important to detect suspicious items and unusual activities. Outlier detection has a significant impact on many applications such as detecting fraud transactions for credit cards and network intrusion detection [1].

Local Outlier Factor (LOF) [2] is the most common local outlier (anomaly) detection algorithm; it is a nearest-neighbor-based algorithm [3]. LOF has become popular because it can detect an outlier without any previous knowledge about the data distribution. Also, it can detect outliers in data that has heterogeneous densities [4, 5]. However, LOF is designed for static data, which means the size of data is fixed. LOF needs very large memory because it requires storage of the whole dataset with the distances' value in memory. Therefore, LOF cannot be applied to a

---

O. Alghushairy (✉)
Department of Computer Science, University of Idaho, Moscow, ID, USA

College of Computer Science and Engineering, University of Jeddah, Jeddah, Saudi Arabia
e-mail: algh5752@vandals.uidaho.edu

R. Alsini
Department of Computer Science, University of Idaho, Moscow, ID, USA

Faculty of Computing and Information Technology, King Abdulaziz University, Jeddah, Saudi Arabia
e-mail: alsi1250@vandals.uidaho.edu

X. Ma · T. Soule
Department of Computer Science, University of Idaho, Moscow, ID, USA
e-mail: max@uidaho.edu; tsoule@uidaho.edu

© Springer Nature Switzerland AG 2021
H. R. Arabnia et al. (eds.), *Advances in Artificial Intelligence and Applied Cognitive Computing*, Transactions on Computational Science and Computational Intelligence, https://doi.org/10.1007/978-3-030-70296-0_81

data stream, where the size of data is increasing continuously [6–8]. To overcome this issue, the Incremental Local Outlier Factor (ILOF) has been developed that can handle data streams [9]. Nevertheless, ILOF also needs to store all the data in the memory. Moreover, it cannot detect the sequence of outliers in the data stream such as unexpecting surges of outliers in several streaming segments. However, it is important to detect the sequence of outliers in a data stream; if they are not detected, many problems may occur, such as network attacks and sensor malfunctions. For instance, the KDD Cup 99 HTTP dataset has several outlier sequences. This dataset contains a simulation of normal data with attack traffic on an IP scale in computer networks for testing intrusion detection systems.

To solve the issues with ILOF, several algorithms have been proposed that are Genetic-based Incremental Local Outlier Factor (GILOF) [10], Memory Efficient Incremental Local Outlier Factor (MILOF) [11], and Density Summarization Incremental Local Outlier Factor (DILOF) [12]. The GILOF demonstrates an efficient performance in the state-of-the-art LOF algorithms in data streams. Our work aims to further improve the performance of the GILOF algorithm by proposing a new calculation method for LOF, called $\underline{L}$ocal $\underline{O}$utlier $\underline{F}$actor by $\underline{R}$eachability distance (LOFR). In LOF, the resulting score is based on the local reachability density, while in LOFR, the resulting score is based on the reachability distance. The newly proposed algorithm is $\underline{G}$enetic-based $\underline{I}$ncremental $\underline{L}$ocal $\underline{O}$utlier $\underline{F}$actor by $\underline{R}$eachability distance (GILOFR). GILOFR includes two stages: (1) the detection stage and (2) the summarization stage. The detection stage detects the outliers and updates the old data information in the memory (window) and determines whether a new data point is an inlier or outlier. The summarization stage summarizes the old data in order to reduce the memory consumption and takes into account the old data density. Thus, GILOFR is able to work in limited memory. This new calculation method (LOFR) is the main contribution of this research paper. In this research, GILOFR was evaluated via a series of experiments by using real-world datasets. The results show that the performance of GILOFR is better than GILOF on several of the datasets.

The remainder of this paper is as follows: The related work is reviewed in Sect. 2; Sect. 3 describes the GILOFR algorithm; Sect. 4 presents and discusses the results of the experiments; and Sect. 5 presents the conclusion of our research.

2 Related Work

Outliers have two categories: global outliers and local outliers. If the data point p_0 is far away from other data points, it is considered a global outlier [13]. Local outlier is a data point that is outlier with respect to its k-nearest neighbors. LOF introduced the idea of local outlier. It is considered a means of density-based outlier detection because it makes a comparison between the density of data points in the dataset and their local neighbors. To determine the local outlier score, LOF assigns a measured degree of the density of data points and their local neighbors [14] (Fig. 1).

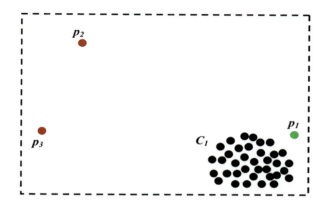

Fig. 1 A two-dimensional example illustrating the outlier's categories, where p_2 and p_3 are global outliers and p_1 is a local outlier

2.1 LOF with Its Extensions in a Data Stream

LOF is a powerful algorithm for detecting local outliers in a static environment. This algorithm aims to calculate the LOF score for every data point in the dataset. The score of LOF indicates whether the data point is considered as an inlier or an outlier. The following steps define the LOF [2]. First, it determines the k-distance (k-dist) of the data point p, where k-dist is the distance between the two data points, p and o, by using the Euclidean distance. Second, it determines the k-nearest neighbor of p. The significance of the k-nearest neighbor of data point p is that each data point q has a distance from data point p that is not greater than the k-dist(p). Third, it determines the reachability distance (Rd) of p in relation to o. In fact, the Rd of a data point p in relation to the corresponding data point o is used when the distance is greater than the k-dist(p). Fourth, it calculates the local reachability density (Lrd) of p. The Lrd of the data point p is calculated by using Eq. (1):

$$Lrd_{MinPts}(p) = 1 \Big/ \left(\frac{\sum_{o \in N_{MinPts}(p)} \text{reach} - \text{dist}_{MinPts}(p, o)}{|MinPts(p)|} \right) \quad (1)$$

Finally, the LOF score of data point p is calculated by using Eq. (2):

$$LOF_{MinPts}(p) = \frac{\sum_{n \in N_{MinPts}(p)} \frac{Lrd_{MinPts}(o)}{Lrd_{MinPts}(p)}}{|N_{MinPts}(p)|} \quad (2)$$

However, the LOF must have a lot of memory to store all of the data points in the dataset. It also needs to keep all distance values between data points. Additionally, it was developed to work on static data and is not applicable for a data stream [6].

ILOF extends the LOF algorithm with respect to the data stream. The idea of ILOF is to update and calculate the score of LOF in a data stream to determine if the newly incoming data point is an inlier or outlier. The method of insertion by ILOF handles the newly incoming data point np by the following two steps. First, it calculates the values of reachability distance (Rd), Lrd, and LOF of np. Second, it updates the values of the k-distance (k-$dist$), Rd, Lrd, and LOF for the existing data points. For example, $k = 3$ and np is inserted. Then, Rd is calculated to the three nearest neighbors of np. After that, Lrd is calculated. The insertion of np might change the k-$dist$ for points that have np in their k-nearest neighbors. K-$dist$ has an effect on Rd. As a result, if k-$dist$ is updated, Rd will be updated too. When Rd is updated, Lrd needs to be updated, which leads to the update of the LOF score. For more detail about ILOF, readers can refer to [9]. However, ILOF still needs to store all the data points in memory in order to calculate the LOF score, which requires a large memory and significant time. The sliding window in [15] is used to detect the change in data behavior. Although the algorithm can detect the change in data behavior, it is difficult for it to differentiate between the new data behavior and outliers.

To overcome the issues of ILOF, three algorithms are proposed for outlier detection for a data stream. The MILOF algorithm summarizes the old data by using k-means [11]. However, this method usually performs badly because clustering old data points by using k-means does not retain the density. The DILOF algorithm shows better performance than MILOF by summarizing the old data using gradient descent. However, gradient descent can get stuck in local minima [12]. GILOF is based on the genetic algorithm (GA) [16], which has demonstrated better performance in several real-world datasets [10]. GILOF summarizes the old data by using a GA, which is a population-based search technique and uses crossover and mutation operators to more widely explore the search space; these are generally better than the simple gradient descent for searching complicated spaces with many local minimums. GILOF contains two stages: the detection stage and the summarization stage. The detection stage aims to detect the outliers and update the old data points when new incoming data points are inserted. The incoming data points are calculated based on the ILOF algorithm alongside a skipping scheme [12]. In the summarization stage, the sliding window is used with a specific size to limit memory usage. Then, the Genetic Density Summarization (GDS) is used to summarize the old half-data points in the window (W) [10].

2.2 Network Intrusion Detection System (NIDS)

The basic purpose of using data mining techniques for intrusion detection in networks is to detect security violations in information systems. To detect an intrusion in networks, data mining processes and analyzes a massive amount of data [17]. Data mining and the knowledge extracted from big data have become more efficient with the new developments in machine learning [18]. Many new algorithms in data mining have been developed that can be applied to datasets to discover clusters, profiles, factors, relationships, predictions, and patterns. Nowadays, data mining algorithms are widely used in different domains, such as business, marketing, laboratory research, weather forecasting, and network intrusion detection [19]. Our work is related to the network intrusion detection system (NIDS). In order to test the effectiveness of the proposed algorithm in NIDS, we used the KDD Cup 99 HTTP service dataset, which is a subset of the KDD Cup 99 SMTP dataset [19]. This HTTP dataset has surprising surges of outliers in several streaming segments [20].

The intrusion detection system is classified by two categories: misuse detection and outlier (anomaly) detection. In outlier detection, it aims to recognize the suspicious items and unusual activity of the network or host. The NIDS monitors and tests the activity in the network, which provides the necessary security for the network of a system [21]. For instance, it detects malicious activities like a denial-of-service attack (DoS) or a network traffic attack. The LOF algorithm has been used in different models and methods for NIDS [12, 22–25]. Despite the researchers' interest in outlier detection, they focus more on global outlier detection in a data stream rather than the local outliers, which have been less studied [26]. Therefore, it is difficult to use this traditional LOF algorithm in data streams for NIDS. However, the ILOF algorithm addressed the limitation of the LOFs in data streams, and other algorithms, such as GILOF, addressed the limitation of time complexity in the ILOFs by summarizing big data streams. The aim of our proposed algorithm GILOFR is to achieve more accurate performance than other state-of-the-art LOF algorithms in data streams for NIDS.

3 Proposed Algorithms

3.1 Local Outlier Factor by Reachability Distance (LOFR)

In this section, we introduce a new calculation method for LOF, which is named LOFR. In fact, LOFR is like LOF [2], except the LOFR does not use the local reachability density when calculating its score. The score of LOFR is based on the reachability distance of the data point p and its nearest neighbors. Below are the key definitions of LOFR; note that the first three definitions are the same as for the LOF [2].

Definition 1 Calculating the k-dist of the data point p

The distance between two data points p and o in a Euclidean is calculated using Eq. (3).

$$d\ (p, o) = \sqrt{\sum_{i=1}^{n} (p_i - o_i)^2} \tag{3}$$

Let D be a dataset. For the data point p, the *k-dist(p)* is the distance between (p) and the farthest neighbor point o $(o \in D)$ with the following conditions:

k is a positive integer, (1) with regard to at least k data points $o' \in D \setminus \{p\}$, it maintains that $d(p, o') \leq d(p, o)$ and (2) with regard to at most k-1 data points $o' \in D \setminus \{p\}$, it maintains that $d(p, o') < d(p, o)$ [2].

Definition 2 Calculating the k-nearest neighbor of p

In this situation, the *k*-nearest neighbors of p are each a data point q whose distance to p is equal or smaller than the *k-dist(p)*. The *k*-nearest neighbor is described in Eq. (2).

$$N_{k-\text{distance}(p)}(p) = \left\{ q \in D \setminus \{p\} \mid d(p,q) \leq k - \text{dist}(p) \right\} \quad (4)$$

Definition 3 Calculating the reachability distance (Rd) of p in relation to o

The *Rd* between two data points such as p with regard to o is defined in Eq. (3).

$$\text{reach} - \text{dist}_k(p, o) = \max \left\{ k - \text{dist}(o), d(p, o) \right\} \quad (5)$$

Some examples of *Rd* when *k*-nearest neighbor = 5 are represented in Fig. 2. When the distance between two data points (e.g., *p2*) and o is not larger than the

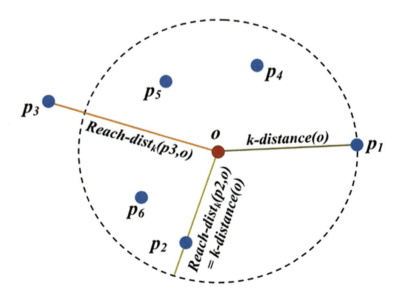

Fig. 2 The reachability distance for different data points (p) with regard to o, when k equals 5

k-dist(o), then the *k-dist(o)* is the *Rd*. On the other hand, when the distance between two data points (e.g., *p3*) and *o* is larger than the *k-dist(o)*, the actual distance between them is the *Rd*.

Definition 4 LOFR score of p

After all the previous steps, the score of LOFR is calculated by using Eq. (6).

$$LOFR_k(p) = \sum_{o \in N_k(p)} \frac{Rd_k(p)}{\left(\frac{Rd_k(o)}{k}\right)} \tag{6}$$

LOFR takes the reachability distance (*Rd*) of data point (*p*) and divides it by the average *Rd* of its neighbors. This method of calculation can provide a little lower "outlierness" score than LOF.

3.2 Genetic-Based Incremental Local Outlier Factor by Reachability Distance (GILOFR)

The main objective of GILOFR is to find the outlierness score in the following instances: (1) when detecting the local outlier with no prior knowledge about the data distribution; (2) when only a small part from the running dataset will be stored in memory; (3) when detecting that the outlier of data point (*p*) has to be finished at current time *T*; and (4) when the algorithm does not have any prior knowledge about future incoming data points when it detects the current outliers.

The designed GILOFR has two stages: detection and summarization. In the detection stage, the ILOFR and skipping scheme are used together to detect an outlier. Note: ILOFR is like ILOF [9], except the ILOFR uses the new calculation that is mentioned above. In the summarization stage, the Genetic Density Summarization (GDS) is applied to summarize the old data points in memory (the window). Algorithm 1 is the GILOFR, which works in the following steps: (1) it determines the window (*W*) size, where the *W* size equals the amount of data points; (2) it uses the LOFR threshold θ to detect outliers; and (3) during the summarizing stage, GDS is applied to summarize the data points. Then, when the new data point *np* occurs, GILOFR detects the outliers by using ILOFR along with the skipping scheme (lines 4–6 in Algorithm 1) [10, 12]. GILOFR will keep detecting outliers and calculating the score of LOFR for every new data point until the present window reaches the *W* size. Thereafter, the GDS algorithm is applied on the window to summarize the 50% of older data points in the window through choosing 25% of data points from that older 50%. After that, the 50% of older data points will be removed from the window, and the chosen 25% will be transferred to the window; this 25% of data points will be combined with the remaining data points in the window (lines 9–13 in Algorithm 1). At this point, the window contains 75% of data points. This 75%

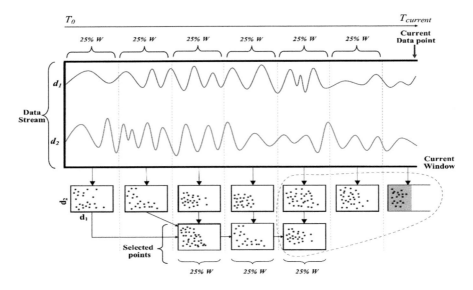

Fig. 3 GILOFR process for a data stream in 2D from time T_0 to $T_{current}$ [10]

of data points will be chosen to combine with the new streaming data points. This process is repeated when the window becomes full again (Fig. 3).

3.3 Genetic Density Summarization (GDS)

GDS depends on the genetic algorithm (GA). A GA is a search algorithm in the evolutionary computation field, which is based on biological evolution [27, 28]. GA includes several components, population, chromosome (or individual), objective (or fitness) function (*Of*), selection, crossover, and mutation [29, 30]. The chromosomes are an array of numerical or binary values that represent the genes evolving and are known as candidate solutions. A population contains a set of chromosomes, which are set randomly as the initial population. The objective function aims to calculate the fitness value for every chromosome in the population. The goal of selection is to find two chromosomes that acquired above-average fitness values. There are many types of selection, which include roulette wheel selection (RWS), linear rank-based selection (RNK), tournament selection (TNT), stochastic universal sampling selection (SUS), and linear rank-based selection with selective pressure (RSP) [31]. Crossover combines the characteristics of the two chromosomes based on a cut-off point to produce new chromosomes. Crossover has several types, as well, such as one-point, two-point, and uniform crossover [32]. Mutation aims to preserve the population diversity and prevent the population from becoming stuck in the local minima. There are also several types of mutation, such as single-point mutation, uniform mutation, and boundary mutation (BDM) [33].

To find the optimal data points, the GDS algorithm is going to summarize *50%* of old data points in the window to balance the density differences between the chosen old *50%* and the new candidates, which are *25%* of data points. The GDS algorithm begins by producing a population, where the population contains random chromosomes assigned with data points. Then, the objective function (*Of*) is executed to evaluate each chromosome. The *Of* is the sum of $a_k(x'_n)$.

$$Of = \sum_{n=1}^{50\%W} a_k\left(x'_n\right) \tag{7}$$

where $a_k(x'_n)$ is the density of x'_n according to its kth-nearest neighbors and x'_n is the *50%* of old data points in the window.

After that, in every generation, GDS executes the selection operation on the present generation. Then GDS executes a crossover operation on two chromosomes. Thereafter, the GDS executes the mutation operation on two chromosomes as well. When GA operations are completed for each generation, GDS stores the best results. After the above steps, the chosen chromosomes are converted into S. After that, S is projected into the binary domain, and the better *25%* of s_n in S is adjusted to 1, and the rest to 0. The GDS is going to choose a data point x'_n in X' if its corresponding s_n in S equals 1 (Fig. 4). Finally, the chosen data points will be presented as Z, which is *25%* of the new data points. For more detail, refer to [10].

3.4 Skipping Scheme

The purpose of the skipping scheme is to keep data points in the dataset in a lower density of the region. Therefore, when the new data points emerge into a new class, that data point will have no effect on the outlier detection accuracy. Because of the benefit of the skipping scheme in outlier detection, it combines with ILOFR to detect any long sequence (series) of outliers [12].

Let $a \in A$; the skipping scheme works by using Euclidean distance in two steps: (1) the average distance $\overline{d_1(A)}$ is computed according to the distance between the data point a with its first nearest neighbor $d_1(a)$ and (2) the distance between a new data point (p) and the latest detected outlier lo is calculated. If the distance between p and lo is less than the distance between p and $\overline{d_1(A)}$, the data point p is considered an outlier [12]. Otherwise, the data point p will be an inlier.

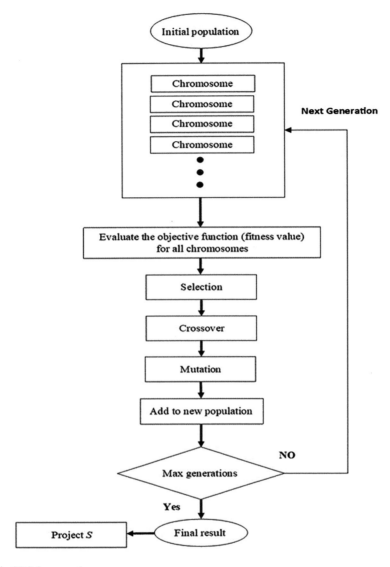

Fig. 4 GDS framework

Algorithm 1: GILOFR

Inputs:	unlimited data streams $P=\{p_1, p_2, ..., p_t, ...\}$,
	The max window size W
	The threshold scores θ of LOFR
	The population size PS
	The number of chromosomes NC
	The number of generations NG

```
1   X ← {}// is the data points in the memory
2   X' ← {}// is the oldest 50% data points in the memory
3   O ← {}// is the detected outliers
4   Skipping_Scheme
5   For each p_t ∈ P do
6       LOFR_k(p_t) ← ILOFR(p_t,o,O,θ)
7       If LOFR_k (p_t) > 0 then
8           X ← X ∪ {p_t}
9           If |X| = W then
10              Z ← GDS(X',PS, NC,NG)
11              Delete the oldest 50%W data points in X
12              X' ← X' ∪ Z
13          End
14      End
15  End
```

4 Empirical Experiment

In this section, the experiment results for GILOFR are presented. We compared the GILOFR results with the results of GILOF, and we compared both algorithms without the skipping scheme (GILOFR_NS, GILOF_NS). The metrics that are used in the experiment are the execution time and the accuracy of outlier detection. For the accuracy of outlier detection, we used the area under the ROC curve (AUC) [34, 35]. All algorithms in the experiment were implemented in C++ on a machine that has an Intel(R) Core(MT) i5-8250U CPU, 250 GB SSD hard disk, 8GB RAM, and Windows 10 (64-bit).

The GILOFR is tested with different window sizes on unnormalized and normalized datasets, described in Table 1. Those datasets are available in the machine learning database repository [36, 37]. The Vowel dataset is modified to the format of the data stream, such as that in [10, 38]. The five percent of uniform noise is added to the Pendigit dataset, so any noised data point will be considered as an outlier [12]. For the KDD Cup 99 SMTP datasets, the same settings as in [39] are used, where the network attacks are the outliers.

The hyperparameters of GILOF were set as in [10], the population size (PS) is 2, the number of generations (NG) is 4, and the selection type is RWS. The crossover used a *two-point* crossover at a rate of 0.7, and the mutation used a *BDM* mutation at a rate of 0.07 for each dataset [40]. For the GILOFR algorithm, we used the same hyperparameters of GILOF, except the number of generations (NG) was changed to 2. The k-nearest neighbors for each dataset were determined as the following for all algorithms: for the Vowel dataset $k = 19$; for the Pendigit dataset $k = 18$; for the KDD Cup 99 SMTP dataset $k = 8$ when the dataset is normalized and $k = 9$ when the dataset is unnormalized for GILOFR and GILOFR_NS while $k = 8$ for GILOF and GILOF_NS; and for the KDD Cup 99 HTTP dataset $k = 9$ when the dataset is normalized and $k = 8$ when the dataset is unnormalized in all algorithms. In addition, we determined ten window sizes for all datasets, $W = \{100, 200, 300, 400, 500, 600, 700, 800, 900, 1000\}$.

4.1 Experiment Results

4.1.1 Accuracy of Outlier Detection

The accuracy of all algorithms was tested by applying them to several real-world datasets (Table 1). Figures 5, 6, and 7 show the AUC of all algorithms for the four datasets. In the unnormalized Vowel dataset, all algorithms were close to each other, but GILOFR_NS was better in most W sizes and demonstrated the highest accuracy result (95.3%). In the normalized dataset, GILOFR_NS was also a little better in most W sizes and demonstrated the highest accuracy result (96.2%).

In the unnormalized Pendigit dataset, all algorithms showed poor results for accuracy. By contrast, for the normalized dataset, all algorithms were close to each other when $W \leq 400$, but GILOF and GILOF_NS were better when $W \geq 500$, and they demonstrated the highest accuracy result (98.7%).

In the unnormalized KDD Cup 99 SMTP dataset, all algorithms were in competition when $W \leq 500$; therefore, there is a specific algorithm that surpasses others in each W size. GILOFR and GILOFR_NS showed better results, and GILOFR_NS demonstrated the highest accuracy result (88%) in $W = 300$. In the normalized KDD Cup 99 SMTP dataset, all algorithms were in competition. Every algorithm showed better accuracy in specific W sizes, while GILOFR_NS demonstrated the highest accuracy result (89%) in $W = 200$.

In the unnormalized KDD Cup 99 HTTP dataset, GILOFR showed better results for accuracy in all W sizes, except when $W = 700$. However, GILOFR demonstrated

Table 1 The real-world dataset characteristics

Datasets	Number of data points	Dimension	Class
UCI Vowel dataset	1456	12	11
UCI Pendigit dataset	3498	16	10
KDD Cup 99 SMTP dataset	95,156	3	Unknown
KDD Cup 99 HTTP dataset	567,479	3	Unknown

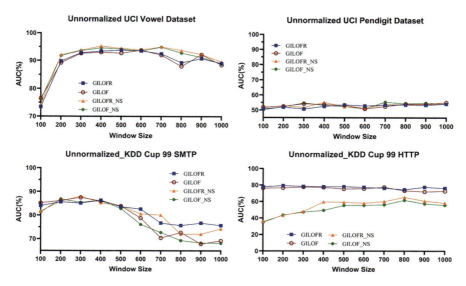

Fig. 5 The comparisons of outlier detection accuracy for unnormalized real-world datasets in all window sizes, where GILOFR and GILOFR_NS demonstrated the highest accuracy result in UCI Vowel, KDD Cup 99 SMTP, and KDD Cup 99 HTTP

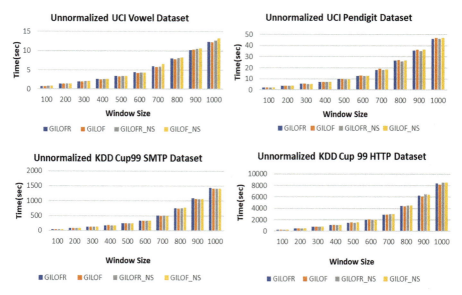

Fig. 6 The comparisons of execution time for unnormalized real-world datasets in all window sizes

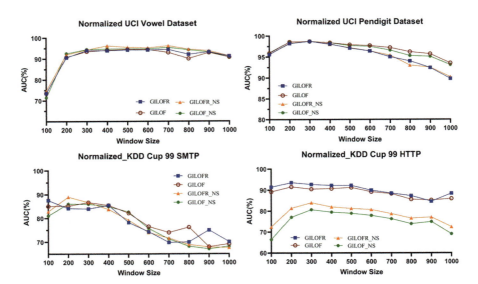

Fig. 7 The comparisons of outlier detection accuracy for normalized real-world datasets in all window sizes, where GILOFR and GILOFR_NS demonstrated the highest accuracy result in UCI Vowel, KDD Cup 99 SMTP, and KDD Cup 99 HTTP

the highest accuracy result (79.3%). Moreover, GILOFR_NS and GILOF_NS showed poor results for accuracy in most W sizes, while GILOFR_NS demonstrated a moderate result for accuracy in $W = 800$ (65.3%). For the normalized KDD Cup 99 HTTP dataset, GILOFR showed better results for accuracy in all W sizes except when $W = 900$. However, GILOFR demonstrated the highest results for accuracy (93.6%). GILOFR_NS and GILOF_NS became better when the dataset was normalized, in which case GILOFR_NS surpasses GILOF_NS in all W sizes.

Finally, the experiment results prove that our new method of calculation can improve the GILOF algorithm, where GILOFR and GILOFR_NS show better results for accuracy than GILOF and GILOF_NS in several datasets, especially in the UCI Vowel and the KDD Cup 99 HTTP datasets.

4.1.2 Execution Time

In Figures 6 and 8, results for the execution time are presented for all algorithms in the unnormalized and normalized datasets, respectively. In the experiments, execution times are measured by seconds. Generally, all algorithms are close to each other in most W sizes. In the UCI Vowel dataset, GILOFR took 0.73–12.3 s, and GILOF took 0.75–12.4 s. In the UCI Pendigit dataset, GILOFR took 1.96–47.19 s, and GILOF took 2.05–47.12 s. In KDD Cup 99 SMTP dataset, GILOFR took 45.09–1442.1 s, and GILOF took 46.2–1420.4 s. In KDD Cup 99 HTTP dataset, GILOFR

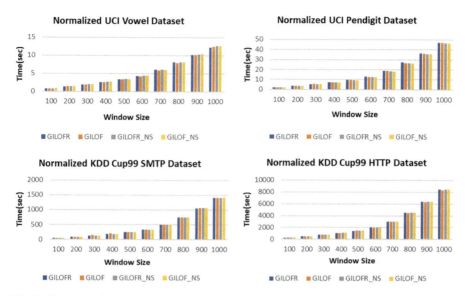

Fig. 8 The comparisons of execution time for normalized real-world datasets in all window sizes

took 269.7–8436.5 s, and GILOF took 274.5–8356.6 s. Moreover, GILOFR_NS and GILOF_NS were very close to GILOFR and GILOF in execution time for most W sizes.

5 Conclusion

The objective of the proposed GILOFR algorithm is to further improve the accuracy of outlier detection in the GILOF algorithm for data stream mining. Like GILOF, GILOFR addresses the limitation of the LOF algorithm in a data stream, but GILOFR further improves GILOF. Accordingly, GILOFR does not need to recalculate all previous steps when a new data point occurs. Additionally, it does not require the storage of all the data points in memory, because it can work under the limited memory. Our new calculation of LOF, which is called LOFR, has a positive impact on the GILOF algorithm and leads to more accurate results. The outcomes of experiments demonstrate that GILOFR and GILOFR_NS are better than GILOF and GILOF_NS for the accuracy of outlier detection in real-world datasets that have sequence of outliers and high-density region, i.e., data points are very close to each other. Specifically, this is true for the KDD Cup 99 HTTP dataset, which includes a simulation of normal data, with abnormal data as attack traffic, on an IP scale in computer networks for testing intrusion detection systems.

References

1. D. Namiot, On big data stream processing. Int. J. Open Inf. Technol. **3**(8) (2015)
2. M.M. Breunig, H.P. Kriegel, R.T. Ng, J. Sander, LOF: identifying density-based local outliers, in *Proceedings of the ACM SIGMOD International Conference on Management of Data*, (2000), pp. 93–104
3. M. Goldstein, S. Uchida, A comparative evaluation of unsupervised anomaly detection algorithms for multivariate data. PLoS One **11**(4), e0152173 (2016)
4. Y. Yan, L. Cao, C. Kulhman, E. Rundensteiner, Distributed local outlier detection in big data, in *Proceedings of the 23rd ACM SIGKDD International Conference on Knowledge Discovery and Data Mining*, (2017), pp. 1225–1234
5. Y. Yan, L. Cao, E. Rundensteiner, Scalable top-n local outlier detection, in *Proceedings of the 23rd ACM SIGKDD International Conference on Knowledge Discovery and Data Mining*, (2017), pp. 1235–1244
6. S. Sadik, L. Gruenwald, Research issues in outlier detection for data streams. ACM SIGKDD Explorations Newsletter **15**(1), 33–40 (2014)
7. O. Alghushairy, X. Ma, Data storage, in *Encyclopedia of Big Data*, ed. by L. Schintler, C. McNeely, (Springer, Cham, 2019). https://doi.org/10.1007/978-3-319-32001-4
8. R. Alsini, X. Ma, Data streaming, in *Encyclopedia of Big Data*, ed. by L. Schintler, C. McNeely, (Springer, Cham, 2019). https://doi.org/10.1007/978-3-319-32001-4
9. D. Pokrajac, A. Lazarevic, L.J. Latecki, Incremental local outlier detection for data streams, in *IEEE Symposium on Computational Intelligence and Data Mining*, (2007), pp. 504–515
10. O. Alghushairy, R. Alsini, X. Ma, T. Soule, A genetic-based incremental local outlier factor algorithm for efficient data stream processing, in *Proceedings of the 4th International Conference on Compute and Data Analysis*, (2020), pp. 38–49. https://doi.org/10.1145/3388142.3388160
11. M. Salehi, C. Leckie, J.C. Bezdek, T. Vaithianathan, X. Zhang, Fast memory efficient local outlier detection in data streams. IEEE Trans. Knowl. Data Eng. **28**(12), 3246–3260 (2016)
12. G.S. Na, D. Kim, H. Yu, DILOF: Effective and memory efficient local outlier detection in data streams, in *Proceedings of the 24th ACM SIGKDD International Conference on Knowledge Discovery & Data Mining*, (2018), pp. 1993–2002
13. E.M. Knox, R.T. Ng, Algorithms for mining distance based outliers in large datasets, in *Proceedings of the International Conference on Very Large Data Bases*, (1998), pp. 392–403
14. I. Souiden, Z. Brahmi, H. Toumi, A survey on outlier detection in the context of stream mining: Review of existing approaches and recommendations, in *International Conference on Intelligent Systems Design and Applications*, (Springer, Cham, 2016), pp. 372–383
15. S.H. Karimian, M. Kelarestaghi, S. Hashemi, I-inclof: improved incremental local outlier detection for data streams, in *The 16th CSI International Symposium on Artificial Intelligence and Signal Processing (AISP 2012)*, (IEEE, 2012), pp. 023–028
16. K.F. Man, K.S. Tang, S. Kwong, Genetic algorithms: Concepts and applications [in engineering design]. IEEE Trans. Ind. Electron. **43**(5), 519–534 (1996)
17. N.A. Azeez, T.M. Bada, S. Misra, A. Adewumi, C. Van der Vyver, R. Ahuja, Intrusion detection and prevention systems: An updated review, in *Data Management, Analytics and Innovation*, (Springer, Singapore, 2020), pp. 685–696
18. R. Sahani, C. Rout, J.C. Badajena, A.K. Jena, H. Das, Classification of intrusion detection using data mining techniques, in *Progress in Computing, Analytics and Networking*, (Springer, Singapore, 2018), pp. 753–764
19. M.K. Siddiqui, S. Naahid, Analysis of KDD CUP 99 dataset using clustering based data mining. Int. J. Database Theory Appl. **6**(5), 23–34 (2013)
20. S.C. Tan, K.M. Ting, T.F. Liu, Fast anomaly detection for streaming data, in *Twenty-Second International Joint Conference on Artificial Intelligence*, (2011)

21. B.R. Raghunath, S.N. Mahadeo, Network intrusion detection system (NIDS), in *2008 First International Conference on Emerging Trends in Engineering and Technology*, (2008), pp. 1272–1277
22. J. Auskalnis, N. Paulauskas, A. Baskys, Application of local outlier factor algorithm to detect anomalies in computer network. Elektronika ir Elektrotechnika **24**(3), 96–99 (2018)
23. T. Ding, M. Zhang, D. He, A network intrusion detection algorithm based on outlier mining, in *International Conference in Communications, Signal Processing, and Systems*, (Springer, Singapore, 2017), pp. 1229–1236
24. Z. Xu, D. Kakde, A. Chaudhuri, Automatic hyperparameter tuning method for local outlier factor, with applications to anomaly detection. arXiv preprint arXiv:1902.00567 (2019)
25. S. Agrawal, J. Agrawal, Survey on anomaly detection using data mining techniques. Procedia Comput. Sci. **60**, 708–713 (2015)
26. M. Salehi, C. Leckie, J.C. Bezdek, T. Vaithianathan, Local outlier detection for data streams in sensor networks: Revisiting the utility problem invited paper, in *2015 IEEE Tenth International Conference on Intelligent Sensors, Sensor Networks and Information Processing (ISSNIP)*, (2015), pp. 1–6
27. A.E. Eiben, J.E. Smith, *Introduction to Evolutionary Computing*, vol 53 (Springer, Berlin, 2003), p. 18
28. A. Ponsich, C. Azzaro-Pantel, S. Domenech, L. Pibouleau, Constraint handling strategies in genetic algorithms application to optimal batch plant design. Chem. Eng. Process. Process Intensif. **47**(3), 420–434 (2008)
29. M. Mitchell, *An Introduction to Genetic Algorithms* (MIT Press, London, 1998)
30. H.G. Goren, S. Tunali, R. Jans, A review of applications of genetic algorithms in lot sizing. J. Intell. Manuf. **21**(4), 575–590 (2010)
31. R. Sivaraj, T. Ravichandran, A review of selection methods in genetic algorithm. Int. J. Eng. Sci. Technol. **3**(5), 3792–3797 (2011)
32. J. Magalhaes-Mendes, A comparative study of crossover operators for genetic algorithms to solve the job shop scheduling problem. WSEAS Trans. Comput. **12**(4), 164–173 (2013)
33. N. Soni, T. Kumar, Study of various mutation operators in genetic algorithms. Int. J. Comput. Sci. Inf. Technol. **5**(3), 4519–4521 (2014)
34. J.A. Hanley, B.J. McNeil, The meaning and use of the area under a receiver operating characteristic (ROC) curve. Radiology **143**(1), 29–36 (1982)
35. A.P. Bradley, The use of the area under the ROC curve in the evaluation of machine learning algorithms. Pattern Recogn. **30**(7), 1145–1159 (1997)
36. D. Dua, C. Graff, *UCI Machine Learning Repository* (University of California, School of Information and Computer Science, Irvine, 2019). [online]. Available: http://archive.ics.uci.edu/ml
37. R. Shebuti, *ODDS Library [http://odds.cs.stonybrook.edu]* (Stony Brook University, Department of Computer Science, Stony Brook, 2016)
38. C.C. Aggarwal, S. Sathe, Theoretical foundations and algorithms for outlier ensembles. ACM SIGKDD Explorations Newsletter **17**(1), 24–47 (2015)
39. K. Yamanishi, J.I. Takeuchi, G. Williams, P. Milne, On-line unsupervised outlier detection using finite mixtures with discounting learning algorithms. Data Min. Knowl. Disc. **8**(3), 275–300 (2004)
40. https://github.com/olmallet81/GALGO-2.0

Variance Fractal Dimension Feature Selection for Detection of Cyber Security Attacks

Samilat Kaiser and Ken Ferens

1 Introduction

Our everyday life is dependent on cyber systems one way or the other. The world of communication uses some form of cyber networks be it a critical infrastructure, telecom sectors, monetary transactions, or mere everyday emails or networking on a social platform. Hence, protecting the cyber network against threats or attacks is important to safeguard the government, economies, information security, and data privacy. But the data that we share are ever growing in terms of volume and complexity. Various machine learning techniques are used for extracting valuable and useful information out of this big data that also contains huge amount of meaningless data.

Over the past decade, data posed perhaps the single greatest challenge in the field of cyber security. The global Internet population growth went from 2 billion in the year 2012 to 4.3 billion in the year 2018. And by January 2019, the Internet reached 56.1% of the world's population. For every minute in 2019, 188 million emails were sent, and 4 million Google searches were conducted. The world uses a staggering 4,416,720 gigabytes of Internet data per minute [1]. According to a big data statistic carried out by IBM in 2017, 90% of all the data in the world back then had been created in the last 2 years. On the other hand, the frequency and damage caused by novel cyber-attacks are increasing with time. Since the cyber-attacks occur simultaneously with usual cyber operations, this adds onto the high volume of data created per day. Such large-scale data poses a challenge in terms of the four data quality dimensions, namely, volume, variety, velocity, and veracity.

S. Kaiser · K. Ferens (✉)
Department of Electrical and Computer Engineering, University of Manitoba, Winnipeg, MB, Canada
e-mail: Ken.Ferens@umanitoba.ca

© Springer Nature Switzerland AG 2021
H. R. Arabnia et al. (eds.), *Advances in Artificial Intelligence and Applied Cognitive Computing*, Transactions on Computational Science and Computational Intelligence, https://doi.org/10.1007/978-3-030-70296-0_82

Traditionally dataset contains many features that carry information of the system activities. However, it also contains redundant and irrelevant features that are the two primary factors which result in large volume and high-dimensional data. High-dimensional data poses several downsides. To begin with, redundant and irrelevant features consume more computational resource. Then, such volume of data negatively affects the performance of machine learning algorithm. Also, they increase the computational time for the learning algorithm. Moreover, such large volume of data will require large storage capacity. Finally, high-dimensional sparse data affects the accuracy of machine learning algorithm due to curse of dimensionality [2]. All these reasons validate the need of dimensionality reduction which can be attained by removing the features that possess no discriminating power. If we can select the discriminative features from the dataset, the machine learning algorithm is expected to achieve optimal result in machine learning model.

Typically, data and/or security analysts do feature selection based on their human intelligence, domain knowledge, and expertise which corresponds to their cognitive ability. However, the human cognitive approach also involves the ability to analyze complexity that exists in the data. This work hence focuses on depicting this cognitive aspect by using variance fractal dimension as a tool to complexity analysis to identify the discriminative features of an Internet dataset. The resultant discriminative features enhance the classification performance of the artificial neural network in terms of detection accuracy and computational time.

This section is ended by an outline for overall structure of remainder of this paper. An overview of feature selection methods and a brief literature review of feature selection in cyber security are provided in Sect. 2. Here cognitive analysis and complexity are also discussed in brief. Section 3 presents our proposed discriminative feature selection method. Section 3 also describes the dataset that was used for this experiment. Next, in Sect. 4, the experiments and results are discussed that summarize the findings and performance of our derived reduced features versus performance for all features of the dataset. Section 5 provides the concluding remarks.

2   Literature Review

Dimensionality reduction techniques play an indispensable role to help improve the learning performance. Feature selection and feature extraction are the two techniques of dimensionality reduction [3]. In this section, we discuss about some related works on feature selection as our research focuses on feature selection. Feature selection in dimensionality reduction is achieved by removing redundant and irrelevant features. It denotes the process of feature subset selection that can describe the specified problem without dropping of performance [4]. There have been many studies on three different evaluation-based categories of feature selection models, namely, (i) filter model, (ii) wrapper model, and (iii) Eembedded model [3]. The filter model performs independent of classifier algorithm while relying on the general characteristics of the training data. In [5], the authors have introduced pairwise correlation analysis-based filter method for feature selection. The correlation between continuous and discrete features was measured by removing weakly relevant, irrelevant, and relevant but redundant features. On the other hand, the wrapper algorithm uses classifiers as a selection process to assess the quality of a given feature subset [6]. Authors in [7] proposed an algorithm that alternates between filter ranking construction and wrapper model iteratively. The model seems promising as it analyzes few blocks of variables that decrease the number of wrapper evaluations drastically. The embedded model for feature selection is directly linked to classification stage training process [8]. An embedded model was used for selection method to a variance-based classification model construction in [9]. The comparative analysis of the experiment was done using real-world benchmark datasets and industrial dataset. The authors used relevance vector machine model based on an automatic relevance determination kernel using variance-based inference for regression in hierarchical prior over the kernel parameter setting. Another feature selection model is hybrid model that merges the filter and wrapper-based methods. Authors in [10] have used a hybrid model to select optimized feature set of protein-protein interaction pairs in. The mRMR filter and a KNN wrapper were used in the hybrid feature selection model.

Researchers have also applied other techniques as selection models. In [11, 12], the authors have identified and analyzed different features that can distinguish between phishing website and legitimate ones. While [11] have developed a computerized tool to automatically extract features, [12] have used Kaiser-Meyer-Olkin test to select features. Feature clustering was used as a selection process using minimal-relevant-redundancy criterion function in [13, 14] used forward selection ranking or backward elimination ranking to determine the feature correlation, and support vector machine function was used for feature ranking. The model was able to select the feature set independent of the classifier used. In [15], the authors have used auto encoder and principal component analysis (PCA) to select low-dimensional features of the CICIDS2017 network intrusion dataset. Later linear discriminant analysis, random forest, Bayesian network, and quadratic discriminant analysis classifiers were used to design an IDS. From the literature, we can observe machine learning and mathematical and statistical analysis are applied for feature selection; however, the threat landscape is ever evolving and still poses a threat when it comes to zero-day attack. Hence, processing of features or attributes still requires manual human involvement to identify the desired features that are discriminative, independent, and informative.

Cognitive analysis refers to the combination of artificial/machine intelligence and human intelligence. Cognitive systems are machines that are inspired by the human brain [16]. Hence, cognitive analysis replicates the human's analysis approach. One of the philosophies used in cognitive analysis is the concept of complexity where complexity can be interpreted as components that are dependent in such a way that it is impossible to distinct the influence of one from another where decomposition of the system into independent components would destroy the whole system [17]. The cognitive complexity measure hence should be based on information theoretic analysis or using computational approaches. Application of fractal dimensions can model the complexity of an object or component where higher fractal dimension would refer to more complexity. Hence, in this experiment, the authors have chosen variance fractal dimension (VFD) to measure the feature complexity.

3 Proposed Methodology

Our approach to variance fractal dimension (VFD) feature selection for the detection of cyber-attacks consists of six consecutive steps: data preprocessing, computation of the VFD, analysis and comparison, selection of discriminative features, application of artificial neural network classifier, and evaluation.

3.1 Dataset Preprocessing

To make the data suitable for the machine learning model, the training data needs to be preprocessed.

(a) Removing missing data: Missing data refers to an attribute's field value that is missing due to the data was not captured or improperly captured. These fields were removed as they do not contribute much.
(a) Encoding of data: Data contains both numeric and non-numeric attributes. The non-numeric attributes were converted to numbers so that mathematical calculations can be performed for the machine learning model.
(b) Normalization of data: Data samples were normalized following the feature scaling process (Fig. 1).

During the preprocessing stage, 3 out of 47 features were excluded due to data irregularities in them and to ease the computation process.

3.2 Computation of the VFD

Fractal dimensions are significant in the context of fractals since it measures the complexity of fractals. This work focuses on identifying features that show significant difference in terms of complexity in normal and attack data. Here, the authors have used variance fractal analysis (VFD) as a tool to measure the complexity of normal and attack samples with respect to each feature. One of the advantages of VFD is that it can be used for real-time application. Further, the unique identification for each class that it provides can be adopted in both segmenting data and extracting features as it is able to emphasize the underlying complexity [18].

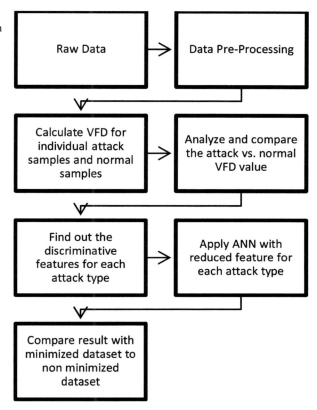

Fig. 1 Proposed framework for variance fractal dimension feature selection method

In this step, VFD values of all 47 features were calculated for normal and individual attack types. First, the normal samples and nine individual attack samples were separated from the dataset, and ten separate subsets of datasets were created. Then we have calculated variance fractal dimension of each feature for eight different attack datasets. Here we excluded the attack type "worm" due to its insufficient number of data samples. After that, we have calculated variance fractal dimension of each feature for the normal dataset. The resultant VFD values of features for individual attacks and normal data will be the basis for the complexity analysis of our proposed feature selection method.

3.3 Analysis and Comparison

For each attack type, a comparative analysis was done for the VFD values of each feature to the VFD values of each feature for normal dataset. This was denoted with V_{diff} in the below equation. To find out the V_{diff} value for individual attacks, we plot a feature v/s VFD graph and further proceed for the complexity analysis. For each attack type, the features that are more significant will have greater V_{diff} value hence discriminative in nature. For each attack type, we established a threshold to identify up to which V_{diff} value we will consider the feature as discriminant.

$$V_{\text{diff}} = |V_{\text{attack}} - V_{\text{normal}}| \tag{1}$$

3.4 Selection of Discriminative Features

Now that we have calculated the V_{diff} values, i.e., the difference between VFDs of features in attack samples with the normal samples, the next question is how to define the threshold for selection of significant features. We have already signified in the previous sections that the higher the difference between VFDs for normal and attack, the greater the discriminative nature of a feature. However, for the algorithm to be effective, we need an objective way to find a threshold to determine which features are significant. In other words, we need to set a threshold for the V_{diff}; any V_{diff} greater than the threshold will be considered important, and the features with V_{diff} lower than this threshold will be considered insignificant. We can utilize the classic midline approach; in this case, we can find an optimum range by finding the midline between the two extrema. Below is the equation for the threshold. The V_{diffth} is the threshold, which is the midline between the maximum V_{diff} and the minimum V_{diff}. Any feature having a V_{diff} value higher than the midline, i.e., features with $V_{\text{diff}} > V_{\text{diffth}}$ value, will be considered as significant. In the example provided in the below figure, the highest V_{diff} is 0.864, and the lowest value is 0.123; hence, our threshold is $(0.864 + 0.123)/2 = 23,423$ (Fig. 2).

The threshold for V_{diff} is given by (2):

Feature	VFD attack 1	VFD Normal	Vdiff (VFD attack1-VFD normal)	
1	.987	.123	.864	Vdiff_max
2	.	.	.	
3	.	.	.	
.	.	.	.	
.	.567	.444	.123	Vdiff_min
47	.	.	.	

Fig. 2 Sample of V_{diff} threshold selection

$$V\text{diff}_{th} = \frac{V\text{diff}_{max} - V\text{diff}_{min}}{2} \tag{2}$$

The discriminative features for each type of attack are maintained in a different list, and this list of significant features is used in later phases to train a neural network.

3.5 Application of Artificial Neural Network Classifier

The concept of the proposed model is to perform human cognitive and complexity analysis and distinguish the significant features of cyber-attack data. After the algorithm has autonomously filtered the significant features based on discriminativeness, they are now are used to train the artificial neural network (ANN). Then we test the neural network which is already trained with reduced features to evaluate the detection performance. This step is performed for each attack type. For the comparison, for each attack type, we use an ANN to detect the attack using a dataset with all the features and then use the same model and dataset with only the selected features. We capture the performance metrics and then compare the outputs.

Artificial neural network is a simple, easy-to-use, well-established method with great learning capabilities and effectiveness in capturing anomalies. ANNs are well known in the field of machine learning for its ability to process and classify information faster. Further, they have ability of self-organization. These are the reasons why ANNs can increase the accuracy and efficiency of intrusion detection model. Our sample ANN used logistic activation function.

Results were compared with the detection performance when used all features of each attack type (Fig. 3).

3.6 Evaluation

The intension of the proposed method of discriminative feature selection as a preprocessing step is to achieve enhanced performance of machine learning classification algorithm. This also gives an efficient solution to the best response that should be taken into consideration with regard to the individual intrusion type. Hence, the ANN classifier performance for each attack type with its reduced number of feature set was fairly evaluated and later compared with the performance of all features of each individual attack. The following metrics were considered to evaluate the performance.

$$\text{True positive rate} = \frac{TP}{TP + FN}$$

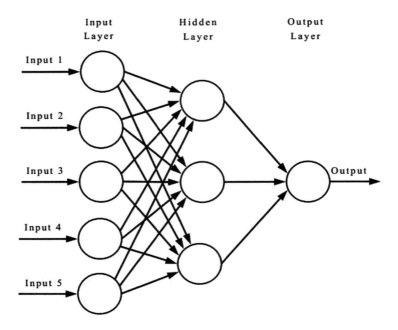

Fig. 3 Structure of a basic artificial neural network model

$$\text{True negative rate} = \frac{TN}{TN + FP}$$

$$\text{False positive rate} = \frac{FP}{FP + TN}$$

$$\text{False negative rate} = \frac{FN}{FN + TP}$$

$$\text{Accuracy} = \frac{TP + TN}{TN + TP + FN + FP} \quad (3)$$

These include accuracy, training time, and false positive rate. Our goal is to obtain a decently high accuracy while lowering the training time.

Variance Fractal Dimension Feature Selection for Detection of Cyber Security Attacks   1037

Table 1 Dataset record distribution and brief attack description [19, 20]

Type of data samples	No. of records or samples	Description
Normal	2,218,761	Natural transaction data
Fuzzers	24,246	Attempting to cause a program or network suspended by feeding it the randomly generated data
Analysis	2677	It contains different attacks of port scan, spam, and html file penetrations
Backdoors	2329	A technique in which a system security mechanism is bypassed stealthily to access a computer or its data
DoS	16,353	A malicious attempt to make a server or a network resource unavailable to users, usually by temporarily interrupting or suspending the services of a host connected to the internet
Exploits	44,525	The attacker knows of a security problem within an operating system or a piece of software and leverages that knowledge by exploiting the vulnerability
Generic	215,481	A technique works against all block-ciphers (with a given block and key size), without consideration about the structure of the block-cipher
Reconnaissance	13,987	Contains all strikes that can simulate attacks that gather information
Shellcode	1511	A small piece of code used as the payload in the exploitation of software vulnerability
Worms	174	Attacker replicates itself in order to spread to other computers. Often, it uses a computer network to spread itself, relying on security failures on the target computer to access it

3.7 Dataset

In our experiment, we have used the publicly available UNSW-NB15 dataset. The dataset was created utilizing the IXIA PerfectStorm tool in the cyber range lab of the Australian Centre for Cyber Security (ACCS) at UNSW in Canberra. A hybrid of normal and abnormal data can be found in the network traffic dataset [19]. Around 2.5 million data samples are distributed in 4 CSV files and consist in 49 different features including class label. Further, detail description of the features can be found at [19, 20]. This dataset contains nine different types of attacks, namely, fuzzers, analysis, backdoors, DoS, exploits, generic, reconnaissance, shellcode, and worms. In this work, we find out the discriminative features for individual attack type. Table 1 shows the distribution of attack and normal samples in the UNSW-NB15 data files and their brief description.

4 Experiments and Results

The experiment was performed using Python engine running on a 64-bit operating system. The data was processed using various Python libraries, i.e., pandas and scikit-learn.

The proposed variance fractal-based feature selection algorithm was executed on the UNSW-NB15 dataset containing both attack and normal samples. As detailed in the above section, first, the variance fractal dimensions of each feature for only the normal data samples were calculated. Then we have calculated the variance fractal dimension of each feature for each of the eight different attack datasets. Hence, we have nine VFD values for every feature, one for only normal traffic and one for each of the eight attack types.

The graphs (Figs. 4, 5, 6, 7, 8, 9, 10, and 11) depict the feature-wise difference of the VFD values of the normal and attacks. The blue line denotes the VFD values of features for the normal dataset, and the orange line denotes the VFD values of features for the attack dataset. The features in the x-axis are organized in increasing order of the difference of VFD values of attack and normal, i.e., with the features with the smallest VFD differences appearing on the left and the ones having the most differences appearing in the right. That is why the blue line and the orange line appear closer in the left and grow apart toward the right. Therefore, the most significant features with the most discriminative characteristics are the rightmost features in the graphs.

Table 2 shows the reduced number of features that are selected as the discriminative features for each attack type.

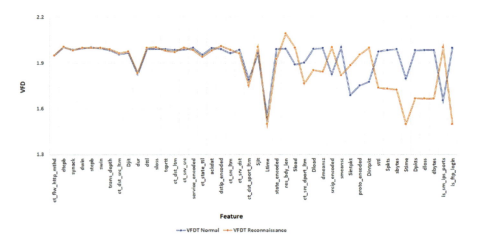

Fig. 4 VFD v/s feature graph for normal and reconnaissance

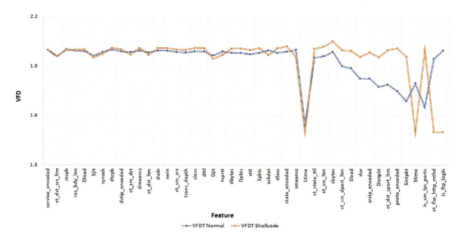

Fig. 5 VFD v/s feature graph for normal and shellcode

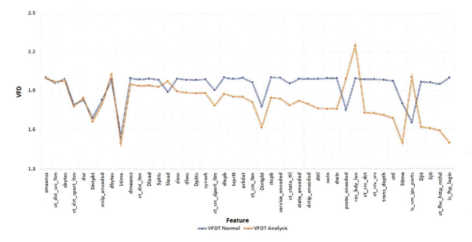

Fig. 6 VFD v/s feature graph for normal and analysis

For example, 11 features have a difference of VFD values greater than the midline threshold for analysis attack, and for exploit, there are only 3 features that have VFD difference greater than the midline threshold.

In order to evaluate the algorithm, we then train and test an ANN using only the reduced number of features. Among the 8 attack types that were considered in the experiment, most of the attacks have attack rows of 10,000 or higher. For these attack types, we construct our data subset by randomly selecting 30,000 normal rows and 10,000 attack rows. And for the attack types like analysis, shellcode, and backdoor, with smaller attack samples in the original dataset, in the range of 10,000 or less, we construct the dataset with 1000 randomly chosen attack rows and 3000

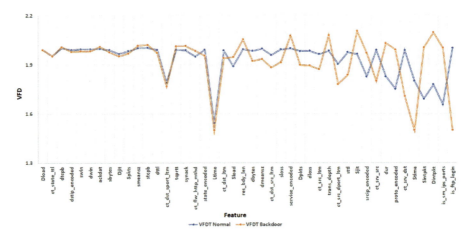

Fig. 7 VFD v/s feature graph for normal and backdoor

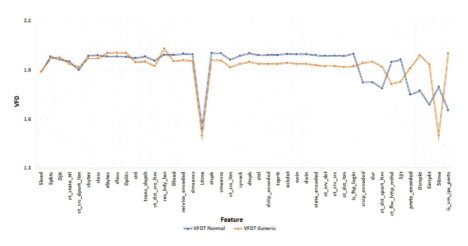

Fig. 8 VFD v/s feature graph for normal and generic

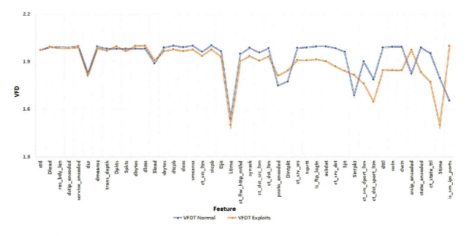

Fig. 9 VFD v/s feature graph for normal and exploits

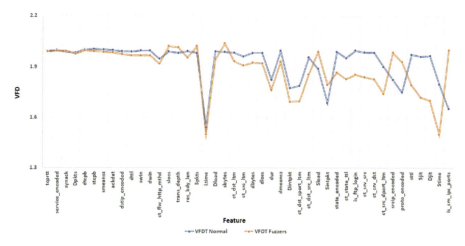

Fig. 10 VFD v/s feature graph for normal and fuzzers

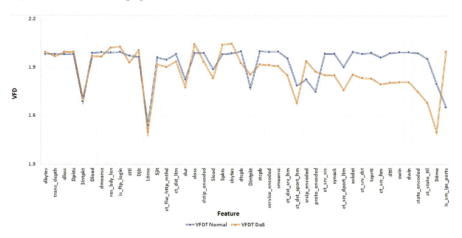

Fig. 11 VFD v/s feature graph for normal and DoS

randomly chosen normal rows. Out of each data subset, we choose 70% for training and 30% for testing. Hence, for the larger datasets, we have 12,000 rows for testing, and for the smaller datasets, we have 1200 rows for testing. Further, the detection performance of minimized dataset was compared with non-minimized dataset.

As shown in Table 3, the accuracy for supervised learning has been very high. Scikit-learn-based supervised learning models are known to provide high accuracy. It is observed that with the decrease in the number of features, the accuracy drops, and the rate of the decrease remains sporadic in nature; however, despite the significant data loss, the accuracy doesn't suffer too much.

Accuracy for shellcode suffers the most at 91.83%, and the accuracy of analysis remains quite high at 99%. For all the other attack types, the accuracy decreases

Table 2 Reduced features of each attack

Attack name	Reduced no of discriminative features
Fuzzers	6
Analysis	11
Backdoors	6
DoS	7
Exploits	3
Generic	4
Reconnaissance	8
Shellcode	6

Table 3 Testing accuracy

Attack type	Accuracy (%)	
	All features	Reduced features
Reconnaissance	99.92%	96.05%
Shellcode	99.72%	91.83%
Analysis	99.79%	99.08%
Backdoor	99.99%	92.92%
Fuzzers	99.59%	96.10%
Generic	99.97%	95.22%
Exploits	99.39%	97.94%
DoS	99.93%	98.04%

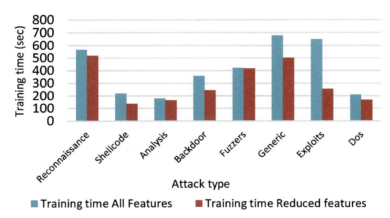

Fig. 12 Training time for attacks with reduced feature and all features

around 1–7%. The results show that losing the attributes was worthwhile; it is because those attributes were non-discriminatory in nature, which is why losing them did not decrease the accuracy substantially. The proposed method reduces the training time taken for reduced features in comparison to the time taken for all features (Fig. 12). Figures 13 and 14 show the confusion matrix for each attack type.

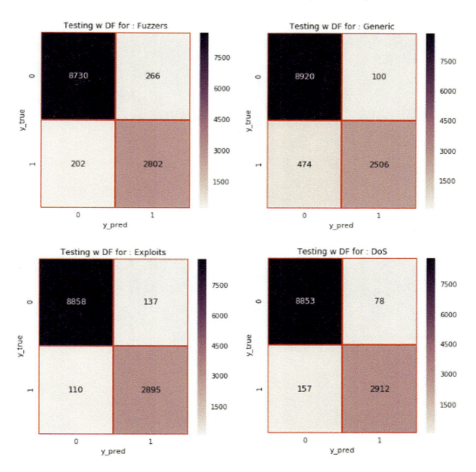

Fig. 13 Confusion matrix for fuzzers, generic, exploits, and DoS

It should be noted that not only the proposed method performs well for the classification problem at hand but it also reduces computational complexity while identifying features that are important to individual distinct attack types instead of an indefinite or collective attack types.

5 Conclusions

In this work, the authors have proposed a variance fractal dimension-based feature selection technique for identifying discriminative features of cyber-attack dataset. The experiment was carried out for eight different attack types of UNSW-NB15 dataset, namely, fuzzers, analysis, backdoors, DoS, exploits, generic, reconnais-

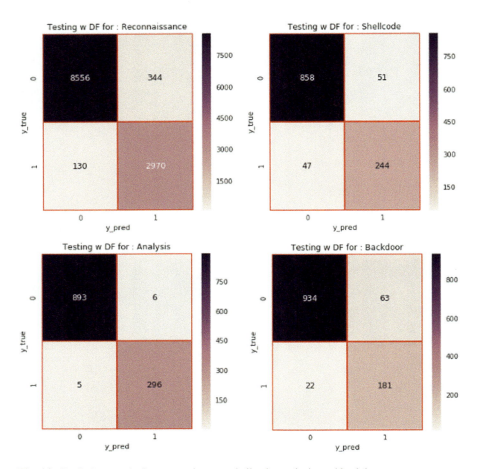

Fig. 14 Confusion matrix for reconnaissance, shellcode, analysis, and backdoor

sance, and shellcode. An artificial neural network was used to compare detection performance with resultant minimized dataset features of each attack type with non-minimized dataset. The experiment with variance fractal dimension as a tool for cognitive complexity-based analysis shows a promising result. It improves the detection performance of the classification algorithm in terms of computational complexity and detection time while giving substantial accuracy rate. However, fractal-based method involves computation time for fractal analysis which motivates the authors to further work in the future to assess a workaround to optimize the computation time in feature selection stage.

References

1. Data never sleeps 7.0 infographic|domo [Online]. Available: https://www.domo.com/learn/data-never-sleeps-7. Accessed 27 Mar 2020
2. H. Bahsi, S. Nomm, F.B. La Torre, Dimensionality reduction for machine learning based IoT botnet detection, in *2018 15th International Conference on Control, Automation, Robotics and Vision, ICARCV 2018*, (2018), pp. 1857–1862
3. V. Bolón-Canedo, N. Sánchez-Maroño, A. Alonso-Betanzos, Feature selection and classification in multiple class datasets: An application to KDD Cup 99 dataset. Expert Syst. Appl. **38**(5), 5947–5957 (2011)
4. M.H. Bhuyan, D.K. Bhattacharyya, J.K. Kalita, Network anomaly detection: Methods, systems and tools. IEEE Commun. Surv. Tutorials **16**(1), 303–336 (2014)
5. S.Y. Jiang, L.X. Wang, Efficient feature selection based on correlation measure between continuous and discrete features. Inf. Process. Lett. **116**(2), 203–215 (2016)
6. R. Kohavi, G.H. John, Wrappers for feature subset selection. Artif. Intell. **97**(1–2), 273–324 (Dec. 1997)
7. P. Bermejo, L. De La Ossa, J.A. Gámez, J.M. Puerta, Fast wrapper feature subset selection in high-dimensional datasets by means of filter re-ranking. Knowledge-Based Syst. **25**(1), 35–44 (2012)
8. H. Fu, Z. Xiao, E. Dellandréa, W. Dou, L. Chen, Image categorization using ESFS: A new embedded feature selection method based on SFS. Lect. Notes Comput. Sci. **5807**, 288–299 (2009)
9. J. Zhao, L. Chen, W. Pedrycz, W. Wang, Variational inference-based automatic relevance determination kernel for embedded feature selection of noisy industrial data. IEEE Trans. Ind. Electron. **66**(1), 416–428 (2019)
10. L. Liu, Y. Cai, W. Lu, K. Feng, C. Peng, B. Niu, Prediction of protein-protein interactions based on PseAA composition and hybrid feature selection. Biochem. Biophys. Res. Commun. **380**(2), 318–322 (2009)
11. R.M. Mohammad, F. Thabtah, L. McCluskey, An assessment of features related to phishing websites using an automated technique, in *2012 Int. Conf. Internet Technol. Secur. Trans.*, (2012), pp. 492–497
12. W. Fadheel, M. Abusharkh, I. Abdel-Qader, On feature selection for the prediction of phishing websites, in *Proc. – 2017 IEEE 15th Int. Conf. Dependable, Auton. Secur. Comput. 2017 IEEE 15th Int. Conf. Pervasive Intell. Comput. 2017 IEEE 3rd Int. Conf. Big Data Intell. Comput.*, vol. 2018, (2018), pp. 871–876
13. J. Martínez Sotoca, F. Pla, Supervised feature selection by clustering using conditional mutual information-based distances. Pattern Recogn. **43**(6), 2068–2081 (2010)
14. S. Zaman, F. Karray, Features selection for intrusion detection systems based on support vector machines, in *2009 6th IEEE Consum. Commun. Netw. Conf. CCNC 2009*, (2009), pp. 1–8
15. R. Abdulhammed, H. Musafer, A. Alessa, M. Faezipour, A. Abuzneid, Features dimensionality reduction approaches for machine learning based network intrusion detection. Electron **3**, 8 (2019)
16. J.E. Kelly, Computing, cognition and the future of knowing. IBM White Pap., 7 (2015)
17. W. Kinsner, System complexity and its measures: How complex is complex. Stud. Comput. Intell. **323**, 265–295 (2010)
18. A. Phinyomark, P. Phukpattaranont, C. Limsakul, Applications of variance fractal dimension: A survey. Fractals **22**(1–2), 1450003 (2014)
19. N. Moustafa, J. Slay, UNSW-NB15: A comprehensive data set for network intrusion detection systems (UNSW-NB15 network data set), in *2015 Military Communications and Information Systems Conference, MilCIS 2015 – Proceedings*, (2015)
20. N. Moustafa, J. Slay, The evaluation of Network Anomaly Detection Systems: Statistical analysis of the UNSW-NB15 data set and the comparison with the KDD99 data set. Inf. Secur. J. **25**(1–3), 18–31 (2016)

A Grid Partition-Based Local Outlier Factor for Data Stream Processing

Raed Alsini, Omar Alghushairy, Xiaogang Ma, and Terrance Soule

1 Introduction

The demand for data is increasing in the big data era. Massive data is generated through various sources such as smart home, sensor, mobile application, communication, and finance, which all lead to an increase in processing the data. One of the challenges in big data processing is how to measure outliers in streaming data. Data streams change every second, and the size is potentially infinite because it keeps increasing constantly. Subsequently, it is difficult to store the entire datasets in the memory and process them with older algorithms [1]. The density-based method is a well-known method used to find the outlier in the multi-density data. Local Outlier Factor (LOF) is currently the most utilized density-based method. It handles the data without assuming any underlying distribution. Moreover, it is capable of finding the dataset in the data with heterogeneous densities [3–5]. However, LOF faces some limitations in data stream processing. First, it works only on static data

---

R. Alsini (✉)
Department of Computer Science, University of Idaho, Moscow, ID, USA

Faculty of Computing and Information Technology, King Abdulaziz University, Jeddah, Saudi Arabia
e-mail: alsi1250@vandals.uidaho.edu

O. Alghushairy
Department of Computer Science, University of Idaho, Moscow, ID, USA

College of Computer Science and Engineering, University of Jeddah, Jeddah, Saudi Arabia
e-mail: algh5752@vandals.uidaho.edu

X. Ma · T. Soule
Department of Computer Science, University of Idaho, Moscow, ID, USA
e-mail: max@uidaho.edu; tsoule@uidaho.edu

© Springer Nature Switzerland AG 2021
H. R. Arabnia et al. (eds.), *Advances in Artificial Intelligence and Applied Cognitive Computing*, Transactions on Computational Science and Computational Intelligence, https://doi.org/10.1007/978-3-030-70296-0_83

1047

that does not change over time and only scans the data at one time to process the whole dataset. Also, for any change in the data points that occurred by adding or deleting data points, LOF needs to be recalculated on the whole dataset. Because of the limitations of LOF, it can't be used in data streams as the size of data streams is potentially infinite and the data are changing over time.

To overcome the limitation of LOF for data stream processing, an incremental version of LOF (ILOF) was introduced [6]. However, it needs the entire data points to detect outliers in a data stream. Therefore, the authors in [3] proposed the MILOF algorithm, which stores a subset of the whole data points by clustering the old data using the k-means algorithm. However, MILOF has a weakness when it clusters the old data, that is, the k-means method cannot maintain the low density [7]. The Density summarizing ILOF (DILOF) algorithms were proposed by summarizing the old data using the gradient descent method and skipping schema [8]. DILOF is by far the best developed method to detect the local outlier in data streams.

To further improve the accuracy of outlier detection and performance in the data stream, we propose a new technique called Grid Partition-based Local Outlier Factor (GP-LOF). The proposed technique has the following characteristics. First, GP-LOF works with limited memory. Therefore, a sliding window is used to summarize the points. Second, a grid method splits the data points for the processing phases. Third, it detects outliers using LOF.

In this paper, GP-LOF is evaluated through a series of experiments with real-world datasets. Based on the experimental result, the GP-LOF algorithm has shown better performance in accuracy and execution time compared with the DILOF algorithm [8]. The rest of this research paper is organized as follows: Sect. 2 gives a review of related works. Section 3 describes the structure of the GP-LOF algorithm. Section 4 explains the experimental result. The conclusion is presented in Sect. 5.

2 Related Work

Several reviews have described the outlier detection method in data mining, such as [9–11]. However, most of the methods can find outliers in a static environment since the data points are available to be measured, and the size is precalculated. When it comes to stream environments, a large amount of data is generated, and the size increases. Therefore, the detection of outliers in the data stream is challenging because it is difficult to scan the data points in a limited amount of memory [12]. Many approaches of outlier detection in the data stream have been categorized, such as distance-based outlier method, density-based outlier method, and clustering-based outlier method [11, 13, 14].

The distance-based outlier method is evaluated depending on the distance between the data points. Authors in [15] introduced the distance-based method as a way to calculate the distance between points and their nearest neighbors. Those with a far distance to the neighbors are counted as an outlier [16]. k-nearest neighbors (kNN) is the most common method used to evaluate the outliers based on their local neighbors. In the data stream, sliding windows have several methods applied in the distance-based model. In [17, 18], the authors utilized a sliding window technique to uncover the global outliers regarding the current window. In [19], the authors enhanced the algorithm in [16] by introducing a continuous algorithm that has two versions to reduce the time complexity and memory consumption. The first version handles multiple values of k-neighbors. The second version involves reducing the number of distances by using the micro-cluster for computation.

The density-based outlier method is based on measuring the densities to their local neighbors. Those densities who are far from their closest neighbors are considered as an outlier. The evaluation in the density-based method is measured by the local density. LOF is an example of the density-based model that uses (kNN) to detect the data points by using the local reachability density. LOF has a high detection accuracy and has several proposed methods in improvement, such as in [20, 21]. The author in [22] used the LOF to distribute data points into several grids. Since LOF works in a static environment, an incremental local outlier factor, i.e., ILOF, is presented using the LOF in the data stream. All the previous extensions in the LOF required the entire data points to be calculated to get the LOF score, which is not necessary for the ILOF technique since it can handle new incoming data points.

The clustering-based outlier method is an unsupervised method that processes data points by dividing the data into groups based on their distribution [23]. The aim of the method is first to cluster a data point and then detect the outlier [24–26]. Some algorithms used the small cluster by representing a small number of points as an outlier, while other methods used the threshold to find the outlier. Several methods used the cluster in the detection of the outlier, such as partition cluster model, density cluster model, and grid-based cluster model [11]. The authors in [27] proposed using the cluster in a high data dimension data stream. k-means is used in the data stream to split the data into several segments to process. Authors in [28] generated histograms for the clusters in the streamed data, which they used later for data mining and in detecting outliers. In [29], the authors proposed an algorithm that works on outlier detection on a data stream that can determine if any point is an outlier within any time. The accuracy of detection depends on the availability of time. In [30], the author used the data stream arriving time to learn and determine the normal behavior of the current period.

3 Methodology and Methods

3.1 Local Outlier Factor (LOF)

LOF is an unsupervised approach in the density-based outlier algorithm to search the anomaly based on the local density with respect to the neighbors. LOF evaluates the data points according to a degree of measurement, i.e., the outlier factor regarding the density of the local neighbors. The definition of LOF was described in [2, 31] and illustrated as:

- Definition (1): k-distance of data point *pt*

  Any space between two data points *pt* and *o* is computed by using the Euclidean distance in *n*-dimensional space.

$$dist\ (pt, o) = \sqrt{\sum_{i=1}^{n} (pt_i - o)^2} \tag{1}$$

Given a point *pt* in a dataset *D* and *k* is a positive integer, the *k-distance(pt)* is defined based on the distance between the furthest distance *o (o ∈ D)* and point *pt* with the following constraint:

- For at least *k* data point $o' \in D \setminus \{pt\}$, it keeps that dist *(pt, o')* ≤ dist *(pt, o)*.
- For at most *k-1* data point $o' \in D \setminus \{pt\}$, it keeps that that *dist (pt, o')* < *dist(pt, o)*.

- Definition (2): *k-nearest neighbors of pt*

  Given the same condition as *k* defined in definition (1), the *k-nearest neighbor of pt* can be described as any data point *q* whose distance does not exceed the *k-distance(pt)*. The following equation defines the *k-distance neighbor of pt* as:

$$N_{k-distance(pt)}\ (pt\ ) = \{\ q \in D \setminus \{pt\ \}|\ dist\ (pt, q) \le k - distance(pt) \tag{2}$$

- Definition (3): *Reachability distance (Reach-dist (pt)) with respect to o*

  When *k* has a positive integer, the *reachability distance of point pt with any point o* can be described in Eq. (3):

$$Reach - distance_k\ (pt, o) = \max\ \{\ k - dist(o), dist\ (pt, o)\} \tag{3}$$

Figure 1 represents several data points when *k* = 6. According to Definition (3), any point under the *k*-neighbor is computed by the *k-dist(pt)*. However, if the actual distance is far from the *k-dist(o)*, it will be calculated as an actual reachable distance as in data point pt_6.

- Definition (4): *Local reachability density of LRD (pt)*

  It is the inverse of the reachability distance of *pt*. It computes the *Lrd* by having a minimum number of points Min-points and the size. The size is calculated according to the reachability distance. The following equation describes the local reachability density as:

$$Lrd_{\text{Min}-points}(pt) = 1/\left(\frac{\sum_{o \in N_{Minpoint}(pt)} Reach - dist_{\text{Min}-point}\ (pt, o)}{|\text{Min} - point(Pt)|} \right) \tag{4}$$

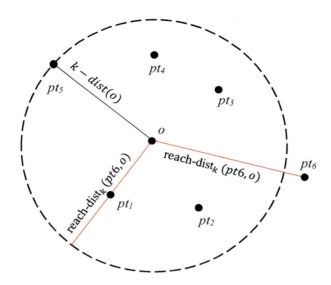

Fig. 1 Several data point (pt) to calculate the reachability distance with respect to o, $k = 6$

- Definition (5): *Local Outlier Factor (LOF) of pt*
  In order to calculate the *LOF (Pt)*, each definition must be followed to obtain the LOF score, as in Eq. (5).

$$LOF_{\text{Min}-point}(P_t) = \frac{\sum_{p \in P_{\text{Min}-point}(pt)} \frac{Lrd_{\text{Min}-point}(o)}{Lrd_{\text{Min}-point}(pt)}}{\mid P_{\text{Min}-point}(pt) \mid} \quad (5)$$

The LOF generates the score according to the proportion of the local reachability density of *pt* and the minimum point neighbors of *pt*. A threshold score θ will be used to determine whether *pt* is an outlier or not.

3.2 Proposed Algorithm: Grid Partition-Based Local Outlier Factor (GP-LOF)

In this section, we will explain the Grid Partition-based Local Outlier Factor (GP-LOF) algorithm. The main objective of the GP-LOF is to find the outlier by the following characteristics: no prior knowledge of the data distribution, a part of the dataset is stored in the memory, and applied the LOF algorithm to detect the outlier. GP-LOF includes three phases to identify the outlier as a preprocessing phase, processing phase, and detection phase, as shown in Fig. 2. Algorithm 1 explains how the GP-LOF method operates as follows: First, the GP-LOF algorithm begins by collecting the data points in *Pre-Processing Window (PPW)* that has specific

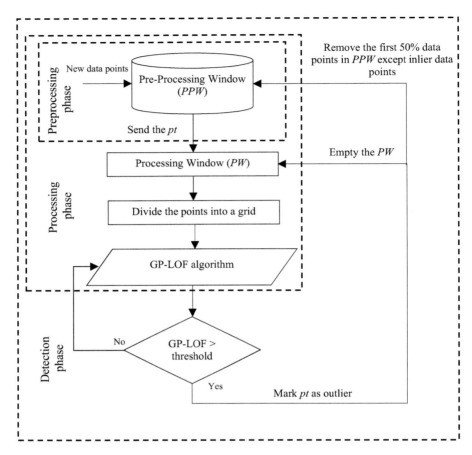

Fig. 2 Framework of the GP-LOF algorithm in the data stream

window size *ws*. Then, once the size in the *PPW* is complete (lines 3–6), the first half of the data points in *PPW* is selected and then sent to the *Processing Window (PW)*. The grid technique is applied in the *PW* to divide data points in the grid index *i*. After the data points are partitioned into the grid, GP-LOF calculates the LOF and gets the result depending on a predefined threshold θ. The GP-LOF ensures any points that exceed the threshold θ are removed from the *PW* (lines 7–17). Each phase in the GP-LOF is described as follows:

A Grid Partition-Based Local Outlier Factor for Data Stream Processing

Algorithm 1: GP-LOF

Input: LOF threshold θ
Infinite data streams points $P = \{p_1, p_2, ..., pt\}$
Pre-Processing Window PPW
Processing Window PW
Number of grids Ng
Grid G

1 **Init** $PPW \rightarrow \{\}$ // is representing the
  Pre-Processing Window
2 **Init** $PW \rightarrow \{\}$ // is representing Processing Window
3 **For each** $pt \in P$ **do**
4   **If** $PPW\ (pt) < PW\ (pt)$
5   **Add** pt to $PPW\ (pt)$
6   continue.
7   **else**
8   **Add** 50% $PPW\ (pt)$ *to the PW*
9   **For** PW **do**
10     $GP\text{-}LOF\ (pt) \leftarrow G(Ng, pt)$
11   **For** every $GP\text{-}LOF\ (pt)$ **do**
12   **If** the $LOF_k\ (pt) > \theta$ then
12   pt is an outlier
13   **End**
14   **End**
15   Empty the PW
16   Remove the first 50% data points from PPW
       except inlier data points
17   **End**
18 **End**

3.2.1 Preprocessing Phase

In this phase, a PPW is used to collect and store the data points that arrives in the stream. Then, once the ws of the PPW is filled with data points, the GP-LOF selects the first 50% of data points from PPW to be processed in the PW.

3.2.2 Processing Phase

The processing phase begins once the data point in the PPW is moved to the PW. Next, PW divides the data points into a grid, which is a number of regions. Then, GP-LOF computes the LOF score for each grid index i. After that, the LOF scores for each i are used to find the LOF score for all data points belonging to the grid.

Then, the LOF results will forward to the detection phase. To compute the outlier score of data points, the *PW* ensures the following steps:

- Divide the dataset's dimensions equally into the grid.
- Allocate each data point p into a grid index i.
- In every grid index i, compute the LOF score for all data points using the LOF algorithm [2].

3.2.3 Detection Phase

Once the LOF result is received in the detection phase, the GP-LOF method will scan all data points and select the data points that are greater than the threshold to be outliers. Then, it empties the *PW* and removes the first 50% of data points in the *PPW* except inlier data points as in Algorithm 1. Then, when any new data point is arriving in a stream, the preprocessing phase starts again to collect them.

4 Experiment Procedures

4.1 Datasets Used in Experiments

This section provides the experimental results on the GP-LOF algorithm. The GP-LOF results are compared with DILOF results in various datasets. The GP-LOF algorithm is implemented in Java by a machine that runs in Intel® core (MT) i7-4940MX CPU, 16GB RAM, 1 TB SSD hard disk, and Windows 10 (64-bit) operating system. The same machine implements the DILOF algorithm in C++, and the source code is available in [8]. For the DILOF setting and hyperparameter, the reader can refer to [8].

Both methods are measured under two metrics in the accuracy of the detection of the outlier and the execution time. In particular, the area under the ROC curve (AUC) is used in the first category as [32, 33]. The efficiency of the AUC is evaluated by applying the true positive rate (TPR) and a false positive rate (FPR) with a scale of the threshold $t = \{0.1, 1.0, 1.1, 1.15, 1.2, 1.3, 1.4, 1.6, 2.0, 3.0\}$. The evaluation of both TPR and FPR rate is tested according to each threshold value to obtain the accuracy rate. Both GP-LOF and DILOF methods are tested with different window sizes *ws* through a real-world dataset, as in Table 1.

Table 1 Real-world dataset

Datasets	Number of data points	Dimension	Class
UCI Vowel dataset	1456	12	11
UCI Shuttle dataset	58,000	9	10
KDD Cup 99 SMTP dataset	95,156	3	Unknown

All these real-world datasets can be obtained from the Machine Learning database repository at UCI in [34]. For the UCI Vowel dataset, it can be obtained from [35]. The UCI Shuttle dataset is provided in [34]. Both the UCI Vowel dataset and the UCI Shuttle dataset have been modified to be suitable for data stream format like in [35]. In the KDD Cup 99 SMTP, we did the same configuration as [36] to be an outlier. SMTP is a subset from KDD Cup 99 dataset that developed to test the intrusion detection in the network. In the SMTP service, it is possible to show some changes in distribution within the streaming series [37].

For the experimental setup, we set k to be 19 for the UCI Vowel dataset. For the rest of the UCI Shuttle dataset and KDD 99 SMTP, we set k to be 8. Both GP-LOF and DILOF have the same setup of the window size ws for their validation. In the DILOF, we setup the ws in the summarization phase to be $w = \{100, 200, 300, 400, 500\}$. For the GP-LOF method, we setup the ws of preprocessing windows $PPW = \{100,200,300,400,500\}$.

4.2 Experiment Discussion

4.2.1 Accuracy of the Outlier Detection

The accuracy of detecting the outlier is tested under the AUC through a series of experiments with different window sizes. Figures 3, 4, and 5 and Table 2 represent the accuracy between the GP-LOF and DILOF for the UCI Vowel dataset, UCI Shuttle dataset, and KDD99 SMTP dataset.

In the UCI Vowel dataset, DILOF has high accuracy in the size of the window $w = 100$ at 72.02% and in $w = 200$ at 89.038% compared to the GP-LOF algorithm, which takes 43.00% and 45.49%. However, when the size of the window

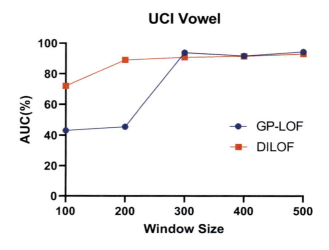

Fig. 3 GP-LOF shows consistently higher accuracy in all the window size compared to DILOF algorithm in UCI Vowel real-world dataset

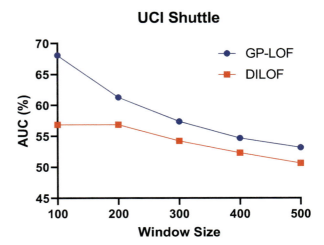

Fig. 4 GP-LOF shows consistently higher accuracy in all the window size compared to DILOF algorithm in UCI Shuttle real-world dataset pared to DILOF algorithm in UCI Shuttle real-world dataset

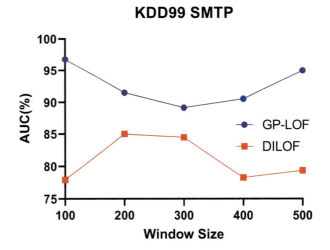

Fig. 5 GP-LOF shows consistently higher accuracy in all the window size compared to DILOF algorithm in the KDD99 SMTP real-world dataset

is increasing, GP-LOF has better accuracy. For example, when the size of the window reaches $w = \{300\}$ or $\{500\}$, the GP-LOF outlier accuracy is better than DILOF. Both algorithms have a similar accuracy rate when the size of the window reaches $w = \{400\}$.

For the UCI Shuttle dataset, we noticed the GP-LOF algorithm in the window size $w = \{100\}$ reaches an accuracy rate of 68.07% compared to DILOF at 56.85%. However, both algorithms' accuracy has been reduced when the size of the window is increased. GP-LOF has higher accuracy in detecting the outlier compared to DILOF in all window sizes $w = \{100, 200, 300, 400, 500\}$.

In the KDD99 SMTP dataset, there is a variance detection accuracy between GP-LOF and DILOF. For example, when the window size $w > 300$, there is a

A Grid Partition-Based Local Outlier Factor for Data Stream Processing

Table 2 Accuracy result for the real-world dataset between GP-LOF and DILOF algorithms

Window size	Vowel		Shuttle		SMTP	
	GP-LOF	DILOF	GP-LOF	DILOF	GP-LOF	DILOF
100	43.0016	72.037	68.07932	56.8506	96.79145	77.9116
200	45.49206	89.0384	61.25319	56.8338	91.46393	85.0027
300	93.8462	90.825	57.36905	54.2211	89.15365	84.484
400	91.68576	91.463	54.65677	52.303	90.50625	78.2639
500	94.36069	92.9275	53.1342	50.6228	94.98262	79.3717

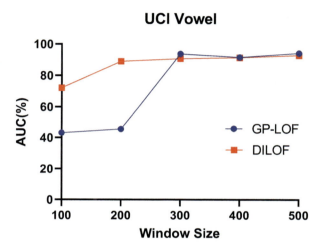

Fig. 6 GP-LOF execution time in the UCI Vowel real-world dataset for all window size has better performance than the DILOF algorithm

significant difference between GP-LOF and DILOF algorithms. The GP-LOF rate keeps increasing, while the DILOF rate has a lower accuracy when the size of the window keeps increasing.

4.2.2 Execution Time

Figures 6, 7, and 8 and Table 3 represent the execution time for the UCI Vowel, KDD SMTP 99, and UCI Shuttle datasets and between GP-LOF and DILOF algorithms. We notice that the GP-LOF execution time is always much lower than the DILOF algorithm, even when the size is increasing. In the UCI Vowel dataset, GP-LOF takes 1.17 seconds, while the DILOF execution takes 4.32 seconds. This is because GP-LOF divides the points into several grids, which reduces the execution time. The same thing occurs for KDD99 SMTP and UCI Shuttle datasets. In KDD99 SMTP, GP-LOF execution in the $w = 500$ takes 56.396 seconds while in DILOF takes 260.544. For the UCI Shuttle dataset, the GP-LOF execution takes 15.714 seconds, which is longer than DILOF at 12.99 seconds. When we increase the size of the window to 500, GP-LOF execution takes 37.022 seconds, while DILOF takes 163.071.

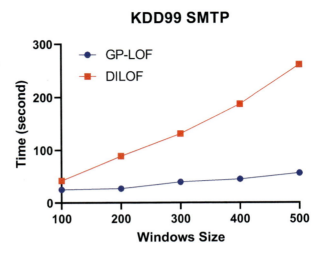

Fig. 7 GP-LOF execution time in the KDD 99 SMTP real-world dataset for all window size has better performance than the DILOF algorithm

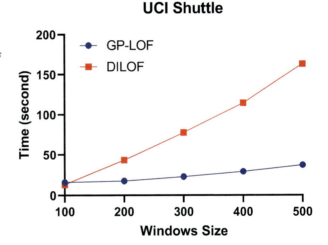

Fig. 8 GP-LOF execution time in the UCI Shuttle real-world dataset for all window size has better performance than the DILOF algorithm

5 Conclusion

This research paper aims to improve the efficiency of algorithms for detecting local outliers in data streams. LOF is one of the algorithms that detect outliers in static data, but it has limitations when dealing with data streams. First, it consumes a lot of memory as the whole dataset needs to be stored in the memory, which offer is not feasible. Next, it needs to process the whole dataset since any change in the data requires that the LOF be recalculated from the beginning, which is not applicable in data streams as the data is changing. We propose a novel algorithm called Grid Partition-based Local Outlier Factor (GP-LOF), which overcomes the

A Grid Partition-Based Local Outlier Factor for Data Stream Processing   1059

Table 3 Execution time for the real-world dataset between GP-LOF and DILOF algorithms

Window size	Vowel		Shuttle		SMTP	
	GP-LOF	*DILOF*	*GP-LOF*	*DILOF*	*GP-LOF*	*DILOF*
100	0.91	0.737971	15.714	12.996	24.91	41.7741
200	0.882	1.552	17.428	43.253	26.95	88.7286
300	0.903	2.32008	22.512	77.3791	39.296	131.205
400	1.01	3.24402	28.845	114.771	44.879	186.835
500	1.179	4.32603	37.022	163.071	56.396	260.544

two limitations of LOF in data stream processing. Our experimental evaluations demonstrate that GP-LOF has better performance in several real-world datasets than the state-of-the-art DILOF algorithm.

References

1. S. Sadik, L. Gruenwald, Research issues in outlier detection for data streams. ACM SIGKDD Explorations Newsletter **15**(1), 33–40 (2014)
2. M.M. Breunig et al., Lof, in *Proceedings of the 2000 ACM SIGMOD international conference on Management of data – SIGMOD 00, 29*, (2000), pp. 93–104
3. M. Salehi et al., Fast memory efficient local outlier detection in data streams. IEEE Trans. Knowl. Data Eng. **28**(12), 3246–3260 (2016)
4. Y. Yan et al., Distributed local outlier detection in big data, in *Proceedings of the 23rd ACM SIGKDD International Conference on Knowledge Discovery and Data Mining*, (2017)
5. Y. Yan, L. Cao, E.A. Rundensteiner, Scalable top-n local outlier detection, in *Proceedings of the 23rd ACM SIGKDD International Conference on Knowledge Discovery and Data Mining*, (2017)
6. D. Pokrajac, A. Lazarevic, L.J. Latecki, Incremental local outlier detection for data streams, in *2007 IEEE Symposium on Computational Intelligence and Data Mining*, (2007)
7. A.K. Jain, Data clustering: 50 years beyond K-means. Pattern Recogn. Lett. **31**(8), 651–666 (2010)
8. G.S. Na, D. Kim, H. Yu, Dilof, in *Proceedings of the 24th ACM SIGKDD International Conference on Knowledge Discovery & Data Mining*, (2018)
9. S. Rajasegarar, C. Leckie, M. Palaniswami, Anomaly detection in wireless sensor networks. IEEE Wirel. Commun. **15**(4), 34–40 (2008)
10. V. Chandola, A. Banerjee, V. Kumar, Anomaly detection: A survey. ACM Comput. Surv. (CSUR) **41**, 15 (2009)
11. H. Wang, M.J. Bah, M. Hammad, Progress in outlier detection techniques: A survey. IEEE Access **7**, 107964–108000 (2019)
12. O. Alghushairy, X. Ma, Data Storage, in *Encyclopedia of Big Data*, ed. by L. Schintler, C. McNeely, (Springer, Cham, 2019)
13. M. Gupta et al., Outlier detection for temporal data: A survey. IEEE Trans. Knowl. Data Eng. **26**(9), 2250–2267 (2014)
14. C.C. Aggarwal, *Outlier Analysis*, 2nd edn. (Springer, Cham, 2015)
15. E.M. Knorr, R.T. Ng, Algorithms for mining distance-based outliers in large datasets. Algorithms for mining distance-based outliers in large datasets, in *Proceedings of the 24rd International Conference on Very Large Data Bases*, (1998). Available at: https://dl.acm.org/doi/10.5555/645924.671334.

16. P. Thakkar, J. Vala, V. Prajapati, Survey on outlier detection in data stream. Int. J. Comput. Appl. **136**(2), 13–16 (2016)
17. F. Angiulli, F. Fassetti, Detecting distance-based outliers in streams of data, in *Proceedings of the sixteenth ACM conference on Conference on information and knowledge management – CIKM 07*, (2007)
18. D. Yang, E.A. Rundensteiner, M.O. Ward, Neighbor-based pattern detection for windows over streaming data, in *Proceedings of the 12th International Conference on Extending Database Technology Advances in Database Technology – EDBT 09*, (2009)
19. M. Kontaki et al., Continuous monitoring of distance-based outliers over data streams, in *2011 IEEE 27th International Conference on Data Engineering*, (2011)
20. J. Tang et al., Enhancing effectiveness of outlier detections for low density patterns, in *Advances in Knowledge Discovery and Data Mining Lecture Notes in Computer Science*, (2002), pp. 535–548
21. A. Chiu, A.W.-C. Fu, Enhancements on local outlier detection, in *Seventh International Database Engineering and Applications Symposium, 2003. Proceedings*, (2003), pp. 298–307
22. M. Bai et al., An efficient algorithm for distributed density-based outlier detection on big data. Neurocomputing **181**, 19–28 (2016)
23. R. Alsini, X. Ma, Data streaming, in *Encyclopedia of Big Data*, ed. by L. Schintler, C. McNeely, (Springer, Cham, 2019)
24. C.C. Aggarwal et al., A framework for clustering evolving data streams, in *Proceedings 2003 VLDB Conference, 29*, (2003), pp. 81–92
25. F. Cao et al., Density-based clustering over an evolving data stream with noise, in *Proceedings of the 2006 SIAM International Conference on Data Mining*, (2006)
26. S. Guha et al., Clustering data streams: Theory and practice. IEEE Trans. Knowl. Data Eng. **15**(3), 515–528 (2003)
27. C.C. Aggarwal et al., A framework for projected clustering of high dimensional data streams, in *Proceedings 2004 VLDB Conference, 30*, (2004), pp. 852–863
28. C.C. Aggarwal, A segment-based framework for modeling and mining data streams. Knowl. Inf. Syst. **30**(1), 1–29 (2010)
29. I. Assent et al., AnyOut: Anytime outlier detection on streaming data, in *Database Systems for Advanced Applications Lecture Notes in Computer Science*, (2012), pp. 228–242
30. M. Salehi et al., A relevance weighted ensemble model for anomaly detection in switching data streams, in *Advances in Knowledge Discovery and Data Mining Lecture Notes in Computer Science*, (2014), pp. 461–473
31. O. Alghushairy et al., A genetic-based incremental local outlier factor algorithm for efficient data stream processing, in *Proceedings of the 2020 the 4th International Conference on Compute and Data Analysis*, (2020)
32. J.A. Hanley, B.J. Mcneil, The meaning and use of the area under a receiver operating characteristic (ROC) curve. Radiology **143**(1), 29–36 (1982)
33. A.P. Bradley, The use of the area under the ROC curve in the evaluation of machine learning algorithms. Pattern Recogn. **30**(7), 1145–1159 (1997)
34. D. Dua, C. Graff, *UCI Machine Learning Repository [http://archive.ics.uci.edu/ml]* (University of California, School of Information and Computer Science, Irvine, 2019)
35. C.C. Aggarwal, S. Sathe, Theoretical foundations and algorithms for outlier ensembles? ACM SIGKDD Explorations Newsletter **17**(1), 24–47 (2015)
36. K. Yamanishi et al., On-line unsupervised outlier detection using finite mixtures with discounting learning algorithms, in *Proceedings of the sixth ACM SIGKDD international conference on Knowledge discovery and data mining – KDD 00, 83*, (2004), pp. 275–300
37. C. Tan, K.M. Ting, T.F. Liu, Fast anomaly detection for streaming data, in *Proceeding of the 2011 Twenty-Second International Joint Conference on Artificial Intelligence*, (2011)

A Cognitive Unsupervised Clustering for Detecting Cyber Attacks

Kaiser Nahiyan, Samilat Kaiser, and Ken Ferens

1 Introduction

The latest cyber threat detection platforms utilize either the perceivable known signatures or heuristic-based behavioral analysis to determine the presence of potential threats. Therefore, it is imperative that the human traits and capabilities are an innate attribute required to analyze and differentiate the true positives from the large pool of normal traffic, and hence this implies that we need to improve the cyber defense tools to integrate human-like cognition capabilities. This claims further credit since humans are the perfect learners, especially when it comes into learning new things. Therefore, in this work, we are incorporating cognitive approach to isolate the attack from the normal traffic.

The traditional machine learning algorithms are performing single-scale analysis. The single-scale approaches used in analyzing autonomous intelligent systems and natural cognitive processes are good for all patterns and processes that are scale independent or single-scale in nature. Such approaches are not adequate for the systems that are inherently scale-free, like network traffic [1]. Hence, if a multi-scale approach is required to capture the local characteristic of the system that contains the information of any underlying complex behavior, the proposed solution will be using multi-scale complexity analysis using variance fractal dimension. The unsupervised nature of the algorithm enables it to be effective not only for known attacks but also for unknown 0-day attacks.

Focusing on the key motivation of cognition-inspired adaptive learning, this study also tries to adopt significant properties of early-stage human learning. It has

---

K. Nahiyan · S. Kaiser · K. Ferens (✉)
Department of Electrical and Computer Engineering, University of Manitoba, Winnipeg, MB, Canada
e-mail: Ken.Ferens@umanitoba.ca

© Springer Nature Switzerland AG 2021
H. R. Arabnia et al. (eds.), *Advances in Artificial Intelligence and Applied Cognitive Computing*, Transactions on Computational Science and Computational Intelligence, https://doi.org/10.1007/978-3-030-70296-0_84

been studied that the ability of humans to identify and recognize objects develop shortly after birth [2, 3]. A very significant skill that they develop during this time to support object perception is gaze control, that is, the ability to direct gaze toward informative or distinctive regions of an object, like edges and contours, as well as to shift gaze from one part of the object to another [4–6]. While focusing on the regions, one of the key processes that they use to identify distinctive and informative regions is by using statistical learning, which is a vital process that infants use to assess their surroundings [7]. Hence, the perfect approach to imitating human learning is segment-wise focus and extraction of statistical features, with complexity analysis as the final layer of discriminative analysis, which are the exact building blocks for the proposed algorithm.

2 Background and Related Work

For years, the authors have explored the significance of statistical learning in human learning, memory, intelligence, and inductive inference. Major contributors in the field of cognitive science, from Helmholtz [8], Mach [9], and Pearson [10], and continuing through Craik [11], Tolman [12], Attneave [13], and Brunswik [14], have all stressed that the brain performs statistical computations in one form or the other and an important use of it is establishing the statistical regularities of the environment predictively and to adapt behavior to future events [15]. It is important to note that "all learning could be regarded as the internalization of environmental regularities" [15], and an important fundamental basis of human knowledge is knowing what is normal and what is abnormal; in other words, humans constantly perform a statistical modeling of the environment and establish a normal environment and have ways to compare things experienced to the normal environment. It is this strong sense of the customary that capacitates the human mind to identify something that is out of the ordinary, sometimes even in subconscious state. And for these reasons, such principles must be effective in detecting cyber intrusions. This perhaps is one of the reasons why statistical features have been widely used in cyber intrusion detection datasets.

Korczyński et al. [17] used statistical features from headers to classify encrypted traffic. They created statistical features of the information embedded in SSL/TSL headers on traffic of the applications and applied Markov chain fingerprinting classification of application traffic and attained satisfactory true positive rate. Authors in [17] designed a method called Statistical Protocol Identification to identify the different types of communication (voice call, SkypeOut, video conference, chat, file upload, and file download) within Skype traffic, including VoIP call, conference calls, file download/upload, and normal chatting. In [18], the authors described a network intrusion detection system (NIDS) that detects infiltration or attacks from within encrypted traffic in real time. The NIDS has two parts; the first one clusters the data and forwards to the second engine, and the second engine identifies the attack pattern. The authors used the detected anomalies to identify the botnet masters. Korczyński and Duda [19] proposed using stochastic fingerprints based on Markov chains for identifying application traffic in SSL/TLS sessions. The method uses payload statistics from SSL/TLS headers of 12 representative applications such as Twitter, Skype, and Dropbox were used.

In our previous research [20], we have used UNSW and AWID dataset and presented an unsupervised attack separation technique using traffic segmentation and higher-order statistical feature extraction measures to group the traffic segments into two distinct groups or sections using k-means algorithm. There the result was two clusters, one normal and one attack; however, both clusters needed to be manually examined to identify the attack cloud. Here in this study we are presenting an automated approach to identify the attack cluster. For this approach, we are using cognitive technique elaborated below.

Cognition, in a literary sense, means knowing, perceiving, or conceiving as act. In the computer science language, it can be described as a state of mental process intervening between stimuli and output response, and the processes can be described as algorithms that lend themselves to scientific investigations [21]. Cognitive systems tend to be aware of the environment consisting of machines and human beings, where they base their efforts on the knowledge of the physical world and work toward their goals [22]. In systems theory, complexity means many interacting elements with many degrees of freedom, whose individual behavior could not be traced back or predicted [22]. Such systems exhibit self-organization (emergence), thus leading to new features that did not exist before and were not programmed to evolve. The author in [22] has categorized objects, systems, and structures into two distinct types, one containing any order or pattern in their behavior and the other class that obeys no pattern at all and displays stark randomness.

Past research has shown great results when fractals were applied into neural network and better learning was attained in comparison with traditional learning methods [23–25].

3 Proposed Methodology

When working with large datasets, it is helpful to divide the dataset into smaller fractions, which can be analyzed individually, and a separate synopsis can be extracted from these events, which, as described in the latter literature, will be used to construct the higher-order feature vectors. For our algorithm to be fast and responsive, we would need to have a much-reduced dataset. With that intent, in our approach [26], we club several network events into a single representation, based on their adjacency in time, i.e., if the events occur consecutively in time, we club these events into a single representation. In our proposed system, this job is done by the component called sampler. Our sampler divides the large dataset into time steps, fragmenting the dataset into smaller sections based on time. In terms of files and the data rows, after the clubbing, each row of the reduced dataset will now contain information representing several data rows from the original dataset.

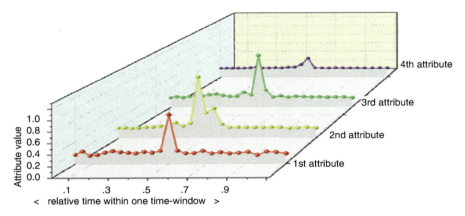

Fig. 1 Time step elaborated

The approach is depicted in Fig. 1. Let us imagine a dataset having five attributes; one of them is the timestamp, which means the time when the values of the four attributes were captured, much like our dataset. In Fig. 1, we plot these four attributes against time. The leftmost point represents the first timestamp within this time-window or time step, and the rightmost point represents the last captured value of the attribute within the time-window. Notice that there are 30+ points in the red time-window. This means the 30 rows had a value of Stime which corresponds to the first time-window, and hence, the entire time-window these 30 rows will be represented by a single time-window, and after we extract higher-order statistical features for the entire time-window, that will represent the entire 30+ points.

These small segments or time steps are then processed, and higher-order feature vectors are constructed from them. Each of the time steps consists a few rows of the original dataset. These small segments or network data flows are now processed to extract features from the statistical analysis of the group. The purpose is to extract key information from this group of events and decrease the amount of data that needs to be processed. Now that the data has been segmented, the feature extraction portion of the algorithm extracts the below statistical properties for each of the attributes within the group. It is meaningful to say that these statistical attributes capture all the details from the sample. Therefore, essentially the 30+ points mentioned in Fig. 1 will be represented by a single row.

Now, feature extractor, the next component in our model, extracts the below-mentioned statistical higher-order features from each of the time steps presented. For all the rows or events within every time step, we select each feature and calculate the higher-order features like median, mode, variance, maximum, minimum, standard deviation, and standard error of mean. And the final feature to be extracted will be the count of rows in each time step. Hence, the reduced dataset has one row per time step, which decreases the number of rows significantly, but simultaneously

A Cognitive Unsupervised Clustering for Detecting Cyber Attacks

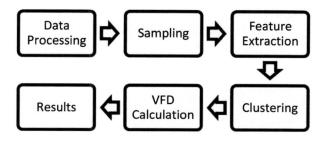

Fig. 2 Proposed method flowchart

the number of columns increases eight times. The reduced dataset is then fed into the unsupervised k-means classifier. The entire process model is depicted in Fig. 2.

Our unsupervised algorithms need to make inferences without any training or reference of the known or labeled examples. K-means clustering algorithm is a good choice for the unsupervised model; it is an np-hard iterative refinement technique that aims to perform k partitions within the n samples presented. It is surprisingly versatile, and hence it is frequently applied in many fields including computer vision, protein enhancement, vector quantization, and computational geometry.

Now that two distinct clouds of items are separated using k-means, the next step is to identify the attack cloud and the normal cloud. In this study, we try to find an automated approach in selecting the attack cloud. Let us re-iterate that every item in the cloud represents a time step (with a start time and an end time) and the higher-order features extracted from the rows or events happening in that time step. In the attack cloud, there should be more items, i.e., time steps that have attack in them, and in the normal cloud, most of the items should have no attack occurring within the time step.

To achieve this task of attack cluster identification, we utilize the concept of complexity; by using which, we could compute the variance fractal dimension trajectory value of items in each of the two clusters, and the cluster with greater number of items having higher VFD value will be identified as the attack cloud. In order to save time and resources, we can also skip doing this for all the items in the clusters and just do it for some randomly chosen items. Hence, the algorithm picks an item randomly, explores into the main dataset, and finds the VFD value at that corresponding time. Once it is done for 10 items in each cluster, algorithm sorts the 20 values from high to low and picks the cluster having highest representation in the top 15 items, as the attack cluster.

4 Experiments and Results

The accuracy of the algorithm has been presented using metrics like accuracy, precision, recall, true positive rate, false positive rate, and f-measure in Table 1. Figure 3 shows the AUC ROC curve. For an unsupervised algorithm, the results obtained were quite satisfactory. The confusion matrix has been presented in Fig. 4.

Table 1 Results obtained

Metric	Proposed cognitive unsupervised clustering algorithm	Basic unsupervised clustering
TPR %	84.93	16.58
TNR %	89.65	63.75
Precision %	89.31	6.06
Accuracy %	87.27	57.92
F-measure %	87.23	8.9

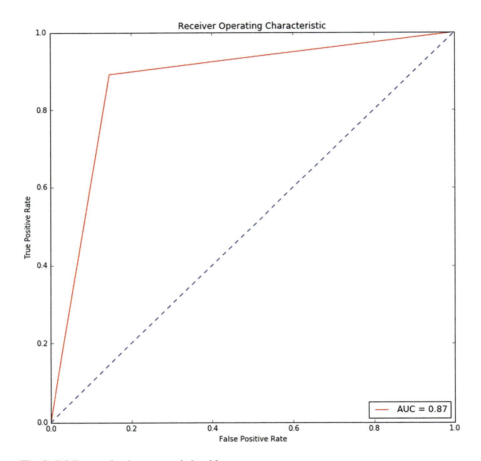

Fig. 3 ROC curve for the proposed algorithm

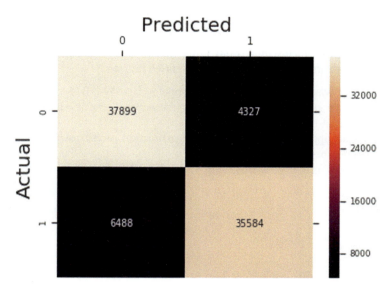

Fig. 4 Confusion matrix

As we can see, the algorithm is able to correctly classify normal as normal and attacks as attacks for majority of the cases (35,582 out of 42,226 for normal and 37,899 out of 42,072 for attack) which eliminates a major load for human analysts. The human analysts can only focus on the attack time steps for further analysis and ignore the normal time steps.

5 Conclusions

As depicted by the results, despite being an unsupervised clustering method, our proposed algorithm was able to achieve a moderately high rate of accuracy in detecting the attack from the normal events. In the VFD calculation section of the algorithm, we utilized the concept of complexity analysis and randomly chose items from attack cloud and normal cloud and calculated the variance fractal trajectory dimension from the original dataset. The difference in the value could accurately detect the attack cloud. The unsupervised clustering technique utilizing statistical higher-order features of network flows was able to bifurcate the normal mundane network traffic from the attack-prone traffic. The intelligent feature extraction approach was able to extract information discriminative enough to enable the classification with an accuracy of 87.27%.

6 Future Work

The study was performed within limited time, resource, and scope. Below are some propositions for future work that can further enrich this study.

(a) This study was tested on the widely popular UNSW-NB15 dataset. Similar study can be performed on a lab with live network traffic and synthetically generated real-time attacks.
(b) The similar approach can be used in classification problems outside network traffic and cybersecurity.
(c) Instead of using k-means clustering, other deep learning methods can be used and compared.

References

1. W. Kinsner, It's time for multiscale analysis and synthesis in cognitive systems, in *Proceedings of the 10th IEEE International Conference on Cognitive Informatics and Cognitive Computing, ICCI*CC 2011*, (2011), pp. 7–10
2. R.L. Fantz, A method for studying early visual development. Percept. Mot. Skills **6**(1), 13–15 (1956)
3. A.m. Slater, Visual perception in the newborn infant: Issues and debates. Intellectica. Rev. l'Association pour la Rech. Cogn. **34**(1), 57–76 (2002)
4. M. Haith, *Rules that Babies Look by: The Organization of Newborn Visual Activity* (Psychology Press, 1980)
5. G.W. Bronson, The scanning patterns of human infants: Implications for visual learning. Monogr. Infancy (1982)
6. G.W. Bronson, Infant differences in rate of visual encoding. Child Dev. **62**(1), 44–54 (1991)
7. J.R. Saffran, R.N. Aslin, E. Newport, Statistical learning by 8-month-old infants. Science **274**, 1926+ (1996)
8. Hermann von Helmholtz – Wikipedia [Online]. Available: https://en.wikipedia.org/wiki/Hermann_von_Helmholtz. Accessed 20 Mar 2020
9. Ernst Mach – Wikipedia [Online]. Available: https://en.wikipedia.org/wiki/Ernst_Mach. Accessed 20 Mar 2020
10. Karl Pearson – Wikipedia [Online]. Available: https://en.wikipedia.org/wiki/Karl_Pearson. Accessed 20 Mar 2020
11. Fergus I. M. Craik – Wikipedia [Online]. Available: https://en.wikipedia.org/wiki/Fergus_I._M._Craik. Accessed 20 Mar 2020
12. Edward C. Tolman – Wikipedia [Online]. Available: https://en.wikipedia.org/wiki/Edward_C._Tolman. Accessed 20 Mar 2020
13. Carolyn Attneave – Wikipedia [Online]. Available: https://en.wikipedia.org/wiki/Carolyn_Attneave. Accessed 20 Mar 2020
14. Egon Brunswik – Wikipedia [Online]. Available: https://en.wikipedia.org/wiki/Egon_Brunswik. Accessed 20 Mar 2020
15. H. Barlow, The exploitation of regularities in the environment by the brain. Behav. Brain Sci. **24**(4), 602–607 (2001)
16. C. Kolias, G. Kambourakis, A. Stavrou, S. Gritzalis, Intrusion detection in 802.11 networks: Empirical evaluation of threats and a public dataset. IEEE Commun. Surv. Tutorials **18**(1), 184–208 (2016)

17. M.K. Korczyński, A. Duda, Classifying service flows in the encrypted skype traffic, in *2012 IEEE International Conference on Communications (ICC)*, (IEEE, 2012), pp. 1064–1068
18. P. Vahdani Amoli, T. Hämäläinen, G. David, A real time unsupervised NIDS for detecting unknown and encrypted network attacks in high speed network, in *2013 IEEE International Workshop on Measurements & Networking (M&N)*, (IEEE, 2013), pp. 149–154
19. M. Korczyński, A. Duda, Markov chain fingerprinting to classify encrypted traffic, in *IEEE INFOCOM 2014-IEEE Conference on Computer Communications*, (IEEE, 2014), pp. 781–789
20. K. Nahiyan, S. Kaiser, K. Ferens, R. Mcleod, A multi-agent based cognitive approach to unsupervised feature extraction and classification for network intrusion detection, in *International Conference on Advances on Applied Cognitive Computing (ACC)*, (CSREA, 2017), pp. 25–30
21. E.D. Reilly, A. Ralston, D. Hemmendinger, *Encyclopedia of Computer Science* (Wiley, Chichester, 2003)
22. W. Kinsner, Complexity and its measures in cognitive and other complex systems, in *Proc. 7th IEEE Int. Conf. Cogn. Informatics, ICCI 2008*, (2008), pp. 13–29
23. L. Zhao, W. Li, L. Geng, Y. Ma, Artificial neural networks based on fractal growth, in *Lecture Notes in Electrical Engineering*, vol. 123, (LNEE, 2011), pp. 323–330
24. E. Bieberich, Recurrent fractal neural networks: A strategy for the exchange of local and global information processing in the brain. Biosystems $66(3)$, 145–164 (2002)
25. E.S. Kim, M. Sano, Y. Sawada, Fractal neural network: Computational performance as an associative memory. Prog. Theor. Phys. $89(5)$, 965–972 (1993)
26. K. Nahiyan, Cognitive unsupervised clustering for detecting cyber attacks 2020 [Online]. Available: http://hdl.handle.net/1993/34747. Accessed 16 Jul 2020

A Hybrid Cognitive System for Radar Monitoring and Control Using the Rasmussen Cognition Model

James A. Crowder and John N. Carbone

1 Introduction: Radar Cognition Models

Radar systems provide essential missions in contested environments spanning from long-range surveillance to air marshalling[1] and weapons queuing. Performance and affordability of these radars are paramount [1]. Performance encompasses the inherent ability to execute the mission and overcoming sophisticated threats and adversarial tactics: technical capability and system availability. A cognitive radar system (CRS) solution enabled by modern architecture techniques and technologies will advance the capabilities through extreme adaptability and facilitate a new, advanced system architecture that supports radar with interferometry and electronic warfare (EW).

One possible non-AI approach would be to construct a database of Objects of Interest (OoIs) distributions identified for previously encountered geographical areas [2]. However, a database of this type would suffer from many issues:

1. The distribution is strongly dependent on the transmit beam width, the time-band width product, and grazing angle. Hence, utilizing a database like this imposes

---

[1] Air marshalling is visual signaling between **ground** personnel and pilots on an **airport, aircraft carrier,** or helipad.

J. A. Crowder (✉)
Colorado Engineering, Inc., Colorado Springs, CO, USA
e-mail: jim.crowder@coloradoengineering.com

J. N. Carbone
Department of Electrical and Computer Engineering, Southern Methodist University, Dallas, TX, USA
e-mail: john.carbone@forcepoint.com

© Springer Nature Switzerland AG 2021
H. R. Arabnia et al. (eds.), *Advances in Artificial Intelligence and Applied Cognitive Computing*, Transactions on Computational Science and Computational Intelligence, https://doi.org/10.1007/978-3-030-70296-0_85

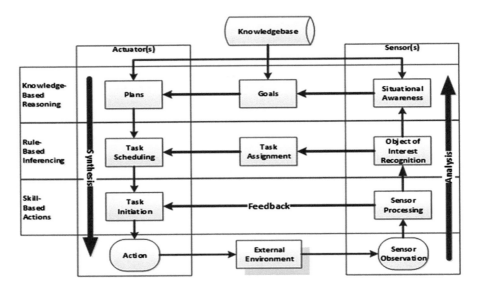

Fig. 1 Cognitive radar system (CRS) architecture

necessary trade-offs between real-time emission controls and the validity of the previous measurements captured in the database.
2. In a long-term temporal sense, the geography may be non-stationary, and human influence can change geographic features and/or introduce discrete clutter into the measurements.

This drives us to design a cognitive radar system that can adapt to present conditions, based on possibilities from past measurements. One of the major questions to be answered during the initial research for this project was which cognitive model was appropriate to design and prototype a CRS. There are many models of human cognition which could be applied against the problem of providing limited cognitive functions to a radar design [3]. During the initial research, it was decided that the initial design, prototype, and testing would entail a hybrid approach to a CRS, with cognitive functions driving a policy-based decision engine that defines changes to the radar parameters and usage, based on the analysis performed within the cognitive processing framework of the CRS [4]. Figure 1 illustrates the overall high-level concept for the CRS architecture.

2 The Rasmussen Cognition Model

This high-level architecture shown in Fig. 1 utilizes a model of human cognition called the Rasmussen Cognition Model (RCM) [5]. The RCM more closely resembles the human decision-making process than other human cognition models

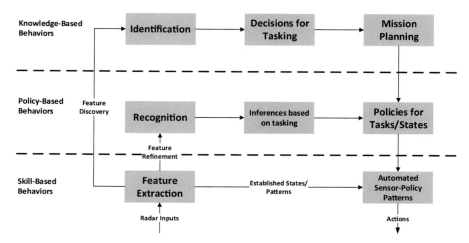

Fig. 2 Rasmussen Cognition Model (CRM) information flow for the CRS

[6]. The purpose of decisions is to achieve some required goal. In the case of a CRS, the system should decide to take a course of action in response to an event, either in normal operations or in a situation that warrants immediate actions. Decision-making is a dynamic analytical process consisting of a series of linked activities and not just a discrete action. Information or courses of action resulting from the analytical processes within a CRS will drive policies that dictate follow-up actions, either by the CRS itself or by radar operators. Decisions stem from mental processes, based on what is currently known, observations that are currently under consideration, and feedback that drives the cognitive process to adapt and learn. The RCM incorporates three levels of abstraction to the cognitive process: knowledge-based processes, rule-based processes, and skill-based processes. Combined, these form the RCM [7]. Figure 2 illustrates the high-level RCM that will be utilized for the proposed CRS. The RCM will provide the CRS with a naturalistic decision-making process that is based on the say human operators use their experience to make decisions in operational settings.

We utilize fuzzy metrics, explained below, to provide decision-making methods, based on the RCM which have the following characteristics essential to real-time mission-critical operations:

1. Ill-structured problems or observations
2. Dynamic, uncertain mission environments
3. Information environments where situations and cognitive cues may change rapidly
4. Cognitive processing that proceeds in iterative action/feedback loops (the OODA loop) illustrated in Fig. 3
5. Time-constrained mission situations
6. High-risk situations (imminent threats)

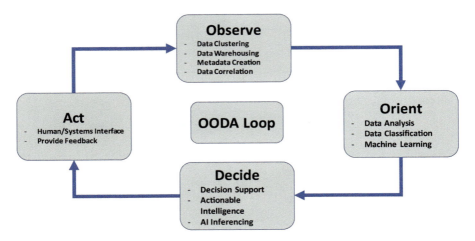

Fig. 3 The OODA loop: Observe, orient, decide, and act

Support for the OODA loop necessitates the need for a hybrid, policy-driven, adaptive CRS. The design of the fuzzy metrics and subsequent policy-driven decision system, we feel, will provide the autonomous Sense, Learn, and Adapt (SLA) technologies for autonomous machine learning and self-adaptation technology with the CRS, shortening the OODA loop. The CRS will store, retrieve, and continuously learn from information based on previously executed radar "missions" creating and modifying scripts that can be used by the CRS as situations "similar to" previous radar missions (in a fuzzy sense) are required and/or encountered.

Scripts that lead to the satisfaction of mission goals are reinforced through the feedback process, while those that interfere with mission goals or mission success will have their weightings reduced. This SLA approach alleviates the problem of realizing a time-varying mapping of neural structures, commonly referred to as the "sequential learning problem [7]." Classical neural system tends to forget previously learned neural mappings when exposed to new or different types of data environments. The SLA approach for the CRS cognitive engine uses the scripts to retain learned data/information. Upon intake of new data environments or continued scanning of the radar system, the CRS cognitive engine provides:

1. Situational recognition: looking for familiar patters or experiences.
2. Option evaluation: looking for relevant cures, providing expectancies, identifying plausible answers (fuzzy metrics), and suggesting typical responses (script selection).
3. Cognitive simulation: does the selected outcome (script) appear workable or not before any action is taken?

Applying these descriptions to the RCM, we get definitions for each level of the Rasmussen model applicable to CRS, and they are:

A Hybrid Cognitive System for Radar Monitoring and Control Using... 1075

- *Skill-Based Behavior* – Trained neural networks (deep learning) that have been trained to recognize features important to mission parameters (e.g., recognizing periscopes). These take place without conscious control of the radar system since they are automated and represent highly integrated patterns of cognitive behavior. Improvement in performance is achieved through feedback control and continuous learning across multiple missions.
- *Rule-Based Behavior* – This entails the composition of a sequence of scripts, each detailing a plan for how to handle a given situation (stored rules) that either have been created initially or are derived empirically through experience and interaction with the CRS. Here, performance of the CRS is goal-oriented with structured feedback control through stored scripts. Here, CRS control is driven by purpose rather than cause in that control (scripts) is selected from previous successful experiences. Script addition/augmentation reflects the functional properties which properly constrain the radar's parameters for successful mission outcomes.
- *Knowledge-Based Behavior* – During unfamiliar situations, when faced with an environment for which there has been no training for the deep learning networks, nor are there scripts to handle the given situation, the CRS move goal-controlled and knowledge-based control, which is essentially operator interaction (HMI). Courses of action are considered and provided to the operator, and their effect is tested against the mission. In doing so, knowledge is gained that can be imparted down through the CRS to the cognitive engine, resulting in new learning and development of new scripts for new situations.

3 The CRS Overall Architecture

Our CRS RCM model is illustrated in Fig. 4. Our use of fuzzy metrics within the RCM adaptive cognitive radar decision analytics forms a knowledge-aided approach, based on past measurements and similar geographical and transmit parameters.

In these terms, the best estimate of the current distribution (current Objects of Interest) is now based on previous and current measurements. Once the current fuzzy distribution is determined, an optima threshold (defuzzification) is derived. Machine learning is then utilized to derive this defuzzification threshold based on the measured data and current knowledge base. Here q^- is the fuzzy metric sample mean, and $\sigma^\wedge$ is the fuzzy metric sample standard deviation. Here we utilize the Extended Ozturk Algorithms (EOA) to find the expected value of the summation of the magnitude of ordered fuzzy statistics multiplied by appropriate weighting factors.

The point associated with fuzzy metric distribution j with weighted shape parameter v is found by computing:

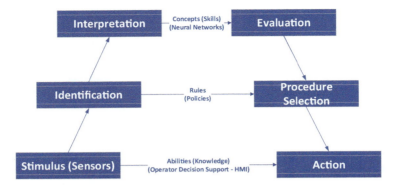

Fig. 4 The CRS RCM hierarchical observation-to-action flow

$$X_{EOA,j}(v) = E\left[\sum_{i=1}^{N} w_i \left|z_{j,(i)}(v)\right|\right]$$

where w_i is the weighting function. We can train a series of neural networks to classify the weighted sum of ordered fuzzy metric statistics (WSOS), which is trained to recognize:

$$X_{WSOS,j}(v) = E\left[\sum_{i=1}^{N} w_i q_{j,(i)}(v)\right]$$

Each Object of Interest is "learned" from previous data collections and is in a library, allowing a threshold to be calculated associated with each Object of Interest. The radar may then form the same fuzzy test statistic from measured data and find the closest (in a fuzzy sense) Object of Interest in the library matching the fuzzy sample test statistic. The associated threshold is then used as a detection threshold for each Object of Interest that is detected by the currently operated radar. The CRS system architecture is illustrated in Fig. 5.

Recognizing that any single radar measurement of an Object of Interest is, at best, random in terms of the cross-sectional scintillation, it cannot be predetermined where the object will be observed and at what angle of observation will be relative to the Object of Interest. However, for given object geometries, the use of higher-order spectrum, in combination with Renyi's mutual information theories, may provide useful algorithms in creating training data sets and in processing radar data sets for Object of Interest (OoI) determination. Since scintillation is manifested as fluctuations in the amplitude of the received radar pulses from an OoI, caused by a shift in the effective reflection point (there are other causes) [8], such fluctuations can be manifested slowly, showing up in consecutive scans, or manifest rapidly, showing up in pulse-to-pulse measurements. Scintillation is utilized in

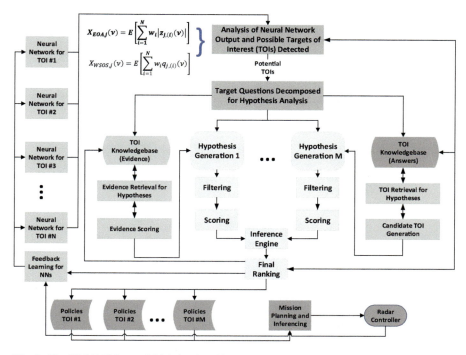

Fig. 5 The CRS RCM overall high-level architecture

target modeling, appears especially at seaside level, and can be used training neural nets, in combination with higher-order stochastic derivative processing and the use of mutual information to determine the "most possible" target being scanned [9, 10].

4 Neural Network and Fuzzy Metric Processing for the CRS

Figure 6 illustrates a high-level block diagram view of the overall CRS architecture shown in Fig. 5. The hybrid cognitive radar concept discussed for this project was to utilize machine learning to drive a set of procedures or scripts which would be used by the CRS system to further refine radar scans once it is perceived by the system that an Object of Interest has been observed by one or more of the on-board radar systems [11]. Referring to Fig. 6, radar scintillation data that has been captured for various OoIs is utilized to train one or more neural networks. The radar data is transformed utilizing stochastic derivative algorithms that is described and derived below. The output from the set of neural networks is fed to a set of fuzzy metric algorithms that determine the level of mutual information the neural network output contains relative to known OoI signatures [12]. The fuzzy membership function

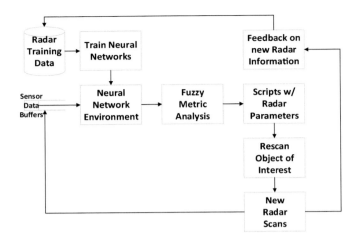

Fig. 6 High-level CRS processing diagram

that indicates the highest mutual information drives the system to speculate (create hypotheses) on the OoI(s) currently being scanned.

Utilizing the hybrid cognitive approach, this drives the next step, shown in Fig. 6, to invoke scripts which have been previously created and tested that modify parameters within the radar system to refine the scans, assuming it is looking for the OoI(s) the system believes it is seeing. This kicks off new set of radar scans which are then processed and compared to the OoI hypotheses from the original scans. The output and information are then fed back into the system, possibly to retrain the neural networks [13]. In this way, the system continually leans and improves its ability to recognize and characterize OoIs. The next two sections provide more details of this process the overall CRS. Once the neural networks are trained with appropriate training data, the CRS is available for use. Figure 6 highlights the portions of the overall architecture illustrated in Fig. 5 for this discussion. Ongoing radar scan data are formatted, transformed, and fed into the neural networks to determine if they detect any OoIs that the networks have been trained to detect and characterize. The output of the neural networks indicate which potential OoIs have been recognized by the networks. This information is fed into a set of mutual information fuzzy membership functions to further refine the potential OoI(s) that are being tracked by the CRS. Figure 7 illustrates this process [14]. The process illustrated in Fig. 7 requires the use of higher-order stochastic derivatives to distinguish differences between the scintillation patterns of OoI scans.

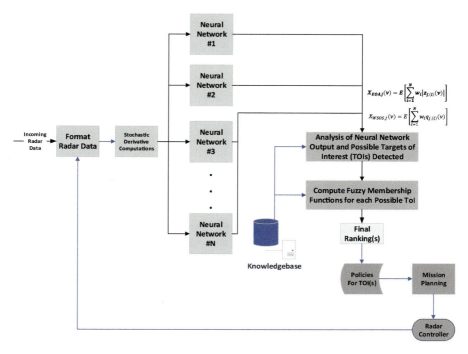

Fig. 7 The CRS neural network data characterization and processing

5 Higher-Order Spectral Theory for CRS

The fundamental properties of stochastic data are important to the detection and characterization of Objects of Interest (OoIs) for use in cognitive radars. Radars scans represent dynamical stochastic processes, since the signals are put through non-linear, stochastic processes imposed by the earth's environments [15]. However, the individual parametric changes within the radars can be directed using hybrid cognitive processing algorithms that understand the subtleties of higher-order stochastic derivatives on the radar return signals. We believe these higher-order stochastic derivatives can be utilized to differentiate OoIs for real-time cognitive radar use. The problem is that stochastic derivatives are computationally challenging in a processing environment. What follows is a derivation of the stochastic derivative, ending in an approximation of the stochastic derivative utilizing an L_2-space stochastic derivative matrix that greatly reduces the computational complexities of stochastic derivatives and produces an $O(n)$ process, based on a discretized generalization of the Levy process [16, 17].

We start with a stochastic integral representation for an arbitrary random variable in a general L_2-continuous martingale space as an integrator for the process. Then, in relation to this, we will define a *stochastic derivative*. Through the derivative, it can be seen whether the proposed random variable admits "or can be characterized" by the derived *stochastic integral* representation [18]. This allows us to show that the stochastic derivative determines the integrand for the stochastic integral which serves as the best *L_2-approximation* for the variable and allows us to establish a discrete *$O(n)$* approximation [19], called a *stochastic derivative matrix*. We start by introducing a stochastic integration scheme in *L_2-space*:

$$H = L_2 (\Omega, \Im, P)$$

for the class of real random variables ξ:

$$\|\xi\| = \left(E|\xi|^2 \right)^{1/2}$$

and then introduce an *H-continuous* martingale process for integration, $\eta_t, 0 \leq t \leq \mathrm{T}$, with respect to an arbitrary filtration of the process [20]:

$$\Im_t, \quad 0 \leq t \leq \mathrm{T}$$

The integrands are considered as elements of a certain *functional L_2-space* of measurable stochastic functions [21]:

$$\phi = \phi (\omega, t), \quad (\omega, t) \in \Omega \times (0, T]$$

with a norm:

$$\|\phi\|_{L_2} = \left(\iint_{\Omega \times (0,T]} |\phi|^2 P (d\omega) \times d[\eta] (\omega) \right)^{1/2} = \left(E \int_0^T |\phi|^2 d[\eta]_t \right)^{1/2}$$

which is given by a *product-type* measure:

$$P (d\omega) \times d[\eta]_t (\omega)$$

associated with a stochastic function $[\eta]_t, 0 \leq t \leq T$, having monotone, right-continuous stochastic trajectories, such that:

$$E (\Delta [\eta] |\Im_t) = E \left(|\Delta\eta|^2 |\Im_t \right)$$

for the increments $\Delta[\eta]$ and $\Delta\eta$ on intervals:

$$\Delta = (t, t + \Delta t] \subseteq (0, T]$$

In particular, for the *Levy process* $\eta_t, 0 \leq t \leq \mathrm{T}$, as integrator:

A Hybrid Cognitive System for Radar Monitoring and Control Using...   1081

$$\left(E\eta_t = 0,\ E\eta_t^2 = \sigma^2 t\right)$$

the deterministic function [22]:

$$[\eta]_t = \sigma^2 t$$

is applicable. Those functions having their permanent $\mathfrak{I}_t\text{-}measurement$ values $\phi^h \in H$ on the $h\text{-}partition$ intervals:

$$\Delta = (t, t + \Delta t] \subseteq (0, T] \quad \sum \Delta = (0, T] \quad (\Delta t \le h)$$

have their stochastic integral defined as:

$$\int_0^T \phi^h d\eta_s \overset{def}{=} \sum_\Delta \phi^h \cdot \Delta \eta$$

Broken into the partition intervals associated with signal environments, it is assumed that [23]:

$$E\left(\phi^h \Delta \eta\right)^2 = E\left(\left|\phi^h\right|^3 \cdot E\left(|\Delta \eta|^2 | \mathfrak{I}_t\right)\right) =$$
$$E\left(\left|\phi^h\right|^2 \cdot E\left(\eta\right) | \mathfrak{I}_t\right) = E\int_\Delta \left|\phi^h\right|^2 d[\eta]_S < \infty$$

which yields:

$$E\left(\int_0^T \phi^h d\eta_S\right)^2 = E\int_0^T \left|\phi^h\right|^2 d[\eta]_S$$

where the *integrands* φ are *identified* as the limit:

$$\phi = \lim_{h \to 0} \phi^h$$

For the functional signal space, which we will define as an $L_2\text{-}space$ $\left\|\phi - \phi^h\right\|_{L_2} \to 0$, the corresponding *stochastic integrals* are defined as limits:

$$\int_0^T \phi d\eta_S = \lim_{h \to 0} \int_0^T \phi^h d\eta_S$$

in the measure space of H, with:

$$\left\|\int_0^T \phi d\eta_S\right\| = \|\phi\|_{L_2}$$

Given data from the same or similar radar scintillation patterns (or related stochasto-chaotic phenomena), we have a system (or condition) that exhibits a set of stochastic variables $\hat{\xi}_i$ and factors that act on these variables that can be a function of these variables [24]:

$$z = g\left(\hat{\xi}_1, \hat{\xi}_2, \hat{\xi}_3, \ldots, \hat{\xi}_m\right)$$

where the factors are atmospheric, magnetic, etc. based on their paths of the radar scans. Thus, we now have a new stochastic variable that can be gathered and constraints determined to track whether this new stochastic variable converges in probability [25]:

$$\hat{\xi}_z = f\left(g\left(\hat{\xi}_1, \hat{\xi}_2, \hat{\xi}_3, \ldots, \hat{\xi}_m\right)\right)$$

The density and distribution functions (pdf and PDF) of $\hat{\xi}_z$, in terms of the pdfs and PDFs of $\hat{\xi}_1, \hat{\xi}_2, \hat{\xi}_3, \ldots, \hat{\xi}_m$, can be obtained through the designation of distribution D_z, where:

$$D_z = \left\{\left(\hat{\xi}_1, \hat{\xi}_2, \hat{\xi}_3, \ldots, \hat{\xi}_m\right) : g\left(\hat{\xi}_1, \hat{\xi}_2, \hat{\xi}_3, \ldots, \hat{\xi}_m\right) \leq z\right\}$$

while noting that:

$$\left(\hat{\xi}_z \leq z\right) = \left\{g\left(\hat{\xi}_1, \hat{\xi}_2, \ldots \hat{\xi}_m\right) \leq z\right\} = \left\{\left(\hat{\xi}_1, \hat{\xi}_2, \ldots \hat{\xi}_m\right) \in D_z\right\}$$

so that:

$$F_z(z) = P\left(Z \leq z\right) = P\left\{\left(\hat{\xi}_1, \hat{\xi}_2, \ldots \hat{\xi}_m\right) \in D_z\right\}$$

which gives us:

$$F_z(z) = \iint_{D_z} f_{\hat{\xi}_1, \hat{\xi}_2, \ldots \hat{\xi}_m}\left(\hat{\xi}_1, \hat{\xi}_2, \ldots \hat{\xi}_m\right) d\hat{\xi}_1 d\hat{\xi}_2 \ldots d\hat{\xi}_m$$

We are exploring if the data contained in a radar scan scintillation environment increases or decreases in stochastic convergence, i.e., if the entropy increases or decreases. We can use stochastic filtering, through looking for the stability of, or lack of, a stochastic and/or stochasto-chaotic process [9]. We want to know if the data environmental processes support a stochastic environment [10]. Data not consistent across stochastic boundaries would produce a lack of stochastic stability or would increase the entropy of the stochastic convergence within the data environment.

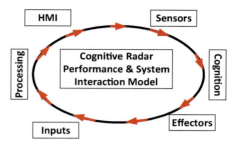

Fig. 8 The CRS test process

6 Next Steps

Now that we have established the methodology and algorithms for the CRS, the next steps involve the following:

1. Modify radar models to allow changing parameters within the radar system from one scan to the next, based on scripts for the given set of Objects of Interest.
2. Create sample scripts that would be utilized to modify radar parameters and scans, based on initial outputs of the neural networks and fuzzy metric analysis.
3. Provide measures of effectiveness (MOEs) to assess the CRS methodology.

Figure 8 illustrates this test process. Simulated radar scans of Objects of Interest are performed (sensors), and the machine learning and fuzzy metric algorithms assess the radar signals (cognition) [26]. Based on the output of these algorithms, the radar parameters are adjusted, and new radar scans are taken (effectors), these scans (inputs) [27]. These new inputs are again put through the cognition algorithms (processing) and the results displayed for the operator (HMI).

For the mutual information calculations discussed above, Figs. 9 and 10 illustrate this process. In Fig. 9, the output values from the neural networks, which represent the higher-order stochastic derivatives of the radar scintillation data, are used to populate the membership function created for OoI #1. The values are normalized from 0 to 1, since this is a fuzzy membership function. The average value from the membership function represents the average mutual information between the output of the neural networks and the information in the knowledgebase for OoI #1. As you can see from Fig. 9, some of the values are very high, but most are near zero on the right side of the curve. This results in a low mean value or low mutual information value, and we would conclude that the radar scan was not from OoI #1.

Figure 10 illustrates an example of a high fuzzy membership, and therefore mutual information, for another OoI, in this case tagged OoI #2. As can be seen from Fig. 10, all of the values are close to the peak of the membership value, producing a high mean membership value or high value of mutual information between the neural network outputs and the stored knowledgebase information for OoI #2. If the membership values were those indicated in Figs. 9 and 10, the system would

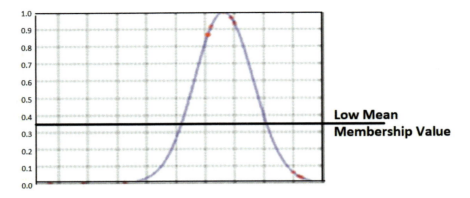

Fig. 9 Example of low fuzzy membership OoI #1

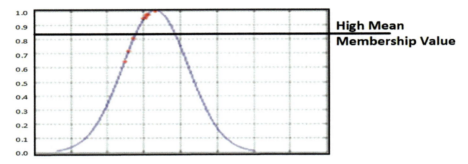

Fig. 10 Example of high fuzzy membership OoI #2

conclude that it is most likely that the scan was of OoI #2 and the scripts that correspond to OoI #2 would be invoked, the radar parameters tuned for OoI #2, and a new radar scan taken. The results of the scan would be provided to the operator, and any new information gathered from these scans would be fed back into the system for assessment of neural network retraining.

7 Conclusions

The derivations described here provide a mathematical basis for the cognitive elements of the CRS. We believe this will provide a functional hybrid cognitive engine for future radar systems. Future work will capture the work being done to improve the radar models, providing the abilities for parametric changes driven by scripts, which are invoked based on the output of the cognitive algorithms described here. Some of the steps that will be required for the future are:

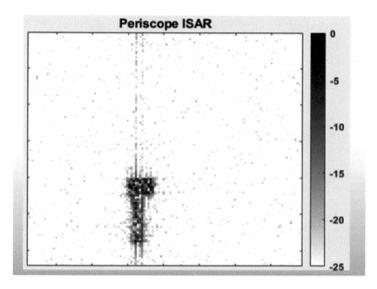

Fig. 11 Example of simulated training data for a periscope

1. Create/simulate/obtain training data to train the neural networks on a given set of Objects of Interest (see Fig. 11).
2. Train the neural networks with the training data.
3. Establish a knowledgebase of scintillation patters and higher-order moment information for the given set of Objects of Interest.
4. Establish the fuzzy membership functions appropriate for the Objects of Interest.

References

1. A. Bar, E. Feigenbaum, *The Handbook of Artificial Intelligence*, vol I (Stanford University, 1981)
2. J. Crowder, J. Carbone, S. Friess, *Artificial Cognition Architectures* (Springer-Verlag, New York, 2013). ISBN 978-1-4614-8072-3
3. R. Thomason, Logic and artificial intelligence, in *Stanford Encyclopedia of Philosophy*, ed. by E. N. Zalta, (1995)
4. J. Ender, S. Brüggenwirth, Cognitive radar – Enabling techniques for next generation radar systems, in *16th International Radar Symposium (IRS)*, (IEEE, 2015), pp. 3–12
5. J. Rasmussen, Skills, rules, and knowledge; signals, signs, and symbols, and other distinctions in human performance models. IEEE Transactions on Systems, Man, and Cybernetics, SMC-13(3) 1983. IEEE
6. G. Klein, Naturalistic decision making. Hum. Factors **50**(3), 456–460 (2008)
7. M. McCloskey, N. Cohen, Catastrophic interference in connectionist networks: The sequential learning problem. Psychol. Learn. **24**, 109–165 (1989)

8. E. Byron, *Radar: Principles, Technology, Applications* (Prentice Hall, Upper Saddle River, 1992). ISBN 0-13-752346-7.
9. J. Crowder, Cognitive systems for data fusion, in *Proceedings of the 2005 PSTN Processing Technology Conference, Ft. Wayne, Indiana*, (2005)
10. J.A. Crowder, The continuously recombinant genetic, neural fiber network, in *Proceedings of the AIAA Infotech@Aerospace-2010, Atlanta, GA*, (2010)
11. J. Crowder, J. Carbone, *Methodologies for Continuous, Life-Long Machine Learning for AI Systems* (International Conference on Artificial Intelligence, Las Vegas, 2018)
12. M. Maahn, U. Löhnert, *Potential of Higher-Order Moments and Slopes for the Doppler Spectrum for Retrieving Microphysical and Kinematic Properties of Arctic ICE Clouds* (Institute for Geophysics and Meteorology, University of Cologne, Cologne, 2017)
13. E. Fremouw, A. Ishimaru, Intensity scintillation index and mean apparent radar cross section on monostatic and bistatic paths. Radio Sci. **27**(4), 529–543 (1992)
14. J.A. Crowder. Derivation of Fuzzy Classification Rules from Multidimensional Data. NSA Technical Paper CON_0013_2002_001 (2002)
15. A. Faruqui, K. Turner, Appl. Math. Comput. **115**, 213 (2000)
16. J. Crowder, T. Barth, R. Rouch. Learning Algorithms for Stochastically Driven Fuzzy, Genetic Neural Networks. NSA Technical Paper, ENIGMA_1999_002 (1999).
17. F. Baudoin, D. Nualart, Equivalence of Volterra processes. Stochastic Process Appl. **107**, 327–350 (2003)
18. J. Cresson, S. Darses, Plongement stochastique des syst'emes lagrangiens. C. R. Acad. Sci. Paris Ser. I **342**, 333–336 (2006)
19. H. Föllmer, *Time reversal on Wiener space. Stochastic Processes—Mathematics and Physics (Bielefeld, 1984) 119–129. Lecture Notes in Math. 1158* (Springer, Berlin, 1986)
20. Rozanov, Y. 2000. On differentiation of stochastic integrals. Quaderno IAMI 00.18, CNR-IAMI, Milano.
21. W. Schoutens, J. Teugels, Lévy processes, polynomials, and martingales. Commun. Statist. Stochastic Models **14**, 335–349 (1998)
22. R. Michalski, Inferential theory of learning as a conceptual basis for multi-strategy learning. Mach. Learn. **11**, 111–151 (1993)
23. L. Brillouin, *Science and Information Theory* (Dover, New York, 2004)
24. J. Crowder, J. Carbone, Recombinant knowledge relativity threads for contextual knowledge storage, in *Proceedings of the 12th International Conference on Artificial Intelligence, Las Vegas, NV*, (2011)
25. C. Stylios, P. Groumpos, Fuzzy cognitive maps in modeling supervisory control systems. J. Intell. Fuzzy Syst. **8**(2), 83–98 (2000)
26. J. Carbone, J. Crowder, Transdisciplinary synthesis and cognition frameworks, in *Proceedings of the Society for Design and Process Science Conference 2011, Jeju Island, South Korea*, (2011)
27. Crowder, J. 2004. Multi-Sensor Fusion Utilizing Dynamic Entropy and Fuzzy Systems. Proceedings of the Processing Systems Technology Conference, Tucson, AZ

Assessing Cognitive Load via Pupillometry

Pavel Weber, Franca Rupprecht, Stefan Wiesen, Bernd Hamann, and Achim Ebert

1 Introduction

Cognitive load, understood as the amount of working memory resources dedicated to a specific task, determines a person's problem solving ability in terms of effectiveness and efficiency [18]. Best task performance is achieved based on a balanced, productive cognitive load level that avoids mental "under-challenge and over-challenge". A software system that is able to dynamically adapt task difficulty based on a person's experienced cognitive load in real time can have great impact on a variety of applications, e.g., learning, driving, and high-performance working environments such as that of pilots.

Many methods have been used to measure cognitive load, such as subjective self-reported measures and analytical approaches [2, 7], and objective psycho-physiological measures, e.g., electroencephalography (EEG), functional magnetic resonance imaging (fMRI), heart rate, blood pressure, skin temperature, and eye activity [21]. Many of these methods have the disadvantage of being intrusive and depending on non-portable equipment. This has been tested only in controlled laboratory settings, and they require complex data analysis. Considering these limitations and the increasing accuracy and affordability of eye tracking

---

P. Weber (✉) · S. Wiesen · A. Ebert
Department of Computer Science, University of Kaiserslautern, Kaiserslautern, Germany
e-mail: weber@cs.uni-kl.de; wiesen@cs.uni-kl.de; ebert@cs.uni-kl.de

F. Rupprecht
insight.out GmbH - digital diagnostics, Kaiserslautern, Kaiserslautern, Germany
e-mail: franca.rupprecht@insio.de

B. Hamann
Department of Computer Science, University of California, Davis, CA, USA
e-mail: hamann@cs.ucdavis.edu

© Springer Nature Switzerland AG 2021
H. R. Arabnia et al. (eds.), *Advances in Artificial Intelligence and Applied Cognitive Computing*, Transactions on Computational Science and Computational Intelligence, https://doi.org/10.1007/978-3-030-70296-0_86

systems, analysis of pupillometry data for extracting cognitive features has become increasingly feasible and common. The possibility of turning a smartphone, a tablet, or a webcam into an eye tracker emphasizes its real-world and real-time applicability [11, 17].

While task-evoked pupillary response (TEPR) was found to be a reliable measure directly corresponding to working memory [6], it does not distinguish between pupillary reflex reactions to light changes and reactions induced by cognitive effort. The only published algorithm that claimed to successfully separate light reflexes from dilation reflexes is the index of pupillary activity (IPA). The IPA was published broadly and openly, in contrast to the patented index of cognitive activity (ICA) that was used in a wide range of studies.

We present a validation of the IPA by applying it to an experiment with finely granulated levels of difficulty of cognitive tasks and compare the results with traditional TEPR metrics, namely the percentage change of pupil diameter (PCPD). First, we analyze a participant's performance for the different levels of task difficulty. Specific hypotheses and expectations have guided our efforts. We expect reaction times to increase and accuracy to decrease with increasing difficulty. Next, we calculate the PCPD and analyze its peaks and magnitudes. We expect both the peaks and magnitudes to increase with increasing difficulty. Finally, we calculate the IPA for each trial, expecting it to also increase with increasing difficulty level. In conclusion, we discuss limitations of the proposed method and provide incentives for future work.

2 Background and Related Work

2.1 TEPR

The correlation between pupil diameter and problem difficulty has already been noted in the 60s [8]. In short-term memory tasks, it was observed that the pupil dilated during the presentation phase and constricted during the recall phase. The peak pupil diameters were found to be directly related to the number of items presented [10]. Other studies found the raw pupil diameter to be not comparable across participants and proposed the PCPD as a metric of interest [9, 12]. It is computed in regard of a certain baseline, typically an average value over a given amount of seconds of pupil diameter data measured before the experiment.

The main problem with this measure is that the changes in pupil size for the most part cannot definitely be attributed to either lighting conditions or actual cognitive effort. It was found that changes in pupil diameter size evoked by light reflexes can be described as large (up to a few millimeters), while those evoked by cognitive activity happen to be relatively small (usually between 0.1 and 0.5 mm) and rapid [1]. Those are however very loose ranges and cannot be directly applied to reliably distinguish the cause of the pupillary reflex.

2.2 ICA and IPA

In the early 2000s, Marshall developed the ICA that seems to be able to distinguish the pupillary reflexes [13]. The ICA uses wavelet analysis to compute the rate of occurrences of abrupt discontinuities in the pupil diameter signal. The assumption is that low IPA values (i.e., few abrupt discontinuities per time period) reflect little cognitive effort, while high values indicate strong cognitive effort. Although the algorithm itself is proprietary and its implementation is undisclosed, it has been used in a variety of studies and is claimed to be reliable across sampling rates and different hardware platforms [3, 4, 20].

Since there is no independent verification of the ICA, another research group has developed their own version of the algorithm, using clues in different papers and the patent manuscript: the IPA [5]. The IPA uses discrete wavelet transformation—similar to the ICA—but differs in the choice of wavelet, thresholding approach, and extrema detection method. The algorithm itself is disclosed in the paper, making it possible for other researchers to reproduce every step of it. A multi-level wavelet decomposition with a *Symlet-16* mother wavelet is used to separate low-frequency components (level-1 detail coefficients), corresponding to light reflexes, and high-frequency components (level-2 detail coefficients), triggered by cognitive activity. The modulus maxima are used to find local extrema in the level-2 coefficients. Those maxima are then compared to a so-called universal threshold, denoted by $\sigma\sqrt{(2\log n)}$. All maxima above this threshold are considered an abrupt discontinuity.

3 Method

3.1 Study Design

The present study was a within-subjects eye tracking experiment based on a simple memory span task. Memory span tasks are used to determine a user's working memory capacity (WMC). With each new trial, the participant is presented with an increasing number of items and then asked to recall them. The WMC is the longest number of sequential items that the user can correctly recall. For a typical young adult, the WMC is 7 ± 2 [15]. In our adapted version, the number of digits was not successively increased but randomized. We were not so much interested in the participant's WMC but rather in inducing different levels of intrinsic cognitive load that somewhat reflect non-ideal real-world conditions.

The independent variable is the number of digits presented—which represents the inherent task difficulty for the trial. The dependent variables are reaction time, answer correctness, and the pupil diameters measured during the whole experiment. From the signal of the latter, we calculated PCPD as well as IPA values for each trial.

3.2 Participants

The study was conducted as part of a HCI lecture with a sample of 34 international students. Data of 4 participants had to be discarded due to difficulties with eye tracker calibration, giving a final sample size of $N = 30$ (16 female, 14 male) with age ranging from 22 to 45 ($\mu = 26 \pm 4$).

3.3 Apparatus

The laboratory was setup in a clean and neutral office with two windows covered by roller blind, and two double flux fluorescent ceiling lights. The inside lighting condition fluctuated between 160 and 192 Lux during the experiment period of two weeks, depending on the weather. A PupilLabs Pupil Core eye tracking system was used to acquire pupillometry data for both eyes at a rate of 120 Hz. The corresponding Pupil Capture software as well as a specially developed JavaFX application was executed on a customary windows machine with a 1080p monitor and standard mouse and keyboard. Brightness and contrast of the display were constant.

3.4 Procedure

After filling out a simple demographic questionnaire, the participant was asked to put on the head-mounted eye tracker. We then started a marker-based calibration sequence since it yielded the best confidence values of the system. Once the confidence level was stable and high enough, the actual digit span task was started. Instead of being asked to recall the digits, the participant was shown a composite number sequence with the same number of digits. This composite sequence could contain built-in errors; hence, the participant was asked whether it corresponds exactly to the single digits that were shown before. The answer was given by pressing the left arrow key for *no*, the right arrow key for *yes*, respectively. The participant had the opportunity to get acquainted with the task by performing six training trials with 3 to 5 digits. The training was followed by five blocks of 24 trials each, with sequences of 3 to 10 digits, resulting in a total of 120 trials per participant (15 for each difficulty level). Each block took around 3.6 ± 0.3 minutes to complete. The participants were encouraged to take a short break after each block.

4 Results

4.1 Analyses

Since the relationship of the difficulty levels cannot be assumed to be linear, but definitely to be monotonic, Spearman's rank coefficient ρ was used to calculate correlation. In addition, repeated measures of analyses of variance (ANOVA) were used to find significant effects. Cohen's parameter d was calculated to assess the effect sizes and variation between-group means, emphasizing respective significance through pairwise T-tests. Descriptive statistics for all dependent variables are summarized in Table 1, while effect size and significance are shown in Table 2. All training data were excluded from the analyses. All analyses were conducted in Python, more specifically *Pandas* and *Pingouin*.

4.2 Task Performance

Task performance was measured in terms of reaction time and answer correctness/accuracy. Reaction time was expected to increase with increased task difficulty, while correctness was expected to decrease with increased task difficulty, i.e., the presumption was that more difficult tasks would take more time to complete and were more likely to be answered wrongly. Our analyses confirmed this presumption. Figure 1 shows a clear linear trend for both observations.

The correlation between task difficulty and reaction time was moderately positive (0.54, $p < .001$), and the one between task difficulty and answer correctness was weakly negative (-0.22, $p < .001$).

ANOVA revealed significant effects of both: task difficulty having impact on reaction time, $F_{(7,203)} = 69.62, p < .001, \eta^2 = 0.706$, and on correctness, $F_{(7,203)} = 23.67, p < .001, \eta^2 = 0.449$. This proves both our hypotheses

Table 1 Statistic results for dependent variables of the experiment; effect of different task difficulty levels

Digits	Reaction time μ and σ (s)	Accuracy μ and σ (%)	PCPD peak, μ and σ (%)	Magnitude μ and σ (%)	IPA μ and σ (Hz)
3	1.25 ± 0.60	97.3 ± 16.1	3.3 ± 11.8	12.9 ± 5.8	$1.134 \pm .378$
4	1.37 ± 0.70	96.2 ± 19.1	3.4 ± 11.6	13.4 ± 6.7	$1.155 \pm .350$
5	1.74 ± 0.92	93.5 ± 24.6	3.9 ± 11.5	14.2 ± 6.5	$1.199 \pm .360$
6	2.00 ± 1.23	90.6 ± 29.1	3.6 ± 13.3	15.1 ± 9.8	$1.183 \pm .351$
7	2.56 ± 1.53	87.1 ± 33.5	4.6 ± 12.8	15.9 ± 8.1	$1.216 \pm .346$
8	2.75 ± 1.75	82.4 ± 38.1	4.2 ± 11.6	15.4 ± 6.1	$1.239 \pm .352$
9	3.11 ± 2.18	80.6 ± 39.5	4.9 ± 12.5	16.7 ± 7.7	$1.222 \pm .331$
10	3.00 ± 2.06	75.3 ± 43.1	4.8 ± 12.1	16.8 ± 7.4	$1.236 \pm .336$

Table 2 Effect sizes and significance between task difficulty level groups

	Reaction time		Accuracy		PCPD peak,		Magnitude		IPA	
Pair	d	p	d	p	d	p	d	p	d	p
(3,4)	.074	***	−.034	−	.009	−	.062	−	.061	−
(4,5)	.229	***	−.082	**	.044	−	.101	**	.126	**
(5,6)	.156	***	−.088	**	−.021	−	.126	**	−.049	−
(6,7)	.343	***	−.111	**	.076	−	.109	*	.096	*
(7,8)	.114	**	−.138	**	−.031	−	−.072	*	−.065	−
(8,9)	.225	***	−.053	−	.062	−	.168	***	−.050	−
(9,10)	−.064	−	−.169	***	−.015	−	.008	−	.040	−

Significance: *$p < .05$, **$p < .01$, ***$p < .001$

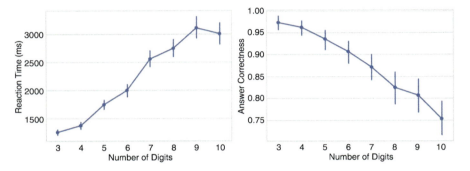

Fig. 1 Average reaction times and answer correctness per number of digits

regarding task performance, i.e., reaction time increases and accuracy decreases with increasing difficulty.

Table 2 shows that the effects of task difficulty levels on reaction time are highly significant, except for the differences between 7 and 8 digits (being significant) and between 9 and 10 digits (not being significant). Concerning answer correctness, the difference between 9 and 10 digits is the only highly significant one. All others are significant, except for the differences between 8 and 9 digits and between 3 and 4 digits. Even though the effect sizes are not large, these results support our paradigm of performing testing with finely granulated task difficulty levels.

4.3 Pupil Dilation

We used the median pupil diameter of the training sequence as baseline for calculating the PCPD. PCPD values for each trial were aggregated to the actual variables of interest, i.e., pupil dilation peaks and magnitudes between pupil dilation valleys and peaks. Figure 2 illustrates the average values for the two metrics considered. The trends of both graphs match our expectations.

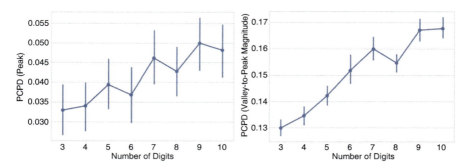

Fig. 2 Average PCPD peaks and valley-to-peak magnitudes per number of digits

The correlation between PCPD peaks and task difficulty was weakly positive (0.059, $p < .001$), with a significant effect $F_{(7,203)} = 4.144$, $p < .001$, $\eta^2 = 0.125$. This proves our first TEPR-related hypothesis that PCPD peaks increase with task difficulty. However, pairwise T-tests revealed that none of the between-groups effects were significant.

The second correlation of interest is the correlation between PCPD valley-to-peak magnitudes and task difficulty. This one is stronger, but it is also only weakly positive (0.225, $p < .001$). ANOVA revealed a significant effect: $F_{(7,203)} = 22.645$, $p < .001$, $\eta^2 = 0.438$. This result proves our second TEPR-related hypothesis that PCPD valley-to-peak magnitudes increase with task difficulty. Regarding the effects between difficulty level groups, 5 out of 7 pairs showed a significant effect of at least $p < .05$, see Table 2. The only pairs that showed no significant effect were between 3 and 4 digits and between 9 and 10 digits.

4.4 Abrupt Discontinuities

Our next goal was aimed at determining whether the same behavior holds for the more sensitive metric that is said to distinguish between pupil dilation reflex and light reflex, the IPA. We calculated IPA values for every commenced second of a trial and averaged values.

Figure 3 shows that the correlation between IPA and task difficulty is indeed weakly positive (0.105, $p < .001$) with a significant effect ($F_{7,203} = 16.327$, $p < .001$, $\eta^2 = 0.36$). Table 2 reveals, however, that only two effects of difficulty level differences are significant, the one between 4 and 5 digits and the one between 6 and 7 digits with the latter only being $p < .05$. Nevertheless, the general significance for the effect of task difficulty on IPA, shown by the ANOVA, is not questioned.

The second plot in Fig. 3 shows another interesting detail about the IPA. Following cognitive load theory, the highest experienced cognitive load should occur during the stimulus presentation phase and the lowest during pauses, while the

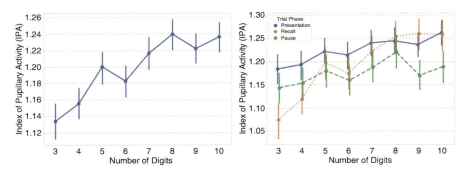

Fig. 3 Average IPA per number of digits and trial phase

mental load experienced during the recall phase should not have a direct correlation to the task difficulty but should follow the same trend as the task inherent difficulty; therefore, the mental load should be low for simple tasks and increase significantly with higher difficulty levels [19]. This behavior—together with the weak positive correlation between IPA and PCPD magnitude (0.112, $p < .001$)—confirms our last hypothesis that the IPA is an indicator for cognitive load that increases with task difficulty.

5 Conclusions

The results of our statistical analyses validate our specific hypotheses, as stated in the Introduction. We have shown that finely granulated difficulty settings can have significant impact on task performance. We have shown that impact on task performance is reflected in the magnitude of pupil dilation amplitudes. Finally, we have ascertained that this behavior is also substantiated by the values calculated with the relatively new IPA algorithm. Our experiments have demonstrated that the IPA correlates with traditional TEPR metrics, even in finely granulated task difficulty settings. While the authors differentiated between three difficulty settings and found no significant effect between the easy and control tasks, we found two significant effects that occurred when increasing the number sequence by just one digit, additionally to the general significant effect of task difficulty on IPA revealed by ANOVA. Effect size and significance levels for the different difficulty levels were not as high for measured IPA data relative to measured TEPR data. In summary, our findings validate the IPA. However, we found the unmodified IPA algorithm to be sensitive to sampling rate and signal length, resulting in very different recognized discontinuity counts. The chosen symlet-16 mother wavelet therefore seems not to be universally applicable; hence, our IPA values differ from the ranges reported by the authors.

These findings underline the need for further investigation to ensure smooth utilization of the IPA. Our experimental design considered only one task, i.e., the digit span task. Since the digits were all shown at the center of the screen, we did not consider eye tracking measures such as fixation and saccade. We have not yet analyzed eye blink frequency and latency. The authors of the IPA removed all data points in a 200 ms window before the start and after the end of a detected blink. Our method, in contrast, uses cubic spline interpolation to reconstruct the signal (see [14]). While the cubic spline approach seems to work well, it would seem of interest to determine how well it performs in a more realistic experimental setting.

Concerning possible future research directions, it would certainly be useful to further examine the validity of the IPA through a battery of tests. Unlike the ICA there is the possibility to do so independently. Users of the IPA would benefit from a large-scale comparison of different sample rates, wavelets, and coefficient resolutions, possibly resulting in proper usage guidelines. The algorithm also remains to be tested under varying light conditions, like it was done for the ICA [16], as well as under different tasks and modalities. It may also be of interest how the measured cognitive load value relates to the relative performance of individuals. These various research directions must be supported by the results of other valid cognitive feature extraction methods.

Acknowledgments This research was partially funded by the German Research Foundation (DFG) within the IRTG 2057 "Physical Modeling for Virtual Manufacturing Systems and Processes".

References

1. J. Beatty, Task-evoked pupillary responses, processing load, and the structure of processing resources. Psychological Bulletin **91**(2), 276–292 (1982). https://doi.org/10.1037/0033-2909. 91.2.276
2. A. Cook, R. Zheng, J. Blaz, Measurement of cognitive load during multimedia learning activities. Cognitive Effects of Multimedia Learning, 34–50 (2008). https://doi.org/10.4018/978-1-60566-158-2.ch003
3. V. Demberg, Pupillometry: the index of cognitive activity in a dual-task study, in *Proceedings of the Annual Meeting of the Cognitive Science Society*, vol. 35, pp. 2154–2159 (2013)
4. V. Demberg, A. Sayeed, The frequency of rapid pupil dilations as a measure of linguistic processing difficulty. PLOS ONE **11**(1), e0146,194 (2016). https://doi.org/10.1371/journal.pone.0146194
5. A.T. Duchowski, K. Krejtz, I. Krejtz, C. Biele, A. Niedzielska, P. Kiefer, M. Raubal, I. Giannopoulos, The index of pupillary activity: Measuring cognitive load *vis-à-vis* task difficulty with pupil oscillation, in *Proceedings of the 2018 CHI Conference on Human Factors in Computing Systems - CHI '18* (ACM Press, Montreal QC, Canada, 2018), pp. 1–13. https://doi.org/10.1145/3173574.3173856
6. E. Granholm, R.F. Asarnow, A.J. Sarkin, K.L. Dykes, Pupillary responses index cognitive resource limitations. Psychophysiology **33**(4), 457–461 (1996). https://doi.org/10.1111/j.1469-8986.1996.tb01071.x

7. S.G. Hart, Nasa-task load index (NASA-TLX); 20 years later, in *Proceedings of the Human Factors and Ergonomics Society Annual Meeting*, vol. 50(9), 904–908 (2006). https://doi.org/10.1177/154193120605000909
8. E.H. Hess, J.M. Polt, Pupil size in relation to mental activity during simple problem-solving. Science **143**(3611), 1190–1192 (1964). https://doi.org/10.1126/science.143.3611.1190
9. X. Jiang, M.S. Atkins, G. Tien, R. Bednarik, B. Zheng, Pupil responses during discrete goal-directed movements, in *Proceedings of the 32nd Annual ACM Conference on Human Factors in Computing Systems - CHI'14* (ACM Press, Toronto, Ontario, Canada, 2014), pp. 2075–2084. https://doi.org/10.1145/2556288.2557086
10. D. Kahneman, J. Beatty, Pupil diameter and load on memory. Science **154**(3756), 1583–1585 (1966). https://doi.org/10.1126/science.154.3756.1583
11. K. Krafka, A. Khosla, P. Kellnhofer, H. Kannan, S. Bhandarkar, W. Matusik, A. Torralba, Eye tracking for everyone (2016). arXiv:1606.05814 [cs]
12. J.L. Kruger, F. Steyn, Subtitles and eye tracking: Reading and performance. Read. Res. Q. **49**(1), 105–120 (2014). https://doi.org/10.1002/rrq.59
13. S. Marshall, The index of cognitive activity: Measuring cognitive workload, in *Proceedings of the IEEE 7th Conference on Human Factors and Power Plants* (IEEE, Scottsdale, AZ, USA, 2002), pp. 7–5–7–9. https://doi.org/10.1109/HFPP.2002.1042860
14. S. Mathôt, J. Fabius, E. Van Heusden, S. Van der Stigchel, Safe and sensible preprocessing and baseline correction of pupil-size data. Behav. Res. Methods **50**(1), 94–106 (2018). https://doi.org/10.3758/s13428-017-1007-2
15. G.A. Miller, The magical number seven, plus or minus two: Some limits on our capacity for processing information. Psychological Review **63**(2), 81–97 (1956). https://doi.org/10.1037/h0043158
16. L. Rerhaye, T. Blaser, T. Alexander, Evaluation of the index of cognitive activity (ICA) as an instrument to measure cognitive workload under differing light conditions, in *Congress of the International Ergonomics Association* (Springer, 2018), pp. 350–359. https://doi.org/10.1007/978-3-319-96059-3_38
17. K. Semmelmann, S. Weigelt, Online webcam-based eye tracking in cognitive science: A first look. Behav. Res. Methods **50**(2), 451–465 (2018). https://doi.org/10.3758/s13428-017-0913-7
18. J. Sweller, Cognitive load during problem solving: Effects on learning. Cognitive Science **12**(2), 257–285 (1988). https://doi.org/10.1016/0364-0213(88)90023-7
19. J. Sweller, Cognitive Load Theory: Recent Theoretical Advances (2010). https://doi.org/10.1017/CBO9780511844744.004
20. J. Vogels, V. Demberg, J. Kray, The index of cognitive activity as a measure of cognitive processing load in dual task settings. Front. Psychol. **9**, 2276 (2018). https://doi.org/10.3389/fpsyg.2018.02276
21. R.Z. Zheng, K. Greenberg, The boundary of different approaches in cognitive load measurement strengths and limitations, in *Cognitive Load Measurement and Application* (Routledge, 2017), pp. 45–56. https://doi.org/10.4324/9781315296258-4

A Hybrid Chaotic Activation Function for Artificial Neural Networks

Siobhan Reid and Ken Ferens

1 Introduction

Chaos theory has become an integral concept in the modeling and analysis of the human brain [1]. Chaos describes the seemingly random, unpredictable, and irregular behavior in a dynamical system that is fully deterministic [2]. A dynamical system is chaotic if it is sensitive to initial conditions, topologically dense, and topologically transitive [3, 4].

Many researchers have investigated the existence of chaos in the brain. A study examining the synapses in the somatosensory cortex of rats determined the synapse's behavior was either fixed, periodic, or chaotic depending on the frequency of the stimulus [5]. The study revealed that individual synapses produce bifurcation patterns. Other studies have found chaotic attractors in the electrical activity of rats' brains [6] and chaotic activity in squid axons when stimulating pulses were applied [7].

Typical ANNs are extremely oversimplified [8] and do not reflect the chaotic nature of the brain. We propose a new ANN which incorporates chaos. The chaotic neurons in the ANN use two AFs: the sigmoid function and the logistic map function. The combination of the two functions will be referred to as the hybrid chaotic activation function (HCAF). The objective of the HCAF is to more accurately represent the complex and chaotic behavior of the brain.

S. Reid · K. Ferens (✉)
Department of Electrical and Computer Engineering, University of Manitoba, Winnipeg, MB, Canada
e-mail: Ken.Ferens@umanitoba.ca

© Springer Nature Switzerland AG 2021
H. R. Arabnia et al. (eds.), *Advances in Artificial Intelligence and Applied Cognitive Computing*, Transactions on Computational Science and Computational Intelligence, https://doi.org/10.1007/978-3-030-70296-0_87

2 Related Work

Commonly used non-chaotic AFs include the sigmoid function, the hyperbolic tangent function (tanh), and the rectified linear unit (reLU) [9]. Each of these functions offers advantages and disadvantages. A disadvantage of the sigmoid and reLU functions is that they are not zero-centered, which causes their gradients to be all positive or all negative values, making the learning process occur in a zigzag manner. Furthermore, the tanh and sigmoid functions can encounter an issue referred to as the vanishing gradient; the derivatives of these functions become very small, slowing the learning process. This occurs because the outputs are saturated; sigmoid only outputs values between 0 and 1, and tanh only outputs values between −1 and 1 [10]. This issue can be resolved by using the reLU function, which is non-saturating. A disadvantage of the reLU function is that it can create "dead neurons" that are never used. Dead neurons occur when many of the network's activation values are zero. They can inhibit the ANN's learning process. Modified versions of reLU have been introduced to prevent this issue; they have small negative slopes to avoid zero-valued activations. These disadvantages can decrease the ANN's performance.

Researchers have attempted improving AFs by modeling them to represent the complex nature of the brain. In [11], a neuron is presented which contains multiple AFs. The first AF is referred to as the inner synaptic function, which is a multi-input, multi-output function. The second AF is referred to as the soma (cell body) AF; it receives the inputs from the inner synaptic function and non-linearly maps them into one output value. This model provides another way of adding non-linearity into a network and presents the concept of separating a neuron into multiple components.

Chaotic AFs have also recently been introduced into ANN's and have been shown to accelerate the learning process. In [12], researchers proposed using the chaotic Duffing map equation in a Hopfield network for pattern recognition. The network showed improved performance, benefitting from the non-linear nature of the Duffing map equation. In [10], the logistic map and the circular map were used as the AF in a network. The AF significantly reduced the number of iterations needed for the network to converge. It was argued that the improvement was due to the chaotic AF's ability to avoid saturation and its resemblance to biological brain behavior. The logistic map function has also been effective in decreasing the convergence time of other optimization problems. In [13], a chaotic simulated annealing model was proposed which incorporated the logistic map function. The model converged quicker due to it implementing a chaotic walk in the solution space, which guaranteed that candidate solutions would not be repeated, as is the case in the conventional random walk simulated annealing.

W. Freeman was one of the first neuroscientists to model biological brain behavior using chaos theory. He theorized that chaos is the mechanism which allows neurons to rapidly respond to stimuli [14]. Freeman modeled the behavior of perception neurons in a rabbit's olfactory system. He proposed the rabbit's neurons are in a low-level chaotic state when no stimuli are applied. When a stimulus is applied and recognized, the neurons converge towards an attractor and display periodic or fixed stable activity. When a stimulus is applied that has not been seen before, the neurons move to a high-level chaotic state. This enables the system to produce a new pattern of behavior instead of attempting to converge to a previously learned pattern. Freeman hypothesized that this is what allowed the rabbits to identify new odors. Chaos provides the system with what Freeman referred to as an "I don't know" state, in which the system searches for new solutions. Although aspects of Freeman's model have been criticized, recent evidence supports the overall notion that the brain behaves chaotically [7].

3 Chaotic ANN Architecture

In this paper, we present a model of neurons in a high-level chaotic state, as described by Freeman [14]. This is done by using the HCAF, which generates a unique chaotic activation value to send to each neuron in the following layer. This provides a population-based search technique, which allows neurons to search a large solution space and prevents the ANN from falling into a local minimum.

Figure 1 illustrates the proposed ANN architecture for solving the XOR problem. In this model, the neurons in the inner layers use the HCAF, and the final layer's neuron uses the sigmoid AF. The upper subscript (k) represents which layer the component belongs to. The first lower subscript (i) represents which neuron the weight or bias is connected to in the subsequent layer. The second lower subscript (j) represents which neuron the weight is connected to in the previous layer.

The following equations are used to calculate the inputs and outputs of the neurons:

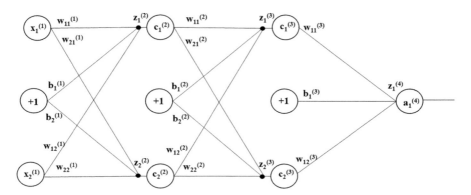

Fig. 1 Proposed ANN architecture. $x_i^{(k)}$ input feature, $w_{ij}^{(k)}$ weight, $b_i^{(k)}$ bias, $z_i^{(k)}$ sum of weighted inputs, $c_i^{(k)}$ activation value(s) of a neuron produced by the HCAF, $a_i^{(k)}$ activation value of a neuron produced by the sigmoid AF

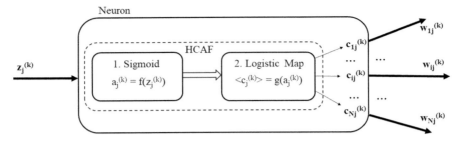

Fig. 2 Proposed chaotic neuron

$$z_i^{(2)} = \sum_{j=1}^{N} x_j^{(1)} w_{ij}^{(1)} + b_i^{(1)} \quad (1)$$

$$z_i^{(k)} = \sum_{j=1}^{N} c_{ij}^{(k)} w_{ij}^{(k)} + b_i^{(k)} \quad (2)$$

$$<c_i^{(k)}> = \left[c_{1i}^{(k)}, \ldots, c_{Ni}^{(k)}\right] = HCAF\left(z_i^{(k)}\right), \text{ where } N \text{ is the number of neurons in the following layer} \quad (3)$$

$$a_i^{(k)} = sigmoid\_AF\left(z_i^{(k)}\right) \quad (4)$$

3.1 Hybrid Chaotic Activation Function (HCAF)

The HCAF uses a combination of the concepts presented in [10, 11]. The HCAF uses a logistic map function to generate chaos, similar to [10]. However, it also incorporates the concept presented in [11], where the neuron is separated into multiple components. This model differs from [11] by having the multi-output AF at its output to send different activation values to each neuron in the following layer.

Figure 2 illustrates the architecture of the proposed chaotic neurons using the HCAF. Input received by the neuron is sent to the sigmoid function. The output from the sigmoid is passed to the logistic map function. The logistic map outputs a different activation value for each neuron in the following layer.

3.1.1 Sigmoid Function

The first function used in the HCAF is the sigmoid function. This function is defined as [9]:

$$a_i^{(k)} = f\left(z_i^{(k)}\right) = \frac{1}{\left(1 + e^{-z_i^{(k)}}\right)} \tag{5}$$

The function accepts the sum of weighted inputs from the neurons in the previous layer and maps the output to a value between 0 and 1. The output value represents the excitement level of the neuron, 0 being the lowest and 1 being the highest. This step is essential because the logistic map function only accepts inputs between 0 and 1.

3.1.2 Logistic Map Function

The second function used in the HCAF is the logistic map function. This function is typically used to represent the evolution of population over time and is given by the equation [15]:

$$x_{n+1} = r\ (x_n)\ (1 - x_n) \tag{6}$$

The x_n term is usually interpreted as the percent amount of a population and ranges between 0 and 1 inclusively [15]. When x_n is 1, the population is at its theoretical maximum, and when x_n is 0, the population is extinct. r is the growth rate of the population, defining how much the population increases over time. The term $(1 - x_n)$ represents environmental constraints which prevent the population from increasing past its theoretical maximum.

The interpretation of the logistic map equation can be modified so it can be applied as the AF in an ANN. The new definition is as follows: the population variable, x_n, now represents a neuron's excitement level; r represents an excitatory rate in a neuron; and $(1 - x_n)$ represents the theoretical constraints of a neuron. The logistic map function generates an array of activation values equal to the number of neurons in the following layer. A single neuron sends a different activation value to each neuron in the following layer. The first activation value is calculated using the output from the sigmoid function:

$$c_{1j}^{(k)} = r\ a_i^{(k)}\left(1 - a_i^{(k)}\right), \tag{7}$$

and the preceding activation values are calculated recursively using the following equation:

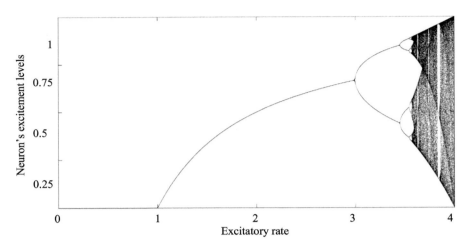

Fig. 3 Bifurcation diagram of the logistic map

$$c_{ij}^{(k)} = r \left(c_{i-1,j}^{(k)} \right) \left(1 - \left(c_{i-1,j}^{(k)} \right) \right) \tag{8}$$

The chaotic activations from the neuron can be represented as an array:

$$< c_j^{(k)} > = \left[c_{1,j}^{(k)}, c_{2,j}^{(k)}, \ldots c_{N,j}^{(k)} \right] \tag{9}$$

The activation values depend on the excitatory rate, r [3, 15]. A rate below 1 will cause the activation levels to converge towards 0. A rate between 1 and 3 causes the activation values to converge towards a fixed point. A rate between 3 and 3.5 causes the activation values to converge towards a periodic solution. Rates between 3.5 and 4 produce chaos, where the activation values will fluctuate between many different values, appearing to be random. A growth rate smaller than 0 or larger than 4 will cause the output to become unbounded, and the results will no longer be in the range between 0 and 1. Figure 3 illustrates the possible values that can occur given a specific excitation rate.

3.2 Backpropagation

The weights in the network are updated using the following equation:

$$w_{ij}^{(k)} = w_{ij}^{(k)} - \alpha \frac{\partial E}{\partial w_{ij}^{(k)}}, \tag{10}$$

where α is the learning rate and E is the mean square error cost function, given by the equation:

$$E = \frac{1}{2} \left(a_1^{(4)} - y \right)^2, \tag{11}$$

y is the output label corresponding to the given input features. The partial derivative for a chaotic neuron's first output can be computed using Eq. 12. The partial derivatives for the neuron's subsequent outputs can be computed using Eqs. 13 and 14.

$$\frac{\partial c_{1i}^{(k)}}{\partial z_i^{(k)}} = \frac{\partial \left(r \, f \left(z_i^{(k)} \right) \left(1 - f \left(z_i^{(k)} \right) \right) \right)}{\partial z_i^{(k)}} = \frac{-r \left(e^{z_i^{(k)}} - 1 \right) \left(e^{z_i^{(k)}} \right)}{\left(e^{z_i^{(k)}} + 1 \right)^3} \tag{12}$$

$$\frac{\partial c_{ij}^{(k)}}{\partial c_{i-1,j}^{(k)}} = \frac{\partial \left(r \left(c_{i-1,j}^{(k)} \right) \left(1 - \left(c_{i-1,j}^{(k)} \right) \right) \right)}{\partial c_{i-1,j}^{(k)}} = r - 2r \left(c_{i-1,j}^{(k)} \right) \tag{13}$$

$$\frac{\partial c_{ij}^{(k)}}{\partial z_j^{(k)}} = \left(\frac{\partial c_{ij}^{(k)}}{\partial c_{i-1,j}^{(k)}} \right) \left(\frac{\partial c_{i-1,j}^{(k)}}{\partial c_{i-2,j}^{(k)}} \right) \cdots \left(\frac{\partial c_{1j}^{(k)}}{\partial z_j^{(k)}} \right)$$

$$= \left(r - 2r \left(c_{i-1,j}^{(k)} \right) \right) \cdots \frac{-r \left(e^{z_j^{(k)}} - 1 \right) \left(e^{z_j^{(k)}} \right)}{\left(e^{z_j^{(k)}} + 1 \right)^3} \tag{14}$$

4   Experiments and Results

The proposed model was tested by solving the XOR problem. Each test consisted of 50 trials. In each trial, the model was run until the mean square error reached a cut-off value of 0.01. The average number of iterations and the average execution time of the 50 trials were then recorded. The excitatory rate of the HCAF was set to 4 to produce chaotic outputs; this is when the neurons are in a learning state. The weights were initialized to random numbers between -1 and 1.

Table 1 Results

AF	r	HCAF number of outputs	Number of iterations	Time (s)
Sigmoid	N/A	N/A	2937.7	0.1122
HCAF	4	1	137.7	0.0156
HCAF	4	Equal to the number of neurons in the next layer	71.3	0.0090

Table 1 provides a summary of the results. The chaotic ANN was also tested with the HCAF only outputting one activation value. An ANN with the same architecture using only the sigmoid AF took an average of 2937.7 iterations and 0.1122 s to converge. The mean convergence rate of the chaotic ANN with $r = 4$ was 71.3 iterations, and it took an average of 0.0090 s to execute. The chaotic HCAF prevented the ANN from becoming trapped in local minimums, allowing it to converge significantly faster.

These results are encouraging for future work. A potential issue that may arise in larger ANNs is the "exploding gradient" problem, where the partial derivatives become very large during backpropagation when the HCAF outputs many values. Potential solutions may include limiting the number of outputs out of the HCAF or implementing gradient clipping. Future work will also include developing an adaptive excitatory rate to control the behavior of the logistic map and the basin of attractors.

5 Conclusions

This paper presents a novel model of an artificial neuron, which incorporates a hybrid chaotic activation function (HCAF). Also, this paper presents a corresponding novel interpretation of how the proposed artificial chaotic neuron behaves and communicates with other neurons in the network. The proposed chaotic neuron behaves in one of the three main states, which implement various degrees of exploration and exploitation: (1) In the most excited states, where the input excitation rate is between 3.5 and 4, the chaotic neuron produces a multiplicity of different chaotic outputs; it sends a different excitation output to each neuron in the next layer. In this way, the neuron is interpreted as promoting exploration by sending different excitation signals to the neurons in the next layer of the ANN. (2) For lower input excitation states, where the input excitation rates are between 3.0 and 3.5, the exploratory signal is lessened, and exploitation is introduced, informing the next layer of neuron to traverse a periodic path in the solution space, which represents a combination of exploration and exploitation. (3) In the lowest input excitation state, where the excitation rates are between 1.0 and 3.0, the chaotic neuron sends the same excitation signal to all neurons in the next layer and, thus, sends an exploitation signal, informing all neurons in the next layer to continue on their current paths in solution space.

With this novel model, HCAF, and corresponding novel interpretation of a chaotic neuron, this paper has shown that the hybrid chaotic ANN converged on average 12.5 times faster compared to an ANN using the sigmoid function alone.

5.1 Future Work

Future work includes applying and testing the model on larger problems; further investigating the idea of a multi-output AF; determining how the multiple activation values should be distributed to the following layer; and experimenting with adaptive rate parameters to control the excitement levels of a neuron.

Acknowledgments This research has been financially supported by Mitacs Accelerate (IT15018) in partnership with Canadian Tire Corporation and is supported by the University of Manitoba.

References

1. G. Rodriguez-Bermudez, P.J. Garcia-Laencina, Analysis of EEG signals using nonlinear dynamics and chaos: A review. Appl. Math. Inf. Sci. **9**(5), 2309–2321 (2015)
2. W. Kinsner, *Fractal and Chaos Engineering: Monoscale and Polyscale Analyses* (OCO Research, Inc, Winnipeg, 2020)
3. S. Elyadi, *Discrete Chaos* (Chapman & Hall/CRC, New York, 1999), p. 137
4. R.L. Devaney, *An Introduction to Chaotic Dynamical Systems* (Addison-Wesley, Massachusetts, 1989)
5. M. Small, H.P.C. Robinson, I.C. Kleppe, C.K. Tse, Uncovering bifurcation patterns in cortical synapses. J. Math. Biol. **61**, 501–526 (2010)
6. A. Celletti, A.E.P. Villa, Determination of chaotic attractors in the rat brain. J. Stat. Phys. **84**, 1379–1385 (1996)
7. H. Korn, P. Faure, Is there chaos in the brain? II. Experimental evidence. C. R. Biol. **236**(9), 787–840 (2003)
8. A.M. Zador, A critique of pure learning and what artificial neural networks can learn from animal brains. Nat. Commun. **10**(3770), 1–7 (2019)
9. Y. Wang, Y. Li, Y. Song, X. Rong, The influence of the activation function in a convolution neural network model of facial expression recognition. J. Appl. Sci. **10**(5), 1897 (2020)
10. A.N.M.E. Kabir, A.F.M.N. Uddin, M. Asaduzzaman, M.F. Hasan, M.I. Hasan, M. Shahjahan, Fusion of chaotic activation functions in training neural network, in *7th International Conference on Electrical and Computer Engineering, Dhaka, Bangladesh*, (2012)
11. W. Youshou, Z. Mingsheng, A neuron model with trainable activation function (TAF) and its MFNN supervised learning. Sci. China **44**(5), 366–375 (2001)
12. M. Daneshyari, Chaotic neural network controlled by particle swarm with decaying chaotic inertia weight for pattern recognition. Neural Comput. Applic. **19**, 637–645 (2009)
13. D. Cook, K. Ferens, W. Kinsner, Application of chaotic simulated annealing in the optimization of task allocation in a multiprocessing system. Int. J. Cognit. Inf. Nat. Intell. **7**(3), 58–79 (2015)
14. C. Skarda, W. Freeman, How brains make chaos in order to make sense of the world. Behav. Brain Sci. **10**, 161–195 (1987)
15. G.L. Baker, J.P. Gollub, *Chaotic Dynamics: An Introduction* (Cambridge University Press, New York, NY, 1996)

Defending Aviation Cyber-Physical Systems from DDOS Attack Using NARX Model

Abdulaziz A. Alsulami and Saleh Zein-Sabatto

1 Introduction

Cyber-physical system (CPS) is a system that enables interaction between physical components to exchange information through a communication network system. CPS allows stand-alone physical devices to communicate with each other and communicate with the physical world [1]. Many industries have been using CPS in different fields because the benefits of CPS are recognized. CPS improves important factors in industries such as performance and quality [2]. Recently, aviation industries have been concentrating on integrating cyber systems into their aircraft. The reason is that CPS improves the efficiency and the safety of the aircraft. Also, using CPS allows pilots to avoid a collision early, especially when two or more aircraft get closer to each other [3]. Even though CPS brings powerful features, it can also be exposed to various security vulnerabilities, which can introduce serious threats to the physical components of aircraft [4]. Unauthorized access to the network system in the aviation cyber-physical systems (ACPS) could result in mechanical failures since attackers can provide false commands to the physical components of an aircraft. Also, attackers can exhaust the network system with so many requests to fail it. Distributed Denial-of-Service (DDOS) attacks are very well-known attacks in cybersecurity that aims to overwhelm the victims (servers) with so many requests to make them unable to use network resources [5]. Servers are flooded by so many requests that are needed to be served at one time. Consequently, when servers are overloaded with a high number of requests, they cannot handle incoming requests from clients. As a result, any incoming useful

---

A. A. Alsulami (✉) · S. Zein-Sabatto
Department of Electrical and Computer Engineering, Tennessee State University, Nashville, TN, USA
e-mail: aalsula3@my.tnstate.edu; aalsula3@tnstate.edu; mzein@tnstate.edu

© Springer Nature Switzerland AG 2021
H. R. Arabnia et al. (eds.), *Advances in Artificial Intelligence and Applied Cognitive Computing*, Transactions on Computational Science and Computational Intelligence, https://doi.org/10.1007/978-3-030-70296-0_88

information packets will be dropped out due to the capacity of the buffer becoming fully occupied. A DDOS attack is considered as the main threat to an essential fundamental security aspect, which is the availability of the network. The reason is that the DDOS attack's main task is to increase the network traffic to disturb servers [5].

This paper proposes an approach to defending the aviation cyber-physical systems (ACPS) from DDOS attack using a nonlinear autoregressive exogenous (NARX) model to forecast packets that were dropped when the network system failed due to a DDOS attack. In other words, the input signal, which is a command signal to a flight surface of an aircraft, e.g., elevator, will be predicted whenever a network system reports a high number of utilizations compared with the normal utilization number of the same network.

The design of the network system and the simulation of the DDOS attack was reported in [5, 6] using the SimEvents library, which is a Simulink library. However, defending the network system from a DDOS attack was not addressed in both references. SimEvents can simulate network system using discrete-time event methodology. Not only this, but also it can measure the performance of the network system such as server utilization, departure packets, and blocked packets [7]. If the utilization number of any network system exceeds the normal utilization number, then that means the network system is under DDOS attack.

The aircraft model used in this research is based on the Transport Class Model (TCM) developed by NASA. The TCM is a full-scale model that simulates the dynamics of commercial aircraft using the state-space representation model [8].

The rest of the paper organized as follows: Section 2 represents the implementation of ACPS and the NARX model. Section 3 discusses NARX training and validation. The simulation of the conducted experiment and discussion about the results are presented in Sect. 4. Section 5 includes the conclusion of the research work.

2 System Implementation

This section explains the implementation of the systems that were used in this research, i.e., the aviation cyber-physical system (ACPS), network system, Distributed Denial-of-Service (DDOS) attack, and nonlinear autoregressive exogenous (NARX) model.

2.1 ACPS

The main components of the ACPS are illustrated in Fig. 1. The main physical system used in this research is the Transport Class Model (TCM), which is a commercial aircraft model developed by NASA for the research community. The

Defending Aviation Cyber-Physical Systems from DDOS Attack Using NARX Model 1109

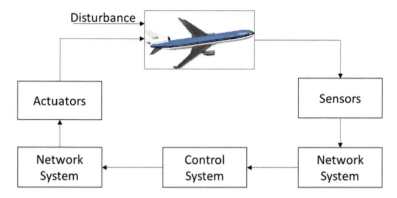

Fig. 1 The overall structure of ACPS

TCM considered a full-scale simulation that simulates the dynamics of commercial aircraft using state-space representation. The original TCM model did not include any communication network involved between its different components. To transform the aircraft model into a CPS, communication networks were added to its components. Hence, sensors detect responses of the aircraft model and then transmit it to the control system through a communication network system. The control system receives the sensor signals as inputs and processes it to generate the desired command outputs, which are the controller signals. Finally, the controller signals are also transmitted by a network system to the aircraft actuators to generate the desirable performance of the aircraft TCM model.

2.2 Communication Network System

The communication network system was developed using SimEvents, which is a MATLAB Simulink library that can be used to model a communication system based on discrete-time events. SimEvents provides some criteria that are used to evaluate the performance of the network characteristics such as packet loss, throughput, latency, and utilization. There are two network system models designed in this research. The first model simulates the normal network system, and the second model simulates a network system with a DDOS attack.

2.2.1 Normal Network System

The design of the normal network system is shown in Fig. 10. It refers to a simulation of an unattacked network system. This model is used to calculate the server's utilization when the network system is not under a DDOS attack. There are two clients which transmit the elevator and the throttle commands of the aircraft sent by the pilot. These two input signals are converted from the time domain to the discrete-event domain because of the SimEvents library. The Time-Based Entity Generator is used to generate packets based on intergeneration times. Intergeneration times refer to the time between two generated packets [7]. The distribution of the packet is constant, and the period is equal to 2. The queue's capacity is equal to 25, and the average utilization number of the two servers under normal conditions is 50%.

2.2.2 Network System with a DDOS Attack

The designed communication network system with a DDOS attack is shown in Fig. 11. The model consists of an attacker, two zombies, and an entity to represent security failure/repair of the server. Basically, servers are forced to only listen to the attacker; therefore, the attacker floods the servers with so many requests. Furthermore, the two legitimated clients turned to zombies because they are assumed to be compromised by the attacker to collaborate on overwhelming the two servers. The attacking event is generated randomly using a uniform distribution, and the time-based entity generator is based on constant distribution. The failure/repair entity is used to simulate the server failure and repair from the DDOS attack. The event-based for the failure/ repair entity is randomly generated using a uniform distribution, and the time-based entity generator used is based on a constant distribution. The average utilization number of the two servers under the DDOS attack is 95%, which is much higher than the average utilization number in the normal case, which was set to 50%.

2.3 NARX

NARX is a dynamic neural network that can be used with a nonlinear system to predict the next value of the input signal. NARX model can be defined using Eq. (1):

$$y(t) = f\left(y(t-1), y(t-2), \ldots, y(t-n_y), x(t-1), x(t-2), \ldots, x(t-n_x)\right) \tag{1}$$

where $y(t)$ is the predicted output, $f(.)$ is the nonlinear mapping function, and $x(t)$ is the input. n_y is the number of delay for the output, and n_x is the number of delays for the input.

There are two types of NARX architecture: (1) series-parallel architecture and (2) parallel architecture. The two architectures are shown in Figs. 2 and 3, respectively.

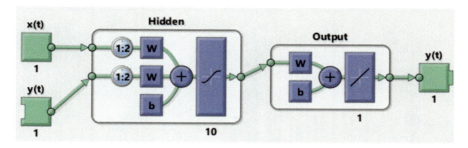

Fig. 2 Series-parallel architecture

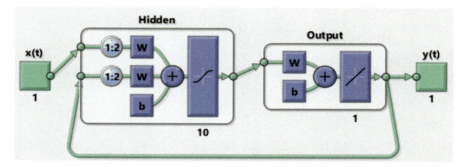

Fig. 3 Parallel architecture

This research uses the series-parallel architecture, which is also called the open-loop neural network because it has two advantages. (i) It uses the true output for the feedforward network, which is more accurate than the feedback value. (ii) It has a pure feedforward neural network design; therefore, the backpropagation training technique can be used.

3 NARX Training and Validation

The training of the data and the training validation are discussed in this section.

3.1 NARX Training

The NARX network consists of an input layer, an output layer, and a hidden layer. The input layer consists of two nodes: the first node receives the input, and the second node receives the true output [9]. The hidden layer consists of ten nodes,

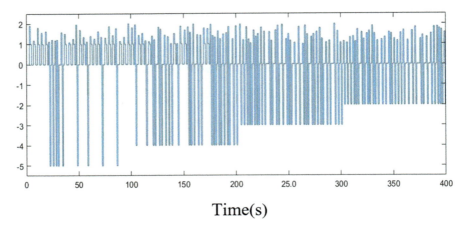

Fig. 4 Elevator signal with DDOS attack

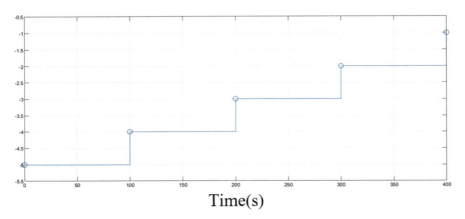

Fig. 5 Normal elevator signal

and the number of delays is 2. The output layer consists of one node. The algorithm used for training the data is Levenberg-Marquardt. The NARX network was trained with one input signal data, which is the elevator, and the size of the data is 40,000. Figure 4 shows the elevator signal when the network system was under a DDOS attack. The normal elevator input signal is shown in Fig. 5. The difference between the two elevator signals can be observed between the two figures, so in Fig. 5, some packets were dropped when the network was under attack.

The output of the NARX model is shown in Fig. 6, which is identical to the normal elevator control signal.

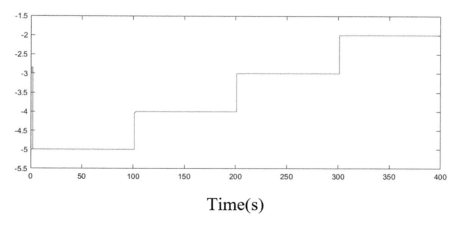

Fig. 6 NARX output

Fig. 7 NARX training performance

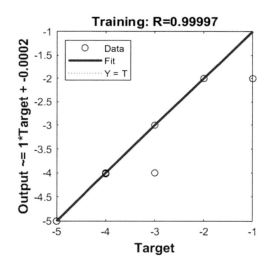

3.2 Training Validation and Testing

The training of the NARX was validated and tested, so 70% of the target data was used for training, 15% for validation, and 15% for testing. Figures 7, 8, and 9 show regression plots of the network output with the target for the training, validation, and testing, respectively. The value of R is very close to 1 with training, validation, and testing, which means that the NARX output and the target data have a linear relationship. Therefore, it can be said that the training, validation, and testing data shows a perfect fit.

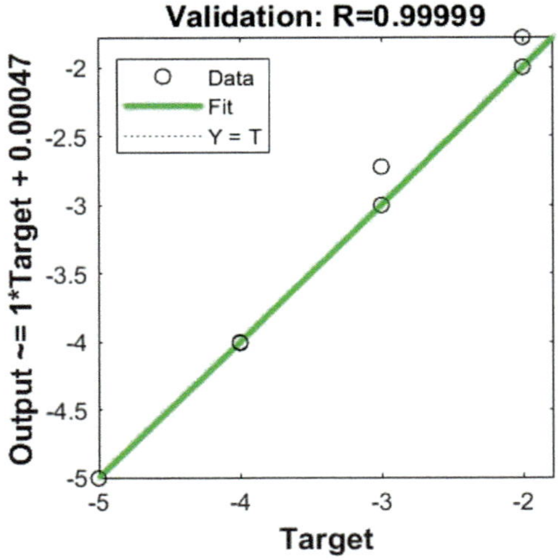

Fig. 8 NARX validation performance

Fig. 9 NARX test performance

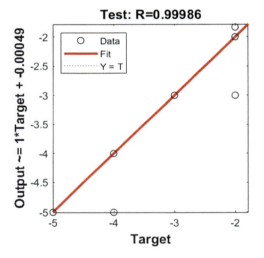

4 Simulation and Result

The simulation of the experiment and results are discussed in this section.

4.1 *Simulation*

The experiment of this research was performed using the Simulink/MATLAB environment. The aircraft model used is the TCM to simulate the dynamics of a

Defending Aviation Cyber-Physical Systems from DDOS Attack Using NARX Model

Table 1 Trim conditions of the aircraft

Flight control	Pilot command	Initial values
Elevator	Staircase signals	−3.0976 deg
Throttle	0%	29.8%

Table 2 Trim conditions of the aircraft

State variables	Trim condition
u	485.69 ft/sec
w	38.72 ft/sec
q	0 rad/sec
$\ominus$	0.079 rad/sec

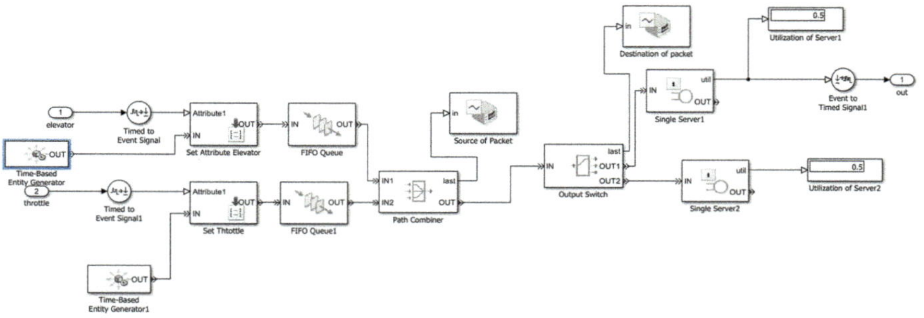

Fig. 10 Communication network system without DDOS attack

commercial aircraft. The performance of the aircraft was tested when the aircraft was at the cruise phase. Therefore, only two flight control signals were used: the elevator and the throttle commands. Table 1 represents the values of the pilot command and the initial values of the flight controls. We only tested the longitudinal model of the TCM. The longitudinal model consists of five state variables: (u) which is the velocity in x-direction, (w) which is the velocity in the z-direction, body pitch rate (q), and Euler angle for pitch ($\ominus$). The trim conditions for the state variables are shown in Table 2. We also tested the performance of the angle of attack (AOA) of the aircraft.

Two communication network systems were integrated into the TCM model. The first network model represents the normal network model without a DDOS attack, and the purpose of this model is to calculate the utilization number of the servers under normal conditions. The second network model represents the network model under a DDOS attack, and the utilization number was also calculated. Finally, the two numbers are compared to detect the DDOS attack, so if the two numbers of the same server are not equal, that means the network is under attack, and the threat level is high. As a result, the system blocks the attacked network system and receives signals from the NARX model (Figs. 10 and 11).

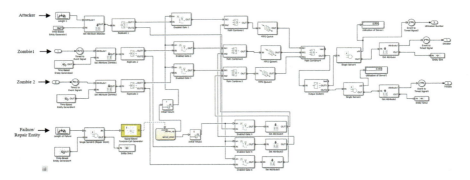

Fig. 11 Communication network system with DDOS attack

4.2 Test Result

The response of the aircraft velocity, which is the state variable (u) under the DDOS attack, is shown in Fig. 12. The blue line represents the TCM performance, and the orange line represents the reference model performance representing normal aircraft performance. We can observe that the TCM does not follow the reference (normal) model response. Also, the TCM model response is not stable. However, Fig. 13 shows the response of both models under a DDOS attack, but with using the NARX neural network to predict the control signal values when packets were dropped due to communication network attack. It can be observed that the system was successfully defending from the DDOS attack. Figures 14 and 15 show the response of the aircraft speed on the z-direction, the state variable (w), which confirms that the NARX network accurately eliminated the effect of the DDOS attack. Also, Figs. 16, 17, 18, 19, 20, and 21 show that the NARX network provided the desired results for q, ⊖, and AOA of the aircraft.

5 Conclusions

In conclusion, this paper proposed, developed, and implemented an approach to defending aviation cyber-physical system (ACPS) from DDOS network attack. The simulation results found that the NARX model successfully forecasted the dropped packets caused by the DDOS attack. Therefore, the NARX model can be used with real-time ACPS to eliminate the effect of a DDOS attack on aircraft performance, hence improving the safety of aircraft flights.

Defending Aviation Cyber-Physical Systems from DDOS Attack Using NARX Model 1117

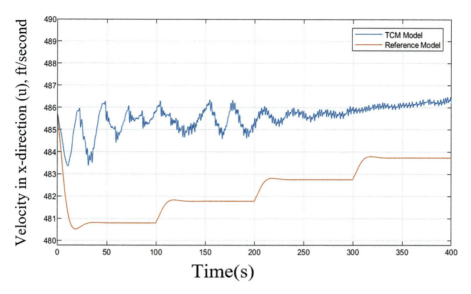

Fig. 12 Aircraft speed performance under DDOS attack

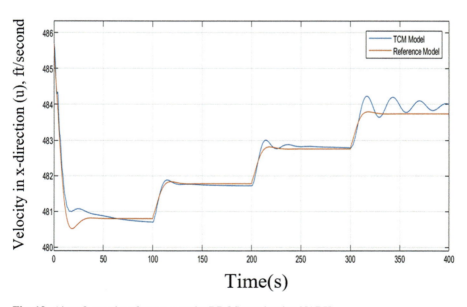

Fig. 13 Aircraft speed performance under DDOS attack using NARX

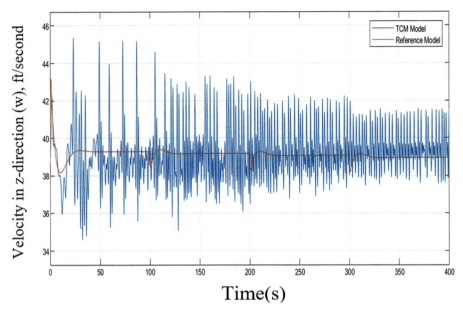

Fig. 14 State-w performance under DDOS attack

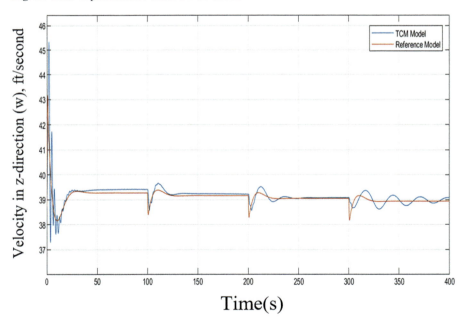

Fig. 15 State-w performance under DDOS attack using NARX

Defending Aviation Cyber-Physical Systems from DDOS Attack Using NARX Model 1119

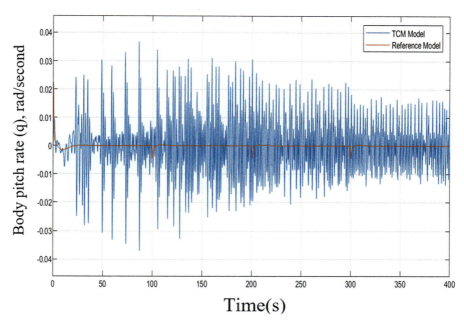

Fig. 16 State-q performance under DDOS attack

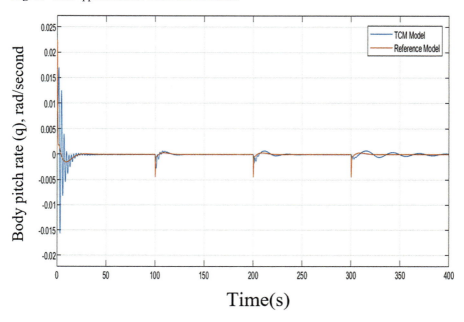

Fig. 17 State-q performance under DDOS attack using NARX

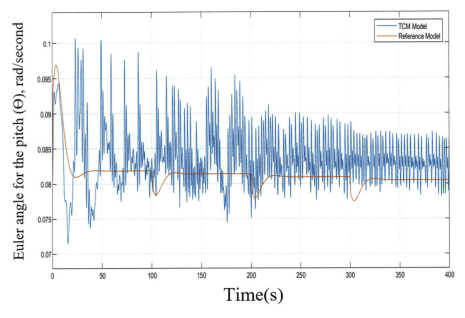

Fig. 18 State-Θ performance under DDOS attack

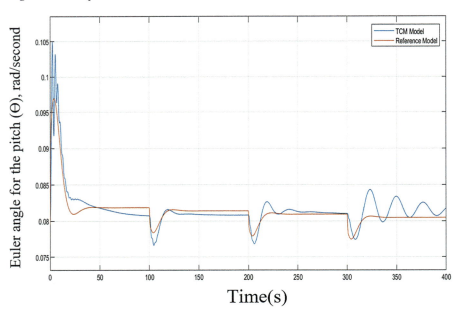

Fig. 19 State-Θ performance under DDOS attack using NARX

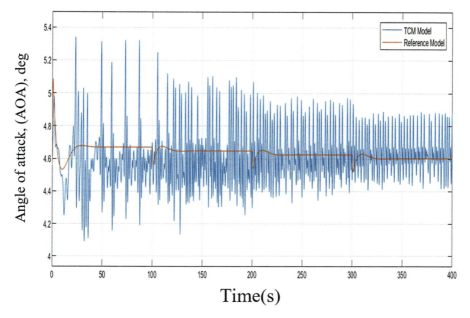

Fig. 20 Aircraft-AOA performance under DDOS attack

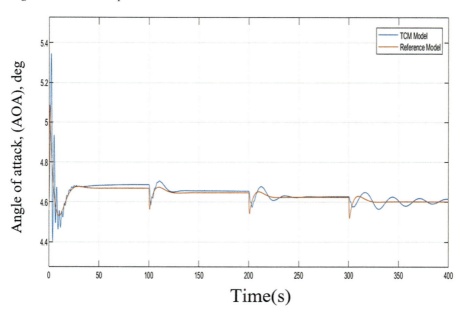

Fig. 21 Aircraft-AOA performance under DDOS attack using NARX

References

1. R. Alur, *Principles of Cyber-Physical Systems* (MIT Press, London, 2015)
2. M. Bhrugubanda, A review on applications of cyber-physical systems. Int. J. Innovative Sci. Eng. Technol. **2**(6), 728–730 (2015)
3. A. Platzer, *Logical Foundations of Cyber-Physical Systems* (Springer, Cham, 2018)
4. A.Y. Nur, M.E. Tozal, Defending cyber-physical systems against DoS attacks, in *2016 IEEE International Conference on Smart Computing, Louis, MO, USA*, (2016)
5. A. Bala, Y. Osais, Modelling and simulation of DDOS attack using SimEvents. Int. J. Sci. Res. Network Secur. Commun. **1**(2), 5–14 (2013)
6. G. Khazan, M.A. Azgomi, A distributed attack simulation for quantitative security evaluation using SimEvents, in *2009 IEEE/ACS International Conference on Computer Systems and Applications, Rabat, Morocco*, (2009)
7. Mathworks.com. 2020 [online] Available at: https://www.mathworks.com/help/pdf_doc/simevents/simevents_gs.pdf. Accessed 10 May 2020.
8. R.M. Hueschen, Development of the transport class model (TCM) aircraft simulation from a sub-scale generic transport model (GTM) simulation, in *NASA, Hampton, Virginia*, (2011)
9. Mathworks.com. 2020 [online] Available at: https://www.mathworks.com/help/pdf_doc/deeplearning/nnet_gs.pdf. Accessed 10 May 2020

Simulated Annealing Embedded Within Personal Velocity Update of Particle Swarm Optimization

Ainslee Heim and Ken Ferens

1 Introduction

In certain cases, the inspiration for an algorithm comes from behaviors that occur in the natural world. In the case of particle swarm optimization, the influence of a particle's movement originated from observing the way a flock of birds or school of fish travel as a group successfully. For simulated annealing, the process is off of the cooling temperature of a liquid to a solid to form an optimally strong structure. This paper will combine these two methods to create a hybrid algorithm. Basic particle swarm optimization (PSO) uses multiple particles to search through a solution space by moving based on their own inertial force, an influence from the personal best position they have ever occupied, and the global best position of all the other particles in the swarm. A disadvantage of basic PSO is that it can become trapped in local optimal solution prematurely. This paper conjectures that the addition of simulated annealing (SA) to guide the movement of a particle's own inertial force may be used to escape this premature local minimum by offering the particle a chance to move around the solution space more freely early on and slow this movement toward the end of the algorithm.

A. Heim · K. Ferens (✉)
Department of Electrical and Computer Engineering, University of Manitoba, Winnipeg, MB, Canada
e-mail: Ken.Ferens@umanitoba.ca

© Springer Nature Switzerland AG 2021
H. R. Arabnia et al. (eds.), *Advances in Artificial Intelligence and Applied Cognitive Computing*, Transactions on Computational Science and Computational Intelligence, https://doi.org/10.1007/978-3-030-70296-0_89

2 Related Work

PSO was created by social psychologist James Kennedy and electrical and computer engineer Russell C. Eberhart in 1995. PSO simulates the natural swarming of birds as they look for food [1]. The algorithm follows three rules for the particles: separation to avoid collision, alignment to match the velocity of a neighbor, and cohesion to stay near neighboring particles.

The simulated annealing (SA) algorithm is a global optimization method in that it has a built method for escaping local minima. Simulated annealing mimics the natural annealing process of thermal-dynamical systems, such as the annealing of metals. In the annealing of metals, the metal is first raised to a very high temperature, where the solid becomes liquid, and the atoms are perfectly free to move throughout the substance. At the initial high temperature state, the atoms explore the solution space attempting to find positions, which collectively represent the lowest energy of the substance at that temperature, which is called thermal equilibrium. This initial movement might cause the atoms to move into temporary positions that have higher energy, with the rational being that lower energy states will eventually be found. Thus, the atoms are able to escape local minima and try to find other better minima. Once thermal equilibrium has been attained, the temperature is lowered by a very small amount. The lowering of the temperature causes the atoms to find new positions to once again find the lowest energy state at the new temperature and once again reach thermal equilibrium. The degree to which atoms are able to move around the substance is given by Boltzmann's probability function, which is indirectly proportional to the change in energy state and directly proportional to the temperature. The lowering of temperature accompanied by thermal equilibrium repeats, for many iterations. As the temperature is lowered, so is the range of motion of the atoms within the substance, and, therefore, the atoms will be concentrating on the current local minima. In this way, the atoms will find local minima and attempt to dig deeper (exploit) the local minima to find the optimal position at the current temperature. This procedure repeats until room temperature has been attained, and the substance will have found the optimal energy state and become a very rigid solid. The initial algorithm that was developed in 1970 by M. Pincus which later became known as simulated annealing was first used to find the global minimum of functions [2].

This paper embeds SA into PSO, with the aim to overcome PSO being trapped into local minima by allowing a probability of a worse solution being accepted in the hopes it leads to a better solution later on, such as the case where PSO can fall prematurely into a local minimum as described in [5]. To do this, several experiments were designed. The basic PSO which is implemented in the first experiment of this project and adapted for the second experiment consists of three steps: (1) Evaluate the cost of each particle. (2) Update the individual and global best position and cost. (3) Update the position of each particle according to a combination of influences. The second experiment uses swap velocity operators for discrete PSO to apply to the Travelling Salesperson Problem. The technique

Simulated Annealing Embedded Within Personal Velocity Update of Particle... 1125

follows the algorithm outlined in Ch. 4-Part 3 of Traveling Salesman Problem [3] with a slight adaptation which will be outlined in section (3). PSO has a habit of becoming trapped in a premature local minimum; SA will be embedded into the velocity update of PSO to try and prevent this from happening. SA is an algorithm based on the cooling process of solids, the idea being that as the temperature cools, the particles rearrange themselves in optimal positions for that level of temperature. At the beginning, the temperature is high, allowing lots of movement; the temperature is lowered periodically with the particles settling into a stable state for each temperature. Near the end of the process, the temperature is much lower allowing for less movement in the particles. The particle moves based on its probability of motion which is given by Boltzmann's probability function in Eq. (1) [4].

Basic PSO

Initialize swarm best position $x^{sb} \Leftarrow$ random(lob,upb)
for each particle i = 1,..., I do % I = the number of particles
    Initialize the particle's position $x^i \Leftarrow$ random(lob, upb)
    Initialize the particle's best-known position to its initial position $x^{pbi} \Leftarrow x^i$
    if $f(x^i) > f(x^{sb})$ then
        update the swarm's best known position: $x^{sb} \Leftarrow x^i$
    Initialize particle's velocity: $v^i \Leftarrow$ random(-|upd − lob|, |upb − lob|).
for each iteration k=1..K do:
    for each particle i=1,... I do
        for each dimension d = 1,... D do
            Pick random numbers: $r_p, r_g \Leftarrow$ random(0,1)
            Update the particle's velocity:
            $v^i_{k+1} = w v_k^i + c_1 r_p (x_k^{pbi} − x_k^i) + c_2 r_g (x_k^{sb} − x_k^i)$ (2)
            update the particle's position: $x^i_{k+1} = x_k^i + v^i_{k+1}$

        if $f(x^i_{k+1}) > f(x^{pbi})$ then
            update the particle's best known position: $x^{pbi} \Leftarrow x^i_{k+1}$
            if $f(x^{pbi}) > f(x^{sb})$ then
                update the swarm's best known position: $x^{sb} \Leftarrow x^{pbi}$

---

3 Description of the Experiments

3.1 Objective

The overall goal of this paper is to improve PSO with the addition of SA; this was tested in two different ways: The first experiment is focused on basic PSO where the solution space is continuous and performance is evaluated on two benchmark functions, the Griewank function and the Rastrigin function [6]. The second experiment applies a similar addition of SA to PSO for TSP where the solution is discrete and the velocity operators become swap operators between a combination of influences. The performance of the second experiment is evaluated by lowest cost (best route) found by the algorithm in a comparable amount of iterations.

3.2 Experiment 1: Basic PSO

Basic PSO will be evaluated on performance and compared to the embedding of SA within personal velocity update of PSO, using the Griewank and the Rastrigin benchmark functions. Both functions have a global minimum at [0 0] with a cost of 0 in 2D space. The Griewank function will be evaluated on the hypercube $x_i = [-600, 600]$. The Rastrigin function will be evaluated on the hypercube $x_i = [-5, 5]$. The two algorithms will be run for a comparable amount of iterations and their solutions evaluated with the expectation that the PSO with embedded SA will perform better with its capacity to avoid a premature local minimum.

The decision was made to embed SA into the velocity update equation by modifying the term that represents the particle's inertial movement. The standard velocity update equation for basic PSO has three terms: the first relates to the inertial force of the particle, the second is a velocity influenced by the personal best position of the particle, and the third is a velocity influenced by the current global best position of all particles in the swarm and is shown in Eq. (2). For basic PSO, the pseudo-code can be found in Fig. 1.

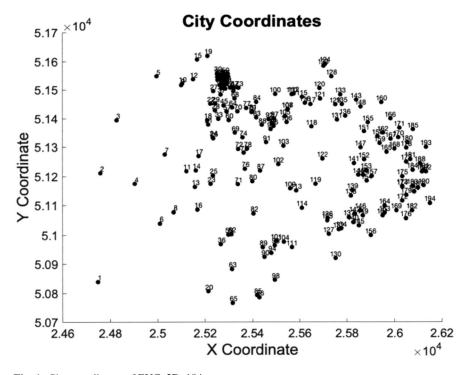

Fig. 1 City coordinates of EUC_2D_194

In the SA embedded within the personal velocity update of PSO version, the inertial velocity term is replaced with a scaled randomly generated velocity term to simulate random movement of particles in SA. This change allows the particles to move around the solution space with higher energy at the start and lower energy at the end of the iterations while still accepting influences from the personal best and global best solutions. The possibility of moving a particle based on the randomly generated velocity is accepted based on Boltzmann's probability function; the probability of moving to a position which offers a worse solution is higher at the beginning in the hopes it will lead to a better solution in the future iterations. As the energy is lowered, the particle is not likely to accept worse solutions and becomes more influenced by the personal and global best positions. The pseudo-code for SA embedded PSO is shown to the right.

SA Embedded Within Personal Velocity Update of PSO
Initialize swarm best position $x^{sb} \Leftarrow$ random(lob,upb)
for each particle i = 1,..., I do % I = the number of particles
    Initialize the particle's position $x^i \Leftarrow$ random(lob, upb)
    Initialize the particle's best-known position to its initial position $x^{pbi} \Leftarrow \alpha^i$
    if $f(x^i) > f(x^{sb})$ then
        update the swarm's best known position: $x^{sb} \Leftarrow x^i$

for each iteration i=1..coolingLoops do:
    for j =1,...equilibruimLoops do:
        for each particle i=1,... I do
            for each dimension d = 1,... D do
                generate random velocity $v^i \Leftarrow$ random(-|upd – lob|, |upb – lob|).
                Move the particle: pos_temp = $x_k^i + v^i$
                Calculate the cost: cost_temp = cost(pos_temp)
                If (cost_temp < cost_current):
                    x_k^i = pos_temp
                    cost_current = cost_temp
                If (cost_temp > cost_current)
                    If (p > random(0,1)) % p = probability
                        x_k^i = pos_temp
                  cost_current = cost_temp
                Pick random numbers: r_p, $r_g \Leftarrow$ random(0,1)
                Update the particle's velocity:
                $v^i_{k+1} = c_1 r_p (x_k^{pbi} - x_k^i) + c_2 r_g (x_k^{sb} - x_k^i)$
                update the particle's position: $x^i_{k+1} = x_k^i + v^i_{k+1}$
                reduce the temperature for the next cooling loop:
            tCurrent = $frac$ * $tCurrent$ %fractional reduction per cycle
        if $f(x^i_{k+1}) > f(x^{pbi})$ then
            update the particle's best known position: $x^{pbi} \Leftarrow x^i_{k+1}$
            if $f(x^{pbi}) > f(x^{sb})$ then
                update the swarm's best known position: $x^{sb} \Leftarrow x^{pbi}$

3.3 Experiment 2: Discrete PSO

In experiment 2, SA embedded within personal velocity update of PSO will be tested to see how it can improve PSO for the discrete TSP. Using PSO to solve the TSP involves modifying the velocity update equation; the new equation creates swaps between city routes based on the particle's own inertial movement, the order of the cities in the best route the particle has ever found, and the order of the global best route that has been found. The algorithm was based on the PSO for discrete optimization problems' pseudo-code that can be found in Ch. 4-Part 3 of Traveling Salesman Problem [3] with a slight modification to allow for the possibility of three city swaps occurring instead of only a single swap/movement for each particle for each iteration as well as a linear adjustment of probabilities based on iterations. The probability of making a swap from a randomly generated swap (used to represent the particle's individual motion) is high at the beginning and reduced each iteration. The probability of making a swap to match the personal best route and the global best route is low at the beginning and increased with each iteration. The data set used will contain 194 different cities, each with 2D coordinates called EUC_2D_194. The pseudo-codes for both versions are shown below.

Discrete PSO
Initialize swarm best route $x^{sb} \Leftarrow$ randomly perturbed city route
Initialize starting probabilities p2 and p3
for each particle i = 1,..., I do % I = the number of particles
    Initialize the particle's route $x^i \Leftarrow$ randomly perturbed city route
    Initialize the particle's best-known route to its initial route: $x^{pbi} \Leftarrow x^i$
    if $f(x^i) > f(x^{sb})$ then
       update the swarm's best known route: $x^{sb} \Leftarrow x^i$
for each iteration k=1..K do:
   for each particle i=1,... I do
     %swap one
     if ($p1 > random(0,1$)
       generate a random swap in the particle's current city route: $x^i([A\ B]) = x^i\ ([B\ A])$
     %swap two
     if $(p2 > random(0,1))$
       choose a city in the particle's best route
       adjust the particle's current route so the chosen city is now in the same position in both x^i and x^{pbi}
     %swap three
     if $(p3 > random(0,1))$
       choose a city in the swarm's best route
       adjust the particle's current route so the chosen city is now in the same position in both x^i and x^{sb}

     if $f(x^i_{k+1}) > f(x^{pbi})$ then
     update the particle's best known position: $x^{pbi} \Leftarrow x^i_{k+1}$
       if $f(x^{pbi}) > f(x^{sb})$ then
         update the swarm's best known position: $x^{sb} \Leftarrow x^{pbi}$
   update probabilities: p1 = p1 - reduction
     p2 = p2 + increase
     p3 = p3 + increase

Simulated Annealing Embedded Within Personal Velocity Update of Particle...

Discrete PSO SA Embedded Within Personal Velocity Update
Initialize swarm best route $x^{sb} \Leftarrow$ randomly perturbed city route
Initialize starting temperature
Initialize starting probabilities p2 and p3
for each particle i = 1,..., I do % I = the number of particles
      Initialize the particle's route $x^i \Leftarrow$ randomly perturbed city route
      Initialize the particle's best-known route to its initial route: $x^{pbi} \Leftarrow x^i$
      if $f(x^i) > f(x^{sb})$ then
          update the swarm's best known route: $x^{sb} \Leftarrow x^i$
for each coolingLoop i=1..I do:
    for each equilibriumLoop j=1,...J do:
        for each particle d=1,... D do
            %swap one
            temp1 = x^i
            generate a random swap in the particle's current city route: *temp1([A B]) = temp1([B A])*
            if (*f(temp1) < f(x^i)*)   %cost is lower
               update particle's current route: $x^i \Leftarrow$ temp1
            if (*f(temp1) > f(x^i)*)   %cost is higher
               if (*p1 > random(0,1)*)   %accept the new route based on probability p1
                  update particle's current route: $x^i \Leftarrow$ temp1
            %swap two
             if *(p2 > random(0,1))*
               choose a city in the particle's best route
               adjust the particle's current route so the chosen city is now in the same position in both x^i and x^{pbi}
            %swap three
             if *(p3 > random(0,1))*
               choose a city in the swarm's best route
               adjust the particle's current route so the chosen city is now in the same position in both x^i and x^{sb}
    if $f(x^i_{k+1}) > f(x^{pbi})$ then
        update the particle's best known position: $x^{pbi} \Leftarrow x^i_{k+1}$
        if $f(x^{pbi}) > f(x^{sb})$ then
            update the swarm's best known position: $x^{sb} \Leftarrow x^{pbi}$
    reduce the temperature for the next cooling loop: tCurrent = *frac* * *tCurrent* %fractional reduction per cycle
    reduce probabilities: p2 = p2 + increase
               p3 = p3 + increase

SA was embedded into the same part of the velocity update equation as it was in experiment 1. The particle will generate a random swap between two of the cities in its current route. The particle will keep that swap if it offers a better solution. If the swap generates a worse solution, the particle will keep the swap based on Boltzmann's probability function. The probability of accepting a worse swap is higher at the beginning and lower at the end in the hopes accepting a worse solution will lead to a better solution in future iterations. This allows the particles to explore the solution space early on and settle into better solutions as the temperature cools. The personal best and the global best routes and costs are updated in the cooling loop, allowing the particles to reach a state of equilibrium for each temperature setting. The probabilities of accepting a swap from either the personal best route or the global best route begin at the same low probability as the discrete PSO and are increased with each iteration.

4 Experiments and Results

Basic PSO was tested with and without the addition of embedded SA in the personal velocity update equation. Two benchmark functions were used to evaluate performance. Both versions of PSO were run for 100 iterations for basic PSO and 100 cooling loops for basic PSO with SA for a total of 20 times for each benchmark function. The results are shown below (Tables 1 and 2).

Basic PSO – Griewank
Best solution cost: 0
   Best solution position: 1.0e−08*
   [-0.645034136315891 0.290458002714703]
   Average solution cost: 0.024054049060007
   Elapsed time for 20 trials: 0.348972 seconds

Basic PSO with SA – Griewank
Best solution cost: 0.009872034404375
   Best solution position:
   [-6.278219327564698 -0.004770265559432]
   Average solution cost: 0.010807125697256
   Elapsed time for 20 trials: 3.324738 seconds

Table 1 Griewank benchmark function solutions

Trial	Basic PSO	Basic PSO with SA
1	0.007396040334115	0.009872034404375
2	0.059178177055109	0.009944880361684
3	0	0.007797560564081
4	0.029584161212070	0.003068793177938
5	0.009864672061006	0.014981283774201
6	0.000301169875875	0.000030641286944
7	0.019991824059437	0.009864719033155
8	0.026453171484132	0.000023757872225
9	0.046835018587874	0.007602361354831
10	0.019719489248185	0.007481923705905
11	0.007396041798466	0.008103287270781
12	0.007396040334115	0.007398846043541
13	0.066584072420795	0.012153116491699
14	0.039458688195417	0.046921250644809
15	0	0.028153962459320
16	0.007396040334115	0.007396433547362
17	0.000000016379735	0.000686523559684
18	0.007396040334115	0.007422058885127
19	0.106358541667201	0.007398416827175
20	0.019771775818382	0.019840662680286

Simulated Annealing Embedded Within Personal Velocity Update of Particle... 1131

Table 2 Rastrigin benchmark function solutions

Trial	Basic PSO	Basic PSO with SA
1	0	0
2	0	0
3	1.989918114186580	0
4	0	0.000000020363494
5	0	0.000000052359297
6	0.994959057093290	0.000705884201434
7	0.994959057093290	0.000000655092638
8	0.994959057093290	0
9	0	0.994959077959351
10	0.994959057093290	0
11	0	0.994959178597057
12	0	0.994959604828733
13	0	0
14	0	0.994959239252545
15	0	0.000002150911431
16	0	0.994959238234877
17	0.994959057093290	0.000038005966822
18	0.994959057093290	0.000000339698211
19	0	0
20	0.994959057093290	0.000000425306816

Basic PSO – Rastrigin

Best solution cost: 0
  Best solution position: [0 0]
  Average solution cost: 0.447731575691980
  Elapsed time for 20 trials: 0.331011 seconds

Basic PSO with SA – Rastrigin

Best solution cost: 0
  Best solution position: [0 0]
  Average solution cost: 0.248777193638635
  Elapsed time for 20 trials: 3.376214 seconds

Basic PSO was able to find the global minimum in the Rastrigin benchmark function. Basic PSO found a cost of zero 2/20 trials, while basic PSO with SA found a cost of zero 0/20 trials for the Griewank benchmark function test. Basic PSO with SA found a better solution more frequently than basic PSO for the Griewank benchmark function as its average solution cost was half that of basic PSO. Neither basic PSO nor basic PSO with SA were able to find the true global minimum for the Griewank benchmark function as can be seen by the best solution coordinates; the cost of zero would be due to the solution's close proximity to the true global minimum. The time elapsed for basic PSO with SA was greater for both benchmark functions. When basic PSO was tested on the Rastrigin benchmark function, it found a cost of zero 12/20 trials compared to basic PSO with SA which found a cost of

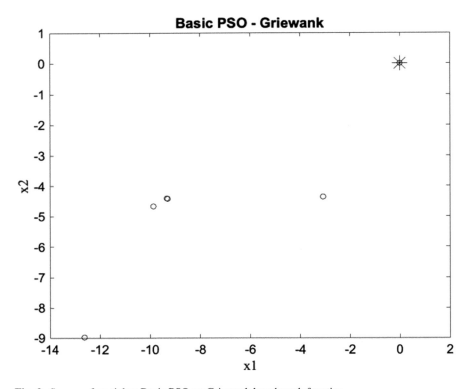

Fig. 2 Swarm of particles: Basic PSO on Griewank benchmark function

zero only 7/20 trails. It is important to note that although basic PSO found the global minimum more frequently than basic PSO with SA for the Rastrigin benchmark function, basic PSO with SA had an average solution cost of approximately half as basic PSO, showing that it found a better solution more frequently than basic PSO if not the true solution for the Rastrigin benchmark function (Figs. 2, 3, 4, 5).

Below are images of a swarm from each of the four trials from experiment 1.

The following graphs show the solution cost for each iteration of the four tests (Figs. 6, 7, 8, 9).

As can be seen in Figs. 6 and 8, the cost of basic PSO converges faster than basic PSO with the addition of SA.

4.1 Experiment 2: Discrete PSO

Discrete PSO was tested on the EUC_2D_194 data set which can be seen in Fig. 1. Discrete PSO was run for 210,000 iterations, while discrete PSO was SA was run for 2100 cooling loops with 100 equilibrium loops for a total of 20 trials. Discrete

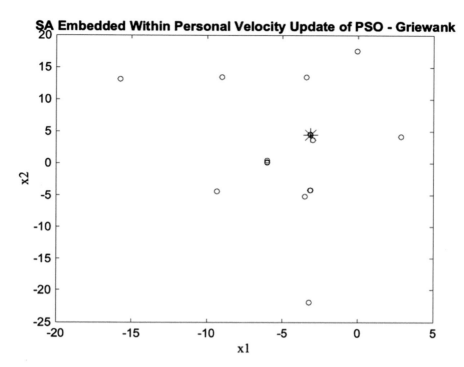

Fig. 3 Swarm of particles: Basic PSO with SA on Griewank benchmark function

PSO was also tested at 420,000 iterations for 20 trials. The results are shown in Tables 3 and 4.

Discrete PSO
Best solution cost: 5.820872239832236e+04
  Average solution cost: 6.126688209539304e+04
  Elapsed time for 20 trials: 589.777172 seconds

Discrete PSO with SA
Best solution cost: 1.628915987648256e+04
  Average solution cost: 1.893068910916914e+04
  Elapsed time for 20 trials: 1066.169095 seconds

Discrete PSO – 420,000
Best solution cost: 5.461175898185149e+04
  Average solution cost: 5.662261292643110e+04
  Elapsed time for 20 trials: 1193.436249 seconds

The addition of SA embedded in the velocity update equation of discrete PSO found significantly better solutions than discrete PSO when applied to the TSP. The best solution found by discrete PSO with TSP had a cost that was 3.6 times

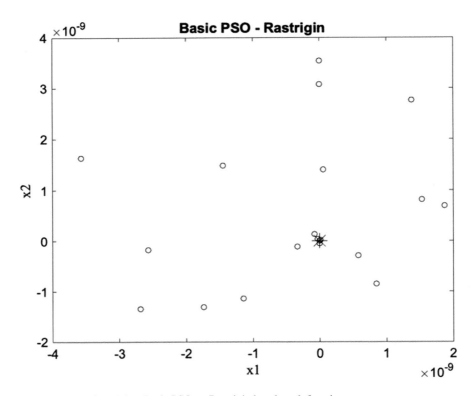

Fig. 4 Swarm of particles: Basic PSO on Rastrigin benchmark function

lower than discrete PSO. The average solution cost was 3.2 times lower for discrete PSO with SA. The run time for discrete PSO with SA was close to double the run time for regular discrete PSO although the improvement in solution makes the trade-off desirable. When studying the solution per iteration graph which can be seen in Fig. 12, discrete PSO showed that with more iterations, a better solution could be found as the best solution per iteration does not appear to have reached a limit. To investigate this, the amount of iterations was doubled, and the results can be seen in Table 4. Allowing for twice the amount of iterations lowered the best solution cost and the average solution cost and had a more comparable run time to the discrete PSO with SA but did not offer solutions as desirable as discrete PSO with SA. Increasing the number of iterations to 10,000,000 resulted in a run time of 1761.842491 seconds and a cost of 3.469141239422924e+04. This trial offered a better solution than previous discrete PSO for TSP tests but still a higher cost than discrete PSO with SA and a longer run time.

An example of the best solution cost vs iteration as well as the best city route for each test is shown below (Figs. 10, 11, 12, 13).

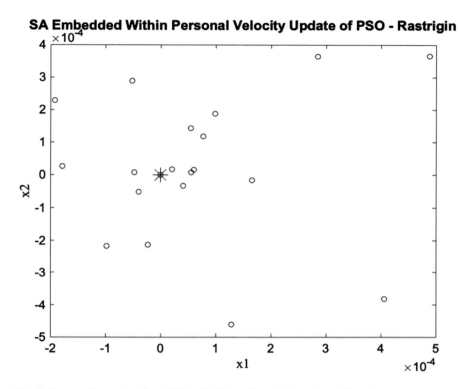

Fig. 5 Swarm of particles: Basic PSO with SA on Rastrigin benchmark function

As can be seen in Figs. 10 and 12, the distance appears to reach a limit in the discrete PSO with SA. The city route for discrete PSO with SA has less overlaps compared to the discrete PSO city route in Fig. 11.

5 Conclusions

The addition of SA in the velocity update equation of PSO showed improvements of results in all trials. The most significant being when it was added to discrete PSO. Embedding SA created longer run times in all trials but offered a better solution making this trade-off acceptable. In the case of basic PSO, the addition of SA offered a better solution more of the time with the average solution being lower than basic PSO. In discrete PSO, the addition of SA was significant. The cost of the solution was 3.2 lower than discrete PSO on its own.

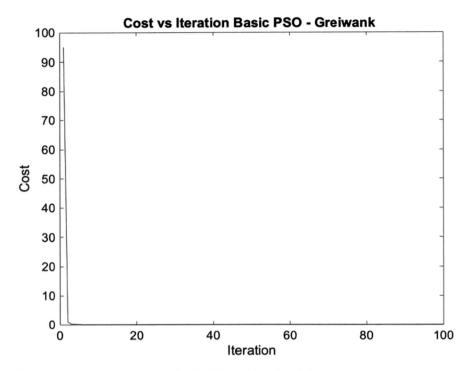

Fig. 6 Cost vs iteration for basic PSO for Griewank benchmark function

5.1 Future Work

Future work in this area for basic PSO with SA should involve adjusting the probabilities of influence from global and local best particles to find a probability that results in the best solution instead of randomizing the influences for each iteration. For discrete PSO with SA, adjusting the probabilities of accepting a swap from either the personal best city route or the global best city route should be investigated. In the experiments done, the probabilities of accepting either of these swaps were the same; making them different could potentially offer improved results if the probability for global best was increased at a faster rate than the personal best probability. The way the probabilities were increased should also be investigated as they were increased linearly depending on the amount of iterations and a different way of adjusting the probabilities could offer better results.

Acknowledgments This research has been financially supported by Mitacs Accelerate (IT15018) in partnership with Canadian Tire Corporation and is supported by the University of Manitoba.

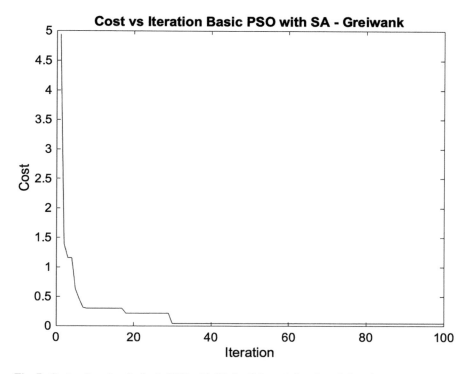

Fig. 7 Cost vs iteration for basic PSO with SA for Griewank benchmark function

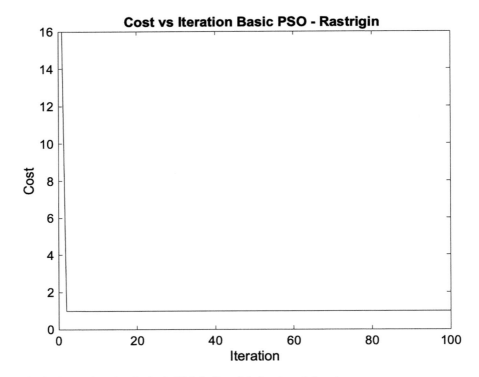

Fig. 8 Cost vs iteration for basic PSO for Rastrigin benchmark function

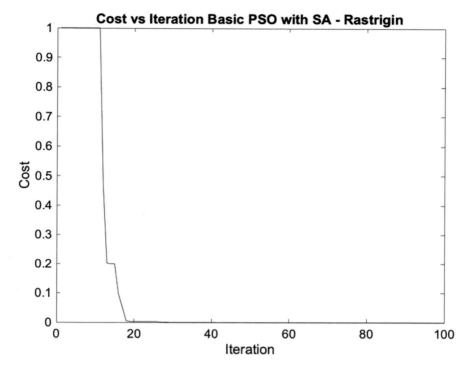

Fig. 9 Cost vs iteration for basic PSO with SA for Rastrigin benchmark function

Table 3 Discrete PSO

Discrete PSO [1.0e+04*]	Discrete PSO with SA [1.0e+04*]
6.117259415644601	1.861170219967018
6.126776161548571	1.855793476663093
6.163652948384005	2.020739111973698
5.926436593501144	1.868242672584329
6.408906599402301	1.841618809846635
5.975725093555228	1.913865285071508
6.223374492625993	2.129286171768821
5.993013822634150	1.839198190809456
5.897761604487618	1.812635374585229
6.342614562636565	1.837781002851906
6.060328490058327	2.059384218531958
6.049795476930349	1.796337132635699
5.969964119206752	1.958043165208614
6.181610736986666	1.962677189080013
6.291910625359307	1.628915987648256
6.412692747576831	1.976489518627333
5.962395249900829	1.816343994842385
6.341269795847896	1.776583340887138
6.267403414666711	1.916967659071302
5.820872239832236	1.989305695683901

Table 4 Discrete PSO at twice the previous amount of iteration

Discrete PSO at 420,000 iterations [1.0e+04*]
5.738039571858577
5.633245079722503
5.635463730831607
5.634043921172185
5.564665291978528
5.666143611908147
5.461175898185149
5.563097950138999
5.659965698380288
5.705695063923603
5.653239307312491
5.853543964745004
5.789023930435751
5.714626268728632
5.684348723019610
5.887953625653828
5.705134095464535
5.500088117247375
5.494979751273196
5.700752250882187

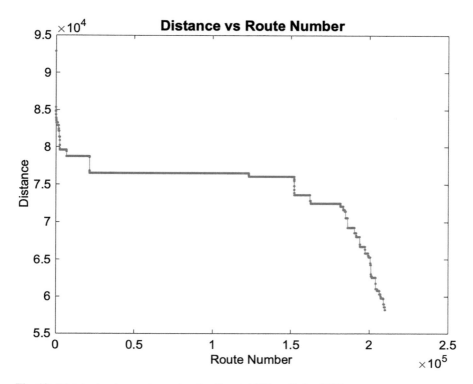

Fig. 10 Distance/cost vs route number for discrete PSO applied to TSP

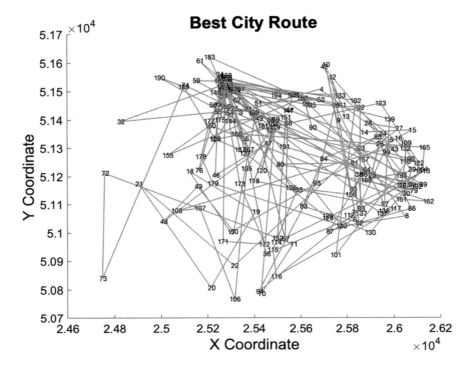

Fig. 11 Best city route map for discrete PSO applied to TSP

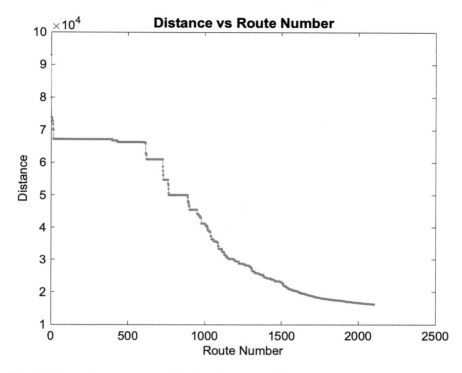

Fig. 12 Distance/cost vs route number for discrete PSO with SA applied to TSP

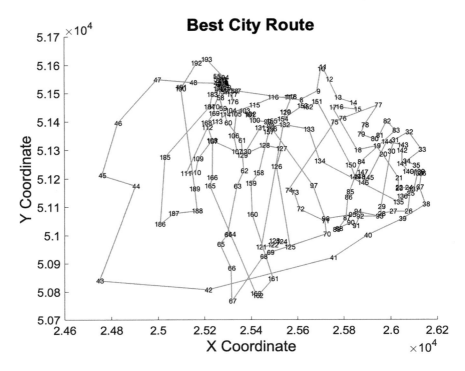

Fig. 13 Best city route map for discrete PSO with SA applied to TSP

References

1. J. Kennedy, R.C. Eberhart, Particle swarm optimization, in *Proceedings of the IEEE International Conference on Neural Networks, Perth, Australia*, (1995), pp. 1942–1948
2. M. Pincus, Letter to the editor—A Monte Carlo method for the approximate solution of certain types of constrained optimization problems. Oper. Res. **18**(6), 1225–1228 (1970)
3. F. Greco (ed.), *Travelling Salesman Problem* (In-teh, Croatia, 2008)
4. J.D. Hedengren, Optimization techniques in engineering, 5 April 2015 [Online]. Available: http://apmonitor.com/me575/index.php/Main/HomePage
5. M. Li, W. Du, F. Nian, An adaptive particle swarm optimization algorithm based on directed weighted complex network. Math. Probl. Eng. **2014**, 1–7 (2014)
6. S. Surjanovic, D. Bingham, Virtual library of simulation experiments: Test functions and datasets (2013). Retrieved May 29, 2020, from http://www.sfu.ca/~ssurjano

Cognitive Discovery Pipeline Applied to Informal Knowledge

Nicola Severini, Pietro Leo, and Paolo Bellavista

1 Introduction

A number of artificial intelligence methods have effectively proliferated into numerous sectors, for example, computer vision, speech recognition, and natural language processing in a broad range of analysis confirm its impact on boosting innovation for all kinds of organizations [1]. Multiple industries are likewise reinforcing their internal R&D capabilities by bringing together the previously segregated internal data sources in combination with external datasets and apply a range of AI methods to generate insights and accelerate the discovery process.

The discovery process is a specialized contribution of AI to accelerate the traditional R&D and IBM research; this is often referred to as Cognitive Discovery [2]. Cognitive Discovery wants to represent a set of methodologies based on the use of AI techniques that aim to reverse, eventually, the standard R&D process, which always starts from an opportunistic discovery of a human being toward an approach in which it is the machine that directs this process.

According to the classic R&D approach, after the opportunistic discovery, a broad spectrum of simulations and experiments opens up, which inevitably leads

---

N. Severini
Alma Mater Studiorum, Bologna, Italy

IBM Active Intelligence CenterQ1, Bologna, Italy

P. Leo (✉)
IBM Active Intelligence Center, Bologna, Italy
e-mail: pietro_leo@it.ibm.com

P. Bellavista
Alma Mater Studiorum, Bologna, Italy

© Springer Nature Switzerland AG 2021
H. R. Arabnia et al. (eds.), *Advances in Artificial Intelligence and Applied Cognitive Computing*, Transactions on Computational Science and Computational Intelligence, https://doi.org/10.1007/978-3-030-70296-0_90

to very high costs and typically ends with the accumulation of a huge quantity of technical knowledge. In some cases, it could be useful to reverse this process by generating insights extracted with natural language understanding and other techniques from the already-accumulated internal and external technical knowledge to better focus and optimize new simulations and experimental activities and, ideally, automatically generate new discovery hypothesis. According to this vision, several industry R&D processes can benefit from this transformation and will have the chance to access to this acceleration modality that leverages the huge quantity of technical knowledge that is growing at a very high rate.

In fact, the motivation of Cognitive Discovery is greatly linked with the increased volume of public as well as proprietary technical knowledge, made available in the form of publications and technical documents/reports. For instance, close to half a million papers in the field of materials science were published in 2016 alone. A range of information extraction (IE) methods have been applied to this task providing a broad spectrum of applications [3]; for instance, IE applied to biomedical literature counts a large number of works [4], as well as works that apply IE to patents [5]. These kinds of documents hold highly complex information, also known as "dark" information: technical plots and diagrams, tables, and formulas are just a few examples.

Cognitive Discovery aims to automate the extraction of knowledge also from dark sources, including the reasoning related to this knowledge, and to store this rich body of knowledge for application to future scientific or technological questions [6]. One clear opportunity that our project is focused on is to leverage and expand the Cognitive Discovery capability to process also "informal sources" of technical content such as the one that is exchanged along informal channels such as physical or web meetings, phone calls, and technical discussions in general that could be a potential source to capture insights about the business practical experience.

The main difference between processing "formal sources" with respect to "informal sources" is in the intrinsic difference between the raw data that the two scenarios consider. In general, the kind of knowledge shared during meetings is heterogeneous, and inherently there is a need to apply multimodal information extraction methods able to act on textual, visual, and audio signals.

Additionally, informal sources of technical data, such as the content we share during a technical meeting, include much more "noise" than that present in well-formed documents such as a scientific paper or a patent. During the meeting, the content is not only well organized, and, growing with the number of participants, typical we could have quite a lot a mix of themes, deviations from the agenda, out of topics, and so on.

2 Methods and Materials

The motivation that is driving our project is to provide in perspective a business tool to automatically formalize the knowledge exchanged during technical meetings by

Cognitive Discovery Pipeline

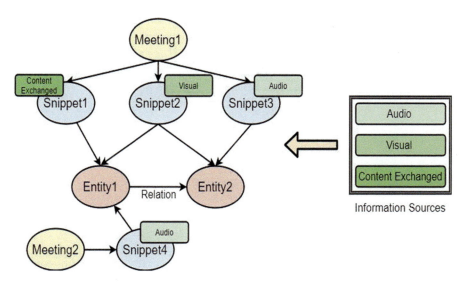

Fig. 1 Graph architecture

creating a "meeting conceptual map." A meeting conceptual map will be in the form of a knowledge graph that will be centered around "meeting snippets" as well as meta-data extracted from the theme. Meeting snippets will be extracted by a number of meeting information sources including the vocal, the visual, and the content that is exchanged among participants as represented in the graph topology reported in Fig. 1.

The graph nodes will include also the main discussion "entities" that will be connected to snippet nodes; eventually, each entity node will be connected to other entity nodes based on a certain semantic relationship that holds among them derived from the snippets. This graph structure provides a means to perform queries such as "give me all the snippets referring to a certain entity" or simply as "what were the main discussion points during the meeting on Monday?" or "which is the evolution of themes during our past months meetings?" as well as allow much more complex queries and analysis providing also a modality to "follow" an argument cross meetings. The multimodal information extraction pipeline our project is working on is represented in Fig. 2 and eventually aims to populate the meeting knowledge graph.

Specifically, in our design, we are considering multiple meeting observation points as well as assuming the presence of multiple and synchronized raw data channels such as the audio, the video recording, as well as material that is exchanged during the meeting among participants such as slides. First of all, we normalize all raw data channels by extracting and representing all kinds of data into a textual form; then, on the extracted textual data, we apply a Snippetizer component that aims to extract portions of text that are particularly important in terms of content. On

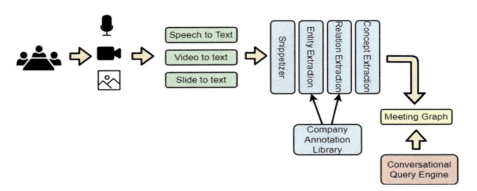

Fig. 2 Project pipeline

the snippets, an entity extraction and a relation extraction components are applied, respectively, to extract the main entities and then connect the entities with semantic relationships. The last two components, in a real scenario, are typical trained to with a specific company annotation strategy. Two enrichment components are also used to better characterize extracted entities such as a concept extraction component that aims to score the main concepts and a Wiki cation component to semantically expand information related to recognized entities by associating them to external information such as a Wikipedia page. This pipeline eventually creates a knowledge graph which can then be queried with a conversational query engine. The latter can be implemented through different methods or technologies; in the case of our prototype, it will be a component capable of processing the natural language queries provided by the user, translating them into the specific language of the knowledge graph, and returning the result.

For the prototype system, just the meeting audio stream is acquired, and a speech-to-text component is used to transcript it. To implement this component, the IBM Watson speech-to-text service was used [7]. The output of this service is a json file containing a list of blocks of text, the division of which is determined according to the pauses of the speakers. Furthermore, the output is free of punctuation; therefore, the data processing steps were two: concatenate the various text blocks of the json file and add punctuation through the use of a python library [8]: it is a recurring bidirectional neural network used to predict punctuation in text without punctuation.

The next step is to extract entities from the text built in the previous step. To do this, we used spaCy [9], a powerful and extensible open-source python library which offers many NLP services such as named entity recognition. For each entity extracted via NER, an entity node is created in the KG (which will be connected with the snippet nodes that we will see later).

One of the main steps for IE is the extraction of relationships between entities; this task is of fundamental importance because in addition to giving us a greater level of knowledge, it also allows us to extract those insights discussed in the previous chapters, for example, we may find a common link in seemingly unrelated meetings. This task was accomplished through Watson's NLU service [10].

Then we have the component that makes the text snippets; this component is highly dependent on the specific use case: it aims to extract only the important parts from the overall text returned by the speech-to-text component. Being the text extracted from an oral speech, much of the content will be pure "noise" so you will have to discard it. The component implemented here is based on a clustering of entities extracted from the text. The extracted snippets will be those portions of text that are particularly important in terms of extracted entities. Specifically, the Jenks natural breaks optimization has been implemented on an array of integers that represent the indices of the sentences within the overall text that contain at least one entity. Taking an example, assuming that the integer array is [1,2,5,7,21,23,24], the end result will be this: [1,2,3,4,5] and [21,22,23,24]. As we can see from the example, in addition to applying the separation in two clusters, we have also inserted the missing indexes within the cluster; this is because we also have the readability constraint of the snippets: a snippet consists of a sort of paragraph that can be consulted depending on your needs; therefore, the final result will be a portion of text that is legible and without missing parts.

The final step is that of "Wiki cation" which consists in adding to each entity node the related link to the Wikipedia page if the latter exists. This step is useful as it enriches the knowledge extracted, thus giving the possibility to the user who makes a query to consult Wikipedia if he wants to deepen the topics covered in a discussion.

As far as the technology used for the graph is concerned, Neo4j [11], an open-source software for managing the graph database, has been chosen which allows you to query the graph using a language called cypher. This language is very similar to sql, therefore not very accessible; to ensure that this product can be used by anyone, a query engine has been developed that can analyze and interpret the user's natural language query.

As regards the conversational query part of the graph, a query engine has been built that can translate the user's natural language query into a cypher query. The approach used is based on the fact that only two elements are needed to perform this translation:

- Search Elements: what the user wants to get from the query, for example, a list of snippet nodes or meeting nodes or a list of entities.
- Filter Elements: The elements that are used to filter the results, for example, if we want to obtain the snippets where we talk about certain entities, the latter will be the filter elements. Depending on the specific implementation, these elements may also be properties of some types of graph nodes.

These two elements are extracted through a grammatical analysis of the dependency parser of the query, a tool that allows you to obtain a hierarchical structure of a sentence in order to understand the dependencies of the words, but given the immense variability of natural language, this method must be coupled with sentence similarity techniques in order to apply these grammatical rules to a set of known patterns: first of all, a query library was defined built through the concatenation of linguistic patterns and graph elements, such as "give me all the [x] where we talk about [y]," where x are the search elements and y are the filter elements. Subsequently, we define a vector space of this library through word embedding techniques. Thanks to this, we can transform the user query into the relative vector of the library vector space and let the user choose one of the three most similar queries in the library and then apply the grammar rules on this to extract the two elements needed for the translation into query cypher, i.e., search elements and filter elements.

3 Preliminary Results

The purpose of this paper is not to show a complete and perfectly functional solution, but highlight our research directions and show the software architecture behind these solutions and therefore show their potential with a prototype.

Given the illustrative purposes of this project, a python application was the deployment choice. Below is the represented UML diagram that describes the application; as you can see, the pattern strategy has been used in order to be able to implement increasingly sophisticated pipeline blocks in the future, without changing anything in the code but only by extending the existing one. This pattern will also allow you to make changes runtime, particularly useful in case you want to create multilingual pipelines (Fig. 3).

The green component is the main class that takes care of running the pipeline; the yellow components are the abstract strategies that expose only the abstract methods common to any implementation; finally, the red components are the actual implementations described above. The idea behind this architectural choice is to be able to improve performance by implementing increasingly sophisticated components.

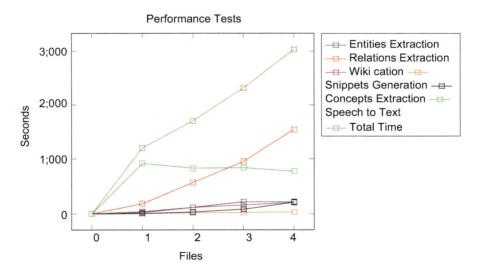

Cognitive Discovery Pipeline

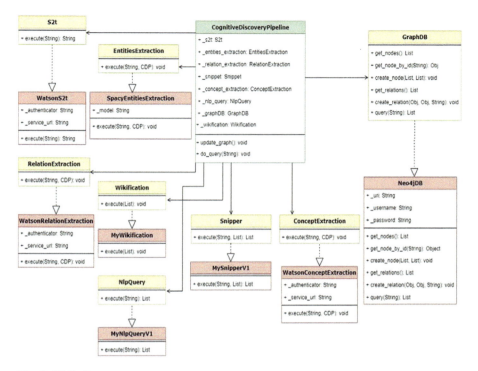

Fig. 3 UML diagram

The figure on the side reports a first set of performance tests for each block of the pipeline. In particular, the entire pipeline was run incrementally on four half-hour podcasts. Two considerations are necessary: The first is the fact that each cognitive discovery solution is designed for deployment on the cloud, mainly for scalability; in this case, instead, the deployment was done on a single core. The second is that the aim of these results is to show what the computational criticalities are. As we can see, the block with the least scalability is that of extracting relations, which did not surprise us since it has the greatest computational load, both in terms of service execution and in terms of graph controls.

4 Conclusion

This paper proposed a preliminary work that aims to develop an architectural model for extracting insights from "informal" technical knowledge sources such as the one produced during meetings by a group of people. The work has the objective of laying the foundations for the realization of complete software solutions. Given the importance of having a system capable of processing this type of knowledge, we have defined a software architecture designed on the concept of expandability;

a prototype has also been developed capable of showing the peculiarities and the critical points of solutions of this type.

5 Acknowledgments and Funding

The work was done by N. Severini, and it was partially funded by IBM for an internship program to develop at the IBM Active Intelligence Center his master degree thesis. P. Leo gave strategic directions and supervised the work for IBM, and P. Bellavista supervised the work for the University of Bologna. We thank G. Camorali, V. Saturnino, and M. Gatti and in general all members of the IBM Active Intelligence Center team in Bologna (Italy) for assistance with comments that greatly improved the work.

References

1. I.M. Cockburn, R. Henderson, S. Stern, *The Impact of Artificial Intelligence on Innovation: An Exploratory Analysis* (University of Chicago Press, 2018), pp. 115–146. https://doi.org/10.7208/chicago/9780226613475.001.0001, http://www.nber.org/chapters/c14006
2. IBM Research, Cognitive discovery: The next frontier in R&D, https://www.zurich.ibm.com/cognitivediscovery/
3. M. Grida, H. Soliman, M. Hassan, Short text mining: State of the art and research opportunities (2019). https://www.thescipub.com/abstract/, https://doi.org/10.3844/jcssp.2019.1450.1460
4. A. Holzinger, J. Schantl, M. Schroettner, C. Seifert, K. Verspoor, Biomedical text mining: State-of-the-art, open problems and future challenges (2014). https://link.springer.com/chapter/10.1007/978-3-662-43968-516
5. L. Aristodemou, F. Tietze, The state-of-the-art on intellectual property analytics (IPA): A literature review on artificial intelligence, machine learning and deep learning methods for analysing intellectual property (IP) data (October 2018). https://www.sciencedirect.com/science/article/pii/S0172219018300103
6. P.W.J. Staar, M. Dol, C. Auer, C. Bekas, Corpus conversion service: A machine learning platform to ingest documents at scale [poster abstract]. (2018). https://doi.org/10.13140/RG.2.2.10858.82888
7. Speech to Text – IBM Cloud API Docs, https://cloud.ibm.com/apidocs/speech-to-text/speech-to-text. Accessed on 03/23/2020
8. O. Tilk, T. Alumae, Bidirectional recurrent neural network with attention mechanism for punctuation restoration, in *Interspeech 2016*, (2016)
9. M. Honnibal, I. Montani, spaCy 2: Natural language understanding with Bloom embeddings, convolutional neural networks and incremental parsing. To Appear 7(1), 411–420 (2017)
10. Natural Language Understanding – IBM Cloud API Docs, https://cloud.ibm.com/apidocs/natural-language-understanding/natural-language-understanding. Accessed on 03/23/2020
11. Neo4j Graph Platform – The Leader in Graph Databases, https://neo4j.com/. Accessed on 03/23/2020

Index

A
AAD, *see* Advisor Agent (AAD)
AAN, *see* Analyst Agent (AAN)
ABSA, *see* Aspect-based sentiment analysis (ABSA)
Abstraction, 789
ACCS, *see* Australian Centre for Cyber Security (ACCS)
ACPS, *see* Aviation cyber-physical systems (ACPS)
Activation functions, *see* Chaotic activation function
Act-STEM skill knowledge teaching technology
  CT, 789
  direct service audiences, 793
  Fun-Joy, 789
  goals, 793
  HE-STEM, 789
  K-14 grades, 790
  manageable problems, 789
  modeling-centered inquiry, 793
  PBL, 793
  pilot projects, 789
  professional development experiences, 793
Act-STEM training modules, 794
AD, *see* Axiomatic design (AD)
Adaptive boosting (AdaBoost), 807
Adaptive training, 211, 216
ADS, *see* Data Steward Agent (ADS)
Advanced encryption standard (AES), 776
Advanced metering infrastructure (AMI) data
  cost-aware policy, 224
  experiment results

drift threshold, 230–232
environment, 230
performance comparison, 232, 233
online learning system, 224
problem description, 224–226
proposed system
  architecture overview, 226–227
  incremental learning scheme, 227–229
two-way communication, 223
Adversarial neural networks
  block cipher mode, 427–428
  ciphertext-only attacks, 426–427
  cryptography, 425
  experiments and results, 430–434
  GAN, 426
  neural architecture, 428–430
Advisor Agent (AAD), 1001
AE, *see* Autoencoder (AE)
AES, *see* Advanced encryption standard (AES)
AF, *see* Atrial fibrillation (AF)
Affective NPC (ANPC), 695–710, 712, 713
Agglomerative hierarchical clustering
  data mining technique, 869–870
  performance evaluation
    monthly prediction, 872–874
    weekly analysis, 874, 875
  proposed work
    data preprocessing, 870
    hierarchy-based cluster analysis, 871, 872
    time sequence, 870–871
  related work, 870
  RTA, 869
AGI, *see* Artificial general intelligence (AGI)

© Springer Nature Switzerland AG 2021
H. R. Arabnia et al. (eds.), *Advances in Artificial Intelligence and Applied Cognitive Computing*, Transactions on Computational Science and Computational Intelligence, https://doi.org/10.1007/978-3-030-70296-0

1154   Index

AHP, *see* Analytic hierarchy process (AHP)
AI, *see* Artificial intelligence (AI)
AI-aided computational thinking, 802
AI based DPLF models
    forecasting model construction, 811–813
    input variable configuration, 808–811
    MAPE, 813
    MSE, 813
    normalization, 812
    PCA and FA, 813
    Python environment, 813
    superiority, 813
    TSCV, 808
AIE, *see* Artificial intelligent entity (AIE)
AIN, *see* Interface Agent (AIN)
AlexNet
    evaluations, 144
    illustration images, 143, 144
    ILSVRC, 141
    object recognition, 141–143
Algorithms, 789, 952, 958–960
ALV, *see* Autonomous land vehicle (ALV)
AMI data, *see* Advanced metering
        infrastructure (AMI) data
Analog VLSI Implementation of Neural
        Systems, 729
Analyst Agent (AAN), 1002, 1008
Analytic hierarchy process (AHP), 608, 609,
        611–613
Angle of attack (AOA), 1115, 1116, 1121
ANN classifier applications, 1035
ANNs, *see* Artificial neural networks (ANNs)
Anomaly detection, 1049
    classification, 511
    data training, 506
    experimental results, 836–838
    forecasting model configuration, 836
    input variable configuration, 834–835
    intrusion detection system, 1015
    IQR, 834
    MEMS, 833
    model configuration, 835–836
    RAD, 510
    smart grid, 833
    time-series data, 833
ANPC, *see* Affective NPC (ANPC)
APF, *see* Artificial psychosocial framework
        (APF)
Application-Specific Integrated Circuit
        (ASIC), 757, 772
Approximation
    accuracy, 568, 569
    activation function, 68
    DL, 561

experimental validation, 566–567
exploration, 328
futurework, 567
PES, 112
preliminaries, 562–563
previous arts, 563
proposed method, 563–566
Shroedinger, 509
stochastic derivative, 1079
Arabic language, 46–51, 53, 54
Archie search engine interface, 914
ARE, *see* Reasoner Agent (ARE)
ARIMA, *see* Auto Regressive Integrated
        Moving Average (ARIMA)
Arrhythmia annotation, 751
Arrhythmia ECG models, 753
Artificial emotions, 486, 489
Artificial general intelligence (AGI)
    AI, 74–76
    ANN, 79–80
    data science, 73, 76
    discussion, 83–84
    DL, 80–83
    domain hierarchy, 74
    ethics, 84–85
    ML, 76–79
    weather forecasts, 73
Artificial intelligence (AI), 74–76, 729, 772,
        775, 807, 883, 951, 952, 1145
    cyber (*see* Cyber security)
    data mining, 841
    discussion, 845–848
    healthcare, 29
    implementation, 578–579
    learner-centric interpretation, 211
    methodology, 953
    non-linearity and complexity, 499–504
    proposed work
        algorithm, 845
        block diagram, 842, 843
        correlation coefficient, 844, 845
        dataset, 843–844
        missing values filling, 844
    related work, 842
    results, 845–848
    techniques
        classification, 30
        cutting-edge, 127
Artificial intelligent entity (AIE), 999, 1008
Artificial neural networks (ANNs), 79–80,
        807, 1035
    artificial neurogenesis, 413–421
    decision-making process, 167
    massively parallel systems, 382

Index

neural network pruning, 410–413
nonlinear statistical models, 188
processor-based AI systems, 388–403
scaling methods, 382–388
single-processor performance, 381
SWaP, 409
training, 32
Artificial neurogenesis
algorithms
ANG overview, 416
critical connection search, 416
extreme member search, 414–416
analysis, 420–421
experiments
accurate architecture, 419–420
minimum priming cycle count, 418
scaling factor x, 418–419
process, 1007
Seed Network, 413
Artificial psychosocial framework (APF),
697–702, 704, 706–710, 712, 713
Artist-artist relationship, 653, 654, 656, 660,
662
ASIC, *see* Application-Specific Integrated
Circuit (ASIC)
ASIC chips, 771
Aspect-based sentiment analysis (ABSA), 312,
313
Aspect category, 313–320
Association of sets, 644, 648
Atrial fibrillation (AF), 750
Attention
detection, 818
experiment, 822, 823
maps, 160, 164, 166–168, 479
mechanism, 819
proposed method, 819–821
semantic segmentation, 818–819
Attention-based LSTM network, 813
Attributes (features)-opinion pairs rules, 721
Audiovisual content platforms, 917–919, 921
Audiovisual works, 917, 918, 920
Australian Centre for Cyber Security (ACCS),
1037
Auto-diagnosis, 743
Autoencoder (AE)
dimension reduction, 268–269
grayscale images, 102
VAE (*see* Variational autoencoder (VAE))
Automatic construction, 951
Automatic opinion-oriented classifiers, 721
Autonomous cars, 369
pilot, 372
procedural trajectory, 372–374

straightening and smoothing trajectory, 374
track mapping and reconstruction, 371–372
Autonomous land vehicle (ALV), 775
Autonomous systems, 1008
Autonomous vehicles, 775
Auto-regressive and moving average models,
890
Auto Regressive Integrated Moving Average
(ARIMA)
ACF and PACF, 891, 892
forecasting, 890
four-month prediction, 891, 893, 894
heavy data preprocessing, 894
hyperparameter values, 891, 892
statistical model, 891
Auxiliary sentences, 313, 317, 319, 320
Availability, 1049, 1071, 1108
Aviation cyber-physical systems (ACPS)
components, 1108
DDOS network attack, 1116
structure, 1108–1109
TCM, 1108–1109
Axiomatic design (AD), 501–504, 511, 512,
520
AXI-to-MBUS wrapper, 777

B

Basic PSO
Boltzmann's probability function, 1127
city coordinates of EUC, 1126
disadvantage, 1123
Griewank function, 1125, 1126, 1130
pseudo-code, 1126, 1127
Rastrigin function, 1125, 1126, 1130, 1131
with SA
Griewank function, 1130–1133, 1136,
1137
Rastrigin function, 1130–1132, 1134,
1135, 1138, 1139
SA embedded personal velocity update,
1126, 1127
solution space, 1125
standard velocity update equation, 1126
Bayesian network (BN) models
credit risk prediction, 461
Ensemble classifiers, 466
experimental setup
dataset, 466–468
evaluation, 468–469
formal definition, 463
Naïve Bayes, 463
related work, 462–463
results and discussion

Bayesian network (BN) models (*cont.*)
  error analysis, 471–472
  prediction evaluation, 469, 470
  structure learning methods
    genetic search, 464
    hill climbing, 465
    k2 method, 464–465
    simulated annealing, 465
    tree augmented Naïve Bayes, 463–464
    Ugandan financial institution, 461
BDM, *see* Boundary mutation (BDM)
Bidirectional encoder representations from
    transformers (BERT) model, 312,
    313, 316–320, 567
Bi-directional long short-term memory
    (Bi-LSTM), 898, 899, 901
Big data, 76, 78, 129, 502, 715, 883, 885, 886,
    1014, 1029, 1047
  analysis, 715
  data mining technique, 869–870
  performance evaluation
    monthly prediction, 872–874
    weekly analysis, 874, 875
  processing, 1047
  proposed work
    data preprocessing, 870
    hierarchy-based cluster analysis, 871,
    872
    time sequence, 870–871
  related work, 870
Bi-LSTM, *see* Bi-directional long short-term
    memory (Bi-LSTM)
Bi-LSTM-LSTM model
  experimentations, 901
  mid-term forecasting applications, 902
  model loss, 902
  results, 901
Bitcoin
  advantages
    anonymity, 907
    noninflationary, 907
    resilience, 907
    secure information, 907
    transaction costs, 907
  challenges, 907–908
  cryptocurrencies, 909
  cryptocurrency payments, 910
  data centers, 909
  electronic system, 905
  GHOST, 909
  incentives, 909
  increase speed, 909
  off-chain transactions, 909

  peer-to-peer payment system, 905
BLE, *see* Bluetooth Low Energy (BLE)
BLE mesh network, 760
BLE mesh system
  host/server, 761
  humidity range, 763
  LSB, 763
  mesh-topology method, 761
  MSB, 763
  network connection, 763
  self-designed boards, 761
  self-manufactured PCB boards, 762
  sensors, 761
  task execution, 768, 771
  temperature range, 764
Blockchain
  applications, 906
  characteristics, 905
  cryptographic puzzles, 905, 906
  energy, 910
  ICOs, 906
  off-chain transactions, 909
  structure, 906
  transportation, 910
  trust, 905
  wallets, 906
Blood cell classification
  algorithmic model, 29
  depth, 30
  GA-enhanced D-CNN model, 31, 36–40
  hyperparameters, 30
  and Kaggle data set
    chromosome representation of CNN, 35
    CNN model, 36
    data preprocessing, 35
    GA, 34–36
    ImageNet ILSVCR competition, 33, 34
  pretrained models, 30
  proposed model's performance, 31
  WBC, 31–32
Bluetooth Low Energy (BLE), 758
BN models, *see* Bayesian network (BN)
    models
Body-consciousness approach, 964
Boltzmann's probability function, 1124, 1125,
    1127, 1129
Boundary mutation (BDM), 1019
Brain tumor
  BraTS dataset, 90
  diagnosis, 89
  discussion, 98
  GBM, 89
  materials and methods

Index 1157

data acquisition, 90–91
data preprocessing, 91
3D deep learning algorithms, 91–92
results
DeepMedic base model, 93–97
Brain wave processing, 348
BraTS dataset, 90, 91, 93, 94
Breast cancer
diagnosis, 851
malignant, 851, 852
mammography, 851
performance evaluation, 855
proposed work, 852–854
related work, 852
Building cluster information, 808, 809

C
CAD, *see* Computer aided vision (CAD)
Caffe, 715
CAGR, *see* Compound Annual Growth Rate (CAGR)
Calories, 857, 858, 861
Cancer
breast, 843, 851–855
colon, 631
diagnosing, 629
lung, 33
skin (*see* Skin cancer)
CAT, *see* Cognitive appraisal theory (CAT)
CCES data, *see* Congressional Election Study (CCES) data
CDR, 864, 868
Cell viability imaging, 202–204, 207, 208
Chaos theory, 1097
Chaotic activation function
ANNs, 1097
Chaos theory, 1097
non-chaotic AFs, 1098
rabbit's olfactory system, 1098
Chaotic ANN architecture
HCAF, 1099
proposed chaotic neuron, 1100
Charged Hierarchical Space (CHS), 954
Chinese language, 在NP上(zài NP shàng)
cognitive approach, 936
cognitive linguistics, 936
geometric scheme, 937
image scheme (*see* Image schemes, 在NP上(zài NP shàng))
locative spatial situation, 936
spatial prepositions, 936
topological scheme, 936
topological type, 937

Chromosome diagnostics
classification, 619–621
feature extraction, 619–621
proposed classification model
chromosomal abnormalities-related disease diagnosis, 628–630
feature extraction method, 623–628
genetic disorder detection, 630–637
scalable multi-layer neural network, 630–637
spectrum analysis, 630–637
CHS, *see* Charged Hierarchical Space (CHS)
CICIDS2017 network intrusion dataset, 1031
Ciphertext-only attacks, 426–427, 432
Classic midline approach, 1034
Classification
algorithm, 1044
binary problem, 130
D-CNN algorithmic model (*see* Deep convolutional neural network (D-CNN))
linear functions, 172
supervised, 67–69
technique, 883
CLF, *see* Conscious life form (CLF)
Climate change, 449
Cloud computing *vs.* edge computing, 758
Clustering algorithm, 952
Clustering-based outlier method, 1049
Clustering performance
CF, 239
contingency matrices, 69
learning objects dataset, 243
nonmetric data, 269
patterns and anomalies, 77
prototype, 236
spatial voting, 513
Clustering techniques
analysis, 242–243
analyzed techniques, 237–239
evolution of information, 235
example of analysis, 243–247
experiments, metrics, 239–240
literature review, 236–237
online unsupervised learning techniques, 236
results, 241
social computing, 958–959
CNNs, *see* Convolutional neural networks (CNNs)
Coefficient of similarity of sets of association, 644, 648
Cognition, 975, 1063
Cognition-inspired adaptive learning, 1061

Cognition models
  radar, 1071–1072
  Rasmussen, 1072–1075
  RCM (*see* Radar cognition models (RCM))
Cognitive analysis, 1031
Cognitive appraisal theory (CAT), 696, 697,
    699, 703, 712
Cognitive complexity, 1031
Cognitive complexity-based analysis, 1044
Cognitive discovery
  AI methods, 1145
  components, 1151
  dark information, 1146
  dark sources, 1146
  data processing, 1148
  extraction of knowledge, 1146
  filter elements, 1149
  formal sources, 1146
  graph nodes, 1147
  IBM approach, 1145
  IE (*see* Information extraction (IE))
  informal sources, 1146
  internal and external technical knowledge,
      1146
  meeting conceptual map, 1146–1147
  meeting snippets, 1147
  motivation, 1146
  Neo4j, 1149
  project pipeline, 1147, 1148
  prototype system, 1148
  query library, 1150
  raw data channels, 1147
  R&D approach, 1145, 1146
  search elements, 1149, 1150
  Snippetizer component, 1147–1148
  speech-to-text component, 1148
  text snippets, 1149
  UML diagram, 1150–1151
  Watson's NLU service, 1149
  Wiki cation, 1148, 1149
Cognitive load, 1087
  method
    apparatus, 1090
    participants, 1090
    procedure, 1090
    study design, 1089
  outcomes
    abrupt discontinuities, 1093–1094
    analyses, 1091
    pupil dilation, 1092–1093
    task performance, 1091–1092
Cognitive radar system (CRS)
  architecture, 1072
  neural network data, 1079

OoI scans, 1078
  processing diagram, 1078
  RCM hierarchical observation, 1076
  RCM model, 1075
Cognitive science, 1062
Cognitive systems, 176, 477, 481, 936, 960,
    1008, 1031, 1063
Cognitive technology, 954
Cognitive unsupervised clustering
  accuracy, 1065
  attack cluster identification, 1065
  AUC ROC curve, 1065
  confusion matrix, 1065
  datasets, 1063, 1064
  feature extractor, 1064
  higher-order features, 1064, 1066
  k-means classifier, 1065
  network events, 1063
  sampler, 1063
  statistical analysis, 1064
  timestamp, 1064
  VFD calculation, 1066
  VFD value, 1065
Color selectivity, 207, 208
Common STEM teaching method, 788–789
Communication network system
  with DDOS attack, 1110, 1115, 1116
  normal network system, 1109–1110
  SimEvents, 1109
  TCM model, 1115
  without DDOS attack, 1115
Complex intelligent algorithms and systems,
    775
Complexity
  analysis, 1034
  cross discipline engineering, 501
  cyber research, 499
  managing, 502–504
  mathematical formalisms, 500
  transdisciplinary approach, 501
Compound Annual Growth Rate (CAGR), 729
Computational resource, 1030
Computational thinking (CT), 789, 799
  advantages, 799
  AI-aided, 802
  complex problem solving, 800
  components, 800
  drawing skill development, 799
  Fun-Joy project, 799
  integrating skills, 799
  K-12 education, 799
  learning progressions, 799
Computation time, 30, 33, 207, 551, 758, 759,
    765–767, 769, 1044

Index

Computer aided analysis, 112, 851, 890
Computer aided vision (CAD), 898
Computer programming, 787
Computing assisted system, 724
Computing efficiency, 384
Computing hardware device, 757
Concept drift, 223–225, 227–232
ConceptNet, 723, 724
Confusion matrix, 1065, 1068
Congressional Election Study (CCES) data, 303
Connectors, 644–646
Conscious life form (CLF), 963, 968
Consciousness, 964, 968, 969, 976
Constituency parsing
  dataset, 53–54
  dense input representation, 49–50
  long sentences
    with maximum split points, 57
    with minimum split points, 56, 57
  neural network-based approaches, 45
  parse tree, 46
    generator model, 50–51
  short sentences
    with maximum split points, 56
    with minimum split points, 55–56
  survey of relatedwork, 47–49
  syntactic distance-based model, 46
  workflow, 52, 53
Constraint solving, 525, 526
Contact networks, 644, 647
Contacts, 644, 645, 647
Control system, 764
Conventional educational method, 788
Conventional machine learning methods, 813
Conventional PSO, 983–985
Conventional time-series forecasting methods, 811
Conversation partner, 345
Convolutional neural networks (CNNs), 134–135, 730
  AlexNet (*see* AlexNet)
  D-CNN model (*see* Deep convolutional neural network (D-CNN))
  facial landmarks (*see* Facial landmarks)
  image information, 21
  LSTM, 132
  model, 36
  objects and features, 19
  person re-identification, 825–830
  profiling layers, 682
    convolution layer, 683–684
    fully connected layers, 684
  retinal pictures, 865

single-layered, 130
skin cancer, 147–153
traffic signs, 82
Copyright
  audiovisual content platforms, 917, 918, 921
  audiovisual works, 917, 918
  countries, 917
  creation, 916, 917
  intelligent recommendation system, 918
  literary and artistic works, 915
  monopoly, 916
  originality, 917, 920
  protection, 917, 918, 920
  search engines, 920
  trademarks, 917–919
Coronavirus (Covid-19)
  adjacency relationships, 271
  differential interaction modules, 272–273
  host-cell
    identification, 267–268
    specific regulatory networks, 272
  human-adaptive, transmutability, 264–266
  human-to-human transmission, 263
  hybrid
    statistical model, 270
    unsupervised clustering, 269–270
  interaction networks, 273–274
  learning-based methods, 267–268
  mammalian and avian hosts, 264
  ML, 265–266
  next-generation sequencing technology, 265
  nonmetric similarity measurement, 269
  prediction, 267
  pseudo-temporal ordering, 271–272
  SARS-CoV-2, 264, 265
  stacked autoencoders, 268–269
  target interaction modules, 272
CPS, *see* Cyber-physical system (CPS)
Credit risk assessment, 461–464, 473
Cross-input neighborhood differences
  dataset, 826
  disjoint camera views, 825
  experimental results, 829, 830
  learning-based, 825
  model, 826–828
Crowds
  combining with GAs, 358–360
  crowd interactions, 355, 357
  design of public spaces, 353
  evolution models, 355
  GAs, 360–361
  hand-collected dataset, 357–358

1160 Index

Crowds (*cont.*)
  record of success, 354
  results
    discussion, 365–366
    time series, 363–365
    trained *vs.* control, 362–363
  structured/unstructured crowds, 354–356
Cryptocurrencies, 905, 907, 909
Cryptocurrency payments, 910
CT, *see* Computational thinking (CT)
Cyber-attack dataset, 1043
Cyber-attacks, 908, 1029
Cybernetics, 924
Cyber networks, 1029
Cyber-physical system (CPS)
  ACPS (*see* Aviation cyber-physical systems (ACPS))
  benefits, 1107
  efficiency and safety, 1107
  physical components, 1107
Cyber security, 1029, 1030, 1067, 1107
  data ingestion, 505–506
  ML and
    discussion, 519–521
    high dimensionality and complexity, 511–512
    high fidelity fusion, 518–519
    information theory and information theoretical methods, 510–511
    ITM, 512–516
    KR, 517–518
    physical representation, 516–517
    supervised and unsupervised, 509–510
    value, characteristics and limitations, 507–509
Cyber systems, 1029
Cyber threat detection, 1061
Cypher, 1149, 1150

D
Daily peak load forecasting (DPLF)
  AI techniques, 807
  energy consumption, 807
  historical electric load data, 810
  KMA, 810
  nominal scale, 809
  1D data, 809
  2D space, 809
  variable configuration, 808
Dark information, 1146
Data analytics, 73, 267, 477, 507
Data and text mining technology, 721

Data mining, 883
  Isle Royale annual moose populations, 184
  isolated moose populations, 183–198
  KNN, 190
  medical domain, 847
  multiple regression
    first maximal model, 186
    reduced parameter maximal model, 186–187
  neural networks, 188–189
  regression trees, 187–188
  results
    constant population assumption, 194
    KNN1, 196
    KNN2, 196, 197
    KNN3, 197, 198
    KNN4, 198
    multiple regression, 194, 195
    neural network, 195, 196
    overview, 192–194
    regression tree, 195
  simulation, 190–192
  techniques, 183
  vital methods, 885
  wolf population, 184–185
  wrangling, 185
Data science (DS), 73, 74, 76, 251, 258, 699
Data Steward Agent (ADS), 1001
Data stream, 1047
  mining, 1014, 1015, 1025, 1049
  processing, 1059
Day-ahead peak load forecasting, 814
D-CNN, *see* Deep convolutional neural network (D-CNN)
DDOS attacks, *see* Distributed Denial-of-Service (DDOS) attacks
Decision-making process
  AI, 578–579
  descriptive similarity, 572–574
  implementation, 575–578
  NATO, 571
  structural similarity, 573–574
Decision-making trial and evaluation laboratory (DEMATEL), 608, 609, 612–613
Decomposition, 789
Deep convolutional neural network (D-CNN)
  algorithmic model, 29
  depth, 30
  GA-enhanced, 36–40
  hyperparameters, 30
  and Kaggle data set

Index

1161

chromosome representation of CNN, 35
CNN model, 36
data preprocessing, 35
GA, 34–36
ImageNet ILSVCR competition, 33, 34
pretrained models, 30
proposed model's performance, 31
WBC, 31–32
Deep embeddings
    exploration domain, 60–61
    methodology
        embedding techniques, 66, 67
        knowledge graph construction, 62–66
        naive vectorization, 61–62
        testing protocol, 67–68
    results
        supervised classification task, 68
        unsupervised discovery task, 69
    semantic graph representations, 60
    tactics identification, 59
Deep learning (DL), 80–83, 883
    AMI data (*see* Advanced metering
        infrastructure (AMI) data)
    brain tumor (*see* Brain tumor)
    breast cancer, 851–855
    classifier, 132–133
    constituency parsing, 45–57
    Covid-19 (*see* Coronavirus (Covid-19))
    definition and relatedwork
        image/video captioning, 19–20
        text summarization, 20
    difference with traditional method, 130
    embeddings (*see* Deep embeddings)
    experiments, 24–26
    image/video captioning, 17
    methodology
        document to title process, 22–23
        video to document process, 21–22
    proposed system, 17, 18
    record pair representation, 130–132
    skin cancer, 147–153
    text summarization, 17
DeepMedic base model
    learning neural network, 90
    residual connections, 92
    survival prediction, 94–97
    3D U-net neural network, 94
Deep Metric Learning (DML), 765
Deep neural networks (DNNs), 730, 813
    BraTS dataset, 90
    diagnosis, 89
    discussion, 98
    GBM, 89
    materials and methods

        data acquisition, 90–91
        data preprocessing, 91
        3D deep learning algorithms, 91–92
    results, DeepMedic base model, 93–97
    VA, 667
Deep reinforcement learning, 82, 103
Defects classification, 545, 550–553, 555
Defects detection
    proposed algorithms
        defect classification, 550–553
        evaluation criteria, 554–555
        experiment results, 555–558
        phase, 546–550
        preprocessing phase, 546
    related work, 544–545
Demand response (DR)
    electrical customers, 877
    grid operators, 877
    HAs, 878
    housing societies, 877
    implementation, 878
    LSTM (*see* Long short-term memory
        (LSTM))
    MAS, 878, 880
    RAs, 878
    real-time forecasting model, 877
    renewable energy, 877
    RTP, 878
    smart appliances, 878
    smart home, 878
DEMATEL, *see* Decision-making trial and
    evaluation laboratory (DEMATEL)
Dempster-Shafer theory, 884, 885
Denial-of-service attack (DoS), 1015
Density-based outlier detection, 1012
Density-based outlier method, 1047, 1049
Density cluster model, 1049
Density Summarization Incremental Local
    Outlier Factor (DILOF), 1012, 1014,
    1048, 1056
DES, *see* Discrete-event simulation (DES)
Descriptive similarity, 572–574, 641, 642,
    648
Design-under-test (DUT), 780
Device free human sensing, 174
Diabetes, 841–844, 846, 857
Dialog systems, 642, 648
Digital coins, 905
DILOF, *see* Density Summarization
    Incremental Local Outlier Factor
    (DILOF)
Dimensionality reduction techniques, 1030
Direct execution, NNEF, 715
Discrete-event simulation (DES), 924, 925

1162   Index

Discrete PSO
   Boltzmann's probability function, 1129
   EUC_2D_194, 1128, 1132
   pseudo-codes, 1128
   with SA, 1133, 1134
     run time, 1134
     SA embedded personal velocity update,
       1129, 1133
     TSP, 1134
   testing, 1133, 1140
   TSP (*see* Travelling salesperson problem
     (TSP))
Discriminative features, 1030, 1035
Disposed system, 759–760
Distance-based outlier method, 1048
Distributed computing, 382, 510
Distributed Denial-of-Service (DDOS) attacks
   ACPS, 1116
   aircraft-AOA performance, 1116, 1121
   aircraft velocity, 1116, 1117
   communication network system, 1110,
     1115, 1116
   elevator signal, 1112
   NARX
     aircraft-AOA performance, 1116, 1121
     aircraft speed performance, 1116, 1117
     state-q performance, 1116, 1119
     state-w performance, 1116, 1118
     state-$\ominus$ performance, 1116, 1120
   network system with, 1110, 1116
   network traffic to disturb servers, 1108
   SimEvents library, 1108
   Simulink library, 1108
   state-q performance, 1116, 1119
   state-w performance, 1116, 1118
   state-$\ominus$ performance, 1116, 1120
   without communication network system,
     1115
   without normal network model, 1115
Distributional models, 955
Distributional semantics, 955
DL, *see* Deep learning (DL)
DML, *see* Deep Metric Learning (DML)
DML based face recognition system, 765
DNNs, *see* Deep neural networks (DNNs)
Domain randomization, 158
   attention maps, 160
   experimental setup
     agent architecture, 162–163
     design, 163–165
     learning environment, 160–162
   results
     attention maps, 166–167
     training in simulation, 165

     transfer to real world, 165–166
   task-specific experience, 157
   transfer learning, 159–160
DoS, *see* Denial-of-service attack (DoS)
DPLF, *see* Daily peak load forecasting (DPLF)
DPLF model construction, 812–813
DR, *see* Demand response (DR)
Drawing, 797
DS, *see* Data science (DS)
DUT, *see* Design-under-test (DUT)

E
ECG, *see* Electrocardiogram (ECG)
Edge computing, 757, 758
   CCTV, 667
   DDN, 667
   evaluation, 673–676
   low-cost video system
     system objectives, 670–671
     velocity based configuration adaptation
      algorithm, 671–673
   MEC, 667
   VA systems, 668
Edge detection
   camera-buffer-vga data path, 777
   design and interconnection, 778
   design architecture, 776
   image processing algorithm, 776
   implementation (*see* Sobel engine)
   IP, 776
   Sobel algorithm, 778–780
   SoC architecture, 777–778
   software level implementations, 776
   video processing platform, 776
Edge devices, 568, 636, 739
Edge network, 760–761
Education system, 964, 965
EEG, *see* Electroencephalography (EEG)
Electricity, 889, 890
   load and consumption, 889
   power plants, 897
Electrocardiogram (ECG), 743
   arrhythmia detection and classification, 743
   based arrhythmia test, 751
   Database, 750
   generation, 745
   signal classification, 746
   signal generation, 745
   signal waveform, 748
Electroencephalography (EEG)
   acquisition of source, 346
   evaluation experiment
     accuracy, 349

Index

method, 348–349
normalization, 346–348
proposed technique, 345–346
pseudo extension, 348
Electronic code book (ECB), 427–428
Electronic system, 905
Elevator signal, 1112
Embedded systems, 757, 758
EMI-mw, *see* Energy-matter-informational
model of the world (EMI-mw)
EMI objects, 964–972
Emotion detection
acquisition of source, 346
evaluation experiment
accuracy, 349
method, 348–349
normalization, 346–348
proposed technique, 345–346
pseudo extension, 348
Encryption
adversarial neural networks, 425
block cipher mode, 427–428
ciphertext-only attacks, 426–427
cryptography, 425
experiments and results, 430–434
GAN, 426
neural architecture, 428–430
Energy consumption, 807
Energy efficiency, 437, 592, 595, 690, 807,
833, 878, 976
Energy, matter and information (EMI),
964–972
Energy-matter-informational model of the
world (EMI-mw), 965–969,
972–973, 975, 977, 978
Ensemble classifiers
boosting and bagging, 466
A*n*DE method, 466
Entity recognition, 1148
Entity resolution (ER)
contributions, 128
cutting-edge technique, 127
DL method, 130–133
experiments and results
CNN, 134, 135
count combining MLP, 136–138
embedding combining MLP, 136, 137
LSTM, 135, 136
real-world Cora data, 138, 139
TF-IDF combining MLP, 138
information system, 127
problem statement, 128–130
problem types, 128

related work, 128–130
Equalization of elements, 643–645, 853
ER, *see* Entity resolution (ER)
Escaping local minima, 1124
Ethics, 84–85
EUC_2D_194, 1126, 1128, 1132
Evaluation metrics, 330
implementation details, 329–330
Event-based keyframing
adaptive learning, 212–214
AI, 211
elaboration and repair, 219–220
event
recognition, 217–218
representation, 217–218
insights, 214–215
keyframes, 218–219
keyframing, 218–219
systems requirements, 212–214
Event recognition, 216–219
Event representation, 216–220, 339
Evolution
biological, 1018
characteristics, 247
host-cell interaction, 270
NN and GA, 355
strategies, 108–109
viral transmutability, 265
Explanation, 4, 46, 165, 239, 301, 302, 324,
455, 514, 791, 865
Extreme gradient boosting (XGB), 813
Eye tracking, 1087, 1089

F

Facial landmarks
attention mechanism, 819
detection, 818
experiment, 822, 823
proposed method, 819–821
semantic segmentation, 818–819
squeeze-excitation network, 818
Factual knowledge, 790–791
Feature clustering, 1031
Feature extraction method
chromosome classification model
adaptive thresholding, 624
chromosomes-image binarization, 624,
625
gray level extraction, 625–626
histogram information, 626–628
medial axis estimation, 625
skeletonization, 624, 625

Feature scaling, 812
Feature selection, 283–284
  algorithms, 278
  and classification algorithms, 294
  cyber security, 1030
  dimensionality reduction, 1030
  domain-driven, 60
  embedded model, 1031
  evaluation-based categories, 1030
  and feature extraction, 1030
  filter ranking construction, 1031
  hybrid model, 1031
  learner-based, 467, 468
  pair-wise correlation analysis-based filter method, 1031
  prediction performance comparison, 468
  proposed framework, 1033
  subset selection, 1030
  VFD (*see* Variance fractal dimension (VFD) feature selection)
Feedforward neural network, 1111
Field-programmable gate array (FPGA), 730, 757, 776, 782
  deep learning operations, 682
  design architecture, 778
  resource
    cost, 782, 783
    scheduling module, 683
    runtime rescheduling, 686–687
Filter elements, 1149
First-order Markov chain, 744
FIS, *see* Fuzzy inference system (FIS)
5-Fold cross-validation method, 724
F-Measure, 724
FNNs, *see* Fuzzy neural networks (FNNs)
Food-types
  MATLAB, 857
  performance evaluation, 861, 862
  proposed work
    algorithm-multilayer perceptron NNs, 858–860
    multilayer perceptron NNs, 860–861
  related work, 858
Forecasting
  electricity, 890
  high-density traffic issues, 354
  resource management and decision support, 450
  training-data generation, 807–814
Forecasting process, 889
  computational time, 897
  electricity consumption, 897
  long-term electricity demand, residential user, 898

Four-plane semantics (FPCS) model, 975, 977
FPGA, *see* Field-programmable gate array (FPGA)
Friedman test, 813
Fuel resources, 897
Functional products, 721
Fundus eye image, 864, 865, 868
Fun-Joy project
  abstraction, 800
  AI-supported Geometrical and Computational Methods, 800, 802
  algorithms, 800
  CT, 799
  decomposition, 800
  K-8th grades, 802
  nature computing phenomena, 799, 802
  participatory approach, 801
  pattern recognition, 800
  Realistic Drawing & Painting, 801, 803
  research methods, 800
  research questions, 799–800
  STEM learning, 799
  strategic curricular design, 802
  visual art skill knowledge, 802
  vital habits and skills, 800
Fuzzy inference system (FIS), 930, 931
Fuzzy neural networks (FNNs), 807
Fuzzy techniques, 928, 931
Fuzzy theory, 608

G

GA, *see* Genetic algorithm (GA)
GANs, *see* Generative adversarial networks (GANs)
GANs-based fake hidden state generation, 748–749
GANs+LSTM decoder, 750
Gated recurrent unit (GRU), 808
Gaussian functions, 744
Gaze control, 1062
GBM, *see* Gradient boosting machine (GBM)
GDS, *see* Genetic Density Summarization (GDS)
General-purpose CPU, 757
Generative adversarial networks (GANs), 743, 746
  accuracy, 3
  agent model, 104
  architecture, 3
  background, 4–5
  backpropagation algorithm, 102
  car racing experimental results, 107–108

Index

1165

controller model (C), 105, 107
discriminator, 5, 106
Doom RNN, 108–109
evolutional strategies, 108–109
experiments
    experiment 1, 8
    experiment 2, 9
    experiment 3, 9–13
    GiniIndex, 6
    predictors, 6, 7
generator, 5–6
human imagination, 83
imaginary fashion models, 82
MDN-RNN (M) model, 105–107
neural architecture, 426
other work, 4–5
reinforcement learning, 101
related work, 102–103
results, 14
VAE (V) model, 104, 105
Genetic algorithm (GA), 34–35, 360–361, 983,
        1014, 1018
algorithm segments, 369
application, 375–376
autonomous car pilot, 372
AUTOPIA, 370
backpropagation, 355
benchmark procedural trajectory, 372–374
car's trajectory, 369
differential modules, 273
evolution models NN, 355
experimental results and discussion,
    377–378
GILOF, 1014
human movement in crowds (*see* Crowds)
hybrid model, 377
hyperparameters' optimization, 33
NN, 370
non-uniform key framing, 376
settings, 375
structure learning methods, 467
track E-Road, 370, 371
trajectory
    application, 375–376
    hybrid model, 377
    non-uniform key framing, 376
    settings, 375
    straightening and smoothing, 374
Genetic-based Incremental Local Outlier
    Factor (GILOF), 1012, 1014
Genetic-based Incremental Local Outlier
    Factor Reachability distance
    (GILOFR)
accuracy, 1022–1024

BDM mutation, 1022
data points, 1017, 1025
datasets, 1021
data stream, 1018
execution time, 1024–1025
GDS, 1017
GILOF hyperparameters, 1021
LOFR, 1012
objectives, 1017, 1025
outlier detection, 1021, 1022
performance, 1012
stages, 1012, 1017
Genetic Density Summarization (GDS), 1014
chromosomes, 1018, 1019
crossover, 1018, 1019
data point, 1017, 1019
density differences, 1019
framework, 1020
GA, 1018
mutation, 1019
selection, 1018, 1019
Genetic diseases, 620, 634
GHOST, 909
GILOF, *see* Genetic-based Incremental Local
    Outlier Factor (GILOF)
GILOF_NS, 1023–1025
GILOFR, *see* Genetic-based Incremental Local
    Outlier Factor Reachability distance
    (GILOFR)
GILOFR_NS, 1023–1025
GitHub repository, 716
Glaucoma screening, 863, 864
K-means algorithm, 866–867
performance evaluation, 866–867
proposed work
    detection of MA, 865
    preprocessing, 865, 866
related work, 864
GLCM, *see* Gray level co-occurrence matrix
    (GLCM)
Global Internet population, 1029
Global optimization, 273–275, 1124
GP-LOF, *see* Grid Partition-based Local
    Outlier Factor (GP-LOF)
GP-LOF execution, 1057, 1058
GPU, *see* Graphical processing units (GPU)
Gradient boosting machine (GBM), 813
Graph evaluation, 657
Graphical processing units (GPU)
and CPU, 55
edge server hardware information, 689
and FPGAs, 682
MEC, 668
Graph nodes, 1147

1166   Index

Graph processing, 66
Gray level co-occurrence matrix (GLCM), 546, 550, 551
  K-means algorithm, 866–867
  performance evaluation, 866–867
  proposed work
    detection of MA, 865
    preprocessing, 865, 866
  related work, 864
Grid-based cluster model, 1049
Grid Partition-based Local Outlier Factor (GP-LOF)
  datasets, 1054–1055
  detection phase, 1054
  execution time, 1057
  hyperparameter, 1054
  KDD Cup 99 SMTP, 1055
  objective, 1051
  outlier detection, 1048
  outlier detection accuracy, 1055–1057
  phases, 1051
  PPW, 1051–1053
  PW, 1052–1054
  real-world dataset, 1054, 1055, 1057
  ROC curve, 1054
  summarization phase, 1055
  UCI Shuttle dataset, 1055, 1056
  UCI Vowel dataset, 1055
Griewank function, 1125, 1126, 1130–1133, 1136, 1137
GRU, *see* Gated recurrent unit (GRU)

H
Hacking, 908
Handwritten digit recognition, 730
Hardware architecture, 771
HAs, *see* Home agent (HAs)
HDL, *see* Verilog hardware description language (HDL)
Hedonic products, 721, 722
HE-STEM, *see* High Efficiency Skill Knowledge Learning with Computational Methods for STEM (*HE*-STEM)
Heterogeneous
  differentiation, 491
  proposed scheduling algorithm, 682–689
  strategies, 491–497
  structured records, 129
High-dimensional data, 1030

High Efficiency Skill Knowledge Learning with Computational Methods for STEM (*HE*-STEM), 789
Higher-order spectral theory
  deterministic function, 1081
  OoIs, 1079
  radar scan scintillation environment, 1082
  stochastic derivative, 1079
  stochastic integrals, 1080, 1081
Higher-order statistical feature extraction measures, 1063
High fuzzy membership, 1084
Hill climbing method, 465
Home agent (HAs), 878–879
Homogeneous, 301, 308, 491–497, 653
Host-cell interaction, 270, 273, 275
Human activity recognition, 173, 277, 294
  experimental results, 285–293
  medical applications, 277
  proposed methodology
    for elderly, 284–285
    feature selection, 283–284
    ML algorithms, 279–283
  related work, 278
  wearable sensors, 277
Human-adaptive transmissibility, 264–266
  adjacency relationships, 271
  Covid-19, 264, 265
  differential interaction modules, 272–273
  host-cell
    identification, 267–268
    specific regulatory networks, 272
  human-to-human transmission, 263
  hybrid
    statistical model, 270
    unsupervised clustering, 269–270
  interaction networks, 273–274
  learning-based methods, 267–268
  mammalian and avian hosts, 264
  ML, 265–266
  next-generation sequencing technology, 265
  nonmetric similarity measurement, 269
  prediction, 267
  pseudo-temporal ordering, 271–272
  SARS-CoV-2, 264, 265
  stacked autoencoders, 268–269
  target interaction modules, 272
  viral transmutability, 265
  WHO, 264
Human cognitive approach, 1030

Index 1167

Human knowledge, 1062
Human-like cognition capabilities, 1061
Human motion recognition
  contributions, 172–173
  encoder-decoder paradigm, 171
  experimental result
    dataset, 178–179
    supervised learning, 179
    unsupervised learning, 179–180
  motion/activity, 171
  proposed method
    preliminary, 176
    projection functions tuning, 178
    semantic auto-encoder adaptation,
      176–178
  related work
    auto-encoder, 175
    supervised learning, 173–174
    unsupervised learning, 174
  researchers, 172
Human sensing, 174, 180
Hybrid
  car
    detection system, 679
    trajectory, 369–379
  crowd-sourced data, 679
  edge server and cloud, 680
  heterogeneous accelerated edge
      environment
    scheduling hardships, 681–682
    task graph complexity, 681
  performance evaluation
    cloud offloading, 692
    performance metrics, 689
    proposed scheme impact, 690–692
  proposed scheduling algorithm
    assumptions and constraints, 684
    phase 1, 685–686
    phase 2, 686–687
    phase 3, 687–689
    profiling layers in CNN Network,
      682–684
  statistical model, 270
  unsupervised clustering, 269–270
  video surveillance system, 680
Hybrid chaotic activation function (HCAF)
  architecture, 1100
  chaotic ANN, 1104
  logistic map function, 1101
  mean square error cost function, 1103
  network, 1102
  sigmoid function, 1101

Hybrid evolutionary algorithm
  PSO (see Particle swarm optimization
      (PSO))
Hybrid feature selection model, 1031

I
IBM approach, 1145
ICOs, see Initial coin offerings (ICOs)
Ideological Influence Index (III), 958, 961
IE, see Information extraction (IE)
ILOF, see Incremental Local Outlier Factor
      (ILOF)
Image color processing, 201, 204
Image generation, 525–541
ImageNet Large-Scale Visual Recognition
      Challenge (ILSVRC), 141
Image processing, 149–150
  algorithm, 776
  data availability, 148
  dataset, 148–149
  evaluations, 144
  FPGA demonstration, 784
  medical, 898
  methodology
    CNN, 150
    results, 151–153
    VGG-Net, 150–151
  object recognition, 141–143
  UV light, 147
  visual examination, 147
Image schemes, 在NP上(zài NP shàng)
  landmark object (LM)
    place, 937
    SURFACE, 939, 941–944
    topological type LINE, 939, 940
  location, 938
  oscillatory motion, 939
  preposition, 940
  spatial configurations, 944
  trajector (TR)
    HOLE, 944
    LM, 937
    PLATE type, 942, 944, 948
    SUPPORT, 940, 941, 948
Image/video captioning, 17, 19–20, 24
Implementation decisions, 571, 575–578
Incentives, 909
Incremental Local Outlier Factor (ILOF),
      1012, 1014, 1048
Index of pupillary activity (IPA), 1088, 1089
Informal learning environment, 799

1168 Index

Information extraction (IE)
  biomedical literature, 1146
  extraction of relationships between entities,
      1148
Information theoretic methods (ITM), 500,
      507, 509
  high fidelity object/data relationship
      modeling, 515
  novel data characterization, 512–513
  spatial constructs, 515–516
  SV, 513–514
Initial coin offerings (ICOs), 906
Input variable generation, 811–812
Inspection systems, 543, 545
Instance-based learning, 957
Insufficient public-school visual art education,
      798–799
Integrated reporting of wildland-fire
      information (IRWIN) dataset, 451,
      452, 454, 455, 457
Integrated SoC, 783
Intellectual properties (IPs), 776
Intelligent algorithm, 952
Intelligent Software Agent (ISA)
  architecture, 999
  cognitive system roles, 1003
  services, 1002, 1004
Interaction network, 272–276
Interactive model predictive control, 776
Interface Agent (AIN), 1002
Internet, 913, 914
Internet data collection method, 952, 959
Internet-of-Things (IoT), 757, 910
Interquartile range (IQR), 834, 837
Intrusion detection, 1062
IoT-Edge-Server based embedded systems
  architecture design, 759
  BLE, 758
  cloud network, 761
  disposed system, 759–760
  edge computing, 758
  edge network, 760–761
  improved data privacy, 758
  platforms, 758
  two channel computation options, 759
  virtual layers, 759
IPF, *see* ISA processing framework (IPF)
IPF SATP
  communication
    classes, 1006
    ecosystem, 1006
  process
    adaptive behavior, 1004, 1005
    collective thinking, 1004, 1005

self-adaptive system roles, 1004
  self-organizing topical maps, 1004
  systems-level collective thinking, 1004
IP rights, 915
IPs, *see* Intellectual properties (IPs)
IQR, *see* Interquartile range (IQR)
IRWIN dataset, *see* Integrated reporting of
      wildland-fire information (IRWIN)
      dataset
ISA processing framework (IPF)
  artificial entity, 1000
  ISA archetypes
    AAD, 1001
    AAN, 1002, 1008
    ADS, 1001
    AIN, 1002
    ARE, 1001, 1008
    mandatory services, 1002, 10003
  ISA service, 1002, 1003
  real-time autonomous systems, 1000
  systems-level, 1001
Isle Royale National Park (USA), 194–198
  annual moose populations, 184
  data sets, 185
  population dynamics, 186
ITM, *see* Information theoretic methods (ITM)
IXIA PerfectStorm tool, 1037

J

Joy emotion classifier
  English/Chinese word segmentation
      facility, 722
  online Chinese music reviews, 725
  opinion mining, 722
  pre-processing system, 722
  products/services, 722
  prototype, 723–724
  SVM, 722
  system framework, 722, 723
  trigger/no trigger, 723

K

Kaggle data set, 34–36
  chromosome representation of CNN, 35
  CNN model, 36
  data preprocessing, 35
  ImageNet ILSVCR competition, 33, 34
Kaiser-Meyer-Olkin test, 1031
KDD Cup 99 HTTP dataset, 1012, 1015,
      1022
KDD99 SMTP dataset, 1056
Keras, 715

Index 1169

Keyframing
adaptive learning, 212–214
AI, 211
elaboration, 219–220
event recognition, 217–218
event representation, 217–218
insights, 214–215
and keyframes, 218–219
repair, 219–220
systems requirements, 212–214
Khronos NNEF parser, 716
KMA, *see* Korea Meteorological
Administration (KMA)
K-means, 238, 241–243, 837, 842
algorithm, 1014, 1048, 1063
AP, 269
classifier, 1065
clustering, 68
data mining techniques, 842
performance evaluation, 866–867
types, 545
unsupervised machine learning, 238
K-nearest neighbors (KNN), 76, 78, 90, 190,
960, 1012, 1016, 1048, 1050
big data, 885, 886
distance rejection, 884
D-S theory, 884, 885
KNN1, 196
KNN2, 196, 197
KNN3, 197, 198
KNN4, 198
k value selection, 885, 886
mean error, 887
pattern, 885
t-SNE, 885–887
KNN, *see* *K*-nearest neighbors (KNN)
Knowledge based systems
replay, 256–257
StarCraft domain knowledge, 254–256
Knowledge graph
construction, 62
domain-specific semantic graph, 62–63
specification graph, 64–66
meeting conceptual map, 1147
reasoning, 323
Knowledge levels taxonomy
conceptual, 790
factual, 790
metacognitive, 790
procedural, 790
Knowledge relativity (KR), 517–518, 520
Knowledge relativity threads (KRT), 516, 517,
520

Korea Meteorological Administration (KMA),
810

L
LAMDA, 236, 238, 241–243, 246–249
Language implicit link processor (LILP), 959
Language processing, 958
Latent Dirichlet allocation (LDA), 236–238,
241–243, 248
Latent spaces, 3, 5, 6, 9–13, 15, 746
LDA, *see* Latent Dirichlet allocation (LDA)
Learning methods, 17, 40, 62, 103
deep learning classifier, 132–133
record pair representation, 130–132
structure
genetic search, 464
hill climbing methods, 465
k2 method, 464–465
simulated annealing, 465
traditional method, 130
unsupervised, 180
Learning models, 30, 31, 33, 41, 52, 112, 120,
133, 139
illustration image, 143
machine, 6, 1030
zero-shot, 174
Least squares Monte Carlo (LSM) model, 928
LED, *see* Light-emitting diode (LED)
Levenberg-Marquardt, 1112
Light-emitting diode (LED), 760
LightGBM, *see* Light gradient boosting
machine (LightGBM)
Light gradient boosting machine (LightGBM),
813
LILP, *see* Language implicit link processor
(LILP)
Linear rank-based selection (RNK), 1018
Linear rank-based selection with selective
pressure (RSP), 1018
Load forecasting
configuration, 836
experimental results, 836–838
input variable configuration, 834–835
IQR, 834
model configuration, 835–836
time-series data, 833
Local Outlier Factor (LOF)
data points, 1048, 1050
density-based method, 1047
DILOF, 1048
GP-LOF (*see* Grid Partition-based Local
Outlier Factor (GP-LOF))

1170

Index

Local Outlier Factor (LOF) (*cont.*)
   ILOF, 1048
   kNN, 1050
   limitations, 1047
   local outlier detection algorithm, 1011,
      1013
   LRD, 1050
   memory, 1011, 1013
   MILOF, 1048
   outliers detection, 1058
   pt, 1051
   reachability distance, 1050
   score of data point, 1013
   static data, 1011
   static environment, 1049
   steps, 1013
Local Outlier Factor by Reachability distance
      (LOFR)
   calculation method, 1012
   data point p, 1015, 1016
   Euclidean, 1015
   k-nearest neighbor of p, 1016
   Rd, 1016
   score, 1015, 1017
Local reachability density (LRD), 1050
Location-based advertising
   analysis results
      AHP, 611–612
      DEMATEL, 612–615
      descriptive statistics, 611
   research backgrounds, 607–608
   research methodology
      analysis approach operation, 609, 611
      evaluation dimensions and factors, 609
      hierarchical evaluation framework, 609,
         610
LOF, *see* Local Outlier Factor (LOF)
LOFR, *see* Local Outlier Factor by
      Reachability distance (LOFR)
Logic programming
   datalog queries, 253–254
   encoding domain knowledge, 252–253
   FOL statements, 251
Logistic map function, 1101
Long-range BLE mesh system, 770
Long-range dependency, 20
Long short-term memory (LSTM), 114–117,
      729, 743, 744, 746, 808
   building block, 20
   chemistry dynamics simulations, 113
   data, 899
   direct dynamics, 112
   electricity, 898
   electricity prices, 880

experimental results, 120–124
flowchart, 118
GRU units, 19
HAs, 878–879
model, 117–120
model loss, 900
predicting load, 880
prediction-correction algorithm, 113–114
prediction of DL, 112
RAs, 879–880
real residential smart meter data, 898
real-time data, 880
RELu activation function, 899
results, 899, 900
simulated MAS results, 880
speech and handwriting recognition
   applications, 82
VTST and PES, 111
Low Binary Pattern Histogram (LBPH)
   lower computation time, 765
   Open-CV library, 764
   Raspberry Pi 3, 764
Low-cost video system
   CCTV, 667
   DDN, 667
   evaluation, 673–676
   MEC, 667
   system objectives, 670–671
   VA systems, 668
   velocity based configuration adaptation
      algorithm, 671–673
LRD, *see* Local reachability density (LRD)
LSM, *see* Least squares Monte Carlo (LSM)
      model
LSTM, *see* Long short-term memory (LSTM)
LSTM and GAN based arrhythmia generator,
      744
LSTM and GAN based ECG signal generator,
      746, 753
LSTM decoder, 753
LSTM encoder and decoder, 744, 747–748
LSTM-RNN, 890

M
Machine intelligence, 74, 1031
Machine learning (ML), 76–79, 500, 715, 718,
      1029
   AI-enabled ensemble, 335
   algorithms, 4
   applications, 883
   behavior recognition problem, 336–337
   data/performance trade-off curve, 335, 336
   discussion, 519–521

Index

DL method, 338
ECG models, 746
efficacy and utility, 67
electricity consumption datasets, 889
event sequence matching, 338–340
high dimensionality and complexity, 511–512
high fidelity fusion, 518–519
and high-powered virtual machine services, 30
human activity (*see* Human activity recognition)
information theory and information theoretical methods, 510–511
ITM, 512–516
KR, 517–518
multi-model, 507
physical representation, 516–517
political
    culture (*see* Political culture)
    participation (*see* Political participation)
results and comparisons
    data requirements, 340–341
    flexibility, 341
    validation issues, 341–342
Spec Graph representation, 64
supervised and unsupervised, 509–510
supervised techniques, 59
thematic text data analysis, 960
time series data, 889
value, characteristics and limitations, 507–509
Magnetic resonance imaging (MRI), 89–91, 96
Mammography, 851
MAPE, *see* Mean absolute percentage error (MAPE)
Markov chain fingerprinting classification, 1062
Markov chains, 516, 531–534, 744, 1062, 1063
Markov decision process, 325–327
Markov WFC (MkWFC) model
    architecture, 532–533
    meta-data configuration, 533
    model usage, 534
    overview, 531–532
MAs, *see* Microaneurysms (MAs)
Massive data, 1047
Mass media sources, 952
Master bus (MBUS), 777
Mathematical ECG models, 744–745
Matlab, 739
Matrix-based Sobel filter, 779
MBUS, *see* Master bus (MBUS)

Mean absolute percentage error (MAPE), 813, 814
Mean-squared error (MSE), 813
Medical image analysis, 149
Medical image processing, 890, 898
Medical imaging, 82, 89
Meeting conceptual map, 1146–1147
Meeting snippets, 1147
Memory Efficient Incremental Local Outlier Factor (MILOF), 1012, 1014
MEMS, *see* Micro electro mechanical systems (MEMS)
Mentality, 964
Mental model of educational environments (MMEE)
    body-consciousness approach, 964
    cognition, 975
    complexness relations, 972, 974
    consciousness, 964, 976
    cultural representation, 964
    ecological thinking, 965
    educational institutions, 964, 965
    EMI-mw, 965–969, 975
    EMI objects, 964–966
    EMIs, 964–966
    FPCS, 975, 977
    non-classical semiotic model, 975
    PNCC, 971–973
    PNC-text, 975–978
    subject-environment-PNC-text, 977
    temporoformation, 964
    transition types, 970–971
Metaheuristics, 592, 597–603, 981
Metasearch engines, architecture, 914, 915
Microaneurysms (MAs), 865
Micro electro mechanical systems (MEMS), 833
Military conflict situation assessment
    definitions and procedures, 644–647
    future works, 647–649
    methodology, 641
    military
        conflict, 641
        domain, 642–643
    proposed methodology, 643–644
    related works, 643
MILOF, *see* Memory Efficient Incremental Local Outlier Factor (MILOF)
MILOF algorithm, 1048
Mini-loss functions, 748
Mission planning and scheduling
    comparisons
        energy consumption rate effect, 597–598

Mission planning and scheduling (*cont.*)
  minimum time between surveils, 596–597
  number of targets, 599, 600
  number of UAVs, 600–601
  random, 590
  random best sensor, 590–591
  total energy effect, 598, 599
  mapping events, 589–590
  UAVs-based, 583, 584
    metaheuristic, 592
  value-based heuristics
    max value, 591
    max value per energy, 591–592
    max value per time, 591
MIT-BIH arrhythmia database, 750
MIT-BIH database, 751, 753
MIT-BIH data set, 744
MkWFC model, *see* Markov WFC (MkWFC) model
ML, *see* Machine learning (ML)
MLP, *see* Multilayer perceptron (MLP)
MLP implementation
  accuracy analysis, 736
  accuracy comparison, 738–739
  double precision, 736, 737
  input image, 734
  network with weights and biases, 731–734
  neurons, 734
  signal-precision, 737
  strength, 734, 737, 738
MLP neural network
  design architectures, 730, 731
  handwritten digit recognition, 730
  implementation (*see* MLP implementation)
  low-cost design, 730, 739
  network training, 730
MMEE, *see* Mental model of educational environments (MMEE)
MNIST hand-writing training, 717
MNIST handwritten digit data, 730
Mobile
  advertising models, 607
  applications, 73, 1047
  motivations, 611
  sensor data, 173
Mobile-edge computing
  car detection system, 679
  crowd-sourced data, 679
  edge server and cloud, 680
  heterogeneous accelerated edge environment
    scheduling hardships, 681–682
    task graph complexity, 681

performance evaluation
  cloud offloading, 692
  experiment environment, 689
  performance metrics, 689
  proposed scheme impact, 690–692
proposed scheduling algorithm
  assumptions and constraints, 684
  phase 1, 685–686
  phase 2, 686–687
  phase 3, 687–689
  profiling layers in CNN Network, 682–684
video surveillance system, 680
Modern educational environments, 963
Modern search technology, 913
Modified National Institute of Standards and Technology (MNIST), 730, 731
Modified social spider algorithm (MSSA)
  case study, 441–443
  constraints
    system, 438–439
    thermal unit, 439–440
    wind unit, 440–441
  in economic operation, 437
  energy sources, 437
  objective function, 438
  proposed method, 441
  simulation results
    only thermal units, 443, 444
    thermal and wind units, 444–446
Money laundering, 908
Monte Carlo simulation, 928
MRI, *see* Magnetic resonance imaging (MRI)
MRS, *see* Music recommendation systems (MRS)
MSA, *see* Multiagent system (MSA)
MSBUS SoC architecture, 777
MSE, *see* Mean-squared error (MSE)
MSSA, *see* Modified social spider algorithm (MSSA)
Multi-agent reinforcement learning
  heterogeneity, 491
  methods, 492–495
  results, 495–497
  source, 492
Multiagent system (MSA)
  autonomous agents, 878
  DR application, smart grids, 878
  heterogenous components, 878
  LSTM (*see* Long short-term memory (LSTM))
  retail market, 880
Multi-dimensional networks, 373, 508, 510, 826

Multigraphs
  CF, 651
  data collection and preprocessing, 653–654
  discussion, 661–662
  method, 654–658
  MRS, 651
  network graph-based method, 652
  related work, 652–653
  results, 658–661
  three node importance measurements,
    663–664
Multilayer perceptron (MLP), 132–133, 730,
    852, 858, 861, 862
  algorithm, 858–860
  network architecture, 79
  NNs, 860–861
Multilingual methods, 951, 956
Multiple regression
  first maximal model, 186
  reduced parameter maximal model,
    186–187
  results, 194, 195
Multi-scale approach, 1061
Multi-scale complexity analysis, 1061
Multi-swarm approach, 982
MusicBrainZ, 653, 661, 662
Music consumer behaviors, 725
Music recommendation systems (MRS)
  CF, 651
  data collection and preprocessing, 653–654
  discussion, 661–662
  method, 654–658
  network graph-based method, 652
  related work, 652–653
  results, 658–661
  three node importance measurements,
    663–664

N

Naïve Bayes (NB), 807
Na NP$_{loc}$, image schemes
  CLF, 938
  landmark object (LM)
    place, 937
    preposition, 937, 938, 947
    SURFACE, 941–944
    topological type LINE, 939, 940
  NCRL, 945, 947
  oscillatory motion, 939
  possessive particle, 938
  progressive marker, 938
  superessive marker, 938
  trajector (TR)

    HOLE, 944, 945, 948
    LM, 937
    multidirectional movement, 946
    reference level, 946
    SUPPORT, 940, 941
  方(fāng), 947
  在NP上, 946
NARX, *see* Nonlinear autoregressive
    exogenous (NARX)
National Corpus of the Russian Language
    (NCRL), 945, 947
Natural language processing (NLP), 20, 45, 47,
    57, 75, 83, 132, 174, 1148
NB, *see* Naïve Bayes (NB)
NB classification, 463
  classification algorithms, 286, 845
  prediction evaluation, 469
  techniques, 279
  tree augmented, 463–464
NCRL, *see* National Corpus of the Russian
    Language (NCRL)
Neo4j, 1149
Network intrusion detection system (NIDS),
    1014–1015, 1062
Network with weights and biases
  hidden layer neurons, 732–734
  output layer neurons, 734
  Python, 732
  single-precision floating point numbers,
    731
  table biases, 734
Neural Network Exchange Format (NNEF)
  code generations, 718
  computational graph, 716
  files, 715–718
  handling tools, 718
  interpreter, 716–718
  operations, 717
  prototype implementation, 716, 718
  standard specification, 716
  training tools, 718
Neural networks (NNs), 188–190, 354, 952,
    959, 960, 1075, 1076
  ANN (*see* Artificial neural networks
    (ANN))
  artificial neurogenesis, 413–421
  Bi-LSTM, 899, 901
  CNN (*see* Convolutional neural networks
    (CNNs))
  combining with GAs, 358–360
  computational model, 729
  crowd interactions, 355, 357
  CRS, 1077–1079
  design of public spaces, 353

Index

Neural networks (NNs) (*cont.*)
  DNNs (*see* Deep neural networks (DNNs))
  electricity, 890
  emerging applications, 729
  evolution models, 355
  food-types, 857–862
  frameworks, 715
  GAs, 358–361
  hand-collected dataset, 357–358
  image processing, 890
  impacts, 729
  LSTM, 20, 890, 894, 898
  LSTM-RNN, 890
  massively parallel systems, 382
  medical image processing, 890, 898
  personality dynamics, 898
  processor-based AI systems, 388–403
  pruning, 410–413
  PV based data, 890, 898
  Python, 890, 899
  real-world applications, 729
  record of success, 354
  results
    discussion, 365–366
    time series, 363–365
    trained *vs.* control, 362–363
  RNN, 898
  scaling methods, 382–388
  single-processor performance, 381
  structured/unstructured crowds, 354–356
  SWaP, 409
Nexys 4 FPGA, 782
NIDS, *see* Network intrusion detection system (NIDS)
NLP, *see* Natural language processing (NLP)
NNEF, *see* Neural Network Exchange Format (NNEF)
NNs, *see* Neural networks (NNs)
Nonlinear autoregressive exogenous (NARX)
  advantages, 1111
  DDOS attacks
    aircraft-AOA performance, 1116, 1121
    aircraft speed performance, 1116, 1117
    state-q performance, 1116, 1119
    state-w performance, 1116, 1118
    state- performance, 1116, 1120
  definition, 1110
  neural network, 1116
  open-loop neural network, 1111
  output, 1112, 1113
  parallel architecture, 1110, 1111
  series-parallel architecture, 1110, 1111
  testing, 1113–114
  training, 1111–1113

validation, 1113–114
Non-minimized dataset, 1041, 1044
Non-player character
  ANPC, 696
  APF, 697
  background information, 698–699
  CAT, 696
  character makeup, 701–702
  emotions, 696
  formal grammar, 706–709
  implementation, 709–712
  NPCs, 695
  overview, 698
  priori setup, 700–701
  psychological evaluation, 702–705
Normalization, 5, 30, 38, 62, 80, 92, 102, 149, 812
  linear, 347
  non-linear, 348
Normal network system, 1109–1110
np-hard iterative refinement technique, 1065

O

Object detection, 18, 21, 159, 620, 667–671, 673, 692, 776, 818
Object of Interest (OoI) determination, 1076, 1079
Object recognition, 141–143
  evaluations, 144
  illustration images, 143, 144
  ILSVRC, 141
Observe, orient, decide, act and learn (OODAL) loop, 478–482
  commercial applications, 477
  SANDI
    agent-based processing, 487
    emotional contexts, 482–484
    inference architecture, 487
    self-organizing emotional maps, 487–489
    synthetic nervous/limbic system, 486
Online learning, 224, 226, 230, 232
OODAL loop, *see* Observe, orient, decide, act and learn (OODAL) loop
OODA loop, 1074
Open-loop neural network, 1111
Operational constraints
  system, 438–439
  thermal unit, 439–440
  wind unit, 440–441
Opinion mining, 722
Optimal projection function, 171, 172, 174–178

Index

Optimization, 327, 923
  algorithm/activation function, 30
  chromosome representation, CNN, 35
  global, 275
  hyperparameters, 33
  standard-response, 450
Organizational cybernetics, 924
Outlier detection, 1011, 1048
Outliers, 1012
Overlapping WFC (OWFC)
  improvements, 531
  iterative entropy resolution, 529
  overview, 527–528
  tradeoffs, 528

P

Pairwise matching, 128, 130, 132, 133, 139
Paradigmatic time sequence, 874
Parallel algorithms
  automatic traditional-inspection systems, 543
  defect detection, 543
  proposed algorithms
    defect classification, 550–553
    defect detection phase, 546–550
    evaluation criteria, 554–555
    experiment results, 555–558
    preprocessing phase, 546
  related work, 544–545
Parse tree generator model, 50–51, 54, 55
Particle swarm optimization (PSO)
  applications, 981
  basic PSO (see Basic PSO)
  bird flocks, 981
  chaotic implementation, 982
  description, 1124
  disadvantages, 982
  discrete PSO (see Discrete PSO)
  evolutionary metaheuristics, 981
  GA, 983
  hyperparameters, 982
  local exploitation, 981
  multi-swarm approach, 982, 983
  performance, 982
  population diversity, 983, 996
  premature local minimum, 1125
  schooling fish, 981
  tribal coefficient, 981
Partition cluster model, 1049
Part of speech (POS), 49, 53, 724
Part-of-speech tagger SINICA CKIP, 724
Passive infrared (PIR), 760
Path-finding process, 325

Pattern identification
  analysis, 242–243
  analyzed techniques, 237–239
  evolution of information, 235
  example of analysis, 243–247
  experiments, metrics, 239–240
  literature review, 236–237
  online unsupervised learning techniques, 236
  results, 241
Pattern recognition, 789
PCA, see Principal component analysis (PCA)
PCG, see Procedural content generation (PCG)
Peer-to-peer payment system, 905
Pendigit dataset, 1021, 1022
Performance analysis, 520
Personality, 696, 698, 700–702, 706, 708–712, 890, 917, 974
Personalized ECG signal, 753
Personalized GANs (PGANs), 746
Personalized recommender system
  CF-based methods, 323
  dimensional embeddings, 324
  experiments
    dataset, 328, 329
    evaluation metrics, 329–330
    explainable recommendation, 331
    performance evaluation, 330–331
  knowledge graph reasoning, 323
  literature review, 324–325
  methodology
    Markov decision process, 325–327
    Q learning, 327–328
    user-item-entity graph, 325
Personnel deployment
  data
    aggregation, 452
    cleansing, 451–452
    collection, 451
  discussion, 456–458
  exploratory analysis, 453–454
  methods, 453–454
  related work, 450–451
  results, 455–456
  USDA Forest Service, 449
Person re-identification
  CCTV-based surveillance technology, 825
  dataset, 826
  disjoint camera views, 825
  experimental results, 829, 830
  learning-based, 825
  model, 826–828
PGANs, see Personalized GANs (PGANs)
Pharmaceutical Industry, 927–932

1176   Index

Photovoltaic (PV), 890, 898
PIR, *see* Passive infrared (PIR)
PIR sensor, 764
PNCC, *see* Post-non-classical concept of culture (PNCC)
PNC-text, *see* Post-non-classical text (PNC-text)
Political culture, 298–299
  data and methods, 303
  hypotheses, 302
  regional, 299–301
  results, 304–307
  variables, 303, 304
Political heritage, 298
Political participation, 301–302
  data and methods, 303
  democratic systems, 297
  hyper-partisanship and dynamic change, 298
  hypotheses, 302
  people's attitudes, 297
  results, 304–307
  variables, 303, 304
Political science, 298
POS, *see* Part of speech (POS)
Post-non-classical concept of culture (PNCC), 971–973, 977
Post-non-classical text (PNC-text), 968, 971–972, 975–978
PPW, *see* Pre-Processing Window (PPW)
P,Q,R,S,T, *see* P-wave, the QRS complex and the T-wave (P,Q,R,S,T)
Predator-prey pursuit, 492, 493
Predictions, 889
Pre-installed arrhythmia classifier that, 753
Pre-Processing Window (PPW), 1051–1053
Principal component analysis (PCA), 811, 812, 1031
Prioritization policy, 923–925
Priority management
  DES, 924, 925
  organizational cybernetics, 924
  organizations, 923
  prioritization policy, 923–925
  VSM, 924
Probability density function, 751
Procedural content generation (PCG), 525, 526
Processing Window (PW), 1052–1054
Processor-based AI systems
  accelerators, 394–395
  communication-to-computation ratio, 389
  computing

    benchmarks, 389–391
    parameters, 400–401
    quantal nature, 398–400
  general considerations, 388–389
  high-speed bus(es), 397–398
  layer structure, 396–397
  rooflines of ANNs, 400
  timing of activities, 395–396
  training ANNs, 401–403
  workload type, 391–394
Products, 721
Pruning, 282, 409–413, 421, 750, 957
Pseudo-code, 778, 779, 985, 1126–1128
Pseudo data, 348
PSO, *see* Particle swarm optimization (PSO)
Psychosocial framework
  ANPC, 696
  APF, 697
  background information, 698–699
  CAT, 696
  character makeup, 701–702
  emotions, 696
  formal grammar, 706–709
  implementation, 709–712
  NPCs, 695
  overview, 698
  priori setup, 700–701
  psychological evaluation, 702–705
Public K-12 education, 798
Pupillometry, 1090
PV, *see* Photovoltaic (PV)
PW, *see* Processing Window (PW)
P-wave, the QRS complex and the T-wave (P,Q,R,S,T), 745
Python, 891, 1148, 1150
Python implementation, 17

Q

QE, *see* Quantization error (QE)
Q learning
  algorithm structure, 482
  optimization, 327
  softmax exploration, 328
  top-N recommendations, 328
Quantization
  and computational geometry, 1065
  matrix multiplication, 561
  self-organizing map, 201–208
Quantization error (QE), 205–206
  distributions, 207
  SOM-QE, 201–204, 207, 208

Index

R

Race and politics, 298, 301–303, 308, 370–374, 376–379
RAD, *see* Registry anomaly detection (RAD)
Radar scintillation data, 1077
Radar systems, 1071
Random Access Memory (RAM), 761
Random forest (RF), 749, 807
Rasmussen cognition model (RCM), 1072–1073
  knowledge-based behavior, 1075
  rule-based behavior, 1075
  skill-based behavior, 1075
Raspberry Pi, 757
Rastrigin function, 1125, 1126, 1130–1132, 1134, 1135, 1138, 1139
Rd, see Reachability distance (*Rd*)
R&D approach, 1145, 1146
Reachability distance (*Rd*), 1014
Realistic Drawing & Painting with Geometrical and Computational Methods (*Fun-Joy*), 789
Real-object image (ROI), 144, 546, 547, 549, 550, 555, 557, 852, 854, 855
Real options-based R&D valuation model, pharmaceutical sector
  commercialization cash flows estimation, 930–931
  cost estimation, 931
  drug development
    clinical trials, 929, 931
    commercialization phase, 930
    fuzzy techniques, 931
    LSM model, 928
    post-marketing surveillance, 929
    preclinical testing phase, 928, 929
  features, 927
  FIS, 930
  fuzzy techniques, 928
  innovations, 928
  investment selection techniques, 927
  LSM model, 928
  pharmaceutical laboratories, 928
Real-time decision making, 775
Real-time systems, 1008
Reasoner Agent (ARE), 1001, 1008
Recommender systems, 914–915
Recurrent neural network (RNN), 730, 898
Register-transfer (RT)-level design, 780
Registry anomaly detection (RAD), 510
Regular Research Paper, 1058
Reinforcement learning (RL), 158–159

ML, 101
  reward-based system, 77
  supervised word prediction, 20
Residential electricity usage, 898
Resizing, 35
Resource management, 923
Retailer agent (RAs), 879–880
Retail market, 880
RF, *see* Random forest (RF)
RF arrhythmia classifier, 749–750
RF classification, 76, 290, 837
  algorithm, 847
  arrhythmia classifier, 749–750
  classifier, 853–855
  risk for diabetes, 842
RL, *see* Reinforcement learning (RL)
RNK, *see* Linear rank-based selection (RNK)
RNN, *see* Recurrent neural network (RNN)
Road-edge detection demonstration, 782
Robotics, 17, 157, 776
ROI, *see* Real-object image (ROI)
Root mean square error (RMSE), 899
Roulette wheel selection (RWS), 1018
R–R time series, 744
RSP, *see* Linear rank-based selection with selective pressure (RSP)
Russian language, *na* NP_{loc}
  causative movements verbs, 939
  cognitive approach, 936
  cognitive linguistics, 936
  geometric scheme, 937
  image scheme (*see Na* NP_{loc}, image schemes)
  locative spatial situation, 936
  spatial prepositions, 936
  spatial relations, 935
  topological scheme, 936
  topological type, 937
RWS, *see* Roulette wheel selection (RWS)

S

SANDI
  agent-based processing, 487
  emotional contexts, 482–484
  inference architecture, 487
  self-organizing emotional maps, 487–489
  synthetic nervous/limbic system, 486
SATP, *see* Systems AI Thinking Process (SATP)
SATP artificial neurogenesis, 1006–1007
Scalability, 908

1178 Index

Scalable surveillance system, 771
Scaling rule
    Amdahl's law, 383–384
    Gustafson's law, 384–387
    modern scaling, 387–388
Scanning electron microscopy (SEM) images,
        201, 202, 207
Scheduling
    assumptions and constraints, 684
    hardships
        heterogeneity in accelerators, 681–682
        resource limitation, 682
    mission techniques, 589–592
    phase 1
        execution time, 685
        fastest resource, 686
        subdeadlines, 685–686
        time-consuming path, 685
    phase 2
        runtime rescheduling task 2, 687
    phase 3
        re-assigning, 688–689
        re-computing subdeadlines, 688–689
    profiling layers, 682–684
    task graph complexity, 681
    UAV, 583
Scikit-learn-based supervised learning models,
        1041
Search elements, 1149, 1150
Search engines, 914, 918, 920
Seed Network, 410, 413–421
Segmentation
    brain tumor (*see* Brain tumor)
    human, 562, 568
    image, 159
    semantic, 817–822
    super-frame, 19
Self-driving car, 775
Self-organizing maps (SOM)
    FSOM, 489
    prototype, 205–206
    SOM-QE, 201, 208
    training and data analysis, 206–207
Semantic analysis
    models, 956
    optimal keyword weights, automatic
        classification, 956
    supertopics, 956–957
    techniques, 956
    visualization, 955
Semantic clustering techniques
    social computing, 958–959
Semantic segmentation, 818–819
    attention mechanism, 819

experiment, 822, 823
facial landmarks detection, 818
proposed method, 819–821
Semantic vector spaces (SVS)
    artificial intelligence tool, 953
    computational linguistics, 953
    data structures, hierarchy, 954
    search engine, 952
    SMART information, 952
    text data, 952
    TOEFL, 953
Sentiment analysis (SA)
    ABSA, 312
    computational analysis, 311
    customer decision making, 311
    dataset and annotation
        annotation schema, 314–315
        composition, 315, 316
        guidelines for the reviews, 314–315
    global optimization method, 1124
    methodology
        models, 316–318
        task definition, 316
    PSO, 1126, 1134
    related work, 313
    results and discussion, 318–320
    sentiment classification, 722
    TABSA, 312
Sequence matching, 338–342
Server-side execution
    back-and-forth data transmission, 766
    cloud server request, 767
    computation time, 767, 769
    face recognition system, 766, 771
    HTML, 766
    LBPH algorithm, 766
    LBPH and DML algorithms, 767, 769
    nodemon app, 765
    python scripting environment, 766
    recognition time, 765
Sets of statements, 643, 644
Shallow neural network (SNN), 813
Short-term load forecasting (STLF), 807
Sigmoid function, 1101
Sigmoid neural network, 731
Signatures/heuristic-based behavioral analysis,
        1061
SimEvents library, 1108–1110
Simple Mail Transfer Protocol (SMTP), 765
Simulating annealing (SA)
    cooling temperature, 1123
    escaping local minima, 1124
    global minimum of functions, 1124
    particle's own inertial force, 1123

Index

PSO (*see* Particle swarm optimization (PSO))
thermal equilibrium, 1124

Simulation
baseline set of randomized scenarios, 593–594
data collection, 190–191
habitat, 191
initialization, 192
moose characterization, 191–192
parameter sweeps, 594
results
discussion, 601
mission scheduling techniques comparisons, 596–601
related work, 602–603
upper bounds on performance, 595–596
system analysis, 190–191
wolf characterization, 191

Simulation-based training, 336

Simulation-to-reality
attention maps, 160
domain randomization, 158
experimental setup
agent architecture, 162–163
design, 163–165
learning environment, 160–162
results
attention maps, 166–167
training in simulation, 165
transfer to real world, 165–166
RL, 157–159
task-specific experience, 157
transfer learning, 159–160

Simulink library, 1108, 1109, 1114

Single-scale approaches, 1061

Situation assessment
definitions and procedures, 644–647
future works, 647–649
methodology, 641
military
conflict, 641
domain, 642–643
proposed methodology, 643–644
related works, 643

Skill knowledge, 787, 788, 802, 803
cognition, 791
feedback, 792
hand/muscle dependent, 792
levels, 792
mastery, 792
modeling-centered inquiry, 793
problem-solving capability, 792
quintessential factor, 789

staged process, 792
STEAMS domain, 793
students learning, 787, 792

Skin cancer
data availability, 148
dataset, 148–149
methodology
CNN, 150
image preprocessing, 149–150
results, 151–153
VGG-Net, 150–151
UV light, 147
visual examination, 147

Skipping scheme, 1019–1020

SL, *see* Supervised learning (SL)

Slave bus (SBUS), 777

SMART, 913

Smart appliances, 878

Smart grids, 807, 878

Smart home, 878

SMTP, *see* Simple Mail Transfer Protocol (SMTP)

Snippetizer component, 1147–1148

SNN, *see* Shallow neural network (SNN)

Sobel algorithm, 778–779

Sobel edge filter, 778

Sobel engine
design, 780
Mentor Graphics ModelSim—Intel FPGA, 782
resource cost, FPGA, 782
RTL-level design, 780–781
simulation, 782

SoC, *see* System-on-Chip (SoC)

SoC architecture, 777

Social computing, 958–959

Social media data collection method, 959

Social networks, 951, 952, 959

Social relations, 700–702, 704, 706, 708–713

Social spider algorithm (SSA)
case study, 441–443
constraints
system, 438–439
thermal unit, 439–440
wind unit, 440–441
in economic operation, 437
energy sources, 437
objective function, 438
proposed method, 441
simulation results
only thermal units, 443, 444
thermal and wind units, 444–446

SoC interfaces, 776

SoC peripheral I2C master, 777

Softmax
    accuracy, 568, 569
    activation function, 68
    DL, 561
    experimental validation, 566–567
    exploration, 328
    futurework, 567
    IoT devices, 561
    preliminaries, 562–563
    previous arts, 563
    proposed method, 563–566
Solar energy consumption, 899
SOM, *see* Self-organizing maps (SOM)
spaCy, 1148
Spatial relations (SR), 935
Spatial skills, 798
Spawn child initialization, 989
Specification graph (spec graph), 64, 65, 69
Spectrum analysis, 345, 346, 630–637
Squeeze-excitation network, 818–823
SSA, *see* Social spider algorithm (SSA)
SSL/TLS sessions, 1063
StarCraft
    AI approaches, 251
    example data analyses
        build order identification, 258–259
        future work, 260
        state estimation, 259–260
    knowledge representation, 254–257
    logic programming, 252–254
State-of-the-art DILOF algorithm, 1059
State-space representation model, 1108, 1109
Statistical features, 1062
Statistical learning, 1062
Statistical models, 1062
    ARIMA (*see* Auto Regressive Integrated
        Moving Average (ARIMA))
    electricity consumption datasets, 889
    Python, 891
    time series data, 889
Statistical Protocol Identification, 1062
STEAM learning, 798
Steel production, 543, 544
STEM disciplines, 788, 797
STEM education, 788, 789, 799
STEM learning, 789
STLF, *see* Short-term load forecasting (STLF)
Stochastic derivative matrix, 1080
Stochastic universal sampling selection (SUS),
    1018
Streaming data, 1047
Structural similarity, 340, 573–574, 642
Structured crowded area, 354–355, 365

Supervised learning (SL), 48, 173–174, 179,
    508, 511, 722
    accuracy, 1041
    machine translation, 20
    and reinforcement learning, 77
    results, 179
    Scikit-learn-based, 1041
Support vector machine (SVM), 76, 173, 278,
    280, 282, 294, 346, 348, 510, 555,
    722–724, 862, 960
Support vector regression (SVR), 807
Surveillance
    cameras, 361
    frequency, 594
    system-wide performance, 584, 585
    time value, 586
    value, 587
    video, 680
Surveillance system
    algorithms, 764
    DML, 765
    LBPH, 764–765
    security, 764
    server-side execution, 765–767
SUS, *see* Stochastic universal sampling
    selection (SUS)
SVM, *see* Support vector machine (SVM)
SVM machine learning method, 723
SVR, *see* Support vector regression (SVR)
Symmetric key encryption, 426
System-on-Chip (SoC), 757, 776
Systems AI Thinking Process (SATP)
    artificial neurogenesis, 1006–1007
    autonomous AI entities, 1008
    ISA IPF processing infrastructure, 1008
    real-time systems, 1008
Systems-level AI process control, 999, 1000
Systems theory, 1063

T
Tactics identification, 59–61, 67, 251, 336, 500
Targeted aspect-based sentiment analysis
    (TABSA), 312
    ABSA, 312
    computational analysis, 311
    customer decision making, 311
    dataset and annotation
        annotation schema, 314–315
        composition, 315, 316
        guidelines, 314–315
    global optimization method, 1124
    methodology

Index

1181

models, 316–318
  task definition, 316
PSO, 1126, 1134
related work, 313
results and discussion, 318–320
Target entity, 311, 312
Task-evoked pupillary response (TEPR), 1088
Taxation, 908
TCGA-GBM dataset, 90
TCM, *see* Transport Class Model (TCM)
t-distributed stochastic neighborhood
      embedding (t-SNE)
  classification procedures, 884
  Dempster-Shafer theory, 884
  distance-weighted k-nearest-neighbor rule,
    884
  immense level, 884
  KNN (*see* $k-$nearest neighbors (KNN))
  low-dimensional embedding, 884
  non-probabilistic classification schemes,
    884
  samples, 884
  spectral clustering, 883, 886
Temporoformation, 964
TensorFlow, 715
Terrorist Threat Degree Index (TTDI), 958,
    960, 961
Testing protocol
  supervised classification task, 67–68
  unsupervised discovery task, 68
Test of English as a Foreign Language
    (TOEFL), 953
Text data processing, 951, 952
Text summarization, 17, 18, 20–27
Texture Synthesis, 525, 526, 529, 532, 534
Thematic text data analysis, 960
Thermal-dynamical systems, 1124
Thermal equilibrium, 1124
3D ECG trajectory, 744
Threshold, 1034
Tiling WFC (TWFC) models
  architecture, 529
  comparison, 538
  improvements over OWFC, 531
  meta-data configuration, 529–530
  and MkWFC, 536
  outputs, 537
  tradeoffs, 531
Time-based entity generator, 1110
Time management, 923
Time-series cross-validation, 808, 811
TNT, *see* Tournament selection (TNT)
TOEFL, *see* Test of English as a Foreign
    Language (TOEFL)

Topological models, 236, 248
Tournament selection (TNT), 1018
Traceability
  analysis, 242–243
  analyzed techniques, 237–239
  evolution of information, 235
  example of analysis, 243–247
  experiments, metrics, 239–240
  literature review, 236–237
  online unsupervised learning techniques,
    236
  results, 241
Trademarks, 917–920
Traditional cloud computing architecture,
    761
Traditional data analysis tools, 715
Traditional machine learning algorithms,
    1061
Training data generation, 808
Training methods, 951
Trajectory optimization
  accuracy, 112
  algorithm, 114
  straightening and smoothing, 374
  tracking, 259
Transfer learning, 8, 20, 82, 83, 148, 159–160,
    167, 174, 853
Transport Class Model (TCM)
  ACPS, 1108–1109
  aircraft model, 1114–1115
  elevator and throttle commands, 1115
  longitudinal model, 1115
  physical system, 1108
Travelling salesperson problem (TSP),
    1124–1125, 1127–1128, 1143
  discrete PSO, 1133, 1134
    best city route map, 1134, 1135, 1142,
      1144
    distance/cost *vs.* route number, 1134,
      1135, 1141, 1143
  SA to PSO, 1125
Trend analysis, 870, 872, 874
Tribal PSO implementation, 983
  benchmark functions
    LANG and EGG, 995, 996
    objective function evaluations, 994,
      996
    parameter tuning, 994–996
    performance metrics, 993, 995
    run time, 996, 997
    σ value, 992–994
    tribal fitness, 993
  conventional PSO, 983–985
  local and global minima, 986

Tribal PSO implementation (*cont.*)
  social coefficient, 985
  spawn behavior
    decision variables, 987–989
    particle fitness, 987
    spawn child initialization, 989
  test functions
    benchmark functions, 989, 991, 992
    benchmark performance and evaluation, 990
Triggered joy, 724
Trinitarian World Information Model and Methodology (TWIMM), 965
Trusted analytical models, 619, 621, 623, 630, 635, 637
TSCV, *see* Time-series cross-validation (TSCV)
t-SNE, *see* t-distributed stochastic neighborhood embedding (t-SNE)
TSP, *see* Travelling salesperson problem (TSP)
TTDI, *see* Terrorist Threat Degree Index (TTDI)
TWFC models, *see* Tiling WFC (TWFC) models
TWIMM, *see* Trinitarian World Information Model and Methodology (TWIMM)
2D matrix, 814
Types of transitions, 970–971

U
UAVs, *see* Unmanned aerial vehicles (UAVs)
UC, *see* Unit commitment (UC)
UCI Pendigit dataset, 1024
UCI Vowel dataset, 1024
Ugandan
  BN models (*see* Bayesian network (BN) models)
  SA (*see* Sentiment analysis (SA))
UML diagram, 1150–1151
Unattacked network system, 1109
Unfair competition, 918–920
Unit commitment (UC)
  case study, 441–443
  constraints
    system, 438–439
    thermal unit, 439–440
    wind unit, 440–441
  in economic operation, 437
  energy sources, 437
  objective function, 438
  proposed method, 441
  simulation results

    only thermal units, 443, 444
    thermal and wind units, 444–446
Universal threshold, 1089
Unmanned aerial vehicles (UAVs)
  example scenario, 584
  number of, 600, 601
  system model, 585–586
    characteristics, 586–588
    problem statement, 589
    surveillance value, 587
    target, 586–588
  system-wide performance, 585
  UAV-based metaheuristic, 592
Un/semi/supervised learning, 20, 22, 48, 76, 77, 173–174, 179–180, 237, 508, 1041
Unstructured crowded area, 353–356, 358, 359, 363, 365, 366
Unstructured records, 127–129, 139
Unsupervised attack separation technique, 1063
Unsupervised classification
  CD4 T-cell, 202
  color-coded cell viability image, 202, 203
  materials and methods
    data analysis, 206–207
    images, 204
    QE, 205–206
    SOM prototype, 205–206
    SOM training, 206–207
  results, 207, 208
  SEM, 202
  SOM-QE, 201, 202
UNSW and AWID dataset, 1063
UNSW-NB15 dataset, 1037, 1043, 1066

V
VA, *see* Video analytics (VA)
VAE, *see* Variational autoencoder (VAE)
VAE and PGAN models, 746
Variance-based classification model construction, 1031
Variance-based inference, 1031
Variance fractal dimension (VFD), 1032
Variance fractal dimension (VFD) feature selection
  accuracy analysis, 1041
  ANN classifier applications, 1035
  comparative analysis, 1033–1034
  computational complexity, 1043
  computation process, 1032–1033
  cyber-attacks detection, 1032
  dataset preprocessing, 1032

Index

discriminative feature selection, 1034–1035
evaluation, 1036–1037
non-discriminatory attributes, 1042
Python libraries, 1038
UNSW-NB15 dataset, 1037, 1038
values, 1038
Variational autoencoder (VAE), 105, 833–838
anomaly detection model, 834
combination agent, 107
experimental results, 836–838
forecasting model configuration, 836
GAN, 104
input variable configuration, 834–835
IQR, 834
model configuration, 835–836
time-series data, 833
Variational transition state theory (VTST), 111
V_{diff} threshold selection, 1034
Vector space models, 953, 958
Verilog HDL, 780
VFD, *see* Variance fractal dimension (VFD)
VFD comparative analysis, 1033–1034
VFD computation, 1032–1033
VFD *vs.* feature graph, 1039, 1040
VGA-enabled interface, 739
VGA-interfaced monitor, 778
VGG-Net, 150–151
Viable System Model (VSM), 924
Video analytics (VA)
CCTV, 667
DDN, 667
evaluation, 673–676
low-cost video system
system objectives, 670–671
velocity based configuration adaptation
algorithm, 671–673
MEC, 667
systems, 668
Video captioning, 17–24
Video processing platform, 776
Viral transmutability and evolution, 265, 273, 275
Virtual currency, 908
Vision and Change report's modelling, 798
VLSI technology, 729
Vowel dataset, 1021, 1022
VSM, *see* Viable System Model (VSM)
VTST, *see* Variational transition state theory (VTST)

W

Wallets, 906
Watson's NLU service, 1149
Wave-based ECG dynamic model (WEDM), 745
Wave function collapse (WFC)
experiment, 535–536
Markov Chain model
architecture, 532–533
meta-data configuration, 533
model usage, 534
overview, 531–532
Maxim Gumin's WFC repository, 527
minimum entropy heuristic, 526
overlapping model
overview, 527–528
tradeoffs of OWFC, 528
OWFC and TWFC, 526
PCG, 525, 526
results
constraint identification, 539
constraint usage, 539–540
performance, 540–541
TWFC and input sample, 536, 537
TWFC and MkWFC outputs, 538
TWFC model, 538
tile-based model
architecture, 529
improvements over OWFC, 531
meta-data configuration, 529–530
overview, 528–529
tradeoffs, 531
WBC, *see* White blood cells (WBC)
Wearable sensors, 277, 294
WEDM, *see* Wave-based ECG dynamic model (WEDM)
Weighted feature map, 822, 823
WFC, *see* Wave function collapse (WFC)
White blood cells (WBC), 29, 31–32, 38, 40, 41
Wiki cation, 1148, 1149
Wildfire containment
forecasting system, 450
U.S. wildfire data, 451
Wind units, 440, 443–446
WIPO, *see* World Intellectual Property Organization (WIPO)
WMC, *see* Working memory capacity (WMC)
Workflow, 47, 49, 52, 53
dataset, 53–54
exploratory data analysis, 253
neutral node, 117
Working memory capacity (WMC), 1089
World Intellectual Property Organization (WIPO), 915
World model, 913
World Wide Web, 914

X

XGB, *see* Extreme gradient boosting (XGB)
XGBoost, 4, 453, 455, 457, 458
Xilinx Nexys 4 FPGA, 776

Z

Zero-shot learning (ZSL)
  coding-based, 174
  human cognitive system, 176
  human motion recognition (*see* Human
    motion recognition)
  image classification, 175
  projection function, 172
  semantic auto-encoder, 178
ZSL, *see* Zero-shot learning (ZSL)